# 1 MONTH OF
# FREE
# READING

## at

## www.ForgottenBooks.com

By purchasing this book you are
eligible for one month membership to
ForgottenBooks.com, giving you
unlimited access to our entire
collection of over 1,000,000 titles via
our web site and mobile apps.

To claim your free month visit:

www.forgottenbooks.com/free893803

ISBN 978-0-265-81615-8
PIBN 10893803

# CANADIAN ELECTRICAL NEWS

## STEAM ENGINEERING JOURNAL

OLD SERIES, VOL. XV.—No. 6.
NEW SERIES, VOL. V.—No. 1.

JANUARY, 1895

PRICE 10 CENTS
$1.00 PER YEAR.

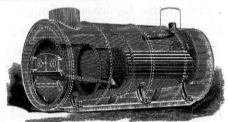

# INDEX

CANADIAN

# ELECTRICAL NEWS

— AND —

## STEAM ENGINEERING JOURNAL

## VOL. V.

1895 :
C. H. MORTIMER, Publisher
TORONTO-CANADA

CANADIAN

# ELECTRICAL NEWS

AND

## STEAM ENGINEERING JOURNAL.

| VOL. V. | JANUARY, 1895 | NO. 1. |

### MONTREAL STREET RAILWAY COMPANY'S NEW BUILDING.

WE present herewith a sketch of the Montreal Street Railway Company's new building in course of construction at the corner of Place D'Armes Hill and Craig streets. It will be remembered that a few weeks ago the iron roof of this building suddenly fell in, killing several of the workmen. The coroner's jury, after a lengthy investigation, found that the accident was due to negligence on the part of the architect and superintendent of construction. These persons will accordingly be brought before the criminal courts. Meanwhile another architect has been employed, under whose direction the construction of the building is being completed.

### LOSS OF POWER.

IF you happen to go into an engine room at any time and find that there is a leak past the cylinder or of the air pump, if you are using a condenser, you may be very sure that you are losing power. While the trouble may not be the cause of any danger, it shows that there has been some mistake in the setting out of the packing ; that the adjustment is bad ; that it has been in use for too long a time, or that the internal surface of the cylinder is cut. If the last is the case, you have no other remedy than that of re-boring the cylinder. In the other instances, the packing may be renewed or the piston may be taken out by removing the cylinder head, as in the case of locomotives. The leaking of piston packing may be detected when the exhaust is continuous instead of intermittent, although a leaky valve will also produce the same results ; or it will be made to appear by opening the cylinder cocks and noting whether they blow steam on the back stroke. A twofold loss is entailed by this defect ; steam is blown away uselessly and back pressure is increased, putting a greater load upon the steam that does the pushing.—Dixie.

### MONTREAL JUNIOR ELECTRIC CLUB.

Nov. 27.—Mr. E. W. Sayer gave an essay on "The Difference between the Cost of Gas and Electric Light Plant."

Dec. 3rd.—Mr. W. T. Sutton read a paper on "Arc Lamps."

Dec. 10th.—Mr. Morrison read a paper entitled "Is Electricity in its Infancy?"

Dec. 17th.—Mr. W. T. Sutton gave a general review of past subjects.

The above meetings were held in the club rooms, No. 6 Richmond ave.

Dec. 20th.—A semi-public illustrated lecture on "Electricity", was delivered by the President in Welcome Hall.

### MOTOR-GENERATORS FOR TELEGRAPH PURPOSES.

THE Great North Western Telegraph Company have completed arrangements for the installation of a dynamo plant in their Montreal office. It is to be of the type known as the motor-generator, and when completed will displace over four thousand cells of gravity battery.

The plant will consist of thirteen machines. The generator side will give the following voltages : 6, 20, 80, 160 and 330 volts respectively. The 6 and 20 volts are used for local work and short lines ; the 80 and 160 volts for ordinary lines, and the 330 volts for quadruplex work. Each group consists of two machines of opposite potentials with a spare machine to be used as a "relief."

The adoption of dynamo currents will effect a great saving in space and materially increase the efficient working of the Company's lines.

The machines are being built by the Toronto Electric Light Company, under the direction of Supt. A. B. Smith, who so successfully installed the plant in Toronto at the head offices of the Company some two years ago.

### THE INTERIOR FRICTION OF OILS.

PETROFF, who has occupied himself very extensively with the examination of lubricants, has investigated, says the Scientific American, the interior friction of oils by means of an apparatus invented by himself, and has given his results in tabular form and graphically by a series of curves. According to his results, the degree of transparency of lubricants, the refining process, viscosity, flash point and fire point, give no basis for estimating the degree of interior friction, though all are of importance.

If two oils which at the same temperature possess different interior frictions be mixed, the mixed product will yield a characteristic curve corresponding to that of an oil the qualities of which lie between those of the two opponents. Consequently the excessive friction of any thick lubricant may be reduced by mixing with it small proportions of solar oil, pyro-naptha, or kerosene, or any oil possessing low interior friction. But this addition can be useful only when the added product does not separate to any great extent.

The addition of such light oils can, of course, be easily detected through the flash point and the fire point. The addition of various resinous materials increases friction in the machinery and in the lubricant itself. These products have also an injurious chemical effect upon the metallic surfaces subjected to friction.

It was also frequently observed that samples of the same oil that were received in the factory at different times did not yield the same characteristic curve, although filling all requirements,

## PROPOSED UNDERGROUND SEC-TIONAL ELECTRIC RAILWAY AT BELLEVILLE.

WE are indebted to the Belleville Sun for the accompanying illustrations and particulars of an under-ground electrical railway system which is to be constructed in that city. It is known as the " E. M." system, of which Mr. James F. Mc-Laughlin, of Philadelphia, is said to be the inventor. The system is described as follows :

The method of distribution consists of a main tube of cast iron of the same height as the rail, made in sections of fourteen feet each, which are bolted on the ties midway between the track rails. In the centre of this tube is placed the main line conductor, which is dependent upon the amount of power used. This rod is continuous and carries the current. It is insulated from the iron tube by a bituminous cement.

The current is distributed to the car motor by means of surface sectional conductors, in conjunction with a flexible controlling device mounted between the two car motors. The controller carries two collectors in the form of flat metal wheels, which are in constant contact with the sectional surface conductor. The electricity connection is made without the use of switches, magnets or automatic devices, the actual method as being kept secret, because of additional European patents which have been applied for.

There is no possibility of "grounds" or short circuits between the rails and the surface conductors, for the reason that there is no current in the conductor except the section directly under the car. This conductor is a strip of rail one inch wide and four and a half inches high. It is laid in sections of varying lengths, which are insulated one from the other by blocks of wood or asphalt a foot wide.

The electric current is supplied to each one of these sections of conductor through a switch-box, but the circuit, of course, is not completed until the trolley wheels, of which there are two on each car, pass over the section of conductor and pick up the electric current for the motor. Thus, this rail, one inch wide, which lies between the tracks, flush with the streets, is never charged with electricity so as to give a shock to a man or animal, except the one section directly underneath the car. It would be impossible, therefore, for a man or child to receive an electric shock severe or otherwise, along the line of this road.

The moment the copper trolley wheels have left a fourteen-foot section of the conductor—or it may be an eight foot section or an eighteen foot section—that section, for all practical purposes so far as street traffic is concerned, is dead. Carriages and horses and men and women crossing the streets have nothing to fear from it. This was shown by Inventor McLaughlin at his shop when he invited the visitor to place his hand on the conductor rail and on the car rail of his model. The current was turned on, but there was no shock. Even a better test was made with a couple of incandescent lamps, attached to copper hooks. There was no electricity in the tracks to light the lamps, but when a workman struck the two sharp copper hooks underneath the car in its passage along the miniature tracks the lights burned brightly.

One of the greatest difficulties in the operation of the trolley system has been in switching, but this seems the easiest thing of all in connection with the E. M. railway system. Because of the two trolley wheels, one in front and one in the rear of the trucks, it is possible to eliminate the conductor rail for a distance of several feet where this is found necessary, at curves, crossings and switches. The two trolley wheels are so far apart that one of them is always bound to be upon the rail, even when the other finds no conductor beneath it.

Telegraph poles are now made of paper pulp in the United States. The pulp is mixed with borax, tallow, and other things, then cast in a wooden mould having a solid core, which makes the pole hollow. The cross arms are to hold the insulators and telegraph wires are attached to pieces of wood fixed in the sides of the pole. What with their lightness and durability, these poles have some advantages over wooden ones.

A new belt fastener recently patented in England, consists of a metal plate adapted to extend across the meeting edges, the plate having one straight side and at the other side a series of spurs arranged in pairs longitudinally opposite, the spurs of each pair being at equidistant points from the transverse center of the plate and arranged in advance of the proceeding in both directions, so that each pair will penetrate the belt at different point',

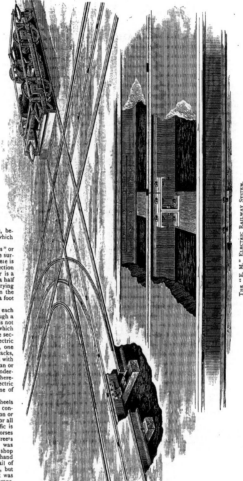

THE "E. M." ELECTRIC RAILWAY SYSTEM.

## A B C OF ELECTRICITY.

### BY H. BRECK, JR.

Abbreviations used : E—Electricity.  +or P = Positive.  — or N = Negative.

WHAT electricity is it is impossible to say, but for the present we may safely say it is a kind of invisible something which pervades all substances. At the present day, however, electricians are able to keep it under perfect control and put it to many uses. So much for the practical view of this science ; but from a theoretical standpoint, supposing all substances—the earth and all on it—as having a certain quantity of electricity upon them, as long as everything has its proper share of electricity, no electrical manifestation is apparent ; but supposing we give a body more than its usual share or take away any from it, leaving it without its proper share, then it is that we become aware of the effects of what we call electricity.

When a body has more than its proper share of electricity, it is said to be in the state of positive electrification ; conversely, if it has less than its usual share, it is said to be in the negative state of electrification.

I may here explain the term "potential." When a body is in the + state of electrification, its potential is said to be higher than that of the earth, so, if a body is in the — state, its potential is lower than the earth, whose potential is reckoned as zero.

To make this clearer, for illustration let us take a number of men. So long as each of them has an equal amount of wealth, there is no thought of poverty amongst them. Suppose one man was to get more than his usual share, he would be considered rich as compared with the rest of them ; when each body has

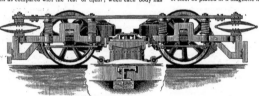

THE "E. M." SYSTEM—METHOD OF MOUNTING CONTROLLER ON CAR TRACK.

its proper share of electricity, the presence is not noticed, but let one body get more than its share, and then its potential is raised.

Suppose one man to lose some of his wealth, then he is considered as poor amongst his fellows ; so if a body loses some of its share of electricity, then also it is reckoned as compared to the earth whose potential is reckoned as zero. The reason the earth's potential is reckoned as zero is for the same reason that we measure the heights of mountains or the depths of mines from the level of the sea.

When a body has a potential higher than that of the earth, it is said to be at a + potential ; conversely, when its potential is lower than the earth's, its potential is —. Although, bear in mind, two bodies may be at a + potential, it is very clear that they may be at different potentials to one another.

Whenever two bodies are at different potentials, the current tends to "flow" from the one at a higher potential to that of a lower potential ; or if a body be at a higher or lower potential than the earth, electricity tends to "flow" between the two until the potentials are equalized. If the difference of potential be very large, electricity will "flow" through anything, even air, as is the case with the lightning flash. With electricity at a low potential, some substances offer a comparatively high resistance to the current. The difference between a conductor and insulator is of one degree only, for whether a body will or will not allow the current to pass depends merely on the difference of potentials of the bodies between which the current tends to pass. There are three principal properties of a current, namely, heating, magnetic and chemical.

*Heating.*—If we pass a strong current through a fine wire, the wire will become heated in proportion to the strength of current passed.

*Magnetic.*—When electricity passes along a wire, it sets up invisible circular lines of force round the wire, and if the current is strong enough the wire will pick up iron filings.

*Chemical.*—If a current of electricity be passed through certain chemical solutions, these will be split up into their component parts.

Let me now divert from electricity and turn your attention for a while to magnetism. What magnetism is it is impossible to say, but we may speak of every particle of iron or steel as being a perfect magnet with a north and south pole.

The earliest form of magnet known to the ancients was lodestone, or "leading stone," as they termed it, as they used it in early times for a compass. It is a brownish-colored stone, dug up from the earth, and from which iron may be extracted, and naturally possesses the magnetic power of attracting iron.

A magnet may be easily told by dipping the piece of iron or steel into iron filings ; if these adhere the iron or steel is a magnet. The poles of a magnet may be discovered by suspending the magnet by a fine thread, and the end which points toward the north is termed the north pole of the magnet, the other the south pole.

There are two kinds of magnets—permanent and electro-magnets. A permanent magnet is one that when once magnetised will retain that power. An electro-magnet is one that will only retain its power while a current of electricity is passed through a coil of wire (insulated of course) wound around it.

The air space surrounding a magnet is filled with invisible lines of space, and is called the magnetic field. You cannot have a magnet with but one pole. If a magnet be broken, each end will form a pole of a magnet. If a magnetic piece of iron or steel be placed in a magnetic field, it will turn so that its own lines of force are parallel with and in the same direction as the lines of force of the magnetic field. From this we have the law of "lines of force," viz., that neighboring lines of force tend to turn themselves so that they be parallel with one another and in the same direction.

The molecular theory of magnetism may be spoken of as follows :—

Suppose we mix a lot of small magnets together, what would be the effect? Simply this : each north pole would unite itself with the south pole of a neighboring magnet, and vice-versa, on account of the first magnetic law, that like magnetic poles *repel* (north and north), unlike magnetic poles *attract* or unite (north and south). When these poles attract one another there is very little free magnetism. The molecules of a piece of unmagnetised iron or steel may be looked upon as in this state, each magnetised molecule completing its magnetic circuit through the neighboring molecules ; the iron or steel as a whole exhibits, therefore, no free magnetism.

Suppose we magnetise the iron or steel, we then arrange its already magnetised molecules so as to give free magnetism by causing their lines of force to act according to the law stated previously. This causes, as you know, the molecules to set themselves in straight lines so that their own internal lines are in the same direction as the lines of force passing through them. So, excepting at the ends of the bar, all the north poles come opposite the south poles, and no free magnetism is apparent, but at one end of the bar we have a row of north poles and a row of south poles ; at the other end, consequently, we have free magnetism.

As to the use of *electro-magnets*, I need not say much, excepting that if any person has occasion to study the practical application of electricity, he will find that this form of magnet is by far the most useful electrical device, as every form (or very nearly every form) of electrical instrument or machine depends for its action on some form of an electro-magnet.

In this paper I hope I have made the description of magnets and the theory of magnetism clearly understood ; but, however, if any reader fails to understand any part, I will be glad to hear from him, either through the columns of this paper or by letter. My address can be had from the editor, who will furnish it to those who desire it.

## THE ONLOOKER.

JUST how rapidly history is made in the present day is best realized when one has occasion to examine into some special subject. The Onlooker thought of this when, in conversation with Mr. Rutherford, chief engineer of the Canadian General Electric Co., he learned of the growth of electric light for domestic purposes. "There are many towns throughout the country now," said Mr. Rutherford, "where electric light takes the place altogether of gas, coal oil or the old tallow candle. Just at the present time we are completing a plant for the town of Mattawa, the capacity of which is 1000 lights. The order was at first for a 500 light plant, but the change was made almost immediately to double the size. When we remember that Mattawa has a population of only 1700, and is a lumbering town, where the large majority of buildings are not the most pretentious, we can see that almost everyone in the place will have his home, no matter how humble it may be, lighted by electricity. In towns where gas is used, the electric light is gradually supplanting it, and where there is no gas it has largely the control." The advance from the tallow candle to the electric light has been made within the memory of the younger men on the stage of life to-day. The Onlooker is not going to give his age away, but it seems only as yesterday since the old home was lighted with the tallow candle, and he can remember when a boy and occasion made it necessary to take a visit to the eastern section of Toronto, that he would go a long step out of his way to escape the delightful odor that came from a tallow candle factory located then in what was known as east Toronto, though to-day one must travel much further east to reach the section so described in the municipal map. To-day the tallow candle—can the bright boys of the public schools tell anything about it? And yet it was to those of a quarter of a century ago what the rude torches prepared by dipping sticks of papyrus or rushes into pitch, and coating them with wax, was to the ancient Greeks and Romans of centuries past. It seemed but a short step from the tallow candle to the kerosene lamp. This method of lighting, however, had little more than fairly become a fixture in the home, and the fears of the good housewife from a kerosene explosion had been overcome, before gas, in the larger towns and cities at least, had become the more common method of domestic lighting. It is hardly to be anticipated that either gas or electricity can drive out kerosene and petroleum. They occupy a wide field in mechanics and arts. But as a method of giving brilliancy to the home after nightfall, their doom would seem to be sealed. Now gas has to fight for its place as an illuminator of towns and cities, and the odds are entirely in favor of its rival, electricity.

     x    x    x    x

On another page of this issue of the ELECTRICAL NEWS is to be found some account of the several railroad undertakings in the construction of which Mr. William T. Jennings, the well-known engineer, has taken a prominent and practical part. The two great railways of Canada, the Grand Trunk and Canadian Pacific, now cover the Dominion so completely from Atlantic to Pacific, that it is hardly to be expected, that in the time of the present generation, at least, there can be any further extensive railroad building. "The trend of railroad construction to-day," said Mr. Jennings, "is in the direction of covering the interior of the country, especially in Ontario, with a network of short and light electric roads that will enable the farmer and small manufacturer to get the products of their industry to the leading markets and shipping centres with less difficulty and cost than is necessary in driving, or the occasional train from one of the regular roads that may, perchance, be passing these points. I look for a large development in the next few years in this direction. Illustration is found of this in the Galt and Preston road, and which is likely to be soon extended to Hespeler. Also in Hamilton and Grimsby, and not only for pleasure-seeking, but likewise as a convenience to that portion of country, the Niagara Falls Park and River Railway, now so well known to everyone." It has occurred to the Onlooker that an intelligent development along these lines will have a healthy bearing on the general progress of the Province. A lamentable lesson from the recent census was the phenomenal growth of a number of the leading cities of Ontario at the expense of the smaller towns and rural sections. This condition is common to all progressive countries, though none the less regrettable, and carrying with it obvious evils. Manufacturers, who have been the life and heart of moderate-sized towns, have been pulling up sticks and locating in the cities, because of the increased shipping facilities there offered. Now if the development that Mr. Jennings hints at, and which is already taking practical shape, can be made the means of holding manufacturers in the smaller towns, who will say that a great good will not have been accomplished for the country as a whole? Those who, like the Onlooker, have their interests centered in the cities, feel a natural pride in the growth of these places, and selfishly congratulate themselves on these conditions, but it is to be remembered that if the big cities owe their growth simply to a sapping of the life blood of the smaller communities, that growth will be of an unhealthy character, and the time will come when a severe reaction will set in. Toronto, Hamilton, London, Peterboro' and Kingston cannot live within themselves. They are going to prosper only as the large and beautiful country that constitutes the province of Ontario prospers. Not the least of the blessings that electricity has brought to the present age will be that of serving to give to the smaller communities the opportunity, not alone of holding their own in the world of commerce and agriculture, but to put on increased strength in these particulars.

     x    x    x    x

It is no easy matter for any one concern in this age of progress to hold for any length of time a particular advantage in mechanical equipment over another. The Onlooker, last month, in describing the new engine of the Toronto Street Railway Co., quoted Mr. Ross, engineer, as saying that this was the only engine in Canada not belted and with a water jacket. Mr. Ross, of course, was speaking out of his own experience, as mechanical superintendent of the Laurie Bros., and which has been far from limited. He had not known then that in the main station of the Incandescent Light Co., of Toronto, according to its superintendent, Mr. Milne, there has been in use for the past fifteen months a compound vertical 750 H.P. engine with generator, direct-driven and also bearings water jacketed. No one will be more pleased than Mr. Ross and the Onlooker to make this further remark as to completeness of equipment in engineering progress in Canada.

## HARTFORD v. THE BELL TELEPHONE CO., THE TORONTO ELECTRIC LIGHT CO. AND THE HOLMES PROTECTION CO.

READERS of the ELECTRICAL NEWS will doubtless remember the particulars of this case. Suit was brought against the above named companies by a Mrs. Hartford, who claimed $25,000 damages for injuries received as the result of coming in contact with a live wire, dangling in the street. Mr. Justice Rose, before whom the case was heard, gave immediate judgment in favor of the Bell Telephone Co., but reserved his decision as regards the other defendants. A few days ago he gave judgment in favor of these defendants also, in the following terms:—

At the trial I gave judgment for the defendants the Bell Telephone Company and Wheeler, reserving for further consideration the question of the liability of the remaining defendants or of either of them.

The Electric Light Co. had a right to have its wires where they were and if there has been a reasonable use of the powers of the Company, if there has been no negligence, then there is no liability; see the National Telephone Company v. Baker, L. R. Q Cb. Div. 193 P. 168; Howard vs. St. Thomas, 19 O. R. 719.

The negligence charged is that putting up its wires under a crossing wire, it did not adopt sufficient safeguards against the wires coming into contact. The suggested safeguards were: separating the wires by a greater intervening space, placing a guard wire above the electric light wires, a better composition for insulating such wires. The evidence would not justify a finding that there was any defect in the material or composition used to insulate the wires but the weight of evidence was much the other way.

The remaining grounds must be considered in the light of the facts. What was the cause of the accident? Undoubtedly on the evidence the falling of the branch or branches from the tree upon the crossing wires put up by the Holmes Co., the branches having been cut by boys whether as trespassers or not did not appear.

Admitting for the sake of argument, that the workmen of the Electric Light Company, knew or should have known that the wire passed among the branches of the tree and that they should have reasonably anticipated that in the course of nature, some natural force such as the wind, or as the result of decay, the branch might fall upon the wire, that the result would naturally be that having regard to the span of such crossing wire and its fastenings and the smallness of the space between the electric light wire and the crossing wire, it, the crossing wire, would sag so that the wires would come into contact and that against such a not improbable occurrence it was the duty of the company to provide either by having a greater space between the wires or a guard above the electric light wire, it is sufficient to say that the accident here was not due to any such a state of facts.

The question in my opinion narrows itself down to this: should the act of the boys in cutting the branches of the tree have been contemplated and guarded against by the Electric Light Company as something likely to occur and reasonably to be anticipated.

After much consideration, which the peculiarly sad condition of the plaintiff in a high degree demands, I am not able to answer such a question in the affirmative, even on the admissions or assumptions above made.

This particular danger was caused by the act of another without the defendants' knowledge or consent; that other was a "responsible third party." As I look at it, there was no neglect of duty on this defendants' part, for there was no duty to guard against such an act. It is not a case of negligence on a defendants' part which was "insulated" by the act of an interven ing third party—see Howard v. St. Thomas—but it is a case of a wire in apparently a safe and secure position, where it would not, so far as we know, have become dangerous but for the act of another over whom the defendants had no control.

A duty may have existed to guard against other possible causes of danger, but such duty, if it existed, was not this duty, and so here there was no neglect of duty and hence no negligence.

This in effect also disposes of the question of the liability of the Holmes Company. Something was made of the fact that the crossing wire was a disused wire, but I cannot see how that makes any difference. I must on this question assume that the Holmes Company had a right to erect the wire; that it was allowed to remain in position although not in actual use, did not make it more but rather less a source of danger. The duty to watch such wire and prevent its becoming loose and dangerous no doubt remained, but on the evidence it was in as safe and secure a position and condition at the time of the accident as when in use. Mr. Cameron cited United Electric Ry. Co. v. Skelton, 14 S. W. Rpter. 863, in support of his argument on this point, but beyond establishing continued responsibility for a disused wire, it does not, as I think, help the plaintiff. The elements of knowledge of the facts, except on which the judgment there is founded, is lacking here.

But there seems to me to be another difficulty in the plaintiff's way. Assuming that the managers of both the Electric Light and the Holmes Protection Co. knew the position of the crossing wire, and how it was fastened, and that it passed along the branches of the tree in question, I do not see how it could be fairly found as a fact that it should have been reasonably anticipated that a blow on the wire at the tree would cause the sagging which here happened. It may be that such a result should reasonably have been anticipated and therefore guarded against, but I do not think on the evidence I could so find the fact.

On the whole I think the action must be dismissed with costs.

The following additional authorities may be referred to: Viners Abridged "Actions B" p. 215; Box v. Jubb, L. R. 4 Ex. Div. p. p. 789; Beven on Negligence p. 1000; Ahern v. Oregon Telephone & Telegraph Co. 33 Pac. Rep. 463. Ward v. Atlantic & Pac. Tel. Co. Am. EL. Cases Vol. 1, p. 259.

Had I found for the plaintiff, and had the husband here joined, I should have assessed the damages at $8,000. For the assistance of the Court, if my opinion should hereafter be held to be erroneous, I asess the damages for the wife at $6,000.

## FIRE HAZARDS FROM TROLLEY WIRES.

THE Inspector for the Board of Underwriters, in Toronto, having represented to the Board the danger of allowing the use of current from trolley wires in buildings insured by the Association, it was suggested that the opinion of well known experts be asked. Accordingly the following letter was forwarded by Secretary McLean to Mr. J. J. Wright, of the Toronto Electric Light Company, Mr. F. A. Badger, City Electrician, Montreal, Mr. W. E. Davis, Toronto Street Railway, and their own inspectors.

The replies as printed below will be interesting, as showing how the matter is viewed from the different standpoints:—

TORONTO, November 12th, 1894.

DEAR SIR:—I have been instructed by the Toronto Board to ask for your opinion as to the danger of an electric ground circuit when high pressure and a ground return are used for power or light. Does this danger exist when the current is taken from the trolley wires? Your reply will much oblige.

Yours truly,
ROBT. McLEAN, Sec'y.

[COPY].

TORONTO ELECTRIC LIGHT CO.

TORONTO, Nov. 16th, 1894.

ROBT. McLEAN, Secretary Board of Underwriters.

DEAR SIR:—In answer to yours of the 13th, re grounded electric wires, I may say that from the very inception of the electric lighting business, the danger of a grounded circuit has been recognized and its use discountenanced. Electrically, it can be worked all right, but the opposition to its use has always come from the insurance fraternity and other parties interested in the security of property from danger of fire. From our standpoint, we shall be very glad indeed if the insurance companies can see their way to

relax their rules on this subject. We have spent many thousands of dollars in copper wire for return circuits, where we could easily have used the ground for a return. The moment the companies allow a ground return we shall at once ground all our power circuits and thus double the capacity of our distributing plant. At the same time, I am free to say, that the risk will be greater. Where there is the slightest dampness, the current will have a tendency to leave the wire and seek the ground, and anything that will burn that lies in its path will be set on fire. This tendency to leave the wire does not exist in the same degree on a metallic circuit. There is also a much greater danger of ignition from lightning with a grounded circuit. The network of wires all over the city form a huge lightning rod. The lightning always seeks the nearest passage to the earth, so that wires grounded through motors or lights are just a picnic for a flash of atmospheric electricity—they form a lightning conductor of the best quality. On the other hand, a metallic circuit insulated from the ground offers no inducement for the lightning to travel on it at all, and consequently it rarely does so. I wrote your Inspector some time ago, when I found that the Railway Co. were putting in grounded wires, and claimed the same right. That letter has not yet been answered. I did not press for an answer as I understood the Association took the stand that is universally taken in England and the States, that it could not be permitted. I want the Association to understand, however, that while we shall continue to observe their regulations in the future as we have done in the past, we shall claim the same privileges that are extended to others, and that if the other companies are allowed to use grounded wires, that we shall do the same. In reply to your query, does the danger of grounded circuits exist when current is taken from trolley wires? I answer emphatically, "yes," and always will as long as the ground is used for the return circuits. There is absolutely no difference between "trolley" or any other electrical current, as far as grounded circuits are concerned.

Yours very truly,
(Sgd.) J. J. WRIGHT.

[COPY.]

OFFICE OF CITY ELECTRICIAN.

MONTREAL, Nov. 15th, 1894.

ROBT. McLEAN, ESQ., Secretary Toronto Board of F. U.

DEAR SIR:—Replying to yours of the 13th inst., regarding the use of trolley circuit for light and power, I will say that while it may be possible in a given case (every precaution being taken to guard against accident,) yet I do not believe that in general practice such precautions would be fully attended to, and we would not sanction nor assume any responsibility for the use of such currents for the purposes mentioned.

Yours, etc.,
(Sgd.) F. A. BADGER.

[COPY.]

THE TORONTO RAILWAY COMPANY.

TORONTO, Nov. 13th, 1894.

ROBT. McLEAN, ESQ., Secretary Underwriters' Association.

DEAR SIR:—Yours re danger from high voltage and grounded circuit to hand. There is no greater danger from high volt system than low—provided the proportion of care taken in installation is equal to the increase in pressure. More fires occur throughout the country from low volt—heavy ampere systems—than others.

The pressure from the trolley wire is 500 volts—not by any means high voltage. All extensive electric light and power systems are more or less grounded—the more dangerous owing to the fact that it is often a partial ground only, and that the locality of same is unknown. A partial ground is liable to overheat the wire and cause combustion, whereas a dead or actual ground would instantly melt the fuses—provided, of course, the other wire was at some point also grounded.

The only danger from current to installations from trolley wire, in my opinion, is from lightning. This I consider small, inasmuch as, when lightning strikes the trolley wire it invariably comes to the power house, taking the path of least resistance, blowing circuit breakers and passing to earth through lightning arresters. We have besides about 60 lightning arresters on the line throughout the city and one on every car. We have experienced over 40 thunderstorms and have never had any damage whatever, either in cars or buildings, otherwise than blowing fuses and the melting of lightning arresters.

We admit that there is danger from such circuit when not properly protected. Where such protection is ample, I believe there is no more danger than from the ordinary overhead circuit. Our experience in Toronto will bear out the above statement.

Respectfully yours,
(Sgd.) W. E. DAVIS.

TORONTO, Nov. 15th, 1894.

ROBT. McLEAN, ESQ., Secretary, Toronto.

DEAR SIR:—In reply to your letter of the 12th inst., I can only repeat what I have stated before, namely, that the use of a ground return for electric lighting and power has been universally prohibited by all Boards of Underwriters, ever since the earliest days of electric lighting. There is today more than ever good and substantial reasons for a strict enforcement of this rule.

That there is a decidedly increased hazard by the introduction of such systems for commercial purposes, is beyond dispute.

At the convention of electrical inspectors held in Chicago in August of '93, the rule bearing on this subject was amended to make it as emphatic as possible, and reads as follows: "Lightning and power from railway wires must not be permitted under any pretence in the same circuit with trolley wires with a ground return, nor shall the same dynamo be used for both purposes except in street railway cars, electric car houses, and their power stations."

I might say the rules of the International Underwriters' Association have been universally adopted in the United States and Canada.

Yours,
A. B. SMITH.

PUBLISHED ON THE FIRST OF EVERY MONTH BY

### CHAS. H. MORTIMER,

OFFICE : CONFEDERATION LIFE BUILDING,
*Corner Yonge and Richmond Streets,*

TORONTO, - CANADA.
Telephone 2362.

NEW YORK LIFE INSURANCE BUILDING, MONTREAL.
Bell Telephone 2299.

***ADVERTISEMENTS.***

Advertising rates sent promptly on application. Orders for advertising should reach the office of publication not later than the 15th day of the month immediately preceding date of issue. Changes in advertisements will be made whenever desired, without cost to the advertiser, but to insure proper compliance with the instructions of the advertiser, requests for change should reach the office as early as the 22nd day of the month.

***SUBSCRIPTIONS.***

The ELECTRICAL NEWS will be mailed to subscribers in the Dominion, or the United States, post free, for $1.00 per annum, 50 cents for six months. The price of subscription may be remitted by currency, in registered letter, or by postal order payable to C. H. Mortimer. Please do not send cheques on local banks unless 25 cents is added for cost of discount. Money sent in unregistered letters must be at senders' risk. Subscriptions from foreign countries embraced in the General Postal Union, $1.50 per annum. Subscriptions are payable in advance. The paper will be discontinued at expiration of term paid for if so stipulated by the subscriber, but where no such understanding exists, will be continued until instructions to discontinue are received and all arrearages paid.
Subscribers may have the mailing address changed as often as desired. *When ordering change, always give the old as well as the new address.*
The Publisher should be notified of the failure of subscribers to receive their papers promptly and regularly.

***EDITOR'S ANNOUNCEMENTS.***

Correspondence is invited upon all topics coming legitimately within the scope of this journal

THE "CANADIAN ELECTRICAL NEWS" HAS BEEN APPOINTED THE OFFICIAL PAPER OF THE CANADIAN ELECTRICAL ASSOCIATION.

## CANADIAN ELECTRICAL ASSOCIATION.

### OFFICERS :

PRESIDENT :
K. J. DUNSTAN, Local Manager Bell Telephone Company, Toronto.

1ST VICE-PRESIDENT :
A. B. SMITH, Inspector Canadian Board Fire Underwriters, Toronto.

2ND VICE-PRESIDENT :
C. BERKELEY POWELL, Manager Ottawa Electric Light Co., Ottawa.

SECRETARY-TREASURER :
C. H. MORTIMER, Publisher ELECTRICAL NEWS, Toronto.

EXECUTIVE COMMITTEE :
L. B. McFARLANE, Bell Telephone Company, Montreal.
GEO. BLACK, G. N. W. Telegraph Co., Hamilton.
T. R. ROSEBRUGH, Lecturer in Electricity, School of Practical Science, Toronto.
E. C. BREITHAUPT, Berlin, Ont.
JOHN YULE, Manager Guelph Gas and Electric Light Company, Guelph, Ont.
D. A. STARR, Electrical Engineer, Montreal.
J. J. WRIGHT, Manager Toronto Electric Light Company.
J. A. KAMMERER, Royal Electric Co., Toronto.
J. W. TAYLOR, Manager Peterboro' Carbon Co., Peterboro'.
O. HIGMAN, Inland Revenue Department, Ottawa.

## MONTREAL ELECTRIC CLUB.

### OFFICERS :

| | |
|---|---|
| President, W. B. SHAW, | Montreal Electric Co. |
| Vice-President. H. RITCHIE, | Royal Electric Co. |
| Secretary, JAMES BURNETT, | 19 Shuter Street. |
| Treasurer, L. M. PINOLET, | 4251 Dorchester Street. |

Committee of Management : H. BROWN, JAS. DOUGLAS, H. R. NORTON.

## CANADIAN ASSOCIATION OF STATIONARY ENGINEERS.

### EXECUTIVE BOARD :

| | |
|---|---|
| President, J. J. YORK, | Board of Trade Bldg, Montreal. |
| Vice-President, W. G. BLACKGROVE, | Toronto, Ont. |
| Secretary, JAMES DEVLIN, | Kingston, Ont. |
| Treasurer, DUNCAN ROBERTSON, | Hamilton, Ont. |
| Conductor, E. J. Philip, | Toronto, Ont. |
| Door Keeper, J. F. CODY, | Wiarton, Ont. |

TORONTO BRANCH NO. 1.—Meets 2nd and 4th Friday each month in Room D, Shaftesbury Hall. Wilson Phillips, President ; T. Eversfield, Secretary, University Crescent.

HAMILTON BRANCH NO. 2.—Meets 1st and 3rd Friday each month, in Maccabee's Hall. Jos. Langdon, President ; Wm. Norris, Corresponding Secretary, 211 Wellington Street North.

STRATFORD BRANCH NO. 3.—John Hoy, President ; Samuel H. Weir, Secretary.

BRANTFORD BRANCH NO. 4.—Meets 2nd and 4th Friday each month. C. Walker, President ; Joseph Ogle, Secretary, Brantford Cordage Co.

LONDON BRANCH NO. 5.—Meets in Sherwood Hall first Thursday and last Friday in each month. F. Mitchell, President ; William Meaden, Secretary Treasurer, 533 Richmond Street.

MONTREAL BRANCH NO. 1.—Meets 1st and 3rd Thursday each month, in Engineers' Hall, Craig street. President, Jos. Robertson ; first vice-president, H. Nuttall ; second vice-president, Jos. Badger ; secretary, J. J. York, Board of Trade Building ; treasurer, Thos. Ryan.

ST. LAURENT BRANCH NO. 2.—Meets every Monday evening at 43 Bonsecours street, Montreal. R. Drouin, President ; Alfred Latour, Secretary, 306 Delisle street, St. Cunegonde.

BRANDON, MAN., BRANCH NO. 1.—Meets 1st and 3rd Friday each month, in City Hall. A. R. Crawford, President ; Arthur Fleming, Secretary.

GUELPH BRANCH NO. 6.—Meets 1st and 3rd Wednesday each month at 7:30. p.m. C. Jorden, President ; H. T, Flewelling, Secretary, Box No 8.

OTTAWA BRANCH, NO. 7.— Meets 2nd and 4th Tuesday, each month, corner Bank and Sparks streets ; Frank Robert, President ; F. Merrill, Secretary, 352 Wellington Street.

DRESDEN BRANCH NO. 8.—Meets every 2nd week in each month ; Thos. Merrill, Secretary.

BERLIN BRANCH NO. 9.—Meets 2nd and 4th Saturday each month at 8 p. m. W. J. Rhodes, President ; G. Steinmetz, Secretary, Berlin Ont.

KINGSTON BRANCH NO. 10.—Meets 1st and 3rd Tuesday in each month in Fraser Hall, King Street, at 8 p.m. J. Devlin, President ; A. Strong, Secretary.

WINNIPEG BRANCH NO. 11.—President, Chas. E. Robertson ; Recording Secretary, L. Brandon ; Financial Secretary, Arthur Harper.

KINCARDINE BRANCH NO. 12.—Meets every Tuesday at 8 o'clock, in the Engineer's Hall, Waterworks. President, Jos. Walker ; Secretary, A. Scott.

WIARTON BRANCH NO. 13.—President, Wm. Craddock ; Rec. Secretary, Ed. Dunham.

PETERBOROUGH BRANCH No. 14.—Meets 2nd and 4th Wednesday in each month. S. Potter, President ; C. Robison, Vice-President ; W. Sharp, engineer steam laundry, Charlotte Street, Secretary.

BROCKVILLE BRANCH No. 15.—W. F. Chapman, President ; James Aitkens, Secretary.

CARLETON PLACE BRANCH No. 16.—W. H. Routh, President ; A. M. Schofield, Secretary.

## ONTARIO ASSOCIATION OF STATIONARY ENGINEERS

### BOARD OF EXAMINERS.

| | |
|---|---|
| President, A. E. EDKINS, | 139 Borden st., Toronto. |
| Vice-President, R. DICKINSON, | Electric Light Co., Hamilton. |
| Registrar, A. M. WICKENS, | 280 Berkeley st., Toronto. |
| Treasurer, R. MACKIE, | 28 Napier st., Hamilton. |
| Solicitor, J. A. McANDREWS, | Toronto. |

TORONTO—A. E. Edkins, A. M. Wickens, E. J. Phillips, F. Donaldson.
HAMILTON—P. Stott, R. Mackie, R. Dickinson.
PETERBORO'—S. Potter, care General Electric Co.
BRANTFORD—A. Ames, care Patterson & Sons.
KINGSTON—J. Devlin (Chief Engineer Penetentiary), J. Campbell.
LONDON—F. Mitchell.

Information regarding examinations will be furnished on application to any member of the Board.

### VOL. V.

THE present number of THE ELECTRICAL NEWS marks the commencement of the fifth yearly volume. During the coming year each number of the journal will consist of 28 pages instead of 24, as hitherto, and we shall be glad to further increase its size from time to time to keep pace with the future development of the important industry which it strives to represent. We recognize also that quality is of greater importance than quantity, and shall aim to make THE ELECTRICAL NEWS, especially from a Canadian standpoint, of the greatest possible interest and value. Every reader is invited to assist, by forwarding information which would tend to make the journal increasingly useful. We are indebted to the friends who have sent us contributions from various parts of the Dominion, and trust that during the coming year their example will be largely followed by others. The editor doesn't profess to know it all, and will at all times welcome information and suggestions.

The year 1894 was marked by great electrical development in Canada, especially in electric railway construction, notwithstanding the severe commercial depression which prevailed in this and all countries. Towards the close of the year there was observable in business circles a feeling of greater hopefulness, which bespeaks more encouraging and satisfactory conditions during 1895. In the hope that such will prove to be the outcome, we extend to every reader of THE ELECTRICAL NEWS our best wishes for a happy and prosperous new year.

FEBRUARY 19th and 21st are the dates chosen for the next annual meeting of the National Electric Light Association, to be held at Cleveland, Ohio.

THE latest application for membership in the Canadian Electrical Association comes from Mr. W. A. Foster, Electrical Engineer to H. H. the Sultan of Johore, India. Having read in the technical press a report of the proceedings of the convention in Montreal, Mr. Foster writes the Secretary of the Association from Singapore, as follows : " Having for several years been connected with the telegraph service of the Canadian Pacific in various parts of Canada, and a Canadian by birth, I should be glad to become a member of your Association. Since leaving the Canadian Pacific Telegraph Service, I have for the past five years been actively employed in electrical engineering, having installed several large lighting plants in this and other districts."

THE boiler explosion at Essery's saw mill near Orangeville, Ont., referred to in our last issue, is but one of many that have resulted from the same cause. The boiler was a second-hand one, about thirty years old, and should have found its place in the scrap heap long ago. The engineer of the Orangeville waterworks, who made an examination of the boiler at Essery's request, said it should not be run at more than 40 or 50 pounds pressure. Mr. Edkins, boiler inspector, from an examination of the boiler subsequent to the accident, came to the conclusion that 20 pounds was the safe limit of pressure at which it could be operated immediately prior to the explosion. To what extent the boiler was properly handled is apparent from the testimony of one of the employees, that twenty minutes before the accident he noticed that the steam gauge indicated a pressure of between 110 and 115 pounds.

ON another page will be found illustrations and a description, both copied from the Belleville Sun, of a street railway system to be introduced in Belleville. It is spoken of as the E. M. system, and the inventor's name is McLaughlin. It is a sectional contact railway, and the description quoted is simply a description of sectional contact railways in general. It states incidentally that the collector wheels are maintained in constant contact with the sectional surface conductor, and current is supplied to the latter from the main conductor through the switch board ; also that a "secret" controller does it all. Now the durable construction of the switches, and reliable methods of operating them, and insuring constant contact of collectors, are the only points in a sectional contact railway that any inventor can claim. They are what constitute a "system." Any so-called system which does not explain the methods of dealing with these essentials is no more a subject for serious criticism than is a patent medicine advertisement. It is an impertinence for an inventor to obtrude his "secrets" on public attention. And the offence is aggravated by column after column of trash about the "deadly trolley," "the death dealing fluid," extravagant claims of economy, personal descriptions of the inventor, his career, and the number and sexes of his children. This kind of newspaper article is out of date, long out of date. It belongs to the earliest and worst days of electrical development, when it was almost a slur on a man's character to call him an electrician. Electrical engineering now-a-days is business, and to merit confidence should be treated in a business-like manner.

ONE of the signs of a revival of business from the depression of the last two years, is the number of electric railway schemes which are now being mooted. Some are appearing for the first time, and some reappearing after an enforced retirement. If there is a quick return of prosperity, there is danger of a boom, with its accompanying construction of unprofitable lines, whose inevitable failure will bring loss to the shareholders and undeserved discredit on electric railways in general. It is of course plain to all that such lines mean a waste of capital. But there is another method of wasting capital which is not so plain. In many parts of Canada, particularly in western Ontario, there are steam railways in abundance connecting all the smaller towns through branches and junctions. But unless on some main lines, these towns are poorly served. The cost of running a train is too great to permit a frequent service, and the inconvenience to traffic of being forced to accommodate itself to a few opportunities at long intervals tends to minimize the development of travel. The inhabitants grumble at the railway which

cannot give them any better steam service except at a loss, and readily welcome an electric railway which promises them relief. Since an electric railway is most economically run when its total traffic is uniformly diffused in small lots throughout the whole running time, it will probably take the bulk of the passenger and small parcel traffic from the parallel competing steam road. It will also in many cases build up and increase traffic which the steam road could not obtain, and it is quite probable that the electric railway may pay handsomely. But it is none the less true that a large portion of the investment in it is wasted capital, so far as the community at large is concerned. The steam railroad with its existing tracks, buildings and organization, might have served the town just as well or better, by undertaking the local electric service, with merely the additional investment required for power, cars and trolley wire. The public benefits by competition—but only up to a certain point. When the business competed for is not enough to pay the expenses of competition, the loss, directly or indirectly, falls on the community. Where a projected electric parallels an existing steam road, it means in general that it is to do work which the steam road could do at less cost. And if the steam road for any cause will not undertake it, the communities to be served, the investing public, and electrical interest in general, would probably all be better off if the electric company simply leased running rights over the steam road, instead of sinking capital in a new and parallel permanent way of its own. Of course operating, or leasing operating rights, by the steam roads, presupposes that the latter are alive to their own interests, and are prepared to deal in a spirit of fairness with the communities they serve, whose interests are interdependent with their own. Unfortunately the record of Canadian railways does not tend to inspire much confidence in their judgment in such matters, and if they cannot or will not see their opportunity, a parallel road is the only solution ; but it is none the less, to the extent of the invested capital, a tax on the resources of the country in general, and a source of justifiable alarm to foreign investors.

THE well known saying of the French general who witnessed the charge of the Light Brigade at Balaklava was, "it is magnificent, but it is not war." On looking at some of the large central stations in America, and seeing a large part of the plant lying idle for twenty-two hours out of the twenty-four, in readiness to tide over a brief heavy local period, European engineers must frequently be tempted to parody the phrase and say, or at least think, "it is magnificent, but it is not business." With more abundant and cheaper capital, generally greater competition and national habits of permanent construction, the fundamental conditions in Europe are not strictly parallel to those on this side of the Atlantic ; and such a criticism, though natural in one imbued with European methods, may in many cases be too sweeping when worked down to details. But it must be confessed that there are some grounds for such a criticism which might profitably receive more consideration. The tendency of American engineering practice is to run in grooves. This is a natural result of the system of machine manufacturing, whose developments is so distinctively American; and justly considered one of the triumphs of American mechanics. Added to this is the national habit of rapid business expansion designed to skim the cream off the opportunities for making money which are so lavishly afforded by the growth of a naturally rich and diversified country. The wonderful development in any one branch due to such enterprise, is dazzling, but it tends to unwise self-satisfaction and to blind the observer to the merits of the steadier advances which are being continually made by the less wholesale methods pursued in Europe—advances to which the American engineer is glad to turn for lessons, when the inevitable time comes that something more than the cream of the business must be worked to maintain profits. The American engineer owes the large direct connected dynamo to the labors of his European confreres, and at least two of the largest central stations in the United States are now availing themselves of European experience in the use of storage batteries as an adjunct to central station plants. In a station load curve there will always be one summit much higher than the others—generally about 6 p. m. The first object of every central station is to add customers till the height of this summit equals the full capacity of the plant. But during most of the running time the load is much less than the maximum, and at these times the surplus power may charge an auxiliary storage plant. This plant may be used to carry an increased maximum load, beyond the capacity of the dynamos and steam plant. In this case the investment, depreciation and expense of storage plant required would be compared with the same items for additional generating plant to do the same work. Or the storage plant may be used to carry the whole load of the station during the light load period, allowing the generating plant to be shut down altogether, saving wages and fuel, &c. For such a use as this last the interest and depreciation on battery plant investment would be compared with the running expenses saved. In the case of smaller stations which run only at night, a day load might be carried by an auxiliary battery plant, which could be charged between midnight and morning. Such uses of the storage battery have been common in European practice for years, but it is only recently that they have been applied to central stations in America. The largest plant of the kind in the United States is that put in last spring by the Edison Illuminating Co., of Boston, and it is now reported that a large increase of this plant has been ordered.

IT is a noticeable characteristic of our neighbors in the United States that they suffer themselves to be inconvenienced or imposed upon in their collective capacity, by any individual or corporation of sufficient impudence to profit by their national weakness in resisting organized extortion. We say weakness advisably, for although their forbearance is no doubt partly due to their equally noticeable national good naturedness and sense of humour, it has its roots in the eclipse of that sturdy civic virtue which resists any oppression on principle, even though resistance involves greater temporary individual inconvenience than would submission. A good nature which finds sufficient general relief from continual imposition by making a joke of it, is not so estimable as it appears if the imposition is no joke to the humble or weaker members of the community, whose unheard blessings would fall upon any "crank" or "kicker" who would refuse to see the joke. Amongst our near relations across the Atlantic, this sturdy civic virtue shines with possibly too undimmed radiance. We in Canada are, in this respect, certainly more like our neighbors than our relations, but we have from the latter a wholesome leaven which causes us to rise more quickly than the former. The leaven has been working in Montreal recently, where an unregarded civic ordinance against over crowding the street cars was invoked by a public-spirited individual, and the Street Railway Co. was fined $1.00 and costs in two cases. This drew forth a plaintive protest from the company that they really could not help it. People would crowd on in "a most discourteous fashion." Their conductors could not be expected to "beat them off with a club, or take them by the collar and jerk them off into the street." People in Montreal are so unruly and so bent on making themselves uncomfortable that if the ordinance is to be enforced, the company "will have to engage and train sluggers and bullies" to keep the people from crowding. This sort of nonsense is simply adding insult to injury. People are not such fools as either to do or believe such things. In the whole letter there is only one semblance of an argument. It says, "extra cars do not meet the requirements, as everybody wishes to get on the first car and will not wait for the next, which is only half filled." Of course they will not, when they have no assurance that the next car will not be just as bad as the first. Let people know that they must wait until they can get the seat they pay for, and they will wait. If there are not enough cars to handle the business, people will walk off with their fares in their own pocket, instead of in the company's, and then kick till a proper number of cars are put on. This is the true reason of the wail from which we have quoted. In England the licensed number of passengers is posted in the car, and when that number is complete, the car stops only to let people off. Do the same thing here and fine the company for every extra passenger carried, and there will be no more tears over the gentle feebleness of the conductors and the untamable fierceness of the passengers. Perhaps, even, in time, the passengers will learn that when they pay their fare they pay for the right to get and keep a seat, and not for the privilege of exercising their politeness by giving up a seat to a lady, who herself ought not to be subjected to the crowding and jostling from which even a seat will not shield her. There are emergencies,

times of extraordinary traffic, or unavoidably irregular cars. At such times it may be in the public interest to permit overcrowding. Let the police magistrates have power to dismiss charges of overcrowding when they think it excusable. This would avoid all hardships to the railways from a cast-iron rule, and the public will be content to leave its interests to the magistrates, instead of as at present in the hands of the other interested party, the street railway itself. If the corporation whose attitude towards the public meets with condemnation, cannot find ground upon which it can defend its position and policy, it had better attempt no defence.

## THE BELL TELEPHONE COMPANY'S NEW EXCHANGE AT OTTAWA.

IN the December number of THE ELECTRICAL NEWS some particulars were published concerning the Bell Telephone Company's new exchange at Ottawa. The following additional facts in connection therewith have since come to hand :—

The new Ottawa building is 80 feet deep, has a frontage of 32 feet and is 3½ stories high. The basement and first story have a brown stone front ; the upper stories are brick with brown stone trimmings ; the woodwork inside is of clear pine with natural wood finish.

The rear of the basement is used for the linemen's quarters, store room for line material, also hot water heating apparatus ; the front part and all of the first floor is rented for offices.

On the second floor are the manager's office, general office, and long distance tool office, operators' cloak room, battery room and instrument store room. The top floor is one large room the full size of the building, and is used entirely for the switchboard and distributing rack.

The switchboard consists of 14 sections of improved 100-point metallic standard switches. These switches are set in the middle of the room to allow a multiple switch being put in when necessary, without having to move the present board. The distributing rack is made of iron, and is known as the Ford or "number four" rack, having sneak and carbon arresters attached, which are known as the "number seven" arrester.

All the lines enter the building through an underground system. The lead cables are spliced in the basement to Akonite cables, which extend to the top floor where the cables are formed up and connected to the arrester ; this method does away with the iron terminal generally used, and saves considerable valuable floor space.

There are over 4,000 lineal feet of conduit in the underground systems, having a capacity for over 1,800 miles of wire. There are at present about 575 miles of wire in use underground, and 200 miles of wire in aerial cables ; the latter cables are spliced to the underground cables in the nearest manhole to the lead they are on.

The ultimate capacity of the exchange and underground system is 3,000 subscribers ; there are about 1,100 at present.

The return ground system is used on the pole routes ; the rest of the system is metallic. The return wires of all the pairs in the cables are connected in the return ground at the cable poles, which makes the entire system nearly as good as straight metallic circuits.

The change from the old to the new system was successfully made on the evening of November 7th, in about three minutes.

The City Council of Halifax have threatened to ask the Legislature to cancel the charter of the Halifax street railway on the ground of inefficient service. The solicitors of the company state that representatives of the various interests concerned in the road are conferring together with the expected result that the disputes and litigation between them will be amicably settled and the electric substituted for the present unsatisfactory service in Halifax. The Council have deferred action awaiting the outcome of the present negotiations.

The New Westminster and Burrard Inlet Telephone Co. has purchased the Postal Telegraph Co.'s line between Vancouver and Seattle, W.T., which has been converted into a telephone line. The line, which is 160 miles in length, is now in first-class working order for telephonic purposes. Besides Seattle, Tacoma and other Sound cities will shortly be connected, while public offices have been established at the following places : Westminster, Clover Valley, Blaine, West Ferndale, Whatcom, Fairhaven, Samish, Brownsville, Wash., Mount Vernon, Stanwood, Walker's, Maryville, Everett, Snohomish and Botholl.

## CHARACTER SKETCH.

### WM. T. JENNINGS, C.E.

*" Because I have neglected nothing."—Nicholas Poussin.*

IT does appear, if one takes a broad view of the world's history, either from the standpoint of the philanthropist or that of the man of business, that greater credit to the world's progress cannot be given to any class of men, than to those whose vocations have led them to open out uninhabitable sections of country and by an iron band to belt continents and connect far distant districts with each other.

So far as the Dominion of Canada is concerned, one of the men, who has played an active part in the construction of its public works, is Mr. Wm. T. Jennings, who was born in Toronto, May 19th, 1846. Mr. Jennings was educated at the Model Grammar School and Upper Canada College in his native city, and commenced his professional career as an engineer in 1869. He was then under Mr. Molesworth, and his first work was to survey the swamp lands of Grey and Bruce for drainage improvements. From 1870 till 1875 he was on the engineering staff of the Great Western Railway, which he left to enter the service of the Dominion Government. From that time forward Mr. Jennings' chief work, perhaps, has been in the line of railroad construction. Several important surveys and construction works in the Northwest, British Columbia and other parts of the Canadian Pacific Railway were made by him while in the service of the Government, the Construction Co., and the C.P.R. Co. From 1886 to 1890 he had charge of the surveys and constructions for the C.P. R. in Ontario, and early in 1890 was appointed Engineer for the City of Toronto.

What his activities amounted to during his years of railroad construction find illustration, in a measure, in the Canadian Pacific and the success that has attended this railway in opening out the extensive and fertile northwest territories, and proving the medium of spanning the Dominion out to the Pacific coast.

Whatever work Mr. Jennings has undertaken has been marked with thoroughness and complete mastery of his profession. Nowhere was this more apparent than during the two years he occupied the position of City Engineer in Toronto. The basis of the present arrangement between the city and the Street Railway Co. had its origin with

WM. T. JENNINGS, C.E.

Mr. Jennings, and not only the excellent character of the road, but also the revenue the city receives from it in the shape of percentages and mileage tax was the thought of the late City Engineer. Toronto is proud of its splendidly paved streets, and this work was planned by Mr. Jennings during his occupancy of the office of City Engineer. It was a special study with him to give to Toronto a system of pavements that should not alone be attractive as public thoroughfares, but that would possess endurance and lasting qualities. Negotiations between the city and Bell Telephone Co., which resulted satisfactorily for Toronto, were also undertaken while Mr. Jennings was in the service of the city. He had planned a system of underground wires in connection with the electric light and telephone contracts, and had he remained in office would likely have brought these plans to completion. During these two years the esplanade matter came up for consideration, and Mr. Jennings had to fight with his known determination the strong railroad corporations.

His official connection with the city did not cover as great a length of time as citizens, desirous of the well-being of Toronto, would have liked. The fact is that Mr. Jennings was too independent and fearless an officer to suit many who at that time occupied the position of aldermen. They quickly learned that he was a man who was master of his business and could not be dictated to, or used by those who had their own little schemes to carry out. His wide experience, outside of that hemmed in by the bounds of a single municipality, had given him a thorough knowledge of human nature, and contractors, as well as aldermen, understood that they

were dealing with a man who would permit of no injustice, much less anything approaching crookedness. Of his own free will, and to the regret of the better elements in the council, after occupying the position for about two years, he resigned, owing to the aldermanic body breaking faith with him by changing the by-law under which he took office.

The past two years of Mr. Jennings professional career have been employed, to a large extent, in the construction of various short lines of railway throughout the province. He has come actively to the front as the engineer of several electric roads recently completed, and that are likely to be the forerunner of many others. He points with natural pride to the Niagara Falls Park and River Railway, in the construction of which he was chief engineer. He served the new Hamilton and Grimsby Electric Railway as consulting engineer, and the line extending from Galt to Preston was built under his superintendence. During the past twenty years the various undertakings with which Mr. Jennings has been connected, represent a total of not less than $35,000,000. On another page of the ELECTRICAL NEWS is discussed the question of short line electric roads suggested by Mr. Jennings' experience and observations on this phase of railroad building.

Personally Mr. Jennings is popular with all who have his acquaintance or friendship. He is ever ready, out of his wide knowledge of public affairs, to impart information to those honestly wanting information ; at the same time he is too busy to rest comfortably under the intrusion of those who hold to a person in his position much the same relationship that the editorial bore does to the journalist in his sanctum. His career furnishes a capital illustration of the saying of Owen Feltham : "That man is but of the lower part of the world that is not brought up to business and affairs."

Mr. Jennings is a member of the Canadian Society of Civil Engineers, the Institution of Civil Engineers, the American Society of Civil Engineers and the American Association for the Advancement of Science. In religion he is a Presbyterian, being a son of the late Rev. John Jennings, D. D., for many years pastor of the Bay St. United Presbyterian Church, in his time one of the best known and ablest clergymen of Toronto.

### PUBLICATIONS.

"The President and Directors of the Canadian General Electric Company wish you a Happy and Prosperous New Year" is the kindly inscription borne by a card addressed by the Canadian General Electric Company to their friends. The ELECTRICAL NEWS heartily reciprocates the cheery sentiment.

An attractively gotten up card reached us the other day, on which was printed the following : "All comparisons are odious, yet we must say that, among our customers, ONE only has first place in our esteem. We have (at great expense) secured a portrait (in colors) of our honored friend. WE PRESENT IT TO YOU AS A MODEL. Study the features ; see in this genial face how CANDOR, JOY, MIRTH and BEAUTY are tempered with WISDOM, JUSTICE AND PEACE. Merit demands recognition. What we say of the original of this apotheosis of art is only justice. Our New Year's wish is, may pleasant relations long continue between our esteemed friend and yours truly." This is followed by the name of the American Electrical Works, Providence, R. I., and its various branches, including the Eugene F. Phillips Electrical Works, Montreal. The reader wonders what it is all about, and is prompted by curiosity to unfasten the little envelope attached to the center of the card, when his own face reflected in a tiny mirror reveals to him the secret. This is decidedly the most original New Year souvenir that has come under our notice.

The Seaforth Electric Light Co. are considering the advisability of extending their lighting system to Egmondville.

It is said to be the intention of Mr. J. Ashley, to establish at Gananoque, Ont., works for the manufacture of electrical machinery.

The citizens of Stratford will vote on a by-law on the 7th inst., to authorize the council to expend the sum of $16,000 on a municipal lighting plant.

It is reported that Mr. S. R. Break, Manager of the London, Ont., Street Railway Co., has been appointed Assistant Secretary of the Detroit Street Railway Co. Mr. Break enjoys the well-deserved respect of the citizens of London, and his removal from that city will be much regretted.

## CENTRAL STATION TYPES.

### By Geo. White-Fraser.

### NO. 2—SMALL MUNICIPAL PLANT, ORILLIA.

MUNICIPAL versus Private Ownership of Electric Lighting Plants is a question which undoubtedly has two sides, both of which have had a good deal of consideration ; and a quantity of statistics on the subject have been collected, chiefly by the champions of private ownership. The writer has followed the discussion throughout, and has had the pleasure of hearing Mr. Francisco speak on it personally, and cannot help being struck by the fact that all the arguments against the economy and expediency of municipal ownership seem to be founded on the *possibility* of aldermanic corruption ; the *possibility* of political partizanship influencing appointments ; and the *possibility* of the same undue influence being excited to keep in office incompetent persons to the detriment of economical management. Mr. Francisco and other writers draw attention to the reports from several municipal plants in the States, in which electric lighting is said to have cost nothing, or some absurdly small figure ; and in refutation of the implied argument in favor of municipal economy, states that in all these cases the salaries of the engineers, the cost of the coal, &c., have been put down to the waterworks or some other municipal department, so as to show a good economy for the electric lighting account.       •

This expose is by many taken as a proof that, for the towns cited, municipal lighting is a failure as regards economy, but while it certainly throws light on the eccentricities of municipal book-keeping, in the absence of any definite data, it still leaves open the question of the economy of municipal lighting, because expensive lighting is not a necessary consequence of poor accounting. In fact if the waterworks account is thereby not so greatly increased as to rouse an outcry, it is obvious that the two municipal departments work together more efficiently than either of them singly. For in all large towns, the engineering staff that is required for the pumping machinery can equally well attend to the electric light engines ; one thoroughly competent chief engineer is sufficient for both plants ; one accounting department controls the finances of both. The mere fact of the waterworks department being able to shoulder the electric lighting expenses, itself might be cited as one illustration of the practical advantages of municipal control. It seems to the writer that the question of corporate management of electric lighting should not be regarded simply *per se*. Whether a city arc lamp costs a little more or less than a private one is no doubt of great interest, but of itself it proves nothing except that the City Superintendent is or is not fully competent to handle electricity. The question is : " How do the various departments affect each other as regards *efficiency* and *general* economy ? " Is the whole level of efficiency raised or lowered by the addition of an electric department ? Obviously it is raised. The whole engineering and business staff must benefit by the accession of another official who, as Lord Kelvin timely says, must by profession be as much of a mechanical as of an electrical engineer. It is impossible to draw a clear line between civil, mechanical and electrical engineering, as these branches of the profession overlap in many places, so it may well be that in municipal affairs, as in others, three heads are better than one. To follow the same line of argument a little further : It needs not much demonstration to prove that the book-keeping as well as the engineering staff will be working at higher efficiency, and therefore relatively more economically. Then as to the possibility of " boodle." It stands to reason that any unnecessary expenditure on behalf of " boodle account," must be in the first cost of the plant, and therefore can only increase the yearly operating expenses by the interest on the excess. This interest, divided by the number of lamps, is not going to very appreciably raise the yearly cost per lamp, because the amount of boodle that a plant can stand is limited, and bears a certain loose proportion to its capacity. There is no really valid reason why political partizanship should not influence civic appointments, so that the whole argument seems to reduce itself to a weighing of facts on the one side and possibilities on the other. The writer is not extenuating bribery and corruption, but merely taking things as they seem to be and considering their practical bearing. The question, however, should not be regarded as a whole and so attacked. If carefully considered, it seems to divide up naturally into portions which have their own peculiar features. The general efficiency of a plant increases, as a rule, with its size and the nature of its service ; so that in a large city electric lighting in connection with an electric railway may quite possibly be more cheaply done by a private plant than as a civic enterprise. And in towns of any size, where a reasonable sum has been paid, or considerable responsibilities undertaken by a private company in return for a franchise, it seems almost a breach of faith for the civic authorities to undertake their own lighting, unless of course the sum charged by such private company be exorbitant.

As communities become smaller, the arguments against their undertaking their own lighting seem to have less and less practical application, and a limit is soon reached when the question becomes one, not of expediency, but almost of necessity. The operating expenses of an electric station can be placed under two heads : fixed and variable. The latter—coal, carbons, repairs, &c.—decrease directly in proportion to the size of the plant ; the former—salaries of electrician, engineers, clerks, linemen, &c.—decrease to a certain extent in the same proportion, but eventually arrive at a point where further reductions can only be made at the cost of imperilling the life of the plant. I assume, of course, that the fallacy that electrical machines and good engines require no experienced care, is quite exploded. Statistics show that in all successful stations the fixed expenses are to the variable, in the proportion of, roughly, 2 to 1. It is therefore quite evident that a private plant that has to pay dividends as well as operating expenses, requires a certain minimum amount of business for its support, and small towns therefore seem to have no alternative but to do their own lighting, or go without. Between these alternatives they of course will choose for themselves, but as the very large majority of towns in Canada are below the " business limit," it will be interesting to watch what results are worked out in one or two typical cases.

There is yet again the case of the small town, which, although beyond the size limit, and therefore offering a fair field for private enterprise, is yet sufficiently near to it to make investors cautious, although such a community may reasonably undertake its own lighting. The problem is greatly simplified if it has already decided the general question of municipal ownership, by owning its waterworks ; for then the two can be operated together under almost the same management, by the same boilers, the same engineers and help, in the same building, and the employees of the two departments can be almost interchangeable. As a result of this system, not only will light be cheap, but water will also cost less, relatively, and there will be a greater number of experienced practical engineers employed around the works, to the obvious benefit of all the machinery.

It would be easy to point out the many direct and indirect advantages accruing from such combination of departments, from the standpoint of pure theory ; but a citation of actual results obtained in one typical case will be much more convincing, and is not far to seek. Orillia is a town of about 4,700 population, and may therefore be taken as representing the class above referred to. To the Mayor, Mr. Geo. Thomson, is due the credit of having combined two municipal enterprises—waterworks and electric lighting—with street and private. There always had been arc-lighting for the streets, but as the people desired house lights too, and there were no offers from private individuals, the matter was at once taken in hand by the Council, with the result that, for compactness and economy, Orillia has a little plant that seems hard to beat. To fully appreciate its advantages, it is well to state that the waterworks plant had been in operation for some time previously, requiring one boiler, one pumping engine, one fireman, one engineer—all working during a part of each 24 hours, under the management of a committee of the Council. The electric light plant was installed in the waterworks building, and required extra—one boiler, two engines and dynamos, one fireman, one engineer, also all working part of the 24 hours, and managed by the same committee. The extra work thrown upon the town clerk is said to be " great," perhaps may necessitate the employment of an extra clerk. So far the advantages may not be evident, but every little consideration will show that the electric light and the waterworks service allow of each boiler working almost continuously for 24 hours a day, while the duplication of boilers gives a good chance for thoroughly cleaning and overhauling them alternately. The two engineers can take night and day spell, week about alternately ;

the one " off" for the day can look after general repairs, cleaning, &c., and so the plant is better looked after. The two firemen also relieve each other week about, the one who is " off" during the day doing all the line work—trimming the arc lamps, of which there are 40—and any little repairs. The electric lighting books, being under the charge of the town clerk, add nothing to the municipal expenses, and even if an extra clerk is necessary, the probability is that his time will be divided between the electric lighting and the general municipal books. The whole municipal accounting, therefore, will undoubtedly be better and more expeditious, and although the price at which lights are sold is very moderate, the income derived therefrom will be sufficient to secure first class advice as to general management.

Evidence of proper attention to business principles is to be seen in every detail of the combined plant. Excellent machinery was purchased in the first place—general electric alternating dynamos and apparatus, Wheelock engines, Northey condenser—and installed properly by the various makers, who were thereby made to prove their guarantees. Well qualified engineers were engaged to run it, and given fair salaries. The wisdom of this policy is often not appreciated (impractical employers preferring cheap men to expensive good men, and putting down the greater cost of repairs to faulty machinery), but is at once made plain in this plant, not only by its thoroughly workmanlike appearance, but by the arrangements made for keeping it in perfect working order. The engines are periodically "indicated"; boilers examined and properly cleaned ; temperature taken of feed and condensing waters ; arc lamps brought in, in turn, off the line for cleaning and repairs, and many other little attentions paid that second rate engineers would neither think of nor be competent to undertake, but which greatly conduce to long life and efficiency.

It is greatly to be regretted that no station records are kept which would enable the cost of production per K.W.H. to be

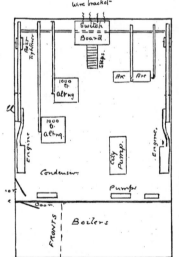

known, but figures are given below which will prove of great interest to many station men, and which certainly show a most satisfactory result. The accompanying diagram gives the general arrangement of the plant. The switchboard being on a raised platform, is a very convenient and sightly feature ; while the pumps, being in the engine room, are thus under the engineer's immediate supervision. Another excellent plan : The two

halves of the shafting are connected by a clutch, which permits of any desired combination of engine and dynamo.

The electric light employees are, with their salaries : One engineer at $700 per annum ; one fireman at $1.25 per day. It has already been explained that the two firemen take line work week about, with their firing shift, so the expense of a trimmer is saved ; and as the cost of installing lamps in houses is paid for by the consumer, outside help can be employed for that purpose, entailing no expense on the station. Wood for fuel costs $1.75 per cord, and is consumed at the rate of 5 cords per night in winter and 3 in summer. The operating expenses of the entire plant, arc and incandescent, are, therefore, taking no account of such items as oil, waste, repairs, &c., of which no data are given :

| | |
|---|---:|
| Engineer at $700 | $ 700 |
| Fireman at $1.25 per day | 460 |
| Fuel (7 summer and 5 winter months) | 2,435 |
| Carbon estimate | 570 |
| **Total** | **$4,165** |

The capacity of the incandescent plant is 2,000 16 C.P. lamps, of which nearly 1,400 are already installed and running ; and 40 2,000 C.P. arcs. The prices per lamp vary from 30c. per month for residences to 80c. for hotels, being based on the number of hours of illumination ; and taking one-third of the number at 60c., and the remainder at 30c., which certainly allows a margin, gives a yearly income of $5,468.22. This leaves a balance of $1,303, against which must be placed the miscellaneous items, unspecified lamp renewals being at the consumer's expense. This balance will remain a net profit, and will probably be enough to pay the salaries of the waterworks engineer and fireman. Or it can, if preferred, be drawn upon to pay interest on the bonds sold to provide the cost of the electric installation. Then it would appear, that working the incandescent plant at two-thirds of its rated capacity, brings in a sufficient income to defray its own expenses, pay off its own debt, and do the town lighting *free ;* besides all the indirect advantages indicated above. It is perfectly obvious that the remaining third of the capacity will bring in a net income that may be used, as municipal income, in streets or public buildings, market house, &c., and will be a still further advantage to the public. The writer believes that the advantages of municipal ownership, as regarded from the purely theoretical standpoint, are fully realized in the above case, and are equally attainable in every other case when accompanied by caution, enterprise and public spirit.

## THE WELSBACH GAS BURNER.

THE process employed in manufacturing the mantles used in the Welsbach incandescent gas burners, which are now so well known in London, is one of considerable interest and delicacy. These mantles, upon which depends the distinctive character of the Welsbach light, are a sort of elongated hood, which hangs over the burner and becomes brilliantly incandescent in the heat of the flame. They are made of fine cotton netting, carefully washed in order to render it chemically clean. The first step is to soak these cotton mantles in a complex fluid containing salts of a number of earthy metals. They are then dried, and, after undergoing one or two intermediate operations, are strongly heated in a Bunsen flame, by which the cotton is burnt away. The skins or films which are left behind consist of the substances dissolved in the fluid just mentioned, and are so fragile as to crumble to dust at a touch. To make them sufficiently strong to bear handling they are dipped in collodion, and after being trimmed with scissors wetted with methylated spirit, are packed in boxes for distribution. The light obtained as a result of this treatment is comparatively rich in actinic rays, so that it is quite possible to take good photographs by means of it, and it is claimed that less gas is consumed to produce a given amount of light than in other forms of burner, and that less heat is developed. The company which owns the patents for this country has lately brought out a novel form of glass chimney, which consists of a ring of elliptical glass rods held in position by brass bands at the top and bottom. This is stated to be less liable to break than the ordinary chimney, and moreover, to yield a more diffused light.

It is proposed to extend the Montreal Park & Island Railway to the village of St. Laurent.

## THE CANADIAN GENERAL ELECTRIC COMPANY'S WORKS AT PETERBORO'

THE Peterboro' factories of the Canadian General Electric Co. (Ltd.) are in extent by far the largest in the Dominion, and for completeness of appointment and variety of output are probably unsurpassed in America. An examination of the ground plan of the works as given below reveals at a glance the great care in arrangement and minute attention to detail with which every possible facility has been provided for handling the immense output of the shops through the various stages of manufacture from its entrance as raw material to its final shipment as finished apparatus.

The factories are located in an admirably situated enclosure of some forty acres, in the south-west corner of the town. Switches from the main track of the Grand Trunk and C. P. R. lines extend through the grounds and provide for the rapid and economical handling of all materials used. The three hundred horse-power required for the different shops is furnished by individual motors conveniently located, receiving their current from an isolated power house, which is in itself a model of central station construction. The three principal buildings, the main machine shop, the tube and carpentering shop and the wire shop, have an aggregate floor space of some 75,000 square feet,

on a scale which should for some time at least meet all possible developments of the business. In the wire works provision is made for turning out insulated wire of every description from standard weatherproof or C. C. rubber covered to flexible cord or armoured cable.

A recent departure of the company is into the field of car building, and the cars so far completed show a solidity of construction combined with quiet elegance of finish which it would be difficult to excel.

In so brief a sketch as this, it is impossible to do more than touch on the many and interesting features presented by the factories of the Canadian General Electric Co., but to anyone interested in the enormous development of electrical industries in the last few years, a day's visit to the company's works at Peterboro' will prove an instructive and fruitful experience.

## QUESTIONS AND ANSWERS.

W. R. R., Stayner, Ont., writes : 1. What is the maximum voltage of electric current that a Blake telephone transmitter will stand without effect on its action ? 2. Why will it not stand any amount of current ? 3. What is meant by a multiphase alternating dynamo ?

ANS. 1.—The best results are obtained from a Blake transmitter

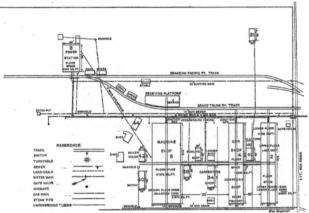

GROUND PLAN OF GENERAL ELECTRIC COMPANY'S WORKS, PETERBORO', ONT.

and there are besides some fifteen smaller buildings devoted to various purposes.

As might naturally be expected the most striking feature of these factories is the great variety of the work required to be be turned out. A recent visit to Peterboro' showed in a hurried run through the works over how wide a range of work their operations actually extended. On the main floor of the machine shop, were in course of construction at the same time the three huge 1,000 horse power double unit generators for the Montreal Street Railway Co.; incandescent dynamos, alternating and direct current, coming through literally by the dozen ; "Wood" arc apparatus, which the company has adopted as standard ; mining apparatus, drills, generators and locomotives ; street railway motors of the well known C. G. E. 800 type ; direct connected units complete with their engines ; transformers; snow sweepers, etc. The galleries are devoted to the manufacture of the smaller parts of the heavy machinery and to the lighter class of apparatus and supplies: Type "K" controllers ; Thomson recording watt-meters ; indicating instruments of all kinds ; arc lamps ; sockets and lamp bases ; porcelain rosettes and cutouts, and the thousand and one other minutiae which go to make up a complete electrical system. In the lamp works, recently removed from Hamilton, the different departments are thoroughly equipped for the manufacture of incandescent lamps of the most approved type, and

when the current in the primary circuit is from 1-2 amp. A pressure of from 1-2 volts will cause such a current to flow through an ordinary Blake transmitter and induction coil. 2. If the voltage is increased more current will flow, the carbon button at the point of contact with the platinum spring will be burnt, thus blurring the articulation. If still more current is sent through, the insulation of the wire of the induction coil will be destroyed and possibly the wire itself fused. 3. A dynamo supplying several circuits with alternating current, of the same frequency on all circuits, but differing in phase by predetermined amounts in different circuits.

W. B. S., Montreal, writes : I note L. O' C.'s remarks re broken wires near commutator, and may say that I have had similar trouble with a motor having a Gramme ring armature which was manufactured by a United States firm. I tried the effect of soldering on a piece of heavier wire and leading it to the commutator, but even this broke in the middle, the soldering of the pieces both at armature wire end and commutator end remaining intact. As I made sure that the motor was right every other way, I fear there must be some additional cause as well as vibration to account for this, and would (like friend L. O' C.) be glad of enlightenment.

ANS.—Staying the leads by weaving them together is a better preventative than stiffening them individually with an auxiliary wire. The latter plan by the added weight to the wire may sometimes increase the tendency to injurious vibration. Staying by weaving is generally found to overcome the trouble.

## DIRECT CURRENT MOTOR AND DYNAMO DESIGN.*—I.

### BY GANO S. DUNN.

I wish to refer to several features of direct current motor and generator design, and not to pretend to cover the whole subject, and I trust this will not be without interest to you.

The first point I wish to bring out is how very imperfect are our means of rating electrical machinery. We have good means of rating other machinery, but we do not have good means of rating electrical machinery. The steam engine is rated by the diameter of its cylinder and length of stroke and its speed. It is not sold by horse power, because an engine which could develop 100 horse power 3-10ths cut-off would develop 200 horse power at 6-10ths cut-off. In electricity the problem is similar, but we have no way of rating motors by the diameter of cylinder and length of stroke as they have in steam engine practice, and we are continually suffering on this account. Water wheels are not sold by horse power. You do not buy a 50 horse power water wheel ; you buy a 28-inch water wheel and run it at so many feet head, and then figure out how much power you can get from it, and if is not enough, you buy a 30-inch wheel.

The limits of loading an electric motor are really two only ; namely, you have reached the load of a motor when it begins to heat so much that it is dangerous to continue to run it, and you have also reached its limit of load when it sparks so badly that if you run it longer it will destroy its commutator. There are builders to-day who build a motor which will have a very high efficiency at one-third load, but which will spark if it is overloaded, and there are other makers who build a machine which

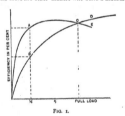

FIG. 1.

has not a high efficiency at light loads, but which will stand three and four times normal load without sparking. Now both of those motors are listed in the catalogues, are charged for and are known as, let us say, 10 horse power motors.

If you buy a 10 horse power motor without knowing something about these facts—it is possible that you will be very much disappointed. If it is agreed that we should call a motor that runs at about 6 horse power, but is capable of running at 10 horse power continuously without dangerous overheating, but will never be called upon for more than 10 horse power,—if we should agree to call such a motor a 10 horse power motor, very well. But then we ought to call the motor which is used in crane service, which is called upon for 30 or 40 horse power, for periods of ten minutes at intervals, something more than, or different from, 10 horse power, and the builder ought not to be obliged to make one motor fill one user's requirements and another user's requirements, when their requirements differ so much. Suppose we sell two 10 horse power motors to ordinary customers. We may hear a report from one of them that this motor is not a good motor ; it does not give 85 per cent. efficiency at one-third load. While the other one will say this motor is not a good motor ; it sparks very badly when loaded to 300 per cent. of its normal load for a few moments. In view of these facts horse power is not a fair way to rate machines.

The lines between these kinds of motors are not so distinctly drawn as I have pointed out, but for the purposes of illustration I have shown here two curves, A and B, Fig. 1. A is the curve of a motor of the kind first mentioned above. It is expected to run at 6 horse power average, can give 10 horse power all day long if you require it, but must not give any more, and has a

*A lecture delivered before the New York Electrical Society, November 27, 1894.

very high efficiency at light loads. Such a motor is an ordinary power motor for running a printing or machine shop. The other motor, whose curve is shown at B, is a motor whose efficiency at light loads is very low ; it is not much more than half of the former. But its efficiency at full load is exactly equal, and at overload is even higher than this other motor. Motor B is much superior for its work to the former motor, because whenever it does run it runs at three times its rated load, and motor A, whenever it runs, as I will show later, runs at only one-third its rated load. Put motor B on the load that A is built for, and A on the load that B is built for, which is what you are likely to do, and you will not secure the best results.

This A curve machine is the type of ordinary power motor running machinery by a belt. The B curve machine is a type of railway motor or motor for operating cranes, or for other intermittent and very heavy work. These two conditions affect the design of a motor very much, and the way they affect it is this : At the point C both motors have the same efficiency. That means that the losses in both motors are equal. But why, then, does curve A differ from B ? The difference is in the distribution of the losses. The losses in a motor are of two kinds, fixed and variable. The fixed losses are the losses due to field current,

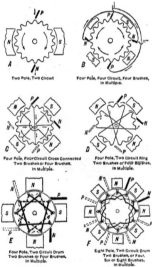

Two Pole, Two Circuit

Four Pole, Four Circuit, Four Brushes, in Multiple.

Four Pole, Four Circuit Cross Connected Two Brushes or Four Brushes, in Multiple.

Four Pole, Two Circuit Ring Two Brushes or Four Brushes, in Multiple.

Four Pole, Two Circuit Drum Two Brushes or Four Brushes, in Multiple.

Eight Pole, Two Circuit Drum Two Brushes, or Four, Six or Eight Brushes in Multiple.

FIG. 2.

hysteresis and eddy currents, brush friction and bearing friction. The variable losses are the losses due to armature resistance and the losses due to commutation. Now motor A is a machine with a small magnetic circuit using comparatively few lines of force. Therefore the wire that is wound around the magnetic circuit to keep it excited, need not be very great, and therefore will take but a small amount of current to energize it. The iron in the armature will be small in quantity, and therefore the hysteresis and eddy currents will be low, and the bearings will be rather light because this motor is never called upon for more than rated load. Therefore the fixed losses are light. The variable losses, such as armature resistance and commutation losses, may be high in this machine without detriment. The reverse of these conditions is true in machine B. These points are of great importance, but they have not received attention.

As a result of tests made on about 200 power motors, at the instance of Mr. H. L. Lufkin of the Crocker-Wheeler Co., it was found that these 200 motors on actual commercial circuits did not average one-third of their load. The readings were taken in this manner. It is customary wherever a motor is installed to put in a meter that measures the current taken. Now if this were a 10 horse power motor and ran 10 hours per day, it ought to have 100 horse power hours for every day it runs. The meter readings were taken and the average was between 25 and 30 per cent. of the power that the motor could have given according to what was stamped on its name plate. Now if we build a motor with the kind of an efficiency curve B, and use it for a purpose such as I have just spoken of, where a motor averages only one-third load, it is practically equivalent to taking a motor of only one-half the efficiency.

Another thing with regard to the rating of motors is the speed. Other things being equal, a high speed motor is cheaper to build than a low speed one. It is a smaller machine operated at greater activity. If brought to the same speed it would have less power. A motor, to be compared carefully, ought to have these points determined :

1. At what portion of its load does it commence to spark?
2. At what portion of its load will its temperature rise abnormally?
3. What is its efficiency at various points?
4. At what speed does it run?

Reduce all these things to a common basis and you have a fair method of comparing motors.

Too much importance has heretofore been attached to full load efficiencies. Motor engineers boast that their machines have 90 per cent. efficiency, and some other maker's machines have only 88 per cent. efficiency. They speak of the highest efficiency that the machine is capable of, or of the efficiency of which they are most proud. Now that may not be the efficiency which the customer wants to know. There are many cases, where a motor of 85 per cent. efficiency would take less current

FIG. 3.

to run a printing office than a motor of 90 per cent. efficiency, for this reason : At point D, Fig. 1 is a motor of 90 per cent. efficiency. But the printing office runs at only about one-third the power of its motor, and would therefore realize the efficiency at B ; whereas at point E here is a motor which has 85 per cent. efficiency, and if the printer bought that he would realize the efficiency at A, which is much higher than that at B.

The next feature to which I wish to call your attention is the most recent methods of winding. I cannot go into this more than slightly, but it might be well to give some idea of what they are. In winding bi-polar machines, windings are all on the same principle ; they are two-circuit. As the current leaves the brush, in diagram A Fig. 2,* half goes down one side of the armature and half down the other and joins the bottom brush and goes out. Now there is not much room for modifications in that winding. It can be made ring or drum, and that is practically all. But when we come to the multi-polar machines there is a great variety of windings which we can use. If we imagine B, Fig. 2, to be a 4-pole machine, with an armature a ring in which the wire is wound around and around, then the current may be taken off this armature by brushes P, N, P, N. These are the field magnets of alternate polarity and each field-magnet generates a portion of the total current. This is called 4-circuit, because there are four circuits from which the current is collected in the armature.

The earliest multi-polar machines were wound in this fashion, but there are some objections to it. If one pole is stronger than

*From "The Practical Management of Dynamos and Motors," by F. B. Crocker & S. S. Wheeler.

its neighbor, then when it sends its current to be collected it will be under greater pressure than the one coming from the other pole, and if there is sufficient difference will neutralize it and send a reversed current through its winding. This causes heating, diminishes output and reduces the efficiency of the machine. Now it is very easy for one magnet to become weaker than another one in an actual machine. Suppose, for instance, that the bearings of the machine wear. The armature will settle down and will be nearer the bottom magnets than it will be to the top, and they will make stronger currents than the top ones will. Then again, if the windings on the magnets are not very carefully made, and one has a few more turns of wire than the other, it will be stronger, and there are a great many ways that the effect of inequality of electromotive forces generated under the poles will produce very bad results.

To overcome this, windings were invented which are called 2-circuit windings. The object of the 2-circuit winding is that the wire shown in D, Fig. 2, instead of going along always under the weak pole, between every section cuts across and goes under a strong pole and comes back, and the winding under the strong pole cuts across every other section to the weak, so that by the time the brush is reached, the windings have each had acting upon them both the weak and strong poles, and the result has been to make equal electromotive forces and produce no trouble.

In actual winding these connections are not made on the armature in this form of two-circuit winding, but are made inside the commutator. This commutator is of the kind shown in Fig. 3. Every bar is connected to the bar directly oppositeto it by a bird-wing form of connecter. The connecter is shaped in the following way : It starts down, goes over in the form of a bird-wing, goes inside about an inch, and after it has reached the under layer of windings it makes the rest of its circuit until it gets down to the bottom. The armature is wound the same as a 4-circuit armature, but when its ends are put into the commutator, these cross connecters have the effect of making the alternate poles generate part of the electromotive force, as shown in the diagram. The armature L have described is but one kind of a 2-circuit armature, but there are other kinds that are used just about as much. The other kind of armature is not wound in ring form, but is like that shown at E in Fig. 2. Now the armature I have just described was wound as a cell, which was connected to another on the other side. There will be no short circuiting due to unequal voltages. (A model of this kind of an armature was exhibited.) I will not describe farther the 2-circuit windings, since they are all on the same principle.

There is another advantage in the 2-circuit winding besides stopping the interaction of parts of the armature against itself ; and that is, that for very high voltage machines, if we had a 4-circuit armature as previously described in B, Fig. 2, the wire under each pole would have to generate full voltage. In a 500

FIG. 4.

volt machine, each pole would have to generate 500 volts, so that the little current it contributed would be at the same pressure as the other current. That necessitates fine wire. If there are only two circuits each pole contributes only half the voltage and the wire may be of twice the size, which is a very important consideration. When we get into very large machines, however, then the wire gets too big for us to handle in 2-circuit machines and we go back to the 4-circuit windings.

The next form of winding that I wish to describe is one in use by the General Electric Co. and known as a double winding. The reason for a double winding is this : It is a rule among electricians that the voltage between any two bars of the commutator must not be greater than, say, 20 volts. If it is much greater in a large machine, any little accident—a piece of carbon —would make a flash, and, once started, the whole commutator would blaze and there would be a short circuit and great trouble. To keep that down the voltage between the bars must be kept down low.

The voltage generated by a machine is proportional .to the size of the magnets and the number of the windings. Now as the magnets get large the windings need be fewer in order to generate the same voltage. As our machines get larger and our magnets get larger we only need a few turns of the heavy conductors on the armature to give us, say, 110 volts for lighting. We soon reach the point where, say, 10 turns on the armature would be enough to give us 110 volts. That would give us 10 sections to our commutator, and as half the sections are on one side, and half on the other, that would give us only five sections of the commutator among which to distribute 110 volts and that would be more than our limit. Now what are we to do? We cannot increase the number of turns because we could get more voltage than we wanted. We cannot decrease the size of the magnets because then we would have sparking and armature reaction and other troubles. There are two things for us to do; one is, to use the multi-polar machines, which have a number of small magnets each contributing to the total current, or to use the form of winding shown in Fig. 4. Now in this diagram there are ten coils and ten commutator bars belonging to each winding. When we use the windings in conjunction we have the voltage due to ten coils and the commutator bars due to twenty, and that is achieved in this manner. The windings are wound just as if they belonged to two different machines. One winding is entirely disconnected from the other and it generates, say, its 110 volts with its few commutator bars—too few to be used by itself. The other winding is entirely separate and it generates its 110 volts with few commutator bars—too few to be used by itself. But when we put these two windings together, then we have still 110 volts, but we have double the current, because each of the windings contributes some, and we have double the number of commutator bars. Now that is in reality just like two dynamos. The brushes are made so wide that they cover at least two of the commutator bars, so that each winding is free to contribute its share of the current, no matter what the position of the armature is. We have shown this for a bi-polar machine. But if we go into multi-polar machines we will soon reach a size where we are agin face to face with it, and we would have to increase our poles and would soon catch up with this difficulty again, and this method of double winding is an excellent one for overcoming it. It has also the advantage that the collection at the brush is better.

## A FEW TELEPHONE FIGURES.

There are now in operation in the United States alone more than half a million miles of telephone line, bringing into speaking relations over 250,000 telephonic subscribers, and employing in daily service over 600,000 telephones, by means of which 600,000,000 messages are transmitted annually. These figures, given on the authority of Mr. Arthur V. Abbott of the Chicago Telephone Company, most graphically portray the remarkable proportions which Prof. Bell's invention has assumed. The earliest application of the telephone necessitated a wire extending from each station to every other one with which communication was desired. How impractical a method this is, however, for covering a territory of any magnitude can be seen without much difficulty. In both New York and Chicago, for example, about 10,000 subscribers have telephonic communication. The most compact system of underground circuits needs about four square inches for every 100 lines, so that, to unite each of the 10,000 subscribers with the remaining 9,999 would require a space of more than a yard square simply to contain the necessary conductors. No present city street could afford the required room for the subways. If communication were thus attempted, each subscriber, according to Mr. Abbott's figures, would require nearly 200 miles of cable, and should the distribution be undertaken by means of aerial wires, pole lines 1,000 feet high would be required to accommodate the necessary circuits. It is to the impossible complexity of such a system, which became apparent even in the earliest telephonic days, that the telephone central station and the telephone switchboard owe their origin and development, accomplishing to-day results whose convenience and importance are all but lost sight of in the busy whirl of existence, and can perhaps be best appreciated only by comparison with the meagre telephone facilities of a dozen years ago.—Cassier's Magazine.

## SPARKS.

It is said to be the intention of the London Street Railway Co. to again ask permission from the Ontario Legislature to extend their lines to Spring bank.

The new power station of the Petrolea, Ont., Light, Heat & Power Co., will shortly be completed. The motive power will consist of a 125 h. p. engine and boiler.

The construction of an electric railway between the town of Collingwood and the village of Nottawa, is said to be receiving the favorable consideration of local capitalists.

The Central Telphone Co., Limited, has been incorporated with a capital of $3,000 to construct lines along the route of the Nova Scotia Central Railway, from Bridgewater to New Germany, N. S.

Mr. Myles, President of the Hamilton, Grimsby & Beamsville Electric Railway, is reported to have announced that the earnings of the road for the first two months of its operation reached upwards of $4,000.

The Seaforth Electric Light Co., which recently purchased the municipal lighting plant, and have since erected a power station, are operating 800 inc. and 75 arc. lights, and are said to have one of the best equipped plants in Ontario.

Mr. H. S. Thornberry, of the Toronto Electrical Works, would like to secure the address of manufacturers of wood cases for electrical bells and such like apparatus. It is a singular fact that be has not found in any of the electrical journals the name of a manufacturer in this line.

The St. John, N. B., Electric Railway Co. have threatened to sue the city for damages for having been compelled to desist running their cars until all the wires blown down by the recent storm were repaired. The company contended that some of their lines could be operated with safety under the existing circumstances.

The town of Richmond, Que., with a population of 3,000, has two electric light systems, two telegraph offices, two telephone exchanges, a burglar alarm detective line, and is to be the headquarters of the Richmond Water Power & Manufacturing Co. In this town, at least, it cannot be said that electricity is in its infancy.

Mr. H. W. W. Kent, manager of the N. W. and B. I. Co., has gone to the Kootenay country and will first establish an exchange at Three Forks. The different mines in the neighborhood will be connected. Nelson and Kaslo have exchanges at present and these will be inspected by Mr. Kent and possibly improved. He expects to be away a couple of months or so.

Mr. T. Viau, chief promoter of the proposed electric railway at Hull, Que., has made arrangements for the construction of a line of electric railway from that city to Aylmer and Gatineau Point. He is debarred from building a line to Chelsey and the Quyon. This, however, will not block the scheme. The charter for electric lighting and heating has met with no opposition in the Legislature. It is the intention to proceed with the construction of the system at an early date.

The Citizens' Light & Power Co., who for three years past have supplied electric light for Montreal Harbor and the suburbs of St. Henri and Cote St. Antoine, have erected a large new power station in St. Henri, in which is being installed a steam plant of 1,200 h. p. capacity. This plant will be used to operate the lights for the above-named corporations, as well as to furnish power for the operation of the electric railway which the Standard Light and Power Co. propose to construct to Lachine.

Mr. Wm. Bayley, a ratepayer of the city of Vancouver, recently obtained from the courts a rule nisi, calling upon the Corporation to show cause why the by-law for raising $100,000 for the purchase of an electric lighting plant, passed on the 8th of October last, should not be quashed on the ground that the by-law was never duly carried out by the ratepayers in accordance with the provisions of the statute, that it did not receive a three-fifths majority of the votes of the ratepayers, and, further, was ultra vires of the Corporation. The case turned on the point whether or not a three-fifths majority was necessary for the passage of the by-law. Mr. Justice Drake, before whom the case was argued, decided that the by-law is valid.

The officers of the Ottawa Electric Co. are as follows: Board of directors, Hon. F. Clemow, Hon. E H. Bronson, T. Ahearn, J. W. McRae, C. Berkeley Powell, G. P. Brophy, Geo. H. Perley, D. Murphy, Wm. Scott; president and general manager, T. Ahearn; secretary-treasurer, G. S. Macfarlane (formerly with the Standard Co.); general superintendent, A. A. Dion (Chaudiere Co.); chief accountant, D. R. Street (Chaudiere Co.). The Line and Construction Department will be in charge of W. G. Bradley (of the Ottawa Co.) and the power houses will be in charge of Mr. John Murphy (Chaudiere Co.) The auditors are: Messrs. Archer Bayley and Redmond Quain, formerly auditors of the Chaudiere Company. The working staff has already been largely reduced. The office of the new company will be at the corner of Sparks and Elgin streets, the premises formerly occupied by the Standard Company. The building formerly occupied by the old Ottawa Company as a power house, and which was vacated two years ago, is being converted into an electrical repair shop, which will be in charge of Mr. P. Beard, formerly power house foreman for the Standard Company. The use of the incandescent electric light has become very general, and the nightly output of current is larger than in any other place in Canada. Upwards of 100 electric motors, of various sizes, are in daily operation in Ottawa. These are used for machine shops, planing mills, passenger elevators, and in one instance a steam engine manufactory is provided with power from an electric motor.

# ELECTRIC RAILWAY DEPARTMENT.

### HAMILTON RADIAL RAILWAY.

HAMILTON, Canada, is to be the center, says the Street Railway Review, of a great electrical railway system which is being constructed by the Hamilton Radial Railway Company. The accompanying map shows the points which the system will embrace. There will be 227 miles of road, made up as follows : Hamilton to Toronto, 39 miles ; Hamilton to Guelph, 29, (to Fergus, 21) ; Hamilton to Galt, 23, (to Waterloo, 35) ; Hamilton to Brantford, 22, (to. Woodstock, 48, and to Port Dover, 49) ; Hamilton to St. Catharines, 33, (to Niagara Falls, 45, and to Buffalo, 67). The road will be up to steam road requirements, only the lines to Waterloo and Fergus being equipped with electricity.

This system of railways is designed to make the city of Hamilton the greatest commercial center of the Niagara peninsula, which includes a large fruit growing and farming region. To collect and bring the produce of this section to one point was the object which has engrossed the attention of John Patterson, who is about to see this plan perfected. Three of the branches will be built and equipped as steam roads, as a portion of the line from Niagara to Woodstock will be used as a link for a fast train service from New York to Chicago. The plans of the company contemplate the shortening of the distances between many of the towns, as compared with that of roads now in existence, in some cases from one-half to one and one-half miles, and in others one-third to one-half the distance is saved.

PROPOSED LINES OF THE HAMILTON RADIAL RAILWAY SYSTEM

Another advantage over existing lines will be the frequency of trains. In one place for instance, where, under present conditions, it is necessary for a man to rise at 5 a.m. to reach Hamilton and return the same day, the Hamilton Radial Railway Company will have cars running every hour, making the trip in thirty to forty-five minutes. The system will also reach one-half the coal consuming population of Ontario, and will supply transportation to a population of about 700,000 persons, or about one-third of the entire population of the province.

The territory is served by the Grand Trunk Railway and and small branches of the Canadian Pacific and the Michigan Central Railroads, but as will be seen on the map, they do not cover the field as the Radial Company proposes to do. The prospects for revenue are good. Bonds are issued with ten years' interest paid up, making the stock the only claim on the net earnings for that period. It has been estimated that it is only necessary to earn $1,400 per mile to make 7 per cent., and this is such a small amount to be expected from a territory so rich in resources, that the most sanguine expectations of the people interested in the enterprise are likely to be exceeded.

Work of construction has been delayed for some time on account of the large expense. But now the cheapness with which a first-class road can be constructed as compared with the cost a few years ago, has removed that obstacle. The eyes of all persons interested will be turned towards this road as the work of construction progresses, for particular attention will be paid to the development of high speed on its electric lines. Plans are being made and figures prepared with the expectation of handling 50-mile transmissions commercially, using the three-phase system at 20,000 volts pressure. At present the plans have not progressed sufficiently for details to be presented, but engineers are confident this prodigious feat can be accomplished.

The Hamilton Radial Railway Company is incorporated with $2,000,000 capital, being owned largely by Boston and Eastern capitalists. It is reported that the Canadian Pacific Railway Company is also interested. The company will probably use the charter of the Canadian Pacific for a line from Hamilton to Woodstock, with a promise of the usual Dominion subsidy of $3,200 a mile, and the charter of the Niagara Central from Hamilton to Niagara Central with a bridge franchise. Recently was purchased the St. Catharines & Niagara Central Railway which runs twelve and one-half miles on the way of the system.

It is expected that the city of Hamilton will give a bonus of $400,000 on account of the benefit the road will be to the city.

### ECONOMY IN STEEL RAILS.

A PROCESS for re-rolling steel rails has recently been patented by Mr. E. W. McKenna, late assistant general superintendent of the Chicago, Milwaukee & St. Paul Railway. By careful tests, the Engineering News tells us, Mr. McKenna ascertained that a rail wears out by deformation of the head and not by actual loss of metal. Sections of 60-pound rail, removed from the track and weighed, were found to have lost only 0.117 to 0.135 pound per yard in ten to fourteen years of service. He, therefore, devised a process by which the worn-out rails are heated in a furnace and the deformed heads are re-rolled into the proper form. The loss of transverse section in this re-rolling is about two pounds per yard, of which part is gained as elongation, leaving the net loss only one-half pound per yard for oxidation and the same amount for loss in crop ends. By this process, which is said to be inexpensive to carry out, a steel rail of 75 pounds or heavier can be renewed from five to fifteen times before its section is so greatly reduced as to make it necessary to scrap it. If Mr. McKenna's invention proves a practical success, the rail mills will have to close up or turn to other products. The demand for rails for renewals is what the rail mills have been chiefly relying on since speculative railway building came to an end."

### A FACTORY BURNED.

Editor ELECTRICAL NEWS.

SIR,—The Whitney Electrical Instrument Co.'s works at Sherbrooke, Que., were damaged by fire to the extent of about ten thousand dollars, on Saturday night, Dec. 29. Insurance, five thousand. Canadian orders will be promptly attended to at their Penocock, N. H., factory until the works are rebuilt.

C. E. SHEDRICK, Supt.

The Coaticook Electric Light Co., of Coaticook, Que., has dissolved, and a new company composed of Fritz E. Lovell and Moodie B. Lovel formed ; style unchanged.

## HAMILTON, GRIMSBY AND BEAMSVILLE ELECTRIC RAILWAY.

IN response to a request for information, Mr. A. J. Nelles, manager of the above railway, has kindly furnished the ELECTRICAL NEWS with the following particulars of the freight and passenger tariffs in use by the company :

"Our freight tariff is based upon one cent per mile for 100 lbs. first-class, using the Canadian classification for classifying goods. Our passenger tariff is based upon one and a half cents per mile with usual proportions for return tickets. We also issue books of 50 and 100 tickets at a reduced rate. School, workingmen's, and commutation tickets are scaled as to distance ; milk, per can.

I have adopted a ticket that after considerable thought and experience with a service extending 18 miles and on which cars are expected to pick up passengers at every house and charge accordingly, fills the bill exactly—viz., a 1000 mile coupon book, which we sell at ten dollars. We have the road miled off, and we take off a coupon for each mile travelled by the passenger, but not less than five coupons for any single trip. This fills the bill, and puts each and every customer upon an equal footing. The " way freight " is going to be quite a problem to solve, as we will in the fruit season have fruit from every farm, which will be moved by special car several times each day."

### TRADE NOTES.

A company is being formed in Perth for the manufacture of car and locomotive wheels.

The Dartmouth Electric Co., Dartmouth, N. S., are enlarging their plant, and have ordered a 125 horse power engine and boiler from the Robb Engineering Co. The engine will be a Robb-Armstrong tandem compound and the boiler a Monarch Economic with Adamson flanged furnace.

The Stanstead Electric Co., Stanstead Que., who have been running their station by water power, have decided to put in an auxiliary steam plant, and have placed their order with the Robb Engineering Co. for a 100 horse power Robb-Armstrong engine and Monarch Economic boiler.

A local company has been formed at Belleville, Ont., to be known as S. A. Lazier, Sons & Pringle, to manufacture for Canada the E. M. Electric Railway System, as description of which appears elsewhere in this number of the ELECTRICAL NEWS. The company are said to have secured the sole right to the E. M. patent for Canada.

The Dodge Wood Split Pulley Co., of Toronto recently received an order from an Australian machinery firm for 600 pulleys of various sizes. The pulleys manufactured by this company are said to be on sale in every important city in Europe, South and Central America and Australia, as well as throughout the Dominion of Canada.

Mr. C. F. Gildersleeve, of Kingston, states that the promoters have under consideration the proposal to operate the projected Kingston and Ottawa Railroad by electricity. No decision has yet been reached.

Notice is given that application will be made to the Legislature of Quebec for an Act to incorporate the Quinze Electric Co., to produce and sell electric light, heat and power, and to build, lease and operate electric railways in the vicinity of the Quinze Rapids, in the county of Pontiac. The proposed capital of the company is $50,000. The promoters are : John Bryson, of Fort Coulonge, lumberer ; James B. Klock and Robert A. Klock, both of Klock's Mills, in the Province of Ontario, lumberers ; James T. MacDougall, of Klock's Mills aforesaid, agent ; and John Malcolm MacDougall, of the city of Hull, advocate.

Negotiations are in progress between the promoters of the Hamilton Radial Railway and the City Council of Hamilton, with regard to the amount of assistance which the city should grant to the enterprise. The railway originally asked for $400,000 in the shape of stock, to be purchased by the city. This proposition was afterwards amended, and the amount of stock reduced to $300,000, together with right of way for single track on Cannon street. The citizens of Hamilton seem to prefer that the city should grant the company a straight bonus of $200,000 to $250,000, while, on the other hand, the Trades and Labor Council have passed a resolution in favor of the city taking stock in the enterprise and having representation on the directorate of the company.

## CANADIAN ASSOCIATION OF STATIONARY ENGINEERS.

Note.—Secretaries of the various Associations are requested to forward to us matter for publication in this Department not later than the 20th of each month.

### HAMILTON ASSOCIATION NO. 9.

HAMILTON, Dec. 20th, 1894.

Editor ELECTRICAL NEWS.

I am very glad to report that Hamilton No. 2 are having very interesting meetings of late. At our last instruction meeting a very good course was taken by the special committee ; instead of the reading of papers a series of discussions were indulged in. The first discussion arose on a question asked by a member of the committee as to which is the most advantageous, a dome or a dry pipe in a steam boiler. The question was discussed for some time, and among the different opinions a great many good points were struck.

A question on boiler feed was also taken up, as well as the proper blow-off for a boiler. These two last questions always give rise to plenty of discussion, notwithstanding they come up so often. There always appears to be something new to be gained out of a discussion by a body of practical engineers.

We also had with us Bro. Mitchell, of London, who is always ready to take part in any discussion that has a tendency to forward the objects of the C. A. S. E.

WM. NORRIS,
Cor.-Secretary.

CARLETON PLACE, Dec. 26th, 1894.

Editor ELECTRICAL NEWS.

SIR,—I herewith send the names of the officers of Branch No. 16, C. A. S. E., and a brief outline of the Association :—President, G. H. Routh ; Vice-Pres., Jos. McKav ; Sec., A. M. Schofield ; Fin. Sec., Hy. Derrer ; Treasurer, James Doughery ; Conductor, J. M. Murphy ; Doorkeeper, David Welsh ; Trustees, A Nichols, A. McCallum, J. D. Armstrong.

This Association was organized Nov. 6th, 1894, by Provincial Deputy Edkins. We have had one meeting each month thus far, but intend to meet twice a month in the future. We have 17 members, and prospects of more. We are going to take up a course in Mechanical Mining Science. The president has papers on this subject, and will give the members questions at one meeting and they will have till the next to answer them. After each one who wishes to take part has had his say on the question, the president will fully explain it. In this way I think the meetings will be both instructive and pleasant. I am of the opinion that Branch No. 16 will be a successful association. Hoping you will find space for these few remarks on Branch No. 16.

A. M. SCHOFIELD,
Rec. Sec., No. 16, C. A. S. E.

## WARNING

# CANADIAN GENERAL ELECTRIC CO.
### (LIMITED)

MANUFACTURERS OF

## ELECTRIC RAILWAY APPARATUS

Multipolar Generators

Bi-polar Generators

C. G. E. 800 Motors

Type K Series-parallel Controllers

Overhead Appliances

Car Bodies

Snow Sweepers

## ELECTRIC LIGHTING AND POWER APPARATUS

Alternating Incandescent Dynamos

Monocyclic Alternating Dynamos

Edison Direct-Current Dynamos

Wood Arc Dynamos

Direct Connected Dynamos

Direct Current Motors

Alternating Induction Motors

Oil-insulated Transformers

Recording Watt Meters

Arc Lamps

Electrical Supplies
OF ALL KINDS

## CANADIAN GENERAL ELECTRIC CO., Ltd.

### 65 Front Street West

### TORONTO, ONT.

## IMPORTANT TO TELEGRAPHISTS.

THE following important order has just been issued by the United States Postmaster-General to all appointed telegraphists. The question, it is understood, have been drawn up by Mr. Preece, Electrician-in-Chief to the Post Office :

"Telegraphists recommended for promotion will, for the future, be examined in the following subjects :

1. Crossing and looping wires with facility and certainty.
2. Tracing and localising faults in instruments.
3. Tracing and localising permanent and intermittent earth, contact and disconnections on wires.
4. Methods of testing the E. M. F. and resistances of batteries and a general knowledge of the essential features of the various descriptions of batteries.
5. System of morning testing, both as regards sending and receiving currents, with the necessary calculations in connection with the same.
6. Making up special circuits in cases of emergency.
7. Joining up and adjusting single needle, single current, and double current Morse, both simplex and duplex, and Wheatstone apparatus.
8. Fitting a Wheatstone transmitter to an ordinary key-worked circuit.
9. A general knowledge of the principles of quadruplex and multiplex working.
10. Measuring resistances by Wheatstone Bridge."

THE Telpher Cable Way across the Devil's Dyke was opened by the Mayor of Brighton on the 13th of October. On either side of the Dyke has been erected a light lattice-work tower. Over the tower is hung a steel wire rope, having its ends anchored to the ground, as in the case of a suspension bridge. At intervals along the rope are attached steel anchors, and resting on the tips of the arms of these ropes are about 2 ft. apart, and on them run the 4 wheels, from the axles of which are suspended the cars or cages, which seat eight people, four on each side sitting back to back. The cars are hauled to and fro by a small endless wire rope, worked by 4½ h. p. Crossley oil engine. The cars have two grippers, each grasping the endless driving rope, and when it is desired to reverse the direction in which the car is travelling, one gripper is released and the other tightened. The driving rope runs on the arms of the anchors which carry what is really the platform of the suspension bridge, and unless proper wheels are provided for this rope to run on, it will quickly cut off the arms.

### SPARKS.

Observations and experience tend to prove that to secure the best results in both transmission and endurance, the diameter of pulleys for wire ropes should be from fifty to a hundred times the diameter of the ropes.

One of the most remarkable discoveries of modern times, says Metal, and one whose use it is hardly possible to foresee at the present time, is the fact that vacuum practically cuts off the transmission of heat. Radiant heat has long been supposed to follow laws similar to light and to be transmitted through space in the same way. The fact that heat does not radiate from a high vacuum works a revolution in one department at least of theoretical science.

In a recent number of Engineer a course of interesting applications of a dynamo to a centrifugal pump was shown. The pumps are shown in series. That is each centrifugal delivers into the one next beyond it and so on, there being four in all. The dynamo is placed on the same shaft as the four pumps, with two on each side. The water is delivered to a height of 157 feet. The plan seems to have been devised and the work directed by M. Dumont, of Paris.

Ball bearings have recently been perfected to a wonderful degree. It is reported that a street car, which was equipped with the latest inventions in ball bearings, that would do away almost entirely with friction, was drawn a distance of several hundred feet by a single man tugging gently at three strands of ordinary sewing thread attached to the car. Perhaps a more interesting experiment was that of a carriage manufacturer in the west, who put a another style of ball bearings upon the wheels of a large coach to which four horses' were ordinarily hitched. Then he took a trained dog and harnessed and hitched him to the pole, when the dog drew a huge coach easily around the yard.

In view of the increasing number of fly-wheel accidents, Power asks if it would not be well in planning power plants, especially electric plants, to keep the vital and dangerous portions out of the plane of the engine pulleys? In the Lowell accident had not a large separator stood directly in the line of the wheel and received the impact of several large pieces, one of the boilers would have inevitably been unseated, with what additional damage and loss can only be surmised. Fortunately the flying fragments of a ruptured wheel will become confined to a narrow vertical plane, and this plane should be so situated with regard to the surroundings as to involve the least danger in case of accident.

In selecting indicator springs it is desirable to have them suited to the conditions of pressure and speed of engine. A good practice says the Stationary Engineer, is to use a stronger spring, for engines operated at a greater number of revolutions, than is the usual practice or as is generally determined by rules which were made before high speed engines became such an important factor as they are at the present time. A card received at this office a short time ago shows a greater initial than boiler pressure. The boiler pressure is 75 lbs. spring 40 and the height of diagram above the atmospheric line indicates 80 lbs. initial pressure ; but then the lines of the card were so heavy that it would be difficult to make an average. A stronger spring would have overcome, this difficulty and would have given a much closer approximation to the actual conditions existing. It is seldom necessary to make the height of card as much as 2"; a little less than this, say 1¾" is much better, especially on high speed engines.

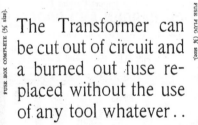

# CANADIAN ELECTRICAL NEWS

## STEAM ENGINEERING JOURNAL

OLD SERIES, VOL. XV.—No. 6.
NEW SERIES, VOL. V.—No. 2.

**FEBRUARY, 1895**

PRICE 10 CENTS
$1.00 PER YEAR.

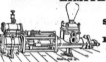

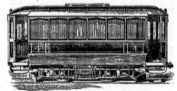

CANADIAN

# ELECTRICAL NEWS

AND

## STEAM ENGINEERING JOURNAL.

| VOL. V. | FEBRUARY, 1895. | NO. 2. |

### THE FIRST CANADIAN ELECTRICAL JOURNAL.

THE fact is probably not very widely known that the publication of a journal devoted to the electrical interests was commenced in Montreal eleven years ago. Singularly enough it was called the " Canadian Electrical News."

For the benefit of our readers who are for the first time made aware of the existence of a Canadian electrical journal antecedent to the present one, and who may be curious to know what it looked like, we herewith present a fac-simile of the title page of the initial number reduced to one-quarter the original size, with the names of the publishers, Messrs. Hart Bros. & Co., and of Mr. John Horn, the editor, appearing thereon. The publication of an electrical journal at this early period, when the electrical industries, with the exception of the telegraph, were literally " in their infancy," was a task the magnitude of which the promoters of the enterprise seem not to have fully considered. The required amount of support in the shape of advertisements and subscriptions was not forthcoming, and with the sixth semi-monthly number, the paper ceased publication.

Mr. John Horn, a well-known telegrapher of the early days, is still a resident of Montreal, and takes an active interest in everything pertaining to the application of electricity.

The following paragraphs, culled at random from the six copies of our predecessor, which happily have found their way into our possession, will doubtless have an interest for the "old timers" in the electrical field, and possibly also for those of more modern date :

Mr. Hugh C. Baker is Sup't of the Western Division of the Bell Telephone Co.

Mr. C. P. Sclater is Secretary-Treasurer of the Canada Bell Telephone Company in this city.

The Toronto branch of the Bell Telephone Co. of Canada is under the able management of Mr. Hugh Neilson.

Mr. George Black is the manager of the G. N. W. at Hamilton, Ont., a position he has ably held for many years.

Manager W. Y. Soper, for the Canada Mutual at Ottawa, is one of the best men that company has in its service.

The Great North-Western Telegraph Company operate nearly 35,000 miles of wire, and are constantly adding thereto.

L. B. McFarlane is the Superintendent of the Eastern Division of the Bell Telephone Co. of Montreal, whose usefulness is constantly extending.

The Ball Electric Light Company have removed their head office from London to Toronto, and have also reduced their number of directors to three.

The Canada Electric Light Company, of Toronto, have had their tender of 6½c. per light for 50 lights, recommended to the City Council for acceptance.

Mr. Erastus Wiman, President of the G. N. W., has promised to distribute in New York 15,000 programmes for the St. Jean Baptiste festival next June.

The Bell Telephone Co.'s exchange office at Toronto, which occupies the southern wing of the Mail building, was burned out on the morning of the 24th May.

Mr. H. P. Dwight, the General Manager of the G. N. W., Toronto, positively refused to produce the telegrams passing between the politicians in the bribery case now going on in this city.

We hear the trial test of elec. tric lights by the two companies at Toronto is giving great satisfaction; The three months experiment, we have no doubt, will end in a contract for their continuous use.

The number of telephone subscribers in the principal towns of the Dominion up to January, 1884, was only 3,256. It will thus be seen there is plenty of room for an extension of this exceedingly useful invention.

The average cost of a well-built telegraph or telephone line, say of No. 8 galvanized wire, with 35 poles to the mile, including the cost of setting the poles, stringing the wire and transportation, is about $165 per mile.

To the uninitiated we would state for electric light purposes the wire is a pure copper, No. 6 size, with insulation of double braided cotton coated with white lead, making it impervious to the weather and absolutely safe to handle, especially in connection with the automatic cut-outs.

Wm. Mackenzie, one of Montreal's leading stock brokers and financiers, is noted as a keen, shrewd and cautious business man. He is also Secretary of the Stock Exchange. Many years ago he was an expert operator in the old Montreal Company's service.

The Toronto Electric Light Co. has for officers: Messrs. E. Strachan Cox, President ; John T. Beckwith, Vice-President ; Robt. Myles, Secretary-Treasurer, and Mr. S. Hamburger as General Manager. We hope later on to say something more of the active operations of this Company.

The electors of Magdalen Island, in the county of Gaspe, and of the Island of Anticosti, in the county of Saguenay, vote by means of the telegraph at election of members for the House of Commons during periods when communication between the island and the railroad is wholly interrupted in consequence of the closing of navigation by the ice.

The Royal Electric Company is about to be incorporated by the following gentlemen : Richard W. Elmenhorst, President of the St. Lawrence Sugar Refinery ; Thomas Davidson, Manager North British Mercantile Insurance Company ; E. S. Clouston, Manager of the Bank of Montreal ; Gilbert Scott, of Wm. Dow & Co. ; James Crathern, of Crathern & Caverhill ; M. Lee Ross, H. E. Irvine, George R. Robertson, and J. Carsie Hatton, Q.C.

We have the pleasure to place before our readers a description of a Single Pole Quadruplex, which has just been invented by Superintendent B. B. Toye, of the Great North Western Telegraph Company's service at Toronto. We understand the patents have been applied for, as we are confident from the serviceable appearance of the innovation, that this new Canadian invention will be adopted everywhere in America and Europe.

We should like to see an electrical society formed in this city from amongst the employees of the various companies. We feel quite satisfied much good would result from the monthly lectures on, and discussion of, subjects affecting the general interest of all. New York, Chicago, Cincinnati and other cities have such associations, and there is no reason why we should not organize. We shall be glad to receive any suggestions.

C. F. Sise, the General Manager of the Bell Telephone Company, is an excellent executive business officer, and has brought the extension of this organization to a very high state of efficiency. The general use of the telephone at Montreal compares very favorably with the other large cities of the universe. On the 1st of this month 1,090 Bell instruments were in use in this city, and by the 1st of May it is expected the number of subscribers will have increased to 1,200. The Company have a large and extensive staff of employees.

The Royal Electric Company prices average from $400 to $5,000 for each machine and from $65 to $80 for each lamp for street lighting. Lamps will burn some 7½ hours and some 14 hours. A single lamp is 1,200 candle power, or equal to 125 gas burners of 16 candle power each. The cost of lighting a square area by 100 gas burners consuming 6 feet per hour, which would be 600 feet per hour, or for six hours burning 3,600 feet at $2.50 per 1,000 feet, would amount to $8.85 and giving 1,600 candle power only. The electric lamps will light the same space for six hours at $1.50, giving 6,000 candle power, or lighting 4,400 candle power greater than gas for $7.35 less money.

Leonard Henkle, inventor and electrician, of Rochester, N. Y., says that he has negotiated for the purchase of land on the Canada side of the river and for power from the great Horseshoe Fall for the lighting of sixty-five American and Canadian cities, connected by means of underground cables with electric lights generated at Niagara. The plans are all drawn for ten hydraulic engines of 200,000 horse-power each, and gigantic machinery. That Henkle himself means business is attested by the fact that he will soon open an office on the Canadian side of the river, and endeavor to complete arrangements with capitalists, whom he expects will furnish $22,000,000 for the undertaking.

The suggestion in our last with regard to forming an Electrical Society in this city has been warmly commended by various members of the profession. The society should be in working order before the advent of the British Association for the Advancement of Science, which will meet at Montreal in August, 1884. If a few influential gentlemen will call a meeting at such time and place as would be most convenient for those desiring to attend same, an organization would be sure to result. Who will make this first move? Remember Sir William Thompson and Mr. W. H. Preece are coming, and they should not fail us unprepared to receive them. Some noted American members of the profession are also coming.

Mr. F. N. Gisborne, F.R.C.S., the able and energetic Superintendent of Government Telegraphs and Signal Service of the Dominion, delivered a very interesting lecture on the "Origin and Development of Electrical Science" at Ottawa, under the auspices of the Literary and Scientific Society, on the 21st of last month. The chief feature of the lecture consisted of colored drawings expressly prepared by him for those uninitiated in electrical science, thus enabling all present to understand the general principles upon which dynamos, cable testing and duplex working, &c., &c., are based. The lecture was delivered without notes in a clear and pleasant off-hand manner. Not the least interesting part was that of describing the introduction of the electric telegraph into Canada in 1846, when Mr. Gisborne was an employee of the Montreal Telegraph Company, and who opened the first office in Quebec. The discourse drew forth frequent applause and was a most comprehensive survey of the whole field of modern electrical application.

## A WIRE-WOUND FLYWHEEL.

The following description of a heavy flywheel composed of wire appears in the American Manufacturer:—"Amongst the most recent and novel applications of wire, perhaps none has greater interest to the mechanical world than that presented by the new wire flywheel lately erected at the Mannesmann Tube Company's Works, Germany. Heavy flywheels driven at high velocities obviously present dangers of breaking asunder from the great centrifugal force developed. The wheel at the factory mentioned consists of a cast iron hub or boss, to which two steel plate discs or checks, about 20 ft. in diameter, are bolted. The peripheral space between the discs is filled in with some 70 tons of No. 5 steel wire, completely wound around the hub, and the tensile resistance thus obtained is far superior to any casting. This huge flywheel is driven at a speed of 240 revolutions per minute, or a peripheral velocity of about 2.8 miles per minute (250 ft. per second, approximately), which is nearly three times the average speed of any express train in the world. The length of wire upon such a constructed flywheel would be about 250 miles."

## THE "WOOD" ARC SYSTEM.

THE Canadian General Electric Co. have, within the past year, placed on the market a complete line of arc apparatus of the well known "Wood" type. The selection of the "Wood" machine as their standard for arc lighting was only arrived at after a careful consideration of the requirements of the best arc practice and of the relative adaptability to the same of the different leading arc systems owned or controlled by the Company.

In the "Wood" dynamo it has been aimed to combine certain essential features of design and construction by which it is claimed a distinct superiority has been gained. The most salient point in the machine is its simplicity in design, combined with solidity of construction. To the design of the armature, always the weak spot in an arc machine, special attention has been paid. The insulation and ventilation are of the best, and the coils are easily and separately removable in case of damage. The automatic regulation is it is claimed the most perfect attainable, variations ranging from full load to one lamp, being handled instantly without sparking or perceptible change in the candle power of the lamps. A recent improvement has been the equipping of these dynamos with the standard Edison self-oiling ring bearing.

The lamps used with the system are the Company's improved standard C. K. single carbon lamp which with a ½ carbon is claimed to be admirably adapted for all night lighting, and for those cases in which a double carbon lamp is required. The standard "Wood" type is manufactured. Both of these types of lamp are designed to secure the greatest simplicity compatible with steady and reliable service. A simple clutch feed is used and the number of moving parts reduced to a minimum. The minor accessories of the system, ammeters, lightning arresters, cut-outs, hoods, hangerboards, etc., have all been carefully worked out with a view to combining in one system the various features which experience has shown to be essential to the satisfactory operation of a modern arc lighting plant.

K

WOOD ARC LAMP.

## ALTERNATORS IN PARALLEL.

A striking illustration of the tendency of alternators to keep in step, when they have once been synchronized and connected in parallel, occurred a short time ago at one of the power houses of the Ottawa Electric Company.

Two 750 light Westinghouse a. c. dynamos, with surface wound armatures, driven from the same countershaft, by separate pulleys, were furnishing current to a circuit of about 1200 lamps.

The machines ran in this manner for several days without anything unusual occurring. One afternoon, shortly after starting up for the evening run, the ammeter, which was connected between the machines, suddenly showed a higher reading than the ammeter on the line. This indicated that one of the dynamos was carrying all the load and was also sending some current through the other. The circuit was removed and both machines shut down. The switches between them were opened and it was then attempted to start again. As soon as the machinery was set in motion the cause of the rise in current between the dynamos became quite apparent—one of them and the pulley on the driving shaft to which it was belted stood still. The key had slipped out of the driving pulley and the dynamo, which had now become a motor, was driving the pulley instead of being driven by it. But for the change in the ammeter reading the accident would not have been noticed—at least not until the heating of the armature carrying the load had attracted the attention of the dynamo tender.

## OVERCOMING TROUBLE WITH A GRAMME RING ARMATURE.

DURHAM, ONT., Jan. 15, 1895.

Editor ELECTRICAL NEWS.

DEAR SIR,—I see in your last issue that W. B. S., Montreal, has had the same trouble with a Gramme armature that I had, so I thought I would let you know the cause of the trouble here. You evidently do not understand the spot where the breaking takes place. The leads are of course stayed by weaving with cord or some other substance. These leads are styled the commutator web. It is the projecting wires connecting the armature to the web that break. These wires are about one inch long, and are really a part of the armature. However, I found the trouble here was caused by the expansion bolts that hold the armature in its place getting loose. This of course would allow the armature to give a little, and the commutator and web holding firm, the consequence was that the whole or a large part of the strain of driving the armature was on the short wires, consequently breaking them. The only wonder to me was that more of them did not go. I took off the armature and tightened these bolts, making sure I got it true ; since then I have had no trouble. I think the bolts were loosened by the vibration of the machine. I think W. B. S. will find that this is the trouble with his motor, and if after tightening the expansion bolts he makes sure the lock nuts are tight, will have no further trouble.

Yours truly,

L. O'CONNOR.

## MUNICIPAL CONTROL OF ELECTRIC LIGHTING.

RENFREW, Jan. 14th, 1895.

Editor ELECTRICAL NEWS.

SIR,—I have gone very carefully over Mr. Fraser's interesting article in your issue of January, and fail to exactly understand the financial part at the close. At the prices he mention, that customers in Orillia pay for incandescent lights, the plant should be even more productive than he makes it to be. Thirty cents per month per lamp would be $3.60 per year per lamp, and an ordinary dwelling house with say 20 lights, would thus pay $72 per annum. Then the hotel keeper who pays 80 cents per month per lamp, would be $9.60 per annum per lamps and if he had fifty lamps (which is by no means a large number for a fair sized hotel) it would cost the proprietor 50 × $9.60, or $480.00 per annum for his lights. I cannot get such prices here, and am fully persuaded in my own mind that very few electric light men can.

I fail also to see just how he gets the yearly income of $5,468.22, as from the data he gives I make the income to be $6,720 per annum, thus :—

½ of 1400 = 466⅔ × 60c. × 12mos. = $3,360.00
⅔ of 1400 = 933½ × 30c. × 12mos. = $3,360.00
or an annual income of           $6,720.00

Doubtless there has been some mistake in putting down the figures. Mr. Fraser no doubt will be able to explain. Would Mr. Fraser also state what the customers pay for the arcs ? This is evidently a pretty good paying plant.

Please state also how you come to reckon $570 for carbons as an expense against incandescent lighting ? Why charge all the amount paid to the engineer and fireman to incandescent work ? A portion of these items are justly chargeable to arc, etc.

Yours, etc.,

A. A. WRIGHT.

## MUNICIPAL CONTROL OF ELECTRIC LIGHTING.

CITY HALL, QUEBEC, January 16th, 1895.

Editor ELECTRICAL NEWS.

SIR,—I have read with much interest under the heading of "Central Station Types" in your paper, CANADIAN ELECTRICAL NEWS, Mr. G. White-Fraser's views on the comparative cost of electric lighting by and at the cost of a municipal body, as against the same service performed by a company ; and quite agree with the writer that for a small town, and especially where the water supply is done by pumping and there is a small addition to the pumping plant and building—say a dynamo or two, an extra boiler and an engine, with a single extra hand in the way of an electro-mechanical engineer—the needful can be done at a slight advance on the first and yearly cost of pumping works alone. In large cities it is very much as on board of "His Majesty's ship Pinafore," where the admiral had to accommodate and care for his "sisters and his cousins and his aunts." Look at the "Tammany Hall" for proof of what I say : that in all such concerns, where, under municipal management, every councillor and alderman—in payment I suppose of giving his services free for the enlightenment of his fellow citizens—must have a finger in the pie, in the shape of a member of his family or some relation or friend or other appointed to swell the already overdone list of useless and ignorant hands altogether unsuited to the requirements of the occasion. Our corporation on several occasions wanted to take over our gas works and our Lower Town street railway, both of which concerns were and have been coining money ; the first 33%, the other 22%. I as persistently opposed this, as there would have been no end to "dead heads" and other scoundrelism ; but what I did and do advocate, sir, in the most forcible manner, is that on account of the sacrifices imposed upon a city in thus giving up its ways to telegraph and telephone and electric posts and wires, to street railways, to gas companies, etc., to dig up and impede, to the great inconvenience of citizens—what I say, I do recommend, is that all such companies be made to pay, not a fixed sum, not so much per post, not so much out of its profits which can not be got at, being always hidden under a bushel, but a percentage on gross receipts. Look at Toronto with its 12½ per cent. on its city tramways gross receipts—nearly $100,000 annually to that city. See how much richer Quebec would now be, if since the establishment of our gas works in 1847 we had had our share of the 33% profit which the company, not daring to own to, for fear of popular outcry, used to pay 8% dividends on the $100 shares, and advertised this in all the city papers to lead people to believe the company was only doing middling well ; while the remaining $25 out of the $33 was put down to capital—½ share per share per annum, one share additional every four years—a share that did not cost a cent, and on which the 8% dividend continued to be paid, or on each four yearly increase of an extra $100 share. Similarly with our street railway company we should have, now more than 20 years ago, stipulated for a percentage of its receipts, which, as I have said were 22% profit. A fixed sum to be paid by any company is absurd, and against the interests either of the one or the other. No company, no municipality can know, foresee exactly what profits a company will make on any plant and service. The fixed sum which our telephone and electric company pay here as a tax was in the beginning too high for the profit of the respective companies. At present these fixed sums are too low. They are to be increased, and the only honest and fair way of doing so—fair to the city and to the company—is a percentage on the gross receipts or on the profits, if these can be got at, which it is difficult to ascertain ; whereas there is more difficulty in blinding people as to gross collection, which there are many ways to arrive at. I would therefore incline, except as Mr. Fraser says for small communities and therefore small plants, and when a company does not care for the slight profits to be made—and especially where there are pumping works and the same staff answer for both—that the electric lighting of such a small place even as Quebec, with its only 570 lights, continue with the present company, which gives entire satisfaction ; but that on renewal two years hence, of our five years term of contract, we do stipulate for a share in the concern, and I would also advise my friend, the City Engineer of Topeka, Kansas, who for the last three years has been putting forth his best endeavors to reduce cost per arc lamp below $90, and has not yet nor will he succeed in doing so, until some cheaper process has been discovered of producing light, to wed his corporation to the better plan of doing the thing by contract with a company, paying the city a good percentage on the venture.

C. BAILLAIRGÉ,
City Engineer, Quebec.

A great deal has been published in engineering journals about scale in boilers, and yet very little has been said about the accumulation of it in feed and blow-off pipes. There are men who maintain that scale can not accumulate in pipes in which the water is circulating constantly, but cases cited will show how fallacious such opinions are. As a matter of fact, these pipes often fill up in a remarkable way, the deposit choking them to such an extent that it becomes a source of positive danger.

## CHARACTER SKETCH.

### ROBERT F. EASSON.

"What is the use of a child? It may become a man !"—Franklin.

FROM smaller beginnings, sometimes, than a grain of mustard seed, grow great things. When Galvani discovered that a frog's leg twitched when placed in contact with different metals, it could scarcely have been imagined that so apparently insignificant a fact could have led to important results. Yet, as Samuel Smiles has remarked, therein lay the germ of the electric telegraph, which to-day binds the intelligence of continents together, and seems destined speedily to "put a girdle round the globe."

A sketch of one who has given nearly half a century to the telegraph business, as has been the case with Mr. Robert F. Easson, of the Great North-Western Telegraph Co., is suggestive of the early trials and triumphs of telegraphy. Mr. Easson cannot go back, except historically in memory, to the rude methods of signalling adopted by the Roman generals who spelled words by means of fires on different substances. The beacon lights on the mountain tops in ancient days were, however, the precursor of the electric telegraph of to-day in all its completeness. Nor 50 years ago were the methods in vogue as crude as those described by Bishop Wilkins, when the custom of conversing at a distance with three lights or torches at night so used as to indicate the 24 necessary letters of the alphabet, was in somewhat general use.

Robert F. Easson entered the service of the Montreal Telegraph Co., at Toronto, as messenger, in 1849, a period only four years later than that which marked the opening of the first telegraph business in which the Morse patents were used. He has, therefore, good claim to rank among the pioneers of that method of telegraphy which in late years has girdled the globe. From messenger boy in 1849 Mr. Easson was promoted, with little delay, to the position of book-keeper. In 1852 he was an operator in the Toronto office, and the records show that he was one of the first operators to read by sound. Chicago became his place of residence a year later, where he was engaged by the late Mr. Ezra Cornell, who was largely interested in western telegraphs, to work as an operator in the telegraph office in that city. A little later he was sent to Laporte, Ind., In that journey the iron horse had not cut his way through the country as he has done to-day, and Mr. Easson's drive from Laporte to Plymouth was right through the bush. Here he, in company with E. B. Stevens, of Chicago, opened a telegraph office, Mr. Easson remaining at Plymouth for about a month, teaching a young Hoosier to telegraph and then installing him in charge of the office.

Spending three or four months after leaving Plymouth in Logansport, Ind., in the fall of 1853 he had again returned to Chicago. For two years from this date he remained in the Chicago office and worked the old Speed-Cornell line, which ran through the woods and along the highways from Chicago to Detroit. The telegraph in those days had not cut any large figure in the commercial world. The entire staff of operators in the Chicago office did not exceed half a dozen, and when Mr. Easson left Chicago in 1855 there were not more than a dozen operators in that now great city. And yet some progress had been made in the decade from 1845 to 1855. The first telegraph line constructed on the Morse plan in 1845 was operated between Washington and Baltimore. The advertised tariff was, "for every four letters—one cent." The receipts for the first four days were one cent and the total income for the first ten days $3.09½.

In 1855 the subject of our sketch returned to Canada and re-engaged with the Montreal Telegraph Co., at Toronto, as an operator. His career from that date on was one of steady progress, and in 1859 we find him adding to his duties as an operator those of associated press agent at Father Point, Que.,

ROBERT F. EASSON.

where the Canadian ocean steamers were intercepted and their news sent on by telegraph in advance of the steamers' arrival at Quebec. Once more he returned to the Toronto office, where he worked the Montreal circuit until about 1864, when he was appointed chief operator, occupying this position until about 1880. Mr. Easson's fitness for the position of press superintendent was shown in the fact that when the Great North-Western Telegraph Co. assumed control of the Montreal and Dominion Telegraph Companies, he was appointed to take charge of the news-gathering branch of the business, which had developed most successfully under the direction of General-Manager H. P. Dwight. This position Mr. Easson occupies at the present time. His talents and training eminently fit him for the work. Of this every newspaper man throughout the country is prepared to testify; and no man is better or more favorably known, either in Canadian telegraph circles, or by the newspaper men of the Dominion, than "R. F. E.," the familiar business nom de plume of Mr. Easson.

Every one who has come in contact with Mr. Easson is prepared to bear testimony to his genial, kindly and courteous character. Among those who know him more intimately he is regarded as one of the best friends that any man can possess.

Robert F. Easson hails from the land of the hills and heather, having been born in Prinlaws, Scotland, June 14th, 1838, though he came to Canada when quite young ; and attached as he is to the land of his birth, he is ever proud of the land of his adoption, and is one of its most worthy citizens.

## NATIONAL ELECTRIC LIGHT ASSOCIATION CONVENTION.

FOLLOWING are the titles of papers to be read at the annual convention of the National Electric Light Association convention which will meet in Cleveland, Ohio, on the 19th, 20th and 21st inst.: "The Storage of Energy Essential to Central Stations, How it May be Accomplished and the Economies Resulting," by Nelson W. Perry. Professor Langley of the Case School of Science, and Professor Stine of Armour Institute, Chicago, will take part in the discussion. The topic, "How to Light Large Cities," will be discussed by Frederic Nicholls, Charles R. Huntley, Frank H. Clark, J. Frank Morrison, T. Carpenter Smith, George A. Redman, E. F. Peck, and others ; "Some Economies in Electric Light and Power Stations," by Professor Edward Weston ; "Arc Carbons and The National Electric Light Association Standard of Light," by L. B. Marks ; "The Monocyclic System," by Dr. Louis Bell ; "The Correct Method of Protecting Electric Circuits," by W. E. Harrington ; "The Evolution of Arc Lighting Machines," by C. N. Black. E. A. Leslie's paper, read at the Buffalo meeting, and entitled "The Operation of High Tension Currents Underground from a Physical and Financial Standpoint," will be taken up and discussed.

---

Mr. W. J. Clark, electrician, Trenton, Ont., has been committed for trial on a charge of having incited one J. J. Cooley to burglary, for the purpose of obtaining the cypher of the Brush Electric Co., in connection with the Toronto boodle inquiry.

The Stormont Electric Light & Power Co. will apply to the Ontario Legislature for an act to ratify and confirm an agreement made on the 18th January last, for the purchase of the Cornwall Gas Works, and for power to operate the same, and to increase the capital stock of the company.

Three-hundred subscribers of the Bell Telephone Co. had their connection with the main office cut off by the recent fire, and about a dozen telephone instruments were destroyed. As speedily as possible, the company intend to place their wires underground in the vicinity of the recent fire.

The Richmond Industrial Company, of Richmond, Que., is seeking incorporation to acquire the real estate, machinery and franchise of the Richmond Water Power and Manufacturing Company, Ltd. The capital stock of the company is to be $100,000, divided into 1,000 shares. The applicants are :—Messrs. Leonard Thomas, Melbourne ; William Ewans Jones, Richmond ; John Matthew Nunns, Henry Autisell Allen, Melbourne ; Kelzer A. Cummings, Francis Henry Nunns, Coaticook.

## THE ONLOOKER.

An ancient writer has said : "I would do what I pleased, and doing what I pleased I should have my will, and having my will should be contented ; and when one is contented, there is no more to be desired ; and when there is no more to be desired, there is an end of it." The Onlooker thought of these words as he learned of the several suggestions made by Mr. Hamilton MacCarthy in a paper on "The Aesthetic Unity of the Fine Arts," read at a late meeting of the Canadian Institute. This well-known sculptor would have the people of Toronto "do away with the dangerous and unsightly trolley," because of the manner in which it mars the beauty of the city. The suggestion shows how directly the practical spirit of the age will come in contact with the sentimental. It would be a delightful thing for Toronto, and other metropolitan cities, if they could be planned on the lines of the artist and sculptor. But if these things were given the people, how much else would they be forced to forego? No boon is without its drawbacks. The telegraph, telephone, and electric system of transit are not an unmixed blessing, but who would be without them ? The unsightly trolley, however, has done not a little to give to Toronto an aesthetic and artistic character. Were it not for this method of rapid transit the city would not be dotted with so many pleasant homes, laid out with skill and taste. The people by means of this quick method of propulsion are enabled to get out to the country, that spot which it is said God made, whilst man made the city. The suggestion is more practical that another method of electric locomotion should be adopted in place of the overhead trolley system. Science is working to-day on the conduit system and some of the brainiest electricians have not abandoned the hope of a successful storage system for the largest cities. · When these come Mr. MacCarthy's desires will have been met, and the advantages of the present will not have been sacrificed.

× × × ×

When one talks of a "penny wise and pound foolish" policy, it is a mistake to suppose that the old adage can be applied only to domestic expenditures. The unsuccess in many lines of manufacture, and the trouble that comes to municipalities is frequently due to the narrow manner in which investments made are viewed. The other day the Onlooker was talking to the representative of a large electrical supply house, who had recently furnished an electric plant for an eastern municipality. Pursuing a cheese paring policy those interested had thought it wise to divide up the purchase, securing one part of the plant from one concern, and part from another, thereby saving, as they believed, a trifle. When everything was supposed to be ready for operation, it was discovered that things would not work satisfactorily. There was a fault somewhere, and after no little trouble it was ascertained that associated with machinery of the better class, there had been placed that of a commoner kind, and as a result the plant would not work. Someone had to be blamed for the trouble, and the man who did the best work came in for his share of it. Let it be said for this municipality, that finally they saw for themselves the mistake that had been made, and a plant perfect in all its details was secured and everything now goes on lovely. It may not be a matter of serious concern if the good housewife makes the mistake of buying shoddy for cloth or cotton for silk. It is not a killing affair, though it may drain her pin money a little. But it is hardly excusable when hard-headed business men, and especially those whose work is in the line of mechanics, allow themselves to be influenced in the securing of plant, that their experience and good sense ought to tell them can never give satisfaction. The poorly made machine, at the best, will give out in a little time, and the labor, expense and trouble of replacing the broken machinery has to be undertaken. An investigation made some time ago by an inspector of a boiler insurance company showed that the larger percentage of accidents caused by the bursting of boilers was due to the owners being satisfied with something cheap. The cheap and nasty, as a shop term runs, may not hurt anyone, when this rule is applied in the purchase of an article for individual wear, or the home, but the "cheap and nasty" in machinery may mean the loss of life, and the wreckage of valuable property. Moreover, when one sets up a shop or mill filled with machinery, he does not do this for the day only. Thousands of dollars are not invested in mechanical and electric plants as one might invest in a toy. The idea of permanency ought to be foremost in every such purchase. These ends can only be attained when the purchaser gets far away from the penny wise and pound foolish policy that controls too often men in their individual, as well as in their corporate, capacity. Better be sure than sorry.

× × × ×

What is to be the future of the electric railway, is a subject that is commanding wide attention among press and people everywhere. The progress of electric propulsion has been on so wide a scale and taken altogether of so satisfactory a character, that those who indulge in prophecy are hardly to be kept within bounds in the optimistic pictures they paint of the future of this rapidly developing power. All this is cheering to the men who have given the best of brain and brawn to a consideration of this question. It occurs to the Onlooker, however, that it is worth while to get away from the heights a little and look at this matter in a more practical and business-like sense. For this is to be noted that the progress of electric power can only grow as it appeals to, and is capable of satisfying, the commercial needs of a community. Electricity is not only supplementing other methods of lighting, and doing this most successfully. It is, also, cutting into the corner that steam has held since the days of Watt, as the one great power for driving the machinery of the workshop ; and as advancement is made in this direction it is a question whether ultimately, because of its cheapness and simplicity as compared with steam, it will not become the one practical power of the day for mechanical purposes. In neither of these instances, however, does it stir up the elements of opposition, as is the case when it is used as the propelling power of travel between municipality and municipality. Whether the electric road can engage in freight and passenger traffic without injury of the older railroads and all that this involves, is the burning question in electrical fields just now. It is useless for railroads, any more than gas companies, to close their eyes to the fact that electricity, in both cases, is on top. But without withholding one jot or tittle from the capabilities of electrical power, in a consideration of its application to railroading, questions of expediency come up for consideration. Some words of caution were printed in the ELECTRICAL NEWS a short time since, touching the wisdom of running electric roads parallel with the steam railroads now in operation. A Montreal journal has supplemented this thought with the remark that at least one place that helped an electric line to compete with a railroad got badly left. The electric railway had to go to the wall and the older railroad then stepped in and the rates for local traffic were advanced to recoup it for the loss by the competition. In certain parts of the United States similar difficulties have arisen, and it is pointed out how unwise is the tendency, which in some places is being carried to excess, of constructing electric lines in places where, as time will demonstrate, they cannot be operated at a profit. A writer in an engineering journal, discussing this phase of the question, takes the ground that nothing seems more certain than that for their own protection the steam railroad corporations will ultimately be compelled to secure control, directly or indirectly, of all electric street railways which seriously compete with them. In view of this possibility, not to say certainty, it would seem to be the part of wisdom for the municipal forces of towns to exercise a good deal of discretion in granting exclusive perpetual franchise for the occupation of their streets and roads by electrical companies. It is easily understood, with the developments of electricity as a method of railway propulsion, that steam railroads would grow alarmed and difficulties, which time would shortly wipe out, will be magnified. It has seemed to the Onlooker, nevertheless, that with all the confidence that the student of electricity may have in the developments of the future, and he can afford to have big faith, there are yet phases of the question that call for caution and wise judgment. This one thought has suggested the present remarks.

It is said to be the intention of the Canadian Association of Stationary Engineers, to renew their application to the Ontario Legislature at its next session, to pass a law making it compulsory on all engineers in charge of steam plant to pass a qualifying examination.

It is the intention of the Corporation of Port Arthur, owners of the Port Arthur Electric Railway, to install and operate a 12,000 c.p. incandescent light plant by and in connection with the railway plant at an early date. The management of the road for 1895—6 is vested in Mr. N. P. Cooke, chairman of the Light Committee, and Mr. A. M. Gill, Chief Engineer.

## CENTRAL STATION TYPES.

George White-Fraser.

### NO. III.—SMALL PRIVATE PLANTS.

THE small amount of business that is sufficient for the support of a paying electric light company, operating under favorable circumstances, is certainly surprising ; and may well afford matter for serious consideration to small capitalists. At the same time, it must be admitted that the smaller the business, the more favorable must those circumstances be, and the more careful the consideration given to any particular case. The question of design assumes an importance in a small station which is well nigh predominant. In a station of any size, large r small, this question is, of course, a determining factor in the ubsequent costs of operation, and resultant dividends ; but, as first cost of plant and frictional and other losses do by no means diminish in the same proportion as the capacity, it requires very little demonstration that faulty or ill considered design of station and proportions of steam and electric machinery, has relatively a greater effect on a small plant than on a large one, in increasing operating and maintenance expenses, and reducing dividends. There is a very general impression among the non-professional public, and unfortunately also observable among persons connected with electric plants in a non-practical capacity, that the designing and operating of an electric lighting or railway plant is really the most simple thing in the world—that any person who is capable of buying a bag of sugar in the open market, and of exercising his judgment as to its quality, is quite equally competent to solve the various "simple" problems governing the design of an electric station ; and as to operation of plant—" why what more is there to do than to shovel coal into a boiler and close a switch. The whole thing is as easy as rolling off a log." So it is, but it seems possible that the person who has studied rolling off logs, so as to get the greatest amount of exercise with the least personal exertion, and with the least damage to his clothes, will probably roll further and better than one who has devoted most of his life to some other pursuit.

The practice of electrical engineering as a profession is not sufficiently individualized ; the lines of demarcation between it and the other branches of engineering are not distinctly enough laid out. On the one hand, we find purely civil engineers, and even land surveyors, whose technical training certainly does not necessarily qualify them to tackle electrical work ; and on the other hand, mechanical engineers who are very close cousins to electrical men, but yet not actually qualified, are both and all, placed in charge of enterprises which not only involve the use of electricity, but in which electricity is the active principle, so to speak. The natural result of such short-sighted policy is that, while a piece of construction is probably very solidly and well done and calculated to last, it has cost thirty per cent. more than necessary ; and a little criticism will show that the first principles of *electrical design* have been lost sight of, and the plant saddled with the irremediable results of inefficient machinery. All this is bad enough as it is. But when we find the ordinary business man in a small way entering the practical field, and buying a dynamo or engine as he would a new hat or a dog collar, and designing his own electric plant as he would the arrangement of his own store, the thing is raised from the ordinary to the sublime. The chances are that the resultant station or power house will present many valuable features (for all intelligent people can have good ideas, and work them out), but at the same time will not bring in dividends. There appears to be a very considerable misconception as to the *commercial* conditions of electrical enterprises, and also as to those conditions under which electrical enterprises will work to advantage. Any person of intelligence will at once admit that to run a successful dry-goods or grocery business requires experience and a substratum of dry goods or grocery knowledge. Similarly with any other line of trade. It is not merely necessary to have a shop and stock. Assistants are required who must have had some training—a book keeping staff to keep track of everything ; petty pilferings must be detected and stopped, and above all, there must be a *head*, to manage things, who is deeply versed in the details of the business, and who can buy in the best market and make the best use of his goods. The sooner electrical enterprises are placed on the same common sense basis, the better for both owners and consumers.

An electric light station is a store where electric light is sold· The machinery - waterpower or steam and electric - the stock ; the engineer, linemen, &c., are the shopmen, chief engineer is the gentlemany floor walker, and the superintendent the guiding intelligence who, let it be strongly emphasized, must be well up in all the details of the business. The secretary will be the book-keeping department. So far the analogy is followed, for all plants have some kind of engineer and linemen—or "Electrician"—as he is called (!) and a secretary. None but the large ones have a superintendent. It may be pertinent, at this point, to ask :—" Then who *manages* the plant? who is the experienced head, who cares for every department, and sees that the stock is purchased in the cheapest market—that none of it goes to waste or is pilfered, or is sold below its reasonable price? who is the guiding hand of the whole business, which hold the reins? Echo will faintly reply, "Who?" Possibly echo may suggest that the floor-walker (engineer), who is a working mechanic with no technical training whatever, and but little general education, can be head too. Then, will this assistant, whose mental vision has for years been directed mainly towards keeping his bearings cool, and his steam gauge at 80 lbs.—will he be *capable* of choosing the really best market? Is his education sufficient? In a word, is it *possible* that a working mechanic, with a working mechanic's education, and opportunities for acquiring scientific knowledge, can "run" an electric plant, taking care that he does not burn too much fuel, or take too much steam, or waste steam, or waste electricity, or introduce too great frictional or other loads ; and can he choose the really best incandescent lamp to run with his plant, and under its peculiar operating conditions ; and so on through the numberless details that constitute efficient managing? And yet in a majority of medium and small sized plants in Ontario, that class of man is expected to generate good current, and at a price low enough to sell in competition with gas. He does so in many cases—but why? Because he runs with a bountiful water power or burns slabs and other cheap wood fuel. Whatever may be the class of fuel burned, it should be one of the objects of the management to reduce the consumption ; whether it be coal at a high price, or cordwood at $1.00 per cord. An unnecessary consumption of 1 lb. per H. P. H., means an unnecessarily increased fuel bill, and just so much knocked off dividends. But is this kind of care taken? By no means ; nor is the class of mechanic usually found running electric light stations qualified by knowledge or experience to undertake such management.

Then in the accounting department, in all stations, municipal or private, a set of books are kept, and very scientific looking reports are sent in daily by the engineer to the secretary. Therein are set forth the times of starting and stopping ; quality of fuel burnt ; maximum and minimum indication of the ammeter, and various other most imposing looking data. And what meaning has it all to the secretary? He doesn't probably know the difference between an ampere and a set nut, the engineer being likely equally well informed. His books, therefore, are simply a record of salaries, purchases, fuel expense, and income from rental of lights ; and th's service, which probably a clerk in a shoestore could perform equally well, is called "managing an electric lighting station"! Where is the classifying and analyzing of accounts? Where is the close scrutiny into the expenditure of every department? Where is the system by which alone wastes (steam, mechanical, electrical) can be traced to the causal defect, and the proper remedy pointed out? Why should electric lighting business not be managed on business principles? A teamster is not necessarily qualified to run a horse car line, nor a carpenter to conduct an extensive contractors' business. A man who is an excellent hardware man and yet he would run a restaurant to earth, simply because neither his professional training nor his subsequent experience has been in the catering line. Why therefore set a railway or sewerage engineer or land surveyor to design a plant? and why manage it afterwards by a committee consisting of a plain working mechanic and a book-keeper? Is a boiler a kind of nickle-in-the-slot machine, into which you throw fuel and get steam?—an engine room to be regarded as a kind of expensive barrel organ out of which an itinerant Italian can grind a tune just as well as a doctor of music?—while a whole electric lighting system is placed pretty much on the same footing as those glass case toys that are shown by sham seamen, into which a little dirty ruffian drops a cent (perhaps a brass button), and in return the mechanism heaves a rolling ocean, and sets a number of jerky figures into spasmodic action. It doesn't matter who drops the cent, the street boy gets the same results as the educated passer by, and so it is apparently with electricity. The ordinary handy man who has run a saw mill, and can oil a bearing, is believed capable of properly, economically and efficiently managing an electric lighting station.

This question of the original design of a station and its plant is one the great importance of which is not sufficiently appreciated. It will be evident, however, on the slightest reflection, that the efficiency of the system, as a whole, depends ultimately on the preliminary consideration given to the *commercial* conditions under which the plant may be expected to operate. Take the familiar case of a large water power. It may be the best economy to use a large wheel, and waste considerable power in order to reduce line copper; or it may cost so much to develop the power, that less line loss and more copper may be the true economy. Who would think of using an expensive triple compound engine where the fuel was waste slabs, or a simple, single acting, low pressure one, with coal at $12.00? The whole thing comes down to dollars and cents eventually, and to a comparison of increased interest on first cost by using higher class machinery, with the decreased fuel and other expenses to be expected, consequent on its more economical operation.

The *consideration of local conditions* should guide in the selection of a plant, not merely the fact that one engine is cheaper than this or that other one. When a light and fire committee is proposing to purchase a plant, do they take into consideration the probable *load line* of their station, and decide on the relative proportions of dynamo and engine, with the reference to that? They apparently do not. Similarly with private companies and individuals. There is an utter absence of all weighing of conditions, and careful consideration of particular details. The diagram gives the usual nightly load line of a small station, expressed in horse power—that is, the wattage of the electrical output—arc and incandescent. The line at the top shows the H. P. of the engine that runs it, at its normal rating. This diagram shows the winter load in the beginning

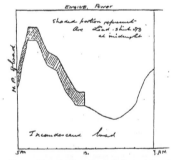

of January; the summer load will naturally be greatly less. It affords a good object lesson to anyone who is capable of reading it. And this sort of thing, which is common, is called "designing a station." The little town of Penetang affords an example of a successful little plant, which is managed with a good deal of common sense and enterprise. Its population is about 2,000, and its electric lighting is done by a private company. This company had the shrewdness to contract with the town to run its waterworks for it. It put up its own building, contiguous to the waterworks engine house, and agreed to run the waterworks plant, not including any of the outside or pipe work, for the same sum it had cost the town for the same service the preceding year. It is hardly necessary to point out that the advantages are to a great extent the same as those obtained in Orillia (described last number) by doubling up the same services. It is a most regrettable circumstance that the above diagram comes

from this very plant; but the very evident loss in efficiency consequent on such questionable proportioning, was to a certain extent offset by the commendable policy of the company in securing the services of a qualified electrician at a regular good salary, to whom was entrusted the responsibility of installation. The policy usually is to give a contract for the installation of so many hundred lamps and the outside line work, trusting more or less to the contractor to do good solid work. Contractors are, unfortunately, contractors all the world over, and propose to make money, consequently work is not as well done in all cases as it is when supervised by a person whose interest does not lie in cheapening labor and lessening expenses. This little plant is good and solid, and does good work. The inefficiency of proportion is further offset by the low price of fuel—$1.25 per cord for slabs—but it will always remain true that first cost was increased by the price of many horse power more than necessary, and that operating expenses will always, of necessity, include the fuel to overcome the frictional load of a very much too powerful engine, operating for a large proportion of its time at a very inefficient length of cut-off.

## RULES OF THE CANADIAN UNDERWRITERS' ASSOCIATION.

THE Canadian Underwriters' Association has recently revised its rules for the installation of electric light and power. No very important changes have been made, but the present rules have been brought up to date and made very comprehensive. They are in harmony with the rules adopted by underwriters universally. The special requirements of the Association are as follows :

A certificate for all new work or changes in old work should be signed by the party installing or controlling any apparatus. The certificate should be sent to the Secretary of the Canadian Fire Underwriters' Association, Toronto.

This certificate is relied upon as a guaranty until the work can be inspected. Permits for the use of light or power may be granted as soon as the certificate is duly filed.

Blank certificates may be obtained by application to the Secretary of the Canadian Fire Underwriters' Association, Toronto.

All work should be inspected before any of it is concealed, and to this end notice of concealed work must be given this Association as soon as work is commenced.

The Canadian Fire Underwriters' Association reserves the right at any time to add to, change or modify the accompanying rules, and to enforce such modifications, changes, etc., as it shall deem necessary for safety; and it will use all reasonable efforts to promptly notify all electric light companies of any such change.

Any additional loading of wires, either in building as a whole, or in any department thereof, without previous notification to the Association such as is required, shall be deemed a sufficient cause for the suspension of any permit previously granted, until the same shall have been inspected and approved by this Association.

This Association reserves the right to disapprove of the use of any wire, switch, cut-out, or any device, or form of material, which it may consider inconsistent with safety from fire risk, even though it may be proposed to instal the same in conformity with these rules.

## THE SMALLEST ELECTRIC BATTERY IN THE WORLD.

In contrast to the very large generators of electricity—battery and dynamo electric machinery—in such common use to-day, it may be interesting to note, says the Manufacturer and Builder, what is perhaps the smallest electric battery ever constructed and no doubt also the smallest generator of electrical or mechanical energy. This battery was constructed some years ago by one of the electricians of the Boston Telephone Company, and consisted of an ordinary glass bead, through which two wires, one of copper and the other of iron, were looped and twisted so as to prevent their coming in contact. The wires acted as the electrodes, and all that was necessary to cause a current to flow was to place a drop of acidulated water in the bead. Certainly such a minute battery furnished but an infinitesimal current, but could be easily used in a delicate telephone; in fact it is said to have actually been used in signaling to a distance of nearly 200 miles,

## COST OF ELECTRIC LIGHTING.

IN view of the general interest which has arisen in the subject of the relative cost of electric lighting under municipal control and by contract, our readers will doubtless be interested in a perusal of the following statement issued by the Toronto Electric Light Co. to the citizens of Toronto :—

as given, the most marked feature is the difference between the estimated cost of operating a civic plant as made by Mr. McGowan, Secretary of the Fire Department, and the offer of the Toronto Electric Light Company to provide lights by contract. Mr. McGowan's estimate is $103.85, as against $74.82 by the Electric Light Company. A careful perusal of the figures will at once explain the apparent anomaly. Mr. McGowan has

## ANALYSIS OF THE DIFFERENT ESTIMATES THAT HAVE BEEN MADE.

| No. | ITEMS. | City Engineer. 1300 Lights. | | R. J. Mc-Gowan. 1300 Lights. | | Bertram. 1300 Lights. | | Chicago Municipal Plant, 1110 Lights. | | Number Toronto Electric Light Co'y Employees for City Lighting. | REMARKS. |
|---|---|---|---|---|---|---|---|---|---|---|---|
| | | No. | | No. | | No. | | No. | | | |
| 1 | Superintendent.................. | .. | .... | .. | $2500 | .. | $2000 | .. | $.... | 1 | [eer, $2108 |
| 1 | Chief Engineer. ............... | .. | $1500 | .. | 2000 | .. | 1200 | .. | }10600 | 1 | Waterworks Engin- |
| 2 | Assistant Engineers .......... | .. | 2000 | .. | 2000 | .. | 1600 | .. | | 2 | Asst. Eng. $1270 each |
| 1 | Electrician ................... | .. | 1500 | .. | 1500 | .. | .... | .. | .... | 1 | |
| 2 | Dynamo tenders .............. | .. | 1500 | .. | 1500 | .. | 1200 | 3 | 8734 75 | 4 | |
| | Oilers ....................... | 3 | 1800 | 2 | 1200 | 2 | 960 | 2 | 4580 70 | 2 | |
| 1 | Dynamo cleaner ............. | .. | .... | .. | 600 | .. | .... | .. | .... | 2 | |
| | Firemen........ ........... | 3 | 2100 | 4 | 2800 | 3 | 1500 | 4 | 12659 92 | 5 | Firemen at W. W'ks, |
| | Helper ·.................. | .. | .... | .. | 500 | .. | .... | .. | 7606 39 | .. | $788 each |
| | Trimmers .................... | 15 | 7500 | 20 | 11000 | 18 | 9000 | 25 | 13367 16 | 24 | |
| | Patrolmen or Inspectors ...... | 2 | 1600 | 7 | 4200 | 2 | 1200 | 6 | .... | 7 | |
| | Foreman of Trimmers and Inspectors.................. | .. | .... | 1 | 800 | .. | .... | .. | .... | 1 | |
| | Horse, wagon and driver for lamp department and keep | .. | .... | .. | 975 | .. | .... | .. | .... | 2 | |
| | Linemen...................... | 2 | 1800 | 5 | 3000 | .. | 1800 | 4 | .... | 5 | |
| | Team of horses, wagon and harness for linemen and keep | .. | 675 | .. | 1000 | .. | 630 | .. | .... | 1 | |
| | Clerk hire and stationery...... | .. | 800 | .. | 800 | .. | 1000 | .. | .... | .. | |
| | Storekeeper .................. | .. | .... | .. | 800 | .. | .... | .. | .... | 1 | |
| | Machinists................... | 1 | 700 | 3 | 2100 | 3 | 1500 | 3 | .... | 5 | |
| | Materials for repairs to dynamos, lamps, engines and boilers ................... | .. | .... | .. | 3000 | .. | .... | .. | }6374 81 | .. | |
| | Maintenance and renewal of tools, repair shop and linem'n | .. | .... | .. | 500 | .. | .... | .. | .... | .. | |
| | Globes ...................... | .. | .... | .. | 1950 | .. | 4000 | .. | 2253 67 | .. | |
| | Coal ........................ | .. | 39420 | .. | 39420 | .. | 16675 | .. | 28509 87 | .. | |
| | Carbons ..................... | .. | 10409 | .. | 14235 | .. | 9031 | .. | 9651 19 | .. | |
| | Oil waste, etc................ | .. | 2000 | .. | 2000 | .. | 1200 | .. | .... | .. | |
| | Interest and depreciation...... | .. | 31020 | .. | 31020 | .. | 20000 | .. | .... | .. | |
| | Insurance .................... | .. | .... | .. | 2100 | .. | 1500 | .. | .... | .. | |
| | Taxes which the city would lose by taking the lighting from private corporation........ | .. | .... | .. | 1500 | .. | .... | .. | .... | $3000 | |
| | Rebates for lights accidentally out which contracting Co. forfeits ................... | .. | .... | .. | .... | .. | .... | .. | .... | $1000 | |
| | Incidentals .................. | .. | .... | .. | .... | .. | 1500 | .. | 2959 47 | .. | |
| | **Cost per lamp per year..** | .. | **$81.78** | .. | **$103.85** | .. | **$59.66** | .. | **\*$96.64** | | |

Cost per lamp per year of lights produced by Municipal Corporations in England, sent by *The Telegram's* special correspondent, November 15th, 1894:

Derby.................................................. £25 per year, or $121 67
Dundee .............................................. 25 " 121 67
Brighton ............................................. 30 " 146 00
Blackpool ............................................ 22 " 107 07
Manchester .......................................... .. " 131 40

OFFER OF THE TORONTO ELECTRIC LIGHT CO., $74.82 PER LAMP PER YEAR.

\* Without interest or depreciation.

For the purpose of comparison the various estimates made by the city officials, and also the estimates made by Mr. Bertram, are tabulated on this opposite page with each item, so that it may be taken by itself and conclusions drawn. There are also given figures of the cost of street lighting by municipalities in England, which appear to be much higher than contract figures in this country, though supplies and labor, especially coal, are supposed to be much cheaper in England than here. The average cost when run by the municipality in England is $125.56 per lamp per year. A detailed statement is also given alongside the estimates for civic plant in Toronto of the cost of the municipal plant in Chicago, which is the only one of any size on this continent. It is said that the service in Chicago is mostly on underground wires. This is true, and the cost for interest would thereby be somewhat increased, though the cost of operating would, if anything, be less. But the figure given above is from the City Clerk's books and is less for labour and material only, no charges for interest or depreciation, and amounts to the sum of $96.64 per lamp per year; with interest and sinking fund added, the amount would be $169.00 per lamp per year.

With respect to the estimates of Toronto officials and others

made a fair and honest presentment and has based his figures for labor on the salaries and wages now paid by the city for similar services. Without taking up too much space one or two items selected from the whole will illustrate this. For instance, Assistant Engineers are figured by him at $1,000 a year each. The Assistant Engineers at the waterworks receive over $1,000 each. It is not likely that assistant engineers in one branch of the city's service would be content with less wages than their co-labourers on another branch, especially as the electric light work would be the more exacting of the two. Mr. Bertram only estimates assistant engineers at $800 each, which from a municipal standpoint is a fallacy. But does any one suppose that a private service or employer pays $1,000 for assistant engineer, or anything like it? Firemen the same. City firemen receive over $700 a year each at the waterworks, and Mr. Gowan is perfectly justified in estimating that sum. Any individual or company can hire all the first-class firemen they want for $500 to $550 per year, and so on through the entire list. The figure of $74.82½ offered by the Toronto Electric Light Co. is cost price to them, and only made because they would rather keep their plant in operation than have it destroyed, and as shown above it will be cheaper for the city to accept the offer rather than pay higher wages and have men in their own employ.

Mr. McGowan's report has been sneered at by the advocates of municipal control, but it has not yet been shown that one single item would cost the city less than his estimate. A comparison of a few of the items in the different estimates will be instructive and conclusive to a fair investigator.

COAL.—This is estimated by the City Engineer, using the same kind of coal as now used at the waterworks, and which if best for the waterworks should be best for the electric lighting service at $39,420 per year for 1,300 lights. Mr. McGowan accepts the engineer's figures and allowed the same $39,420. The city of Chicago uses for only 1,110 lights, $28,509.87 —and coal is cheaper in Chicago than in Toronto— yet Mr. Bertram says that the coal for 1,300 lights will only cost $16,575. This is manifestly an error, and either arises from an erroneous calculation of hours run, or an error in figuring. An ordinary engineer can calculate this, and to prove the statement the calculation is here given. One 2,000 c. p. arc light takes, including friction of engines, shafting, belting and dynamos and loss on line, one horse power. The number of hours per year run, according to the lighting schedule now in use, are 3,932, equalling 3,932 horse-power hours per light. This multiplied by 1,300, the number of lights, equals 5,111,600 ; multiplied by 2½ pounds of coal per horse power hour, which is a very low figure, and which is given by Mr. Bertram himself, amounts to 6,390 tons. Now as the fires have to be banked every morning, or new fires built and steam got up at night, this would add at least 3 tons per day making a total of 7,485 tons. This, at the present contract price of $4.34 per ton, delivered, would amount to $32,484, or very near up to Engineer Keating and McGowan's figures, and over double the amount of Mr. Bertram's estimate. *This item alone would bring up Mr. Bertram's estimate to the contract price of the Toronto Electric Light Company,* or very near it.

TRIMMERS.—A good man can trim in all weather about 60 city lights per day, taking the year round, this average is not exceeded in any city where plant is operated municipally or otherwise. 1300 divided by 60 equals very nearly 22 men. This supposes them to work seven days a week, 365 days a year, no Sundays, no holidays—a dog's life—yet the City Engineer estimates fifteen ! Mr. McGowan is nearer the mark with twenty ; as a matter of fact the Toronto Electric Light Company employ on 1000 city lights, eighteen men, and give each of them a week's holiday in the year without the loss of pay.

INSPECTORS OR PATROLMEN.—These men are not, as has been erroneously supposed, intended to inspect other men. It is a technical name for men whose duty it is to cover the city during the night to look after the lamps and re-light or re-trim any that may be out, to report faulty lamps to be taken to the works for repairs the next day and replaced. It is a dangerous occupation and an unpleasant one, especially during bad weather. The men have to be trained and supplied with special appliances. It would be utterly impossible or unprofitable to attempt to train the whole police force, and have each man carry round the insulated stool necessary for safety, yet neither estimate of the Engineer or Mr. Bertram allows for sufficient men. They each allow 2 men without horses, and expect them to walk over 250 miles of city streets. The 7 men of the T. E. L. Co., some with horses, manage to see the whole of the present lamps every night. If the police were allowed to be used and the items of fines for lights out abolished, the Electric Light Company could afford to reduce their estimate by the number of men required for an efficient service.

GLOBES.—The City Engineer allows nothing at all for breakage of globes and renewal, which is a serious item in the catapult season. Chicago, for 1,110 lights cost $2,253.00 per year. The experience of the Toronto Company more than confirms this, and a duty of 30 per cent. has to be met besides, though Mr. McGowan allows only $1,950 in his estimate. Another item which costs over $1,000 is put down in neither estimate given, that is renewals of ropes, on which the wear and tear is great as they are continually exposed to the weather.

TAXES.—The Toronto Electric Light Co. will have to pay for taxes next year at 16 mills, 6,160 dollars, more than half of which is on plant used in city lighting, and which the city will lose on the establishment of a civic plant.

DEDUCTIONS.—Although the Toronto Electric Light Co. employ seven inspectors for night work, in place of two as estimated by Mr. Bertram, lamps are occasionally unavoidably out. The deduction last year amounted to over $1,000, which the city would lose in the event of establishing its own plant, and for which no allowance is made in any estimate.

GUARANTEE.—It is guardedly said by Mr. Bertram that he is prepared to guarantee a certain price. Mr. Bertram knows, as well as the most dull witted citizen, that if the city instal a plant and pay a quarter of a million dollars for it, *that they will run it themselves* and not farm it out—it would be stupid to do so—and therefore his imaginary guarantee is quite safe.

CONCLUSIONS.—That the figures given by Mr. McGowan as the cost of civic management are quite correct, and are proved by the cost of municipal plants in Chicago, U. S.,Derby, Manchester and other English towns, where the average cost is in excess of his estimate. That the Toronto Electric Light Co. being a private concern, can run the lights for considerably less, and

offer to do so for 74.82 per year each and take the responsibility for deductions for lights out, suits for damages and depreciation of plant. It is manifestly in the interests of the tax payers that a contract should be awarded at the price offered, rather than $300,000 of their money should be sunk in a civic lighting experiment.

## MOONLIGHT SCHEDULE FOR FEBRUARY.

| Day of Month. | Light. | Extinguish. | No. of Hours. |
|---|---|---|---|
| | H.M. | H.M. | H.M. |
| 1...... | P. M. 11.00 | A. M. 6.20 | 7.20 |
| 2...... | " 11.10 | " 6.20 | 7 10 |
| 3...... | " 12.00 | " 6.20 | 6.20 |
| 4...... | .......... | " 6.20 | } 4.50 |
| 5...... | A. M. 1.30 | .......... | |
| 6...... | " 3.00 | " 6.20 | 3.20 |
| 7...... | No light. | No light. | .... |
| 8...... | No light. | No light. | .... |
| 9...... | No light. | No light. | .... |
| 10...... | P. M. 6.00 | P. M. 8.00 | 2.00 |
| 11...... | " 6.00 | " 9.10 | 3.10 |
| 12...... | " 6.00 | " 10.30 | 4.30 |
| 13...... | " 6.00 | " 11.40 | 5.40 |
| 14...... | " 6.00 | A. M. 1.00 | 7.00 |
| 15...... | " 6.00 | " 1.00 | 7.00 |
| 16...... | " 6.00 | " 2.00 | 8.00 |
| 17...... | " 6.00 | " 3.10 | 9.10 |
| 18...... | " 6.00 | " 4.10 | 10.10 |
| 19...... | " 6.00 | " 5.50 | 11.50 |
| 20...... | " 6.00 | " 5.50 | 11.50 |
| 21...... | " 6.00 | " 5.50 | 11.50 |
| 22...... | " 6.00 | " 5.50 | 11.50 |
| 23...... | " 6.00 | " 5.50 | 11.50 |
| 24...... | " 6.00 | " 5.50 | 11.50 |
| 25...... | " 6.00 | " 5.40 | 11.40 |
| 26...... | " 6.10 | " 5.40 | 11.30 |
| 27...... | " 7.00 | " 5.40 | 10.40 |
| 28...... | " 8.00 | " 5.40 | 9.40 |
| | | Total, | 200.10 |

## PERSONAL.

Mr. Wm. T. Jennings, of whom a character sketch appeared in the ELECTRICAL NEWS for January, has been elected third vice-president of the Canadian Society of Civil Engineers.

Messrs. A. E. Edkins and A. M. Wickens have been appointed representatives of the Canadian Association of Stationary Engineers on the Board of Management of the Toronto Technical School.

Mr. Frank Pitcher, of Stanstead, Que., has recently received the appointment of demonstrator of electrical engineering in McGill University, Montreal. Mr. Pitcher is an honour graduate of the University, and has had considerable practical experience as an electrical engineer.

Mr. Chas. W. Hagar, on vacating the position of manager of the Royal Electric Co., was presented by the officials and employees of the company with a complimentary address, accompanied by a handsome office desk and other fixtures. Messrs. F. Duffy and W. Darlington made the presentation.

Mr. L. M. Pinolet, late Treasurer of the Montreal Electric Club, left Montreal a few weeks ago for Newark, N. J., where he has accepted a position as assistant in the laboratory of the Moore Electric Mfg. Co. Mr. Pinolet has numerous friends in Canada, who will be glad to hear of his future prosperity.

It is understood that Mr. W. E. Davis, electrician of the Toronto Street Railway Company, will shortly leave Toronto to assume a position with the new company of which Mr. H. A. Everett is the promoter, at Detroit, Mich. Rumor has it that Mr Davis is also about to enter into a life partnership with one of Toronto's fair daughters.

Mr. Charles E. A. Carr, late private Secretary to Mr. H. A. Everett, of the Toronto Railway Company, has been appointed manager of the London Street Railway, vice Mr. S. R. Break, who is to take the management of Mr. Everett's new company at Detroit. Mr. Carr, who has entered upon the duties of his new position, is said to be the youngest street railway manager on the continent.

The township council of Waterloo has passed a by-law in behalf of the Berlin & Preston Street Railway.

The Court of Revision of the City of Hamilton placed an assessment of $85,000 on the underground mains of the Hamilton Gaslight Co. The company appealed against this assessment, arguing that the gas mains should not be assessed while the wires and poles of the electric light and telephone companies were exempt. Judge Muir held that gas mains were real estate, and as such were not exempt from assessment.

PUBLISHED ON THE FIRST OF EVERY MONTH BY

### CHAS. H. MORTIMER,

OFFICE: CONFEDERATION LIFE BUILDING,
*Corner Yonge and Richmond Streets,*

**TORONTO,      -      -      CANADA.**
Telephone 2362.

NEW YORK LIFE INSURANCE BUILDING, MONTREAL.
Bell Telephone 2299.

*ADVERTISEMENTS.*
Advertising rates sent promptly on application. Orders for advertising should reach the office of publication not later than the 25th day of the month immediately preceding date of issue. Changes in advertisements will be made whenever desired, without cost to the advertiser, but to insure proper compliance with the instructions of the advertiser, requests for change should reach the office as early as the 22nd day of the month.

*SUBSCRIPTIONS.*
The ELECTRICAL NEWS will be mailed to subscribers in the Dominion, or the United States, post free, for $1.00 per annum, 50 cents for six months. The price of subscription may be remitted by currency, in registered letter, or by postal order payable to C. H. Mortimer. Please do not send cheques on local banks unless 25 cents is added for cost of discount. Money sent in unregistered letters must be at senders' risk. Subscriptions from foreign countries embraced in the General Postal Union, $1.50 per annum. Subscriptions are payable in advance. The paper will be discontinued at expiration of term paid for if so stipulated by the subscriber, but where no such understanding exists, will be continued until instructions to discontinue are received and all arrearages paid.
Subscribers may have the mailing address changed as often as desired. *When ordering change, always give the old as well as the new address.*
The Publisher should be notified of the failure of subscribers to receive their papers promptly and regularly.

*EDITOR'S ANNOUNCEMENTS.*
Correspondence is invited upon all topics coming legitimately within the scope of this journal

THE "CANADIAN ELECTRICAL NEWS" HAS BEEN APPOINTED THE OFFICIAL PAPER OF THE CANADIAN ELECTRICAL ASSOCIATION.

## CANADIAN ELECTRICAL ASSOCIATION.

**OFFICERS:**
PRESIDENT:
K. J. DUNSTAN, Local Manager Bell Telephone Company, Toronto.

1ST VICE-PRESIDENT:
A. B. SMITH, Inspector Canadian Board Fire Underwriters, Toronto.

2ND VICE-PRESIDENT:
C. BERKELEY POWELL, Manager Ottawa Electric Light Co., Ottawa.

SECRETARY-TREASURER:
C. H. MORTIMER, Publisher ELECTRICAL NEWS, Toronto.

EXECUTIVE COMMITTEE:
L. B. McFARLANE, Bell Telephone Company, Montreal.
GEO. BLACK, G. N. W. Telegraph Co., Hamilton.
T. R. ROSEBRUGH, Lecturer in Electricity, School of Practical
Science, Toronto.
E. C. BREITHAUPT, Berlin, Ont.
JOHN YULE, Manager Guelph Gas and Electric Light Company,
Guelph, Ont.
D. A. STARR, Electrical Engineer, Montreal.
J. J. WRIGHT, Manager Toronto Electric Light Company.
J. A. KAMMERER, Royal Electric Co., Toronto.
J. W. TAYLOR, Manager Peterboro' Carbon Co., Peterboro'.
O. HIGMAN, Inland Revenue Department, Ottawa.

## MONTREAL ELECTRIC CLUB.

OFFICERS:
President, W. B. SHAW,         Montreal Electric Co.
Vice-President, H. O. EDWARDS.
Sec'y-Treas., CECIL DOUTRE.         81A St. Famille St.
Committee of Management, T. F. PICKETT, W. GRAHAM, J. A. DUGLASS.

## CANADIAN ASSOCIATION OF STATIONARY ENGINEERS.

**EXECUTIVE BOARD:**
President, J. J. YORK,         Board of Trade Bldg, Montreal.
Vice-President, W. G. BLACKGROVE,         Toronto, Ont.
Secretary, JAMES DEVLIN,         Kingston, Ont.
Treasurer, DUNCAN ROBERTSON,         Hamilton, Ont.
Conductor, E. J. PHILIP,         Toronto, Ont.
Door Keeper, J. F. CODY.         Wiarton, Ont.

TORONTO BRANCH No. 1.—Meets 2nd and 4th Friday each month in Room D, Shaftesbury Hall. Wilson Phillips, President; T. Eversfield, Secretary, University Crescent.

HAMILTON BRANCH No. 2.—Meets 1st and 3rd Friday each month, in Maccabee's Hall. Jos. Langdon, President; Wm. Norris, Corresponding Secretary, 211 Wellington Street North.

STRATFORD BRANCH No. 3.—John Hoy, President; Samuel H. Weir, Secretary.

BRANTFORD BRANCH No. 4.—Meets 2nd and 4th Friday each month. C. Walker, President; Joseph Ogle, Secretary, Brantford Cordage Co.

LONDON BRANCH No. 5.—Meets in Sherwood Hall first Thursday and last Friday in each month. F. Mitchell, President; William Meaden, Secretary Treasurer, 533 Richmond Street.

MONTREAL BRANCH No. 1.—Meets 1st and 3rd Thursday each month, in Engineers' Hall, Craig street. President, Jos. Robertson; first vice-president, H. Nuttall; second vice-president, Jos. Badger; secretary, J. J. York, Board of Trade Building; treasurer, Thos. Ryan.

ST. LAURENT BRANCH No. 2.—Meets every Monday evening at 43 Bonsecours street, Montreal. R. Drouin, President; Alfred Latour, Secretary, 306 Delisle street, St. Cunegonde.

BRANDON, MAN., BRANCH No. 1.—Meets 1st and 3rd Friday each month, in City Hall. A. R. Crawford, President; Arthur Fleming, Secretary.

GUELPH BRANCH No. 6.—Meets 1st and 3rd Wednesday each month at 7:30. p.m. C. Jorden, President; H. T. Flewelling, Secretary, Box No 8.

OTTAWA BRANCH, No. 7. — Meets 2nd and 4th Tuesday, each month, corner Bank and Sparks streets; Frank Robert, President; F. Merrill, Secretary, 352 Wellington Street.

DRESDEN BRANCH No. 8.—Meets every 2nd week in each month; Thos. Merrill, Secretary.

BERLIN BRANCH No. 9.—Meets 2nd and 4th Saturday each month at 8 p.m. W. J. Rhodes, President; G. Steinmetz, Secretary, Berlin Ont.

KINGSTON BRANCH No. 10.—Meets 1st and 3rd Tuesday in each month in Fraser Hall, King Street, at 8 p.m. J. Devlin, President; A. Strong, Secretary.

WINNIPEG BRANCH No. 11.—President, G. M. Hazlett; Recording Secretary, W. J. Edwards; Financial Secretary, Thos. Gray.

KINCARDINE BRANCH No. 12.—Meets every Tuesday at 8 o'clock, in the Engineer's Hall, Waterworks. President, Jos. Walker; Secretary, A. Scott.

WIARTON BRANCH No. 13.—President, Wm. Craddock; Rec. Secretary, Ed. Dunham.

PETERBOROUGH BRANCH No. 14.—Meets 2nd and 4th Wednesday in each month. S. Potter, President; C. Robison, Vice-President; W. Sharp, engineer steam laundry, Charlotte Street, Secretary.

BROCKVILLE BRANCH No. 15.—W. F. Chapman, President; James Aitkens, Secretary.

CARLETON PLACE BRANCH No. 16.—W. H. Routh, President; A. M. Schofield, Secretary.

## ONTARIO ASSOCIATION OF STATIONARY ENGINEERS.

BOARD OF EXAMINERS.
President, A. E. EDKINS,         139 Borden st., Toronto.
Vice-President, R. DICKINSON,         Electric Light Co., Hamilton.
Registrar, A. M. WICKENS,         280 Berkeley st., Toronto.
Treasurer, R. MACKIE,         28 Napier st., Hamilton.
Solicitor, J. A. McANDREWS,         Toronto.

TORONTO—A. E. Edkins, A. M. Wickens, E. J. Phillips, F. Donaldson.
HAMILTON—P. Stott, R. Mackie, R. Dickinson.
PETERBORO'—S. Potter, care General Electric Co.
BRANTFORD—A. Ames, care Patterson & Sons.
KINGSTON—J. Devlin (Chief Engineer Penetentiary), J. Campbell.
LONDON—F. Mitchell.

Information regarding examinations will be furnished on application to any member of the Board.

A SUCCESSFUL working test of Gray's teautograph is reported to have taken place on the 15th of December over the long distance telephone line between Paris and London—a distance of 312½ miles—23 miles consisting of submarine cable and 5½ miles of buried conductors. Eighteen words were transmitted in 36 seconds. The writing as received in London is said to have been perfectly legible, though of ragged appearance.

THE first preliminary meeting of the Executive Committee of the American Street Railway Association, in connection with the annual convention of the Association to be held in Montreal next autumn, will take place at the Windsor Hotel, Montreal, on the 27th inst. It is expected that at this meeting a local committee will be appointed to further the arrangements for the convention. Mr. Granville C. Cunningham, manager of the Montreal Street Railway, is a member of the Executive of the Association, and is deeply interested in the success of the Montreal meeting.

IN Montreal, where the auer light first made its appearance in Canada, and where it came largely into use, its defects are becoming understood, and its popularity is said to be now on the wane. The opinion is freely expressed that after the burners have been for a time in use, the consumption of gas is almost if not quite as great as with the ordinary burner, while the color of the light changes from white to green. It is further stated that the atmosphere is heated and deprived of its oxygen to quite the same degree as when gas is burned in the ordinary way. The opinion which once prevailed that the new light would seriously affect, if not supersede, the incandescent electric light, no longer obtains. The fact that during 1894 the number of incandescent lamps in use in the city of Montreal increased by 13,000, is the best possible proof that the incandescent electric light is growing in popular favor.

IN response to our request for his opinion as to the soundness of the views expressed in the ELECTRICAL NEWS by Mr. George White Fraser on Municipal control of electric lighting in towns, Mr. J. J. Wright, manager of the Toronto Electric Light Co., writes : "The best answer I can give to the article in question is the recommendation of West Toronto Junction's Mayor to the Council, that if they can get their principal streets lighted by the company supplying Toronto at a reasonable rate, they close up their station and dispose of the plant if possible." We publish also expressions of opinion on this subject from Mr. A. A. Wright, of Renfrew, and Mr. Chas. Baillairgé, City Engineer of Quebec. The subject is an interesting and important one, and its thorough consideration at the present time is opportune and calculated to prove beneficial alike to electric lighting companies and municipal corporations. We therefore invite a free expression of opinion.

AS the result of recent interviews with a number of Canadian electric manufacturing and supply companies, it is learned that, considering the commercial depression which has prevailed for two or three years past, a very satisfactory business was done during 1894. The development of electric railways was somewhat hindered last year by the prevailing financial stringency, but a fair amount of work in this line was done, and the prospect is that many new enterprises will be carried out during the present year. Electric lighting, principally incandescent lighting, continues to grow at a surprising rate. The representative of one of the electrical companies expressed to the writer his opinion that this was due in some measure to the high price of coal oil in Canada. It is a well-known fact that the best grades of American oil, which is the kind most largely used, costs nearly four times as much in Canada as in the United States, the price here being 29 cents per gallon, as against 8 cents per gallon in the United States. It is not surprising that with so little difference in cost, the public should largely adopt incandescent electric lighting. In Montreal we learn that the incandescent lamp has come into use, not only in the residences of wealthy citizens, but also in the houses of persons of moderate means. The use of electricity for power has kept pace with its use for other purposes, and the number of electric motors manufactured and sold during the year was surprisingly large. Owing to the disposition of some of the smaller manufacturers to cut prices, there is little or no profit to the manufacturer in this line. This, however, is a condition of things which will probably adjust itself in the near future. It is impossible to construct and sell at the prices now accepted by these manufacturers, a motor which will combine in a satisfactory degree efficiency with durability. As soon as this fact becomes realized, as no doubt it shortly will, those who are responsible for reducing prices to the point where a satisfactory machine cannot be produced, will have to give place to the manufacturer who offers a good machine at a fair price.

OF special interest to Canadians, as being the invention of Mr. T. L. Willson, originally of Hamilton, is a new process for producing cheaply and conveniently an illuminating gas which gives a very brilliant and white light. The gas is nothing but the well-known acetylene, and the invention consists in the method of generating it. Commercially it depends upon the production of calcium carbide cheaply and in large quantities, and Mr. Willson discovered the means of doing this whilst working on the production of aluminium by an electric furnace. This furnace is a species of huge arc lamp. A carbon rod connected to one pole of a dynamo dips into a carbon crucible connected to the other pole, and the most refractory substances may be melted by the intense heat due to the current passing between rod and crucible. Subjected to this intense heat a mixture of coal dust and powdered burnt lime is converted into calcium carbide. The process of using this for gas making is simplicity itself. Water is decomposed by calcium carbide and pure acetylene produced. Calcium carbide is a solid easily transported, and it will be readily seen that an extremely simple gas machine, to slowly drop water on the solid carbide, is all that is required for gas making. Such a machine would be excellently adapted to private gas making. The point lies in the cost of the calcium carbide. One ton of carbide is obtained from a mixture of 1200 lbs. coal dust and 2000 lbs. burnt lime. The figures given for cost are :—

| | |
|---|---|
| 1200 lbs. fine coal dust | $2.50 |
| 2000 lbs. powdered burnt lime | 4.00 |
| 180 electrical horse power for 12 hrs., using water power | 6.00 |
| Labor, &c. | 2.50 |
| Cost of 2000 lbs. calcium carbide | $15.00 |

These figures seem very close, but even if we double the total for the selling price, making the carbide cost $30 per ton to the consumer, the results are striking. One ton of carbide will produce 10500 cub. ft. of acetylene, which when burnt with a sufficient supply of air, gives the same illumination as 100,000 cub. ft. of ordinary illuminating gas of 22 to 25 candle power per 5 foot burner. Thus the cost of carbide sold at $30 per ton is only 30 cents for material, giving light equal to 1000 cub. ft. ordinary gas. In addition, the light given is of a much whiter quality. If this invention fulfils its present promise it is difficult to say what effect it will have on other illuminants. One point concerning it should be mentioned. Acetylene has a very disagreeable smell. The odour from the imperfect combustion of ordinary gas is due to the formation of acetylene. A leaky gas pipe would be intolerable.

THAT there is a great deal of ignorance displayed and nonsense talked about the "deadly trolley," is perfectly patent to the majority of sensible men, even though they are without the technical knowledge which would enable them to see exactly where the errors and fallacies come in. But amongst these same men there is undoubtedly a feeling, which sometimes finds expression, that although there is no real danger to the public to be apprehended now, the present safety is largely because the overhead construction work is new, and that with lapse of time there will naturally come a general deterioration in the strength of the overhead construction which will constitute it a serious menace to life and property. This feeling is honest and therefore deserves serious and respectful treatment. Such points as that the 500 volts of a trolley system are not really dangerous to life, it is not necessary to consider in this connection. Those who share the feeling referred to concede that, be that as it may, there is at any rate no danger when the overhead construction is kept in its proper place—overhead. What they fear is that in time it may tumble down. It is to this point and to these men that we address ourselves, and ask them to look at the matter from the point of view of how it would affect the street railways. The trolley wire is a vital point of the railway, and that it should be maintained in the best possible condition is of the utmost importance to the railway as a business concern. A car may break down and it can be pushed home by the next car. A dynamo or steam engine in the power house may give way—it can be put out of service for repairs. But with the trolley wire down, cars and power house are both useless ; everything stops until it is repaired. That the railway is liable for damages caused, is the smallest guarantee the public has against neglect by the railway to avoid chance of damage. There is the greater guarantee that every minute's stoppage means a steady loss of income which can never be recouped, besides the prospective loss which always follows an unreliable service. The public has therefore the best guarantee against neglect by a corporation—however soulless and bodyless—that the consequences immediately touch that tenderer point than their consciences, their pocket. The object of the railway is to make money, and when well managed with that object, its first business interest is to keep the overhead work from tumbling down. Deterioration is inevitable, but not being built like the "wonderful one horse shay" it will not all give way together. A weak point develops first, showing faulty design or construction ; with good business management it is promptly repaired, and not only it, but all other points in the system which have the same original defect. Thus before they have developed their inherent weakness, the defect is remedied by improved construction due to this experience. Applying this to all points in a well managed prosperous road, it follows that the construction improves with time, and that the longer the overhead work stays up, the stronger it becomes, through the repairs dictated by the railway's business interests. We have said a prosperous road. If not prosperous, repairs may be skimped and the management may be bad, and like other "lame ducks" it may become a nuisance to the community.

IN consequence of the announcements that two large lines in the Eastern States are each to try working a branch on the trolley system, the customary unbalanced forecasts are being made of the time—always in the near future—when the steam locomotive will have disappeared from the trunk lines of the country, driven out by the trolley. This sort of thing is nonsense, but nonsense when often repeated is infectious, and it is just as well to keep steadily in mind what are the conditions which will make the trolley more economical than the steam locomotive. So shall we be better able to resist the assaults of the promoter, who armed with his charter and its stocking and bonding privileges, may seek our franchises and bonuses, sell us his watered stock, and leave us the poorer by his gains and the wiser by his incompetence or dishonesty. And these conditions in brief depend entirely on the volume of traffic per mile to be handled. If a road has one train a day it is obviously cheaper to use a locomotive to haul that train, even if the locomotive burns coal wastefully, than it would be to maintain a trolley line and a power station or series of power stations, in order to handle the transient load of the passing daily train. On the other hand it is equally obvious that steam locomotives could not perform with equal economy, such constant services as is required on a large street railway. Somewhere between these two extremes lies the critical point where the whole cost of transport by steam locomotives equals the whole cost by trolleys. With greater than this critical traffic, trolleys are best. With less, they are unprofitable. Improvements in the economy of steam locomotives force this critical point up to a greater traffic volume. Improvements in the efficiency of generators, line and motors ; reduction in their cost ; the ability to use for electrical work the natural waterpowers ; all operate to drive the critical point down to a lower traffic volume. Where is this critical point for the road in question ? That is the essence of the whole matter put in a nutshell. A proved saving in coal, for instance, by the trolley is only a part, and it may be an insignificant part, of the question at issue. . It might well occur that if the trolley system got coal for nothing, it would still be unprofitable as compared with steam locomotives wastefully using dear coal. The whole cost must be considered, including interest and depreciation charges on plant. It will then be found to hinge on the volume of traffic per mile. The ELECTRICAL NEWS will rejoice in all legitimate developments in the use of electricity, and legitimate developments are best advanced by sitting down hard on inflated or unfounded speculations.

## THE CARE AND MANAGEMENT OF DYNAMOS.*

### BY H. BUCK.

IN considering this subject to-night, my friends, I don't intend to go into the theoretical part or the construction of dynamos, but I intend giving a few pointers to any engineer who may some time or other have a dynamo or small electric plant under his charge.

The first thing that comes under our notice is the place for the dynamo to rest. It should be a dry, cool place, and free as possible from dust and, unless direct driven, should allow for a belt of proper length. The foundations also must receive our attention, and this is a most vital part if we wish to have as little trouble as possible. Concrete or stone may be used to advantage, or brick with cement as mortar, having a large stone placed as a top.

For large dynamos the bolts which hold the machine in place should be long enough to reach right down to the bottom and fix into iron plates built in, but for small dynamos these bolts may be set in place by lead in holes in the stone top of the foundation. If long holes are left in the foundation for the bolts, these holes should be filled in with cement after the bolts are in position. The bed for the dynamo must be quite level, and the armature shaft set properly parallel with the driving pulley.

Supposing our dynamo to be in position and the wiring all completed and everything in readiness to start, see that your connections are all secure and carefully examine your dynamo before starting it for the first time. Clean your journals and bearings, then see that your lubricators are all filled with oil. In oiling the parts of a dynamo (i.e., filling the oil cups) use nothing but copper oil cans. Turn your armature around slowly by hand to make sure that no loose wire or waste is attached to it, and that nothing catches. Then clean your commutator with the finest sand paper, and be sure that no copper dust is lodged between the bars. A good stiff brush will remove it.

See that the brush-holders are adjusted correctly and that the catches

*Paper read at open meeting of Kingston Association No. 10, C.A.S.E.

that hold the brushes from the commutator when not running, are in order.

Be sure your brushes are trimmed correctly. Some brush manufacturers supply their dealers with a special tool to guide the file in trimming the brushes, so if the engineer is fortunate to possess one of these he will have no difficulty.

Adjust the brushes firstly, so that they protrude at the proper length from the holders ; secondly that they bear in a proper position, and that they press with a firm but moderate pressure on the commutator. After you have the brushes properly adjusted, raise them off the commutator so that the hold-off catches hold them. You can tell if the brushes are properly adjusted on the commutator by two arrows or some significant mark which the makers place on the dynamos. If these are not on the commutator, the rule is that the brushes bear on the directly opposite bars of a commutator, in a two-pole machine, but in a four-pole machine they bear on the bars that are a quarter of the circumference apart.

Start your engine, then release the catches of your brushes on the dynamo, close the switch on your dynamo, (if the dynamo is shunt-wound it will at once excite itself with the main switch open), look to your brushes and see if they spark. If there is any sign of sparking knock the brushes forward or backward until the non-sparking position is found. Then close your main switch and light your lamps.

Dynamos should be oiled every day and constant attention given to the brushes to see if they require to be fed forward or trimmed. No oil should be used on the commutator, (except arc light machines with special commutators), but use only vaseline applied with a cotton cloth (never use waste). Sparkless running is a matter that cannot be too strongly impressed on the engineer who has a dynamo under his charge. A dynamo, if properly attended to, will soon assume a beautiful chocolate-brown colored commutator surface, but the commutator, even of a good machine, may be ruined in a very short time by careless handling. If you allow the brushes to press too heavily your commutator will soon have ridges all around it and, if too light, the vibration will cause them to spark and the result will be, your commutator will be soon worn away in patches at the edges of some of the sections, and will lose its roundness of outline. If this ever occurs, there is only one remedy. Take off all the brush holders and amuse yourself by turning it down or filing it true ; but this should occur very seldom. It is well for the engineer to acquaint himself with the positions of all cut-outs in his station, so that if any fuses "blow out" he can place them in himself and not have to send for an electrician to do so for him. If he follows these directions he will have no trouble. In every cut-out there are two binding-screws that the fuse connects ; he should loosen these and take out the burnt pieces of fuse wire. Place one end of your fuse wire under one binding screw and tighten it (not too heavy or you will cut through your wire). With the end of your screw-driver push the other end of your fuse wire under the remaining binding screw. If your circuit is all right it will not burn out. If it does, you have either lamps turned on or else a ground or short circuit. The two latter will always blow your fuse but the former only sometimes (if the wire be very fine, say 1 ampere.)

In closing, let me impress upon the engineer in charge of electrical machinery to keep a " cool head." If anything goes wrong, do not get excited and lose your wits, but do everything quickly (if needed) and carefully. Again, keep all your machines thoroughly clean ; dust your dynamos every day, and clean your brushes and commutators, where possible, every day ; and last but not least, have a place for everything and keep everything in its place, so that if anything should occur that you would need any tool, you would know exactly where to find it without having to search for it.

## PUBLICATIONS.

In the February Arena, Henry Wood, writing on " The Dynamics of Mind," claims that as a matter is now held to be instinct with life, so thoughts are as much dynamic forces in life as any other of the phenomena of nature,—electricity or magnetism, for instance.

The proceedings of the International Electrical Congress held in Chicago during the progress of the World's Fair in 1893, edited by Max. Osterberg, have been published in a volume of 500 pages by the American Institute of Electrical Engineers. A photogravure reproduction of portraits of the official delegates of this Congress forms a fitting frontispiece to the book. The thanks of the electrical fraternity is due the American Institute of Electrical Engineers for having undertaken and so successfully carried out this important work.

Application has been made for a charter for the Burlington and Lake Shore Electric Railway Company, to construct and operate an electric railway from a point within or near to the city of Hamilton, in the County of Wentworth, to the village of Burlington, in the County of Halton, via the south side of Burlington bay, and across Burlington Beach, or via the north side of Burlington bay, or both, with power to extend the same along the north or south shore, or both, of Lake Ontario. Mr. Maitland Young, of Hamilton, is the organiser of the company.

## CANADIAN ASSOCIATION OF STATIONARY ENGINEERS.

Note,—Secretaries of the various Associations are requested to forward to us matter for publication in this Department not later than the 20th of each month.

### CARLETON PLACE ASSOCIATION NO. 16.

A very interesting meeting of this association was held on Jan. 5th. Three applications for membership were received. On the 19th January a special meeting was held to consider the advisability of renting a room to be kept always open for the exclusive use of the members, and to be supplied with suitable reading matter, etc. It has been decided to carry the idea into effect. The association is reported by A. M. Schofield, Recording Secretary, to be in a prosperous condition.

### MONTREAL ASSOCIATION NO. 1.

The annual dinner of the above association was held in the rooms of the Society, 662½ Craig street, on the evening of Feb. 2nd. There were present nearly one hundred persons, among the visitors being Mr. E. A. Edkins, Provincial Deputy for Ontario. Following the repast the following toasts were given : " Queen and Country ;" Steam Engineering," responded to by H. Cooper ; " Manufacturing Industries," responded to by Messrs. Fisher and Green ; " Sister Societies," responded to by A. E. Edkins and J. Marchand, president of St. Lawrence No. 2 ; " Brotherhood of Locomotive Engineers," responded to by Mr. Spencer ; " The Executive Council," responded to by Past President T. Ryan and Provincial Deputy O. E. Granberg ; " License and Inspection Law," responded to by Bro. A. E. Edkins, and " The Press." Recitations and songs were given by Bros. Nuttal, Edkins, Joly, O'Neil, Weaver and Nadin. Bro. Driscoll presided at the piano.

### KINGSTON ASSOCIATION NO. 10.

A largely attended and most interesting open meeting of this Association was held in Fraser Hall, Kingston, on the evening of Jan. 15th. The proceedings were presided over by Mr. R. Ring, who stated that the meeting had been called for the purpose of making the public acquainted with the benefits conferred by the association upon engineers and steam users. During the two years that the Association had been in existence it had made about fifty members. Mr. B. W. Tolger and Mr. J. H. Breck were elected honorary members. The former advocated a law making it compulsory on those in charge of steam engines to obtain certificates of competency. Mr. Breck gave an address on " My First Knowledge of Steam." The balance of the programme was as follows : Paper on " History and Objects of the Executive," by J. Devlin ; paper on " Principles of Pipe Boilers," by Mr. Doris ; reading, H. Breck ; paper on " Care and Management of Boilers," by R. Charlton ; paper on " The Arc Lamp," by F. Simmons ; paper on " The Pump," by H. Hoppers ; song, Mr. Manning ; paper on " Steam Heating," by W. Little ; address on " The Order," by Mr. Wickens ; paper on " The Dynamo, its Care and Management," by H. Buck ; paper on " Belting and Lacing," by C. Asselstine ; reading, Mr. Gilmore ; paper on " Line Shafting," by A. Strong ; paper on " The Engineer's Duty," by A. Donnelly ; " Paper on " The Indicator," by H. Youlden.

### WINNIPEG ASSOCIATION NO. 11.

The regular meeting of the above association held on Jan. 25, in room 19, Grain Exchange, was most enthusiastic and well attended.

There was no lecture, as the installation of officers and other business took the time and attention of the members.

The following is a list of the officers installed by the District Deputy, Bro. C. E. Robertson : President, Bro. G. M. Hazlett ; Vice-President, Bro. John McKechnie ; Recording Secretary, Bro. W. J. Edwards ; Financial Secretary, Bro. Thos. Gray ; Treasurer, Bro. J. Stuart ; Conductor, Bro. W. F. Brown ; Door Keeper, Bro. Robt. Sutherland ; Trustees, Bros. Harper, Douglas and Whyte.

A very interesting paper was read from the Kingston Association, showing the work done by the C.A.S.E. there.

Regarding the progress and future prospects of the Winnipeg Association, the District Deputy, C. E. Robertson, writes the ELECTRICAL NEWS as follows : " Everything points to a very successful term, as the officers appointed are young and active, and showing great energy in the work of advancing this society. They have arranged to make themselves entirely independent by getting a meeting room of their own, and are in dead earnest to push the society."

### MONTREAL JUNIOR ELECTRIC CLUB.

DEC. 31.—Paper on " Storage Batteries," by R. H. Street.

Jan. 7.—Paper on " Transformers for Medical Use," by W. T. Sutton.

Jan. 14.—The annual election of officers took place on this date, resulting as follows :—President, E. W. Sayer ; Vice-President, Wm. T. Sutton ; Treasurer, R. H. Street ; Secretary, E. A. Brissette.

Dec. 21.—Paper on " Induction Coils," by E. W. Sayer.

### MONTREAL ELECTRIC CLUB.

THE membership of the Club continues to grow, and its meetings are of a very interesting character. Additional interest would attach to the meetings, however, if a considerable proportion of the younger members would make an effort to overcome the diffidence which apparently prevents them from rising to their feet to express their opinions on the matters which come up for consideration. Just in proportion as these members seek to overcome their natural aversion to speaking in the presence of their fellow members will they derive personal benefit from their connection with the Club, and be the means of adding to the interest and value of the work which the society aims to perform on behalf of its members.

The following motion, introduced by Mr. Geo. H. Hill, has given rise to considerable discussion. Some of the members seem to think that the matter to which it refers is one which is beyond the scope of the society, while others believe that, unless the Club takes cognizance of such matters, it will be difficult for it to show proof of its *raison d'etre*. The resolution is as follows : " In the opinion of this Club the laws governing inspection of electric light wiring in Montreal are unsatisfactory, and that a fixed set of rules for such should be drawn up by competent persons, and insurance companies be called upon to enforce same, and make their rating accordingly."

The Secretary of the club writes : " Kindly insert the following notice in your valuable paper : The third annual meeting of the Montreal Electric Club was held on Jan. 7th. The Secretary's and Treasurer's report for the past year showed the Club to be in a flourishing condition. The membership is very good and there is a good balance in the treasury. The following were elected to office for the ensuing year : W. B. Shaw, Pres.; H. O. Edwards, Vice-Pres.; Cecil Doutre, Sec'y-Treas.; T. F. Pickett, W. Graham, J. A. Duglass, Committee of Management.

I am requested by the members of the Club to thank you most sincerely for your past kindness in publishing the papers read before the Club, and also Club notices."

### TRADE NOTES.

The Sterling Co., of Chicago, have appointed Darling Bros., of Montreal, as Canadian agents for the Sterling Safety Boiler.

The Canadian General Electric Co., Ltd., have concluded an arrangement with the National Carbon Co., of Cleveland, by which they become sole Canadian agents for this well-known brand of arc carbon.

Mr. James Hardman, formerly representative in Toronto of Messrs. Robin & Sadler, leather belting manufacturers, Montreal, has transferred his services to Messrs. Goodhue & Co., manufacturers of leather belting, Danville, Que. Mr. Hardman's headquarters are at 90 Bay St., Toronto.

Messrs. Robin, Sadler & Haworth, manufacturers of leather belting, who were among the sufferers by the recent Toronto fire, have advised their customers that the loss of their warehouse and factory on Jordan street, will in no way affect their business, as they have engaged other premises on the same street, and orders addressed to them as formerly will be promptly supplied from their factory in Montreal.

A meeting to complete the organization of the Packard Electric Company, and to decide upon the future location of the company, will be held during the present month. The authorized capital of the new company is $300,000. It is understood that two thirds of this amount has already been paid up. The company propose to engage in the manufacture and sale of all kinds of electrical supplies. There is a probability that the headquarters of the Company will be located in Ontario.

## STEAM ENGINEERING.*

STEAM ENGINEERING has assumed such vast proportions as an agent of modern progress and civilization, that it has given birth to a profession the scope and functions of which are not yet very clearly defined. The engineer's duties, in the performance of his daily routine, involves the application of the laws of nature in various ways, to understand and explain which requires a wide range of scientific knowledge. While there are found to be in the profession men whose intelligence and acquirements would shed lustre on any calling, there are others who by their disregard of correct rules show that they are laggard in the acquisition of that real knowledge so essential to men in their profession. This is to be regretted, in view of the vast amount of property and the great number of lives entrusted to their care, both on sea and land.

Whenever an attempt is made to induce engineers to qualify themselves for their calling, it is met with the old-time question regarding the relative merits of theoretical and practical engineers and the comparative value of theory and practice. The practical man who has no theoretical knowledge scoffs at the theorist, and the latter sneers at the former. It requires very little experience on the one hand, and not much study on the other, to show that each is equally important, only in different ways. Both parties to the controversy should know that theory and practice make perfect. Theory with practical experience will, without doubt, enable men to excel in whatever work they may undertake. Therefore, it should be the highest ambition of engineers to combine theory with practice and prove the one by the other. This object may be effected by devoting a portion of their leisure hours to study, and by pursuing a systematic course of self-culture. Engineers whose early training has been neglected, and who are now debarred from the advantages of a good education, need have no cause for despondency, because the extra exertion and effort required to educate themselves will confer advantages of their own, which the routine work of a school cannot develop. Of course there are men in this, as in all other callings, who will fail, however much they may try to accomplish in the way of educating themselves. This arises from the fact that, though all men morally may be equal, intellectually they never can be. Consequently the ability of men to educate themselves varies in proportion to the amount of natural intelligence they possess. But in any case, study gives quickness of apprehension, enables a man to profit by all the recorded experience of others, develops accuracy, and, if properly directed, teaches men to qualify facts, make proper deductions and reason logically. The knowledge acquired from the study of books is of inestimable value to the young engineer ; without it he can never be thoroughly qualified for the duties of his profession, since he will be lacking in certain definite information which can only be obtained from the use of books, and wanting which he is almost sure to be narrow-minded, very slow to receive new ideas and to estimate the proper value of old ones. Such persons, if occupying positions in which they exercise authority, are very apt to become intolerant of other people's opinions, and to assume that all knowledge begins and ends with them. One of the common excuses for ignorance is the old stereotyped expression, "I am too old to learn." Now, if this expression is made in sincerity, it is a great mistake, as it is a false pride which neglects an opportunity to learn because it came late in life, and it is a false fear which shrinks from an effort on account of its difficulty. One fact is very important and ought to be considered in this connection, viz., that knowledge throws light upon itself, and that it is the first step only that must be taken gropingly, as it were, in the dark ; the bugbears in such cases, like shadows, disappear the moment they are boldly approached. Truths are, in the main, simple and easily to be understood, and are daily being brought more within the grasp of the most ordinary comprehension by means of good books, which may be had at a trifling cost.

It is frequently stated by members of this calling that they are no "book engineers," which statement betrays their ignorance of the manner in which some of the most valuable books on the steam engine originated. They were written by engineers of experience, who wished to advance their profession, and who thought that, if their predecessors could commence their studies in their younger days, they themselves might advance and improve still further, leaving the benefit of their experience to posterity. The art would therefore advance with the age.

As much information may be learned in a few weeks from the works which they have left us as took them years of observation and trial to ascertain. Most of the abuses connected with steam engineering have arisen from two causes, viz., avarice and ignorance ; avarice on the part of owners of steam engines and boilers, who entertain the idea that cheap steam engines and boilers might be managed by a class of men who are willing to work for very low wages ; and ignorance on the part of those who claim to be engineers, but who are only men of all work, or at best mere laborers in the treadmill of routine. It is evidently one of the greatest mistakes connected with the use of machinery to entrust its care and management to persons of inferior judgment, as a compe-

tent engineer, who could command good wages, would probably save three times the difference by his judgment and skill in its proper maintenance. If engineers wish to raise the standard of their profession to what it ought to be, and command remunerative compensation for their services, they may do so by educating themselves, and not otherwise. It will not do for them to shrug their shoulders and say, "I am a practical man," and reject theory, for it is a well-known fact that such men have become a nuisance in every branch of mechanics, being the least progressive, least enlightened and most stubborn in the assertion of their views, because their minds are cramped and will not allow of either the substitution or admission of ideas different from their own—however crude and primitive they may be. No man is practical unless he proves practice by theory and theory by practice, and attaches no importance to statements not sustained by facts.

One of these self-styled "practical" engineers was asked, "What is a vacuum ?" His answer was that he thought it was foul air.

While I have endeavored to show you that book knowledge is very essential to young engineers, we must not forget the practical part as well. It is a common thing to see men parading good certificates, simply because they are good mathematicians or theorists, while they have little or no practice. While this class of men should receive their due reward, it does seem unjust, in the awarding of certificates, to place them above men who have spent the greater part of their lives, and who have shown by their industry, truthfulness and sobriety, that they are reliable in every respect. These are nice points to decide, especially when it has to be done by one man, without, perhaps, very much practical experience.

The question is often asked, "Should an engineer be a machinist ?" The answer I would give to that question is, "Not necessarily so." There is no need of a man learning two trades in order to follow one. Besides, experience has shown me that while a machinist may be the best judge of things that may transpire in relation to the repairing of steam machinery, he is nevertheless frequently less careful, less reliable, less ingenious than one who never learned a regular trade. An engineer should be possessed of natural talent, and should be ingenious and able to discover any defect that may take place in the machinery under his charge, and should be able to take up the lost motion, and take apart and put together any of the different parts of his engine.

The class of men known as engineers have conferred upon mankind the greatest boons, and the monuments which display their conceptions are indestructible as the firmament or the ocean. It cannot be said of the engineer, as has been said frequently of the lawyer or doctor, that if mankind could do without him it would be well for the human race.

*Paper by Mr. Gilmour, read on Jan. 15th at an open meeting of Kingston Association, No. —, C.A.S.E.

---

## ROYAL ELECTRIC CO.

A number of important changes are taking place in the management of the Royal Electric Co., of Montreal. Mr. C. W. Hagar, who for ten years past has occupied the position of secretary and manager of the company, has resigned to enter into business on his own account. The directors of the company passed a resolution expressing their appreciation of services during the period he was connected with the company.

Mr. W. H. Browne, late general manager of the United Lighting Companies, of New York City, has been appointed general manager of the Royal Electric Company. He had barely entered on his duties when seen by a representative of the ELECTRICAL NEWS, and could therefore say but little regarding his future course of action. He intimated, however, that an energetic policy would be pursued, and that the public would have no reason to doubt the existence of the company of which he is the head. While the company would make an effort to secure a fair share of the electrical business, there was no expectation that they would get all the business; and he hoped to enjoy pleasant relations with business competitors.

Referring to the new factory now in course of construction, Mr. Browne said he hoped to see the building completed and the manufacturing plant removed into it before the first of April. In his opinion, the new factory when completed, would offer every facility for turning out work in the most expeditious and economical manner. He believed that the Stanley apparatus which his company has secured the right to manufacture in Canada, was fully up to date, and consequently they were starting out under most favorable auspices. Mr. J. A. Kammerer, who for several years has been the company's agent in Ontario, will occupy the position of general sales agent under the new régime.

---

The electric power station and plant at Merrickville, Ont., were destroyed by fire on January 10th.

## FORETHOUGHT VS AFTERTHOUGHT.

BY W. H. WAKEMAN.

IT is said of some men that their "foresight is hindsight" and their "forethought always comes afterward." This is not a very handsome expression, but it answers the purpose very well in describing the characters referred to. When one of these men is put in charge of a steam plant, there is trouble almost continually, and the plant is frequently shut down, that his hindsight may be made use of and his lack of forethought made prominent. Such a man never makes it his business to inspect the lacings in his main belt at short intervals to see that it is in good order, but allows it to run as long as possible, and when all the machines in the factory are running, thus bringing a heavy strain on the nearly worn out lacing, it fails and the whole factory is shut down for about an hour while a new lacing is put in ; or perhaps a part of the lacing gives way first and the belt is thrown to one side of the pulley, is caught by the floor or wall and badly torn, making it necessary to get a new piece and put it in, and as the job must be done in a hurry, there is no time to properly scarf, cement and rivet it, so that it is laced on, and ever afterwards there are two lacings to care for instead of one. It does not really need to be a very large factory to make such a shut down cost as much as is paid the engineer for a week's work, consequently a man who watches such things and avoids the shut down saves his employer many dollars.

It is a good plan to draw in pieces of old lacing over the new simply to protect the lacing which holds the belt together from wear as it runs over the pulleys. These pieces will then wear out first and so give warning, when they may be renewed and the others kept intact.

Such a man as forms the subject of this article, does not remove small accumulations of sediment from his sight-feed oilers, but waits until the dirt is about half an inch deep in them and the oil passages choked up with it, and as the bearings are not oiled, hot boxes are the result. He is then not slow in applying some heroic remedy and boasting of his skill in curing the evil. The flange joints in his cast iron main steam pipe are leaking drops of water while his engine is shut down, but he has not foresight sufficient to enable him to know that unless they receive proper attention, the packings will be blown out and it will be necessary to shut down to renew them.

If the packing around his piston rod begins to leak, he simply screws up the nuts which hold the gland in place, and when it leaks again he repeats the process, but does not heed the warning that new packing is needed, until some morning after starting up he finds that he can no longer stop the hiss of steam in this way, consequently throughout the entire day, at each revolution of the engine it sounds as if it were about a hundred geese in the engine room, and visitors and employes are not slow to take note of it and rate him accordingly.

This man has an injector in his boiler room which formerly worked very well, but of late it will break occasionally, and frequently he finds it difficult to make it start as it should. This tells him that it is becoming coated with scale on the inside. He should have foresight to enable him to determine that in a short time it will become so filled up as to make it useless, but he lacks this most desirable qualification, and when his pump is being repaired the injector refuses to work and he can not feed his boilers. To cover up his blunder he advances the idea that no injector will last long anyway, and that they fail without giving warning, when the truth is that they do give such warning, but he either does not understand the story they tell, or is too indifferent to profit by it. It matters little which it is, as the result is the same in either case.

With a man in charge who lacks foresight, when the girth seams on the under side of his boilers commence to leak, he does not look ahead and calculate what the result will be if this leakage continues, but proceeds to calk up the leaky seams, and continues the same practice that caused the trouble in the first place. He can not foresee that if he fills a hot boiler with cold water, severe contraction will be the result, or that if he feeds cold water into the bottom of a boiler while under steam pressure, the cold water will settle to the bottom and cause the seams to leak.

His boiler is badly scaled and he introduces some scale resolvent to remove it, but does not possess sufficient foresight to enable him to see that if his remedy is of any value whatever, it will throw down a large quantity of scale which will lodge on the parts immediately over the fire and prevent the water from coming in contact with the iron, the consequence being burned plates and leaky seams.

If a small hole appears in the blow-off pipe, he puts a slip patch over it to stop the leak temporarily, but does not have orethought enough to show him that if corrosion has weakened he pipe in one place it soon will be in others ; but when this pipe fails and his boiler room is filled with clouds of steam and the boiler is unceremoniously emptied of its contents, his afterthought has a chance to secure a prominent position.

If an oil agent offers him a commission on all of the oil that he buys of a certain kind, he repeats the old axiom that "a bird in the hand is worth two in the bush," without taking into consideration the fact that he has made a wrong application of it. He can not see into the future enough to discover that he will soon be no longer a free man, but will be under obligations to those from whom he has taken bribes, forgetting that all of these deals are brought to light sooner or later and always to the disadvantage of those who are concerned in them. The engineer who is capable of getting out of scrapes in short order, often passes as a hero, while the unassuming engineer who is thoughtful, and by his thoughtfulness keeps out of scrapes, attracts but little attention and frequently fails to get as much credit as is really his due. When he leaves a situation where he has had but little trouble, and where shut-downs were few and far between, and is replaced by a man whose forethought comes afterward, the difference is often plainly to be discerned without the aid of a magnifying glass.

There is one more point which I wish to mention, as follows : When a man takes charge of a steam plant, he should have foresight enough to study out the characteristics of his employer, know just what his ideas are as far as possible, and then govern himself accordingly. By this I do not mean that he should sacrifice any of his own opinions or ideas which are proven to be correct, for this is not at all necessary, but he should adapt himself to circumstances and by skillful management of affairs, secure the respect and confidence of his employers.

## PRACTICAL NOTES.

NOTHING helps the introduction of a new machine or device among practical mechanics more than simplicity of design and the absence of numerous joints and pieces, which tend to shorten the life of the machine as well as impair its efficiency. Joints are good things to avoid where possible, as the inevitable wear is followed by lost motion, which affects the accuracy of the machine.

It is a bad practice to put an over-loaded belt down out of sight, especially where there is any inflammable material. The slipping of a belt on its pulley from overload is a good heat producer, especially if the belt hooks happen to stop in contact with the pulley. The writer saw a case of this kind several years ago, and the streams of sparks that came from that pulley rim would have done credit to a Chinese pin-wheel. Such occurrences are dangerous, and precaution should be taken to render them impossible.—Machinery.

A very bad habit in mills where there are large driving belts, is shifting belts with a square stick, no regular shifters being used. The result of this is the belts are more or less injured on the edges. All heavy machines should have shifters to act so that they shift the belt over steadily, not putting too much strain on the driving belt too suddenly. Two pieces of gas pipe just large enough to revolve on round iron supports, for shifters, will, lessen the friction on the edges of heavy belts as these pipes revolve while the belt is being shifted. It effects a great saving in long driving belts ; in fact, any belt at all, leather or rubber.

The transmission of power by ropes has been largely resorted to in England, the preference being given to what is known as the Lambeth cotton rope, which is made of four strands, the center or core of each strand being bunched and slightly twisted, the outside of the strand having a covering of yarns that are firmly twisted. The four strands are further laid with a core in the center to form a rope and twisted in the same way as any four-stranded rope. In this way a rope is formed possessing extreme flexibility, and the fibers will not break by bending when run on pulleys, the rope also standing elongation or stretching some 12 inches in a length of 50 inches before breaking.

## WATER TUBE BOILERS.*

### By M. R. DAVIS.

IT is not my intention to discuss the merits of one particular boiler, but what I wish to do is to bring more fully before you the rise and growth of the coil boiler, and if possible to point out a few of the more important points, as gained from experience.

About Aug the 6th, 1791, the first patent was issued in this country for a water-tube boiler, and was taken out by a Mr. Rumsey. This we may fancy was rather a crude affair, but it was soon followed by the Barlow boiler in 1793, which, together with the Fulton, was put in use.

This was the pioneer of a great variety of boilers of this class. The records of the patents in this line alone would be a revelation to the average reader.

The year 1807 brings to light another boiler of this class built by a Mr. Stevens, of Hoboken, N. J., who became interested in pipe boilers and built one to put in a steam yacht. This boiler was a ft. long, 15in. wide, and 12ft. high, and consisted of 81 1in. pipes. We are not told how this boiler was made, but we may be allowed, from the proportions given, to judge that it was not of the type we would care to use to-day. We have very little in the history of the pipe boiler to interest us for nearly one hundred years, when in 1878, Mr. Roberts, who has become famous for his great success in this line, built his first boiler at Red Bank, N. J. Mr. Roberts felt the need of something better than the market had yet supplied, and by untiring effort has succeeded in giving the world at least a stepping-stone on which to build what the engineering world calls to-day the water-tube boiler. This boiler stands to-day side by side with our best pipe boilers.

But we need not stop with this type. Look at the market to-day, and compare this class of boilers with the shell boilers. Where do they stand as regards variety? We are told, and there is every reason to accept it as a fact, that there have been patents issued on nearly 200,000 pipe boilers—coil or water tube boilers as you choose to call them.

Among the great variety we will name only a few of the most important types. We have the Roberts, Babcock, Root, Sterling, Heine, Alma, Sebury, Hazelton, Ward, Yerrow, Thorneycroft, Belleville, Mosher, Clark, Warrington, Worthington and many others. These boilers belong to a type which in the past few years has largely increased in this country, and not without reason, especially for marine service. A few of the above mentioned boilers are used only for stationary purposes.

American engineers are almost alone in the employment of water-tube boilers. Aside from the Yerrow and Thorneycroft used in England, they are practically unknown outside of American steam vessel—with one exception only. French engineers have recognized their value and have them of very high power in many deep water ships. Water tube boilers are absolutely safe from explosion, safe from disaster even if ordinary care be taken, and they stand abuse better—so far as immunity from costly repairs is concerned—than the shell boiler. A burned sheet in a shell boiler is a serious matter; a burned pipe in a water tube boiler, if such a thing should happen, is of no moment whatever. A fitter with a pair of pipe tongs can take out the old and put in the new one in a few hours—minutes sometimes —according to the location; but it takes days to repair a shell boiler. Water tube boilers stand getting up steam quickly much better than shell boilers. The latter are hot and cold in the wrong places, but a water tube boiler gets hot all over at once; all parts go together and expansion strains are reduced to nothing. The water in them circulates like the blood in a man's body—up one side and down the other—and it keeps on circulating all the while. There is no dead water in a water tube boiler; it is a constant wash of water over hot metal from start to finish.

You might ask me the question: What is the difference between one water tube boiler and another? or in other words, What kind of boiler is the best to use? Since there are so many kinds, there are several ways of answering this question; and if you will follow me a little way in the matter I will try and point out a few types that are most popular.

We will commence with the pipe boiler, or that class made with the joints screwed together—made from lap welded pipe—for no other kind should be used. This class is made in a great variety of shapes, and I may say here that they are like watches: if you get the right kind you will have something reliable, and if you get a poor one you would be better without it.

This class of boiler has been in use longer than any other class of water tube boiler, but the merits of the pipe boiler has been seriously affected by the use of bad material and unskilled workmanship. Pipe boilers have been made by men who knew little or nothing about a boiler or the hardships it had to stand; and this cause, if no other, has been the means of spoiling the merits of a first class maker. I could call to mind instances where pipes have split with only a few days use—or weeks at longest—while good pipes, together with good workmanship, will stand the hardships of years.

One fault with some of the pipe boilers to-day lies in the fact that they have their cross-section or up-take pipes too flat over the fire. Such being the case and small pipes being used, they are very apt to sag, and in a short time something will happen which I will leave you to guess.

Some of our pipe boiler makers have recognized this fact and are remedying it by using short pipes and not driving the water so far through a short pipe—if, indeed, it goes through at all.

Among the water tube class we have a far greater variety and a far superior boiler. You may say, "How is that? I thought pipe and water tube boilers were the one thing!" Not so; water tube boilers have no screwed joints; all screws are expanded into the main drums. This is one of the reasons why the water tube boiler is superior to the pipe boiler, and a great many pipe boiler makers call their boilers "water tube," because it sounds better, and is better.

The water tube boiler is doing to-day what the pipe boiler never has done, or, indeed, never can do. The water tube boiler stands to-day at the head of all others in the eyes of the mercantile world. You say, "Why don't capitalists put them in use then?" I told you a little while ago that unskilled mechanics have placed this boiler where it is, but it has a bright future. Look over the commercial world to-day. What do engineers think about it? A few years ago some of them said, "Oh, it's no use; you can't depend on it; it will fail you when you need it most." It may fail you, but it won't blow up and kill all hands in the factory or on the ship; it will simply stop and you can see at once where the trouble lies. Neglect a shell boiler and see where you or the boiler will go.

I have seen pipe and water tube boilers red hot from top to bottom and cold water pumped into them, without even affecting them to any extent.

Other points about water tube boilers which commend them for general use in the manufacturing world, are, that they are light; take up very little room; are easy to manage; quick to get up steam, and easy to repair. You can put them down in the cellar or up in the garret; distribute them through a building and take up less space than with any other boiler.

For marine use this boiler stands at the head of all others to-day. It has been adopted by the principal naval powers of the world, after having been subjected to the most severe tests.

A number of ocean steamers have been equipped of late with water tube boilers, and the result has been entirely satisfactory.

Comparing the weight of a water tube boiler of the Babcock & Wilcox make with one of the Scotch boilers of same power, we have for the Babcock boiler 25 lbs per sq. ft. of heating surface, while in the Scotch boiler we have from 75 lbs. to 115 lbs. per sq. ft. of heating surface. Thus you see there is a great advantage in favor of the water tube boiler. We may use other comparisons which will convince still further that the water tube boiler as a marine boiler commends itself for use. For one comparison we shall take the Ward boiler, as representative of the water tube type, while the data for the fire tube or shell boiler will be representative of good modern practice, and in each case boilers of a size suitable for large steamers will be assumed:

Weight per sq. ft. heating surface of boiler without water, 27 lbs and 13 lbs. Weight per sq. ft. of heating surface of water in boiler, 15 lbs. and 1¾ lbs. Weight per sq. ft. of heating surface of boiler with water, 42 lbs. and 14½ lbs.

It appears, therefore, that water tube boilers of this type or of such types as may be fairly represented by this boiler, are about one-third of the weight of fire tube boilers of equal heating surface. The water tube boiler is seen to be of about one-half the weight of the fire tube or shell boiler, while the amount of water contained in the former is inconsiderable in comparison with that contained in the latter. We could dwell on this point much longer, but as time will not allow we will simply draw your attention to a few points not yet mentioned. We would refer to such boilers as the Yerrow, Mosher (improved marine type), Seabury, Hazelton and a few others which have no bent tubes, but are made with thin steel tubing only. These boilers will stand hard service much better than that type made with bent pipes or tubes, and why? Because expansion and contraction are much more equally distributed, and over-straining is not so apt to take place when crowded. For example, we will take the Robert and the Thorneycroft; also the Mosher yacht type. All those are made with bent tubes, and have a tendency to straighten when crowding takes place. It has been found that in a certain boiler of the above class, the steam drum was raised by the straightening of the tubes. You can easily prove what I say to be correct by taking a rubber hose; bend it and turn a pressure of steam or water through it, and you will find at once the action of the tubes in some water tube boilers when under pressure.

What do you think is the secret of the success of some of the modern fast steamers? While admitting the improvement in models, we say it is principally due to the boiler. They get more power with less weight with this boiler than with the old type—therefore, less displacement and greater speed. Look if you will at the Hornet, the Daring, the Ferret, the Hazard, all of which have attained a speed of 32 miles or more an hour; this was never done till water tube boilers came into use. Up and down our own shores we find such boats as the Fireen, Norwood, Buzz and a few others, which claim a speed of 28 to 32 miles an hour. Even at Kingston we have found the great advantage of using the water tube boiler.

I might go on discussing the water tube boiler, dwelling on many points, such as thickness of tubes, corrosion, cleaning and repairing, circulation and distribution of heat, but will differ, doing so to a future occasion.

========================

A considerable difference of practice appears to prevail among engineers in the matter of packing the stems of Corliss valves. We have seen the packing pulled out of the stuffing boxes of these stems, and found that in some cases as few as two pieces were used, and yet the stems did not let steam. We saw a valve removed a few days since and about a dozen pieces of packing had been removed before all was out. This, it seems to us, is a waste of packing, and might produce a considerable amount of friction on the stem. If the box is so deep that it requires an extra amount before the gland can take hold, it would pay to fill up that extra space with something else.

*Paper read before Kingston Association No. 10, C.A.S.E.

## CHIMNEY 100 FEET HIGH STRAIGHTENED.

THIS corner is a view representing a brick chimney 100 feet high, which was pulled back into position from a point 28½ inches out of plumb. It is 9½ feet square at the bottom, 5½ feet square at the top, with a central flue three feet square. The estimated weight is 206 tons, and the chimney stands on a foundation 14 feet deep. The soil is affected by the rise in a river, and although two similar chimneys has been already built in the vicinity, no trouble had been experienced. When measurements were first taken the chimney was found to lean about 16 inches, and a few days later was 22 inches out of line. No particular change was noticed for four months, when it was found that the chimney was 28½ inches out.

A scaffolding was built about the chimney, and 42 feet above the stonework and 4½ feet below the center of gravity of the brickwork were placed eight oak timbers 6 inches by 10 inches by 10 feet. The timbers were used to spread the bearing of wire ropes over as large a section as practicable. Around the timbers were placed wire ropes, to which was fastened another wire rope 2½ inches in diameter, having eyes at each end, the lower eye being connected with a system of eleven pulleys secured at a point 78 feet distant and opposite the direction in which the chimney leaned. Cables with turnbuckles were placed at right angles to the main cable and a guard cable was placed in the rear. The earth was excavated on the high side of the foundation nearly half way around to the bottom, a depth of 13 feet, and the main cable put under strain by the pulleys. In three weeks the chimney was straightened 4 inches. A post hole digger eight inches in diameter was used to excavate eleven holes around the trench, which relieved the pressure of the earth, and by the following morning the chimney had moved back in place eight inches. By tightening the rope three times a day and digging additional post holes when necessary, the chimney was brought back to place in a few weeks. The holes were filled in with fine broken stone and gravel thoroughly rammed. The illustration is reproduced from Engineering News.

---

## SPARKS.

It is reported that another electric light plant is to be installed at Renfrew, Ont.

It is reported that Frank Bryden of Toronto will install an incandescent lighting plant at Napanee.

Mr. F. Anderson has been appointed electrician of the Summerside, P. E. I., Electric Light & Power Co.

At Sorel, Que., the Bell Telephone Co. is reported to have reduced the rental of its instruments to $15.00 per year.

Mr. M. W. Corbitt has commenced business in Montreal under the registered title of the Montreal Electrical Supply Company.

The London Electric Co. has contracted to erect and operate 274 2,000 c. p. lamps. 260 of these lamps have been put in operation.

Mr. A. Rowan, of the Customs Department, St. John, N. B., has been appointed government inspector of electric meters in that locality.

The Thompson Electric Co.'s tender of $1,000 for a 50 arc light plant for Toronto Island, has been accepted by the city council of Toronto.

T. Belanger, of Montreal, has invented a three rail surface system which is designed to take the place of the overhead trolley electric system.

A by-law recently submitted to the ratepayers of Stratford, authorizing the expenditure of $16,000 for a municipal plant, was defeated by a large majority.

It is said to be the intention of the Cookshire, Que., Machine Works Company, to go out of the electric lighting business and to dispose of their electric plant.

Application for incorporation is being made by the Dundas County Telephone Company, for the purpose of operating a telephone system at Chesterville, Ont.

Mr. P. S. Archibald, chief engineer of the Intercolonial Railway, discovered in a recent visit to Manitoba, that nearly all the small towns in the North-West are lighted by electricity.

Mr. H. Calcutt, of Ashburnham, has applied for a patent for a combined water tube and tubular boiler, which it is claimed will effect a large saving in fuel, and where necessary will increase the boiler capacity.

Application for incorporation is being made by the Midland Electric Light & Power Co., of Midland, Ont. The capital stock of the company is to be $10,000.

The Crown Pressed Brick Co., of Ormstown, Que., have submitted to the Council of Huntingdon, Que., a proposition to light the streets of that place with electric light.

The London Street Railway Company has made a new offer for an electric franchise. It embraces seven tickets for 25 cents, and workingmen's tickets eight for 25 cents.

The Hamilton, Grimsby and Beamsville Electric Ry. Co., have been granted permission by the railway committee of the privy council, to cross the tracks of the G. T. R. Co. in Hamilton.

It is reported that in consideration of the sum of $20,000 in cash, Mr. W. S. Williams will relinquish his stock in the Montreal Park and Island Railway Co., and all other claims against the company.

The Toronto Street Railway Co. have suggested to the City Council that an electric fire-engine be purchased, and have offered to supply free of cost the electricity required to operate the same within reach of their conducting wires.

It is reported that the owners of timber limits in the locality are considering the question of constructing an electric railway between Lake Temiscamingue and Lac des Quinze, by means of which to get lumber more easily to the Temiscamingue.

The Montreal Incline Railway Co., represented by Messrs. F. B. McNamee and Wm. Mann, and the Montreal Street Railway Co., are each endeavoring to secure authority from the City Council to construct and operate an electric railway through Mountain Park.

The Town Council of St. Boniface has under consideration a by-law, which has received its first reading, for the establishment of an electric light plant. If the by-law should pass the Council, operations will be commenced about the first of March.

The recently organized Ottawa Porcelain and Carbon Company have purchased a site for a new factory on Elgin street. The erection of the factory, which is to be 200 feet long, 150 feet wide, and three stories high, is to be commenced early in the spring.

The railway committee of the privy council has granted to the Montreal Street Railway Co., the privilege of extending their line along Etienne street, and of crossing the G. T. R. tracks, on condition that the company shall bear the cost of maintaining the necessary semaphores, gates and lights at the crossing. This will afford electric railway communication to the residents of Point St. Charles.

At a recent meeting of the town council of Kincardine the following resolution was adopted : That the council feel it their duty to tender their thanks to Mr. L. A. Campbell, foreman of construction for the Canadian General Co., on the incandescent electric light work for the very kind and courteous way in which he has at all times treated the committee and council, and for the very skilful manner in which the work appears to have been done, as evidenced by the satisfactory way in which the plant was started and is now working.

The annual report of the Montreal Telegraph Co., presented at the annual meeting of the shareholders held in Montreal on Jan. 10th, showed : Assets—Telegraph lines, $1,625,890 ; telegraph cables, $33,487.39 ; offices and equipment, $212,500 ; real estate in Montreal, Ottawa, Quebec and Toronto, $279,946.46 ; cash, other real estate, accounts receivable, etc., $101,853.19. Total, $2,253,677.04. Liabilities—Shareholders' capital, $2,000,000 ; dividend No. 122, payable 15th January, 1895, $40,000 ; unclaimed dividends, etc.. $1,794.75. Total, $2,041,794.75. Excess of assets over shareholders' capital, $151,823.85. Contingent fund, $60,058.44. Total, $2,253,677.04. Mr. Joseph H. Joseph's motion that 3 per cent. of the available assets be distributed among the shareholders was ruled out of order. A motion by Mr. John Crawford that the directors consider the advisability of dividing among the shareholders the present contingent fund, was adopted. The old board of directors was re-elected, Mr. Andrew Allan was re-elected President.

The third annual meeting of the shareholders of the Toronto Railway Company was held on the 16th of January. The annual report of the President and Directors for 1894 shows a net profit of $250,965, against $212,859 in 1893. The directors desire to explain their policy in expending their surplus earnings in the improvement of the property by stating that in the operation of a modern street railway they have felt it to be of the utmost importance that it should be sufficiently provided with electrical and steam power, cars, car houses, machine shops, tools and machinery, so that in the extensions of its system the bonds of the company, which can be issued at $35,000 per mile, would provide sufficient capital for its requirements. During the present year the directors have completed the purchase of the Toronto and Mimico Electric Railway and Light Company's property, and paid for it out of the surplus earnings of the company. The net earnings of this property have paid 5 per cent. upon its cost, and a surplus of $2,502. The statistical statement shows the gross earnings for 1894 to have been $958,370, an increase of $58,138 over 1893. The operating expenses were $517,707, or $19,890 less than in 1893, leaving the net earnings $440,663, an increase of $78,028. Twenty-two million, six hundred and nine thousand, three hundred and thirty-eight passengers were carried in 1894, as against 21,215,010 in 1893 The operating expenses were 54 per cent. of the earnings, as against 59.07 per cent. the previous year. During the year $7,973 was paid out for injury and damages.

# ELECTRIC RAILWAY DEPARTMENT.

## PROPOSED LONDON AND WESTERN ONTARIO ELECTRIC RAILWAY.

NOTICE has been given that H. A. Everett, E. W. Moore, T. H. Smallman, Greene Pack, S. R. Break and C. S. Ivey will make application at the next sitting of the Legislature of Ontario, to incorporate the London and Western Ontario Electric Railway Company, with power to construct electric railways as follows :

(1) London to St. Thomas and Port Stanley, also St. Thomas to Aylmer ; (2) London to Deleware, Glencoe, also from Deleware to Mt. Brydges and Strathroy ; (3) London to Lucan, St. Mary's and Stratford ; (4) London to Dorchester, Ingersoll, Woodstock, and also Ingersoll and Tilsonburg.

As will be seen by the accompanying map, this scheme proposes to make London the distributing point for Western Ontario, and these lines run through and tap one of the best sections of country in Ontario. The population served will be about 200,000 people, and the line to Port Stanley includes connection with the line of steamboats to Cleveland. This is one of the largest schemes of the kind proposed in Canada, and further developments will be eagerly looked for.

## LEGAL.

GAUTHIER VS. THE MONTREAL STREET RAILWAY COMPANY.—The plaintiff in this case claimed $335 damages by reason of being knocked down and dragged along a considerable distance by a car of the company defendant, on the 7th October, 1893. He alleged that the injury was caused by the fault and negligence of the defendants. The defendants pleaded, first, a demurrer to one of the allegations of the declaration, to the effect that the company habitually ran its cars faster than is permitted by the by-law regulating the matter. The defendants further pleaded want of notice of action as required by the company's charter, and that the notice given was irregular. They further alleged that there was no negligence or fault on their part. The court maintained the demurrer on the ground that the fact, if proved, that the company habitually ran its cars faster than allowed by the by-law was not of itself a good reason for holding the defendants liable to damages in this case. The plea of want of notice was rejected, because notice had been given, and the defendants had not shown in what respect it was irregular. On the merits of the action, however, the court was of opinion that the plaintiff had failed to prove that the injury was due to any fault or negligence on the part of the defendants. Although there was some evidence that at the time plaintiff was

struck by defendant's car, the car was moving at too high a rate of speed, yet such excess of speed did not appear to have been the determining cause of the injury suffered by the plaintiff. On the contrary, the weight of evidence went to show that the injury was attributable to the fact that the plaintiff hurriedly endeavored to cross the tracks of defendants in front of the approaching car at a moment when, had he used the most ordinary prudence, and looked where he was going, he must have seen that such an attempt would result in his being struck by the bar. The action was therefore dismissed.

## HAMILTON RADIAL RAILWAY.

We present herewith an illustration of a proposed passenger depot to be erected in Hamilton by the Hamilton Radial Railway Company. The proposed site of the building is Cannon

PROPOSED DEPOT HAMILTON RADIAL RAILWAY

street, with frontages also on James and McNab streets. The building is to be constructed of brown sandstone, and is to contain commodious waiting and dining rooms.

## SPARKS.

An agitation is said to be on foot at Owen Sound for the construction of an electric railway between that town and Meaford.

Mr. G. H. Campbell, the manager, is authority for the statement that considerable extensions will be made this year to the Winnipeg Electric Company's lines.

Mr. S. R. Break, the popular manager of the London Street Railway Company, has received the appointment of General Superintendent of the new electric railway at Detroit, and will enter upon the duties of his new position at once.

Incorporation is being sought for the Kingston and Gananoque Electric Railway Company to construct an electric railway between Kingston and Gananoque, with power to extend the same east as far as Brockville, and north to Westport.

The ratepayers of Burchton held a meeting recently and appointed a committee to obtain signatures to petitions for the extension of the Ottawa Street Railway to Hintonburgh, Skead's Mills, Birchton, Britannia, Deschenes and Aylmer.

The Toronto & Scarboro Electric Railway has passed into the hands of the Toronto Railway Co. The shareholders of the suburban line have accepted in exchange for their property Toronto Railway stock. It is understood that the new management will construct loop lines via Blantyre and Victoria Park.

At the annual meeting of the Hamilton, Grimsby and Beamsville Railway Company, recently held, it was decided not to extend the line to Grimsby Park at present, but to endeavor to secure a site for a park on the lake shore between Stoney Creek and Winona. The following officers were elected :— Directors, Messrs. C. J. Myles, Thomas W. Lester, John Hoodless, A. H. Myles, Walter Glover, John Gage and Robert Ramsay. Mr. C. J. Myles was re-elected president ; Mr. T. W. Lester, vice-president ; and Mr. Adam Rutherford, secretary-treasurer.

At the recent annual meeting of stockholders of the Galt & Preston Railway, a statement was presented showing the total cost of the works to the present time to be $59,537.61, which is somewhat below the estimate. The earnings of the road to December 31st—being a few days over five months —was $5,980.27, which is considerably in excess of expectations. The running expenses for the same period, including interest, was $4,152.22. The directors have been instructed to apply for authority to increase the capital stock of the company to $100,000. In the event of this authority being given, the extension of the road to Hespeler will be undertaken in the spring. The following gentlemen have been elected as the officers of the company for the ensuing year :—Messrs. Thomas Todd, President ; R. G. Cox, Vice-President ; Lutz, Secretary-Treasurer ; Directors, H. McCulloch, D. Spiers, F. Clare, Preston J. D. Moore, M.P.P.; Auditors, J. M. Duff, J. M. Irwin.

## INCANDESCENT LIGHTING ON THE METER SYSTEM COMPARED WITH THE CONTRACT SYSTEM.*

### BY G. L. COLE.

In taking up the subject of incandescent lighting on the meter plan, I shall state my own experience from the date of my going to the city of Beloit in September, 1893. At that time I found my customers divided, some on the contract and others using meters. Now the question arose, whether I should furnish this current on the European, meter plan—pay for what you receive —or the American plan by contract—take all you can get. Two questions came up for consideration in changing from contract plan to meter system. First—A considerable outlay in meters. Second—Would the additional returns warrant the change?

To compensate me for the extra investment in the meters, I charge meter rental, which pays a good percentage. As showing the additional profit in the meter over the contract system, I will give an example. I had a customer on the contract plan who was paying $3.50 per month for the electric current in his residence. I changed him to meter basis and his bills, run on the meter plan, for several months, were $53.49, an average of $7.64 per month. There is a good deal of satisfaction to feel that what current the customer uses I am receiving pay for, and not is wasted, as is generally done on the contract plan. When I furnish current on the meter plan and the customer leaves his residence for a short time, I make a minimum charge of $1 per month in addition to meter rental, while if this service were furnished on contract the usual cry would be raised, "We have not used the lights." Unless a discount were made dissatisfaction would arise.

Another advantage was that I could use a smaller transformer to supply the customer on meter plan than I could on contract plan, for I felt assured that the customer would not use at any one time the entire number of wired lights unless for special occasions, when we could temporarily install a larger transformer. To illustrate: For a factory wired with 95 lamps I only use a 50 light transformer, and the bills run at 20 cents per 1,000 watts, $459.10 for eight months, an average of $57.38. Another example: We have 180 lights in one factory; these are only used until 6 P. M., so we put up a 50 light transformer, and from the same transformer we connect an additional 40 lights for a lodge room. Both of these customers are on the meter basis, and I will say that the income on the investment has been very satisfactory to me; and this manner of furnishing the current has given entire satisfaction to the consumer, he pays for what he uses and no more. Another advantage is the large number of lights we can wire from a medium sized dynamo. I feel assured the customer will turn off all the lights he can possibly spare to keep his bills down as low as possible. I now have 3,300 incandescent lights wired, an average increase of 151 lights per month.

My dynamos are one of 750 lights and one of 1,000 lights. When I purchased the plant I found the entire number of incandescent lamps wired was 1,029, and the average daily load was 550. It is my opinion, based on former experience, that where there were 825 incandescent lamps wired entirely on the contract plan the average daily load was 450. This shows that over one-half the number wired were being used. On the meter plan I can calculate on about one-third the number burning that I have wired, or about 850 on an average for 10 hours.

Still another advantage is the profit in wiring residences, for I find that people will put in more lights when they are to be used on the meter system than they would when on the contract

*Address at the meeting of the Northwestern Electrical Association, Milwaukee, January 6-18, 895

system, where the basis of furnishing current is so much per lamp. It follows that our customers feel better satisfied to have their houses thoroughly equipped with lamps from garret to cellar and in the stables as well, knowing that after they are once installed they are no expense to maintain unless they are used.

The following points are self-evident in favor of the meter: The station receives pay for current furnished; the customer pays only for the current received; service is more satisfactory to the customer because it is more complete, and investment under the meter system commands a larger field than is possible to obtain under the contract system. It stimulates economy on the part of the customer and induces study on the part of the owner. The meter system is eminently fair to both parties, and soon after installation commends itself to the public.

As the result of being caught in the driving belt of a dynamo, Mr. A. Ross, superintendent of the Oxford, N. S., Electric Light Station, recently sustained severe injuries.

On the occasion of his removal to assume the management of the Bell Telephone Company's exchange at Orillia, Mr. E. H. Farrow, who for five years has been manager of the Stratford exchange, was waited upon by a number of the subscribers and presented with an address expressing appreciation of the efficiency of the service under his management and regret at his removal, together with best wishes for his future success. Mr. Farrow's successor is Mr. John E. Bull, manager of the exchange at Orillia.

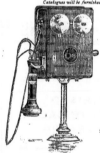

## SPARKS.

A satisfactory dividend has been declared by the Hamilton Street Railway Co.

The passengers carried by the Kingston Street Railway Co. are said to number 8,000 per week.

The city council of Hamilton gives notice of its intention to apply to the legislature to dissolve the perpetual charter of the local gas company, and to authorize the corporation to own and operate electric railways.

Temporary telephone communication has been established between Three Rivers and Nicola, Yamaska, Drummond and Athabaska, Que. Early in the spring a cable will be laid across the river providing permanent communication.

A dispatch from Nanaimo, B. C., states that an eastern syndicate has bonded the charter for a tramway between Wellington and Nanaimo, and that the city will be asked to guarantee bonds to the amount of $59,000 for the construction of the road.

Judge McDougall has reserved his decision in the case of the appeals of the Toronto Incandescent Electric Light Co., and the Toronto Electric Light Co., against the assessment of nearly one-million dollars imposed by the city on the plant of these companies.

Application is to be made to the legislature to incorporate the Delaware, Parkhill & Lobo Electric Railway Co,. to construct and operate an electric railway between Delaware, Komoka, Poplar Hill, Fernhill and Parkhill, to Port Franks on Lake Huron.

The officers elect of the British Columbia Marine Engineers' Association are as follows:— President, W. Cullom; vice-president, James Lauderdale; secretary-treasurer, A. Goddyn; board of directors, Alex. McNiven, James McArthur, John McGraw, R. McGill, Charles McKechnie.

Messrs. Colquhoun & McBride, of Berlin, will apply to the Ontario legislature for the incorporation of the Grand Valley Railway Co., which proposes to construct and operate a steam or electric railway from Berlin, southerly to Brantford, and north-westerly to Listowel, westerly to Stratford, or northerly to Elora.

The contract for carrying the mails between the Toronto post office and the railway station will expire on the 31st of March. It is said to be the intention of the Toronto Street Railway Co. to tender for the contract, with the object of using electric mail cars similar to those which have been in use for some time past in Ottawa.

The divisional court at Vancouver has allowed the appeal of Mr. Bailey, a ratepayer of that city, against the Vancouver electric light by-law. This by-law, it will be remembered, authorized the expenditure of $100,000 for the purchase of a city lighting plant, and the erection of suitable buildings for the same. Under this decision of the divisional court the by-law is quashed.

The Canadian Marine Engineers' Association, have elected their officers as follows:—President, O. P. St.John, re-elected by acclamation; 1st vice-president, E. J. O'Dell; and vice-president, J. S. Adam; council—J. D. Banks, R. J. Garsall, W. Harwood, J. Findlay, R Hughes; treasurer, J. H. Ellis, re-elected by acclamation; auditors—S. A. Mills, D. L. Foley; inside guard, I. Hopkins.

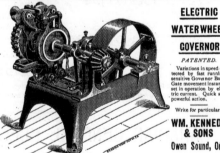

CANADIAN ELECTRICAL NEWS

STEAM ENGINEERING JOURNAL

OLD SERIES, VOL. XV.—No. 6.
NEW SERIES, VOL. V.—No. 3.                    MARCH, 1895                    PRICE 10 CENTS
$1.00 PER YEAR.

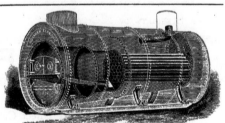

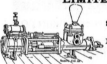
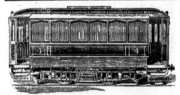

CANADIAN.

# ELECTRICAL NEWS

AND

## STEAM ENGINEERING JOURNAL.

VOL. V.                    MARCH, 1895                    NO. 3.

### THE DODGE PATENT SPLIT FRICTION CLUTCH AND CUT-OFF COUPLING.

THE utility of friction clutch pulleys for power transmission has been fully demonstrated by long and continuous service, and their advantages over the belt destroying shifter are so numerous and obvious that one wonders why their already extensive use is not universal. Even in the matter of first cost the clutch equipment is not greatly in excess of that of tight and loose pulleys, when the extra pullevs and double widths necessary for the drivers are considered.

By placing the clutch pulley upon the driving shaft, the belts and all the auxiliary shafting connected or controlled by the clutch are thrown out of action, saving belting, power, oil and danger from hot bearings and pulleys. Amongst the comparatively new clutches on the market is the Dodge Split Clutch, manufactured by the Dodge Wood Split Pulley Company. This clutch is made for service as a cut-off coupling, or may be used in connection with pulleys, gears, sprockets, rope sheaves, friction or hoisting drums, and various other power connections. Its simplicity is

DODGE CLUTCH WITH SHIFTER.

readily appreciated by mechanics who have ever had any experience with clutches of more or less complicated mechanism and those having a large number and variety of parts.

The friction disc is made of iron with perforations therein, through which hardwood friction blocks are fastened, presenting two surfaces of end grain for frictional contact. This disc is a

CLUTCH ON SPROCKET WHEEL.

part of the extended sleeve or portion of the clutch connected to the pulley, or whatever driving appliances may be used, and runs loose on the shaft where the clutch is located at the driven end of the transmission. The friction connection is made through two finished cast iron plates, one of which is keyed to the shaft, and which are thrown in contact with the wood filled disc by throwing in a sliding collar which works loose on the shaft, through the thrust of the collar actuating the toggle levers which operate four draw-bolts, forcing the friction plates to contact with the friction disc—this connection operating the pulley or transmission wheel in conformity with the moving shaft. One of the main difficulties existing in the various styles of clutches is the lack of clearance between the friction disc and plates; this trouble is entirely obviated in the Dodge clutch, the clear-

ance being large and instantaneous, actuated by powerful coil springs which separate the plates quickly upon a withdrawal movement on the sliding collar. Two levers are used with four points of contact on the plates, there being no loose or rattling joints ; the levers are made solid in one piece and have carefully finished fulcrum points on the outer or loose friction plate.

The Dodge Wood Split Pulley Co., appreciating the trade demand for a simple, quick acting clutch, with all possible points of advantage considered, have incorporated the split or separable feature as being one of the most important and quickest of appreciation by consumers. The advantages in a split clutch are manifold, they are easier and quicker to adjust to shaft or repair, and effect quite a saving in time and labor. None of the shafting or other equipment need be disturbed in placing the clutch in position. When this point is fully understood and appreciated we bespeak a more rapid change from the old tight and loose pulley ideas to the modern plan of machine driving. It is the expense of the split clutch as formerly made, as well as the trouble and expense involved in putting on solid clutches that has kept many manufacturers from making the changes long ago. This clutch is put on the market at about the same price as any other first-class clutch, but having the split feature to its credit. The Dodge split clutch is particularly adapted to service with gears, sprockets and other connections, and the only necessary features of these appliances over the regular goods is the large bore necessary to fit the extended sleeve. This sleeve is separate from the friction disc and may be easily detached for repairs without handling any portion of the clutch mechanism. For ordinary service the sleeve is lined with genuine babbit and fitted with compression grease cups to insure continuous efficient lubrication. The pulley is clamped over the sleeve and keyed securely. An improvement of very great practical importance is the patented separable or detachable hub, which bears to the clutch the

SHOWING SPLIT SLEEVE.

PATENT INTERCHANGEABLE SLEEVE.

same relation that the Dodge and Philion bush bears to the pulley. It enables the manufacturers to carry finished clutches in stock to be furnished with hubs as ordered also from stock—or at most with delay of only a few hours. It also enables the owner to keep a clutch on hand, and at the expense of a new hub use it in a shaft of different size or as a cut-off coupling, as he may desire.

Many patents have been taken out on the special features, and if the numerous favorable expressions of mechanics and engineers go for anything we feel safe in predicting a large demand

SECTION OF CLUTCH AND PULLEY SHOWING EXTENDED SLEEVE.

for the Dodge patent split clutch. Many shifting devices are shown ; one is adjusted to the floor stands used for shaft supports, and another a plain geared apparatus mounted independently and operating through a rock and pinion.

The company issue a handsome catalogue covering their specialties, and are pleased to mail same free to any one interested. Address, Dodge Wood Split Pulley Co., 68 King Street West, Toronto.

### THE AMERICAN STREET RAILWAY ASSOCIATION.

THE Executive Committee of the American Street Railway Association held meetings on the 27th and 28th of February, at the Windsor Hotel, Montreal, to arrange for the preliminaries of their next annual convention, which will take place in that city on the 15th and 18th of October next, inclusively. There were present Messrs. Joel Hunt, President, Atlanta, Ga.; W. Worth Bean, 1st Vice-President, St. Joseph, Mo.; D. G. Hamilton, St. Louis, Mo.; Granville C. Cunningham, Montreal, Que.; and W. J. Richardson, Secretary-Treasurer, Brooklyn, N. Y.

The chief feature of the convention will be an exhibition of the different electrical appliances used in operating an electric railway. Negotiations are now in progress to admit the exhibits coming from the States into Canada free of duty, a privilege which will no doubt be accorded, as was done for the electrical exhibition held some three years ago in Montreal.

At the last annual convention held in Atlanta there were 1100 delegates present, but the number that will attend the next convention will no doubt be increased, and eclipse all other conventions of the Association previously held.

### ELECTRICITY AND ECONOMY.

THE nature of electrical generation and dynamo working is such that only sufficient amount of current required to do the work in is used, so its economy is at once obvious. In factories where the machinery is working intermittingly, and liable to great fluctuation, the economy of driving by electricity is even more marked, as the electric current can be switched on or off with the greatest ease and rapidity, after which crossed belts and fast and loose pulleys appear a heavy and clumsy, not to say unscientific, method of utilizing power.—Manufacturers' Gazette.

### CANADIAN ASSOCIATION OF STATIONARY ENGINEERS.

*Note.—Secretaries of the various Associations are requested to forward to us matter for publication in this Department not later than the 20th of each month.*

#### HAMILTON ASSOCIATION NO. 2.

Mr. Wm. Norris, the Secretary, writes that while not many new members are being received into the above Association, the organization is otherwise making progress. The open meetings are being continued, and are proving most beneficial to the members. At the last of these meetings Bro. Brice gave an instructive talk on Electricity, touching upon many points with which engineers having electric plants require to be familiar. The Association is about to set about the arrangements for the annual dinner.

### BELL TELEPHONE COMPANY.

At the annual meeting of the above company, held on Feb. 27th, reports showed that the gross revenue for the year crossed a million, being $1,012,839, as against, for 1893, $961,174, an increase of $51,665. The working expenses for 1894 amounted to $729,611, as against, for 1893, $724,791, an increase of $4,820. The net revenue for 1894 was $283,227, as against, in 1893, $236,383, an increase of $46,844. A dividend of 8 per cent. was paid and $50,000 added to the contingent account, while $10,698 was placed to the credit of revenue account. The company's assets over liabilities are given at $921,429. The old board of directors was re-elected as follows :—Messrs. C. F. Sise, president ; G. W. Moss, vice-president ; W. H. Forbes, Hon J. R. Thibaudeau, John E. Hudson, Robert Archer, Robert Mackay, Wm. R. Driver, Hugh Paton.

At a special meeting of shareholders it was unanimously decided to float $600,000 worth of 5 per cent. bonds, redeemable in 30 years, the money raised to be expended in the erection of a new building at Montreal and for other purposes.

### INJURY TO BOILERS BY GREASE.

It has often been observed that small quantities of grease in combination with deposits lead to boiler accidents. This compound gets deposited on the plates, and the most violent water circulation is sometimes insufficient to remove it. The plates, in consequence, get overheated and accidents result. The introduction of grease inside the boiler should be avoided, especially where the water from the condenser is used for feeding the boiler, by the use of a sufficiently large feed-water filter. The Berlin Boiler Inspection Society had the following case brought under its notice : Two single-flued boilers, 4 feet 8 inches diameter, 23 feet long, flues 18 to 22 inches diameter, pressure 12 atmospheres, were used to generate steam for a 150 horse-power engine with surface condenser. The installation had only been at work since July, 1893. A considerable portion of the flue of the left boiler had collapsed. This could not be attributed to shortness of water. On examination it was found that nearly all over the boiler a fatty brown slime had been deposited, which, being placed on a red-hot iron, burst into flame. The feed-water pump got its water from a large open tank over which a small filter was placed. The condensed water was led to this filter in order to have the grease removed. Unfortunately, the arrangements were so bad that a considerable portion of the grease found its way into the boiler. A similar case was recorded by Mr. Abel at the last meeting of the Markisch Society for Testing and Inspecting Steam Boilers. Four boilers, the feed water of which was heated by the exhaust steam from a Westinghouse engine, after being in use about six weeks, were so damaged that one boiler had to be completely removed ; the other three had to receive extensive repairs. An examination showed that the flues were covered with a deposit of fatty slime. An analysis of this showed that about 52 per cent. of it consisted of mineral oils and paraffine, and 27 per cent. of animal fat. It is strongly advised, therefore, that feed water shall always be filtered so as to remove any oils or grease.—Scientific American.

It seems probable that an agreement will shortly be reached between the London Street Railway Co. and the City Council of London, under which the present street railway system will be transformed into an electric system.

## CHARACTER SKETCH.

### K. J. DUNSTAN.

MANAGER TORONTO BRANCH BELL TELEPHONE CO.

*"Patience is the finest and worthiest part of fortitude, and the rarest too."—John Ruskin.*

THE Hello of the telephone is no longer an object of curiosity with the people of Canada. We have come to accept it as a convenience of modern life that cannot be done without. But what of the men who were associated with this great invention in its earlier days, and to whose enterprise, business skill, and back of that, inventive genius, patiently exercised, we enjoy this system of communication to-day?

Mr. K. J. Dunstan, whose portrait is here presented to ELECTRICAL NEWS readers, does not make any claim to being the inventor of the telephone, but few men in Canada have been more intimately associated with its progress. Of him, it may be said, that he was in at the start of the telephone in Canada and has grown up with the business, by this means possessing an intimate relationship with telephone affairs.

Mr. Dunstan was born in Hamilton, where the first telephone exchange in the Dominion had an existence. In fact, before the telephone proper was known, he was associated with Mr. Baker, who had established in Hamilton what was known as the District Telegraph Co., employing a system of signaling, which preceded the birth of the telephone. When Professor Alexander Graham Bell brought his experiments to a successful issue, and the Bell Telephone was evolved into a reality, Mr. Dunstan became manager of the Hamilton branch. This was in the year 1878. During the years that he was in charge of the Hamilton office he also travelled considerably throughout the province, engaging in the work of inspection, and otherwise promoting the growth of the Bell telephone. In 1891 Mr. Dunstan removed to Toronto, becoming manager of the exchange in the Queen City.

When one goes back to 1878, a short period in the growth of any great invention, and contrasts these beginnings in telephoric history with the conditions existing to-day some idea is realized of how quickly, in late years especially, the people have learned to appreciate the telephone. Like every new invention they fought shy of it at first, and in Ottawa and St. John, N. B., where the telephone

MR. K. J. DUNSTAN.

to-day is as great a necessity as anywhere else, it was almost impossible in the early years of the business to get the people of these cities to evidence any practical appreciation of Mr. Bell's invention. The first canvass in Ottawa was a complete failure and no subscribers could be secured. A two weeks' canvass in St. John resulted in the enrolling of one subscriber. These cities to-day have well equipped exchanges, and in business and domestic circles the Hello of the 'phone is familiar to all classes of people.

The growth of the Toronto exchange is naturally a matter of satisfaction to Mr. Dunstan. There are to-day in this city 4,500 telephones, distributed among houses and private residences. Some of the defects that were worrisome to subscribers a few years ago, have been almost entirely abolished, owing to the degree of perfection attained in the construction departments of the work. The Toronto system is nearly altogether worked by metallic circuits, and to a greater extent than is the case in any other city in Canada. A larger proportion of underground work exists in Toronto than in any city of the same size in America. It is difficult to imagine the headquarters of any exchange to be more comfortably and completely appointed than is the case with the Toronto exchange, in the new building on Temperance St. It had been hoped that the new switch board, in course of construction for the exchange would have been completed before this date, when the entire business would be conducted under the roof of the new building. Within probably two months at the outside this end is likely to be attained. This switch board, Mr. Dunstan claims, will be the largest installation board in the

world, and will possess many new and valuable features. Three hundred persons are employed in the Toronto exchange.

To Mr. Dunstan, the general growth of the telephone business, throughout the entire Dominion, is a matter of keen interest. Some figures on this point will be interesting. On 31st Dec., 1894 there were 350 telephone exchanges in Canada using Bell instruments. These were represented by 32,485 subscribers, distributed as follows : Business places, 21,733 ; residences, 10,621 ; public pay stations, 131 ; and in addition 528 private line subscribers. For this service 34,595 miles of wire is in use on over 300,000 poles, besides underground conduits and house top fixtures.

An interesting development in the use of the telephone, and this has been largely within the past few years, is the long distance line. In 1877 it was quite a novelty when a line was placed between the residences of Messrs. Baker and Cory, of Hamilton. Then, hardly Professor Bell himself looked forward to the time when conversation would be carried on daily between points up to 700 miles apart. Yet within the past decade long distance telephoning has developed to this extent, and is growing most rapidly, rendering it somewhat difficult to say what successes may yet be scored in this department of the work. Mr. Neilson, of the Toronto exchange, thinks there is no doubt as to the possibility of building lines that would work say from Quebec to Sarnia ; the only question is whether wires of that length in Canada would pay. There is good reason to believe that time will solve satisfactorily that question. A great difficulty in the early days of long distance telephoning was the continual presence of induction, making conversations difficult. This trouble, however, has been quite successfully overcome by the almost general use to-day of metallic circuits and copper wire. The long distance telephone of the Bell Telephone Co. of Canada alone comprises 13,091 miles of wire on 5,361 miles of poles, and gives the means of verbal communication between the subscribers to their 300 different exchanges, and also to 262 other places, where they have no exchanges but only toll offices. The great convenience it is to telephone subscribers to be able to communicate with subscribers in other towns, what it means in a business way, and socially, scarcely needs so much as a suggestion to a people who have been so quick to appreciate in a general way the benefits of a telephone service in their own communities.

Mr. Dunstan is not the man to assume any large share of credit for the widespread growth of the Bell telephone system in Canada. He, of course, is pleased with its success in the exchange, directly under his own supervision, and it would be a strange thing if he did not feel a very vital interest in its general progress, watching as he has been able to do its growth from year to year since its first start in 1878 in the city of Hamilton.

In everything that pertains to electrical matters the subject of our sketch takes an intelligent interest. He has been an active member of the Canadian Electrical Association since its commencement, and at the meeting in Montreal in Sept. of 1894 the highest gift in possession of the Association was conferred upon him when he was elected president. Active and energetic in any work he undertakes, and possessed of a measure of geniality and courtesy that is not common to all men, he makes an excellent executive head of one of the most important commercial and scientific organizations of the Dominion.

M. Armagnat notes the various methods for electrically igniting gas in gas engines, and points out the defects and remedies. He points out that the connections usually adopted by having connecting the frame of the engine with both the primary and secondary of an induction coil, is not the best, and often better results are obtained if there is no connection between the two coils ; also, that with a reversal of the connection of the two poles of one of the coils, when a common return wire is used, the sparks may become greater or less.

## OUR GOVERNMENT TELEGRAPHS—THEIR SCOPE AND SPECIAL FUNCTIONS.

FROM time to time, it may be found, reference is made with more or less particularity to the Government Telegraph Service in press articles, dealing alike with the colonization and internal development of the country, and with the coast navigation and our marine interests in general ; the implication being thereby conveyed that the Government lines are of considerable extent, and that their ramifications serve some general and useful purposes beyond the sphere of operation of the commercial systems of the telegraph companies.

As a matter of fact, the Government lines do occupy a special field, even in the midst, as it were, of a telegraphic environment ; and a ready explanation of this is afforded by the consideration that the several districts or sections of the country in which the Government lines are operated are too sparsely settled for the creation of sufficient traffic to warrant the enlistment of private enterprise, while at the same time the character of the service afforded is of such importance to the general welfare, that the Government is warranted in providing the requisite facilities for its performance.

Such being the case, it is self-evident the telegraphs operated by the Government are not calculated to be tangibly remunerative in so far as a comparison of the revenue with the expenditure figures is concerned, but that their utility must be reckoned from some such basis as that of the post office, whose receipts and expenditures have but a secondary place in respect of the manifold interests that are fostered by its maintenance. At the same time it is instructive to look into the matter from the material standpoint, as affording some definite knowledge of what the actual cost of telegraph construction and maintenance is, and to that end reference may be had to the blue-books (the published annual reports) of the Department of Public Works—in conjunction with which the Government telegraphs are administered by the Hon. J. A. Ouimet—for figures on these points which will be found to compare favorably with those to be had in connection with any other of the existing telegraph organizations.

As it was only so far back as 1880 that the Government telegraphs, as such, were established, it may prove interesting to revert to that period for an insight of what brought about the creation of the service. It appears there had been for a long time considerable agitation going on because of the dangers of navigation in the Gulf of St. Lawrence ; so many wrecks involving losses of life and property were happening along the coast and on the inland shores, that it seemed imperative something should be done to render that particular locality less hazardous, and to this end it was conceived the establishment of telegraphic communication round about there would at least minimize the liability to total wreck of vessels cast ashore, as any such casualty could be reported and steamboats promptly dispatched to the rescue. So favorably was this view entertained by those most directly interested, that even the marine insurance companies advanced the probability of a very material reduction in rates if the telegraph connections were made. The scheme was discussed by Government in a very comprehensive " Report on the Advisability and Necessity of Establishing a Submarine Telegraph System for the River and Gulf of St. Lawrence "—printed by order of Parliament at Ottawa, 1876. And again it was strenuously advocated in " A Pamphlet Compiled and issued under the auspices of the Boards of Trade of Montreal and Quebec : Telegraphy with the Coasts and Islands of the Gulf and Lower River St. Lawrence and the Coasts of the Maritime Provinces—Its Relation to Shipping, Fisheries and Signal Service."—Quebec, 1879. Thus the project took on a broader aspect and assumed more imposing proportions ; but it was not all realized at once. In that same year (1879) a " proposed annual grant " of $15,000 was voted by Parliament " for the purpose of establishing telegraphic connection with the Island of Anticosti and the Magdalen Islands and Bird Rock." It was found, however, that this amount was insufficient to induce any company to undertake the work, so the Government decided to capitalize the proposed annual grant, and at the session of 1880 a vote of $200,000 was accordingly obtained for the purposes of construction. The cables were then contracted for and laid, land lines were built to complete the connections, and the whole of those have since been maintained and operated by the Government,

along with other extensions and systems elsewhere established in the meantime, under annual appropriations voted by Parliament for their continuance ; and in the meanwhile, too, the projects dealt with in the pamphlet above mentioned have materialized. Not only are all of the more important lighthouses electrically connected—thus affording the advantage sought of ready communication in behalf of vessels in distress—but regular systems of reporting in the interests of the meteorological service and of the fishing industry have been put in operation. And daily reports are communicated to the marine newspapers, of the passages in and out of the Gulf of vessels observed from the several stations, with any of which messages can be exchanged, if need be, by the use of flags and semaphores displayed in accordance with an international code of signals that has been generally adopted and is now familiar to all seafarers the world over. It is in this way that news is obtained and telegraphed of the appearance and whereabouts of ice at certain seasons in the waters of the Gulf, which can thus be avoided by outgoing vessels that would otherwise be exposed to dangers and delays of serious moment in consequence of its presence.

In view of all this, it is appreciative that the representations that were made to bring about the existence of these telegraphs, are shown to have been more than amply justified by the circumstance mentioned in one of the annual reports of the Department (1890), that it was found the predicted probable reduction in insurance rates had actually amounted to 50 per cent. in the interval since 1876, when the statistics were under discussion, and that all this was due to the telegraphs and other features that had been introduced to render the navigation of the Gulf so much less hazardous than it used to be. Whence, it is readily conceived that these telegraphs fill an actual need that must have been hard felt before their establishment.

Turning now to the inland lines ; it is found that the Government is not unmindful of what is best calculated to render our outlying regions inviting to prospective settlers. Wherever postal and telegraphic facilities are at hand no one can be said to be actually isolated, and as it is obviously desirable for the common weal that all parts of the country should be in touch, the Government's intervention to this end appears to be calculable where there are sufficient interests concerned to warrant it. Hence, wherever there has been a manifest effort on the part of the people in any section to build up and cultivate a business intercourse, it seems the Government has been found ready to lend its co-operation in the way of affording those reasonable facilities for the development of their resources.

In British Columbia the coal mines and the gold fields are reached, so that the produce can be readily listed in the markets, the trades promoted and the common interests of the people engaged in such industries advanced ; while incidentally the settlers along the route of the telegraph lines are alike directly benefitted by being thus brought into touch with the trade centres. The same is true of the conditions in the North West Territory, where widely separated towns are connected by long stretches of telegraph over prairies that will ere long be dotted with settlements whose very creation may in no small measure be contributed to by the existence of such facilities for intercourse with the outside world. As it is, the old adage about straws showing the way the wind blows would find apt application to these inland telegraphs, as the successive changes they undergo afford an object lesson of the rapid settlement and substantial progress that is being made in the country.

As the railways and the active operations of commerce become extended over territory occupied by the Government telegraphs, these latter are absorbed or superseded by the private enterprises whose development has been hastened by their use, while elsewhere new extensions are called for, to be in the same way withdrawn in due course when they shall have served the same useful purpose for which they were designed. In this way a good many changes have occurred in the course of the past few years, but, as might be expected, the Government service as a whole is gradually but steadily increasing in its proportions.

At the present time the Government telegraph service comprises a number of systems, aggregating 2,451 miles of land lines and 206 miles of submarine cable, with a total of 148 offices.

The lengths and whereabouts of these several lines ; the years of their construction, and the volume of paid traffic handled in each instance, will be found in the accompanying table. It will be noted that the total volume of the paid traffic is by no means insignificant.

The tariffs imposed by the Government in connection with the inland lines are about the same as those of the telegraph companies, and on the Gulf lines the tolls are very reasonable. For example, any office on the north shore of the St. Lawrence can be reached from Murray Bay, where connection is made with the G. N. W. Tel. Co.'s line, for 25c.; or for offices within 100 miles apart, 15c.; and any office on Anticosti Island can be reached from Murray Bay or from Gaspe for 50c. The tariff to the Magdalen Islands from North Sydney, C. B., is 50c., and the local rate on the Cape Breton lines is 25c., and the same on the Magdalen Islands.

The figures of the blue book for 1892-93—the latest that has been issued by the Department—may be taken to represent the yearly cost of the service. For the 12 months covered by that report the expenditure upon the whole was about $49,000, and the revenue collected amounted to a little over $9,000.

The operation on the whole appears to be carried on in a very systematic and agreeable way. A book of rules for the guidance of the employees generally was issued some time ago, after the working of the lines for several years had developed what was most advantageous and best calculated to facilitate the business of the service, and a standing invitation in furtherance of the same object is given in its introduction :

"Agents are requested to communicate to the General Superintendent, any suggestions they may have to make affecting the improvement of facilities for transacting business in their respective localities. Any suggestions calculated to enhance the efficiency of the service will be gladly received and carefully considered."

It is pleasing to note this, as it affords an evidence of intended encouragement to co-operation, and a desire on the part of the management to cultivate personal interest in the work.

Under the regulations that have thus been established, it appears the divers conditions and requirements found to exist in the different sections of the country can be conveniently dealt with through the offices of several district superintendents, acting under the direction of the head office at Ottawa, in which is vested the direct management and control of the entire service. This important office, for which he was eminently well qualified, was very ably filled by the lamented Mr. F. N. Gisborne, from the beginning of the service until his death in August, 1892, since when these affairs have been attended to in a manner which is understood to be highly satisfactory to the Department, by Mr. D. H. Keeley, who was associated with Mr. Gisborne as Assistant Superintendent from 1882 onwards.

In the list of the District Superintendents, which is given hereunder, there will be found the names of several widely-known and able telegraphers, whose valuable services, it is gratifying to know, the Government has at its command.

Inland Lines : District Superintendents—Messrs. E. Pope, Quebec, for North Shore, Escuminac and Quarantine ; James Wilson, Vancouver, B.C., for Barkerville and Cape Beale ; F. C. Gamble, C.E., Victoria, B. C., for Comox system ; H. Gisborne, Quappelle Station, N.W.T., for Edmonton and Wood Mountain ; J. McK. Selkirk, Leamington, Ont., for Pelee Island.

Gulf Lines : D. C. Dawson, St. John, N. B., for Meat Cove, Mabon and Cape Sable ; C. C. Seely, Grand Monar, N. B., for Bay of Fundy system ; H. Pope, South West Point, for Anticosti ; A. Le Bourdais, Grindstone, for Magdalen Islands ; E. H. Tetu, Pentecost, Q., for North Shore, east of Bersimis.

As part of the equipment of the service, the Government SS. "Newfield" was at the outset provided with the necessary appliances for picking up and relaying the cables, and is made available for the work of repairs in the Gulf when needed. The incidental electrical work in connection with the ship's operations has been personally attended to for several years past by the now acting General Superintendent. It almost invariably happens that when a break-down occurs in a cable, its whereabouts is unknown, and electrical tests have to be made to determine the location of the trouble.

In the course of a recent visit to the head office at Ottawa, there were shown some very interesting samples and specimens of the different kinds of apparatus and materials used in the construction and equipment of the lines ; also the requisites for the various kinds of instrument and battery tests incidental to telegraph maintenance; one of the latest acquisitions being a machine for the determination of the properties of wire intended for line construction. With such appliances as these the management is enabled to deal with the material needs of the service in an intelligent way, thus ensuring a proper equipment of the lines in order that they may be rendered as reliable as possible ; and it may safely be assumed that so long as this end is had in view the interests of the Government in respect of these useful and important telegraphs will be well conserved and looked after.

Thus much for our Government telegraphs. The service is seen to be an admirable institution ; its management is evidently in good hands, and the policy of its perpetuation unquestionable.

LINES OF THE GOVERNMENT TELEGRAPH SERVICE.

| LOCATION. | Year of Construction. | LENGTH IN MILES. | | | Number of Offices | Sent* Messages per annum. |
|---|---|---|---|---|---|---|
| | | Land Lines | Cables | Total. | | |
| **Newfoundland.—** | | | | | | |
| Port au Basque to Cape Ray. | 1883 | 14 | | 14 | 2 | |
| **Nova Scotia.—** | | | | | | |
| North Sydney to Meat Cove... | 1880 | 151¼ | 1 | 152¼ | 12 | 5700 |
| Barrington to Cape Sable...... | 1883 | 16 | 1¾ | 17¾ | 3 | 450 |
| Mabon to Cheticamp........... | 1887 | 63 | | 63 | 7 | 2000 |
| **New Brunswick.—** | | | | | | |
| Eastport to Campobello, Grand Manan and Whitehead Id... | 1880 | 34 | 10¼ | 44¼ | 7 | 600 |
| Chatham to Point Escuminac... | 1885 | 42 | | 42 | 5 | 750 |
| **Quebec.—** | | | | | | |
| Magdalen Islands to Meat Cove. | 1880 | 83 | 55½ | 138½ | 9 | 500 |
| St. Paul's Island to Meat Cove. | 1890 | 3 | 20 | 23 | 2 | 50 |
| Anticosti to Long Point........ | 1890 | 9 | 21 | 30 | } o | 500 |
| Anticosti to Gaspe............. | 1880 | 242¼ | 44¼ | 286½ | | |
| Murray Bay to Pt. Esquimaux. | 1887 | 456¾ | 39¼ | 496 | 35 | } 18400 |
| Bay St. Paul to Chicoutimi..... | 1881 | 92 | | 92 | 6 | |
| Grosse Isle to Quarantine...... | 1885 | 48 | 4¾ | 52¾ | 7 | 3400 |
| **Ontario.—** | | | | | | |
| Pelee Island to Leamington.... | 1888 | 24 | 8¾ | 32¾ | 7 | 500 |
| **North West Territory.—** | | | | | | |
| Quappelle to Edmonton and St Albert.................. | 1883 | 607½ | | 607½ | 14 | 4200 |
| Moosejaw to Wood Mountain.. | 1885 | 90½ | | 90½ | 2 | 250 |
| **British Columbia.—** | | | | | | |
| Ashcroft to Barkerville......... | 1879 | 276½ | | 276½ | 8 | 2000 |
| Victoria to Cape Beale........ | 1890 | 118 | | 118 | 6 | 250 |
| Nanaimo to Comox............ | 1892 | 81 | | 81 | 6 | 2000 |
| Total..................... | | 2451¾ | 206½ | 2658¼ | 248 | 41550 |

*Meteorological and signal service messages and fishery reports are handled free of tolls, and are not included in the count.

## TRADE NOTES.

The storage batteries manufactured by the Ballard Electric Works, 43 Adelaide St. W., are said to be giving satisfactory service where a current of small voltage is required.

Messrs. Roe & Graham, Ottawa, report the following recent sales of water wheels :—60" to the E. B. Eddy Co., Hull, Que. ; 36" to J. Oliver & Son ; 48" to the McKay Milling Co., Ottawa ; 36" wheel to F. M. Pope, Robinsonburg, Que.

The Canadian General Electric Co. have closed a contract with the Winnipeg Street Railway Co. for a 600 H.P. direct-connected railway generator. This machine will be of the Company's standard multipolar type, similar to the 1200 H.P. generator recently installed by them for the Toronto Railway Co.

The Montreal Railway Co. have within the last few days placed an order with the Canadian General Electric Co. for 80 motors of their new C.G.E. 1200 type, together with a large additional order for C.G.E. 800 motors, with which they have had, during the recent severe weather in Montreal, a most satisfactory experience.

Mr. Wm. T. Bonner, of New York City, has succeeded Mr. E. C. French as General Agent for Canada for the Babcock & Wilcox Co., manufacturers of the well known water tube steam boilers of that name. The Babcock & Wilcox boilers are now built in Canada, large shops having been fully equipped with special tools and other necessary plant for handling orders of any size. Mr. Bonner will continue the principal Canadian office at 415 Board of Trade Building, Montreal.

The Robb Engineering Co. has received the following letter from Principal Grant, of Queen's University :—"The Mining Institute of Ontario held its quarterly meeting here last week, and we took that occasion of formally opening the Mining Laboratory ; and your engine and boiler were both voted satisfactory. As a Nova Scotian I was delighted that we had so much of our machinery from Nova Scotia, and as this is the only mining laboratory in Canada, I was delighted that you had contributed to its equipment."

PUBLISHED ON THE FIFTH OF EVERY MONTH BY

## CHAS. H. MORTIMER,

OFFICE : CONFEDERATION. LIFE BUILDING,
*Corner Yonge and Richmond Streets,*

TORONTO,   -   CANADA.
Telephone 2362.

NEW YORK LIFE INSURANCE BUILDING, MONTREAL.
Bell Telephone 2299.

***ADVERTISEMENTS.***
Advertising rates sent promptly on application. Orders for advertising should reach the office of publication not later than the 15th day of the month immediately preceding date of issue. Changes in advertisements will be made whenever desired, without cost to the advertiser, but to insure proper compliance with the instructions of the advertiser, requests for change should reach the office as early as the 22nd day of the month.

***SUBSCRIPTIONS.***
The ELECTRICAL NEWS will be mailed to subscribers in the Dominion, or the United States, post free, for $1.00 per annum, 50 cents for six months. The price of subscription should be remitted by currency, in registered letter, or by postal order payable to C. H. Mortimer. Please do not send cheques on local banks unless 25 cents is added for cost of discount. Money sent in unregistered letters will be at senders' risk. Subscriptions from foreign countries embraced in the General Postal Union, $1.50 per annum. Subscriptions are payable in advance. The paper will be discontinued at expiration of term paid for if so stipulated by the subscriber, but where no such understanding exists, will be continued until instructions to discontinue are received and all arrearages paid.
Subscribers may have the mailing address changed as often as desired. *When ordering change, always give the old as well as the new address.*
The Publisher should be notified of the failure of subscribers to receive their papers promptly and regularly.

***EDITOR'S ANNOUNCEMENTS.***
Correspondence is invited upon all topics legitimately coming within the scope of this journal.

THE "CANADIAN ELECTRICAL NEWS" HAS BEEN APPOINTED THE OFFICIAL PAPER OF THE CANADIAN ELECTRICAL ASSOCIATION.

## CANADIAN ELECTRICAL ASSOCIATION.

### OFFICERS :

PRESIDENT :
K. J. DUNSTAN, Local Manager Bell Telephone Company, Toronto.

1ST VICE-PRESIDENT :
A. B. SMITH, Inspector Canadian Board Fire Underwriters, Toronto.

2ND VICE-PRESIDENT :
C. BERKELEY POWELL, Manager Ottawa Electric Light Co., Ottawa.

SECRETARY-TREASURER :
C. H. MORTIMER, Publisher ELECTRICAL NEWS, Toronto.

EXECUTIVE COMMITTEE :
L. B. McFARLANE, Bell Telephone Company, Montreal.
GEO. BLACK, G. N. W. Telegraph Co., Hamilton.
T. R. ROSEBRUGH, Lecturer in Electricity, School of Practical Science, Toronto.
E. C. BREITHAUPT, Berlin, Ont.
JOHN YULE, Manager Guelph Gas and Electric Light Company, Guelph, Ont.
D. A. STARR, Electrical Engineer, Montreal.
J. J. WRIGHT, Manager Toronto Electric Light Company.
J. A. KAMMERER, Royal Electric Co., Toronto.
J. W. TAYLOR, Manager Peterboro' Carbon Co., Peterboro'.
O. HIGMAN, Inland Revenue Department, Ottawa.

## MONTREAL ELECTRIC CLUB.

OFFICERS :
President, W. B. SHAW,   -   Montreal Electric Co.
Vice-President, H. O. EDWARDS,   -
Sec'y-Treas., CECIL DOUTRE,   -   81A St. Famille St.
Committee of Management, T. F. PICKETT, W. GRAHAM, J. A. DUGLASS.

## CANADIAN ASSOCIATION OF STATIONARY ENGINEERS.

EXECUTIVE BOARD :
President, J. J. YORK,   -   Board of Trade Bldg, Montreal.
Vice-President, W. G. BLACKGROVE,   -   Toronto, Ont.
Secretary, JAMES DEVLIN,   -   Kingston, Ont.
Treasurer, DUNCAN ROBERTSON,   -   Hamilton, Ont.
Conductor, E. J. Philip,   -   Toronto, Ont.
Door Keeper, J. F. CODY,   -   Wiarton, Ont.

TORONTO BRANCH No. 1.—Meets 2nd and 4th Friday each month in Room D, Shaftesbury Hall. Wilson Phillips, President ; T. Eversfield, Secretary, University Crescent.

HAMILTON BRANCH No. 2.—Meets 1st and 3rd Friday each month, in Maccabee's Hall. Jos. Langdon, President ; Wm. Norris, Corresponding Secretary, 211 Wellington Street North.

STRATFORD BRANCH No. 3.—John Hoy, President ; Samuel H. Weir, Secretary.

BRANTFORD BRANCH No. 4.—Meets 2nd and 4th Friday each month, C. Walker, President ; Joseph Ogle, Secretary, Brantford Cordage Co.

LONDON BRANCH No. 5.—Meets in Sherwood Hall first Thursday and last Friday in each month.  F. Mitchell, President ; William Meaden, Secretary Treasurer, 533 Richmond Street.

MONTREAL BRANCH No. 1.—Meets 1st and 3rd Thursday each month, in Engineers' Hall, Craig street. President, Jos. Robertson ; first vice-president, H. Nuttall ; second vice-president, Jos. Badger ; secretary, J. J. York, Board of Trade Building ; treasurer, Thos. Ryan.

ST. LAURENT BRANCH No. 2.—Meets every Monday evening at 43 Bonsecours street, Montreal. R. Drouin, President ; Alfred Latour, Secretary, 306 Delisle street, St. Cunegonde.

BRANDON, MAN., BRANCH No. 1.—Meets 1st and 3rd Friday each month, in City Hall. A. R. Crawford, President ; Arthur Fleming, Secretary.

GUELPH BRANCH No. 6.—Meets 1st and 3rd Wednesday each month at 7:30 p.m. J. Fordyce, President ; J. Tuck, Vice-President ; H. T. Flewelling, Rec.-Secretary ; J. Gerry, Fin.-Secretary ; Treasurer, C. J. Jorden.

OTTAWA BRANCH, No. 7.— Meets 2nd and 4th Tuesday, each month, corner Bank and Sparks streets ; Frank Robert, President ; F. Merrill, Secretary, 352 Wellington Street.

DRESDEN BRANCH No. 8.—Meets every and week in each month ; Thos. Merrill, Secretary.

BERLIN BRANCH No. 9.—Meets 2nd and 4th Saturday each month at 8 p. m.  W. J. Rhodes, President ; G. Steinmetz, Secretary, Berlin Ont.

KINGSTON BRANCH No. 10.—Meets 1st and 3rd Tuesday in each month in Fraser Hall, King Street, at 8 p.m.  J. Devlin, President ; A. Strong, Secretary.

WINNIPEG BRANCH No. 11.—President, G. M. Hazlett ; Recording Secretary, W, J. Edwards ; Financial Secretary, Thos. Gray.

KINCARDINE BRANCH No 12.—Meets every Tuesday at 8 o'clock, in the Engineer's Hall, Waterworks. President, Jos. Walker ; Secretary, A. Scott.

WIARTON BRANCH No. 13.—President, Wm. Craddock ; Rec. Secretary, Ed. Dunham.

PETERBOROUGH BRANCH No. 14.—Meets 2nd and 4th Wednesday in each month. S. Potter, President ; C. Robison, Vice-President ; W. Sharp, engineer steam laundry, Charlotte Street, Secretary.

BROCKVILLE BRANCH No. 15.—W. F. Chapman, President ; James Aitkens, Secretary.

CARLETON PLACE BRANCH No. 16.—W. H. Routh, President ; A. M. Schofield, Secretary.

## ONTARIO ASSOCIATION OF STATIONARY ENGINEERS.

BOARD OF EXAMINERS.
President, A. E. EDKINS,   -   139 Borden st., Toronto.
Vice-President, R. DICKINSON,   -   Electric Light Co., Hamilton.
Registrar, A. M. WICKENS,   -   280 Berkeley st., Toronto.
Treasurer, R. MACKIE,   -   28 Napier st., Hamilton.
Solicitor, J. A. MCANDREWS,   -   Toronto.
TORONTO—A. E. Edkins, A. M. Wickens, E. J. Phillips, F. Donaldson.
HAMILTON—P. Stott, R. Mackie, R. Dickinson.
PETERBORO'—S. Potter, care General Electric Co.
BRANTFORD—A. Ames, care Patterson & Sons.
KINGSTON—J. Devlin (Chief Engineer Penetentiary), J. Campbell.
LONDON—F. Mitchell.
Information regarding examinations will be furnished on application to any member of the Board.

___

MR. D. McFarlan Moore, of Harrison, N. J., is reported to have attained very encouraging results in the direction of phosphorescent lighting. "The line which Mr. Moore has marked out for himself," says the Electrical Engineer, "contemplates the introduction of phosphorescing glow lamps on continuous or alternating current circuits of ordinary potential, with the addition of but the simplest auxiliary apparatus." The details of the method by which are achieved the results which Mr. Moore has demonstrated, are withheld for the present, pending the securing of patents.

___

AT the meeting of the Executive Committee of the Canadian Electrical Association held since the publication of our last issue, quite a number of new members were elected. The members of the Association resident in Ottawa are looking forward with a great deal of interest to the convention of the Association to be held in that city next autumn, and are determined that if possible it shall eclipse anything of the kind held, in the past. The history of the Association thus far has been marked by steady progress, and we have strong faith that it will accomplish greater things in the future.

___

WE hear of two companies being formed, one at Ottawa, the other at Hamilton, to engage in the manufacture of storage batteries. The latter is being formed to manufacture an American battery which is especially designed for the propulsion of electric cars. The inventor of this battery claims that by practical tests he has demonstrated its ability to drive a car one hundred and twenty five miles without recharging. This is the kind of battery that electric railway projectors and managers are looking for. Apart from the possibilities for railway use, the demand for storage batteries for other purposes is we believe sufficient to support at least one Canadian manufactory, provided it can produce a battery that in point of durability, efficiency and cheapness, will compare favorably with those now being imported.

THE paper on "Electricity for Architects," by Mr. John Langton, printed in the present number, was prepared with the view of making clear to the minds of a non-technical audience the meaning of electrical terms and the methods of operating electrical currents. We re-print the paper in the belief that it will prove valuable to students of electricity to whom a thorough understanding of the underlying principles of the science is of the greatest importance.

THE attention of owners and operators of electric lighting stations throughout Canada is directed to the paper on "Central Station Dividends" by Mr. Cecil Doutre, of Montreal, published in part in this number of the ELECTRICAL NEWS. The author demonstrates very clearly the means by which profits can be made or lost, and emphasizes particularly the necessity for skilled superintendents, firemen and engineers, without which the best business manager is powerless to achieve satisfactory results. In this connection an article appearing in the present number, calls timely attention also to the fact that electric lighting might be made more profitable than it often is, if business principles were applied to the operation of lighting stations to the same extent as to electric railways, for example. These are facts which those interested in electric lighting enterprises, whether as investors or managers, cannot afford to disregard.

ELECTRIC railway and lighting companies are finding it necessary to expend large sums of money in defending the increasing number of actions for damages brought against them by persons who adopt the methods of the black-mailer to extort money to which they know themselves not entitled. Managers of electric companies would welcome at as early a date as possible, judgments of the Canadian Courts which would define the liability of such companies in actions for damages and serve as precedents for the settlement of future cases. If, as would appear to be the case, the law as it stands at present does not afford sufficient protection from the assaults of the blackmailer, an effort should be made to have it properly amended. Electric companies, we take it, expect to make compensation for injuries resulting from carelessness on the part of their management or employees, but they should not be allowed to become the easy prey of the blackmailer.

ON the evening of Feb. 25th, and less than half an hour after the employees had left the building, a small upright boiler in a soda water factory on Sherbourne street, Toronto, exploded, completely wrecking the building and knocking out the windows and otherwise damaging surrounding residences. Luckily the absence of employees from the factory and of foot passengers from the sidewalks opposite the building, avoided more serious results. Portions of the outer shell of the boiler were found imbedded in the frozen ground in the yards of the locality, having cut their way through the felt and gravel roof of the factory. The force of the explosion is sufficient evidence that it was not caused by low water. Had five hundred pounds of gunpowder exploded inside the building, it could not have exerted a more destructive force. It is supposed that the fire in the boiler had not been securely banked for the night, and that it had burned up, causing a rise in steam pressure to at least 300 lbs., and that the safety valve was not in proper working-order. This occurrence should suffice to dispel the prevalent notion that while proper inspection and skilled supervision are required for steam-plants of large capacity they can safely be dispensed with in connection with small plants. It should likewise direct the attention of the municipal authorities to the necessity for an ordinance to prohibit the locating of factories in the center of thickly populated residential districts, like the one in which this accident occurred.

THE eyes of electricians and manufacturers have recently been turned to the operations of the Cataract Construction Co., for the transmission of power from Niagara Falls to Buffalo and other points. Much speculation has been indulged in as to whether power thus transmitted could be sold to the consumer at a less figure than it could be supplied to him from a local steam plant. The promoters of the transmission scheme are now confronted with another obstacle which was probably unforeseen, and which bids fair to seriously affect enterprises of this character. It is stated that since the proposal to transmit power from Niagara Falls has taken tangible form, the owners of manufacturing sites in the vicinity of Buffalo, which, from their location, would make the cheap power available, have jumped to ten times their former price—or from $300 to $3,000 per acre. This advance is believed will fully offset any advantage which the Buffalo manufacturer is likely to derive from cheap power. Exactly the same principle applies, says the American Machinist, to comparisons made between manufacturing plants located in Buffalo and similar plants located in Cleveland or elsewhere. If in each case the annual rental value of the land occupied, or the interest on its actual value be considered as a part of the cost of production—as it should be—it will, we think, be found that things are nearly balanced, i. e., that whatever advantages a given manufacturer may have over another in the matter of cheap power will be balanced or nearly so by the cost of the privilege of using it ; and this is one of the reasons why manufacturers in Hartford, Providence, Cleveland or Cincinnati need not especially fear the competition of manufacturers who get power at much lower cost elsewhere, unless the owner of the land upon which the factory stands—who may be also the manufacturer occupying it —is willing to ignore its value in making up his accounts ; something which he does not usually do if he is wise.

THE use of the gas engine in connection with electric lighting plants is not receiving the attention that it should in Canada, where so many towns have gas and electric plants combined under the same management, and where small and medium sized isolated plants are so frequent in the larger cities. Gas companies neglect their own product, and use steam engines to operate dynamos, when the most superficial investigation into the subject would have probably modified their designs, to their great advantage. A gas company should consider this question, not so much with a view of determining whether it is cheaper to produce a gas light or an incandescent light of equal candle-power ; but in order to determine whether gas consumed in a burner to produce light, or in a cylinder to produce power, is the more efficient method of consumption. Most such combined plants sell the two forms of light at the same, or what are intended to be the same, figures per candle power. If the electric light is more popular than the gas, less gas is consumed, less is produced, and the general efficiency of the gas plant lowered, and vice versa. This simply means that the income from the total lighting (whether by gas or electricity) has got to pay interest on capital sufficient for two plants. If gas companies were to make greater efforts to popularize their electric light, and used gas engines instead of steam engines to run their dynamos, the efficiency of their electric plant would rise ; and that of their gas plant might be kept fairly constant, because the consumption of gas for power would increase with the extension of the electric service. Presumably all gas companies know how much gas they distil from a certain quantity of coal ; and if they do not know how many pounds of coal they burn in their steam boilers per horse power hour, it argues very poor management. There are gas engines in use to-day in Europe and America, whose makers guarantee their performance at so little as 15 cubic feet of gas per h. p. h.; and plenty of them are guaranteed at from 20 to 25 cubic feet per h. p. h. for reasonably large sizes. A manager should therefore be able to compare the price of the coal burnt per h.p. h. to raise steam, with the price of the gas necessary to produce one h.p. h. in a gas engine, and in connection with this he can make another calculation as to the comparative efficiencies of gas consumed in a burner and in a cylinder. Thus, a 16 c.p. gas burner will require on the average 5 cubic feet of gas per hour. One thousand cubic feet of gas will therefore give 200 16 c.p. hours. The same quantity of gas consumed at the rate of 25 c. f. per hour will give 40 h. p. h. This at 75 per cent. electrical efficiency, and using 4 watt lamps, gives 349 16 c.p. hours as against 200 using gas in burners. A gas engine would have to consume 44 cubic feet per horse per hour before the efficiencies of the two methods of consumption would become the same, on the assumed data. Here is a matter which is well worth the attention of gas managers. Lighting is too much regarded from the standpoint of the gas company or of the electric company, and not from a purely unprejudiced commercial one ;

and thus it is that lighting companies of all sizes and using all kinds of methods are suffered to manage themselves according to most unbusiness-like principles, and hence fail of success.

THAT electric lighting is an industry requiring special study, not only of its practical, but also of its commercial features, is not so well recognized as it should be. Practical electricity is a most fascinating science, but commercial electricity has also its problems, the study of which often indicates practice that from a commercial point of view is very paradoxical. The spending of an extra amount of money, for instance, in running a system of "secondary mains," utilizing the transformer secondary pressure of 100 volts, instead of the alternator primary pressure of 1,000 volts, will in certain cases be an actual economy. Economy can be observed in the preliminary design of a plant, and subsequently in its operation ; but a seemingly economy in the design may be the direct cause of an increase in subsequent operation expenses, much greater than the interest on the sum saved by such economy. This proposition is not evident to the untrained intelligence of the non-professional "electrician." The efficiency of large transformers is greater than that of small sizes ; the drop of voltage between full and light loads is less in the former than in the latter ; and as a rule the price per light is less in the large sizes. Consequently, in a block of houses and stores, where the lights are fairly numerous, the placing of a large transformer at each corner, and connecting all these in multiple or to secondary mains, whence are tapped off the branches into the various houses, will, under certain conditions to be examined in their purely commercial bearing, be a very superior method to running primary mains, and putting on a small transformer for each customer. The increased cost of secondary mains at 100 volts over primaries at 1,000, may well be compensated by the less cost of large transformers per light, and the saving effected in labor and supplies by putting up only four, instead of a number of small ones ; and thus, the great advantage of a much better regulation of pressure is obtained at small cost. The desirability of a constant pressure at the lamps is established by the facts that the life of a lamp is reduced about 15 per cent. for every 1 per cent. increase of voltage above that at which it is intended to be run ; while its candle power decreases about 5 per cent. for every 1 per cent. decrease of that voltage. In the former case the consumer has to buy a greater number of lamps, because they burn out too quickly ; in the latter he pays for light that he does not get. The variation of load on a transformer will always be great, probably ranging from full load at about 7 o'clock on a winter's night, to less than a quarter at 3 a.m. next morning ; the variation of pressure on the lamps will therefore also be great, but greater if small transformers be used, than large ones ; and the efficiency of small will be less than that of large sizes, as the following figures will show :

| Capacity. | ¼ load. | Full load. |
|---|---|---|
| 12 lamps...... : .......... | 84 per cent.............. | 94 per cent. |
| 90 " .......... .... | 94 " " | ........... 97 " " |

A comparison of costs per light of the two sizes is also instructive :—

5 light transformer.... . ......................$3.60 per light.
90 " " ............ ............. .....$1.05 " "

The efficiency alone of the larger size, at light load, is a considerable saving, to say nothing of the decreased cost. It is points like these that constitute station management, and that help towards dividends.

### PREVENTING DECAY IN TELEGRAPH POLES.

A FRENCH engineer has arrived at the conclusion, according to the Railway Review, that the principal seat of decay in telegraph poles is in the ten or twelve inches immediately below the surface of the ground. To protect them he proposes to excavate the soil to this depth around the pole, to clean it thoroughly from soil and all decaying wood, and then to give it a coat of hot tar. The pole is then surrounded by a sleeve of glazed earthenware, which is made in semi-cylindrical halves, to facilitate the putting in place. The annular space between this sleeve and the pole is then filled with some dry material, which is finally capped with a waterproof layer of asphalt or some similar material. The cost is said to be about 50 cents per pole.

### THE NEW BRANCH TELEPHONE EXCHANGE AT QUEBEC.

ON February 4th a branch exchange of the Bell Telephone Company was opened in Quebec to take in the subscribers from the St. Roch district. The building, which is situated at the corner of Caron and Charest streets, was built by the Bell Telephone Company expressly for the purpose, and is of white brick with stone foundations, one storey high, with a basement. In this basement are the heating apparatus and the batteries for the operators, transmitters, and other circuits of the switch which require battery power. The cables from outside are also brought into the building here from a pole in the yard, which is placed close to the wall of the building, and pass through the floor to iron terminals, on which are mounted the standard combination lightning arresters, consisting of a heat coil and a carbon plate air space arrester.

The operating room is almost 51 feet long by 21 feet wide, and has light on three sides. Portions are reserved for the operators' cloak room, talking booths, a public counter, and desk room for a clerk.

The exchange is at present equipped for four hundred subscribers, with an ultimate capacity of two thousand. The switch is of the divided type, having one section of multiple board for the trunk operator, on which to give connections called for from the main office, and also some local connections, the remainder of the operator's positions having only the usual jacks and drops. Each operator has charge of one hundred lines. The various parts of the board are of the latest design, such as open jacks, self restoring drops, and the new combination ringing and listening key, such as are being used on the new Toronto switch. From the arresters on the cable terminals, switchboard cables are carried to one side of a Hibbard rack ; the other side being connected to the multiple jacks in the first section. From here other cables are run to the connecting boards on the various switches, and thence to the jacks and drops. Chairs of special design are provided for the operators, and both gas and electric light are provided as illuminants. On the whole, this is one of the most modern and best equipped offices the company have.

### NATIONAL ELECTRIC LIGHT ASSOCIATION CONVENTION.

THERE was a larger Canadian representation than usual at the National Electric Light Association Convention held in Cleveland on the 19th, 20th and 21st of February. Among those who attended from Canada were Messrs. John Langton, Frederic Nicholls and J. K. Kammerer, of Toronto ; J. W. Taylor, Peterboro'; Frank Badger, Montmorency, Que:, and J. A. Corriveau, of Montreal. The total attendance at the Convention numbered about three hundred. "It was the first of the American Conventions that I had attended," said one of the Canadian delegates, "and I was so disappointed that I am not in the least anxious to attend another. I had polished up my mental apparatus to the greatest possible degree, in anticipation of coming in contact with the brightest intellects of the continent. Judge of my surprise to find the discussions of the tamest character, and largely monopolized by sales agents of the various manufacturing companies for advertising purposes. One whole day was taken up with the discussion of the monocyclic system. I was surprised that the President of the Association made no effort to limit the discussion so as to allow time for the consideration of other subjects of general interest to the electrical fraternity. Mr. Nicholls' paper on "The Lighting of Large Cities," for example, received no discussion whatever. I consider that our Canadian Electrical Association can give the National Association pointers as to how to manage a successful Convention The Canadian Association is noticeably superior in point of the ability of its members to present intelligibly their views upon the subjects up for discussion, and in their knowledge and observance of parliamentary practice."

### MONTREAL JUNIOR ELECTRIC CLUB.

Jan. 28th—Paper on "Review of Past Papers," by H. H. Morgan.

Feb. 4th—Paper on "Electric Bells, Batteries and Push Buttons," by T. W. Sutton.

Feb 11th—Paper on "Street Railway Trolley System," by R. H. Street.

Feb. 18th—Paper on "Incandescent Lamps," by E. W. Sayer.

## MUNICIPAL ELECTRIC LIGHTING.

TORONTO, Feb. 16th, 1895.

Editor ELECTRICAL NEWS.

SIR,—May I request you to insert the following in answer to Mr. A. A. Wright's questions re Orillia municipal plant, and Mr. J. J. Wright's reference to action recently proposed by West Toronto Junction?

First, Mr. A. A. Wright's calculation of income is based on 1400 lights. I gave "nearly 1400" as the number installed and running. The exact number was 1340. I calculated the income on this figure, and allowed 15 per cent. discount for prompt payments. This discount should have been alluded to, but was inadvertently omitted.

The schedule of rates in Orillia is as follows : (a) ordinary store lighting, 60c. per month per 16 c. p. lamp ; (b) stores remaining open every night, 70c. per 16 c. p. per month ; (c) Hotels, all night lights, 80c. per light ; 12 o'clock lights, 70c.; bed-room and dining-room lights, 30c. per light per month, etc.

Mr. Wright will see that these prices are fair and attainable in most towns that have not the advantage of profiting by the competition of two rival electric plants, under which undesirable conditions I believe he himself operates. I say "undesirable" conditions, because, having been a central station manager myself, I know just how he feels ; from the standpoint of public interest, I should regard those conditions as highly proper and satisfactory. To return—there are no "commercial" arcs, and I cordially agree with Mr. Wright in saying that Orillia "*is* very evidently a pretty good paying plant." I am delighted to find that I have carried conviction to the mind of a central station man as to the economy with which the operations of at least one municipal plant are conducted, and I hope to produce several more such arguments.

As to my charging the $570 for carbons against the incandescent plant, readers of the article in question will, on re-perusal, notice that the print reads (just above the table of expenses) : "The operating expenses of the entire plant, *arc* and incandescent, are therefore as follows, etc." The carbons, therefore, are properly charged. In stating that "a portion of these items (engineer and fireman) are justly chargeable to arc account," Mr. Wright shows us that his accounting system is inexorable in its adherence to philosophical principles, but I ask his pardon for suggesting that he has missed the point I made, which was distinctly shown half way down the last paragraph of the article : "Thus, it would appear, that working the incandescent plant at two-thirds its rated capacity, brings in a sufficient income to defray its own expenses, pay *off* its own debt and do the town lighting *free*." It is evident, therefore, that these items were charged wholly to the incandescent plant, in order to show that its income is sufficient to defray all arc expenses, and leave the town arc lighting a clear gain.

I really feel like a criminal, in this discussion, as it is evident that however good may be my intentions, I am not on the "Wright" side. I cannot even feel confident that Might is against Wright. Mr. J. J. Wright's strongest argument is "that Toronto Junctions mayor has recommended the council to get the lighting done by the Toronto Company, if that company will do it at a reasonable figure." This amounts to "it is wrong for Toronto Junction to do its own lighting if any other party can do it more cheaply." Does anyone doubt this? Has anyone stated the opposite? It is an unassailable truism, whose roots spring out of eternal fact. But can this truism be taken as a proof that municipal ownership is expensive? My article contains plenty argument, and is quite open to criticism ; let us have a few such criticisms from Mr. J. J. Wright. We have had all sorts of theories, and exposés, and not a little honest invective, and assumptions from all sorts of persons. Let us relegate all that, and also truisms, to the rear, and have solid arguments. Besides, I claim an argument taken from Toronto Junction cannot apply to towns situated like Orillia. Toronto Junction can simply tie on its wires to those of the Toronto Lighting Company, and I should be the last person to say that a 50 light plant can be run as cheaply as a 1500 light plant. But Orillia has not got a 1,500 light plant at its doors, nor have a hundred other provincial towns of the same size, so what are they going to do? My article was descriptive of a particular case, governed by peculiar conditions. Any criticism founded on completely different conditions cannot apply. Thanking you for your space.

I remain, yours very truly,

GEORGE WHITE-FRASER, E. E.

## MOONLIGHT SCHEDULE FOR MARCH.

| Day of Month. | Light. | Extinguish. | No. of Hours. |
|---|---|---|---|
| | H.M. | H.M. | H.M. |
| 1...... | P. M. 9.30 | A. M. 5.42 | 8.10 |
| 2...... | " 10.40 | " 5.40 | 7.00 |
| 3...... | " 11.00 | " 5.40 | 6.40 |
| 4...... | " 12.00 | " 5.40 | 5.40 |
| 5...... | ............ | " 5.40 | } 3.20 |
| 6...... | A. M. 2.20 | ............ | |
| 7...... | " 3.30 | " 5.40 | 2.10 |
| 8...... | " 3.30 | " 5.40 | 2.10 |
| 9...... | No light. | No light. | .... |
| 10...... | No light. | No light. | .... |
| 11...... | No light. | No light. | .... |
| 12...... | P. M. 6.10 | P. M. 9.20 | 3.10 |
| 13...... | " 6.10 | " 10.30 | 4.20 |
| 14...... | " 6.10 | " 11.50 | 5.40 |
| 15...... | " 6.10 | A. M. 1.00 | 6.50 |
| 16...... | " 6.10 | " 2.00 | 7.50 |
| 17...... | " 6.20 | " 3.00 | 8.40 |
| 18...... | " 6.20 | " 4.00 | 9.40 |
| 19...... | " 6.20 | " 4.30 | 10.10 |
| 20 ..... | " 6.20 | " 5.00 | 10.40 |
| 21...... | " 6.20 | " 5.00 | 10.40 |
| 22...... | " 6.20 | " 5.00 | 10.40 |
| 23...... | " 6.20 | " 5.00 | 10.40 |
| 24...... | " 6.30 | " 5.00 | 10.30 |
| 25 ..... | " 6.30 | " 5.00 | 10.30 |
| 26...... | " 6.30 | " 5.00 | 10.30 |
| 27...... | " 7.00 | " 5.00 | 10.00 |
| 28...... | " 8.00 | " 5.00 | 9.00 |
| 29...... | " 9.00 | " 4.50 | 7.50 |
| 30...... | " 10.00 | " 4.50 | 6.50 |
| 31...... | " 11.00 | " 4.50 | 5.50 |
| | | Total, | 205.10 |

## REMOVAL OF BOILER SCALES.

THE great bulk of the solid matter deposited from the feed water, remarks the Locomotive, may be removed by frequent and judicious blowing. It cannot all be removed in this manner, however, for where the plates are hot more or less of it is sure to bake on, forming the hard, stony layer known as "scale."

The commonest components of scale are carbonate of lime (limestone) and sulphate of lime (gypsum): Carbonate of lime seldom forms a stony scale. It may collect in large masses and do serious injury to the boiler, but the deposits which it forms are usually lighter and more porous than the corresponding deposits of the sulphate of lime.

Most substances are more soluble in hot water than in cold ; but carbonate of lime is a notable exception to this rule, for, although it is somewhat soluble in cold water, in boiling water it is almost absolutely insoluble. It follows from this fact that when feed water is pumped into a boiler, the carbonate of lime it contains is precipitated in the form of small particles as soon as the temperature of the water reaches the neighborhood of 212 degrees. These particles are whirled about for a considerable time in the general circulation, and if the circulation is good they do not usually settle until the draft of the steam is stopped for some reason—as for instance, in shutting down for the night, or in banking the fires for the noon hour.

The best time to remove this sediment by blowing is, therefore, just before starting up at one o'clock, or after the boiler has stood idle for an hour or so at night, or just before beginning work in the morning ; for at these times the carbonate deposit has settled into a kind of mud at the bottom of the boiler.

Sulphate of lime differs from the carbonate in being more soluble in hot water than in cold ; and it is, therefore, not deposited in the same way. The sulphate deposit is formed at those points where the evaporation (and consequent concentration of the solution) is most rapid, that is, in contact with the shell, the tubes and the back head. Being deposited practically in contact with the iron, it forms a hard adherent coating, which often resembles natural stone so closely that nobody but a skilled mineralogist could tell the difference between them. The best way to treat water containing sulphate of lime is to convert the sulphate into carbonate, and remove the carbonate thus formed by means of the blow-off, as already described. This can be done without injury to the boiler by the use of soda ash, which is a crude carbonate of soda.

## STATION DIVIDENDS.*

### By C. Doutre.

In taking into consideration the advisability of putting up a plant of any description, one or two things have got to be looked into very carefully before going ahead with the work—the first, cost of the plant, and secondly, the cost of supplying electricity to consumers. Now, gentlemen, it is not my intention to dwell upon either one or the other, but to try and explain why some plants do not give as fair a return as might be expected for the amount of money invested, and how to apply the remedy. In the majority of cases the reason is found to be ignorance on the part of either the manager or the party who has charge. If a station is not properly managed a satisfactory service cannot be expected, and a poor service is not conducive to gaining new customers or keeping old ones. And again, the operating expenses of a poorly managed plant must necessarily be more than if it were managed otherwise. Sometimes we find plants which are well installed, the apparatus is very good, the superintendent knows his business, local conditions are such that the plant ought to pay a fair return, and yet every year there is a balance on the wrong side of the books. The manager does the buying and manages everything. He enters into contracts which he ought to know would result disastrously to the company. A prospective customer comes along who probably lives in an isolated spot half a mile from the nearest circuit. He wants his house wired, or has it wired and would like to have light ; he probably has 15 to 20 lights, and his yearly account would not average more than 3 to 4 dollars a month. The cost of putting up the line and the possible income from same never come into consideration at all ; the manager agrees to light his house, and at the end of the year they find, although they have so many more customers, and the station output has increased 10%, there is still a deficit. A coal merchant offers coal at 5% cheaper price than what they are paying. No trial order is given, no sample offered to the superintendent or engineer for their opinion as to whether the article in question is of an inferior grade to what they are using ; a yearly contract is entered into, and at the end of the year it is found that instead of the consumption of fuel per electric H.P. being in the neighborhood of 3 lbs., it is nearer 4.

A plant managed in such a way will never pay. The superintendent or electrician in charge may be a first-class man in every respect, and although he is doing his best to run the plant as economically as he can, so as to make it pay, he is handicapped to such an extent by the lack of executive ability and judgment on the manager's part, as to be able to accomplish nothing further than to reduce the balance which, as I said before, is sure to be found on the wrong side of the ledger. Sometimes we find just the reverse state of affairs—the manager is an exceedingly shrewd business man ; manages everything to the advantage of the company and does everything to promote its welfare. He is unfortunate enough to engage a superintendent, who, although coming fairly well recommended, does not understand his business. The plant is comparatively new and has only been in operation a year or so. The difference between operating expenses and receipts for the year was so small as to indicate that something was radically wrong. So the directorate or proprietor decides that a change of management would be beneficial to the company's interests. As I said an A1 manager is secured who engages a man entirely unsuited for the work there is in hand. The plant is in bad condition, and there is plenty of room for improvement. Steam pipes are not protected with any suitable covering ; exhaust steam is going to waste ; cold water is pumped directly into the boilers. Boilers are not properly fired, and are cleaned when it becomes apparent to the fireman that he has to work twice as hard to keep steam up. The dynamo room is pretty much in the same condition—oil is everywhere except in the proper place ; loose coils of wire are lying around the floor ; carbons everywhere ; incandescent lamps, good and bad, in places where they are easily broken ; machines not properly cleaned, and the remainder in keeping with the above. The engineer, who knows his business, recommends certain changes and improvements—for instance, that the steam pipes be covered with some suitable covering. He explains that the loss due to convection and radiation is considerable, which we will figure out later ; also that by pumping the water directly into the boiler before passing it through some suitable heater, involves a great loss. The superintendent prefers getting along without the above improvements. His policy in regard to the management of the plant is the same throughout. He thinks that if the expenditures are kept low the company will think him a shrewd superintendent, and that his position will be enhanced. In the meantime the coal pile is diminishing at an extraordinary rate ; what was a small mountain in the morning is a mole-hill at night. The engineer in a short time looks upon the superintendent with one closed eye, and any suggestions or improvements which he may deem advisable to be carried out he keeps to himself. The superintendent displays the same knowledge and ability in regard to the dynamo room and other departments. Lamps are broken carelessly ; no account is kept of the same ; every-

*Paper read before the Montreal Electric Club.

one has access to the stock room ; material is taken for which an account should be given, but is never rendered. Here again, we have a parallel case, except that the positions are reversed—the manager on one side working hard for the company ; buying to advantage ; entering into paying contracts ; looking after the company's interest ; seeing that the office staff are doing their duty, and that the office and other departments under his immediate supervision are run as economically as is consistent with efficient work. What he has gained for the company has been lost in the station through the inefficient manner in which it has been operated, and through the false economy on the part of the superintendent.

Now suppose we take an ordinary every-day plant—one which you can find almost anywhere—say of 600 H.P. capacity, using non-condensing engines and operating the Edison 3 wire system ; and let us see if we cannot effect a saving by making a few alterations. We will start in the boiler room. Now, gentlemen, I think you will agree with me, when I state that the loss in the furnace is greater than in any other part of the system ; therefore, great care ought to be exercised to see that the boilers are handled efficiently. I think the " corpus delicti " of most of the non-paying stations would, on investigation, be found to be right in the boiler room. Firing is a science which few can master, and very few fixed rules or principles can be laid down for the efficient firing of any boiler, as boilers vary exceedingly in construction, and every different grade of coal has its peculiarities. The fireman is the person under whose immediate care the boiler properly comes, and his duties, from not being generally understood, are apt to be undervalued, although they call for more knowledge than is generally supposed. It is not too much to say that a really good fireman is an almost invaluable man, and that he saves his wages to his employer more than twice over by the care and economy which he exercises.

As stated before, the main object is to convert the heat energy of coal into electrical energy. To do this we must get as complete a combustion in our furnace as possible. The firing is only done properly when the fuel is consumed in the best possible way—that is, when no more is burned than is necessary to produce the amount of steam required and to keep the pressure uniform. Now, to attain this end, complete combustion must be attained in the furnace, and this is going on when the fuel is burning with a bright flame evenly all over the grate. Before considering the conditions we must have to attain this end, it may be of interest to find out exactly what the word combustion means. We find combustion is an energetic chemical combination of oxygen with some other substance, accompanied by light and heat. The substance with which it combines is called "the combustible," or when combustion takes place in an ordinary furnace, it is called "fuel," as for instance, coal, wood, oil, etc. The products of perfect combustion are water (steam) and carbonic oxide, and to insure it, a sufficiently high temperature and a sufficient supply of oxygen is necessary. The first step towards effecting the combination of any gas is, to ascertain the quantity with which it will chemically combine and the quantity of air required to supply the amount of oxygen. Much of the apparent complexity which exists on this head arises from the disproportion between the relative volumes or bulk of the constituent atoms of the several gases, as compared with their relative weights. For instance, an atom of hydrogen is double the bulk of an atom of carbon vapor ; yet the latter has six times the weight of the former. Again, an atom of hydrogen is double the bulk of an atom of oxygen, yet the latter is eight times the weight of the former, and so on. Coal must be distilled into gas before it can be properly burned, and in order to do this, hot air must be introduced at a temperature which will not cool the gases below the igniting point. A large supply of hot air is needed whenever a fresh lot of coal is thrown into the furnace, as this when first introduced generates a large amount of gas. If insufficient air is admitted, imperfect combustion will take place and therefore waste. An air space in the grate bars must be preserved and arrangement made so that fresh supplies of hot air are introduced above the fire. In nine cases out of ten this air is supplied through the medium of a strong chimney draught, and in regard to same I quote Prof. Rankin, the highest and best authority on the subject. In the 11th edition of his classical work on "The Steam Engine," (page 272), Prof. Rankin shows that with a chimney the best possible draught should be produced by a temperature in that chimney of 600° Fah. above the temperature of the external air, and then he says : "It appears that under no circumstances can it be necessary to expend more than one-fourth of the latent heat of combustion for the purpose of producing a draught by means of a chimney." Continuing Prof. Rankin says : "When the draught is produced by means of a blast pipe or of a blowing machine, no elevation of temperature above that of the external air is necessary, and with a forced draught less air is required for dilation, consequently, a higher temperature of the fire, a more rapid conduction of heat through the heating surface and a better economy of heat than there is with a chimney draught."

Now Prof. Rankin distinctly states that under no circumstances can it be necessary to expend more than one-fourth of the total heat of combus.

tion for the purpose of producing a draught by means of a chimney. Now one-fourth of the total heat of combustion means considerably more than one-fourth of the fuel, because it does not include the unburnt fuel nor the fuel escaping as combustible gases.

With chimney draughts the experiments of the U. S. navy show that the ordinary furnace requires about twice the theoretical amount of air to secure perfect combustion. Prof. Schwackhoffer, of Vienna, found in the boilers used in Europe an average excess of 70% of the total amount passing through the fire, or over three times the theoretical amount, was used. A series of analyses by Dr. Behr on the escaping gases from a well-known make of boilers with chimney draught, show the excess of air to be 42 per cent. of the whole quantity. A series of 12 tests made by the same, with artificial blast, gave an average excess of only 22 per cent. (which was almost a saving of 50%) of the whole quantity, and in a few cases none at all, with only traces of carbonic oxide. So by putting in a mechanical device for creating a draught, we can safely figure on a saving of 20 per cent.

Now what percentage will we gain by putting in a water heater? We are carrying a pressure of 95 lbs., having a temperature of 328° Fah. The best of heaters do not heat the water to above 200° Fah., and as the water has to be heated from the normal temperature to that of steam before evaporation can take place, some arrangement ought to be made so that this should not be done at the expense of the fuel, which should be utilized in generating steam. As stated, the temperature of steam at 95 lbs. pres. is 328°, and if we take 60° as the average temperature of feed, we have 268 units of heat per lb., which, as it takes 1.151 heat units to evaporate one pound of water from 60°, represents 23 per cent. All of this heat, therefore, which can be imparted to the feed water, is so much saved, not only in fuel, but in capacity of boiler. As we heat our water to 200° before pumping it into the boiler, we save 140 heat units, which represents 12.5 per cent. of the fuel saved.

Now let us find out what we are losing through the steam pipes being unprotected. Say we have 300 sq. ft. of superficial area of high pressure piping. The steam has a temperature of 328°, the surrounding air 75°; difference in steam and air 253° Fah. In experiments on bore wrought iron pipe, conducted at Cornell University under the direction of Prof. R. C. Carpenter, it was found that the loss due to convection per square ft. per hour in a four inch pipe at a difference of temperature of 277°, was 425 B.T.U., the loss due to radiation 340 B.T.U., making a total 765 B.T.U. At given difference of temperature, the loss due to convection per sq. ft. per hour would be 381 B.T.U., loss due to radiation 305 B.T.U., total 686. As we have 300 square feet of piping, the loss would amount to 205,800 B.T.U. per hour, or 205 lbs. of coal—equivalent to 6.8 H.P. at the switch board. Supposing the average load per hour was 300 indicated H.P., and we were running 24 hours a day. Assuming that it was necessary to burn 4 lbs. of coal per H.P., let us see what we have saved so far. We were burning 28,800 lbs., or 14.40 tons per day. By the addition of the mechanical draught we save 20%, or 2.88 tons per day; water heater 12½%, or 1.45 tons per day; by covering our pipes with one inch of magnesia and one inch of hair felt we save 492 lbs. per day. Taking these figures together we find that we have saved 8.352 lbs., or 4.2 tons a day. I think a fair average price for coal would be about $3.00 per ton, so the saving in plain figures would be $12.60 per day, or $4,599.00 for the year; and instead of the consumption of coal per H.P. being 4 lbs., it would be 2.84.

So much for our engines and boilers. Some stations derive quite an income from the exhaust steam. A concern with which I was connected heated three large office buildings, having a total of 1,340 rooms, for which they received the sum of $1,400, and if I remember rightly the back pressure was only 1¾ lbs. Any company using non-condensing engines can, by the simple addition of a back pressure valve, utilize their exhaust steam for this purpose, if they can find a sale for the same.

Before going on to the dynamo room and other departments, I would like to say a few words in regard to a furnace which was invented by Monsieur De Linet. Prof. De Linet, who was a lecturer in the Ecole Polytechnique, left Paris in 1870 to avoid conscription in the war, and started business as a boiler maker in London. The old lectures returned to his mind and he decided to put them into effect. This he did, and wherever they have been put in operation the whole series has been changed. By a system of expanding fans and strong mechanical draught, any substance, whatever its nature be, can be consumed. In experiments which have been performed with this boiler, sludge containing 40 per cent. of moisture has been operated on successfully. The Linet Electric Light Co., of Halifax, as the name implies, use the above-mentioned boilers, with such success that they can produce light at the rate of 13s. per 16 C.P. lamp per annum. They dispose of all the town refuse, and anything they can get in the way of refuse from manufacturers, etc. I may mention that from 4 lbs. of carbon lining from gas retorts, which has hitherto been difficult to dispose of, they have produced as much power as can be obtained from one pound of coal in the ordinary boiler. These boilers are operated with the above success

in Halifax, Harrogate and in numerous other places. There is no doubt that there is being wasted every day valuable power producing substances, and no one at this moment can attempt to gauge the vast economy which it is possible to attain in regard to the generation of power, as seems to be foreshadowed by the working of this invention; and it is not too much to expect that the time will come when, by the utilization of every waste product, the economy of living will meet the increased demands for the comfort of the masses, and will go some way towards turning the luxuries of to-day into the necessities of the future.

*(To be Continued.)*

## A NOVEL ELECTRIC FURNACE FOR HEATING IRON STRIP.

THE variety of ways in which the electric current can be applied for heating purposes seems to be steadily on the increase. Thus the heat can be applied directly to the body by passing the current through it either by the direct current or by the alternating current as in Prof. Thomson's welding processes. It can also be subjected to the arc as in the Benardos and like processes. Again it can be heated under water as in the recently developed Hoho process; and finally, the object to be heated can be brought into contact with another body previously heated by the passage of the current.

It is an application of the last mentioned kind that has recently been made in Montreal, where the Montreal Electric Co., by order of Jas. W. Pyke, Canadian representative for Messrs. Siemens & Co., have lately installed a furnace for heating iron strip used in making horse-shoe nails, at the rolling mills of Messrs. Peck Benny & Co., Montreal.

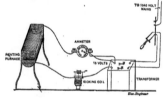

ELECTRIC FURNACE FOR IRON STRIP.

The accompanying engraving shows the arrangement. The local lighting company's mains are run in and deliver current at 1040 volts alternating to the Siemens transformer which reduces it to 12½ or 15 volts as required. The secondary flexible cables pass through an ammeter built on the dynamometer principle, also through a kicking coil which has a movable core, and thence to the furnace.

The furnace consists, says the Electrical Engineer, of a carbon tube, 24 inches long, with a bore of 1 inch, and walls ½ an inch thick; the tube is covered over with sand. The current at 15 or 12½ volts passes through this tube, bringing it to a white heat. 500 amperes is the usual current allowed, but at starting it is increased a little to hasten the heating up of the carbon which decreases its resistance, and the current is then lowered to normal. It is intended for continuous feed, and will heat five feet of strip per minute. There was some doubt that the heat could be got up, but this has been conclusively proved by the experimental apparatus described above.

## ADVANTAGES OF WOOD PULLEYS.

The practical advantages of wooden over iron pulleys are briefly summarized by Power and Transmission as follows:—

Saving in power by better traction surface for belt, 33 per cent.

Saving in weight, 70 per cent.

Reduced size and weight of shafting, hanging, etc.

Corresponding saving in power required to overcome friction in bearings, etc.

Gain in safety speed limit, 400 per cent.

Saving in time in putting on or off the shafts.

Saving in time in procuring pulley when wanted.

Saving in avoiding mutilation or distortion of shafting.

All these considerations become augmented in importance as speeds are increased, and are, therefore, of especial interest to all who are operating electrical machinery.

## ELECTRICITY FOR ARCHITECTS.*

### By JOHN LANGTON, TORONTO.

THE engineer or architect will find that, though the quantities dealt with in electrical work may be new to him, the ideas involved are largely the same as he is already familiar with in other branches of physical science. There is a common impression to the contrary, and this no doubt is partly due to the peculiar names of the practical electrical units; volt, ampere, &c.

These are merely arbitrary names, agreed upon by international convention, to shortly express compound units, and so avoid the repetition of cumbersome phrases. It is, for instance, as if it were agreed to call the ordinary British unit of fluid pressure a Newton, so that we might say shortly, but with perfect definiteness, 75 Newtons, instead of 75 pounds pressure per square inch above the atmosphere. Electrical units are in this manner named after eminent men of science. The Volt, after Volta, the discoverer of the galvanic battery; the Ohm, after Ohm, the discoverer of Ohm's law; the Ampere and Culomb, after the French physicists of the same names; the Farad, after Faraday, the Henry after Joseph Henry, and the Watt, the unit of power, appropriately named after James Watt.

In the common commercial uses for electricity for light and power, which the architect has ordinarily dealt with, the main ideas and phenomena present close analogies to the familiar facts of the pressure and the flow of water—so much so that the readiest way of getting a very fair general idea of commercial electricity is to consider the facts concerning it as being illustrated by the similar facts of hydraulics.

I will endeavor briefly to present this analogy, but I must ask you to remember that it is of course nothing but an analogy. I do not mean to imply that an electric current is in any sense a flow of a material fluid, merely that the results are very much as if it were a kind of fluid.

Referring to Table 1:—The 1st column gives the names of the electrical units which measure the four electrical quantities involved in ordinary light and power work. The 2nd column shows the letters by which they are symbolized in formulae. The 3rd column gives the electrical quantities which the units designate. The 4th column states the general ideas involved; and the last column the hydraulic quantities which are analogous to the electrical quantities in the 3rd column.

TABLE 1.

| NAME OF ELECTRICAL UNIT. | SYMBOL. | ELECTRICAL QUANTITY. | IDEA. | ANALOGOUS HYDRAULIC QUANTITY. |
|---|---|---|---|---|
| Volt. | E | Dif. of Potential. Electromotive force (E. M. F.) | Pressure. | Head or pressure of water. |
| Ohm. | R | Resistance. | Wasteful Resistance. | Friction of pipes and channels. |
| Ampere. | C | Current. | Rate of Flow. | Flow per second. |
| Watt. | W | Power. | Rate of doing work. | Power. |
| | | | 746 Watts = 1 Horse Power. | |

OHM'S LAW.

Current = Pressure ÷ Resistance.

$$\text{or Amperes} = \frac{\text{Volts}}{\text{Ohms}}$$

In Symbols, $C = \dfrac{E}{R}$

or, $R = \dfrac{E}{C}$

or, $E = CR$

ELECTRICAL POWER.—(Kilowatt = 1000 Watts.)

Watts = Volts × Amperes.

In Symbols, W = EC

or, $W = C^2R$

or, $W = \dfrac{E^2}{R}$

As the flow of water is due to difference of level or head, so is a current of electricity due to a difference of electric potential. And in both cases the amount of the flow through some path provided for it, is dependent—1st, on the head or pressure which causes it; and 2nd, on the frictional resistance which the provided path opposes to that flow, and in both cases the work expended in overcoming this resistance appears as heat. With the same resistance, the greater the pressure the greater the flow. With the same pressure, the greater the resistance the less the flow.

In electrical work, the relation between the pressure, the resistance and the current is a very simple one, and is expressed by Ohm's Law, which is that "The current is equal to the pressure divided by the resistance," or that the current is equal to the ratio of the pressure to the resistance. This is a definite numerical statement that the number of the amperes is equal to the number of the volts divided by the number of the ohms. For instance, 100 volts applied to the ends of a wire whose resistance is 50 ohms, will produce in the wire a current of 2 amperes. With 1000 volts and 500 ohms, the current would still be 2 amperes. And the same current of 2 amperes would be produced with 10 volts and 5 ohms.

Now as to the unit of power. In a fall of water, the weight of the water in pounds multiplied by the number of feet fall or head, is its energy—that is, its capacity for doing work—in foot pounds. And consequently, the rate at which this energy is developed, the rate of doing work, that is to say the *power* of the fall, is measured by the rate of flow multiplied by the head. In mechanical units 33,000 foot-pounds per minute is 1 horse power. Whether the flow is 33 pounds per minute under 1000 ft. head, or 1,000lbs. per minute under 33 ft. head, or 33,000 lbs. per minute under 1 foot head, the power in each case is the same, namely: 33,000 foot pounds per minute, or 1 horse power.

Similarly in electrical units, power is measured by the current multiplied by the E. M. F. The watts equal the volts multiplied by the amperes. 1000 volts and 10 amperes, 100 volts and 100 amperes, 10 volts and 1000 amperes, all give the same power, namely: 10,000 watts or 10 kilowatts.

Since it is the same thing that is measured in both cases,—power—there must be a definite numerical relation between the electrical and mechanical units, which is that 746 watts equals 1 horse power.

In buying electric power at a rate of, say, 4 cents per horse power hour, it is a very simple matter to calculate the cost of the current consumed.

\* Paper read at the fifth annual convention of the Ontario Association of Architects.

Power circuits supply current at a constant pressure. 250 volts is one usual pressure. If the current used is 3 amperes, the watts are 3 × 250 = 750 watts; practically 1 horse power, and costing 4 cents per hour. On a 125 volt circuit, a 6 ampere current would mean 1 horse power. So that for a general rule, multiply together the average amperes, the volts and the hours. Divide the product by 746 and the quotient will be the horse power hours consumed.

The resistance of an electrical conductor is analogous to the hydraulic friction in pipes; but whilst the mechanical friction in pipes varies according to the most complicated rules, the resistance of conductors is fortunately governed by very simple laws. It depends only on the material, the area of cross section and the length of the conductor. And fortunately again a cheap metal, copper, is one of the best conductors. It is second only to silver, which is better still, but only by a small percentage. The resistance of iron is between 6 and 7 times that of copper. Copper is therefore universally used for wiring, and we need only consider the effects of area and length, which are that the resistance of a wire varies directly as the length and inversely as the area.

Suppose the resistance of a wire 1 foot long and of a certain cross-sectional area is 1 ohm; if a feet long it would be 2 ohms; if 10 feet, 10 ohms. Another wire of 10 times the sectional area and 1 foot long would be 1-10 of an ohm, or if 10 feet long, 1 ohm. There are plenty of published tables of the resistance per foot of copper wires, and by the aid of this simple relation between resistance, area and length, they may be extended to any actual case. The question of resistance is of direct concern to architects in the wiring of buildings, but its bearing will perhaps be plainer after considering as briefly as possible the three systems of lighting in general use:

I. The constant current system, which is generally used for lighting by arc lam s.

II. The constant potential direct current system, used for incandescent lamps and for motors.

III. The alternating current system, used for incandescent lamps.

These can all be very well illustrated by analogous hydraulic systems, and for this purpose I have prepared diagrams 1, 2 and 3.

Take first diagram 1—the constant current system. Here the pump represents the dynamo, which maintains a steady flow circulating round the

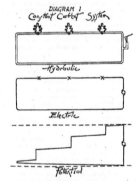

DIAGRAM 1
Constant Current System

Hydraulic

Electric

Potential

main pipe. At intervals in the main pipe are stop cocks, and round each of them a by-pass consisting of a long pipe, which opposes a high frictional resistance to the flow, so that the pump must exert greater pressure to maintain the same steady flow. These by-passes represent the lamps, in which the whole work done in forcing the current through them against their resistance, appears as heat, raising the temperature of the carbons to such a degree that they give out light. It is obvious that except for the constant resistance of the main pipe, a resistance which is made small, the work the pump must do increases directly with the number of the by-passes the flow must traverse; and, since the flow is constant, it follows that this increased work is due entirely to increased pressure.

Underneath the hydraulic diagram is that of the corresponding electric system, representing a dynamo and arc lamps. The amperes are the same no matter how many lamps are burning, but the dynamos must generate about 45 volts for each lamp burning.

Underneath this again, I have drawn a diagram of potentials, which shows graphically how the difference of potentials established by the action of the dynamo is consumed in different parts of the circuit. It is as if the dynamo were a pump raising water to a height, from which it flows down through the channels offered by the different parts of the circuit back to the pump, which again raises it to retrace the same course, maintaining a constant circulation.

Starting at the highest potential made by the dynamo, the potential gradient falls gently, owing to the slight consumption of volts required to overcome the small resistance of the main conductor. When it reaches the first lamp there is an abrupt fall, by the amount of the volts consumed in forcing the current through the high resistance of the lamp. Then follows a gentle grade to another abrupt fall at the next lamp; and so on to the last lamp, from which there is a last gentle grade back to the dynamo through the return wire.

When the same current passes through one lamp after another the lamps are said to be connected in series.

Turning to Diagram 2—the constant potential direct current system—the hydraulic diagram is an ordinary waterworks system, with the addition that all water used is discharged into a main return pipe which leads back to the pump, and from which the pump draws its supply.

The function of the pump is to maintain a constant difference of pressure between the mains. Each cross pipe from main to main takes whatever its own frictional resistance will allow the main pressure to produce. And it is obvious that as more cross paths are opened, the increased work the dynamo must do, is due to the increased flow; the pressure remains constant.

In the electric diagram underneath, the high friction cross tubes are replaced by incandescent lamps, and the water motors by electric motors.

The potential diagram shows that the pressure in the mains cannot be quite constant, since there must be some fall of potential in the main conductors, and the amount of this fall is less when there are fewer lamps burning, i.e., when the total current is less. The potential gradient is

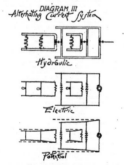

DIAGRAM II
Constant Potential Direct Current

Hydraulic

Electric

Potential

greatest near the dynamo where the conductor has to carry the total current, but the grade gets flatter and flatter as the current diminishes, by the amperes subtracted at each cross part. These potential gradients are repeated in the reverse order in the return wire.

The form of gradient drawn is that for a conductor of uniform size. But if the conductor is reduced at each cross path so that its area always bears the same proportion to the amperes carried, the potential gradient would be the same throughout, and a continuation of the first grade starting from the dynamo, as is indicated by dotted lines.

When lamps are connected side by side, so that each takes its own separate current, the lamps are said to be connected in parallel, or in multiple.

There remains the alternating current system shown by Diagram 3. The pump in Diagrams 1 and 2 produces a flow always in the same direction, representing a direct current of electricity. In Diagram 3 the pump is re-

DIAGRAM III
Alternating Current System

Hydraulic

Electric

Potential

placed by a movable piston in the main pipe, which being oscillated from one side to the other, produces a pressure first in one pipe and then in the other, with an accompanying back and forth flow which will vary in amount according to the number of cross paths open. This represents the primary circuit, to which the lamps are *not* connected. The object of an alternating system is to save in the cost of main conductors by transmitting power with a small current at a very high pressure. But for convenient and safe use, this power must be transformed into a larger current at a lower pressure, in a secondary circuit to which the lamps are connected. It is the peculiarity of the alternating system, that this can be done with very little loss and without any moving mechanism. The actual means by which this is effected in the "transformers" or "converters," are beyond the scope of this paper, but a simple mechanical contrivance in the hydraulic system will give us analogous results. A sliding piston in the small primary cross pipe and another piston in the large secondary pipe are connected by a bar pivoted in the middle, so that the two pistons oscillate together and move equal distances. If the area of the secondary piston is, say 10 times that of the primary, a secondary pressure 1-10 that of the primary will balance the contrivance, whilst a secondary flow of 10 times the primary is produced by any oscillation.

The electric diagram shows the alternator and primary circuit, feeding two separate secondary circuits through two transformers.

The potential diagram for both primary and secondaries is similar to diagram 2.

I may say parenthetically that the hydraulic analogy goes still further. The effects of the inertia of water represent excellently those of the electric quantity, self induction, whilst, if the pipes are made elastic, the results would be very similar to those due to electric capacity. But as inductance and capacity are inappreciable in such work as an architect will ordinarily have to deal with, I have not considered them in this paper.

In wiring for incandescent lamps the object is to maintain as nearly as possible a constant difference of potential between the terminals of the lamps. But when, at different times, there are are at different points on the same line, different numbers of lamps burning, it is impossible that the volts at all points should be always the same. And the question is, what variation is permissible?

An incandescent lamp is simply a carbon wire of high resistance, which the current passing through it heats to incandescence. But the resistance of the carbon decreases as the temperature increases. Therefore a rise or fall in the volts causes more than a proportionate rise or fall in the amperes, and the consequent heat and temperature. Also, the light given out increases much more rapidly than the temperature. Roughly speaking a variation of 1% from the rated volts of the lamp, up or down causes a variation of 5% in the light. And it is informally agreed that a% total variation in volts at any lamp constitutes good and satisfactory regulation. This variation is of course reckoned between the maximum and minimum load—not between full load and no load. If a group of lamps is fed by a wire direct from the constant volts of the source of supply, and the lamps in the group are always all turned on or off together, the drop of volts in the wire may be any amount, and yet because it will be always the same, there will be no variation in the volts when the lamps are burning.

But besides minimizing the variation in the volts at any one lamp, it is desirable that the lamps in the same building should all get about the same volts, so that the same class of lamp may be used interchangeably throughout the building. This is the object of the feeder system of indoor wiring, in which feeder wires carry the current to convenient points in the main wires to which the lamps are connected. Diagram 4 shows graphically the effect on the distribution of volts.

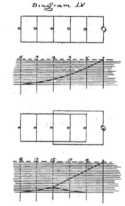

Diagram IV

The consolidation of experience in wiring, into practical rules, embodying safe practice as regards danger from fire, has been performed by the Fire Underwriters Associations. Their rules and regulations are published, and give in detail the minimum standard of safe wiring. The wiring must be at least as good as the rules prescribe, to avoid trouble with fire insurance. Among other things the rules prescribe the greatest amperes different sized wires may carry, and here is a point where they are not necessarily a sufficient standard to guide the architect. The Underwriter's aim is a single one, safety from fire; the architect has an additional aim, good and unvarying light; and our previous considerations concerning the resistance of wires, will show how these objects are not simultaneously attained.

The work spent in forcing a current through a wire generates heat in the wire, and the degree the temperature will rise to depends upon the relation between the rate at which the heat is generated and the rate at which it is got rid of. Now the rate at which heat is generated, is the rate at which work is spent in forcing the current through the resistance of the wire; i.e., it is the watts spent in the wire. And from Table 1, watts=(amperes)²× ohms. But we have seen that a long thick wire and a short thin wire may have exactly the same ohms resistance. With the same current the heat generated is the same in both. How about the resulting temperature? In the long wire the heat generated is spread over a greater length. The heat generated per foot run is less, and, the long wire being thicker, it also presents greater surface per foot run for cooling by radiation, convection or conduction. Obviously the short thin wire gets much the hotter. The safe carrying capacity of wires, both exposed and cased, laid down in the Underwriters' rules, is really a statement of the heat generated per foot that the wires can get rid of without becoming dangerously warm. This is independent of the length of the wire. The heat generated per foot remains the same, but the longer the wire, the greater the total heat and the greater the total volts lost in forcing the same amperes through the resistance of the greater length. But the total drop in volts is what the architect must limit in order to get good and steady light. He must therefore be guided by the total resistance, and the ampere capacity of the Underwriters' rules is useful to him only as setting the inferior limit to which he may reduce the size of the wire.

With regard to the different classes of wiring—exposed wiring on cleats or knobs is best for factories, but for domestic work it is only suitable in stores or houses where hanging kerosene lamps, or exposed gas pipes stapled to the walls and ceilings, would be the alternatives. What is called concealed work, in which rubber covered wires are run between floor and ceiling, and brought out only where brackets or hanging fixtures are to be placed, is probably the most suitable for the ordinary run of dwellings. Where once in, there is little liability of its being disturbed. But in larger buildings, warehouses, stores and such like, where there are likely to be changes in partitioning off the space, and in the distribution of the lights, to suit different tenants, or the same tenant at different times, wires run in mouldings are by far the most convenient. They are not only most readily accessible for changes of wiring, but the wiring is much less liable to damage by workmen in other trades making repairs. The principal objection is the unsightliness of lines of mouldings straggling all over the ceiling to wherever a lamp is needed. But moulding work has frequently many advantages of convenience to recommend it. Architects, I believe, are always looking for a "motive" in design. Mouldings can be made of any section, and perhaps they might be used to panel off the ceiling in some decorative pattern, suitable to any probable distribution of lights, giving a wide choice of paths for wiring and points for outlets. But this touches the artistic side of architecture, which is beyond my province. The best class of wiring is interior conduit work, in which buildings are piped with strong non-metallic, waterproof and poorly combustible tubing, and wires subsequently drawn into. This is particularly suitable to such work as the best class of office buildings. And where it would be too expensive to carry throughout a building, it may sometimes be used for the main lines to centres of distribution

When incandescent electric lights were first introduced they were distributed on wall brackets and hanging fixtures in the same manner as gas jets, as if this were the natural arrangement of lights, instead of having originated in the necessity of keeping gas jets within reach for ease in lighting, and in keeping them away from walls and ceilings for fear of fire. This force of habit for some time prevented, and in a measure still prevents full advantage being taken of the possibilities offered by electric lights for getting better illumination with the same amount of light. For a desk light or a reading light we cannot do better than replace the shaded oil lamp or gas drop.light by a shaded electric lamp, but for the general illumination of a room the incandescent light can, in general, do much better. The illumination we perceive depends not only on the amount of light reflected from an object, but also on the amount of the reflected light the eye takes in, and with lights a little above the level of the eye, we are always partially dazzled by them, and our perception of surrounding objects is indistinct compared with what it would be if the lights were out of sight. To get the best illumination for the quantity of light, a room should be lighted as a picture is lighted for exhibition. Electric lamps high up near the ceiling remove the dazzling effect and at the same time give a more generally diffused light, especially if there are light tinted walls and ceilings to reflect the light without much loss.

For lighting large rooms, arc lamps are used to a considerable extent on the continent of Europe, constructed so that they throw all their light on a white or light tinted ceiling, the room being thus lighted entirely by reflection from the ceiling. The result is a diffused light as shadowless as diffused daylight.

Now with regard to uses of arc and incandescent lights. Diagram V is copied from one by Prof. Nichols, of Cornell, embodying the results of experiments made by him, and is instructive in illustrating the difference of light from different sources in quality, as opposed to quantity.

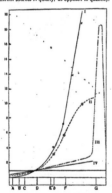

This diagram represents the brightness of different parts of the spectra of the electric arc, clear daylight and clouded daylight, in comparison with the same colours in the spectrum of an incandescent lamp. The brightness of the latter is taken as the standard in all parts of the spectrum, and is represented in the diagram by the horizontal line at the height 1. The other spectra are reduced to the same brightness at the yellow line D, and their brightness in other regions of the spectrum is shown by curves.

Curve I represents daylight on a cloudless summer day.
Curve II represents daylight under a densely clouded sky.
Curve IV is from the lime light.

The abrupt rise and almost immediate descent again of curve III indicates a narrow but very bright band of light in the violet end of the arc spectrum, which accounts for the value of the arc light in photography, and also explains the predominant bluish tinge of its light.

The curves show how far all artificial illuminants fall short of equalling the quality of clear daylight, which latter must always be our standard of perfect white light. Even the light of a very dull day is a better all round light than an arc light or lime light of equal general brightness.

Lights are usually rated by candle power, and this is gauged by the relative blackness of shadows thrown on a white ground. Candle power there-

fore merely measures the ability, to distinguish between black and white. For this purpose the yellow rays are much the most effective; but the blue and violet rays are the most useful for showing the distinction between colors, that is, for bringing out the colors of natural objects. And this being so, the curves show that the incandescent lamp gives us the most of that kind of light wanted, for reading or writing, whilst the arc light gives a closer approach to the effect of daylight upon colors. In addition to positive utility, the feeling of suitability has a value. The incandescent lamp, with its warm yellowish red glow, gives a cheerful and cosy air to a small room, where an arc lamp would be simply garish. In a large hall or store, to which the idea of coziness is inappropriate, the same light that makes a small room cheerful may give only an impression of dullness, whilst the arc light would give an agreeable effect of brilliancy. Of two large stores side by side, one lighted by arcs and the other by incandescents, the arc lit store has in general a more attractive air of being brilliantly lighted, although, measured by candle power, its actual illumination may be decidedly inferior.

Diagram VI is a curve of current consumed in one trip of an electric eleva-

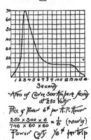

Diagram VI

Area of Curve 500 sq in taken as being at 250 Volts

Price of Power 6¢ per H. P. Hour

$$\frac{250 \times 500 \times 6}{746 \times 60 \times 60} = \frac{1}{6} \text{ (Nearly)}$$

Power Cost ⅙¢ per trip

tor. The time in seconds is measured horizontally, and the amperes vertically. This is a diagram from an actual elevator, running 250 feet per minute in a three-storey building. I have taken several such diagrams from different elevators and they all have the same general form as the one shown. I have chosen this one of an elevator having a short run, in order to better bring out the effect on the current consumption of frequent stops. The peak at the beginning of the curve shows the large current required to start the car and accelerate it to full speed, in comparison with the small current needed—from 7th to 14th, second in diagram—to keep the car in motion after speed has been attained. Nevertheless, even in this unfavorable case, the curve shows the very small cost of power per trip.

For the diagram given the cost is only ⅙ cents per full trip one way, at the rate charged in Toronto for very intermittent use of current, which is 50% higher than the regular Toronto meter rate for power supplied to an elevator in constant use.

Hitherto the application of electricity to architecture has owed little to architects. Trade competition has forced in electric power to take the place of power from other sources which had been already applied to elevators, pumps, ventilating fans, &c.; but beyond this little has been done. The convenience of, electric power has certainly led to the extension of mechanical ventilation, with its steady displacement of a fixed volume of air per minute, independent of the degree of dryness or temperature which makes ventilation by natural draught so variable. In ventilation architects have been fairly quick to utilize the opportunity afforded them. But in the larger problems of architecture, the possibilities of applying electricity seem to have received little or no attention. An illustration of what I mean is what might conceivably, though not probably, be the effect of cheap electric power in modifying the design of dwelling houses. If it were desirable to use elevators it would not be very difficult to devise perfectly safe methods of operating them without trained attendance, and a constantly used elevator would have almost as great an effect in modifying house planning as the substitution of stairs for ladders. Supposing such a use of elevators were practicable; whether it would be desirable, whether it would really add to the comforts and conveniences of life, nobody can say so well as the architect, who alone is trained to appreciate at their proper value all points bearing on such a question. And that is the point I wish to bring out by the illustration. Whether in the future electricity is to play any part in modifying architecture; whether it be of any real assistance to the architect in dealing with the particular problems of his profession, depends principally on the architect himself. The electrical engineer must co-operate in devising working details, but if the applications of electricity to architecture are ever to be more than superficial, the initiative must come from within, not from without.

## AN ILLUSTRATION OF POWER.

If it were possible, says the Polytechnic, to place 300 oars on each side of the ocean steamer Campania, making 600 oars altogether, each worked by three men, there would be 1,800 men at work at one time. As they could not work continuously for twenty-four hours, but only for a total of eight hours each man, divided into four watches, it would be necessary to have a crew of 5,400 men to man the oars. If six men could develop 1 H.P., the total horse power developed by the 600 oars handled by 1800 men would be but 300, as against 30,000 in the Campania, or the same power would require the employment of 180,000 oars and a crew of 558,000 men to manipulate them.

The American Bell Telephone Company has appealed to the United States Circuit Court of Appeals against the decision of Judge Carpenter declaring the Berliner patent, No. 463,569, void.

## THE ONLOOKER.

BETWEEN fifteen and twenty companies, asking for powers to operate electric railways, have made application to the Ontario Legislature this session for incorporation. The fact is evidence of the important part electricity will occupy in the railways of the near future. The Onlooker heard comment on this point by experts in engineering, like Mr. Wm. Jennings, C. E., and Mr. Rutherford, chief engineer for the Canadian General Electric Co., (Ltd.) It is hardly likely, should letters of incorporation be granted to all the companies applying, that they will enter actively into the work of construction at once. Mr. Jennings felt assured, however, that certain of the projects now contemplated would be entered upon this season, and the Onlooker at the time he chatted with Mr. Rutherford, found that he was then in communication with the promoters of one of the railway companies included in the list. The extension of several roads already in operation will likely be prosecuted as soon as the weather permits. These will almost certainly include an extension of the Galt and Preston road to Hespeler, and certain work on the Guelph road. The Onlooker took occasion last month to point out where possible mistakes might be made in a too rapid development of the electric railway. The very fact, however, that a cautionary signal had to be raised, adds emphasis to the leading position electricity will play in the railroads of the future. When one contrasts the railroad of less than half a century ago, with its imperfect and faulty construction, and entire barrenness of the thousand and one conveniences the travelling public enjoy to-day, some conception may be realized of how great a work has been accomplished in a short time. With electricity, than which no science is forging to the front with so great rapidity, clamoring for the foremost place as a method of railway propulsion, what will another half a century bring forth—nay another decade? Electricity is already coming into general use for lighting cars. It is used to no small extent for heating cars; and at Baltimore, an electric plant is rapidly approaching completion that is to provide power for an electric motor of sufficient capacity to move heavy trains through a tunnel half a mile in length. A recent engineering authority in noting this circumstance, significantly asks : "If this can be accomplished may we not expect to see electricity supersede steam as the motive power of the locomotive?"

x   x   x   x

The third, and in many respects, the most serious of the recent Toronto conflagrations, which resulted in the destruction of the large seven-storey building of Mr. R. Simpson, and other adjacent buildings on Sunday morning last, has caused the people to renew the enquiry, what has been the cause of these fires? At the time of the Globe fire two months ago, and that of the Osgoodby building and several large warehouses a few days later, it was suggested, as a possible explanation, that electricity had, in some way, proven the primary cause of these fires. This suggestion is being offered with increased intensity as, not only a possible, but reasonable, explanation of the causes that led to the fire Sunday, March 3rd. Some of the daily papers have been very persistent in the effort to fix the cause of the fire in some way on the feed wires of the companies from whom power was obtained to light the Simpson building. The Onlooker would wish that every effort might be exercised to get at the bottom of this last fire, and also of the two previous ones. But don't let Mother Grundy take the place of common sense and fact. It has been said that the Simpson fire started in the basement where the feed wires came in. Let this statement be disposed of at once. The feed wires entered at the front of the building. The fire broke out in the rear. Another theory put forth is that possibly power from certain wires might be conveyed to a gas pipe and combustion take place there. The Onlooker has made careful enquiry on this point, and the unreserved opinion of experts who knew the building, and who are able to speak on a matter of this kind on general principles, say plainly that such a thing could not occur. An inquest on the Simpson fire is to be held without delay. The Onlooker sincerely trusts that it will result in an intelligent and satisfactory explanation of the trouble. It is hardly anticipating the finding of that body, however, to say that a multitude of absurd theories regarding the connection between electricity and the destruction of valuable property by fire will be thoroughly exploded.

It would be unwise to say that there is no risk whatever from fire in the use of electric lights. There are certain dangers. But compared with the use of gas, for example, not to say anything of less modern methods, these are infinitesimally small. The Onlooker has before him at this writing a statement showing that no less than 284 fires were caused in 1894 in London, Eng., by what is termed gas explosions, but not a single explosion produced by electric light. In a conversation with Mr. A. B. Smith, electrical inspector for the Underwriters' Association of Toronto, this fact was mentioned, and Mr. Smith believed, from his knowledge of the subject, and, perhaps, no one in Canada can speak more intelligently on this question, the London figures would be found to apply, relatively, to all other cities—a fact worth remembering when it is remarked that where fifteen years ago hardly a single commercial incandescent lamp was in existence, to-day 12,000,000 are used in the United States and Canada, and throughout the world there are nearly 25,000,000 incandescent and electric light circuits. The security and safety to be found in the incandescent system of lighting can readily be named. Matches are unnecessary. How many fires are constantly occurring through the careless disposition of the lighted match after it had been used to light a gas jet ? As a covered light, there is no danger from the various inflammable oils, gases and dusts that are found in the air, and which, where the open gas jet is used, have frequently produced serious trouble. The one thing necessary to give the most perfect protection to incandescent lighting is care, experience, and perfect work in the construction and placing of the plant and wires. This department of work, Mr. Smith remarked to the Onlooker, had in the Dominion attained a very high degree of perfection.

x   x   x   x

The Onlooker has had pointed out to him the danger that comes from the burning of the fusible cut-outs, thereby throwing out sufficient sparks to ignite any inflammable or combustible matter that might be near by, and the manager of a large concern in Toronto that instanced two cases of the kind. In the particular building named, Mr. Smith said some of the older methods of construction was still in use. At the present time fuses were protected in such a way by a plate that when they gave out the sparks were confined to the enclosure, and could not possibly do any harm. Saving this much, the Onlooker at the same time would not want to excuse faulty construction. The future of electricity as a method of lighting is going to rest in no small degree on the character of the work performed from this day out. A writer in the current number of the Engineering Magazine has said, perhaps no element among those which entered into the causes of fire and fire losses has caused more discussion and difference of opinion than electricity, since its introduction for light and power during the past decade. And this writer proceeds to show wherein certain dangers of fire will come from electricity. With the quarterly report of the Electrician for the Chicago Fire Underwriters' Association before him he discusses the result of fires alleged to have been caused by "grounds;" by the return current passing through the earth; a result of contact between conductors carrying electricity for light and power ; and wires intended to carry only "battery" currents. Candidly admitting that fires have occurred in all these ways, and in other ways, he states the whole question when he says : "Anyone who cares to study the causes of fire started by electricity will be convinced that a very large portion of these fires are from defects, which would never have existed had the work of installation been done by men having a proper knowledge of the subject." In almost so many words this is the position taken by Mr. A. B. Smith, in his conversation with the Onlooker. Beyond any doubt, electricity is the coming method of lighting, alike for commercial and domestic purposes. The extent and rapidity of its growth will be retarded if such suggestions as come from Mr. Cabot, author of the article in the Engineering Magazine, from Mr. Smith, Mr. Rutherford, and other electrical experts are ignored. Contrariwise putting into practice the suggestions there made will give to this method of lighting full, complete and quick supremacy over all others.

Mr. C. F. Beauchim, of Montreal, is endeavoring to organize a telephone company in Quebec.

## CHEAP ELECTRIC POLE FOUNDATIONS.

IN the construction of the Negaunee and Ishpeming electric railroad and of the Marquette and Presque Isle road in Michigan, some difficulty and expense was anticipated in setting the poles where it was necessary to locate them in the edge of the lake in shallow water and soft mud. Pile driving was not practicable, coffer-dams were too troublesome and slow, iron caissons were too costly, and ordinary dredging was impossible. Finally the best empty oil barrels that could be procured were purchased, and after removing both heads were driven down solidly to about water level. Then the soft mud and clay was scooped out as much as possible and the pole set up on end in the barrel and worked down by hoisting and racking as far as it would go, usually 5 or 6 feet. A 2" x 10" stick was then driven with mauls each side of it, their flat sides against the pole and their bottom ends beveled to draw apart. When they were all in place the barrel was pretty well filled up by them. The pole was securely guyed and the mud again scooped out of the barrel and the water pumped down as low as possible, when about 30 pounds of dry cement was thrown in the bottom, and on top of that enough concrete, one-half small stone, was put in to fill up the barrel. After it had set a few days the guys were removed and the pole remained very firm and rigid. Some that were cut out showed excellent solid concrete extending nearly a foot below and around the barrel, as shown in the figure, when the dry cement had been pushed out and mixed with the surrounding sand. When very small barrels were used it was found better to drive the 2x10 sticks before setting the post. When the poles were used for span work they were battered excessively, up to 1 in 3, but for bracket work they were battered about 1 in 10. The cost was : Barrel, 60 cents ; cement, 50 cents ; lumber, 75 cents ; labor setting barrel, 50 cents ; driving posts, 40 cents ; setting and concreting pole. 40 cents ; total, $3.25. About 1 mile of poles were set thus in 1890, and are reported to have given complete satisfaction ever since. The above description has been prepared from the notes of Alexander Thompson, C. E., then resident engineer of the work, who devised the arrangement.—Engineering Record.

CHEAP ELECTRIC POLE FOUNDATIONS.

## LOOK TO THE BOILER ROOM.

THE enormous coal consumption per unit of output in many electric lighting and power plants is cause for general comment, especially since a recent committee report brought the wide variations of efficiency prominently into notice.

Theories innumerable are advanced to account for the difference between the fuel burned per horse-power in driving the dynamos, as compared with other service ; a favorite conclusion seeming to be that compound engines are not satisfactory when working through other than narrow ranges of power variation.

While this is probably true of a great many of the engines used for such work where the cylinder proportions and general make-up are no credit to those responsible for the designs, yet there is little doubt but that one main cause of the trouble must be looked for in another direction. A glance through the power houses discloses the fact that many of them are run on an easy-going basis, no attempt seemingly being made to maintain proper discipline among the attendants, each of whom shifts for himself without let or hindrance from the directing authority.

A genuine fireman, thoroughly trained in the principles of his profession, would blush with shame at the sight of these boiler departments.

In place of clean grates, giving a bright glow beneath, the bars are masked by clinkers, and the ash pits yawn without a ray of light to show what is going on within.

The air wheezes as it forces a passage through the refuse, instead of passing in with that rustling sound that tells of free combustion.

At frequent intervals, between the discussion of politics, or other matter foreign to the work in hand, an individual, whose only claim to being a fireman lies in his ability to heave carbon against the back of the furnace, rises from an ottoman of coal and canvas, and throwing open a furnace door, leisurely proceeds to shovel in a half ton or so of fuel, after which work of art, with no attempt to level the fire, or clear the grate, he throws himself upon his couch for another restful season.

Under the too common management of such places, there seems to be not the slightest incentive offered a man to properly attend his boilers.

An enormous grate and heating surface, and immense chimney, are relied upon to maintain steam, when a few first-class men at the fires would make fewer boilers do better work.— American Machinest.

## PERSONAL.

Mr. J. R. H. White has been appointed Secretary-Treasurer of the Montmorency Electric Light & Power Co., of Quebec.

Mr. Geo. A. Cox has been elected a director of the Toronto Railway Company to fill the place on the board made vacant by the resignation of Mr. J. W. Leonard.

Mr. Chas. A. E. Carr, the new manager of the London Street Railway, was a passenger on the Grand Trunk train which came into collision with a train going west from Toronto last month. Mr. Carr fortunately escaped uninjured.

Mr. J. W. Taylor, who has been manager of the Peterborough Carbon and Porcelain Co. since its organization, has resigned to accept the management of the Company recently organized at Ottawa to engage in the same line of manufacture. Mr. Taylor has already entered on the duties of his new position.

Mr. Frederic Nicholls, 2nd Vice-President and Managing Director of the Canadian General Electric Co., has been elected 1st Vice-President of the National Electric Light Association of the United States at the Convention recently held in Cleveland, Ohio. This international recognition of Mr. Nicholl's personal and business standing is the more pleasing in view of the fact that this will be his fifth year on the Executive Board of the Association, in which last year he filled the office of 2nd Vice-President.

Mr. J. P. Sparrow, for the last three years on the staff of the Canadian General Electric Co., as foreman of construction, has removed to New York, where he has been appointed to an important position on the staff of the Edison Illuminating Co. Mr. Sparrow, while in this country, has had charge of some important installations for the Canadian General Electric Co., such as the Niagara Falls Park & River Railway, the Brantford Street Railway and the power plant for the Dominion Government at the Canadian Sault.

A large number of the contemporaries of Mr. W. E. Davis, in the service of the Toronto Street Railway Company, on the eve of his departure for Detroit, to superintend the construction of the electrical system of the Detroit Railway Co., entertained him at a smoking concert at the St. Charles Restaurant, and presented him with a handsome gold watch, bearing the following inscription : " Presented to W. E. Davis by the employees of the electrical department of the Toronto Railway Company, in token of respect and esteem, on his retiring from the position of electrical engineer, Feb. 12, 1895." Mr. Thomas Graham, chief of the stores department, occupied the chair, and Mr. A. G. Horwood made the presentation. Mr. Davis was married a fortnight ago to Miss Meta Gallon, of Parkdale, Toronto. After the wedding breakfast a special car conveyed Mr. Davis and his bride to the Union Railway depot, where they embarked on a tour to Washington and Old Point Comfort.

Electric conduits made of small. vitrified stoneware sewer pipe laid in concrete on brick subways have been introduced in England. The conductors employed are naked copper strips which rest on insulators, the pipes themselves, each two feet long, resting on porcelain bridges. In order to join the conductor with a house connection, a special length of pipe is furnished, which is U-shaped for a part of its length. The house lead is taken out through an insulated removable cover forming the top of this part of the pipe. The joints are made on the Doulton system, and require no cement.

## SPARKS.

A semi-annual dividend of three per cent. has been declared by the Canadian General Electric Co.

Incorporation has been granted to the Mica Boiler Covering Co., of Toronto, with a capital stock of $50,000.

The number of telephone subscribers to the Windsor, Ont., exchange has increased in eight years from 54 to 800.

It is reported that the electric light is to be introduced at Eganville, Ont., by Messrs. John Childerhose & Sons.

The central station of the Victoria, B. C., Electric Light Co., was damaged by fire to the extent of $15,000 on the 26th February.

Negotiations are in progress with a view to the absorption of the St. John, N. B., Gas Light Co., by the St. John Street Railway Co.

The Hamilton Street Railway Company is said to have made an order forbidding employees from entering a saloon at any time.

Mr. D. C. Dewar has succeeded to the management of the Bell Telephone Company's Exchange at Ottawa, vice Mr. T. Ahearn resigned.

Application for incorporation has been made by the Milton Electric Light and Power Co., of Milton, Ont., with a capital stock of $15,000.

It is reported that an effort is being made to construct an electric street railway at Stratford, with radial lines extending out to neighboring villages.

A very successful "At Home" of the officers and members of the Canadian Marine Engineers' Association and their friends, was held in Toronto recently.

The annual meeting of shareholders of the Ontario Mutual Telegraph Company will be held at the company's offices in Montreal on the 28th instant.

The council of the town of St. Henri has granted permission to the Merchants' Telephone Co., to erect poles and string wires within the limits of the municipality.

The contract for carrying the mails between Hamilton and Bartonville, Stony Creek and Fruitland, has been secured by the Hamilton, Grimsby & Beamsville Railway.

The authority of the Ontario Legislature is asked to allow the Hamilton Iron and Steel Co. to construct a tramway to connect the works of the company with its quarry property.

The announcement is made that the exclusive rights and privileges of the Hull Electric Railway Co. have been purchased, on behalf of an Ontario syndicate, by Mr. Peter Ryan, of Toronto, and that construction will be commenced in the spring.

The London, Ont., Gas Co. has during the past year added machinery for the purpose of cheapening the production of gas, and is said to be about to make a big cut in price to meet the competition of the electric lighting.

It is reported that the first electric railway to be built in Muskoka, is being projected at Huntsville, and that it will span a difficult portage a mile in length, thus forming a connecting link in the navigation of the lakes of that vicinity.

Mr. C. G. Ballantyne, a native of Peel County, Ontario, has been granted an exclusive electric franchise for an electric railway in Honolulu, in recognition of services which he rendered the government in the recent battle with the rebels.

Incorporation has been applied for by the Peterboro' & Chemong Park Railway Co., with a capital stock of $100,000, to construct an electric railway from the town of Chemong Park, and to any other points which may be thought desirable.

At the annual meeting of the Sarnia Gas & Electric Light Co., held recently, the old Board of Directors was re-appointed. The Directors at a subsequent meeting elected Mr. Thos. Kenny as President, and Mr. Wm. Williams, as manager and secretary of the company.

Application is to be made to the Ontario Legislature to incorporate the Guelph Railway Co. to construct and operate a surface street railway in the city of Guelph, under the agreement entered into on the 7th of August last, between the corporation of Guelph and Mr. Geo. Sleeman. Power is also asked to extend the railway outside the city limits.

Mr. R. C. Cowan has been exhibiting in Montreal a snow sweeper, the invention of a Boston man named Callet. The sweeper can be operated by electricity or by horses, and by means of large brushes making 3,000 revolutions per minute, is said to be capable of perfectly cleaning the street of snow and dirt, which is carried by a Sturtevant blower through a funnel into carts.

At the annual meeting of the Guelph Light & Power Co. held recently, the annual report showed that notwithstanding the general depression, a fair business had been done. This applies also to the power distribution branch of the company's business. Mr. Guthrie, the president, stated that the success of the company is largely due to the careful and efficient management of Mr. John Yule. The old Board of Directors was re-elected.

A snow-brush for trolley cars, the invention of Mr. R. G. Olmstead, of Hamilton, is said to have been successfully tested on the Hamilton, Grimsby and Beamsville electric road. It can be fastened to the front of an ordinary trolley car and will cost only about $150 each, whereas the large plows in use now cost over $6,000. The brushes are of steel and rest on each rail. They are geared from the axle but rwolve four times as fast. There is also a small revolving fan underneath the car which prevents the snow banking up more than three inches between the rails.

The proposal made by one of the local papers that the town of Amherst, N.S., should purchase the plant of the Canada Electric Co., and operate the same, does not meet with favor. It is pointed out that in view of the necessity for a large expenditure in the near future for sewerage purposes, the town is not in a position to add the sum of $60,000 to its debt for the purpose mentioned.

The electric power houses at Ottawa have recently experienced much difficulty owing to the formation of anchor ice at the falls of the Chaudiere. The street railway company had great difficulty in keeping their cars running, and some of the streets were temporarily without light. This difficulty is experienced to some extent every year, but owing to the severity of the frost, is said to have been greater than usual the present winter.

The Hamilton, Valley City & Waterloo Railway Co., are applying for a charter. The capital stock is placed at $650,000, and the shares at $100 each. The promoters are: The Rev. Dr. Burns, Messrs. John Hoodless, J. E. O'Reilly, A. McKay, M. P., F. A. Carpenter, Thomas Ramsay, James F. Smith, R. H. McKay, C. J. Myles, W. N. Myles, H. C. Fearman, Wm. Andrews, Guelph; Thomas Bain, M. P., Dundas; E. J. Powell, London.

A three days' Convention of agents of the Stanley Electric Manufacturing Co. was held last week at Pittsfield, Mass., the headquarters of the Company. The object was to have the agents look over the extensive plant in Pittsfield, and discuss and familiarize themselves with all questions and departments connected with the work. It is proposed to hold similar conventions once a year. Among those in attendance were: W. S. Hine, J. B. Wallace and E. L. Barr, of Chicago; M. D. Barr and T. E. Theberth, of New York; Wm. H. Browne and J. A. Kammerer, of Montreal; Fred. P. Barnes of Boston and Wm. C. Whitner of S. C.

Mr. E. A. C. Pew, one of the leading promoters of the Lake Erie aqueduct project states that the canal will require to be only eight miles in length, connecting the upper part of the Welland River with the lower part of the Jordan river. The depth of the canal will be 23 feet. It is claimed that by means of this canal it will be possible to generate 300,000 h. p., half of which it is proposed to transmit for use in the city of Hamilton, and the other half to operate an electric railway from Toronto Junction to Bartonville, there connecting with the Hamilton, Grimsby and Beamsville road.

"The Niagara Falls and Lundy's Lane Street Railway Co., Limited," is seeking incorporation for the purpose of operating lines of street railway in the municipalities of the town of Niagara Falls, the village of Niagara Falls and the village of Stamford, and to connect with the Niagara Falls Park & River Railway Co.'s lines. The capital stock of the company is placed at $50,000. The promoters are :—Henry Charles Symmes, contractor; James Alfred Lowell, gentleman; Joseph Gibbons Cadham, clerk; Luther Richardson Symmes, mechanical engineer, all of Niagara Falls South; and Henry David Symmes of St. Catharines.

Tests have recently been made to ascertain the comparative efficiency of heating by electricity and by combustion. Masses of metals were heated to a red heat by combustion methods, and by electric current. In the first series of experiments a platinum rod was heated. The results showed that less than 5 per cent. of the thermal energy produced in the flame was transferred to the bar, whereas 90 per cent. of the electrical energy appeared as heat. In a second series of experiments an iron bar was used, and the result in the former case gave 75 per cent. of thermal energy transferred, and 88 per cent. in the latter case. These experiments show that for some purposes electrical energy has important claims.

The Ontario Legislature is asked to grant charters as follows: To Henry A. Everett, Edward W. Moore and Greene Puck, of Cleveland, Ohio, and T. H. Smallman, S. R. Brock and Charles H. Ivey, of London, as the London & Western Ontario Electric Railway Company; Hamilton Valley, City & Waterloo R.R ; Hamilton, Burlington & Lake Shore Electric R.R.; Brantford, Port Dover & Galt Radial Electric R.R.; St. Thomas Radial Electric R.R. Company; London & Springbank Electric R.R ; London Radial Electric R.R.; Toronto, Hamilton & Niagara Falls Electric R.R. Company; Georgian Bay Ship Canal & Power Co.; Hamilton & Lake Erie Power Co. The Stormont Electric Light & Power Company apply for an act to ratify and confirm a certain agreement. The Hamilton Radial Electric Railroad and the Hamilton & Dundas Street Railway Co. ask for amendments to their acts of incorporation.

The car barns and cars of the Halifax Street Railway Co. were burned a few days ago. The property destroyed was insured for $30,000 in favour of the Nova Scotia Power Co., who control the common stock. The Bank of British North America has garnisheed the insurance under a judgment which they hold against the Power Co. for $47,000. The opinion prevails that the railway will come under the control of the syndicate represented by Mr. Henry M. Whitney, which is at present seeking a charter from the legislature to operate an electric road on the streets of Halifax, as well as in other towns throughout the Province. The city council of Halifax has passed the following resolution in relating to this application : resolved, that with reference to the bill now before the legislature proposing to charter a company to construct and operate an electric railway in the streets of the city of Halifax, the city council hereby protests against the legislature granting the valuable franchises of the city of Halifax, unless provision is made in the act for fully remunerating the city therefor, such remuneration to be not less than five per cent of the gross earnings of the road, in addition to taxes, and that the road be constructed on such conditions as the city Council shall approve of.

# ELECTRIC RAILWAY DEPARTMENT.

## CENTRAL STATION TYPES.

### NO IV.

#### By Geo. White-Fraser.

##### INTERURBAN ELECTRIC RAILWAY.

ELECTRICITY has a very great deal to contend against— violent enemies and injudicious. friends. The writer had a man in his office one day, discussing a new method of propulson or traction, and in reply to a comment on one piece of mechanism, got the answer, "Well, I haven't thoroughly worked up that little point yet, and don't quite see my way to it, but I guess electricity will manage it somehow." And so electricity has to "manage it somehow" in all kinds of adverse conditions, against prejudice, ignorance, neglect and abuse. It says a great deal for the inherent vitality of electrical enterprises generally, that they are so popular, for it seems impossible to find a business which is less thoroughly understood and which is run with less educated intelligence. The conditions which contribute to the success of an electrical enterprise—commercial, practical, preliminary or operative—are completely lost sight of, if ever they did receive consideration, and we find small central stations in every part of Canada that actually " run" themselves, for all the attention that is given them from a business point of view. So long as the receipts are somewhat higher than the operating expenses the owners are satisfied, and the interior economy of the plant is not thought of, much less understood. It seems hard to understand why the shrewd business methods that achieve success in other branches of manufacture, should not be applied to the production of electric current, and in the management of enterprises involving its use. The object of all the machinery and apparatus in a lighting station being the incandescence of the little 25c bulb, what percentage of the men who are responsible for the business management of such stations know anything of the commercial conditions under which that bulb will produce the best dividend? How many of them study "lighting" from a broad standpoint, as a legitimate business? There are very few stations where it can be said that a pound of coal at the boiler door, can be traced through its various changes, until it finally bursts out into light in the lamp, and that every heat unit can be accounted for. Until this can be done, the station must be said to run itself, and all claims of strict economy are worth nothing, for it is only by means of watching this pound of fuel throughout its course that its performance can really be checked and its little delinquencies remedied.

If careful scrutiny and the most rigid application of business principles are important in electric lighting enterprises, they can truly be said to be the very life of an electric transportation business. An enterprise having for its object the transportation of large numbers of passengers and quantities of freight, involves the consideration of so many main and side issues, so many commercial as well as practical questions, that as a rule we might expect better results to be worked out in such a case, than in one where it is simply a question of light. The necessity for the careful consideration of such matters as route, rights of way, terms of franchises, contracts for freighting and such purely commercial questions, would seem naturally to lead to a more thorough investigation into, and therefore a much better and more comprehensive grasp of the purely practical features of the transportation. In fact, in the latter class of enterprise, "transportation" is recognized to be the aim and object, and the broad principles underlying the successful prosecution of such a business are more thoroughly understood, "electricity" being properly regarded as a means only, whereas, in the former class, this generalization is not made; there is no study of the broad, fundamental principles of "lighting," and consequently, not the same success. In this connection it is a somewhat significant fact that, although cases are not infrequent, of electric light enterprises earning no dividends, we do not hear of electric railways failing of success in that respect—the reason being that one set of enterprises is managed on business principles, whilst the other is not.

While electricity seems to offer a better solution of the various practical problems encountered in the working out of an urban,

or interurban, transportation scheme, neither electricity nor any other force can make business, the possibility of which must be pre-existent ; and a very instructive example of how such a business may be worked up is afforded by the Galt & Preston Street Railway Co., using electricity in connection with a steam dummy as its motive power.

This railway runs between the towns of Galt and Preston, a distance of a little over four miles, and is going to be extended to Hespeler, another stretch of about four miles. The power house, being at Preston, will thus be at about the centre of the track. The C. P. R. runs through Galt, and the G. T. R. has stations at Galt, Preston and Hespeler. At Preston are situated several flourishing businesses : furniture factories, foundries, woollen mills, grist mills, &c.; and in Galt are the best educational establishments and high class stores of the surrounding country: The fact that the C. P. Ry. does not touch Preston was at once taken advantage of by the directors of the Electric Railway Co., who arranged to afford to Preston a competitive outlet by running in connection with the C. P. R. themselves. Goods can be booked at any station on the C. P. R. to or from Preston, and between Galt and the latter place are handled, in the C. P. R. cars, by the G. & P. railway, who use for this purpose a steam dummy engine. An amount of freight averaging 110 tons a month, is handled in this way in direct competition with the G. T. Railway ; the advantage to consumers being that the G. & P. Railway has switch tracks right into its patrons' yards, whereas the Grand Trunk deposits goods at its own depot. A special attempt was made to induce business by getting the large factory men in both termini interested in the road, and with great and evident advantage. Here then is a good foundation for dividends. Special inducements are held out to capture certain classes of business : school tickets at greatly reduced rates are sold to scholars, and special arrangements made for their comfort. Commutation tickets are also sold, either for local use in Galt only, or for use between termini. A special feature is the freighting business for lighter packages. Two cars only are run on this road, one of which is a double-truck, another car, divided into compartments, one for passengers and the other for small freight. This question of small freight is one of particular interest in all interurban railways, which may not be so favorably situated as is the G. & P. Railway Co., with respect to railway co-operation. It is a class of business that is capable of very considerable extension under enterprising management, and productive of great profit.

The accounting department, which is in the hands of Mr. W. H. Lutz, the Secretary of the Company, is so managed that any disbursement can be at once traced to its cause, and, by the voucher system, is without any difficulty debited under its proper heading, whether " operation," " maintenance," or other. The practical part of the electric system is under the superintendence of Mr. A. Lea, who is well qualified to care for it, by both technical knowledge and actual experience. It was originally proposed by a gentleman interested in the undertaking to do without an electrical superintendent, for the sake of economy, and to let the " electrician " of the local arc plant take a " look over the machinery now and again." The company is greatly to be congratulated on having escaped this peril. That sort of economy is what generally results in a beautiful scrap heap.

The practical features of this enterprise are just as admirable as its business arrangement. The track is 56 lb. Tee. rails, laid to a gauge of 4 ft. 8½ in.; placed on ties, spaced 2 ft. centre to centre, with 8 in. face. It is single bonded, with, I believe, no supplementary return. The route is in general undulating, with slight curves and grades ; one 100 ft. curve being on a 5% grade 1200 feet long. A No. 0000 (four 0) feeder runs along the track and is tapped to the trolley wire every five or six poles. The power house itself is very compactly laid out, containing two boilers, two compound Wheelock engines, so arranged on the same shaft that they may be clutched together ; and as yet, one Westinghouse 175 ampere generator. The two cars have each two 30 H.P. Westinghouse motors, mounted on Taylor trucks, with cars built by the Ottawa Car Co., and heated by electric heaters. The series parallel controller is used. The engines are each rated at 130 H.P. The proper and economical proportioning of the various parts of a power house is always a more difficult matter with a small plant or road than with a large one, and a much greater margin must be allowed in a small one. Thus the proportioning in this plant—130 H.P. steam, 117 H.P. generator and 120 H.P. in motors, is probably as good as any with only two cars, which may both be starting on a grade at once. Especially is this so in summer, when a load of 240 persons has been carried on one motor car and a trailer. On a road

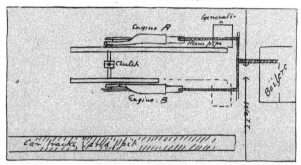

running a greater number of cars, a much greater economy could be effected at equal efficiency.

The steam plant is, so arranged that the feed water from the hot well is heated by the exhaust steam from the low pressure cylinder, on its way to the condenser. It is run into the hot well from the condenser at a temperature of about 90°, and is raised to about 112° in the heater.

The two cars are housed right in the power house—a track (with pit) having been run up along one side of the engine room, but it is the intention of the Company to build a regular car barn when their extension is completed.

The two cars are served by three sets of conductors and motormen; motor and overhead repairs being done by the set that is "off," who are paid extra for this service. One trackman is constantly employed on the rails, &c.

This road has not been running long enough to arrive at any satisfactory figures as to the cost of running per car mile, for maintenance, fuel, &c., or to give the results of experience with its various kinds of apparatus; but under its very capable business and practical management, no doubt these figures will work out well in comparison with data obtained from larger and more established roads.

It is greatly to be regretted that, even in this otherwise satisfactory plant, injudicious prelimininary design should be able to affect subsequent working. The general disposition of the boilers, engines and dynamos will, by reference to the diagram, be seen to be such as to necessitate much more piping than good design would call for, as well as a very clumsy method of changing the generator from one engine to the other.

Belting back from the engine to the generator has the effect of pushing the latter away into a corner, while the piping is unnecessarily lengthened, with all the consequent loss in fall of pressure, condensation, &c. There is plenty of room to belt forward, and the cylinders might then have been almost against the boiler room wall. And the method of changing engines is humorous in its simplicity. In order to run the generator off engine B, it is necessary to stop engine A, close the clutch by dropping a pin in the hole indicated, take off the connecting rod of engine A entirely, and then start up B, which thus has to run not only the generator but also A's flywheel. The fall in pressure between the boiler and the high pressure cylinder is 9 lbs., due entirely to this injudicious design.

The Mattawa Electric Light and Power Co. are applying for power to increase their capital stock to $30,000.

The Hamilton Street Railway Company has hit upon a method of increasing traffic during the winter season, corresponding to the attraction of suburban parks in summer. With the object of attracting crowds of skaters to the bay, the company keep clear of snow a wide strip of ice, which is illuminated by the rays of a search light mounted on the roof of the power house.

# CANADIAN GENERAL ELECTRIC CO.
## (LIMITED)

Authorized Capital, $2,000,000.00.
Paid up Capital,    $1,500,000.00.

### HEAD OFFICE:
## 65-71 Front Street West
## TORONTO, ONT.

### Branch Offices and Warerooms:

| 1802 Notre Dame St. | - | MONTREAL. | 138 Hollis Street | - | HALIFAX. |
| 350 Main Street | - | WINNIPEG. | Granville Street | - | VANCOUVER. |

PETERBOROUGH WORKS—LARGE DOUBLE UNIT FOR MONTREAL RAILWAY, ASSEMBLING
PRIOR TO SHIPMENT.

## MULTIPOLAR RAILWAY GENERATORS

### Belt Driven or Direct Connected

Our Railway Generators have established themselves as the standard for economical
and satisfactory operation, and are used exclusively by the following leading railway companies in Canada:—

**Montreal Railway    -    6,000 Horse Power**
**Toronto Railway    -    4,500 Horse Power**
**Niagara Falls P. & R. Ry, 1,200 Horse Power**

...... ALSO AT ......

KINGSTON, PETERBOROUGH, BRANTFORD, ST. CATHARINES, LONDON,
PORT ARTHUR, ONT.; WINNIPEG, MAN.; VANCOUVER, WESTMINSTER, B.C.,
YARMOUTH, N. S., etc.

## SPARKS.

. A charter of incorporation is being applied for by the Ingersoll Electric Light & Power Co., of Ingersoll, Ont.

A very enjoyable sleighing party was held by the night staff of the Bell Telephone Co, Toronto, on the 12th of February.

The electric welding machine recently installed in Gillies' Carriage Factory at Gananoque, by the General Electric Co., is said to be giving good satisfaction.

The electric lighting plant at Hespeler, Ont., has been purchased by a Mr. Skinner. The purchaser has not decided whether to operate the plant or remove it.

Application for incorporation is to be made for the Hamilton Storage Battery Co., Limited, with a capital stock of $10,000. The applicants are Dr. Stark, Dr. Osborne, W. D. Long, G. H. Bisby, Geo. Lowe, Jos. Farrell and H. E. Copp.

The earnings of the Toronto Street Railway Co. for February were $69,000, working expenses 59 per cent., or 10 per cent. less than during the same month last year. The net earnings for the month showed an increase of $7,200 above those of February, 1894.

Petitions are being circulated in Stratford, asking the council not to enter into a five years' agreement for electric street lighting, with the Stratford Gas & Electric Lighting Co , until the price of gas is reduced, or until tenders have been received for the lighting franchise.

The Winnipeg Electric Railway Company have distributed along their lines old horse cars fitted up with stoves and otherwise made comfortable as waiting rooms In such a climate as that of Manitoba, such a provision will doubtless be appreciated at its true value by passengers.

A report of the city electric inspector of Victoria, B. C., shows that it cost the municipality last year $97.95 per annum for each arc light, running five and one-quarter hours per day. It is claimed that when the improvements which are now being made to the plant are completed, the cost will be reduced to $80.00 per lamp per year, including interest and sinking fund. The municipality operate 300 arc lamps.

The City Engineer of Toronto, has recommended that under the requirements of clause 38, of the agreement between the Toronto Railway Co. and the City, the carrying capacity of ordinary closed cars be limited to fifty per cent. above their seating capacity (allowing a space of 18 inches on the seat for each person) and that open cars be limited to their seating capacity. He further recommends that in order to avoid disputes, a notice should be posted in each car, stating the exact number of passengers it is allowed to carry.

A number of Ottawa men are said to be endeavoring to secure control of a tramway line at Kingston, Jamica. The line is at present operated with mules, and extends through the city and seven miles outside to a large hotel which was erected at the time of the Jamaica exhibition, and which is said to have proved a very successful enterprise. Mr. C. Berkeley Powell has gone to Jamaica on behalf of the Ottawa syndicate to see what can be done. It is understood to be the intention to substitute electricity for animal power, should the syndicate get control of the present line, the franchise for which expires in October next, and endeavor to secure a franchise for a new line, if the present one cannot be got; also to construct electric roads to supply communication between the sugar plantations on the island.

## DUMMY ENGINE WANTED

Second-hand, in good order; cylinders 10 by 14 or larger. Would rent for one year or buy.
W. N. MYLES, H. & D. R. Co.,
Hamilton, Ont.

## FOR SALE

25 Arc Dynamo, Reliance system, with wire, lamps, pulleys, cross-arms, pins, etc.; cheap for quick sale. Write for prices  THOMAS. R. FOSTER, Tara, Ont.

The series of moonlight tables for 1895, now appearing in the ELECTRICAL NEWS were compiled by the Cleveland Carbon Co., Cleveland, Ohio.

Notice is given that application will be, made by the Hamilton Radial Electric Railway Company to the Legislature of Ontario for an Act to amend the Act of incorporation, by providing that the said company shall have authority to operate the Guelph and Berlin branches of the said railway by steam as well as by electric power and increasing the bonding powers of the company in respect of such portions of the said railway as may be constructed for operation by steam power.

At a recent meeting of the Hubbell Primary Battery Co., of the organization of which at Ottawa, mention was made in a recent number of the ELECTRICAL NEWS, the following gentlemen were elected directors of the company:—Messrs. J. W. McRae, Dr. A. A. Henderson, N. C. Sparks, H. B. Spencer, Archie Stewart, J. R. Trudeau, S. M. Rogers and E. F. Hubbell. At a subsequent meeting of the directors, Dr. Henderson was elected president, F. C. Sparks, vice-president, and E. F. Hubbell, secretary-treasurer. Satisfactory tests of the Hubbell Battery are said to have been made, and the directors express their intention of proceeding immediately with its manufacture.

# CANADIAN ELECTRICAL·NEWS

## STEAM ENGINEERING JOURNAL

OLD SERIES, VOL. XV.—No. 6.
NEW SERIES, VOL. V.—No. 4.
APRIL, 1895
PRICE 10 CENTS
$1.00 Per Year.

CANADIAN

# ELECTRICAL NEWS

AND

## STEAM ENGINEERING JOURNAL.

| VOL. V. | APRIL, 1895 | No. 4. |

### THE INCANDESCENT LIGHT CO. OF TORONTO.

WE illustrate and describe herewith the central station of the above company. The building was erected in 1889 and the foundations rest on good solid clay. It does not cover the entire lot, which has a frontage of 100 feet and a depth of about 175 feet, enough ground having been left for future extensions.

The boiler room occupies a space at the rear or west end of the building 56 feet wide by 65 feet long, and is separated from the engine and dynamo room by a solid brick wall. There are eight boilers—six return flue and two Babcock & Wilcox water tube, the former being rated at 175 h. p. each and the latter at 400 h. p. each, making a total of 1850 h.p. Each boiler is provided with a separate damper and in the breeching is placed another damper controlled automatically.

Coal is delivered direct from the teams into the coal hole, and descends by gravitation to boiler room floor, and is then wheeled to the various boilers as required. It is carefully weighed and each fireman keeps accurate account of the amount of coal burned during his run. The ashes are removed by an endless belt with steel buckets driven by a 2 h. p. motor, and are raised a height of 20 feet and dumped into a chute which conveys them outside the building.

There are three feed pumps, viz., two duplex and one tandem compound. The feed water pipes are so arranged that water can be taken from either of the main pipes indicated on ground floor plan (Fig. 1), and also from the well outside of the building. The drips from all the engines run into this well, and whatever heat is got from same is utilized. The water is pumped through Wainright exhaust steam heaters and enters the boilers at about 200° Fah. Water is supplied from the city mains by meter measurement. There are two smoke stacks, the positions of which are shown on Fig. 1, their heights being 100 feet and 150 feet respectively. At present the two water tube boilers are connected to the latter, which is of sufficient capacity for at least 2,500 h. p. in boilers.

The engine and dynamo room is 56 feet wide by about 115 feet long. Fig. 1 shows the general layout of the plant and Fig. 2 shows a general view taken from one end of the room. The engines are six in number, viz., two Straight Lines, three Armington & Sims, and one Lake Erie Cross Compound. Owing to the station being situated fully a mile from the lake, the plant is entirely non-condensing. The figures show the arrangement so plainly that a detailed description is unnecessary.

The generators are of the Edison type throughout—the two 250 k. w. direct connected being the latest addition, having been made by the Canadian General Electric Company at Peterboro'. These machines have been running right along since they were installed over eighteen months ago, and running at times up to 300 k. w., have given the best possible satisfaction, and are certainly a credit to the builders and also to those in charge for the care that appears to have been taken in keeping them in their present first-class condition. It may be mentioned that these were the first of this kind made at Peterboro'. The dimensions of the vertical engine are : 19 × 38 × 22 inch stroke, 130 revolutions per minute.

The switchboard shown in Fig. 4 is divided into three sections —all switches on the positive side of system in one section, all negative switches in another, and the neutral in another. The position of same is shown in Fig. 1 and Fig. 2. It is arranged that two different pressures can be run, but owing to the peculiarity of the load, and the distribution of same at times, three different pressures are run by dividing the auxiliary bus. We described in our February, 1894, issue, the manner in which Mr. Milne, the general superintendent, arranged the switchboard for running at 500 volts for street railway purposes, together with the three-wire system. It might be wel for our eaders to refer to that again and take in Fig. 4 in conjunction with same, seeing we have now a more general description of the plant. Fig. 3 represents diagrammatically the generators and their connections to switchboard—the generators, commencing from the left hand side, being connected in pairs to their respective engines, and are : 1st pair, two 250 k. w.; 2nd and 3rd pairs, 100 k. w. each, and the remaining three pairs 80 k. w. each. By tracing out the connections it will be clearly seen how the three-wire system was run in connection with the 500-volt system. Recording volt meters are used which record the pressure at the junction boxes throughout the city. A careful record is kept of the ampere output, which is plotted on a chart and kept for reference.

All current is sold by meter. The dayload is composed chiefly of motors, there being almost 1000 h. p. connected to system. The total number of lights connected is about 25,000. The station

THE INCANDESCENT LIGHT CO. OF TORONTO—FIG. 2.

has run day and night since it commenced, February, 1890, without interruption. It is a model of cleanliness, and everything appears to be in the best order, and speaks well for those in charge. The service given by the company is of excellent character.

Among the directors of the company may be mentioned : W.

any four-stranded rope. In this way a rope is formed possessing extreme flexibility, and the fibres will not break by bending on each other when run on pulleys, the rope also standing elongation or stretching some twelve inches in a length of fifty inches before breaking.

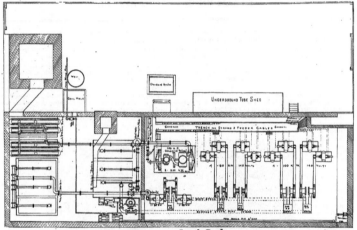

GROUND PLAN

THE INCANDESCENT LIGHT CO. OF TORONTO—FIG. 1.

D. Matthews, President ; W. R. Brock, Vice-President ; H. P. Dwight ; Frederic Nicholls, Managing Director.

Mr. James Milne, the general superintendent, in point of natural ability and education, is undoubtedly one of the most efficient electricians in Canada to-day.

### TRANSMISSION.

THE transmission of power by ropes has been largely resorted to in England, the preference being given to what is known

### ROPE DRIVES.

For rope drives, the common rule is not to make the diameter of the smallest pulley less than thirty times the diameter of the rope, and even larger than this is to be preferred. For wire rope it should be still more, and from 50 to 100 times the rope diameter is the common practice when these are used. Excepting for very long transmissions, the wire is seldom used in regular driving, as its weight is objectionable, and its advantages are not enough to make it popular over manilla or cotton ropes.

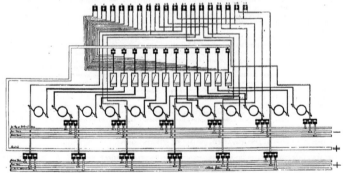

THE INCANDESCENT LIGHT CO. OF TORONTO—FIG. 3.

as the Lambeth cotton rope, which is made of four strands, the centre or core of each strand being bunched and slightly twisted, the outside of the strand having a covering of yarns that are firmly twisted. The four strands are further laid with a core in the centre to form a rope and twisted in the same way as

The Bell Telephone Company has entered an action against the Montreal Street Railway Co., claiming the sum of $27,000 damages by reason of interference with the working of their lines, by disturbance from inductive currents from the railway company's lines. The railway company's defence is that they have not gone beyond the rights conferred on them by their charter and their contract with the city. The case is still in progress.

## LIGHTING AND ELECTRIC LIGHTING AS A BUSINESS.

BY GEORGE WHITE-FRASER, E.E.

"APPLICATIONS for the position of electrician and fireman to operate the waterworks and electric light systems, are invited, etc." This is an advertisement that no doubt has been seen by every one in a weekly paper recently ; and has electricity, the baby of yesterday and the giant of to-day, the science of Kelvin, Hertz, Helmholz, Edison, turned out to be such a quack that the duties of "fireman and electrician" can be efficiently performed by one man? It is almost difficult to decide from what standpoint to criticize this combining of services, whether from that of economy or from that of practical engineering, theory or practice. From whichever side it be looked at so many weak points are presented that it really seems a waste of energy to attack it at all. The fact, however, that the municipal authorities of a certain reputable town propose to operate the combined departments in the manner indicated, opens up such a vista of disastrous mismanagement that for the credit of electricity as an illuminant, and for the electrical engineering profession at large, some efforts should be made to bring about a better comprehension among those interested in lighting stations, of the principles of lighting as a business, and more especially of the economics of electric lighting.

If these principles and economics received any study at all, a most radical and comprehensive change would shortly be observed in the methods followed in the operation of every electric plant in Canada, always excepting some few honorable examples of thoroughly up-to-date practice in the large towns and cities. As an example of the violation of one of the most thoroughly established principles of lighting—most common in electric light service—why is almost every incandescent lamp in Ontario installed on the "flat rate" basis instead of on the meter basis? Has any one ever heard of a

THE INCANDESCENT LIGHT CO. OF TORONTO—FIG. 4.

gas company contracting with consumers at so much per month per burner? The thing seems so perfectly obvious that one is disinclined to weaken the position by descending from generalities to particulars, but at the same time the methods of managing stations are so very crude that one is forced to conclude that no argument will have any weight unless it is copiously illustrated. "A" has ten lamps in his house, and he pays a certain definite sum per month for them, whether he burns them or not. There are any number of men who would burn them all so as to "get their money's worth." Anyway, since A does not benefit himself by turning out a lamp that he really does not require, the chances are that he will leave it burning, either from forgetfulness or from meanness. This means reduced profit to the supplier. Put a meter in A's house and watch. Do you think he will leave his bedroom without shutting off the lamp? Not much. In this case the unnecessary light touches his pocket, not the supplier, and he economizes. One advantage of the meter basis applies more particularly to electric lighting and will be referred to later. This important matter seems to receive no consideration at all, although it lies at the threshold of the "dividend" question.

Light is a manifestation of energy, and is accompanied by more or less heat, as the optical efficiency of the transformation of stored into radiant energy is less or more perfect. In proportion as the lengths of the waves set up in the ether are greater or less, so will the transformation of stored energy produce more heat or more light. Now, coal being the form of stored energy most frequently used, it is perfectly plain that whether more heat or more light is produced depends entirely on how the coal is burnt. In the furnace of a boiler, for instance, the design may be such that a great deal more heat goes up the chimney in the form of unconsumed gases than is utilized to raise steam ; or, and more likely, the ignorance of fireman, engineer and superintendent as to the theory of combustion and its applications causes the same waste. No sooner, therefore, do we begin to study the economics of electric lighting than we find a fruitful source of unnecessary expenditure at the very first step. It might be supposed that the means taken to guard against the introduction of waste would be commensurate with the liability towards such waste. Therefore, where quantities of fuel are to be consumed to produce steam we might expect to find such consumption being watched by a person who, to a certain extent, knew what was going on, and was capable of detecting, at least, large wastes and of remedying them. At this important strategical point we generally find placed an ordinary laborer, who gets $1 or $1.25 per day, and who knows as much about combustion as combustion knows about him. If an engineer be employed he is rarely better than a fair mechanic who can make little repairs and do odd jobs. So that, actually, at the point where a very large waste can easily occur "false economy" places the very man who is least capable of checking it.

It is a commercial principle that a "business" is to be treated as an individual, with whom a debit and a credit account is kept, and the comparison of these accounts shows profit or loss. In a business involving the manufacture of a commodity the general accounts are divided into two main divisions—1, the cost of production, and 2, the cost of distribution and sale. There will be one complete set of books kept by the factory, and a separate set kept by the warehouse. In the former set can be traced every stage in the manufacture, from purchase of raw material to delivery at warehouse door of finished product, and great importance will be attached to returns showing (a) the quantity of such product in comparison with the quantities of raw material used ; (b) the cost of the product (for wages, materials, etc.) at various points in its progress toward completion. In the latter will be set forth the cost of handling the material from its delivery by the factory authorities up to the moment it is finally disposed of to a consumer. The factory will be debited with the cost of the raw material, wages, cost of power, etc., and credited with the amount of its output at a price sufficient to wipe out the debit. The warehouse again will be debited the cost of the finished product it receives from the factory, salaries of clerks, general expenses of distribution and management ; and it will be credited with the receipts on sale of goods. In this case there really are two separate businesses—a manufacturing and a distributing. It is just as necessary for the factory to know the amount of its output in manufactured goods as it is for the warehouse to know exactly what quantity of goods have been disposed of. If the warehouse does not know how many articles have been sold for a certain sum, it cannot tell whether they have been sold above or below cost price. If the factory does not know how many articles have been manufactured out of a certain quantity of raw material, and with a certain expenditure for wages, etc., it cannot tell how much those articles have cost each ; it cannot tell whether half the raw material has not been pilfered by operatives ; in fact it simply is not in a position to do business against competition in the open market, and a business conducted on such lines would very rapidly go to the wall. Moreover, the arrangement of the books would be such that returns might be made showing the most profitable line of goods; the direction in which other lines might be pushed, that one particular goods should no longer be manufactured ; and that improvements in the material or design of another are necessary in order to hold the market. In fact the principles on which such a business is conducted are such that every detail can be examined and kept in order like the works of a watch, and every "tick" sent forth by the great commercial timepiece of the world's market is promptly echoed within its little case, and its little hands move "on time."

An electric light or power business differs in no respect from the above. Electric current is manufactured and sold for various purposes. Raw material—coal—is used, wages paid, current produced, distributed and disposed of. It really includes two businesses—the generation of current and its distribution and sale—the former purely practical, the latter more or less commercial. A diagram will clearly set forth the analogy.

ELECTRIC BUSINESS.

COMMERCIAL BUSINESS.

I am afraid, however, that this analogy is purely a paper one, judging from actual practice, for I should greatly like to hear of five stations in Ontario that can tell what their output is. I do not mean that take periodic readings of their ammeters and so construct a sort of curve which gives them a fairly approximate notion of it, but that can say that as a result of a month's running their output has been so many kilowatts and can prove it. The writer knows of a very nice little plant, the engineer of which makes a report every morning to the secretary, giving with great accuracy several details, among them being set down the highest and the lowest reading of the ammeter. This is the nearest approach he has seen to keeping any account of output, and he entirely fails to see the use of such report.

It must be evident, on the most cursory consideration, that a station operating in this way really cannot know what its output is. It is working purely in the dark and has no more chance of success than the ordinary commercial business that keeps no accounts of goods between the initial stage of raw material and the final stage of cash from customers. It simply knows that a certain sum has been paid out for coal, wages, etc., and that another sum has been received from customers for light. But for how much light? Is it any wonder at all that complaints are frequent that electric lighting does not pay? How can it pay under the extraordinary mismanagement it receives? And in proof of the statement that I have made that such stations do not know their output, I give here the actual results of a night's run, taking hourly readings of the ammeter, and compare them with the report furnished next day to the secretary. Averaging the hourly ammeter readings gives 22.57 amperes. The maximum reading was 39, the minimum was 11. Of course these latter figures are given in order to arrive at an average which in this case would be figured $\frac{39+11}{2} = 25$ A. These two approx-

imations differ by very nearly $8\frac{1}{2}$ amperes, or ten per cent. of the higher one. Reducing this to horse power hours, for a run of 14 hours at the initial voltage of 1040, shows that whatever amount of fuel has been consumed to generate power has generated 49-horse power hours less than as calculated from the reports, and that consequently the consumption of coal per h.p.h. has been greater than as calculated. Assuming that this difference of 49 h.p.h. is kept up through the year, which is after all fairly allowable, shows that the station has been credited with over 17,000 h.p.h. that it has not generated. What kind of management is this ?

Another instance of how the perfect and comprehensive ignorance of principles, or blind disregard of their application, contributes to an actual considerable loss came under the writer's notice. A hotel man paid $45 a month for his lights on the "flat rate" basis, and considered he could save money by putting in a meter. Next month his charges by meter record were $63. Of course he objected, but the meter was right. In this case the station was losing 18 good dollars every month by supplying him on the flat rate. Had the management known really how much current was being consumed, by keeping any reliable record of station output, the change from flat to meter rate would have originated with them instead of with the customer.

Returning to our diagram, and confining ourselves to the factory at first, here system reigns, ensuring strict economy. A pound of raw material is followed through its various changes ; the various ingredients that go towards the finished product are well known—their proportions, the cost of each, the necessary wastes that occur during the successive processes, are clearly recognized and very carefully kept down to their proper limits, and an exact biography can be written of a copper kettle, giving quantity and price of pig copper, waste in smelting, polishing, modeling, etc., and resultant cost price, cost price during its various stages, and suggestions as to effecting economies. Does this analogy hold in an electric station? Barring the few bright exceptions alluded to, is there one station in Ontario that weighs every shovel of coal put into the furnace? Of course it is not necessary to weigh smaller amounts than wheelbarrow loads, but the idea is an accurate knowledge of the exact number of pounds burnt during a night's run? If he neither weighs his fuel nor knows his exact output, what station manager will be bold enough to say that his current costs him such and such a figure? It is pure guess-work and therefore quite unreliable.

*(To be Continued.)*

## TRADE NOTES.

George White & Sons, of London, have fitted out the saw mills of Gow & McLean, Fergus, and George A. Patrick, Delaware, with new internal fired boilers, and " clipper " engines.

The announcement is made that the business heretofore carried on by the Dominion Electric Co., Ltd., and the Packard Lamp Co., Ltd., of Montreal, has been acquired by the Packard Electric Co., Ltd.

The Canadian General Electric Co. have closed a contract with the Kingston, Portsmouth and Cataraqui Ry. Co. ; for additional equipment, including a 200 Kilowatt generator and several G. E. 800 equipments, with forty foot car bodies.

The Ottawa Porcelain & Carbon Co., has been incorporated with a capital stock of $100,000, to manufacture carbon and porcelain goods for electrical purposes. Mr. J. W. Taylor, formerly manager of the Peterboro' Carbon Co., is the manager of the new company.

The Montreal Electric Co., agents for the Fensom Elevator Co., of Toronto, report having changed over freight and passenger elevators in Nordheimer's Building, Montreal, from hydraulic to electric. The freight is a 6 k. w. motor and belted, while the passenger is a 7½ k. w. motor direct connected. The circuit is a 250 volt one.

In October last, the Kay Electric Works, of Hamilton, called a meeting of their creditors and compromised their liabilities at 33⅓ cents on the dollar. It was hoped that this compromise would enable them to discharge their liabilities, and place themselves in a position to carry on their business successfully. These expectations, however, have not been realized as the company have just made an assignment of their estate. The creditors have accepted the offer of Mr. J. L. Job for the purchase of the assets. The business will be continued under the same name.

## CHARACTER SKETCH.

### JAMES GUNN,

SUPERINTENDENT TORONTO STREET RAILWAY CO.

Whate'er thy race or speech, thou art the same ;
Before thy eyes Duty, a constant flame,
Shines always steadfast with unchanging light,
Through dark days and through bright.—The Ode of Life.

IT has been said, that an honest man is the noblest work of God. In an age when indifference, to the smallest, as well as the largest, duties of life, is a crying evil, one may well wish to apply the term "honest" in one particular direction of life. The men who can be relied upon to do the work placed in their hands are none too many. As one has said, a man is already of consequence in the world when it is known that he can be relied upon ; that when he says he knows a thing he does know it ; that when he says he will do a thing, he can do, and does it.

In Mr. James Gunn, superintendent of the Toronto Street Railway Co., we have one, who, by faithfulness to duty, has attained to a high position in his chosen calling. When in 1869 he identified himself with the street railway of Toronto there was nothing very tempting, from a purely material point of view, in the position assumed. The street railway of that day with its few odd horse cars, and a somewhat imperfect route on Yonge and Queen streets, carrying a few hundred passengers daily, was a small affair, contrasted with the railway of to-day, so splendidly equipped as an electric system, and covering nearly all parts of the city. But, as secretary at that time, Mr. Gunn did his duty no less faithfully than when he became a leading official of the mammoth concern of to-day.

The subject of our sketch was, to a large extent, to the manner born, so far as his experience in dealing with men is concerned. His birthplace was Banniskirk, in the parish of Halkirk, near Spittle Hill, Scotland. He has not to this day lost the burr of his Scottish home, as a few minutes conversation with him, readily proves. His father was a road contractor in Scotland, and the son worked with him, keeping track of the men's time, and in other ways mixing up with them, and having to assist in the handling of large numbers of men. After a time he left his native parish and found his way to Edinburgh,

MR JAMES GUNN.

where he engaged in mercantile life, holding a position for a considerable time with Christie & Son, large military tailors.

From the time he first entered the activities of life he had an ambition to settle in America. "Westward the course of empire takes its way" had no meaningless inspiration for him, and in 1867 he broke away from the ties of Scotia's land and took sail for America, coming direct to Canada, and locating almost immediately on his arrival in the city of London. There he was engaged for two years with Mr. Charles Dunnett, tanner and leather merchant. It did not seem to be much chance for him to push ahead in the position obtained. He had formed an acquaintance with Mr. E. W. Hyman, a capitalist of the Forest City. Mr. Hyman had obtained an interest in the Toronto Street Railway, of that day, through having made certain advances to Mr. William T. Kiely, who was then part owner of the railway. Mr. Gunn came to Toronto at the suggestion of Mr. Hyman, partly, no doubt, to watch his interests in the company, and at the same time accepting the position of secretary.

Thus it was that Mr. Gunn's connection with the Toronto Street Railway, and which through its various changes, has remained unbroken, was commenced in May, 1869. He can look back from the present with a large share of interest at the operations of the road in those early days. Mr. W. T. Kiely was general manager of the road and Mr. Gunn was secretary, two important officers. But, as with the beginnings of every business, the men who were in at the start could not be too particular in choosing the work that came to them. Mr. Kiely and Mr. Gunn

in those days had to take their turns in relieving the conductors of the road. They did not always please the public any better than the ordinary conductor. It was not an uncommon thing for Mr. Gunn to be sitting at his desk and have parties come into the office and lay complaints of some irregularity, and on enquiry to learn that the conductor who had blundered was the general manager, Mr. W. T. Kiely. It would then become the duty of Secretary Gunn to inform Conductor Kiely, that if this kind of thing occurred again he would get his walking ticket. At another time General Manager Kiely would be in charge of the office and a complaint would be made against Conductor Gunn, who, on reporting at the office, would be informed that his position was in jeopardy if, as a conductor, he could not meet the public wants in a more efficient manner. These were a few of the pleasantries of the business that in those early days came to the present superintendent of the Toronto Street Railway Co.

Historically it may be of interest to remark that prior to the days of Wm. T. and Geo. W. Kiely, who afterwards became so intimately associated in street railway matters in Toronto, the business of public propulsion in the city was in the hands of Mr. Alexander Easton, who, with others, operated the first street railway in Canada, in Toronto, in the year 1861. The construction of the Yonge street line was commenced on the 26th day of August of that year and it was opened to the public on the 11th day of September following. The Queen street line was commenced on the 16th day of October, 1861 and opened to the public on the 2nd day of December of the same year. The average number of passengers carried daily on Yonge street at that time was 1270 and on Queen st. 688. A few years later the street railway property fell into the hands of the Bowes' estate, and in 1869 it came into the possession of the Kiely Bros. In 1873 Frank Smith, now Sir Frank, senator, became a large shareholder in the concern, buying out Mr. Wm. T. Kiely for some $250,000, and the property, as is generally known, remained practically in the hands of Sir Frank Smith and Geo. W. Kiely up to the time that it was taken over by the city in March, 1891.

The history of negotiations at that time are so recent as to be clear to the memory of almost everyone. The amount paid to the owners, as per the award, was $1,453,788, an illustration of how valuable the franchise had become in those thirty years. From March until September 1891, when the present company took hold of the business, it remained in possession of the city. During the interim Mr. James Gunn was constituted general manager, with extraordinary powers of management and heavy responsibilities, a high tribute to his ability and sterling integrity. The heavy receipts of the road were all deposited in the name of James Gunn, and he personally, without any countersigning by any city officer, signed all cheques for disbursement.

With something like 80 miles of track to-day, and the whole of this, practically, worked by the electric system, the Toronto street railway will always have a very lively interest for electricians.

Mr. Gunn remains one of the chief officers under the new management. He is known as superintendent, being the man in charge of the practical operations of the entire system. His wide and intimate knowledge of the system from its commencement peculiarly fits him for this position, and it is needless to say that the work is being performed with that same high regard to duty that led to success in the several important positions he has previously held in connection with the company during the past eighteen years.

There have been of too busy a character to permit Mr. Gunn to enter at all largely into the work that comes to citizens outside of their regular business. He has done well the work that was beside him, and this, after all, is the highest compliment that can be paid to any man. Mr. Gunn is a Presbyterian, and an active and efficient member of the board of management of the Westminster Presbyterian church. Personally he is possessed of the social characteristics of the Scottish people and has a host of warm friends in the city of his adoption.

## QUESTIONS AND ANSWERS.

"A Subscriber" writes: I notice that when we are running our dynamo that the outside brush will sometimes get red hot. They are carbon brushes. I have done almost everything I can think of to try and remedy it, but she will do it. I notice our lamps burn out very quickly; the fuse never burns out, but the lamps go very quickly. Is it the fault of the lamp or the machine?

ANS.—It is impossible, without examination, to assign the real cause for lamps burning out. They may be poor lamps; or more probably, the dynamo pressure is not maintained constant. The potential indicator may be out of adjustment, reading too low; or the transformer system may be such as to allow a very large variation in pressure between full and light load. Again it is very likely that the wiring system is so faultily designed, as to leave too high a pressure at the lamps. Try a change of lamp make; if that does not answer examine the potential indicator as to its correctness, and go over the wiring system. As to the brush trouble, examine the machine; tighten all connections; and especially see that the "outer" brush makes good connection with the brush holder, and that there is no poor contact anywhere. It is very unlikely that the trouble is caused by either machine or lamps—much more probably in the attention it receives.

J. J. C., Kemptville, Ont., writes: Would you kindly through your valuable ELECTRICAL NEWS describe the latest process of producing Calcic Carbide (T. L. Wilson process); also a description of the electrical furnace used in its production. I feel confident this would interest your readers as this discovery will affect electric lighting, should the right to manufacture same be covered with patents.

ANS.—There are various kinds of electrical furnaces. The main thing seems to be to place a mixture of lime and carbon—say anthracite coal—within the influence of a powerful electric arc. The exact nature of Mr. Wilson's process has not yet been made public. When it is, we shall be glad to lay it before our readers.

J. L. M., Orillia, Ont., writes: Some time ago I wrote you re. making a bar magnet, and followed your directions, but when the steel bar was placed on end on the field magnet of a Ball system arc dynamo and left there for five hours it was no more magnetized than when it was put on. Can you explain this, or can you give me further instructions?

ANS.—You have made a mistake somewhere. Take a file and place it in a similar position on the dynamo. If it is strongly magnetised your material or hardness is not right. If the file does not mag-. netize, the position on the dynamo or the dynamo itself is to blame. Five minutes will do as well as five hours.

W. R. R. writes: 1. What is the reason telephonic communication cannot be carried on through the Atlantic cable? 2. Where can I get a description of the duplex telegraph? 3. What is "cross section of the armature wire?" 4. Are there telephones which transmit and produce speech without losing its original loudness, in practical use? If not, why not?

ANS.—1. Retardation of current is too great owing to large static capacity of cable. 2. In any modern work in telegraphy. 3. Cross section of armature or any other wire is the area or surface of the end of the wire if cut across, usually measured in circular mils. 4. There are no telephones in practical use to produce loud speech. There is no particular demand for them. If there was, a louder speaking telephone than the ordinary instrument would doubtless be forthcoming.

E. V. B., Stayner, Ont., writes: Would you be so kind as to inform me where I can purchase the non-arcing metal used in a form of lightning arrester described in a recent number of the ELECTRICAL NEWS?

ANS.—So-called non-arcing metal is made by manufacturers of lightning arresters in their own apparatus, but so far we do not know of any such metal being placed on the market as such.

"Fireman" asks: Will you kindly give me a receipt for a boiler compound that will remove scale and not injure the boiler. We use Lake Ontario water which is very hard on a boiler. State what quantity of the compound to use.

ANS.—There are numberless substances used to prevent scale. One of the best is caustic soda, but too much should not be used, as it is liable to attack the brass seats of check valves and stop valves, and corrode them somewhat. Vegetable substances, as sumach and slippery elm bark, or any bark or wood, such as oak chips that contain tannic acid, are also very good and quite harmless.

## LEGAL.

DIXON v. TORONTO.—This was an action for $5,000 damages brought against the City of Toronto, the Toronto and Mimico Railway and Light Co., Ltd., and the Toronto Electric Light Co. by one Dixon for injuries alleged to have been sustained by coming in contact with an electric light pole while riding on the upper deck of an electric car on the Mimico Electric Railway. The case was heard before the Hon. Mr. Justice Robertson and a Jury in the Court of Assize at Toronto. Following are the terms of the judgment:—This action against the corporation of the City of Toronto is dismissed so far as said corporation is affected by it—is dismissed without costs; and as against the Toronto Electric Light Co. this action is also dismissed with costs; and the jury find a verdict for the plaintiff against the defendants—the Toronto and Mimico Electric Railway and Light Co. Ltd., and assess the damages at $1000. I order judgment to be entered for the above sum of $1000 with full costs of the action to be entered against the Toronto and Mimico Electric Railway and Light Co., Ltd. on and after the 3rd day of the next sittings of the divisional court.

## THE WESTON FLUE SCRAPER.

THE attention of the engineers of Canada is invited to the advertisement on another page and to the accompanying illustration of the Weston Flue Scraper. This scraper has been pronounced by some of the most competent American engineers to be, and is guaranteed to be capable of fulfilling all demands. Being made solely of malleable iron, the blades will always keep a good cutting edge, and the action of the coil spring keeps

THE WESTON FLUE SCRAPER.

them to their work, preventing them from riding over the scale and dirt.

Any tension required can be put on the spring by judicious adjustment of the set nuts, thus making it adaptable to the strength of the operator and the amount of scale and soot on the flue.

Attention is also requested to the "shearing cut," which is an exclusive feature of this scraper and which it is claimed makes it the easiest running in the market.

## PERSONAL.

Mr. F. E. Lovell, of the Coaticook Electric Light Company, Coaticook, Que., was recently married in New York city to Miss Jean Norton, formerly of Coaticook, and recently of Jackson, Mich.

The sudden death is announced of Mr. Francis Northey, who for sixteen years past held the position of engineer at the Hamilton water works pumping station. Mr. Northey was a brother of Mr. Thos. Northey, of Toronto.

The town of Hespeler has given a five years contract for street lighting to James Fenwick, of Preston, at 17 cents per light per night for 300 nights per year.

The earnings of the Montreal Street Railway Company for the six months from September 30 to April 1 amounted to a total of $462,262.63, making a net increase over the same six months of the year previous of $92,714.46.

## MUNICIPAL VERSUS PRIVATE OWNERSHIP OF ELECTRIC LIGHTING PLANTS.

Editor CANADIAN ELECTRICAL NEWS.

DEAR SIR,—I was much interested in the article on municipal ownership of electric lighting plants by Mr. G. White-Fraser, which appeared in the January number of the NEWS, as well as in the discussion which followed.

As Mr. Fraser in this article, as well as in a communication which appeared in the Toronto Mail over his signature some time since, criticises the paper which I presented at the Montreal convention of the Canadian Electrical Association, I beg leave to reply.

In that paper I argued and endeavored to show from returns given that in towns large enough to support a fair sized plant, the municipal lighting can be supplied cheaper and will be performed better by a private company than by the corporation itself. I also expressed the opinion that it is unjust for a municipality to establish a fighting plant in opposition to an existing company, provided the company are acting fairly, and that in small places, or where there is no other plant, a municipal plant may be justified on the ground of public necessity. I am pleased to see that my argument is supported by such eminent men who are able to pass an independent opinion, as Mr. Baillargé, of Quebec, as well as by a large number of engineers who·have had experience with municipal stations. Mr. Fraser takes exception to my argument, though he expresses himself as in accord with my opinions otherwise.

In towns where the lighting plant can be operated in connection with some other branch of the public service, such as a water pumping station, it is quite evident that a saving can often be effected in the total original investment as well as in running expenses, although in cities where the machinery is running the full twenty-four hours each day—where the load is large enough that all the machinery can be operated at its maximum efficiency, and where all the men employed have their time fully occupied, I do not see that anything can be gained by combining these two departments. In the smaller towns there may be other considerations, too, which weigh in favor of a municipal plant. In such places, if they have a private plant, it is usually carried on in connection with some other business and does not receive the necessary attention to details to give a good service. If, at such places, the plant is owned by the town, the public take a considerable interest in it and the committee in charge pride themselves on having everything in good order ; moreover, the chances for "boodle" are for various reasons very small. A number of such plants have been installed of late. In Bracebridge, for example, the town bought out a private plant which had been running for some time, overhauled it and enlarged it somewhat. It is now run in connection with the water pumping station, and is giving such good satisfaction that it has already been found necessary to further increase its capacity.

The combined pumping and lighting station at Orillia is a good one in point of construction ; it is well arranged throughout, and it is quite probable that this plant can under the circumstances be operated very efficiently and economically. The figures which are given concerning it are very alluring, but I do not find it stated anywhere that they have been realized. No account is taken of insurance, depreciation, interest and taxes ; moreover, some of the men employed seem to have very long working hours, and private lighting companies usually find it necessary to have a man patrol the street circuits during lighting hours.

In his communication to the Mail, Mr. Fraser claimed that a municipal plant is not required to pay rates and taxes. Arguing from the same premises, I suppose he would claim that when a man happens to own the house he lives in he is not required to pay anything in the way of rent.

When we come to consider the large towns and cities the question assumes a somewhat different aspect.

In the argument whether a corporation or a private company can operate a large plant more economically, I repeat what I stated in my Montreal paper, that the burden of proof lies with the advocates of municipal ownership and not with the side of private control, as Mr. Fraser would have us suppose, and it is therefore only necessary on the part of the latter to point out possible defects in the arguments put forward by the champions of municipal ownership. It must certainly be admitted that the possibility of aldermanic corruption, the possibility of political partisanship influencing appointments, and the possibility of the same undue influence being exerted to keep in office incompetent persons to the detriment of economical management, are not imaginary but are real difficulties in the way, examples of which are not far to seek.

I agree with Mr. Fraser that the whole argument seems to reduce itself to a weighing of facts on the one side and possibilities on the other. Considering the question only from the commercial point of view, it must be decided by the relative efficiency and economy of municipal plants and private plants. Now, of the facts on the one side, the principal one is that the municipal plants which have so far been established in this country and the United States have not yet demonstrated their ability to supply a better and cheaper light than is furnished by the private plants. The possibilities on the other side, as pointed out by Mr. Fraser himself, are those above stated. From the commercial standpoint this is really all there is to be said.

E. CARL BREITHAUPT.

Berlin, March 30, 1895·

## CANADIAN ASSOCIATION OF STATIONARY ENGINEERS.

Note,—Secretaries of the various Associations are requested to forward to us matter for publication in this Department not later than the 20th of each month.

### HAMILTON ASSOCIATION NO. 2.

Editor ELECTRICAL NEWS.

At the last two meetings of Hamilton No. 2, considerable business of importance was transacted, and some very interesting discussions indulged in. Among the most important was one in which the requirements of an engineer for different plants and the kind of men required for engineers was considered. A discussion took place upon the appointment of representatives to wait upon the Ontario Government in connection with the Engineers' license bill. The members were very much pleased with the report of the delegates upon their return home.

We intend holding our annual supper on the eve of Good Friday. As usual, it will take place at the Commercial Hotel, and it is not necessary for me to say that all attending will have a good time.

WM. NORRIS,
Cor.-Sec.

A deputation representing the Canadian Association of Stationary Engineers recently waited on the Ontario Government, and requested the aid of the Ministry in support of the bill to make compulsory the obtaining of license certificates by stationary engineers. In view of the opposition to the measure on the ground that the Association are seeking to become a close corporation, the deputation expressed their willingness to leave the licensing in the hands of the Government, and offered on behalf of the Association to furnish an examining board so as to render the operation of the act inexpensive. The deputation was composed as follows :—Messrs. James Devlin, Kingston ; J. T. Smart, Peterboro' ; Arthur Ames, Brantford ; Geo. Gilchrist, Guelph ; Robert Mackie and Duncan Robertson, Hamilton ; Fred. Donaldson, Ottawa ; John Graham, Dresden; Wm. Vaughan, Berlin ; A. E. Edkins, A. M. Wickens, M. V. Kuhlman and O. B. St. John, Toronto. They were introduced by Mr. Crawford. The spokesmen were Messrs. Edkins, Robertson, Devlin and St. John.

Mr. James Devlin, Secretary of the Kingston Association of the C. A. S. E., had the misfortune recently to break his leg while stepping off an electric car.

## MONTREAL JUNIOR ELECTRIC CLUB.

February 25.—Paper on "Treatment of Suffering by Accidental Shock," by R. H. Street.

March 4.—Paper on "New Electrical Discovery in Primary Battery," by E. W. Sayer.

March 11.—Paper on "Telephones," by W. T. Sutton.

March 18.—Paper on "Dynamos," part 1st, by I. Turner.

The following are the officers elect of the London Street Railway Company for the present year : President, H. A. Everett; vice-president, E. W. Moore ; manager and treasurer, Chas. E. A. Carr ; secretary, S. R. Break ; assistant secretary, Chas. Curry ; superintendent, I. H. Deharte.

PUBLISHED ON THE FIFTH OF EVERY MONTH BY

## CHAS. H. MORTIMER,

OFFICE : CONFEDERATION LIFE BUILDING,
Corner Yonge and Richmond Streets.

TORONTO, - - CANADA.
Telephone 2362.

NEW YORK LIFE INSURANCE BUILDING, MONTREAL.
Bell Telephone 2299.

*ADVERTISEMENTS.*

Advertising rates sent promptly on application. Orders for advertising should reach the office of publication not later than the 15th day of the month immediately preceding date of issue. Changes in advertisements will be made whenever desired, without cost to the advertiser, but to insure proper compliance with the instructions of the advertiser, requests for change should reach the office as early as the 22nd day of the month.

*SUBSCRIPTIONS.*

The ELECTRICAL NEWS will be mailed to subscribers in the Dominion, or the United States, post free, for $1.00 per annum, 50 cents for six months. The price of subscription should be remitted by currency, in registered letter, or by postal order payable to C. H. Mortimer. Please do not send cheques on local banks unless 25 cents is added for cost of discount. Money sent in unregistered letters will be at senders' risk. Subscriptions from foreign countries embraced in the General Postal Union, $1.50 per annum. Subscriptions are payable in advance. The paper will be discontinued at expiration of term paid for if so stipulated by the subscriber, but where no such understanding exists, will be continued until instructions to discontinue are received and all arrearages paid.
Subscribers may have the mailing address changed as often as desired. *When ordering change, always give the old as well as the new address.*
The Publisher should be notified of the failure of subscribers to receive their papers promptly and regularly.

*EDITOR'S ANNOUNCEMENTS.*

Correspondence is invited upon all topics legitimately coming within the scope of this journal.

THE "CANADIAN ELECTRICAL NEWS" HAS BEEN APPOINTED THE OFFICIAL PAPER OF THE CANADIAN ELECTRICAL ASSOCIATION.

## CANADIAN ELECTRICAL ASSOCIATION.

**OFFICERS :**

PRESIDENT :
K. J. DUNSTAN, Local Manager Bell Telephone Company, Toronto.

1ST VICE-PRESIDENT :
A. B. SMITH, Inspector Canadian Board Fire Underwriters, Toronto.

2ND VICE-PRESIDENT :
C. BERKELEY POWELL, Manager Ottawa Electric Light Co., Ottawa.

SECRETARY-TREASURER :
C. H. MORTIMER, Publisher ELECTRICAL NEWS, Toronto.

EXECUTIVE COMMITTEE :
L. B. McFARLANE, Bell Telephone Company, Montreal.
GEO. BLACK, G. N. W. Telegraph Co., Hamilton.
T. R. ROSEBRUGH, Lecturer in Electricity, School of Practical Science, Toronto.
E. C. BREITHAUPT, Berlin, Ont.
JOHN YULE, Manager Guelph Gas and Electric Light Company, Guelph, Ont.
D. A. STARR, Electrical Engineer, Montreal.
J. J. WRIGHT, Manager Toronto Electric Light Company.
J. A. KAMMERER, Royal Electric Co., Toronto.
J. W. TAYLOR, Manager Peterboro' Carbon Co., Peterboro'.
O. HIGMAN, Inland Revenue Department, Ottawa.

## MONTREAL ELECTRIC CLUB.

OFFICERS :

President, W. B. SHAW, Montreal Electric Co.
Vice-President, H. O. EDWARDS.
Sec'y-Treas., CECIL DOUTRE, 81A St. Famille St.
Committee of Management, T. F. PICKETT, W. GRAHAM, J. A. DUGLASS.

## CANADIAN ASSOCIATION OF STATIONARY ENGINEERS.

EXECUTIVE BOARD :

President, J. J. YORK, Board of Trade Bldg., Montreal.
Vice-President, W. G. BLACKGROVE, Toronto, Ont.
Secretary, JAMES DEVLIN, Kingston, Ont.
Treasurer, DUNCAN ROBERTSON, Hamilton, Ont.
Conductor, E. J. Philip, Toronto, Ont.
Door Keeper, J. F. CODY, Wiarton, Ont.

TORONTO BRANCH NO. 1.—Meets 2nd and 4th Friday each month in Room D, Shaftesbury Hall. Wilson Phillips, President ; T. Eversfield, Secretary, University Crescent.

HAMILTON BRANCH No. 2.—Meets 1st and 3rd Friday each month, in Maccabee's Hall. Jos. Langdon. President ; Wm. Norris, Corresponding Secretary, 211 Wellington Street North.

STRATFORD BRANCH No. 3.—John Hoy, President ; Samuel H. Weir, Secretary.

BRANTFORD BRANCH No. 4.—Meets 2nd and 4th Friday each month, C. Walker, President ; Joseph Ogle, Secretary, Brantford Cordage Co.

LONDON BRANCH No. 5.—Meets in Sherwood Hall first Thursday and last Friday in each month. F, Mitchell, President ; William Meaden, Secretary Treasurer, 533 Richmond Street.

MONTREAL BRANCH No. 1.—Meets 1st and 3rd Thursday each month, in Engineers' Hall. Craig street. President, Jas. Robertson ; first vice-president, H. Nuttall ; second vice-president, Jos. Badger ; secretary, J. J. York, Board of Trade Building ; treasurer, Thos. Ryan.

ST. LAURENT BRANCH No. 2.—Meets every Monday evening at 43 Bonsecours street, Montreal. R, Drouin, President ; Alfred Latour, Secretary, 306 Delisle street, St. Cunegonde.

BRANDON, MAN., BRANCH No. 1.—Meets 1st and 3rd Friday each month, in City Hall. A. R. Crawford, President ; Arthur Fleming, Secretary.

GUELPH BRANCH No. 6.—Meets 1st and 3rd Wednesday each month at 7:30. p m. J. Fordyce, President ; J. Tuck, Vice-President ; H. T. Flewelling, Rec.-Secretary ; J. Gerry, Fin.-Secretary ; Treasurer, C. J. Jorden.

OTTAWA BRANCH, No. 7. —Meets 2nd and 4th. Tuesday, each month, corner Bank and Sparks streets ; Frank Robert, President ; F. Merrill, Secretary, 352 Wellington Street.

DRESDEN BRANCH No. 8.—Meets every and week in each month ; Thos. Merrill, Secretary.

BERLIN BRANCH No. 9.—Meets 2nd and 4th Saturday each month at 8 p. m. W. J. Rhodes, President ; G. Steinmetz, Secretary, Berlin Ont.

KINGSTON BRANCH No. 10.—Meets 1st and 3rd Tuesday in each month in Fraser Hall, King Street, at 8 p.m. J. Devlin, President ; A. Strong, Secretary.

WINNIPEG BRANCH No. 11.—President, G. M. Hazlett ; Recording Secretary, W, J. Edwards ; Financial Secretary, Thos. Gray.

KINCARDINE BRANCH No 12.—Meets every Tuesday at 8 o'clock, in the Engineer's Hall, Waterworks. President, Jos. Walker ; Secretary, A. Scott.

WIARTON BRANCH No. 13.—President, Wm. Craddock ; Rec. Secretary, Ed. Dunham.

PETERBOROUGH BRANCH No. 14.—Meets 2nd and 4th Wednesday in each month. S. Potter, President ; C. Robison, Vice-President ; W. Sharp, engineer steam laundry, Charlotte Street, Secretary.

BROCKVILLE BRANCH No. 15.—W. F. Chapman, President ; James Aitkens, Secretary.

CARLETON PLACE BRANCH No. 16.—W. H. Routh, President ; A. M. Schofield, Secretary.

## ONTARIO ASSOCIATION OF STATIONARY ENGINEERS.

BOARD OF EXAMINERS.

President, A. E. EDKINS, 139 Borden st., Toronto.
Vice-President, R, DICKINSON, Electric Light Co., Hamilton.
Registrar, A. M. WICKENS, 280 Berkeley st., Toronto.
Treasurer, R. MACKIE, 28 Napier st., Hamilton.
Solicitor, J. A. McANDREWS, Toronto.

TORONTO—A. E. Edkins, A. M. Wickens, E. J. Phillips, F. Donaldson.
HAMILTON—P, Stott, R, Mackie, R. Dickinson.
PETERBORO'—S. Potter, care General Electric Co.
BRANTFORD—A. Ames, care Patterson & Sons.
KINGSTON—J. Devlin (Chief Engineer Penetentiary), J. Campbell.
LONDON—F. Mitchell.

Information regarding examinations will be furnished on application to any member of the Board.

THE Canadian Controller of Customs has decided that electricity brought into Canada from the United States is subject to a duty of 20 per cent. This matter came up for decision at Washington a year or two ago, and if we mistake not, electricity was declared to be a manufactured product, and as such subject to duty.

MUCH sympathy is being expressed with Mr. Nikola Tesla, the brilliant electrical inventor, in consequence of the destruction by fire on March 13th of all the apparatus which he was accustomed to employ in his experiments. Unfortunately there was no insurance on the property. Mr. Tesla lost no time in vain regrets, but immediately set to work to replace what had been destroyed.

THE Supreme Court has reserved its decision in the case of the appeal of the Toronto Railway Company to be granted a refund of $56,000 paid as duty on steel rails imported into Canada for use in the construction of their system. Item 173 of the Customs Act exempts from duty steel rails over 25 pounds in weight for use on railway tracks. The Exchequer Court has decided that this item does not apply to street railways. The present appeal is from this decision. The case appears to hinge on whether an electric road is a "tramway" or a "railway." The extent of mileage and nature of construction of the suburban electric roads built during the last two years should be sufficient evidence upon which to conclude that they are in the truest sense entitled to be classed as railways. A proof of this is to be found in the electric railway bill now before the Ontario Legislature, the author of which, in introducing it said that "it was intended that the bill should bear the same relation to the electric 'railways' that the general act does to the steam railways."

A SECOND meeting of the executive of the American Street Railway Association to complete arrangements for the Montreal Convention is announced to take place at the Auditorium building, Chicago, on the 10th inst. It has been decided that the convention shall extend over four days. viz., Oct. 15th, 16th, 17th and 18th. The meetings will be held in the hall of the Windsor hotel. The Victoria Rink, opposite the hotel, has been secured for the use of exhibitors of electrical appliances. Exhibits from the United States will not be required to pay duty, but a reasonable charge will be made for space in the Victoria Rink. Messrs. Granville C. Cunningham and E. Lusher, of the Montreal Street Railway Company, are devising means for the entertainment of the members of the Association during their stay in Montreal.

IT is said to be the intention of the Chicago City Railway to lay this spring thirty miles of track with cast welded joints. The testimony of the electrical engineer of the Cleveland Electric Railway on 1,000 feet of continuous track laid between two and three years ago with hot rivetted joints, is, that it is as good as when put down, and has required no repairs. Experiments with hot rivetted, driven bolt, and electrically welded joints in various parts of the United States and in New South Wales, seem to demonstrate that a track laid in this manner is not thrown out of alignment by contraction and expansion due to extreme changes of atmosphere. The statement of the engineer of the Cleveland Railway on this point is, that notwithstanding that during the past winter the thermometer frequently ranged at 10 below zero for days in succession, only six out of three thousand electrically welded joints gave way. The result of the adoption of this method on an extensive scale in Chicago will be looked for with much interest by street railway managers.

A BILL has been introduced in the Ontario Legislature by Mr. Bronson, and has received its second reading, designed to regulate the manner in which electric railways shall be constructed and operated. The bill debars directors of an electric railway company from any interest, direct or indirect, in any contract for construction. It provides that plant and material shall be paid for in cash, that the original issue of stock must be sold for cash, and subsequent issues by tender to the highest bidder, that all contracts for construction for leasing rolling stock or power, shall be subject to a two-thirds vote of the shareholders. Dividends are limited to 8 per cent. above a fair charge for working expenses. The limit of fare is placed at five cents for a distance of three miles and under, a provision which may interfere with present arrangements on the H. G. and B. and other suburban roads on which a system of graded fares has been put in operation. Sunday service is forbidden on roads extending more than one mile beyond the limits of any city, town, or incorporated village. Companies are to be allowed to purchase and improve pleasure parks, but such parks must not be opened on Sunday. These are a few of the more important features of the bill, which is a lengthy measure. It is to be hoped that before such a bill is placed on the statutes, its provisions will have undergone the most careful revision, and that an opportunity will be afforded representatives of the electric railway interests of the country to submit their views on the subject. The greatest care should be exercised to avoid hampering the development of electric railway enterprise.

WE are pleased to observe that the Bill recently introduced by Mr. German, the member for Welland, in the Ontario legislature, with the object of making electric light, telegraph and telephone wires and poles assessable as personal property, has been withdrawn. Evidently the author of the Bill did not foresee the injurious results which would follow the putting into operation of the measure ; and it is creditable to him that when he became possessed of larger information on the subject, and was able to see what the effect of the measure would be, he decided upon its withdrawal. Immediately after the Bill was introduced the Canadian Electrical Association issued a circular calling the attention of electric lighting companies throughout the country to the Bill, and the injurious effects which must follow should it be placed on the Statute books of the province, and urging them to lay the facts before their representatives in the legislature so that the Bill might not become law. In this action the Association has shown itself to be alive to electrical

interests. In addition to the influence thus exerted, a deputation of fruit growers from Mr. German's own constituency waited on the Committee of the House and urged that the Bill should be thrown out, pointing out the advantage which at present is derived from telephone communication throughout the country districts, and showing that if telephone poles and wires were taxed in the manner proposed it would be the means of doing away with the present convenient and valuable means of communication, and of again placing fruit and agricultural producers in the isolated conditions with which they were surrounded previous to the introduction of these modern facilities. The presentation of facts resulted in the withdrawal of the support of the Patron element from the Bill, and sufficed to show Mr. German that the measure was likely to result in serious injury to the very class whose interests he was especially seeking to serve. So far as electric light companies located in towns and villages are concerned, it is a well-known fact that very few of them are more than self-sustaining under present conditions, and there is required to be exercised the closest economy in order to make ends meet ; consequently, they are not in a position to be burdened with additional taxation. It should be borne in mind that the property of an electric light, telephone or telegraph company differs from that of a stock of dry goods, for example ; the latter can be sold and removed to another locality without suffering in value ; but electric light poles and wires cannot be so removed after having once been erected in the streets of a municipality. The cost of removing poles would be as much as their value, and the same would apply to wires. The only proper method of assessing these companies is to base the assessment upon the earnings of the company, and in the event of legislation being again proposed, we trust this fact will be borne in mind.

THE class of man usually found operating electrical machinery is such as to cause a doubt as to the advisability of any further improving the efficiency of dynamos and lamps. It seems to be of very little use to raise the efficiency from 95 to 96 per cent. by alterations in design, improvements in material, and by the expenditure of valuable time and money when the person responsible for its subsequent operation is not competent to either appreciate the saving or to continually watch that the conditions of operation are such that the higher efficiency may affect results. For instance, 3 watt lamps are, broadly, more efficient than 4 watt lamps, but the condition of efficiency is a more perfect pressure regulation. A 3 watt lamp will, generally speaking, suffer more by a varying pressure than a 4 watt. Pressure varying directly as speed it would appear that it is very questionable economy to use 3 watt lamps and supply them from a dynamo run by an ordinary saw mill engine. The writer has seen this done, likewise been told that those lamps were "no good," although the agent claimed them as "more economical." There appears to be very little advantage in manufacturing dynamos or transformers with a very high percentage of efficiency when the actual economy effected by their use instead of low priced low efficiency machines, is not recognized. What is the actual advantage of high efficiency, and how may it be measured? The actual advantage is a direct saving of money, and it can be measured by the increase in dividends. The efficiency of a dynamo being the percentage of the mechanical power required at the pulley to turn the armature, given out in the form of electrical power by the armature, measured at the brushes, it follows that two or more machines can be directly compared. Take two, of efficiencies 94 and 95 per cent. This means that of 100-horse power delivered at each pulley one dynamo, will return 94-h.p. and the other 95-h.p. Neglecting losses in engine, belting and boiler, and assuming 5 lbs. coal per h.p.h., the machine with lower efficiency will require 5 lbs. of coal per h.p.h. more to run it than the other. Taking 10 hours of operation every night for 365 nights, with coal at $3.75 the short ton, we find that the low efficiency machine costs $33.75 per horse power per year more to run it than the high efficiency. This is 10 per cent. on $337. The high efficiency machine therefore might cost $337 more per h. p. than the low and yet be actually no more expensive as an investment. The same remark applies to transformers, lamps, etc. The efficiency of a boiler is largely a matter of firing, and this depends entirely on the fireman. In a 50-h.p. plant, running 10 hours each night, 365 nights, coal as above, 5 lbs. per horse power hour, the consumption of a quarter of a pound per h.p.h. more than necessary, means $427 spent per annum which might be saved. What fireman at $1 a day knows enough to save this? What "superintendent and electrician" at $50 a month knows that this is being wasted? And yet, very few small plants, managed by inexperienced persons, can say that they are not wasting a good deal more?

THE statement was recently made to the writer by the owner of a medium sized arc plant, that "Arc plants do not pay   Communications received from owners of arc plants all over Canada show that not one of them is paying anything. I don't believe there's any money in electric lighting, anyway." Whether this be a fact or not, with regard to electric lighting generally, with regard to the particular plants communicating their sad experience it seems to indicate either ridiculously low prices for light or conspicuously poor management. If the former, then it shows a very considerable "margin of ignorance" in the consideration given to the preliminaries, especially commercial ; if the latter, it points to the same ignorance of the principles of what now is almost a science—" station management "; and in either case it emphasizes the fact that electric lighting has emerged from the stage in which all that is necessary to be known about it can be learned from the pages of the daily press. As an illustration of this ignorance factor, the same gentleman whose opinion is given above stated that he was "operating at very high economy ; the electrician's time, and therefore salary, was equally divided between a gas plant and the arc plant ; and the only other employee was a trimmer, and he was burning coal at the rate of 4 lbs. per horse power hour." Under all these favorable conditions why doesn't that plant pay? Is it possible that these conditions are not so very favorable after all? Is it possible that a man, who is not even a scientific gas maker, may not be the very most competent person to become "electrician" of an arc plant? or that a gentleman whose business is hardware does not necessarily know "all about" the economical generation and utilization of the electric current? It seems quite impossible, but then it doesn't pay. The stated coal consumption of 4 lb. per horse power hour seems to indicate not only remarkable—quite remarkable—economy, but also very capable, skilled management, and as such is quite incompatible with ignorance or carelessness. In fact it is such a very remarkable figure as to call for some analysis. The plant is one of about 50 200-c.p. arcs ; engine simple, non-condensing, belted to dynamo, boiler being return tube type. The engine, therefore, must develop about 50-h.p. Such a small engine will certainly require nearly 40 lbs. steam per h.p.h. This will require for evaporation from 100° feed to 80 lbs. pressure—44280 h.u., supposing that everything is in first-class order and the fireman thoroughly skilled in his work. As 4 lbs. of coal is stated to be the consumption, each pound of coal must have the calorific value of $\frac{44280}{4}$ h.u. = 11070 h.u., (which is the value of really very high-class, expensive coal) supposing that every h.u. is rendered available for evaporating purposes. Assuming that such a very high percentage as 80% of the h.u. of the coal are made use of, we find that at least 55380 heat units will be required to raise the necessary amount of steam, and if we assume the calorific value of the coal used to be 9000 h.u. per pound (which seems reasonable) we get as the amount of coal required to raise steam for the 50-h.p. engine—assuming all machinery to be in good condition, no loose contacts on lamps, and everything as it would be in a first-class station—6.15 lbs. How the figure 4 lbs. was arrived at hardly appears, but it would be pretty safe to say that 7½ to 8 lbs. are used. Does all this ignorance, not only of theoretical possibilities, but of actual practice and results, not point to reasons why plants do not pay dividends? It seems almost superfluous to point out that it is because electric lighting is not generally recognized as a business possessing its special features and problems, each requiring special and careful study.

Negotiations are said to be in progress for the absorption of the St. John, N. B., Gas Light Company, by the St. John Street Railway Company. A two-thirds vote of the shareholders will be required to consummate the deal.

To make filaments of electric incandescent lamps more brilliant, says Invention, Messrs. Chaney and Depouy soak the fibre in a solution of the nitrates of magnesia, zirconia, and lathania, and the effect is said to be analogous to that of the Welsbach lamps.

## GAS ENGINES.

REFERRING to the article in our March number on the subject of gas engines, in this issue we call attention to those manufactured by The Fried Krupp Grusonwerk, of Magdeburg, Germany.

These engines, made of three different models B, C and D (the accompanying cut represents model B), are designed for electric lighting, industrial and other purposes. Those for electric lighting are provided with patent cut-off gear, and are so regulated that they are constantly supplied with the same quality of gas mixture, the quantity of which is automatically varied by the governor.

In some gas engines for electric lighting that are offered for sale only the quantity of gas is altered by the regulator according to the load and consequently the explosive mixture is not constant and therefore sometimes uneconomical.

In the Krupp cut-off gear engines, however, the air supply is also influenced by the governor, so that whatever may be the

THE KRUPP GAS ENGINE.

quantity of mixture in the motor there is always at the intake valve and before the igniting valve an easily inflammable and under all circumstances perfectly constant mixture.

In the cylinder itself the remaining gases in the compression chamber, which are always equal in quantity, have a more diluting action on the pure fresh combustion charge in the cylinder in the case of small charges than in the case of large, and therefore the mixture burns slower and consequently more unfavorably in the case of small charges. In order to obtain a more favorable combustion in the case of small charges their ignition is made earlier than in the case of large charges. The governor accomplishes this automatically, so that in all circumstances a proper combustion takes place in the engine, regularity and economy being thereby secured even when only half loaded.

The gas consumption is more favorable the larger the engine. With full load it reaches 0.45 cub. in., and for the smallest engines about 0.7 cub. in. per brake h.p. an hour. The cylinders are made of the best chilled cast iron and are slid into the cooling jacket. Crank shaft, connecting rod, etc. are made of steel; axle boxes and bushes are made of phosphor-bronze or are lined with special bearing material.

The engines with automatic cut-off gear are made in three models, viz. (1) one cylinder horizontal engines model C up to 12-h.p; (2) one cylinder horizontal engines model B from 16-h.p. upwards, and (3) two cylinder engines model D from 40-h.p. upwards.

The representatives in Canada for The Fried Krupp crusonwerk are Messrs. Jas. W. Pyke & Co., Montreal, to whom all communications for particulars should be addressed. They have imported a small sample engine and made arrangements to have it erected for testing purposes at the establishment of Mr. Samuel Fisher, 57 St. Sulpice street, Montreal, who is also acting as local agent for Montreal and district. Those who have already examined this engine admit that it is a fine piece of workmanship and runs with remarkable smoothness. Mr. Fisher will be pleased to show the engine working, and give all information regarding it to interested parties.

The town of St. Louis Du Mile End, has entered into a contract with the Citizens' Light & Power Co., to light the streets of the town, and has also entered action against the Montreal Street Railway Company for laying their tracks within the limits of the municipality. The company claim that they were obliged by the city by-laws to lay down the tracks, and will hold the city of Montreal responsible for any damages which may be incurred.

## THREE-PHASE PLANT AT ST. HYACINTHE, P. Q.

THE town of St. Hyacinthe, at which place is located the first three-phase plant installed in Canada, is on the Portland line of the Grand Trunk Railroad, about thirty-five miles from Montreal. A branch line of the Canadian Pacific also reaches the town, and a new line of railway called the United Counties passes through it, connecting it with the town of Sorel on the west and Iberville on the south. The population is at present about 11,000, which is rapidly increasing. A fine water power on the Yamaska river is utilized to operate the granite mills owned by Feodore Boas & Co., manufacturers of woolen goods, and several other factories. There has been for some time, however, a demand for more power than was available in the town itself, and in the fall of 1893 the transmission of power from the Rapid Plat, 4½ miles below the city, was first discussed. In February, 1894, this power was acquired by Mr. A. M. Morin, and in April of that year a company called La Cie des Pouvoirs Hydrauliques de St. Hyacinthe was formed to improve and distribute it for motive purposes and lighting in the town. Work was commenced at once on the water power, and in July contract was closed with the Canadian General Electric Company for the necessary electrical apparatus for the plant. A

tal shaft by means of clutch coupling. The shaft is divided into two sections connected by a Hill cut-off clutch, two wheels being geared to each section. The main driving pulleys, which are four in number, and each provided with Hill clutches, are placed on an extension of this shaft under the dynamo room. Hand wheels controlling each of the four wheels and the four clutch pulleys are placed in a convenient position in the dynamo room, so that the entire operation of the plant can be absolutely controlled from the switchboard. Two electric governors, one for each pair of wheels, are connected to controlling mechanism, which is also placed in the dynamo room. It is intended to connect a tachometer to the shafting which will at all times indicate the speed.

### ELECTRICAL PLANT.

The electrical equipment of the power house, installed throughout by the Canadian General Electric Co., Ltd., consists at present of three of their type A. T. 12-150-600 standard three-phase alternators, each having a capacity of 150 k. w. at 2,500 volts. They are compound wound in the same manner as that company's single-phase high periodicity alternators of the Thomson-Houston type, the commutator, however, being in three sections to accommodate the three-phase current. The

THREE-PHASE PLANT AT ST. HYACINTHE, P. Q.

thorough investigation was made of the different systems of electrical transmission, and the President of the company, Mr. Louis Cote, and their consulting engineer, Rev. Father Choquette, visited a number of power transmission plants before their decision was reached. The contract for the water wheels and shafting was awarded to the James Leffell Co., of Springfield, Ohio, who also furnished plants for the installation of the wheels.

### WATER POWER PLANT.

The power had formerly been utilized for the operation of a grist mill and woolen factory on one side of the river, and for a small grist mill on the opposite side. No change was necessary in the dam, although it will be possible by raising this to greatly increase the power available. The canal leading to the mill was almost entirely reconstructed and deepened, so that its capacity is nearly three times that of the old canal. In addition to this a long tail-race was excavated, greatly increasing the head, which is now about 17 feet. The water is led directly to the wheels, which are four in number, of the Leffell Co.'s Sampson type, 50 inches in diameter, and running at a speed of 100 revolutions per minute. These wheels are on vertical shafts and placed in wooden penstocks with separate gates. At the top of the vertical shaft is placed a crown wheel 6′ 2″ in diameter, having 78 iron wood teeth. This is geared to a pinion, 24 9-16 inches in diameter, having 26 teeth, and connected to a horizon-

separate exciting current is supplied by two 6 k. w. standard Edison dynamos, either one of which is capable of exciting the fields of all three machines.

The periodicity of these alternators is 60 cycles per second, this having been adopted in place of the old standard of 125 cycles, as it has been found from experience that motors operate very much more satisfactorily the lower the periodicity, and this number was decided upon as being more suitable for the combination of motors, arc lamps, and incandescent lamps, the steadiness of the latter being affected when the periodicity is much further reduced. The current from these machines is led to the centre panel of the switchboard, as shown in the accompanying cut, and is there connected to the main bus bars in multiple through three high potential triple pole switches. On this panel is also placed a current indicator and potential indicator for each machine, together with the phase indicator by which the machines are thrown together. The feeder panel is to the right and is equipped with three current indicators, one for each leg of the system, a ground detector, lightning arresters and feeder blocks. On the left are the three station transformers and the exciter current indicators and switches.

### LINE.

The distance between the power plant and the town, as stated above, is 4½ miles. The line consists of four number ∞ B & S bare copper wires placed on double petticoat insulators. The poles are all of cedar; 30 feet in height above the ground, and a double set of cross arms, pins and insulators are placed on each pole. Only three of the wires normally are in use, the fourth being kept as a spare in case of accident. The line is of the most solid and substantial construction throughout, and has been built with the object of providing amply for any addition to the lines which may be required at any future time.

### SECONDARY DISTRIBUTION.

The primary wires are brought to the centre of distribution in the town, and from this point primary mains extend over the district which is to be furnished with light and power. The greater portion of the lighting is from a four-wire system of secondary mains fed by banks of transformers at suitable points. This system combines the economy of both the three-wire and three-phase systems and insures a uniform potential at all points. All large motors will be connected to separate banks of transformers, only the smaller sizes being operated from the secondary mains.

### THE COMPANY.

The directorate of the company includes the names of nearly all the prominent business men of the town. Mr. Louis Cote, the president, is well known as the inventor of several important labor saving machines for shoe manufacturing. Mr. Payan, vice-president, is a member of the firm of Duclos & Payan, tanners and manufacturers of leather. The construction work and wiring has been done under the supervision of Mr. R. Duperouzel, superintendent of the Hydraulique Company, to whom much credit is due for the manner in which he has carried out an installation having so many novel features.

### RECENT CANADIAN PATENTS.

Canadian patents have recently been granted for the following electrical and steam engineering devices :—

No. 47,796, for gas engine, to Frank S. Mead, Montreal.

No. 47,832, for automatic railway car protector, to Chas. Gletner, Cincinnati, Ohio.

No. 47,834, for feed water heater, to Stirling L. Bayley, Chicago, Ill., and John W. Dowd, Toronto, Ont.

No. 47,848, for car fender, to Wm. Hofmeister, and W. F. Madaus, New York City.

No. 47,850, for a closed conduit electric railway, to James Francis McLaughlin, Philadelphia, Pa. The inventor's claim for this device, which is herewith illustrated, is as follows :—

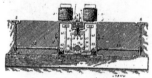

No. 47,850.—Closed Conduit Electric Railway.

1st. In an electric railway, the combination with the closed conduit, of a main or supply conductor housed therein, a sectional working conductor composed of sections in the conduit and exposed sections seated in the top of the conduit and electrically connected to the sections in the conduit, and magnetically operated switches, pivoted to the main conductor, formed of switch plates in operative relation to the underground sections of the working conductor, and with armatures closed to the top of the conduit. In an electric railway, the combination with a closed conduit provided with a central longitudinal groove or trough along its top, or a main or supply conductor housed in the conduit, a working conductor composed of sections in the conduit and exposed sections seated in the trough and electrically connected to the sections in the conduit, and magnetically operated switches pivoted to the main conductor and formed with switch plates in operative relation to the underground sections of the working conductor and with armatures close to the top of the conduit between the trough and sides of the conduit.

No. 47,858, for electric switch or circuit breaker, Frank Stevens and Robert Rodwell Kesteven, Philadelphia, Pa.

No. 47,859, for an electric motor, to Jas. H. K. McCollom, Edwin Krickmore, Thos. Ed. B. McCollom, Melville B. R. Gordon, John W. Sweetnam and Thos. W. Hector, Toronto, Ont. The following description of this motor, an illustration of which appears herewith, is abstracted from the statement of claim of the inventor :—

The combination with a stationary armature comprised of a ring having a series of sections arranged equi-distant around it between the toothed projections, the sections having a plurality of coil layers arranged in multiple and suitably connected to the corresponding sections of the commutator, of an arc-shaped field magnet supported and magnetically insulated from the main shaft of the motor by a disc as shown and for the purpose specified. The combination with the ring-shaped stationary armature composed of a series of sections arranged as specified and having arms secured to the ring.

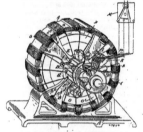

No. 47,859.—Electric Motor.

and extending inwardly and terminating in hubs, which form bearings for the main shaft of the motor, of a stationary commutator secured in one of the hubs, rotating brushes and arc-shaped field magnets supported upon and magnetically insulated from the shaft by a suitable disc and designed to rotate in unison with the commutator brushes, which are suitably supported and derive current from electrically insulated rings through the contact brushes resting on such rings and connected to the main circuit.

No. 47,861, for a trolley, to Carl Ast, Gorlitz, Prussia, Ger.

No. 47,863, for a multiphase motor, to the Canadian General Electric Co., Toronto, Ont., assignee of Louis Bell, Chicago, Ill. The accompanying illustration and statement of claim will serve to explain the invention :—

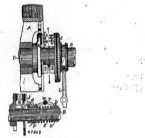

No. 47,863.—Multiphase Motor.

In an electric motor having a closed-circuited armature-winding, the combination with resistance in circuit with said winding and carried therewith upon the armature shaft, said resistance being divided into sections, of automatic switching mechanism also mounted on said armature shaft and responsive to the speed thereof, whereby successive sections of the resistance are short-circuited as the speed increases.

No. 47,867, for car brake, to Geo. Hill Kinter, Geo. D. Teller and Geo. Tait, Buffalo, N. Y.

No. 47,892, for telephonic relay, to Geo. Gilmour, Douglas, Isle of Man.

No. 47,908, for telephone annunciator and call bell, to Fred. G. Warrell, Philadelphia, Pa.

No. 47,909, for telegraphic transmitter, to Frank F. Howe, Marietta, Iowa.

No. 47,935, for an electric railway, to the Thomson-Houston National Electric Co., Portland, Maine.

No. 48,030, for a steam generator, to Chas. Wm. Vanderberg, Wellington, B. C.

It is understood that a satisfactory understanding has at last been reached between the London Street Railway Co. and the City Council, in accordance with which the railway system of the city will immediately be transformed to the trolley system. The company is said to have secured property adjoining that of the General Electric Company at the foot of Bathurst street, on which to erect a power station and car barns. The situation is a very desirable one, water for condensing purposes being obtainable from the river Thames.

## DYNAMO DESIGN.*

### BY E. B. MERRILL.

#### PRELIMINARY.

BEFORE proceeding with the design of a dynamo, it is necessary to ascertain as definitely as possible what is required of it. It is to be built for a given purpose. It is a generator, perhaps, to do lighting or electro-plating, or to supply a power circuit, or it may be a motor to run a fan, a lathe, or a workshop, an elevator or a street car ; a generator for a constant pressure or a constant current circuit ; or a motor for fixed, or for variable speeds, to be used on parallel or series circuits. We must obtain information on these points :

(1) What circuit is it to supply or be supplied from ?

(2) What power is required of it ?

(3) What are the general conitions of use—such as the position that it is to be suspended in, the limit of floor space, the necessity for protection from dust, water, iron filings, and other sources of injury, the mode of driving, etc.

From the answer to the first question we determine one of the factors of the power. It will decide the pressure to be obtained between the terminals if for parallel working, as for incandescent lighting, and for motors on constant pressure circuits, or the current to provide for in series circuits ; it will help us in settling the question of open or closed coil armatures, and of series, shunt, or compound fields. The degree of constancy necessary in the pressure of generators and in the speed of motors will determine whether they are to be shunt or compound wound. We shall know also whether special design and regulation will be necessary to effect alterations of electromotive force in generators, or of speed in motors.

From the answer to the second, we determine the other factor of the power, for, as we know electrical power is the product of current and electromotive force, and we have only to divide the required power—reduced to Watts—by the one given to obtain the other ; usually we are given the pressure and require to determine the maximum current needed. As power is lost in the various windings of a dynamo, this also should be considered. Any such heat losses may be calculated just as they would be if the energy were being used in the external circuit, armature and series field coils as in series circuits, and shunt fields as in parallel circuits. If $C$ is the total armature current and $E$ the generated electromotive-force, driving the current in the generator or resisting the current in the motor,

$$E \; C = P \; (watts)$$

gives the total transformation of energy, mechanical to electrical in the generator or electrical to mechanical in the motor. The power given to or received from the external circuit is

$$E \; C \mp C^2 r \mp c_1{}^2 \; r_1 = E_1 \; C_1 = P_1$$

the negative signs applying to the generator and the positive ones to the motor.

$r$. = resistance of armature or armature and series field winding.

$r_1$ = resistance of shunt field winding.

$c_1$ = current in shunt field winding.

$E_1 = E \mp Cr$, the pressure between the terminals.

$C_1$ = current given to or received from external circuit.

The answer to the third question will affect the general design, particularly the selection of the type.

There is an almost endless variety of different types of continuous current machines ; they may all, however, be classed as bipolar and multipolar. The bipolar machines may be divided into those having single or multiple magnetic circuits, and these, again, as having one or more exciting coils. Multipolar machines, in practice, never have more than one magnetic circuit per pair of poles, and often this is abridged ; that is, two or more pairs of poles may have parts of their circuits in common. Multipolar machines may, therefore, be classed as having one exciting coil—half as many coils as poles, or as many coils as poles, etc.

As to the effect of the power of a machine on the type, one may say, generally, that, for small powers, bipolar machines are preferable on account of their simplicity and economy in construction, while, for large powers, multipolar machines are most economical.

The common types of armature are the drum and ring

* Part of a paper read before the Engineering Society of the School of Practical Science, Toronto, March 20, 1895.

(Gramme). In bipolar machines for small armature diameters, the drum winding is the more economical in length (i. e., resistance) of conductor, and the core more simple in construction ; while, for large diameters and shorter cores, the ring type gives more economy in winding, and offers much better opportunity for ventilation. Ring winding is done in distinct sections, so that it is much easier to insulate it and to replace portions of it than in the case of drum winding, in which the sections all overlap. In ring winding there is one active conductor in each loop ; in drum winding there are two.

In multipolar machines, both rings and hollow drums are used ; and for these we have the ring, and lap and wave windings, with as many parallel circuits as poles, excepting for wave winding, for which there are always two. The cores of armatures may be smooth or slotted. The advantage, on the whole, lies with the smooth surface. The teeth in the slotted core form good driving horns, and somewhat decrease the magnetic reluctance of the circuit ; they also allow the heat to escape more rapidly from the core ; but their additional cost of construction, the heating of the pole pieces, due to the unequal distribution of magnetism, and the trouble they cause in sparking, tell heavily against them.

Let us now, to fix our ideas, suppose that we are to design a dynamo for constant pressure. In the first place we know that the terminal pressure for a generator is less, and for a motor is greater, than the generated electromotive force by a product $Cr$ ; so that if $Cr$ is small, as it must be for economy, the same machine generates nearly the same electromotive-force as a generator or as a motor with terminal pressures the same, and, therefore, as other conditions are unaffected, the speed must be nearly the same also, so that a generator and a motor to be used in the same circuit may be considered very nearly as the same machine.

#### THE ARMATURE.

As the electromotive-force of a dynamo is generated in the armature, and the whole current used must flow through it (excepting for the shunt field of a motor), so that both factors of the power affect it directly ; and as the field provided by the magnetic circuit is only for the use of the armature, and must, therefore, be designed to suit it, the consideration of the armature is evidently the first and most important part of the design of a dynamo ; so that, after deciding the number of poles to be used, we should proceed to determine what is required of the armature, and select the type.

We know approximately the electromotive-force that it must generate $(E = E_1 + Cr$, of which $E_1$ is known, and $Cr$ is small, and need not be considered at this stage), and the maximum current that must flow through it, allowing a percentage for the shunt circuit (see at end of paper). The total armature current is divided between two parallel paths in bipolar, and between two, four, or more parallel paths in multipolar dynamos, so that the current that the armature conductor has to carry is fixed, i. e., one-half, one-fourth, etc., of the total current, and from this, knowing the safe carrying capacities of conductors, we can fix the cross-section necessary. From 400 to 800 circular mils per ampere for copper gives the common range of practice, the lower values when the machine is run intermittently, or when there is good ventilation, and the higher values for continuous running, or when the ventilation is poor.

The armature of the common types of dynamos consists of the arrangement of a number of conductors (usually copper wire) on a core which is a good magnetic conductor, which is attached to a shaft, and revolves in a magnetic field, so that the conductors cut through the lines of that field as they pass from the poles of the dynamo across the gap,—composed of space and clearance and space occupied by the conductors and their insulation—to the core. Now, the cutting by a conductor of 10₈ lines per second produces in it an electromotive-force of one volt, so that the total electromotive-force produced in an armature is given by the equation

$$E = \frac{c \; v \; l \; \mathfrak{B}_a}{10_8}$$

where $\mathfrak{B}_a$ = the average number of lines per square inch in the field.

$l$ = length of field in inches being cut by the conductors, i. e., about the length of the conductor on the face of the armature or the length of the armature core.

$v$ = velocity of the conductors cutting through the field in inches per second, $i.\ e.$, the peripheral velocity of the armature. The electromotive-force produced in each conductor is therefore $\dfrac{v\,l\,\mathfrak{B}_a}{10^8}$

$c$ = number of effective conductors in series; $i.\ e.$, the number of those in series which are cutting the field of strength $\mathfrak{B}_{as}$, and which are therefore within the polar arcs.

The cross-section of the magnetic field is approximately the same as the area of the pole face, and the distribution $\mathfrak{B}_a$ is fairly uniform in a good design. As a means of decreasing the exciting power necessary for a given total magnetic flux, the intensity of magnetization, $\mathfrak{B}_a$, of the gap spaces is taken quite low, especially for smaller machines, as compared with the limits of saturation of the poles and armature core. The value in practice increases with the capacity of the machine, and is about 50 per cent. higher for wrought iron or steel than for cast-iron pole pieces. For bipolar machines, with cast-iron pole pieces, $\mathfrak{B}_a$, ranges from 15,000 for 1 kwt. capacity to 30,000 for 300 kwt.[*]

$v$ as an easily-produced factor of the electromotive-force should be as high as possible. It is limited, however, by mechanical and electrical considerations, such as strain in the moving parts, vibration due to irregularities of balance, friction in the bearings and air friction in the clearance space, eddy currents and hysteresis losses which increase with the speed. The hysteresis losses depend on the number of reversals per second and the intensity of magnetization of the core. In practice, the peripheral velocities of drum armatures range from 25 to 50 feet per second, increasing with the capacities ; and those for ring armatures—on account of their better ventilation and the better hold of the conductors—reach double that amount.

We must now select values for $_a$ and $v$. We then have $c$ and $l$ to determine, having the relation $cl = \dfrac{10^8\,E}{v\,\mathfrak{B}_a}$ between them.

As a question of internal resistance, $c$ and $l$ may nearly balance each other, and it becomes a question then as to whether altering $l$ increases or decreases the idle wire necessary in the particular winding used. But there are other conditions which limit the fixing of $c$, a consideration of which will help us in making the adjustment. $c$ is the number of conductors in series—in one of the two or more parallel armature circuits—which are actually cutting the magnetic field at a given instant, at any of its two or more pole faces ; its ratio to the total number of conductors on the cutting surface is approximately that of the polar arcs to the periphery of the armature space. The length of the polar arc, we find in practice, ranges between 50 and 100 per cent. of the total circumference, and usually, lies between 70 and 80 per cent. ; questions of magnetic leakage, sparking, etc., affect its length. Fixing this ratio now also fixes the relation between $c$, the *effective*, and the total number of *active* armature conductors.

Again, the number of conductors around the circumference, and the number of layers, with a knowledge of the space required for insulation (between conductors and core, between layers, between sections, and between individual conductors), and that needed for driving horns, will fix the circumference required for the armature and the depth of winding.

The radiating power of the armature fixes the limit of depth of conductors. The greatest current density allowed in practice, that is, the ampere turns per inch circumference, is fixed at about 800 or about 2,500 per inch diameter. The average would be about 600 or 1,900 per inch diameter, which corresponds to a rise of 125° to 140° F. The depth of winding varies, in practice, with the diameter of the armature ranging with drum armatures from .25 to .8 inch for diameters of from 2 to 30 inches.

Another question affecting $c$ is the number of sections that there are to be in the armature and in the commutator. Many sections increase the difficulty of winding and insulating, and increase the size and cost of construction of the commutator, while few sections give greater losses in the coils short-circuited under the brushes, causing sparking, and give greater variations

[*] We are indebted for these figures, as well as for others that follow, to the valuable articles by A. E. Wiener, entitled "Practical Notes on Dynamo Calculation," which have been running for some time in the *Electrical World*.

in the total electromotive-force generated. Thirty-six divisions cause only a variation of one-fifth of one per cent., so that this would be plenty for steadiness ; but, besides this, the self-induction in the short-circuited coils require that the number of loops per section should be kept down, which would both tend to decrease the total number of conductors and to increase the number of sections. From 40 to 60 sections is good practice for pressures up to 300 volts on bipolar machines.

For high pressures, the effect of self-induction in the coils comes still more into play in increasing the number of sections, as also the necessity of keeping the pressure between adjacent segments low enough not to maintain an arc across the insulation between them.

Let us, then, decide upon a number of sections for the commutator. Each commutator bar will begin one coil and end another, so that the number of armature and commutator sections will be the same.

We may now select a value for $l$, obtained from a similar type of machine in practice, and obtain $c$ and the total number of conductors ; then select the nearest number to this which will give the chosen number of sections, and correct the assumed value of $l$. Knowing now the number of conductors per section, we can decide, from considerations above given, and from convenience in winding and insulating, the form that the sections will take and how they will be placed and wound. The circumference of the armature is readily deduced from this, and, therefore, the diameter. We may now see if the length chosen and the diameter deduced bear reasonable proportions to each other ; if not, a new value of $l$ may be selected.

The selection of the type of armature need not be finally made until after a preliminary calculation of the dimensions, though, as we have seen, a higher $v$ is allowed for ring than for drum armatures, which should be considered.

*(To be Continued.)*

## "A BOOSTER."

AT the request of the Canadian representatives for Crompton & Co., London, Eng., the Montreal Electric Co. have lately installed at the Royal Victoria Hospital, Montreal, a "booster" in connection with their storage battery plant. The "booster" allows them to light the house circuits at the same time as they are charging the battery. The accompanying diagram shows the principle :

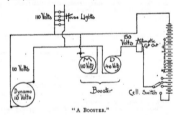

"A BOOSTER."

Heretofore it was necessary to run a dynamo for charging alone, speeding it up to the required voltage, this of course being too high to allow of their running their lamp circuits at the same time.

The "booster" is arranged so that it can also be switched in on their three wire bus-bars ; it then supplies the boosted current from one circuit when the motor is fed from the other. The batteries can also be discharged either on their two wire or three wire bus-bars. The amount of current furnished to the batteries is 50 amperes normal.

The Common Pleas Divisional Court recently dismissed an action brought by a workman against the Hamilton Street Railway Co., for injuries received by coming in contact with an electric car while engaged in lining in rails for the new track. The court decided that there was no statutory duty on the part of the motor-man to sound the gong at the place where the accident happened. and that unless the presence of danger was apparent to him in time to prevent the collision, he was not to blame.

## TRANSMISSION OF POWER BY BELTS.*

By Geo. Fowler.

I venture to say that there are few appliances so much abused and neglected as the one under consideration, namely, the old and tried friend of all shops and factories, the belt. We find it stretched out of all resemblance to its former self. We see it laced in a slipshod manner with perhaps half the lace holes torn out, giving opportunity for the belt to catch against the fingers of the shifter and finally tear out and come down on somebody's head. When we go into a shop or factory and see the belts in the condition described, we are pretty sure to find a shop where the time of attending to the shafting, hot bearings and attendant ills would make a big item in the accounts if it was counted on the list of running expenses. But this kind of a shop never keeps much account anyway, and guesses at the charges to be made for work, with the result of losing money.

It is not idle capital to have belts running slack and doing less work than they might possibly be made to do, for it is much better to have the capital invested in this way than to have delays, cut boxes, and the annoyance that follows in the wake of all unsatisfactory machinery and parts in the whole establishment. It is a pleasure to see a nicely running belt, to go into an engine room and see the great driving belt that is running the whole of a great plant and doing it without apparent effort, the belt running so loose as to give a sag to the upper half, and the lower half running straight as a line. This is a sure sign that the journals are running cool and everything is going along nicely.

I do not wish it understood that everything in this paper is original with me; on the contrary, some of it is borrowed from the best engineering practice in the country. I have been very generously assisted by the several belt manufacturing companies, who gave me good hints on the use of belting. I have also studied such works as Morin's, Cooper's, Nicholson's, Thurston's, and out of these I have taken and adopted several valuable rules and formula.

There are few engineers who have not been frequently in want of information or readily applicable formulæ, upon which they could place reliance, giving the power which, under given conditions and velocity, is transmitted by belts without unusual strain or wear, therefore I believe it is well to study the experiments which are given in the works of the different authors, and acknowledge and adopt formula therefrom, and apply it to daily use. But in doing so we must be careful, because, notwithstanding the existence of this mathematical and experimental information, the numerous tables that have been given by mechanical engineers appear to have had only that kind of a basis which has come from guessing that an engine, or a machine, either the driving or the driven, with a belt of given width, was producing or requiring some quantity of power which might be expressed in foot-pounds generally without any stated arc of pulley contact. For instance, one writer says that a single leather belt one inch wide, running 1000 feet per minute, will transmit .76 horse power; another asserts .93 horse power; another claims one horse power; another makes out 1.33, and still another figures it out to be 1¼, and so on, thus producing conflicting testimony.

The rule which I have acknowledged and adopted may be thus expressed: An ordinary single leather belt one inch wide, with a velocity of 600 feet per minute, will transmit one horse power. After an examination of different text books, I find that General Morin's data gives us the clue to the truth of this rule, and also that it is supported by other good authority. Morin says: "Belts which are designed for continuous service may be made to bear a tension of .551 lbs. per .00155 square inches of section, which enables us to determine the breadth according to the thickness." This is equal to 355 lbs. per square inch of belt leather, and is also equal to about one tenth of the breaking strength of the same as given by Mr. Rankine and other good authorities. Cooper in his works says if we substitute 330 lbs. for 355 lbs. per square inch, we strike the component part of a horse-power and deduce the following: one square inch of belt leather at a velocity of 100 feet per minute will transmit one horse power with safety, and from these data get the rule: The denomination of the fraction which expresses the thickness of the belt in inches, gives the velocity in hundreds of feet per minute at which each inch of width will transmit one horse-power; and as the ordinary thickness of a single leather belt is generally about ⅙ of an inch, we simply multiply the denominator of this fraction by 100 and get the 600 feet at which a single strap one inch wide should run to transmit one horse power.

No rules can be given that will apply to all cases—circumstances and conditions must and will modify them. Belts, for instance, for machines which are frequently stopped and started, and shifting belts, must be wider to stand the wear and tear and to overcome the starting friction, than belts which run steadily and continuously. The breaking strength per inch width of belts when made from good ox hide, well tanned, has been determined as follows:

| | |
|---|---|
| In the solid leather.................. | 675 lbs. |
| At the rivet holes of splices....... | 362 " |
| At the lace holes.................. | 210 " |

*Paper read before Toronto No. 1, C. A. S. E., February 8th, 1895.

Engineers are often required by their employers to put up new shafting, pulleys and belts for the purpose of doing an additional amount of work which may be stated in horse power, and the matter of proper dimension of same, such as size of shaft, diameter and speed of pulley, width of belt, etc., are left to the judgment of the engineer. I have no doubt that a majority of the members of this association are perfectly competent to oversee such work, but to those whose practice along this line has not been very extended, and who may be called upon at any time to take such matters in hand, I offer the following information, which is taken from standard works and may be relied on for everyday use:

The safe working tension is assumed to be 55 lbs. per inch of width, which is equal to a velocity of about 50 square feet per minute per horse power, which is safe practice.

Now let C = circumference in inches of pulley,
D = diameter in inches of pulley,
R = revolutions per minute,
W = width of belt in inches,
H = horse power that can be transmitted by the belt.

Then, to find the horse power that a single belt can transmit, the size and speed of pulley and width being given, the formula would be:

$$\frac{C \times R \times W}{144 \times 50} = H, \text{ or } \frac{C \times R \times W}{7200} = H,$$

or we may still further simplify the process by substituting D for C and divide the constant 7200 by 3.1416, which is the proportion of circumference to diameter. The formula would then be $\frac{D \times R \times W}{2300} = H$.

The transmitting efficiency of double belts of average thickness is to that of single belts as 10 is to 7, therefore for double belts the formula would be $\frac{D \times R \times W}{1575} = H$.

The horse power to be transmitted, and the size and speed of the pulley being given, to find the width of belt required:

For single belts—$\frac{H \times 2300}{D \times R} = W$. For double belts—$\frac{H \times 1575}{D \times R} = W$.

The horse power, speed of pulley, and width of belt being given, to find the diameter of pulley required:

For single belts—$\frac{H \times 2300}{R \times W} = D$. For double belts—$\frac{H \times 1575}{R \times W} = D$.

The horse power, diameter of pulley, and width of belt being given, to find the number of revolutions required:

For single belts—$\frac{H \times 2300}{D \times W} = R$. For bouble belts—$\frac{H \times 1575}{D \times W} = R$.

In the rules I have assumed that the belts are open, the pulleys of equal diameters, and the arc of contact is the semi-circumference. If, however, the pulleys are of different diameters and the arc of contact is less than the semi-circumference, the rules must be modified accordingly. The width of a belt required for any work depends on three conditions: 1st, the tension of the belt; 2nd, the size of the smaller pulley and the proportion of the surface touched by the belt; 3rd, the speed of the belt. The average strain under which leather will break has been found by many experiments to be 33,200 lbs. per square inch of cross section. In use on pulleys, belts should not be subjected to a greater strain than one-tenth their tensile strength, or about 330 lbs. to the square inch of cross section. This will be 55 lbs. average strain for every inch in width of single belt ⅙ of an inch thick. The strain allowed for all widths of belting (single or double) is in direct proportion to the thickness of the belt. This is the safe limit, for if a greater strain is attempted the belt is likely to be overworked, in which case the result will be an undue amount of stretching, tearing out at the lace holes, and damage to the joints.

The working adhesion of a belt to the pulley will be in proportion both to the number of square inches of belt contact with the surface of the smaller pulley, and also to the arc of the circumference of the pulley touched by the belt. This adhesion forms the basis of all right calculation in ascertaining the width of belt necessary to transmit a given horse power. A single belt ⅙ of an inch thick, subjected to the strain which I have given as a safe rule (55 lbs. per inch in width) when touching ½ of the circumference of the pulley, will adhere ½ lb. per square inch of the surface contact; or if the belt touches ¼ the circumference of the pulley, the adhesion will be ¼ lb. per square inch of contact, and so on.

Mr. Evan Leigh, C.E., of Manchester, Eng., gives the following rule for finding the horse power that any given width of double belt is capable of driving: Multiply the number of square inches of belt contact on the smaller pulley by one-half the velocity of the belt in feet per minute and divide the product by 33,000, and the quotient will be the horse power. Mr. Leigh also gives a rule for finding the proper width of double belt for any given horse power: Multiply 33,000 by the horse power required and divide the product, first by the length of contact in inches on the smaller pulley, and again by one-half the speed of the belt, the quotient will be proper width of belt.

Now, if these rules (which the author devised some 20 years ago) can be compared with the single straps as at present used in mills, it will be found that they considerably overshoot the mark ; yet single belts, being so much weaker and more liable to stretch than double ones, ought to have less strain upon them. The secret of wide double driving belts running so mysteriously long without attention will at once be seen, when it is considered that single belts are generally made to do two or three times more than they ought to do for their width and speed.

For existing establishments where it is not convenient to alter the speed of shafting or size of drums, in driving machines with single straps, the following rule will come nearer to actual practice : Multiply 33,000 by the horse power required and divide the product, first by the length in inches covered by the belt on the smaller pulley, and again divide by the speed of the belt in feet per minute ; the last quotient will be the proper width for a single belt.

This, and more than this, is what single belts are made to do when driving machinery. Comparatively, then, the strong double belts, working as per first rule, have exceedingly light work, which can be done with great ease while running in a slack state. Hence their durability, and the nearer a user of belts can approach the rule given for double belts, the longer his straps will last.

To determine the strength and size of a belt, find first the amount of labor to be performed by it. This labor is its tension with velocity. If a belt passes over a 3 foot pulley which makes 100 revolutions per minute, its velocity will be: 100 x 3 x 3.1416=942.48 revolutions per minute. Now, if this belt is to transmit 2 horse power, its tension on

the pulling side will be : $\dfrac{2 \times 33,000}{942.48}$ =70 lbs. In this case it is assumed

that one side of the belt is slack ; if this is not the case (which in the average of practical instances may be depended on), the tension on the following side of the belt is subtracted from the above. We here see of how much more service the horizontal belt is than the vertical one, for it increases the tension by its own weight and also by the arc of contact. In most of these cases we may neglect the width of the pulley in the calculations of friction ; for the strength of the belt, if sufficient to stand the tension, makes the belt wide enough for adhesion. In all cases it is advisable to make the belt sufficiently wide. No other loss arises from too wide a belt than that of first cost. If a belt is too narrow or the arc of contact too short, the tension must be increased in order to afford sufficient adhesion to the pulleys.

Short belts are very disadvantageous and so are vertical ones ; they always require more tension than either long or horizontal ones. Those which are too narrow will stretch, in consequence of which tension and adhesion are diminished.

The adhesion of leather upon smooth surfaces is greater than upon rough surfaces, and for this reason pulleys ought to be made perfectly sound and smooth. Frequently we see the surface of pulleys convex in order to prevent the running off of the belt, but this convexity must be very small, or it will diminish the adhesion.

It is of great importance that a belt should be of such a length that it will adhere to the pulley enough to prevent it from slipping without the necessity of putting on the belt so tight as to wear the bearings. Every belt, to run easy and well, should be so slack when running that the slack side should run with a wavy, undulating motion, without any tension except on the working side ; and when belts will so run without slipping on the pulleys, they wear for a great length of time, for although a belt may be heavily loaded, yet if at every revolution it can have an opportunity for relief from its tension so as to contract back to its natural texture, it will prevent it from breaking by the stress upon it. But if it be kept constantly strained to its greatest extent on both sides of the pulleys it will wear but a short time and will soon be destroyed.

## A NEW REDUCING GEAR FOR INDICATOR USE.

THE readers of mechanical publications will have observed that in late years much attention has been paid to the details connected with

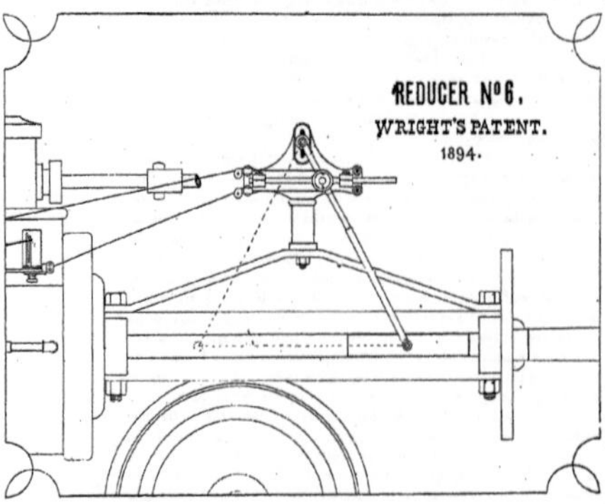

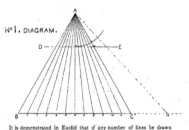

**REDUCER Nº 6.**
**WRIGHT'S PATENT.**
**1894.**

the use of the steam engine indicator. All of the usual forms of reducing gear have been condemned, some from error in principle and some from error developed in use. Many engineers have contented them-

selves with a pine lever swinging on a wood screw, and some apparently thought that any backward and forward movement was good enough to communicate the motion of a piston to the paper barrel of an indicator.

The development of the compound engine was probably the first event which directed the close attention of engineers to this subject. A certain amount of possible error may be tolerated in engines of small size, but in the case of engines of such size and power that a difference of one pound in the mean effective pressure made a difference of 75 or 100-horse power in the work done, it became evident that accuracy in every detail was essential in order to arrive at reliable conclusions.

Following the application of the steam engine for generating electric currents, the electrical engineer in establishing the relations between the first mover and electric energy, required a precision and durability which at first was considered impossible of attainment. It must be acknowledged that the makers of indicators have kept abreast of the demands, while the reducing gear remains a poor complement.

The pantograph and lazy-tongs were in favor for a long time, but now it would be a rare thing to hear an observant engineer say a word in their favor. As a geometrical question they are correct in principle but short lived in use, losing truth from the effects of wear in the numerous joints, and in best condition they are unfit for high speed in revolutions.

In the improved device, herewith illustrated, the reduced reciprocating motion for operating the paper barrel of an indicator is obtained from a sliding rod, which is actuated by a lever swinging on a fulcrum, the other end following the motion of the cross-head or other suitable point. The conditions under which a swinging lever must act on a sliding rod in effecting a true reduction of the motion of a piston is best illustrated by the diagram.

**Nº 1. DIAGRAM.**

It is demonstrated in Euclid that if any number of lines be drawn from a point A to the line B C, then any line D E drawn parallel to B C cuts all these lines in the same proportion or in constant ratio. In the diagram, A represents the fulcrum of the lever, B C the motion of the crosshead during a stroke, A B, A1 and so on to A C, successive positions of the lever during a stroke, and D E the axis of the slide rod which must be parallel to B C. The part of the line D E, between the lines A B and A C, is the reduced motion on the line D E, corresponding to the stroke B C.

To put the above in a practical form the first condition is, that during the stroke the lever is free to alter in length, as required by the varying distance between the fulcrum and the driving point on the crosshead. This is accomplished by a rod sliding in a tube, the common telescopic connection.

**Nº 2.**

fig. 1.
fig. 4.
fig. 2.
fig. 3.
fig. 5.

Figures 1 and 2 are front and side view of the lever. The pin (Fig 4) connects the tube end of lever with the crosshead, the fulcrum pin (Fig. 5) passing through the eye on end of rod. In Fig. 3 the fulcrum is intermediate.

The second condition is that to maintain a constant ratio in all positions between the length of the lever and the distance between the fulcrum and the axis of the slide rod, the point on the lever which actuates the slide rod must shift its position during a stroke and always be on the line D C or some line parallel to it. This is effected by a sliding contact of the lever, with a cylindrical piece called the rocking slide carried in a case on the slide rod.

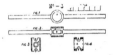

Cut No. 3 is the slide rod ; Fig 1, a front view, showing the circular case A ; Fig. 2, a plan ; Fig. 3, an end view ; and Fig. 4, a section of the case.

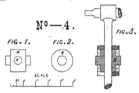

Cut No. 4 is the rocking slide R. Fig. 1 is a side and Fig. 2 an end view. Fig. 3 is a section of the case and rocking slide, with the fulcrum end of the lever in position.

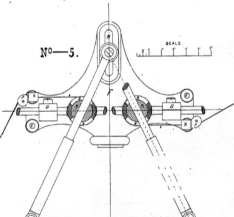

Cut No. 5 is the frame F, carrying all the moving and stationery parts in position, with the lever in two positions and case and rocking slide in section. As seen in cut 5, the slide rod is held in position by the slides D D as shown in cuts 4, 5 and 6, the rod of lever passes diametrically through the hole H in the rocking slide, and the slots B B in the case. The lever is free to slide in the rocking slide ; the slots permit the required angularity of the lever during a stroke, and the consequent roll of the rocking slide in the case.

Referring to diagram, cut 1, all points on the rod of lever when working describe arcs with A as a centre. The point K on the lever coincided with the line D E when the lever stood in position A6. In the position A C the point K is at L, the rod having slid the distance L M through the rocking slide and case, and the point on the lever actuating the slide rod has been transferred on the lever from L to M, and in all positions is found on the line D E with the axis of the rocking slide intersecting the same point.

In this manner the geometrical conditions to communicate a true reduction of the motion of a piston during a stroke to a sliding rod is complied with, and the velocity ratio between crosshead and slide rod is constant and invariable. On diagrams taken by this motion the points

of admission, cut-off, exhaust and compression are accurately laid down, the form of the expansion curve is not distorted and can with confidence in the results be compared with a hyperbolic, a saturation, or an adiabatic steam curve.

The fulcrum is adjustable in the slot N, cut 5. By this means long or short diagrams can be made as desired.

It is preferable, but not necessary, that the lever at mid-stroke be at a right angle to the piston rod. In diagram, cut 1, the line A12 is cut by D E in the same ratio as the others.

In conjunction with the adjustment of the fulcrum in the slot N the lever may be of any convenient length, but in its position of least length should not be less than ⅞ of the stroke. In cases where extreme variations in the length of stroke are to be met the tube of lever is made in two or more pieces, buckled together in the ordinary manner.

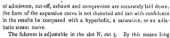
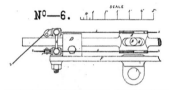

Cut No. 6 is a detail of the string operation. Strings S S may be led from the ends of the slide rod, but preferably from the case. Referring to Fig. 3, in cut 4, it will be seen that the case is wider than the contained rocking slide, and that from circumferential grooves cut on the exterior string holes are drilled to the interior where it is not covered by the rocking slide. The strings are held by knots in the interior of the case, and may if required be led in the groove to a point where it leaves the case, at a tangent parallel to the slide rod. This parallelism is constantly maintained on the frame F by pulleys P turning on a stem that is held in standards X set in the holes H in the frame, the axis of the stem and the groove of sheave coinciding with the line of the string. When properly set, the strings after passing the sheaves go direct to the indicator at any angle. In this manner four or more indicators can be operated at the same time from one reducing gear. In the case of engines which should not be stopped reserve strings may be reeved in empty holes and used if a working string breaks. In cases where the strings would be objectionably long, and errors or irregularities be introduced by its deflection or unsteadiness, the slide rod is lengthened to the extent required with a piece of cold drawn tube buckled to the end of the slide rod.

It will be observed that the device between the operating strings and the crosshead is a combination of three rocking and four sliding contacts, and contains in a straight line the principle of the pantograph, with true action limited to parallel lines.

It worked steady, and diagrams were of uniform length at slightly over 400 revolutions per minute, the highest it has been applied to. It is in use in the largest steam using establishments in this country, and is peculiarly adapted for lighting and power stations and locomotives.

There are four permanently erected on engines in the lighting and power stations of the Royal Electric Co., Montreal, where they have worked day and night as required for the last five months.

The elevation of design No. 6 shows it erected on a locomotive with a four bar slide. Manufactured at the Reliance Works, Darling Bros., Montreal.

The Dunnville Electric L'ght Co., Limited, have decided to erect a power station, and to install an incandescent lighting plant of 1,000 lamps capacity. The last annual statement of the company shows a surplus of $7,117, equal to 13 per cent on the capital.

The municipality of St. Laurent, has granted a thirty years exclusive lighting and railway franchise to the Montreal Park & Island Railway Co. The company guarantee to have an electric road in operation by the end of the present year. St. Laurent is one of the largest municipalities on the island of Montreal, embracing an area of over 54 square miles

# ELECTRIC RAILWAY DEPARTMENT.

### AN ELECTRIC SHUNTING LOCOMOTIVE.

Whatever opinion may be held of the applicability of electricity to heavy railway work at the present time, there can be no question that it is eminently suitable for light service. We have on several occasions emphasized its advantages for light shunting work or for conveying goods cars to the main lines from manufactories or other points of supply situated at some distance therefrom ; and in the issue of this journal for September, 1893, we described two electric locomotives constructed for such service in America. At that time there was, we believe, no similar service performed in this country ; but we are now able to give some particulars regarding a plant that for some months has been constantly employed at the textile machinery works of Messrs. Tweedales and Smalley, Castleton, from whom we learn that the entire installation has worked admirably. Although the service is very light, it is hardly necessary to point out that, were it required, a locomotive of much greater power could be provided ; and wherever locomotive power is needed at private works possessing a stationary power plant, there is probably no method of obtaining motive power more conveniently and economically than by the addition of an electric generator and the construction of an overhead wire system. The interest of the arrangements now to be described arises chiefly from the hint which they convey of possibilities, for the locomotive is designed to draw a loaded wagon not exceeding twenty tons weight at a speed of about two miles an hour. It is used for shunting wagons on a siding connecting the boiler house and delivery stores of the textile machinery works with the main line. The electric generator, which is of the patent Manchester type, is designed for an output of 100 volts, 54 amperes, at 1,100 revolutions per minute, and is driven off the main shafting in the works. It is fitted with fast and loose

ELECTRIC SHUNTING LOCOMOTIVE.

pulleys, so that it can be stopped when not required. The current is conveyed from the generator along the top of the engine-house to the overhead wires, and returns through the rails, which are bonded with copper strips and rivets.

The locomotive, which has been designed and constructed by Messrs. Mather and Platt, Limited, of the Salford Iron Works, somewhat resembles an ordinary goods wagon. It is fitted with coil spring buffers of the standard height and centres, axles boxes, and guides, and hand screw breaks with wooden break blocks bearing on the car wheels, which are 28 inches in diameter. The locomotive is roofed in with galvanized corrugated iron carried on wrought iron pillars. These continue through the roof and carry the collector bars.

The driving motor is also of the Manchester type, and is mounted on a cast iron bed-plate which slides on cast iron brackets bolted to the framing of the car. The motor is fitted

with a vulcanized fibre pinion with steel end plates of 21 teeth, which gears with a cut cast iron wheel of 72 teeth on the gudgeon shaft, on which is keyed a chain pinion of seven teeth, driving a chain wheel of 22 teeth. The latter is split into two halves, and is fitted to one axle of the locomotive. A sand box is provided, and the car is fitted with a controlling switch, resistance box for starting and regulating the speed, and a reversing switch. The weight of the locomotive is 3 tons, 0 cwt., 9 qrs.

The system of collectors on the locomotive lends itself particularly well to the requirements of this line, as there are many points, curves, and crossings. The system consists of two wrought iron bars placed about 6 ft. apart, one of which is always rubbing on the under surface of the overhead wire. This system is controlled by Messrs. Mather and Platt's patents, and has lately been used on an extended scale on the Douglas and Laxey Electric Tramway, of which a description was given in a recent issue of this journal.

As the locomotive is only required for two or three hours each day at odd times, it will be at once seen that a very great saving is effected by using an electric locomotive in place of steam, as the generator is simply started whenever it is required. As an ordinary labourer looks after the entire plant, including an overhead electric travelling crane, supplied by Messrs. Mather and Platt, to lift three tons, and worked from the same generator dynamo as the locomotive, the cost for attendance is also small.—Railway World.

### SPARKS.

Several extensions will be made to the lines of the Ottawa Street Railway Co., during the present year.

A movement is said to be on foot to extend the Peterborough Electric Street Railway to Lakefield, a distance of nine miles, using the G. T. R. Co.'s track.

Mr. Fowler, of Carleton Place, who is the chief promoter of the Perth & Lanark Electric Railway scheme, states that the enterprise will shortly assume practical shape.

The Metropolitan Railway Co., of North Toronto are seeking power from the Legislature to operate their system on Sunday, for the alleged purpose of carrying milk to the city. It is said to be the intention of the company to extend their line during the present year, and to actively engage in the handling of freight for the Toronto market.

As foreshadowed in our last number, a company to be called the Halifax Electric Tramway Co., Limited, has been incorporated to acquire the ownership of the Halifax Street Railway system. The promoters of the company are: H. M. Whitney, of the Dominion Coal Co., G. B. M. Harvey, Boston ; James Ross, Montreal ; M. Dwyer, David McKeen, J. Y Payzant, Allan Haley, Thos. Lynch, A. Burns and W. C. Ross, Q. C. Nova Scotia.

The New Westminster & Bying Inlet Telephone Co. of Vancouver, B. C., of which Mr. H. W. Kent is manager, have secured control of the telephone system at Rossland, B. C. It is the intention to extend the system to all the leading mines in the neighborhood, and there is a possibility that a line will be built in the near future to Northport. A new building is under construction, and is about ready for occupation. The local management will be in the hands of Mr. G. A. Smith.

## ENGINEERING NOTES.

It is of course necessary to have a set of heavy fire tools in every boiler room for the purpose of handling heavy fires, but there should also be a set of light tools there, for in many places the latter may be used to advantage during a large portion of the time, thus saving much labor on the part of the fireman. Do not compel him to use a hoe that weighs 75 pounds, more or less, to draw the ashes out of the ash pits, when a much lighter one will answer every purpose.

When buying gaskets with which to pack man-hole or hand-hole covers on steam boilers, be careful to select those that are soft and tough, and not too thin, for the inside of the heads where these are to be used, and also the covers themselves are frequently anything but smooth and true, and the gaskets must "fill the gaps" as it were.

It is a good idea to have extra man-hole cover guards on hand, so that if one is broken on Sunday or some holiday when it may be difficult to procure another, no loss of time will be necessary. Especially should this be attended to in plants that are located at a distance from foundries and machine shops.

In case of accident to the feed pump, or any part of the boiler which makes it necessary to reduce the temperature at once, it is much better to cover the fire with damp ashes or fresh coal, rather than to attempt to draw it, for when a fire is disturbed it gives out an intense heat for a few minutes.

It is a good idea to be as economical as possible in the use of oil, but it does not pay to attempt to run an engine with an insufficient quantity of cylinder oil, for not only will the cylinder be ruined, but you will use extra oil enough to much more than pay for all the cylinder oil needed.

Always have a sight feed oiler located where it will drop oil on the piston rod as it travels back and forth, for it lessens friction, saves wear on the rod, and makes the packing last much longer. This applies to both fibrous and metallic piston rod packing.

In laying out the holes in a belt for the lacing, do not get them too near together, for while this practice makes the finished lacing stronger, it makes the belt weaker on account of the large amount of material cut away in making the holes.

After cleaning boilers do not screw up the nuts on the man-hole and hand-hole covers any tighter than is necessary, for you may break the guards or dogs that hold the covers in place, and cause yourself much trouble.

When wiping up the engine be constantly on the watch for loose set-screws, keys, nuts and pins, for by attending to this simple matter, many an expensive shut down has been avoided.

When fitting grate bars to a furnace do not make them too tight a fit, for expansion by heat must be provided for, or else the bars or furnace will be ruined.

When setting a boiler, pieces of common steam pipe, say about one inch in diameter, should be built into the outside walls in such a way that they will allow the air in the space between the two walls to escape when the heat expands it, and also allows it to enter this space when the boiler cools off.

Try gauge cocks often and keep them in perfect order, for you cannot tell how soon the gauge glass will leave you in the lurch, unless you have them to fall back on.

Asbestos packing for valve stems and similar purposes is much improved for use by oiling it well with cylinder oil before putting it into place.—Power and Transmission.

## PUBLICATIONS.

The March "Arena" is a good representative number of this alive and progressive magazine, which, whatever may be said of contemporary literature in general, is certainly showing no decline in vitality and virility with the progress o the New Year.

Students of telegraphy will be interested in knowing that a useful work on their behalf has lately been published at 29 Ludgate Hill, London, E. C., entitled "The Telegraph Guide to the New Examinations in Technical Telegraphy," by James Bell, A. I. E. E. The questions therein will be found useful by the student in testing his knowledge. The price of the book is 1 s. 6d.

"The Engineers' Annual" is the title of a hand-book for Canadian marine engineers, issued by the Canadian Marine Engineers' Association. The name of Mr. O. P. St. John appears on the title page as having been the compiler of the book, which we regard in itself as a guarantee of the accuracy and value of the contents. The book is meeting with a ready sale at $1.00 per copy.

### MOONLIGHT SCHEDULE FOR APRIL.

| Day of Month. | Light. | | Extinguish. | | No. of Hours. |
|---|---|---|---|---|---|
| | | H.M. | | H.M. | H.M. |
| 1 | P. M. | 11.20 | | ........ | } 5.20 |
| 2 | | | A. M. | 4.40 | |
| 3 | A. M. | 1.20 | " | 4.40 | 3.20 |
| 4 | " | 2.00 | " | 4.40 | 2.40 |
| 5 | " | 2.40 | " | 4.40 | 2.00 |
| 6 | " | 3.00 | " | 4.40 | 1.40 |
| 7 | No light. | | No light. | | .... |
| 8 | No light. | | No light. | | .... |
| 9 | No light. | | No light. | | .... |
| 10 | P. M. | 7.00 | P. M. | 9.20 | 2.20 |
| 11 | " | 7.00 | " | 10.30 | 3.30 |
| 12 | " | 7.00 | " | 11.40 | 4.40 |
| 13 | " | 7.00 | A. M. | 1.00 | 6.00 |
| 14 | " | 7.10 | " | 1.40 | 6.30 |
| 15 | " | 7.10 | " | 2.30 | 7.20 |
| 16 | " | 7.10 | " | 3.00 | 7.50 |
| 17 | " | 7.10 | " | 3.30 | 8.20 |
| 18 | " | 7.10 | " | 4.00 | 8.50 |
| 19 | " | 7.10 | " | 4.20 | 10.10 |
| 20 | " | 7.10 | " | 4.20 | 10.10 |
| 21 | " | 7.10 | " | 4.10 | 10.00 |
| 22 | " | 7.20 | " | 4.10 | 9.50 |
| 23 | " | 7.20 | " | 4.10 | 9.50 |
| 24 | " | 7.20 | " | 4.10 | 9.50 |
| 25 | " | 7.20 | " | 4.10 | 9.50 |
| 26 | " | 8.00 | " | 4.00 | 8.00 |
| 27 | " | 9.10 | " | 4.00 | 6.50 |
| 28 | " | 10.20 | " | 4.00 | 5.40 |
| 29 | " | 11.00 | " | 4.00 | 5.00 |
| 30 | " | 11.20 | " | 4.00 | 4.40 |
| | | | Total, | | 170.10 |

Where the valve stems of Corliss or other engines are fitted with stuffing boxes for fibrous packing, it should be renewed at intervals and not be allowed to remain in place until it becomes so hard that it will no longer do the work for which it was intended.

# CANADIAN GENERAL ELECTRIC CO.
## (LIMITED)

THE problem of successful transmission of power to distances of from five to twenty-five miles or over has been solved by the introduction of

# THE THREE PHASE SYSTEM

More than twenty-five plants of this description, aggregating thousands of horse power in capacity, have been installed within the past eighteen months with uniformly satisfactory results.

# ALTERNATING REDUCTION MOTORS
### SIMPLE! COMPACT! DURABLE!

These motors, after a most satisfactory preliminary experience extending over two years, have now been standardized in sizes from 1 to 150 horse power, and are placed on the market with the fullest confidence in their ability to meet the most exacting requirements of electric power service. **They are equal in starting, torque and efficiency, and superior in regulation to the best shunt wound direct current motors.**

In operation they require a minimum of attention, having no starting box and being without brushes, commutators or moving contacts of any kind.

## SPARKS.

Work is proceeding on the construction of a telephone line from Moncton to Hopewell Cape. N. B.

A company is said to be in progress of formation, to supply electric light to the Parish of St. Rumuxid, Que.

The Mattawa Electric Light & Power Co., Limited, has been empowered to increase its capital stock from $10,000 to $30,000.

It is said to be the intention of the Kingston Light, Heat & Power Co., to make considerable additions to their steam and electric plant.

The town council of Magog, Que., are negotiating with the Dominion Cotton Mills Company, to supply electric light for the streets of that place.

An offer of $175,000 for the St. Catharines & Niagara Central Railway, is said to have been made recently on behalf of the Hamilton Radial Railway Co.

The Water & Light Committee of the city council of Vancouver, B. C., have resolved to call for tenders for both incandescent and arc lighting, for terms of five, seven or ten years.

It is said to be the intention of the Welland Electric Light & Power Co., to remove their plant to the American side of the river, for the purpose of obtaining cheap power from the hydraulic canal.

Mr. J. H. Eckert, formerly local manager of the Bell Telephone Co., at Brantford, has been transferred to Windsor, and Mr. D. Roberts, of St. Catharines has been appointed local manager at Brantford.

Incorporation is being sought for by the Sault Ste. Marie Pulp & Paper Co. Amongst other things, the company ask for power to construct and operate an electric railway or railways. The capital stock is placed at two million dollars.

It is reported that the necessary equipment is being purchased for the construction of a new metallic telephone circuit between Montreal and Toronto, which will be in every way the equal of the famous long distance line between New York and Chicago.

The Niagara Falls, Ont., Electric Light Co., is said to be asking for a ten years franchise, in consideration of which they will agree to furnish arc lights at $28 50 per annum, instead of $35.00 as at present. The company's present contract with the municipality has yet three years to run,

Mr. J. C. Mullin was presented with a purse of gold by his friends in Ottawa, previous to his departure for Valparaiso, Chili, where he proposes to engage in the electrical business. Mr. Mullin was foreman of the Ottawa Car Works.

Incorporation has been granted to the Victoria, Vancouver & Westminster Railway Co., for the purpose of constructing a line from near Garry Point on the Fraser river, through Richmond South, Vancouver and Burnaby to Westminster.

A charter of incorporation has been granted by the Ontario Legislature to the London Radial Electric Railway Co. which proposes to build lines to Woodtock, St. Thomas, Port Stanley, Aylmer, Strathroy, Delaware, Lucan, St. Mary's and other places.

The Canadian Locomotive & Engine Co., of Kingston, are understood to be negotiating with one of the largest electrical supply companies in Chicago, for the establishment of a Canadian branch at Kingston. The company have also secured the right from a German firm to manufacture gas engines.

The annual report of the Niagara Falls Park & River Railway Co., for the year 1894, shows the recel ts to have been $62,481, as compared with $58,064 the previous year. The operating expenses, (including $10,000 for advertising), were $42,994, as compared with $34,196 in 1893. The number of passengers carried during the year was 479,710.

The Berlin and Waterloo Street Railway is to be immediately converted to the electric system. Mr. E. Carl Breithaupt, of Berlin, has been appointed consulting engineer for the work, and is at present preparing the necessary plans and specifications. A contract has been made with the Berlin Gas and Electric Company to supply the power for the operation of the road.

Robert Conroy and John C. Nelson, Aylmer, Que., Wm. Jackson, Dennis, of Ottawa, will apply for incorporation under the name of " The Deschene Electric Company," for the purpose of constructing and operating works for the production of electricity for light, heat and power, and to distribute and sell the same in the town of Aylmer, and in other places in the county of Ottawa, and in the township of Nepean and the city of Ottawa, in the province of Ontario. The headquarters of the company is to be at Aylmer, and the capital stock to consist of $60,000, divided into 600 shares of $100 each.

The promotors of the Hamilton Valley City & Waterloo Railway are said to have had much of the necessary surveying for the line done, and arrangements made to commence construction immediately, should the charter for which application has been made to the Legislature be granted the company.

Application has been made to the Ontario Legislature for the incorporation of the Niagara Falls Electric Street Railway Co , with a capital stock of $250,000 to construct a street railway and to supply electricity, for light, heat and power in the municipalities of the town of Niagara, the village of Niagara and the town of Stamford, Ont.

The corporation of Bracebridge have recently exchanged the 1,000 light alternator installed by the Canadian General Electric Co. last fall for a 2,000 lighter of the same type. This increase was necessitated by the phenomenal increase of their private lighting, the number of incandescents now installed in the town, with a population of about 2,000, being something over 1,800.

The Mattawa Electric Light and Power Co, have decided to establish an alternating power service, and for this purpose have purchased from the Canadian General Electric Co., a one hundred horse-power generator of their new monocyclic type. The power will be transmitted from the company's power station at McCools' Mill, a site to the town of Mattawa, a distance of some three and one half miles.

Mr. C. H. Stickles has presented to the City Council, of Victoria, B. C., a legenthy report on the present condition of the new electric lighting plant. The report goes to show the the lighting system has in many respects been poorly constructed and loosely managed, and that owing to the fact that the council expended $7,000 above the estimate on a site for the power station, the capacity of the station has been reduced by 65 lights, and it has been necessary to retain two systems of lighting—the Ball and Wood—instead of one. The report says no system has been hitherto followed in handling supplies, which has resulted in unnecessary waste and destruction of property, and suggestions are made to remedy this defect.

# CANADIAN ELECTRICAL NEWS
## STEAM ENGINEERING JOURNAL

OLD SERIES, VOL. XV.—No. 6.
NEW SERIES, VOL. V.—No. 5.

MAY, 1895

PRICE 10 CENTS
$1.00 Per Year.

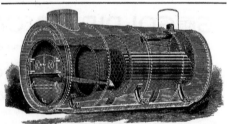

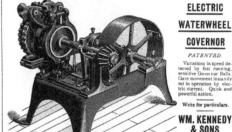

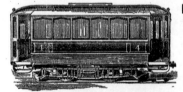

CANADIAN

# ELECTRICAL NEWS

AND

## STEAM ENGINEERING JOURNAL.

| VOL. V. | MAY, 1895 | NO. 5. |

## CHARACTER SKETCH.

### CHAS. E. A. CARR.

MANAGER AND TREASURER LONDON STREET RAILWAY CO.

"Either I will find a way or make it."—Norseman Motto.

" To let you know that I live"—a favorite expression, which might properly be employed as a motto by him—affords an index to one phase of character in the personality of Manager Carr, which explains the phenomenally rapid rise of this bright young Canadian, to the trusted and responsible position which he holds to-day, and which, all who know him well, regard as a stepping-stone to higher honors in the broad field of railway enterprise. Mr. Carr is not to the manner born. It cannot be said that he grew up with a railway, or in a railway office. His advancement is due to sheer natural merit and adaptability, and the intuitive recognition of these qualities by Mr. H. A. Everett, of Cleveland, O., the well-known street railway promoter, whose interests are identified with great railway corporations in nearly every large city in the United States and Canada.

Sir John A. Macdonald was wont to say that his greatest discovery was Sir John Thompson. It was Mr. Everett who discovered Mr. Carr, while the latter was employed as a clerk in the City Engineer's office in Toronto, and Mr. Everett claims him as one of his ablest and most trusted associates to-day—one of his best finds.

Mr. Carr was born Feb. 8th, 1870, a little over 24 years ago, on his father's farm near Barrie, in Simcoe County, Ontario, and he received his early education in the public schools of that town. When sixteen years of age he came to Toronto and passed through a three years' course in one of the leading commercial colleges. An initial experience in practical office work in the " Mail Order " department of the T. Eaton Co. lasted but a few weeks, when he accepted a position in the office of W. T. Jennings, then City Engineer of Toronto.

Mr. Carr's active career may be said to date from that time. His genial frankness and manly sociability soon made him one of the most popular attaches of the office, and a conscientious fidelity to duty, combined with an exceptionally high order of ability, won him the confidence and generous approbation of his superiors in office. During his stay of nearly three years at the City Hall, Mr. Carr acquired a thorough general insight into routine work in one of the most important branches of the municipal service, and a practical knowledge of civil engineering.

The investigations for the transfer of the Street Railway from the Frank Smith Company to the city, which was chiefly carried on through the City Engineer's office, was brought on while Mr. Carr was connected with the department, and first brought him into familiar association with street railway matters. He was one of the party of City Hall attaches delegated to make the memorable midnight demand for the surrender of the railway to the city on April 30, 1891. Later in the summer of that year,

MR. CHAS. E. A. CARR.

when the railway was turned over by the city to the present company, Mr. Carr was assigned to checking over the stock, tools and equipment, which went with the transfer.

In January, 1893, Mr. Carr left the employ of the city to accept the position of private secretary to Mr. H. A. Everett, then vice-president and general manager of the Toronto Railway Co., and shortly afterwards the greater portion of the Toronto Railway system was converted from horse to electric power. During this time and through 1894, Mr. Carr acquired the practical experience in the installation and operation of an electric railway, which he is now putting to good account in the London Street Railway system.

His appointment as manager and treasurer of the London Street Railway Company was made Feb. 11, this year, and he is now engaged on the conversion of that railway from horse to electric power, negotiations having been going on between the Company and the city of London for some months.

The new system in London will cover over 13 miles of streets within the city limits, and will also include an extension of the line to Springbank, a delightful summer resort on the River Thames, some three miles below the city, popularly known as the Water Works Park, and which is visited by many thousands every season.

In railway circles Mr. Carr bears the distinction of being the youngest railway manager in America, and his success affords a present illustration of the prominent place occupied by young men in business, as well as in many other walks of life. But it is with Mr. Carr, as with everyone who would succeed in life, be they young or old, a definite and earnest purpose must follow them in everything that they do. Those who best know this young street railway manager know that " he lives," and those whose business dealings are kept in touch with his tactful energy and enterprise, are quick to perceive why he has made a success, even in his short life-time, of whatever he has put his hand to.

Mr. Carr is married and an active member of the Methodist church, and though but twenty-four years of age, has already won a measure of self-made financial success, which would be reckoned remarkable, gauged even by modern ideas of affluence.

A NEW contrivance for scraping boiler tubes has been introduced in Australia. Hitherto the appliances in general use have been worked by spring expansion, and are soon rendered useless. The new scraper works on a hinge, which is closed as the cleaner is pushed into the tube. In the act of drawing out, the hinge is opened, and two disks, which can be gauged to fit any tube, are thrown out. These disks fit so closely to the inside of the tube that their passage causes the removal of all dirt and scale. It is claimed that the new apparatus is cheap, that it cleans the tubes more effectively than any other scraper, that no brushes are required with it, and that it pulls soot, etc., before it, instead of pushing it into the fire-box at the back end of the tubes. It is stated that the scraper has been found especially serviceable on board large steamers.—The Age of Steel.

## DYNAMO DESIGN.*

By E. B. MERRILL.

*(Continued from April number.)*

### THE MAGNETIC CIRCUIT.

WE now come to the consideration of the magnetic circuit which produces the fields to be cut by the armature conductors. There are a great many different types of magnetic circuits. Dynamos are classified by the forms and arrangements of these. They can only be utilized for dynamos by conductors cutting through gaps in them. The rest of the circuit is, therefore, only of use as it provides the fields. Electric current circulating in a continuous direction, in coils which surround the material of the magnetic circuit, is necessary to produce and maintain the magnetic flux. The excitation is proportional to the product of two terms, the current and the number of times it encircles the flux or to $NC$. Only the current involves energy, so that we are able to reduce the energy waste by increasing $N$.

As they are inter-related we may consider together the questions : How do we adjust the proportions of the magnetic circuit? what excitation will be needed for it? and how shall this be provided?

They are, in fact, the discussion of the equation

$$N C = \frac{10}{4\pi \times 2.54} \, \phi \, \frac{l_1}{S_1 \, \mu_1} + \frac{l_2}{S_2 \, \mu_2} + \frac{l_3}{S_3 \, \mu_4} + \&c.$$

where $l_1$, $l_2$, $l_3$, etc., are the average lengths of the magnetic paths in the different parts of the magnet circuits.

$s_1$, $s_2$, $s_3$, etc., the corresponding areas of cross section.

$\mu_1$, $\mu_2$, $\mu_3$, etc., the corresponding permeabilities.

$\phi$ the magnetic flux forced through the circuit by the ampere turns $N C$.

In this equation we have to fix the quantities $\phi$ and for each part of the circuit the term $\frac{l}{S \mu}$. To fix $\phi$ we have already the intensity of the field $\mathcal{B}_a$, its width, which is taken as the length of the pole face, and its depth is calculated from the ratio of polar arc and the circumference of the armature ; so that we have the area of the pole face, which is taken as the cross section of the field $S$ and therefore $\phi = S \, \mathcal{B}_a$ is determined.

We will now consider the terms $\frac{l}{S \mu}$ for (1) the air gaps, (2) the armature core, (3) the pole pieces, (4) the magnet limbs, and (5) the yoke or connecting pieces.

For the air gaps, we have already settled $S$ ; $\mu$ is unity, and $l$ is the distance between armature core and the pole face, which is made up of depth of winding around and clearance. The depth of winding has already been settled. The clearance varies with the diameter, in practice, between 1·32 and 7·16 of an inch. The larger distances are for slotted armatures, being found necessary to prevent sparking ; it should be as small as possible, but there should be safety assured, for the surface of the armature, from touching the pole face. The smallness of $\mu$ makes this the most important term in the calculation. The main reason for making $S$ large or $\mathcal{B}_a$ small is now apparent.

Let us now deal with the armature core. In the first place, it has to be well laminated, for the reason that iron is a good electrical conductor ; so that if the core were made of solid iron, this, cutting the magnetic lines which pass through it, would have the same effect as though conductors on the surfaces were short-circuited, which would waste power if it did nothing worse. The current that would flow in it would be in the same direction as in the conductors ; the lamination, therefore, is to effect discontinuity in this direction. If the lamination is too thick, there will still be formed circuits in it sufficient to cause serious loss. The range of practice seems to be between 10 and 80 mils in thickness. Special insulation is not required between the plates ; the coating of oxide formed by heating the iron is sufficient.

The radial depth of the core is fixed so that, after allowing for air space in the lamination, the total cross-section is sufficient to keep the value of $\mathcal{B}_a$ in it well within the limits of saturation. In bipolar machines the total flux has two paths to take about the centre. In multipolar machines, for each magnetic circuit, it has but one path in the armature. The $S$ for the armature is now fixed, since we know the length and have corrected it for lamination already. We must determine $\mu$ from tables giving the relation between $\mathcal{B}_a$ $\left( = \frac{\phi}{S} \right)$ and $\mu$. $l$ is the average length

of magnetic path through the armature core, and may be estimated. The hysteresis losses in the core are proportional to the number of magnetic reversals and to the 1.6th power of the intensity of magnetization ; for this reason $\mathcal{B}_a$ should be lower as the speed increases, to keep down the heat and the heat losses.

The quantities, $l$, $s$ and $\mu$ are readily estimated for the pole pieces—the pole face has already been fixed. The general design of the pole should be such as to prevent the unequal distribution of the field. They are often made of cast iron, especially in smaller sizes, when the intensity of the magnetic flux carried by them is not great, and therefore the permeability is large. .

The magnet limbs, on the other hand, should be of the best annealed wrought iron, for the cross-section, as it affects the cost of winding, as well as the weight of metal, should be a minimum. It should also be as nearly circular as possible, as this has the least circumference for a given area. The limbs are usually run pretty well up to saturation, so that $\mathcal{B}_a$ and therefore $S = \frac{\phi}{\mathcal{B}_a}$ can now be fixed. For the present, the value of $l$ will have to be estimated. This may be done from comparing similar machines. It is decided later, when we find the space required for winding. $\mu$ is fixed by $\mathcal{B}_a$. If the dynamo is to have field regulation for electro-motive force or speed over any considerable range, then the value of $\mathcal{B}_a$, chosen should correspond with the field needed for maximum pressure or minimum speed, so as to keep the field below saturation.

The cross-section of the yoke or other connecting pieces between the limbs should, at least, be as great as the latter, if of the same kind of iron. It is better to have it somewhat larger, so as to bring $\mathcal{B}_a$ and $\mu$ down. If of cast iron, the value of S, being decided by $S = \frac{\phi}{\mathcal{B}_a}$, would be considerably larger, as the permissible $\mathcal{B}_a$ would be much smaller. It is again the length of the average path of the magnetic lines (not the length of the yoke over all).

We now have all the data for calculating the ampere turns $N C$ necessary to produce the field for the armature conductors to cut. We should find, however, that if we took this value and designed the fields according to the cross-sections, etc., above obtained, and provided windings and current accordingly, that the useful flux that we should actually obtain would perhaps be only ¼ or ⅔, perhaps, even less, of the amount calculated upon.

The explanation is this : air is a magnetic conductor—not a good one, but still it has conductance, and magnetic lines, instead of passing around, and keeping within the bounds of the circuit, run out from the exciting coils, in more or less wide paths through the air, constituting-magnetic leakage. The part of the total flux that does not go through the armature is considerable.

If we were to take a practical example of the magnetic circuit, and calculate the ampere turns, or the magnetomotive force necessary for each part of it, we should see that by far the greatest term would be that for the air gaps ; so that if we consider the magnetomotive force about the magnetic circuit in the same way that we do electro-motive-force about the electric circuit, we see that the drop is proportional to the resistances. The air gap is to the magnetic circuit in very much the same relation that the space between plates suspended in acidulated water is to an electric circuit. The reluctance of the air gap is greater than that of the rest of the circuit, and, therefore, the greatest drop of magnetomotive force takes place over it. Wherever we have difference of magnetic potential in a magnetic conductor, we shall find magnetic lines. The pole pieces have great difference of magnetic potential. There are, we may say, two magnetic conductors between them, the air gaps and armature core as one, and the remaining possible paths as the other. Most of the lines will follow the former path, but only in proportion to its magnetic conductivity as compared with the other. In the same way there will be leakage between the limbs, and between the pole pieces and the yoke; there will be very little between the ends of the yoke. In all cases given the machine the magnetic conductivity of the air spaces is perfectly definite and can be ascertained, and, therefore, the proportion of the lines in a given case that leak through the air and those which are used in the armature can be ascertained.

As all the lines (or nearly all of them) will have to pass through the iron within the exciting coils, and through the yoke, we shall have to increase the cross-sections of these as calculated, if we wish to keep the value of $\mathcal{B}_a$ and $\mu$ the same—adding sufficient to them to take the leakage of the rest of the circuit at the same densities $\mathcal{B}_{a1}$ thereby retaining the flux $\phi$ for the armature.

Doing this will be seen not to affect the value of $NC$ calculated; the increased flux is simply proportional to the increased conductance. The ratio of the total to the useful flux is called the co-efficient of leakage, and ranges from 1.1 in large machines of good design to 2 for very small ones.

The determination of the conductances of the air circuits is rather troublesome, so that if it is not necessary to have very accurate results at first a value for the co-efficient of leakage may be assumed by comparing those of similar types of machines; and an allowance may be made for increasing or decreasing the excitations when the machine is tested.

There are two more items to be considered in providing ampere turns, which, in bad construction or in bad design, may become of considerable importance; they are the effects of joints and the demagnetizing action of the armature. The former becomes of importance when there are too many joints in the circuit, or when their surfaces are not perfectly even and smooth. The latter is due to a certain number of turns in the armature between the pole horns, which actually surround the magnetic circuit, and have a current in the opposite direction to that of the magnet. It is due to the lead given the brushes to prevent sparking. The effect, evidently, varies with the armature current, i.e., with the load. It can be allowed for, for some particular load, or it may be counteracted by compounding.

If the field and speed can be kept constant in a generator, then the electro-motive-force generated will be constant; but the terminal electro-motive-force will drop as the load increases, as we see from the relation $E_1 = E - Cr$. Now, as it is the terminal pressure that must be kept constant, since the speed cannot be very well increased with the load, the machine is compounded; that is, a winding is provided on the magnet which, by taking the armature current (or that, less the shunt field current), is designed to produce an additional flux, which will increase the electro-motive force generated by the amount $Cr$ over as wide a range of load as possible.

A motor is compounded to maintain constant speed when run on a constant pressure circuit. The series winding acts against the shunt, decreasing the counter electromotive force by the part $Cr$ in the relation $E_1 = E + Cr$ as the load increases. As the action of the armature turns, due to negative lead in the motor is opposed to that of the field turns, just as it is in the generator; it may be made use of in the design of the motor to maintain constant speed instead of providing a special series winding.

We may now decide the relations between the turns and current in the fields. We have for the shunt field.

$$NC = k_1$$

Where $k_1$ is the calculated value, we have also the relations

$$C = \frac{E}{R} \qquad \text{and}$$

$$R = k_2 \frac{L}{s}$$

Where $L$ is the total length, and $s$ the cross-section of the wire used, and $k_2$ is the resistance of unit length of copper conductor of unit cross-section. If $L$ is in feet and $s$ in circular mils, then $k_2$ is the resistance of one foot of wire of one mil diameter= 10.381 ohms at 75° F. Again $L = k_3 N$ where $k_3$ is the average length of a turn. Now, if $k_3$ can be considered nearly constant for a fairly wide range of turns of a given wire, then knowing the excitation required must fix the gauge of wire to be used, for we derive from the above equations the relation

$$S = \frac{k_1 k_2 k_3}{E}$$

$k_3$ is estimated at first from the diameter of the spools on which the wire is to be wound, and is corrected by trial, as the space to be occupied by the windings is determined.

We may now select a value for the current to be used, which must be within the safe carrying capacity of the gauge of wire determined; and calculate the number of turns. The adjustment should be made by balancing running loss against cost of construction. As the field losses are rather a matter of absolute than of relative cost, we find that much larger percentages of the total output are used in field circuits in small than in large

machines. They range from about 15 to .015 per cent.

In adjusting the excitation due to the series field of the compound winding, the principal necessity is the knowledge of the properties of the magnetic circuit above the degree of its magnetization by the shunt coils, if for a generator, or below, if for a motor, because if lines are added to those already in the circuit or are taken away, it affects the value of $\mathcal{B}_a$ and $\mu$ throughout, so that the ratio of the increase of $\mathcal{B}_a$ per increase of the magnetizing force must be known.

## LARGE ELECTRIC TRANSMISSION IN NORWAY.

A SYNDICATE has lately applied to the Norwegian government for a concession for an electric power transmission for some 20,-000 horse power from the Raanaas waterfall at Sorum and the Fossum waterfall at Askin, to the town of Christiania. It is understood that the plan of the installation is made by an English electrical engineer, and approved of by Lord Kelvin. It comprises the transmission of 20,000 horse-power to Christiania from the above two falls, but it is proposed to commence with a transmission of 10,000 horse-power. The tension is not to exceed 10,000 volts. The length of the line from the Fossum fall is close upon 25 miles; it passes through Spydelizerg, Tomter, Ski, and East Aker. The other, the Raanaas line, is about a couple of miles shorter, and passes through Sorum, Skedsmo, and East Aker.

At the borders of the borough of Christiania it is intended to build a distribution station, where the high tension current will be transformed and distributed to the various parts of the town by means of underground cables, the proposed tension varying from 100 to 400 volts. The installation cannot only supply convenient and cheap power for the larger and smaller industries of the town, but it is also under contemplation to supply the whole town of Christiania—or at least the part of the same which is not lighted from the central electric station—with electric light at a cheap price. The lighting of the public thoroughfares will, according to the plan, be effected by arc lamps of 2,000 and 1,000 candle-power, according to the importance of the street, the price will be much cheaper than that charged by the present central station.

There will be several large turbines at each waterfall, which are some of the most important in Norway, and the high tension will be generated either direct or by transformers. It is proposed to use naked wires.

The installation can, it is calculated, be completed in the year 1895, although much will depend upon the water level of the waterfalls, as the works in connection with leveling, damming, etc., require low water. The cost is calculated as 6,000,000 kr., or about 350,000l.—Engineering.

## THE DIFFICULTY AT THE C. G. E. WORKS ADJUSTED.

It is understood that a satisfactory adjustment has been made of the difficulty which arose at the General Electric Company's works at Peterborough a fortnight ago, as the result of the employees being required to sign a certain agreement. Some of the features of the agreement to which objection was taken by the employees have been modified by the company, and as a result harmony has been restored and the works will resume operations.

## PERSONAL.

Mr. John Langton, electrical engineer, Toronto, recently spent two months in the interest of Eastern States capitalists in visiting mining properties in Colorado, Arizona and elsewhere. He reports that business conditions across the border do not as yet exhibit anything like their former activity.

We chronicle with regret the sudden death from pneumonia, of Mr. Geo. M. Phelps, manager of the New York Electrical Engineer. The sad event occurred on the 11th of April. Mr. Phelps was widely known and most highly esteemed. The position made vacant by the death of Mr. Phelps has been filled by the appointment of Mr. T. C. Martin, one of the editors of the paper.

Mr. W. J. Richardson, Secretary of the American Street Railway Association, died at his home in Brooklyn, N. Y., on the 26th of April. He was for many years connected in an official capacity with the Brooklyn Street Railway Company, on whose system occurred the recent great strike. This strike is said to have had much to do with bringing about his death. Mr. Richardson had occupied for many years the position of Secretary of the American Street Railway Association, and his loss will be deeply felt by that organization, more especially at the present time when arrangements are in progress for the annual convention to take place in Montreal in the autumn.

## LIGHTING AND ELECTRIC LIGHTING AS A BUSINESS.

By Geo. White-Fraser, E. E.

II.

AGAIN, how many superintendents, and so-called "electricians," have the faintest idea of what goes on between the moment when coal or wood is thrown into the furnace and the other moment when the lamp key is closed and the lamp bursts forth into incandescence? They will all tell you that the fuel raises steam, which runs the engine, which turns the dynamo, which generates current (somehow), which lights a lamp. And then they go on to say that they burn 45 tons of coal per month, and get $470 per month on the average for rental of lights, and that therefore their business pays, or does not pay, as the case may be. And this, together with the cleaning of the machines, the tightening up of the lines, etc., is all the management the station receives.

Now, the generation and utilization of electricity, in connection with steam power, involves several very complicated transformations. I am perfectly well aware that it is the opinion of most Canadian station owners (an opinion which they endorse by their practice) that there is nothing whatever complicated about them, and that a perfectly inexperienced person can run an electric business. Still, I have used the word complicated, and I mean it. There are the following transformations: The potential energy of coal into the actual energy of steam, acting both by direct pressure and expansively. The utilization of this steam energy to produce first linear and then rotary motion. Up to this point we have mechanical energy. Here comes in the transformation of mechanical to electrical energy. We next have the transmission of electrical energy, and lastly the utilization thereof. The utilization involves either the retransformation of electrical to mechanical energy—by motors—or the transformation of electrical to radiant energy in the the incandescent or arc lamp. Perhaps this is not a complication; again, perhaps it is. Each person can judge for himself. Every one of these transformations, utilizations and transmissions is, and must be, accompanied by some waste of energy. Of the total heat in the coal, some goes up the chimney, some is used in heating the draft air, some in heating furnace walls, bars, grates, &c., and boiler shell, and the remainder in heating the water. Of the total heat in the steam generated, some goes towards running the engine—the rest is wasted in heating piping, cylinders, &c. Of the total theoretical power of the engine, a considerable portion is required simply to turn itself over against its own friction and inertia. Quite a considerable percentage is lost, between the engine and dynamo, by belt slippage. In the dynamo itself a still further loss takes place; yet another on the transmission lines and transformers, and in the last transformation—from electrical to radiant energy—such a tremendous loss occurs that one authority states that only 5% of the electrical energy given to an incandescent lamp is converted by it into light. The losses in lamps, wires and dynamos are unalterable, being dependent on design; those in belting, shafting, engine, piping and boiler are partly the result of necessity, but can to a great extent be kept down.

What does our diagrammatic factory do? It watches every process in manufacture, and accounts for every pound of material throughout; recognizes some wastes or losses as inevitable, others as extravagant, and remedies them. Does our electric factory do this? No; it shovels coal into the boiler and is satisfied when the engine turns round. In blissful ignorance it doesn't weigh raw material (coal), buying it by the car load, and dumping it somewhere handy. It doesn't, by careful experiment, determine which is actually the cheapest coal to use in its peculiar condition as to draft, &c. Oh no! It trusts in Providence and keeps on shovelling coal. The turning over of the the engine is a stage in the manufacture. And seeing that the modern steam engine is the result of the concentration of great minds, and the application of scientific principles, it might be worthy of receiving some attention. The prime object of its being is to convert the energy of the steam into motion, linear and rotary; a special feature being the utilization of both the direct force and the expansive force, in such proportions as to attain the highest efficiency, i.e., the greatest force with the least loss of energy. As everyone knows—even our hypothetical "electrician"—this is done by cutting off steam at a point in the stroke and allowing it to act expansively afterwards. And

all engines designed for electric service allow a variation in the point of cut-off from say a quarter to over a half, this variation being effected through the governor, eccentric and valves. All machinery is liable to fall out of adjustment; it must wear, and it requires attention. Any deviation from perfect adjustment means loss of energy, therefore, waste; and as it is caused by misadjustment, such waste is remediable. In how many stations operated by steam, is a card taken off the engine once a year? How, otherwise, can the operation of the engine be watched? It is quite possible for an admission or exhaust to open or close too late or too early, and every slight loosening of a nut will produce very appreciable results. And yet this matter is neglected, as is apparently almost every other matter connected with electricity, and hence it is that a comprehensive statement may be made, that "electric lighting stations run themselves."

Now, if electric lighting be not a business, then let that be clearly understood. But if it be such, then let electric business be conducted on business principles. I think it is evident to any one who has had the opportunity of watching the methods followed generally, that seventy-five per cent. of the electric stations in Canada neither know their consumption of fuel, nor the amount of their electrical output, with any degree of accuracy; have no person in charge who is in any smallest detail competent to manage; keep no checks whatever on any of their apparatus, and have not the faintest idea as to the interior economy of the plant. If the plant pays it is purely good luck, and they accept their dividends without question. If it doesn't pay, then "There's no money in electric lighting anyway." Their $1 a day "Fireman and Electrician," may, by injudicious use of the draft, waste pounds of fuel; the engine may through some slip in adjustment of cut-off, use 5% more steam than necessary; there may be a most healthy ground on the lines. No matter— "Electricity don't pay!" I should like to hear of the business that would pay under these conditions. Let all these precautions be taken and checks made, and see the improvement. Every station will know within a fraction of a cent, how much it has cost to produce a kilowatt hour; how much it has been sold for; where losses may be diminished, gains increased; and a report can be prepared, showing in an intelligent business way the working of the station, and the directors, or other authorities, will understand clearly what they are doing.

Among the many matters requiring consideration at the hands of electrical investors, as preliminaries, the one of capital investment is very prominent. The purchase of land, building, machinery and construction expenses, involve a capital outlay that, large or small, excessive or reasonable, once made has got to stand. The plant must pay its own proper operating expenses, and is expected to pay interest on the investment. The larger the investment, the less evidently, the percentage of interest. It is impossible to raise this percentage by decreasing operating expenses, beyond a certain limit, and it is equally impossible to do so by decreasing the investment once made. Hence the importance of keeping it down in the first place is obvious. This can be done by either providing cheap and nasty machinery, or by most carefully considering the entire conditions under which the plant will operate, and balancing cheapness against efficiency. The cheapest is often the most expensive; conversely, the most expensive is sometimes—generally—the least so. Why? Because the more expensive machine, by its being more carefully built, with better materials, and greater attention to scientific principles, will save in cost of operation, a greater sum than the interest on the difference in price between the more and the less expensive. True economy, therefore, is only attained at the cost of patient investigation.

There should be unity in all things, even in electric lighting, although it seems not to be generally admitted. An electric station is intended to perform a certain service—to give a certain quantity of current. It is a condition of all lighting that more is required at certain periods than at others, and so electrical machinery must be capable of supplying the highest demand, as well as the lowest, or average. Therefore, although in a 1000 light town, it is found that for ten hours of lighting only 300 lights are going; during 7 hours there must be the machine capacity to supply the whole 1000 lights during the three hours they are required. The dynamo capacity being fixed, there is no doubt as to the necessary size, but the engine capacity varies with the cut-off, and is generally rated at some particular point, say ¼ or ⅜ cut-off, with a range of variation from ⅛ to ⅝. The engine therefore may be chosen, so that it will be powerful enough for the 1000 lights at the maximum admission of the ⅝ cut-off. By so doing the rated capacity of the engine is less than that required, the actual capacity sufficient, and the price less than if one were purchased whose rated capacity at ½ cut-off were sufficient.

All machinery has a most efficient rate of working. A boiler will require less coal to be consumed to evaporate 1 lb. of water into steam, at a certain rate of supplying steam, than at any other rate. An engine will require least steam when a certain definite relation exists between the admission of steam into the cylinder, and the expansion thereof. A dynamo will waste less power when running a large number of lamps than a small number. As a rule, machinery is so designed that it will work at the highest efficiency when it is being worked at its full rated capacity. Therefore, a boiler rated at 100 horse power, will not furnish steam to a 50 h.p. engine, at anything like the same economy. It is very bad economy, and unwise precaution, to allow too large margins, for thereby an element of waste is introduced, and no useful purpose gained. On the other hand, too small margins are worse. Here is just where the question comes in, of the proper proportioning of the various apparatus that go towards the production of electric current. In any actual case there is always a happy mean—avoiding on the one hand excessive margins, which increase not only first cost, but also operating expenses; on the other, a mistaken economy that allows not enough. It has been said that in the most ordinary objects around us—the leaves on the trees, the beetle's wing, &c.—there is abundant evidence of design, and fitness for purpose. Nature allows sufficient material and no more—a beautiful harmony and symmetry being the distinguishing characteristics of her works. Nature would chuckle over a few of the the electric light stations in Canada. In some of these the only unity of design seems to have been to purchase in strictly the cheapest market. I have seen a 150 h. p. engine bought to run a 600 light incandescent dynamo, because it was second-hand, cheap. It would be very interesting to arrive at how much money this station is losing by the inefficiency of this proportion, remembering that the dynamo runs probably only 100 lights for half the time. Nature has not had much to do with these stations. An electric light man comes along and a dynamo is bought from him. An engine "big enough to run the dynamo" is duly put up, and a boiler. They have to have these things and so they get them. The engine man makes more out of a large engine than a small one, and so he recommends a good large one—two if they will stand it; similarly with the boiler man. And so the station grows; a source of waste and expense; is managed by a "Fireman and Electrician"; and "electric lighting doesn't pay!"

One word as to the use of meters. Dynamos are always wired for more lamps than their rated capacity, because all the lamps will never be going at the same moment. If a man pays by the meter, he will be very careful not to be using more at any moment than he actually requires. This has been found by experience to allow of a still greater margin of overload, and this is tantamount to a reduction of capital investment relatively.

This article will be seen by a great many to whom all the above is well known, also to a greater number who have not made electric lighting their study and profession for years. The former will, I hope, pardon the repetition; the latter will, I sincerely trust, come to see that electric lighting as a business is worthy of study; that it is not the simple affair they have hitherto regarded it, and that by the application of business methods, and in that way only, can they expect to succeed in it.

## DAMAGE TO CHIMNEYS BY LIGHTNING.

An investigation was recently carried on in Germany, by C. Carlo, upon the subject of the damage done by lightning to chimneys, both with and without lightning conductors. From a study of twenty-four cases, he draws the following conclusions:

1. Lightning very seldom strikes a chimney in such a way as to leave any perceptible effect.

2. The damage done by lightning to chimneys is in most cases inconsiderable; only in one case was a chimney actually destroyed, and in four cases only was the damage so great that it was necessary to pull the chimneys down.

3. Lightning strikes chimneys both with and without lightning conductors; the latter appear, however to be struck oftener than the former. Of the cases reported on, two were with and fifteen without lightning conductors; in four cases it was not definitely known whether a conductor was in position or not.

4. In low, marshy grounds, lightning flashes seem to occur more often than in high and dry neighborhoods.

5. In one case only has lightning struck a steam boiler so as to necessitate repair.

## WATER HAMMER IN STEAM PIPES.

Recently numerous explosions of high-pressure steam pipes led a German engineer to call special attention to the great danger from water hammer. To prove this experiments were undertaken to show the high pressure in a pipe when water hammer occurs. A pipe 12 inches in diameter, ⅜-inch thick and 21 feet long, blank-flanged at one end, was partially filled with water, and at the other end steam supplied through a three-inch pipe. Three pressure gauges at equal distances were screwed to the pipe and one in the black flange. When steam of five atmospheres, 73 pounds per square inch, was admitted suddenly above the water, the pressure gauges indicated respectively pressures of 426 pounds, 114 pounds, 199 pounds and 114 pounds per square inch, or nearly 30, 8, 14 and 8 atmospheres. When steam entered slowly again above the water, hardly any concussions and abnormal pressures were noticed. Steam was then admitted through a valve 2 inches in diameter, and the steam, at a pressure of 5 atmospheres, now entered below the water. The concussion was so violent that the threads of four of the nuts were shorn off, the fourth gauge placed there was crushed, though the gauges were designed for a maximum pressure of 2,133 pounds per square inch, while the other gauges indicated pressures of 483, 385 and 923 pounds per square inch. The end of the pipe bulged considerably. On a new trial the first three gauges registered 313 pounds, 185 pounds and 853 pounds per square inch, the fourth, refitted, over 2,130 pounds; a rent of eight inches in length formed, starting about four inches from the far end. This damaged part was then cut off, the pipe closed again, and the water level lowered to 6¾ inches; pressures of 498 pounds, 498 pounds and 853 pounds per square inch were then observed. The water level was then raised to 10 inches and steam turned on again; this time two bolts broke in the end plate, and fissures formed near the middle of the pipe. In al. cases air and water were thrown out through the air and water outlets. This occurred always in sudden rushes, after an interval of 15 seconds, when steam was turned on fully, and of several up to four minutes, when the valve was only partially opened, to one-fifth in the last instance. Only part of the water was forced out; a minimum of 3 inches always remained in the pipe. The experiments prove that the blow did not begin before the steam had condensed and the water had acquired the respective temperature. The different indications of the gauges seem to show that the blow was propagated in waves, which affected the pressure gauges according to their positions. The maximum pressure observed was 30 times higher than that of the steam which caused the concussion. If we consider that the steam inlet had only one-thirtieth of the area of the steam pipe, that steam pressures of three times the intensity of those experimented with are actually used on shipboard, and that part of a pipe might, under circumstances, be entirely filled with water, we must admit that these lodgments of water may lead to most disastrous consequences.—Engineer and Iron Trades Advertiser.

## QUESTIONS AND ANSWERS.

H. Bros., Kincardine, Ont., write: We want to build a brick chimney 80 feet high, square chimney. What size should the base be for this chimney, and should cement or mortar be used.

ANSWER.—It will depend largely upon the nature of the soil upon which the chimney is to be built, as to the dimensions at the base. If built on clay soil, allow about three tons to the square foot; if on sand, about four tons to the square foot. You should take into consideration also, the thickness of the walls. Unless the chimney requires to be rapidly constructed, the use of cement mortar is not necessary, a good quality of ordinary mortar will be sufficient.

## SOME NOTES ON THE GOVERNING OF STEAM ENGINES, PARTICULARLY WHEN COUPLED TO DYNAMOS.*

By G. L. ADDENBROOKE.

THIS is a subject which has always been in my mind in connection with central station electric lighting ; and the fact that it has been brought into prominence by Mr. Swinburne, in his simple but lucid paper before the British Association, shows that the question is now beginning to attract the attention which it deserves.

In order to understand the exact position of matters at present, as regards the governing of the engines used for other than marine purposes, it is advisable to look back a little at the history of the steam engine governor, and the purposes to which it has been chiefly applied up to the present date, premising the fact that so far electrical engineers have not paid any very great amount of attention to the problem, as they have been occupied by other considerations of more immediate importance, and have been generally content in this particular to accept the practice of industrial engineers, and to rely on their judgment as having the greater experience.

What really caused careful attention to be paid to the construction of governors and valve-gear was the application of steam engines to driving spinning machinery, the rise of mechanical weaving and spinning and the best modern type of steam engines being almost contemporaneous. For a loom to work properly regular speed is essential, or the shuttles will not throw properly, and uneven strains are produced in the material. Not only, therefore, has the best steam practice been usually devoted to the construction of engines for driving textile factories for the sake of economical working, but at the same time the utmost pains have been taken to design valve and governing gear which would admit of the engine doing its work at a constant speed whatever the load. The latter condition was essential, because one engine was only employed for driving at a time, though the load would constantly vary as different machines were thrown in and out ; and at times a part of the factory might be altogether closed if work in one of the departments was slack.

The engineers who devote themselves chiefly to the construction of mill engines have for many years been able to meet these conditions and to supply plant which would readily retain a speed within 2 per cent. or 3 per cent. of the normal, whatever might be the load. It is true that large fly-wheels and comparatively slow speeds have hitherto been chiefly in vogue for this class of plant, but this is a somewhat minor point.

Such being the state of affairs when electric lighting first became practical, electrical engineers naturally turned to the constructors of mill engines for advice and help ; the result was the application of the mill engine, with rope driving, to turning dynamos, a system which has worked well wherever it has been applied, while for freedom from breakdowns and exactness of working this method has probably not yet been surpassed. Such plants, however, occupy a great deal of space, are rather expensive, and there is a certain loss of power in rope-driving. Consequently the high-speed engine has been developed to get over these difficulties, so far, chiefly of the single-acting type ; but now that the conditions of running of such engines is better understood, and confidence in high speeds is increasing, there are indications that similar engines, but double-acting, will gradually come to be preferred.

Our cursory glance at the history of the application of governors to engines has now made it manifest that the best practice hitherto has been in the construction of governors to produce perfectly even motion whatever the load. Further, it must be noted that such engines have usually been engaged in supplying energy direct in the form of motion to lines of shafting, and to machinery which it is required to run at a constant speed.

Let us now turn to the driving of dynamos by means of steam engines. The proper methods of governing such engines will naturally depend on the purposes to which the current produced is to be applied. If for the transmission of power or the running of arc lights in series, very little alteration of existing methods is required ; but if the engine is used for lighting incandescent lamps direct, whether for a private installation or in a public supply station—which latter is, of course, the most important

case, and the one to which it is proposed to devote most attention—then it is desirable to clearly define the limiting conditions of the problem ; so that, knowing exactly what is required, we may be in a position to judge how far existing appliances meet the conditions and where they are defective, and thus arrive at a definite idea of the directions in which alterations and improvements are required.

Now in a central station the ultimate object is the incandescing to a certain point of carbon filaments, to procure from which the standard amount—16 candles from a 16 C. P. lamp—the pressure or voltage must be steady and exact, with one per cent. on either side of the normal, any further variation either way making a great difference in the light.

Perhaps, in order that we may present to ourselves a clear idea of the conditions, an entirely mechanical analogue may be useful. Let us suppose that the engine room in a central station is the engine room of a large mill which is worked by several steam engines, driving by means of pulleys and friction clutches on to a line of counter shafting. Imagine that these clutches are not very well designed, and that there is always a certain amount of slip in them, not more than 6 or 7 per cent. at light loads, but increasing considerably at heavy loads. It is, of course, required to keep the mill machinery running evenly within 2 or 3 per cent. If the load is a light one not only will the engine or engines be lightly loaded but there will be very little slip in the clutches ; on the other hand, as the work done in the mill increases and becomes heavy not only will the load on the engines increase, but, having to keep the mill machinery revolving at a constant speed, the speed of the engine will also have to be increased to make up for the extra motion lost in the clutches, the friction of which will absorb energy, which will be dissipated as heat as the losses in dynamos are. Under these circumstances it is clear that it will not be sufficient if we have a governor which keeps the engines running always at the same rate ; we must over-govern to some extent, that is, we must not only have the speed of the engine under control, but we must also have an arrangement by which the engine shall go a certain percentage faster at heavy loads than at light. We may diminish the difference of slip at different loads by improving the efficiency of our friction clutches, but the fact remains that it exists to some extent and that it must be compensated for.

In an electric-lighting central station the dynamo may be regarded as the friction clutches in the above simile. Besides a certain loss of power greater at heavy loads than at light, and which may be likened to the loss in friction in the clutches, there is also a loss of motion, increasing with the load, due to self-induction and the back action of the armature current on the fields, which weakens them and necessitates the speed being increased to obtain the same voltage ; this is like the slip in the friction clutch. If not corrected by outside influences it may amount to from 10 to 20 per cent.

In the central station there is the further point that at the hours of heavy load it is necessary to run at a higher speed altogether than at light loads in order to keep up a higher general pressure in the mains to allow for increased loss in them.

Consequently, in driving dynamos in central stations for lighting, we have not only to secure regular running or absence of sudden variations of speed at all loads, but we have got to provide, taking the light load as a basis, that as the load increases the driving speed shall be increased in order—

1. To increase the pressure on the mains.

2. To overcome the increased armature resistance and provide increased excitation of the field magnets.

3. To make up for the slip or loss in pressure caused by the increased back action of the armature on the magnets and increased self-induction, which necessitates a higher rate of driving, though it does not directly mean loss of energy.

It is true that it is not absolutely necessary to increase the speed of the engines ; it may be kept constant at all loads as in mill engines, and the requisite variations may be produced by the use of resistances or by varying the excitation of the dynamos, by means of resistance or the use of a separate exciting engine.

## PUBLICATIONS.

The Arena for May is an excellent number, and the high character of its contents is one of the encouraging literary signs of the times.

*Abstract from the London Electrician.

## THE ONLOOKER.

THE question of electric lighting continues a leading one with the people of Toronto. The Onlooker has no desire to add one more to the army of disputants. The economic side of the subject has been touched with a good deal of vigor by some writers. Let a word be added on this point. Because the cost per light in one place may be shown to be $5.00 less than in another, some writers have concluded that money is actually saved the municipality. Quality and quantity come into the question here, as much as with the good housewife, when she does her Saturday's shopping. Consideration needs to be given to the amount of work done by each lamp. Stating the cost does not state the real value. The Onlooker has been interested in an article in which are furnished valuable statistics on electric lighting by an expert. He has pointed out that as a matter of fact Buffalo pays $127.75 per year for a lamp, while South Norwalk pays less than half that amount ; but Buffalo gets almost twice as much light. The question suggested is, Would the light which South Norwalk receives be sufficient to satisfy Buffalo, even if the latter could have it at the lower price, and if not, is the excess of lighting which Buffalo gets costing that city more than the same excess of light would cost South Norwalk if the latter were to increase output, plant and equipment, to a point where it could furnish a supply of light equal to that necessary to supply Buffalo? The Onlooker stops here, but commends this thought to the tyros who are telling where Toronto can secure electric light at a much greater reduction than some of the figures that are offering to-day. A penny saved is a penny made, sometimes, but not always.

x x x x

It is always the case when a "boom" is on—it matters little whether it is in real estate or electricity—the average man is prepared to take a good man chances, and the more uncertain the project the greater are likely to be the chances taken. If he does not know all about it he is ready to assume that he understands the whole situation, and often rushes in where angels would fear to tread. This thought is nicely brought out in a recent article in the ELECTRICAL NEWS on interurban electric railways, by Mr. Geo. White-Fraser. The Onlooker had the same thought presented to him a short time since, when discussing electrical projects with a well-known engineer of Toronto. This person drew attention to the injustice often done the experienced engineer by the manner in which his designs, which had been prepared by request and at the cost of much thought, are altered afterwards, or entirely ignored, by the directors or officers of electrical enterprises, while the suggestions of a novice, who may have a purpose to serve in selling a plant, are accepted, because, for the moment, it may seem a few dollars can be saved. The case was cited of an electric railway in western Ontario, where the engineer's plans, as to the laying of the road, were untouched, and the work of a road-man, who thought he knew all about the laying of tracks, was accepted. The road was built ; the cost was less than if the plans of the engineer had been followed. But what of the work? It was not long before a large part of the road had to be taken up and relaid on a scientific and common-sense basis. By the way, the Onlooker is informed, that the criticism of Mr. White-Fraser as to the disposition of the boilers, engines and dynamos in the Galt & Preston power house, could not have been made had the plans of Mr. Wm. Jennings, C. E., engineer for the project, been followed ; but after Mr. Jennings had done his work the directors of the road, and they certainly had a right to do as they pleased in this matter, viewing the question from at least one standpoint, fitted the machinery to certain buildings in their possession, and in cutting the coat to the size of the cloth, the designs of the engineer had to be discarded. To quote another old saw, it is possible to be penny wise and pound foolish in the building of electric railways, as in the matter of electric lighting.

----

Two open double truck cars with accommodation for 150 passengers are being constructed for summer traffic on the Kingston, Portsmouth & Cataraqui Electric Railway.

The Royal Electric Company recently recovered from Mr. Walbank, the architect, who planned and supervised the erection of a chimney stack for the company's works, the sum of $1,281.04, on account of the defective manner in which the work was performed. The architect in turn brought suit against the contractor, from whom he secured damages.

## ECONOMY OF SMALL v. LARGE TRANSFORMERS.
### BY F. G. PROUTT.

IT is to be supposed that every company operating an electric lighting system, operates it for the purpose of making money, and perhaps the best way to make money is to save it "A penny saved is a penny earned," is true in every case. In very many instances of plants operated on the alternating system, a separate transformer is used for each and every customer, and no attention whatever paid to that the core loss in a small transformer is enormous when compared with the loss in a larger one, a fact which is very clearly shown by the accompanying curve.

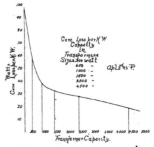

We will take as an illustration, what is perhaps an extreme case. With 1,000 volts applied to the primary and the secondary open circuited, the watts core loss in a 300 watt transformer was found to be about 29, or at a rate of about 97 watts loss per K. W. of capacity. The core loss in a 4,300 watt transformer, was found to be only 88 watts, or at a rate of 19.5 watts per K. W. capacity. Now fifteen 300 watt transformers have the same capacity as one 4,500, whilst the loss in the former would be 435 watts, and in the latter only 88, a saving at all times, whether the transformer is running fully loaded, or only partly loaded, or on open secondary circuit of 347 watts, or enough energy to maintain six 16 c. p. lamps.

While it would perhaps be hard to find such an extreme case, there are plenty of instances where two or three or four transformers might be replaced by one of a larger size, necessitating perhaps the putting up of a little larger secondary wire for a few feet, but the slight addition in cost for a small amount of wire is nothing when compared to the saving of energy—not for a day or a week, but for as long as the customers have to be supplied with light.

A little explanation of the diagram may be necessary. The ordinates represent not the losses in the transformers, but the losses per K. W. capacity of the transformers ; and the abscissae represents the capacity of the transformers. The transformers tested are of a very well known and reliable make, and range in size from 300 to 4,500 watt. Heavy ordinates have been drawn from points on the curve to the abscissae below which show clearly the size of each transformer tested and the point it takes on the curve.

It will be well to remember that even a little saved each day amounts to something in a year. If we can so arrange our transformers as to make a saving of one kilowatt, it means $50 per annum on the coal bill alone, allowing 4 lbs. of coal per h. p. hour, and a twelve hour run each day in the year, with coal at $4 per ton, and in electric lighting as in all other business, what is saved can be very conveniently used in paying dividends.

Malden, Mass., April 3rd, 1895.

----

Notice is given of application to be made for incorporation of the "New Light Co., Ltd.," to carry on an illuminating business and manufacture illuminating apparatus. The headquarters of the company is to be at Montreal. The names of the applicants are.—William Robertson, merchant, Angus M. Thom, merchant, Joseph A. E. Whyte, merchant, William D. Aird, merchant, of Montreal, and Geo. W. Booth, of Toronto, merchant ; the first three of whom shall be the provisional directors of the company.

PUBLISHED ON THE FIFTH OF EVERY MONTH BY

**CHAS. H. MORTIMER,**

OFFICE : CONFEDERATION LIFE BUILDING,
*Corner Yonge and Richmond Streets,*

**TORONTO,    -    -    CANADA.**
Telephone 2362.

NEW YORK LIFE INSURANCE BUILDING, MONTREAL.
Bell Telephone 2299.

*ADVERTISEMENTS.*

Advertising rates sent promptly on application. Orders for advertising should reach the office of publication not later than the 15th day of the month immediately preceding date of issue. Changes in advertisements will be made whenever desired, without cost to the advertiser, but to insure proper compliance with the instructions of the advertiser, requests for change should reach the office as early as the 22nd day of the month.

*SUBSCRIPTIONS.*

The ELECTRICAL NEWS will be mailed to subscribers in the Dominion, or the United States, post free, for $1.00 per annum, 50 cents for six months. The price of subscription should be remitted by currency, in registered letter, or by postal order payable to C. H. Mortimer. Please do not send cheques on local banks unless 25 cents is added for cost of discount. Money sent in unregistered letters will be at senders' risk. Subscriptions from foreign countries embraced in the General Postal Union, $1.50 per annum. Subscriptions are payable in advance. The paper will be discontinued at expiration of term paid for if so stipulated by the subscriber, but where no such understanding exists, will be continued until instructions to discontinue are received and all arrearages paid.

Subscribers may have the mailing address changed as often as desired. *When ordering change, always give the old as well as the new address.*

The Publisher should be notified of the failure of subscribers to receive their papers promptly and regularly.

*EDITOR'S ANNOUNCEMENTS.*

Correspondence is invited upon all topics legitimately coming within the scope of this journals

THE "CANADIAN ELECTRICAL NEWS" HAS BEEN APPOINTED THE OFFICIAL PAPER OF THE CANADIAN ELECTRICAL ASSOCIATION.

## CANADIAN ELECTRICAL ASSOCIATION.

OFFICERS:
PRESIDENT:
K. J. DUNSTAN, Local Manager Bell Telephone Company, Toronto.

1ST VICE-PRESIDENT:
A. B. SMITH, Inspector Canadian Board Fire Underwriters, Toronto.

2ND VICE-PRESIDENT:
C. BERKELEY POWELL, Manager Ottawa Electric Light Co., Ottawa.

SECRETARY-TREASURER:
C. H. MORTIMER, Publisher ELECTRICAL NEWS, Toronto.

EXECUTIVE COMMITTEE:
L. B. McFARLANE, Bell Telephone Company, Montreal.
GEO. BLACK, G. N. W. Telegraph Co., Hamilton.
T. R. ROSEBRUGH, Lecturer in Electricity, School of Practical Science, Toronto.
E. C. BREITHAUPT, Berlin, Ont.
JOHN YULE, Manager Guelph Gas and Electric Light Company, Guelph, Ont.
D. A. STARR, Electrical Engineer, Montreal.
J. J. WRIGHT, Manager Toronto Electric Light Company.
J. A. KAMMERER, Royal Electric Co., Toronto.
J. W. TAYLOR, Manager Peterboro' Carbon Co., Peterboro'.
O. HIGMAN, Inland Revenue Department, Ottawa.

## MONTREAL ELECTRIC CLUB.

OFFICERS:
President, W. B. SHAW,           Montreal Electric Co.
Vice-President, H. O. EDWARDS,    -
Sec'y-Treas., CECIL DOUTRE,     -     81A St. Famille St.
Committee of Management, T. F. PICKETT, W. GRAHAM, J. A. DUGLASS.

## CANADIAN ASSOCIATION OF STATIONARY ENGINEERS.

EXECUTIVE BOARD:
President, J. J. YORK,        Board of Trade Bldg, Montreal.
Vice-President, W. G. BLACKGROVE,   -     Toronto, Ont.
Secretary, JAMES DEVLIN,       -     Kingston, Ont.
Treasurer, DUNCAN ROBERTSON,     -     Hamilton, Ont.
Conductor, E. J. Philip,      -     Toronto, Ont.
Door Keeper, J. F. CODY,      -     Wiarton, Ont.

TORONTO BRANCH NO. 1.—Meets 2nd and 4th Friday each month in Room D, Shaftesbury Hall. Wilson Phillips, President; T. Eversfield, Secretary, University Crescent.

HAMILTON BRANCH No. 2.—Meets 1st and 3rd Friday each month, in Maccabee's Hall. Jos. Langdon, President; Wm. Norris, Corresponding Secretary, 211 Wellington Street North.

STRATFORD BRANCH No. 3.—John Hoy, President; Samuel H. Weir, Secretary.

BRANTFORD BRANCH No. 4.—Meets 2nd and 4th Friday each month. C. Walker, President; Joseph Ogle, Secretary, Brantford Cordage Co.

LONDON BRANCH No. 5.—Meets in Sherwood Hall first Thursday and last Friday in each month. F. Mitchell, President; William Meaden, Secretary Treasurer, 533 Richmond Street.

MONTREAL BRANCH NO. 1.—Meets 1st and 3rd Thursday each month, in Engineers' Hall, Craig street. President, Jos. Robertson; first vice-president, H. Nuttall; second vice-president, Jos. Badger; secretary, J. J. York, Board of Trade Building; treasurer, Thos. Ryan.

ST. LAURENT BRANCH No. 2.—Meets every Monday evening at 43 Bonsecours street, Montreal. R. Drouin, President; Alfred Latour, Secretary, 306 Delisle street, St. Cunegonde.

BRANDON, MAN., BRANCH No. 1.—Meets 1st and 3rd Friday each month, in City Hall. A. R. Crawford, President; Arthur Fleming, Secretary.

GUELPH BRANCH No. 6.—Meets 1st and 3rd Wednesday each month at 7.30. p. m. J. Fordyce, President; J. Tuck, Vice-President; H. T. Flewelling, Rec. Secretary; J. Gerry, Fin.-Secretary; Treasurer, C. J. Jorden.

OTTAWA BRANCH, No. 7 — Meets 2nd and 4th Tuesday, each month, corner Bank and Sparks streets; Frank Robert, President; F. Merrill, Secretary, 352 Wellington Street.

DRESDEN BRANCH No. 8.—Meets every week in each month; Thos. Merrill, Secretary.

BERLIN BRANCH No. 9.—Meets 2nd and 4th Saturday each month at 8 p. m. W. J. Rhodes, President; G. Steinmetz, Secretary, Berlin Ont.

KINGSTON BRANCH No. 10.—Meets 1st and 3rd Tuesday in each month in Fraser Hall, King Street, at 8 p.m. J. Devlin, President; A. Strong, Secretary.

WINNIPEG BRANCH No. 11.—President, G. M. Hazlett; Recording Secretary, W. J. Edwards; Financial Secretary, Thos. Gray.

KINCARDINE BRANCH No 12.—Meets every Tuesday at 8 o'clock, in the Engineer's Hall, Waterworks. President, Jos. Walker; Secretary, A. Scott.

WIARTON BRANCH No 13.—President, Wm. Craddock; Rec. Secretary, Ed. Dunham.

PETERBOROUGH BRANCH No. 14.—Meets 2nd and 4th Wednesday in each month. S. Potter, President; C. Robison, Vice-President; W. Sharp, engineer steam laundry, Charlotte Street, Secretary.

BROCKVILLE BRANCH No. 15.—W. F. Chapman, President; James Aitkens, Secretary.

CARLETON PLACE BRANCH No. 16.—W. H. Routh, President; A. M. Schofield, Secretary.

## ONTARIO ASSOCIATION OF STATIONARY ENGINEERS.

BOARD OF EXAMINERS.
President, A. E. EDKINS,     -     139 Borden st., Toronto.
Vice-President, R. DICKINSON,   -   Electric Light Co., Hamilton.
Registrar, A. M. WICKENS,    -     280 Berkeley st., Toronto.
Treasurer, R. MACKIE,      -     28 Napier st., Hamilton.
Solicitor, J. A. McANDREWS,   -     Toronto.

TORONTO—A. E. Edkins, A. M. Wickens, E. J. Phillips, F. Donaldson.
HAMILTON—P. Scott, R. Mackie, R. Dickinson.
PETERBORO'—S. Potter, care General Electric Co.
BRANTFORD—A. Ames, care Patterson & Sons.
KINGSTON—J. Devlin (Chief Engineer Penetentiary), J. Campbell.
LONDON—F. Mitchell.

Information regarding examinations will be furnished on application to any member of the Board.

THE negotiations which are understood to have been in progress of late between the General Electric and Westinghouse Electric Companies of the United States, with the object of arriving at a satisfactory adjustment of their patent disputes, are declared to have come to a stand-still.

THE entrance of women into business pursuits for which formerly men only were thought to be adapted, is one of the features of this evolutionary age. There are indications that ere long the position of the street car conductor will be invaded by the New Woman. Several applications from young women for positions as conductors are said to be in the hands of Superintendent Folger, of the Kingston, Portsmouth and Cataraqui Electric Railway. As usual the enterprising Canadian girl is the first to enter upon an untried path. Managers of suburban pleasure lines especially may possibly see in the lady conductor a means of increasing their passenger traffic and of still further popularizing electric railways in general.

IN view of the recent decision of the Canadian Controller of Customs, imposing a duty of twenty per cent. on electricity generated in the United States and transmitted to Canada, the New York Electrical Review wonders "just how the 'juice' will be measured." "Probably," it says, "at so much per H.P. for strong currents, but how the quantity used in transmitting telegrams will be taxed is evidently a refinement of calculation not yet attempted." It will be interesting also to notice what action the United States Government will take in the case of the company which is generating electricity at St. Stephen, N. B., and transmitting the same into the United States. Thus far, the American customs authorities have not attempted to collect duty from the company. The fixing of a rate of duty would seem to be a much easier task than determining the method of measuring the current and calculating duty on the same.

A FEW errors crept into the description of the central station of the Incandescent Light Co. of Toronto, printed in the ELECTRICAL NEWS for April. In the fifth line below cut in first column, first page, it should read "30 feet". On twentieth line, second column, of first page, "Size of vertical engine, 19 × 38 × 32 inch stroke." In last line on first page "25,000" should read "35,000."

ON all sides come reports of new electric light, railway and power enterprises, to be carried out in the near future. If these prospects should materialize, they will give rise to a demand for electrical apparatus, such as has not been experienced in Canada in any previous year in the history of the industry. In keeping with the brighter outlook, the faces of the manager and sales agents of the various electric manufacturing companies are marked by an unusually cheerful expression.

It is a consequence of the "Infancy" theory, that electrical enterprises do not receive that careful consideration and attention without which success is impossible. When it is clearly understood by electrical investors that the proper methods for the construction and operation of electrical machinery have attained a certain degree of finality, and that they can be formulated on thoroughly scientific principles, then a direct experience in electricity will be a necessary qualification for the appointment to positions in power houses, &c. It is not yet recognized as it should be, that electricity is a distinct business, to be studied carefully and intelligently.

A BY-LAW to authorize the expenditure of $277,000 for an electric lighting plant will shortly be submitted to a vote of the property owners of Toronto. It is very doubtful whether any saving could be effected in consequence of the ownership and control of the generating plant being vested in the city instead of a private company. It is altogether certain, we believe, that the saving, if any, would be so very trifling as not to warrant the city in entering upon such an extensive undertaking, involving so large an outlay. For these reasons, as well as the fact that aside from this expenditure it will be impossible to avoid increasing in a material degree this year the present rate of taxation, we believe the by-law is doomed to be defeated. Whether or not it will ultimately be to the city's advantage to own and operate its own lighting plant, will be much better understood say five years hence than it can possibly be to-day, when the business is to some extent in a transition state.

THE cost of steam production has declined in recent years in consequence of the introduction of devices for creating artificial draft, whereby it is possible to use coal screenings and refuse of various kinds instead of fuel of better quality and higher price. It will be interesting to notice what effect the increased demand will have on the price of the cheaper grades. The opinion obtains with some engineering authorities that the result must be to increase the price of these materials. On the other hand, coal for domestic purposes, shows an inclination to drop in price as the result of the lessened demand existing for steam purposes. The manufacture of artificial fuel for manufacturing purposes has already been commenced in Canada, and will doubtless prove to be an important factor in the future in keeping down the cost of steam production to a point to harmonize with the economical methods which are being adopted in every branch of manufacturing in the present day.

A CORRESPONDENT "J. M.," writes the ELECTRICAL NEWS as follows : "How does the enclosed application strike you? The question arises was he a good man? The copy is to my mind an improvement on the original." The application referred to reads as follows : "I have leave to inform you I have been A engine-here for over 20 years and has been a tester and prover of the—engine and is a good machinest to." Whether the author of the above application be a capable engineer or not from the standpoint of practical experience, it is certain that he would have been much more capable had he given more attention to self-education. An application which on its face so clearly displays the lack of even a rudimentary education, within the power of every man to acquire, is doomed to fail. In nine cases out of

ten the person to whom it is addressed will not consider it necessary to enquire further into the qualifications of the applicant. There is a lesson here for every engineer who desires to improve his position. If he is not so fortunate as to have had educational opportunities in his youth, he may nevertheless pursue a course of self culture which will to a large extent make good the deficiency. The man who neglects to make this effort must expect to be counted out of the race for advancement.

WHY is it that there are so few high-speed engines in use throughout the Dominion for electric lighting? There are many very good makes, and they are specially adapted for use in smaller lighting stations, where the load, which varies greatly between maximum and minimum, is not great even at the maximum. The two types, high and low speed, have each their peculiar advantageous features, which adapt them more especially to peculiar conditions, and a choice between them, for any individual case, which is made on purely personal preference for one or the other, is not always certain to secure the best results. The fact is that every case should be considered on its merits, and selection made accordingly. In small plants, of say 500 to 1000 incandescent lights, the sudden extinguishing of 50 lights, as in the case of a church, large store, or other considerable building, relieves the engine instantly of between 10 and 5 per cent. of its load. This has the tendency to quicken the engine speed, and consequently to temporarily raise the electrical pressure at the lamps. This excess of pressure, as is well known, has a very injurious effect on the life of the lamps, which will be more marked in proportion as the regulation of the engine is slow. This regulation is affected directly by the variations in speed, and will therefore be very much more rapid in a rapidly revolving wheel. It is evident therefore, that for a small-sized plant, where any fluctuation will very greatly vary the percentage of the load, a high-speed engine has considerable advantages. This is not the only one, however. They have generally a very considerably greater range of cut-off, thus giving a much greater flexibility of service. For instance, a high speed engine rated at 100 H.P., at ¼ cut-off, and giving a range of ½ to ⅛, will, according as the cut-off is varied, and at the same steam pressure, give either 170 H.P. or about 65 H.P. Now, although it is not equally economical at these abnormal points of cut-off, as cards taken would show, it is plain that an engine that will increase its power to meet the heavy demand at about 8 or 9 o'clock, and that will just as amiably shut itself off to suit the light-running conditions at 1 o'clock in the morning, all the time regulating its speed very closely, has its strong features, and another point in its favor is that shafting, with its friction, can be done away with. All shafting requires an expenditure of energy to turn it round, and it makes very little difference whether the load be heavy or light, that is, the power absorbed by its own friction, is practically constant, and it is obvious that the smaller the load, the greater is the percentage required for the shafting only. So that towards the middle of the night, when the lights are dropping off, perhaps 25 or 30 per cent. of the total power developed by the engine is used solely by the shafting. This is quite worth saving. There is also the purely commercial aspect of the question. High speed engines are sold at practically the same figure as the low speed ; they do away with one large main belt, belt tightening apparatus, and a quantity of shafting and standards, and so on. They take up less room, and altogether they can very well claim to have their merits considered in every case where a new installation is contemplated.

THE usual journalistic statement that "electricity is in its infancy yet," has been so persistently made, that the public, who somewhat naturally regards the public press as its guide, has come to accept that dictum, regardless of the numberless evidences of approximate perfection that are to be seen on every hand. If a generator with an electrical efficiency of 98%, and a commercial efficiency of 94% or 95%, is an "infant," what an efficiency we may expect when that puny infant attains full growth! Surely, we may confidently look for 300% or 400%. But when will the public learn that electricity is not in "its experimental state; " thatit is a force whose actions are perfectly understood, easily controllable, and as capable of being chained to the service of man, as is steam or water power. This ignorance of the true state of affairs directly affects the interest of the public through their pockets ; and it reacts indirectly on the electrical industries through the odium cast on the operation of electrical enterprises, either by reason of the poorness of ma-- chinery ignorantly purchased, or the unsatisfactory results obtained when employing unqualified persons to run it. It more especially affects that portion of the public who have small sums of money to invest, and who are continually on the watch for "snaps." This class is very much at the mercy of a fluent patent attorney, who can talk glibly of "largely increased effi- ciency"; improvements in "design of magnetic circuits," and consequent decrease of cost of manufacture and operation and so on ; and invest their money in appliances that have probably been tried and found wanting, in the very early days of electri- cal investigations, twenty and thirty years ago. The extraordi- nary conceit of some professional inventors, is not less remark- able than the credulity of those who invest their money in the crude productions of an uneducated mechanic. The methods for the generation and utilization of current are so perfectly un- derstood, and the proper design of the mechanical and other ap pliances necessary, has been so carefully investigated by persons in all civilized countries, having the advantage of scientific edu- cation as well as practical experience, and access to the most complete sets of tools, instruments, &c., and backed by plentiful capital, that it is improbable in the extreme that a person who never had other than ordinary every day education, and whose practical experience in electrical matters has been that of an ordinary mechanic, is going to effect any revolution in methods or any appreciable improvement in mechanical, magnetic, or electrical design of machinery. Electricity is not now, what it was when T. A. Edison, then a newsboy, started inventing in the telegraphic field. In those days almost anything was an im- provement. The field for original investigation was so large, and the explorers so few, that intelligence and industry was al- most the whole equipment required by the prospector. We have entirely changed that now-a-days. The system on which the young electrical engineer is now educated, seems to show not only the immense amount of work that has been necessary to evolve electrical science to the point it has reached, but also it clearly indicates that an incomparably more complete equip- ment is necessary before an intending investigator can consider himself competent to enter the field at all. An illustration of this whole souled conceit, on the part of an inventor, and guile- less innocence on that of the investor, was recently manifest- ed very strikingly in the case of a new motor, that was said to operate on a new principle, being a "magnetic motor," and the usual claims were made for it—more efficient, lighter, stronger, cheaper—going through the whole list to which we are so well accustomed. The "prominent capitalist" was also there, and altogether, everything went happily, the inventor being cordially congratulated on having produced such a marvel, and the only wonder was that all the scientific electricians in the world had not been able to forestall him. It was the unhappy fate of the writer to have to point out that the "new" principle was uncom- monly old, that the motor was the exact reproduction of one that was a new design in 1837, just 58 years ago, when electri- city really was an infant, and that therefore—to go no further in- to its peculiarities—it was not *likely* to be a better one than those built to-day, on the principles ascertained during a course of investigation which this 1837 motor was but a low down rung on the ladder. Now, in the first place, had the intending investors not been in the habit of regarding electricity as in its infancy, they would have very much more carefully and thorough- ly investigated the merits of this wonderful machine, before put. ing their money in, instead of afterwards, and in the second place, had the inventor had less conceit, and posted himself in the smallest degree, as to the developments of electrical machinery, he would probably have set to work to acquire the rudiments of electrical knowledge, which he very evidently lacked, and in course of time, he too would probably have discovered to his wonder, that to effect improvements in electrical machinery to- day, requires not only vast knowledge of electrical, magnetic and mechanical principles, but close study of electrical history, a highly trained mind, and—capital.

The Welland Electric Light Co. are enlarging their buildings and putting in additional plant.

## THE GAS ENGINE AND THE STEAM ENGINE JOINED.

REFERRING to the discussion of the gas engine question, a prominent engineer remarked in conversation that he failed to see why the gas engine and the steam engine should not be compounded, so to speak. He put the case something in this manner : In the gas engine one of the problems is to keep the cylinder reasonably cool, and in the steam engine to keep the cylinder hot. Now, suppose we have a gas engine running and jacket its cylinder with water, which is then used for boiler-feed water, thus saving the heat which is now thrown away. Then take the exhaust from the gas engine through the jacket of the steam cylinder, and, if necessary, as it probably would be, add a heating chamber for the steam to pass through just before reaching the cylinder, so that more heating force could be em- ployed. Two such engines adapted to each other would pro- bably mean a relatively small gas engine and a steam engine large enough to carry all the load in case the gas engine refused to work for any reason, and, arranged in this way, each would supplement the other so far as the proper distribution of heat is concerned. The proposition is a novel one, and there is a chance to do some thinking over it. Possibly someone may be so situated as to make it easy to try the plan and let us know the results.—American Machinist.

## ZINC WILL PREVENT SCALE.

IT is a well-known fact that zinc slabs suspended in steam boilers prevent the formation of scale, and large quantities are used annually for this purpose. The following directions will enable one to use it successfully. The proportions necessary to insure complete protection are one square foot of zinc to fifty square feet of heating surface in new boilers, which may be diminished after a time to one in seventy-five or even one in one hundred square feet.

Merely placing the zinc in trays, hangers or strips will not insure metallic contact, and the action of zinc to prevent corosion under such circumstances will be weak and limited. The best method of fixing the zinc is to place a number of studs in the sides of the furnaces and combustion chambers, and to bolt on to these studs the zinc plates, which should be about 10x6x1 inches. It is important to see that the contact surfaces are clean and bright and the nut screwed close to the zinc to exclude the water and deposits from the contact surfaces, thus compara- tively insulating them and preventing the galvanic action. Otherwise the zinc is acted upon as a solvent that renders the water innocuous or non-exciting.

## LONG DRIVING BELTS.

A VERY bad habit in mills where there are large driving belts, is shifting belts with a square stick, no regular shifters being used. The result of this is the belts are more or less injured on the edges. All heavy machines should have shifters to act so that they shift the belt over steadily, not putting too much strain on the driving belt too suddenly. Two pieces of gas pipe just large enough to revolve on round iron supports for shifters will lessen the friction on the edges of heavy belts, as these pipes revolve while the belt is being shifted. It effects a great saving in long driving belts, in fact any belt at all, leather or rubber.

The Vancouver Street Railway Co. have purchased at auction from the trustees of the debenture holders, the Westminster and Vancouver Tram- way. The price paid was $280,000. The line will be operated in connection with the city lines. The City Council and the Street Railway Co. have come to an agreement with regard to new tracks to be laid down, the com- pany to have the lease of the streets for five years at a nominal rental of $1 per annum, and the city to have the option of purchasing tracks at a valuation at the end of that period. Should the city not exercise this option, the company is to have the lease of the tracks for a further period of nine years, and so on from time to time for the unexpired period of the franchise for the streets granted the company for its original contracts. After the first five years the company is to pay a rental according to their earnings in ex. cess of $5,000 per mile per year.

## ELECTRIC LIGHTING FOR LARGE BUILDINGS.

By Geo. White Fraser, E. E., Toronto.

There have, until recently, been but two general methods employed for the electric lighting of large buildings. 1. The renting of the current from the electric company· operating a central station; and 2. The installation in the building itself, of a private generating plant. There can be no rules laid down, of general application, pointing to the adoption of either method preferentially to the other; every case must be considered on its merits. Such consideration will reduce the question to its lowest terms, viz., a comparison of the costs of lighting by the rival methods. The variable quantities that will enter into this reduction as governing factors are: The price of current supplied from the central station per lamp hour, or per kilowatt hour on the one hand, and on the other, the cost of generating the current required in the building, using a private generating plant. This latter cost will be the total of such items as coal and water, etc., engineer's salary, little repairs, depreciation of plant and interest on capital expended in purchase of machinery, etc.

It is evident that local data will greatly influence the selection of method. For instance, a building full of dingy offices, and employing already an engineer and using quantities of steam, for heating, might very reasonably be expected to effect an economy in lighting by using a private engine· or dynamo; while a church using a number of lights periodically, would not. These are, of course, extreme cases, illustrations merely, of the general principle, but it is a great mistake to assume in all cases, that because there is a central station operating, it must be better and cheaper to rent current instead of generating it.

The object of this article is to introduce to the owners and designers of large buildings a modification of each of the two general methods described above, that have the merit of presenting very interesting features from an engineering point of view, and of promising well from that of the owner.

The first modification is the use of a gas engine instead of a steam engine, to run a dynamo in a private installation. This plan presents many very advantageous features. No boiler is required, with its coal dust, dirt, and chance of explosion. Less ground space is taken up. Gas engines are very largely used in Europe, and in the States for working dynamos, and many of them are absolutely guaranteed to use only 15 cubic feet of gas per horse power per hour. In connection with this, there is a very interesting and suggestive calculation easily made. Thus:-One thousand cubic feet of gas in Toronto costs the consumer $1.10, for lighting purposes, and 90c. for power and heating purposes. If burnt in a 5 foot burner, this thousand feet will give 3200 candle power hours. If consumed in a gas engine requiring 25 feet per horse power per hour, this 1,000 feet will give 40 horse power hours. These 40 H. P., if used to run a dynamo, assuming 75% electrical efficiency and 4 watt lamps, will give 5584 candle power hours. The Toronto Incandescent Light Co. sells current at the net rate of 6 cents per 160 candle power hours, barring special arrangements.

Reducing these prices down to their equivalents per 1,000 candle power hours, gives:

Gas in burner...........costs, 34.375 cents per 1,000 c. p. h.
Gas Engine and Dynamo " 16.177 " " " " "
Tor. Inc. Co. . " 37.500 " " " " "

Now, one 16 candle power incandescent lamp burning for 62½ hours is equivalent to 1,000 c. p. h., so it can be seen how, in a large building requiring extensive illumination, it may very easily be actually less expensive to run a private dynamo, operated by a gas engine, than to rent either gas or current. So far, there is a difference, in favor of the gas engine, of over 100 per cent., but no account has yet been taken of two considerable items, salary of attendant, and interest and depreciation. These will of course greatly increase the cost as calculated above, but to what extent depends entirely on the size of the installation; there seems to be, however, a considerable margin to come and go on.

·The second modification is the use of a storage battery located in the building, and kept constantly charged by wire from the central station. I am aware that the advantages of this plan are open to criticism, and would have to be the result of special arrangements with the central station. I am also aware that "Central Station Management" as understood in most Canadian stations, except the really large ones, is not the science that it has become in the States and in Europe, and that as a fact, so little attention has been given to its study, that managers do actually not know those conditions of operation which are productive of best financial results to themselves. At the same time I am convinced that a little persevering effort on the part of interested persons would bring about a state of affairs favorable to the economical operation of the modification I have just suggested.

The position may be briefly and clearly described thus :-The best and most paying load for a central station is a *constant* one, one that doesn't jump from 0 at 5 o'clock p.m. to 500 H. P. at 8 p. m., but that remains fairly steady all the time. This is just the very kind of load that a storage battery gives, and moreover, it presents another very attractive feature to the central station as follows: A building is wired for say 500 lights; it will use these 500 lights all at the same time two or three times a year, but the machinery in the central station has got to have this 500 light capacity consuming interest and suffering depreciation all the year round, although it earns money only three or four times a year. A storage battery, however, that can be charged at the rate of 20 amperes continuously, will give out when required the whole current for the 500 lights, and keep them going for four or five hours. Thus the central station will only have to provide generating machinery sufficient for 20 amperes, which will be earning money all the time, instead of sufficient for 500 amperes which will be idle most of the time. Consequently it is to the obvious interest of the central station to encourage the use of the battery, and in order to do so it will lower price of current. A very well known authority on electric matters, has stated that it would pay central stations to sell current (for use with storage battery) at 4 cents per kilowatt hour in order to get the constant load that is so advantageous. This figure is equivalent to 16 cents per 1000 candle power hours, and to this must again be added interest of and depreciation on accumulators. Here again particular conditions will affect the total per 1000 c. p. h., but as in the former modification a considerable margin is left . I have no doubt, that central stations would see the advantage of lowering their rates on such loads, if the case were properly presented, and advantage to both parties would result.

There being no particular installation to study, it is impossible to be other than general in this article; it was intended merely to point out to parties having large buildings to light, or to supply elevators for, that there are several ways of doing it, each of which is best only under ·certain favorable conditions, while under different conditions it should give place to some more efficient method, the actually best method being, determined only after due consideration of the peculiar conditions.

## THE S. K. C. SYSTEM IN BROOKLYN.

On Friday evening, March 22nd, says Electricity, an informal reception was given at the station of the Citizens' Electric Illuminating Company in Brooklyn, the guests being invited especially to see the S. K. C. two-phase system in operation.

A 600 k. w. generator supplied current, operating both a 3 h. p. and a 10 h. p. motor, in connection with a bank of 250 110 volt 16 c. p. incandescent lamps and twenty-eight arc lamps. The system worked to perfection, and nothing but favorable criticism was heard.

There was a large attendance of Brooklyn people, and a goodly delegation of New York men who had not been fortunate enough before to see the practical workings of the system.

The Citizens' Company, until recently, had been unable to make a feature of incandescent lighting owing to the patent situation. Now, however, it is freed from further annoyance, and will no doubt be a stronger factor than ever before in the local lighting situation in Brooklyn.

It is said to be the intention of the new electric light company organized at Gananoque, to operate on the underground system. They are said to have already in hand orders for 500 lights.

It is proposed to convey current for electric lighting purposes from the Chaudiere Falls to Notre Dame de Levis. These falls are 95 feet high, and are said to be capable of generating 500,000 H. P

## THE MONOCYCLIC SYSTEM.*

### BY DR. LOUIS BELL.

IT is the purpose of this paper to call your attention to the various methods of central station distribution, involving motor service on the alternating current system, and more especially to a modified single-phase alternating system, which lends itself very readily to a very simple and straightforward distribution of lighting without sacrificing the excellent motor service which makes the true polyphase systems so desirable. For all around central station work, the lighting service is of most fundamental importance, and convenience and economy must, in a vast majority of cases, be the first consideration. (The author here reviews the different systems of electrical distribution indicated by the accompanying diagrams.)

It is instructive to glance over this list of alternating lighting systems to see their relative complexity and advantages. It is especially noteworthy that any and every method of distribution that saves copper introduces in some form or other the question of balance. This is the price we pay for reduction in cost of conductors. It has not seriously interfered with the use of the Edison three-wire system; in fact, those most familiar with that system were the first to make light of the difficulty; nor do I think it stands as a valid objection to the use of the polyphase systems in cases where they are desirable, as none of them are more sensitive in the matter of balance than the Edison three-wire system which is now in such extensive and uniformly successful use. We may further note that in each of the alternating systems where a great saving of copper is accomplished, a fourth wire is necessary, at least if both lights and motors are to operated; in each case, however, of trifling size.

Having now looked over the field in general, we may pass to the more minute consideration of the somewhat striking electrical peculiarities of the monocyclic system; peculiarities which, although they do not involve any particular complexity, are yet of decided interest.

The general principle of this system is well shown in this diagram. So far as the main work of the generator is concerned, its winding is closely similar to that of any well-designed alternator. The armature is of the ironclad type, and the winding is made in machine wound coils which are invariably insulated, and can be very readily slipped into place. There is, however, upon the armature, a second set of coils of cross section equivalent to that of the main coil, but composed of comparatively few turns, so that the room taken up on the armature is very small, and owing to the shallowness of the slots necessary to accommodate this second or teaser coil, the output of the machine, considered as a single-phase generator, is not affected. This teaser coil is located with reference to the main coil as shown in the diagram. Its place on the armature is midway between the other coils, and the electromotive force generated is in a direction at right angles to that of the principal coil. It is evident now that if we connect the terminals of the main coil, we shall get an electromotive force compounded of the two, and in some intermediate direction. In general, by varying the proportions of the two coils, and hence their electromotive forces, we could obtain a resultant electromotive force between their terminals, having any angle we pleased with either the main or the teaser coil. If, then, wires are taken upon the line from the terminals of the main coil, and also from the teaser coil, we can obtain from the main line three electromotive forces, two of which are symmetrically situated with reference to the E. M. F. of the teaser coil, and bear to it any phase relation that we please. One of the most convenient arrangements, and that which is most generally adopted, involves such a relation of the electromotive force of the teaser coil to that of the main coil, that we shall have on the line three electromotive forces approximately 60 degrees apart. In other words, the resultant E.M.F.'s between teasers and main coils are each 60 degrees from the E. M. F. of the teaser coil. Such an arrangement is that shown in the diagram. Under these circumstances, it is clear that if one of these electromotive forces were reversed, either in transformers or anywhere in the translating devices, the result would be three electromotive forces 120 degrees apart, one of them having been turned through an angle of 180 degrees. Meanwhile, the relation between the power wire, which is connected to the teaser

*Abstract of paper read before the National Electric Light Association at Cleveland, Ohio, 1895.

coil, and the outside wire has no effect upon the electromotive force between these outside wires, since the electromotive force of the main coil itself does not interact with the power wire, except in so far as a portion of it may act with the power wire to form a resultant phase, and electromotive force for running motors. Consequently, so far as lights are concerned, the two outside wires behave precisely like the leads from any other alternating generator, while so far as motors are concerned, we have the power of getting our three electromotive forces 120 degrees apart, and hence have the same magnetic effect as with a three-phase system. The arrangement of lights and motors with this device is clearly shown in the diagram. It is evident that we can take from the outside wires of the monocyclic system either arc or incandescent lights anywhere and to any extent the capacity of the machine permits, working for the incandescents either two or three wire distribution at option. For a motor, two transformers are connected anywhere we please, one between the power wire and each of the outside wires. At this point the resultant phases come into play and the necessary reversal of one of the electromotive forces is accomplished by the very simple and obvious device of reversing one of the sec-

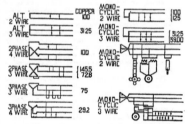

ondaries, as shown in the diagram. To the secondary circuit thus constituted, we can connect a standard induction motor which will start and run as well as if connected to a regular three-phase system, or instead of reversing one of the transformer secondaries we may accomplish the same virtual reversal of the electromotive force by reversing one of the coils in the motor itself. We therefore have a system which, so far as lights are concerned, is a simple alternating system; so far as motors are concerned, the dynamical equivalent of a polyphase system. With a differently proportioned formation we could place upon the secondaries two electromotive forces 90 degrees apart if necessary, and then run two-phase motors instead of three-phase motors, if there were any object in so doing. Such an arrangement, however, would be less desirable than that of the quasi three-phase system, for the reason that without gaining anything in the motors we should have to generate a larger electromotive force in the teaser coil, and hence take up more room on the armature with it, perhaps enough to have an effect upon the output of the machine considered as a single-phase generator. It is sufficiently evident that the method shown would not be the only way of getting the same result. For example, in this second diagram a somewhat different arrangement is shown accomplishing precisely the same end. Here our object is to operate secondary mains on the Edison three-wire system, and in connection with them to run motors at any point we please. A large transformer, to which the secondary is connected on the three-wire system, is, therefore, installed, and the secondary mains distributed in any manner we please. A second and small transformer, proportioned to the total amount of motor service desired, is connected as shown in the diagram, and the power wire leading from it is taken through the whole or part of the three-wire system. The device is analogous to the arrangement of the generating coils themselves, and the result is the ability to operate a standard induction motor by connecting it anywhere on three-wire system to the two inside wires and to the power wire. Such an arrangement as this is immensely convenient in distributing power and light in cities where, for example, it is desired to establish an extensive system of secondary mains through Edison tubes, or other convenient underground distribution. It is, furthermore, interesting to know that one is not confined to the use of either two or three-phase induction motors, since a monocyclic generator connected to the primary circuit makes an excellent synchronous speed without the assistance of a starting motor, in this particular being vastly superior to the pure single-phase synchronous machine. But, it may be asked, how about this power wire? In case, for example, of a transmission over a considerable distance before the distributing point is reached, must the power wire be part of the transmission system? In answer, I need only call your attention to the fact that the essential point of the monocyclic system, so far as motors are concerned, is the establishing of an electromotive force bearing the same relation to the system as is borne by the teaser coil of the generator. Consequently, in case of a transmission plant, the main generators at the distant station may be simple single-phase machines, the subsidiary electromotive force being furnished by a synchronous motor or similar device at any point in the system. So we might readily have an extensive transmission with a monocyclic machine in the sub-station of such size as is necessary to furnish current for what motors may be upon the system. The power wire would then run only to the sub-station. Another interesting peculiarity of the monocyclic system, and one which is not without importance in case of an extensive power distribution, is the following :

Under ordinary circumstances, induction and synchronous motors are wound so that the counter electromotive force affects the system in a perfectly symmetrical manner, and the current flows over all the wires with some degree of symmetry in response to the demands of the motors on the system. It is customary, however, in the monocyclic system, to employ motors so wound as to throw a high counter electromotive force into the power wire when the motor is at speed and loaded, thereby reducing the normal current carried over the power wire to a purely nominal amount, and this can evidently be done without sacrificing much in the matter of starting, since at the start all the counter electromotive forces in the motors are zero. We have, then, a motor system of a type really peculiar to the monocyclic system, in that each motor will start under the same conditions of impressed electromotive force as if it were a polyphase motor, while, when at speed and loaded, it would be operating virtually as a single-phase machine. If, however, it were overloaded so that it would tend to slow down or stop, sufficient energy would flow over the power wire to bring it back to speed, just as if it were a polyphase machine. This is only one of various interesting ramifications in the system when developed to meet special conditions.

The connections shown in the diagram, however, are those of the most direct applicability and probably which would be most extensively used for central station service. At this point it may be appropriate to ask what is the advantage of such a system. It evidently secures exceedingly marked advantages in the ability to operate the lights on existing circuits or with the simplest possible kind of distribution, and at the same time to run at any point in the system synchronous motors or induction motors of well-tried and familiar types. The question can be readily answered; in fact, the question is almost obvious. The price which we have to pay for this advantage is the installation of the power wire, which necessarily adds something to the weight of copper in the system and to the trouble of installation. Under all ordinary circumstances the power wire need be of trivial cross-section compared to that of other wires, since, as a rule, the energy required for operating the motors from the given central station is small compared with the total capacity of the stations ; and further, it is worth remarking that the monocyclic motors, either synchronous or induction, will run perfectly well if the power wire is disconnected after the power wire is at speed, operating them as single-phase machines. It is, of course, well known that the single-phase synchronous motor gives admirable results, and it is also true that a single-phase induction motor can be constructed of excellent efficiency and other electrical properties. The only material difficulty is to get the motor started with a good torque. The monocyclic connection enables this to be accomplished. After the motor is at speed the power wire becomes no longer necessary to successful operation, so that in spite of the necessary existence of the power wire it is easy to see that the additional amount of

copper is not likely to be burdensome in central station operation. It would hardly ever be necessary to install a power wire of more than one-fourth the joint cross-section of the others, as given in the diagram, and generally a much smaller wire will suffice.

I have thus endeavored in a brief space to give a good working idea of the monocyclic method of combined light and power distribution which has been devised, and is urged upon the public as specially adapted for the work of central station distribution on account of its unique simplicity in practice. As to what has already been done in the installation of such apparatus, a considerable number of these monocyclic generators are in daily use in central stations, for the most part operating over circuits already established and displacing the higher frequency alternators which had been previously used. They are giving excellent results, and the operation of the motors, wherever employed, has been highly satisfactory.

## MOONLIGHT SCHEDULE FOR MAY.

| Day of Month. | Light. | Extinguish. | No. of Hours. |
|---|---|---|---|
|  | H.M. | H.M. | H.M. |
| 1...... | ........... | A. M. 4.00 | } 3.20 |
| 2...... | A.M. 12.40 | ........... | |
| 3...... | " 1.10 | " 4.00 | 2.50 |
| 4...... | " 1.30 | " 4.00 | 2.30 |
| 5...... | " 2.00 | " 4.00 | 2.00 |
| 6...... | No light. | No light. | .... |
| 7...... | No light. | No light. | .... |
| 8...... | No light. | No light. | .... |
| 9...... | P.M. 7.30 | P. M. 9.30 | 2.00 |
| 10...... | " 7.30 | " 10.30 | 3.00 |
| 11...... | " 7.30 | " 11.30 | 4.00 |
| 12...... | " 7.30 | A. M. 1.00 | 5.30 |
| 13...... | " 7.40 | " 1.10 | 5.30 |
| 14...... | " 7.40 | " 1.40 | 6.00 |
| 15...... | " 7.40 | " 2.00 | 6.20 |
| 16...... | " 7.40 | " 2.20 | 6.40 |
| 17...... | " 7.40 | " 2.40 | 7.00 |
| 18...... | " 7.40 | " 3.00 | 7.20 |
| 19...... | " 7.40 | " 3.20 | 7.40 |
| 20 .... | " 7.40 | " 3.40 | 8.00 |
| 21...... | " 7.40 | " 3.40 | 8.00 |
| 22...... | " 7.50 | " 3.40 | 7.50 |
| 23...... | " 7.50 | " 3.40 | 7.50 |
| 24...... | " 7.50 | " 3.40 | 7.50 |
| 25 .... | " 7.50 | " 3.30 | 7.40 |
| 26...... | " 9.00 | " 3.30 | 6.30 |
| 27...... | " 10.00 | " 3.30 | 5.30 |
| 28...... | " 10.40 | " 3.30 | 4.50 |
| 29...... | " 11.00 | " 3.30 | 4.30 |
| 30...... | " 11.10 | " 3.30 | 4.20 |
| 31...... | " 11.40 | " 3.30 | 3.50 |
| | | Total, | 148.20 |

## EXPERIMENTS WITH SUPER-HEATED STEAM.

IN giving the results of their protracted experiments with saturated and super-heated steam, the Alsace Union of Boiler Owners say that, theoretically, it has never been denied that super-heated steam should give a higher efficiency than saturated, yet no experiments were undertaken with super-heated steam. Subsequently, however, after numerous trials, the oldest engine even was found capable of being safely used with super-heated steam, and not only without injury, but more economically than with saturated. It is also declared by the union that in installing a super-heater care is essential that the advantages gained are not lost either by less perfect combustion or by greater radiation losses—the cost of the super-heater not to exceed, of course, the saving obtained in coal consumption ; the super-heater to be connected with the boiler, so that both can be fired from the same furnace ; and after leaving the super-heater, the gases should come in contact with the heating surface of the boiler, and, lastly, with the heating surface of the economizer. Further, these experiments showed that the use of super-heated steam does not exclude the use of steam jacket. Though both super-heating and steam jackets were used, yet condensation in the high-pressure cylinder occurred. The use of low-pressure, seven and one-half atmospheres, did not give such good results as the use of high pressure, eleven and one-half atmospheres.

## INSTRUCTIONS TO BOILER ATTENDANTS.

THE Manchester Steam Users Association of England, has issued a revised edition of its " Instruction to Boiler Attendants."

In forwarding these instructions to its members, the Association says :

"These instructions have been drawn up with much care, it being desired to make them as complete and educational as possible. There are so many points affecting the safety and proper treatment of boilers, that it was found impossible to compress the instructions into a small space. In boiler and engine rooms, height of wall space is more generally available than width, and, therefore, the sheet was made long and narrow, rather than short and wide. If hung up so as to be about two feet from the floor, it can easily be read from top to bottom.

" It is desirable that the sheet should be mounted, and the best plan of doing this will perhaps be to have a board about ¾ in. thick built in three or four widths and stiffened by a batten at each end, the joints being grooved and tongued. On this board the sheet might then be pasted, and varnished for preservation. In most cases it might be well to have this done by a bookbinder.

" When mounted, the sheet should be placed in a good light, and where the boiler attendants can have convenient access to it. They should be encouraged to study and master its contents. Much of the information contained therein will be of service daily, and not merely on the occurrence of an emergency."

GETTING.UP STEAM.—Warm the boiler gradually. Do not get up steam from cold water in less than six hours. If possible, light the fires over night.

Nothing turns a new boiler into an old one sooner than getting up steam too quickly. It hogs the furnace tubes, leads to grooving, strains the end plates, and sometimes rips the ring seams of rivits at the bottom of the shell. It it a good plan to blow steam into the cold water at the bottom of the boiler, or to open the blow-out tap, and draw the hot water down from the top.

FIRING.—Fire regularly. After firing, open the ventilating grid in the door for a minute or so. Keep the bars covered right up to the bridge. Keep as thick a fire as quantity of coal will allow. Do not rouse the fire with a rake. Should the coal cake together, run a slicer in on top of the bars and gently break up the burning mass.

Repeated trials have shown that under ordinarily fair conditions, no smoke need be made with careful hand-firing. Alternate side firing is very simple and very efficacious.

CLEANING FIRES AND SLAKING ASHES.—Clean the fires as often as the clinkers render it necessary. Clean one side at a time, so as not to make smoke. Do not slake the clinkers and ashes on the flooring plates in front of the boiler, but draw them directly into an iron barrow and wheel them away.

Slaking ashes on the flooring plates corrodes the front of the boiler at the flat end-plate, and also at the bottom of the shell where resting on front cross wall.

FEED-WATER SUPPLY.—Set the feed valve so as to give a constant supply, and keep the water up to the height indicated by the water-level pointer.

There is no economy in keeping a great depth of water over the furnace crowns, while the steam space is reduced thereby, and thus the boiler rendered more liable to prime. Nor is there any economy in keeping a very little water over the furnace crowns, while the furnaces are rendered more liable to be laid bare.

GLASS WATER GAUGES AND FLOATS.—Blow through the test tap at the bottom of the gauge hourly, as well as through the trap in the bottom neck, and the trap in the top neck twice daily. These taps should be blown through more frequently when the water is sedimentary, and whenever the movement of the water in the glass is at all sluggish. Should either of the thoroughfares become choked, clean them out with a wire. Work the floats up and down by hand three or four times a day to see that they are quite free. Always test the glass water gauges and the floats thoroughly the first thing in the morning before firing up, and at the commencement of every shift.

It does not follow that there is plenty of water in the boiler because there is plenty of water in the gauge glass. The passages may be choked. Also, empty gauge glasses are sometimes mistaken for full ones, and explosions have resulted there-

from. Hence the importance of blowing through the test taps frequently.

BLOW-OUT TAPS AND SCUM TAPS.—Open the blow-out taps in the morning before the engine is started, and at dinnertime when the engine is at rest. Open the scum tap when the engine is running, before breakfast, before dinner, and after dinner. If the water is sedimentary, run down ½ in. of water at each blowing. If not sedimentary, merely turn the taps round. See that the water is at the height indicated by the water-level pointer at the time of opening the scum tap. Do not neglect blowing out for a single day, even though anti-incrustation compositions are put into the boiler.

Water should be blown from the bottom of the boiler when steam is not being drawn off, so that the water may be at rest and the sediment have an opportunity of settling. Water should be blown from the surface when steam is being drawn off, so that the water may be in ebullition and the scum floating on the top. If the water be below the pointer, the scum tap will blow steam ; if above the pointer, the scummer will miss the scum.

SAFETY VALVES.—Lift each safety valve by hand in the morning before setting to work, to see that it is free. If there is a low-water safety valve, test it occasionally by lowering the water level to see that the valve begins to blow at the right point. When the boiler is laid off, examine the float and levers and see that they are free, and that they give the valve the full rise.

If the safety valves are allowed to go to sleep, they may get set fast.

OPENING DRAIN TAPS AND STEAM PIPES.—If the boiler is one of a range, and the branch steam pipe between the junction valve and the main steam pipe is so constructed as to allow water to lodge therein, open the drain tap immediately the boiler is laid off, and keep it open until the boiler is set to work again. If the main steam pipe is so constructed as to allow water to lodge therein, open the drain tap immediately the engine is shut down, and keep it open till the engine is set to work again.

If the water is allowed to lodge in the pipes, it is impossible to blow it out under steam pressure without danger. Attempting to do this frequently sets up a water-hammer action within the pipes, and from this cause several explosions have occurred. The only safe plan is not to let the lodgment occur, or to shut off the steam before opening the drain taps.

SHORTNESS OF WATER.—If the boiler is found to be short of water throw open the fire doors, lower the dampers, ease the safety valves, and set the engine going, if at rest, so as to reduce the pressure. If the boiler is one of a series, shut down the junction valve. If there is reason to conclude that the water has not sunk below the level of the furnace crowns, and they show no signs of distress, turn on the feed and either draw the fires quickly, beginning at the front, or smother them with ashes or anything ready to hand. If there is reason to conclude that the water has sunk below the level of the furnace crowns, withdraw, and leave the safety valves blowing. Warn the passers by from the front.

EASING THE SAFETY VALVES.—If either the construction of the boiler or the character of the feed water is such as to render the boiler liable to prime, the safety valve should be eased gently.

TURNING ON THE FEED.—From experiments the association has conducted, it appears that this is the best thing to do in nearly every case, especially where the feed is introduced behind the firebridge, as it would tend to restore the water level, and at the same time to cool and reinvigorate the furnace plates. While, however, the experiments showed that showering cold water onto red-hot furnace crowns would not, as has been generally supposed, lead to a sudden and violent generation of steam which the safety valves could not control and the shell could not resist, it is thought that if the furnace crowns were very hot and just on the point of giving away, the generation of a few additional pounds of steam might turn the scale and lead to a collapse. Thus it might be wise to turn on on the feed in some cases and not in others, according to the extent to which the furnaces are overheated, and this it is difficult to ascertain. Under these circumstances a hard and fast rule, applicable to all cases, cannot be laid down, and therefore, having regard to the safety of the fireman, the advice to turn on the feed, as a general rule is confined to those cases where the water has not sunk below the level of the furnace ground.

DRAWING THE FIRES.—This ought not to be attempted if the furnace crowns have begun to bulge out of shape.

It is an extremely responsible task to give any recommendation with regard to the treatment of a boiler when short of water and working under steam pressure, that shall be applicable to every case under every variety of circumstance. A boiler attendant has no right to neglect his water supply and allow it to run short ; nor has he a right to charge the fires without making sure that the furnace crowns are covered. Should he neglect these simple precautions it is impossible to put matters right without some risk being run. A boiler with hot fires and with furnace crowns short of water is a dangerous instrument to deal with, and the attendant who has done the wrong must bear the risk. The best advice the association can give the boiler attendants on this subject is, do not let shortness of water occur. Keep a sharp look-out on the water-gauge.

USE OF ANTI-INCRUSTATION COMPOSITIONS.—Do not use any of these without the consent of the association. If used, never introduce them in heavy charges at the manhole or safety valve, but in small daily quantities along with the feed-water.

Many furnace crowns have been overheated and bulged out of shape through the use of anti-incrustation compositions, and in some cases explosions have resulted.

EMPTYING THE BOILER.—Do not empty the boiler under steam pressure, but cool it down with the water in ; then open the blow-out tap and let the water pour out. To quicken the cooling the damper may be left open, and the steam blown off through the safety valves. Do not, on any account, dash cold water on the hot plates. But in case of an emergency pour cold water in before the hot water is let out, and mix the two together so as to cool the boiler down generally, and not locally.

If a boiler is blown-off under steam pressure the plates and brickwork are left hot. The hot plates harden the scale, and the hot brickwork hurts the boiler. Cold water dashed on to hot plates will cause severe straining by local contraction, sometimes sufficient to fracture the seams.

CLEANING OUT THE BOILER.—Clean out the boiler at least every two months, and oftener if the water is sedimentary. Remove all the scale and sediment as well as the flue dust and soot. Show the scale and sediment to the manager. Pass through the flues, and see not only that all the soot and flue dust has been removed, but that the plates have been well brushed. Also see whether the flues are damp or dry, and if damp find out the cause. Further, see through the thoroughfares in the glass water gauges and in the blow-out elbow pipe, as well as the thoroughfares and the perforations in the internal feed dispersion pipe and the scum pipe are free. Take the feed pipe and scum troughs out of the boiler if necessary to clean them thoroughly. Take the taps, if not asbestos packed, and the feed valve to pieces, examine, clean and grease them, and, if necessary, grind them in with a little sand. Examine the fusible plugs.

All taps, whether asbestos packed, or metal to metal, should be followed in working, especially when new. The gland should be screwed down as found necessary so as to keep the plug down to its work, otherwise, it may rise, let the water pass, and become scored.

PREPARATION FOR ENTIRE EXAMINATION.—Cool the boiler and carefully clean it out as explained above, and also dry it well internally. When the inspector comes, show him both scale and sediment as well as the old cap of the fusible plug, and tell him of any defects that manifested themselves in working, and of any repairs or alterations that have been made since the last examination.

Unless a boiler is suitably prepared, a satisfactory entire examination cannot be made. Inspectors are sent at considerable expense to make entire examinations, and it is a great disappointment when their visits are wasted for want of preparation.

PRECAUTIONS AS TO ENTERING BOILER.—Before getting inside the boiler, if it is one of a series, take off the junction valve handwheel, and if the blow-out tap is connected to a common waste pipe, make sure that the tap is shut and the key in safe keeping.

From the neglect of these precautions, men working inside boilers have been fatally scalded.

FUSIBLE PLUGS.—Keep these free from soot on the fire side and from incrustation on the water side. Change the fusible metal once every year, at the time of preparing for the association annual entire examination.

If fusible plugs are allowed to become incrusted, or if the metal be worked too long, they become useless, and many furnace crowns have been rent from shortness of water, even though fitted with fusible plugs.

GENERAL KEEPING OF BOILER.—Polish up the brass and other bright work in the fittings. Sweep up the flooring plate frequently. Keep ashes and water out of the hearth pit below the flooring plates. Keep the space on the top of the boiler free, and brush it down once or twice a week. Take a pleasure in keeping the boiler and the boiler house clean and bright, and in preventing smoke.—The Safety Valve.

## CANADIAN ASSOCIATION OF STATIONARY ENGINEERS.

Note.—Secretaries of the various Associations are requested to forward to us matter for publication in this Department not later than the 20th of each month.

### HAMILTON ASSOCIATION NO. 2.

The annual dinner of the above Association was held at the Commercial Hotel on the evening of April 18th. There was a good attendance of members and visitors—about seventy in all. As usual, Mr. Maxey, the proprietor of the Commercial, distinguished himself by the manner in which the dinner was prepared and served. On the removal of the cloth the President, Bro. Joseph Langton, took the chair, and addressed a few well-chosen remarks of welcome to the visitors, after which the toast list was proceeded with, the first toast being that of "Her Majesty the Queen."

The toasts to "The Governor-General" and "The Dominion and Local Legislatures" were responded to by Mr. Alex. McKay, M. P.; "The Army and Navy," by a song—"Rule Britannia," by Prof. Cline, and a trio by Messrs. Maxey, Cline and Leroy Mallard ; Alderman McKennan responded to the toast "The Mayor and Corporation"; "The Executive Head" was responded to by Bros. Edkins, Wickens and Blackgrove, of Toronto ; Mr. Watson of Kingston, Bros. Clapperson, Holden and Brice, of Hamilton, responded to the toast "Our Manufacturers"; "Our Sister Associations" brought responses from Bros. Phillips and Huggett, of Toronto ; Bro. Pettigrew, of the Board of Education of Hamilton, responded on behalf of "The Learned Professions"; The President, Bro. Joseph Langton, and the Corresponding Secretary, Bro. Wm. Norris, replied in a suitable manner to the toast "Hamilton No. 2," proposed by Bro. A. E. Edkins, of Toronto ; Bros. Robertson and Stott, of Hamilton, responded for "The Ladies," and Mr. Mattie of the Toronto Globe, and Mr. R. J. Robb of the Hamilton Spectator, for "The Press."

The proceedings of the evening were enlivened by songs from Prof. Cline, Messrs. Martin, Maxey, Leroy Mallard, Thomas and Edkins.

Letters of regret were read as follows :—From S. S. Wrightman, M. P., J. T. Middleton, M. P. P., Hon. J. M. Gibson, and Mayor Stewart, of Hamilton ; the Ottawa, Montreal and Kingston Associations, C. A. S. E., and the editor of the ELECTRICAL NEWS. The evening throughout was a most enjoyable one.

The members of this Association express their disappointment at the fate of the Engineers' Bill before the Ontario Legislature at its last session. They think it peculiar that a matter of so much importance should have been so lightly dealt with, and are in hopes that the leaders in the movement for legislation will carefully consider the most effective manner in which to bring the bill again before the Government.

The Berlin Gas Co. have closed a contract with the Canadian General Electric Co for a 65 Kilowatt generator from which they will furnish power for the Berlin and Waterloo railway.

The Town Council of Listowel, Ont., has invited tenders for lighting the town with 14 arc lights of 1,500 to 2,000 candle power. The plant is required to be in operation by the first of July.

Mr. F. S. Mead, of Quebec, has recently been granted a United States patent for a gas engine ; Mr. G. E. Smith, of Ridgetown, for a machine for bending rails, and Mr. C. H. Waterous, Brantford, Ont., for a steam boiler.

## WATER WHEELS—A QUESTION OF EFFICIENCY.

It was but a few years since when the old testing flume at Holyoke, Mass., was in charge of Mr. James Emerson that the manufacturer of turbine wheels who could not boast of a reliable (?) test that showed in useful effect a percentage of the power of the water from 80 to 85 per cent. was considered decidedly behind the times, and while it is a fact that some of those same wheels that gave such remarkable results at the testing flume when put into actual use in the mills under ordinary working conditions, proved to be less effective than many others which made no such claims, and had no other test than that of actual use. The writer was frequently scouted at by certain parties ten years ago when he asserted that, notwithstanding all the scientific tests that had been made, and the flaming circulars that had been issued by prominent water-wheel makers, that in every-day practice and under ordinary conditions where one wheel could be found that would yield 80 per cent. in useful effect ten would be found that would yield less than that amount. Now it transpires from the statements of Mr. J. McCormick, who has heretofore been considered by a certain class of hydraulic engineers as the "king of water-wheel designers" that the writer was correct. In a recent article by that gentleman with reference to the "Hercules," which had been by many considered the perfection of turbines, and has claimed a power of from 85 to 90 per cent., he says ; " In the present testing flume with Mr. Emerson's apparatus those same wheels that had for several years been upon the market and guaranteed to yield from eighty-five to ninety per cent. in useful effect *required material alterations to bring them up to an efficiency of 85 per cent.";* from which we may reasonably infer that those wheels which were rated at from eighty-five to ninety per cent. were in reality nearer from seventy-five to eighty. Again Mr. McCormick further says : " Wheels made from the same pattern which had produced from eighty-eight to ninety per cent. in the old flume (under the Emerson administration of course) would only yield a little over eighty per cent. in the new one " (under the charge of Mr. Hershel). In consequence he says "he was compelled to change the *vital patterns* of the Hercules wheels of all sizes, both right and left hand, after which they were tested and brought up to efficiency of *eighty-eight per cent. or better.*" Now if it required those changes in order to bring the wheel up to eighty per cent. or a trifle better (he does not say how much better) the reader is at liberty to form his own opinion as to the efficiency of those wheels before those *vital parts* were changed. Now while all will admit that Mr. McCormick, one of the most experienced water-wheel designers in the country, has probably devoted more time and study to this subject than any other man, and while we do not for a moment question either his honesty or integrity, yet one thing is self-evident that the former tests made with the same instruments by Mr. Emerson, whether intentionally or not, certainly deceived him as well as the public, and we cannot but admire his honesty in coming out and practically admitting such to be the case ; and further that either much of the former testing or the latter is a humbug. This only confirms what has often been stated that there is but little confidence to be placed in those tests, no matter how honestly they may be conducted, unless the identical wheel that has been subjected to the test is obtained, for the miller has no assurance that the next wheel of the same size will give the same efficiency. To those who have examined some of the wheels that have been fitted up for that purpose, it is unnecessary to say that they are fitted up with the utmost care and perfection. Everything is as perfect as mechanical skill can render it and the surface of the buckets are scraped smooth, and in some cases polished, and it is strange if such wheels do not give better results than the common every-day wheel where much less labor and care is bestowed. The fact is with the present condition between the different manufacturers and the low price that wheels are being sold, if the same labor were expended upon every wheel that is expended upon some of the test ones, the price would not cover the cost of production.

There is no doubt but all water-wheel makers endeavor to obtain as smooth castings as possible and fit up their wheels as well as prices will admit, but no extra work is found in the average wheel. Now when we consider that the efficiency of a wheel depends much upon the resistance offered to the water in passing through the buckets, it is evident that with those which are perfectly smooth the efficiency will be greater than those which are not. That many of the millers have but little confidence in those tests is well illustrated in the reply of an old miller to an agent of a certain wheel who was boasting of the high percentage of power that there wheel had shown at Holyoke testing flume. "Young man," said he, "I want a good wheel and one that will do the work. I don't care a —— for all your tests ; if you want to put a wheel in my mill and hitch it on to my mill-stones, and if it will grind on an average of — bushels of corn per hour, I will keep it and pay for it ; it not you will take it out at your own expense."

Another question upon which there is an honest difference of opinion, is the economy in wheels of small diameter and high speed, over those of larger diameter and less speed, both using the same quantity of water. There is no question but within a few years the tendency of all water-wheel makers has been toward wheels of small diameter with large openings, and that many of the present style of wheels of from eighteen to twenty inches in diameter use the same quantity of water that was formerly used by wheels of from twenty-five to thirty, while the speed is increased in proportion to the diameter, and all that has been said upon the subject of small turbines, not one *good argument* has been offered in their favor. It is true that it costs less to make a small wheel both in stock and labor, and by enlarging the openings so that the same quantity of water may pass through it, and figuring the power from the quantity of water discharged, may enable the manufacturer to realize a greater profit, and this may be one of the reasons, but if so, it is no argument in its favor in an economical point. The high speed is one of the objections to small wheels. If a turbine wheel was like a steam engine and could be placed in a shop where it could be looked after every day and kept in perfect order, the case would be different. An engine to run one hundred and fifty revolutions per minute, must needs have an engineer to look after it, and must be oiled and wiped at least once a day in order to keep it up and in good working order, while the poor little water-wheel is buried in some out-of-the-way place in a flume or penstock where it is practically inaccessible except when the mill is stopped and the water drawn off (which is not very often) and there left to its fate to run anywhere from three to eight hundred revolutions per minute without oil and comparatively no care, with mud, sand, stones, and other rubbish to contend with, and is expected to perform its daily work day after day and week after week without a murmur, and the only wonder is, that they last one-half as long as they do ; and the faster they run the sooner they wear out. Now this, if nothing else, furnishes one good argument in favor of wheels of larger diameter and there is no question but a wheel thirty inches in diameter with a capacity for using the same amount of water as one of fifteen and run at one-half the speed, would be as efficient, if not more so, and much more durable. The thirty-inch wheel would cost a few dollars more probably in the first instance, but as the cost of the wheel is but a small item in the outfit when durability is taken into consideration, there is no question but the larger would be the cheapest in the end. There are other points in favor of large wheels with small openings which may be discussed in a future article.—C. R. Tompkins, M. E., in *Milling*.

## NOISE FROM A GAS ENGINE.

Among the various engineering investigations which for some time have engaged the attention of mechanical experts is that having in view some ready method for deadening the objectionable noise made by the puffs from the exhaust of the gas engine, but only an indifferent amount of success has hitherto attended these efforts. The most recent contrivance of the kind is a device described in a French journal, and claimed to be simple, efficient and inexpensive. Briefly, a pipe split for a distance of about two metres is attached to the end of the exhaust, with the split end upward, and beginning at the lower end of the cut, which may best be made by a saw, dividing the pipe into two halves, the slotted opening is widened out toward the top until it has a width equal in extent to the diameter of the pipe. Under this arrangement the puff of the exhaust spreads out like a fan, and the discharge into the open air takes place gradually, the effect produced depending somewhat on the flare of the tube.

# ELECTRIC RAILWAY DEPARTMENT.

## UNDERGROUND RAILWAY SYSTEM AT WASHINGTON.

THE Metropolitan Street Railway Company, of Washington, D. C., who a few years ago are said to have dropped $360,000 on experiments for the propulsion of street cars by storage batteries, are now turning their attention to a metallic subway system, substantially on the lines of the Buda Pesth road. By a recent Act of Congress the company are compelled to adopt some underground system on their Ninth street line, and to have the same in operation by August next. The method of construction employed is thus described by a correspondent of the New York Electrical Review :—

"The line now under construction extends from the Potomac River to the northern boundary of the city, and is a double track road four miles long. It has about a dozen modern curves, and its maximum grade is three per cent. It intersects the busiest section of the city and crosses three cable lines.

The construction is, in the main, identical with that of the ordinary cable lines. The conduit, however, is only 24 inches deep, while most of the cable conduits are 39 inches. The track is laid with 83-pound steel rails, seven inches deep. The electric conductors consist of angle irons in 27-foot lengths, weighing 23½ pounds per yard, with a conductivity equal to 400,000 circular mils copper. They will be hung from porcelain insulators, four inches in diameter, placed 13½ feet apart. These insulators and their iron castings are set in place from the surface of the street and can be readily and separately removed and replaced. The conductor irons will be slipped into the conduit through narrow pockets in the slot-rail 500 feet apart, and snaked along to their places. They will be fitted with expansion joints and flexible copper bonds. Any section can be removed without disturbing anything else. In designing the road quite a number of new features have been devised by Engineer Connett and patented by the company. Provision has been made for easily substituting a cable should it for any reason ever be found desirable. Ample drainage for the conduit will be provided.

A 10-way terra cotta conduit of the Lake pattern will be laid along the track to carry the necessary feeder cables. Wire made by the General Electric Company will be used, and 35,000 feet of 1,000,000 circular mils will be required. It will have a covering of five-thirty-seconds india rubber and one-eighth lead with an insulation resistance of 750 megohms per mile. The manufacturers guarantee it to last five years.

## SUGGESTIONS FOR REDUCING ACCIDENTS IN BROOKLYN.

The Advisory Committee appointed by Mayor Schieren, of Brooklyn, prior to the strike, to investigate the trolley system, as operated in Brooklyn, especially regarding the speed, made its report last month. The committee consists of Peter T. Austen, John Giff, Ditmar Jewell, William H. Nichols and R. S. Walker. The report is as follows :

In compliance with the request contained in your esteemed communication of December 29, 1894, in which you asked the undersigned to act as an advisory committee to investigate "the question of speed of the trolley cars, and a fender, or system of fenders, for them, and in general, the proper regulation of the trolley systems of this city," your committee has held a number of meetings, at which there have been present the officials of all the trolley roads, as well as representative motormen and conductors, and an opportunity has also been given the public to be heard.

After careful consideration of the information thus gained, together with considerable personal investigation of the practical details of the trolley system, we beg to submit the following recommendations for the regulation of the trolley system, which we believe will afford the best service to the public at the least risk to life :

1. The speed of trolley cars should not exceed ten miles an hour.

2. Every car should be provided with a device giving an audible signal when the car exceeds ten miles an hour.

3. Passengers should not be allowed to ride on the front platforms, and both gates of the front platforms should be kept closed when the cars are in motion.

4. The gates on the track side of the rear platforms should be kept closed.

5. Cars on all lines crossing main thoroughfares on which there are car tracks should come to a full stop before crossing. Cars on main thoroughfares must be kept under perfect control and run at a reduced speed at such crossings.

6. All cars should be provided with reliable fenders which should be approved of by a commissioner of experts.

7. In case of an accident through the negligence of a motorman, the motorman should be held criminally responsible therefor.

8. It being the opinion of this committee that accidents have occurred which have been due to the use of intoxicants by employes, we earnestly recommend that the companies provide at their respective depots comfortable waiting rooms for the men, where tea and coffee may be obtained at reasonable cost.

9. The tracks should be kept sufficiently sanded where needed.

10. As cars are often wilfully and unnecessarily obstructed by traffic wagons, we advise that the law that makes it a misdemeanor wilfully to obstruct, hinder or delay the passage of any car running on a street railway be rigidly enforced.

11. We consider that the present overcrowding of the cars is indecent and a fruitful source of inconvenience, delay, and danger, and we therefore strongly recommend that the number of passengers carried on any car should not exceed its seating capacity by more than 50 per cent. We are aware that the enforcement of this rule will necessitate an increase in the number of cars, but we consider that the public is entitled to proper and decent accommodation.

Mayor Schieren, it is said, will take action on the report at once. It is believed the Board of Aldermen will be called upon to adopt an ordinance in conformity with the recommendations made by the Advisory Committee.

## AN OLD ELECTRIC RAILWAY PATENT.

ON February 24, 1881—that is to say, 14 years ago—Profs. Ayrton and Perry took out their well-known patent, No. 783, for "improvements in electrical conductors applicable to electric railways and to electric signalling, and in apparatus connected therewith." The invention, it is stated, was "designed for the better and more convenient insulation of the electrical conductor for the transmission of power in electric railways, and for furnishing an automatic block system without the use of signalmen, and, further, by that method of insulation to give an easy and simple indication of the position of a train or carriage at any point of its travel." The inventors state that neither the third rail nor overhead conductor have been found to be satisfactory, and that their idea is to bring a sectional insulated conductor into temporary contact at stated intervals with a thoroughly well insulated conductor. The method by which this required contact was insured was by making use of the pressure of a contact brush on the car, or of the actual wheels to depress a stud on to the end of the insulated cable ; the connection once made continued so long as the car remained on the section, and was broken electro-magnetically in the case of long sections as the car quitted the section. This device lent itself to the registration of the stud-depressing car, and to an automatic block being placed upon the following car if it tried to enter a section upon which there was already a car. The specification was a very lengthy one, and contained five claims—for the division of an elastic conductor rail into short insulated sections, the use of contact boxes of the kind described, the automatic and graphic method of showing the position of a car, the division of the conductor rail into comparatively long insulated sections, with automatic devices for disconnecting the conductor rails as the car or train left the section, the absolute block system.—The Electrician.

## SPARKS.

. It is said to be probable that the Toronto Junction electric railway will be extended to Islington.

An electric street railway for Amherst, N. S., is said to be among the possibilities of the present year.

Incorporation has been granted to the Hamilton Storage Battery Co., with a capital stock of $10,000.

The construction of lines of electric railway from Gananoque to Mallorytown and Lynn, Ont., is talked of.

It is proposed to construct immediately and have in operation by August, eight miles of street railway in Halifax.

The name of the Galt & Preston Street Railway Co. has been changed to the "Preston & Hespeler Street Railway Co."

Application has been made by the Selkirk Elec tric Co. for authority to increase their capital stock from $10,000 to $25,000.

The general meeting of the American Institute of Electrical Engineers is to be held at Niagara Falls commencing Tuesday, June 18th.

The president of the Waterdown, N. Y., Electric Railway is said to be one of the promoters of the Kingston and Gananoque Railway Co.

The Board of Trade of Kincardine have under consideration the question of the construction of an electric railway from Kincardine to Teeswater.

Mr. Sleeman, the principal promoter of the proposed Guelph Electric Ry., has purchased an acre of land on Gordon, Surrey and Wellington streets, on which to erect a power station.

Mr. J. M. Turner, of Detroit, late superintendent of construction of the Windsor Electric Railway, is said to be preparing plans for the proposed Windsor, Amherstburgh and Lake Erie Railway.

The Vankleek Hill Electric Light Co. is reported to have gone out of business. The plant is said to have been purchased by Col. Wm. Higginson, who will continue, and endeavor to extend the business.

It is said to be the intention of the Montreal Park and Island Railway Company to in future use in winter snow-ploughs made of a number of rotating scoops, followed by a rotating broom—a device which is said to give excellent results where heavy snow-drifts are to be encountered.

Mr. J. R. Roy, Chief Engineer of the Montreal Park and Island railway, is at present engaged in making a survey of the proposed extension of the company's lines from Mile End to St. Laurent and Cartierville, a distance of over seven miles. It is the intention to construct this extension during the present summer.

The Huntsville and Lake of Bays Transportation Co. propose to establish an electric railway on the Portage between Lake of Bays and Peninsula Lake, Muskoka, and to manufacture and distribute electricity for light, heat and power. The capital stock of the company is $25,000.

An incandescent lighting plant is being installed at St. Stephen, N. B.

The plant of the Brussels, Ont., Electric Light Co , is being offered for sale.

The promoters of the Peterboro' and Chemong Park railway Co. are J. Hendry, Thos. Cahill, G. W. Halton, Robt. Fair and T. A. S. Hay. The capital stock of the company is $100,000.

The Quebec Street Railway Co. are asking from the Council the privilege of substituting for the trolley system the electric storage battery system, provided permission is granted them to run through Dalhousie and St. Andrew streets.

Mr. Geo. H. Campbell, manager of the Winnipeg Street Railway Company, has recently visited the eastern provinces and Minneapolis for the purpose of securing information with a view to making improvements in the Winnipeg system.

For the first six months of its operation, the Hamilton, Grimsby and Beamsville Electric Railway carried 69,851 passengers and a large amount of freight. The company is said to be making arrangements to establish a free fruit market at Hamilton.

The Nanaimo Electric Light, Heating and Power Co., Ltd., has been incorporated, to carry on a general electric business at Nanaimo, B. C. The incorporators are: Thos. J. Jones, Jos. Hunter and Albert Lindsay. The capital stock is $100,000, 5,000 shares of $20 each.

The city of Victoria, B. C., proposes to appoint a city electrician at a salary of $125 per month, to give his whole time to the service of the corporation. Mr. R. B McMicking, who has been acting for the city, is unable to give his whole attention to the duties of the position.

The Grand Trunk Railway have awarded the contract for the electrical apparatus required for the new depot to the Canadian General Electric Co. The plant, which will be of the very latest type, includes two fifty Kilowatt direct-connected multipolar generators and a most complete equipment of station apparatus.

Capt. Carter, the promoter of the Oshawa Electric Railway, has arranged with Messrs. Ahearn & Soper, of Ottawa, to construct and equip the road. Construction will begin almost immediately. Arrangements have been made with Mr. E. S. Edmonson, of the Electric Light Co., to supply the necessary power, for which purpose a special dynamo will be installed.

The amalgamation of the St. John Gas Co. and Street Railway Co., foreshadowed in a recent number of the ELECTRICAL NEWS, has been consummated. The basis of amalgamation as to capital stock is as follows: Street Railway Co., $600,000; Gas Co., $400,000. Each company keeps its own book debts and is paid for its supplies by the new company; which in the case of the Gas Company equals $33,000 in addition to the $400,000. A proviso was adopted, however, to the effect that of the nine directors of the new company four shall be shareholders of the present gas company. The new order of things went into effect on May 1st.

The Milton Electric Light and Power Co., Ltd., Milton, Ont., have closed a contract with the Canadian General Electric Co. for a 500 light alternating plant.

The T. Eaton Co. are adding a 1,000 light direct connected generator to their electric plant and have closed a contract with the Canadian General Electric Co. for the complete equipment.

The Toronto Ferry Co. are making extensive additions to their electric plant. They are installing Wood arc apparatus of 125 light capacity, manufactured by the Canadian General Electric Co., and the Goldie and McCulloch Co. engines.

The London Street Railway Co., which was recently granted a charter for the construction of a line to Springbank, is having the route surveyed and will push on the construction of the road as rapidly as possible. The solicitors of the railway company and of the city are having frequent conferences with regard to the street railway by-law. It is said that the matters in dispute are being adjusted, and the question will shortly be in shape to be again submitted to the Council.

### TRADE NOTES.

Mr. R. E. T. Pringle, of Montreal, has arranged with J. W. Gibbony, of Lynn, Mass., to act as sole agent for the Gibbony & Thompson Alternating Telephone System.

The Robb Engineering Co. are getting out a new line of engines to meet the demand in some quarters for slow speed automatic machines, on which their governor will be used.

A new partnership has been formed between V. Thomas, W. Ness Norman, W. McLaren and Charles Bate, under the name of T. W. Ness & Co,, to manufacture and deal in electrical supplies at Montreal.

Application is being made for incorporation by the J. C. McLaren Belting Co., of Montreal. The applicants are : David W. McLaren, Mrs. A. Cummins Walker, Alexander Walker, B. S. Sharing, Joseph Ryan and G. W. McDougall, all of Montreal.

The Canadian General Electric Co. have been awarded the contract by the Galt & Preston Railway Company for the equipment of their extension to Hespeler. The contract covers a 100 Kilowatt generator, two passenger cars and one freight car, together with the necessary overhead construction.

We regret to announce the assignment of Messrs. J. Ross, Son & Co., manufacturers of wire for electrical purposes, at Montreal. The liabilities of the company are placed at $13,000. We trust that a satisfactory arrangement can be made with the creditors so that the business of the company may be continued.

Messrs. John Starr, Son & Co., of Halifax, N. S., advise us that they are meeting with excellent success with their "Unique" Telephones. They are said to be giving universal satisfaction wherever used, and a large number have been put out both in Canada and the United States. As an evidence of their superiority, the Imperial War Department at Halifax have adopted them after thorough competitive tests and are now equipping their forts with the Messrs. Starr's standard instruments.

The Packard Electric Co., Limited, of Montreal, announce that owing to increased business they have found it necessary to seek larger manufacturing premises. These have been found at St. Catharines, Ont., to which city the company have removed their business, and where all future correspondence should be addressed to them. They expect to have their new factory thoroughly equipped and in operation in a short time. In the meantime, they are prepared to fill all orders promptly from stock.

The Berlin and Waterloo street railway is being converted from horses to electricity. The Canadian General Electric Co. have the contract for the necessary equipment.

# CANADIAN GENERAL ELECTRIC CO.
## (LIMITED)

THE problem of successful transmission of power to distances of from five to twenty-five miles or over has been solved by the introduction of

# THE THREE PHASE SYSTEM

More than twenty-five plants of this description, aggregating thousands of horse power in capacity, have been installed within the past eighteen months with uniformly satisfactory results.

# ALTERNATING INDUCTION MOTORS
### SIMPLE! COMPACT! DURABLE!

These motors, after a most satisfactory preliminary experience extending over two years, have now been standardized in sizes from 1 to 150 horse power, and are placed on the market with the fullest confidence in their ability to meet the most exacting requirements of electric power service. **They are equal in starting torque and efficiency, and superior in regulation to the best shunt wound direct current motors.**

In operation they require a minimum of attention, having no starting box and being without brushes, commutators or moving contacts of any kind.

# THE
# ROYAL ELECTRIC COMPANY

*MONTREAL, QUE.*　　　　　　*Western Office:* **TORONTO, ONT.**

Have just completed new manufacturing building providing additional floor area of 40,000 square feet, and have secured the sole right for the manufacture and sale in the Dominion of Canada of the celebrated

# S. K. C. Two Phase
# Alternating Current System

AS MANUFACTURED BY THE

*STANLEY ELECTRIC MANUFACTURING COMPANY,*
**PITTSFIELD, MASS., U. S. A.**

Acknowledged to be the only complete and perfected system by which light and power can be supplied from the same generator and circuit.

## GENERATORS

Have no moving wire, no collectors, no brushes.
Greatest Efficiency, Extreme Simplicity, Best Regulation.

## MOTORS

Self-starting, simple, efficient.　Have no commutators.
Superior in many ways to direct current motors.

## TRANSFORMERS

The Stanley Transformers are standard.
All others are compared with them.
They are the most efficient, best regulating and safest.

All S. K. C. apparatus made from drawings, patterns and details of construction as used by The Stanley Electric Manufacturing Company, Pittsfield, Mass.

THE MANUFACTURE WILL ALSO BE CONTINUED AND EXTENDED OF :

## ARC DYNAMOS　RAILROAD GENERATORS
## ARC LAMPS　RAILROAD MOTORS

### Direct Current Generators and Motors
### Station Equipments and Instruments
### Switchboards, Wire, Electrical Appliances

CORRESPONDENCE SOLICITED FOR

Electric Lighting, Railway, Manufacturing and Mining Work, Isolated Plants, Central Stations, Long Distance Transmission for Light and Power.

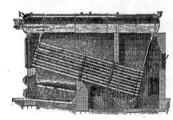

# CANADIAN ELECTRICAL NEWS
## STEAM ENGINEERING JOURNAL

OLD SERIES, VOL. XV.—No. 6.
NEW SERIES, VOL. V.—No. 6.

JUNE, 1895

PRICE 10 CENTS
$1.00 PER YEAR.

---

CANADIAN

# ELECTRICAL NEWS

AND

## STEAM ENGINEERING JOURNAL.

| VOL. V. | JUNE, 1895 | No. 6. |

### THE BRINTNELL ELECTRIC MOTOR.

MR. Archibald H. Brintnell, of Toronto, has invented and patented an electric motor which so far departs in construction from the standard types at present in use that the accompanying illustration of the device, together with the inventor's claim, will doubtless be examined with interest by our readers :—

The object of the invention is to devise a simple slow-speed motoi, incapable of being burned out, and which may be reversed at full speed and with a full potential current, without damage to the motor, and which also may be instantaneously stopped at any desired point, without the necessity of using any brake, and it consists essentially of a series of double field magnets

A-A are the vertical standards in which are journalled the main shaft B. C are a series of double magnets supported in a circular open frame, D, of non-magnetic metal. The horse magnet core, e, of each magnet, C, is laminated as shown and straddles the outer rim, D', while the inner ends of the magnets extend into notches, d, in the rim, D', and are concentrically flush with the interior surface of such inner rim. C' are the coils of the magnet C, which aie supported in the openings D', in the circular frame D.

E-E are two rings substantially Z shaped in cross-section, which extend around the magnets and are secured to the ring D, close to the horse-shoe cores of same on each side by screws,

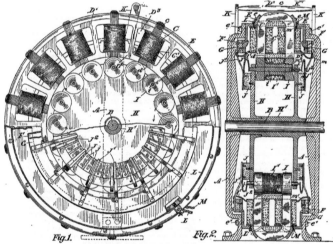

Fig. 1.    Fig. 2.

THE BRINTNELL ELECTRIC MOTOR.

held in a circular non-magnetic frame, supported on suitable standards and having the inner ends of the pole of the magnets flush with the inner circle of the frame and a number of spool armatures journalled free to rotate in bearings in supporting discs secured to the main shaft of the motoi and in direct successive contact with the poles of the magnets, a commutator of peculiar construction being provided to throw the current into both magnets and spool armatures, so as to cause a pull upon the spool armatures in succession, and rotate the shaft as hereinafter more particularly explained.

Fig. 1 is a side elevation, showing the upper half of one side of the commutator and face rings broken away.

Fig. 2 is a vertical section through the motor on a line with the shaft.

In the drawings like letters of reference indicate corresponding parts of each figure.

e. The rings are secured at the outer face by screws, e', to the standards, A. By this means the frame, D, is maintained in position. F-F are supplemental rings secured within the inner edge of the ring, E, and G-G are rings of insulating material suitably secured to the rings, F.

H-H are two discs preferably of non-magnetic metal secured to or foiming part of the hub, H', which is keyed to the shaft. I are a series of spool armatures having coils, I', and spindles, i, which aie journalled near the outer periphery of the discs, H-H, and extend through the same. The spindles, i, are made of non-magnetic metal. The ends of the spool armatures are preferably recessed as shown.

The spool armatures and magnets are arranged in diametric pairs. The commutator is comprised of two rings, J-J, of insulating material, each situated to the outside of the discs, H-H, and a plurality of brushes, J'-J', corresponding in number to the magnets and a plurality of brushes, J', corresponding in number to the armatures and situated directly opposite to them.

The brushes, J', extend from the insulating ring, F, to the ring, J, and are bent at their outer ends so as to form spring ends. The brushes, J', at one side press upon the ends of the spindles, i, and at the outer side press upon insulated pins, i', in the opposite end of the spindles, i. The insulating rings, J, of the commutator, are provided with a number of dead metallic sections, j, extending between the spring contact brushes, J' which dead sections are designed to prevent sparking as commutator revolves within the brushes, J', and to give a smoother running surface.

K-K is a double arm secured to the rings, F-F, and connected by a cross rod, K'. By means of this double arm, K, the position of the brushes, J', are changed either to increase or decrease the speed of the motor, stop it, or to reverse the direction of rotation as hereinafter described.

The wiring is as follows: We will suppose that L is the positive wire and M the negative. The current comes in to each diametric pair of coils (see Fig. 2) successively, it being necessary that two of the brushes, J', are over two of the brushes, J', each brush being diametrically opposite the other. The first brush, J', to the left hand side of the standard, A, is shown partially over the first brush, J'. The brush diametrically opposite is also in the same position. The second brush, J', to the left is also shown partially over its corresponding brush, J', and its diametric brush will also be in the same position circumferentially. In fact, two pairs of diametric brushes will always be in circuit, so that their corresponding magnets will be exerting a pull upon the armatures. The current would consequently pass in each case from the positive wire, L, in upon the wire, l, through one of the coils, C', of the magnets, C, the wire, l, brushes, J' and J', through the spindle, i, and coil, I, of the spool armature, I, thus magnetizing such spool armature out through the wire, m', pin i', opposite brushes, J' and J', wire, m', through the other coils, c', and out by the wire, m, to the negative wire, M.

It will consequently be seen that the pull upon the diametric spool armatures, I, will be such as to cause the discs, H, to rotate in the direction indicated by arrow. Each successive diametric pair of brushes, J', pass on to the brushes, J', serving to continue the rotation of the discs in the same direction, and the result of the pull of the magnets upon the spool armature being to rotate them in their bearings in the opposite direction.

The nearer the brushes, J', are placed circumferentially to the magnet core the less will be the speed, and the farther away the greater, except that there is a maximum distance in which they must be placed, and that is indicated as closely as possible in the drawings.

To reverse the direction of rotation of the motor, it will necessarily be seen that the brushes J', attached to the ring, G, may be moved by the double arm, K, so as to place them in the same relative position circumferentially on the opposite side of the cores.

It will be noticed in the construction of my machine that there is no air space between the armatures and poles of the magnet, but that they rotate directly thereon, and consequently I am enabled by the construction shown to always maintain a very strong pull between the spool armatures and magnets. As soon as the armature comes opposite magnet, of course the current is thrown out, and both magnet and armature are demagnetized.

By bringing around the brushes, J', opposite to the magnets the machine may be stopped practically instantaneously, and held from rotation by the magnetism being retained in the field coils and armatures as the current is still on. This, it will be readily seen, is very important when my motor is employed to run elevators, or for other purposes in which it is necessary to stop the machine instantaneously and hold it in any desired position. No brake, therefore, would be required, as the impetus is checked immediately.

From the construction adopted it will also be seen that the direction of rotation may be reversed at full speed, with a full potential of current, without burning out or otherwise damaging the motor.

CLAIMS.

1. An electric motor comprising a plurality of magnets circularly arranged in a suitable frame, so that the poles touch and are circumferentially flush with a common circle, and a number of spool armatures journalled in bearings in discs secured to the main shaft and caused to rotate in their bearings by the pull exerted by the poles of the magnets with which they come in direct and rolling contact as and for the purpose specified.

2. The combination with a plurality of horse shoe magnets supported in a circular frame of non-magnetic metal, and with their poles extending through notches made in the sides of the inner rim of the frame, so that the ends of the poles are circumferentially flush with the inner rim, of a series of spool armatures journalled in discs secured to the main shaft and means whereby a current is thrown into one or more diametric pairs of magnets and armatures successively as and for the purpose specified.

3. The combination with a circular non-magnetic frame D, provided with a plurality of laminated horse shoe magnets, with coils and the rings, E, secured to the outer ends of the magnets and supported in the frame, and the rims, F and G, provided with spring contact brushes, J', of the spool armatures, I, provided with spindles, i, which are journalled in the discs, H-H, secured to the shaft, B, the insulating rings, J-J, the brushes, J' secured to the rings, J, and commutating means whereby the current is carried simultaneously and successively through the co-acting magnets and armatures, as and for the purpose specified.

4. The combination with a circular non-magnetic frame D, provided with a plurality of laminated horse shoe magnets with coils and the rings, E, secured to the outer ends of the magnets and supported in the frame, and the rims, F and G, provided with spring contact brushes, J', of the spool armatures, I, provided with spindles, i, which are journalled in the discs, H-H, secured to the shaft, B, the insulating rings, J-J, the brushes, J' secured to the ring, J, and the main current wires, L and M, connected by a wire, l, to one coil, and the current being carried by wires, l', brush, J', brush, J', spindle, i, coil I, wire, m', pin, i', brushes, J' and J', wire m', coil, C', wire, m, to negative wire, M, successively for each diametric co-acting magnets and armatures, as and for the purpose specified.

## ELECTRIC LIGHTING IN TORONTO.

JUDGING by the overwhelming defeat of the by-law to authorize the purchase of a municipal electric lighting plant for the city of Toronto, the property owners regarded the tender of the Toronto Electric Light Co. as being so low that it would be unwise for the city to go into the lighting business in the uncertain hope of being able to effect a saving of a few hundred dollars per year. The pros and cons of the question were discussed in detail through the public press for months prior to the submission of the by-law, so that the verdict of the ratepayers cannot be said to have been reached without full and deliberate consideration of all the arguments for and against municipal lighting. We believe the decision to be a wise one so far as it relates to Toronto and cities of large population. It is to be hoped that after all the worry to which the Toronto Electric Light Co. have been subjected for a year past, they will now be given a five years' contract on the basis of their tender.

---

An electric light plant is to be installed at Brighton, Ont.

The Packard Electric Co., who recently removed their business from Montreal to St. Catharines, have purchased the Neelon mill property, and are fitting up the mill building in a way to meet the requirements of their manufacturing business. The property consists of about ten acres of land, a valuable water power, the capacity of which is estimated at about 700 h. p., in addition to wharves, storehouses, etc. It is understood that the company bought the property at a bargain, and have been offered a considerable advance on the price paid since closing the purchase. The mill building, which is to be used as their manufactory, is a most substantial structure, the outer walls being of stone faced inside with brick, and plastered with cement. The walls in the basement are upwards of four feet in thickness, while the wood framing of the building is equally substantial in character. The basement floor is of concrete, making a foundation for machinery so substantial as to be almost noiseless. The Packard Company are putting up a separate building in which the cementing will be done, thus preventing the possibility of fire destroying the main building. The company expect to have their new works in operation by the first of July, and contemplate engaging in some new lines of manufacture. We hope to be able to furnish additional particulars concerning their new manufactory in a future issue.

## RECENT CANADIAN PATENTS.

Canadian patents have recently been granted for the following electrical devices:—

No. 48,407, for an electric heater, to John Emory Meek, New York City. The inventor's claim for this device, which is herewith illustrated, is as follows:—

ELECTRIC HEATER.

In an electric heater, an electric resistance fabric, consisting of a conducting wire, woven with non-conducting strands, preferably asbestos, said non-conducting strands preferably forming the warp of said fabric, and having said wire preferably woven as or with the filling thereof, so that the fabric may be cut parallel with said filling strands into pieces of suitable size for use as cushions, pads, blanks, garments and for other heating purposes, each piece so cut from the fabric being in itself a complete electrical heater when connected in circuit.

No. 48,433, for a rail cleaner, to Robert Leslie, Toronto, Ont. The claim for this device, which is herewith illustrated, is as follows:—

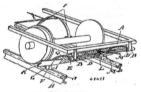

RAIL CLEANER.

A track cleaner consisting of a rock shaft, bearings for the rock shaft, means for allowing the rock shaft a side motion in its bearings, a shoe carrying arm mounted on the rock shaft and so arranged as to have an independent side motion, and also to move sideways in conjunction with the rock shaft, means for operating the rock shaft, a cleaning shoe connected to the end of the shoe carrying arm, and a coulter connected to the inner side of the shoe and arranged to scrape the side of the rail, substantially as specified. The combination with the car truck of an arm depending from each side of the car truck in close proximity with the rail, a cleaning shoe carried by each of the depending arms, and a rail connected to the said arms and extending across the under side of the car truck.

No. 48,482, for an electric controller, to James Parmelee, New York, assignee of Elmar A. Sperry, Cleveland, Ohio.

ELECTRIC CONTROLLER.

Claim.—In an electric controller, a series of contacts, an element of the controller moving to and fro over said series of contacts, a co-operating contact bone upon such an element, an air-discharge nozzle upon one side of the said co-operating contact, a source of fluid pressure as bellows O, means connected with said moving element whereby said source shall inhale during one, and discharge during the other of the to and fro movements of said moving element, and a duct from the said source to the discharge nozzle.

## ONTARIO ASSOCIATION OF STATIONARY ENGINEERS.

The annual meeting of the O.A.S.E. was held in Brantford on May 27th. The meeting took place in the hall of Brantford No. 4, C.A.S.E.

There were present engineers from London, Hamilton, Kingston, Galt, Ayr, Guelph, Toronto and many other places, including a number of Brantford engineers.

President A. E. Edkins called the meeting to order sharp at 11 a. m. and proceeded with routine business. After minutes of last meeting were read and approved the president addressed the meeting as follows:

### PRESIDENT'S ADDRESS.

On this the occasion of our fifth annual meeting I feel it my duty to thank the members and officers of this Association for having shown their confidence in me at last year's meeting in London by electing me to fill the chair of this Association, and the Board for another year, when I was unavoidably detained from attending the meeting myself.

The business to be brought before you on this occasion will consist chiefly in the appointment of members to the Board to replace the retiring members whose time has expired. In connection with this matter I deem it scarcely necessary for me to impress upon you the necessity of selecting members to fill vacancies on the Board who are energetic in Association work, as the success of this institution must to a great extent depend on the perseverence and energy of the members on the Board.

I take this opportunity of congratulating you on the fact that during the past two years no complaint has been made to the Board respecting any certificate holder, either for incompetency, intoxication, or any other action unbecoming an engineer, and as a result we have not been put to the painful necessity of revoking any certificates for such causes.

The membership of the Association has not increased so rapidly during the past year as it did in the years preceding it, which is quite natural and to be expected, owing to the fact that in our cities and towns, all engineers who feel inclined to pass an examination (without being compelled to do so by law) have taken out certificates, therefore our hope for membership in the future must be in the direction of engineers who are in charge of plants in the smaller towns and villages.

I do not mean to infer by this that our membership in the cities will not increase, for I certainly expect it will do so materially, but not in the same proportion year by year during the next four years as it has in the past.

During the years of its existence the Association has not increased in membership to the extent that it should. When we consider the fact that (at a very moderate estimate) there are 4,000 stationary engineers in active service in the province, we must be disposed to consider that our strength numerically is far from what it should be.

I hope within the next four years to see the membership of the Association reach 2,000, and I hope that with this end in view, not only the members of the Board, but also the individual members of the Association will interest themselves in the work of increasing our membership and sphere of usefulness.

As a member of the Association I shall do all in my power to place the membership of the Association at the limit of which I made mention, irrespective of the fact whether I may continue to be a member of the Board or not.

I am sorry that at the last meeting it was decided to alter the day of meeting, as in my opinion, the 24th of May being a holiday, was a much better time, owing to the fact that many engineers who would and could attend on that day are this year prevented from doing so, as they have to be on duty, and as some of them have expressed themselves to me, they are debarred from attending the meeting and taking part in the proceedings.

During the last session of the Local Legislature, as most of you are aware, we presented a bill to provide for the inspection of steam boilers, and the examination of engineers in charge.

The bill passed its first and second readings and was referred to a special committee, by whom it was thrown out, and a resolution passed to the effect that the committee, while accepting and endorsing the principle involved in the bill, took exception to the machinery provided for its working in the event of its becoming law.

This proposed legislation was introduced by Mr. Crawford, one of the members for the city of Toronto, who, while he advised us strongly to get one of the government supporters to champion our cause, very generously informed us that if we insisted he would accede to our wishes and do the best he could.

There was, to my mind, several things about the bill calculated to raise opposition. For instance, the appointment of a chief inspector and a staff of assistants, who might do this, that and the other thing, but whose duties were by no means clearly defined. I think you will agree with me, that a measure of that kind would be very apt to be looked upon by many members (of the opposition side of the house, at least,) as a scheme to' provide soft snaps for some of the supporters of the government, who had arrived at the end of their useful period.

As far as the engineers of the province are concerned, I do not think they are very anxious for a government inspection law, but they would like to see some standard set up for the qualification of engineers of land engines,' and I think we shall stand a better chance of securing moderate legislation 'governing the latter, than any measure dealing with both. The provinces of Manitoba and Quebec have both passed laws of this kind, which are as unique in some ways as they are useless, and I trust that when an engineers' law is passed by our legislature, it will be something useful and beneficial to all concerned.

In connection with this legislation it has been necessary to spend some of the Association's money, and if we have no other return to show for it, we can, at least, say that it has kept the Association before the public, and I have no doubt but that you will endorse the disbursement.

It is gratifying to us to know that the Association's certificates are being generally recognized by manufacturers and steam users as a guarantee that the holders thereof may be depended upon as being reliable and sober engineers. We have had evidence of this many times during the past year, when men who held certificates received appointments to positions in preference to men who were not registered, even where the latter were fully equal to the former in point of ability, etc. In several cases engineers have come from a distance to pass their examination, in order to secure positions, where none but a certificated man would be employed. This is encouraging, and it is the duty of the Association and of the Board particularly to see to it, that no certificate is issued to any person who has not had the necessary experience required by the act, and in the event of any person applying for examination who has not had that experience, or who is in the habit of using intoxicating liquors to excess, or who, for any other reason is unfit to shoulder the responsibilities of a position as an engineer, and the facts are known to a member of this Association, it becomes that member's duty to the Association and to himself, to see that the chairman of the Board is at once informed of the facts.

At a previous meeting of this Association, I suggested to you the advisability of reducing the annual renewal fees for certificates, which suggestion you saw fit to act upon, and consequently reduced the fees from $1.50, $1.25, and $1.00 to $1.25, $1.00 and 75c. for 1st, 2nd and 3rd class respectively. This reduction has been the means of inducing a larger number of members to keep up their renewals, and has, I believe, on the whole proved beneficial to the Association. Believing, as I do, that the fees for renewals should be made as low as possible consistent with the proper management and economical administration of its affairs, I would respectfully suggest to you the advisability of further reducing the renewals to $1.00, 75c. and 50c. for 1st, 2nd and 3rd class respectively. This further reduction will necessitate a slight alteration in our by-laws. I refer to sec. 3, art 5, which reads as follows, viz : " During the life of a certificate the holder thereof may apply for one of a higher grade which will be granted to him on his passing the necessary examination and paying the amount for renewal charges for such higher grade." It will be necessary for this section to read in such a way that the applicant should pay the difference between the examination fee for the grade of paper he holds, and the fee for the higher grade required.

The registrar's and treasurer's report will be submitted to you during the day, when you will be afforded an opportunity of hearing something of the numerical, and also the financial standing of the Association. I have before made reference to the fact that considerable money has been spent in connection with our efforts to secure legislation, as we had a representative constantly at the Parliament buildings, who was well known in the House, and who did most excellent work in pushing our claims before the notice of the members. I have also much pleasure in stating that during the session of the House, the members of the Board received very practical help from the officers and members of the different associations of the C. A. S. E., and in my estimation the interests of these two Associations are so identical that I am sometimes inclined to feel it a pity that there should be two organizations in place of one Association incorporated by act of the Dominion Parliament, and having all the powers and privileges of the two Associations combined. The Ontario Association is of course an outcome of the C.A.S.E., and might almost be called a branch of it, but I am inclined sometimes to think that one Association instituted as I have said, would prove much more effective and beneficial.

I am pleased to be able to report that during my terms of office in this chair, and so far as I am aware, only one loss by death of a member has been reported to the Association, that one being Mr. J. Campbell, engineer Parry Sound waterworks. The first intimation I had of Mr. Campbell's death was when, having occasion to write to him on business for the Association, I received a letter from his wife informing me that he had been dead over a month. Mr. Campbell was of Scotch nationality, a thoroughly practical engineer, an ardent member of this Association, and a genial and good living man, and I am sure that his wife and family have the sincere sympathy of this Association in the sad loss which they have sustained.

There is one other matter which I would like to bring to your notice, and in my estimation it is one of considerable moment

to us as steam engineers. I refer to the detestable action of some manufacturers of engineers' brass goods and fittings, who in order to save a few cents worth of metal, are in the habit of coning out the squares of blow-off cocks, and otherwise reducing their strength, until they are positively dangerous to handle or move under steam pressure. Several cases of narrow escapes from scalding of engineers have come to my notice from such causes, and I have here a specimen for you to inspect which will speak for itself, and I think you will all agree with me that any manufacturer of brass goods who would deliberately turn out such trash, especially in these days of constantly increasing pressures, when the tendency should be to increase rather than decrease the strength, is worthy of a place in penitentiary, and I hope this Association will pass a resolution severely condemning such perfidy on the part of manufacturers who for the sake of a few cents worth of metal would turn out articles that will endanger our lives while in the discharge of our daily duties.

In conclusion I would ask you all to kindly give your strict attention to the business to be brought before you, and to refrain from all useless discussion, so that the business of the Association may be done with despatch, and the meeting not unnecessarily prolonged.

Then came the registrar's (A. M. Wickens) report, which showed that 450 certificates had been issued, the income for the year from all sources amounting to $699.10, with the disbursements at $83.71 less, leaving a total balance in the treasurer's hands of $145.00.

The Treasurer's report (by Treasurer R. Mackie) was then received, showing the Association to be in good position financially. The registrar's and treasurer's report were received and referred to the auditors, Messrs. Ames, Sutton and Wells.

The appointment of special committees was then proceeded with, and the meeting adjourned at 12 noon for dinner.

At 1 p.m. the meeting was called to order by the President, and the different committees were then given two hours in which to do their work.

At 3 p.m. the meeting was again called to order and the reports of the auditors, Committee on Good of Order, Expenses, etc., were received and adopted after some discussion.

The election of members of the Board to fill the positions of retiring members was then proceeded with. The retiring members were given out by the registrar as follows: Messrs. R. Dickinson, Hamilton; S. Potter, Peterborough; F. Robert, Ottawa; Ames, Brantford. Messrs. T. Elliott, Hamilton; W. Phillips, Niagara Falls; Thos. Wensley, Ottawa, were elected to the board, being new members, and Mr. A. Ames, of Brantford, was re-elected by acclamation.

The election of officers was then proceeded with and resulted as follows :—A. Ames, Brantford, President; F. G. Mitchell, London, Vice-President; A. E. Edkins, 139 Borden street, Toronto, Registrar; R. Mackie, Hamilton, Treasurer.

Galt was selected as the next place of meeting. Kingston and Toronto were both nominated.

The renewal fees for certificates were reduced to $1.00, 75c. and 50c. for 1st, 2nd and 3rd class respectively, to take effect Jan. 1st, 1896.

The report of the Committee on the Strength of Boiler Appliances took the form of the following resolution :—

Resolved: That this Association express their unanimous condemnation of the practice of some of the brass goods manufacturers in reducing the thickness of their brass goods, more especially in blow-off cocks, as in practice many of them are not of the requisite thickness for strength, and we further resolve and pledge ourselves to use only those that we know are of sufficient strength to be perfectly safe in their operation.

The discussion on this resolution brought out samples of defective and weak blow-offs that were positively dangerous to use. The resolution was adopted.

After a vote of thanks to the retiring officers, and the members of the Brantford C. A. S. E. for courtesies extended, the meeting adjourned at 4.45.

Two miles of grading have been completed on the extension of the Galt and Preston railway, and a distance of three quarters of a mile has been laid with rails and ties. It is expected that the new line will be in running order by the 1st of July.

The Three Rivers Iron Works Co. has been incorporated with a capital stock of $100,000 for the manufacture and sale of iron, steel and brass goods, and the supply of electric light, heat and power. The promoters of the company are :—Antoine Aime Charlebois, gentleman, and Alphonse Charlebois, contractor, of Quebec ; W. Duncan, of Three Rivers, mechanical engineer, and George Duval, barrister, of Ottawa.

## QUESTIONS AND ANSWERS.

"Constant Reader" writes: 1. What is the loss of pressure in a 3-inch pipe, 100 feet long, at 100 lbs. pressure per square inch, supplying steam to an engine doing 50 horse power? 2. What would be the loss in the same pipe supposing there were four elbows and two globe valves in it.

ANSWER: 1. Loss of pressure in steam pipes is due to three causes: (a) The pressure required merely to overcome friction in the pipe. (b) The pressure required to produce the required discharge of steam through a pipe of given diameter. (c) That loss of pressure which is due to the difference in temperature of the steam at opposite ends of a long pipe, caused by radiation of the heated pipe surface into the surrounding atmosphere. With the data given, the combined losses of pressure due to the two first causes will be, with a straight pipe, about one-third of a pound per square inch; the globe valves will make no difference practically; the entire drop due to friction of pipes and bends and velocity through same would be one-half a pound per square inch. The drop due to difference in temperature is likely to be more appreciable, but with the data given it is not possible to more than approximate to it. Assuming 100 lbs. pressure at boiler; 100 feet of 3-inch pipe, bare pipes, and an atmospheric temperature of 60°—there would be a difference of temperature of about 10° Fahr., corresponding to a loss of pressure of about 15 lbs. per square inch. What the actual drop is, depends on the material with which the pipe is covered, and upon the actual temperature of the outside air.

"Fireman" writes: Please explain to me how to find the mean pressure on this card, also what horse power is the engine if the card on the other end is the same.

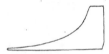

Steam, 72½; revolutions, 94; spring, 40; cylinders, 13"x30".

ANSWER.—The mean steam pressure of an indicator card can best be found by running a planimeter round the card. If you have not such an instrument, then divide up the horizontal line representing the length of the stroke, into a considerable number of equal parts; draw from each point of division a vertical line cutting the steam line. Then add the lengths of the verticals between the stroke line and the steam line together, and divide by the number of division. To this result add the vertical dis-

tance between the stroke line and the atmospheric line (which you have not shown), and the figure you get will represent mean pressure in lbs.

## SPARKS.

The City Council of St. John, N. B., will call for tenders shortly for electric streeet lighting.

Mr. M. Hutchison has been appointed superintendent of the electric light plant at Victoria, B. C.

A meeting of shareholders of the Guelph Railway Company will be held in that city on the 17th inst. for the purpose of organization and other business transaction.

A collision of two trolley cars occurred on the Galt and Preston street railway about midnight on June 1st, by which a motorman named Jenkins had his leg broken. Although about two hundred passengers were on board the cars at the time of the accident, all escaped uninjured.

At the annual meeting of shareholders of the Ottawa Electric Company held on the 3rd of June, the gross earnings for the year were reported to be $152,000, and a dividend of 8 per cent. was declared. The company have now in operation 48,000 incandescent and 430 arc lights, also 89 motors. The board of directors were re-elected, namely: T. Ahearn, president; Hon. E. H. Bronson, vice president; G. H. Brophy, William Scott, C. Berkeley Powell, G. H. Perley, J. W. McRae, D. Murphy, P. Clemow, R. Quain and A. Bayley, auditors.

## THE STATUS OF THE GAS ENGINE.*

Mr. Chairman and gentlemen: The subject before the meeting to-night is the present status of the gas engine, and, as I understand, my sole duty here this evening is to state the case for discussion.

The status of the gas engine, as the status of every other thing whatever in the world, depends entirely upon the surrounding conditions. A plant exists merely because the sun shines and the seed has been planted. Unless the surrounding conditions are favorable the plant cannot exist, and the amount of its growth and the character of its growth depend entirely upon the surrounding conditions. The gas engine follows this general law, in that its status to-day is what it has been forced to be and what it has been allowed to be by the general commercial and engineering conditions of the world. The status of the gas engine, therefore, is the status of the power question of the world, and the importance which the gas engine takes in the field depends as much upon the condition of other forms of power as it does upon the condition of the gas engine itself. Now practically all other power that is in use is steam power. I do not mean in saying this to lose sight of the tremendous amount of water power that is in use in this country or the world to-day. But water power is restricted to certain localities and the gas engine is not. It comes into competition only with those forms of power which are applicable to almost any locality or set of conditions; and, aside from gas power, steam power is almost the only other occupant of that class of prime movers.

The condition of steam power engineering has very rapidly altered during the past few years, at any rate within the past ten or fifteen years. The steam engine started off as the forerunner of the factory. Perhaps the factory started first on water power. But practically the factory and the steam engine grew up together; and the idea that we, nearly all of us, have carried in our minds of a commercial manufacturing plant driven by any power was a set of tools driven by a central motor, through the medium of a line of shafting and a number of belts. That scheme of industrial works grew in size and importance until it had reached tremendous proportions and in certain lines of industry it still survives and will continue to survive in the future in still larger sizes, and on still more important a scale. But for the vast majority of industries, those species of manufacturing which involve a varied number of processes and departments, and involve the production of a comparatively complex commodity—such, for instance, as the building of steamships or railway cars, or the complicated machines involving wood work and iron work and steel forging work, and all that sort of thing—the modern factory is a collection of factories. The various factories may be merely departments, merely various rooms in one building or on one floor. But at any rate the factory consists of a large number of departments, and those departments are quite distinct. They often run entirely independently, have separate foremen or superintendents, run different hours and have different classes of labor. In some, labor is a small item and power is larger. In others, power is a small item and labor is the principal feature. All these varied conditions make it almost impossible for such an establishment to be driven by one central prime mover. In the first place, distance of transmission comes in, and consequent losses. In the second place come in varying conditions. Every engineer knows that no piece of apparatus can work well under varying conditions. We have therefore seen grow up in the last few years the modern industrial works in which the power has to be transmitted quite a distance, and subdivided among a large number of different sorts of tools. This, of course, was first attempted by shafting and belting or by rope drives. But it is evident that loss by transmission in all various transitions and distances of transmission are very great. Consequently we have seen other schemes tried. First came the subdivision of the steam engine itself into a large number of units, and we have only to look about us in our large factories to see steam-driven plants where the power is furnished from a central boiler plant to anywhere from ten or a dozen to seventy different engines scattered all over the works—sometimes twenty or more in one room. Sometimes the losses in steam transmission are under average conditions less than for shafting and belting. Of course, no cast-iron

* A statement of the subject by Mr. Reeve, before the American Society of Mechanical Engineers at their monthly meeting April 10, 1895.

rule can be given for all conditions. But the subdivided steam plant has come in to stay. Of later years, superseding the subdivided steam plant, came in first compressed air and then electrical transmission and subdivision of power.

I have given this resumé of the changes of the power question to show that the problem of the present is economical transmission of power, not economical development of power. Of course economical development at the original point of production is of great importance, but it is vastly more important to transmit it economically, because the losses in transmission can easily exceed the largest losses possible in production.

Now those of you who have spent any time on the problem of the subdivision of power or its transmission have seen that none of the systems heretofore provided satisfy the question. They all involve tremendous losses in transmission. They all involve heavy first cost, heavy expenditure for generating plant, transmission plant and redeveloping plant at the other end. For instance, in electrical transmission, if your total works need a thousand horse-power, besides your thousand horse-power of boilers you must follow with a thousand horse-power of steam engine, a thousand horse-power generator, your mains for carrying the electric power, and then on top of that a thousand horse-power of motors. That is, of course, losing sight of all small factors and percentages of loss. Now the present status of gas power in this country or any other, and also of its immediate promise for the future, depends upon this statement of the problem in this part—that gas power offers the ideal solution for the sub-division and transmission of power, the mechanical difficulties for the time being lost sight of. In other words, let us suppose that a large industrial plant, requiring say 1,000 H. P., subdivided into say 50 different units, which are utilized at various points in different buildings on different floors at different speeds for different hours during the day, under different conditions of varying and steady load—suppose that in such a plant as that we install a 1000 H. P. gas generator or its equivalent, and then lead from it or from the holder to a large number of large central gas mains—it is evident that the first cost is away below that of any system with the possible exception of the subdivided steam plant, where our boiler plant corresponds to our generating plant, our steam mains to our gas pipes, and our steam engines to our gas engines. But there is one big advantage which a gas plant has over a steam sub-divided plant in point of operation, and that is that in the steam sub-divided plant, no matter how large a proportion of the load be off, the central generating plant must be run, and the fixed charges of running cannot be altered. Steam must be kept up, the boilers must be kept hot, the stack, if there be one, must be kept hot, and the labor must be there to take care of the whole matter. I have myself tested one factory in which for a large proportion of each day the efficiency for the transmission of the power between the boiler plant and the work was 5 per cent., simply because they had to keep the whole plant going in order to move one small department. With the gas plant that entirely disappears. Your gas generator works for a certain number of hours a day on whatever load or capacity is best suited to produce maximum economy, and as we all know there is only one point in capacity in which any apparatus can work at minimum economy. During those certain number of hours a day the generator makes gas and stores it in the holder. The generating plant is entirely unconscious of the consumption of the power, provided it be large enough to fill all demands. The consumers of power, the foremen of different departments, are as unconscious as is the generating plant of the consumption of power. They simply know that all they have to do is to turn on their gas and start their engine. They may run twenty-four hours in a day while the gas generator runs eight, provided the total maximum production of the gas generator is large enough to cover the whole output of power. The comparison between such a gas plant and any plant relying upon transmission and sub-division of power, by compressed air or by electricity, or by any scheme wherein the power is first developed by the steam engine and then converted into another form and then converted back again —a comparison between the gas power plant and any such plant as that is really hardly possible from the economical standpoint. The operation of any such gas plant would be incomparably more economical than that of the compressed air or the electrical or the hydraulic system of transmission of power.

You will notice that I have entirely left out of the question the mechanical side of it, which I purposely wished to do. But for the merely commercial side there is an absolutely unlimited field for the development of power and its transmission and sub-division in industrial works by means of the gas generator and the gas engine.

The next big argument for any plant of that sort for division of power is that in every industrial works—the supply of power or steam—the supply of power is made to correlate with the other portions of the work. To be more plain, perhaps, in nearly all of our large textile mills in New England, and in a great many other forms of industry, the steam is used as much, and sometimes more, for heating purposes and boiling and dyeing as it is for power. In fact the big promoter of steam power in New England, where it has proved an indispensable auxiliary of water power, is the fact that the steam had to be had anyhow. That is true also of compressed air plants. There are a great many forms of factory where compressed air is indispensable for

blowing, furnishing draft, cleaning, and innumerable purposes to which it can be applied, and in a great many plants compressed air is used to transmit and sub-divide power where no other system would be tolerated, simply because the compressed air has to be there anyway, and the compressed air mains have got to be there, and they might a great deal better use it for power. The same thing applies to electricity. Those factories relying on the electric current entirely for light may often bring the power question into an entirely secondary importance compared with light. If they have got to have their central engines and generators and mains for the production and distribution of light, and if they need light more than power, then of course the electrical transmission of power is the thing, without any regard to the general arguments against. But this same factor of the correlation of the system of transmission of power and the system of transmission of other forms of energy needed in the works applies also to gas. In fact the gas producer has reached its present state of perfection largely owing to the fact that gas is the most economical form of fuel for a large number of industrial devices—metallurgical, for glass works, and for a large number of sorts of cooking and heating and baking, where exact temperatures and exact control of temperatures have to be had. In all these plants the power may again become secondary to other purposes in the factory ; and in those plants where generators have already been installed for the purpose of supplying fuel gas, the gas engine follows as a natural sequence.

The reply to this side of the discussion—and you will remember that I am merely stating the discussion—the reply to this side of the discussion is, that the mechanical difficulties have not yet been overcome ; that the gas engine to-day, after having had spent on it the best energy, or of the best energy, in the line of mechanical engineering which the world has been able to produce for some thirty years, is still more crude in a great many mechanical features than was the steam engine of a century ago. It is still very heavy. In nearly all of the devices only one impulse is received by the fly-wheels for every two revolutions ; in the case of the single acting engines only one is received ; so that the fly-wheels are heavy. The regulation, as a rule, is accomplished by simply dropping out a certain proportion of the impulses instead of varying their strength ; and the necessity for the ignition of the charge in a minute fraction of a second has led, until very recently, to extreme uncertainty in the matter of ignition and also on the question of perfection of combustion. That last is not so marked a feature, because even with poor combustion, the gas engine is an exceedingly economical prime mover. But the mechanical difficulties still stand in the way of the accession of the gas engine to the proper field in which it belongs—that is, the universal factor for the production and transmission and subdivision of power in industrial works.

At the risk of being considered rather superficial in skimming over this subject, I will take one step into the future and say that while gas engines have hitherto been almost entirely run with illuminating gas, yet, already there has been considerable done in the way of supplying gas engines with special producer gas from special producers built just for that purpose. There has also come upon the field—I will just mention it—the incandescent gas burner, that has just begun to attract wide attention as being an established fact. These two coupled together—first that the gas engine can be run much more economically upon producer gas, not illuminating gas, and second, that there is a means attained of producing illumination by a non-illuminous gas—lead us to surmise—perhaps I should not state it as a surmise, I will state it as a hypothesis, which I hope to hear discussed—that the near future will see the distribution of energy— all energy—which is derived from coal in the form of a non-illuminous cheaply produced fuel gas—that this gas will be relied on entirely for power, for lighting where gas lighting at all is permissible, and where it is not, where the electric light is needed, that electric light will be produced through the medium of gas engines, and that this same gas will be used for all sorts of purposes—heating, domestic heating and cooking, and industrial heating of all sorts.

I have tried to make as brief a statement as I could of the engine problem as it appears to me to-day, not on the basis of the condition of the gas engine itself, but as a statement of its possibilities, the demands which are going to be made upon it in the near future, and what the near future may bring forth in the way of powerful auxiliary to aid in the adoption of the gas engine as the universal prime mover.

The St. John Railway Co. are considering two propositions for lighting the city. One is to light the entire city with 2,000 c. p. lights, and purchase at a valuation the city lighting plant in the north end. The other is to light the eastern portion of the city only, in the same manner as has been done heretofore.

Messrs. W. McLea Walbank and Thos. Pringle of Montreal, will apply for permission to construct works on the north shore of the river St. Lawrence at Lachine Rapids, for the utilization of the water power from the rapids. Plans of the proposed undertaking have been submitted for the approval of the Minister of Public Works.

Tenders are invited until the 10th inst. at noon, for an electric plant with a capacity of 1,000 16 c. p. lights, and four motors of an average of 9 h. p. each, for the Montreal General Hospital. Contractors are required to supply full plans and specifications with their tenders. Tenders are to be submitted to Jas. Paton, Secretary of the Committee of Management.

## VISUAL SOUND.

### By W. E. Irish.

It is well-known that vibrations caused by the utterances of words, musical notes or other sounds made near a telephone, or transmitter diaphragm, or stretched drum head, or any other flexible disc held firmly at its periphery, leaving its centre free, will vibrate in unison with the sounds conveyed to it.

There can be no doubt as to this fact in the mind of anyone who has ever examined and studied the "Bell" telephone. The vibrations of the diaphragm at the transmitting end of a pair of "Bell" phones varies with every sound, but for the same sound the vibrations remain a fixed quantity; therefore, it is necessary for the diaphragm at the recurring end to make exactly the same number of vibrations and in exactly the same tune to reproduce the same sound, which set in vibration the transmitting diaphragm.

This being the case, it only remains for some one to devise a simple means whereby, instead of reproducing the sound as in a telephone, these vibrations are caused to be recorded in ink on paper, so as to give a distinct character or figure for every different sound affecting the diaphragm. When he has done this, he will also have discovered a key to a universal language, and will have given to the world one of the greatest labor-saving devices ever enjoyed by man.

It does not appear to be and really is not such a very great problem to solve. Such a device would, in addition to being the key to a universal language and great labor-saver, also be the key to a universal system of recording our own thoughts and the thoughts of others by means of the same simple sound characters as in writing, printing, telegraphing; but most important of all, it would be the means of solidifying the thoughts of our ablest men as fast as they could utter them. Through its agency our children would acquire more useful knowledge in two years with far less labor than they now do in five years. The letters of the alphabet and orthography as now known would be entirely abolished, and correct pronunciation and true natural and visual sound character substituted. The student would be helped by hearing the teacher pronounce the word, which the instrument would at the same time describe in sound characters.

Such a machine would write the whole of Webster's dictionary with a little over fifty different characters, all of which could be more readily learned than the spelling of fifty of the simplest words by means of the alphabet. These characters would convey more readily to the mind and senses their import, than is possible by any other means. The future generation would not be mentally taxed to reason out why if P L O U G H spells plough, D O U G H must spell dough. The arbitrary characters which we use to express our thoughts visually are by no means limited to twenty-six or to several hundred, for the reason that there are in many instances over thirty accepted different types of character representing the same letter.

The first lesson book for children should start out thus: A stands for ass, symbolical of the man who introduced the first variation in the formation of letters, for variation sake alone, making the teacher's task perplexing, the pupil's much more irksome and difficult, and the general reader's vexatious.

As true sound figures cannot be changed without altering their meaning, we may hope the characters, if they ever should come into general use, will be permitted to retain their natural form.

This machine would save the author, editor, reporter, student and commercial man a great amount of the drudgery of the pen, and their brightest ideas and most valuable time. They would simply have to talk their best thoughts to the machine to have them accurately recorded at unlimited speed, in characters which will not require transcribing to be understood by others. A person having a knowledge of these characters would be able to read German, French, Spanish, or any other language recorded by the machine, perfectly and with correct pronunciation, although the meaning might be quite unintelligible to him. It would be a great help to the traveller in a foreign land, as well as to the student learning a foreign language.

In the study of music these characters would be of immense advantage. Records of pieces played by our most able masters, with feeling such as notes alone cannot give, will represent the music sheets of the near future, and all newspapers, books and other printed matter will be published in these characters on one-fifth the amount of paper, and with larger and more distinct type and spacings.

A knowledge of these characters will be acquired without any special effort or mental training, as while the learner listens to a 'phone he will see the characters representing the words he hears within, and while speaking he will see his own words recorded. Such an instrument would be found an able automatic secretary, a trusty automatic stenographer and a reliable automatic typewriter, always ready and never tired or in the way. Sound characters would be less complex and fewer in number than any arbitrary system of shorthand. By aid of such an apparatus telephones would be brought to their fullest measure of usefulness, and all messages would be recorded without additional trouble. It would become a requisite in every office, school and home, the tutor, helpmate, slave and friend of everyone from baby to great grandpapa, and the idiot to the scholar; even to the deaf or the blind, or the dumb, or the armless, it would be a comfort and a blessing, as all could use it to advantage.

We have but to call to mind the phonograph to remember that attempts have been made in this direction, but the stylus of the phonograph makes indentations in a soft medium, such as wax, from which the sounds may in a measure be reproduced. These indentations, however, have never been visually deciphered, and it is doubtful whether they ever will be. If a magnified section at the centre could be obtained, showing the variations in the depth, etc., of the indentations, it might then be possible to read them, but such a plan would have no useful application. In the successful sound-writer the stylus or pen must be free to vibrate, discharging a continuous jet of ink after the manner of the pen in Thompson's syphon recorder or in the telephonograph.

### MOONLIGHT SCHEDULE FOR JUNE.

| Day of Month. | Light. | Extinguish. | No. of Hours. |
|---|---|---|---|
|  | H.M. | H.M. | H.M. |
| 1...... | P.M. 11.50 | ............ |  |
| 2...... | ............ | A. M. 3.30 | } 3.40 |
| 3...... | A.M. 1.00 | " 3.40 | 2.40 |
| 4...... | " 1.10 | " 3.40 | 2.30 |
| 5...... | " 1.40 | " 3.40 | 2.00 |
| 6...... | *No light. | No light. | .... |
| 7...... | No light. | No light. | .... |
| 8...... | No light. | No light. | .... |
| 9...... | P. M. 8.00 | P. M. 11.00 | 3.00 |
| 10...... | " 8.00 | " 11.30 | 3.30 |
| 11...... | " 8.00 | " 11.50 | 3.50 |
| 12...... | " 8.00 | A. M. 12.20 | 4.20 |
| 13...... | " 8.00 | " 12.40 | 4.40 |
| 14...... | " 8.00 | " 1.00 | 5.00 |
| 15...... | " 8.00 | " 1.20 | 5.20 |
| 16...... | " 8.00 | " 1.30 | 5.30 |
| 17...... | " 8.00 | " 2.00 | 6.00 |
| 18...... | " 8.00 | " 2.30 | 6.30 |
| 19...... | " 8.00 | " 3.00 | 7.00 |
| 20...... | " 8.00 | " 3.40 | 7.40 |
| 21...... | " 8.00 | " 3.40 | 7.40 |
| 22...... | " 8.00 | " 3.40 | 7.40 |
| 23...... | " 8.00 | " 3.40 | 7.40 |
| 24...... | " 8.00 | " 3.40 | 7.40 |
| 25...... | " 9.00 | " 3.40 | 6.40 |
| 26...... | " 9.30 | " 3.40 | 6.10 |
| 27...... | " 10.00 | " 3.40 | 5.40 |
| 28...... | " 10.30 | " 3.40 | 5.10 |
| 29...... | " 11.00 | " 3.40 | 4.40 |
| 30...... | " 11.00 | " 3.40 | 4.40 |
|  |  | Total, | 136.50 |

The Consolidated Railway and Light Co., of New Westminster, B.C., is said to have under consideration the construction of a tramway, to connect New Westminster with Vancouver and Steveston. The distance between the two cities, via Steveston, is about twenty-six miles. The same company is reported to have made an offer to the city council of New Westminster to light the streets of that city at twenty-five per cent. less cost than at present. It is understood that the construction of the railway is contingent upon assistance in the shape of a bonus from the city of New Westminster and the municipality of Richmond. It is proposed to utilize the water power at Seymour Creek, and to instal an electric and power plant at a cost of about $200,000, and thus operate the whole system from a power house established at Seymour Creek.

PUBLISHED ON THE FIFTH OF EVERY MONTH BY

### CHAS. H. MORTIMER,

OFFICE : CONFEDERATION LIFE BUILDING,

*Corner Yonge and Richmond Streets,*

**TORONTO,** — — **CANADA.**

Telephone 2362.

NEW YORK LIFE INSURANCE BUILDING, MONTREAL.

Bell Telephone 2299.

---

*ADVERTISEMENTS.*

Advertising rates sent promptly on application. Orders for advertising should reach the office of publication not later than the 15th day of the month immediately preceding date of issue. Changes in advertisements will be made whenever desired, without cost to the advertiser, but to insure proper compliance with the instructions of the advertiser, requests for change should reach the office as early as the 2and day of the month.

*SUBSCRIPTIONS.*

The ELECTRICAL NEWS will be mailed to subscribers in the Dominion, or the United States, post free, for $1.00 per annum, 50 cents for six months. The price of subscription should be remitted by currency, in registered letter, or by postal order payable to C. H. Mortimer. Please do not send cheque on local banks unless 25 cents is added for cost of discount. Money sent in unregistered letters will be at senders' risk. Subscriptions from foreign countries embraced in the General Postal Union, $1.50 per annum. Subscriptions are payable in advance. The paper will be discontinued at expiration of term paid for if so stipulated by the subscriber, but where no such understanding exists, will be continued until instructions to discontinue are received and all arrearages paid.

Subscribers may have the mailing address changed as often as desired. When *ordering change, always give the old as well as the new address.*

The Publisher should be notified of the failure of subscribers to receive their papers promptly and regularly.

*EDITOR'S ANNOUNCEMENTS.*

Correspondence is invited upon all topics legitimately coming within the scope of this Journal.

THE "CANADIAN ELECTRICAL NEWS" HAS BEEN APPOINTED THE OFFICIAL PAPER OF THE CANADIAN ELECTRICAL ASSOCIATION.

---

## CANADIAN ELECTRICAL ASSOCIATION.

### OFFICERS :

PRESIDENT :

K. J. DUNSTAN, Local Manager Bell Telephone Company, Toronto.

1ST VICE-PRESIDENT :

A. B. SMITH, Inspector Canadian Board Fire Underwriters, Toronto.

2ND VICE-PRESIDENT :

C. BERKELEY POWELL, Manager Ottawa Electric Light Co., Ottawa.

SECRETARY-TREASURER :

C. H. MORTIMER, Publisher ELECTRICAL NEWS, Toronto.

EXECUTIVE COMMITTEE :

L. B. McFARLANE, Bell Telephone Company, Montreal.

GEO. BLACK, G. N. W. Telegraph Co., Hamilton.

T. R. ROSEBRUGH, Lecturer in Electricity, School of Practical Science, Toronto.

E. C. BREITHAUPT, Berlin, Ont.

JOHN YULE, Manager Guelph Gas and Electric Light Company, Guelph, Ont.

D. A. STARR, Electrical Engineer, Montreal.

J. J. WRIGHT, Manager Toronto Electric Light Company.

J. A. KAMMERER, Royal Electric Co., Toronto.

J. W. TAYLOR, Manager Peterboro' Carbon Co., Peterboro'.

O. HIGMAN, Inland Revenue Department, Ottawa.

---

## MONTREAL ELECTRIC CLUB.

OFFICERS :

President, W. B. SHAW, - - Montreal Electric Co.

Vice-President, H. O. EDWARDS.

Sec'y-Treas., CECIL DOUTRE, - 81A St. Famille St.

Committee of Management, T. F. PICKETT, W. GRAHAM, J. A. DUGLASS.

---

## CANADIAN ASSOCIATION OF STATIONARY ENGINEERS.

EXECUTIVE BOARD :

President, J. J. YORK, - Board of Trade Bldg. Montreal.

Vice-President, W. G. BLACKGROVE, - Toronto, Ont.

Secretary, JAMES DEVLIN, - - Kingston, Ont.

Treasurer, DUNCAN ROBERTSON, - Hamilton, Ont.

Conductor, E. J. PHILIP, - - Toronto, Ont.

Door Keeper, J. F. CODY, - - Wiarton, Ont.

TORONTO BRANCH No. 1.—Meets 2nd and 4th Friday each month in Room D, Shaftesbury Hall. Wilson Phillips, President ; T. Eversfield, Secretary, University Crescent.

HAMILTON BRANCH No. 2.—Meets 1st and 3rd Friday each month, in Maccabee's Hall. Jos. Langdon, President ; Wm. Norris, Corresponding Secretary, 211 Wellington Street North.

STRATFORD BRANCH No. 3.—John Hoy, President ; Samuel H. Weir, Secretary.

BRANTFORD BRANCH No. 4.—Meets 2nd and 4th Friday each month, C. Walker, President ; Joseph Ogle, Secretary, Brantford Cordage Co.

LONDON BRANCH No. 5.—Meets in Sherwood Hall first Thursday and last Friday in each month. F. Mitchell, President ; William Meaden, Secretary Treasurer, 533 Richmond Street.

MONTREAL BRANCH No. 1.—Meets 1st and 3rd Thursday each month, in Engineers' Hall, Craig street. President, Jos. Robertson ; first vice-president, H. Nuttall ; second vice-president, Jos. Badger ; secretary, J. J. York, Board of Trade Building ; treasurer, Thos. Ryan.

ST. LAURENT BRANCH No. 2.—Meets every Monday evening at 43 Bonsecours street, Montreal. R. Drouin, President ; Alfred Latour, Secretary, 306 Delisle street, St. Cunegonde.

BRANDON, MAN., BRANCH No. 1.—Meets 1st and 3rd Friday each month, in City Hall. A. R. Crawford, President ; Arthur Fleming, Secretary.

GUELPH BRANCH No. 6.—Meets 1st and 3rd Wednesday each month at 7:30. p.m. J. Fordyce, President ; J. Tuck, Vice-President ; H. T. Flewelling, Rec. Secretary ; J. Gerry, Fin.-Secretary ; Treasurer, C. J. Jorden.

OTTAWA BRANCH No. 7. — Meets 2nd and 4th Tuesday, each month, corner Bank and Sparks streets ; Frank Robert, President ; F. Merrill, Secretary, 352 Wellington Street.

DRESDEN BRANCH No. 8.—Meets every and week in each month; Thos. Merrill, Secretary.

BERLIN BRANCH No. 9.—Meets 2nd and 4th Saturday each month at 8 p.m. W. J. Rhodes, President ; G. Steinmetz, Secretary, Berlin Ont.

KINGSTON BRANCH No. 10.—Meets 1st and 3rd Tuesday in each month in Fraser Hall, King Street, at 8 p.m. J. Devlin, President ; A. Strong, Secretary.

WINNIPEG BRANCH No. 11.—President, G. M. Hazlett ; Recording Secretary, W. J. Edwards ; Financial Secretary, Thos. Gray.

KINCARDINE BRANCH No. 12.—Meets every Tuesday at 8 o'clock, in the Engineer's Hall, Waterworks. President, Jos. Walker ; Secretary, A. Scott.

WIARTON BRANCH No. 13.—President, Wm. Craddock ; Rec. Secretary, Ed. Dunham.

PETERBOROUGH BRANCH No. 14.—Meets 2nd and 4th Wednesday in each month. S. Potter, President ; C. Robison, Vice-President ; W. Sharp, engineer steam laundry, Charlotte Street, Secretary.

BROCKVILLE BRANCH No. 15.—W. F. Chapman, President ; James Aitkens, Secretary.

CARLETON PLACE BRANCH No. 16.—W. H. Routh, President ; A. M. Schofield, Secretary.

---

## ONTARIO ASSOCIATION OF STATIONARY ENGINEERS.

BOARD OF EXAMINERS.

President, A. AMES. - - Brantford, Ont.

Vice-President, F. G. MITCHELL - London, Ont.

Registrar, A. E. EDKINS - 139 Borden st., Toronto.

Treasurer, R. MACKIE, - - 28 Napier st., Hamilton.

Solicitor, J. A. MCANDREWS, - Toronto.

TORONTO—A. E. Edkins, A. M. Wickens, E. J. Phillips, F. Donaldson.

HAMILTON—F. Stott, K. Mackie, T. Elliott.

BRANTFORD—A. Ames, care Patterson & Sons.

OTTAWA—Thomas Wesley.

KINGSTON—J. Devlin (Chief Engineer Penetentiary), J. Campbell.

LONDON—F. Mitchell.

NIAGARA FALLS—W. Phillips.

Information regarding examinations will be furnished on application to any member of the Board.

---

THE proposed exhibition at St. John, N. B., the coming autumn, has been named Canada's International Exhibition. The dates chosen for this Exhibition are September 24th to October 4th. The Board of Management is composed as follows :—W. C. Pitfield, president ; A. S. Law, vice-president ; Ira Cornwall, vice-president ; Jas. Reynolds, treasurer ; J. C. Mitchell, secretary ; Chas. A. Everett managing, director.

---

THE Property Committee of the Toronto City Council have very properly refused to consider the request of the Toronto Island Residence Association, that a clause be inserted in the patents for the water lots prohibiting all kinds of vehicular traffic, and especially the operation of a street railway on the Island. Even the persons who presented the request will probably be pleased a few years hence that it did not receive favorable consideration, as it is well nigh certain that an electric railway on the Island will be an accomplished fact within a few years, and will tend to enhance the already great popularity of the Island as a pleasure resort.

---

THE United States Circuit Court of Appeals at Boston, on May 18th, reversed the decision given by Judge Carpenter on Dec. 18th, 1894, declaring the Berliner telephone patent to be invalid. The reasons upon which this decision is founded have as yet not been made public. The Electrical Engineer, commenting on the decision, remarks that while it complicates the situation, it does not appear to narrow the telephonic field of opportunity in any essential particular, as recourse can always be had to the magneto telephone, which had already been brought to a high state of perfection until its use was rendered unnecessary by the decisions of Judge Carpenter and that of the U. S. Supreme Court in the Bate case. The opinion prevails that the Government will not ask the Supreme Court to review the case.

THE Executive Committee of the American Street Railway Association held a special meeting in New York City a fortnight ago, to further arrangements for the Annual Convention to be held in Montreal. It was decided that the convention should open on Tuesday, the 15th of October, and continue for four days. Mr. G. C. Cunningham, manager of the Montreal Street Railway Co., who is a member of the committee, was present, and assured the committee that the Association would receive a hearty welcome from the citizens of Montreal. Col. John N. Partridge, of Brooklyn, N. Y., was elected secretary pro tem., to fill the vacancy caused by the recent death of the secretary of the Association, the late Wm. J. Richardson.

IT now appears that calcium carbide, from which acetylene gas is manufactured, will cost at least $160 per ton exclusive of the cost of the air and water tight drums, in which it is required to be packed for shipment. It will be remembered that when first this new competitor of the electric light was announced, the cost of calcium carbide was placed at $30 per ton, and electric lighting concerns were in consequence somewhat startled at its advent. The knowledge that the cost will exceed five times that sum should be sufficient to relax the electric light man's fears, at least, until the next wonderful discovery shall be announced. The fate of such discoveries thus far has served to strengthen the conviction that the electric light has come to stay.

THE truth of the wise man's saying that "A little knowledge is a dangerous thing" was very well illustrated recently in the proposal made by the "electrician" of a small lighting plant to add to the capacity of the station by the purchase of a new alternating dynamo, which should be run in parallel with the old alternator. Thus boldly stated, the boldness of the proposition does not strike one; but when investigation shows that, in the first place, the speeds of the two machines were such as to produce quite different periodicities, and that, in the second, the old machine was surface wound, non-compounded, while the new one is ironclad and compounded, it makes one wonder what would have been the result of this parallel running. Suppose them to have had the same frequency, they certainly could not be expected to have anything like the same e.m.f. curves, and the combining of the two would have given a resultant curve that would have been interesting to trace, if not conducive to high efficiency. But add to this dissimilarity the further condition of different frequencies, and we have something sublime. Let anyone draw a diagram for himself, both of assumed curves and of connection to feeders, and it will be plain that while the machines are in step (which they will be periodically) everything will be all right; but this condition of affairs will be of very short duration, and by far the larger part of the time the alternators will be more or less short-circuited through each other, until, periodically, they will be actually dead short-circuited! The "electricians" of small plants can amuse themselves with direct current machines, if they like; but they should leave alternators and induction apparatus severely alone.

THE trolley is charged with breaking seriously into the traffic hitherto controlled by steam and horse-power. The bicycle, on the other hand, is proving quite a formidable competitor to the trolley. In Denver, Col., competition from the "bike" has been so great that the wages of motormen have been reduced, and the schedule of the railway cut down at certain times of the day, and when the weather is fine. The statement is that there are 10,000 bicycles in daily use in Denver, a city of about 110,000 population. It is figured, that if each wheel takes only ten cents a day from the receipts of the street railway, a loss of $1,000 a day in gross receipts takes place, or $315,000 per year. This is, perhaps, an exceptional case, for in Toronto, where wheelmen are supposed to be on top, the maximum census has placed the number of bicycles in use at not more than 10,000. But even in the larger cities, where competition is less likely to be felt, there is a noticeable interference with the daily business of the trolley. It was stated by a daily paper, that the Toronto Street Railway was seriously considering what method could be adopted to meet the competition of the bicycle that is already recognized as a factor here. One suggestion was that cars be constructed to carry bicycles, so that those planning a trip to some of the out-lying districts might have their wheels carried for them. The suggestion hardly seems a practical one to wheelmen. One thing is certain that electrical power for railway propulsion has come to stay, and users of steam must count it as a leading element of the future. Equally so, the bicycle has come to stay, and street railway companies, and those controlling lines running out into suburban districts, have got to take the wheel into consideration in their plans for the future.

THERE is so great a difference between various makes of arc carbons, as regards their light giving qualities, that every arc plant should make careful experimental tests to determine which is the best for its peculiar conditions, before deciding in favor of any one in particular. Carbons, to be thoroughly satisfactory, and economical in use, must possess a number of features, all of which require a very great deal of attention, during the various processes of manufacture, and the finished article represents years of close study and the result of experiments costing hundreds of thousands of dollars. It has been shown by a high authority recently, that the light giving qualities of several carbons of different makes varied during an experimental test among themselves, by so much as 50%, that is, that of two carbons tested under the same conditions of current, pressures, etc., one would give twice as much light as the other. If the test had also shown that the light, and life, varied inversely, then there would be no advantage in using the higher efficiency carbon (unless, of course, customers were grumbling as to the amount of light), but they did not show any such relation, as holding between carbons of different makes. On the contrary, they showed that comparing make with make, they were distinctly carbons that were better, commercially and electrically, than others that would give a better light, and in proportion would last longer than others. A careful study of the reasons for such differences, led to the very intelligible conclusion that that carbon to the manufacture of which was given the most scientific attention, was the best all round. It is unnecessary to more than suggest that complex processes of manufacture necessarily impose a higher cost on the finished product, and every intelligent station manager should decide for himself after experiments, which is the best for himself. Cheapness is by no means desirable in electrical apparatus, and there are many station managers who believe (and act on their convictions) that a $700 carbon is more than $100 dearer than an $800 one. The electrical machinery and apparatus and methods of to-day are the outcome of the knowledge gainer during the last 25 or 30 years of experimental work, and modern methods of station management are based on the experience gained, often very dearly purchased, during a similar length of time. As each detail is more carefully studied, new directions are discovered in which economy is possible, and it behooves every intelligent station manager to keep himself thoroughly abreast of the advance taking place along the whole line of progress.

AN unfortunate accident occurred at the power-house of the Toronto Railway Company some days ago. The electrician, from Chicago, in charge of the new Siemens-Halske generator, that had been placed in position some three weeks previous, undertook to scrape the machine with a knife. The result was that the knife was drawn in and the current short-circuited; the machine was badly damaged, and severely burned the attendant. The accident threw on to the generator of the Canadian General Electric Company, which has been doing excellent work for the past six months, so great a strain that a portion of the foundation gave way, and throwing some parts of the machine out of gear, considerable injury was done to it. Traffic on the street railway was for a short time stopped and during repairs there was some restriction in the number of cars running. The Siemens-Halske machine had not passed out of the hands of the electrician of the firm, who had come on from Chicago to furnish the usual tests. When the accident occurred and the electrician in charge was disabled, in response to a telegraphic message, Mr. O. S. Lyford, chief electrician of the manufacturers, came to Toronto to superintend repairs. Within a week the generator was again in running order, and that of the Canadian General Electric Company some days before. In conversation, Mr. Lyford remarked that so rapid has been the growth of electric power for railway propulsion, lighting and manufacturing purposes, that those interested have not realized the necessity there is for experienced management in handling the business. Only a week before the more serious accident mentioned some trouble had occurred with the Siemens Halske generator through the carelessness of an employee in undertaking to perform some work, which, had the necessary experience been possessed, would have been executed in a proper and skillful manner. Since the removal of Mr. Davis, formerly electrician of the Toronto Railway Company, to Detroit, there has been no expert electrician in charge of the Company's business. The trouble of the past week would seem to indicate the unwisdom of this course, and yet the Toronto Railway Company is not alone negligent in this important particular. The writer was informed by Mr. Lyford that trouble was constantly occurring in establishments where electricity was used from the fact that skilled help was not employed. The impression is entertained, that if a man is a master mechanic it is not difficult for him to add to his knowledge a perfect understanding of electrical engineering. Experience is proving this to be a great mistake. A comparison of the methods pursued by the steam engineer in the management of an engine or locomotive with those of the majority of men in charge of electrical machinery, shows a very marked contrast. The engine or locomotive is cleaned and inspected every day, and if only a slight defect has become manifest, it is put in repair on the shortest possible notice. We fear that nothing like the same carefulness is exercised in the operation of street railway and other electrical machinery.

---

THE necessity for careful calculation of every feature of an electric lighting installation, was never more strikingly illustrated nor the danger of guessing more plainly shown, than in a case coming recently under the notice of the writer. It was proposed to add to the capacity of a station, and to extend the wiring into several new directions, and this necessitated running a new pair of mains, and the establishment of new centres of distribution. The old wiring system had been calculated for 5% loss, and the new machine was to be overcompounded for 10%. It may be incidentally mentioned that it was first proposed to run these two machines—the first an old type uncompounded, and the second, a quite modern, high-classed, compounded alternator—in parallel, only for some reason the electrician in charge hesitated and the plan was not carried out. However, the wiring plan was arranged, and was more or less a copy of the old 5% system. This included a pair of No. 2 mains, from the dynamo to the main distributing point, and this size was adopted for the new system to carry the same amperage, the same distance as in the old, the reason being that this was the size "in the other system that had always given good satisfaction." So that because No. 2 wire was right for 5% loss, it would be all right for 10% loss. The expression for the proper size of wire to carry any desired current, A any required distance, D with an E. M. F. of E., with any percentage of loss P, may be reduced to the following :

$$Size = C \times \frac{100 - P}{P}$$ C being a constant depending on A, and D, and E, which expression shows that the size will vary, very nearly, inversely, as P. Taking this particular case into consideration, we find from the above, that $\frac{size\ for\ 5\%}{size\ for\ 10\%} = \frac{19}{9}$ (the general expression being) $\frac{size\ p}{size\ P} = \frac{(100 - p)\ P}{(100 - P)\ p}$ ("P" and "p" being the two percentages.) So that the new mains should only have been $\frac{9}{19}$ ths the size of the old ones, instead of the same size. Following this up with the particular data used, results in finding at the primary transformer terminals, about 42 volts more than there should be, and, taking the ratio of transformation into account, we should get 4% too high voltage at the lamps. Every one knows what the effect of this will be. This is probably a very exceptional case of ignorance or carelessness —let us hope so at least—but it serves to indicate the results of insufficient care in the calculation of a wiring system.

In connection with this above calculation, comes the very pertinent question, "Why was 5% loss used in the first-case, and not 10%?" The distance of transmission was over a mile, and the expression shows that less than half the weight of copper might have been used, with an appreciable saving in first cost.

A very easy calculation further shows that taking into account the slightly increased fuel consumption required by the 5% greater initial E. M. F., the annual saving in interest and depreciation by so reducing the capital expenditure, is quite appreciably greater than the cost of the fuel increase required. Great stress has been laid in the columns of this paper on the necessity for carrying the strictest economy into every detail of the design and operation of electric lighting and power plants, and the instance here cited clearly shows the results of such rule of thumb methods of working. Each wiring plan should be calculated for itself, and every volt drop accounted for between the brushes and the lamp, in order that the terminal voltage may be as required.

---

## THE PROPOSED BELL TELEPHONE COMPANY BUILDING, MONTREAL.

MENTION was made in the ELECTRICAL NEWS some months ago of the fact that the Bell Telephone Company had purchased a block of land bounded by St. John, Notre Dame and Hospital streets in Montreal, as the site for a magnificent new fire-proof building, to be erected in the near future for the company's use. Since this announcement was made the architect of the proposed structure, Mr. Maxwell, has been busily engaged in designing the building, and the plans are now sufficiently advanced for us to be able to give our readers some idea of what the structure will be like. The building will have a frontage of 35 feet on Notre Dame street, 108 feet on St. John street, and 98 feet on Hospital street, and will be six storeys in height. It will be constructed of terra-cotta ashlar and pressed brick of a light salmon tint, with terra-cotta trimmings, and ornamented in the same shade of color, and will be made as thoroughly fire-proof throughout as possible. With this object in view, no stone or wood will be employed in the exterior of the building, which will be constructed entirely of steel, brick and terra-cotta. The windows of the rooms in which the telephone apparatus is located will be protected by rolling steel shutters. There will be fire escapes leading to the roof from the operating room and from every story to the basement, from whence a fire-proof passage will afford exit to the street. In addition to this there will be a stand-pipe for fire purposes, with hose on each floor.

The floors and halls of corridors will be laid with mosaic. The main entrance will be on Notre Dame street. A corridor ten feet wide and eighteen feet high, panelled in marble, will extend right through the building to Hospital street.

The ground floor of the building will be fitted up as a bank, and will be entered from the corner of Notre Dame and St. John streets. The greater part of the building will be occupied by the Bell Telephone Company, the general offices of the company being located on the second floor, and the operating room on the fourth floor. This room will be thirty-two feet wide by one hundred and twenty-eight feet long, and will be splendidly lighted by four windows on each side, as well as by sky-lights. Opening off the operators' room will be a recreation and lunch room for the operators, with steel lockers for each operator, accommodation being provided for 125 operators.

The third floor will be occupied as a battery room. The second and fifth floors will be fitted up as business offices. The building will have two elevators with a speed of 350 feet per minute, operated by electricity.

The cost of this magnificent building will be in the neighborhood of a quarter of a million of dollars. It is expected that the structure will be roofed in before the close of the present season, and will be in possession of the company by the beginning of May, 1896.

---

It is said that the Hamilton Electric Light & Power Co. will refuse to accept the city's offer of a two years' contract at $90.00 per light. The company offer to enter into a five years' contract at $91.25 per light.

The following gentlemen have been elected as the Board of Directors of the International Radial Railway Co., which is seeking a charter from the Dominion Government : Alexander Burns, M. A., D. D., Alexander McKay, M. P.; Thos. Bain, M. P., Dundas; John Hood, less, Thos. Miller, M. D., J. E. O'Reilly, master in chancery ; Peter D. Crerar, M. A., F. A. Carpenter, W. N. Myles, Thomas Ramsay, R. H. McKay, of McKay Bros.; Ald. A. H. McKeown, J. F. Smith, William Andrews, Guelph, and E. J. Powell, London.

## NEW NON-MAGNETIC ARC LAMP.

### BY W. Æ. IRISH.

This lamp for either constant current series, alternating or constant potential circuits from any generator, is governed by the thermal wire, lever and counteracting spring as usual, but instead of a clutch or rack, it employs a screw feed with more sensitive mechanism.

The screw feed is a new, simple and valuable feature. It retards the fall of the carbon so as to prevent over-feeding, and it is not affected by a dirty carbon rod. The carbon rod carries at its upper end a traveller which engages in the deep square cut thread of a long screw, the threads having a pitch of about half an inch. The traveller simply carries the rod, which is free to turn for convenience of trimming.

The object of the screw is to maintain a slow, regular and continuous feed, so as to obtain an absolutely steady light in place of the intermittent feed and constantly varying c. p. and e. m. f. incidental to most of the clutch and gear lamps. For constant current series circuits the lamps are equipped with a simple variable shunt resistance and an automatic switch which also acts as the cut-out.

This device tends to maintain the e. m. f. and current of the line constant in opposition to the contrary tendency of the lamp, for as the resistance of the arc increases the resistance of the shunt will correspondingly decrease, and vice versa. By employing this automatic cut-out and variable shunt, there is practically no burning out of the shunt coils or opening of the circuit. The longer the arc, the greater will the current be passing through the shunt and the less its resistance, as the resistance shunt will automatically decrease in proportion to the increased resistance of the arc.

In the event of the carbons breaking, the lever of the cut-out will act and by rapid steps vary the shunt resistance until the lever reaches the line terminal and makes a short circuit, when the lamp and shunt will be completely cut out of circuit and the line wire bridged across. As there is no breaking of the circuit within this lamp, there is no arcing or interference to prevent the perfect working of lamps and circuit.

On starting a newly trimmed lamp employing this automatic switch, the current will, when the hand switch is open, have two paths—one through the lever of cut-out and a small resistance, the other through the carbons.

The resistance of the line through the cut-out is made a trifle higher than the resistance of the lamp circuit, so that the automatic device may respond more promptly, and by opening the line force the whole of the current through the lamp. Here again the current finds two paths, one through the carbon and the other through the variable resistance shunt.

The chimney top is furnished with a cross arm and insulated hanger. The cross arm carries the wires leading to the lamp, the whole making a good substitute for a hanging board. The lamp top, chimney and cross arm may be considered as a separate part from the lamp proper. The top carries the hand switch and the automatic variable resistance, contact plugs, terminals and wires.

The wires pass through insulating and water-tight bushings to the terminals. To this upper part the lamp proper may be attached in a simple, handy and rigid manner, while the upper part is hanging, so that a defective lamp can be removed and replaced without interfering with wires, switch or circuit. To lower the globe for trimming it is simply necessary to release the clamping screw in globe holder, when the globe may be let down until its top is in line with the lower carbon holder, at which position it will be held by a flexible wire cord. When the globe is replaced the wire is coiled back into the base of lamp by the tension of a spring. To remove the globe entirely it is simply necessary to unscrew the thumb nut at the bottom so' as to release the wire from the holder. The cover is readily lowered or removed by partly unscrewing the two thumb screws at the side. A glass tube encircles the carbons to increase the length of their life and keep down the resistance of the arc by excluding cold air. This lamp for all circuits is bug and waterproof, and being practically sealed, needs no spark arrester.

A branch of the Canadian Marine Engineers' Association is in existence at Halifax, and meets every Friday evening in Temple of Honor, in that city. Mr. W. B. Parkes is the secretary.

## SPARKS.

An electric railway between Barrie and Allandale is being talked of.

Messrs. Dobson & Co. have become the owners of the Cannington Electric Light Co.

Permission has been granted the citizens of Vancouver to lay a telegraphic cable to the American side of Puget sound.

The Gananoque Electric Light Co. are reported to be replacing their incandescent circuits by wires of larger cross section.

Mr. P. A. Lavin, of Galt, has accepted a position in connection with the new electric road now under consideration at Oshawa.

The Toronto Street Railway Co., have rented to campers thirteen old horse-cars, which will be used as sleeping tents at Victoria Park.

The contract for steel rails for the Guelph Electric Railway has been given to Mr. Geo. Baker, of Hamilton, representing the Illinois Steel Co., of Chicago.

The receipts of the Toronto Street Railway Co., on last Queen's Birthday, were $5,700, being $1,700 in excess of the corresponding day of the previous year.

The Montreal Park & Island Railway have ordered a number of new cars for use on their several lines. Work on the St. Laurent extension will be commenced immediately.

A by-law has passed its second reading in the City Council of St. Thomas, to grant a franchise for the construction of an electric street railway through the streets of that city.

Messrs. Thos. W. Ness, Norman Westwood McLaren and Chas. Bate have been registered proprietors of Ness, McLaren & Bate, dealers in electrical supplies, Montreal.

The Ottawa Street Railway Co. have concluded an agreement under which the company's lines will be extended to Hintonburg. The work is expected to be completed by the first of July.

The proprietors of the recently organized Kay Electric Mfg. Co., of Hamilton, are adding new machinery to their manufactory, and will in future manufacture transformers and alternating machinery for lighting purposes.

The Montreal Street Railway Co. recently inaugurated an all-night service. It is reported that the enterprise of the company has been thus far poorly rewarded, and unless the traffic improves the service will probably be withdrawn.

At the adjourned annual meeting of the Toronto Street Railway Co. the following directors were re-elected : Wm. McKenzie, Jas. Ross, Geo. A. Cox, Jas. Gunn and H. A. Everett. The statement of earnings show a net gain for April, '95, of $12,116.77.

Messrs. W. J. Thompson, O. Burke, J. MacW. Telfer, J. J. O'Donohue, E. Latimer, K. A. McRae and W. A. P. Byrch, have passed the examination for juniors in electricity and magnetism at the Toronto Technical School. Mr. W. Hahn passed the senior examination.

It is announced that the Cataract Construction Co., of Niagara Falls, will not undertake to deliver power even to local consumers until their second large dynamo shall be ready for operation. One dynamo is now in position and working order. It is expected that the company will be in a position to furnish power by the first of August.

Arrangements have been made under which the Hamilton, Grimsby & Beamsville Electric Railway will make connection with the Hamilton Steamboat Co., for the transfer of passengers and freight. There will be four freight deliveries daily at Bartonville, Stoney Creek, Fruitland, Winona, Grimsby, Smithville, Grimsby Park and Beamsville.

The construction of the Oshawa Electric Street Railway is being proceeded with by Messrs. Ahearn & Soper, of Ottawa. Mr. E. W. D. Butler is overseeing the work on behalf of the Rathbun Co., of Deseronto, who are the owners of the franchise. The road will run from the railway depot through the town, with branch lines to the principal manufactories, and will be principally used for the transportation of freight.

A syndicate represented by Mr. Beemer have made definite proposals to the city council of Quebec, for an electric railway franchise through the principal streets of the city and St. Roch ; the system to be connected with the Quebec, Montmorency & Charlebois Railway. The syndicate offered to pay to the city three per cent. of the gross receipts for two years, and four per cent. for the balance of thirty years, in consideration of a franchise being given them for the period mentioned.

Mr. J. S. Larke, who recently took up his residence in Australia, as Commercial Agent for Canada, has submitted to the Sydney Board of Commerce a scheme for construction by Great Britain, Canada and Australia of the proposed Pacific cable. It is proposed to borrow eight million dollars, the estimated cost of the work, on the credit of the countries interested, at two and one-half per cent. Mr. Larke estimates that the interest, capital and working expenses can be paid in twenty-five years out of half of the estimated business, at two shillings a word. The scheme is said to have been well received.

## LEGAL DECISIONS.

TORONTO STREET RAILWAY CO. v. GOSNELL.—A decision of the Supreme Court of Canada led the Toronto Street Railway to expect that they might have the rule there laid down, that a person driving a team across a steam railway track is obliged to look both ways before he crosses a track, to see if a car, is coming, applied to persons driving across their tracks in Toronto. The Supreme Court, however, did not apply the same rule to cases of electric cars in cities, and where an action was brought for damages arising in consequence of injury by a trolley car colliding with the respondent's team, and the applicants contended that he did not see if the car was coming before he started across the track, and that his own negligence, therefore, led to the accident, the respondent was held entitled to retain the verdict against the company.

The Supreme Court has dismissed the appeal of the Toronto Street Railway Co. from the decision of the lower courts, under which the company were held to be liable for damages for injuries caused by the accumulation of ice and snow on the sides of the street, part of which would come from the railway tracks, and part from the sidewalks. The court also dismissed the appeal of the company in the case of an employee named Bond, who was awarded damages by the lower courts for injuries sustained while engaged in coupling cars on the company's lines. The court held that the buffers 'of the cars, which the respondent was coupling together when injured, were constructed on different levels, and so over-lapped as not to prevent the cars from coming together, and that this condition of affairs was sufficient proof of negligence on the part of the company, in addition to the fact that the company had received notice that buffers were necessary for the protection of their employees.

## PERSONAL.

Mr. T. Langlois has been appointed electrical engineer for the municipality of Maisonneuve.

Mr. Alexander Moffatt, electrical engineer, of New York, was married recently to Miss Madelin Spratt, daughter of Mr. Robt. Spratt, of Toronto.

Mr. J. C. Grace, Secretary-Treasurer of the Toronto Street Railway Co., has resumed his duties after an absence of several months in the South, induced by ill health.

Mr. Sam J. Heenan, formerly with the Thomson-Houston Co., and the Brooklyn Street Railway Co., has been appointed superintendent of the Yarmouth, N. S., Street Railway.

We regret to announce the sudden death at Dereronto, on the 19th of April, of Mr. E. C. French, who until recently was the Canadian agent at Montreal for the Babcock and Wilcox Co.

Mr. G. White Fraser has been appointed consulting electrical engineer for the Guelph Electric Railway and for the corporation of the town of Collingwood, in connection with the new lighting plant to be installed at that place.

Mr. Eckley B. Coxe, a portrait and sketch of whom as president of the American Society of Mechanical Engineers, appeared in this journal coincident with the society's convention in Montreal, died at his home in Drifton, Pa., on May 13th, of pneumonia.

Mr. Cecil Doutre, secretary of the Montreal Electric Club, has been appointed electrical engineer for the Richelieu and Ontario Navigation Co. He is at present busily engaged in superintending the installation of new lighting plants on six of the company's steamers. All the boats running between Hamilton and Montreal are to be lighted by electricity.

## TRADE NOTES.

Two 300 h. p. tandem engines have been ordered from the Robb Engineering Co., by the Halifax Electric Street Railway Co.

The Richelieu & Ontario Navigation Co. has ordered five Robb-Armstrong engines for electric lighting purposes on their steamships.

The Halifax Electric Street Railway Co. has ordered from the Robb Engineering Co., two 300 horse-power tandem compound engines.

The Toronto Electrical Works announce that they have commenced the manufacture of the McIntosh Current Controller for regulating electric light currents for medical purposes.

The Ingersoll Rock Drill Company have added another branch to their extensive works at St. Henry, in which they are at the present time manufacturing electric railway specialties. They have completed a contract with Albert & J. M. Anderson, of Boston, who are perhaps the largest manufacturers and designers of this class of machinery, for

the exclusive manufacturing right in Canada, and our electric lines can now procure this class of machinery and effect a saving of the duty. Their advertisement appears in one of our advertisement columns.

The Canadian General Electric Co. has ordered three 100 horse-power Robb-Armstrong engines with extension base and outboard bearing for direct connected dynamos. One of these is to be placed in the building of the T. Eaton Co., and the others in the Union Station of the G. T. R., Toronto.

A handsomely printed catalogue, describing in detail the Robb-Armstrong engines, simple, compound, tandem compound and cross compound, with tables of dimensions and H. P. capacity at various initial pressures and speed at the most economical point of cut-off, has just been issued by the manufacturers, the Robb Engineering Co., Amherst, Nova Scotia.

The new manufacturing works of the Ottawa Carbon and Porcelain Co. are rapidly nearing completion. The buildings have a frontage of 300 feet alongside the Canada Atlantic Railway Depot. The carbon and porcelain departments will each occupy a distinct portion of the building. The kilns are cylindrical in shape and forty feet high. The works when completed will be adapted in the best possible way to the perfect and economical production of the products to be manufactured, while the shipping facilities are first-class. The work is being done under the careful supervision of the general manager of the company, Mr. J. W. Taylor.

## SPARKS.

The New Brunswick Telephone Co.'s lines are being extended from Moncton to Albert.

Mr. James Bonfield expresses his intention to establish a second electric light plant at Eganville, Ont.

A charter of incorporation is being applied for by the Lindsay Light, Heat and Power Co., of Lindsay, Ont.

The Montmorency Electric Light and Power Co. are adding to their plant two power generators, with a combined capacity of 600 h. p.

The City Council of Hamilton invite tenders until the 7th inst. for lighting the city with about 400 arc lights for a term of five years from the 1st of September next.

The foundations for the Mattawa Electric Light and Power Co.'s new power station are under construction. The station building will be constructed of brick, and it is expected will be operated both night and day for the supply of electric light and power.

It is understood that negotiations are still going on between the Canadian Locomotive and Engine Co., of Kingston, and the Siemens-Halske Co., of Chicago, with the object of having the Siemens-Halske electrical machinery manufactured by the Locomotive Company. The matter will no doubt be settled at an early date.

The Tagona Water and Light Co., of Sault Ste. Marie, have purchased a monocyclic plant from the Canadian General Electric Co., for supplying incandescent and arc lights and power to the town. The arc lighting will be done by 60 Helios alternating lamps operated from the low pressure mains.

There are at present running in France no less than 328 electric light supply stations. This number, although considerably in excess of those installed in Great Britain, does not utilise, in the aggregate, such a large number of horse-power as the English stations, the greater number being quite small works, and many of them run by means of water-power.

A very satisfactory statement was presented at the recent annual meeting of shareholders of the Northwest Electric Co., at Winnipeg. The following board of directors was elected for the ensuing year: G. H. Strevel, J. M. Graham, G. A. Simpson, J. A. McArthur and H. Cameron. G. H. Strevel was re-elected president and H. Cameron re-elected manager and secretary.

At the first meeting of the Board of Directors of the St. John Railway Co., which recently amalgamated with the St. John Gas Co., the following officers were elected: Mr. James Ross, president; J. Morris Robinson, vice-president, and F. W. Warren, secretary-treasurer. The president, vice-president and Mr. H. H. McLean were appointed an executive committee to take general charge of the business of the company.

It is learned from the Ottawa papers that a very satisfactory test was made a few days ago in that city of the Hubbell Primary Battery, manufactured by the Hubbell Battery Co., of Ottawa. Mr. Geo. A. Whistler, expert electrician for the United States Navy, Mr. Geo. H. Hill, electrician for John Forman, of Montreal, and Mr. Dion, superintendent of the Ottawa Electric Light Co., have given their testimony to the efficiency of this battery, the latter gentleman stating that its capacity per pound is 50 per cent. greater than of most storage batteries with which he is acquainted. The manufacturers propose lighting a C. P. R. and C. A. R. passenger coach in the near future as a proof of the efficiency of this battery.

## ELECTRIC LIGHTING BY THE MUNICIPALITY AND BY CONTRACT.

THE advisability or otherwise of municipal ownership and control of electric lighting is engaging the attention of the City Council of St. Thomas, Ont. The City Engineer, Mr. J. H. Campbell, was recently commissioned by the Council to visit a number of American cities where municipal control of electric lighting prevails, and report as to the advantages and disadvantages of the system. Mr. Campbell has kindly furnished the ELECTRICAL NEWS with a copy of his report, which is a lengthy document, and contains much that is interesting on a subject which at present occupies considerable public attention. For the information of our readers we print a synopsis of the report as follows :—

To the Chairman and Members of No. 3 Committee of the Council of the City of St. Thomas :

GENTLEMEN,—In compliance with a resolution passed at a regular meeting of your Committee held on the 28th of February last, and confirmed by the City Council at its meeting held on the 6th of March, requesting me to prepare a plan of the city showing the location of the gas and electric street lamps now in use, also the location and number of electric lamps required for properly lighting the streets of the city, together with a report of the description and cost of a municipal electric street lighting plant, utilizing as far as practicable the buildings and power of the municipal waterworks system, an estimate of the cost of operating and maintaining such a plant, together with such further information as may be of service in considering the question of street lighting, I beg to submit as follows :

PLANS.

The accompanying plan shows the location of the number of electric lamps required to efficiently light every portion of the city. The radiating power of each lamp of 2,000 candles is shown by a red circle surrounding the lamps, which at a glance will show how completely the territory will be lit. It will be seen from the plan that on account of the irregular manner in which the streets have been laid out and the irregular intersections of the streets, that more lamps are necessary than would be required if the blocks were uniform and the intersections regular. For this reason I find that a less number of lamps cannot be so disposed as to cover the territory and do the lighting as efficiently as would be expected from a municipal plant. The location of the present gas and electric lamps are also shown.

DESCRIPTION OF PROPOSED PLANT.

1. One of the present waterworks boilers to be used to generate the necessary steam. A wing to the present waterworks building, containing 1400 square feet of floor surface, to be built at the south-east angle of the boiler room and to extend southerly to the line of the south end of the filter room, to provide space for engines, dynamos and necessary apparatus storage room, to have a stone foundation, brick walls and a slate roof.

2. One 120 horse-power tandem compound condensing engine and condenser, capable of operating 158 regular arc lamps of 2,000 c. p. each.

3. One boiler feed water pump and heater.

4. Two dynamos of a capacity of 60 lamps of 2,000 c. p. current, to measure 9.6 amperes and fifty volts for each lamp, with provision for an additional dynamo to be provided with switches, by the use of which the lamps can be burned at 2,000, 1,600 or 1,200 c. p.

5. One hundred arc lamps.

6. Two circuits of Number 4 and 6 B and S insulated copper wire.

7. Line poles to be placed 125 feet apart, to be iron on Talbot street, and cedar, straight, planed and painted, on all other streets.

8. Lamps to be suspended in the centre of the street and wherever practicable at street intersections.

9.

| ESTIMATED COST OF THE PLANT. | |
|---|---:|
| Building.......... ......... | $1,000.00 |
| 120 horse power engine and condenser.......... | 2,800.00 |
| Heater and feed water pump.......... | 400.00 |
| Foundation for engine.......... ......... | 200.00 |
| Foundation for dynamos.......... | 300.00 |
| Boiler connections.......... ........... | 75.00 |
| Two 60 light dynamos, 2,000 c. p........ ..... | 3,200.00 |
| 100 arc lamps, hoods, globes, suspenders, complete | 3,500.00 |
| 121,000 feet of insulated wire circuit.......... | 1,870.00 |
| 68 iron circuit poles.......... ......... | 1,360.00 |
| 472 cedar circuit poles.......... ......... | 2,832.00 |
| Cross arms.......... ......... | 150.00 |
| Insulators.......... | 100.00 |
| Insulator pins.......... | 25.00 |
| Labor on circuit.......... | 1,000.00 |
| Publishing by-law and incidental expenses.......... | 188.00 |
| Total.......... | $19,000.00 |

10. If the plant is to be constructed for street lighting only, the direct system is preferable, but if it is your intention to do commercial lighting, it would be advisable to adopt the alternating system, as both arc and incandescent lights can be furnished by the same machine, otherwise you would be obliged to install arc and incandescent dynamos and separate circuits. If the direct system is adopted public buildings could be lit with arc lights and opaque globes.

I have estimated for iron poles on Talbot street only. If you wish to use a greater number of these, the additional cost will be fourteen dollars each.

There being a sufficient quantity of water in the creek wasted during the greater part of the year, I have estimated on using a low pressure engine and condenser, which will prove a great saving in the cost of fuel, as the condenser can be worked about eight months in the year.

All the lamps will be 2,000 c. p., but the plant will be provided with automatic regulators so that the lamps, or any circuit of lamps, can be immediately reduced to 1,600 or 2,000 c. p. if required, during certain nights in the year, or after a certain hour of the night, thus effecting a further saving in the cost of fuel.

MAINTENANCE.

Operating the plant from the Waterworks station will be advantageous in the matter of site and buildings, machinery, operating staff, floor-room, use of water for condenser; while to a small extent it presents some economy in fuel. This will, however, be discounted by its greater distance from the centre of electric distribution, and the extra wiring necessary.

2. The staff shall, in addition to the men at present employed at the Waterworks, consist of one engine and dynamo tender, at a salary of $700 per year, and two line-men and carboners, at $600 and $480 per year respectively.

3. One of the chief operating expenses is that of fuel. The amount of coal required, as determined by the American National Electric Light Association, averages, in a well equipped station, five pounds per I.H.P., (746 volts). On a basis of ninety per cent. mechanical efficiency for the engine and the same for the dynamo, one pound of coal will produce 120.85 watt hours. The estimate which I have prepared provides for two generators working ten hours per day each, producing 9.6 amperes under 2,500 volts pressure, making a total daily output of 480,000 watt hours. At 120.85 watt hours per pound of coal, therefore, the amount of coal necessary is 3.971 pounds per day.

4. The item of depreciation on electrical apparatus has been variously estimated at from six to fifteen per cent. of the cost of the plant. A committee of the United States Congress placed it at ten per cent., and good authorities in the United States and Europe state that this is a safe margin. In my opinion ten per cent. should be allowed for depreciation of the electrical apparatus and six per cent. for the remainder of the plant.

| ESTIMATED COST OF OPERATION. | |
|---|---:|
| Coal, 3,971 lbs. per day, at $3.75 per ton.............. | $2,719.25 |
| Carbons.......... | 560.00 |
| Salary of engineer.......... | 700.00 |
| Salary of two carboners at $40 and $50 per month...... | 1,080.00 |
| Cylinder oil.......... .............................. | 75.00 |
| Dynamo oil.......... .............................. | 7.00 |
| Engine oil.......... .............................. | 30.00 |
| Management.......... .............................. | 200.00 |
| Waste and sundries.......... .............................. | 35.00 |
| Cost of operation per year.......................... | 5,006.25 |
| Cost per lamp per year.......................... | 54.0625 |
| Cost per lamp per night.......................... | 14.81 |
| Interest on investment, $19,000, at 5 per cent... | 950.00 |
| Depreciation on $7,542 at 6 per cent.................. | 452.52 |
| Depreciation on boiler, $1,500, at 6 per cent............ | 90.00 |
| Depreciation of $11,458 at 10 per cent............ | 1,145.80 |
| Insurance.......................................... | 10.00 |
| Total interest and depreciation per year .......... | 2,648.32 |
| Total cost per annum............................. | 8,054.57 |
| Total cost per lamp per annum..................... | 80.5457 |
| Total cost per lamp per night......................... | 22.06 |

Should you determine to reduce the number of lamps to 75 in order to lessen the annual outlay, the cost of the plant and the total annual cost of lighting would be reduced, although the cost per lamp per annum would be slightly increased. From the experience of other cities I am convinced, however, that it would be false economy to do so, as under municipal ownership demands would be made for more lamps, which would ultimately reach the proposed number, entailing expensive alterations and additions, while portions of the plant would be found inadequate.

$7,975 of the cost of the proposed plant is taxable, and if owned by a company, would, at the present rate, 16 mills, amount to $127.60 per annum. Under municipal ownership this would in all probability be lost to the city, in which event it should be included in the cost of maintenance.

Following the above is a description of municipal lighting plants at Norwalk, Conn., under the operation of which the average cost per lamp per year burning from dusk until 1:30 or 2 a.m., is given as $64.53¾. A plant at Marblehead, Mass., is also described, which having been in operation but a few months, affords no data. Municipal plants at Dunkirk and Jamestown, N. Y., and Aurora, Ill., are likewise described. In the case of the latter the cost per lamp per year is stated to be $53.65.

The following schedule for lighting in a number of Canadian and American towns and cities is embodied in the report :

| | COST OF LIGHTING IN CANADIAN CITIES. | | | |
|---|---:|---:|---:|---|
| City | No. of Lamps. | Cost per night. | Cost per annum. | No. of nights allowed for moonlight. |
| Barrie............. | 33 | $0.28 8-10 | $70.00 | 72 |
| Kingston ......... | 100 | 30 | 81.00 | 96 |
| Brockville......... | 29 | 35 | 94.15 | 96 |
| Hamilton......... | 348 | 28 | 102.20 | 96 |
| Brantford......... | 35 | 23 | 83.95 | none. |
| Guelph............ | 90 | 24½ | 70.00 | 72 water and steam. |
| Stratford......... | 73 | 18 | 49.50 | 90   do. |
| Belleville......... | 30 | 35 | 94.55 | 96   do |

| City. | No. of Lamps. | Cost per night. | Cost per annum. | No. of nights allowed for moonlight. |
|---|---|---|---|---|
| Peterborough...... | 77 | $0.25 | $75.00 | 65 water and steam. |
| Chatham......... | 56 | 23½ | 65.80 | 85      do. |
| Galt............... | 39 | 22 | 66 00 | 65 |
| Ottawa............ | 331 | 23 2-10 | 64 96 | 85 waterpower. |
| Woodstock........ | 56 | | 56.00 | to midnight. |
| Owen Sound...... | 30 | 37 7-10 | 90 00 | 72 water power. |
| London............ | 226 | 25 | 81.96 | 60      do. |
| St. Thomas....... | 31 | 28 | 102 20 | 96 |
| Windsor........... | 108 | 14½ | 52 93 | none. |
| Pembroke......... | 13 | | 65 00 | muon to 10 o'clock. |
| Port Hope........ | 32 | | 45.00 | 65 water power. |
| Ingersoll......... | 35 | | 62.50 | to midnight. |
| Renfrew .......... | 7 | | 91.25 | to midnight.. |
| Perth............. | 26 | | 56 73 | midnight water. |
| Wallaceburg ..... | 17 | | 75 00 | midnight. |
| Winnipeg......... | 121 | | 102.00 | moonlight all night. |
| Sarnia............ | 30 | | 80.00 | moon all night. |

AMERICAN CITIES LIGHTED BY PRIVATE CONTRACT.

| City. | No. of Lamps. | Cost of each per annum. | Schedule. |
|---|---|---|---|
| Allentown, Pa................... | 114 | $100.00 | All night. |
| Auburn.................... | 150 | 87.50 | All night. |
| Burlington. Ia................ | 120 | 100.00 | Moonlight. |
| Camden, N. J................ | 119 | 146.00 | All night. |
| Canton, O.................. | 120 | 70.00 | All night. |
| Columbus, Ga............... | 40 | 126.00 | All night. |
| Dallas, Tex................ | 125 | 93.85 | All night. |
| Kalamazoo, Mich............. | 103 | 95.00 | Moonlight. |
| Leavenworth, K............. | 87 | 96.00 | All night. |
| Michigan City, Ind............. | 100 | 75.00 | Moonlight. |
| Springfield, Mo.............. | 70 | 114.00 | All night. |
| Wilkesbarre, Pa............. | 99 | 120.00 | All night. |
| Texarcanna, Ark............ | 31 | 160.00 | All night. |
| Danville, Ill................ | 80 | 80.00 | As ordered. |
| Jacksonville, Ill............. | 71 | 96.00 | All night. |
| Streator, Ill................ | 60 | 96 00 | All night. |
| Komoka, Ind................ | 56 | 100.00 | All night. |
| Loganport, Ind............. | 85 | 100.00 | All night. |
| Arkansas City, Ark........... | 35 | 72.00 | To 12 P.M. |
| Fort Scott, Kan............. | 75 | 80.00 | Moon, 1 P.M. |
| Queensbury, Ky............. | 32 | 110.00 | Moonlight. |
| Augusta, M................ | 58 | 76.33 | 9 hours. |
| Bath, M................... | 31 | 125.00 | To 1 P.M. |
| Grand Rapids. Mich.......... | 120 | 100 50 | All night. |
| Lansing, Mich............. | 100 | 100.00 | Moon all night. |
| Bellaire, O................ | 52 | 90 00 | Moon all night. |
| Tremont, O................ | 70 | 70 00 | All night. |
| Killsborough, O.............. | 63 | 70 00 | Moon all night. |
| Lebanon, Pa............... | 50 | 80 00 | To 12 P.M. |
| New Castle, Pa............. | 50 | 80.00 | All night. |
| S. Bethlehem, Pa............ | 55 | 81.82 | Noon to 12 m. |
| Houston, Tex.............. | 92 | 150.00 | All night. |
| Parkersburg, Tex. ........... | 59 | 102.00 | All night. |
| Port Huron, Mich............ | 88 | 100.00 | Moonlight. |
| Buffalo.................. | 1880 | 127.50 | All night. |
| Cincinnati................ | 254 | 84.90 | All night. |
| Detroit.................. | 1279 | 136.00 | All night. |
| Binghampton.............. | 256 | 109.50 | |
| Syracuse................. | 648 | 109.50 | |
| Cleveland................ | 241 | 88.67 | |
| Lowell, Mass............. | 325 | 127.75 | |
| Worcester, Mass........... | 410 | 127.75 | |
| Toledo.................. | 600 | 100.00 | |
| Pittsburg................ | 1594 | 96.00 | |
| Albany.................. | 600 | 144.17 | |

quoted in favor of municipal control, and of D. E. Price, of Paris, Ill., in favor of the contract method.

The report concludes as follows :

The statistics regarding lighting by contract were obtained from various electrical authorities and engineering journals, which have, at considerable expense, collected reliable data for use in discussing the lighting question.

The prices given in the table for lamps lit by private companies are doubtless correct, for their prices were fixed by contract, and can therefore be relied on, as showing the whole cost to the various cities for street lighting. But in the figures given as the yearly cost per lamp, where plants are operated by a city, we are at once confronted by the question whether the estimates represent the entire cost to the city. In general the chief items of error, where error exists, are first, confusion in regard to interest and depreciation on the investment, and second, where the plant is operated as an adjunct to the Waterworks station, or where private lights are furnished from the same plant, the charging to the street lights a larger or smaller sum than their fair proportion of the total expenses. In the list of Canadian cities, Windsor is the only one operating its own plant, but the cost given, fourteen and one-half cents per lamp per night, is so low compared with other cities, operating under similar circumstances, as to convince me that only the cost of operation is given. From their financial report of 1894, I find that the total expenses for the year amounted to $6,762.28. Of this amount $342 85 was apparently expended in repairs, which sum should be included in the depreciation on the plant ; $500 was expended in the purchase of a lot ; in all $842.85, which should be deducted from their cost of operation. The cost would then be as follows :

| | |
|---|---|
| Running expenses............................... | $5,919.43 |
| Interest on cost of plant, $18,000, at 5 per cent,........ | 900.00 |
| Depreciation on cost of plant at 8 per cent. .......... | 1,440.00 |
| Total cost for 1894. .................. | $7359 43 |

This amount is equal to $68.14 per lamp per annum or 18⅜ per cent. per lamp per night. The lamps are only 1000 candle power. In considering these tables it is to be pointed out also that a first-class electric lighting plant for which contracts can be let to-day, is not only better and more efficient than the first-class built five years ago, but it also represents a smaller expenditure. It will be noticed that there are wide differences in prices paid for street lighting, both by municipal plants and by contract. To account for this would necessitate an analysis of each plant, but in general the divergence will be accounted for by the differences in the cost and quality of coal, in the power (water or steam) employed, and in the management ; in the number and kind of lamps and in the kind of generating apparatus, etc.

It is apparent that municipal lighting plants are yearly increasing in number, although there are some reports of failures. The reason for this increase is perhaps due in many cases to the private companies themselves, which, after obtaining a valuable franchise, have no other interest than to curtail the cost of operation at the expense of good service ; or it may be due to the corporations themselves insisting on a light at a rate which does not afford a fair profit. If the city would secure its light from a private company at the lowest possible price, it must make that company's investment secure ; first, by making its franchise for street lighting an exclusive one, and second, by agreeing on some adjustment between the city and

## MUNICIPAL LIGHTING IN AMERICAN CITIES.

| City. | Population. | When Installed. | Cost of Plant. | Power. | No. of Lamps. | Candle Power. | Cost per lamp per year. | Schedule. | Do you do Commercial Lighting. | Should a municipality own its own plant. |
|---|---|---|---|---|---|---|---|---|---|---|
| Aurora, Ill......... | 20,000 | 1886 | $17.986.12 | Steam. | 75 | 1200 | $ 58.00 | All dark and cloudy. | No. | Yes, by all means. |
| Titusville, Pa...... | 8,000 | 1887 | 9,000.00 | do. | 58 | 2000 | 57.00 | | No. | Yes, by all means. |
| Jamestown, N. Y.... | 18,000 | '91 & '95 | 62,000.00 | do. | 271 | 1200 | 48.00 | All night, every night. | No. | Yes, every time. |
| Ashtabula.......... | 8,300 | 1891 | 80,000.00 | do. | 70 | 2000 | Expenses nearly paid by com. lights. | Till 3 a.m. except moon. | 4,200 inc. | Yes. |
| Madison, Ind....... | 8,800 | 1886 | 20,907.00 | do. | 86 | 2000 | 66.00 | Moonlight. | No. | Yes. |
| Metropolis, Ill .... . | 7,000 | 1892 | 51.000.00 | do. | 35 | 2000 | 58.00 | ................... | Yes, | No. |
| St. Peters, Minn.... | 4,000 | 1892 | 15,000.00 | do. | 45 | 2000 | 45.00 | Moonligh. | 800 inc. | Yes, by all means ; franchise under no circumstances. |
| Batavia, Ill ........ | 12,000 | 1889 | 40,000.00 | do. | 120 | 2000 | Com.lights pay, | Moonlight. | Yes. | I do. |
| Bowling Green, Ky.. | 7,800 | 1888 | 15,000.00 | do. | 72 | 2000 | 50.27 | Moonlight. | No, sir. | Yes. |
| West Troy........ | 12,900 | 1893 | 39,000.00 | do. | 113 | 2000 | 66.00 | None. | No. | If managed properly, yes. |
| Marshalltown, Ia.... | 8,900 | 1889 | 18,000.00 | do. | 64 | 1200 | 12.00 | Moonlight and 3 a.m. | No. | Our experience has been satisfactory. |
| Elgin. Ill......... | 20,000 | 1887 | 24,000.00 | do. | 110 | 2000 | 64.93 | Moonlight, all night. | No. | Yes. |
| Dunkirk, N. Y...... | 10,000 | 1887 | 19,000.00 | do. | 75 | 2000 | 43.45 | 8 hrs. 6 min. per night. | .... | |
| Mariette, O........ | 8,200 | 1889 | 18,000.00 | do. | 110 | 2000 | 35.23 | All night. | No. | Depends. |
| Chambersburg, Pa.. | 7,800 | 1890 | 31,000.00 | do. | 144 | 2000 | 64.39 | Moonlight. | No. | Yes. |
| Madison, Ga..... .. | 5,000 | 1891 | 24,000.00 | do. | 40 | 2000 | 58.00 | Moonlight, all night. | Yes, if best light wanted. | |
| Decatur, Ill........ | 17,000 | 1885 | 40,000.00 | do. | 105 | 2000 | 100.00 | Dark nights. | No. | Yes. |
| Little Rock, Ark.... | 25,800 | 1884 | 35,000.00 | do. | 210 | 2000 | 44.88 | M'nlight and dark nights. | No. | Yes, by all means. |
| Marquette, Mich.... | 9,000 | 1889 | 70,000.00 | water. | 100 | 2000 | 50.04 (int. and dep.) | All night, every night, except bright moon. | Yes. | We consider it advisable. |
| Painesville, O...... | 7,800 | 1885 | 12,000.00 | Steam. | 81 | 2000 | 40.00 | Moonlight, all night. | No. | Yes. |

NOTE.—Except where mentioned in the above list the cost per lamp per annum does not include interest on the investment or depreciation in the value of the plant.

All the statistics relating to municipal lighting were collected by me either personally or in answer to a circular addressed to the mayors of the various cities.

The opinions of the mayors of Ashtabula, Ohio, Aurora, Ill., Marshalltown, Iowa, Marquette, Mich., Titusville, Pa., and the Commissioner of Water and Light, Chambersburg, Pa., are

company at the termination of the franchise, which shall be equitable to both parties.

In the general rules of the Fire Underwriters is the following clause with reference to operating an electric light plant from the Waterworks station : " The pumping station must be an independent and separate first-class building, unexposed, and shall not be used for an electric light station or other purposes." But as a separate building is to be erected for the electrical apparatus, no doubt satisfactory arrangements can be made. A number of American cities are operating the electric light and waterworks systems together, from six of which I have the following reports :

| City. | Population. | No. of Lamps. | Cost per Lamp. | Cost of Plant. |
|---|---|---|---|---|
| Bloomington, Ill. | 20,400 | 220 | $51.60 | $73,000 |
| Madison, Ga. | 5,000 | 40 | 58 00 | 24,000 |
| Dunkirk, N.Y. | 10,000 | 75 | 43 45 | 19,000 |
| Marquette, Mich. | 9,000 | 100 | 50.04 | 70,000 |
| Mariette, O. | 8,800 | 110 | 35 23 | 18,000 |
| Metropolis, Ill. | 7,000 | 40 | 75 00 | 51,000 |

Reasons advanced in favor of municipal ownership :—

1. In the construction of a municipal station the municipality saves the profit that a private company would expect to make.

2. A city is generally in a position to borrow money at a lower rate of interest than a private company.

3. Most of the plants owned by private companies in the large cities were put in some years ago, when the cost of electrical appliances was greater than it is to-day, while their efficiency was less.

4. In the construction of a private plant capitalists will not risk investing money unless they are assured of good dividends, while all the municipality requires is a sufficient return to pay the cost of running, including interest on the outlay, with a margin for renewals.

5. A company operates usually under a franchise which is not exclusive, and under a contract which is made for a short term of years, making it inadmissable to venture capital unless the returns are so large as to warrant taking the risk.

6. There is reason for believing that a city can, by letting contracts publicly, erect more cheaply than a company which obtains the machinery by private contract.

Reasons advanced in opposition to municipal ownership :

1. In American cities, where state politics enter largely into the Council Chamber, it is urged that the situation is not always favorable to the operation of a municipal plant, but no valid reason has been advanced why, under the government of a Canadian city, a municipal system of lighting should not be as economically managed as a private plant.

2. A company furnishing light, heat and power being given a secure franchise, can, by a slight additional investment, make the capacity of their apparatus sufficient to do the municipal lighting at a much less cost than could a plant erected for street lighting only.

3. The possibility of great advancement in electric generators and transmission in the near future.

There are certain financial responsibilities which attend the ownership of a municipal plant, which are deserving of careful consideration. Among these I include, (a) the liability of loss from fire, (b) destruction of poles and circuit storm, (c) danger from a broken wire or defective insulation, and the possibility of costly litigation resulting, (d) damages from lightning, instances of which I found where as many as three armatures were burned out in the course of a year; with us this would cost $1,500; (e) the probability of great advancement in electric generation and transmission in the near future.

A company must estimate on a margin for these risks and responsibilities, and very often no doubt this margin is over-estimated ; but in order to determine fairly what the cost to the city will be, and in comparing this cost with the figures of private companies, a reasonable allowance should be made and added to any estimated cost of operation and maintenance.

After going through some of the largest manufactories of electrical appliances in America, and examining machinery now being developed, and looking into recent theory and practice, I am much impressed with the fact that our means of utilising the electric current has by no means, as some affirm, safely passed the experimental stage, but on the contrary the electric domain is changing every day ; old appliances ceaselessly give way to new ; methods of production, distribution and utilization vary yearly, and we may soon expect it to thrust a revolutionary force into the lighting and power system, and for this reason I do not feel it a position to estimate this margin.

On visiting a number of lighting plants owned by companies and municipalities considered to be modern, without exception I found a large floor space occupied by engines, shaftings, pulleys, belting and dynamos, presenting a giddy maze ; and except a few plants driven by water wheels, gas engines and compressed air, this is universally the case. Now each step from consuming coal in the furnace to the glow of light is attended with great loss. The energy is frittered away through this complication of machinery and belting

It is argued that the standard make of electric generators of to-day possesses a mechanical efficiency of about 95 per cent., that is, that every one hundred horse-power required to turn the armatures will be given back in the form of 95 horse-power worth of current—the other five per cent. is taken up in mechanical and electrical losses—and that, therefore, it is not obvious that any extensive improvement can be made in the generating apparatus. But actual tests prove that barely five per cent. of the current produced manifests itself, as light in an incandescent lamp. One great object to be attained, is the consolidation of the parts of machinery to prevent as far as possible the great loss of energy, and a notable stride in this direction is in the machine now being put on the market, in which the engine and dynamo is combined, yet leaves a complication of mechanism with a dynamo in which some part is not utilized every moment. With the oscillator, the recent great invention of Tesla, fly-wheel generator balls, eccentric valves, belting, etc., are thrown aside. The piston of the engine is the only

part doing work, and to this is attached the armature, which, instead of revolving between magnets, is simply darted in and out of the field of force. It is evident that such a machine must have an economy far beyond anything at present in use, and it is possible that when perfected, from its simplicity and economy, it will render valueless the machinery of to-day.

Efforts of electrical engineers are now being directed towards the conservation of energy in the transmission of the electrical currents. When this is accomplished, and it would seem to be a matter of a short time, the vast power of Niagara and others of a similar character will be turned to account, and the present system of lighting revolutionized by companies of central distribution furnishing cheap power for this and like purposes. The Niagara company has already, I understand, contracted with the city of Buffalo at a rate of fifteen cents per lamp per night, and negotiations are in progress with other cities, and even this price will be reduced as the amount of work increases.

A new arc lamp is now being tried, which promises to make an innovation in electric lighting. It may be said to be a combination of the incandescent and arc principles. It is claimed that with this lamp, carbons of ordinary length will burn from one hundred to two hundred hours instead of ten or eighteen as now. The current required to operate the lamp is only from four to five amperes as distinguised from seven to ten commonly used. The labor and carbons, if it meets the expectations of the promoters, will effect a saving in the cost of an ordinary plant.

A new illuminant, a gas called acetylene, has recently been discovered and is attracting much attention. Its illuminating power is 240 candles, whereas a like quantity of London gas is only sixteen candles. What the cost of this is likely to be has not yet been ascertained, although it is claimed that it will be very low.

I have enumerated these phases of electrical development to show the necessity of a fair allowance for these trifles.

## THE TWO-PHASE SYSTEM.

To the Editor of Electricity.

DEAR SIR,—In an editorial in your issue of April 10th you say in reference to the new Westinghouse shop : "Aside from all this, these new works unboubtedly constitute the most complete electrical shops in the world of any kind, and the only ones using the two-phase currents for all operations." As to the first part of this statement it is a matter of individual judgment, and we have no comment to make ; but we are certainly astonished at the latter part, as a reference to your own files would prove its inaccuracy. Our shops have long been operated by two-phase currents, and entirely so. There is not even a temporary use of direct current for elevators and cranes. The whole work is done by two-phase currents. Neither is our apparatus of such peculiar design that it can be operated only at an abnormally low frequency. All the apparatus in our shop is supplied from the central station of the town by a generator using the standard frequency of 16,000 alternations. It is not necessary for us to limit ourselves, therefore, saying that our apparatus could also be used at Niagara. It can be used in connection with any central station, and is being used in connection with many. We are surprised above all that you should undertake to decide the legal question as to the right to use two or three-phase currents for transmission purposes. Surely an anti-monopoly journal need not start in to create a monopoly in advance of the decision of the courts. Had these statements appeared in a journal known to be biased, or to be affected editorially by its advertising columns, we should have passed them over in silence as of no importance. In the militant advocate of fair trade, however, they carry weight. We trust that, your attention having been called to their erroneous nature, in your usual spirit of fairness you will rectify them.

JOHN F. KELLY.

Stanley Electric Manufacturing Co., Pittsfield, Mass.

The above appeared in Electricity, and the editor makes the following comment : We thank Mr. Kelly for correcting us in making too sweeping a statement, as the Westinghouse shops are not "the only ones using the two-phase currents for all operations." More than a year ago Electricity published a comprehensive write-up showing the application of the S. K. C. two-phase system in the Stanley works at Pittsfield, which, we believe, was the first extensive installation, and which has worked with perfect success. In regard to our remarks as to the legal questions involved, we made no attempt to anticipate the decisions of the courts, discussing merely the Westinghouse and Monocyclic systems, as a careful reading will show. We stated, what we knew to be a fact, that the General Electric people had *acknowledged* their Monocyclic system to be an infringement of the Tesla patents.

## THE TIME SYSTEM OF THE TORONTO ELECTRIC LIGHT CO.

UNIFORM and standard time has now come to be looked upon as a necessity. To meet the growing demands for this the Toronto Electric Light Co. have inaugurated a system of time distribution from their station on the Esplanade by means of which electrically driven clocks are located wherever desired in any part of the city, giving absolutely accurate and uniform standard time corresponding with the time signal of the Toronto observatory. The system is an entirely new one, the invention of Mr. J. J. Wright. The necessary apparatus—transmitters, switchboards, etc., as well as the secondary clocks, were constructed in the machine shop of the Electric Light Company. Many of

## CANADIAN ELECTRICAL ASSOCIATION.

THE Executive of the Canadian Electrical Association have commenced the necessary preparations for the approaching annual convention to be held in Ottawa in September next. The exact date will be fixed in the course of a few days. It is expected that the convention will take place in the early part of the month. Advices from Ottawa state that the members of the Association resident in that city, are enthusiastically interested in making the coming convention the best that has yet been held under the auspices of the Association. Several features in the line of entertainment have already been announced. His Excellency the Governor-General has very kindly placed at the disposal of the Executive his electric launch for a trip on the Rideau. Mr.

TIME SYSTEM OF THE TORONTO ELECTRIC LIGHT CO

the difficulties that have heretofore been met in the distribution of time signals have been successfully overcome in this system, and clocks are now in use in the various manufacturing establishments, hotels, offices, and on the streets for the use of the railway company. The cost per year to subscribers is nominal, and the system bids fair to become exceedingly popular. The illustration is from photographs of some of the apparatus and shows different styles of clocks. In our next issue we propose to give details of the service, illustrated by drawings, which we think will be of considerable interest to our readers. Patents have been applied for covering the main points of the system.

Soper will give a garden party at his residence, and the Ottawa Electric Railway Co., have offered to do everything possible to make the event a success. A local committee will be appointed at an early date to complete arrangements for the entertainment of the visitors. A meeting of the Executive will shortly be held to arrange for the necessary papers and otherwise to provide an attractive programme.

A Halifax dispatch states that the amalgamation of the Halifax Street Railway Co. and the Halifax Gas Co., is expected to be an accomplished fact in the near future.

# ELECTRIC RAILWAY DEPARTMENT.

### THE BERLIN AND WATERLOO STREET RAILWAY.

THE conversion of the Berlin and Waterloo Street Railway from horses to electricity was successfully completed on the 18th ultimo. The overhead construction was done under the supervision of Mr. J. Lawson, foreman of the Canadian General Electric Co., with Mr. E. C. Breithaupt as Consulting Engineer. The car equipments are of the C.G.E. type, and the necessary power is furnished by a 60 K.W. Edison bi-polar generator, recently installed by the Berlin Gas Co. for the purpose.

### THE GUELPH ELECTRIC RAILWAY.

MR. GEO. SLEEMAN has commenced active operations on the construction of his new electric railway in Guelph, and has awarded the contract for the entire electrical equipment and cars to the Canadian General Electric Co. About 4½ miles of the road will be constructed this year, extending from the Agricultural College to the Cemetery in one direction ; a second line running from Mr. Sleeman's brewery to the Grand Trunk and C.P.R. depots. The power house and car sheds adjoining will be a solid stone structure of handsome design, with pits in the latter extending the full length of the tracks. The site selected is an excellent one, near the brewery, provided with an ample supply of water for condensing.

The track will consist of 56 lb. T rails, laid on squared cedar ties, and the overhead structure will be for the most part cross-suspension. The rolling stock at the start will comprise three closed and two open cars, each of the former being equipped with two C.G.E. 800 motors, mounted on Blackwell trucks. The car bodies are to be 21 feet long, with removable vestibules and interior fittings of the most elegant description.

The generating plant will consist initially of a cross-compound Wheelock engine, driving a one hundred kilowatt generator of the Canadian General Electric Co.'s multipolar type.

Altogether the Guelph road promises to be when completed at once a credit to the enterprise of Mr. Sleeman and a model of electric railway construction.

### THE INTERURBAN ELECTRIC RAILWAY.*
##### BY S. H. SHORT.

STUDENTS in political economy deplore the tendency of our people to congregate in or immediately around cities, but the practical street railway man sees right here one of the greatest fields for his work, with the best promise . of great profits, for which he is always looking. Every city of more than 50,000 inhabitants has, just beyond its limits, within a radius of fifty miles, a large population more or less dependent on it for business and supplies. These hangers on are more apt to form their own little communities and villages than to be entirely isolated, and these settlements are universally upon roads which furnish a comfortable means of access to the common centre. The steam railroads have long since catered to these "outside citizens" with local trains running at short intervals during the busiest portion of the day ; but this service, being expensive and limited, has not quite satisfied the American people, who want to go just when they please, as rapidly as possible and for as small a sum as may be. Into this ready made business the interurban railway companies must step, with all the accommodation of a street railway, cheap fare, frequent stoppages and more frequent trains, greater cleanliness and even higher speed than the steam roads.

This service will be found to be of mutual benefit to both city and surrounding country. It has often been proved in street railroading that an increase in the number of trains has brought out an immediate and entirely disproportionate increase in the travel, and this can be easily accounted for. If it is possible for one to reach the business centre easily and quickly from a quiet home on the outskirts, where there are no city taxes, noises or nuisances, he will be very apt to purchase such a home and be come a regular patron of the road which accommodates him. The city has the benefit of his money and trade, the country of

*Abstract of Paper read before the Cleveland Electric Club, May 22, 1895.

his beautified home and the increased value of the surrounding property, and the railway of his goings to and fro, and as we are none of us disinterested, all three can rejoice. That the electric railway, when properly constructed and operated, is best adapted for this interurban service is fast being acknowledged by all classes of railroad men, and the eagerness of capital to invest in such undertakings is growing daily. I therefore propose to discuss the various points developed in this new branch of travel, from the side of the practical working, making an effort to care for both the comfort of the passengers and the assets of the company.

After selecting the country through which an interurban railway is to be run, the point of next greatest importance is, where the road-bed should be constructed. There are two possible locations, one along the existing common highway, connecting the termini of the road, the other over a private right of way similar to a steam railway. Each has a number of points in its favor, and generally local conditions must be considered in making a decision. The latter is most tempting in its freedom from obstructions consequent upon the general travel of a highway, the power to establish grades and curves, and chiefly the liberty to maintain a high rate of speed. It will be most advantageously used when the road is intended to operate between a few stations of considerable distance apart. But in most cases it will undoubtedly be best to adopt the public highway, as along this road will be found the greatest number of people, and upon them the success of the road depends. These highways are usually arranged with two roadways, one well macadamized and the other a dirt road. On either side of these is often found a considerable space, before the fence is reached, which is used for drainage purposes, and with very little work one of these spaces may be prepared for the track.

In the construction of the road I would advocate as few grades (never exceed 5 per cent., or 250 ft. per mile) and curves as possible, owing to the high rate of speed which it will be necessary to maintain, but always keeping near the highway.

The road bed itself should be prepared according to the best received practice in steam railroad construction—ties close together—a sixty or seventy pound rail and stone or gravel ballast, with the rails far above the surface of the road, which keeps them clean and dry, affording good traction and a smooth rail. As much care should be taken in holding the rails at the joints and upon the ties as though steam passenger trains were to be operated on the line.

When a single track road is constructed, sidings should be provided, not only for the passing of trains, but occasional ones where cars may be left for light freight or market produce. These sidings should be provided with the latest and best automatic switching devices, or lock switches, with signal targets and lamps attached. Where very high speed is to be maintained, or the trains to be very frequent, a double track throughout should invariably be constructed, for, while the outlay at first may seem very large, the greater convenience, safety and efficiency obtained will soon repay the additional cost.

The overhead trolley wire with the ordinary under running trolley is undoubtedly the only method to be considered for supplying the car with current. The wire should be of large size, certainly not less than No. 00 B. & S. It should be held with flexible supports, the best construction being a cross wire attached to poles on both sides of the roadway. Where this is not possible a single wire pole with brackets may be used, but some flexible support between the trolley wire and the arm must be provided in order that the trolley may be obliged to lift the weight of the wire at all times, and the wire be free to follow all slight movements of the trolley arm.

The poles should be put far enough from the cars to make it impossible to reach them from windows or platform. These same poles can be used to support the supply wires, and telephone and signal lines for the convenience of those operating the road.

Regular stopping-places should be provided along the railway line, as it would be out of the question to stop wherever signalled. The intervals between these stops should be as great as possible without too seriously inconveniencing the people, as the time consumed in stopping and starting would make large inroads on the schedule time of the trip. Shelter should always be provided at these stations, and they can be lighted from the trolley wire.

When the interurban road enters large cities with existing street railway lines, it is generally possible to make a traffic arrangement whereby trains can be run directly into the heart of the city. This should be done, as it avoids discomfort and the delay of a transfer of passengers.

The rolling stock equipment for the interurban electric railway requires most careful and intelligent consideration. The requirements differ so greatly from ordinary street railway customs that we have little to guide us save ordinary steam railway practice. I will, therefore, endeavor to place before you information gained from the use of electric motors for tractive purposes, and apply it to the new conditions met with in this class of work. In almost every case the interurban road enters into competition with some steam railway, and is, therefore, called upon for the same or better service ; so high speed is the first requirement.

The cars should be built like those found most practicable for passenger coaches on steam roads. The body should be about forty feet long—fifty, including the platforms—mounted on double swivel trucks which have thirty-three inch or thirty-six inch wheels, with wide tread and deep flanges. The wheels should all be of the same size, and the trucks centrally pivoted so the weight will be equally distributed on all the wheels. The trucks should be provided with both elliptical and spiral springs, as on all high speed coaches. Such a car will seat fifty passengers, hold 100, weigh, empty, about twelve tons, and about twenty tons when filled with passengers and fitted with its electrical equipment of two fifty horse-power motors. This car will run at a speed of thirty-five miles an hour on the level, and make a schedule time of twenty-five miles, including stops and slow speed on curves and grades.

The motors should be mounted one on each truck. The controlling apparatus should consist of the ordinary series parallel controller, worked by hand, and an air brake outfit of the regular Westinghouse type, with an air compressor operated by a separate motor automatically controlled by the air pressure.

Believing it to be absolutely necessary to use air brakes on these high speed cars and trains, we have designed an air compressor with a special small motor to be placed in the motorman's cab. When the air pressure reaches a point below normal in the main reservoir, the current is automatically turned on, and the motor pumps in air till the pressure is restored ; then the current is cut off from the motor. This air pressure may be made to operate a signal whistle, and to force water from a tank under the car into wash basins, closets and drinking faucets.

The train should be lighted thoroughly with incandescent lamps, and the interior finish should be good quality and attractive. The entire train is heated by electricity directly, or probably more economically by hot water obtained by immersing resistance coils in water tanks provided for the purpose.

Trains of more than two cars, of heavier cars, or requiring higher speed, must have larger motors, preferably of the gearless type, but for such work as specified, the single reduction gear motors are best. All motors should be as nearly spring mounted as possible, to save destructive blows upon rail joints and wheels.

Engineers, in dealing with problems in interurban railways, should exercise the greatest care in making out specifications for cars, motors and generating plants, as there has been very little data obtained from larger motors than the so-called twenty-five horse power machine, and their performance in ordinary street railways. I take the liberty to present a number of facts gleaned from our own experience, and all conformable with regular practice, which may form a basis for all preliminary calculations.

The standard motor will give its full rated brake horse power for one hour without heating more than 150 degrees above the surrounding air. This motor will not run continuously doing this work without danger to its insulation, but it may temporarily, in starting or accelerating a train, take 100 per cent. overload without injury or sparking, while the efficiency at this extreme overload will remain high. Such are its peculiarities.

Our last problem is that of the distribution of power to our railway system. It is seldom practicable to have more than one power station, so it is always best to place the station as near the centre of the line as possible, and feed in both directions. Where the line is short and traffic not heavy this can easily be done, but when long, several methods may be used to maintain a constant E. M. F.

The "booster" system is one which may be used where fuel is inexpensive, and consists in placing a small series machine on the long feeders, whose capacity is just large enough to supply the losses and maintain a constant E. M. F. on the distant trolley. This may be carried to any extent, but it is not economical, as the extra watts are all destroyed by the feeders and do no useful work.

The alternating current method, by which the feeders are charged with a pressure of 2,000 volts or more, and rotary transformers used at intervals to reduce the pressure to 500 volts and change the current to a direct one, is very good, but the regulating apparatus is very complicated.

A more direct and practicable method, we believe, is to have an armature wound for both 500 and 1,500 volts, with a commutator on each side. This needs no separate exciter, and can be so designed as to regulate for the drop on the trolley feeders as the load comes on, thus maintaining absolutely the full E. M. F. so necessary for high speeds.

I have recently designed machines of this character and believe they will give better results than any other method proposed.

I give below a table comparing the amount of copper necessary in the feeders for a fifty-mile road, running trains of two cars equipped with 100 H. P. motors, on a half hour headway, eight trains in operation at all times, when fed from a central station at 100 volts by the old method, and with 1,500 volts using sub-stations with these direct current transformers.

RELATIVE AMOUNT OF COPPER REQUIRED WITH AND WITHOUT ROTARY TRANSFORMERS.

Figures based on a 50-mile road, double track, 8 64-ton trains, each equipped with 2 100 H. P. motors, schedule time 25 miles an hour. R of return ⅓ of total.

| 1 CENTRAL STATION. | 1 CENTRAL STATION. | 1 CENTRAL STATION. | 1 CENTRAL STATION. |
|---|---|---|---|
| Ordinary 500 Volt Distribution. | 2 Rotary Transformers 1,500 Volts Transmission. | 4 Rotary Transformers 1,500 Volts Transmission. | 6 Rotary Transformers 1,500 Volts Transmission. |
| Copper = 100. | Copper = 48.7. | Copper = 23.3. | Copper = 13.9. |

The ideal system of transmission, however, is an alternating current on the feeders at high pressure, stationary transformers at regular intervals between feeder and trolley lines, reducing the current to 500 volts, and an alternating street railway motor on the cars. But the last is not yet forthcoming, and while it is possible to operate induction motors successfully, their tendency to synchronism renders them exceedingly low in efficiency when run at varying speeds, and the controlling devices become very complex.

However, we await the development of the motor, which will enable us to extend our interurban railways into transcontinental lines speeding from ocean to ocean at 100 or more miles an hour.

---

A successful trial trip was recently made over the newly constructed electric road between the towns of Waterloo and Berlin.

A proposal has been made for the construction of an electric railway from Port Hope to Bewdley, on Rice Lake, a distance of nine miles.

Notice is given that the Canada Switch Mnfg. Co., Limited, will ask for authority to extend the powers of the company so as to enable them to engage in the manufacture, purchase and sale of all kinds of railway, electrical, chemical and contractors' supplies.

The London City Council have passed a by-law and draughted an agreement under which the street railway system of that city may be changed to the electric system. A special general meeting of the shareholders of the street railway company will be held on the 6th instant, to ratify the by-law and agreement, and to take into consideration and decide upon changing the motive power to electricity, and to authorize the increase of the capital stock of the company to $250,000 and the issue of debentures for this amount.

## PERSONAL.

Mr. G. W. Moss, Vice-President of the Bell Telephone Co. of Canada, died in Montreal on May 22nd, at the age of fifty-nine years. The deceased gentleman came to Canada from Lincoln, England, upwards of 30 years ago, and was for many years prominently identified with the dry goods trade. He was held in highest esteem by the business community. He had been in ill health for nearly 20 years. At a special meeting of the directors of the Bell Telephone Co., held on the 27th ultimo, the following entry was ordered to be made on the minute books of the company : " The directors record the death, on the 22nd May, 1895, of their honored friend and associate, Geo. W. Moss. Mr. Moss became a director of the company in the year 1881, and served continuously and faithfully for fourteen years Since May, 1890, he has filled the office of vice-president. He was a man of remarkable intelligence, of good business judgment, and most generous impulses. He took a most lively interest in the welfare of the company and in the plans for its future development, in all of which his advice was of great value to his associates on the board, where his place will not be easily filled. The directors place upon their minutes this tribute of their affectionate remembrance of Mr. Moss and of sincere sympathy with his family in their bereavement "

## PUBLICATIONS.

One of the most valuable papers in the June Arena is " A Review of the Brooklyn Street Railway Strike," by G. Emil Kichter.

The Canadian General Electric Co. have in press a new and most elaborate illustrated catalogue of their railway, lighting and power apparatus.

It is reported to be the intention of the C. P. R. to construct an electric tramway from Three Forks to Sandon and Cody Creeks, B. C.

Mr. T. Silvene, of Victoria, B. C., a locomotive engineer on the E. & N. Railway, has secured a patent for an air brake, which is declared by railroad men to be an improvement on the Westinghouse air brake. Mr. Silvene has already received several handsome offers for the purchase of his patent.

A pleasure park has been purchased by the Kingston Street Railway Co. on the outskirts of that city. A new branch line leading thereto went into operation on the Queen's Birthday. The Kingston Street Railway is in the hands of an enterprising company, who furnish an excellent service, and it is the boast of the citizens that the car equipment of the road is the finest in Canada.

## FOR SALE

25 Arc Dynamo, Reliance system, with wire, lamps, pulleys, cross-arms, pins, etc.; cheap for quick sale. Write for prices. THOMAS. R. FOSTER, Tara, Ont.

## SITUATION WANTED

ADVERTISER HAS HAD SIX YEARS' EXperience in management of rail and tramway traffic (passenger and commodity) and would take charge of this department, or sole charge of new or other electric line ; energetic and capable ; age 32 ; understands how to work up business ; best of references both as to ability and character. Address Box 24, ELECTRICAL NEWS.

# CANADIAN GENERAL ELECTRIC CO.
### (LIMITED)

Authorized Capital, $2,000,000.00.
Paid up Capital,   $1,500,000.00.

### HEAD OFFICE:

## 65 FRONT STREET, WEST,  -  -  TORONTO, ONT.

**BRANCH OFFICES:**

| 1802 Notre Dame St. | - | MONTREAL. | Main Street | - | - | WINNIPEG. |
| 138 Hollis Street | - | HALIFAX. | Granville Street | - | - | VANCOUVER. |

### Factories: Peterborough, Ont.

50 KILOWATT SET.

# DIRECT CONNECTED GENERATORS
## For Railway and Lighting Service
—— MEAN ——

Economy in Space
Economy in Power
Economy in Repairs

Sizes from 2½ to 1500 Kilowatt capacity.

## SPARKS.

An electric lighting company is being promoted at Sussex, N. B. If it is decided to form the company, the plant will be got ready for operation by the first of September.

The town of Hastings is to have incandescent light. A local company has taken the matter up and closed a contract for the necessary plant with the Canadian General Electric Co.

Incorporation has been granted to the Ingersoll Electric Light and Power Co., Limited, with a capital stock, of $45,000. The company consists of:—Messrs. Stephen Noxon, John Gayfer, Alfred E. Gayfer, Henry Richardson and George E. Gayfer, of Ingersoll.

A special committee appointed by the Quebec city council to consider and report on the offers of the syndicate represented by Mr. Beamer, for the running of electric cars through the streets of the city, recommend that the company be required to make a deposit of $5,000 with the city treasurer on a thirty years' contract, that for the first three years the company be not required to pay any percentage of receipts to the city; that for the subsequent twenty-two years, the company pay to the city four per cent. on gross receipts, and five per cent. for the remainder of the term of the contract, the franchise not to be an exclusive one, and the contract to be renewable every five years after the expiration of the thirty years. for which it is originally made; the company to have three months in which to commence the construction of extensions of lines in any of the suburbs which may hereafter be annexed to the city; the company's work-shop to be located in the city of Quebec; cars to run between 6 a.m. and midnight, or longer at a speed not exceeding eight miles per hour; and the company is to be responsible for any damages resulting from negligence or otherwise. In winter time, the company may, if it chooses, substitute sleighs drawn by horses. A model of the proposed cars is to be submitted for approbation to the City Engineer, and he will fix the number of persons to be carried in each. All expenses of laying the rails, &c., are to be borne by the company. The first track is to be completely finished by January 1, 1897, and the whole line by June first, 1898. The price of a single ticket for adults is fixed at 5 cents, including transfers, and 10 cents after 10 p.m. Reduced rates are to be given to children. Six tickets to be given for 25 cents, and twenty-five for $1. Each infraction of any of these stipulations is to be punishable in the Recorder's Court by a fine not exceeding $40.

# CANADIAN
# ELECTRICAL NEWS
## STEAM ENGINEERING AND JOURNAL

OLD SERIES, VOL. XV.—No. 6.
NEW SERIES, VOL. V.—No. 7.

JULY, 1895

PRICE 10 CENTS
$1.00 PER YEAR.

CANADIAN

# ELECTRICAL NEWS

AND

## STEAM ENGINEERING JOURNAL.

Vol. V.                    JULY, 1895                    No. 7.

## IMPROVED WATER WHEEL.

THE accompanying illustrations show a water wheel of improved pattern which is being manufactured by Mr. Robert Graham, of Ottawa. The manufacturer lays great stress on the fact that this wheel is one solid casting, and that the buckets are so shaped that 84 per cent. of power can and is guaranteed, being, it is claimed, the highest efficiency yet reached. The wheel being solid, the annoyance and expense of buckets getting loose and falling out is avoided. The gate is as easily worked as a steam valve ; one-half turn of hand wheel turns the water full on or off, so that any ordinary governor will secure a perfectly steady power, no matter how fluctuating the machinery driven may be.

The manufacturer claims for this wheel superiority in the following points :

(1) The ease with which the gate of largest size wheel can be closed under any head of water, as the pressure of water does not in the least affect the working of the gate.

(2) The mode of applying and shutting off the water is so scientific that one half turn of the hand wheel does the whole work, making it just as easy to start, stop and govern as any steam engine.

(3) The construction is such as to almost entirely overcome delay and breakage of machinery by sticks getting into the wheel.

(4) The absence of complicated attachments. Five pieces

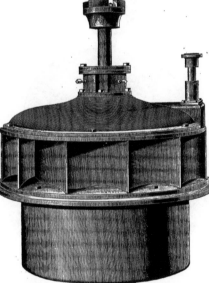

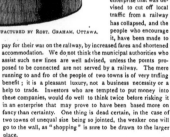

IMPROVED WATER WHEEL, IN CASE, MANUFACTURED BY ROBT. GRAHAM, OTTAWA.

include the whole of wheel case, either with cylinder or register gate.

The wheel and case, as well as the mode of producing the same, are covered by patents.

Further information may be obtained by addressing the manufacturer.

## ELECTRIC LINES AND RAILROADS.

THOSE who keep track of the companies formed, or projected, as they appear in the official gazettes, says the Financial Chronicle, of Montreal, must have been struck by the number of enterprizes organized, or proposed, for providing the towns westward of this city, with communication by an electric car service. If

this continues much longer, we shall be able to reach the western boundary of Ontario without travelling by either of the existing railways, as we could go on from one point to another by the local electric cars. Although this class of service has advantages, and pleasures, we are inclined to think it is in danger of being overdone. It will entail drawbacks, which may develop into more serious proportions than the promoters of these lines seem to forsee. They are reckoning upon the diversion of the great bulk of the passenger traffic between the two towns they are severally proposing to connect by an electric service, away from the railway which is now available. Were this done, the railway so injured would certainly make reprisals, they would cut off a number of trains from stopping at such points and reduce freight handling facilities where local competition for passengers had reduced their earnings; almost certainly also, they would enhance freight rates to and from such points. Considering how all the towns from here to the Detroit River have been built up or developed by railway connections, and how necessary they are for business facilities, it seems somewhat ungrateful, as it must be also unwise, to antagonize the railway which has done such invaluable service to the country. One such enterprise that was devised to cut off local traffic from a railway has collapsed, and the people who encourage it, have been made to pay for their war on the railway, by increased fares and shortened accommodation. We do not think the municipal authorities who assist such new lines are well advised, unless the points proposed to be connected are not served by a railway. The mere running to and fro of the people of two towns is of very trifling benefit ; it is a pleasant luxury, not a business necessity or a help to trade. Investors who are tempted to put money into these companies, would do well to think twice before risking it in an enterprise that may prove to have been based more on fancy than certainty. One thing is dead certain, in the case of two towns of unequal size being so jointed, the weaker one will go to the wall, as "shopping" is sure to be drawn to the larger place.

## CURRENTS OF ELECTRICITY.

WHEN a difference of potential exists in two places connected by a conductor, or a series of conducting bodies, the electricity will seek to equalize the field and a current will flow between the two points. The difference of potential may be due to several causes, but whatever the cause the current will flow. The two places may be a few inches apart, as in a wire from a primary battery, or miles apart, as in a transmission system, or in the currents of the earth or air. They may be connected by a small copper wire, the pipes of the waterworks or gas company, the rails of a street railway, or by a combination of a large number of conducting bodies. This conductor may be of large capacity, of low resistance, or may be a poor conducting medium.

To designate the character of a current, and also a current with reference to its origin, various terms are used. Battery current, dynamo current, earth current, etc., are terms used to designate the current from a battery, a dynamo, currents flowing through the earth on account of a difference of potential at different points, etc. In the production of electricity for lighting and power, two kinds of current are generated, distinguished by the direction in which they flow. The continuous or direct current, which flows continuously in one direction, and the alternating or reversed current, which alternates the direction of its flow, are the currents used, and these two kinds include a number of classes. Both currents are further designated by the voltage; thus we have a 500 volt direct current, 110 volt alternating current, or any other volt direct or alternating current.

The alternating current may alternate the direction in which it flows ten thousand times a second, or twenty-five times a second; this is called the frequency of its alternations. The alternating currents are also distinguished by the number of phases in a given period. When alternating currents are generated by transmission considerable distances, so that a current of higher voltage is used on the transmission lines than can be used for lighting and other purposes, the transformer is used for securing a current of higher or lower voltage. The current flowing in the primary wire of the transformer is called the primary current, and that in the secondary wire the secondary current.

A constant current is an unvarying current. Although the voltage may vary, the amount of current does not change. In series arc lighting systems the current is universally constant. A constant potential current is a current whose voltage is constant, as found in the multiple incandescent lighting system.

One of the great discoveries made by Faraday was that of induction or induced currents. While experimenting with electricity and magnetism he found that if he took a wire, joined the ends and moved it rapidly in front of a magnet, a current would be induced in the wire. This action of the magnet is called electro-magnetic induction. The current is called the in-

IMPROVED WATER-WHEEL, REMOVED FROM CASE.

duction or induced current and it is upon the principle discovered by Faraday that all dynamo electric machinery is based. If we take a coil of wire and a bar magnet and pass the magnet rapidly through the coil, a current will be induced in the coil, or if we move the magnet into the coil and then withdraw it we will have a current flowing first in one direction, and as we withdraw it, in the other. The more rapid the motion the stronger the current.

This discovery was soon followed by another of equal importance—that a current of electricity whose strength is changing in one conductor could induce a current in another conductor forming a closed circuit, and that a current brought near a conductor and then removed would induce a current in the second conductor. Suppose we have two coils of wire, the terminals of one connected to the terminals of a primary battery and the terminals of the second connected to a galvanometer. If these coils are placed near together with a battery current flowing through the first there will be no current detected in the second. If, however, we place a switch in the circuit of the first and open and close it rapidly, the galvanometer will show a current in the second coil. If we close the switch and make the coil approach and recede from the second coil, the second coil will have an induced current. Upon this principle are based the alternating current transformer and similar devices.—Electrical Industries.

## TESTING MOISTURE IN STEAM.

A METHOD of testing the amount of moisture in steam has been discussed by the Institution of Engineers and Shipbuilders, Scotland. The principle in this case, more particularly applicable to marine engines, consists in comparing the saltness of the steam with that of the water in the boiler. The test, as explained, is carried out by means of nitrate of silver, and the reaction is so delicate that, with only 1 per cent. of salt in the boiler, 1 per cent. of priming water can be accurately determined to the second decimal. To one part of salt boiler water there is added 100 parts of pure condensed water, and into this is poured a small quantity of concentrated solution of yellow chromate of potash ; then a nitrate of silver solution containing about 1.10 per cent. of this salt is slowly added. With each drop the salt water turns locally orange red, but this color disappears at first ; later on, when all the salt has been acted on, the whole fluid changes color from pale yellow to orange. The quantity of nitrate solution is noted, and then the experiment is repeated on the condensed steam from the engine, undiluted with distilled water. The ratio of the quantities of nitrate of silver solution used in the two tests expresses the amount in per cent.

MR. Theophile Viau, the chief promoter of the electric railway scheme between Hull and Aylmer, states that the construction of the road will certainly be commenced this summer, probably within the next month.

## UTILIZATION OF NIAGARA FALLS POWER ON THE CANADIAN SIDE.

WE reproduce herewith from a recent issue of the New York Electrical Engineer, a diagram showing the method to be adopted for the utilization of the power of the Niagara Falls on the Canadian side of the river. This plan is said to have been approved by the Ontario Government. The work will be carried on by a company which is in close affiliation with the Niagara Falls Power Co., on the American side.

By reference to the plant, it will be seen that it is proposed to erect two power houses, each having a capacity of 125,000 horse power, and fed by a separate canal. The water discharged by the turbines will be carried through tunnels 300 to 800 feet in length respectively to the outlet.

The construction of the works on the Canadian side of the river will be very much less expensive than those on the American side, where an enormous amount of money has been expended.

## WHY LIGHTNING ARRESTERS SOMETIMES FAIL.

The failure of lightning arresters is too often due to careless installation. It may be instructive, says Alexander J. Wurtz in

## THE TORONTO RAILWAY COMPANY'S ELECTRICIAN.

Editor ELECTRICAL NEWS.

SIR,—In your last issue you make some remarks on the care of apparatus at the power house of the Toronto Street Railway Co., which in my opinion are somewhat unfair to the men in charge of the machinery. I do not think you would do any man an intentional injustice, and therefore ask you to look more closely into the matter, and if you find that injustice has been done, to make amends in as public a manner as the original reflection was made.

You state by inference that since Mr. Davis was removed, that there has been no expert electrician in charge of the company's business. This may be so, depending of course on your idea of what an "expert electrician" is. I understand that the same man is in charge of the dynamos who has been in charge since the establishment of the railway plant, and if Mr. Davis had been nominally in charge, as he was previously, it would not have prevented some fool "electrician" putting a knife into the dynamo.

As the dynamo in question was still in the hands of the contractors, and it was one of their own men who did the damage, it looks a little as though the contractors wanted to unload the

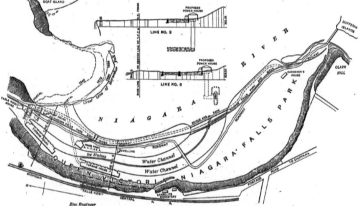

PLAN ADOPTED FOR TWO POWER HOUSES AND TWO CANALS.

an article in Practical Science, to note several examples :

(1.) One plant is reported as having, for better protection, connected two arresters in series. This was probably done with the idea that if a little was good more would be better.

(2.) A large bank of station arresters was grounded to an iron bolt, about two feet long, driven into dry sand.

(3.) Line arresters were grounded by pushing the ground wires into the earth.

(4.) Line arresters were grounded on iron poles, which were themselves set in Portland cement.

(5.) An annual inspection of automatic lightning arresters developed the fact that the arresters were nearly all burned out— in other words that the line was left unprotected.

(6.) The ground plate of a bank of arresters was thrown into a neighboring stream, which subsequently changed its course, leaving the ground plate high and dry.

(7.) The ground plate of a bank of station arresters was laid on the rock bottom of a neighboring stream.

(8.) In a large number or cases a portion of the ground wire is wound into a fancy coil (choke coil).

And so on, the list might be indefinitely extended, each such case forming a source of complaint that the arresters "fail to protect." But, when these curious mistakes are located and properly remedied, the complaints cease.

failure of their machine upon innocent shoulders (a most common occurrence with some manufacturers, I am sorry to say), and take advantage of the removal of the nominal head to find an excuse for their own shortcomings. I do not know that I should interfere in a matter not immediately concerning myself, but the natural instinct of British fair play which, by the way, should pervade the columns of a Canadian journal, moves me to say a word on behalf of Mr. McCullough, the skilful and painstaking electrician of the Toronto Railway Co., to whose ability is due the satisfactory service that has been given to the street railway patrons of this city.

Yours, &c.,

J. J. WRIGHT.

TORONTO, June 12, 1895.

The Street Railway Co. of Brantford, Ont., has been granted permission to lay down 1,200 feet of additional track on Colborne street, provided the company agree to place fenders on all cars.

The Toronto Railway Co have recently placed an order with the Canadian General Electric Co. for twenty motors of their new C. G. E. 1,200 type. These motors, as their name indicates, have a draw bar pull of 50% in excess of the G. E. 800 motor, and are intended for high speed and heavy service. They are to be in operation in time for the Industrial Exhibition, and will no doubt prove of considerable assistance in handling the large crowd which have to be provided for on that occasion

## SOME THOUGHTS UPON THE ECONOMICAL PRODUCTION OF STEAM WITH SPECIAL REFERENCE TO THE USE OF CHEAP FUEL.

THE following is an abstract of a lengthy paper on the above subject, by the late Eckley B. Cox, ex-President of the American Society of Mechanical Engineers, and reprinted from Power :

Mr. Eckley B. Cox, the author, is a prominent Pennsylvania coal miner and merchant, as well as a mechanical engineer, and has given special attention to the utilization of the smaller and cheaper grades of fuel. His paper, which covers nearly eighty printed pages, is a very thorough and original consideration of the use of fuel, but its appreciation involves careful persual and study. The conception derived from the author's abstract presentation unfortunately was not sufficient to draw out the discussion that the paper deserved, but its value is sure to be appreciated on more intimate acquaintance.

When the temperature of the furnace reaches a point sufficiently high to melt the ash or earthy matter contained in the coal, clinker is formed and this clinker will absorb some of the carbon, so that the percentage of "ash" taken from the ash-pit will be frequently much higher than the legitimate ash shown by an analysis of the coal. Our author says : "While too much stress cannot be laid upon the importance of reducing the carbon in the ash to a proper minimum, yet this must not be pushed too far, because a low percentage of ash may be obtained by firing very slowly and gently, thus reducing very greatly the efficiency of the boiler ; or by allowing such excess of air to enter the furnace that the loss due to the increase of free oxygen, nitrogen and steam in the stack, may far outweigh the advantages obtained by the low percentage of ash. I have had a number of specimens of ash analyzed, and find that they contain between nine and sixty per cent. of carbon, although it is not impossible to find ash with a greater or less per cent. of carbon than these." The importance of this loss is dependent upon the percentage of ash in the fuel, as well as the percentage of carbon in the ash, and a table is given showing percentage of the fixed carbon of the fuel lost when the carbon in ash produced is from 1 to 74 per cent., and ash in fuel from 9 to 18 per cent., varying from 11 per cent. with the lowest to 71.05 per cent. with the highest of both values. "In many boiler tests sufficient attention is not paid to this subject. The experimenter simply deducts what he takes out of the ash-pit from the fuel fired, calls the difference 'combustible' and uses the figures in calculating the amount of water evaporated per pound of combustible. Now what we want to know really is how much water is evaporated per pound of combustible in the fuel actually put into the furnace. It is of no advantage to obtain a high evaporation per pound of combustible by throwing away a good deal of good fuel with the ashes."

In order to show specifically the effect of the different losses Mr. Cox presents tables, one showing the effect of burning a hundred pounds of anthracite coal of average analysis with exactly the theoretical amount of air, and the other assuming that double the amount of air theoretically required was used, both with variable conditions of stack temperature. In the first case the per cent. of total heat of fuel lost (except that due to radiation, varies from 11.098 with the stack gases at 400° to 19.108 at 800°, while with twice the theoretical quantity these values for corresponding temperatures are 17.521 and 32.612. These losses are reduced to money value in other tables, using buckwheat coal costing 50 cents at the mines and $2.50 at the furnace, and pea coal costing $1.50 at the mines and $3 50 at the furnace.

From these tables it can be seen, first, how the economy in using cheap fuel decreases as the distance from the mine increases, in other words how important a factor the cost of transportation is in the choosing of a fuel. Second, how much more important it is, the more valuable fuel you use and the further you are from its point of production, that you pay attention to avoidable causes of loss. Thus when using buckwheat coal, with double the quantity of air, and at a stack temperature of 800 degrees, instead of the theoretical quantity of air, and a stack temperature of 400 degrees, the loss is $0.5379 per ton, while with pea coal the loss is $0.7530. So that if you were burning $10,000 tons of pea coal in a year, you could afford to spend to bring about the better result, a sum of money on which $7,530 would give a fair interest and profit, while if you were

burning buckwheat coal, you could only afford to spend an amount from which $5,379 would do so. At the mines, buckwheat coal with double the quantity of air, and at a stack temperature of 800 degrees, instead of the theoretical quantity of air, with stack at 400 degrees, the loss per ton would be $0.1075, while with pea coal the loss would be $0.2825, so that at the mines you could only afford to spend in improving your plant, the sum on which $1,075 and $2,825 would give a fair interest and profit. The same remarks will apply to carbon in the ash, etc.

Thus far we have assumed that the combustion in the furnace was complete, that is to say that all the hydrogen was burned to water, and that all the carbon was burned to carbonic acid gas. This, however, is generally not the case, with the more modern methods of burning coal by forced draft, particularly if a steam jet is used. In the latter case, it is not uncommon to find 6 to 14 per cent. of carbonic oxide and from 2 to 6 per cent. of hydrogen in the stack gases. Of course all the heat units of the hydrogen that go up the stack are lost, and 69.62 per cent. of all the heat units of the carbon which is burned to carbonic oxide are lost. Carbonic oxide is almost always formed unless the bed of coal in the furnace is thin. The air on reaching the lower part of the bed of incandescent fuel converts the carbon into carbonic acid, part of which in passing through the incandescent carbon is converted into carbonic oxide, the latter to a greater or less extent burns in the furnace, uniting with the free oxygen in the gases that have traversed the bed of fuel. Carbonic oxide will not unite with the oxygen unless the temperature is pretty high, variously estimated from 1,150 degrees to 1,500 degrees F., so that if the gases are carried to the furnace too rapidly and reach the cooler parts of the boiler before all the carbonic oxide has been burned, it will pass out of the stack unconsumed.

There is a velocity below which there is loss in the efficiency of the boiler, that is, in the amount of water evaporated per square foot of heating surface, and there is a velocity above which the percentage of unconsumed carbonic oxide and possibly hydrogen becomes so great as to diminish very much the amount of water evaporated per pound of fuel. They had found that in some cases where the passage of the gas through the boiler was very much obstructed and interfered with, it was necessary, to get the best results, to produce a suction in the stack; while in other cases they found that by closing the damper they could burn less coal, reduce their blast and evaporate more water per pound of coal and per hour. The reason was that a part of the carbonic oxide which would otherwise have been burnt, passed up the stack unconsumed because it was carried so rapidly to the cooler part of the boiler that it could not unite with the free oxygen in the gases and burn.

The analysis of the coal, of the gas, and of the ash, where proper arrangements have been made, or where there is a laboratory at the works, is a comparatively inexpensive operation compared with the great value of the results obtained. The indications given by the analysis of the coal he considers of greater value than that of the calorimeter, although where analyses are made from time to time the calorimeter would be a very valuable adjunct. The calorimeter, under the most favorable circumstances, can only give the ash and the heat units developed by a given weight of coal. While two coals might give the same number of thermal units, one might be much more economical than the other on account of different percentages of water present. In the calorimeter test all the heat that goes to make steam is returned by the condensation of steam formed, whereas if the water is evaporated in the furnace it goes up the chimney and is lost. Where the same character of coal is treated, or where the tests are checked up from time to time by chemical analysis, data can be obtained which obviate the necessity of frequent chemical analyses.

In regard to the methods of obtaining the furnace draft, he said, there are three in general use—the chimney, the fan, and the steam jet. The advantage of the chimney is, first, that it allows the deleterious gases to escape at a point considerably above the surface, and therefore, where they are the least likely to cause trouble. Secondly, when once properly built it can be maintained at little or no expense, and if sufficiently large and high, the variations of temperature, etc., can be compensated for by regulating the damper. On the other hand, it is a very uneconomical heat engine where the heat used can be saved, but where there are no means of utilizing it, it is of little consequence how economical or uneconomical it is. The steam jet has the advantage of costing very little to put in and keep in repair. Its disadvantages are, first, it requires a very large amount of steam to run it, and, secondly, it introduces a large amount of water, or steam, into the fire, all of which has to be heated and carried up the chimney, and so far as it condenses in the ash pit has to be reconverted into steam. A still greater disadvantage is the fact that unless very carefully managed there is a large development of carbonic oxide, hydrogen, and marsh gas, due to disassociation of the water, which has a tendency to carry off a great deal of heat in the stack.

The fan is more expensive to install and may cost more to keep in order, but where the arrangements can be made to utilize the heat in the stack gases, it is more economical so far as heat units used are concerned. It has one great advantage—it is possible to at all times obtain the exact blast necessary to produce the best results in the furnaces, which is very important.

He was of the opinion, though advancing it with some reservation, that particularly for the finer anthracites the best results will be obtained by blowing the air by means of a fan through the coal, and, either by a suction fan or chimney, drawing the furnace gases through the boiler, etc., in such a way that there is practically no plenum or vacuum in the furnace or under boiler—or rather a very slight plenum, sufficient to prevent the inflow of any air where it is not wanted.

" In what precedes I have not discussed the question of testing the boiler by means of weighing coal, weighing the ash, and weighing the water, as tests are ordinarily made. While nobody appreciates more highly than I do the advantage of such tests, for in our investigations we often make them, I wish to call attention to the fact that the analysis of the coal, the ash, and stack gases, and the determination of the stack and ash temperatures, give us a means of quickly and cheaply determining practically how much water we are evaporating per pound of coal, when you have once thoroughly tested your apparatus and checked up the results obtained by analysis with the results obtained by the ordinary test. There are several advantages in this manner of testing. We in this way keep track of the coal we are buying, and every large purchaser of coal ought, from time to time, to analyze it. The analysis of the test is not only important for the test, but is very valuable from another point of view Every manufacturer using large quantities of coal should know how much money he is throwing away in his ash-heap. Stack temperatures, if they are below 800° (and they always should be), can easily be determined by the mercurial pyrometer, which can be done in a few minutes by a person of ordinary intelligence, who knows enough to read a thermometer. But the great advantage of this system of keeping track of what you are doing with your plant is, it can be and should be done when the boilers are being fired in the ordinary way and by the ordinary men. The taking of the samples of the coal and ash does not in any way interfere with the operation of the plant, nor does the taking of the samples of the gases and the temperature of the stack ; it can be done at any time and as often as you please. If your man is firing badly, or better at one time than another, the composition of the gases will show it ; if your boiler tubes are coated, either inside or outside, your stack temperatures will give you a very good indication that you ought to clean them, provided that you systematically watch what is going on.

"It seems to me, coal should be kept as dry as possible ; not exposed to the air, if it can conveniently be avoided. Where coal has been exposed to the air and wet it is advisable to get it in the boiler house and allow it to dry as much as possible before using it. It should be handled carefully so as to avoid breakage. The boiler plant should be well covered in, so as to prevent water, snow, or ice from getting upon it. The boiler house should be kept as warm as it can be and comfortable, not, of course, by heating the building with steam or fuel, but by so constructing it that there will be a minimum loss by radiation from it, and should be as free from currents of air as possible. No water, moisture, or steam should be allowed to get into the coal, even under a patent, if it can be avoided.

"While it is a self-evident proposition that the lower the temperature at which the stack gases are discharged into the atmosphere, under the same circumstances, the greater must be the amount of water evaporated per pound of fuel, yet it is not so

certain that we understand the best method of reducing these temperatures. The heated gases from the furnace come first in contact with the boiler, and are generally discharged into the stack after leaving the boiler. Under these circumstances their temperatures must be above that necessary to generate steam at the pressure maintained in the boiler—the higher the pressure of steam the higher must be the temperature at which the stack gases leave the boiler. If, however, an effort is made to reduce their temperature to a minimum, a large portion of the heating surface of the boiler nearest the stack will evaporate very little water, the difference in the temperature between the gases on one side of the iron plates and the water on the other being so slight. In the more modern plants the gases, after passing through the boiler, are used to heat the feed-water, which, entering the feed-water heater or economizer at a very much lower temperature than the steam, is able to absorb heat from the gases, which the boiler is not.

"A still further economy is ideally possible by utilizing a portion of the remaining heat of the gases to warm the air which is taken into the ash-pit. While I am not really an expert in this matter, yet I venture to make a few suggestions as to what appears to be the line in which efforts for further economy should be made. I would try to have the highest possible temperature of combustion in the furnace, protecting the boiler and furnace from the direct radiation of the fire as well as I could, and would allow the gases to escape from the boiler at the highest temperature consistent with the heat contained in them, above the temperature which would be of use for this purpose, all the rest being absorbed by the feed-water ; that is to say, if I were feeding 100 pounds of water per minute, I would try to increase the temperature of the gases, leaving the boiler until the thermal units in them were about what could be utilized in heating the feed-water to the temperature at which it could be placed in the boiler, without allowing the temperature of the gases escaping from the feed-water heater or economizer to be above that at which they could do useful work. My reason for this is the following : The higher the temperature at which the gases pass from the boiler, the greater will be the evaporation per square foot of heating surface, and also the higher the temperature of the feed-water pumped into the boiler will be. Consequently the greater the evaporation of water per dollar invested in the boiler. To obtain the best results, the feed-water heater should be arranged so that the water passes through the heater in the opposite direction to the gases ; that is, the hottest gases should come in contact with the hottest water, and as they are cool they should come in contact with the cooler water. If the gases leaving the feed-water heater were allowed to pass either around the outside or through a series of metal pipes of sufficient length, and the air supplied to the furnace was carried through these pipes or around them, much of the heat still contained in these gases above the temperature of the atmosphere might be abstracted and returned to the furnace, and the loss of heat from the stack gases be largely diminished. There are two objections to this, the cost of the apparatus, and its maintenance ; and as there would be no heat available for chimney draft, it would be necessary to have a suction fan to draw the gases out, or to have a strong forcing fan. While it is probable that in most cases it would not pay to utilize the heat of the escaping gases in this way, yet where coal is expensive it is possible that satisfactory results might be obtained."

## HAMILTON, GRIMSBY AND BEAMSVILLE ELECTRIC RAILWAY.

The management of this road are making eleven round trips per day, and business both in passengers and freight is steadily increasing. In January, the number of passengers carried was 12,000 ; this number has now increased to 15,000 per month. The freight traffic amounts to upwards of 200 tons per month.

The company issue single, return or commutation tickets, working men's and school children's tickets, and a 1,000 mile ticket for $10.00. They are arranging for the establishing of a fruit market in Hamilton where fruit will be sold to dealers and consumers, and connection has been made with the steamboat lines by which freight from Toronto and Montreal is received and discharged over the line.

Mr. A. J. Nelles, the manager, has shown himself to be well qualified to develop the business of the road to the greatest possible extent, having devised a number of original schemes for this purpose. The experience which he gained as freight clerk on the G. T. R. railway is proving serviceable to him in his present position.

PUBLISHED ON THE FIFTH OF EVERY MONTH BY

**CHAS. H. MORTIMER,**

OFFICE: CONFEDERATION LIFE BUILDING,
*Corner Yonge and Richmond Streets,*

TORONTO,　　-　　-　　CANADA.
Telephone 2362.

NEW YORK LIFE INSURANCE BUILDING, MONTREAL.
Bell Telephone 2299.

*ADVERTISEMENTS.*

Advertising rates sent promptly on application. Orders for advertising should reach the office of publication not later than the 25th day of the month immediately preceding date of issue. Changes in advertisements will be made whenever desired, without cost to the advertiser, but to insure proper compliance with the instructions of the advertiser, requests for change should reach the office as early as the 22nd day of the month.

*SUBSCRIPTIONS.*

The ELECTRICAL NEWS will be mailed to subscribers in the Dominion, or the United States, post free, for $1.00 per annum, 50 cents for six months. The price of subscription should be remitted by currency, in registered letter, or by postal order payable to C. H. Mortimer. Please do not send cheques on local banks unless 25 cents is added for cost of discount. Money sent in unregistered letters will be at senders' risk. Subscriptions from foreign countries embraced in the General Postal Union, $1.50 per annum. Subscriptions are payable in advance. The paper will be discontinued at expiration of term paid for if so stipulated by the subscriber, but where no such understanding exists, will be continued until instructions to discontinue are received and all arrearages paid.

Subscribers may have the mailing address changed as often as desired. *When ordering change, always give the old as well as the new address.*

The Publisher should be notified of the failure of subscribers to receive their papers promptly and regularly.

*EDITOR'S ANNOUNCEMENTS.*

Correspondence is invited upon all topics legitimately coming within the scope of this Journal.

THE "CANADIAN ELECTRICAL NEWS" HAS BEEN APPOINTED THE OFFICIAL PAPER OF THE CANADIAN ELECTRICAL ASSOCIATION.

## CANADIAN ELECTRICAL ASSOCIATION.

**OFFICERS:**

PRESIDENT:
K. J. DUNSTAN, Local Manager Bell Telephone Company, Toronto.

1ST VICE-PRESIDENT:
A. B. SMITH, Inspector Canadian Board Fire Underwriters, Toronto.

2ND VICE-PRESIDENT:
C. BERKELEY POWELL, Manager Ottawa Electric Light Co., Ottawa.

SECRETARY-TREASURER:
C. H. MORTIMER, Publisher ELECTRICAL NEWS, Toronto.

EXECUTIVE COMMITTEE:
L. B. McFARLANE, Bell Telephone Company, Montreal.
GEO. BLACK, G. N. W. Telegraph Co., Hamilton.
T. R. ROSEBRUGH, Lecturer in Electricity, School of Practical Science, Toronto.
E. C. BREITHAUPT, Berlin, Ont.
JOHN YULE, Manager Guelph Gas and Electric Light Company, Guelph, Ont.
D. A. STARR, Electrical Engineer, Montreal.
J. J. WRIGHT, Manager Toronto Electric Light Company.
J. A. KAMMERER, Royal Electric Co., Toronto.
J. W. TAYLOR, Manager Peterboro' Carbon Co., Peterboro'.
O. HIGMAN, Inland Revenue Department, Ottawa.

## MONTREAL ELECTRIC CLUB.

OFFICERS:

President, W. B. SHAW,　　-　　Montreal Electric Co.
Vice-President, H. O. EDWARDS,　-
Sec'y-Treas., CECIL DOUTRE,　-　81A St. Famille St.
Committee of Management, T. F. PICKETT, W. GRAHAM, J. A. DUGLASS.

## CANADIAN ASSOCIATION OF STATIONARY ENGINEERS.

EXECUTIVE BOARD:

President, J. J. YORK,　　Board of Trade Bldg., Montreal.
Vice-President, W. G. BLACKGROVE,　-　Toronto, Ont.
Secretary, JAMES DEVLIN,　-　Kingston, Ont.
Treasurer, DUNCAN ROBERTSON,　-　Hamilton, Ont.
Conductor, E. J. Philip,　-　Toronto, Ont.
Door Keeper, J. F. CODY,　-　Wiarton, Ont.

TORONTO BRANCH No. 1.—Meets 2nd and 4th Friday each month in Room D, Shaftesbury Hall. Mr. Lewis, President; S. Thompson, Vice-President; T. Eversfield, Recording Secretary, University Crescent.

HAMILTON BRANCH No. 2.—Meets 1st and 3rd Friday each month, in Maccabee's Hall. E. C. Johnson, President; W. R. Cornish, Vice-Pres.; Wm. Norris, Corresponding Secretary, 211 Wellington Street North.

STRATFORD BRANCH No. 3.—John Hoy, President; Samuel H. Weir, Secretary.

BRANTFORD BRANCH No. 4.—Meets 2nd and 4th Friday each month, C. Walker, President; Joseph Ogle, Secretary, Brantford Cordage Co.

LONDON BRANCH No. 5.—Meets in Sherwood Hall first Thursday and last Friday in each month. F. Mitchell, President; William Meaden, Secretary-Treasurer, 533 Richmond Street.

MONTREAL BRANCH No. 1.—Meets 1st and 3rd Thursday each month, in Engineers' Hall, Craig street. President, John J. York, Board of Trade Building; first vice-president, J. Murphy; second vice-president, W. Ware; secretary, B. A. York; treasurer, Thos. Ryan.

ST. LAURENT BRANCH No. 2—Meets every Monday evening at 43 Bonsecours street, Montreal. R. Drouin, President; Alfred Latour, Secretary, 306 Delisle street, St. Cunegonde.

BRANDON, MAN., BRANCH No. 1.—Meets 1st and 3rd Friday each month, in City Hall. A. R. Crawford, President; Arthur Fleming, Secretary.

GUELPH BRANCH No. 6.—Meets 1st and 3rd Wednesday each month at 7:30, p m. J. Fordyce, President; J. Tuck, Vice-President; H. T. Flewelling, Rec. Secretary; J. Gerry, Fin.-Secretary; Treasurer, C. J. Jorden.

OTTAWA BRANCH, No. 7.—Meets 2nd and 4th Tuesday, each month, corner Bank and Sparks streets; Frank Robert, President; F. Merrill, Secretary, 352 Wellington Street.

DRESDEN BRANCH No. 8.—Meets every and week in each month; Thos. Merrill, Secretary.

BERLIN BRANCH No. 9.—Meets 2nd and 4th Saturday each month at 8 p. m. W. J. Rhodes, President; G. Steinmetz, Secretary, Berlin Ont.

KINGSTON BRANCH No. 10.—Meets 1st and 3rd Tuesday in each month in Fraser Hall, King Street, at 8 p.m. J. Devlin, President; A. Strong, Secretary.

WINNIPEG BRANCH No. 11.—President, G. M. Hazlett; Recording Secretary, W. J. Edwards; Financial Secretary, Thos. Gray.

KINCARDINE BRANCH No. 12.—Meets every Tuesday at 8 o'clock, in the Engineer's Hall, Waterworks. President, Jos. Walker; Secretary, A. Scott.

WIARTON BRANCH No. 13.—President, Wm. Craddock; Rec. Secretary, Ed. Dunham.

PETERBOROUGH BRANCH No. 14.—Meets 2nd and 4th Wednesday in each month. S. Potter, President; C. Robison, Vice-President; W. Sharp, engineer steam laundry, Charlotte Street, Secretary.

BROCKVILLE BRANCH No. 15.—W. F. Chapman, President; James Aitkens, Secretary.

CARLETON PLACE BRANCH No. 16.—W. H. Routh, President; A. M. Schofield, Secretary.

## ONTARIO ASSOCIATION OF STATIONARY ENGINEERS.

BOARD OF EXAMINERS.

President, A. AMES,　　-　　Brantford, Ont.
Vice-President, F. G. MITCHELL　-　London, Ont.
Registrar, A. E. EDKINS　-　139 Borden st., Toronto.
Treasurer, R. MACKIE,　-　28 Napier st., Hamilton.
Solicitor, J. A. MCANDREWS,　-　Toronto.

TORONTO—A. E. Edkins, A. M. Wickens, E. J. Phillips, F. Donaldson.
HAMILTON—P. Stott, R. Mackie, T. Elliott.
BRANTFORD—A. Ames, care Patterson & Sons.
OTTAWA—Thomas Wensley.
KINGSTON—J. Devlin (Chief Engineer Penetentiary), J. Campbell.
LONDON—F. Mitchell.
NIAGARA FALLS—W. Phillips.

Information regarding examinations will be furnished on application to any member of the Board.

---

AMERICAN electrical journals are protesting against the manner in which electrical goods are being slaughtered in price, and particularly incandescent lamps, which one dealer at least offers to sell at "two for a quarter." The cost of manufacturing a medium grade lamp in Germany, where labor and material at least are much cheaper than in America, has been shown to be 11¼ cents. It is estimated that in America the cost of production, exclusive of marketing, will exceed 12 cents. It is therefore certain that lamps cannot be sold at a profit under 20 cents at least. In Canada the price has come near this limit. It is not improbable that the rise in price of materials will add to the expense of manufacture in case of electrical supplies generally, to such an extent as to necessitate a stiffening of prices of the manufactured products. The wisdom of the policy of selling at prices so little above the actual cost of production is open to question.

---

WE had occasion some months ago to question the soundness of the statements contained in an article in the Belleville, Ont., Sun, concerning the conduit electric railway which it was proposed to construct in that city. The article was full of glowing generalities concerning the system, but nothing came in answer to our request for something definite in the way of particulars of the method of construction and operation. It is not surprising to learn, therefore, that Messrs. Lazier & Sons, who are the principal promoters of the road, have fallen back upon the trolley system. Contracts for equipment have been awarded, and construction is being proceeded with. We hear of others who are experimenting with systems designed to dispense with the use of overhead wires, but their success or failure remains to be proven. It is quite probable that the trolley system may eventually be superseded by one which will be equally efficient, less expensive and more sightly, but the experiments in this direction have thus far not reached the point where railway projectors and municipal authorities would be warranted in undertaking to construct electric roads on other than the prevailing method.

THE railroad committee of the New York Legislature has recommended the adoption of a Bill requiring every person or corporation operating a street-surface railroad within the State, to equip every car with a guard fender or sweep, on or before the 1st of September next, and fixing a penalty of $25.00 a day for each car used without such device, the fines thus incurred to be applied to the improvement of the streets of the city in which the railroad company operates.

IN another column will be found a communication from Mr. J. J. Wright, manager of the Toronto Electric Light Co., in which exception is taken to some of the statements appearing in the article in the ELECTRICAL NEWS for June touching the recent accident at the Toronto Railway Company's power station. It should be said in answer to Mr. Wright's objections that the statements published were based on information procured by a representative of this journal at the power station from what was believed to be reliable and impartial sources, and he is loth to believe that either the officials of the railway company or the representative of the Siemens-Halske Company were guilty of wilful misrepresentation. The fact that Mr. Davis was the railway company's electrician in name only, while the actual duties of the position were being discharged by Mr. McCullough, is one with which we, in common with other outsiders, could not be supposed to be familiar. We are pleased to learn from Mr. Wright that the machinery of the Company's power-house is under such competent supervision, and to give proper credit to Mr. McCullough for his ability. In fact, we fail to see why, if he satisfactorily discharges the duties of a position which calls for the highest skill of the trained electrician, the company should have allowed another person to wear the honors.

A PERUSAL of the following extract from an article in the Western Electrician regarding the result of the municipal lighting experiment in Chicago, confirms the wisdom of the decision recently given by the taxpayers of Toronto on this question : "Depending upon the published statements of interested or irresponsible persons, taxpayers have come to believe that the city was pursuing an economical policy in establishing a plant for electric street lighting. It is not at all difficult to understand how this impression was gained, although it is quite remarkable that the public should so long remain in ignorance of the true condition of affairs. It can only be explained by the fact that there was implicit confidence in the ability of Supt. Barrett to fulfil the promises he made which induced the city to adopt the policy of lighting its streets, and that he has thus far evaded making a complete statement of the result of this experiment. Such a statement would show that instead of fulfilling the promises made for it, and thereby effecting an actual saving for the taxpayers, the present municipal lighting system of Chicago has been a very expensive and unsatisfactory experiment. Sup. Barrett has been given ample opportunity to carry out his agreement to provide an economical lighting system for the city streets, but he has failed to show any advantage gained for the taxpayers. He has been urged repeatedly to present data in support of his claims that he was furnishing light at a price named by himself as the actual cost of production, but he has never been able to prove his case. In fact the whole subject of the cost of this project to the city has been shrouded in mystery, and thus far it has been utterly impossible to secure from the city electrical department any detailed statement that would be at all reliable. But while it is true that Sup. Barrett has positively refused to enlighten the electrical fraternity upon the cost to the taxpayers of his favorite scheme, and has encouraged the belief that the project was a profitable one for the city, electrical experts of established reputation have made it their business to investigate the subject thoroughly, and they have proved conclusively that Sup. Barrett's statements are based upon figures that are not complete, and that they are therefore neither fair nor honest, but misleading and injurious to the electrical business. Sup. Barrett cannot plead ignorance of these facts, for he has been confronted with them, and he has been urged to disprove the charges of bad faith made against him or to admit the error of his statements. He has done neither, but he has repeatedly promised, or rather threatened, to publish figures that would convince the world of the truthfulness and the justice of his claims and cover his critics with confusion. For six years the fraternity has waited, occasionally reminding him of these promises, but he has ignored all demands."

THE necessity for better comprehension of electrical principles by those placed in charge of electrical machinery, and for the employment of a better class of men, properly qualified, in power houses, was never more emphatically shown than recently in a small town where the dynamo was run in connection with a saw mill, and by the same "engineer." The owner was heard to remark that there was "no money in electric lighting," and that all the profits were more than eaten up by the constant repairs to machinery. Enquiry elicited the facts that the saw mill engine was used to run the dynamo; the "electrician" of the plant was also the engineer, and that they "used up" an armature every three months or so,—the idea being that the armature is consumed in the production of electricity as a tallow candle is consumed in its own special way. The engineer was "all right." Of course, the engineer always is all right ; it is always the electrical machinery that is no good. A suggestion that perhaps a saw mill engine is not the best adapted to electric lighting service was met with a polite sneer. The owner (a lawyer, by the way) knew better than that, and the suggestion was regarded as rising out of the hope of selling a new engine. He also considered that the "engineer," having run a saw mill, and "been around machinery" all his life, should know all about machinery, and was, ipso facto, a competent electrician. This state of affairs is fortunately not very common. The combination of a bigoted owner, an ignorant engineer, and an engine not adapted to electric service, is enough to bring discredit to any electrical enterprise, but, fortunately, luck seems to have so far befriended small plants. At the same time, plants should be run with judgment, and some experienced person employed as dynamo tender, or electrician, instead of the mechanic usually engaged. It is certainly true that, as Lord Kelvin said, an electrical engineer must be nine-tenths a mechanical engineer; but it by no means follows that an ordinary mechanic, who has merely picked up what he knows by having shoved coal under a boiler, and cleaned an old engine, is in any smallest detail qualified to call himself "electrician." And that is the mistake that is made in a very large number of lighting plants. It is perfectly true that it does not take a very high intelligence to carbon a lamp, or clean a dynamo, but because a man can do these he is not therefore an "electrician," and it takes something more than this before a man can properly manage a power house. In the States and in Europe it is recognized that to properly acquire a thorough knowledge of electricity, some years of special technical education following the ordinary education of the schools, is necessary; and that the curriculum must include special courses in mechanics, mathematics, and general physics. Then again, in these old-fashioned countries, a qualification for the manager of a lighting or power house embraces not only a very thorough training in practical electricity, but also a very intimate knowledge of mechanics, of steam, of book-keeping, even of elementary chemistry; in fact a candidate must be very comprehensively educated, and remarkably capable. Here in Canada, however, we have left behind us these antiquated ideas; our youth are so naturally gifted that technical education is not necessary for them; they are so versatile that they are equally competent as farm hands, engineers, carpenters, lawyers and so on, and any man who can look after a 60 H. P. brewery plant and keep it going, is just as qualified to design and operate an entire electric power plant as a School of Practical Science graduate. It is all these impractical ideas that cause a state of affairs described quite recently to the writer by a prominent American electrical engineer, in the words "Canada is a regular graveyard of old out-of date apparatus." He might equally truthfully have said that the methods of operating are as inefficient as the machinery used.

THE numerous different makes of steam and electrical machinery, the variety of somewhat similar apparatus for attaining the same end, and the absolutely incompatible claims of manufacturing companies and their agents, all depreciating each other's makes, and fearlessly asserting their own to be without an equal, imposes a most embarrassing responsibility on the purposing investor in electrical apparatus. At this date, a company going into the electric lighting business can choose between at least a dozen different makes of electrical machinery, most of which are manufactured in Canada,—at least nine different steam engines, some high, some low speed, with all their varieties of simple, cross and tandem compound, triple expansion; while the choice offered in switches, lamps, sockets, motors, brushes, wires, etc., is large. The details of design or construction on which the rival makes base their claims for absolute superiority, are, of course, just as numerous as the rival makes themselves. One well-known and excellent maker says that every other known make is simply and perfectly wrong in mechanical and electrical design combined, while his own, by a peculiar arrangement of armature and field magnets, attains that measure of perfection that is permitted to mere mortals. Another over-compounds his fields, and "begs to draw your earnest attention to the fact" that thereby the regulation of voltage is rendered automatic and perfect; a third says that instead of compounding his *alternator* fields, he takes a rectified portion of the alternating current, and therewith compounds his *exciter*, attaining the same regulation with a great deal less wire; while a fourth asks you scornfully what you want with "automatic" regulation anyway? Give him a smart, live dynamo tender, who will mind his business, and that's a great deal better and more reliable than any automatic regulation; so he doesn't compound at all, and sneers at those who do. And then a new-comer laughs at them all, because they revolve their armature through a stationary field, whereas he has struck out a line for himself, and does the opposite, claiming as an advantage the ease of replacing armature coils, and the fact that he has no moving wire. It is impossible to more than enumerate such varieties, as surface wound versus ironclad armatures; disk versus drum versus gramme ring windings; cast iron versus wrought iron versus steel, for pole-pieces; carbon versus copper for brushes, etc., *ad infinitum*. And similarly with steam engines and boilers. Everyone of these rival makes wants a purchaser's order, and will go great lengths to get it. The first object, of course, is to impress him with the clear value of one particular machine, and to run down all others; so the agent talks about magnetic circuits, and freedom from hysteresis and wasteful eddy currents; will touch lightly on the mechanical design, and fill the unfortunate purchaser full of technical phrases; and will, discreetly, appeal to him as judge on these matters, as to whether it is not so! How many of the non-professional public are able to say whether it really is, or is not so? The public, as a rule, has enough to do to keep itself posted on matters connected with its own business, and if it could once be brought to see that an electrical machine is not a mere question of iron, and copper, and day labor, it would probably begin to think that electricity is best managed by electricians, and that a business man, shrewd and intelligent though he be, is not and cannot be final judge of the comparative merits of machines, that must not only conform to mechanical principles, which after all are fairly obvious, but also to magnetic and electrical principles, which certainly cannot be considered obvious. The distinct points in an electrical machine, wherein it differs from those of all other makes, are of course claimed as improvements, and the object of all improvements is to save money. Take, for example, two rival methods of compounding—that one that puts a rectified series current round the alternator fields, and that one that takes the same rectified series current, but puts it round the exciter field instead. (The writer has no intention of comparing the two methods, but takes them only as an illustration.) Each one claims superiority. One would require probably more copper than the other, while the other would take a heavier exciter than the one. The point of the respective costs of the two machines is soon settled, and will be obvious to anyone; but is any non-professional person capable of saying whether the methods are equally efficient? and if they are not, which is the better of the two? The maker who compounds his fields will, naturally, say that the other is no good; the other will return the compliment, and the business man who knows nothing whatever about the matter is left to judge between them. Or again, one manufacturer uses Swedish iron in his castings; another uses American. The first claims superiority on account of the better magnetic qualities of his material. How does the purchaser know whether that is a valid claim? He has only got the manufacturer's word for it, and a

manufacturer can hardly be considered a disinterested person, where he has a sale to make. Once again, a manufacturer makes a sale on a guarantee of such and such efficiency. How does the purchaser know whether this guarantee is made good? The manufacturer makes a test himself, and says it is all right. Of course he says so. But ought that to be sufficient for the man who has to pay the money? It seems hard to improve on the plan followed by waterworks companies, or railway companies, and other parties requiring machinery. These engage the services of a competent engineer, to whom are referred all practical questions, his duty being to act as the technical adviser of those unacquainted with engineering matters, and the results are what might be expected. Waterworks are as a rule much better arranged and operated than electric light and power plants, for the simple reason that professional knowledge and experience is admitted to be necessary for a hydraulic engineer; whereas, apparently, any farm hand or master carpenter is sufficiently equipped by nature to tackle the most difficult electrical problems.

## LOOKS LIKE AN ELECTRIC RAILWAY COMBINE.

THE bill to incorporate the International Radial Railway Co., now before parliament, will bear careful scrutiny and watching by everyone desirous of seeing a healthful development in electric railways. Some very estimable citizens, principally from Hamilton, appear as charter members of the Company. These are Messrs. Alexander Burns, Alexander McKay, M. P., John Hoodless, James Edmund O'Reilly, Thomas Miller, F. A. Carpenter, M. P., Peter D. Crerar, Thomas Ramsay, William N. Myles, R. H. McKay, Arthur H. McKeown and James Frank Smith, of Hamilton; Thomas Bain, M. P., of Dundas; William Andrews, of Guelph, and E. J. Powell, of London.

An array of good names, it so happens, is not always a guarantee of the character and real purpose of a project. The names of not a few good business men are identified with the International Radial Railway Co. There are several members of parliament, and there is not wanting a lawyer or two. But let no one be carried off in any project of the kind simply by the names of those high in church courts, business circles, or the bar.

The business men of Canada know something of the power of railway corporations, and especially after the smaller roads had been brought under the control of one or other of the two large railways of this country. With competition crowded out they know how difficult it is to secure freight rates or other concessions that would be helpful to their business, and to the general business, very often, of the whole country. The railway pool, and the joint circular that is an important document in the offices of the Canadian Pacific and Grand Trunk Railways, can hardly be forgotten by business men, and must cause them to think twice when any movement is made to tie up their interests by a possible combine.

Let us look at some of the terms of the proposed bill. It does not limit the motive power to electricity, but it is very plain, as one studies the measure, that it owes its suggestion to the rapid development of electric power for railway propulsion, and especially of recent years in the opening of short electric lines in different parts of the country, proving of inestimable value to the public.

The lines of the International Radial Railway Co. will have their start in Hamilton. It is proposed to construct a line to Waterloo, passing through Galt, Preston and Berlin, with a branch to Guelph. Another line will connect Hamilton and Fort Erie, passing through Wentworth, Lincoln and Welland, with a branch through Dunnville to Lake Erie, terminating at or near the mouth of the Grand River. A third line will connect Hamilton and St. Mary's, passing through Brantford and Woodstock, with a branch to Fort Burwell. Expropriation powers exceeding those provided by the railway act are applied for. The bill empowers the company to bond the road to the extent of $20,000 per mile of single line, and $6,000 per mile additional for double track. Authority is also asked to lease the railway to the Canadian Pacific, the Canada Southern or the New York Central Railway Co., and to acquire the franchise and properties of the Hamilton and Dundas Street Railway, the Galt, Preston and Hespeler Street Railway Co., and the Berlin & Waterloo Street Railway Co. It is proposed that the company shall be given power to enter upon public highways with the consent of the municipal authorities, for the purpose of constructing a telegraph and telephone system.

The scheme is a big one, and calls for a large outlay of capital. That the projectors are likely to use their own means to carry through the undertaking was seriously questioned by Mr. Sutherland when the bill was before the railway committee of the House. He said : "They appear to be procuring the charter simply to secure the territory covered by it." Were it carried out in its entirety it would mean that every electric railway of importance in Western Ontario would pass into the hands of the proposed company. Not one of these would continue to be an independent corporation. New lines would be constructed tending to serve . as useful extensions ; and this much done the directors would then have power to hand this most valuable franchise over to the Canadian Pacific, the Canada Southern, or the New York Central Railway.

Have the people of this country, and especially Ontario, forgotten the efforts put forth some years ago to secure local railway accommodation for their different towns and municipalities ? Some will have a lively recollection of the days of bonusing and the flattering speeches of the bonus-hunter, as he travelled from school-house to school-house and grew eloquent over the benefits that were to accrue to the country by the construction of these local and short-line railways. Where are these railways to-day ? What is the position of the towns that were to become important railway centres, with workshops and many other privileges granted to them ? Every one of these short-line railways has been gobbled up either by the Grand Trunk or the Canadian Pacific, and individuals and municipalities are groaning under the heavy interest and taxes contracted for at the time.

The building of short-line electric railways is one of the hopeful omens of the present day, furnishing a medium of communication where railway accommodation is unknown or very imperfect. To what extent these undertakings should be pushed and on what lines, has been discussed in this journal before, and we have not hesitated to speak words of caution where these seemed to be required. But anything we have said will not alter the fact that if there is to be a development in electric railway building it must come through independent effort. No doubt the steam railways see the competition that will meet them from the electric railways of the future. Be it so. How disastrous would be the result if parliament should sanction any measure that would tend to placing the present electric railways and the many possible, and in some cases, planned roads of the future, in the hands of any one large corporation.

Everyone interested in electrical enterprises and desirous of seeing this new force attain the high commercial purpose and place that it is designed to do, will see at a glance how anything in the shape of a combine would thwart this end. We write as among those vitally interested in the progress of the electrical industry. We can write with equal force as citizens anxious for the best interests of the country, profiting by the experience of the past and the conditions of the present, when entering a vigorous protest against any scheme such as that indicated in the bill of the International Radial Railway Co.

From whatever standpoint the scheme is viewed, the nerve of the promoters is remarkable. It is not shown that the public will be in any way benefited by the extraordinary privileges asked for in this bill. On the contrary every clause points in an opposite direction. And then, having secured about all the rights that it would seem possible to conceive of by the most wide-a-wake railway promoter, the company have the gall to ask that they be permitted to "convey or lease" to the Canadian Pacific, or other company, in whole or in part, any or all rights acquired under the act. Or failing to convey or lease that they may amalgamate with any such company. What was the Railway Committee doing when they allowed this bill to pass safely through their hands? It is now plainly the duty of parliament to see that the measure is not further advanced, and allowed to become law.

The Dundas Telephone Co., have completed their limes from Kemptville to Chesterville. The company has also 100 miles of poles ready for wiring connecting the villages of Morrisburg, South Indian, South Finch, Avomore, Monklands, Metcalf, Moorewood, Vernon, Duncanville, and Kenmore.

## THE ONLOOKER.

AFTER the very successful experiment made in Boston on the Nantasket branch of the New York, New Haven and Hartford Railway on the 21st June, where an electric locomotive reached a speed of 80 miles an hour, the question is being asked with renewed force, are steam roads doomed? An electric locomotive built to haul Baltimore and Ohio trains through the tunnel in Baltimore, made its first run on the 30th ult., and the trial there proved so complete a success, that it is anticipated the locomotive is likely to go into permanent service from that day on. The Boston experiment was critically watched by the officials and attachees of the road. The work of changing the motive power of the road from steam to electricity was done under the direction of Col. H. Heft, formerly president of the Bridgeport Traction Co., but who recently was engaged by the consolidated road to superintend their electric work. The engines are of the Green-Corliss build, and especially designed for the work they are given to do. The shaft is 18 inches and the fly wheel, which weighs 64,000 pounds, is 18 feet in diameter. The condensers are so piped that the engines can work with or without them. They are arranged to regulate from no load to a maximum of 1,420 horse power. They can also be stopped by simply pressing one of several buttons. The two generators especially built for the line, run at a speed of 110 revolutions a minute and are guaranteed to develop 1,500 horse power each. The armatures, instead of being built up of wire, are made of copper disks, each insulated from the other The generators are 10 feet high and the armatures 8 feet in diameter. The switchboard is arranged with two main generator panels. The voltage is 700 volts. Save for a single feature, it is said, there is not the slightest resemblance to the steam locomotive. That feature is the cowcatcher at both ends of the locomotive cars. It is situated underneath the platforms instead of projecting beyond the body of the locomotive as with the steam locomotive. The wheels are about the size of the largest wheels used on steam cars. But the axles are considerably heavier to withstand the strain of the electric gearing. The Nantasket Beach Branch was chosen for the experiment for the reason that within its limits are concentrated most of the difficult problems which will have to be met. The curves are many and sharp, and the grade steep. Mr. John Patterson, of Hamilton, who is one of the projectors of the Niagara, Hamilton and Pacific Railway Co., who have within the past ten days secured a bill of incorporation from the House of Commons, has been to the States investigating the experiment spoken of, with the view of applying electric power to anticipated extensions of this road. To the Onlooker he said, as far as one can judge, the Boston experiment has been a complete success, and certainly if time proved the lasting character of the experiment it must go a long way to cause electricity to supplant steam for railway purposes.

## CONCERNING FIRES.

IT is very generally argued, that when a boiler is being heavily worked, a thick fire is absolutely necessary, but from some experiments lately made, the opinion appears to be an erroneous one. As to the economy of the two, some maintain that heavy fires give the most economical results ; but this, also, is questionable. Valuable information on the subject has recently been brought out by the results of two evaporative tests, says the Mechanical World. They were made on a 72-inch return tubular boiler, having 1,000 3½-inch tubes, 17 feet in length. The heating surface amounted to 1,642 square feet, and the grate surface to 36 square feet, the ratio of the two being 45.6 to 1. On the thick fire test, the depth of the coal on the grate varied from eight to twenty inches, being heaviest at the rear end, and lightest at the front end. On the thin fire test, the depth was maintained uniformly at about six inches. The difference in the results, as appears from the figures given, indicates an increased evaporation, due to thin fires amounting to 15.6 per cent.

The annual meeting of the Canadian Electric Light Co. was held in Montreal recently, at which the following directors were re-elected ; Messrs. R. McLennan, Toronto ; Adolphe Davis, Henry Hogan, Robert Bickerdike, John D. McLennan, Cleveland, O.; C. C. Claggett and F. S. McLennan, At a subsequent meeting of the directors, Mr. R. McLennan was elected president, Mr. Davis, vice-president, and Mr. F. S. McLennan, secretary-treasurer.

## THE TIME SYSTEM OF THE TORONTO ELECTRIC LIGHT CO.

THE accompanying illustrations and particulars will serve to explain the above system, of which brief mention was made in the ELECTRICAL NEWS for June :

The object of the invention is to devise an electric time indicator which will accurately indicate time at one or more distant stations synchronously with a master clock at the central station, and it consists essentially of a distant station clock mechanism responsive only to currents of alternately changed polarity ; a transmitter in circuit with the distant station clock and adapted when set in motion to change regularly the direction of the currents received from a dynamo or other source of current ; a master clock in circuit with a device controlling the transmitter and adapted to make and break the circuit at regular pre-determined intervals, and thus start and stop the transmitter, the current in the master clock circuit being generated by a battery or source of current of low potential substantially, as hereinafter more particularly described.

Figure 1 is a skeleton view, showing the general arrangement of the apparatus. Figure 2 is a perspective detail of the transmitter.

In Figure 1, A is the master clock arranged in circuit with transmitter controlling electro-magnet, B, and the battery, C. One wire, D, of the circuit is always in contact with the revolving disk, E, connected to the main arbor of the master clock, A. The other wire, F, of the circuit is connected to one end of the platinum spring, G, the other end of which lies in the path of the two projections shown on the disk, E. As this disk revolves once a minute, the circuit is made and broken regularly every half minute.

In the main circuit are arranged the dynamo or other source of current, I, the transmitter, J, and the distant clock, K. In this main circuit, H and L are wires of the line circuit, and M and N the wires of the dynamo circuit. The mechanism of the distant clock, K, is only responsive to currents of alternately changed polarity, and the transmitter is so arranged that each time it is set in operation by the master clock, the polarity of the current flowing to the distant clock is reversed, and the mechanism of the clock thus put in operation.

In Fig. 2, the construction of the transmitter, J, and the transmitter controlling electro-magnet, B, is shown in detail. O is a commutator suitably journalled and P a drum, around which is wound the cord or wire, Q, which is fastened thereto and also connected to a clock weight, not shown in the drawings. The commutator and the drum are connected by the gearing, R-R-R, as shown, so that a rotary motion will be given to the commutator when it is set free by the master clock. S is a fly, connected to a shaft, geared to the other end of the commutator, as shown in the drawing. Suitable provisions for winding the drum are also made, but as this is an ordinary clock-work train, further detailed description is unnecessary. T-T are arms connected to the near end of the spindle of the commutator, O, and which engage alternately with the end of the armature, U, of the electro-magnet, B. This armature is pivoted at V, and is provided with a weighted tail, X, as shown, to normally retain its point in a position to engage with one of the arms, T. An adjustable stop, W, is connected to the frame of the transmitter, T, so as to limit the upward motion of the armature. Thus the transmission of a current through the wires, D and F, of the master

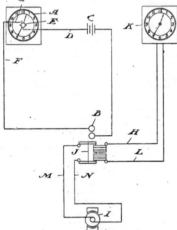

TIME SYSTEM, TORONTO ELECTRIC LIGHT CO.—FIG. 1.

clock circuit will draw down the armature and leave the commutator, O, free to revolve, till the other arm, T, comes in contact with the armature, U, which has been returned to its normal position on the cessation of the current by the weighted tail, X.

The commutator, O, is constructed of non-conducting material and is provided at each end with metal rings, a-b. c and d are metal sections, connected respectively with the rings, a-b, but insulated from one another. e and f are metal sections, arranged diametrically opposite to the sections c and d, insulated from one another, but connected respectively by the wires, g and h, to the sections, b and a. Bearing on the rings, a and b, are the brushes, i and j, provided with weighted tails and pivoted in the usual manner on the brush standards, k and l, to which are connected respectively the wires, M and N, from the dynamo. M are a series of brushes pivoted on the brush standard, n. The ends of these brushes are in position to make contact with the sections, c and e, when the commutator revolves. o are a series of brushes pivoted on the brush standard, p. The ends of these brushes are in position to make contact with the sections, d and f, when the commutator revolves.

These brush standards are connected respectively with the line wires, H and L, of the line circuit. When the commutator is in the position shown in Fig. 2, no current will flow through the line circuit, but when the commutator is released by the energizing of the electro-magnet, B, the sections, e and f, of the commutator will come in contact with the brushes, m and o, and a current will then flow through the wire, M, standard, l, brush, j, ring, b, wire, h, commutator section, e, brushes, m, standard, n, line wire, L, distant clock, K, line wire, H, standard, P, brushes, o, commutator section, f, wire, g, ring, a, brush, i, standard, k, and wire, n.

When the sections, e and f, have passed from underneath the brushes, m and o, the other arm, T, on the end of the commutator spindle has come in contact with the end of the armature, U, and the commutator becomes stationary. The next time the electro-magnet, B, is energised and the commutator released, commutator sections, c and d, come into contact with the brushes, m and o, and the course of the current is then through the wire, M, standard, l, brush j, ring, b, commutator section, d, brushes, o, standard, P, wire, H, distant clock, K, line wire, L, standard, n, brushes, m, commutator section, c, ring, a, brush, i, standard, k, and wire, N. The commutator is again stopped as before, when the commutator sections, c and d, have passed from underneath the brushes, m and o. It will thus be seen that the revolution of the commutator sends currents o falternating polarity to the distant clock, K, which is made to respond only when receiving such currents.

The construction of the mechanism of this clock will be readily understood. q is a polarized steel armature connected to a spindle, r, journalled on the frame, s, in a vertical position, in a manner similar to the balance wheel of a common clock. t is a worm formed on the spindle, r, which engages with the worm wheel, u, on the arbor, v, journalled as illustrated. w and x are the hands of the clock which are driven from this arbor in a similar manner to that in every-day use in ordinary watches and clocks, and which will be readily understood by one conversant with such matters.

From this construction it follows that the revolution of the armature, q, will impart the proper motion to the hands, w and

x. a¹ is an electro-magnet suitably supported on the frame, s, with its poles, b¹, in close proximity to the circular armature, q. The wires, H and L, of the line circuit are connected to this electro-magnet, as shown, so that when a current passes through them, the electro-magnet is energized and its poles attract the poles of the armature, q, having an opposite polarization. It is evident that as long as the current flows through the electro-magnet, a, in the same direction, no further motion of the armature, q, is possible, but when a current is sent through it in a reverse direction, the polarization of the electro-magnet is changed and, as before, opposite poles of the armature, q, are attracted, and another half revolution of the armature takes place. c¹ is a spring connected, as shown, to the frame, s, with its lower end lying in the path of the cross-bar of the armature, q, so as to permit of its motion in one direction only.

The features which the inventor considers particularly important are—First, placing the master clock in an auxiliary circuit so that its office is merely to set in operation a transmitter controlling the flow of current in the main circuit. This enables the use of a current of very low potential in the auxiliary circuit, so that no trouble can occur through the carbonizing of the contacts on the master clock, and as these contacts may be made very light, no disturbance of the time-keeping qualities of the

## THE WORKING SURFACE OF A PULLEY.

IT has taken considerable time to settle the question in regard to belts made of leather, as to which side should run next to the shaft wheels, if, indeed, it has been settled, for even now it is rehashed occasionally by saw mill men, etc., says an exchange. It is always a pleasure to see the best side of a belt stand out whenever a new belt is to be set in motion, and good looks go a long way on all such occasions.

In spite of all tests that have been made on leather belting, nothing has ever been said of the extra cling that the flesh gets by being easily squeezed into every depression on the face of the pulley, which the grain side has a tendency to bridge over. This seems to follow the law of friction where the particles of one material interlock themselves with those of another. Pulleys covered with leather and wheels made of hardwood of all kinds have given much greater driving power from the same grasp of belt than the handsomely polished metal pulleys have done, though this latter class of wheels has all the advantages that are to be derived from atmospherical influences.

But the fine imperfections on the true surface, which are the real gear teeth of friction, are not there in the abundance found in the material that is more closely allied with the belting itself. Everything would seem to indicate that a driving wheel is fin-

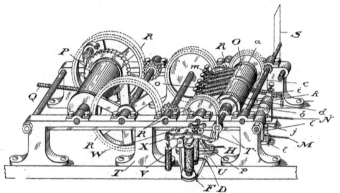

TIME SYSTEM, TORONTO ELECTRIC LIGHT CO.—FIG. 2.

clock can occur, such as might happen with the stronger and heavier contacts required, if the high potential current in the line circuit also passed through the master clock.

If, by the collection of dust on the contact points or by vibration of the contact spring, a double contact takes place in the master clock, no trouble is caused, as the first contact has energized the transmitter controlling electro-magnet, and released commutator, and the second contact produces only a vibration of the armature of the electro-magnet, which will not in the slightest affect the transmitter.

The second point is the alternate reversal of the currents in the main circuit. If two contacts are made on the same sections of the commutator through dust or in any other way, the first contact alone will operate the distant clock, the second contact producing a current in the same direction, which, as already described, is not competent to produce any effect, as the current must be reversed before a further revolution of the armature in the distant clock can take place.

In practice, the wires, H and L, leading from the transmitter would be connected to a switch-board, suitably arranged so that the current from the transmitter can be distributed to any desired number of circuits of preferably about fifty clocks to the circuit. A single wire might in many cases be sufficient for the distant clock-station, as suitable ground connections might be substituted for the return wire.

ished in the wrong direction when a covering of leather adds so much to its driving capacity.

The teeth of gear wheels are not cut lengthwise, and this gives all the hold that its strength will allow to the turn of a pulley, with the finishing cut taken crosswise and ground on a polishing wheel, herringbone fashion. This may not be appreciated in the machine shop, but the object to be obtained is the very one that a draw file is used for, namely, to pitch the minute grooves found on every surface in the right direction.

The corporation of Collingwood have closed a contract for a 1,000 light alternator and a 140 h.p. slow speed engine, the electrical part of the work going to the Canadian General Electric Co., Ltd., and the steam plant to the John Abel Co., of Toronto.

An establishment has for some time been in operation at Pribram, Austria, where tests of fuel are carried out on a sufficiently large scale to enable the heating efficiency of any kind of fuel to be practically determined. In carrying out this process ten tons of the fuel to be tested are divided into two lots of five tons each, a separate and distinct trial being made of each lot, the results obtained being checked one against the other. The tests, which are carried on day and night continuously, are made on a mild steel boiler; the grate is of the stepped type, with a total surface of 23.6 square feet; the heating surface of the boiler is 624 square feet, and that of the feed water heaters 356 square feet; the chimney stack is 106 feet high, the inside diameter at the bottom being 3.6 ft., and that at the top 2 ft. 6 in., there being good draught for the boiler. Special care is requisite that the boilers and accessories be thoroughly cleaned out after each test.

## INSPECTION OF ELECTRIC LIGHT IN CANADA.

As most of our readers are no doubt aware, an Act has been passed by the Dominion Parliament providing for the inspection of electric light, and provision is now made for the inspection of electric light and current the same as for gas. The organization has recently been completed by Mr. O. Higman, of the Inland Revenue Department, Ottawa, who is said to have framed it on the lines of the greatest possible economy in connection with the gas inspection department. Below is given a list of the staff, together with the inland revenue divisions over which each inspector has charge :

John Williams, London—Inland Revenue Divisions of London, Windsor and Stratford.

D. McPhie, Hamilton—Inland Revenue Divisions of Hamilton, Brantford and St. Catharines.

J. K. Johnstone, Toronto—Inland Revenue Divisions of Toronto, Guelph and Owen Sound.

Wm. Johnson, Belleville—Inland Revenue Divisions of Belleville, Kingston, Prescott, Cornwall and Peterborough.

H. G. Roche, Ottawa—Inland Revenue Divisions of Ottawa and Perth.

A. Aubin, Montreal—Inland Revenue Divisions of Montreal, Three Rivers, Terrebonne and Joliette.

M. Levasseur, Quebec—Inland Revenue Division of Quebec.

A. F. Simpson, Sherbrooke—Inland Revenue Divisions of Sherbrooke, St. Hyacinthe, St. Johns and Sorel.

A. Rowan, St. John, N. B.—Inland Revenue Divisions of St. John and Chatham.

A. Miller, Halifax, N. S.—Inland Revenue Divisions of Halifax, Yarmouth, Pictou, Cape Breton, Charlottetown, Prince Edward Island.

The act provides for the testing of incandescent lamps for candle-power, and a sample of each style of lamp furnished to customers must be passed by the inspector. For the testing of meters a scale of prices varying from 75 cents to $3.50 for each test has been arranged, the test to be made once every five years, the cost to be paid by the company ; but should the company or consumer desire a test oftener than provided by the act, the fee must be paid by the party at fault.

## QUESTIONS AND ANSWERS

"Fireman" writes : In handling an automatic cut-off Wheelock condensing engine, should the throttle valve be partly closed or open, according as there is full load or only part load on ? Is it in the interest of economy or otherwise to run an automatic cut-off engine with partly closed throttle ? The engine in question has a fair load to carry.

ANSWER.—With an automatic engine the throttle valve should always be left full open, no matter how small the actual load at the moment may be. It is not in the interest of economy or anything else to try and control the engine by the throttle valve, as the object of the automatic valve is to make this unnecessary, as the very name " Automatic " indicates.

## PERSONAL

Mr. Wm. McKenzie, President of the Toronto Street Railway Co., is at present on a visit to England.

Mr. R. O. King, son of Mr. R. W. King, of Georgetown, Ont, a student in electricity at McGill University, was recently awarded the gold medal and a two years' scholarship.

Mr. John Starr, one of the leading men in the trade in Canada, was in New York last week. Mr. Starr sold the first T.-H. apparatus in France, and is now a general dealer in supplies at Halifax.—Electricity, N. Y.

At St. Gregory's church, Montreal, Miss Gertrude Sise, second daughter of Mr. C. S. Sise, Vice-President of the Bell Telephone Co., was recently married to Mr. Ernest Nash, son of the late Captain Nash, of Ottawa.

The marriage is announced of Mr. Stephen L. Coles, associate editor of the Electrical Review, and Miss Sallie E. Field, of New York City. The happy couple enjoyed a fortnight's trip to the Thousand Islands and Montreal. Our congratulations are extended to Mr. and Mrs. Coles.

## PUBLICATIONS.

The initial number of the Electrical Journal, to be published monthly by Geo. P. Lowe, at San Francisco, has reached our table. It bears evidence of careful and efficient editing and management, and will we trust meet the expectations of the publishers.

## CANADIAN ASSOCIATION OF STATIONARY ENGINEERS.

Note.—Secretaries of the various Associations are requested to forward to us matter for publication in this Department not later than the 20th of each month.

### TORONTO ASSOCIATION NO. 1.

THE above Association at its last regular meeting elected the following officers :—Pres., Bro. Lewis; Vice-Pres., Bro. S. Thompson; Rec.-Secretary, Bro. T. Eversfield (by acclamation); Fin.-Secretary, Bro. Butler; Treasurer, Bro. A. M. Wickens; Conductor, Bro. Mose; Doorkeeper, Bro. Slute; Co Secretary, Bro. Huggett; Trustees, Bros. Fowler, Phillip and Huggett.

### MONTREAL ASSOCIATION NO. 1.

Montreal, No. 1 has recently elected the following officers : President, John J. York, (re-elected); First Vice-President, J. Murphy; Second Vice-President, W. Ware; Secretary, B. A. York; Treasurer, Thomas Ryan, (re-elected); Financial Secretary, H. Nuttal, (re-elected); Corresponding Secretary, H. Thompson; Conductor, P. J. Mooney, (re-elected); Doorkeeper, W. McHalpin; Trustees, John H. Garth, George Hunt and J. G. Robertson.

### HAMILTON ASSOCIATION NO. 2.

At the regular meeting of the above Association held on the 21st June, the following officers were elected for the ensuing term : President, E. C. Johnson ; Vice-President, W. R. Cornish; Corresponding Secretary, Wm. Norris, (acclamation); Financial Secretary, A. Nash, (acclamation); Treasurer, W. Nash, (acclamation); Conductor, Wm. Jones ; Doorkeeper, A. Vollick, (acclamation); Trustees, P. Statt, R. Mackie, R. C. Pettigrue.

The Corresponding Secretary reports the Association to be in a flourishing condition, and much interest is taken in the meetings by the members.

J. F. Philbir, of Rat Portage, was recently killed by coming in contact with an electric wire.

Mr. Wm. Johnson, of Belleville, has been appointed divisional inspector of the district for electric light.

A brick-maker named Harry Dent, of Bracondale, was recently killed on the Toronto and Suburban Electric Railway.

A movement is on foot for the construction of a telegraph line from the 150 mile house to the Forks of Quesnelle, B. C.

Forty men employed on the Springbank section of the London Electric Railway recently struck for an advance in wages.

Five directors of the Bell Telephone Co. have recently been engaged in a tour of inspection of the lines in the eastern townships of Quebec.

Mr. P. R. Randall has applied to the Township Council of Hope for permission to construct an electric railway on a portion of Rice Lake gravel road.

The Toronto Mineral Wool Manufacturing Co. has been incorporated, with a capital stock of $24,000, to manufacture mineral wool and boiler coverings.

Canadian patents have recently been granted to Mr. F. S. Smead, of Montreal, for a gas engine, and to Mr. W. J. Still, of Toronto, for an electric motor.

Work has been commenced on the extension of the Outremont branch of the Montreal Park & Island Railway. The extension will be double tracked and will be three miles in length, extending from the present terminus at Cote des Neiges to Westmount.

Mr. John A. Seely, of the firm of Beldon & Seely, New York, was recently in Montreal and accompanied Mr. Corraveau over the Montreal Park & Island Railway. Mr. Seely considers there is a splendid future for electric railways on the Island of Montreal.

The contract for supplying the ties and trolley poles for the electric railway at London, Ont., has been awarded to Messrs. Kernahan, Webster & Ferguson, of that city. About 40,000 cedar ties and 1,500 trolley poles be required. The trolley poles must be not less than 30 feet in length.

The first transmission of electrical power generated by the monster dynamo of the Niagara Falls power house took place on June 28th, when power was transmitted to the Pittsburg Reduction Co. The test was eminently successful. It is understood the amount of power transmitted exceeded 2000 horse power.

Letters' patent have been issued incorporating the Dominion Electric & Manufacturing Co., with a capital of $50,000, and headquarters at Montreal. The applicants are : Messrs. C. F. Sise, Robert McKay, Hugh Paton, J. R. Thibaudeau, Robert Archer, C. P. Sclater and L. B. McFarlane, all of Montreal.

Mr. F. S. Pearson, of the Dominion Coal Co., who is also interested in the Halifax electric street railway, states that all the material for the road has been contracted for with the exception of the cars. Active work on the construction will commence this month, and part of the road will be in operation in August. The electric power station will be equipped with the latest improved machinery.

## SPARKS.

Mr. John Childerbose is installing an electric light plant at Eganville, Ont.

The Kingston & Gananoque Electric Railway Co. has obtained incorporation.

The new electric cars for the Kingston electric railway will seat sixty people.

The Brantford Electric and Power Co. are offering for sale 80 bonds of $500 each.

The Canada Paper Co. are installing an electric light plant at Windsor Mills, Que.

The large lumber mill now being erected at Whitney, Ont., will be lighted by 300 electric lights.

The prospect is said to be favorable for the construction of an electric street railway at Barrie, Ont.

The owners of the Jeffery mine, Danville, Que., are seeking a franchise to run electric cars from the mine to the depot.

The village of Aylmer, Que., is considering a proposition from Mr. Conroy to light the streets of the town by electricity.

A survey has been made of the site for the power house of the Consolidated Light & Tramway Co., New Westminster, B, C.

The City Council of Quebec have accepted Mr. Beemer's proposition to construct an electric railway along the streets of that city.

The Canadian General Electric Co., of Peterboro', have just completed the second car for the Galt and Preston Street Railway.

The Selkerk Electric Co., Ltd., Selkerk, Man., has applied for letters patent to increase the capital stock from $10,000 to $25,000.

The earnings of the Toronto Street Railway for the month of June last are said to be $2,400 below that of the same month last year.

The Montreal Street Railway Co. will run a night service of refrigerator cars between the cattle market, the abattoirs and the meat market.

Work has been commenced on the extention of the Mimico Street Railway to Long Branch. Grading and tie laying are being proceeded with.

The town of Berlin, Ont., has granted the Bell Telephone Co. a five years' franchise in return for a free fire alarm system and three telephones.

The bill empowering the Thousand Island Railway Co. to construct electric lines from Gananoque to Kingston and Brockville is said to have been dropped.

Mr. George C. Robb, chief engineer of the Boiler Inspection and Insurance Co., has recently made a favorable report on the boilers in the City Hall, Hamilton.

The electric light contract between the town of St. Marys, Ont., and Mr. L. H. Reesor expires this fall, and a special meeting of the council is announced to discuss the making of a new contract.

The long-distance telephone line between Toronto and Montreal is expected to be completed about the end of September. The length of the line will be about 400 miles, which will be the longest in Canada.

General Riley, Consul General at Ottawa, has obtained a charter on behalf of an American syndicate to build an electric road about twenty miles in length, in the province of Quebec, from St. Rami to Napierville.

The ratepayers of the village of Huntingdon, Que., have granted a franchise to the Stadacona Water, Light & Power Co. for the construction of an electric light system. The work is to be completed by November 1st next.

At a recent meeting of the directors of the Bell Telephone Co. of Canada, Mr. Robert Mackay was elected vice-president and Mr. Charles Cassels a director, to fill the vacancy on the Board caused by the death of Mr. Geo. W. Moss.

Mr. Forsyth, projector of the Hamilton Radial Electric Railway, is said to have been successful in financing his scheme, and surveys of the line from St. Catharines to Schaw station, on the C. P. R., and to Toronto, have been completed.

The Executive Committee of the Toronto City Council has recommended that the tender of the Toronto Electric Light Co. be accepted for lighting the streets for five years, from January 1st, 1896, at 20½ cts. per light per night, or $78.81½ cts. per light per annum.

Herbert Cottrell, Newark, N. J., has received a United States patent and a Canadian patent on a telephone without electrodes, which operates on the principle of an electrical shunt, having a path of high resistance and of low resistance, whereby the necessary variations in the current may be produced.

An electric crossing alarm, invented by Mr. John Phillips, has been used by the Grand Trunk Railway at Brockville, and is said to have given satisfaction. Mr. Phillips is engaged in making improvements on it, which will cause a bell to ring automatically from the time the train comes within 40 rods of the crossing until it has passed.

The incorporation is announced of the Niagara Falls Electric Street Railway Co., the capital stock being placed at $125,000. The company is composed of Messrs. Alex. Manning, Hume Blake, Z. A. Lash, and P. A. Manning, of Toronto, and Mr. Chas. Black, of Welland. It is proposed to operate an electric street railway within the limits of Niagara Falls, and to build works for the production and distribution of electricity for light, heat and power.

Several changes have recently been made in the staff of the St. John Street Railway. Mr. C. D. Jones, superintendent, and Mr. A. R. Bliss, electrician, have retired from the service of the company, and Mr. H. Brown, electrician ot the old Gas and Electric Light Co., becomes electrician of the railway. The office of superintendent has been abolished

Mr. G. H. Campbell, manager of the electric street railway at Winnipeg, has recently let contracts for extensive improvements to the railway system. Some of the machinery has already arrived, and is being installed by a staff of experts. The power house is to be extended by a large addition, which will have brick walls and iron roof, absolutely fire proof. A large chimney 150 feet high will also be erected. These improvements will cost upwards of $60,000.

Mr. W. M. Kyle, of Toronto, is said to be interested in a scheme for the construction of an international electric railway belt line at Niagara Falls, Ont. The trolley line will start from the foot of Bridge street, and extend to the old horse car line in Drummondville. It will then run along Lundy's Lane to Falls View, back over the route of the old line to the Michigan Central Railway station, thence down the hill to the Canadian end of the Suspension Bridge. Mr. Kyle states that the right of way for the Canadian portion of the road has been secured.

Col. Lazier and his associates of the Belleville Traction Co., have commenced active operations in the construction of the road. The Canadian General Electric Co. have been awarded the contract for the cars and the entire electrical apparatus. The generating plant will consist initially of a 100 Kilowatt generator, and the total road will start up with three motor cars equipped with C. G. E. 800 motors and type K controllers. The overhead construction will be of the usual type, the Belleville people seeming to have for the present at any rate, abandoned the idea of utilizing the E. M. system, regarding which they were somewhat enthusiastic about a year ago.

## TRADE NOTES.

The J. C. McLaren Belting Co., of Montreal, have been granted incorporation.

The Bennett & Wright Co., of Toronto, have been incorporated, to manufacture boilers, furnaces, etc. The capital stock is placed at $98,000.

The Amherst Boot & Shoe Mfg. Co. are building a power house and will put in a 40 horse power Robb-Armstrong engine and Monarch Economic boiler.

Messrs. Ahearn & Soper, of Ottawa, representing the Westinghouse Company, have been given the contract for the overhead construction of wires, etc., for the Belleville electric railway.

The Victoria Granite Co., of St. George, N. B., has ordered a Robb-Armstrong automatic engine and a Monarch economic boiler from the Robb Engineering Co., of Amherst, N. S.

The Golden Lode Mining Co., of South Uniacke, N. S., has recently put in a duplex compound condensing pump of the Northey pattern. This pump will deliver water from a vertical depth of 400 feet.

The Canadian General Electric Co. have been awarded the contract for the entire electrical equipment of the re-organized Halifax Street Railway system. The motors will be of the usual C. G. E. 800 type with type K controllers.

The dissolution is announced of Messrs. Roe & Graham, manufacturers of water wheels, mill machinery, etc., Ottawa, Ont. Mr. Roe retires and the business will be continued by Mr. Robert Graham, whose announcement appears in our advertising pages.

Messrs. Ahearn & Soper, of Ottawa, recently placed the following motors, viz; One 5 h. p. motor for the Ottawa Electric Plating Works ; one 1 h.p. motor for the Carling Brewing Co.; one ½ h.p. motor for O. Robert ; one 1 h.p. motor for Batterton Bros.; one 1 h.p. motor for Frotheringham and Popham, stationers.

Allgemeine Elektricitats-Gesellschaft (The General Electric Co.), Berlin, Germany, has recently issued a handsomely illustrated catalogue, having reference to the overhead trolley system for street railways, and containing descriptions of important roads equipped by this system. The agents for this firm in Canada are Munderloh & Co., of Montreal.

In commemoration of the 4th of July, the American Electrical Works, of Providence, R. I., of which the Eugene F. Phillips Electrical Works. Montreal, is the Canadian branch, sends out to its friends a miniature skyrocket, in which is inclosed a small American flag accompanied by this sentiment : "Our best wishes that the day may be an enjoyable one to you. There is but one American flag and but one American Electrical Works."

Last week our representative called at Mr. David Starr's comfortable offices in the Board of Trade Building, Montreal, and was pleased to find him doing a very good business as consulting engineer and general electrical contractor. Mr. Starr has a number of good things on hand, and has been quietly, for the past year, working up a very good business. His long connection as general agent of the Royal Electric Co., has made him conversant with the trade all over Canada as well as the United States. He has a number of clients, in the way of central stations, etc., that he buys for exclusively, and who pay him a brokerage. A visit was also paid to Mr. Starr's repair shop, where he is employing a number of expert armature winders, and has a lot of work on hand, consisting of repairs to generators of different systems. His factory is specially fitted up for re-winding the Thomson-Houston arc dynamos.

# ELECTRIC RAILWAY DEPARTMENT.

## SPECIAL TRACK WORK FOR ELECTRIC STREET RAIL-WAYS, ESPECIALLY REFERRING TO THE MON-TREAL AND TORONTO SYSTEMS. *

### BY E. A. STONE, MA.E., A.M. CAN. SOC. C.E.

SPECIAL track work should be good substantial construction, with the greatest care paid to the designing of the parts which wear most rapidly. It is most important that track, especially in the central parts of a city, should require renewal as seldom as possible, for such renewals are very expensive, apart from the actual cost of the new track work, as traffic is interrupted, causing great inconvenience and sometimes loss of business to the public, and generally demoralising a whole route of cars, and sometimes the greater part of the entire system. Special work should be made in such a manner as to cause the least possible obstruction to vehicles, no part rising above the level of the paving more than is avoidable ; the necessary recesses, grooves, etc., should be as narrow and shallow as possible, to prevent wheels of vehicles from catching. Flat surfaces should have a rough top, to prevent horses from slipping upon them. All pieces should be finished so as to facilitate the paving, no long, unnecessary projections being left on bolts, etc. The curves should be of as great a radius as the width of the streets will allow. The sharper the curve the greater is the wear on the track and wheels of cars, the slower the rate of motion, the more power required to drive the cars, the more uneven the motion and the greater liability to derailment.

The track may be made on longitudinal stringers, on cross ties, or directly on concrete with tie bars connecting the rails. The old tracks of strap rail were laid on stringers, and the rail generally called stringer rail. (Figs. 1 and 2.) The greater part of the new construction is laid on ties, and in many respects is similar to steam track work. A combination of these two methods, consisting of planks laid longitudinally on cross ties, in order to give a more even surface, has been tried, but the results do not seem to have been so satisfactory as were expected. In several streets in Montreal where permanent paving has been laid, the rails have been laid directly on concrete, and bound

together by flat tie bars with threaded ends and double nuts. This, with the concrete between the ties, and paving, makes a very solid bed ; however, it does not seem to have so much elasticity as track laid on ties in macadam.

The rails used in Toronto and Montreal are "Girder" rails. Those first laid had a height of 6½ in. with a flange of 4½ in., while those laid later are 6⅜ in. high with a flange of 5 inches ; the web of the rail is not directly below the centre of the head as in the "Tee" rail, but nearer the gauge line, while a flangeway 1¼ in. wide at the top is provided for by a projecting lip. These rails average 75 lbs. per yard. This type of rail (Fig. 3) is used on all straight pieces and outside rails on curves

_____
* Abstract of paper presented before the Canadian Society of Civil Engineers.

in the special work ; the inside rails are made of a section very similar to this, the principal difference being that the lip is much heavier, being one inch in width at the top and rising 5-16ths in. above the level of the head of the rail ; this provides an efficient guard for the cars in running round a curve; the groove is ¼ in. wider than in the ordinary girder rail. This rail weighs 84 lbs. per yard. (Fig 4.) Another section (Fig. 5) is, however, coming into use, and will no doubt largely replace these sections for special work ; it is the same as the guard rail section, except that the groove is filled up with solid metal to within 9-16ths in. of the top of the head, thus providing a double bearing for the wheels, as both flanges and treads of wheels rest on the metal, so that the cars pass over all points without jolting, and the wear on the least durable parts of special work, viz., points, is greatly diminished. This section gives a rail of 89 lbs. to the yard. (Figs. 1 to 6.) The peculiar sections of these rails, with their thin flanges and webs, and much thicker heads, cause a variable amount of toughness in the section; the head having received

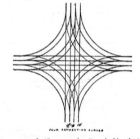

the least amount of rolling proportionally and taking the longest time to cool, is not so tough as the web and flanges. Tests on pieces taken from the guard rail (Fig. 4) have given the following results :—

    Head :—Tensile strength—64,300 lbs. per sq. in.
            Elastic limit—75 per cent. of tensile strength.
    Elongation on 4 in.—3½ per cent. ; reduction in area—2 per cent., with an even and uniform whitish gray fracture, moderately fine grained.
    Web :—Tensile strength—91,250 lbs. per sq. in.
            Elastic limit—75 per cent. of tensile strength.
    Elongation of 4 in.—27 per cent. ; reduction in area—20 per cent., with a fine grained light gray fracture.

The necessity for the increase in the weight of the new rails over the old is made apparent when it is considered that the weight of a motor car averages about 6 tons, while the weight of the old horse cars averaged only about two tons, and whereas horse cars run at the rate of about 6 miles per hour, electric cars frequently have a speed of 15 miles per hour. Tee rail (56 lbs.) is also used lately for this work, but its use is generally confined to macadamised roads in the suburbs, as its height is not suitable for paving purposes (unless raised on chairs), although otherwise quite as efficient. (Fig. 6.) The girder rail being so high admits of block paving, and by the lip on the inside provides a good edge for the pavers to work to, whilst the narrow groove offers a very slight hindrance to vehicles.

In tee rail special work, the inside rail on curves is generally guarded by a second rail being bolted to it, the two rails being held apart by cast iron filling pieces ; the space between these rails is afterwards filled with cement to within an inch from the top, so as to cause as little obstruction to traffic as possible; the guard rail is slightly elevated above the running rail. Frequently rails are used in paved streets of insufficient height to admit of a paving block between the ties and the head of the rail ; when this is the case, the difference in height has to be made up by the use of chairs ; this leads to rather complicated joints, and requires a longer time to lay than the method of direct spiking to the ties.

### MAIN DIVISIONS OF SPECIAL WORK.

Special work may be divided into four classes considered with respect to its use and position when in place, viz. : Intersections, passing sidings, crossovers and turnouts, and miscellaneous combinations.

1. INTERSECTIONS.—By the term intersection is meant the special work placed at the intersection of two or more streets, and may assume an almost endless variety of forms as regards number and direction of curves and the alignment of the main tracks. The work must be so constructed as to guide the cars in whatever direction required, without any other external assistance than the moving of the tongues in the switches by the

begins near the back of the switch, as shown in Fig. 18. If the cars always run to the right (as in Montreal and Toronto) the switch is made left hand, i. e., the p. c. of the curve turning to to the left is in front of the p. c. of the curve turning to the right by the length of the switch (approximately) ; thus a car approaching the siding travels straight along on the tangent past the point of the switch, and is then curved out of its path to the side by the curve in the rail behind, and when leaving the siding runs over the curve of the switch ; this is the best arrangement for such sidings, as it is the simplest, most durable, and causes least delay to the cars.

In the thrown-over siding (Fig. 17) one track is continued straight through, whilst the other is thrown over to one side of it ; this is suitable for single track lines on a side street, or in places where the track is on one side of the street. If cars are

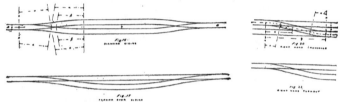

Fig. 16 DIAMOND SIDING

Fig. 20 LEFT HAND CROSSOVER

Fig. 17 THROWN OVER SIDING

Fig. 22 RIGHT HAND TURNOUT

motor men ; the cars must ride as smoothly as possible, i. e., there should be no jolting ; in places where a groove is to be crossed that would cause the car to run unevenly, the floor should be raised so as to give a bearing on which the flanges may run. On double track lines the distance between tracks is usually from four to five feet, but in order that cars may pass one another on the curves, and not be obliged to wait at the ends, this distance is increased to about seven or eight feet to provide ample clearance; this extra width is obtained by striking the curves from different centres, i. e., the curves are not concentric. The practice in Montreal and Toronto has generally been to make the inner and outer curves of the same radius when the apex angle has been nearly 90°, but when the angle varies greatly from a right angle, the outer curve has generally been made sharper than the inner when running round the obtuse angle. When the centre line of street changes direction, or has a "jog" at the intersection, necessitating a plain or reverse curve on the through tracks, the complications increase very rapidly.

2. PASSING SIDINGS.—These are used on single track lines when cars run in both directions ; they may be divided into two classes, viz., diamond and thrown-over sidings.

In the diamond siding (Fig. 16) the track diverges like a Y at either end, so that the centre line between the tracks in the sidings is on line with the centre line of the single track ; this is the form usually adopted on single tracks running through narrow streets. If it is desired that cars should run either to the right or left at these points, the switches of the sidings must be provided with movable tongues ; but if the cars always run in the same direction, they may be guided in the direction required

to be run to either side, switches with movable tongues are necessary ; but if the cars always keep to the same side, the tongues must be provided with springs, or blind switches used. With the latter the problem is not so simple as in the diamond siding, and in order to solve it the main track has a slight reverse curve placed in it extending from the first of the switch to a short distance inside the curve cross ; by introducing this, the general arrangement for the diamond siding holds good. (See Fig. 19). The radius for the curves of passing sidings in Montreal and Toronto is 300 feet to inside gauge line.

3. CROSSOVERS AND TURNOUTS.—Crossovers (Fig. 20), sometimes called connecting tracks, are used on double track lines for the purpose of transferring cars from one track to the other, and consequently are placed at the terminations of regular routes and at points which are made temporary terminii to accommodate special traffic.

Turnouts (Fig. 22) are used when a double track runs into a single track, the centre line of the single track being on line with the centre line of one of the tracks of the double track line.

These crossovers and turnouts, as well as all special work, should change the direction of the car's motion from one line into another with the least amount of resistance possible consistent with the date given ; those in Montreal and Toronto have 75 feet radius curve and about 25 feet of tangent, the latter varying with the distance between tracks ; this gives a crossover of about 60 feet between extreme ends of switches. Crossovers and turnouts are said to be either left or right hand, according to the direction in which they curve from the track, as seen from the switch when looking towards the cross. Fig. 20 shows a right hand crossover. If a crossover of either hand is suitable

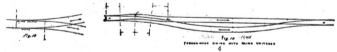

Fig. 18                    Fig. 19
PASSING SIDING WITH BLIND SWITCHES

by a movable tongue held to the proper side by a spring, so that a car facing a switch is always guided to the same side, and a car trailing it compresses the spring, and passes on, the tongue of the switch falling back to its proper position. (See plate Fig. B.) This guiding of the car in one direction, however, may be provided for much more simply by means of a switch without any movable part, commonly called a blind switch. One side of the switch is straight and the other curved; the front of the switch coincides approximately with the end of the curve of the switch, whilst the curve to the opposite side

at a certain point of the line, one of the same hand as the side to which the cars run should be chosen, i. e., right hand crossovers are preferable for systems on which the cars run to the right and left hand, or those in which the cars keep to the left ; this is on account of the fact that cars running always to the right will trail all switches of right hand crossovers and face those of left, so that they cannot possibly take the wrong track in the first case, while they may be suddenly thrown out of their course in the second, and accidents result.

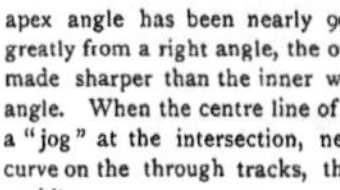

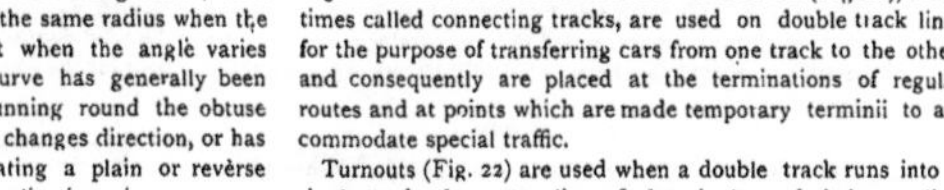

In addition to permanent crossovers it is always necessary to have temporary ones during construction, which are laid directly on top of the paving wherever required. These are so constructed as to be easily and quickly laid in place and readily moved from one part of the line to another by a small gang of men.

4. MISCELLANEOUS COMBINATION.—Besides the work already mentioned, there are several kinds of diamonds made to fill

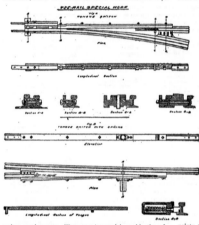

various requirements. There are also special combinations for car houses, etc. The simplest kinds of diamonds are those used when electric lines cross electric lines, and only require the running rails. When an electric road crosses a steam road, the steam road track requires guard rails for greater safety, and the electric line should also be guarded either by an additional rail or plate.

### SUB-DIVISIONS.

Intersections, cross-overs, etc., are composed of several pieces, which may be divided into the following sub-divisions, viz. :—Tongue switches (single and double curve), blind switches, mates (single curve, double curve and combination), curve crosses (single curve, double curve and combination), diamonds (for electric and steam crossings), split switches, stub switches and lengths of rail (curved and straight). (See accompanying illustrations).

1. TONGUE SWITCHES.—The tongue switch is perhaps the most important piece in any combination of special work, as it is subjected to greater and more frequent shocks than any other piece, its duty being to change the direction of the car's motion from one line to another. When made of girder rail, it is constructed of the guard rail section to ensure the perfect guidance of the wheels. When made of tee rail, a guard is formed either by bolting on another piece of rail, or by carrying up the casting on the side to form the required guard. The switch generally consists of four main parts, viz. :—the tongue, a casting and two pieces of rail. The tongue is made of steel, and should be of a substantial size, having a cross section near the point, proportioned to resist violent shocks ; at the same time the point must be rather sharp to ensure the car "taking" it exactly ; if blunt, the car may mount the tongue, and drop again, causing a severe jolt. If the top of the tongue rises above the level of the head of the rail, it is sloped at both ends so as to allow the rise and fall of the car to be imperceptible. The pin must be so placed as to make it impossible for a wheel to touch the tongue behind the pin, and so throw the switch before the back wheels have reached the point. If the tongue were made so long that the distance from the centre of the pin to the tongue point were greater than the wheel base of the cars (about 7 feet), this would be impossible ; this method, however, would necessi-

tate a too expensive switch, and the difficulty is easily overcome by rounding the back of the tongue and placing the pin sufficiently far back. The pin should also be placed so that the wheels do not run over it, and so cause it to become loose, and should be so fastened to the casting that the tongue may easily be removed at any time. The top of the casting on which the tongue slides and the bottom of the tongue should be truly even, as if not, dirt will collect between the two, and after a short time the tongue will tilt when a car runs over it, and may cause the tongue to throw to the opposite side, or the back wheel may strike the point, either of which may be sufficient to throw the car off the track. Single curved switches are those curved only on one side ; double curved switches are curved on both sides.

2. BLIND SWITCHES.—The blind switch is used in place of the tongue switch when cars always run off the curve at that point and never enter it. It closely resembles the mate in general construction. In order that the guidance of the car facing the switch may not altogether depend on the fact that the car will naturally take the straight track in the direction in which it is moving, rather than turn into the curve, a ridge is left along the floor on the straight track which acts as a gauge line, to make it practically impossible for the car to enter the curve.

3. MATES.—The mate is the piece opposite the switch, on which the wheels of one side of the car run while the wheels on the other side are being pulled around by the switch ; its sole use is to provide a surface for the wheels to run upon, and has nothing to do with the change in direction of the car's motion. It is made of two pieces of rail, and sometimes there is a casting. One piece of rail extends over the whole length, and is straight if for a single curve mate, and curved if for a double curved mate ; the other piece is shorter and always curved, the head terminating in a point ; this point should be so designed that the gauge at the point is quite slack, so that a wheel facing the mate may not strike upon it. The width of the point should not be less than $\frac{3}{8}$ inch, as if made sharper it will wear to this. In girder rail the solid floor section makes the best mate, as it provides a wide floor for the wheels to roll upon, and the depth of the floor below

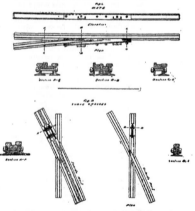

the head of the rail being less than the depth of the flange of the wheel, it quickly wears so as to provide a double bearing for the wheels, so that the point is passed without the wheels dropping heavily upon it. If the mate is not made of the floor section, but of the ordinary girder rail as used on the straight track, or if of tee rail construction, a steel casting is necessary to carry the wheels over the point from the long rail on to the short one; this casting is more efficient if carried up on the inside to provide a guard; for in case of the gauge being too slack, the tongue may have a tendency to jerk the car off the track. This casting must project considerably inside the gauge line of the short rail, the path of the rear wheels on a truck not coinciding with that of the front ones, but lying about ½ inch inside, as may be clearly seen on any worn mate.

4. CURVE CROSSES.—Curve cross is the name given in this work to the piece corresponding to the frog in steam railroad work; it differs considerably from the frog, however: one, at least, of

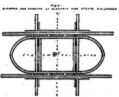

the rails in a curve cross is generally curved to a very sharp curve, whilst the frog is straight on either track; the frog has wing rails, and a wheel crossing a frog runs from one piece of rail across the channel on to another rail, whilst in the curve cross a wheel generally runs the entire length of the cross on one piece of rail, the channel for the flanges being shaped out of the head of the rail. According as one or both rails are curved, the cross is said to be a single or double curve cross.

5. DIAMONDS.—Diamonds are made in various ways, according to the requirements they are to serve. A simple single track diamond for the crossing of two electric lines consists of two

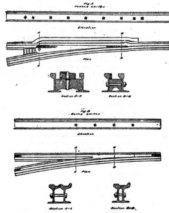

main parts, each part being made of five pieces of rail, one long piece with four short pieces butting up against it, two on each side; the long rail is usually made to form part of the track on the street having the greater amount of traffic. When an electric road crosses a steam road, the diamond is usually all made of tee rail, of the same section as the rail of the steam road. If the rails of the steam road are not to be cut, the diamond is made of three parts (Fig. E), two parts and one inside the steam track, the whole being so constructed as to lift the street car before reaching the rails of the steam track on to the flanges of the wheels, and running across on them to the

other side, and then dropping gradually to the ordinary level again, so that the only place where any jolt can occur to a car while crossing such a diamond is when it crosses the channel of the steam track rails, notwithstanding the fact that the rails of the steam track are not cut to the smallest extent to provide a passage for the flanges of the street-car wheels.

6. SPLIT SWITCHES.—Split switches are used to a comparatively small extent on this class of work. They are more especially adapted to suburban traffic when tee rail is used, rather than crowded thoroughfares of cities. They are especially suitable when cars always run to the same side, when the switch may be made to work automatically by means of a spring, and in this way have been found very satisfactory.

7. STUB SWITCHES.—Stub switches are suitable for yard purposes and sidings only occasionally used; they are cheap, which is always a point in their favor. The use of a stand prohibits their use in city thoroughfares.

8. LENGTHS OF RAIL.—Rails for all special work should be ac-

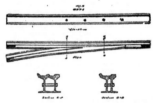

curately cut to the required lengths, and carefully bent to the proper template if for use on a curve, or accurately straightened if required for straight track. If part of a rail is to be straight and the remainder curved, the rail must not only agree with straight edge and template for the required lengths, but it must be tested, to determine whether the straight part is tangent to the curve, for if not, the piece will not fit correctly when placed in the work of which it forms part.

THE DETERMINATION OF NECESSARY SPECIAL WORK.

Having laid down the routes of any street railway system necessary for the accommodation of the present traffic and that of the near future, the special work required becomes apparent. It is most important that curves likely to be required in a few years, but not necessary at the present, should be laid, if at all possible, during construction, as the addition of a single curve

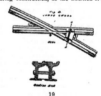

19

to an intersection in some cases necessitates the reconstruction of the greater part of the whole intersection.

SURVEYS.

A careful survey must be made of the intersection of streets requiring special work, and all measurements of lines and angles taken which are necessary to plot with the greatest accuracy the centre lines of the proposed tracks together with the street and curb lines.

These measurements are plotted to a suitable scale (say 10 feet to 1 inch), and the most suitable radii for the required curves determined, which are usually from 40 to 75 feet radius (45 and 50 ft. are most common in Montreal and Toronto.)

The attempt is sometimes made to ease these curves as on steam railroad work; but when it is remembered that the length of most of the curves is about 80 ft., it will be seen how limited the space is in which to attempt anything of the kind; however, an improvement may be introduced by making the switches at the ends of curves of a longer radius than the main part of the curves, such as using 75 ft. radius switches on 45 ft. radius curves; this eases the curves for 10 ft. at each end and meets all practical requirements; any further steps in this direction would seem to lean towards "hair splitting."

It might here be mentioned that although these curves would appear very sharp to engineers accustomed to steam railroad work, yet there is a case on record of a 50 ft. radius curve on a trestle being used on a steam railway, and operated successfully, the speed on it being from 8 to 10 miles per hour. (U.S. Military Railway, Petersburg, Va.; see Trans. Am. Soc. C. E. 1878.) The Manhattan Elevated Railway in New York city has curves of 90 ft. radius.

There should be, if possible, sufficient space between the inside rail of the curve and the curb stone for a vehicle to pass a car easily; this, however, requires very wide streets; if this can not be done, the rail should be at about two feet from the curb stone at the corner, for if at say four feet, there would not be sufficient room for a car and vehicle to pass, but the attempt might be made and an accident ensue. The radii of the curves should also be determined, with a view to sufficient room for the switches; if this is not looked to, special short switches may be required, which is not desirable. The interesting points of the gauge lines should also be carefully observed, as by the slight alteration of a radius, combination pieces of complicated construction and of an unendurable character may often be avoided. The radii having been fixed, the gauge lines alone may be laid down to a large scale (say 4 feet to 1 inch), and the calculations proceeded with.

### WORKING DRAWINGS.

Having completed the calculations for an intersection, the detail drawings for each piece are made, and sent to the shop, together with a print showing the whole intersection with the distinguishing marks of all pieces and lengths of the various rails. A drawing is also made for assembling the work in the street, showing all necessary measurements for laying out the work, together with the position and marks of the various pieces.

### SHOP WORK.

A bill of the rails required and the necessary new prints and references to old ones having been obtained from the drawing office, the manufacture of the work may be proceeded with. The bill of rails required (made out so as to give a minimum amount of scrap) is given into the hands of the man in charge of the rail saw, who proceeds to cut up the rails into the required lengths, marking the length of each and whether required straight or curved upon the web. The rails next, with few exceptions, go to the rail bender, to be either curved to the required radius, or straightened; they next proceed to the "marker off," who carefully marks the necessary lines for all machine work required to be done upon them; he also stamps the rails on the end with their distinguishing marks; the rails afterwards pass on to the machines (milling machines, slotters, shapers, planers, etc.) suited to the work required; they then go to the fitting shop to be assembled according to the drawings.

In a tongue switch the long rail has to be properly curved, and slotted or bent for the tongue to fall into place. The tongue is made of hammered steel, and the turned pin is shrunk in; this is dropped into place, and all measurements checked before being considered ready for the track.

In the blind switch and mate, one rail is planed so as to leave a long notch on one side, while the other rail is planed to a point which fits into the notch; the two are strongly bolted or riveted together and sometimes finished on a planer.

The curve crosses have usually two pieces of rail, one of which has the upper part so shaped at the crossing point as to allow the second one to drop down on the first, and fit accurately into the place allowed for it; while the second has the lower part shaped so as to allow the first rail to pass through, the two rails jointing neatly into one another. Great care is necessary in the fitting to have the angles of the intersection exactly as required; in order to obtain the correct angle, the drawing shows the curves spread at a fixed distance, together with the deflections of the curves at that point; so that this distance is measured along the rails from the intersection point and the deflections marked from the gauge line; the spread is then measured between the points so marked.

### CHECKING.

When an intersection has been made, it is sometimes advisable to have it assembled as a final check before shipping; for this purpose a large piece of ground, as level as possible, is required, and much more than is actually occupied by the work when in place should be available; the tangents of the intersection should be laid out, and a sufficient number of points fixed to actually check the end of each curve. Having laid out the ground, the pieces are assembled, and any errors observed may be corrected; this last step ensures the work being absolutely correct, and is the best check on the work that can be adopted.

### ASSEMBLING IN THE TRACK.

In laying an intersection, it makes a great deal of difference whether the whole space required is graded at once and all traffic stopped, or if only part of the intersection is graded, leaving part undisturbed so as not to interrupt traffic. When the work has to be performed in the latter way, great care is necessary in placing the work, so that the remaining part when laid may fit up to and line accurately with the first part. If it is necessary to lay out a curve, it is generally most easily performed by tangent and chord deflections or by ordinates from a chord. In grading a corner, when an important intersection is to be laid, care should be exercised in excavating to the correct depth and having the grading done evenly, for if the track has to be lifted, say six inches, after being laid, it means very much more than the same lift on ordinary track, as the weight of rail is sometimes enormous as compared with the extent of ground it covers; also, if the work has been carelessly done, and presents a very uneven bed, much more time is necessary to couple up the joints than would have been required had the grading been properly performed. The spacing of the ties for this work should receive more attention than is sometimes given to it, as it is a very important matter. The ties should be the very best available, and spaced more closely than those on the straight track.

The centre lines of tracks for both streets are accurately fixed, and if there is no diamond, the ends of the curves must be found; otherwise this is not essential. If there is a diamond in the intersection, this is laid first, bolted up and lined accurately. The other pieces having been scattered about in their approximate positions, are next drawn to place and bolted together. The rails are then securely spiked to gauge, and lifted (if necessary) to grade, when the intersection may be paved and so completed. If there is no diamond to lay, an end of a curve may be taken as the starting point. To lay the intersection so as to have the through straight tracks in perfect alignment, requires great care, as the joints are usually very close together.

An idea of the amount of rail that may be used in a single intersection, and the consequent amount of labour required to make one, may be formed from the following figures, for one laid at the intersection of St. Lawrence, Main and St. Catherine streets, Montreal (same as Fig. 15). It is built of 75 lbs. and 84 lbs. girder rail (Figs. 3 and 4). It contains 2,150 feet of rail, and has a total weight of about 26 tons. There are 86 built up pieces (switches, mates and curve crosses), and 78 lengths of connecting rails, making a total of 164 pieces in the complete intersection. The extreme length between ends and opposite switches is about 110 feet. The radius of the inside gauge lines of all the curves is 45 feet, and the distance between tracks varies from 4 ft. to 8 ft. 6 in. This intersection, as well as all others in Montreal and Toronto, was made by the Canada Switch Manufacturing Co., Lim., of Montreal.

Such work, when properly constructed and laid, represents a large amount of capital, and deserves much more attention and care than the old cast iron work; but, unfortunately, it seems sometimes to be treated no better. The curves at intersections are necessarily very sharp, and in order to diminish the amount of power required and the wear on the rails (as well as on tires) they require oiling at least once a day for heavy traffic, while the rate at which cars run over special work should be strictly regulated to a low speed. The groove of the rail and the tongue switches require to be constantly cleared of the dirt which inevitably collects, and if not removed causes great inconvenience. The life of such work may be appreciably prolonged by such attention, and when one considers the cost of renewal and the consequent interference to traffic while doing so, it will be readily seen that it pays in the end.

## SPARKS.

The Crossan Car Mnfg. Co., of Cobourg, are shipping two first-class passenger, two baggage and express cars to the Toronto, Hamilton and Buffalo Railway.

The Port Dalhousie, St. Catharines and Thorold Street Railway Co. have purchased additional equipments of the C. G. E. 800 type from the Canadian General Electric Co.

The Lundy's Lane Electric Street Railway Company has been incorporated, with capital stock of $50,000 in $100 shares, the promoters being H. C. Symmes, J. A. Lowell, J. G. Cadham, G. R. Symmes, all of Niagara Falls, and H. D. Symmes, of St. Catharines.

At the present time there are over 850 electric railways in the United States, operating over 9000 miles of track and 23,000 cars, representing a capital investment of over $400,000,000. In 1877 the number of such roads amounted to only thirteen, with scarcely 100 cars.

At a recent meeting of the stockholders of the St. John Railway Co., of St. John, N. B., the following directors were elected : James Ross, J. M. Robinson, R. B. Emerson, James Manchester, J. J. Tucker, H. P. Timmerman, C. W. Weldon, H. H. McLean and William Barnhill. At a subsequent meeting the directors elected James Ross, of Montreal, president ; J. M. Robinson, vice-president, and James Warren, secretary and treasurer. Messrs. Ross, Robinson and McLean were named as the executive of the company.

After a series of delays caused by the inability of the City Council to come to a satisfactory basis of agreement with the London Street Railway Co. for the renewal and extension of their franchise, a satisfactory arrangement has at last been reached, and the electrical equipment of the road is being rapidly proceeded with. The contract for the entire electrical equipment has been awarded to the Canadian General Electric Co. The motor cars, 25 in number, will be constructed in the same manner as those running on the Wilson Avenue line in Cleveland, with cross benches and side aisle, with doors at the centre and rear end of the car, a footboard being provided running the entire length of the car. This latter feature has proved itself of considerable value in handling heavy crowds. The motor equipments for each car will consist of two C. G. E. 800 motors with type K-2 controllers. The entire system is intended to be in operation in time for the Western Fair in September.

### MOONLIGHT SCHEDULE FOR JULY.

| Day of Month. | Light. | Extinguish. | No. of Hours. |
|---|---|---|---|
| | H.M. | H.M. | H.M. |
| 1...... | P.M. 11.20 | A. M.  3.30 | 4.10 |
| 2...... | " 11.50 | .......... | } 3.40 |
| 3...... | .......... | "    3.30 | |
| 4...... | A.M. 12.30 | "    3.30 | 3.00 |
| 5...... | "   1.30 | "    3.30 | 2.00 |
| 6...... | No light. | No light. | .... |
| 7...... | No light. | No light. | .... |
| 8...... | No light. | No light. | .... |
| 9...... | P. M. 8.00 | P. M. 10.30 | 2.30 |
| 10...... | "   8.00 | "  10.50 | 2.50 |
| 11...... | "   8.00 | "  11.10 | 3.10 |
| 12...... | "   8.00 | "  11.30 | 3.30 |
| 13...... | "   8.00 | "  11.50 | 3.50 |
| 14...... | "   8.00 | A. M. 12.10 | 4.10 |
| 15...... | "   8.00 | "   1.00 | 5.00 |
| 16...... | "   8.00 | "   1.00 | 5.00 |
| 17...... | "   8.00 | "   1.40 | 5.40 |
| 18...... | "   8.00 | "   2.30 | 6.30 |
| 19...... | "   8.00 | "   3.30 | 7.30 |
| 20...... | "   8.00 | "   4.00 | 8.00 |
| 21...... | "   8.00 | "   4.00 | 8.00 |
| 22...... | "   7.50 | "   4.00 | 8.10 |
| 23...... | "   7.50 | "   4.00 | 8.10 |
| 24...... | "   7.50 | "   4.00 | 8.10 |
| 25...... | "   8.30 | "   4.00 | 7.30 |
| 26...... | "   9.00 | "   4.00 | 7.00 |
| 27...... | "   9.20 | "   4.00 | 6.40 · |
| 28...... | "   9.50 | "   4.00 | 6.10 |
| 29...... | "  10.20 | "   4.00 | 5.40 |
| 30...... | "  11.00 | "   4.00 | 5.00 |
| 31...... | "  11.00 | "   4.00 | 5.00 |
| | | Total, | 146.00 |

### SPARKS.

The electric light system at Windsor, Ont., is to be enlarged.

Mr. R. Anderson, electrical engineer, of Ottawa, is installing a plant of 600 incandescent lights at Vankleek Hill, Ont., for Col. Wm. Higginson. Mr. Anderson is also installing a 100 light electric plant for the Geo. Matthews Co., Ltd., for their pork packing establishment at Hull, Que.

Mr. Geo. Gillies, of Gananoque, is said to run the only electric welder in Ontario. Power is supplied by the Gananoque Electric Light and Power Co. The articles to be welded are placed end to end in grooves suitable to their shape; two clamps come down fitting on the articles to be welded, which form the connecting link between the clamps and the copper plates underneath. A lever turns on the current and at the same time presses the articles together, and the welding is done almost simultaneously. Articles to the weight of eight pounds can be welded on this machine.

The Ottawa Electric Co., of Ottawa, Ont., are putting in an additional 240 k. w. alternator in their steam power station, where they already have two others of the same power, supplied by Ahearn & Soper. They are also putting in a 200 k. w. D. C. 500 volt generator for stationary motor service, and are engaged in remodelling their motor circuits so as to obtain better distribution. Their workshop is busy in the construction of a large switchboard, which will be 57 feet long, entirely fireproof, and constructed on modern principles. It will be placed in power house No. 1, late of the Standard Co., which will remain the distributing station. The generators in the other stations will be connected to this switchboard, and will be run at the accustomed voltage, the regulation being done at the central station. This company is at present installing a private fire alarm system connecting their stations, work shop and offices with the fire stations in their districts. The above changes were all designed by Mr. A. A. Dion, general superintendent and electrical engineer for the company, and are being carried out under his supervision by Messrs. J. Murphy, superintendent of power stations, W. G. Bradley, superintendent of construction, and W. H. Baldwin, hydraulic engineer. The company are sparing no pains or expense in making improvements so as to maintain a high standard of efficiency throughout the service. Eighty motors aggregating 350 h. p. are supplied with power by this firm, and 450 arc lights and 48,800 incandescent lamps are supplied with light.

# CANADIAN ELECTRICAL NEWS
## STEAM ENGINEERING JOURNAL

OLD SERIES, VOL. XV.—No. 6.
NEW SERIES, VOL. V.—No. 8.

AUGUST, 1895

PRICE 10 CENTS
$1.00 PER YEAR.

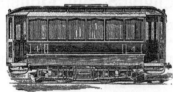

CANADIAN

# ELECTRICAL NEWS

AND

## STEAM ENGINEERING JOURNAL.

| VOL. V. | AUGUST, 1895 | No. 8. |

### THE STREET RAILWAY CONVENTION, MONTREAL.

A representative of the ELECTRICAL NEWS had an interview a few days ago with Mr. Stonewall Jackson, of Montreal, the local secretary of the American Street Railway Association. The Fourteenth Annual Exposition of that association will take place in the Victoria rink, Montreal, the 15th of October next, lasting four days. The officers and executive board are as follows :— President, Joel Hurt, Pres. Atlanta Consolidated Street Railway Co., Atlanta, Ga. ; 1st Vice-President, W. Worth Bean, President St. Joseph and Benton Habour Railway and Light Co., St. Joseph, Mich. ; 2nd Vice-President, John H. Cunningham, Dir. Lynn & Boston R. R. Co., Boston, Mass. ; 3rd Vice-President Russel B. Harrison, Pres. Terre Haute Street Railway Co., Terre Haute, Ind. ; Acting Secretary-Treas., John A. Partridge, Brooklyn Street Railway Co., Brooklyn, N. Y. ; Executive Committee : the President, Vice-President and Hy. C. Payne, Vice-President Milwaukee St. Ry. Co., Milwaukee, Wis. ; Wm. H. Jackson, President Nashville St. Ry. Co., Nashville, Tenn. ; D. G. Hamilton, President, Cass Ave. and Fair Grounds Ry. Co. and St. Louis Ry. Co., St. Louis, Mo. ; Granville C. Cunningham, Man. Montreal Street Ry. Co., Montreal, Que. ; John N. Partridge, President Brooklyn City & Newton R. R. Co., Brooklyn, N. Y.

Exhibition of supplies and manufactures of every nature used in the street railway business will be displayed and electric power is to be provided for the running of machinery which may need it. All machinery, will, if possible, be exhibited in motion. Every precaution will be taken to guard against fire, and a full corps of watchmen will be on duty day and night. The association heartily invites all manufacturers, inventors and street railroads to exhibit their machinery and will make the utmost effort to devote the requisite space to all applicants. All the leading street railway men will attend this exposition, and the directorate will do all in their power to make it the best street railway exposition ever held. For full particulars address, Stonewall Jackson, Local Secretary, 17 St. Sacrament St., Montreal, Que.

---

### LEGAL DECISIONS.

HARTFORD V. BELL TELEPHONE CO., TORONTO ELECTRIC LIGHT CO., ET AL.—The appeal from the judgment of Mr. Justice Rose in favor of the defendants in this case, was dismissed by Mr. Chief Justice Meredith in the Common Pleas Divisional Court, Toronto, in the following terms : "To have entitled the plaintiff to have succeeded against any or either of the defendants, it was incumbent upon her to prove that the defendant or defendants sought to be made liable, had been guilty of some wrongful or negligent act which was the proximate cause of the injuries received by her, and in respect of which the action was brought. On both branches of the case the plaintiff, in the view of my learned brother Rose, failed upon the facts, for he has by his findings of fact acquitted each of the defendants of the wrongful or negligent acts charged against them, and has found that, even if the defendants were guilty of the wrongful or negligent acts alleged to have been committed by them, those acts were not the proximate cause of the injury and damage to the plaintiff for which she sues. There was, we think, evidence which fully warranted the learned judge's findings. . . . Upon the facts of the case it was properly found (for it appeared upon the plaintiff's own case, and there was no evidence to the contrary) that the wires of the Electric Light Co. and of the Holmes Co. were brought into contact owing to the conduct of a boy who chopped off a branch of a tree which stood near the wires of the Holmes Co. and the Bell Telephone Co., between Portland street, where the contact, as I have mentioned, took place, and the building to which the fire was communicated, and the branch, falling upon the wires adjacent to and somewhat below it, brought the Holmes wire down upon and into contact with the Electric Light Co.,s wires, and, but for the boy's act, the negligence of the companies, if negligence there were, would have produced no damage to the plaintiff. It appears to me, therefore, that, according to both the principle acted upon in the Howard case (22 S. C. R. 147), 'the wrong and the damage are not sufficiently conjoined or concatenated as cause and effect to support an action,' and, as put in the Howard case, the negligence and the injury are insulated by the intervening act of the boy—the causal connection between the negligence and the damage being broken by the interposition of independent responsible human action, and the plaintiff's case, therefore, failed. There are probably other difficulties in the way of the plaintiff's recovering, but it is unnecessary to refer to them. The result is that the judgment of my brother Rose is right and must be affirmed, and the appeal from it dismissed with costs. I do not feel, however, that I should part with with the case without expressing the hope that some provision of law may be enacted that will place under proper governmental or municipal supervision and control the vast network of wires which is to be found in a city like Toronto, and may at any moment become the cause of serious injury to life and limb, as well as to property, and for requiring companies and others, whose disused or 'vagrant' wires may become a source of danger, to remove them.

---

The gross earnings of the Toronto Street Railway Company for July amounted to $93,049.94.

John F. Payzant has been elected president, and W. B. Ross, secretary, of the Halifax Electric Tramway Co.

Mr. Chas. W. Wasson, of Cleveland, Ohio, has been elected a director of the London Street Railway Co., to succeed the late Hon. Greene Pack.

The electric light inspection branch of the Inland Revenue Department will be self-sustaining, as up to the present $2,000 has been collected in fees.

The first truck with electric motors for locomotive work to be used in Canada was shipped last week to Oshawa by Ahearn & Soper, of Ottawa. It is a specially constructed truck of heavy steel and weighs with the motors something over eight tons. The motors combined have a capacity of 120 horse power. The truck will be used in hauling freight cars to and from the G. T. R. at Ottawa.

Mr. Thos. Ahearn, of Ottawa, has devised a method for preventing variation of E. M. F. occasioned by sudden withdrawal or addition of load in connection with self-excited water driven dynamos. An independently driven water wheel is employed to generate current exclusively for the purpose of exciting the field of the generators. An ammeter is included in each dynamo field circuit, and upon each dynamo is placed a small double throw switch so that in case of accident, the several dynamos could be self-excited by throwing the switch on each dynamo, thereby connecting the armature of each to its own field. The advantages claimed for this method are : Steadiness of voltage, removal of the danger of burning out fields by abnormal armature speed, relief to the driving machinery, removal of fields from the line circuit preventing any possible damage to them by lightning or other cause, the prevention of damage to commutators formerly caused by short circuits upon the line throwing open the circuit breakers and short circuiting the current across the commutator. Considerable time is also saved in throwing in dynamos, which is now done without delay after the circuit breakers are reset. This formerly required a very considerable time in synchronizing fields.

## CANADIAN ASSOCIATION OF STATIONARY ENGINEERS.

Note.—Secretaries of the various Associations are requested to forward to us matter for publication in this Department not later than the 20th of each month.

### TORONTO ASSOCIATION NO. 1.

IN the absence of the President, E. J. Philp, the last regular meeting was presided over by A. E. Edkins, Dist. Deputy. After general business had been disposed of, the newly-elected officers were installed as follows :

Pres., Walter Lewis ; Vice-Pres., Samuel Thompson ; Rec. Sec., T. Eversfield ; Cor. Sec., Jas. Huggett ; Treas., A. Wickens ; Door-Keeper, A Slute ; Conductor, Martin Mose.

Bros. Fox, Huggett and Wickens were appointed delegates to the annual convention.

At a recent regular meeting the following resolutions of condolence were adopted :—

"Whereas, it has pleased our Heavenly Father to remove from the family of our esteemed and worthy brother, W. G. Blackgrove, two of his children, therefore be it resolved that we deeply sympathise with our brother, his wife and family in their sad bereavement, but commend them to that allwise Supreme Ruler who, though sometimes inscrutable in His dispensations, yet doeth all things well. And be it further resolved, that these resolutions be spread on the minutes of this Association, and a copy of same be sent to the bereaved family and also to the mechanical press.

GEORGE FOWLER,
GEORGE C. MORRING, } Committee.
WILSON PHILLIPS,

### MONTREAL ASSOCIATION NO. 1.

A representative of the ELECTRICAL NEWS called on the Montreal No. 1 Canadian Association of Stationary Engineers at their handsome and comfortable quarters at 666½ Craig St. They had just completed business, and the secretary, Mr. B. A. York, stated that they had just elected delegates to the annual convention to be held in Ottawa in September next.

The following are the delegates elected : Brothers T. Ryan, J. G. Robertson and E. Valiquet, with alternates, Bros. Hy. Nuttal, Jos. Marchand and J. Murphy. The delegates will be accompanied by Bros. J. J. York, (President C. A. S. E.), Geo. Hunt and O. E. Granberg, members of the C. A. S. E. executive.

An invitation was extended to our representative to attend their annual picnic, Aug. 3rd, and make an inspection of their lodge room, which is the best equipped for the study of steam engineering of any in Canada. On the walls are drawings of Babcock engines and designs of different engine and boiler makers, together with photos of the past and present officers and groups of the members of the association. They have the latest models and appliances, and a library of no small size which they hope to add to from the proceeds of their picnic.

Bro. B. Cowper, chief engineer of the Canadian Rubber Co., presented the association with a model double plunger pump with glass cylinders. They have also a model steam pump with cylinders cut in half, showing working of valves.

Many of the merchants of the city contributed to the prize list for the picnic games, and a goodly number turned out to the Exhibition grounds to enjoy them.

In the drawing for a forty gallon barrel of oil, a little maid of eight summers picked out the lucky ticket. In the lacrosse match the team captained by Mr. Hunt won victory over the team led by Mr. Murphy. The Executive have reason to congratulate themselves on the complete success of the picnic.

### BROCKVILLE ASSOCIATION NO. 15.

AT the last meeting of the above association, the election of officers for the ensuing year resulted as follows :

Pres., W. F. Chapman ; Vice-Pres., Archie Franklin ; Rec. Sec., Wm. Robinson ; Fin. Sec., John McCaw ; Treas., John Grundy ; Conductor, W. S. Baverstock ; Trustees, Ernest Carr, Fred. Andrews and Edward Devine ; Delegate, W. F. Chapman.

The Secretary reports that good work is expected during the next term, the newly-elected officers being all energetic men.

### WINNIPEG ASSOCIATION NO. 11.

AT a meeting of the above association at their hall on the 11th ult., the following officers were appointed :

President, G. Hazlett ; Vice-President, Thos. Gray ; Rec. Sec., J. Sutherland ; Fin. Sec., A. B. Jones ; Treas., R. Sutherland ; Conductor, E. Simpson ; Door-Keeper, J. Harrison ; Trustees, G. Hazlett, C. E. Robertson, Thos. Gray.

The District Deputy, Mr. C. E. Robertson, installed the officers, and in a short address asked them to pay great attention to their work in this association, as it was one of the most important in the world.

All the officers are working engineers, and prospects seem to point to another year's successful work.

CHAS. E. ROBERTSON, Dist. Deputy.

### CARLETON PLACE ASSOCIATION NO. 16.

Editor ELECTRICAL NEWS.

SIR,—The following officers of Branch No. 16, C.A.S.E., were elected for the present term, July 6th :

Past President, Geo. H. Routh ; Pres., Jos. McKay ; Vice-Pres., Henry Derrer ; Rec. Sec., A. M. Schofield ; Fin. Sec., John Hamilton ; Treas., John McFarlane ; Conductor, Thos. Meehan ; Door-Keeper, W. M. Taylor ; Trustees, A. Nichol, J. D. Armstrong, J. M. Hamilton.

Branch No. 16 is making good progress. The membership does not grow very fast, owing to the limited number of engineers in the town. We have met once a week so far, but are thinking of changing our rooms, and meeting only twice a month, doing away with our reading-room for a time. The weather being so warm has interfered with our meetings of late, although in spite of that we are in good shape financially and every other way.

Branch No. 16 wishes the other sister branches every success.

A. M. SCHOFIELD, Rec. Sec.

## ONTARIO ASSOCIATION STATIONARY ENGINEERS.

Editor ELECTRICAL NEWS.

SIR,—I wish to call the attention of certificate holders of the O.A.S.E. who have not renewed their certificates to Sec. 2, Art. 7, of the By-Laws and Constitution, which is incorporated by the Ontario Legislature, and reads as follows :

"The certificates shall be good for one year, and shall remain the property of the Board, and must be returned to the Registrar within 30 days after the holder thereof has been notified so to do."

A notice was sent to every member in January last, yet there are quite a number who have not sent in their renewal fees.

A further notice will be sent out (to all who have failed to renew) in a few days, and after thirty days from date of said notice, all certificates not then renewed will be cancelled and means taken to collect old certificates.

This will entail a good deal of work and expense, but the Board are determined to carry out the act as laid down.

I would also request all certificate-holders who may change their place of residence, to communicate same to me by post card. I remain,

Yours very truly,
A. E. EDKINS, Registrar.

P. S.—Renewal fees are $1.25, $1.00, and 75c. for 1st, 2nd and 3rd class respectively. A. E. E.

## SPARKS.

A liquidator has been appointed for the Victoria Electric Light Co., Lindsay, Ont.

The Burk's Falls (Ont.) electric light plant has arrived in that town and is being placed in position.

The local electric light company's plant at Port Arthur, Ont., has been purchased by the town for $7,000.

The Bell Telephone Co. has served a writ on the Dundas Telephone Co., claiming damages to the amount of $10,000 for alleged injuries to the former's service.

The Stratford Gas Co. has accepted the electric light agreement proposed by the Council. Lights will be supplied on the moonlight schedule at the rate of $57 per lamp.

An order-in-council has been passed under the Electric Light Inspection Act, making the following additional regulations :—All electric light supply meters in use at the time of the passing of the Electric Light Inspection Act shall be presented for verifications as follows :—One-third before 1st December, 1895, one-third before 1st March, 1896, one-third before 1st July, 1896. For every unverified meter found in use after the first day of July, 1896, the owner thereof shall incur a penalty of twenty-five dollars. For every failure or neglect to comply with the provisions of section 22 of the Act in relation to affording the department testing facilities, the contractors shall incur a penalty of fifty dollars. For every failure to procure a certificate of registration as required by section 35 and the payment of the fee established thereof, within thirty days after the first day of July, in each year, the contractor shall incur a penalty not exceeding one hundred dollars and not less than fifty dollars.

## CANADIAN ELECTRICAL ASSOCIATION.

ARRANGEMENTS for the Annual Convention of the Canadian Electrical Association are being pushed forward as rapidly as possible, and are making satisfactory progress. A strong local committee has been appointed at Ottawa, to arrange for the proper reception and entertainment of the members of the Association who may attend the convention. This committee is composed of the members of the Executive resident in Ottawa, with whom are associated Messrs. T. Y. Soper, J. W. McRae, and Thomas Ahearn. The committee is manifesting an enthusiastic interest in the work which has been assigned to it to perform, and the members of the Association can confidently look forward to a convention which will be in every respect the equal if not superior to any which has previously been held. A number of interesting papers on various phases of electrical work have been promised for this convention by persons who are well qualified to write interestingly and instructively upon the subjects which have been assigned to them, or which they have voluntarily chosen.

It is a well-known fact that Ottawa is one of the most interesting cities in the Dominion, and this is particularly true from an electrical standpoint. It possesses one of the greatest water powers in Canada, from which sufficient current is generated for the operation of the city railway and lighting systems. The street railway system has become known far and wide as a model of what such a system should be, and will consequently well repay careful inspection of all its details both of management and equipment.

The dates for the convention have been fixed for the 17th, 18th and 19th of September. The local committee are arranging with the railway companies for reduced rates, and hope to be able to secure a single fare rate. The very satisfactory attendance at the Montreal convention last year gives ground for the expectation that at Ottawa a considerably larger attendance will be witnessed, as the convention will be located mid-way between Ontario and Quebec.

## DEVELOPMENTS IN THE TELEPHONE BUSINESS.

IT is astonishing to observe to what an extent the telephone business has become a part of the World's commercial equipment and to notice what strides have been taken by the telephone system in Canada since Professor Alexander Graham Bell made his first experiments in telephony at Tuttello Heights, on the outskirts of Brantford, in the year 1875.

The business which has now become consolidated under the control of the Bell Telephone Co. of Canada, having a paid up capital of over $3,000,000, consists of upwards of 500 offices and toll stations, with an enormous mileage of wire connections, and notwithstanding all that has been done a large amount of money is still being expended in erecting and otherwise perfecting the system so far as it relates to the successful operation of local exchanges; while in recent years also Long Distance lines have been constructed covering the greater part of the Provinces of Quebec and Ontario.

In all the large cities the Company has shown willingness to adopt the more expensive underground system for its wires, dispensing to a great extent with the over-head pole work on the principal streets in business sections. Toronto, Hamilton, Ottawa, London and Montreal all have systems of underground work more or less extensive. The subways in the latter city are used principally for trunking between exchanges. As has previously been mentioned in this journal a large amount of this work has been done within the past five years in Toronto, where subscribers now enjoy telephone communication over metallic circuits free from all noise and disturbance.

At present there are nearly five thousand subscribers in Toronto divided among four exchanges, consisting of the new Main office on Temperance St., having about three thousand subscribers, and the balance being divided between Yorkville, Parkdale and Toronto Junction Branches.

The underground cables which run to the New Main Exchange enter that building from two directions, passing over and round a roller curve built of iron and steel rods, on which are small cast iron rollers over which the cables pass. The cables are then bent up through holes in the ceiling of the cellar to the floor above and are there attached to iron cable terminals. The cables are protected from abnormal currents by the usual combination,

heat coil and carbon arresters. From the terminals the lines are carried in switchboard cables under a false flooring to one side of the Hibbard distributing rack, from the other side of which similar cables are taken through a shaft up stairs to the large operating room, which is located on the top floor of the building. Here they are connected to an intermediate board fastened on the wall close to the first section of the large switchboard.

The switch itself is of the branch terminal type, with the usual jacks and self-restoring drops. It has an ultimate capacity of 4,200 lines, and is at present wired for 3,600. Ten positions are arranged for incoming trunks from branch offices, and especially designed sections are used for Long Distance work. In front of every operator is a small 10 volt incandescent lamp which lights when an annunciator falls, attracting the attention of the operator and also facilitating supervision by the Chief operator. Small incandescent lamps are also used for dis-connect signals on the inter-office trunk lines. Outgoing trunk lines are equipped with mechanical visual busy test signals. Instead of the usual cam and ringing keys, a combination key is used, consisting of two buttons. Depressing one button cuts in the operator's telephone, while a depression of the other button cuts its out, and enables the operator to ring a subscriber.

The frame work of the switch is of iron veneered with polished cherry. The brass work is of dull finish, and the whole presents a very handsome effect. The board in all its parts, except the cables and wires, was made in the Bell Telephone Co.'s factory in Montreal, being put together and set up here by local employees.

The power plant consists of a two horse power motor-generator for charging the storage batteries, and two-half horse power motor generators for ringing bells—one being kept as a spare—twenty storage cells, 300 amp. hour and fourteen 30 amp. hour cells. The cells are arranged in sets, one set being in use while the other is being charged; suitable switches throw them in and out of circuit. A Weston Standard voltmeter and ammeter are in the charging circuit, while four Weston current indicators are in the discharging circuits.

Attention was called in a former article in this journal to the magnificent fire proof head office building which the Company is now erecting in Montreal. The building will front on Notre Dame, St. John and Hospital Sts. In Ottawa also the Local Exchange has been thoroughly remodeled and now occupies a handsome new building, adapted in every particular to its requirements. A description has already been given in the ELECTRICAL NEWS of the new Exchange lately erected in Quebec.

All along the line the same energetic policy is being pursued. New Exchanges are being opened in small places, and additional Long Distance lines are being erected in different sections of the country.

## A FEW STEAM PUMP CALCULATIONS.

WANTED—A steam pump to deliver 1,000 gallons per minute. Strokes per minute, 40; length of stroke, two feet; steam pressure, 80 pounds; head to pump against, 100 feet; allowance for loss, 20 per cent. A loss of 20 per cent. necessitates calculations for 1,000 gallons + 20 per cent., or 1,200 gallons per minute. This divided by 7.48 gives 160.4 cubic feet of water per minute. Dividing 160.4 by 40 we have 4.01 cubic feet per stroke, and call it 4, omitting the decimal. Dividing again by the length of the stroke (in feet) we get $4 + 2 = 2$ square feet as the area of the pump cylinder, or about 19½ inches for diameter; a pretty large diameter for the stroke, but necessary to meet the requirements, although it would be better to lengthen the stroke to three feet. The head of 100 feet (.434 pounds per foot, but calling it .5, makes an allowance for friction) gives us fifty pounds pressure per sq. inch of piston, and the piston area equals $2 + 144 = 288$ square inches, $288 + 50 = 14,400$ pounds total pressure on the piston to be overcome by steam pressure on the steam piston. Dividing the total load by the steam pressure we have $14,400 + 80 = 180$ square inches for the steam piston plus 20 per cent. loss in the steam cylinder, etc.—$15.25 + 3.05 = 18.3$ inches as steam cylinder diameter. The conditions here given are a little unusual, the hand being low for the pressure used, and the stroke short for the diameter; also the small number of strokes per minute, but the method of calculating is clearly shown and can be done for any selected case.

## THE ONLOOKER.

THE college-bred mechanic is a good deal in evidence in these days of technical colleges and schools for manual training. He is, in some respects, a much discussed individual, for it has not been settled in all minds that schools and colleges can turn out capable mechanics. They may make scholars and professional men, but, as Rudyard Kipling has said, that is another story. Opinion differs whether the course actively pursued of recent years of producing electrical engineers through our colleges and universities is going to give to the electrical industry the strong and efficient men, that this science, with its great development, must require. True, the schooling, if we may use the term, obtained by the electrical student, is usually supplemented by a measure of practical training in some one or other of the large electrical manufacturing companies, but is this sufficient to make a capable electrical engineer? In this particular, important advances have been made within a very few years, for Mr. James B. Cahoon, the electrician in charge of the Expert Department of the Thomson-Houston Co., has said, that a few years ago no special requirements for entrance were required or exacted and the result was that the student class was composed mainly of young boys from sixteen to twenty years of age, who could not overcome their boyish proclivities, and were in for fun more than serious work. To-day, in most of the colleges and universities furnishing an electrical engineering course, the age limit has been raised, and no student is admitted unless he is an engineering graduate of some technical college.

x x x x

But the question is a broader one than that of method and efficiency, as reflected through college and university. Experienced men divide on the question, whether the mechanical engineer or the college-bred electrician is likely to make the most capable electrical engineer. The former has a practical knowledge of mechanics, that is as necessary, in many respects, to the electrical, as to the mechanical, engineer. With this thorough knowledge as a foundation, the contention is that he can add to that an electrical knowledge, and thus equipped, no mere college taught electrician can expect to cope with him. This may seem like treating very slightingly the work that is being done in our colleges and universities, and which, by those whose views ought to count for something, is pronounced timely and capable. It does not seem unlikely, in this case, as in others, that the best results will be secured by striking a happy medium between the two methods. Though prejudice has condemned the educated man in many practical walks of life, opinion, based on experience, shows that education, whether with the mechanical or the professional man, is important ammunition in his possession. If the thoroughly trained mechanical engineer can add to that invaluable capital the electrical training that comes through the student course of the universities of the present day, he ought to develop a measure of strength that would at once place him at the top of the electrical engineering profession. Much of the criticism levelled against mechanical engineers, who undertake to call themselves electrical experts, is due to the fact that these men are not mechanical engineers any more than they are electricians. They are firemen and stokers too often, and doing their work in a bungling manner, the capables and incapables come in together for severe criticism.

x x x x

A conversation the Onlooker had a day or two since with Mr. D. C. McLean, Chief Engineer for the Toronto Street Railway, brought out in bold relief the contrast between the capable and incapable engineer. Mr. McLean is a mechanical engineer, having received a training and experience that is common to few men. He is one of the seventy-four, and only seventy-four, the world over, who have passed an honorary examination in engineering, that in Great Britain holds a parallel position with the degree of B. A. from Cambridge or Oxford. There it is necessary, in order to become an engineer, to be apprenticed, under articled indentures, for a period of seven years, where the training the young man is to receive, rather than the immediate emoluments, is the main consideration. Having put in this length of time in apprenticeship, then in order to obtain a third class certificate he must put in one year's actual experience at sea. To obtain a second-class certificate another twelve months at sea is necessary, and the same length of time is called for

when trying for a first-class certificate. In the latter case the engine of the vessel must be at least 3,000 h. p. and whilst the student-engineer is not actually in charge of the vessel, he has charge of an important watch, so that the responsibility, and the experience that comes from that responsibility, is thrown upon him. Mr. McLean has obtained this experience, besides having had thirteen years' actual experience as a marine engineer. The Onlooker enquired of him why so great emphasis was laid on the training of a marine engineer, and the reply was that only by this means could one become thoroughly equipped in his work. The ordinary experience will make a man what will be termed an engineer, but not as the term is understood in its highest sense. Queried as to his view of the training necessary to become an electrical engineer, Mr. McLean replied that the best authorities were of one view that the perfectly equipped electrical engineer must be nine-tenths a mechanical engineer. "How absurd," said he, "to suppose that because a young man can handle a coil of wire and perform a few mechanical acts connected with an electrical plant, that this makes him a master of electrical engineering. And yet I have seen this kind of thing. I have known those calling themselves electrical engineers to be unable to give an intelligent answer to what was an armature, while it was altogether beyond their comprehension to work out the simplest equation." The study of mechanics in Mr. McLean's opinion embraces so much, that it is impossible for one to become a master of his work except by years of toil. Going over the long list of text books that an Old Country examination calls for it was clearly shown that these could not be mastered except after many years of study. Then, on top of the study there had to be the real experience. It may be that Mr. McLean's ideal of a mechanical and electrical engineer is a difficult one to reach. The very fact that it has been placed high ought to be an incentive for those who would aspire to complete success to endeavor to reach it. One thing seemed very clear to the Onlooker that with a man of Mr. McLean's education and experience at the helm the mechanical and electrical affairs of the Toronto Railway Co. were in strong hands.

## MOONLIGHT SCHEDULE FOR AUGUST.

| Day of Month. | Light. | Extinguish. | No. of Hours. |
|---|---|---|---|
|  | H.M. | H.M. | H.M. |
| 1...... | P.M. 11.40 | ............ | } 4.20 |
| 2...... | ............ | A. M. 4.00 | |
| 3...... | A.M. 12.40 | " 4.00 | 3.20 |
| 4...... | " 1.40 | " 4.00 | 2.20 |
| 5...... | No light. | No light. | .... |
| 6...... | No light. | No light. | .... |
| 7...... | No light. | No light. | .... |
| 8...... | P. M. 7.30 | P. M. 9.30 | 2.00 |
| 9...... | " 7.30 | " 9.50 | 2.20 |
| 10...... | " 7.30 | " 10.10 | 2.40 |
| 11...... | " 7.30 | " 10.30 | 3.00 |
| 12...... | " 7.30 | " 11.00 | 3.30 |
| 13...... | " 7.30 | " 11.30 | 4.00 |
| 14...... | " 7.30 | A. M. 12.20 | 4.50 |
| 15...... | " 7.30 | " 1.00 | 5.30 |
| 16...... | " 7.30 | " 1.10 | 5.40 |
| 17...... | " 7.30 | " 2.10 | 6.40 |
| 18...... | " 7.30 | " 3.30 | 8.00 |
| 19...... | " 7.20 | " 4.30 | 9.10 |
| 20...... | " 7.20 | " 4.30 | 9.10 |
| 21...... | " 7.20 | " 4.30 | 9.10 |
| 22...... | " 7.20 | " 4.30 | 9.10 |
| 23...... | " 7.00 | " 4.30 | 9.30 |
| 24...... | " 7.00 | " 4.30 | 9.30 |
| 25...... | " 8.00 | " 4.30 | 8.30 |
| 26...... | " 9.00 | " 4.30 | 7.30 |
| 27...... | " 9.40 | " 4.30 | 6.50 |
| 28...... | " 10.40 | " 4.30 | 5.50 |
| 29...... | " 11.00 | " 4.30 | 5.30 |
| 30...... | " 11.30 | " ...... | } 5.00 |
| 31...... | " ...... | " 4.30 | |
| | | Total, | 153.00 |

Kamloops, B. C., is about putting in an electric plant of the capacity of 1,000 sixteen candle-power lamps.

It is reported that the Weston Union Telegraph Co., is about constructing a telegraph line to Alaska, via British Columbia.

The following board of directors have been elected by the North-West Electric Co., of Winnipeg: G. H. Streyel, president; J. M. Graham, G. A. Simpson, J. A. McArthur and H. Cameron, manager and secretary.

## RECENT CANADIAN PATENTS.

CANADIAN patents have recently been granted for the following electrical devices :—

No. 48,819, for a closed conduit electric railway, to James Francis Mc-Laughlin, Philadelphia, Pennsylvania, U.S.A., and May, 1895; 6 years.

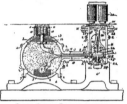

CLOSED CONDUIT ELECTRIC RAILWAY.

In an electric railway, the combination with a closed conduit provided with main and supply conductors, of switch boxes arranged alternately on opposite sides of the conduit and provided with switching mechanism for coupling the main conductor with sections of the working conductor, and two series of electro magnets, on opposite sides of the motor car, in line with the switch boxes, for operating the switches therein by magnetic attraction.

No. 48,838, for a car fender and brake, to William McBeth, Hamilton, and Harriett Belle Lewis, Winona, both in Ontario, Canada, 3rd May, 1895; 6 years.

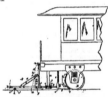

CAR FENDER AND BRAKE.

The combination with an electric or other railway car, of a frame a, b, c, standards d, d, horizontal bar e, e, vertical bangers l, l, provided with openings m, m, adjustable diagonal braces, f, f, with covering a, shaft J, rollers k, rubber tubing b, and cushion p, all constructed substantially as and for the purpose specified. In an electric or other railway car, the combination with a fender or brake shoes, and brake shoe rods, the same constructed to be operated by the fender being pushed against the brake shoe roads, when meeting an obstruction on the track, to apply the brakes on the wheels automatically, substantially as set forth. In an electric or other railway car the combination with a fender A, of brake shoes g, connected by a shaft r, supported by springs s, brake rods, t, t, attached to the brake shoes, brackets u, provided with lugs 4, and spiral spring v, v, to push the brake shoes off the car wheels, and brake rods operated by the rear contact movement of the fender A, against the said brake rod, substantially as and for the purpose specified. In an electric or other railway car, the combination of the fender and brake mechanism, substantially as and for the purpose specified.

No. 48,870, for a furnace grate, to Edward Gurney, Toronto, Ontario, assignee of Henry Truesdell, Hawarden, Iowa, U.S.A., 7th May, 1895; 6 years.

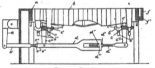

FURNACE GRATE.

The combination of a grate frame, rock shafts provided with alternating lateral arms and also operating arms as at C, a rod d, pivotally connecting arms C, one of its ends being connected to one of said arms C by a screw connection, an operating lever, an adjustable connection between rod d, and the operating lever, and grate bars supported on the lateral arms. The combination of a frame, two sets of bars adapted to move alternately in vertical planes, a stationary coupling bar having journal bearings and rigidly attached at its ends to the frame and interposed between the two sets of

grate bars rock shafts mounted in the journal bearings of the coupling bar, and having means for alternately moving the grate bars vertically, and means for operating the rock shafts, as described.

No. 48,935, for an electric arc lamp, to Peter Kirkegaard, Brooklyn, N. Y., U.S.A., 13th May, 1895 ; 6 years.

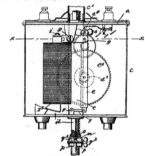

ELECTRIC ARC LAMP.

In an arc lamp, a frame carrying a gear train controlled by an escapement, said frame being pivotally mounted on parallel arms, said arms being pivoted to fixed supports, whereby the frame will always be parallel to a given plane, in combination with an electro magnet located in the shunt circuit of the lamp and attached to the frame, and an armature permanently fixed with respect to the magnet, a carbon holder consisting of a U shaped yoke pivotally connected with the end of the carbon rod, in combination with two jaws pivoted respectively to the arms of said yoke, and means for closing and opening said jaws.

## ELECTRIC WINNIPEG.

A very odd thing, and one that not only astonishes, but startles, the stranger, says a writer in Toronto Saturday Night, is the peculiar electrical condition of everything in Winnipeg during the winter. If you reach out to touch an electric bell, before your finger is within three inches of the enunciator there is a flash of lightning that goes up through your arm and will probably make you jump six feet. If you touch any metallic substance there is a flash of lightning ; when you get into bed the clothing crackles, and one would think that the landlord had provided you with a blanket adorned with fire-flies.

After a while one gets so nervous one is afraid to touch anything. I have stumbled around my room and bruised my shins rather than take chances lighting the gas or turning on the electric current. If you reach for the gas jet, "crack" it goes. If you shuffle your feet along the carpet you can light the gas with your finger. It is really one of the most startling phenomena in the whole northwest. Imagine turning over in one's bed and having the quilts emit sparks, or reaching for a bell and being immediately answered by a flash that is apt to make one howl. I saw my traveling companion, who had grown somewhat careful, wrap his finger up in a piece of paper to touch the bell. He jumped back with a shriek, and the whole paper seemed on fire. The people of the effete east who want to become electrified and have their systems filled with electricity should go up north ; they will get in proper shape and will learn to avoid radiators and every other metallic substance.

The fact remains, however, that in that climate one feels wonderfully hopeful and able to work, and no task seems too hard to be undertaken, and I am firmly convinced that the great men of Canada will be developed there. It is a remarkable fact that Ontario, the most blessed of all the sections of Canada, has developed, proportionately to its population, fewer brainy and energetic politicians and business men than the provinces by the sea or the great western stretch of land with the wonderful electricity in the air and the greatest difficulties of money making in the climate.

Messrs. E. Carl. Breithaupt, Berlin, A. A. Dion, Ottawa, Chas. B. Hunt, London, and E. B. Merrill, Toronto, were representatives of Canada at the annual convention of the American Institute of Electrical Engineers held last month at Niagara Falls.

PUBLISHED ON THE FIFTH OF EVERY MONTH BY

**CHAS. H. MORTIMER,**

OFFICE : CONFEDERATION LIFE BUILDING,
*Corner Yonge and Richmond Streets,*

**TORONTO,**                    **CANADA.**
Telephone 2362.

NEW YORK LIFE INSURANCE BUILDING, MONTREAL.
Bell Telephone 2299.

*ADVERTISEMENTS.*
Advertising rates sent promptly on application. Orders for advertising should
reach the office of publication not later than the 25th day of the month immediately
preceding date of issue. Changes in advertisements will be made whenever desired,
without cost to the advertiser, but to insure proper compliance with the instructions
of the advertiser, requests for change should reach the office as early as the 22nd day
of the month.

*SUBSCRIPTIONS.*
The ELECTRICAL NEWS will be mailed to subscribers in the Dominion, or the
United States, post free, for $1.00 per annum, 50 cents for six months.  The price
of subscription should be remitted by currency, in registered letter, or by postal order
payable to C. H. Mortimer.  Please do not send cheques on local banks unless 25
cents is added for cost of discount.  Money sent in unregistered letters will be at
senders' risk.  Subscriptions from foreign countries embraced in the General Postal
Union, $1.50 per annum.  Subscriptions are payable in advance.  The paper will be
discontinued at expiration of term paid for if so stipulated by the subscriber, but
where no such understanding exists, will be continued until instructions to discontinue are received and all arrearages paid.
Subscribers may have the mailing address changed as often as desired. When
ordering change, always give the old as well as the new address.
The Publisher should be notified of the failure of subscribers to receive their papers
promptly and regularly.

*EDITOR'S ANNOUNCEMENTS.*
Correspondence is invited upon all topics legitimately coming within the scope of
this journal.

THE "CANADIAN ELECTRICAL NEWS" HAS BEEN APPOINTED THE
OFFICIAL PAPER OF THE CANADIAN ELECTRICAL ASSOCIATION.

## CANADIAN ELECTRICAL ASSOCIATION.

### OFFICERS :

PRESIDENT :
K. J. DUNSTAN, Local Manager Bell Telephone Company, Toronto.

1ST VICE-PRESIDENT :
A. B. SMITH, Inspector Canadian Board Fire Underwriters, Toronto.

2ND VICE-PRESIDENT :
C. BERKELEY POWELL, Manager Ottawa Electric Light Co., Ottawa.

SECRETARY-TREASURER :
C. H. MORTIMER, Publisher ELECTRICAL NEWS, Toronto.

EXECUTIVE COMMITTEE :
L. B. McFARLANE, Bell Telephone Company, Montreal.
GEO. BLACK, G. N. W. Telegraph Co., Hamilton.
T. R. ROSEBRUGH, Lecturer in Electricity, School of Practical
Science, Toronto.
E. C. BREITHAUPT, Berlin, Ont.
JOHN YULE, Manager Guelph Gas and Electric Light Company,
Guelph, Ont.
D. A. STARR, Electrical Engineer, Montreal.
J. J. WRIGHT, Manager Toronto Electric Light Company.
J. A. KAMMERER, Royal Electric Co., Toronto.
J. W. TAYLOR, Manager Peterboro' Carbon Co., Peterboro'.
O. HIGMAN, Inland Revenue Department, Ottawa.

## MONTREAL ELECTRIC CLUB.

OFFICERS :
President, W. B. SHAW,         Montreal Electric Co.
Vice-President, H. O. EDWARDS,
Sec'y-Treas., CECIL DOUTRE,    87A St. Famille St.
Committee of Management, T. F. PICKETT, W. GRAHAM, J. A. DUGLASS.

## CANADIAN ASSOCIATION OF STATIONARY ENGINEERS.

EXECUTIVE BOARD :
President, J. J. YORK,    -    Board of Trade Bldg, Montreal.
Vice-President, W. G. BLACKGROVE,    -    Toronto, Ont.
Secretary, JAMES DEVLIN,    -    -    Kingston, Ont.
Treasurer, DUNCAN ROBERTSON,    -    -    Hamilton, Ont.
Conductor, E. J. Philip,    -    -    Toronto, Ont.
Door Keeper, J. F. CODY,    -    -    Wiarton, Ont.

TORONTO BRANCH NO. 1.—Meets 2nd and 4th Friday each month in
Room D, Shaftesbury Hall.  W. Lewis, President ; S. Thompson, Vice-
President ; T. Eversfield, Recording Secretary, University Crescent.

MONTREAL BRANCH NO. 1.—Meets 1st and 3rd Thursday each month,
in Engineers' Hall, Craig street.  President, John J. York, Board of Trade
Building ; first vice-president, J. Murphy; second vice-president, W. Ware ;
secretary, B. A. York ; treasurer, Thos. Ryan.

ST. LAURENT BRANCH No. 2.—Meets every Monday evening at 43
Bonsecours street, Montreal.  R. Drouin, President ; Alfred Latour, Secretary, 306 Delisle street, St. Cunegonde.

BRANDON, MAN., BRANCH NO. 1.—Meets 1st and 3rd Friday each
month, in City Hall.  A. R. Crawford, President ; Arthur Fleming,
Secretary.

HAMILTON BRANCH No. 2.—Meets 1st and 3rd Friday each month, in
Maccabee's Hall.  E. C. Johnson, President ; W. R. Cornish, Vice-Pres.;
Wm. Norris, Corresponding Secretary, 211 Wellington Street North.

STRATFORD BRANCH No. 3.—John Hoy, President ; Samuel H. Weir,
Secretary.

BRANTFORD BRANCH No. 4.—Meets 2nd and 4th Friday each month,
F. Lane, President ; T. Pilgrim, Vice-President ; Joseph Ogle, Secretary,
Brantford Cordage Co.

LONDON BRANCH No. 5.—Meets in Sherwood Hall first Thursday and
last Friday in each month.  F. Mitchell, President ; William Meaden, Secretary Treasurer, 533 Richmond Street.

GUELPH BRANCH No. 6.—Meets 1st and 3rd Wednesday each month at
7:30, p.m.  J. Fordyce, President ; J. Tuck, Vice-President ; H. T. Flewelling, Rec.-Secretary ; J. Gerry, Fin.-Secretary ; Treasurer, C. J. Jorden.

OTTAWA BRANCH, No. 7. — Meets 2nd and 4th Tuesday, each
month, corner Bank and Sparks streets ; Frank Robert, President ; F.
Merrill, Secretary, 352 Wellington Street.

DRESDEN BRANCH No. 8.—Meets every and week in each month; Thos.
Merrill, Secretary.

BERLIN BRANCH No. 9.—Meets 2nd and 4th Saturday each month at.
8 p. m.  W. J. Rhodes, President ; G. Steinmetz, Secretary, Berlin Ont.

KINGSTON BRANCH No. 10.—Meets 1st and 3rd Tuesday in each month
in Fraser Hall, King Street, at 8 p. m.  President, S. Donnelly ; Vice-
President, Henry Hopkins ; Secretary, J. W. Tandvin.

WINNIPEG BRANCH No. 11.—President, G. M. Haslett ; Recording
Secretary, J. Sutherland ; Financial Secretary, A. B. Jones.

KINCARDINE BRANCH No. 12.—Meets every Tuesday at 8 o'clock, in the
Engineer's Hall, Waterworks.  President, Daniel Bentt ; Vice-President,
Joseph Hall ; Secretary, A. Scott.

WIARTON BRANCH No. 13.—President, Wm. Craddock ; Rec. Secretary, Ed. Dunham.

PETERBOROUGH BRANCH No. 14.—Meets 2nd and 4th Wednesday in
each month.  S. Potter, President ; C. Robison, Vice-President ; W.
Sharp, engineer steam laundry, Charlotte Street, Secretary.

BROCKVILLE BRANCH No. 15.—President, W. F. Chapman ; Vice-
President, A. Franklin ; Recording Secretary, Wm. Robinson.

CARLETON PLACE BRANCH No. 16.—President, Jos. McKay ; Vice-
President, Henry Derrer ; Fin. Secretary, A. M. Schofield.

## ONTARIO ASSOCIATION OF STATIONARY ENGINEERS.

BOARD OF EXAMINERS.
President, A. AMES,    -    -    Brantford, Ont.
Vice-President, F. G. MITCHELL    -    London, Ont.
Registrar, A. E. EDKINS    -    130 Borden st., Toronto.
Treasurer, R. MACKIE,    -    -    28 Napier st., Hamilton.
Solicitor, J. A. McANDREWS,    -    Toronto.

TORONTO—A. E. Edkins, A. M. Wickens, E. J. Phillips, F. Donaldson.
HAMILTON—P. Stott, R. Mackie, T. Elliott.
BRANTFORD—A. Ames, care Patterson & Sons.
OTTAWA—Thomas Wesley.
KINGSTON—J. Devlin (Chief Engineer Penetentiary), J. Campbell.
LONDON—F. Mitchell.
NIAGARA FALLS—W. Phillips.

Information regarding examinations will be furnished on application to
any member of the Board.

THE Chicago City Railway Co. has recently established a
school in which their motormen are given instruction which is
calculated to fit them to discharge their duties in an intelligent
and efficient manner.  This is only possible when men have
acquired a knowledge of the method of construction and operation of the apparatus placed in their charge.  The experiment
is one which seems to be in the direction.

IN the present number of the ELECTRICAL NEWS is printed
the decision of Mr. Chief Justice Meredith of the Common
Pleas Divisional Court, Toronto, upholding the judgment of Mr.
Justice Rose, in dismissing the action brought by Mrs. Agnes
Hartford against the Bell Telephone Co., the Toronto Electric
Light Co., the Holmes Protection Co., and Silas Wheeler, to
recover damages for injuries sustained by contact with an electric wire on the streets of Toronto.  The learned judge took the
ground that the direct cause of the injury did not orginate with
the defendant companies, but was due to the act of a boy who
chopped a branch off a tree which stood near the wires of the
Holmes Company and the Bell Telephone Company, and the
falling of the branch, which caused the Holmes Company's
wires to come in contact with the wires of the Electric Light
Company.  While the complainant in this case is deserving of
the greatest amount of sympathy, there can be no question that
the decision given is a just one, and that had the plaintiff been
well advised legal action would not have been brought, at least
as a first resort.  Had a more conciliatory course been adopted
it is by no means improbable that some of the thousands of
dollars which are being expended in law costs might have found
their way into the pocket of the unfortunate victim of this
accident.

SO sanguine is the Street Railway Gazette of the success of the conduit electric street railways now under construction on Lenox avenue, New York, and in the city of Washington, that it advises persons desirous of obtaining franchises for the operation of electric railways by the overhead system to lose no time in securing them. The view is that after the conduit road shall have been shown to be workable, the public will be found much more reluctant to grant permission for the construction of roads on the overhead method.

THE turbine water wheels are about as unique and interesting as any other feature of the great power plant which is about to be set in operation by the Cataract Construction Co., on the American side of Niagara Falls. These wheels, which were designed in Geneva, have six times greater capacity than any turbine wheel previously constructed, being capable of giving 5000 H. P. while running at 250 revolutions per minute, under a head of 136 feet. No higher praise could be bestowed upon the skill of the designers of these wheels than to state that during a recent test their speed varied only 7 per cent. as a result of 3,000 H. P. being suddenly dropped from the load.

THE extent to which electricity will ultimately supersede steam as a motive power on railroads is the subject of much speculation and interest. The ELECTRICAL NEWS has been slow to place confidence in the assertion so fully made, that in a few years the steam locomotive would be superseded by the electric locomotive. On the contrary, there is little room to doubt that under certain conditions and for certain purposes, the electric locomotive will find a place on steam railroads. This subject has received its latest consideration in a paper entitled "The Substitution of Electricity for Steam in Railway Practice," by Dr. Louis Duncan, presented at the recent convention at Niagara Falls of the American Institute of Electrical Engineers. The questions considered by Dr. Duncan in his paper were : (1) Given a railway system at present operated by steam, will it pay to change entirely to electricity, or to make a partial substitution, and how should the change be made? (2) If entirely new lines are to be built, will it pay to equip them electrically? How should they be equipped? The ultimate conclusions arrived at by the author on these points, are as follows : 1st. The tendency of passenger transporation on the steam lines has been in the direction of the greatest electrical economy, while the tendency of the freight transportation has been in the direction of the least electrical economy. 2d. It will not pay any through line with considerable traffic, having two tracks, to equip their main tracks electrically. 3d. With four track roads it will pay to equip all of the tracks electrically unless a considerable portion is through passenger traffic. 4th. It will pay all the larger roads either to equip a number of their branch lines electrically, or to control competing electric lines. 5th. In order to remain on a dividend paying basis it is imperative that most of the two track lines either build additional tracks or control the electric roads that parallel them. 6th. Believing that ultimately all of the traffic will be done by electricity, it is imperative that the managers of steam roads keep constantly in touch with electrical progress.

THE question of the rating of arc lamps seems to require a considerable amount of adjustment. Should an arc lamp be rated according to candle power, or current, or voltage, or how? The rating usually adpoted is according to nominal candle power, that is, a specification says that the lamp shall give a full candle power of nominal 2,000 or 1,500, etc., with a certain amperage. Now, as a matter of fact, this is no specification at all, and refers to no standard whatever. First of all, what is the full candle power of a nominal 2,000 c. p. arc lamp? Next, at what angle is this full c. p. to be measured. It is well known that the intensity of the light from an arc varies according as the eye is above, on a level with or below the arc, and that it is maximum at an angle of 45° below the horizontal. It is well known to those who make any study of electrical matters, that the arc lamp does not give the light, but that its givenly apiece of mechanisim for regulating the distance a part of the carbons, and that it is the carbons that give the light ; that carbons vary among themselves so much that those of different makes may give 50 per cent. difference in light for the same expenditure of energy. So how can the lamp be rated at so many candle power without distinctly specifying the kind of carbon, its size, and the angle of measuremhnt ? Candle power depends directly on the temperature of the crater, and this temperature is nearly inversely proportional to the size of the carbon, the amperage remaining constant, so that any lamp may be adapted to give any candle power (within reasonable limits) by merely varying the diameter of the carbon. However, as the carbon decreases in diameter, it takes a greater pressure to produce the same amperage than with a larger carbon, so that candle power and amperes remaining constant, pressure and diameter must vary inversely according to some proportion. All these general conclusions have been very carefully examined by Professor Ayrton, and experimental results prepared, which show the proportion rather closely that exists between amperes, volts, diameter of carbon, and candle power, and the effect on an actual case of varying one or more of these factors. Starting with the assumption that a certain candle power (at a certain angle of course), is required, and keeping that intensity invariable, he shows that with a certain definite diameter of carbon a certain quantity of energy is required, which, within limits, remains constant, no matter whether the volts be high and the amperes low, or vice versa. Now this gives us a very satisfactory basis where to rate our arc lamps. Instead of specifying them to be of so many "candle power"—a standard which no one can verify, and which depends on so many uncertain factors—let them be rated at so many volts with a certain amperage, and a carbon of such and such clearly defined diameter. These data can be easily verified by experiments, and purchasers of arc lighting plants will have something to go by, instead of nothing as they have now.

A COUNTRY like Canada, when water powers abound, and where manufacturing industries afford outlets for large amounts of capital, is peculiarly well adapted for the development of the electrical methods of power transmission by the use of polyphasal currents. The demand for power is increasing every day ; the supply is practically unlimited in the numerous streams that waste their potential energy in innocous desuetude, and the means for the conversion of this potential energy of water into kinetic energy of rotation are open to all. The Canadian manufacturing interest should be particularly wide-awake in adopting any means for cheapening the cost of their power, and it is somewhat remarkable that the electric motor has not had a greater success in ousting from the factories and other places where power is scattered about in small units, such a very inefficient prime mover as the small steam engine. Central station plants would do well to work up this kind of business. A great source of expense and loss, in such plants, is the fact that for a great part of the time the machinery is being idle, and so earning no money, while if work could be found for it during a longer period a deficit might be converted into a profit. In other words, a higher "load factor" would be an advantage. Now a few motors would go a long way towards helping out returns, and there are very few towns where power is not required to a smaller or larger extent. There are always some people who use a little power, and the fact of it being obtainable would of itself create a demand. A principal reason why electric motors are not more used in factories, saw mills, &c., seems to be the regrettable ignorance on the part of the owners and operators of electric lighting plant of the possibilities of electric transmission, and this ignorance is largely taken advantage of by steam engine manufacturers, who sell small inefficient engines under the very noses of electric light owners. Why do superintendents of central stations not keep themselves up to date in the matter of electrical developments? It might perhaps be too much to expect of an average "electrician" that he should study the more scientific elements that govern the design of new apparatus, but at least he should know that improvement is taking place along certain lines, or that certain novel types of apparatus are being brought out, or that electricity is being used for such and such new purposes. And yet it is very unusual to find a man in charge of a small power house with any general information at all on electrical subjects. Owners are just as bad. The unfortunate results of this carelessness and indolence are observable wherever one goes, and can be seen in almost every detail of the managment. Machinery is purchased because it is cheap, without any reference whatever to its efficiency. Wires are put up by guess work as to size, and the instruments used are frequently curiosities. These matters, we can understand, are passed over by inexperi-enced persons, but why the use of motors for all kinds of in-dustrial purposes is not more pushed by those having power to sell, is a question that can only be answered by reference to the backwardness of electrical knowledge throughout the Dominion. There are many towns in which are situated factories of all kinds that use steam power in small scattered units. These small inefficient steam engines might very well and successfully be replaced by electric motors, that would do the work equally well, and economize much in the saving of condensation in pipes, radiation, &c., to say nothing of the higher efficiency possessed by the electric motor over the small steam engine. Here is a promising field to work in for central station men, who are suffi-ciently enterprising to study what may tend to their own pe-cuniary advantage.

SINCE the article appearing elsewhere in this number relating to the approaching convention of the Canadian Electrical Asso-ciation was put in type, some additional particulars have come to hand from Ottawa concerning the arrangements for the con-vention. The local committee have secured from the railway companies a rate of a fare and a third for delegates attending the convention from any part of Canada. Any member with or without his wife desiring to attend the convention should buy a first-class single fare ticket to Ottawa from the ticket agent where he lives, and at the same time procure from the agent a standard certificate. This standard certificate is in the posses-sion of every ticket agent, and is absolutely necessary in order to buy a return ticket from Ottawa back home at one-third the regular fare. The Ottawa hotels have quoted very low rates for the accommodation of delegates and their wives; in no case will the rate exceed $2.00 per day. We are advised that the local committee are not only making most complete arrangements for the entertainment of the mem-bers of the Association, but are likewise making provision for the ladies. A banquet to members and their wives has been arranged for at the Russell House, at the close of the Convention. This banquet will be of an electrical nature in its appointments and quite unique in all respects. We strongly recommend all members of the Association to visit Ottawa on the occasion of this Convention, and take their wives with them.

. OUR readers will be interested in noting by another column, that the Canadian General Electric Co. have commenced the manufacture of carbide of calcium under the Willson patents. So much has been written and prophesied regarding this new product of the electric furnace and of acetylene gas, the result of its decomposition in water, that the opportunity of obtaining it will be welcomed, if only for experimental purposes. We are not, however, able to share the roseate views of its promoters to the extent of seeing in it an invincible or even a formidable rival of the electric light. It would indeed be the irony of scientific evolution should the electric arc in the end have produced of it-self a rival from which should come a successful challenge of its supremacy as the most brilliant and economical of artificial illuminants. This, however, is far from likely. Already certain limits, undescernible, naturally, in the first glow of an inventors enthusiasm, have been indicated by more recent developments. The earlier representations as to the cost of production of the carbide, have been found unreasonably low, even under the most favorable conditions, and difficulties in the actual use of the gas for illumination, while possibly not insurmountable, have led to the belief that it is more likely to prove of service as an enricher of existing bases, than as an actual illuminating medium. Look-ing at the matter from an electrical point of view, there seems to be no reason to fear, but rather many to welcome, the new dis-covery of Mr. Willson. Carbide of calcium and its most im-portant resultant, acetyline gas, will take their proper rank amongst the contributions of electrical science to the industrial art, not as revolutionary or destructive intruders, but as valuable and now indispensible auxiliaries.

Sherbrooke, Que, capitalists are applying for letters patent to constitute a company to run electric and horse cars in that city, with power to extend their railway to any place in the district of St. Francis.

## THE TELEPHONE AND ITS TROUBLES.

WHEN a telephone was first used on a telegraph circuit, says A. Dolbear, in the Cosmopolitan, it was noticed that hissing and frying sounds could be heard, as well as telegraphic signals of all sorts, all of which had their origin in other electric circuits. Sometimes the extraneous noises were so much stronger than the telephonic speech that they quite overpowered it. The din destroyed the articulation. This was the case when the auto-matic Wheatstone transmitter was employed on a telegraph circuit parallel to a telephone circuit on the same poles. This was at first interpreted as being due wholly to induction, and for business purposes telephone lines were removed as far as practi-cable from telegraph lines. The trouble did not cease. In some cases it was nearly as bad as before ; and then it was apparent that the source of the disturbance was the earth itself. Both circuits made use of it as a part of their systems, and their ground connections were adjacent, oftentimes practically the same. When the telephone ground, as it was technically called, was moved away, there was some relief, but it was found possible to detect telegraph signals from lines separated by miles of earth.

When compared with telegraphic instruments, the telephone is found to be exceedingly sensitive. A sounder requires about the tenth of an ampere to work it properly, a relay, about the hundreth of an ampere ; but a telephone will render speech audible with less than the millionth of an ampere, and is, there-fore, more than ten thousand times more sensitive than a tele-graphic relay. When the earth is made to form a part of an electric circuit, the current does not go in a narrow strip from one ground terminal to the other, but spreads out in a wide sheet, much broader than perhaps most have imagined. Thus, if the grounds be no more than three or four miles apart, the spread-ing earth current can be traced in a sheet as much as two miles wide. If the grounds be still further apart, the sheet will be cor-respondingly wider. This earth current in its course may meet with streams of water, gas and water-pipes, and other conductors better than the earth itself, and these will conduct some of it, but not all. The stronger the current the more it is spread, and a telephone ground connection anywhere in its path will receive its share unavoidably.

In cities and towns employing the trolley railway systems, the rails form part of the circuit. As they lie upon the earth, the earth necessarily conducts away a notable part of the current, no matter how large the rails and good the connection. For in-stance, in Boston, where great pains has been taken to provide ample metallic conductors in rails and return wires, a thousand amperes have been found to return through the earth to the power house, and this is something like ten per cent. of the whole output. How widely such a current may spread may be imagined, and one may compare such a current with the minute one needed for telephonic work. One must remember that a steady current does not effect the telephone at all. It is only when the cur-rent varies in strength above a certain rate, thirty or forty times a second, that it begins to be troublesome. The variations in strength come from the Morse key or its substitutes in tele-graphy, from some types of arc lighting dynamos, from alter-nating dynamos for incandescent lighting, and from the motors in railway work. Though there be thousand amperes in the earth, if the variation be but one ampere, the nine hundred and ninety-nine which are constant are not offensive, hence it does not so much matter how much current is in the earth as how rapidly it varies. There are other currents in the earth due to natural causes, such as lightning, auroras, etc., which have sometimes been destructive to the telephone and its connections. To protect both service and the telephone itself there is one remedy, namely, to cease using the ground as a part of the need-ful circuit, and to provide each instrument with a complete wire circuit. Telephone companies are adopting this method every-where. It is more costly to establish and maintain, but it has been made necessary by the nature of elec-trical action and by the great increase in industrial enterprises within the past ten years.

Iron ore is now smelted by electricity in some parts of Nova Scotia. The new method will likely supplant the old blast furnace process.

The half-yearly dividend of 1¾ per cent. declared by the Toronto Rail-way Company recently, is equivalent to a dividend of 3½ per cent. on $600,000, the amount of the company's original investment.

## SOMETHING ABOUT INJECTORS.

### HINTS ABOUT THEM FOR ENGINEERS AND FIREMEN.

IN some instances it may be found impossible to adjust the injector for the work required, as it may have been especially for a far different pressure than that at which you wish to work it, for the higher the steam pressure used the smaller in proportion must the steam tube opening be, and no injector can be made which will fit all conditions equally well, regardless of advertisements to that effect.

Suppose our injector acts as we have stated before, we immediately know that it is not the fault of the injector, for if it was it would not start at all, unless in rare cases there may be a tube loose, and after the injector has started this may move and alter the relation between the water and the steam supply.

If our injector does not receive steam from the same pipe, the engine does, and the boiler is not forced to such an extent that it lifts the water badly, we may neglect the wet steam cause and look for others. First of all, we will make sure that our water supply is not interrupted by some unknown cause, for this would cause a deficiency of water and the steam would show at the overflow, making the injector break. This water deficiency may be caused by the water valve having a loose disk, which may move on the steam enough to alter the opening for water, and this is a fruitful cause of trouble many times both in steam and water pipes.

Or it may be that a pump in the neigborhood is taking the water at intervals, and at times the lessening of water may be enough to cause a "break" in the injector's working. Other causes which give trouble may be given briefly:

In many instances the pipes leading to the injector are long and small, and often filled with rust and other deposits, and while the injector will start all right it breaks just as soon as it has used the amount of water that is in the pipe, for this acts as a reservoir, supplying water enough for a start, but being soon exhausted.

In a case of this kind it will not do to blame the injector after being sure that there is nothing loose about it, for if it will start it will run until worn out, unless stopped by some outside cause, and this cause must be looked for.

In cases where small injectors are used on large pipes, confusion often arises as follows : The injector will start all right, and after a very short period of operation, will suddenly break and we wonder why. In cases that have come under my notice this has been caused by there not being an opening into the boiler, the check being either stuck or the stop valve shut. The injector starts well enough, but after it gets the large pipes filled and the pressure rises to the limit of the injector, then it breaks. A long pipe between injector and boiler, even if not so large, will have the same effect.

Great difficulty will sometimes be experienced in starting an injector, and one of the most common causes for this is a leaky check valve, allowing hot water from the boiler to come back into the injector and boil the water, or prevent it from condensing the required amount of steam. This can be readily found by care, carefully noticing whether any hot water shows at the overflow when the steam supply is shut off; this will indicate a leaky check valve unless the steam valve leaks, and a little care will soon determine which is the leaky valve.

The checks that give the most trouble are what are termed straight way or swinging checks, which, while very good for some work, are not as good for injector work as the old-fashioned plain check. The reason is this : The passage of water through them wears the side of the seat farthest from the hinge, and in a very short time the check is not tight, and this little leakage back from the boiler makes it hard to start the injector. And if a very slight obstruction becomes lodged near the hinge, the opening at the outer end of the swinging valve is much greater and the leakage is considerable. This is not said to injure any maker of swinging checks, but merely to give my own experience in this class of work.

When you have your doubts as to the quantity of water that can be supplied for injector, just measure the flow by letting it run into a measure of known quantity, and note the time taken to fill the measure. If we have a two-gallon pail, and the water from the supply pipe of the injector will fill it in five seconds, we know that as there are sixty seconds to the minute the pail will be filled twelve times per minute, which is twenty-

four gallons a minute, or 24x60 equals 1,440 gallons per hour.

Then, if the capacity of the injector is only 1,000 gallons per hour at the steam we are carrying, we know we have an ample margin for working. This, of course, is a very large injector, and will supply a large boiler or boilers.

In many cases the injector is made useless by the manner in which the piping is put up, and the writer has found cases where the injector refused to work, in which the supply of water had been cut down to less than half by the man who did the piping screwing all the pipes so far into the valves and elbows as to almost close the openings. This is particularly apt to be the case in the valves and checks; as the brass of which they are made gives so much more than iron fittings that the men do not stop until the pipe refuses to turn with the same force that they apply to iron pipe fittings. A little judgment helps wonderfully in a case of this kind. It is sometimes necessary or convenient to pipe the injector to the same supply and delivery pipes as used by the pump, although it should never be done where both are to be used at the same time, as the pulsating action of pump is very apt to take the water from the injector momentarily and cause it to break.

Where this is done there should be valves so that the pump connections can be shut tightly from the injector and vice versa, particularly in the case of a lifting injector. One instance of this kind was brought to my notice aboard of a little yacht which was being hurriedly fitted for a southern winter cruise, and in which the injector would start nicely and work for a minute perhaps, and then break or fly off, as some call it. The first thought was that there was a piece of wood or waste floating in the water tank in the bow of the boat, and that the action of the water drew it over the pipe and shut off the water supply, as often happens in cases of open tanks. This was not correct, however, as investigation showed that the men who had piped the injector had connected the water supply to the same pipe that supplied the wash basins in the cabins, and whenever the faucets in the cabins were open or leaked the air was drawn into the pipe and into the injector, and caused the break. By piping the two water supplies separately the trouble was remedied, and the boat was ready for her trip in tropical climates.—The Tradesman.

## CALCIUM CARBIDE AND ACETYLENE GAS.

The Canadian General Electric Co. have commenced to manufacture calcuim carbide under the Canadian patents of Mr. T. L. Willson. An electric furnace has been erected at the Peterboro works, under the supervision of Mr. Willson's representative, and a number of orders for the carbide have been filled. A considerable demand for small quantities, principally for experimental purposes, has already arisen. Should the sanguine expectations of the inventor be realized, the use of this product in the manufacture of acetylene gas will in the near future assume immense proportions. For the carbide as a potential source of energy are claimed, as especial advantages, its extremely low first cost of production, as well as its portability and convenient form for transportation. Besides its use in the production of acetylene, it seems likely to be of great commercial value in the production of cyanides and in various other processes of metallurgy.

Acetylene itself is a colorless gas with a penetrating odor. Its specific gravity is 0.91 and it is soluble in water, which, at 64° F. will absorb its own volume of the gas. It can be condensed into a liquid and in that form is readily portable. As an illuminant, properties are claimed for it which should, if justified by the facts, establish it as unmeasurably superior to coal or water gas, and make it easily, the most formidable rival which electric lighting has yet had to encounter. When burned at the rate of five cubic feet per hour, it has produced a light equal to 250 candles, as against an average equivalent of 16 to 20 candles with ordinary illuminating gas. If, however, the results actually attained in practice should fall far short of the possibilities thus indicated, there will remain for acetylene a field of great value as an enricher of ordinary illuminating and fuel gases.

The Willson process for the manufacture of the carbide is a most interesting one, and we hope shortly to present to our readers an account of it as now in operation at Peterboro.

## THE NEW CANAL AT SAULT STE. MARIE.

APPLICATION OF ELECTRICAL MACHINERY FOR OPERATING THE LOCK GATES AND VALVES OF CANAL LOCK.

THE remark merits reflection that at a time when railroads are cutting seriously into the carrying trade by water, there has, perhaps, seldom been greater activity in canal building, and more thought given to projects pointing to the development and expansion of existing waterways, and the opening of new channels of commerce along these lines.

It matters little what part of the world is studied, unusual effort in canal building is discovered. One of the great projects of the past year has been the completion of the Manchester ship canal, providing a direct route between Liverpool and Manchester. Italy has important maritime canal schemes under consideration, and there has lately been completed a notable maritime canal across the high and rocky Isthmus of Corinth. Certain difficulties have hindered the progress of the Nicaragua canal, a scheme in which, at least, one province of the Dominion, British Columbia, is largely interested, as it will be the means of shortening the route between the Pacific Coast

was finally completed with the enlargement of the Lachine canal, to the new dimensions in 1848.

Meanwhile, during the construction of the St. Lawrence canals, the Welland canal, between Lake Erie and Lake Ontario, had been completed and enlarged once. This canal was begun by a private stock company in 1824, after several years investigations by government commissions, and was completed in 1829. It was 27 miles long in 40 locks constructed of wood 110 feet long and 22 feet wide and with 8 feet depth of water on the sills.

After the union of the several Canadian provinces in 1867, further steps began to be taken towards developing the canal system. In 1870 a canal commission was appointed, which reported in 1871, advising an uniform and enlarged waterway with locks 270 feet long, 45 feet wide and 12 feet depth of water on the sills. This depth of water was afterwards increased to 14 feet. In other directions the energy of the government and private parties in canal development has been shown.

BUILDING THE SAULT STE. MARIE CANAL.

In not a few respects, the Sault Ste. Marie canal, in which we are particularly interested at the present time, and which has a special interest to readers of the ELECTRICAL NEWS, marks in several ways new developments and progress in canal building.

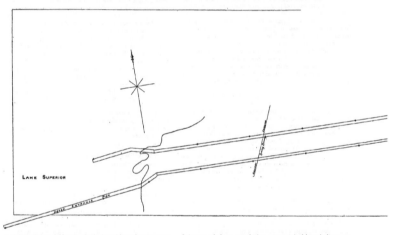

LAKE SUPERIOR

and Great Britain by just one-half; but of its ultimate consummation there can be no doubt.

The development of the Canadian canal system furnishes one of the most interesting chapters in Canadian history. The first practical step towards the construction of these artificial waterways was taken in 1815 and 1816, although they had been talked of long before, and indeed the rapids of the St. Mary's river and some of the rapids of the St. Lawrence had been passed by primitive canals and locks as early as 1798. In 1815 the legislature of Lower Canada voted a grant of money to build a Lachine Canal, and in 1818 a joint commission from Upper and Lower Canada reported in favor of a canal system on the St. Lawrence river, the canals to have a minimum depth of water of 4 feet. Work was begun on the first Lachine canal in 1821, and it was completed in 1825, at a cost of $440,000. The canal was 28 feet wide at the bottom, 48 feet wide at the top and 4½ feet deep, and the locks were of masonry 100 feet long and 20 feet wide. Hardly had the Lachine canal been finished when the Royal Engineer in the charge of the Rideau canal, then being built to connect the Ottawa river at Ottawa with the St. Lawrence at Kingston, urged the government to construct the remaining St. Lawrence canals with longer and wider locks, and with depth of water of 9 feet. This the government decided to do in an act passed in 1832, and the system

It has already been stated that a canal had been built across St. Mary's Island as early as 1798. This canal was built by one of the Northwest fur companies, and, according to such records as can be obtained, was 300 feet long and 45 feet wide, with a lock that raised the water 9 feet, or one-half the total fall at the rapids, so that the remaining height must have been overcome against the current—no great task for the light bateaux of the fur hunters. Between the building of this primitive Sault Ste. Marie canal and the construction of the great work of the same name, which is here illustrated and described, 96 years have elapsed.

The total length of the new canal across St. Mary's Island is 3,500 feet, and the dredged approaches under water at the two ends are about 18,000 feet long, with a depth of water of 21 feet. The essential feature of the work is, of course, the lock by which the 18 feet fall of the Sault Ste. Marie is overcome. This lock is built of masonry, and is 900 feet long between quoin posts, and 60 feet wide, with a depth of water of 20¾ feet on sills on low water. The height of the top of the walls above the floor of the lock chambers is 43¾ feet.

The gates are of wood, composed of white oak and iron truss rods. They are built on what may be called the truss bowstring type. Each leaf of these consists of a quoin (or heel) post formed of three pieces, a mitre (or toe) post formed of two pieces, 3 intermediate vertical frames and the requisite number of horizontal frame trusses spaced and proportioned nearly in accordance with the pressure due to depth sheered with 3 inch pine plank, spiked to the horizontals. Each horizontal frame consists of an upper or upstream chord, bent into a circular arc, a straight chord bar and iron truss rods. The latter are secured in the quoin and mitre posts in the intervals between the horizontals, but form part of the latter in reality.

There are five sets of gates, 2 at the upper or west end, and 3 at the lower end, i.e., a lock and guard gate at each end and an extra or auxiliary lock gate at the lower end for immediate use in case the lower main gate should get injured. Two sets of those gates (the lower main and auxiliary) are 44½ feet in height × 37 feet in width, weighing about 87 tons per leaf. The guard gates are of course to be used only when the lock chamber is being pumped out for examination or repairs.

Water is admitted to the lock chamber by four 8 × 8 ft. culverts, extending under the breast wall and underneath the floor and having openings at their tops. The inlets and outlets to these culverts are closed by butterfly valves 10½ × 8 ft. area, constructed of steel. Both the valves and gates are operated by electric power.

running transversely across the canal. The shafts are used for each set of four valves, one running from the right side-wall chamber to the centre, and carrying two valves, and the other from the left side-wall chamber, also carrying two valves. At the ends of each shaft in the wall chamber is a crank arm of forged steel, its least leverage being 4 ft., to which a vertical draw rod 55 ft. long is attached, and steadied in line by 2 sets of guide rollers. This draw rod is placed in a well in the lock wall, and when moved vertically up or down by the operating mechanism opens and closes the two valves on the shafts to which it is attached, i.e., the valves are operated in pairs.

The operating wire rope cables connect with the gate leaf on hooks secured to the gate near the mitre post, the front or closing cables run to and around a horizontal pulley secured on the mitre sill platform and from thence to a horizontal pulley at the bottom of the well, then under a vertical, then up the including well to another pair of vertical pulleys which gives the diverging angle to the cable which passes to and around one of the sheaves of the travelling pulleys operated by the gate machine, and back to and around the deflecting pulley stationed in the end of the frame, thence again to and around the second sheave of the travelling pulley and thence back to a standard to

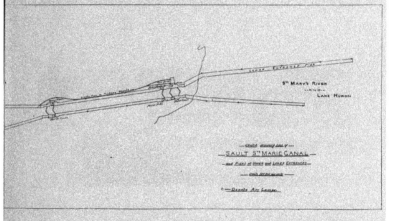

S^{T} MARY'S RIVER
AND
LAKE HURON

— *Sketch showing Line of* —
SAULT S^{TE} MARIE CANAL
— *and Piers at Upper and Lower Entrances* —

⌀ — *Denote Arc Lamps.*

The culverts are constructed entirely of wood, those for filling the lock are 8 × 8 ft. inside, and those for emptying 8 × 10½ ft. inside. In constructing the culverts 12 × 15 inch longitudinal sills were first bolted to the rock foundation, with 1¼ in. round bolts 6 ft. long or over, spaced 6 ft. apart. On the above 12 × 12 in., transverse timbers were laid 6 in. apart, and the interstices filled with Portland cement, concrete and grout, flush with their tops. On the top of this was laid a flooring of two thicknesses of plank, 3 in. and 2 in. thick respectively. The walls between the culverts 2 ft. thick, composed of two thicknesses of 12 × 12 in. timbers, were then built and capped with 12 × 22 in. transverse timbers laid close, having planed joints. This range of cap timbers was then bolted to the longitudinal sills, first laid by bolts 2 ft. apart, extending down through the culvert walls and the transverse timbers ; to give the bolts a good surface for holding down the cap timbers, continuous iron straps were placed crossing the timbers and acting as washers and the nut screwed firmly down on these. Of course these bolts had to be put in while laying the longitudinal sills, and the culvert walls built around them. The cap timbers were then covered with a double flooring of 3 inch and 2 inch plank respectively.

VALVES AND VALVE MACHINERY.—The admission of water into the culverts is controlled by valves. These valves are of steel, and are mounted on horizontal steel shafts, 10 in. in diameter,

which it is secured in a shackle bolt having an adjusting screw.

The back or opening cable, when hooked on, the back of the gate, passes direct to the horizontal pulley in the bottom of the well and from thence as already described for the closing cable passing round its pulleys and being attached to the opposite end of the gate machine. The four turns of the rope around the travelling pulley, which travels 8 ft. 9 inches, causes a travel of 35 ft. to the end attached to the gate leaf, and opens or closes the leaf by a single stroke of the cross head.

In all there are six gate machines, one for each leaf of the upper lock gate, lower lock gate and auxiliary gate. A one storey wooden motor house covers each of the gate machines and its connecting motor. Four of those houses are L shaped, this additional portion being to enclose in the same building the valve machine and its motor.

With this machinery the time required to pass a vessel through the lock going up stream is, after the vessel has taken her place in the chamber, 50 seconds for closing the lower gates, plus 50 seconds for opening the valves, plus 9 minutes for filling the lock, plus 50 seconds for opening the upper gates, or 11½ minutes altogether. As the lock can be emptied in 7½ minutes a vessel can be locked down in 10 minutes.

It may be noted, that both the gate and valve machines are

governed by automatic switches, operated by what may be called cut-off, or adjustable tripping bolts, which push the switch handles over and thereby cut off the current, so that the cross heads will not go beyond the intended point.

The tripping bolts (which push the handles) are adjustable in a slot by a nut and washer on the back of the plate, so as to make them cut sooner or later, or to the point required. These tripping bolts are isolated by 3-16ths of an inch hard rubber sockets, and washers, so as to prevent the current from passing on to the metal of the machinery. Chords run from the switch handles to pulleys on the ceilings, and by these are conducted to the controllers, and the switches are closed by the motor-man pulling the chords without having to leave his position. By this arrangement the danger of damage to the machinery (from the cross head running ablock at the ends of the screws) will be prevented.

### FIRST APPLICATION OF ELECTRICAL MACHINERY.

The machinery which has been described is, as far as we can learn, the first electrical power machinery ever used for operating the gates and valves of canal locks, and that it should for the first time be applied to a lock of this size and importance, indicates the confidence with which this form of power, which was hardly considered manageable a decade ago, is now regarded. For both the old 1881 lock and the new 1800 feet

One turbine will be used for running the generators, the other for running the arc light dynamo and general shop work, but when it is required to pump out the lock, the two wheels can be coupled and used to operate the centrifugal pumps. There are two of these pumps, and they have a combined capacity of 32,000 gallons per minute. The two pumps will lay the lock chamber dry in between 6 and 7 hours.

It should be noted also that near the upper end of the supply pipe there is a 6′ 8″ valve operated horizontally by two Tobin bronze screws, also two 5 ft. valves placed in the supply pipe (and operated vertically by screws of the above named bronze) immediately above the power house, permitting of either the whole of the pipes or of either or both turbines being laid dry when necessary.

It should be further noted that there is a 13 inch turbine water wheel set horizontally at the rear of one of the large turbine cases. This wheel has its water supply from a T shaped pipe placed between and supplied from the 5 ft. supply pipes, and having a valve on each arm of the T, so that in the event of the large turbines having to be stopped, or in the event of one or either having to be laid dry, by the arrangement of these valves on the T's a supply of water can always be obtained for the small turbine wheel, which, by belt to and from a small counter-shaft on the second floor, drives the incandescent light dynamo,

THE LOCK FROM THE WESTERN END.

lock on the American side of the St. Mary's river, hydraulic machinery is used.

The reasons which led to the adoption of electric power on the Canadian lock are stated by Mr. J. B. Spence, Chief Draftsman Department of Railways and Canals, as follows :

As regards economy, I think the difference between electric and hydraulic power will be very trifling, and here the point of economy was not taken into consideration. One of the main objects of using electricity was to overcome the great trouble caused by frost when hydraulic machinery is used. During the closing weeks of navigation the cold is so great that oil has to be used in the hydraulic engines placed on the lock walls, and even then the cold causes the oil to thicken and makes the action of the engines slow and tedious. Of course, frost would not have interfered with hydraulic valve engines placed at the bottom of the lock, but in this case eight engines would have been required, while only four screw power machines are needed with the machinery as designed. These considerations seemed to make it advisable to use electric power throughout, and I have every reason to think that everything will operate successfully when we open for navigation.

Two 45 in. 155 H. P. turbines, equalling a combined power of 310 H. P., supply the power for operating the generators and pumps. These turbines are set horizontally and are supplied with water from the upper level by a 6 ft. 8 in. diameter steel pipe, placed at the back of the lock-wall, just before entering the power house ; this supply pipe divides into two 5 ft. pipes—one for each turbine. The discharge pipes from the turbines are also 5 ft. in diameter. The turbines are placed on the first floor of the power house, and operate by belt a horizontal counter-shaft on the second floor. From this counter-shaft are operated the dynamos and generator on the second floor and the centrifugal pump shafts on the first floor.

so that a full supply of incandescent lights can be obtained throughout the buildings, pump well, etc., supposing that the large turbines are still.

The electrical plant for operating the gates and valves and for lighting the canal and approaches, was supplied by the Canadian General Electric Co., Ltd., of Toronto and Peterboro, under detailed specifications and designs drawn up by the government electrician, Mr. D. Bryce Scott.

The current for power purposes is supplied by two 45 K.W. 500 volt Edison standard 'bi-polar dynamos, either of which is of sufficient capacity for operating under normal conditions.

The lighting plant consists of a No. 7 Wood arc dynamo, having a capacity of 40 2000 C. P. lamps, and a 3 K. W. Edison bi-polar incandescent machine for lighting power house and repair shops.

The switchboard (illustrated) is a beautiful piece of work, and is a great credit to the manufacturers. It consists of three polished black slate panels 7 ft. long by 5 ft. wide and 2″ thick. These are supported by a heavy oak frame of ornamental design. The centre panel carries the instruments and controlling apparatus for the power generators, while on the right is the arc machine panel and on the left that for the incandescent machine.

The mechanical arrangement of the gate and valve mechanism has already been described, and it therefore only remains to give the electrical arrangement. The motors are of the Canadian General Electric Co.'s standard W. P. 50 railway type, and are operated in pairs by means of series parallel controllers classified by the manufacturers as type " K," that is to say, the two motors situated opporite each other on the canal walls and operating one pair of gates, are electrically connected in exactly the same manner as the motors on a street car, the connections across the canal being made by heavily armoured submarine cables, each having 14 conductors and being about 2½″ in

SWITCHBOARD IN POWER HOUSE.

diameter. The valve motors are also connected in pairs in exactly the same manner as described above.

The lighting of the canal and approaches is accomplished by means of a row of arc lamps down each side of the canal, situated at about 300 feet apart. These lamps are double carbon of the standard " Wood " type and are supported by means of iron poles and hoods placed on the top of 40 ft. poles.

In connection with the electric plant a somewhat unprecedented and novel system of an electric regulator is now almost completed and ready to be placed. Recognizing the well-known fact that by using the ordinary electric regulator the generator has to be worked at its full capacity, therefore wearing out the machine unnecessarily, to avoid this Mr. Spence saw, that by giving the large water wheel sufficient work, equal to about three-fourths of its power when running the generator, the amount of current when taken off was but a fraction and almost imperceptible. Therefore he decided to try a system to meet these requirements, and as the end of the water wheel shaft projects over the large wheel in which two large centrifugal pumps are placed, he arranged by placing a mitre wheel on the projecting shaft supported by bridge, and driving a horizontal mitre wheel placed on an upright shaft which extends deep in the well and firmly secured in step. Then on this shaft is placed the propeller wheels of a size to meet the power required, one facing up and the other down, which it is expected will cause no undue strain either up or downwards, and by a tip coupling at the horizontal mitre wheel, and when the regulator is not required, such as when running the large centrifugal pumps, the horizontal mitre wheels can be uncoupled. By this arrangement it is considered that the object will be accomplished.

The contractors for the lock and canal (being section 2), also the lower entrance under water, including the crib work, which form the wharf piers (being section 1), were Hugh Ryan & Co., Toronto, Ont. Mr. M. J. Haney, one of the members of the firm, was the superintendent in charge of the work, and Mr. William Birmingham was the engineer for the above contract-

ors. The contractors for the upper entrance submarine work (being section 3) were Messrs. Allan & Fleming, of Ottawa, Ont. The lock gates were built and placed by the noted gate builder, Roger Miller, Ingersoll, Ont. The contractors for the turbines were William Kennedy & Sons, Owen Sound, Ont. The contract for the gate and valve machines and all pulleys was executed according to departmental detail drawings by the Canadian Locomotive & Engine Co., of Kingston, Ont., and the contractors for the electric plant were the Canadian General Electric Company, of Toronto, Ont.

MR. JAMES BRUCE SPENCE.

For considerable of the data on which this article is based we have to thank Mr. J. B. Spence, Chief Draftsman of the Department of Railways and Canals for the Dominion. These particulars were also in part furnished to the Engineering News, of New York, though revised to date by Mr. Spence when given to the ELECTRICAL NEWS. We cheerfully credit our New York contemporary with the information that we have found it convenient to borrow from its pages.

Mr. Spence is a son of the land of the brown heather and shaggy wood, a native of Kingedward, Aberdeenshire, Scotland. Shortly before reaching manhood's years he came to Canada, choosing the city of Hamilton as his place of abode. After a brief period in that city, in April, 1861, he received an appointment on the civil engineering staff of the old Great Western Railway of Canada, serving under George Lowe Reid, then Chief Engineer of the railway. He remained on the staff of this railway for a consecutive period of sixteen years and left the service in 1876, to accept the position of assistant to the late John Page, then Chief Engineer of Public Works and Canals, for the Dominion. He served under him and his two successors for a period of nearly 19 years. During recent years he has held the post of designing engineer and chief of the draughting staff of railways and canals. It is thus seen that Mr. Spence's experience in Canada covers a period of over 34 years. As James Bruce Spence he is registered a member of the Canadian

MR J. B. SPENCE.

Society of Civil Engineers. To him is due much of the credit for the success that has attended the completion of the Sault Ste. Marie canal. We are pleased to furnish among our illustrations a portrait of Mr. Spence.

A retired farmer of Cote des Neiges, was recently crushed to death under an electric car in Montreal.

## NOTES ON THE RECONSTRUCTION OF A SMALL CENTRAL STATION PLANT.*

### By FRANKLIN L. POPE.

The financial condition of the smaller central station electric lighting plants throughout the country is at the present time by no means satisfactory, and in too many instances cannot even be truthfully said to be encouraging. A survey of the field shows that very few such plants located in towns having less than 10,000 inhabitants are earning more money than is necessary to meet their operating expenses and to provide for indispensable current repairs. In the State of Massachusetts, in which the operations of all electric lighting companies are by law made a matter of public record, it appears from the latest reports that the aggregate liabilities of the 57 companies operating in that State, including stocks, bonds, and floating indebtedness, amounted on June 30, 1894, in round numbers to $14,000,000, nearly all of which stands charged to construction account. The net earnings for the preceding year were $1,000,000, or about 7.1 per cent. on the total investment; a sum obviously quite insufficient to provide for depreciation and at the same time pay a fair dividend on the capital which has gone into the business. But if half a dozen of the larger plants, in cities like Boston, Lowell, Worcester, Springfield, Lynn and Fall River were excluded from the list, the showing for the smaller plants would be even far worse than it now appears.

Many of these small plants were started at an earlier day than could have been justified by any reasonable estimate of the business then in sight, and now find themselves hampered by inconvenient buildings, and with unsuitable machinery, bought at high prices, and encumbered with defective business methods which experience has shown to be wholly inconsistent with the dictates of good judgment.

With the owners of many of these plants, it has become a very serious question whether the easiest way out of the dilemma which confronts them may not be to relegate the entire plant to the junk shop and the scrap pile, and commence over again with new buildings, modern machinery and improved methods of administration. When the necessary capital is readily forthcoming, there can be no doubt that this would often be the wisest course of procedure, but for obvious reasons, it is one which is not always, nor even usually practicable. The alternative is to remodel the existing plant, bringing it as nearly as may be into accordance with the best modern practice, and utilizing so far as possible the old material; a course which at least has the merit of avoiding an undue expansion of the construction account, in most cases already sufficiently burdensome.

Having been called upon during the past year to advise the owners of a plant of the character above referred to, in reference to certain changes which had been suggested as desirable, and having afterward been employed in a professional capacity to design the work and superintend its execution, I have thought that some account of what we undertook to do and how we did it, might not be without interest to the members of the Institute.

The Great Barrington (Mass.) Electric Light Company was organized and commenced business in 1888. The population of the district intended to be served was about 3,000, and most of the expected consumers were located within 2,000 feet of the point decided upon for the station. This was built of wood, in the most inexpensive manner possible, and was placed alongside the railway for convenience in receiving coal, although at the same time the danger from fire was materially increased. The original outfit was an Edison 3-wire, equipped with a pair of 250-light 110-volt dynamos, and the company commenced business with 281 lights on contract of $10 per year each; wiring free. The centre of distribution was 1,800 feet from the station, necessitating over a ton of copper in the feeders alone. Generally speaking, the plant was well laid out, and well built as things went in those days. The two dynamos were belted to a single 80 H. P. Armington & Sims engine. The original cost of the plant was about $16,000. The following year a Schuyler arc plant for street lighting was added, carrying 35 arcs, nominally of 1,500 c. p., which was run from the same engine and boiler. In 1890, the plant was considerably enlarged by the addition of a second arc machine, a Westinghouse 500-light alternator, and a second engine and boiler of the same capacity as the first. An

*A paper, slightly condensed, presented at the Twelfth General Meeting of the American Institute of Electrical Engineers, Niagara Falls, N. Y., June 27, 1895.

80-kw Westinghouse dynamo of more modern type was afterward substituted for the original one.

Upon examining the plant last year, I found the Edison machines carrying on Saturday evenings a maximum load of some 450 lights, while three evenings in the week (with the stores closed) it fell to perhaps half that amount. The two Schuyler machines, with an aggregate capacity of 55 to 60 lights, were carrying about 38 to 40, or an equivalent of that amount, while the Westinghouse machines were seldom as much as half-loaded, carrying a maximum of possibly 500 lights during three or four months of the summer season, and not much more than one fourth that amount the remainder of the year. Necessarily, with so many dynamos of different types, and with such a variable, yet small average output, the consumption of coal was excessive as compared with the light delivered and paid for.

The street lines, according to the usual practice, were of No. 6 B. & S. weather-proof wire; the poles were of cedar, of good size and fitted with pine or spruce cross-arms, with common green glass insulators set upon wooden pins. In consequence of a silly prejudice, which had been fomented among the citizens by interested parties against permitting poles to be set in the streets, the wires, in a very great number of instances, had been attached, by cross-arms or brackets, to the trunks of the immense elm trees with which the streets of the town were shaded; a practice which occasioned an enormous loss of current every wet night, as well as much irregularity in the performance of the lights. The effect on the trees was by no means salutary, while the appearance was as much worse than that of poles in the streets as could possibly be imagined.

The village of Great Barrington extends for the most part along a single broad thoroughfare for a distance of nearly three miles, and the street lighting circuits are consequently very straggling. The 1,500 cp lamps, which were suspended at intervals of 800 to 1,000 feet, were actually of very little service in illuminating the densely shaded streets.

After a careful consideration of the situation, keeping in view the greatest possible reduction of present and future operating expenses, it was determined the wisest course to pursue would be to consolidate the whole service so that it could be supplied by one dynamo, in place of five underloaded ones. In pursuance of this plan it was decided to adopt the two-phase alternating system, at a maximum pressure of 2,100 volts in the primaries, and 105 volts in the secondaries, with a frequency sufficiently low to permit the advantageous use of induction motors if required. It was furthermore decided to abandon the steam plant, and to make arrangements to utilize some one of the excellent water-powers which were available within practicable distances. Under ordinary circumstances, I should have hesitated to recommend the substitution of water-power for steam as the sole source of power for the operation of an electric-lighting plant. Water power is an invaluable auxiliary, and when conveniently available for use in conjunction with steam, may often be made to save a very large coal bill in the course of a year. On the other hand, the excessive fluctuations to which it is subject—which are scarcely realized by those but casually acquainted with the subject—render it in most cases a very uncertain reliance for a business which is compelled to go on, per force, every night in the year, and which cannot suspend operations, as an ordinary manufactory does, if worst comes to worst, for a week or two at a time. Even a water privilege which, during ten months of the year, furnishes twice as much power as is needed, and even more, may be expected to fall off, during one of the extraordinarily dry seasons which occur at intervals of from five to ten years, to one-third its usual amount. In such a case, an electric plant solely dependent upon water-power would find itself in a most undesirable predicament.

In the present instance, the choice of a water privilege finally reduced itself to two sites, one in the town itself, within half a mile of the centre of consumption, and the other at Glendale village, seven miles distant, both situated on the Housatonic river. The privilege first mentioned being already occupied by a woolen factory, only the surplus water was available, but this was known to be quite sufficient for the requirements of the electric company at least nine months in each year, leaving three months to be run by steam. It had the advantage of being close at hand, and was capable of being fitted up at a moderate cost. As to the Glendale privilege, it was necessary to be very sure that the lowest water of a dry summer would give all the power required to run the plant without the aid of steam. Having invariably found the value of a water-power to be greatly exaggerated, not only in popular estimation, but in the opinion of its owners, the matter was investigated with much care.

While negotiations were still pending with the owners of the Glendale privilege and also the one in the village already referred to, overtures were received from a manufacturing company owning a third exceptionally desirable privilege, on the same stream, at an intermediate point considerably nearer than Glendale. This company had only recently completed a new dam, headgates, race-ways, etc., at a very considerable expense, and was willing to lease the complete establishment, including a new McCormick turbine of 325 H.P. and a two-phase Stanley generator of corresponding capacity, at a monthly rental based upon the actual output as measured in kilowatt hours at the dynamo terminals, provided that a certain minimum monthly consumption was guaranteed. With the same volume of water as at Glendale, the fall at this point was 20 feet, assuring at least 417 H.P. at lowest water, during lighting hours. All the hydraulic apparatus and appointments were of the best possible construction, and well calculated to insure absolute permanency of operation.

The minimum rental exacted was somewhat less than the amount of the coal bill of the Great Barrington company for the preceding fiscal year, but while the immediate saving in operating expenses was not large, the acceptance of the proposition would place the company in a position to reduce its rates to consumers, for the reason that its output might be very largely increased without materially augmenting its operating expenses. A lease for a term of years was accordingly closed.

In laying out the plant it was determined to bring the main feeders directly to a distributing station in the village, to be used principally as a convenient headquarters for testing the circuits and controlling the street-lighting service. In laying out the transmission line, a surveyor was employed, and a preliminary line was run directly from the power house to the distributing station. The air line distance was found to be 5.15 miles. With the assistance of the surveyor, the actual line was then staked out, going directly across country, and keeping as near as circumstances permitted to the transit line. About half the distance, the transit line was found to so nearly coincide with existing highways, that the consent of the local authorities was obtained to set the poles along the highway location; the remainder of the route lay principally through uncultivated land of little value, so that a comparatively small expenditure was sufficient to secure a release from all claims for land damages. This enabled the line to be located with long stretches absolutely straight, avoiding all sharp angles; a very important consideration when heavy wires are used. The poles were of selected chestnut with natural butts, usually set five feet in the ground at maximum intervals of 125 feet. Shorter poles were ordinarily 25 feet long and eight inches thick at the small end. Shorter poles were sometimes used on elevations and longer ones in depressions, in order to equalize the strain as much as possible. The insulators used were of the large double-bell white porcelain type (German government standard), and were imported by us from Hagen. The insulator of the top wire is set upon a malleable iron stem 14 inches long, screwed into the top of the pole, which is tapered to five inches in diameter and protected from splitting by driving on a wrought-iron ring. The tapered part of the pole, as well as the top, was given a coating of mineral paint mixed as thick as it could be spread with a brush. The insulator of the second wire is carried on a malleable iron gooseneck, screwed in a five-eighths inch hole bored in the side of the pole, in such position as to bring the wires about 16 inches apart. Another hole was bored on the opposite side of the pole, intended to take the goose-neck of the third wire at some future time, leaving the same interval between the second and third wires. The porcelain insulators are fixed to their iron supports by a packing of oakum placed between the screw threads, which serves to prevent any danger of fracture by expansion or contraction. The line wire is laid in a groove formed in the top of the insulator, except upon the curves and angles, in which case it is tied at the side in a circumferential groove, as is usual in this country. The German method of tying is quite complex, and unnecessarily strong; in case of undue strain if anything

gives way it had best be the tie wire. We therefore devised a simple tie which was easily and quickly applied, and which has so far served an admirable purpose. We were obliged to string the wires during very cold weather; sometimes as cold as eight or ten degrees below zero, and hence it was necessary to strain them very tight. A block and fall and a well-trained horse were used in pulling up, usually six or seven spans of one wire at a time. The hook of the block was always attached to the copper wire, whether bare or insulated, with a chain-knot made of three-quarter inch rope. The feeder wires were of No. 3 B & S soft copper, covered with weather-proof "insulation" along the highway (as a concession to enlightened public opinion), but elsewhere bare. The lengths of wire were joined with McIntyre twisted couplings; the unusual strain we had to put upon them occasionally pulled one apart, and this led us, out of abundant caution, to solder them, although this was done for mechanical rather than for electrical reasons. Only two feeder wires have as yet been strung, providing for a single-phase current from one side of the two-phase generator, but it is the intention to run a third feeder at an early day, which will enable two-phase induction motors to be connected to the same distributing system.

A pair of telephone wires of No. 12 steel were strung below the feeder wires, and these were supported upon small German porcelain insulators on iron goose-necks on opposite sides of the poles. These wires were transposed at intervals of about a mile, in order to eliminate the inductive effects of the alternating current in the feeders. The feeder lines were carried under the railway at an undergrade crossing by placing the insulators upon iron brackets leaded into the stone abutments. The plan of construction above described makes a strong, handsome and durable line, while the insulation of the circuit, even in the worst of weather, is simply faultless.

The system has been planned to deliver the current at the distributing station at a uniform pressure of 2,100 volts. Two distributing centres were fixed upon in the old Edison three-wire network, and at each of these points a pair of large transformers, having a ratio of 20 : 1 were fixed upon a pole, with their respective primaries in series between a pair of branch feeders from the distributing station, and their secondaries were coupled in series in like manner, with the neutral wire between them. None of the consumers on the old Edison system knew when the change had been made to the new service from anything they were able to notice in the behavior of the lights.

The next thing done was to reconstruct the street-lighting system. In place of the 36 arcs of 1,500 nominal c. p. formerly in use we substituted 126 incandescent lamps of 50 volts and 32 c. p. placed in Iona fixtures projecting horizontally from the poles 14 feet above the ground. The lights, as a rule, were fixed upon every alternate pole, but in the business centre, the street being broad, they were placed on each side at intervals of about 250 feet, and staggered, so as not to come opposite each other. A Shallenberger shunt cut-out was applied to each lamp. The usual number of lamps in each circuit was 42, although we have since placed, in some cases, as many as 47 in one series without reducing the brilliancy of illumination sufficiently to be noticeable by any one but an expert. One end of each street-lighting circuit is joined to a special feeder leading to the sub-station, where it is connected with the main feeder through a knife switch. The other end of each lamp-circuit is connected to any conveniently located branch feeder of the regular commercial lighting service. Each lamp-circuit has, or will have, a fuse-block and cut-out inclosed in a weather-proof box at each end, where it joins the opposite feeders. These 32 c. p. lamps, when run at full candle power, furnish a most satisfactory illumination, and give the streets a very attractive appearance. So far as possible, each lamp was located with the aid of a transit and level, so as to get them in absolutely straight lines both vertically and horizontally, a precaution which adds materially to the decorative effect. It is admitted by all that the streets of the town are much more satisfactorily lighted by the incandescents than they formerly were by the arc lamps, while the actual cost to the company is considerably less. The new lamps were cut in, one at a time, on the old arc wires, jumpers being temporarily placed across the terminals until everything was in readiness to discontinue the use of the arc-light machines.

One of the most marked advantages of the series street-lighting system, especially when shunt cut-outs are used, is its great flexibility and convenience. For example, instead of placing from 40 to 45 fifty-volt lamps in one series, we may use 20 to 23 one hundred-volt lamps, or if an odd number be required, less than is necessary to make up a circuit, the deficit may be supplied by adding extra shunt-boxes in series at any convenient point in the circuit, until the pressure has been reduced to the required point. From time to time, as new lights are added, these spare shunt-boxes are one after another brought into use in connection with them. Sometimes, also, we temporarily install extra street lights by connecting them in parallel to the secondary mains of the regular commercial service, ultimately transferring them to new series circuits.

It has been found to be desirable to use a lamp of rather low efficiency for the street-lighting service, as there is always danger of leakage and short circuits from wet boughs of trees and other objects getting into contact with the wires, and thus diverting an abnormal current through some portion of a lamp circuit. In such case, a lamp of high efficiency is pretty certain to be burned out, or at least to have its career of usefulness materially abridged. In this plant, the average consumption of energy in the street-lights, including lamps, lines, shunts and leakage, is found to be about 140 watts per lamp of 32 c.p.

Perhaps the most ticklish part of the whole undertaking was the changing over of the Westinghouse system, which was a 1,050-volt primary and a 52-volt secondary, running at 16,500 alternations. In accordance with the new plan, it was of course necessary to double the pressure both in the primary and secondary circuits, and to substitute 104-volt for 52-volt lamps throughout. A preliminary test of one of the transformers demonstrated, that which perhaps might have been foreseen from theoretical considerations, viz., that a dangerous quantity of heat was developed within a few hours when it was used to convert from 2,000 volts down to 100. In order to utilize, as far as possible, the old transformers, and at the same time avoid the above difficulty, various expedients were resorted to. Wherever a group of consumers was located in one neighborhood, a pair of large transformers were installed, with secondary mains extending from 500 to 600 feet in various directions, these transformers being of course placed in series with each other. Scattering consumers as far as practicable were united in pairs or small groups, and supplied by a pair of small transformers coupled in the same way. The Westinghouse meters, having been originally constructed for a frequency of 16,500 alternations, ran slow when the frequency was reduced to 8,000. The necessary coefficient for correction of the readings was easily ascertained by experiment, and as fast as possible the meters were fitted with new discs, supplied by the Westinghouse Company at a trifling expense, adapted to the lesser frequency.

Of course it will be understood that the reason for resorting to these various shifts and expedients, was merely that we might utilize the old apparatus as far as it could possibly be done, and also that we might carry on the work of reconstruction, for the most part, with the ordinary force of the establishment.

The selection of the best among the many available types of turbines for electric work is a matter which merits far more consideration from a scientific standpoint than it generally receives. Water-wheels, like dynamos and motors, are sometimes sold on commission by agents, and it not infrequently happens that the salesman who makes the largest "claims," especially if he sells his goods the cheapest, carries away the contract. It needs to be said, however, that there is a far greater difference than is often suspected, in the work that different types of wheels will do with a given, and especially a limited amount of water. There are, furthermore, a great many types of wheels in the market, which although as efficient as could be asked for with a full head of water, are very far from being so when the volume of water is reduced, even by a comparatively small percentage. It is but just to say that it is seldom that a turbine makes so favorable a showing, not only in this but in other respects, as the one provided by the company from which we lease our power.

The results of tests made in the testing flume of the Holyoke Water Power Company are worthy of particular note, for the reason that they show a very high percentage of efficiency maintained through a wide range of variation in the quantity of water passing through the wheel; a most valuable characteristic for electric work. When the quantity of water used was diminished from 81.75 to 42.55 cubic feet per second, the percentage of efficiency fell only from 80.99 to 63.9, and what is even more remarkable, it was found that the efficiency remained well above 80 per cent. over a range of variation of discharge from 83.23 to 70 cubic feet per second, or 15.9 per cent. More than one type of turbine which enjoys a high reputation and extensive sale among power-users, will not reach 65 or even 60 per cent. efficiency at "three-quarters gate," while the 33-inch wheel above referred to has been found to give by actual test no less than 78 per cent. under similar conditions.

The turbine carries upon its shaft a driving-pulley 100 inches in diameter, weighing 1,000 pounds, which serves as a balance-wheel. It is also provided with a Replogle electric governor, operated by three cells of gravity battery, which has never failed to do its work quickly and certainly, even under trying conditions.

In carrying out this work, some things have been learned by experience which may be of use to others called upon to advise or to undertake the construction of similar works, and I will therefore venture to summarize some of my conclusions as follows:

1. In considering the advisability of operating an electric plant by water-power, do not on any account neglect to ascertain from authentic sources of information, just how much water can be depended upon during the low stage in an extra dry year, for this is the measure of its value for electric work, except when used as an auxiliary to steam. The ordinary estimates of the commercial value of a water-power are only too apt to prove preposterous exaggerations.

2. If rights-of-way or releases of damages can be obtained without too much trouble and expense, it is better to build the feeder line as directly across country as may be, than to follow a highway. The saving in cost of construction will usually be more than enough to pay for the right-of-way, and on such a route there need be no interference from trees, while many inconvenient angles and much trouble in guying and bracing are avoided. Shorter and stouter poles may also be used; in itself a very important consideration.

3. In electric line construction it is preferable to dispense with cross-arms unless there are more than six wires. The best arrangement is to place one wire on a top-pin and the others alternately on the front and back of the pole, at a vertical distance apart of 12 inches. This construction not only costs less than properly braced cross-arms, but is much less conspicuous and therefore much less objectionable in a public street, is less interfered with by trees, and is far more durable. Much trouble is caused by the decay of cross-arms after they have been exposed a few years to the weather; they split at the ends so that the pin comes out, and not infrequently break in two in the middle, thus fouling the wires.

4. In medium-sized towns and cities, especially in shaded streets, the incandescent lamp may be made to give a better distribution of light for the same money than is possible with the "half-arc" so extensively used, and is much less troublesome to maintain in good working order. My own experience leads me to think that the lamps ought not to be of less than 24 or more than 32 candle-power. Use lamps of low rather than high efficiency, but run them at full candle-power, or even a trifle above. Good street lights, well arranged, and renewed sufficiently often, are the best possible advertisement for any electric company.

5. Use large transformers as far as practicable, placing the consumers within 500 or 600 feet radius upon secondary mains. We have used both two-wire and three-wire mains. The latter plan is certainly to be recommended when the distance approximates or exceeds 500 feet, but for short distances, as for example when distributing within a single block at a pressure of 100 volts or more, it is a question whether the gain in cost of copper over the two-wire plan is of sufficient importance to offset the additional complexity.

6. It was found that raising the voltage in the residence district from 1,000 : 50 to 2,000 : 100 greatly improved the uniformity of distribution, by lessening the potential drop without entailing any corresponding disadvantages. It would seem to be preferable, on every account, to use the higher pressure.

7. One of the most important minor points in the management of a plant is apt to be too much neglected; the maintenance of the insulation of the wires by promptly replacing all cracked

and broken insulators, and by keeping the wires absolutely free from contact with uninsulated objects. The covered wires which lead into the hoods of the street-lamps need to be carefully looked after. —

8. Number all the poles with yellow paint applied with a stencil on a black ground ; and keep a record book of the position , of each one and its distance by the line from the test-station.

9. In selecting a turbine-wheel, consult competent authorities as to the available fall and minimum quantity of water, and when making the purchase do not expect to get a $1,000 wheel for $100. Pay a fair price and insist, not only that the wheel shall be well made in every way, but that it shall be tested by an expert before acceptance. If it does not give an average efficiency of 76 per cent. between half-gate and full-gate, it is not advisable to accept it, inasmuch as you can easily do better, as our own experience proves.

10. I think our experience shows that it is possible to largely increase the net earnings of an old plant without necessarily renewing it throughout, but plenty of time should be taken for considering as well as for execution, in order to secure satisfactory results with a moderate expenditure.

## SCIENTIFICALLY CUT LAMP GLOBES.

AN invention that undoubtedly will be developed into great utility, and that among many other applications, would seem to hasten the adoption of small arc lamps for interior or even desk use, is described in the London Journal in an article on " Holophane Globes," which is the name applied to glass globes that are cut on scientific principles for the proper dissemination of light. It is stated that holophane globes, when enclosing any light of high candle power, such as the Welsbach incandescent gas, or the electric lamp, give the appearance of a vase filled with light, brilliant, yet soft while the actual burner or filament cannot be discerned.

The principle of the holophane globe is readily explained. The interior surface of the globe is formed into vertical grooves, which are so shaped as to spread out horizontally the rays proceeding from every part of the light source. The mouldings on the outer surface of the globe are horizontal, and have the effect of distributing the emergent rays in the vertical sense ; and inasmuch as the light may be required in some instances to be cast downward and in others to be equally dispersed, the angles of the outside grooves are modified accordingly. This is a very different thing from the uncientific cutting seen in ornamental cut glass globes which do nothing for the diffusion of the light.

As for the loss entailed by the reflection and refraction of Holophane globes, it is certified by M. de Nashville to amount in the case of an arc light to from nine to thirteen per cent., and as this observer remarks, there is no other kind of globe in existence capable of realizing such diffusion of light and presenting such uniformity of effect. As the loss of light by transmission through clear glass is from eight to ten per cent., the claim that holophone globes do their special work for about four per cent. of loss, is well established.

## THE ONLY TEST OF MERIT.

THAT the people are quick to appreciate a good thing when they see it, is abundantly shown by the phenomenal record of the Toronto Industrial Exhibition. The Fair which begins on the 2nd of September next, is the seventeenth of the series. It has grown steadily in popularity and yearly attracts increasing numbers which is the best possible proof of its superior excellence. This season the display will be more complete and varied than ever. The number of enteries is unusually large in all departments. Already every foot of space in the building is taken up, though additions and re-arrangements have been made to accommodate the increased number of exhibitors. Great improvements have been made in the accommodations provided, and all arrangements for public convenience are as nearly perfect as possible. An attractive and diversified programme of entertainments is offered. All railways will give low rates and special excursions will be run from many points, presenting an opportunity of which all should avail themselves.

The Brantford Electric Street Railway Co. is inclined to charity. Its gross receipts on August 6th will be given to the public hospital.

## PERSONAL.

Mr F. J. Proutt, Superintendent of the Malden Electric Co., of Boston, and formerly of Bowmanville, Ont., was recently married to Miss Laura J. Yarnold, of Whitby.

We are pleased to notice that Mr. D. H. Keely, of Ottawa, has recently received the appointment of General Superintendent of the Government Telegraph Service of Canada. Mr. Keely was for some years assistant to the late F. N. Gisborne, who was for many years at the head of the Government Telegraph Service of this country. In this capacity Mr. Keely had the most favorable opportunity of becoming acquainted with the requirements of the service, and the means of meeting those requirements in the most satisfactory manner. Since the death of Mr. Gisborne, Mr. Keely has been discharging the duties of General Superintendent in a manner so satisfactory to the public and the Government, as to warrant his permanent appointment, and we have no doubt that he will justify the wisdom of the Government's choice.

## TRADE NOTES.

Rhodes, Curry & Co., Amherst, N. S., have received a contract from the Halifax Electric Railway for fourteen street cars and a $90,000 car house.

The Bell Telephone Company have contracted with the Babcock & Wilcox Company for two of their latest wrought steel type of boilers for their new building now being erected at the corner of Notre Dame and St. John streets, Montreal. While it is not intended to instal the electric light plant at present, the boilers will be abundantly large to furnish steam for the electric light engine whenever wanted, and they will also be built to carry 200 lbs working pressure if desired. The Babcock & Wilcox Company report that their business is very good indeed, their shops at Belleville being well filled with orders for boilers to be delivered during the summer and fall.

The Gooderham & Worts Company, Limited, are just now installing at their new distillery at Toronto, a complete independent water works pumping plant, for the purpose of giving them additional fire protection. This new plant is not intended to furnish all needed fire protection, but rather to supplement the resources of the regular City Water Works. Gooderham & Worts' plant, however, will be very complete and perfect, and the equipment will be first class in every particular. Two large compound condensing pumps of 1,500,000 gallons-capacity each will be used. These pumps to receive their steam from two Babcock & Wilcox wrought steel boilers. The boilers will be of the well-known Babcock & Wilcox Co.'s latest improved type, all pressure parts being constructed of wrought steel ; boilers when completed to be capable of carrying a working pressure of 200 lbs. per square inch. As many of our readers already know, the Babcock & Wilcox Company are now building their boilers in Canada, having equipped large shops at Belleville, Ont., with special tools, patterns, etc., so that they are now prepared to turn out large orders promptly. · The Gooderham & Worts Company are locating their new pumping plant in a handsome new brick building with brick stack, entirely independent from their other works, and the arrangement of the boilers and pumps will be such that the apparatus will always be in readiness for use at a moment's notice.

## SPARKS.

The gross earnings of the tramway companies of Montreal and Toronto average about $4,000 a day.

Charlottetown, N. B., has received a number of tenders for electric light supply, but the contract has not yet been awarded.

The Hubbell Primary Battery Co. have commenced manufacturing their batteries. They are already introduced in the C. P. R. and C. A. R.

The Ottawa Carbon and Porcelain works have commenced grinding coke and carbon. It is expected these works will soon give employment to 100 men.

Dr. Corbett, of Port Hope, proposes to put in a three phase system and new apparatus in his electric light plant. He also proposes supplying power to some local manufacturers.

The town of St. Marys, Ont., having declared incandescent lighting both expensive and inefficient, the Council has decided to advertise for tenders for thirty-two arc lights of 1,000 candle power, each.

The Co-operative Telephone Co. of the counties of Lake St. John and Chicoutimi. Que., with a capital stock of $10,000, with headquarters at Herbertsville, has been formed to build and operate a telephone line

The last annual report of the Ottawa Electric Co. shows 2,192 meter customers, 677 ordinary commercial and 138 monthly accounts, making a total of 3,007 different customers being at the present time supplied with electric light.

Halifax, N. S., is probably the last city of importance to adopt an electric . street railway, but it is at last an assured fact. The company has purchased from the Johnston Steel Co., of Louraine, Ohio, 1,000 tons of rail. The same company has also a contract to furnish the special work necessary for the curves, sidetracks, turnouts, etc., the aggregate cost of which is $30,000.

The Canadian Electric Forging and Smelting Co., of Toronto, seeks incorporation for the purpose of smelting, heating, cooking, and the manufacture of chemicals, by products, gases and electricity ; the manufacture and sale of machinery and construction of necessary plants for all electrical circuits, etc. The capital stock of the company is to be $500,000, divided into 5,000 shares of $100 each. The principal stock holders are from the States of Massachusetts and New York and the Province of Ontario.

# ELECTRIC RAILWAY DEPARTMENT.

## AMERICAN STREET RAILWAY ASSOCIATION CONVENTION.

THE executive committee of the American Street Railway Association has made an arrangement with M. Davis, customs broker of Montreal, for a reduction in custom house charges as follows on exhibits for the Montreal Convention : Warehouse and bond entry $1 ; export bond entry, $1 ; making and procuring consular certificates, $1. When goods to be returned are valued at $50 or more a consular certificate which costs $2.50 must be procured, but this is unnecessary in the case of goods which are valued at less than $50. The fees therefore to be paid for goods under the value of $50 would be $2 ; and $3.50 would be added to that when a consular certificate is required. Shippers should mark goods with their own name, and " Care of M. Davis, Montreal, for exhibition purposes," prepaying the freight, and sending invoices marked "certified correct," and signed. On arrival, Mr. Davis will make warehouse bond entry, and have goods delivered at the Victoria Rink.

When the exhibition is over, the owners of the goods will have to repack them, using preferably the same cases that the goods came in, and they will be returned under the export bond. They must be careful not to make more packages of the goods in sending them out than they had in bringing them in, and it is a distinct advantage to have them in the same cases, so that the marks on these cases may be identified. Consignors must pay all freight and cartage.

The following regulations have been adopted :

Space will be allotted on Aug. 1 to all exhibitors whose applications have been filed with the secretary and accepted on or before that date. Applications for space received and accepted after Aug. 1 will be allotted remaining space, if any, in the order of their acceptance.

The space will be charged for at the rate of 15 cents a square foot, and no space less than 50 square feet will be rented, nor more than 1,000 square feet unless by special arrangement with the secretary.

Space allotted cannot be transferred without permission and must be taken possession of on or before Oct. 9.

Articles placed on exhibition cannot be removed without the written permission of the secretary.

All goods shipped to the exhibition should be plainly marked "Street Railway Exposition, Montreal, Canada." It is advisable to secure a time-limit delivery. Be sure to allow plenty of time for transportation.

On and after Oct. 8 exhibitors and their agents and workmen will be admitted to the building for the purpose of preparing necessary structures. The general reception of articles for exhibition will commence on Oct. 9.

Exhibitors of machinery in operation must have everything in running order, in readiness to start their machinery on the morning of the opening day.

All goods intended for exhibition must be on the premises and properly displayed on or before Monday evening, Oct. 14.

Exhibitors must provide all counter shafts, pulleys, belting, switches, switchboards, etc., necessary for the operation of their machinery.

No platform or other structure must be nailed to the floor or walls.

Exhibitors must not place any sign or circulate advertisements, except such as pertain to their own business (and those only in their own space), without written permission from the secretary.

Electric power will be furnished to those who use power. The charge therefore during the entire time of the exposition will be 45 cents per rated kilowatt of machine actually using current. The minimum charge for power will be $15.

All machinery will, if possible, be exhibited in motion, and should be kept in motion at regular work during the hours 9 to 12 a. m., 2 to 6, and 7 to closing p. m.

Parties desiring to sell and deliver in the building any article whatever, must first obtain a written permit from the secretary for such consideration as may be determined upon.

Any permit to sell may be revoked at any time, at the pleasure of the association.

Every possible precaution will be taken to guard against fire, and a full corps of watchmen will be on duty day and night ; but the association will not be responsible for loss or damage to articles on exhibition, by theft, fire or otherwise.

The association reserves the right to charge an admission fee to the citizens of Montreal should it so determine, but the admission of exhibitors and their agents will be free.

## THE WESTINGHOUSE CONDUIT RAILWAY SYSTEM.

THERE is now on exhibition at the New York offices of the Westinghouse Electric and Manufacturing Company, says the Electrical Review, a model of an underground electric railway system which is attracting considerable interest. It embodies the inventions of Mr. Malone Wheless and Mr. Geo. Westinghouse, jr. The patents taken out by Mr. Wheless were controlled by the Electro-Magnetic Traction Company, of West Virginia, but have recently been acquired by the Westinghouse company. Mr. Wheless' system has been practically tested in Washington, D. C., where a line three-quarters of a mile long was laid last fall on North Capitol street and successfully operated all of last winter. Another line is in operation at the new plant of the Westinghouse company in East Pittsburgh, and it was this line that Manhattan Railway officials recently inspected with a view to its possible adoption on the elevated railways of New York city. The system was originally designed for street surface traffic, but a few modifications will permit its adoption on elevated roads.

The principle of operation is very simple and the construction of the road involves a minimum amount of digging, as it is placed near the surface. The feeding conductors are laid underground at the side of a single track or between double tracks. The feeders are connected at suitable intervals with automatic switches. At corresponding intervals, in the centre of each track, are triple-point contact plates. Under each car are three collector bars which make a sliding contact with the triple-point plates. As the car passes over these plates a storage battery carried on the car automatically operates the switches, and thus the current is thrown from the feeders through the switch to the contact points and on through the collector bars to the car motor. When the car has passed a contact plate the switch automatically breaks the connection with the feeder and the plate remains dead until the passage of the next car. The collector bars are sufficiently long to prevent sparking. It is said that the system is so arranged that overhead trolley lines can be used in suburban districts and the same car run on this underground method in city streets.

Should the Manhattan Railway Company decide to use the Westinghouse conduit system, the triple contact plates will be replaced by a succession of metal bars separated by distances varying from 10 to 20 feet.

## SPARKS.

During the month of June the Galt and Preston electric railway carried 13,000 passengers.

The largest telegraph office in the world is the general post office building, London. There are upwards of 3,000 operators, 1,000 of whom are women. The batteries are supplied by 90,000 cells.

Mr. C. J. Morris, of Montreal, has entered an action against the street railway company of that city for $2,500 damages on account of the death of his child, which was killed by one of the cars which was being backed into the shed.

The Privy Council has granted the Toronto Street Railway special leave to appeal from the decision of the Supreme Court of Canada dismissing the appellant's action to recover the amount paid for custom dues levied on steel rails.

The route is now definitely decided for the Halifax, N. S., electric tramway. The main line of four and ahalf miles, and branches of four miles additional, will be in running order by November 1st. The road, building and rolling stock will cost in the neighborhood of $340,000.

Mr. S. R. Break, superintendent of the Detroit street railway lines, is a resident of London, Ont. He has resigned his position, to which a salary of $2,500 was attached, owing to religious scruples, the duties of his office having made it necessary to transact a certain amount of business on Sun. day, to which he objected.

The heavy blasting on the side of the stone cliffs for the Gorge electric railway at Niagara Falls has been causing havoc in the neighborhood. A recent blast was sent off which tore out several thousand tons of rock, sending it up in the air some 300 feet over on the Canadian side about a quarter of a mile from where the blasting took place. Hundreds of pieces of rock, weighing from ten to thirty-five pounds, dropped like grape shot from their great height on the lawns and roofs of residences on the street facing the river bank. The Niagara Falls Park and River railway have closed their incline railway and promenade, being afraid of accidents to tourists.

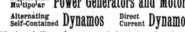

Please mention the CANADIAN ELECTRICAL NEWS when corresponding with Advertisers

## SPARKS.

Cornwall expects to have an electric railway shortly.

A radial electric railway between Sarnia and Florence, to run through Petrolia, Ont., is being discussed.

The Nanaimo Electric Light, Power & Heating Co., have purchased the Nanaimo Electric Light Works. The price paid is said to be $55,000.

The extension of the electric railway between West Toronto Junction and the village of Weston is practically a settled fact. It is understood that the Canadian Electric Co. will carry out the work.

The Niagara Falls and Lewiston, N. Y. Electric Railway, running along the river bed, was formally opened July 19th. On the initial trip one of the cars jumped the track and several persons were injured.

Hintonburgh council has decided to give the Ottawa electric railway the right of way over the Richmond road through the village for twenty years provided the company pay $200 per year for five years and macadamize the road.

At the annual meeting of the Kingston Street Railway Co., recently held, it was decided to increase the capital stock by issuing stock and bonds to the extent of $100,000, to be expended in extending the system to Cataraqui and other points.

The contract between the City of Quebec and Mr. H. J. Beemer, representing the Quebec, Montmorency and Charlevoix railway, for the construction of the city electric railway, has been signed. Work will be proceeded with at once.

The total passenger receipts of the Ottawa Electric Railway Co., for the year ending May 31st, 1895, were $183,394.68, and from mails, rents and other sources, $10,596.68, making a gross revenue of $193,991.36. The total expenses were $122,335.69.

One of the latest undertakings in electrical science is the construction of an electric line for the transportation of passengers, mail and express from Chicago to Buffalo and New York. The run to Buffalo is to be made in four hours and to New York in from eight to ten hours.

At the 26th annual general meeting of the Dominion Telegraph Co., held in Toronto on July 17th, the directors submitted a very favorable report of the year's business. The following gentlemen were elected directors for the ensuing year : Thos. Swinyard, Sir Frank Smith, K. C. M. G., Gen. Thos. T. Eskert, Chas. A. Tinker, A. G. Ramsay, Henry Pellatt, Hector Mackenzie, Thos. F. Clark and Thos. R. Wood. At a subsequent meeting of the newly elected Board, Mr. Thos. Swinyard was re-appointed president, Sir Frank Smith, vice-president, and Wm. Fred. Roper, secretary and treasurer.

CONVENTION · NUMBER

CANADIAN

ELECTRICAL·NEWS

STEAM ENGINEERING JOURNAL

OLD SERIES, VOL. XV.—No. 8.
NEW SERIES, VOL. V.—No. 9.

SEPTEMBER, 1895

PRICE 10 CENTS
$1.00 PER YEAR.

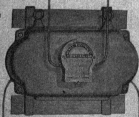

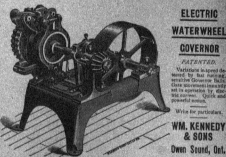

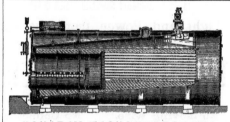
---

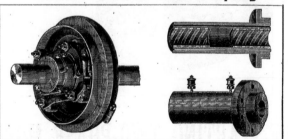
---

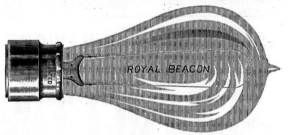

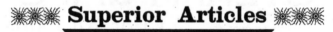

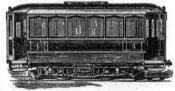

CANADIAN

# ELECTRICAL NEWS

AND

## STEAM ENGINEERING JOURNAL.

| Vol. V. | SEPTEMBER, 1895 | No. 9. |

### ELECTRICAL OTTAWA.

A Description of Some of the Electrical Features of
the Capital City.

IF in selecting a place for their next annual convention the members of the Canadian Electrical Association had been in search of a city combining great natural beauty with unceasing electrical development, they could not have made a more felicitous choice than that of the City of Ottawa. The site of the Canadian capital is one of the most favored spots in the whole Dominion. Built on a commanding elevation on the right bank of the noble river from which it takes its name, the city spreads away to the south, east and west in regular and

Electric Light Company's multiple series system had been used but its many drawbacks prevented its general adoption for house lighting. In 1889 the introduction of the Westinghouse system of distribution gave a great impetus to the lighting industry, and its subsequent growth may be gauged by the fact that the present installation is equivalent to 50,000 16 c. p. lamps, or one lamp for each man, woman and child in Ottawa.

But the electrical feature which gives Ottawa special pre-eminence is undoubtedly its street railway system. In this field, as in that of electric lighting, Ottawa was also a pioneer. When the project was first mentioned grave doubts were expressed on all sides as to the possibility of operating an electric road in Ottawa during the severe winter season, but the promoters of the undertaking had the courage of their convictions, and the uninterrupted service which they have given their patrons shows that their confidence was not misplaced. The electric roads now in operation in all the principal cities and towns in Canada

VIEW OF CHAUDIERE FALLS, OTTAWA.

well built streets; while in the distance to the north, the Lauren-tian mountains form a beautiful and imposing background.

The great water power afforded by the Chaudiere and the Rideau Falls has long since made Ottawa the chief lumbering-mill centre of the Dominion, and in addition to this distinction its people are now predicting that in the near future it will be a great manufacturing and railway centre as well. With its present industrial activity this article does not propose to deal further, than to mention and briefly describe those features of special interest to the electrical world.

It is just ten years since Ottawa took the lead of Canadian cities by introducing a complete system of electric street lighting. Two years later, in 1887, when incandescent lighting was a luxury enjoyed elsewhere almost solely by the owners of isolated plants, a local company was already supplying the stores and dwellings of the citizens of Ottawa with incandescent light. It was not until 1889, however, that the boom in incandescent lighting really took place. Up to this time the United States

are to a great extent the result of Ottawa's foresight and enter-prise.

The Ottawa electric railway has attracted so much attention that a short description of it will not be out of place. The power house is driven entirely by water and is situated close to the Chaudiere Falls. The plant comprises 1-700 h. p., and 2-400 h. p. multipolar, and 1-100 h. p. bipolar Westinghouse generator. The latter machine is driven by a separate water wheel and is used during the day time to excite the fields of the multipolar machines. At night it supplies current for running the mail cars, the workshop motor and for lighting and heating the power house and car sheds. The multipolar generators are driven by six 66 inch turbines which are all belted to the same counter-shaft and are provided with friction clutches so that any or all of the wheels may be run together or separately as the require-ments demand. One of the greatest difficulties that had to be overcome at the power house was the regulation of the speed of the machinery. Several types of automatic regulators were tried, but the variations of load were so great and occurred so frequently that the regulators were discarded after a very short trial and the present system of hand regulation adopted instead. A voltmeter operated by pressure wires from the centre of the city is placed on a small board at one end of the generator room and directly above the water wheels. The man in charge is seated in front of this voltmeter and convenient to a lever attached by an eccentric box to a friction pulley below. The gates are opened and closed and the speed thus controlled by

STREET RAILWAY POWER HOUSE, OTTAWA.

simply moving the lever backwards or forwards. The regulation of the speed by means of this arrangement alone was fairly satisfactory, but since the introduction last winter of a separately driven exciter, the variations in voltage, due to speed variations, have been as small, if not less than those of any closely regulated steam driven plant. The switchboard is situated at the opposite end of the generator room from the pressure indicator, and the entire equipment consists of Westinghouse apparatus, with the exception of a Weston total load ammeter, and a voltmeter of the same make. The generator panels are made up of a rheostat, triple-pole jaw switch, automatic circuit breaker and ammeter; and the feeder panels of a single pole jaw switch, automatic circuit breaker and the Wurts non-arcing lightning arrester. The well-known "Tank lighting arrester" manufactured by the Westinghouse Co. has also been in use for a long time and no accident attributable to lightning has occurred since its introduction. Visitors to the power house are apt to think that the method employed of regulating speed by hand is a very crude arrangement. At first sight there seems to be some ground for this opinion, but when it is explained that the plant is never for a moment from under the eye of a watchful attendant and that it has run for more than four years without a single accident, the visitor will likely conclude that the managers are wise in continuing the present system.

The Street Railway Company's car sheds are situated on Albert street, between Lyon and Kent streets. The sheds are used both for the purposes of barns and repair shops. The offices of the company occupy the central portion of the sheds, fronting on Albert street, and are divided into the offices of General Manager, Secretary-Treasurer, Superintendent and Accountants' offices. In the rear of these offices come in the order mentioned the lavatories, conductors' and motormen's rooms, winding room and machine shop. The remaining portion of the building is fitted up with pits, switches, transfer tables, etc., for the convenient handling and repair of cars.

The manufacturing establishment of the Ottawa Car Co., an outcome of the Street Railway Co., is quite close to the car sheds,

and the excellence of their work is attested to by the fact that they have supplied numerous cars to nearly all the street railways in Canada.

In the winter the snow is removed from the street railway by the well-known Lewis and Fowler sweepers and the Walkaway snow plows. The sweepers clean the tracks and the plows following force the snow back to the sidewalk where men shovel it into large box sleighs; it is then hauled to the river and dumped on the ice. During the winter season street car travel is just as popular in Ottawa as in the summer. The cars are heated by the famous Ahearn heater and are always very comfortable. The same heater is also used to heat the power house, car sheds and offices of the company.

Of the many benefits conferred on the people of Ottawa by the street railway there is none that they appreciate more than the extension of the line to Rockliff Park, which is situated on the Ottawa river just below the city. The ride to Rockliff is a charming one, the scenery being varied by pine woods and rocky bluffs which overhang the river and from which the park derives its name. Every evening during the summer the company provides an open air band concert for their patrons and the park is visited daily by thousands of children for whom innocent amusements are provided.

The Ottawa Canoe Club, whose headquarters are on the river bank at Rockliff, have a novel arrangement for hoisting their canoes out of the river, an electric motor supplied with current from the street car circuit furnishes the motive power.

Next in importance and interest to the street railway, stands the electric lighting company. In 1885 street lighting was begun by the Ottawa Electric Light Co., and in 1887 the Chaudière Electric Light & Power Co. commenced supplying incandescent lamps and motors. Four years later the Standard Electric Co. entered the same field as the C. E. L. & P. Co. Each of these companies carried on its business separately until a little more than a year ago when an amalgamated company was formed under the name of the Ottawa Electric Co. Before handing over the privileges enjoyed by each of the companies the City provided against any increase in the rates of supply and also for reduced rates when the earnings of the company exceeded a certain per cent on the capital stock. Some idea of the extensive nature of the lighting business in Ottawa may be had from a glance at the power houses now operated by the Ottawa Electric Co. The three companies had each their own water power stations, and one of them an auxiliary station driven by steam. The steam station is intended for use only in times of low water

STREET RAILWAY CAR SHEDS, OTTAWA.

and during a portion of last winter when the river was unprecedentedly low it prevented the city from being plunged in almost total darkness. Two 600 h. p. compound condensing engines, fed by six boilers, comprise the driving plant at the steam power house; and the electrical plant consists of three alternators of 240 K. W. each and three 50 arc light machines. In building this power house the company provided for future extensions, and among other arrangements the chimney was made with ample draft capacity for double the present number of boilers; it stands on a cut stone base 24 ft. sq.; it is 120 feet high, is hexagonal in shape, and 9 ft. in diameter at the bottom and 7 ft. at the top. Water for the boilers and condensing purposes is drawn direct from the river, and there is also a connection with the city mains to provide against emergencies. Hard coal and mill wood are used for fuel, the former almost exclusively when the engines are run for four hours or more at a time. Coal is received by rail and dumped direct from the cars into a shed situated alongside the railway track ; while the wood supply is purchased at the neighboring saw mills.

The largest and most important of the water stations is known

POWER HOUSE NO. 1, OTTAWA ELECTRIC CO.

as power house No. 1. There are at present in this power house four 500 h. p. wheels in operation. The water wheels, crown wheels and pinions are in a building projecting from the power house proper, and much of the vibration usually felt in water driven stations is not transmitted to the dynamo room by reason of this isolation. Three-ply leather belts 50 inches wide and 120 feet long transmit the power from the main shafts to the countershafts in the power house proper. All pulleys on the countershaft are provided with friction clutches and the countershafts are so arranged that they may be driven from either one of two waterwheels. The dynamo capacity of this station is equal to 2000 k. w. in alternators, and the continuous current machines for motor work to 250 h. p., to which is being added a multipolar machine of 250 h. p. capacity.

Probably the most interesting feature of this power house will be the switchboard now in course of erection. It is built after the model of the well-known Westinghouse Company's Worlds Fair lighting plant board, modified to suit the requirements of the Ottawa Electric Co. It will consist of 35 panels, each 20″ wide and 11 feet high, and will be divided into dynamo, feeder, and motor sections. The frame work is built of angle iron ($1\frac{1}{2}″ \times 1\frac{1}{2}″ \times \frac{1}{4}″$) and stands on rubber cushions, to ensure steady working of the voltmeters and ammeters. The dynamos in all the other stations will be wired to this switchboard and distribution of the load among these other stations will be done by the switch board attendant. The circuits which formerly were regulated and operated from the power houses belonging to the different companies will all run direct to this board. Each circuit will be supplied with a Stilwell regulator so that its pressure may be varied considerably without changing the pressure of the dynamo at the other stations. A portion of this switchboard already in position presents a remarkably handsome appearance.

Although the arc lighting business has not increased with the

same rapidity as the incandescent, still it has kept up a steady pace, and has now reached very large proportions. Ottawa is perhaps the most thoroughly lighted city in America. Some of the principal streets are not as brilliantly illuminated as the corresponding ones in other cities, but the general lighting of all the streets is more thorough and complete. The growth of the city and the consequent extension of its limits has much more than offset any damage which may have been done to arc lighting by the introduction of the incandescent.

The Ottawa Electric Co.'s motor business is represented by a constant day load of about 300 h. p., distributed among mills, machine shops, printing offices, hotels, and stores. The most important motor installation is at Messrs. Martin & Warnock's Dominion Roller Mills about a mile and a half from the power house. The mill was formerly driven by a 100 h. p. steam engine, which has been replaced by a motor of similar capacity, and runs day and night all the year round except an occasional stop for repairs. For the better regulation of the voltage on the motor circuits, a separately driven exciter is at present being set up at power house No. 1 in a similar way to the installation at the railway power house.

The arc light station of the Ottawa Electric Co. is known as power house No. 2. The present building was erected about three years ago, and the premises adjoining, which were formerly the arc station, are now used by the company as a workshop. The arc station also is driven by water power, the arrangement of the machinery being similar to that at No. 1 power house, except that rope drives are used instead of belts between the main shafts and countershafts.

Power houses Nos. 4 and 5 are situated close to the arc station and supply current for incandescent light and motor work.

The total equipment of the Ottawa Electric Co. in dynamo

POWER HOUSE NO. 2, OTTAWA ELECTRIC CO.

capacity, including apparatus in course of installation, is as follows :

Arc lights, 2,000 c. p................ 550
Incandescents, 16 c. p.............. 35,000
Motors................................. 600 h. p.

Chief among the many isolated plants worthy of mention in Ottawa and vicinity are those of Mr. J. R. Booth and the E. B. Eddy Co. Mr. Booth has a lighting station built on the very edge of the Chaudiere Falls, and his plant of arc and incandescent dynamos which are used in lighting his mammoth lumber mills would meet the requirement of many a good-sized town. The Eddy Co. have also a large outfit, and among other interesting things in connection with the plant may be mentioned an electric welder for joining iron hoops for their patent butter tubs.

The Bell Telephone Company's business has grown with the city, and they have recently moved into their new four-storey

STEAM POWER HOUSE, OTTAWA ELECTRIC CO.

building on Queen Street. The basement and lower floors are used for storage and office purposes, while the operating room is situated in the top storey. All wires enter the building through underground conduits which extend through the central portion of the city. Aerial cables connect with the underground cables at various places and carry the wires to still further distributing points on the poles. Metallic circuits are used altogether, the Telephone Company having presented "the earth" to the Street Railway.

The Ottawa Porcelain & Carbon Company, which is just com-

## LEGAL.

GREEN V. THE TORONTO RAILWAY COMPANY.—A car of the defendents' electric street railway was moving very quickly along a down grade on a street in a city where the plaintiff, who was in the employment of the city corporation, was engaged in his duty of sweeping the roadbed. The motorman did not sound the gong on the car, as was customary, and ran into the

TWO DYNAMO PANELS, OTTAWA ELECTRIC CO.'s NEW SWITCH-BOARD.

defendant, injuring him. Held by the Court of Chancery, that although the defendant had the right of way, the omission to sound the gong or give any warning of the approach of the car was actionable negligence.

The Brantford Electric & Power Company have applied for an injunction to restrain Messrs. Wood Bros. from using the canal water for their mill on the Power Company's water way.

Mr. Thomas Willson, the discoverer of the method of manufacturing acetylene gas, was in Ottawa recently, for the purpose of securing a Cana-

HIS EXCELLENCY THE EARL OF ABERDEEN'S ELECTRIC LAUNCH.

mencing business, is another institution that owes its existence to the success that has attended the many electrical industries in this city. The company's works are situated close to the Canada Atlantic Railway depot, and with their tall kilns and chimneys present an appearance that is sure to attract the attention of the arriving tourist.

dian patent on his invention. Mr. Willson states that he does not expect the new gas to come largely into use until electricity for the production of the calcium carbide can be cheapened. He further states that when the product is ready for popular use, it will be sold like coal; each house will have an apparatus in which by dropping pieces of carbide into water, gas will be made, which will be conveyed through the house in pipes, and be available for use in the same manner as ordinary gas is now.

# CORRESPONDENCE

## A QUESTION OF PRIORITY OF INVENTION.

CHICAGO, ILL., August 26th, 1895.

Editor CANADIAN ELECTRICAL NEWS.

DEAR SIR:—I read with great interest an article in your issue of August, 1895, page 125, concerning a new method of Mr. Thos. Ahearn, of Ottawa, for preventing variation of E. M. F. occasioned by sudden withdrawal or addition of load in connection with dynamos driven by water power.

The method is very good, but as I pointed out in other journals, Mr. Ahearn does not deserve the credit of being the first designer or inventor of this method.

In my paper : "Practical Notes on the Electrolytic Refining of Copper," read before the American Institute of Electrical Engineers, June 6th, 1892, I say : "A few words as to electric generators may not be out place. The author prefers separately excited machines for the reason that they cannot be reversed and other incidental advantages. When water is used as prime mover a good deal of trouble has been experienced in the regulation of the wheels. As a matter of fact, there is no water governor in existence which will regulate so perfectly as the governor of a modern automatic engine under varying loads. By running all the exciters from an independent prime mover (either water or steam) the strength of the fields of the generators will be uniform at all times whether there are fluctuations in the external circuit or not ; the strength of the field of the generators which, with self-exciting machines, is subject to the fluctuations in the external circuit, and is a variable, becomes a constant. The author proposed this arrangement over three years ago for railway and power stations with the very best results."

It will readily be seen that what you please to term " The Ahearn Method" was suggested and tried by me as early as 1889. At the time I first suggested this method, I discussed with my patent attorney the question, whether it was a patentable invention or not, and he stated most positively that it was not. I simply mention this as a matter of record.

You will greatly oblige me by publishing this communication in the next issue of your paper.

Yours respectfully,

J. B. BADT.

---

GOUVERNEUR, N. Y., AUG. 20th, 1895.

Editor CANADIAN ELECTRICAL NEWS.

SIR :—I beg to call your attention to a note on page 125 of your August issue, in which Mr. Thos. Ahearn, of Ottawa, claims he has devised a method for preventing variation of E. M. F. occasioned by sudden withdrawal or addition of load in connection with self excited water driven dynamos, etc. I installed an Edison three wire system some two years ago, consisting of six dynamos of different sizes driven by water, and connected just as you describe in your article, so that one dynamo (or any one) could excite the fields of all, or each be self-exciting, or the fields of the dynamos on one or other side could be excited by one of themselves.

My switchboard was so connected up that it was almost an impossibility for any one to make a mistake, no matter what switches they operated, as the dynamos, resistance boxes, ampere meters, volt meters and switches were all numbered in the most prominent places. A man standing in the centre of the board could see and operate every switch, field and main, and see every volt meter and ampere meter for the six dynamos.

I trust sir you will see that credit is given where credit is due in this matter.

Yours truly,

J. AUG. FARLINGER.

The following note has been received from Mr. Ahearn, whose attention was directed to the above communications :—

" I am not aware that I have made any claims in the matter of dynamo regulation, but at the same time what has been accomplished here is the result of independent work and experience, and so far as I am concerned I have had no knowledge of Prof. Badt's application in the same direction. I have no desire to enter into a controversy over this matter. My only object in stating what has been accomplished here in the matter of regulating dynamos driven by water power, was to give the benefit of our experience to other water driven stations. I am not aware that any other station is using anything of this kind, and, as stated before, what has been done here has not been borrowed from Prof. Badt or any other person."

## QUALIFICATIONS OF AN ELECTRICAL ENGINEER.

TORONTO, Aug. 26th, 1895.

Editor ELECTRICAL NEWS.

SIR,—In your August issue you published some of Mr. McLean's opinions on the subject of the qualifications of an electrical engineer. Mr. McLean will perhaps allow me to heartily endorse those opinions. It is precisely, because, as he says, " a young man who can handle a coil of wire and perform a few mechanical acts with an electrical plant, thinks himself a master of electrical engineering," that such a quantity of bad work is done in Canada. Any person with a little capital, who has run line wires, or served for a few months in a machine shop, goes manufacturing electrical machinery, and starts into business as an electrical engineer, or "expert," as some of these persons call themselves. The consequence is just what one might expect—dynamos that certainly furnish currents ; and that shake themselves to pieces before long ; armatures that burn out ; instruments that are curiosities, and general wiring that is a derision. I know one "mannfacturer," who, I believe is an electrical, mechanical, mining, hydraulic, and every other kind of engineer, besides architect and general scientist, who makes quite a decent living out of the repairs necessary on machinery of his own make. This speaks highly for his goods, equally highly for the public taste. Another "electrician," when it came to the fine point, couldn't calculate his wires, and these are the men who are responsible for the electricity of Canada. Lord Kelvin has said that and electrical engineer is nine-tenths a mechanical engineer ; he did not say that a working mechanic possessing not even a good education is, by virtue of his trade, an electrician.

Mr. McLean is again right in saying that "much of the criticism levelled against mechanical engineers, who undertake to call themselves electrical experts, is due to the fact that these men are no more mechanical than they are electrical engineers." The public will, perhaps, some day discover to their cost, that the man who shovels coal and wipes up the engine is not a mechanical engineer, and the "expert," who juggles with the rheostat, and runs wires by guess work, is not a fully educated electrical engineer. When it is properly understood that in the States and in Europe, some years of scientific education and practical experience are necessary before a man may call himself an electrical engineer, or can be considered competent to take charge of electrical enterprises, then perhaps we shall find the "electrician" of to-day, relegated to his proper status as a kind of superior day-laborer, who will not be permitted to express an opinion on electrical matters, or to sink the whole science of electricity to the level of a trade.

Electrical knowledge should be disseminated broadcast electrical men should co-operate and combine for the purpose of exchanging experience ; linemen, dynamo tenders, lamp trimmers, should be encouraged to read, and everything done to specialize our profession, and to raise the general tone of men following it. There should be examinations for wiremen, tenders, superintendents and the like, and certificates given, and a sure means to all the above ends is to diligently peruse electrical papers, and to strengthen the Canadian Electrical Association by lending it vigorous help. The more specialized the electrical profession, the fewer quacks as described by Mr. McLean, we shall have in it to disgrace it.

Yours,

GEORGE WHITE-FRASER, E. E.

---

It is reported that the hackmen on the Canadian side of the Niagara river have determined to take legal action against the Niagara Falls Park and River Railway Co., with the object of compelling the company to cease operating their road on Sunday. The true inwardness of this action on the part of the hackmen is apparent on the surface. The electric road has been he means of lessening, to a large extent, the profitable business which Niagara hackmen formerly did. It is to be presumed that the present action has been taken by the men who have only entered the hack business in recent years, as those who were formerly engaged in it, at this point, might reasonably be supposed to have accumulated such vast fortunes as would make them indifferent whether any further business came their way or not.

MR. E. A. CARR,
Manager London Street Railway.

MR. JAMES GUNN,
Superintendent Toronto Street Railway.

MR. ROSS MACKENZIE,
Manager Niagara Falls Park & River Railway.

MR. G. C. CUNNINGHAM,
Manager Montreal Street Railway.

MR. A. J. NELLES,
Manager Hamilton, Grimsby and Beamsville Electric Railway

A GROUP OF CANADIAN ELECTRIC RAILWAY MANAGERS

## THE CANADIAN ELECTRICAL ASSOCIATION CONVENTION.

As the present number of the ELECTRICAL NEWS goes to press, the finishing touches are being put to the arrangements for the annual convention of the Canadian Electrical Association, which will open at Ottawa on the 17th inst. The officers of the Association feel confident that this convention will prove to be the most pleasant and profitable, and in every particular the most successful of any which the Association has yet undertaken. The papers, as in former years, are instructive in character, and cover a variety of topics, some of which have not been touched upon at any of the previous conventions.

The illustrated article appearing on front page of this paper will, it is hoped, prove an object lesson to those members of the Association who may not be familiar with the many interesting electrical features of the Capital City. Certainly there is no more interesting place in Canada, from an electrical point of view.

It may be permissible to emphasize what was said in the ELECTRICAL NEWS for August concerning the completeness of the arrangements which have been made by the local committee for the entertainment of visitors to this convention. Nothing has been left undone which would enhance the enjoyment of the occasion, and it only remains for the members and their lady friends to be on hand to participate in the pleasure and profit of the gathering.

The Canadian Pacific Railway is offering a return fare of $7.00 to the Montreal Exposition, on the 16th and 18th inst., good for return until the 23rd inst., and it is understood that members of the Association resident in the vicinity of Toronto, who may wish to attend the Ottawa convention, can purchase a return ticket for Montreal at the price mentioned, and have it marked "via Ottawa." This will lower the fare by more than $3.00 as compared with the rate previously arranged for by the Association, and should render it unnecessary for any member to deny himself the pleasure of attending the convention on the ground of expense.

Following is the program of the convention :—

### HEADQUARTERS—RUSSELL HOUSE.

### BUSINESS PROGRAM.

#### SEPTEMBER 17TH.

11:00 A. M.    Formal opening of the Convention in the Railway Committee Room of the House of Parliament, when His Worship the Mayor will read an address of welcome.
At the conclusion of the address, members and ladies will be shown through the Senate, the House of Commons and Parliamentary Library.

2:30 P. M.    Opening of First Session at Board of Trade Rooms, Elgin Street.
President's Address.
Reading Minutes of last Meeting.
Secretary-Treasurer's Report.
Reception of Reports of Committees on : Constitution, Statistics, Legislation.
General Business.
Presentation of Papers.
Discussion.

#### SEPTEMBER 18TH.

10:00 A. M.    Consideration of Reports of Committees.
Election of Standing Committees for the ensuing year.
Selection of Place of next Meeting.
Election of Officers and Executive Committee.
General Business.
Presentation of Papers.
Discussion.

#### SEPTEMBER 19TH.

10:00 A. M.    Presentation of Papers.
General Business.

### LIST OF PAPERS.

"Some Notes on the Consolidation of Two Systems of Electric Supply,"
A. A. Dion, Ottawa.
"The Telegraph in Canada,"
Chas. P. Dwight, Toronto.

"Suggested Forms for Electric Light Accounting,"
D. R. Street, Ottawa.

"From the Coal Pile to the Meter,"
Jas. Milne, Toronto.

"Some Modern Alternating Current Apparatus,"
H. T. Hartman, Peterborough.

"Non-Interference Diplex Relay,"
"A Percentage Method for Circuit Measurements,"
D. H. Keeley, Ottawa.
"————————,"
J. J. Wright, Toronto.

### SOCIAL FEATURES.

#### SEPTEMBER 17TH.

8:00 P. M.    Members and ladies will be conveyed by special electric cars to view the Chaudiere Falls, the Lumber Mills, and Electric Power Houses. This is a sight which for novelty and interest can scarcely be duplicated outside of Ottawa.

#### SEPTEMBER 18TH.

8:00 P. M.    Banquet to Members and Ladies at the Russell House.

#### SEPTEMBER 19TH.

Immediately after the adjournment, electric cars will be provided to carry members and ladies over the Street Railway Company's lines out to Rockliffe Park and return.

It is anticipated that arrangements will be consummated for members and ladies to run the water slides on a raft of square timber. His Excellency Lord Aberdeen has placed his electric launch at the disposal of members and ladies.

#### RAILWAY AND HOTEL ARRANGEMENTS.

Arrangements have been made with all railways for a reduced rate of one and one-third fare for members and ladies accompanying them. To obtain this concession, members must purchase a first-class ticket, obtaining from ticket agent a Standard Certificate, which will entitle them to purchase at Ottawa a return ticket at one-third the usual fare. This concession is not obtainable prior to 14th Sept. Special hotel rates have been arranged for as follows ; Russell House, $3.00 per day ; Grand Union, $2.00 ; Windsor, $1.50.

## CANADIAN ASSOCIATION OF STATIONARY ENGINEERS.

Note.—Secretaries of the various Associations are requested to forward to us matter for publication in this Department not later than the 20th of each month.

#### TORONTO ASSOCIATION NO. 1.

AT a regular meeting of Toronto No. 1 C.A.S.E., held Aug. 9th, the following resolutions of condolence were passed :

"Whereas, it has pleased Almighty God to remove from our midst the beloved daughter of our esteemed Brother, Edward Dunn, be it therefore resolved, that we do extend to Bro. Dunn and his family our heartfelt sympathy in this their hour of bereavement, and commend them to our all Wise and Supreme Ruler, who doeth all things well, and be it furthermore resolved, that a copy of the above be placed on our minute book and the mechanical press be furnished with the same.

W. G. BLACKGROVE,
G. FOWLER,    } Committee.
T. EVERSFIELD,

#### DELEGATES TO THE OTTAWA CONVENTION.

The following delegates and alternatives have been appointed to the annual convention at Ottawa, by their respective Associations :—

London No. 5—Bro. F. G. Mitchell ; Bro. R. Simmie, alternate.

Toronto No. 1—Bros. Fox, Huggett and Wickens ; Bros. Lewis, Bain and Eversfield, alternates.

Kincardine No. 12—Bro. Jos. H. Walker.

Hamilton No. 2—Bros. R. Mackie, C. Pettigrew and Wm. Norris.

Peterborough No. 14—Bro. A. C. McCallum.

Kingston No. 10—Bros. Sandford Donnelly, Pres., and Harvey Hoppins, vice-president : Frederick Simmons, alternate.

Brockville No. 15—Bro. W. F. Chapman ; F. B. Andrews, alternate.

Wiarton No. 13—Bro. John F. Cody.

Montreal No. 1—Bros. T. Ryan, J. G. Robertson, Elph. Valiquett.

The secretaries of other associations from whom the names of delegates were requested have failed to respond, for which reason we are unable to publish the complete list.

PUBLISHED ON THE FIFTH OF EVERY MONTH BY

**CHAS. H. MORTIMER,**

OFFICE: CONFEDERATION LIFE BUILDING,

*Corner Yonge and Richmond Streets,*

**TORONTO,        -        CANADA.**

Telephone 2362.

NEW YORK LIFE INSURANCE BUILDING, MONTREAL.

Bell Telephone 2299.

*ADVERTISEMENTS.*

Advertising rates sent promptly on application. Orders for advertising should reach the office of publication not later than the 25th day of the month immediately preceding date of issue. Changes in advertisements will be made whenever desired, without cost to the advertiser, but to insure proper compliance with the instructions of the advertiser, requests for change should reach the office as early as the 22nd day of the month.

*SUBSCRIPTIONS.*

The ELECTRICAL NEWS will be mailed to subscribers in the Dominion, or the United States, post free, for $1.00 per annum, 50 cents for six months. The price of subscription should be remitted by currency, in registered letter, or by postal order payable to C. H. Mortimer. Please do not send cheques on local banks unless 25 cents is added for cost of discount. Money sent in unregistered letters will be at senders' risk. Subscriptions from foreign countries embraced in the General Postal Union, $1.50 per annum. Subscriptions are payable in advance. The paper will be discontinued at expiration of term paid for if so stipulated by the subscriber, but where no such understanding exists, will be continued until instructions to discontinue are received and all arrearages paid.

Subscribers may have the mailing address changed as often as desired. *When ordering change, always give the old as well as the new address.*

The Publisher should be notified of the failure of subscribers to receive their papers promptly and regularly.

*EDITOR'S ANNOUNCEMENTS.*

Correspondence is invited upon all topics legitimately coming within the scope of this Journal.

THE "CANADIAN ELECTRICAL NEWS" HAS BEEN APPOINTED THE OFFICIAL PAPER OF THE CANADIAN ELECTRICAL ASSOCIATION.

## CANADIAN ELECTRICAL ASSOCIATION.

### OFFICERS:

PRESIDENT:

K. J. DUNSTAN, Local Manager Bell Telephone Company, Toronto.

1ST VICE-PRESIDENT:

A. B. SMITH, Inspector Canadian Board Fire Underwriters, Toronto.

2ND VICE-PRESIDENT:

C. BERKELEY POWELL, Manager Ottawa Electric Light Co., Ottawa.

SECRETARY-TREASURER:

C. H. MORTIMER, Publisher ELECTRICAL NEWS, Toronto.

EXECUTIVE COMMITTEE:

L. B. McFARLANE, Bell Telephone Company, Montreal.

GEO. BLACK, G. N. W. Telegraph Co., Hamilton.

T. R. ROSEBRUGH, Lecturer in Electricity, School of Practical Science, Toronto.

E. C. BREITHAUPT, Berlin, Ont.

JOHN YULE, Manager Guelph Gas and Electric Light Company, Guelph, Ont.

D. A. STARR, Electrical Engineer, Montreal.

J. J. WRIGHT, Manager Toronto Electric Light Company.

J. A. KAMMERER, Royal Electric Co., Toronto.

J. W. TAYLOR, Manager Peterboro' Carbon Co., Peterboro'.

O. HIGMAN, Inland Revenue Department, Ottawa.

## MONTREAL ELECTRIC CLUB.

### OFFICERS:

President, W. B. SHAW,      Montreal Electric Co.

Vice-President, H. O. EDWARDS,

Sec'y-Treas., CECIL DOUTRE,      81A St. Famille St,

Committee of Management, T. F. PICKETT, W. GRAHAM, J. A. DUGLASS.

## CANADIAN ASSOCIATION OF STATIONARY ENGINEERS.

### EXECUTIVE BOARD:

President, J. J. YORK,      Board of Trade Bldg, Montreal.

Vice-President, W. G. BLACKGROVE,      Toronto, Ont.

Secretary, JAMES DEVLIN,      Kingston, Ont.

Treasurer, DUNCAN ROBERTSON,      Hamilton, Ont.

Conductor, E. J. PHILIP,      Toronto, Ont.

Door Keeper, J. F. CODY,      Wiarton, Ont.

TORONTO BRANCH No. 1.—Meets 2nd and 4th Friday each month in Room D, Shaftesbury Hall. W. Lewis, President; S. Thompson, Vice-President; T. Eversfield, Recording Secretary, University Crescent.

MONTREAL BRANCH No. 1.—Meets 1st and 3rd Thursday each month, in Engineers' Hall. Craig street. President, John J. York, Board of Trade Building; first vice-president, J. Murphy; second vice-president, W. Ware; secretary, B. A. York; treasurer, Thos. Ryan.

ST. LAURENT BRANCH No. 2.—Meets every Monday evening at 43 Bonsecours street, Montreal. R. Drouin, President; Alfred Latour, Secretary, 306 Delisle street, St. Cunegonde.

BRANDON, MAN., BRANCH No. 1.—Meets 1st and 3rd Friday each month, in City Hall. A. R. Crawford, President; Arthur Fleming, Secretary.

HAMILTON BRANCH No. 2.—Meets 1st and 3rd Friday each month, in Maccabee's Hall. E. C. Johnson, President; W. R. Cornish, Vice-Pres.; Wm. Norris, Corresponding Secretary, 211 Wellington Street North.

STRATFORD BRANCH No. 3.—John Hoy, President; Samuel H. Weir, Secretary.

BRANTFORD BRANCH No. 4.—Meets 2nd and 4th Friday each month. F. Lane, President; T. Pilgrim, Vice-President; Joseph Ogle, Secretary, Brantford Cordage Co.

LONDON BRANCH No. 5.—Meets in Sherwood Hall first Thursday and last Friday in each month. F. Mitchell, President; William Meaden, Secretary, Treasurer, 533 Richmond Street.

GUELPH BRANCH No. 6.—Meets 1st and 3rd Wednesday each month at 7:30, p. m. J. Fordyce, President; J. Tuck, Vice-President; H. T. Flewelling, Rec.-Secretary; J. Gerry, Fin.-Secretary; Treasurer, C. J. Jorden.

OTTAWA BRANCH, No. 7.—Meets 2nd and 4th Tuesday, each month, corner Bank and Sparks streets; Frank Robert, President; F. Merrill, Secretary, 352 Wellington Street.

DRESDEN BRANCH No. 8.—Meets every and week in each month; Thos. Merrill, Secretary.

BERLIN BRANCH No. 9.—Meets 2nd and 4th Saturday each month at 8 p. m. W. J. Rhodes, President; G. Steinmetz, Secretary, Berlin Ont.

KINGSTON BRANCH No. 10.—Meets 1st and 3rd Tuesday in each month in Fraser Hall, King Street, at 8 p. m. President, S. Donnelly; Vice-President, Henry Hopkins; Secretary, J. W. Tandvin.

WINNIPEG BRANCH No. 11.—President, G. M. Hazlett; Recording Secretary, J. Sutherland; Financial Secretary, A. B. Jones.

KINCARDINE BRANCH No. 12.—Meets every Tuesday at 8 o'clock, in the Engineer's Hall, Waterworks. President, Daniel Bentt; Vice-President, Joseph Hall; Secretary, A. Scott.

WIARTON BRANCH No. 13.—President, Wm. Craddock; Rec. Secretary, Ed. Dunham.

PETERBOROUGH BRANCH No. 14.—Meets 2nd and 4th Wednesday in each month. S. Potter, President; C. Robison, Vice-President; W. Sharp, engineer steam laundry, Charlotte Street, Secretary.

BROCKVILLE BRANCH No. 15.—President, W. F. Chapman; Vice-President, A. Franklin; Recording Secretary, Wm. Robinson.

CARLETON PLACE BRANCH No. 16.—President, Jos. McKay; Vice-President, Henry Derrer; Fin. Secretary, A. M. Schofield.

## ONTARIO ASSOCIATION OF STATIONARY ENGINEERS.

### BOARD OF EXAMINERS.

President, A. AMES,      Brantford, Ont.

Vice-President, F. G. MITCHELL      London, Ont.

Registrar, A. E. EDKINS      139 Borden st., Toronto.

Treasurer, R. MACKIE,      28 Napier st., Hamilton.

Solicitor, J. A. McANDREWS,      Toronto.

TORONTO—A. E. Edkins, A. M. Wickens, E. J. Phillips, F. Donaldson.

HAMILTON—P. Stott, R. Mackie, T. Elliott.

BRANTFORD—A. Ames, care Patterson & Sons.

OTTAWA—Thomas Wesley.

KINGSTON—J. Devlin (Chief Engineer Penetentiary), J. Campbell.

LONDON—F. Mitchell.

NIAGARA FALLS—W. Phillips.

Information regarding examinations will be furnished on application to any member of the Board.

WITH the aid of subscriptions received from outside sources, the members of the Montreal Association No. 1, C. A. S. E. have the nucleus of an engineering library. This is directly in line with the avowed object of the Association, which is declared to be first and foremost an organization for the purpose of improving by means of education the standard efficiency of engineers. If the Canadian Association of Stationary Engineers has come to stay, as no doubt it has, the officers of the various branch associations throughout the country could not do better than follow the example of their Montreal brethren in this particular.

THE question of the right of electric companies to operate their road on Sunday is likely to come to a test, and be decided at no distant date. In view of the present condition of the law on the subject, it is desirable that such a decision should be reached at as early a date as possible. It is certainly an anomalous state of affairs that street cars should be allowed to run on Sunday in Hamilton, while in Toronto, forty miles distant, they are prohibited from doing so. The matter is one which should be placed on a more consistent basis. The law should declare either that it is legal and proper that street railway lines should be operated on Sunday, or that it is illegal and improper that they should be so operated ; this decision should apply every. where throughout the Dominion. In answer to a deputation representing the Kingston Electric Railway Co., Mr. Harty, a member of the Ontario Government, made the sensible suggestion that the various street railway companies should get together and formulate and present to the government their views on the question of the operation of electric roads on Sunday, when the government would be in a position to consider the question intelligently and take some action in reference thereto.

THE recent mishap to the conduit pipe of the Toronto Water Works should tend to popularize the electric elevator, resulting as it did in stopping the operation of every hydraulic elevator in the city.

NOTWITHSTANDING the rapid increase in the use of the bicycle the receipts of the Toronto Street Railway Co. continue to grow. The total receipts for the month of August were $90,-285.78, of which the city's percentage amounts to $7,222.86, as against $7,013.30 for the corresponding month of last year. This franchise is proving a very valuable one for both the company and the city.

AT the time of this issue the Toronto Industrial Exhibition is in progress, and the same success which has marked it ever since its inception still attends it. The remark so frequently made that it is the same old thing from year to year is not correct, neither is it reasonable to suppose that it can be very different. Still there is marked progress, and every year shows some advance in every department of industry which the exhibition covers. This year, as usual, a large number of our electrical firms make creditable exhibits, showing that the rapidly growing demand for appliances in this branch of science and industry can be fully met in our own country.

The Calumet Electric Railway Co., which was, we believe, the first to give trolley parties, now so popular, has introduced another new idea in the form of trolley funeral trains. A procession of this kind conveyed a funeral recently from West Pullman to Oakwoods cemetery. The forward part of the first car was appropriately draped, and formed the hearse, and the remainder of the train was occupied by the relatives and friends of the deceased. Mr. Farson, the general manager of the road, anticipates some difficulty in overcoming a not perhaps unnatural prejudice against such an innovation, but as our cities grow, and longer distances have to be traversed to the cemeteries, such a system must commend itself from the standpoint of both economy and convenience. Why, for instance, should not a funeral from Parkdale to the Necropolis proceed by trolley instead of with hearse and cabs? It would be a saving of both time and money.

AN Act came into force in New York State on the 1st of September which will tend to protect those who use the telephone for business purposes as fully as their interests are now guarded by the law relating to the telegraph. The Act makes it punishable by a fine not exceeding $1,000, or imprisonment for not more than six months, for any person to wrongfully obtain or attempt to obtain any knowledge of a telegraphic or telephonic message by connivance with a clerk, operator, messenger or other employee of a telegraph or telephone company who wilfully divulges to any one but the person for whom it was intended the contents or nature of any message or despatch intrusted to him for transmission or delivery. The same penalties attach to refusal or neglect of any employee to duly transmit or deliver such message, or the aiding or abetting of any unlawful business or traffic. Someone suggests that this law will settle the question whether a woman can keep a secret.

VOLT-METERS, or pressure indicators, are very important instruments in the operation of lights, and as they are liable to get out of adjustment from various causes, it would be well to have them frequently compared with a standard. A very slight cause may result in throwing a volt-meter out of adjustment. The writer knows of an instance where a small spider, having got into the case in some mysterious manner, left a thread across the space within which the needle swings, which made a difference of about 1 volt in the reading. A little dust will get in and clog the pivot; and more especially, the cheap instruments which are supplied with small plants, although quite good enough to run by, have not been, as a rule, carefully made, or calibrated, and so are quite seriously faulty sometimes. It is the same thing as with a watch. A Waterbury $2.50 will probably keep excellent time for several months, and then simply wear

out, because the material, although worth the $2.50, was cheap. Keep your plant up to the mark; spend a little money on it, and the popularity of the light will repay the cost. Save a $5 bill in small ways, by neglecting to keep it up, and your customers will become dissatisfied and put your lights away.

A SUCCESSFUL test was made a few days ago of a new engine, built at the Grand Trunk shops in Montreal, embodying improvements which give greatly increased power with economy of fuel. There are two cylinders, one of 19 inches and the other of 29 inches, so arranged that the steam is used in the larger one after it has done duty in the smaller. The steam is thus used twice, and in such a way that very little of its force is lost. In the test the new engine drew forty-six loaded cars from Montreal to Brockville, a distance of 125 miles, with a consumption of 4½ tons of coal, an average of 1.66 pounds of coal per car per mile, on an up grade, which is considered by the railway officials a remarkable achievement. An ordinary engine hauling half the number of cars the same distance usually consumes at least five tons of coal. The engine, on the return trip, took fifty-six cars. On its second trip the engine drew 41 and 50 cars respectively. A number of prominent officials accompanied the engine on the trial trip. This compound principle, as applied to locomotives, is covered by patents secured through Sidney Stevens, of Brockville, travelling representative of the Rhode Island Locomotive Works, of Providence, R. I. It seems to promise a revolution in the construction of railway engines.

IT is of great importance to electric lighting stations that the pressure all over the system should be sensibly the same, and equal at all loads. It is, of course, practically impossible to so arrange the system of primary wires and transformers if alternating currents be used, and the mains and feeders if direct currents be used, as that there shall be exactly the same electrical pressure on the lamps, at every distributing point ; but a careful laying out of such a system can always result in obtaining a very small, and comparatively negligible difference. The wiring of a town or building is really a very potent factor in the future operation of the lights, and is deserving of quite as careful calculation as any other feature ; in fact, possibly more money is wasted, and more serious faults made, in the figuring of the mains, feeders, etc., than in almost any other way. This seems a rather bold statement, but a little consideration will shew its force. Consider a very simple case. Lamps placed around the circumference of a circle, with the generator at the center, and a main primary leading from the generator to two points on the circle, at opposite ends of a diameter. The lamps are 110 volts, and the generator gives 125 volts, which allows a drop of 15 volts between generator and lamps. In this case, as the generator is equi-distant from the center of distribution, the mains in every direction may be the same size, to give the same resultant pressure at the lamps; but taking another case, which more closely represents conditions practically obtaining, shews the necessity for careful calculation. Take a square with the corners A, B, C, D; the generator situated at A, and the lamps in bunches at B, C and D. Assume the same number of lamps in each bunch. Then the distance from A to C is almost half as long again as A to B, or A to D. Everybody knows that the size of wire to carry a certain current, with a certain initial pressure, has to be greater as the distance increases, and less as the distance diminishes ; so that the wire between A and C should be larger than the wire between A and B, or A and D. Now, we have above 15 volts that we can drop between the generator and the lamps. Suppose we find that number o wire will carry all the current required, and drop 15 volts over the distance A B ; then it is evident that it will drop more than 15 volts between A and C, so that if we use o wire between A and C, our lamps at C will not get 110 volts, so that they will not be burning at proper candle power, and customers will kick. On the other hand if number o wire suffices to drop 15 volts between A and C, then it will not drop so much between A B or A D, and if we use No. o here, the lamps will have more than 110 volts—will burn quite brilliantly, but will burn out before their time, resulting in another kick. It is quite usual to run a pair of mains direct from the generator, along all the streets perhaps to a distance of three-fourths of a mile from the power house, and to tap transformers on to this wherever lights are required, without any reference whatever to centers of distribution, or the many different pressures resulting at the lamps. Let us consider what this means. Take a line A, F, and divide it into equal parts at the points B, C, D, E; let each of these parts be 200 feet long. The generator, giving 125 volts is at A, and at each of the points B, C, D, E, F is situated a bunch of lamps, requiring say 10 amperes altogether, at 110 volts. So that the generator has to supply altogether 50 amperes. The usual mode of running this wire is as above described, to carry one size all the way from A to F, tapping off at intervals; and if any calculations at all are made, which is doubtful, they are made thus—What size of wire will carry 50 amperes over a distance of 1,000 feet (A to F) dropping 15 volts? The proper size on this basis is 35,000 circ. mils., or nearly No. 5 B & S gauge. Now, if this wire be run the results will be as follows—remembering that the wire has to carry 50 amperes between A, B; 40 between B, C; 30 between C, D; 20 between D, E, and 10 between E, F. At any moment, when all the lamps are burning and the generator voltage is the proper 125, (direct current is assumed, but the theory is just the same for alternating at a higher voltage), the pressure at the lamps at F will be 110, as it should; at E it will be 113; at D it will be 116; at C it will be 119, at B 122 volts. Thus the lamps everywhere but at F, will be burning at far too high pressure, and will burn out in a very short time. Some authorities state that every one per cent. too high voltage reduces the life of the lamp 15 per cent. The above is certainly a rather exceptional case, but in a modified form actually occurred within the writer's experience. Suppose that the wire had been calculated to carry the 50 amperes as far as D, dropping the 15 volts; then the lamps at D would have 110 volts; those at E and F would have too little; while those at B, C would have too much. E would have 108; F, 106; C, 112 and B, 114 volts. Here we have some not up to candle power, and the others burning out. There has been enough shewn to prove that the careful calculation of the wiring of a system is of the highest importance, and that slap-dash methods of guessing at wires are almost sure to be disastrous. It is too usual to conclude that No. 16 will be the proper size for inside wiring, irrespective of the outside pressure. A case came recently under the notice of the writer, where the contractor, out of the honest intention to give good square work for his money, increase d the size of the wires all around, thinking that nobody could complain if he put up No. 6 where he had contracted for No. 8. His honesty, not to say his beautiful ignorance, resulted in burning out lamps in all directions. The fact is that hundreds of dollars are wasted by guessing at wires, and hundreds of lamps are burned out by excessive pressures. In quite a small village the wiring system is as capable of scientific calculation as in a large town; and experience and care can save money in this way as in every other. It would certainly pay electrical men to have their systems overhauled and put in order.

IN view of the approaching convention of the Canadian Electrical Association at Ottawa, and of the American Street Railway Association at Montreal, the effort has been made to make the present number of the ELECTRICAL NEWS one of special character. A largely increased edition has been printed in order that copies may be placed in the hands of every delegate to these conventions, as well as of manufacturing and other concerns who should be interested in acquainting themselves of the progress which is taking place in the applications of steam and electricity. It is hoped that the readers of this number, whether casual or regular, will find the contents of both reading and advertisement pages of more than usual interest. If you are not a subscriber, remember that for only one dollar per year you may enjoy that privilege. If you are not an advertiser, it will pay you to become one.

### NOTES ON AN INTERESTING STEAM PLANT AT MONTREAL.

A representative of the ELECTRICAL NEWS recently called on the superintendent and chief engineer of the, Board of Trade building, Montreal, Mr. J. J York, and it was with much pleasure he saw the neatness and brightness in the power and boiler rooms.

In the pumping rooms are two Davidson steam pumps, 24 × 14 × 24 which run the three passenger elevators. These elevators carry 4,000 passengers daily during the summer months. The pumps pump the water over and over again and only 60 gallons per week are wasted. Two feed pumps (Davidson) are in use and one Otis feed water heater, which receives water from the city mains and delivers it to the boilers at 210° Fr. The only circulating hot water supply system used in the city is used here. It is so arranged that hot water may be had from any of the 170 hot water taps at once, while in other buildings you have to wait till the water arises from the basement. The temperature in the hot water tank is regulated by Power's temperature regulator.

The building is lighted by a private electric plant, supplying light to 1400 lights. The system is divided into ten sections, each connected to one switch board, and so arranged that, in case of an accident, light can be had by connecting to the Royal Electric Co.'s wires. The building is wired for 2% loss at 52 volts, and all wires are rubber covered cable laid in armored conduit.

In the power room are three engines and three generators. The engines are Robb-Armstrong, Class A. 10½ × 12 × 260. The generators are Edison compound wound 30 k. w. A new switch board has just been built in the power room, having Weston volt metres and Edison ametres. New brass railings have just been put around the engines and generators giving them a nice appearance. A 10 h. p. Sprague motor drives a fan which supplies air to the boilers, where hard coal screenings are used for fuel. The motor is controlled by a damper regulator connected with the starting box, so that when steam is up and no air is required the motor and fan will be at rest, and when steam goes down, needing more air for the fires, a weight slowly descends automatically, at the same time pulling a cord attached to the handle of the starting box, which sets the motors and fan again in motion till the required steam is up. This automatic invention does not require any attention from the fireman. This invention originated in the fertile brain of Mr. York and is one he may well feel proud of.

In the boiler room are three boilers, 18 feet long by 5½ feet diameter. They are multitubar and have an average pressure of 90 lbs. Mr. York, after experiments with mixtures of hard and soft coal screenings, has decided to stick to hard coal screenings.

In the kitchen of the restaurant next to the power room is a 2½ h. p. Ball motor which drives a 30″ Blackman fan which can change the atmosphere of the kitchen three times a minute. The kitchen is the best ventilated in the city. The power of the fan was clearly demonstrated to the reporter by Mr. York. He closed all the windows and doors with the exception of one, and then speeded the motor to its utmost capacity. The air was drawn so quickly out of the kitchen that the reporter found it almost impossible to close the open door against the rush of air from outside.

Mr. York is experimenting on an electric heater for a tea-broker of the city. He is making it of German silver insulated with mica and is pleased with results. He says he can heat two quarts of water in ten minutes and not burn out the heater. Mr. York, as President of the C. A. S. E. goes to Ottawa to the convention this month.

### PERSONAL.

Dr. McMaster has been appointed principal of the Toronto Technical School Board, at a salary of $700 per year.

Mr. W. F. Clockenberg, electrician, of Niagara-on-the-Lake, was married recently at Toronto, to Miss Sherrin.

Mr. J. J. Wright, manager of the Toronto Electric Light Co. has also assumed the management of the Hamilton Electric Light & Power Co.

Mr. E. B. Merrill, late principal of the Toronto Technical School, has entered the employ of the Westingbouse Electric Manufacturing Co., Pittsburgh, Pa.

Mr. E. L. Barr, brother to Mr. M. D. Barr, formerly manager of the Canadian Edison Co., has recently been appointed secretary of the Wallace Electric Co., of Chicago.

Messrs. John W. Mackay, president of the Postal Telegraph Co., and C. R. Hosmer, superintendent of the Canadian Pacific Telegraph Co., have recently returned from a trip to Alaska.

At the recent convention of the Association for the Advancement of Science, held at Springfield, Mass., Prof. Galbraith, principal of the School of Practical Science, Toronto, was elected secretary of the Mechanical Science and Engineering section.

## LOCATION OF GROUNDS IN ARMATURES, FIELDS, ETC.*

### By CLARENCE E. GIFFORD.

IF the work can be performed in a very quiet room, two or three cells of battery, a telephone receiver and connecting wires, comprise the necessary apparatus. In some cases two "table binding posts" and a foot or two of No. 18 or No. 20 bright iron wire will be a convenient addition. Where noise will not permit of the use of a telephone, a dead-beat reflecting galvanometer, a milli-voltmeter, or some other form of delicate and rapid working visual indicator, must be used instead. If an armature is to be tested without removing it from the machine, connection with the battery may be made through the brushes, first making certain that the short-circuiting switch is open, if dealing with an arc machine. The points of connection with the battery need not be diametrically opposite, and may be made by the wires being firmly pressed against the commutator by an assistant, if more convenient.

Good electrical contact between metallic surfaces can better be secured by cleaning the same thoroughly with kerosene, which removes foreign matter, and is so fluid that it will in no way interfere with perfect contact, when moderate pressure is applied. Especially when making measurement of resistance of armature sections, it is even advisable to have the surface of the commutator quite wet with kerosene during the operation, as this avoids trouble from grease or dirt which might get on the surface from handling, subsequent to cleaning, and it also prevents the contact points becoming oxidized by any sparks which may occur at the moment of breaking contact. True, the oil is an insulator, but we use it in this case as a detergent simply.

Connection being made between battery and commutator, first determine whether the armature circuit is complete throughout. If the circuit is complete, a click will be heard in the telephone when the two terminals of the same are brought in contact with any two contiguous bars of the commutator, or when contact is broken. If an open circuit exists on either side of the circuit, of course no sound will be heard in the telephone when used on that side, except when connection is made or broken by it between the bars lying on opposite sides of the break. See Fig. 1.

Close any open circuit temporarily by bridging between the two bars with a drop of solder. Two or more breaks can evidently be located by suitably shifting the battery contacts and searching as before. Open circuits will, of course, when an armature continues in work, soon cause burns between the bars

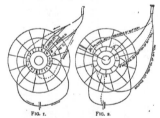

FIG. 1.      FIG. 2.

that will indicate unmistakably their location. Having closed any open circuits, and the battery being connected to two points of the commutator, approximately opposite each other, one terminal of the telephone is connected to the armature shaft, or frame of the machine, and the other terminal is drawn completely around over the surface of the commutator, while the telephone is held to the ear. If only one ground exists, two balancing points, or points giving the least noise in the telephone, will be found.

In an armature of ordinary construction, one of the points so found will be on the bar nearest the real ground, while the other balancing point bears what might be termed a "bridge relation" to the first, being at practically the same potential; the armature itself forming in reality a veritable Wheatstone bridge.

Now, shift the points of battery contact a few bars either way,

* Abstract of a paper read before the American Institute of Electrical Engineers, Niagara Falls, June 26-30, 1895.

and the true ground, if but one exists, will be indicated in precisely the same position as before, while the other balancing point will shift every time the battery contacts are shifted. See Fig. 2.

If two grounds exist, two balancing points will be found, as before, but both points will shift more or less when the battery contacts are shifted, provided the grounds lie on opposite sides of the same battery contact.

In the case of one ground, having determined its location approximately, fix it as closely as may be by making and breaking contact with the telephone terminal on each of the more quiet bars, separately, until by comparison, the two giving the faintest clicks are determined. If your hearing has served you correctly these two bars lie nearest the trouble, the fainter one being the nearer. Prove the non-existence of a second ground by placing

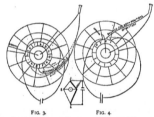

FIG. 3.      FIG. 4.

one of the battery contacts on the first bar to the right of the apparently permanent balancing point just found, and then on the first bar to the left of said point, the other contact being nearly diametrically opposite. This balancing point should still remain unchanged if no other ground exists.

The next step is to connect the battery to these two bars just fixed upon as lying nearest the trouble. The armature still forms a "bridge," the portion included between the two contiguous bars to which the battery is now connected forming the one side, and the remainder of the armature, the other side. See Fig. 3.

One of the telephone terminals is now connected to the shaft as before, and the other terminal again drawn around the commutator. If the balancing point is found, say one-sixth to one-half, the long way round from one battery contact to the other (these contacts being on two contiguous bars), the trouble lies in the coil between those two bars, and the point of trouble divides the coil in the same ratio as the balancing point divides the remainder of the armature, the ground and the balancing point being respectively nearest the same battery contact. If the balancing point falls on the same bar as one of the battery contacts, the ground is located on that bar or on the lead between it and the armature, provided the balancing point is found to be upon the same bar when the battery contacts are both shifted one bar to the right or the left of their original position. See Fig. 4.

If the balancing point appears to be found within three or four bars from one of the contacts, the precaution should be taken to test its correctness by moving both battery contacts one bar toward the balancing point. If the trouble was between the battery contacts when in their previous position, this shifting of the contacts will now throw the balancing point clear around on to the contact which was, in the previous position, farthest away from the balancing point. If, on the contrary, the balancing point remains unmoved by this shifting of the battery contacts, it shows that this balancing point is the point nearest the real ground, and that the ear was deceived in its first supposed approximation, which, with due care, however, is not likely to occur.

If such error had been made, the new point, as indicated, together with first the bar on one side of it, and then on the other, must be tried as points of battery contact; or much better, make a new start with the contacts nearly at opposite sides of the commutator and proceed as before. A single 20,000 ohm ground on a one ohm armature should be located accurately in not over three minutes, in a quiet room. High resistance grounds require more battery and more care. Armatures of very low resistance also offer greater difficulty.

Where two grounds are found to exist, as indicated by the change of location of both balancing points, under the conditions before stated; when the battery contacts are shifted, the following mode of procedure will answer the purpose well, and is simple. Fix the battery contacts at any two points of the commutator nearly opposite each other, preferably at points to be determined by trial, that will cause the balancing points to fall nearly diametrically opposite to each other, and determine and mark the two balancing points, as then shown. Now place the battery contacts on the balancing points just found. If only one ground exists, the two balancing points and one battery contact will all be coincident in one point. If two grounds exist, both balancing points will be shifted from their former position. Open the armature circuit by unsoldering one of the ends of a coil connecting with the lead of the bar that is marked in the first part of this test, as one of the balancing points. Place one of the battery contacts on the armature shaft, and the other on the marked balancing point that is farthest from the point where the circuit has been opened. Next place one telephone terminal on the first bar to the right of the opened wire, and draw the other terminal from the same point, toward the right, over the surface of the commutator. The telephone will be absolutely silent until the moving terminal has just passed the ground nearest to it, and strikes the first bar beyond the same, when it will click. This ground lies in the coil between this first bar giving a click and the one passed just previously, or else in the said previous bar.

The other ground is obviously to be located in a similar manner, by placing one telephone terminal on the bar just to the left of the open wire, and from that point searching toward the left with the other terminal. Only in cases where one ground is of very low, and one of very high resistance, will any difficulty be experienced in locating both accurately before either is removed.

The coils thus indicated may have their terminals unsoldered, when it can be readily ascertained with each, whether the ground be in the coil or in the bar just preceding it.

If scientifically inclined, or if otherwise preferable, the circuit may be opened at a point somewhere midway between the two indicated coils instead of disconnecting those coils, and the exact location of each ground determined as follows : Take a piece of "broom wire" about eighteen inches long, new and clean, screw the ends firmly into two clean, brass table binding posts, and into the other holes of the same posts screw the battery terminals. Have an assistant press the corners of the bases of the binding posts into very firm contact with the two bars that lie at the ends of the indicated coil, observing the directions previously given for securing clean contact. Place one telephone terminal in contact with the shaft, and with the other find the balancing point on the wire. This point will indicate the relative position of the ground in the coil, or commutator bar, as the case may be. If more than two grounds were suspected, the two lying the farthest apart would be approximately located by the first part of the two-ground process, and if these coils were not disconnected before proceeding farther, it would be well to make two openings in the circuit, close to and lying between these outer grounds ; then locate definitely these two extreme grounds, and proceed with the remaining section somewhat as with a complete armature, except that you would commence by connecting the battery to the terminals of this section, and would then bridge the telephone from the shaft to the different portions of the section, and would complete the process by applying the remainder of the two-ground test.

In dealing with a cross connected Gramme ring, an obvious change would be made in the points of application of the battery ; and as many points of apparent trouble would be indicated as there were series of cross-connections.

After location of these points it would be necessary to use the auxiliary wire loop, as before described, between these points, to determine which is nearest the trouble. This fact being determined, it would in case of a single ground (indicated by the permanency of the balancing point) become necessary to remove the cross-connections from two bars before proceeding further.

The auxiliary wire loop would properly be used to complete the process.

The ordinary "closed coil" ring or drum armatures are types to which these methods are directly applicable.

The sections of open coil armatures would receive the same treatment as field coils.

Whenever necessary to deal with wet grounds in testing, it is better to make at least four tests, reversing the battery after each test, and taking the mean of the four determinations.

Field coils, also any wires of uniform cross-section, the extremities of which are accessible, and within a reasonable distance of each other, can of course be easily tested for grounds by soldering or firmly clamping a bare wire of suitable size between the extremities of the conductor to be tested, applying a battery to the junctions, and bridging with a telephone between the bare wire and the object upon which the conductor is grounded. This will give only the location of a single ground, or the "resultant" of two grounds. A "T.-H." rheostat should have the battery connected to the two extremities, and the point of apparent ground determined by bridging with a telephone between the frame and the several contact plates. Then apply the battery to the frame and point of apparent ground, connect one terminal of the telephone with each extremity successively, and search from it toward the center with the other terminal, as in the case of searching for two grounds in an armature.

In determining the location of grounds that are of very low resistance, a good induction coil similar to that used in the Blake transmitter may be used with advantage in connection with the telephone receiver. The receiver is placed in circuit with the secondary of the coil, and the "bridging" is done with the primary. With high resistance grounds the best results are obtained by using the receiver only.

## MOONLIGHT SCHEDULE FOR SEPTEMBER.

| Day of Month. | Light. | Extinguish. | No. of Hours. |
|---|---|---|---|
| | H.M. | H.M. | H.M. |
| 1...... | A.M. 12.40 | A. M. 4.30 | 3.50 |
| 2...... | " 1.40 | " 4.30 | 2.50 |
| 3...... | " 2.30 | " 4.30 | 2.00 |
| 4...... | No light. | No light. | .... |
| 5...... | No light. | No light. | .... |
| 6...... | No light. | No light. | .... |
| 7...... | P. M. 6.40 | P. M. 8.40 | 2.00 |
| 8...... | " 6.40 | " 9.20 | 2.40 |
| 9...... | " 6.40 | " 9.40 | 3.00 |
| 10...... | " 6.40 | " 10.20 | 3.40 |
| 11...... | " 6.40 | " 11.00 | 4.20 |
| 12...... | " 6.40 | " 12.00 | 5.20 |
| 13...... | " 6.30 | A. M. 1.00 | 6.30 |
| 14...... | " 6.30 | " 1.10 | 6.40 |
| 15...... | " 6.30 | " 2.20 | 7.50 |
| 16...... | " 6.30 | " 3.40 | 9.10 |
| 17...... | " 6.20 | " 4.50 | 10.30 |
| 18...... | " 6.20 | " 4.50 | 10.30 |
| 19...... | " 6.20 | " 4.50 | 10.30 |
| 20 ..... | " 6.20 | " 4.50 | 10.30 |
| 21...... | " 6.20 | " 4.50 | 10.30 |
| 22...... | " 6.20 | " 4.50 | 10.30 |
| 23...... | " 6.20 | " 4.50 | 10.30 |
| 24...... | " 8.00 | " 5.00 | 9.00 |
| 25...... | " 9.10 | " 5.00 | 7.50 |
| 26...... | " 10.30 | " 5.00 | 6.30 |
| 27...... | " 11.00 | " 5.00 | 6.00 |
| 28...... | " 11.30 | " 5.00 | 5.30 |
| 29...... | .......... | " 5.00 | } 4.20 |
| 30...... | A.M. 12.40 | .......... | |

Total, 172.30

Since the publication of the article relating to the Sault Ste. Marie Canal, which appeared in our August issue, we have been informed that half the motors required for operating the canal gates were supplied by Messrs. Ahearn & Soper, of Ottawa.

Referring to the article published in the ELECTRICAL NEWS for August, of the electrical apparatus employed in connection with the Sault Ste. Marie Canal, a communication has been received from Mr. J. B. Spence, Government Engineer, stating that he has visited the work since the publication of the article referred to, and by trial has found the valves and gates to open and close (with all caution) in from 45 to 50 seconds. The lock filled in barely 7 minutes, and discharged in barely 5 minutes, being in less time than the estimate made in Mr. Spence's report.

## AN INGENIOUS ELECTRICAL DEVICE.

A NOVEL and interesting application of electric power has recently been made in Ottawa, Ont., in an elevator for boats, constructed by the Ottawa Canoe Club. This club has its headquarters in a prettily designed and well appointed boat-house, built on a ledge of rock and against a high bluff well known as "Rockcliffe," situated on the Ottawa River, a short distance from the city and easily reached by the electric railway.

The Ottawa River rises every spring to a height of 10 to 15 feet above the average summer level, consequently the club house had to be perched quite high up the hill. At this time the lower sills lie some eighteen feet above the water. Under these conditions the transfer of the skiffs and canoes between the boat-house and the water was a difficult, slow and laborious process, that offset to a large extent the many attractions of the place.

This season the Executive decided to overcome this difficulty by the installation of an elevator or "lift," operated by electricity, and Mr. A. A. Dion, of the Ottawa Electric Co., who is the Honorary Treasurer of the club, designed and had constructed under his immediate supervision the apparatus which is now used. Its operation is perfect. The boats are handled quickly and safely, and are less strained than they were when carried by hand. In fact, the lift is a complete success.

The apparatus consists mainly of a framed gangway 10 feet wide by 32 feet long, a skeleton car, an ordinary worm geared hoisting drum, and a 3 K. W. 500 volt Edison motor. The gangway, which is made of two heavy timbers, with lateral ties and braces, forms an inclined plane extending from the lower platform of the boat-house to the water, into which it dips at an angle of about 45 degrees. On the inner face of the

ELECTRIC ELEVATOR FOR BOATS.

side timbers of this gangway, rails of 2 inch angle iron are fastened. The car which runs upon this track consists of an oblong bed frame, from two sides of which arms made of 2 inch by ½ inch iron, looped, extend upwards at an angle of about 90 degrees from the bed-frame. Slings made of 3 inch rubber belting are suspended between the top corners of the bed-frame and the tops of the iron arms, forming two flexible, elastic supports about 9 feet apart, intended to receive the boat. This carriage is mounted on four independent 4 inch cast iron wheels with one inch flanges. There are also two wheels running under the rails to prevent the car leaving the track.

The carriage is attached to the drum by a ¾ inch steel cable running over sheaves to the back of the building, where the hoisting apparatus is placed under the floor. The motor is belted to a countershaft, that is in turn belted to the shaft carrying the worm.

When a boat is to be raised or lowered, the motor is first started on a loose pulley. The drum is started to raise or lower the car by means of a belt shifting lever placed on the edge of the boat-house platform. The car stops automatically at the top and bottom of its allotted course, the lower stop being adjustable to suit the water level.

When the car is at the bottom, its highest part stands about one foot under water, so that a boat may be floated on and off.

When at the top its highest part stands over and inside the edge of the platform, so that a boat may be lifted on and off easily. It will be understood that the boats are carried side on, or at right angles to the line of traction. The speed of the car is about 75 feet per minute. The construction and operation of the apparatus will be more clearly understood from the accompanying photograph, for which the writer is indebted to Mr. J. A. D. Halbrook, a talented amateur photographer and a member of the club.

The automatic stop device is an ingenious modification of the well known arrangement of a nut moving along a screw extending from the drum shaft, and is very positive in its action.

All the machine work was done by Mr. Geo. Low, and the carpenter work by Mr. A. Sparks.

## FIRING STEAM BOILERS.

IF an engineer must hire the fireman, let him look first for a sober man; next see that he is neat, careful and reliable; next ascertain if he wants to learn something new each day. If the man is a "know-it-all" it will not do to take him into the fire room. No matter what his other qualifications may be, he will not prove a financial success. His introduction to the coal pile will mean a considerable hole in the owner's pocket book.

The new fireman, if he understands his business, and especially if he has a new boiler, will start a slow fire. He will be easy on that boiler for a day or two; he will start the fire with wood, if possible, as that fuel can be regulated closer than any other form.

For a medium sized boiler, say 5x16 feet, he will be very lazy in getting up steam the first day. Probably three or four hours will be consumed in getting up the pressure. While this is being done he will have a good look at every seam and every rivet that is within his reach. He will take pains to let the air out of the boiler as soon as the pressure begins to start. This is easily done by leaving a gauge cock or two open, or by raising the safety valve if the lever variety is used.

After the new boiler has been gradually worked up to a pressure, he will let it stand an hour or two, then open the blow-off at surface, and give a chance for all the oil and light dirt to run out. After this the boiler may be put to work in earnest, and if the above directions be followed he will have very little trouble from leaky seams or tubes.—Tradesman.

## BRITISH STREET RAILWAY STATISTICS.

An English paper gives the following statistics of the street railways of Great Britain: Altogether 37 civic authorities and 110 companies own tramways in the United Kingdom. On these tramways 30,528 horses, 564 locomotives, and 4,179 cars are used. The number of men employed is about 20,000. The total number of passengers carried during the year 1894 was 616,872,830. The gross receipts were £3,615,837, and the net profit £758,781, giving a return of 5¼ per cent., or an increase of 1 per cent. on the previous year.

The Condensed Mill Co., Truro, N. S., are putting in a 30 horse-power Robb-Armstrong engine.

## CANADIAN ELECTRICAL ASSOCIATION.

WE print by request of the Executive Committee the following copy of the constitution of the above Association, as recently revised by the Committee on Constitution, in order that members of the Association may be in a position to discuss intelligently its provisions at the approaching convention in Ottawa :

### ARTICLE I.

NAME.—This organization shall be known as the Canadian Electrical Association.

### ARTICLE II.

OBJECT.—The object of this Association shall be to foster and encourage the science of electricity and to promote the interests of those engaged in any electrical enterprise and for discussion and interchange of opinions among its members.

### ARTICLE III.

MEMBERSHIP.—The Association shall consist of active, associate and honorary members. The term Active Members includes all members actually engaged in electrical business. The term Associate includes those interested or actively engaged in any electrical pursuit, and they shall be entitled to attend all meetings of the Association, except those of the Executive, and take part in all discussions, but shall not be entitled to vote or be eligible for office. Honorary members shall be elected by a two-thirds vote of the Association.

### ARTICLE IV.

OFFICERS.—The officers shall consist of a President, 1st and 2nd Vice-Presidents, Secretary and Treasurer, and an Executive Committee, consisting of ten members, five of whom shall act on the Committee for two consecutive years. The President and Vice-Presidents shall be ex-officio members of the Committee. Five shall form a quorum. The office of Secretary and Treasurer may be held by one person.

### ARTICLE V.

FEES.—The annual fee shall be for active members $3.00, associate members $2.00, payable in advance.

### ARTICLE VI.

ELECTION OF OFFICERS.—All officers shall be elected by ballot at a general meeting of the Association. The ballot shall be taken in the following manner :—The Secretary shall read the list of active members alphabetically, and each member shall deposit with the Secretary a slip of paper on which he has recorded his vote, the Secretary checking off his name on the list of voters. Two scrutineers named by the Chairman shall assist the Secretary in counting the votes, and the Chairman shall declare elected the person receiving the majority of the votes cast. In case no one candidate receives such majority on first ballot, another ballot is to be taken, and so on until a clear majority is given in favor of some one candidate. Officers shall hold office until the close of the session, at which their successors are elected, such successors to be elected on the second day of the first general session after the expiry of ten months from day of previous election.

### ARTICLE VII.

ELECTION OF EXECUTIVE COMMITTEE.—Members of the Executive Committee shall be elected by ballot in the following manner, the vote being taken immediately after the election of officers :—Ballot papers containing the names of the ten members of the Executive Committee, five of whom must be re-elected, shall be given the members. The Secretary shall read a list of those entitled to vote, and members, having first marked a cross opposite the names of the five persons selected for re-election, shall deposit the ballots with the Secretary, who, assisted by the two scrutineers named by the Chairman, shall count the vote, and the Chairman shall declare elected the five persons receiving the greatest number of votes. Members shall then proceed to elect the five other members of the Executive, the election being by ballot and the Secretary reading the names as before. Each active member of the Association shall have the right to vote for an active member of the Association, including the retiring members of the Executive, and the vote being counted in the usual way, the Chairman shall declare elected the five persons receiving the greatest number of votes.

### ARTICLE VIII.

PLACE OF MEETING.—Place of next meeting shall be decided by ballot, taken in same manner as laid down for election of officers.

### ARTICLE IX.

VACANCIES IN OFFICE.—Vacancies in office, caused by death or resignation, shall be filled by the Executive Committee to cover the term until the next general meeting of the Association, at which the officers are elected.

### ARTICLE X.

NOTICE OF MOTION.—Permission to introduce any notice of amendment or amendments to this constitution must be granted by a majority of two-thirds of the active members present. Permission being granted, notice may be given and the proposed amendment moved at any subsequent sitting. After discussion the amendment must be submitted to a Committee of five, named by the Chairman. The report of said Committee cannot be considered on the same day on which it is introduced. A two-thirds vote of all active members present shall be necessary for its adoption.

### ARTICLE XI.

Notice of substantive motions is required, and no motion shall be discussed at the sitting at which the notice has been given, but this rule does not apply to merely formal motions, such as motions to adjourn. All reports of standing Committees are to be discussed at a sitting subsequent to the one at which such reports have been received. This rule may be suspended by a vote of two-thirds of the members present.

### ARTICLE XII.

All motions must be duly proposed and seconded, and shall, except those of a purely routine character, be in writing.

### ARTICLE XIII.

No member shall speak more than once, or at a greater length than five minutes, upon any question until all others have had an opportunity of doing so, nor more than twice on any one question without permission of the Chairman, or a majority of the members entitled to vote. The mover of a substantive motion has the additional right to reply.

### ARTICLE XIV.

Questions may be re-considered upon a motion to re-consider being made by a person who voted with the majority, provided such motion is carried unanimously. No discussion of the said question is allowed until the motion for re-consideration has been carried.

### ARTICLE XV.

VOTING.—Every active member present must vote, but any person entering the room after the question has been put by the Chairman may not vote. The Chairman shall not vote except in the case of a tie. Voting by proxy shall not be allowed.

### ARTICLE XVI.

Except where vote is by ballot the chairman will take the sense of the meeting by voice, or by asking members to stand, but on call of five members the Secretary shall read the list of persons entitled to vote, and record the yeas and nays.

### ARTICLE XVII.

An appeal may be taken without debate against the ruling of the chair, a vote of two-thirds being required to reverse the decision.

### ARTICLE XVIII.

The President shall nominate a Committee of three to strike the Standing Committees for the following year and define their respective duties, the report of the Committee being considered at a subsequent sitting to its introduction. The number of Standing Committees must be decided by the Association.

### ARTICLE XIX.

The first person named on any Committee shall act as Chairman until Committee is called together, when they will elect their own Chairman, but the President, in his absence the 1st or 2nd Vice-President, shall be Chairman of the Executive Com. mittee. In the event of the absence of ex-officio members, the Executive Committee shall proceed to elect a Chairman pro tem.

The general order of business at all sessions shall be as follows :

Reading Minutes of last meeting.
Report of Secretary-Treasurer.
Report of Standing Committees.
Election of Standing Committees for following year.
Selection of place of next meeting.
Approximate date of next meeting.
Election of Officers and Executive Committee

Time being allowed for general business and social affairs, at the discretion of Executive Committee or Chairman of meeting. Selection of next place of meeting and election of Officers and Executive Commitee must be on second day of meeting. Order of business may be altered only by unanimous vote of members present.

### ARTICLE XX.

Ten active members of the Association shall be a quorum for business.

### ARTICLE XXI.

Todd's Parliamentary Practice shall be the governing law of the Association in all cases not provided for in its own rules.

### ARTICLE XXII.

DUTIES OF THE PRESIDENT.—It shall be the duty of the President to preside at all meetings of the Association and to call meetings of the Executive Committee, and when requested by the Executive Committee, to call a special meeting of the Association.

### ARTICLE XXIII.

DUTIES OF THE VICE-PRESIDENTS.—The 1st, or in his absence, the 2nd Vice-President, shall act in the absence of the President.

### ARTICLE XXIV.

DUTIES OF THE SECRETARY.—The duties of the Secretary shall be to attend all meetings, take record of all proceedings, and shall perform such other duties as the Executive Committee shall direct.

### ARTICLE XXV.

DUTIES OF THE TREASURER.—The duties of the Treasurer shall be to keep a correct account of all receipts and disbursements in connection with the Association. All checks for disbursements shall be signed by the Treasurer and countersigned by the President, after being approved by the Executive Committee.

### ARTICLE XXVI.

THE DUTIES OF THE EXECUTIVE COMMITTEE.—The Executive Committee shall be the governing body of the Association, shall manage its affairs, pass upon all applications for membership, eligibility of representatives, subject to the constitution, and such special rules or regulations as may be adopted by the Association from time to time.

### ARTICLE XXVII.

DUES.—Dues shall be payable annually on the 1st June, in advance. Members in arrears for dues, other than those for current year, shall not exercise the privileges of membership.

### ARTICLE XXVIII.

The permanent office of the Association shall be in Toronto.

## STEPPED GRATES.

There is one story which comes up to me very often, and it has the special merit of being true, writes Robert Grimshaw, in Power and Transmission. A certain firm drew up plans and specifications for a bridge to cost about a million and a half dollars. When they were done, a certain engineer pointed out where, by a slightly different construction, equal strength, durability, convenience of erection, slightliness, etc., might be had, with a saving of two per cent. "Oh, bother the two per cent," said the designers, "do you suppose that we are going to overhaul and re-make all our calculations and strain-sheets and drawings for a measly two per cent?" "Well," said the critic, considering that two per cent of a million and a half is thirty thousand dollars, I think you could very well overhaul the whole business"

This story is good for any latitude and in any business, and can be especially well applied or considered by large coal users.

What a lovely and useful thing the multiplication table is! Say fifty tons of coal at four dollars a ton, and three hundred working days in the year; and figure up only two per cent on that : $50 + \$4 \times 300 \times .02 = \$1,200$.

How many coal users know or remember that, no matter how well their boilers are set and managed, they can not possibly get more than seventy-five per cent of the heating effect of the coal, and are not likely to get more than seventy-five per cent?

The manufacturer who gets six or seven pounds of dry steam at good pressure per pound of hard coal, is doing better than the average,—and that is not saying much. The little two per cents may be picked up by attention to details,—such for instance, as grates, ash-pits and dampers, and by finding out what combustion rate pays best.

Many manufacturers have tried all sorts of plain grates, and have even ventured into the field of "patent" grates, with rocker arms, fingers and so on ; and those who have properly experimented with rocking grates, suitable to the conditions under which they are applied, have usually found a saving. But the stepped grate, although considerably used and well liked in Europe, where coal costs money (and money is nailed down fast,) is comparatively a stranger in the United States.

The grate should be from 36 to 42 inches long ; and it is better, usually not to exceed 20 inches (or at most 24) in width, (i. e. in the length of the plates forming the steps). When a wider grate is required, there should be two sets of steps side by side. Underneath the slanting part of the grate should be an ash hole, leading to a masonry ash pit, and a similar hole under the dead-plate or ash-plate at the lower or back end, leading to the same pit. The plates may be, if of wrought iron, only about one-third inch to one half inch thick, and they may be so set that each is about three quarters inch to one inch below the other, between centers. As they are set flat, there is more risk of bellying than those bars in ordinary grates, which are set on edge, hence the precaution of having a length of 24, or better yet, only 20 inches. About 45 degrees is a good angle for the grate considered as a whole, but 35 degrees with the horizontal is usually better,—this depending on the kind of coal, some coal sliding more freely and requiring to be held back, and some needing a good deep angle to keep it from banking up instead of moving backwards and downwards. At the back and lower part there should be a slot controlled by a sliding plate, for ashes to drop through and for air to come up through should it be needed ; and this may usually be kept open about an inch and a quarter, although it should be capable of opening six or eight inches to facilitate rapid cleaning, re-making of the fire, etc.

The stepped grate has usually the advantage of requiring but little attention on the part of the stoker. It generally works best when the principal combustion takes place on the lower half ; and this part requires the most slicing and poking from the front and raking from the back. The thickness of the bed depends, of course, on the kind of fuel and on the draft ; the proper amount may be learned in a few days stoking. Usually it is best to have the thickness rather more than twice as great on the upper part of the steps as on the lower,—as, for instance, two inches below and four and a half to five above, for hard coal ; usually thicker for soft.

The smaller the coal the better the stepped grate works ; that is, it works with small coal better than with large ; also it makes a greater saving with small coal over large than the ordinary grate will. It is is not at all suitable for lump and steamer coal, and per contra, it will burn stuff that other grates will utterly refuse to raise steam with ;—for example, such trash as is found for a depth of a foot or so under a coal pile which has been standing for one or two years with constant changes, as on a shipping wharf,—this stuff consisting of from twenty to fifty per cent. of clay, sand, or other incombustible trash.

## OBSTINATE THUMPING.

SOMETIMES an engine which usually runs well develops an obstinate pound or thump, which persists in spite of all the doctoring that can be done to the machine. In vain the engineer will go from the wrist pin to the cross head, and from eccentric to bearing. Even the fly wheel and the manner in which it is keyed upon the shaft will be investigated, to see if the thump is located therein. After all these things have been tried in vain, just give the engine a trifle more compression and note the result. Probably it will cure or make it worse. In the latter case change the valve again and give a little less compression than there was before. In nineteen cases out of twenty the change in compression will do the business. The philosophy of the business is this: The compression is too little or too great to allow the engine to run smoothly over the centre; and at that point the piston gives a "yank," which causes wrist pin and connection and sometimes the main bearing to vibrate to the extent of the lost motion, forming the thump or pound, which is so objectionable to the good engine runner.

## BOILER FEEDWATERS, THEIR TREATMENT.*

BY W. D. JAMESON.

WATER is a wonderful agent produced and given us by nature, and has its advantages and drawbacks ; it is the greatest solvent of all natural or artificial liquids known to chemistry ; it becomes impregnated with all different elements, in one form or other, in which it comes in contact, and absorbs free carbonic acid gas from the air and ammonia from the air and earth. Carbonic acid gas thus formed becomes the life of the water and enables it to take up the otherwise insoluble carbonates of lime, magnesia, etc., holding them in solution as bicarbonate of lime, magnesia, etc.; the colder the water and the heavier the pressure the more gas it contains ; consequently the larger the body of water or the deeper the well, the more heavily impregnated it is with the salts of lime, magnesia, etc.

All natural waters are imbued with the salts of the following mineral bases : lime, magnesia, sodium, potassium, iron, silica and aluminum, combined with carbonate, hydrochloric and sulphuric acids, and sometimes medicinal waters with phosphoric acid, or all of them to a more or less extent, according to the nature of the soil or the conditions in which the water percolates the soil.

The calcium, commonly termed lime, is taken up in the forms of sulphate and bicarbonate ; the magnesia as bicarbonate, sulphate, and chloride ; the sodium and potassium as chloride, sulphate and carbonate ; the iron as bicarbonate. Iron as well as copper is found in solution as a sulphate. The aluminum exists in the water as a sulphate or in suspension as an oxide ; the silica as silicic acid. When we find a water containing sulphate of iron or copper in solution, we generally find free sulphuric acid also.

The salts of lime and magnesia, iron, silica, oxide, etc., are scale forming ingredients ; the sulphate of lime forms a very hard compact incrustation, adhering very tenaciously to the hot metal, is very hard to break up, decompose or dissolve, and, like all sulphates, it is a very staple salt ; it is conveyed into the boiler by the water as a sulphate, and as such enters the scale formation, and is not even soluble in its own acid, and it is impractical to dissolve it with hydrochloric acid except in laboratory work.

The only substances which can be successfully used in the boiler to break up and convert sulphate of lime into a form in which it can be readily washed out, are sugars properly blended, which, when used under the high heat, and the existing conditions of the steam boiler, convert this sulphate of lime into a complex mixture of saccharates and carbonate of lime, and this, in the presence of the tannin matters, is practically converted into tannates of lime.

Carbonates of lime and magnesia enter into the scale formation as such, forming a very compact incrustation, due to the great chemical affinity they have for hot metal, which is also the cause of the adhesive properties of sulphate of lime (gypsum). They can be readily and successively converted into a complex mixture of the tannates of lime and magnesia without any contamination to the steam or injurious effects to the steam receptacle or its connections.

Silica enters the scale formation as such, and also as silicate of magnesia. Sodium salts enter into the scale formation only in small quantities. Being very soluble they remain in solution until the water in the boiler becomes supersaturated, and unable to hold a greater quantity ; these salts then cake on the hottest parts of the boiler, falling out of solution ; this is very dangerous, having been the cause of the burning of a great many boilers in localities where the feed water is highly impregnated with soda salts. They cause internal corrosion, wasting away of the iron, eating through the joints and connections, and are the cause indirectly of one class of corrosion of which I will speak later under another head.

Chlorides of lime and magnesia, found in some feed waters, are very corrosive agents of iron. Being very unstable salts, they readily decompose with the high heat into oxides of lime and magnesia, entering the scale formation as such. The free chloride combines with the hydrogen of the water as a hydrochloric acid, and has a direct corrosive action on the iron. The action of sulphate magnesia is very similar to that of the chloride under the influence of high heat. The sulphates of iron and

* Read at the convention of the Northwestern Electrical Association, Chicago, July 18, 1895.

copper are direct corrosive agents to the iron and boiler connections, and will not enter the scale formation.

It is almost impossible to neutralize sulphates of iron, copper or magnesia in a practical manner. If you do it with soda, and convert the sulphuric acid into sulphate of soda, you get an excess of soda salts, which sets up galvanic action. If you use lime, converting the sulphuric acid into sulphate of lime, you get such large quantities of gypsum that in a short time your boilers will be so full of a hard incrustation that it will be impossible to run them. The only thing which has been half way successful in the handling of soluble sulphates and free sulphuric acid, is a mixture of sugars and starchy matters of a complex organic nature, which have offset the action of the acid by breaking up the acid radical, taking the sulphur and incorporating it with the aid of some of its oxygen into its own organic compositions.

Speaking of sodium and potassium salts, I would ask if it does not look unreasonable to endeavor to treat water for the prevention of the scaling deposits by the use of sodium and potassium salts, yet these salts are, in 99 cases out of a 100, the principal ingredient of the so-called boiler compounds and water purifiers, and it is these salts which cause most of the internal corrosion of steam boilers by their galvanic action.

Internal corrosion is the eating and wasting away of the threads, plates and joints, causing leakage and also causing the boilers and their connections to assume unsafe conditions. Where the corrosion is due to chlorine, free hydrochloric or hydrofluoric acids in the water, we find the pumps and feed pipes eaten through, the submerged parts of the boiler being free from such action on account of these acids readily passing off with the steam, and we get a similar action again in the steam-exposed surfaces of the boiler and the steam piping.

Free sulphuric acid has a very similar action, attacking the feed pipes a great deal more rapidly than the boiler itself ; its corrosive action in the boiler is more uniform and not so much of a pitting and grooving nature ; its action in the steam piping having almost entirely a grooving appearance. Where the deleterious action is due to the presence of an acid, it is called a direct corrosive action, and is generally found prominent in the feed pipes (colder pipes) and in the steam exposed surfaces. Where the corrosion takes place mostly in the submerged parts of the boiler, it is generally an indirect action, due to an excess of salts or too pure a water, coming under the head of galvanic action, termed by electricians electrolysis.

The boiler, as it is generating steam, is also generating a certain amount of galvanic current. The boiler is a galvanic battery in itself, the valves and their brass connections, composed of copper, babbitt, and other alloys, are negative, the iron being positive, forming the negative and positive poles, and under the high heat and other conditions existing in the steam boiler we have a galvanic battery ; not only is copper negative to iron's positive, but the very molecules of the iron in the plates and tubes are negative and positive to each other ; but electrolysis does not take place in the plate because the impurities, or we might say, foreign matter, such as silicon, oxygen and carbon compounds, are not and do not act as conductors between these negative and positive poles ; the water in the natural condition, that is, its chemical affinities and solvent properties, being satisfied with lime and natural salts, will not act as a conductor between these poles, consequently, having no conductor, the battery is not connected by water, but when using distilled water, rain water, or water with an excess of sodium salts, we then have a perfect conductor, the water assuming the position of a battery and of a battery solution, connecting our negative and positive poles, and inciting and generating a galvanic current. We then have a true galvanic battery existing, due to the general make-up and influence in the steam boiler. The purer the water, or the greater the excess of sodium salts, the stronger our galvanic current, the more pronounced our electrolysis.

You well understand that water contains a very corrosive radical in the nature of a hydrate ; the hydrate radical is HO. Water is composed of two atoms of hydrogen and one of oxygen, which is a very strong chemical combination, not readily decomposed except with a soluble metallic base or red hot metal, but in this case, under the influence of the galvanic current, the positive metal, which is iron, exercises a chemical affinity over the water, chemically combining with its hydrate, forming ferric hydrate, taking up the oxygen and part of the hydrogen of the water, freeing part of the hydrogen, which goes off with the steam. This ferric hydrate gradually converts into corresponding oxides, due to the high heat and boiling of the solution, gradually converting into the black magnetic oxide of iron, so named owing to the galvanic action in its manufacture ; its physical properties are that of a black gritty powder found at the bottom of the boiler when washed out, when electrolysis is going on. If you will take a boiler that is pitting from this cause, you generally find zigzag pits and grooves coated over with a baked film, and by tapping these with a hammer you find a reddish brown soft powder underneath, which is the more freshly formed ferric hydrate ; that of a lighter shade is the partly converted oxides, and the few handsful of black gritty powder from the bottom of the boiler, which you can examine after rinsing the other oxides from your hand, you will find to be the black magnetic oxide of iron.

Speaking of electrolysis, which we, from our standpoint, term galvanic action, we believe it truly exists as such, and to prove it consider the large ocean-going vessels and think of the trouble they have from this cause and how and why they treat it. They use tons and tons of zinc to offset this very action, due partially to using too pure a water on account of the hot well system, and further by what salt water they are compelled to use. We all know zinc to be one of the most positive metals known in galvanic battery work ; it is more positive than iron. The zinc put into the boiler assumes the position of the positive pole, consequently it is destroyed in place of the iron by the battery solution in the steam boiler. Its reaction and conversion into its oxide are similar to that of the iron, it being destroyed under the same influences.

Of all the deleterious actions which take place in steam boilers this is the easiest to handle, for you simply need to satisfy that water with some vegetable starch and saccharine matter, and in that way break up your conductor between the negative and positive poles, whether they be brass connections (negative) and the boiler plate and flues (positive) or the molecules of the iron of the boiler plate. It is impossible to set up a galvanic action without the water assuming the position of the battery and acting as the conductor. This same saccharine inert matter in conjunction with tannin extracts will cause these pits and grooves in the iron plate (where the case hardening protective surface of the plate and tube is broken and the raw steel or iron exposed) to heal over, assuming that same case hardening appearance as before. Do not understand me to say that you can fill up the little holes, as that cannot be done, the iron being gone, but the surface of these little zigzag holes and pits will heal over, serving as a protection against the water or the atmospheric oxidation.

Scaling ingredients are converted from crystallizable scale-forming carbonates and sulphates, having a great affinity for hot metal, into non-crystallizable tannates and saccharates of lime and magnesia, being a complex mixture of these with some carbonate, the sodium salts being readily handled in the same manner. This complex mixture of the saccharates, carbonates and partially converted tannates is of an inert nature, having the physical properties of a soft oozy mud, of the same specific gravity as the water, and no affinity for hot metals, neither has it the clay-like properties, but it will readily wash out with the water when cleaning the boiler.

In conclusion I might say a few words relative to the deleterious action of oil in steam boilers. Many of you to-day are running large condensing plants with your hot-well systems, and you are getting oil, with the condensation, into the boilers, possibly 5 to 15 drops per gallon. These oil separators are a good thing, and do, possibly, 50 or 60 per cent. of the work. You often hear of the tubes in a water-tube boiler buckling up and having to be taken out ; you often hear of the bagging of the fire-sheet in tubular boilers. Why is this? The specific gravity of the oil is lighter than that of the water ; the oil does not settle in its natural state. We explain it as follows : The oil coming into the boiler floats on the water ; there is just a sufficient quantity of fresh water coming in to convey salts of lime, magnesia, etc., which are thrown out of solution, chemically combining with the animal oil as insoluble oleates, and mechanically combining with the mineral oils as a heavy mass, both these chemical and mechanical combinations being of a greater specific gravity than the water in the form of little glob-

ules, sinking to the bottom, the great chemical affinity and adhesive properties of this mixture causing them to adhere to the hot metal, and they, being a perfect non-conductor, retarding the transmission of the heat units to the water, concentrating heat in that part of the plate, causing the iron to melt, and the pressure in the boiler forces it down.

Sodium salts, so commonly found in water, or where it is used to counteract this action, saponifies the oil, causing the boilers to foam and carry over into the engines, and should not be used. This defect can be successfully handled with tannin extracts, the tannates forming complex organic compositions with the oils of an inert, light, powdery nature, having no chemical affinity or physical adhesive properties and readily washing out with the water at the opening of the boiler. To prove this go to the tannery and watch the tanner take the hides out of his vat after he is through with the tanning process, and when he lets the liquor run out of the vat you will find two or three scoop shovels full of an inert powder, which readily dries out and is termed pure tannin by the tanning experts. They claim that this is insoluble, and are in want of a solvent so that they can successfully use it for its tanning properties. We do not believe this to be the case, as the tannin in this mixture is, chemically speaking, part of the mixture, and the tannin is satisfied by the fatty matters contained in the hide. We aim to get this same reaction with the oil by pumping into the boiler a properly blended mixture of slippery elm, starches, sugars and tannin extracts.

We have found that we can successfully cope with most of the deleterious actions taking place in steam boilers with vegetable matters, and vegetable matters only, sometimes using from 5 to 10 per cent. of carbonate of soda to partially cut the starches and aid in the action of the sugars, but, correctly speaking, we are vegetarians on this subject, and do not believe that perfect results can be obtained from any other methods known to science.

## THE CARE OF BOILERS.

THE boiler being the vital part of the steam plant, which again is the center of all motion and life in a mill or factory dependent on that form of power, all the skill and attention possible should be directed to their preservation in good order, and at the smallest possible expense consistent with good results. To this end all means proposed should receive the careful consideration of those interested, so that the best plan applicable may be chosen in each place. It is evident that the same method is not practicable under all circumstances, for while the general principles involved are in all cases the same, the working out of these principles necessarily varies. Thus all water derived from wells where the underlying rocks are anything except granite or sandstone, contains a greater or less proportion of solid matter, varying, according to one list in my possession, from as little as 6.7 grains per gallon to as much as 353.8 grains per gallon. In the same localities the water of the streams is likely to partake to a considerable extent of the characteristics of that in the wells. So it may be said that over the greater part of the country it is impossible to procure even comparatively pure water. Even that which falls as rain and snow in inhabited localities contains impurities washed from the air in its descent, although the proportion is so small as not to interfere with its use in boilers, provided it could be obtained in sufficient quantity; but this, from the nature of the case, is impracticable.

Of course not all the solid matter found in well water is of the kind which forms scale. Lime and magnesia are the principal ingredients of scale, with at times a combination of iron and some organic matter, a mixture of iron especially forming a peculiarly hard and obstinate scale. The question of greatest interest to a man in charge of steam boilers is : "How shall I get rid of the scale in my boilers?" The correct answer perhaps smacks of the Hibernian, but I believe it to be : "The best way to remove scale from boilers is not to let it in." After a dozen years of experience with water containing seventeen to twenty grains of solids per gallon, the greater part being of the incrusting kind, I am satisfied that with a little care and the use of moderately good exhaust steam heaters, no trouble need be had with scale in a boiler which is well taken care of.

One great trouble in this matter is that owners are unwilling to allow the firemen reasonable compensation for the extra time required to properly do the work connected with keeping the boilers clean. Some only allow a quarter of a day's pay for the time necessary on Sunday to wash out and clean up generally. It is safe to say that the fireman, unless made of sterner stuff than the majority of the race, does not, on an average, put in much more time than he is paid for. Other owners allow full pay for the day, depending on the engineer and fireman to keep the plant up to the highest condition possible. In one such plant with return tubular boilers, which has been run for fifteen years, with the kind of water just mentioned, no trouble has been had with scale on the boilers for ten years at least; and the heaters are not of the most recent construction either.

Very much depends on the care taken of the heaters as to their efficiency, for if they are allowed to become foul, the accumulation of slush is liable to pass on to the boiler, at least, if the heater is one of the closed variety. While it is a little more trouble to take care of an open heater, as they are generally provided with some kind of a filter which requires some attention to keep in good order, they are, I think, a little more efficient in heating the feed water, while the proportion of steam condensed in the process, being pure water, is also of some advantage. Where the plant is of sufficient size to warrant the expense, or where the water is so hard as to require it as a measure of safety, the addition of a live steam heater of proper size will almost prevent scaling. The water being raised to the temperature of that in the boiler, practically all the incrusting matter is dropped by the water, which is then frequently filtered through a layer of finely-ground coke or similar substance, and so enters the boiler practically pure.—F. Riddel, in American Miller.

## SHAFTING, PULLEYS, ETC.

IN designing a mill or manufacturing plant, says C. R. Tompkins, M. E., one of the most important features, aside from the arrangement for good and sufficient power, is the line of shafting and the necessary pulleys for the purpose of transmitting the power to the several machines to be used. Now, it is just as important that good judgment be manifested in this part of the plant as in any other. The fact is that much needless expense is often caused in the first instance, besides a continual loss of power in the second, by an injudicious selection of the shafting.

A line of shafting unnecessarily heavy, with pulleys and couplings to match, not only involves a greater expense in the first place, whether it is purchased by the pound or foot, but the extra amount of friction on the journals caused by that weight is a factor that should also be taken into consideration. It is a well-known fact that the frictional resistance with all bodies in sliding contact is in direct proportion to the weight pressing them together, so that the weight of a line of shafting with heavy pulleys, no matter what the speed may be, will exert a constant frictional resistance in proportion to the weight.

While there can be no question as to the economy in all cases of using a lighter shaft at greater speed than was formerly the case, still it is not advisable under any condition to go to extremes in either case, for the reason that, with a little forethought and calculation in the first instance, we may avoid either.

As a rule, in all modern mills and factories, the tendency has been toward lighter shafting and pulleys of small diameter, with a corresponding higher speed, and there is no question but much more satisfactory results have been obtained. The shortest and most reliable rule that has been found to obtain the torsional strength of all sizes of shafting, is to multiply the cube of the diameter by 600, and this product by the number of revolutions per minute, and divide by 33,000 for the horse-power. The ultimate torsional strength of a shaft is not the power required to twist it off, but a power not quite sufficient to give it a permanent set.

Now, according to this rule, which has been verified in many cases, a shaft 3 inches in diameter at 200 revolutions per minute should not be required to safely transmit 32 horse-power while by the same rule a shaft of 2 inches diameter of the same quality of iron running at 300 revolutions will safely transmit 43 horse-power. Now, all other things being equal, it is evident that where not over 35 horse-power is required, a 2-inch shaft at 300 revolutions per minute is the most economical. For example, the weight of a line of 3-inch shafting 40 ft. long, without couplings and pulleys, is 955 pounds, while a 2 inch shaft of the

same length weighs 424 pounds, a difference in weight of 531 pounds. Now, the frictional resistance, as before stated, is in proportion to the weight, and without any lubrication it is estimated that it amounts to 25 per cent., but with a good lubrication this may be reduced, according to the best authorities, to 8 per cent.

Now, taking 8 per cent. as the average, we find that with a 3-inch shaft we have a constant frictional resistance of 76.40 pounds to contend with, while on the contrary, the frictional resistance upon a 2-inch shaft amounts to but 34 pounds. Here an important question arises which has been frequently discussed, and that is whether the speed has anything to do with the frictional resistance.

One authority says that "with hard substances and within the limits of abrasion, friction is as the pressure, without regard to surface, time or velocity." In another place the same author states as follows : "A regular velocity has no considerable influence on friction ; if the velocity is increased the friction is greater, but this depends on the secondary or incidental causes as the generation of heat and the resistance of the air."

Now, without entering into a full discussion of this question, if we take the question of speed into consideration, the argument is still in favor of the lighter shaft. We found the frictional resistance in the 3-inch shaft without taking the speed into consideration to be 76.40 pounds. Now, if we multiply this by the speed, as some contend it should be, we have a total resistance of 15,280 pounds per minute to overcome, while with the 2-inch shaft by the same proposition we have 10,200 pounds per minute to overcome, showing a difference in frictional resistance in favor of the 2-inch shaft of 5,080 pounds per minute.

Now, as to the question of pulleys. In order to obtain say 900 revolutions from a pulley driven from a 3-inch shaft at 200 revolutions per minute, it will require a pulley 36 inches in diameter, while the same power and speed may be obtained from the 2-inch shaft at 300 revolutions from a pulley 24 inches in diameter.

Now, in the foregoing argument in favor of lighter shafting and higher speed, the torsional strength of the shaft has only been taken into consideration, and while the torsional strength of a shaft of a certain diameter may be amply sufficient to transmit the required power with perfect safety, still the lateral strength must also be considered. A shaft, no matter what the size may be, in order to fulfill all the conditions of practical use, must possess sufficient lateral strength to stand the pull of the belts, together with the sudden shocks which may be sustained when heavy machines are started suddenly, and for this reason, under peculiar conditions, it may be advisable to use a shaft a trifle larger than the rule calls for. But under ordinary conditions, if the distance between the boxes or hangers is in proportion to the size of the shaft, it will not be found necessary to vary much from the foregoing rule.

One of the most common faults in erecting a line of shafting is in too great a distance between the bearings, and it is often the case that a shaft abundantly heavy is rendered ineffective from this cause, and when a machine is started the shaft springs, so as to cause the belt to slip, unless the pulley happens to be close to the bearing.

While it is good practice in all cases where the conditions will admit to run all heavy pulleys as close to the bearing as possible, still it is not always practical to do so, consequently the size of the shaft and the distance between the bearings should be so calculated that there will be sufficient lateral strength to admit of placing the pulleys upon any part of the shaft between the bearings.

There is no question but as a general rule a shaft that possesses sufficient torsional strength to perform the work, with a moderate allowance for contingencies, will, if the bearings are placed at a proper distance apart, also possess sufficient lateral strength for all practical purposes.

In practical experience it has been found that the most reliable rule for this purpose is to take three times the diameter of the shaft in inches for the distance from center to center of the bearings in feet. Thus a shaft of 2 inches in diameter should be 6 feet from center to center of its bearings. One of 2½ inches would call for 7 feet and 6 inches, while one of 3 inches may be 7 feet, and so on.

## THE STRATTON SEPARATOR.

The following is a letter from Prof. R. C. Carpenter, of Sibley College, Ithaca, N. Y., reporting a test made of the Stratton Improved Separator this year :

"I send you with this letter a short summary of the test which we have made on the Stratton Separator. The results show that the separator is practically perfect, and removes all the moisture which can possibly be taken out by mechanical means. I think we will make another test in which we inject water into the steam pipe, thus increasing the percentage considerably of water in the steam supplied. This latter will not be of great practical interest, but will bring out, of course, the capacity of the separator for extraordinary conditions. If you have no objection I will publish a copy of this report in the next number of the Sibley Journal, and for that purpose would be pleased to have you loan us an electrotype showing vertical section.

THE STRATTON SEPARATOR.

"TEST OF STRATTON IMPROVED SEPARATOR.—For this test the steam pipe leading to the separator was surrounded for a portion of its length with a jacket which could be filled with water to any desired height, the purpose of the water jacket being to condense as great a per cent of the steam as possible. The discharge of steam from the separator was led to a surface condenser, where it was condensed and the amount carefully weighed. The drip of water discharged from the separator was led to a barrel standing on a pair of scales, and accurate weighings were made of the water taken out from the steam by the separator. A throttling calorimeter was placed in the steam pipe directly after the steam left the separator. Pressure gauges were placed either side of the separator. Observations were taken and the results reduced by Messrs. Collins, Hubbard and Thomas of the class of '94. The following is the general summary of the results : The steam supplied to the separator contained moisture, the percentage of which varied from a little over 5 to nearly 21. That discharged from the separator was in every case nearly dry, it containing in every instance less than 1 per cent. of moisture. The separator was worked up to its full capacity, and there was no appreciable reduction of pressure. The summary of the results of different runs is given in the appended table. During these runs the water was kept at a constant height in the separator :

| No. of Run. | Pressure of Steam. | Moisture in Steam Supplied Separator, Per cent. | Moisture in Steam Leaving Separator, Per cent. | Quality of Steam Leaving Separator. Per cent. of Dry Steam. |
|---|---|---|---|---|
| 1........ | 60 | 6.55 | 0.95 | 99.05 |
| 2........ | 61 | 17.2 | 0.94 | 99.06 |
| 3........ | 62 | 15.31 | 0.9 | 99.1 |
| 4........ | 76 | .15.6 | 0.6 | 99.4 |
| 5....... | 61 | 20.9 | 0.8 | 99.2 |

The Goubert Manufacturing Company, New York city, who are sole manufacturers of the Stratton Improved Separator, are represented in Canada by Wm. T. Bonner, 415 Board of Trade Building, Montreal.

## PATENT WATER TUBE HEATER AND PURIFIER.

WE herewith illustrate a patent water tube heater and purifier, manufactured by Laurie Bros., of Montreal. The manufacturers claim for these heaters that when applied as rated they will raise the temperature of feed water to from 210° to 212°, and that the impurities in the feed water that precipitate at boiling point will be deposited at the bottom of the heater, where provision is made for blowing it off.

As will be seen from the cut, the upper tube plate is entirely

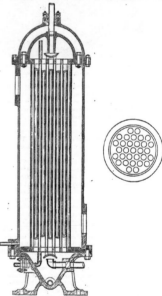

PATENT WATER TUBE HEATER AND PURIFIER.

separate from shell or body of heater, thus providing for free expansion of tubes independent of any other materials used in construction. They are constructed entirely of cast iron and brass, making them almost indestructible.

The water entering at bottom (and being distributed by deflector) passes slowly up the tubes absorbing heat in its passage, to upper chambers, where it is still surrounded by exhaust steam till discharged.

The discharge pipe projects downward into chamber to avoid carrying scum from surface of water into boiler. A scum blow-off is provided, with discharge at bottom of heater as shown in cut.

## ELECTRICITY FOR MINING.

AN important application of the three-phase system of long distance transmission of power has recently been installed and put in operation at the Silver Lake Mines, near Silverton, Col., where power is transmitted a distance of three miles through some of the roughest country in Colorado. It is attracting con-

FIG. 1.—THREE-PHASE GENERATOR FOR LONG-DISTANCE TRANS-
MISSION AT SILVERTON, COL.

siderable attention among mining men, as it is the first three-phase plant established in the Rocky Mountain region.

The success of direct current transmission has been thoroughly demonstrated by transmission plants operating, not only in Colorado, but elsewhere all over the world. The expense, however, inseparable from direct current transmission, precludes the utilization of that system in most places where the distance exceeds a certain limited number of feet. This will be readily understood when it is stated that if 100 H.P. can be transmitted by direct current one mile at 500 volts, 10% being allowed for loss in the line, the copper wire necessary will cost about $2,000, while for the same horse power transmitted by the same system for ten miles it will cost about $200,000. If, however, the three-phase current at 5,000 volts be employed to transmit the 100 H.P. ten miles, the cost will not be more than $2,000. In other words, a given horse power can be transmitted by the three-phase system at 5,000 volts, ten times further than a similar horse power by the direct current system at 500 volts for the same expenditure in copper. As, therefore, the question of dollars and cents is a most prominent factor in all transmission installations, the three-phase system, where long distance transmission is concerned, is the most practical system because the most economical, commercially speaking, and the installation at Silverton is a striking example of this fact.

The Silver Lake group of mines lies about four miles south-east of Silverton, and is situated at an altitude of 12,300 feet above the sea level. They are owned by Edward G. Stoiber. The ore mined carries both gold and silver, is of a comparatively low grade and requires concentration.

Previous to the installation of electricity, the mill, which is situated on the shores of the lake, near the mouth of the mine tunnel, was run by steam. Coal was brought to the steam engine by the zigzag path up the mountain shown in figure 3, and by the time it reached the furnace cost $8.75 a ton. This represented a monthly expenditure of almost a thousand dollars, and the expense proved a burden which went far to eat up the profits of the mine. Reform, therefore became imperative.

The plant is now operated by water power, which is brought from the Animas River, above Silverton, through a 3 x 4 foot flume, 9,750 feet in length, which carries 2,350 cubic feet of water per minute. Flume and trestle are shown in figure 4. One of

the great advantages of electrical utilization is here demonstrated, for it was found less expensive to build this costly two mile flume, running from above Silverton to a spot where the necessary head could be utilized, and then to transmit the electricity back to the mine, rather than to continue to burn coal at the price which it brought at the mouth of the Silver Lake Mines.

The head of water obtained is about 180 feet; this develops on the water wheel shaft 640 H.P. The plant consists of two four foot double nozzle Pelton water wheels, with special buckets, belt connected to two 150 K.W. (200 H.P.) General Electric three phase generators.

The current from these machines is given out at 2,500 volts and is transmitted over a distance of a little more than three miles to the Silver Lake Mill and Mine.

The conductors are No. 3 B & S bare copper wires, one for each branch of the three-phase circuits. These are strung from the power house, shown in figure 2, up the mountain passes and through the rugged forbidding country shown in the illustrations. In one place where a chasm has to be spanned the wires leap from pole to pole a distance of 275 feet. They have been strung with especial care, as befitted the abnormal conditions. At each insulator the wire is run through a short piece of rubber tube, as an extra precaution against leakage. Lightning arresters are placed at each end of the line, and an additional safeguard against possible damage by lightning is provided in the shape of a barbed iron wire, which extends the entire distance of the line along the tops of the poles, and is grounded at every second pole. In this country, where the storms are frequent and the lightning disastrous in its effects, every precaution is taken to frustrate possible damage from atmospheric discharges. The most effective lightning arresters which have been found for this work are those which the General Electric Company itself manufactures.

In the winter time the snow lies thickly on the ground and all intercourse between the mine and civilization is almost cut off. Some idea of the difficulties of the road may be gathered by referring to figure 3, where the zigzag route up the mountain is shown.

Arriving at the mine, the current is supplied to a 100 H.P.

FIG 2.—POWER HOUSE AT THE END OF THE FLUME, SILVERTON, COL.

three-phase induction motor, run directly from the primary circuit. Another 100 H.P. motor, as well as one of 75 H.P., are located beneath the ground, and current is supplied to these at a pressure of 220 volts, the reduction in pressure being effected by step-down transformers. In addition, a 15 H.P. motor runs a pump, raising water from the lake to the mill, and one small 1 H.P. motor operates a blower and the lights for a bunk, office and other buildings, both being connected to the secondaries. The induction motor used is an excellent representation of the latest type of alternating current motor built by the General Electric Co. There are no commutators, collector rings or brushes; the field winding is connected to the circuit, but there is no connection between the armature and any external source of current. The three-phase currents rising and falling in the field windings induce corresponding currents in the armature winding, and the armature revolves. The field armature cores are so completely laminated that all loss from eddy currents is practically eliminated. The speed of the motor is at all loads practically constant. The starting resistance for preventing any excessive current in the armature winding, while gaining speed, is contained in the armature itself, and the handle shown in the cut serves to cut out the resistance when the motor is fully up to speed.

The interest in this mine centres, of course, upon the economy induced by the electrical installation. The power used in the mill and the mines at the present time is more than three times as

FIG. 3.—ZIGZAG TRAIL UP THE MOUNTAIN, SILVERTON, COL.

much as was used when generated by the steam engine previously employed. If, therefore, $1,000 a month or $12,000 per year would be economized by using the same power, an economy of not less than $36,000 a year is effected by the operation of the mine by electricity, and a greater power is available. This has allowed of the operation of machinery inside the mine for the first time.

Before such figures as these any question as to the economy

## EDITOR v. SUBSCRIBER.

DIE MUHLE, a paper printed in Berlin, Germany, indulges in some caustic comments concerning those subscribers who expect everything, and give nothing. It declares that a trade paper is expected to supply information very often outside of its sphere and field; that it must inform manufacturers of the particulars of its own business, even to the extent of showing them how to reduce expenses; must, in fact, be a general advertiser on all subjects within the commercial domain. If, says this paper, the editor should ask where he must get the information, the subscriber is apt to reply, " It is none of your business." The editor thereupon shows precisely what are the duties of the subscriber in relation to the journal he takes, and berates the reader for not supplying him with such facts as may come within his notice, and which, when developed, would probably prove to the mutual advantage of editor and subscriber. There is much force in this. Subscribers have it in their power to extend the usefulness of their trade paper by making suggestions and submitting facts; and he would, indeed, be an indifferent editor who refuses to consider them. Newspaperdom perhaps, puts it in the best possible light by saying that a man who subscribes for a trade paper does so, not only because he is alive to the interests of his trade in general, but because he expects to find in it—and generally does—information and suggestions of value in the conduct of his own business. Such a man reads his paper from end to end, advertisements and all, commenting as he goes along. Many things

FIG. 4.—THE FLUME, SILVERTON, COL.

of an electrical installation, when coal is anywhere near the price it is in this case, becomes irrational.

Joseph Brisbois, of Guelph, Ont., in the employ of the G. N. W. Telegraph Co. as repairer, was run over and killed by a Grand Trunk train recently.

are jotted down on his memo pad, for everyday use, and for enquiring further into on his first visit to the market. And as it is necessary for a man to be wide awake nowadays in order to succeed in business, these are the men who subscribe to their trade journal, and are the advertisers' best patrons.—The Effective Advertiser.

## IMPROVED CORLISS AUTOMATIC ENGINE.

THE accompanying illustration represents a Corliss automatic engine, as manufactured by Laurie Bros., of Montreal, for electrical purposes, with extra heavy fly wheel.

The general design is a modification of what is known as the girder frame engine. The cylinders, frame and pillow block are cast separate and bolted together. The guides (being circular) are bored and end of frame faced at one operation, thus securing perfect alignment with the cylinder. The frame at outer end of guides forms a complete circle, at which point a pedestal is placed, thus forming a very rigid arrangement. The steam cylinder has four valves, two for steam and two for exhaust, independent of each other in action, and placed so close to the bore of the cylinder as to leave the least possible amount of clearance. The steam valves are operated by means of bronze spindles or stems; the exhaust by steel spindles or stems. Either of the four valves may be removed by unscrewing the cap screws that hold the back end bonnet in place. The exhaust valves are located below the cylinder, thereby securing perfect drainage.

These engines are built from extra heavy patterns, and every possible precaution taken to prevent the possibility of accident or derangement.

## MUCH INFORMATION IN A SMALL SPACE.

DROPPING a steel magnet, or vibrating it in other ways, diminishes its magnetism.

It is said that steel containing 12 per cent. of manganese cannot be magnetized.

Flames and currents of very hot air are good conductors of electricity. An electrified body placed near a flame soon loses its charge.

In charging a secondary battery, the charging electro-motive force should not exceed the electromotive force of the battery more than 5 per cent.

Lightning has an electro-motive force of 3,500,000 volts and a current of 14,000,-000 amperes. The duration of the discharge of lightning is 1-200,000 of a second.

IMPROVED CORLISS ENGINE.

The resistance of copper rises about 0.21 per cent. for each degree of Fahrenheit, or about 0.38 for each degree Centigrade.

A lightning rod is the seat of a continuous current so long as the earth at its base and the air at its apex are of different potentials.

The rate of transmission on Atlantic cables is eighteen words of five letters each per minute. With the "duplex" this rate of transmission is nearly doubled.

The effect of age and of strong currents on German silver is to render it brittle. A similar change takes place in an alloy of gold and silver.

A test for the porosity of porous cells consists in filling the cell with clean water and taking the per cent. of leakage. The correct amount of leakage is 15 per cent. in twenty-four hours.

If the air had been as good a conductor of electricity as copper, says Prof. Alfred Daniell, we would probably never have known anything about electricity, for our attention would never have been directed to any electrical phenomena.

A perfect vacuum is a perfect insulator. It is possible to exhaust a tube so perfectly that no electric machine can send a spark through the vacuous space, even when the space is only one centimeter.

For resistance coils, for moderately heavy currents, hoop iron, bent into zigzag shape, answers very well. One yard of hoop iron, ½-inch wide and 1-32 inch thick, measures about 1-100 of an ohm; consequently, one hundred yards will be required to measure an ohm.

Compression of air increases its dielectric strength. Cailletèt found that dry air compressed to a pressure of forty or fifty atmospheres resisted the passage through it of a spark from a powerful induction coil, while the discharge points were only 0.05 centimeter apart.

An accumulator with seventeen plates, ten by twelve inches, is reckoned in horse power hours, equal to about 1 horse power hour. Taking this as a basis, it will require six cells for 1 horse power for six hours, or thirty cells for 5 horse power for the same length of time.

The voltage of a secondary battery must always be equal to or slightly in excess of the voltage of the lamp to be burned. For example, a twenty volt lamp will require ten secondary cells, but ten cells will support more than twenty lamps.

To obtain the length of wire on an electro-magnet, add the thickness of the coils to the diameter of the core outside of the insulation, multiply by 3.14, again by the length, and again by the thickness of the coils, and divide by the diameter of the wire squared.

Blotting paper, saturated with a solution of iodide potassium, to which a little starch paste has been added, forms a chemical test paper for testing weak currents. When the paper (slightly damp) is placed between the terminals of a battery, a blue stain appears at the anode, or wire connected with the carbon or positive pole of the battery.—Boston Journal of Commerce.

## PUBLICATIONS.

Mr. Street of the Ottawa Electric Company, is introducing a new meter ledger and system of keeping electric light meter accounts with customers, which appears to be much superior to the methods generally used. It effects a saving of time and book space. A customer's account, occupying from ¼ to ½ page, shows the meter readings, debits and credits for a number of years, so that the whole history of the account can be seen at a glance as well as the balance owing to the company at any time.

The publishers have favored us with a copy of the Stationary Engineer's Gazetteer of Illinois, being a directory of chief engineers, and engineers in charge, owners of steam plant, etc., etc., within the boundaries of said State.

The September number of the Review of Reviews contains two articles which will be of special interest to electricians, as well as readable by all who watch the world's progress. One is "Nikola Tesla and the Electrical Outlook," the other "Industrial Niagara." Both are extremely interesting.

The Parry Sound Electric Light, Heat and Power Co. are applying for a charter. The company have been granted a five years franchise by the Town Council of Parry Sound, and have agreed to furnish incandescent street lamps of 50 c. p. at $17 per lamp per annum, and to supply lamps for residences at the following prices:—first five, 30 cents each; five additional, 25 cents each; all over ten, 20 cents each. For stores, first two 50 cents each; two additional, 40 cents each; all over six, 25 cents each, with special rates for churches and public halls.

The tender will probably be accepted of the St. Thomas Street Railway Company to light the streets of the city for eight years, from January 6th, 1896, with 90 electric arc lights, 2,000 candle-power, cedar poles on a moon light schedule of 305 nights a year, at 25 ½ cents per lamp per night. The annual cost of lighting under this scheme will be $7,000. The present cost of 31 electric lamps of 1,200 candle-power, and 98 gas lamps, is $5,226, and the cost of their continuance under a proposition from the gas company for a renewal of their contract would be $3,820.

An order has been placed by the Michigan Central Railway Company, with the General Electric Co., of the United States, for two powerful search lights, which will be used to illuminate Niagara Falls. These lights will each have a brilliancy of 100,000 candles, and will be operated with different colored lens. The power to generate the current will be taken from the Niagara river. The effect is expected to surpass anything of the kind to be seen in the world. The only similar attempt at illuminating water falls is that to be seen at the Rhine Schloss, Laufen, Germany.

### SPARKS.

There are female locomotive engineers in Colorado and Kansas.

The project of a Pacific cable is likely to be much advanced by the visit of Mr. Hosmer of the C. P. R. telegraph, and Mr. John W. Mackay, the American millionaire, to the Pacific coast.

The telephone, it is said, is not making much progress in Russia. And no wonder ! Fancy a man going to the 'phone and shouting ; "Holloa, is that you, Divsostkivchsmartvoiczski ? " No, it's Zollemschouskaffirnocknstiffsgowoff, who's speaking?" "Sezmochockiertrjuaksmzyskischohemoff. I want to know if Xlifsromanskeffiskillmajuwchzvastowsksweibierski is still stopping at Dvisostkivchsmartvoiczski."—The Katipo.

The first electric boat on the Rideau canal made the trip from Ottawa a few days ago, in charge of Mr. O. Higman, chief electrician of the inland revenue department of Canada. The Minosee, an Indian word signifying beautiful boat, is thirty-seven feet in length with a beam of seven feet, carrying fifty-two storage batteries, with a four horse-power motor, and when charged she will run about seventy-five miles. The wheels and rudder are attached to the same shaft and so arranged that one man can manage the motor and steer. Her average speed is eight miles an hour. She will accommodate thirty persons comfortably.

Messrs. Wm. Kennedy & Sons, Owen Sound, have an order from the Sault Ste. Marie Water, Light and Power Co. for eighteen water wheels of the latest "New American" type. The wheels are to be 57 inches in diameter and aggregate 5,868 h. p. They are to be used in driving the machinery of the large wood-pulp mill being erected by the above company at the Sault Ste. Marie, Ont.

## TENDERS

### Electric Lighting Plant

Tenders are invited at once for a plant of 1000 light capacity by the undersigned. Information, specifications, &c., can be obtained from the undersigned, or from Mr. Geo. White-Fraser, Consulting Engineer, Imperial Loan Building, Toronto. Tenders must be on forms furnished on application.

D. GRAHAM, SONS & CO.
Inglewood, Ont.

# ELECTRIC RAILWAY DEPARTMENT.

## CANADIAN ELECTRIC RAILWAYS.

WHILST speaking needed words of caution to the Canadian Electrical Association at its annual convention in Toronto in 1893, President J. J. Wright took no optimistic view of electrical possibilities when he referred to the wonderful future ahead of the electric railway. The events of the past two years have furnished much evidence in this direction.

The CANADIAN ELECTRICAL NEWS placed itself in communication some time since with the managers of the various electrical railways in Canada, with the purpose of ascertaining what progress had been made in this direction. We have to thank a considerable number of these for the ready response made to our inquiries, and from the data thus secured we are able to present a fairly full statement of the extent of the electric railways of the Dominion.

Altogether there are about 30 street and suburban railways in Canada. The large majority of these are operated by electric power, though a few still hold to animal power.

Nearly $10,000,000 capital, or to give the exact figures, $9,905,000, is represented in the 12 railways of which we give statistics below. 255 miles of road is covered by these companies, who operate together 387 motors and 144 trailers.

To a large extent a uniform fare is charged by the different street railways of the country. Except in the case of suburban roads, where the fare must be regulated by the length and conditions of travel, five cents is the usual fare, or 6 tickets for 25 cents, 25 for $1.00, with 8 tickets for 25 cents at certain hours of the day, and children's tickets 10 for 25 cents.

The franchise of the different roads varies from 20 to 50 years. The Galt, Preston and Hespeler road, as also Hamilton, have a 20 years' franchise, with Toronto 30, and London, the highest, 50 years.

The most important railway from which an answer to our enquiries was not received at the time of writing, was Montreal, where there is about 70 miles of road, and which, since the adoption of electric power, has attained to remarkable success. The stock of the Montreal Street Railway is viewed in financial circles as one of the best investments on the market, and has been going up with leaps and bounds. No report is received from British Columbia, where there is a road of about 7½ miles length in Vancouver, and another of nearly double that length at Victoria. Aside from the Winnipeg Electric Street Railway, of which full particulars were received from the manager, Mr. G. H. Campbell, there is also a railway about 5 miles length operated by horse power. St. John, N. B., has quite a successful electric railway, and in Halifax and Yarmouth, N. S., there are two roads under operation. Of other roads that have failed to report may be named Belleville, Brantford, Kingston, Peterboro', Port Arthur, St. Catharines, St. Thomas, Waterloo and Quebec. With some of these animal power is still used, and with others, as Kingston, electric power has recently been brought into force. Approximating the mileage of these several roads at 187 miles, which is close to the figure, and adding this to the 255 mileage already noted, it may be said that there is 442 miles of street and suburban railways in Canada, or perhaps in round figures 500 miles.

Individualizing the reports received we take the leading cities first, and the statistics are as follows :

TORONTO.—Capital stock : $6,000,000. Officers : President, William McKenzie; superintendent, James Gunn; secretary, J. C. Grace; comptroller, J. M. Smith. Mileage, 81¼. System in use: General Electric, Westinghouse, and Thomson-Houston. Motors, 166, trailers, 70. Power plant : two 1,600 h. p. engines with multipolar generators coupled direct, 800 K. W. each, working up to 2,000 amperes each at 560 volts; four 620 h. p. and one 430 h. p. engines, with 10 generators driven by belt, 200 K. W. each; output, 4,000 amperes at 560 volts. Capacity of total output, 7,000 amperes at 560 volts. Indicated horse power, 6,110. Period of franchise : 30 years from Sept. 1, 1891.

OTTAWA.—Capital stock authorized $1,000,000; paid up $814,800. Officers : J. W. McRae, president; W. Y. Soper, vice-president; T. Ahearn, managing director; James D. Fraser,

secretary-treasurer ; J. E. Hutchinson, superintendent. Mileage, 23. System in use: Westinghouse. 60 motors. Power plant : one 700 h. p. generator; two 400 h. p. generators; three 100 h. p. generators. Franchise: 30 years from 13th Aug., 1893.

HAMILTON.—Capital stock : $205,000. Officers : B. E. Charlton, president; E. Martin, Q. C., vice-president; J. B. Griffith, secretary-treasurer and manager. Mileage, 22. System in use: Westinghouse. 35 motors and 14 trailers. Power plant : 3 Wheelock, 260 h. p., and one Corliss 260 h. p. engine. Franchise, 20 years.

LONDON.—Capital stock : $250,000. Officers: H. A. Everett, president; E. W. Moore, vice-president; C. E. A. Carr, manager and treasurer; S. R. Break, secretary; Charles Currie, assistant secretary; D. L. D. DeHart, superintendent. Mileage, 25. 60 motors and trailers. Franchise, 50 years.

WINNIPEG.—Capital stock : $300,000. Officers : James Ross, president ; W. Whyte, vice-president ; William McKenzie, treasurer ; F. Morton Morse, secretary; G. H. Campbell, manager. Mileage, 16. System : Edison. 24 motors and 12 trailers. Power plant : one 900 h. p. Corliss engine (Laurie type), one Wheelock 250 h. p. engine, and three Edison generators. Franchise, 35 years.

WINDSOR, ONT.—Capital stock : $500,000. Officers : John Coventry, M. D., president ; Geo. M. Hendrie, vice-president ; William J. Pulling, treasurer ; James Anderson, secretary. Mileage, 10. System in use : Westinghouse. 25 motors and trailer cars. Power plant : two Robt. Armstrong engines, 200 h. p. each, and one Brown-Corliss, 125 h. p. Franchise: 20 years from March, 1893.

SARNIA, ONT.—Capital stock : $50,000. Officers : J. S. Symington, president; H. W. Mills, secretary and manager. Mileage, 4. System, animal power. 9 cars and 21 horses. Franchise, 30 years.

NIAGARA FALLS, ONT.—Niagara Falls, Wesley Park and Clifton Tramway Co. Capital stock : $50,000. Animal power in use; about changing to electric. Mileage, 4. Plant : 8 cars, 20 horses. Franchise, 20 years.

NIAGARA FALLS, ONT.—Niagara Falls, Park and River Railway. Capital stock : $1,000,000. Officers : E. B. Osler, president; William Hendrie, vice-president; R. A. Smith, secretary-treasurer ; W. Phillips, electrician ; H. Rathery, superintendent ; Ross Mackenzie, manager. Mileage, 13½ ; 12 miles double track. System in use: Canadian General Electric. 14 engines, one luggage van with motors, 16 trailers, and 10 observation cars with double trucks, motors, (41 in all). Power plant : one water power station with two 1,000 h. p. wheels, and three 250 h. p. dynamos; large steam station with two 150 h. p. engines, and two 125 h. p. dynamos. Franchise, 40 years. Charter admits of about 27 miles more line being constructed, viz.: to Fort Erie and Niagara-on-the-Lake, and also a railroad on the water's edge from the Falls to Queenston.

HAMILTON, ONT.—Hamilton, Grimsby and Beamsville Electric Railway. Capital stock, $200,000. Officers : C. J. Myles, president : T. W. Lister, vice-president ; Adam Rutherford, sec.-treas.; A. J. Nelles, manager ; C. K. Green, electrician. Mileage, 25. 7 motor cars, 1 trailer, 1 motor freight car, 3 trailer fruit cars. Power plant : 2 Inglis engines, 150 h. p. each, 3 boilers, 2 Westinghouse generators ; Westinghouse electrical equipment.

GALT, ONT.—Galt, Preston and Hespeler Street Railway Co., Ltd. Capital stock, $100,000. Officers : Thomas Todd, president ; R. G. Cox, vice-president ; W. H. Lutz, sec.-treas. ; W. A. Lee, manager. Mileage, 9. System : Westinghouse and Canadian General Electric. 5 motor and 3 trailers. Power plant : 2 Wheelock engines, 125 h. p. each ; 2 generators—1 Westinghouse 85 K. W., 1 General Electric 100 K. W. Power house is located at Preston, being about the centre of the line. Franchise, 20 years. The company have further in the way of equipment a Baldwin steam motor, hauling capacity, 500 tons, with which freight is carried between Preston and Hespeler and the C. P. R. at Galt. The tracks connect with the C. P. R. and are the same guage. There is also an electric freight car, equipped with 2 Canadian General Electric 1200 motors, and also a freight trailer for carrying less than car load lots. The tracks from Galt to Preston are on the highway, and between Preston and Hespeler a private right of way is held. The average freight amounts to about 600 tons per month, and passengers about 20,000 per month. Sidings are run into all the factory yards in Preston and Hespeler, so that freight can be loaded at their own doors. 30 passenger trains are run every day. The freight work is all done at night after 10 o'clock. This is the pioneer Canadian road, combining freight and passenger traffic, and has reached very marked success.

TORONTO JUNCTION, ONT.—Toronto Suburban Street Railway Co. Capital stock, $250,000. Officers: R. Wilson Smith, president; E. P. Heaton, vice-president; R. H. Fraser, secretary and manager. Mileage, 10. System: Edison, General Electric and Westinghouse. 5 motors and trailers. Power plant: 1 100 K. W. B. P. Edison generator, 1 150 h. p. engine. Franchise, 20 years.

### THE COMING STREET RAILWAY CONVENTION.

THE convention of the American Street Railway Association, which assembles at Montreal in October, is exciting much attention from railway men, who look upon it as one of more than usual importance, on account of the questions which are likely to come before it. The association comprises not only the officials but also the electrical and mechanical experts of the leading railway companies of Canada and the United States, and therefore wields great influence, as well as containing the concentrated experience of a large number of able men. There are nearly 1,000 companies in existence, operating 13,000 miles of track, and representing a capital of something like $1,200,000,000, with a yearly earning capacity of from $125,000,000 to $140,000,000.

Referring to the questions likely to come before the convention, a writer in the St. Louis Globe Democrat says:—

Considerable speculation is being indulged in by the street railway men of this and other cities concerning the proposed action of the fourteenth annual meeting of the American Street Railway association, which is shortly to be held at Montreal. It was given out some time ago that the scope of the organization is to be enlarged and its character somewhat changed. At the last annual meeting formal steps were taken to bring about the change. The executive committee was instructed to elaborate plans. This, it is understood, has been done. The committee has not given out what kind of a report it is going to make. However, among some of the propositions to be presented is the establishing of a central bureau of information, which shall be charged with the duty of answering all questions propounded by members on problems affecting their interest. Questions of operating methods, of patents, of legislation, of franchise, etc., etc. The bureau is to be in charge of men well versed in law, patents, insurance and the like. To run a bureau of this kind, the promoters say, will require a great deal of money, that can only be obtained by subscription and by putting up of the dues. Opinion is divided as to the value of the bureau, in relation to its great cost to maintain. The large companies are, it is said, in favor of the scheme, and will insist on its establishment, while the small companies will object thereto for the reason of the large yearly assessment they will be called upon to pay. The annual dues at present are $25, when, under the new arrangements, the dues, it is expected, will be at least ten times that amount. This and the other equally radical changes contemplated is what has started the talk and the discussion in railway circles. Another change that will be made is holding of more executive sessions, and that, hereafter, the proceedings, aside from the reading and discussion of papers, will be secret. Heretofore the annual gatherings have been regarded more as a pleasure outing than a business assembly. The thoughtful ones are begining to see that more unity and a great deal more action is needed on account of the magnitude of the interests at stake, and the complications that are on the increase. There are many troubles and difficulties arising during the course of the year that bother many roads in different parts of the country. In other words, many roads are afflicted with the same annoyances at the same time. If all the troubles were referred to the bureau, much worry and expense would be avoided. This is the

illustration given by the friends of the bureau. They say that the proposition for its establishment will be carried by a large majority, and that half of the benefits to be derived therefrom have not been told, and can only be known by those on the inside. They further state, that once started, its merits will be highly appreciated by all and voted a grand affair and acknowledged that it fills a long felt want.

### THE BERLIN AND WATERLOO STREET RAILWAY.

THIS railway has been in operation for many years as a horse-car road. It is two and one-half miles in length, connecting the two thriving towns of Berlin and Waterloo, and has also a short spur line running to the Grand Trunk depot in Berlin.

The management has for some years been desirous of changing the road to be operated by electricity, but various obstacles have so far hindered the project. The road was originally built for light horse-car traffic; in the main part of both towns the old style of flat rails, weighing about 27 pounds per yard, were put down, while over the remainder of the line a 30 pound steel tee rail was laid. The road-bed was well graded and ballasted and is still in good condition. No very heavy grades are met with, the steepest being 4%, about 300 feet long, and the general being 2½%.

The closed cars in use were of a 12 foot body, and a number of 16 foot open cars were kept for summer use.

To change the road into an electric one, according to the most modern practice, would therefore have necessitated the discarding of all the old material and stock—in fact, building and equipping an entirely new road. The company had for some time past not been making more than expenses, and the necessary funds for such a complete overhauling were not available.

In view of these circumstances, it was decided last winter, on the advice of the Company's consulting engineer, Mr. E. Carl Breithaupt, to utilize the old material in so far as possible and undertake the work of making the change.

The cars were in good condition. Three of the closed cars were altered and equipped as motor cars, a 4½ foot closed vestibule being built on at each end, thus making the car 21 feet over all. One of these was equipped with two Canadian General Electric "800" motors, using series parallel controllers, and the other two were each equipped with one 25 P Westinghouse motor. The two latter are used in the general service and the double motor car is kept as a spare and for special occasions, when one or two trailers can be used with it. Peckham trucks are employed throughout and give excellent satisfaction. The wheel base being 11 ft. 6 in. from centre to centre brings the main part of the car body, which carries the main part of the load, directly over it. The cars thus ride very easy and entirely without any rocking motion.

The road-bed was re-ballasted and re-graded where necessary, but it required very little work, since the road-bed was in good order and the ties were mostly sound and firmly imbedded. The old tracks were used throughout; at rail joints particular attention was given to secure a solid foundation and a rigid joint, and the track was double bonded throughout.

The track construction was the chief point of difficulty. It was feared by the management that the old rails would be too light for the heavier traffic, and to put down a new track using 52 pound rails would have involved too great an expenditure. As they are now used, the engineer expects that the old rails will serve for two or three years at least; at the end of that time it can reasonably be expected that the Company will be able to put down a new track. In the present work the bond wires were made somewhat larger than actually required, so that they can be cut off and used again. The only extra item of expense incurred in the complete reconstruction of the road-bed will therefore be the labor of bonding on the present track—a comparatively small item. The cost of operation will be somewhat increased, for a larger or efficient of traction must be allowed for with these rails, and the cost of track maintenance will also be slightly greater. In these respects this road will furnish some interesting data. It is Mr. Breithaupt's intention to make a series of experiments with special reference to track resistance.

Power is supplied from the electric station of the Berlin Gas Co., a 100 K, W. Edison bi-polar generator being used.

The new road has been in operation since May 18, and so far

the expectations as to increased traffic have been more than realized, and everything has worked well, though the system has already had some severe tests. On the occasion of the annual bicycle meet, July 1 and 2, the traffic was very heavy. A ten minute service was maintained and five thousand passengers were carried on the first day without any serious mishap.

The case furnishes a good example of what can be done in the way of improving street railway properties, which have depreciated somewhat, chiefly from the fact of their being out of date. Horse car traffic is at the present time too slow in any case, and particularly so in suburban and kindred work. It is not claimed that the Berlin and Waterloo road is a model one, embodying all the latest improvements in street railway work, but it has been put on a firm footing and is thoroughly well equipped ; moreover, the work has been done cheaply and no money has been expended in utilizing old material, which will be wasted when this material is replaced.

### THE COMING STREET RAILWAY CONVENTION AT MONTREAL.

WE had hoped to be able to print in this number of the ELECTRICAL NEWS the program of the convention of the American Street Railway Association to be held in Montreal from the 15th to the 18th of October. This cannot be done, however, owing to the arrangements being as yet incomplete, in consequence of which the program cannot be issued for perhaps a fortnight. A meeting of the Executive to further the arrangements was held in Montreal on the 5th inst., but nothing of importance was done. It is probable that a luncheon and drive will be given the visitors by the city council. Montreal can be depended upon to maintain on this occasion the enviable reputation it has gained for hospitality. We present herewith a portrait of the gentleman who will preside over the deliberations of the assembly.

### SPARKS.

Arrangements are said to have been completed for the amalgamation of the St. Thomas Street Railway Co. and the Radial Electric Railway Co.

MR. HENRY C. PAYNE, MILWAUKEE, WIS.
President American Street Railway Association.

The Parry Sound Electric Light, Heat and Power Co. has commenced the construction of its plant. It is expected to be in operation by the 1st November.

Negotiations are in progress for an electric road between Renfrew and Portage du Fort. The distance is eight miles, and water power will generate the electricity.

The promoters of the Aylmer Electric Railway will be compelled to seek an extension of time for the construction of their road, as their charter will expire shortly.

Messrs. Howard, Leamy & Murphy, contractors for the section of the belt line railway extending from Hochelaga to Bout de L'Isle, expect to have the work completed in three months.

It is probable that the Hamilton and Dundas Railway will be changed into an electric road on the termination of the present lease in nine months time. It carries 250,000 passengers a year.

Electric power for manufacturing purposes is now being transmitted from the power house of the Cataract Company at Niagara Falls, to the aluminum works of the Pittsburg Reduction Company.

Mr. F. G. Mitchell will represent No. 5, London, at the convention of the Canadian Association of Stationary Engineers, Ottawa, 24th to 27th Sept., with Mr. Simmie, of the waterworks, as alternate.

The Knechtel Furniture Company, of Hanover, Ont., have purchased a water power at Maple Hill, and are said to be considering the question of transmitting electric power to manufacturies at Hanover,

The prospect of the electric railway being built in Hull this year are not very good, through want of capital. It is expected, however, that the electric light service, for which the same contractor, Mr. Vaiu holds the franchise, will be completed.

The Village Council of Hintonburg are negotiating for the securing of the necessary right of way for the extension of the Ottawa Electric Railway Co.'s lines to that place, and as soon as these negotiations are successfully completed, the extension will be made.

It is reported that Scotch capitalists have sent a representative to Schomberg, Ont., to report as to the business prospects for a railroad, to be operated either by steam or electricity. The locality in question is one of the richest grain growing districts in Ontario.

The earnings of the Montreal street railway for August were $109,316.30. In the corresponding month of 1894 they were $90,202.66, showing an increase of $19,113.66. The largest receipts for any single day were on Saturday, the 17th, when they amounted to $4,404.04.

The Toronto Railway Company has issued £323,000 of first mortgage 4½ per cent. sterling bonds. Of this amount £250,000 was issued in Canada and the remainder in London. The subscription lists closed in Toronto August 2, when about £150,000 had been subscribed.

Chas. McLeod, a lineman employed by the Windsor, Sandwich & Amherstburg Street Railway, was badly injured by coming in contact with a loose wire, one end of which was lying on the trolley wire and the other touching the ground. The strength of the current he received was 500 volts. The doctor says he will recover.

A representative of the Canadian General Electric Co. has been in Kalto, B. C., in connection with a project to supply the city with water by a somewhat novel plan. It is proposed to take water from the lake and force it into the mains by means of an electric pump, the power for which is generated by water on the river a short distance above the city.

Mr. E. Franklin Clements, of the Standard Telephone Co., of New York, is said to be endeavoring to obtain the consent of the Prince Ed. ward Island Government to the construction of a transcontinental telephone system in that province. Mr. Clements is said to be also negotiating for the establishment of an electric street railway to be constructed by American capitalists in the City of Charlottetown.

The Lachine Rapids Hydraulic Company are proceeding with the construction of a dam 4,000 feet in length, for the purpose of generating electric power to be transmitted to the city of Montreal. It is expected that when the work is completed, the company will have 8,000 horse power at its disposal, and the contractors guarantee that the work will be completed, and that the company will be in a position to supply power by the 3rd of May, 1897.

It is proposed to form a company to supply Ashcroft, B. C., with water and electric light. Water is to be brought from the mountains to serve for household purposes and to furnish motive power for the electric plant.

Steps are being taken to incorporate a company to build an electric railway between Detroit and Port Huron. The road will run close to the river and lake St. Clair, through a territory not tributary to the Grand Trunk. Much of the right of way, it is said, has already been secured, and surveys made. The distance is 65 miles, and the cost of the road, power houses and equipments is put at six hundred thousand dollars.

The work of constructing the new electric railway at London, Ont., is being vigorously pushed forward. On one branch of the system a car has already been successfully operated, and an effort will be made to provide electric transit for the visitors to the approaching Western Fair. The construction of a substantial power station from which the system will be operated has been commenced. The line to Springbank has been completed and is in operation.

The Montreal and Toronto Street Railway Companies have recently inaugurated a new idea in the shape of excursion trains, which are handsomely fitted up, and are used for the purpose of conveying social parties and strangers visiting these cities, over the most interesting of the company's lines. This method of seeing the interesting features of cities has become popular in the United States, and it is believed will in time become equally so here. It has the advantage of being very inexpensive as compared with hiring carriages.

## SPARKS.

The Aurora electric light plant has been sold by the Royal Electric Co., to the Metropolitan Lighting Co., who will introduce some radical changes.

R. S. Willison, who claims to have been injured last January by a motor car at Sherbourne and Carlton streets, has filed suit against the Toronto Railway Company for $2,000 damages.

The work of erecting the trunk line for the Bell Telephone Company between Toronto and Montreal is progressing rapidly. It is within a few miles of completion between Toronto and Belleville, and is finished between Kingston and Mallorytown.

W. O. Ogilvie, the great miller, has presented the Winnipeg General Hospital with a pair of steam boilers with fittings, for the purpose of establishing an electric light plant in the hospital, and for other purposes for which steam is required.

The Royal Electric Company have contracted with the City Council of Charlottetown, P.E.I., to supply that city with 65 arc lights of 1,260 candle power each for $73 per lamp per annum. The company also agree to allow the corporation 50 cents per light when the lights are out.

The town of St. Marys recently advertised for tenders for electric light. The only tender received was that of Mr. H. L. Reesor, the present contractor. Mr. Reesor made two offers as follows : to supply 32 arc lights at $45 00 per light, and 24 incandescent lights at $87.00 per year ; or to supply 13 arc lights of 2000 c. p. at $49.00 per light, and 55 incandescent lights of 30 c. p. at $15.00 per light per year. It was decided not to accept either of these offers, but to call a public meeting to discuss the best method of lighting the town.

# CANADIAN GENERAL ELECTRIC CO.
### (LIMITED)

Authorized Capital, $2,000,000.00.
Paid up Capital,     $1,500,000.00.

### HEAD OFFICE:

## 65 FRONT STREET WEST, - - TORONTO, ONT.

**BRANCH OFFICES:**

| | | | |
|---|---|---|---|
| 1802 Notre Dame St. - | MONTREAL. | Main Street - - | WINNIPEG. |
| 138 Hollis Street - | HALIFAX. | Granville Street - - | VANCOUVER. |

PETERBOROUGH WORKS—Part of Main Floor, Machine Shop.

STANDARD ELECTRICAL APPARATUS AND SUPPLIES FOR

RAILWAY LIGHTING POWER AND MINING PLANTS

## TRADE NOTES.

E. S. Stephenson & Co., St. John, N. B., have ordered a 50 horse-power Monarch Economic boiler from the Robb Engineering Co.

The Dominion Coal Co. has placed an order with the Robb Engineering Co. for a 60 horse-power engine to run their machine shop at Glace Bay.

It is the intention of the Mica Boiler Covering Co., of 2 Bay street, Toronto, to remove their business to more extensive premises on Jordan street.

The Halifax Electric Railway Co. have ordered two 300 horse-power Robb-Armstrong engines, in addition to two of the same kind now under construction for them.

Messrs. Patterson & Corbin, of St. Catharines, are furnishing twenty-five cars for the London electric street railway, and four for the Montreal Park and Island railway.

The Canadian General Electric Co. are installing a direct connected engine and dynamo in the Ladies' College at Whitby. The engine is a 50 horse-power Robb-Armstrong.

The Dodge Wood Split Pulley Co., of Toronto, have been given the contract for supplying the split pulleys, and split friction clutch pulleys for the Ottawa Porcelain & Carbon Co.'s extensive new works at Ottawa.

The North Sydney Electric Co. have decided to enlarge their plant and have ordered two dynamos from the Canadian General Electric Co., and a 100 horse-power Robb-Armstrong engine and Monarch Economic boiler from the Robb Engineering Co.

The recently organized firm of Stilwell, Ralston & Co., at Hamilton, Ont., has been succeeded by Messrs. Stilwell & Co., who propose to manufacture incandescent lamps and other electric specialties. The advertisement of the new company appears in the present number.

The Dodge Wood Split Pulley Co., of Toronto, have supplied R. Thackeray, of Ottawa, with a very neatly designed rope drive for the transmission of the power required in the new extension just erected to his extensive planing mills. They have also supplied the required belt pulleys.

The Babcox & Wilcox Company report business as excellent, their orders for June alone exceeding 25,000 h. p. Of this amount they have orders for 6,000 h. p. of their water tube marine boilers from the Plant Steamship Company. All the boilers built by the Babcock & Wilcox Company at their Belleville shops are of wrought steel construction, having a capacity for 200 lbs. working pressure.

In the description published in the ELECTRICAL NEWS for August, of the fire-protection plant of the Gooderham & Worts Co., Ltd., Toronto, mention was omitted of the fact that the pumping engines were supplied by the Northey Manufacturing Co., Ltd., Toronto, and are of their latest type of compound condensing pumps of the "Underwriter" pattern. These pumps are to be finely finished, and will deliver ten or twelve fire streams of much greater efficiency than can be furnished by the ordinary city pressure.

The Royal Electric Company have in operation, in Machinery Hall, at the Toronto Industrial Exhibition, their S. K. C. two phase-alternating current system, and from their two-phase alternating dynamo, are supplying current at the same time for lighting incandescent lamps, arc lamps, and operating motors.

The Dodge Wood Split Pulley Co., of Toronto, have in hand two mammoth rope drives, for the E. B. Eddy Co., of Hull, Que., each drive to have a guaranteed capacity of 700 h. p. The drives are used in the transmission of power from new McCormack water-wheels, being installed for the purpose of increasing the pulp grinding capacity of the company. The E. B. Eddy Co. are of the opinion that the rope drive is a long way ahead of any other means of transmission, especially for heavy work.

The Packard Lamp Co., Ltd., have recently issued a circular announcing that their new factory at St. Catharines is now in full operation, manufacturing lamps and transformers, and is equipped throughout with new and improved machinery, by means of which the product will in the future be of the most superior description. Mr. W. D. Packard has assumed the general managership of the company, with Mr. A. G. Powell as assistant. Prof. Thomas, of the Ohio University, who was chairman of the World's Fair Committee on incandescent lamps, has recently conducted a series of tests of Packard lamps, and concludes his report with the following statement : "Taking economy, maintenance of candle power, and freedom from blackening into account, the results obtained from these lamps are much superior to any heretofore published."

The Canadian Electric Repair Co., has recently been organized under very favorable auspices, and has begun business at No. 623 Lagauchetiere street; Montreal, with Mr. Geo. E. Matthews as manager. Mr. Matthews is well known in the electrical field, he having spent some 20 years in the mechanical and electrical business. He has had the advantage of being an early beginner with the Royal Electric Co., and succeeded in working himself up to be superintendent of the winding department, which position he held for eight years. He was also associated with Mr. David A. Starr, Montreal, for about eight months in a repair business. The above concern will undertake the rewinding of armature fields and transformers, also all other electrical and mechanical repairs required for electrical machinery and apparatus.

We take pleasure in announcing to our readers that the Boudreaux dynamo brush which has met with marked success in the United States and abroad, is now being introduced into Canada, the Boudreaux Dynamo Brush Co. having placed the agency for Canada with R. E. T. Pringle, of Montreal, who is now ready to fill all orders for brushes ranging in thickness from ⅛ to ½ inch, and in width from ¾ inch to 3 inches. The Boudreaux "Foliated" brush, it will be remembered, is made of anti-friction metal rolled into sheets 1/16 inch thick, and folded and refolded until the desired thickness is attained. It differs essentially, therefore, not only in the material, but also in the manner of its application from the old "laminated" brushes. Besides its anti-friction properties, the Boudreaux brush also possesses in an eminent degree the non-sparking property, so that it preserves the commutator from wear, leaving it with a smooth and polished surface. It might be advantageous to consumers to make further enquiries regarding these brushes.

## SPARKS.

The Treasurer of Montreal Association, No. 1, C. A. S. E., acknowledges through the press a large number of subscriptions to the Association's Mechanical Library.

Mr. Jeule Behm, an employee of the Montmorency Electric Power and Light Co,, of Quebec, had his hands severely burned while working at a new switch board.

Several accidents have recently taken place in connection with the operation of the Oshawa Electric Railway, one being the burning out of the coils of the generator armature, and another the blowing out of the cylinder head of the engine at the power station.

At the annual meeting of stockholders of the New Brunswick Electric Telegraph Co., held on the 12th of August, the old board of directors was re-elected as follows:—C. W. Weldon, president; D. C. Dawson, secretary-treasurer; L. J. Almon, J. J. Tucker and D. M. Sutherland.

On the occasion of the civic holiday of the town of Galt, the Galt, Preston and Hespeler Electric Railway carried 1,300 passengers, and this number, it is said, would have been considerably increased had the weather not proved unfavorable during the early part of the day. Regular trips are now being made to Hespeler every half hour.

An invitation has been received to attend the observance of the twenty-fifth anniversary of the founding of the business of the American Electrical Works, Providence, R. I., and in connection therewith the 17th Annual Rhode Island Clam Dinner, tendered to the electrical fraternity at Haute Rieve, on Aug. 17th. It was a matter of regret that we were unable to participate in the pleasure of the occasion.

The Town Council of Trenton, Ont., have awarded to the Brush Electric Co., of Cleveland, Ohio, a contract for lighting the town for the period of ten years. The company will, in addition to supplying incandescent light, also distribute power for manufacturing purposes throughout the town, having at its disposal one of the finest water powers in the Dominion, situated immediately north of the town, and capable of producing 12,000 H. P.

One of the largest shoe manufacturers in the city of Quebec has given in writing to the Montmorency Electric Power Co., his testimony to the advantage which he has obtained by the use in his factory of electric motors, as compared with the steam plant formerly employed. He states that apart from the fact that the motors are instantaneous in operation, smooth running, free from danger, occupy but little space and are less expensive to maintain, he has been enabled since putting them in to manufacture, with the same amount of machinery, from 200 to 300 pairs of boots more than formerly.

Please mention the CANADIAN ELECTRICAL NEWS when corresponding with Advertisers

## SPARKS.

A pair of 250 H. P. wrought steel boilers are being placed in the Bell Telephone Co.'s new building in Montreal, by the Babcock & Wilcox Co.

Charles McLeod, lineman of the Windsor, Sandwich & Amherstburg Electric Railway, was injured by coming in contact with a live wire on Aug. 22nd.

It is said to be the intention of the Kingston Electric Railway to secure permission of the City Council to construct a belt line on the eastern side of the city, similar to the one now in operation on the western side.

A test of the magnetic motor invented by Mr. Brintnell, of Toronto, described in a recent number of the ELECTRICAL NEWS, was recently made at the Hubbell Primary Battery Co.'s works at Ottawa. The test is said to have been so successful that it has been decided to commence immediately the manufacture of the machine.

Some of the leading citizens of the village of Cessaire have purchased a water power with the object of providing the municipality with electric light.

A satisfactory understanding has recently been reached at Ottawa, under which the Berlin and Waterloo Street Railway Co.'s cars will cross the Grand Trunk Railway on King street, Berlin.

Mr. H. Pim, the representative of the Canadian General Electric Co. at Vancouver, B. C., is engaged in promoting the installation of an electric light plant at Nelson, B. C. It is proposed to utilize the power from Kalso Creek.

It is reported that owing to what was considered to be an unreasonable demand on the part of some of the promoters of the Windsor-Selkirk Electric Railway, the conference which was to have taken place between the promoters of the road and St. Paul capitalists has been declared off for the present at least.

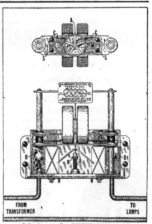

## SPARKS.

The light committee of the Windsor Council have awarded to the Thompson Electric Co., a contract for dynamos, and to Leonard & Sons, a contract for engine and boiler.

The charter has been granted to a local company, with a capital stock of $2,000, to build and operate a telephone line in the counties of St. Maurice, Champlain, and Three Rivers, Que.

A lineman named Alfred Sarazin came in contact with a live wire while at the top of a telephone pole in Ottawa recently, and fell from the pole; fortunately a store awning intervened and broke the fall, thus saving his life.

Mr. D. H. Keely, superintendent of the Government Telegraph lines, is superintending the laying of the new telegraph cable for the Grosse Island quarantine service. The cable extends from the Quarantine Station to Isle Aux Reaux.

Mr. Fraser, Secretary of the Toronto Suburban Street Railway Co., has recently held conferences with the Reeves of the townships of Etobicoke, Markham and York, and with the warden of the County of York regarding the extension of the company's line to the village of Islington.

The New Westminster and Burrard Inlet Telephone Co., has purchased the interest of the local telephone company at Kamloops, B. C., and is about to rebuild and put in good order the system. The arrangements were carried out on behalf of the purchasers by Mr. H. W. Kent, Manager.

In view of the application of the Merchants' Telephone Co. to the Montreal City Council, for exemption from taxes, the Bell Telephone Co. have requested a similar concession, urging the fact that they have been giving a good service at a cost averaging only 11 cents per day to each subscriber, while the average number of calls for each subscriber is twelve, as a reason why their petition should receive consideration.

The control of the Dundas Telephone Co.'s business has recently passed into the hands of the Bell Telephone Co., and new officers have been elected as follows:—W. C. Scott, Toronto, president; W. J. Gilmour, Brockville, secretary; W. Gardener, Winchester, treasurer; directors, W. J. Gilmour, Brockville; S. S. Reveller, Winchester; J. E. McFarlane, Montreal; H. N. Horton, Montreal; W. Gardener, Winchester.

Messrs. French & Hardill, of Stratford, Ont., are said to have invented an improved steam engine, in which the utilization of the steam in the cylinder is effected on a new principle. The most novel feature of the invention is, that there are two piston rods working in opposite directions in the cylinder, the stroke commencing at the centre. A small model of this type of engine has been tested with a 3 H. P. dynamo with satisfactory results.

# The National Electric Co. <sub>EAU CLAIRE, WIS.</sub>

... MANUFACTURERS OF ...

**Bipolar and Multipolar** Power Generators and Motors

**Alternating Self-Contained** Dynamos   **Direct Current** Dynamos

The National Transformers are the best in the market. Our Alternators light the Towns of Oakville, St. Marys, Oshawa, Mount Forest, Palmerston, Grimsby, Port Dover, and hundreds of other towns in Canada and the United States. Ask the companies how they like them. Send for estimates to.

**JOHN W. SKINNER**

500 LIGHT ALTERNATOR.   or apply at 146 YORK ST., TORONTO.   **MITCHELL, ONT.**

---

# Ტᴴᴱ GOLDIE & McCULLOCH Cᴼ. [LIMITED.]

### MANUFACTURERS OF

# Improved Steam Engines and Boilers

## ✦ FLOURING MILLS ✦

*And the Erection of same in the most Complete Style of Modern Improvement.*

## WOOL MACHINERY, WOOD-WORKING MACHINERY, SAWMILL, SHINGLE AND STAVE MACHINERY

### Fire and Burglar Proof Safes and Vault Doors.

---

**Special attention** called to the "WHEELOCK" IMPROVED STEAM ENGINE as being unequalled for simplicity, efficiency and economy in working, and especially adapted for Electric Lighting, Street Railways, etc.

### ⇝ GALT, ONTARIO. ⇜

---

# Ꭷᨆ ROBERT GRAHAM ᨆᎧ

**Iron Founder and Machinist**
Water Wheels, Engines and Mill Machinery a Specialty.        ... OTTAWA, ONT.

### ⇝ THE ⇜

## STANDARD ...
## WATER WHEEL

**M**ADE in sizes from 6 inches to 84 inches diameter. Wheel one solid casting. 84 per cent. of power guaranteed. In five pieces. Includes whole of case, either register or cylinder gate. Water put on full gate or shut completely off with half turn of hand wheel, and as easily governed as any engine . . . . . . . . . . .

Cut showing Wheel Removed from Case.

Write for Estimates, References and Catalogues of the STANDARD WATER WHEEL, also Milled and Rough Gearing of every size and description ; Engines, Mill Machinery and Electric Power Plants ; Latest Improved Band Saw Brazing Tables ; Shears and Gummers ; also Surface Grinder for Shingle Saws.

---

# CANADIAN ELECTRICAL NEWS
## STEAM AND ENGINEERING JOURNAL

OLD SERIES, VOL. XV.—No. 6.
NEW SERIES, VOL. V.—No. 10.

OCTOBER, 1895

PRICE 10 CENTS
$1.00 PER YEAR.

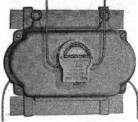

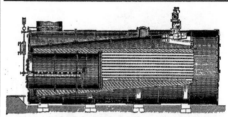

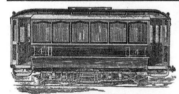

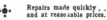

CANADIAN

# ELECTRICAL NEWS

AND

## STEAM ENGINEERING JOURNAL.

| VOL. V. | OCTOBER, 1895 | No. 10. |

### ELECTRIC LIGHT INSPECTION STAFF.

THE accompanying group comprises the electric light inspection staff, with the chief electrician, appointed under the authority of the Electric Light Inspection Act, the scope and aims of which have already been given to the readers of THE NEWS.

The majority of those comprising the group have been engaged since the inception of the Gas Inspection Act in 1875 in

ledge of that science. We are informed by the chief electrician that the work of instructing these gentlemen in the use of the various electrical apparatus with which their offices are equipped was of a very pleasing nature. Their well-trained analytical minds fitted them in no small degree for the new work, and the readiness with which they acquired a knowledge of the instruments and the manner of using them, confirmed Mr. Higman

ELECTRIC LIGHT INSPECTION STAFF.

1. O. HIGMAN, M.Inst.E.E., A.M.Can.Soc.C.E., Chief Electrician, Ottawa, Ont.
2. A. AUBIN, Montreal, Que.
3. D. McPHIE, Hamilton, Ont.
4. A. MILLAR, Halifax, N. S.
5. A. ROWAN, St. John, N. B.
6. J. K. JOHNSTON, Toronto, Ont.
7. JOHN WILLIAMS, London, Ont.
8. N. LE VASSEUR, Quebec, Que.
9. A. F. SIMPSON, Sherbrooke, Que.
10. WM. JOHNSON, Belleville, Ont.
11. H. G. ROCHE, Ottawa. Ont.

the duty of testing gas and gas meters ; and anyone at all acquainted with the delicate nature of these tests, especially the chemical analysis for impurities in the gas, must admit that their training has peculiarly well fitted them for the new work under the Electric Light Inspection Act.

Being warned some years ago that the work of testing electric light meters would be imposed upon them, the gas inspectors throughout the country took up the study of electricity, and have made considerable progress in acquiring an elementary know-

in the opinion he had formed from the beginning that the gas inspectors would do the work satisfactorily and well.

Of Mr. Ormond Higman, the chief electrician, upon whom the work of inauguration has mainly devolved, very little need be said, as his experience in electrical work covers a period of over thirty years. He is a native of Cornwall, England, and entered the service of the Electric & International Telegraph Company at Liskeard in September, 1864. Coming to Canada in 1869, he entered the service of the Montreal Telegraph

Company in October of that year.  In 1873 he was made chief
of the operating staff at Ottawa and manager of the company's
sessional staff in the House of Commons.  For nearly twenty
years Mr. Higman continued in this capacity, also acting as
Inspector of the Ottawa Division.  For five or six years prior
to 1892, in addition to his other duties, Mr. Higman filled the
position of Electric Light Inspector of the Ottawa district for
the Canadian Underwriters' Association.  In September, 1892,
he was invited by the Hon. John Costigan, then Minister of
Inland Revenue, to draft a bill having for its object the inspection of electric lighting ; the measure to be along the lines
of the Gas Inspection Act.  This duty, we are informed, Mr.
Higman performed to the entire satisfaction of the Minister
and the chief officers of the department.  In August, 1893, the
Hon. Mr. Wood, Comptroller of Inland Revenue, sent Mr. Higman to Chicago to represent the Department at the meeting
of the International Congress of Electricians at the World's
Fair.  On presenting his credentials to the committee, Mr.
Higman was chosen to represent British North America in
the Chamber of Delegates—a body composed only of official
delegates from the different governments of Europe and
America—to consider and adopt a system of units of electrical
measure.  He was also made a vice-president of the General
Congress.

During the session of 1894 Parliament passed two electric
bills, one for the purpose of legalizing the international units
adopted at Chicago, the other dealing with the work of inspection.  Mr. Higman's appointment in connection with the
inauguration of the latter followed as a matter of course.  We
believe Mr. Higman is possessed of sound judgment and
understands thoroughly both the nature and standing of the
electric lighting industry in this country ; and in the administration of the new law we are satisfied that he will not only
protect the rights of the consumer, but will conserve, so far
as he is able to do, the best interests of the supply companies.

Mr. Higman is a member of the Institute of Electrical Engineers, London, England, and an associate member of the
Canadian Society of Civil Engineers.

### KIND WORDS CAN NEVER DIE.

THE CANADIAN ELECTRICAL NEWS has issued a special convention number, containing among its special features an illustrated article on the electrical features of the Dominion capital,
the programme of the Canadian Electrical Association convention and the convention of Stationary Engineers.  The NEWS is
always a valuable publication for those interested in electrical
and engineering matters, and this special number is of unusual
interest and very creditable.—Peterborough Review.

We extend our congratulations to our contemporary upon the
very creditable production which it has just issued in the form
of a special convention number of the CANADIAN ELECTRICAL
NEWS, containing an article descriptive of the electrical features
of Ottawa, the programmes of the Canadian Electrical Association convention and the convention of the Stationary Engineers.
The illustrations are good and everything about the journal indicates prosperity and advancement.  C. H. Mortimer, the publisher of the NEWS, is one of the best known men in Canadian
electrical circles and his popularity is deserved.—Western Electrician, Chicago.

### ANNUAL REPORT OF THE ROYAL ELECTRIC COMPANY.

THE eleventh annual report of the Royal Electric Company, of
Montreal, recently published, shows that the net gain on the
fifteen month's operations was $106,209.14, out of which five
quarterly dividends of 2 per cent. each, amounting to $99,900.10
have been declared, the remainder going to swell the balance of
$308,758.98.  The lights on the direct current arc system have
been increased from 1617 to 1666 ; the lights on the alternating
current incandescent system from 40,013 to 53,977, and the
motors from 347 to 688.

Reference is made in the report to the faithful service of Mr.
Charles W. Hagar, who recently retired from the position of
secretary and manager, and a tribute paid to the experience and
business ability of his successor, Mr. Wm H. Browne.

### SPARKS.

The Ottawa electric railway has been extended to the village of
Hintonburg.

An electric road is proposed between Trout Lake and the landing, in
the Kootenay country, B. C.

The street intersection of the electric railway at the corner of Richmond and Dundas streets, London, consists of 171 pieces, and the iron
weighs from 45 to 50 tons.

A company, called the Co-operative Telephone Co., of the counties
of Lake St. John and Chicoutimi, Que., with headquarters at Herbert-
ville, has been formed to build and operate a telephone line.  Capital
stock, $10,000.

There are forty men working in the McLaren and Buckingham mica
mine at present, and the work is being carried on night and day.  A
large vein of very fine amber colored mica has been struck.  A shaft 200
feet deep has been sunk.

William Kyle, charged with attempting to bribe a Niagara Falls,
Ont., councillor in connection with an electric railway franchise, and
Robert F. Segsworth, of Toronto, an alleged associate in the matter,
have been committed for trial on the charge.  They were released on
bail.

The Halifax street railway has been sold by the Sheriff to the new
Electric Railway Co. for $25,000.  The sum of $50,000 has been paid
into the Bank of Nova Scotia, the amount named by the Legislature to
be paid the bond holders of the old company before work on the new
road can be commenced.

While Mr. W. McIntosh, Woodstock, and Mr. Geo. Leacock, Sutton, were stringing wire for the Guelph electric railway from an elevated
truck, the wire broke and both men fell to the ground.  Mr. McIntosh
fell on a dray and had his left leg broken.  Mr. Leacock fell behind the
horse's heels and received a severe shaking up, but had no bones
broken.

Steam jets in furnaces produce destructive effects on the
metal if there is sulphur in the coal.

Heat applied to a solid first expands it, then melts, and finally
converts it into a vapor, if the temperature is sufficiently high.

Sand has been recommended to catch the drippings of oil
tanks or barrels.  The use of sawdust for the purpose is objectionable in that it may cause spontaneous combustion.

After cleaning boilers, do not screw up the nuts on the manhole and hand-hole covers any tighter than is necessary, for you
may break the guards or dogs that hold the covers in place and
cause yourself much trouble.

Prof. Blondel, of Paris, in measuring the total spherical intensity of arc lamps, found that it was nearly trebled by diminishing
the carbons from 21 and 13 mm to 14 and 8 mm diameter, the 13
and 8 presumably referring to the negative carbon.

The state telephone system of Sweden is soon to be connected
with the state telegraph system.  Instead of addresses, the
telephone numbers will be used, the telegraph clerks looking up
the address.  Messages may be telephoned to the telegraph
office and telephoned back, thus dispensing with the greater
number of messenger boys, as in Sweden nearly everyone uses
the telephone.

Within the next year the Illinois Central and the Chicago
and Northern Pacific railroads, both of which are now operated
by steam, will be run by electricity.  It is possible that the latter will use electricity on its entire system.  The Illinois Central
will begin with its suburban service only.  Bids have already
been secured from builders of electrical appliances for the complete equipment of the latter.  It is estimated that the entire
cost will be from $3,000,000 to $4,000,000.  This will include the
substitution of electricity for steam as the motive power in the
company's car shops at Burnside.

Prince Henry of Prussia displayed the courage of the scientific
enthusiast when he stood the other day with tongues of flame
more than a yard long shooting forth from his hands in quiver-
ing zigzags, accompanied by incessant cracklings.  It was in the
lecture room of the scientific society Urania, at Berlin.  He had
offered himself as a subject to Professor Spies to demonstrate
the fact that alternating electric currents of high frequency
passed through a human body, far from causing death, produce
no ill effects.  The Prince declared that he felt no inconvenience
whilst Professor Spies was passing through his body a discharge
of 100,000 alternations a second with a tension of 10,000 volts.

## MR. A. B. SMITH.

WE present herewith to the readers of the ELECTRICAL NEWS a portrait of Mr. A. B. Smith, the newly-elected President of the Canadian Electrical Association. It was fitting that in an Association which aims to represent the various branches of electrical industry in Canada, the first presiding officer, who was representative of the electric lighting interest, should have been succeeded by one identified with the telephone department, and that he in turn is now superseded by a gentleman connected with the telegraphic service. The honors have in every case been well bestowed.

Mr. Smith, the present incumbent of the office, is a native of Montreal, has had a long and varied experience as a practical electrician, and in his capacity as electrical inspector for the Underwriters' Association, has rendered valuable service to the electrical fraternity at large by his common sense dealings with such matters as come under his control. As a youth in 1862 he entered the service of the Montreal Telegraph Company at Montreal. In August, 1869, he was appointed chief operator at Hamilton. In 1872 he was promoted to the position of general inspector, in which capacity he displayed noticeable ability, and in 1859 was further advanced to the position of superintendent of construction for the entire system, the duties of which he has since discharged in a most satisfactory manner.

He was one of the original promoters of the Canadian Electrical Association, and since its organization has been one of the most efficient workers in behalf of its welfare. The honor which has been conferred upon him has therefore been fairly earned. Knowing as we do his adaptability for the position he occupies, we cannot but consider the interests of the Association to be safe in his hands.

## A NEW FUEL SAVING INVENTION.

A TEST has recently been made in this city of an invention, new in Canada, though in use for some time in Philadelphia and some other cities of the United States, by which it is claimed a large saving is effected in fuel for steam boilers, by producing more perfect combustion. The principle on which it is based is that in order to bring about the combustion of the smoke and gases given off by the fuel, a greater degree of

MR. A. B. SMITH, President Canadian Electrical Association.

heat must be created than was required to set free the smoke and gases. This is accomplished by means of gas, produced from oil by means of steam, in a retort attached to the furnace, this gas being forced into the fuel and causing a degree of heat sufficient to ensure perfect combustion. The principle seems to be a reasonable one, and the tests would indicate that the invention will prove a very valuable one.

The test was made with a boiler which supplies power to a number of industries at 109½ Adelaide Street West, Toronto. Prof. Welton, of McMaster University, was present, and as an independent witness watched the experiments. The boiler was first worked for a day of eight hours in the ordinary way, and then for a similar period with the apparatus, the amount of fuel consumed and water evaporated being carefully noted in both cases. The result showed that under the ordinary system the average evaporation per pound of coal was six and six-tenths pounds of water, while with the new apparatus the average was nine pounds evaporated, a gain of 36 per cent. The quantity of oil consumed in producing the gas was only one and a quarter gallons.

The inventor guarantees a saving of at least 15 per cent. in fuel in boilers with the most improved settings. In ordinary boilers the saving will run all the way to fifty per cent., and Prof.

Welton informs us that for half the day when the test was made the saving was even greater. Another test is to be made, and if the results are as satisfactory, a number of users of steam power will adopt the apparatus. The cost is not large, being about $300 for boilers up to 100 horse power, a trifle compared to the saving effected in fuel.

Mr. Teter, the inventor, has great faith in the merits of his system, in which he appears to be justified by the results. Prof. Welton, who is an impartial spectator, is satisfied that it is all that is claimed for it.

## SPARKS.

Mr. E. Rutherford has opened an electrical supply store in Peterboro.

In its mileage of electric railways, Germany stands first, France second, England third.

The Milton Electric Light and Power Co., Milton, Ont., has been incorporated. Capital stock $15,000.

During the week of the operation of the trolley in London, numerous accidents are reported to have occurred.

Sir Henry Tyler, ex-president of the Grand Trunk, is on his way to Peru to experiment with an electric locomotive on a mountainous railway.

The Olympic Mill Company, Olympia, Wash., is operating a small saw mill by electricity, said to be the first of the kind on the Pacific coast. The new mill contains a resaw, a planer, sticker and turning lathe, and is run by a 16 horse power dynamo.

St. Thomas has accepted the tender of the Street Railway Company to light the city, conditional upon its operating the electric street railway.

The Kingston Electric Railway Co. has closed a contract with the Canadian General Electric Company, of Toronto, for six new open cars, to be delivered May first.

The Dominion statistician's figures show that last year the three hundred odd miles of electric railways in Canada carried fifty-seven million passengers.

Mr. Jas. Milne has been appointed lecturer on electricity in the Toronto Technical School. He is general superintendent of the Incandescent Light Company.

The test of the new double cylinder engine on the G. T. R. is declared to be satisfactory, but stronger coupling pins and drawbars will be necessary for such heavy trains.

Arrangements had been made to light Acton, Ont., by electricity, and part of the plant arrived, when a hitch arose, and Mr. Ebbage, who had the matter in hand, declined to proceed. It is likely it will be taken up by someone else.

Brockville will have an electric railway next year. Wm. Henry Comstock, Charles Snow Corsitt, David Spencer Booth, Oliver Kelly Fraser, George Ira Mallory, Wm. Andrew Gilmour and Matthew Munsell Brown will form the company to build it. They are all local men. The capital of the company is $200,000.

Application will be made at the next session of the Dominion parliament for an act to incorporate a company of prominent Canadians, with a capital of $100,000, to introduce into Canada a new process for manufacturing gas for illuminating and fuel purposes. The process is a most interesting one and produces gas for fuel, lighting and power from crude petroleum, water and peat, dispensing with the use of coal in any form.

It is evident that the business of the Peterboro' Carbon and Porcelain Co. has not been profitable, and mis-management is assigned as the cause. A meeting of the shareholders has been held for the purpose of considering the question of liquidating the concern, but it was adjourned without a decision. They owe their banker over $20,000, and without fresh capital, an entire suspension must take place. The paid-up capital appears to have been $40,000.

The Pittsburg Reduction Company has closed a contract with the Niagara Falls Hydraulic Power Manufacturing Company for 3,000 horse power, delivered on the shaft of the turbine, to be placed under the high bank by the Hydraulic Power Company. They will install upon these turbines direct current generators, the current from which will be used for the manufacture of aluminum. The Reduction Co. has now practically got control of all the available cheap power, thus shutting off competition and will use about 25,000 horse power. Their furnaces will be placed under the high bank,

# CANADIAN ELECTRICAL ASSOCIATION

## REPORT OF PROCEEDINGS OF THE RECENT CONVENTION AT OTTAWA.

THE Fifth Convention of the Canadian Electrical Association was opened in the Railway Committee Room of the House of Commons, Ottawa, at 11 o'clock a. m. on Tuesday, September 17th, 1895. Mr. K. J. Dunstan, the President of the Association, occupied the chair.

The following persons were in attendance :—

J. J. Wright, A. M. Wickens, F. C. Armstrong, Jas. A. Baylis, Wm. B. Jackson, T. F. Dryden, J. F. H. Wyse, John C. Gardner, F. C. Maw, A. B. Smith, K. J. Dunstan, J. A. Kammerer, C. H. Mortimer, H. P. Dwight, Chas. P. Dwight, Joseph Wright, E. B. Biggar, Toronto ; Wm. T. Bonner, L. B. McFarlane, D. A. Starr, H. O. Edwards, James A. Burnett, D. W. McLaren, P. W. Atkinson, John Carroll, Montreal ; John Murphy, J. W. Thompson, H. G. Roche, A. A. Dion, Mayor Borthwick, O. Higman, G. F. Macdonald, C. Routh, H. Bott, D. C. Dewar, J. W. Taylor, Warren Y. Soper, C. Berkeley Powell, D. R. Street, Ottawa ; F. W. Harrington, John H. Dale, W. R. McLaughlin, New York, U.S.A. ; H. O. Fisk, J. Knapman, Peterboro' ; W. A. Mackay, C. H. Wright, A. A. Wright, Renfrew ; George Black, Hamilton ; J. A. McCrossan, Rat Portage ; V. B. Coleman, Port Hope ; J. E. Brown, Aylmer ; W. J. Gilmour, Brockville ; E. Carl Breithaupt, Berlin ; R. G. Moles, Arnprior ; F. H. Badger, jr., Quebec ; A. F. Simpson, Sherbrooke.

The President announced that the present meeting was in the character of a formal opening, and that the first business session would be held in the afternoon in the Board of Trade rooms at 2:30 p.m.

### ADDRESS OF WELCOME.

Mayor Borthwick made a speech welcoming the Association to Ottawa, in which he said :

It affords me a great deal of pleasure in having the opportunity of welcoming a body of gentlemen interested in such an important industry, to the city of Ottawa.

We claim to be living in a progressive age, and when we contemplate the rapid strides that have been made in every department that affects the human race, I think there are few who will dispute our claim ; and while you, gentlemen, who are occupied to give all credit and honor to those who are occupied in the development of other branches of knowledge, yet I feel that you will be justified in claiming that you have outstripped all the others in the marvellous achievements that have been made in your own branch during the last few years.

It is in the memory of most of us when the use of electricity was practically unknown, and when first discoveries began to be made the world wondered, but the new discoveries have been so rapid and so great that we have almost ceased to wonder, and content ourselves by asking in a placid manner, "What next ?"

If we had been told a few years ago that we would have the lightning chained down so that we would use it as the motive power to transport us from one part of the city to another, or from one part of the country to another, or if we had been told that we would be able to sit in our own home and hear the voice of absent ones in distant cities, or many of the other things that can be accomplished by the use of electricity, we might have expressed ourselves by saying, " Eye hath not seen, neither hath it entered into the heart of man, the marvellous things that are in store for those who live twenty years hence." Now there is a feeling here that some of the citizens of Ottawa have contributed something in the way of assisting in the development of this great science, and we claim as a city to be keeping fairly abreast of the times. Whether such is the case or not it is for you to judge, but if we are not up to the mark in electrical appliances ; I know that you will find the gentlemen who are associated in this line large hearted and generous, and our citizens know how to entertain strangers, and if we cannot teach you anything in electricity, we are progressive enough to learn ; and I sincerely hope that you will be so electrified by your visit that you will all be pleased to pay us another at an early date. I therefore in behalf of the citizens extend to you a very cordial welcome.

The President, Mr. Dunstan, replied as follows :

On behalf of the Canadian Electrical Association, I wish to thank you very heartily, Mr. Mayor, and through you the members of the Ottawa City Council, for the kindly spirit which has prompted you to present an address the tone of which is so complimentary, the welcome which it extends so hearty, that we would have indeed a high opinion of the merits of our Association and the work which it is performing were we not highly flattered and were we not greatly pleased with the reception which we have met with at your hands.

You have laid stress upon the great progress made in applied electricity within the past ten or fifteen years, and remembering as I do that I am speaking in the city of Ottawa, in the city which is fast earning for itself the reputation of being the electrical centre of Canada, remembering as I do that I have the honor of addressing the first magistrate of that city, I feel that words of mine are unnecessary to add emphasis to your statements in that direction.

You, sir, have no doubt been an eye witness of the great strides made by Ottawa in industrial progress during the past fifteen years, and you cannot but realize the great part played by electricity in such development.

The same thing, sir, holds good, perhaps not to the same extent, but yet holds good in every city, town or village throughout the broad Dominion—villages rescued from Egyptian darkness by electric light, isolation broken down by inter-communication, made possible by the electric railway and telephone, the very world itself made small by reason of the telegraph girdle which encircles it.

You have truly said that the work accomplished in the past is but an earnest of what may be expected in the future, and if this be so and if the future industrial progress of Canada is to a great extent dependent upon the widespread application of electricity in its various branches, then, sir, I think we may fairly claim that an association such as this, in endeavoring to encourage the science of electricity and promote the interests of those engaged in electrical industries, has a mission to fulfil worthy of every encouragement. That great changes, great advancements, will take place I think there can be no doubt, and if in the year of grace 1905 it should so happen that this Association should hold its convention in Ottawa, I can look forward in imagination and fancy I hear the then mayor in his address of welcome contrast the small beginnings of to-day with the vast accomplishments of his time and generation.

You have, Mr. Mayor, kindly expressed the hope that this will not be the last occasion of an annual gathering in your city ; without assuming too much authority, I can safely answer you on that point, in fact I have a suspicion that before voting time to-morrow not a few members will be ready to vote Ottawa as the permanent headquarters of the Association, with annual meetings always here. However this may be, there can be no doubt that all members will leave your hospitable city with the warm desire to return as soon as possible consistent with the aims and objects of the Association. I will not trespass longer upon your time further than to again personally and on behalf of the Association thank you for the kind address which you have read and to say how gratified we will be if you can spare time to attend our sessions, which will be held at the Board of Trade rooms during the next three days. (Applause.)

Mayor Borthwick then stated that should the Association find the Board of Trade rooms not suitable for their meetings, the city would be pleased to place the City Hall at their disposal.

The President then thanked His Worship and stated that they might be glad to avail themselves of his kind offer.

After having examined, under the guidance of Capt. Bowie and Senator Clemow, the objects of interest in and around the Houses of Parliament, a visit was made to the Patent Department, where numerous models of electrical appliances were viewed with much interest.

### AFTERNOON SESSION.

The President took the chair at 2:30 p.m. and opened the convention with the following remarks :

#### PRESIDENT'S ADDRESS.

Gentlemen,—I have to congratulate you on the good 'attendance at this the open sitting of the fifth annual meeting of the Canadian Electrical Association. I believe I am within bounds in stating it to be the largest attendance on an opening day in the history of the Association. This is a matter for congratulation, viewed not only from the standpoint of this particular meeting, but more especially from the broader ground of the strong evidence which it affords of life and vitality in the Association itself. It is generally accepted as a fact that the third and fourth years in the history of a body such as this are most trying and critical. Early enthusiasm for a new organization will influence many persons to join a society and remain members for the first year or two ; fees are paid willingly and with promptness, but if there be no good reason for the organization and continued existence of such a society, or if the seed has been sown on shallow ground, the third and fourth years will show a withering away of the membership roll and the attendance at the annual meetings will gradually grow beautifully less; fees come in slowly and you may know the end is near.

Those who were the original promoters of this Association, and who believed firmly in the wisdom of founding it on a basis broad and wide enough to embrace all persons in the Dominion interested in electrical industry, have watched therefore with no little anxiety its progress during the past two years, half fearing to find a reduced membership roll, diminished attendance and half-hearted interest at the annual gatherings. It is therefore a matter of much satisfaction that the meeting in Montreal last year was such an unqualified success and that indications all point to a Convention of even greater interest and pleasure this year.

The Secretary-Treasurer's report will show the Association to be in a sound financial condition.

The Committee on the Revision of the Constitution will report at an early opportunity, and I would ask the members to give this report most careful consideration. You will all agree that changes should not be made in the Constitution lightly or without good reason, but our present set of rules falls far short of composing a complete and comprehensive Constitution for the proper government of an organization such as this. We have reached now a stage when, assisted by four years' experience, we can draw up and adopt a Constitution not too complex in character yet broad enough to meet all requirements. The rules should afford members the opportunity of accomplishing every legitimate object of the Association and at the same time should provide ample safeguards against hasty, ill-considered action.

The report of the Committee on Legislation will point out the good work done by the Association at the last session of the Ontario Legislature, when an assessment bill was introduced, no doubt with the best of motives, but which, if it had become law, would have laid a very unfair burden upon electrical industries already established, and would have hampered and to a great extent prevented the further extension of the telephone, telegraph and electric light, especially in the smaller towns and villages and throughout country districts. By bringing this measure promptly to their notice, those interested were able to point out to their various representatives in the House the true character and far-reaching effect of the legislation, and I was glad to say the bill was withdrawn.

During the progress of the Convention papers will be read on various subjects, and I cannot point out too strongly the advantage of full discussion, not only as a mark of intelligent appreciation of the labor bestowed on the preparation of the papers, but also because during such discussion new points and fresh ideas are brought out and there is that free interchange of opinion which is of such great value in an Association such as this, comprising as it does so many persons all trying to solve the same problems, all seeking the same results.

Throughout the commercial world of Canada, and to a still greater extent in the United States and other countries, severe depression of trade and great stringency in money has been experienced during the past few years. It could not be expected with reason that those engaged in electrical industries could entirely escape such a storm, but I believe it will be found to be a fact that they have suffered much less than have those engaged in other branches of trade and commerce. More than this could not have been expected—anything less would have been a great disappointment.

At the last session of the Ontario Legislature there were incorporated no less than twelve electric railway companies. This fact alone gives a fair indication of the great activity in that particular branch of applied electricity.

I am well aware that railway construction does not in every case immediately follow the acquisition of a charter. Too often charters are obtained for purely speculative purposes, and legitimate enterprises are blocked by unreasonable demands on the part of speculating incorporators, who do not hesitate to ask heavy compensation for giving up charters never seriously intended to be used by themselves. One of the most important electric railways opened for traffic during the past year is that known as the Hamilton, Grimsby and Beamsville Railway, running from the city of Hamilton to Grimsby, a distance of about 18 miles. It is not only the longest road of the kind in Canada, but is exceptional on account of the large amount of freight it handles.

I am informed that during the three months of June, July and August just passed, it carried 94,164 passengers and handled 559 tons of freight, in addition to 2,917 cans of milk and a large quantity of fruit.

A large fruit market has been established in connection with this railway in the central part of Hamilton. It is needless to point out the many advantages afforded by such a road to the farmers and fruit growers living along its line, and in the villages and towns through which it passes, nor is it necessary to mention the great value to the city of Hamilton in having this market established there, and thus being made the headquarters of the great fruit trade of a large section of the Niagara Peninsula.

On every hand throughout the whole country we now find similar electric roads projected or under construction; towns and villages are being connected together, with the result that this cheap and convenient means of local transit, together with the intercommunication afforded by the telephone, will go far to break down that isolation which makes farm and country life so distasteful to the younger members of the community, and may have a far-reaching effect upon the great problem of how to attach the people to the soil.

In the telephone field the event of most moment has been the opening of the new main exchange belonging to the Bell Telephone Company in Toronto. The switchboard is of the most modern type, complete in every detail, and known technically as a branch terminal board. It has all the novel features, including self-restoring drops, incandescent pilot lamps, automatic disconnect signals, &c. This switch is not only a sample of the most modern type of the multiple switchboard, but is the largest installation of the kind in the world.

The great Niagara Falls power plant will soon provide an object lesson in the transmission of power over long distances, and we should very shortly be in possession of fairly reliable information upon which to estimate the economic distance within which electric energy may be transmitted in successful competition with local plants.

While it must be admitted that electric light plants have been established already in all the cities and larger towns throughout the country—and it is therefore in the enlargement of existing plants rather than in the installation of new systems that increase in that branch of electrical industry must be expected so far as these larger towns and cities are concerned—there is yet the larger field presented by the many smaller towns and villages scattered over this broad Dominion, and perhaps a still larger field existing in connection with isolated plants. In a country also, such as Canada, where there are so many opportunities for obtaining water power from natural falls or by damming up the surplus water of rivers and streams, is it not a fact that there is still room for a large number of installations using water power, the energy developed being available for light and power within economic distances. It has been wisely said that the utilization of energy is a fair test of the progress and civilization of a country, and realizing as we must to what great extent the future industrial progress of Canada is bound up with and dependent upon the growth and development of electricity in all its various branches, we as an Association have a right to feel that in endeavoring to foster and encourage this industry we are working, not only in our own interests, but also for the general advancement of the country.

The question of municipal control of city lighting was fought to an issue in Toronto in a contest remarkable for warmth and energy. Every effort was made on both sides to educate the people in the way they should go to the polls, the result being that the by-law to provide funds to erect the civic plant was defeated by a vote of eight to one.

Last year the Welsbach burner formed a slight unsettling element in the lighting business, there being those who felt that the greater efficiency of this burner would, by cutting down the cost of gas, injuriously affect electric lighting interests. These fears have proved to a very great extent groundless, but we find this year a new disturbing feature, in the form of acetylene gas, but to what extent it will become a live issue yet remains to be seen, as it is too early to predict the commercial outcome of Mr. Willson's cheapened method of production. The gas has defects which may prevent it ever coming into general use, but on the other hand it is possible it may become an important factor. Whatever the outcome, electric light men must face the fact that prices, from competition or other causes, have a downward tendency, and this tendency must be met with improved methods of production. Electric light is of such superior intrinsic value as an illuminant that it is only necessary to maintain a high standard of efficiency, combined with a price not greatly in excess of that charged for gas, in order to control the best market as against competition in any form, but this high efficiency and close economy of production can be obtained only by most skilful business management, combined with technical knowledge and a thorough grasp of details.

We must prepare for every eventuality of our business, and these annual conventions, the communion of men working on parallel lines in various places throughout the country, the interchange of thought, opinion and experience, the rubbing of mind against mind, must tend towards the systematizing of methods and towards placing the conduct of business upon a higher, more scientific, more economical plane.

Indications point strongly to our being on the verge of a "horseless age;" an age when tricycles, carriages and a large proportion of vehicles in general will be self-propelled. Will the motive power be derived from electricity, petroleum, compressed air, or some other source of energy? Tests have resulted so far greatly in favor of petroleum, but electricity has so many advantages, due to freedom from dirt, smell and risk of explosion, that the discovery of a lighter, more economical form of storage battery would enable electricity to control a trade the magnitude of which it is difficult to even estimate. The person who makes this discovery will reap the greatest reward of the age.

In bringing these very inadequate remarks to a close, I wish to thank the members of the Association for the honor conferred upon me when I was elected your presiding officer for the past year, and while I claim a warm interest in the affairs of the Association and a strong desire to build it up, as far as lies in my power, on the broad plan contemplated at its formation, I fully realize that the work actually accomplished has fallen far short of what you might have expected, with all fairness, at the hands of your presiding officer.

The minutes of the last meeting, read by Mr. Mortimer, the Secretary, were confirmed. The Secretary-Treasurer's report was then read, as follows :

### SECRETARY-TREASURER'S REPORT.

I am pleased to be in a position to state that notwithstanding the business depression which we have recently experienced, this Association has made satisfactory progress during the period covered by this report. The success which attended the convention in Montreal last year appeared to give to the organization a new impetus, and I have no doubt that a like result will follow our present gathering.

At the date of my last report, Sept. 19th, 1894, there were enrolled on the books of the Association 99 active and 38 associate members. This number has increased to 169 active members and 41 associates, a total gain in membership during the year of 71. With this membership, and at the present reduced fees, the Association should have a yearly income of $589, while the annual expenses amount to only about $400. In making this statement it is presumed that every member can be depended on to pay his yearly fee when the same shall become due. This expectation has, unfortunately, not been realized in past years, as shown by the fact that the arrears of fees outstanding on the 31st of May last amounted to $321. Your Secretary-Treasurer, acting under the direction of the Executive Committee, has more than once issued accounts to members in arrears, accompanied by special requests for payment. It is therefore to be feared that much of the above stated amount in arrears will prove to be uncollectable. The Executive Committee still have under consideration the question of what action should be taken as regards members in arrears for fees who have failed to respond to repeated requests for payment. In view of the reduction in active membership fee decided upon at last convention, there would seem to be good ground for the hope that prompt payment of fees will be the rule for the future.

Following is a statement of receipts and disbursements for the year ending 31st May, 1895, showing also the condition of the finances of the Association brought up to the present date :—

*Receipts.*

| | |
|---|---|
| Cash on hand June 1st, 1894........................ | $23 30 |
| Cash in bank   "   "   " ......................... | 250 18 |
| 14 active members' fees at old rate, $5.00............ | 70 00 |
| 71 active members' fees at new rate, $3.00............ | 213 00 |
| 14 associate members' fees at $2.00................. | 28 00 |
| Captain Carter.................................... | 2 50 |
| Increased amount paid by associate members to become active.. | 2 00 |
| Exchange on cheque.............................. | 10 |
| | $589 08 |

*Disbursements.*

| | | |
|---|---|---|
| Expenses of Convention at Montreal............... | | $278 76 |
| By cash as per local committee's statement.... | $106 50 | |
| By cash, stenographer......................... | 27 00 | |
| By cash, Canadian Electrical News, printing...... | 92 15 | |
| By cash, Canadian Photo Engraving Co ........ | 52 41 | |
| By cash, express charges on books sent to and from convention................ | 70 | |
| | $278 76 | |

| | | |
|---|---|---|
| Refund to members on account of reduction of fees..... | | $18 00 |
| John Yule.................................... | $ 2 10 | |
| A. A. Wright................................ | 2 00 | |
| J. W. Taylor................................ | 2 00 | |
| F. J. F. Schwartz........................... | 2 00 | |
| W. A. Mackay.............................. | 2 00 | |
| J. A. Kammerer............................ | 2 00 | |
| O. Higman................................. | 2 00 | |
| R. T. Dickenson........................... | 2 00 | |
| E. Carl Breithaupt........................ | 2 00 | |
| | $18 00 | |

| | |
|---|---|
| Postage........................................ | $ 38 50 |
| Stationary and printing, Electrical News............ | 7 75 |
| Grant to Secretary.............................. | 25 00 |
| Statistical Committee, expense account.............. | 25 00 |
| Telegrams and messages.......................... | 79 |
| Exchange on cheques............................ | 1 15 |
| Ribbon for badges............................... | 1 60 |
| Envelopes...................................... | 75 |
| Macrae & Macrae, delivery of 21 copies convention report .... | 11 |
| Blackhall & Co., 50 certificate covers.............. | 4 00 |
| Receipt forms.................................. | 35 |
| Express charges................................ | 35 |
| | $402 31 |

| | |
|---|---|
| Cash in bank May 31st, 1895..................... | 186 77 |
| Cash on hand.................................. | |
| | $589 08 |

*Receipts since May 31st, 1895.*

| | |
|---|---|
| Refund by Statistical Committee for fees............ | $ 23 62 |
| Receipts for fees since May 31st, 1895............. | 173 00 |
| | $196 62 |

*Expenditure since May 31st, 1895.*

| | |
|---|---|
| Postage ....................................... | $ 8 10 |
| Exchange...................................... | 75 |
| Blackhall & Co. (covers) ...................... | 4 00 |
| | $ 12 85 |

| | |
|---|---|
| Cash in bank, Sept. 17, 1895.................... | 84 12 |
| Cash on hand, Sept. 17, 1895................... | 99 65 |
| | $196 62 |

| | |
|---|---|
| Total standing to credit of Association......... | $370 54 |

The Executive Committee have held six meetings during the year, viz. : On Oct. 16th and Nov. 19th, 1894, Feb. 7th, March 19th, June 21st, and Sept. 17th, 1895. At the first of these, accounts in connection with the Montreal convention were passed for payment ; draft of a circular by the Secretary urging members to endeavor to increase the membership approved ; the Secretary instructed to notify Messrs. McFarlane, Powell and Yule of their appointment as a Committee on Legislation, requesting them to communicate promptly to the President

the introduction of Dominion or Provincial legislation affecting the electrical interests.

At the second meeting 25 persons were elected to active membership.

At the third meeting, seven active members were elected ; the Secretary was instructed to invite from each member of the Executive suggestions for needed amendments to the constitution. To the suggestion made by Mr. George Whyte-Fraser, in a letter to the President, that the Association should co-operate with the Underwriters' Association in demanding a proper standard of efficiency on the part of persons engaged in electrical work, the Secretary was instructed to reply that the matter was one which could only be dealt with by the Association as a whole, and suggesting that Mr. Fraser should bring the matter up for discussion at this meeting.

At the fourth meeting the Secretary was instructed to draft a circular to be sent to all electric companies in Ontario re proposed assessment legislation referred to more fully in report of Committee on Legislation.

At the fifth meeting, correspondence was considered from members of the Executive resident in Ottawa relative to arrangements for the present convention. The dates of the convention were fixed, and the following persons were requested to act as a committee to make the necessary local arrangements : Messrs. C. Berkley Powell, O. Higman, J. W. Taylor, J. W. McRae, T. Ahearn and W. Y. Soper. A grant of $100 was voted for the use of this committee ; the Toronto members of the Executive and the Secretary were appointed to arrange for papers, and complete and have printed the program for the convention. Three active and two associate members were elected.

At the sixth meeting of the Executive held immediately prior to the opening of this convention there were elected 17 active and 2 associate members.

C. H. MORTIMER,
Secretary-Treasurer.

Certified Correct, Sept. 17, 1895.

F. C. ARMSTRONG } Auditors.
CHAS. P. DWIGHT }

Moved by Mr. J. A Kammerer and seconded by Mr. Taylor, that the report be received and adopted. Carried.

Mr. Smith enquired the total membership ; the President replied, stating that there were 169 active members and 41 associate members, a gain during the past year of 71, a showing which he was sure would be very gratifying to everyone.

Mr. Kammerer presented the following report of the Committee on By-Laws and Constitution :

TORONTO, Sept. 16th, 1895.
*To the Officers and Members of the Canadian Electrical Association :*

GENTLEMEN : Your Committee on Constitution and By-Laws beg to report that after carefully perusing and digesting the Constitution and By-Laws of the Association as they now stand, they found quite a number of conflicting laws, and also a very large number of important points which were not covered by it. So many changes were considered necessary that your Committee deemed it advisable to undertake a complete and thorough revision, and not to proceed in the usual way of adding to or cutting out portions of the existing Constitution and By-Laws, but to submit to you an entirely new Constitution, in the construction of which they have retained as much of the existing matter as was in their judgment thought suitable, and added new clauses to cover those points which were considered essential for the good government of the Association. This new Constitution, as formulated by your Committee, was published in the September issue of our official organ, the CANADIAN ELECTRICAL NEWS, in order that all members of our Association might become perfectly familiar with the proposed changes, give them careful thought and be prepared to discuss the points on which they may be at variance with what is proposed by your Committee. We beg to submit to you herewith the result of our labors, which we trust will be satisfactory to the members present and beneficial to the Association at large :

### ARTICLE I.

NAME.—This organization shall be known as the Canadian Electrical Association.

### ARTICLE II.

OBJECT.—The object of this Association shall be to foster and encourage the science of electricity and to promote the interests of those engaged in any electrical enterprise and for discussion and interchange of opinions among its members.

### ARTICLE III.

MEMBERSHIP.—The Association shall consist of active, associate and honorary members. The term Active Members includes all members actually engaged in electrical business. The term Associate includes those interested or actively engaged in any electrical pursuit, and they shall be entitled to attend all meetings of the Association, except those of the Executive, and take part in all discussions, but shall not be entitled to vote or be eligible for office. Honorary members shall be elected by a two-thirds vote of the Association.

### ARTICLE IV.

OFFICERS.—The officers shall consist of a President, 1st and 2nd Vice-Presidents, Secretary and Treasurer, and an Executive Committee, consisting of ten members, five of whom shall act on the Committee for two consecutive years. The President and Vice-Presidents shall be ex-officio members of the

Committee. Five shall form a quorum. The office of Secretary and Treasurer may be held by one person.

### ARTICLE V.

FEES.—The annual fee shall be for active members $3.00, associate members $2.00, payable in advance.

### ARTICLE VI.

ELECTION OF OFFICERS.—All officers shall be elected by ballot at a general meeting of the Association. The ballot shall be taken in the following manner:—The Secretary shall read the list of active members alphabetically, and each member shall deposit with the Secretary a slip of paper on which he has recorded his vote, the Secretary checking off his name on the list of voters. Two scrutineers named by the Chairman shall assist the Secretary in counting the votes, and the Chairman shall declare elected the person receiving the majority of the votes cast. In case no one candidate receives such majority on first ballot, another ballot is to be taken, and so on until a clear majority is given in favor of some one candidate. Officers shall hold office until the close of the session, at which their successors are elected, such successors to be elected on the second day of the first general session after the expiry of ten months from day of previous election.

### ARTICLE VII.

ELECTION OF EXECUTIVE COMMITTEE.—Members of the Executive Committee shall be elected by ballot in the following manner, the vote being taken immediately after the election of officers:—Ballot papers containing the names of the ten members of the Executive Committee, five of whom must be re-elected, shall be given the members. The Secretary shall read a list of those entitled to vote, and members, having first marked a cross opposite the names of the five persons selected for re-election, shall deposit the ballots with the Secretary, who, assisted by the two scrutineers named by the Chairman, shall count the vote, and the Chairman shall declare elected the five persons receiving the greatest number of votes. Members shall then proceed to elect the five other members of the Executive, the election being by ballot and the Secretary reading the names as before. Each active member of the Association shall have the right to vote for an active member of the Executive, including the retiring members of the Executive, and the vote being counted in the usual way, the Chairman shall declare elected the five persons receiving the greatest number of votes.

### ARTICLE VIII.

PLACE OF MEETING.—Place of next meeting shall be decided by ballot, taken in same manner as laid down for election of officers.

### ARTICLE IX.

VACANCIES IN OFFICE.—Vacancies in office, caused by death or resignation, shall be filled by the Executive Committee to cover the term until the next general meeting of the Association, at which the officers are elected.

### ARTICLE X.

NOTICE OF MOTION.—Permission to introduce any notice of amendment or amendments to this constitution must be granted by a majority of two-thirds of the active members present. Permission being granted, notice may be given and the proposed amendment moved at any subsequent sitting. After discussion the amendment must be submitted to a Committee of five, named by the Chairman. The report of said Committee cannot be considered on the same day on which it is introduced. A two-thirds vote of all active members present shall be necessary for its adoption.

### ARTICLE XI.

Notice of substantive motions is required, and no motion shall be discussed at the sitting at which the notice has been given, but this rule does not apply to merely formal motions, such as motions to adjourn. All reports of standing committees are to be discussed at a sitting subsequent to the one at which such reports have been received. This rule may be suspended by a vote of two-thirds of the members present.

### ARTICLE XII.

All motions must be duly proposed and seconded, and shall, except those of a purely routine character, be in writing.

### ARTICLE XIII.

No member shall speak more than once, or at a greater length than five minutes, upon any question until all others have had an opportunity of doing so, nor more than twice on any one question without permission of the Chairman, or a majority of the members entitled to vote. The mover of a substantive motion has the additional right to reply.

### ARTICLE XIV.

Questions may be re-considered upon a motion to re-consider being made by a person who voted with the majority, provided such motion is carried unanimously. No discussion of the said question is allowed until the motion for re-consideration has been carried.

### ARTICLE XV.

VOTING.—Every active member present must vote, but any person entering the room after the question has been put by the Chairman may not vote. The Chairman shall not vote except in the case of a tie. Voting by proxy shall not be allowed.

### ARTICLE XVI.

Except where vote is by ballot the chairman will take the sense of the meeting by voice, or by asking members to stand, but on call of five members the Secretary shall read the list of persons entitled to vote, and record the yeas and nays.

### ARTICLE XVII.

An appeal may be taken without debate against the ruling of the chair, a vote of two-thirds being required to reverse the decision.

### ARTICLE XVIII.

The President shall nominate a Committee of three to strike the Standing Committees for the following year and define their respective duties, the report of the Committee being considered at a subsequent sitting to its introduction. The number of Standing Committees must be decided by the Association.

### ARTICLE XIX.

The first person named on any Committee shall act as Chairman until Committee is called together, when they will elect their own Chairman, but the President, in his absence the 1st or 2nd Vice-President, shall be Chairman of the Executive Committee. In the event of the absence of ex-officio members, the Executive Committee shall proceed to elect a Chairman pro tem.

The general order of business at all sessions shall be as follows :

Reading Minutes of last meeting.
Report of Secretary-Treasurer.
Report of Standing Committees.
Election of Standing Committees for following year.
Selection of place of next meeting.
Approximate date of next meeting.
Election of Officers and Executive Committee,

Time being allowed for general business and social affairs, at the discretion of Executive Committee or Chairman of meeting. Selection of next place of meeting and election of Officers and Executive Commitee must be on second day of meeting. Order of business may be altered only by unanimous vote of members present.

### ARTICLE XX.

Ten active members of the Association shall be a quorum for business.

### ARTICLE XXI.

Todd's Parliamentary Practice shall be the governing law of the Association in all cases not provided for in its own rules.

### ARTICLE XXII.

DUTIES OF THE PRESIDENT.—It shall be the duty of the President to preside at all meetings of the Association and to call meetings of the Executive Committee, and when requested by the Executive Committee, to call a special meeting of the Association.

### ARTICLE XXIII.

DUTIES OF THE VICE-PRESIDENTS.—The 1st, or in his absence, the 2nd Vice-President, shall act in the absence of the President.

### ARTICLE XXIV.

DUTIES OF THE SECRETARY.—The duties of the Secretary shall be to attend all meetings, take record of all proceedings, and shall perform such other duties as the Executive Committee shall direct.

### ARTICLE XXV.

DUTIES OF THE TREASURER.—The duties of the Treasurer shall be to keep a correct account of all receipts and disbursements in connection with the Association. All checks for disbursements shall be signed by the Treasurer and countersigned by the President, after being approved by the Executive Committee.

### ARTICLE XXVI.

THE DUTIES OF THE EXECUTIVE COMMITTEE.—The Executive Committee shall be the governing body of the Association, shall manage its affairs, pass upon all applications for membership, eligibility of representatives, subject to the constitution, and such special rules or regulations as may be adopted by the Association from time to time.

### ARTICLE XXVII.

DUES.—Dues shall be payable annually on the 1st June, in advance. Members in arrears for dues, other than those for current year, shall not exercise the privileges of membership.

### ARTICLE XXVIII.

The permanent office of the Association shall be in Toronto.

Respectfully submitted.

J. A. KAMMERER, Chairman.

Moved by Mr. Kammerer, seconded by Mr. J. J. Wright, that the report be now received and that it be discussed tomorrow.

The President : I would like to say in connection with this that printed copies of the report have been distributed, but if any member wants one, copies are to be found on the table. This is a most important matter, and if you will go into it carefully before the sitting to-morrow, it will be a great advantage.

Mr. E. Carl Breithaupt presented the following report of the Committee on Statistics:

REPORT OF THE COMMITTEE ON STATISTICS.

Your committee beg to report as follows:—

The information which it was considered desirable to obtain was principally such relating to central stations for the supply of electric light and power and electric street railways. Detailed information from such stations regarding the original cost of installation, the cost of operation, the volume of the output, the prices realized, and any particular difficulties, or extraordinary circumstances encountered, would doubtless be of great value to every person interested in these branches of electrical work. The members of the electrical fraternity have everything to gain and nothing to lose by a free interchange of ideas and an open discussion of experiences met with; this object is indeed one of the principal motives for the formation of the Canadian Electrical Association. On the other hand, the data referred to, although they are not strictly business secrets, are not of such a nature as a company desires to publish openly. The previous Committee on Statistics had sent out blanks requesting data from central stations; on looking over their returns received, we found that only a small proportion of sheets sent out had come back, while on those which were returned the answers to questions were incomplete and therefore of little value. It was thus evident that the compilation of the desired statistics would involve a considerable expense, and since the funds of the Association did not warrant it, your committee did not proceed in this way.

During the past year, however, the Dominion Government has commenced to gather statistics relating to the electrical industries and it was suggested that the Committee should lend its assistance to this work. Mr. Higman was added to the Committee as advisory member, and separate forms were drawn up by the Committee, one to be sent to electric lighting and power companies, the other to electric railway companies. These returns being the property of the Government, are kept secret. The detailed information concerning each separate company can, therefore, not be presented, but the totals will be published and these will be of undoubted value because they will be complete. Mr. Johnston, the Dominion Statistician, assures us that in another year he will be in a position to give the Association a large amount of information concerning these industries, all properly summarised and tabulated.

All of which is respectfully submitted.

E. CARL BREITHAUPT, Chairman.

Moved by Mr. Breithaupt, seconded by Mr. Armstrong, that the report be received and discussed at the next session.

Mr. L. B. McFarlane presented the following report of the Committee on Legislation:

REPORT OF COMMITTEES ON LEGISLATION.

Shortly after the last meeting of this Association your Committee on Legislation was called upon to act in conjunction with the Executive in an important crisis, which it was pointed out at its organization might arise, and in such case organized effort would prove one of the valuable functions of the Association. Legislation, which, though not intended to be hostile, might, through lack of thorough knowledge of its promoters, result most disastrously to electrical interests.

During the last session of the Ontario Legislature a bill was introduced which, if it had become law, would most assuredly have worked an injustice to many, and would have been burdensome to all who have invested their capital in electrical enterprises, by laying a tax as personalty upon all street equipment of electrical companies.

Immediately upon the introduction of the bill your President called a meeting of the Executive, when the following circular was drafted and sent by the Secretary to all the electric light companies in Ontario:—

TORONTO, March 19th, 1895.

DEAR SIR:—As you are interested in an electric light plant, I am instructed by the Executive Committee of the Canadian Electrical Association to call your attention to an amendment to the Assessment Act which has recently been introduced in the Legislative Assembly of Ontario by Mr. German, and which is contained in Bill No. 91, the object of which is to make electric light poles and wires assessable.

Such an addition to the expense of the electric light business would prove to be a serious burden, more especially in the smaller places where lighting plants are scarcely self-sustaining under present conditions. In such places the electric light has in most instances been established by public-spirited citizens for the public convenience and the welfare of the place, rather than with a view to any profit from the invested capital. The business, as you are aware, is at present sustained only by the exercise of the closest economy, and is consequently not in a position to stand any additional burden.

There is no public demand for the proposed legislation, and it is felt that the measure is one that should be opposed by those whose interests are at stake. I would therefore request you and the stockholders of your company to communicate immediately with the representative of your riding in the Legislature, and any others with whom you are acquainted, and ask them to oppose the bill as being detrimental to the prosperity and development of the electric lighting industry, and as certain to work serious injury and injustice to capital which was invested in the belief that it would continue to remain exempt from taxation. It should likewise be pointed out that even though it may not be possible to defeat the Bill entirely, no increase in taxation should in any event be made until existing contracts with municipalities shall have expired.

As the Bill may come before the Committee of the House within a few days, immediate action in the direction indicated above is necessary.

Very truly yours,

C. H. MORTIMER, Secretary.

As a result of this circular we have reason to believe that a large number of the representatives who compose the House had information placed before them in the most effective way that the proposed legislation would work an injustice, inasmuch as it would enable municipal councils to make a breach of the understanding on which nearly all the electrical companies of the Province were organized; that in many cases where the business is barely self-sustaining under present conditions such taxation would prove so burdensome that it would mean the practical confiscation of the plant; that instead of proving a benefit to the public, the result would be exactly the reverse, as many companies which were organized, not with a view to profit, but as a public convenience, would be compelled to suspend, thus not only causing a loss of the investment, but depriving the country of its advantage. But more important than all, it was clearly shown that the first result of such hostile legislation at the present experimental stage of the practical application of electricity to modern requirements would prevent the investment in the business of capital, which is so important to the development of this great interest, and thus prove of incalculable injury to the commercial interests of the country.

So forcibly were the conditions presented to the legislators that a large majority were convinced of the injustice and inexpediency of the measure, and the mover of the bill himself voluntarily withdrew it at its first appearance in Committee.

The matter of municipal assessment in many of its phases presents a difficult problem, but in no feature is this more apparent than in its application to electrical interests. Your committee would therefore urge, not only upon members of this Association, but upon all who are interested in any way in electrical enterprises, a thorough consideration of the practical application of this question, and especially to see that their representatives in the legislature, who may be called upon at any time to legislate on this subject, have some knowledge of the interests involved, so that they may know the effect of ill-considered changes which may possibly be proposed.

JNO. YULE,
Chairman of Committee.

It was moved by Mr. McFarlane, seconded by Mr. McDonald, that the report be received and considered at the next session.

The President: It would now be in order to consider any general business in connection with the interests of the Association.

Mr. Kammerer: A great many members of the Association, and those connected with electric stations generally, seem to be very much at sea respecting the inspection fee charged by the Inland Revenue Department, that is, the charges for inspecting meters. As Mr. Higman is here we would like to have him explain what the inspection fee is for.

It was understood that Mr. Higman would explain the matter at the next session.

Mr. Kammerer: There is also the matter of the taxation of poles and wires on public highways.

The President: That matter really was covered by the report of the Committee on Legislation. It will be found, as stated in that report, that the bill was withdrawn.

Mr. A. A. Dion then read the following paper:

SOME NOTES ON THE CONSOLIDATION OF TWO SYSTEMS OF ELECTRIC SUPPLY.

BY A. A. DION, M.A.I.E.E.

1. The march of electrical progress has been so rapid within the last few years, such marked advances have been made in the methods of supply and distribution of electrical energy for light and power, that central stations, which six or seven years ago were looked upon as the embodiment of the best and latest practice, are already handicapped in the race for wealth, in view of the many improvements which have been made since that time.

2. The constant and rapid increase in the use of electricity in cities has correspondingly increased the difficulties of distribution at constant potential, and new systems have had to be devised to meet the new conditions. Electric supply companies, whose stations were equipped when distribution at one thousand volts seemed like tempting providence, and small generator units were the rule rather than the exception, now find it impossible to adopt more economical systems of distribution without undue sacrifice of apparatus, and must confine their efforts towards the improvement of their services to changes within the limits of existing pressures.

3. The amalgamation of rival electrical interests, which is not infrequent in these times, brings up another and more difficult problem, that of consolidating various and oftentimes conflicting elements to form a single and uniform system. To do this without throwing any apparatus out of service was the task that the writer was lately called upon to undertake.

4. He does not claim originality for any of the features of the plan adopted, but simply states how it was done, in a particular case, believing that in furnishing each other information regarding work done in our respective fields of action, we best carry out the objects of this Association, and he trusts that some of the members may be benefitted by the discussion which this paper may bring out, if not by the paper itself.

5. The amalgamation above referred to comprised three electric light companies, namely, "The Ottawa Electric Light Company," "The Chaudiere Electric Light and Power Company," and "The Standard Electric Company of Ottawa."

THE OTTAWA ELECTRIC LIGHT COMPANY.

6. This was the oldest company, it having commenced business in 1881, and its operations were confined to arc lighting. It owned a substantial stone power house. The motive power was water, and was transmitted through four vertical turbines operating under a head of sixteen feet. The electrical equipment consisted of eighteen T. H. ten Ampere generators manufactured by the Royal Co. of Montreal, supplying 325 lights for lighting the streets of the city and 95 lamps for private lighting. This company also owned a small workshop for armature and arc lamp repairs.

THE CHAUDIERE ELECTRIC LIGHT AND POWER COMPANY.

7. This company was the next in point of age, it having commenced business in 1887. Its business was confined to incandescent lighting and supplying power for motors. Its first plant was a multiple series system, using the well known U.S. double magnet generators of 25 amperes and 550 volts. The lighting was limited to stores and other public places; five lights were run in series. Each light pendent consisted of two lamps, one above the other. The lower lamp alone usually burned. When, however, it burned out, an electro-magnetic device, con-

tained in the socket, instantly brought the upper lamp in circuit, thereby preserving the continuity thereof.

8. These machines were replaced in 1889 by the Alternating Current Converter system, but were used later for other purposes. The first installation of the latter system consisted of two Westinghouse smooth core alternators of 750 lights capacity each, that were separately excited by small machines of the U.S. type. At the time of amalgamation this company had installed 27,000 incandescent lights and 42-500 volt motors ranging from one and one-half to 20 h. p. and aggregating 320 h. p.

9. This company occupied three power houses, which, for the purpose of this paper we will designate as "a," "b" and "c."

10. "a" was the original power house, and was operated by water. It contained eight 750 light Westinghouse alternators separately excited. From this station eleven pairs of lighting feeders ran to various parts of the city. The switchboard was equipped with indicating instruments of the Westinghouse pendulum type,—one ampere meter for each pair of feeders and one voltmeter for each alternator—Westinghouse compensators, Wurtz non-arcing lightning arresters and a large number of double-throw switches by means of which the feeders and generators were made interchangeable. Some of the longer circuits were supplied with regulators or "boosters."

11. "b" was the next power house to be occupied. It was also a water power station and was built when the daily loads outgrew the capacity of "a." The electrical equipment of "b" consisted of a 1,500 light Westinghouse alternator with smooth core armature and a 120 K. W. alternator with toothed core armature, both separately excited, and a 75 K. W. 500 volt U.S. direct current generator of the upright type. The alternators were separately connected by wires to the switchboard in station "a," some four hundred feet away, and the D. C. generator supplied the motor circuits, two in number, which ran from this station.

12. "c" was a steam power station which had been built in 1893 as an auxiliary, made necessary, on account of periodical diminution of the water power through anchor ice and other causes. No place could be found for the steam plant on the premises of the other stations, therefore it had to be erected some distance away on a water course where an abundant supply of water was available for condensing purposes. Additional electrical equipment had therefore to be provided for this station. The building was a one brick structure with stone foundation 85 ft. by 130 ft. It contained six return tube boilers 14 ft. by 60 inches and a pair of tandem compound condensing engines, rated at six hundred horse power each. These engines were belted through clutch pulleys to a six inch shaft running through the building. Two Westinghouse alternators of 240 K. W. capacity each with toothed armatures were belted to the shaft also through clutch pulleys. They were separately connected by wires to the switchboard in station "a," some two thousand feet distant. In this case pressure wires were run back from the switchboard to the voltmeters in the steam station. Floor and shaft space and stone piers were provided for additional generators.

13. The alternators of this company were run at about 1,100 volts, except those in the steam station, which, owing to their distance from the switchboard, etc., were run at nearly 1,200 volts, when fully loaded, that being their rated capacity. The frequency in every case was about 133 cycles per second. Westinghouse converters—1,000/50 volt—were used, mostly small ones, 1,000 to 2,000 watts and a few of 4,000 watts and 5,000 watts. Over three-quarters of the current output was supplied through meters, the Schallenberger being used exclusively. This company also had a small workshop for re-winding armatures and field coils.

### THE STANDARD ELECTRIC COMPANY OF OTTAWA.

14. This was the junior company, it having commenced business in 1891. It could thus profit by the experience of others, and it had made provision for considerable extensions of the original plant. It occupied a substantial two storey building with a hydraulic plant consisting of four 66 inch turbines operating under a head of twenty-two feet with shafting, clutch pulleys, etc., which made each turbine capable of running the whole station or any part of it. This station contained six separately excited alternators of The Royal Company's manufacture, i. e., one of 5,000 lights capacity, one of 2,000 lights capacity, and four of 1,500 lights capacity, and four sixty horsepower direct current compound wound generators, also manufactured by The Royal Company. The direct current machines were used for the supply of power for motors ; two of them were run in series operating a one hundred horsepower 500 volt motor running an entire flour mill day and night. Another was used to supply 33 250 volt motors ranging from ¼ h. p. to 20 h. p., and aggregating 105 h. p. The other was held in reserve.

The alternators were run at a frequency of about 133 cycles per second. The lighting switchboard was equipped with T. H. measuring instruments and plug panels which made the ten lighting circuits and the six alternators interchangeable. The voltmeters were connected with the centres of distribution by pressure wires, the distribution being made through T. H. and " Royal" transformers—1040/52 volts—52 volt lamps and T. H. wattmeters were used throughout the system.

15. There were 18,000 incandescent lights installed.

### CONSOLIDATION.

16. The plans adopted for consolidating these several systems have not all been carried out at this time. The work is being done in a gradual manner in order to cause no commotion among subscribers, but for the purpose of this paper we will assume that this work has been completed and speak of things as they will be. As a first step towards carrying out the proposed changes, the small work shops above mentioned were merged into a single one in larger and more commodious premises known as the old arc light station, owned by the company and unoccupied at that time. Some additional tools were provided and a foreman competent to superintend any electrical and mechanical work that might be required, was put in charge.

17. For several reasons it was deemed advisable to maintain the arc light service as a department entirely separate from the other branches of the business ; for instance, the hours of lighting are limited, and the men connected with this service in most cases have no connection with the other departments. No changes were made in this station beyond the addition of a 60 light Westinghouse arc light machine, in order to increase the reserve and decrease the liability of impaired service from burn-outs, etc.

18. Each circuit is usually run independently from two generators, of a capacity of thirty-five and twenty-five lights respectively, in series.

19. Three patrolmen drive through the streets of the city during the lighting hours starting up lamps that have gone out and reporting every morning all lamps out, or requiring the attention of the repairer, as well as cases of improper carboning, etc.

20. These patrolmen also answer all fire alarms during lighting hours, and remain on hand at fires in order to cut wires, if necessary, and perform any other duties which may suggest themselves in the interests of the company. The daily reports of these patrolmen are posted in a book kept for that purpose in which the history of any particular lamp in the service can be read at a glance.

21. In the attempt to consolidate the two systems of incandescent lighting it soon became evident that all the feeders must be concentrated at one power house, in order that one station only need be kept running during daylight, and water power being cheaper than coal, that station which had the largest water wheel equipment was the most suitable for a central station. The Standard Electric Company's large and commodious power house best answered the requirements and was selected as the central or distributing station, and the alternators in the other stations were connected, each by a pair of wires, to a central switchboard in this station.

22. In the steam station a 500 volt, direct current, compound wound generator of 250 h. p. was installed as a part of the power system, to take the place of the 500' volt U. S. machine above referred to.

23. The stations a, b and c, of the Chaudiere Company, having become sub-stations, a switchboard panel for each generator was provided in every station. This panel is made of marble set into an iron frame. Each panel contains a T. H. voltmeter connected by pressure wires with the switchboard in the central station, a T. H. ampere meter, alternator field rheostat, main combined switch and cutout, and exciter combined switch and cutout. As these cutouts or fuse blocks, that serve at the same time the purposes of a switch, are also used in the central switchboard, they may be described here.

24. They consist of a block of lignum-vitæ hollowed in the centre so as to form a chamber, air tight but for a small aperture in one side. This chamber contains a fuse of aluminum alloy. The terminals are outside this chamber and fully protected. When a fuse blows the sudden expansion of the air contained in the chamber causes a sudden air blast through the aperture effectually breaking the arc. The terminals extend outward in the form of metallic plugs, which may be inserted in or withdrawn from spring receptacles set in the switchboard. There are no metal parts exposed on the face of these panels from which there is danger of receiving a shock or getting burned.

25. Each generator in the steam station is excited by a separate machine, but each of the exciters is of sufficient capacity to excite any two of the generators.

26. Even the most approved water wheel governors are not sufficiently sensitive or rapid in their action to maintain constant wheel speed under large or sudden changes of load and the speed of water wheels on power service varies to a considerable extent. To prevent wheels racing when a heavy circuit is opened, hand levers were arranged to throw the governor into faster gear with the gate, so as to close it in a few seconds. While this was an excellent feature as a preventive of accidents, a remedy for the more or less continuous variations of voltage in the circuits had to be found, and for this purpose a separate turbine was set up to run dynamos capable of exciting the fields of not only all the direct current generators, but also those of the alternators in this station. The fields will now remain constant, no matter how the speed may vary and the fluctuations of E. M. F. will be materially reduced.

27. The machines used as exciters are one of the 250 volt D. C. generators (run at 125 volts) for the alternators, and two of the 550 volt U. S. machines before referred to (run on a three wire system), for the 250 and 500 volt generators.

28. These exciters are also used to directly supply the motor

(Continued on page 180.)

PUBLISHED ON THE FIFTH OF EVERY MONTH BY

### CHAS. H. MORTIMER,

OFFICE : CONFEDERATION LIFE BUILDING,
*Corner Yonge and Richmond Streets.*

TORONTO,      -      -      CANADA.
Telephone 2362.

NEW YORK LIFE INSURANCE BUILDING, MONTREAL.
Bell Telephone 2299.

---

***ADVERTISEMENTS.***

Advertising rates sent promptly on application. Orders for advertising should reach the office of publication not later than the 25th day of the month immediately preceding date of issue. Changes in advertisements will be made whenever desired, without cost to the advertiser, but to insure proper compliance with the instructions of the advertiser, requests for change should reach the office as early as the 22nd day of the month.

***SUBSCRIPTIONS.***

The ELECTRICAL NEWS will be mailed to subscribers in the Dominion, or the United States, post free. for $2.00 per annum, 30 cents for six months. The price of subscription should be remitted by currency, in registered letter, or by postal order payable to C. H. Mortimer. Please do not send cheques on local banks unless 25 cents is added for cost of discount. Money sent in unregistered letters will be at senders' risk. Subscriptions from foreign countries embraced in the General Postal Union, $1.50 per annum. Subscriptions are payable in advance. The paper will be discontinued at expiration of term paid for if so stipulated by the subscriber, but where no such understanding exists, will be continued until instructions to discontinue are received and all arrearages paid.

Subscribers may have the mailing address changed as often as desired. *When ordering change, always give the old as well as the new address.*

The Publisher should be notified of the failure of subscribers to receive their papers promptly and regularly.

***EDITOR'S ANNOUNCEMENTS.***

Correspondence is invited upon all topics legitimately coming within the scope of this journal.

---

THE "CANADIAN ELECTRICAL NEWS" HAS BEEN APPOINTED THE OFFICIAL PAPER OF THE CANADIAN ELECTRICAL ASSOCIATION.

## CANADIAN ELECTRICAL ASSOCIATION.

**OFFICERS :**

PRESIDENT :
A. B. SMITH, Superintendent G. N. W. Telegraph Co., Toronto.

1ST VICE-PRESIDENT :
C. BERKELEY POWELL, Director Ottawa Electric Light Co., Ottawa.

2ND VICE-PRESIDENT :
L. B. McFARLANE, Manager Eastern Department, Bell Telephone Company, Montreal.

SECRETARY-TREASURER :
C. H. MORTIMER, Publisher ELECTRICAL NEWS, Toronto.

EXECUTIVE COMMITTEE :

GEO. BLACK, G. N. W. Telegraph Co., Hamilton.

J. A. KAMMERER, General Agent, Royal Electric Co., Toronto.

E. C. BREITHAUPT, Berlin, Ont.

F. H. BADGER, JR., Superintendent Montmorency Electric Light & Power Co., Quebec.

JOHN CARROLL, Sec.-Treas. Eugene F. Phillips Electrical Works, Montreal.

K. J. DUNSTAN, Local Manager Bell Telephone Company, Toronto.

O. HIGMAN, Inland Revenue Department, Ottawa.

W. Y. SOPER, Vice-President Ottawa Electric Railway Co., Ottawa.

A. M. WICKENS, Electrician Parliament Buildings, Toronto.

J. J. WRIGHT, Manager Toronto Electric Light Company.

## MONTREAL ELECTRIC CLUB.

OFFICERS :

President, W. B. SHAW,   -   -   Montreal Electric Co.
Vice-President, H. O. EDWARDS,
Sec'y-Treas., CECIL DOUTRE,   -   81A St. Famille St.
Committee of Management, T. F. PICKETT, W. GRAHAM, J. A. DUGLASS.

## CANADIAN ASSOCIATION OF STATIONARY ENGINEERS.

EXECUTIVE BOARD :

President, W. G. BLACKGROVE,   -   -   Toronto, Ont.
Vice-President, JAMES DEVLIN,   -   -   Kingston, Ont.
Secretary, E. J. PHILIP,   -   -   Toronto, Ont.
Treasurer, DUNCAN ROBERTSON,   -   Hamilton, Ont.
Conductor, W. F. CHAPMAN,   -   -   Brockville, Ont.
Door Keeper, F. G. JOHNSTON,   -   -   Ottawa, Ont.

TORONTO BRANCH No. 1.—Meets 2nd and 4th Friday each month in Room D, Shaftesbury Hall.  W. Lewis, President ; S. Thompson, Vice-President ; T. Eversfield, Recording Secretary, University Crescent.

MONTREAL BRANCH No. 1.—Meets 1st and 3rd Thursday each month, in Engineers' Hall, Craig street.  President, John J. York, Board of Trade Building ; first vice-president, J. Murphy; second vice-president, W. Ware ; secretary, B. A. York ; treasurer, Thos. Ryan.

ST. LAURENT BRANCH No. 2.—Meets every Monday evening at 43 Bonsecours street, Montreal.  R. Drouin, President ; Alfred Latour, Secretary, 306 Delisle street, St. Cunegonde.

BRANDON, MAN., BRANCH No. 1.—Meets 1st and 3rd Friday each month, in City Hall.  A. R. Crawford, President ; Arthur Fleming, Secretary.

HAMILTON BRANCH No. 2.—Meets 1st and 3rd Friday each month, in Maccabee's Hall.  E. C. Johnson, President ; W. R. Cornish, Vice-Pres.; Wm. Norris, Corresponding Secretary, 211 Wellington Street North.

STRATFORD BRANCH No. 3.—John Hoy, President ; Samuel H. Weir, Secretary.

BRANTFORD BRANCH No. 4.—Meets 2nd and 4th Friday each month, F. Lane. President ; T. Pilgrim, Vice-President ; Joseph Ogle, Secretary, Brantford Córdage Co.

LONDON BRANCH No. 5.—Meets in Sherwood Hall first Thursday and last Friday in each month.  F. Mitchell, President ; William Meaden, Secretary Treasurer, 533 Richmond Street.

GUELPH BRANCH No. 6.—Meets 1st and 3rd Wednesday each month at 7:30, p'm.  J. Fordyce, President ; J. Tuck, Vice-President ; H. T. Flewelling, Rec.-Secretary ; J. Gerry, Fin.-Secretary ; Treasurer, C. J. Jorden.

OTTAWA BRANCH, No. 7. — Meets 2nd and 4th Tuesday, each month, corner Bank and Sparks streets ; Frank Robert, President ; F. Merrill, Secretary, 352 Wellington Street.

DRESDEN BRANCH No. 8.—Meets every 2nd week in each month; Thos. Merrill, Secretary.

BERLIN BRANCH No. 9.—Meets 2nd and. 4th Saturday each month at 8 p. m.  W. J. Rhodes, President ; G. Steinmetz, Secretary, Berlin Ont.

KINGSTON BRANCH No. 10.—Meets 1st and 3rd Tuesday in each month in Fraser Hall, King Street, at 8 p. m.  President, S. Donnelly ; Vice-President, Henry Hopkins ; Secretary, J. W. Tandvin.

WINNIPEG BRANCH No. 11.—President, G. M. Hazlett ; Recording Secretary, J. Sutherland ; Financial Secretary, A. B. Jones.

KINCARDINE BRANCH No. 12.—Meets every Tuesday at 8 o'clock, in the Engineer's Hall, Waterworks.  President, Daniel Bentt ; Vice-President, Joseph Hall ; Secretary, A. Scott.

WIARTON BRANCH No. 13.—President, Wm. Craddock ; Rec. Secretary, Ed. Dunham.

PETERBOROUGH BRANCH No. 14.—Meets 2nd and 4th Wednesday in each month.  S. Potter, President ; C. Robison, Vice-President ; W. Sharp, engineer steam laundry, Charlotte Street, Secretary.

BROCKVILLE BRANCH No. 15.—President, W. F. Chapman ; Vice-President, A. Franklin ; Recording Secretary, Wm. Robinson.

CARLETON PLACE BRANCH No. 16.—President, Jos. McKay Vice President, Henry Derrer ; Fin. Secretary, A. M. Schofield.

## ONTARIO ASSOCIATION OF STATIONARY ENGINEERS.

BOARD OF EXAMINERS.

President, A. AMES.                        Brantford, Ont.
Vice-President, F. G. MITCHELL   ·   -   London, Ont.
Registrar, A. E. EDKINS   -   -   139 Borden st., Toronto.
Treasurer, R. MACKIE,   -   -   28 Napier st., Hamilton.
Solicitor, J. A. McANDREWS,   -   Toronto.

TORONTO—A. E. Edkins, A. M. Wickens, E. J. Phillips, F. Donaldson.
HAMILTON—P. Stott, R. Mackie, T. Elliott.
BRANTFORD—A. Ames, care Patterson & Sons.
OTTAWA—Thomas Wesley.
KINGSTON—J. Devlin (Chief Engineer Penetentiary), J. Campbell.
LONDON—F. Mitchell.
NIAGARA FALLS—W. Phillips.

Information regarding examinations will be furnished on application to any member of the Board.

---

THE citizens of Ottawa and of Rat Portage are boasting of their enlightenment—there being, it is said, in each of these places, one incandescent lamp in operation for every unit of population.

---

THE question as to the legality of Sunday street cars is to receive an authoritative decision in the courts of Ontario. The Lord's Day alliance secured in May last, a fiat from the Attorney General, to allow the question to be tested, and a few days ago their solicitor, A. E. O'Meara, acting for John Henderson, of Hamilton, who appears as plaintiff, applied for an injunction to restrain the Hamilton Street Railroad Co., and their employees from operating their road on Sunday.  The case will be tried at the Hamilton assizes in November, but whatever the decision then, it will without doubt be carried to a higher court.

---

IT is understood that the Department of Inland Revenue has ordered from James White, of Glasgow, a complete set of Lord Kelvin's standard ampere balances for the Standards Branch at Ottawa.  These instruments will be of the one-ampere type shown on page 24 of James White's catalogue and will be in all respects similar to the set made for the Board of Trade Standardizing Laboratory in London.  They are being made with the view of obtaining great permanency and accuracy, so as to put them beyond any question of dispute.  The set of four instruments will range (a) from 0 to 1 ampere ; (b) from 1 to 5 amperes ; (c) from 5 to 25 amperes ; (d) from 25 to 125 amperes. We understand that when these instruments are placed in position, the Department will be prepared to standardize instruments for the electric lighting industry free of charge, giving a certificate of the comparison.

IT has been hinted quite recently that possibly the Montreal Electric Club would not resume its regular meetings during the coming winter season. Some of the leading spirits in the Club appear to have become in a measure discouraged on account of the apathy of many of the members, and the failure of the local electric manufacturing companies to extend to the Club the recognition and support to which they believe it is entitled. We should indeed regret to witness the winding up of an organization which has shown so much activity and has done so much to awaken and maintain interest in electrical matters in Montreal. Perhaps if those who have so successfully managed the affairs of the Club in the past would stand by the ship a little longer, more favorable conditions would ensue, resulting in a prolonged life of usefulness for the Society.

ONE of the most profitable adjuncts to an electric street railway is a conveniently located park. A number of Canadian roads have invested in property of this character, and so far as we can learn the investment has in every instance turned out satisfactorily. Rockliffe Park, owned by the Ottawa Electric Railway Company, is an extremely popular resort, and has been rendered increasingly so by the erection of a merry-go-round and other means of amusement. It may surprise the managers of other roads, who have not yet gone into this feature of the business, to learn that the merry-go-round at Rockliffe Park, which cost the railway company about $2,500, has produced a revenue of $5,000 during the past season, in addition to increasing very largely the passenger receipts of the road. It is learned that Mohawk Park has been the means of doubling the receipts of the Brantford Electric Railway Company, and that the new park opened recently in connection with the Winnipeg Electric Railway nets the company as much in receipts as the balance of the entire system.

Prof. Alex. Graham Bell, inventor of the telephone, has just returned from Europe, and has apparently got some new ideas in his head, which may materialize into as famous and useful inventions as the telephone. He is interested in flying machines, and believes the day is not far distant when men and women will go flying through the air like birds. What a relief that will be to our congested streets. Balloons and butterfly wings must, however, he says, be discarded. He believes a machine of greater specific gravity than the air to embody the true principle, and says that there are several French designs, one especially called the kilikoptes, which meets his idea. He thinks the introduction of the bicycle will not displace the horse to the extent some people predict, but that a means will be discovered by which that animal can be taken off the ground, and still used as a motive power. When not only men but horses take to flying, what may we not expect. Then as to his radiaphone, he says it is as perfect as the telephone, but impracticable for long distances on account of the rotation of the earth. Prof. Bell has also invented a condenser for providing the Newfoundland fishermen with fresh water at sea. He is a wonderful man.

THT fifth convention of the Canadian Electrical Association, the proceedings of which are fully reported in this number of the ELECTRICAL NEWS, viewed either from the standpoint of instruction or pleasure, was eminently successful. The subject of the papers presented were sufficiently varied to attract the interest of members in every department of electrical effort. It is a matter of much regret that three of the authors, Messrs. Hartman, Keeley and Milne, were unavoidably detained from the meeting, necessitating the reading of their interesting productions by proxy. The absence of these gentlemen had also the effect of lessening discussion. In this connection it might not be out of place to remark, that while the papers presented at the conventions of the Association have usually been of an interesting and instructive character, the discussions upon them have not been as spontaneous and thorough as could be desired. The discussions should be taken part in by every member who holds an opinion on the subject under consideration, and if this were the case, they would prove to be one of the most valuable features of every convention. We look for an improvement in this particular at future meetings. It will be seen that the Association has revised and greatly improved its constitution, thus

placing itself on a sound working basis. The increase in membership during the year is most gratifying, and is due in no small measure to the enthusiastic efforts of Mr. Dunstan, the retiring President. The newly elected President and Executive officers are well qualified to maintain the progress and usefulness of the organization. It would be impossible to speak in too high praise of the efforts of the Local Committee in making such complete arrangements for the entertainment of the visitors. For the successful carrying out of these arrangements, the Committee spared neither time, labor or expense. It is to be hoped that they will feel in some measure rewarded by the success achieved and the expressed appreciation of their work on the part of every member and guest. The opinion is freely expressed that the Ottawa convention has in many respects strengthened the Association and brightened its outlook for a prosperous future. The fact that this youthful organization has already taken first rank with older societies of similar character in other countries, should be a source of pride to Canadians interested in the science of electricity, and should ensure to it their hearty support. The next convention is to be held in Toronto in June, the most beautiful month of the year, and should prove to be worthy of its predecessors and surroundings.

## MOONLIGHT SCHEDULE FOR OCTOBER.

| Day of Month. | Light. | | Extinguish. | | No. of Hours. |
|---|---|---|---|---|---|
| | | H.M. | | H.M. | H.M. |
| 1...... | A.M. | 1.40 | A.M. | 5.10 | 3.30 |
| 2...... | " | 2.40 | " | 5.10 | 2.30 |
| 3...... | No light. | | No light. | | ..... |
| 4...... | No light. | | No light. | | ..... |
| 5...... | No light. | | No light. | | ..... |
| 6...... | P.M. | 5.50 | P.M. | 7.30 | 1.40 |
| 7...... | " | 5.50 | " | 8.10 | 2.20 |
| 8...... | " | 5.50 | " | 8.50 | 3.00 |
| 9...... | " | 5.50 | " | 9.50 | 4.00 |
| 10...... | " | 5.50 | " | 10.50 | 5.00 |
| 11...... | " | 5.50 | " | 12.00 | 6.10 |
| 12...... | " | 5.50 | A.M. | 1.00 | 7.10 |
| 13...... | " | 5.50 | " | 1.10 | 7.20 |
| 14...... | " | 5.40 | " | 2.30 | 8.50 |
| 15...... | " | 5.40 | " | 3.50 | 10.10 |
| 16...... | " | 5.40 | " | .5.40 | 12.00 |
| 17...... | " | 5.40 | " | 5.40 | 12.00 |
| 18...... | " | 5.40 | " | 5.40 | 12.00 |
| 19...... | " | 5.40 | " | 5.40 | 12.00 |
| 20 ...... | " | 5.40 | " | 5.40 | 12.00 |
| 21...... | " | 6.00 | " | 5.40 | 11.40 |
| 22...... | " | 7.00 | " | 5.40 | 10.40 |
| 23...... | " | 8.00 | " | 5.40 | 9.40 |
| 24...... | " | 9.20 | " | 5.40 | 8.20 |
| 25...... | " | 10.20 | " | 5.40 | 7.20 |
| 26...... | " | 11.00 . | " | 5.40 | 6.40 |
| 27...... | " | 11.30 | " | 5.40 | 6.10 |
| 28...... | ......... | | ......... | | } 5.10 |
| 29...... | A.M. | 12.30 | ......... | | |
| 30...... | " | 1.30 | " | 5.40 | 4.10 |
| 31...... | " | 2.30 | " | 5.40 | 3.10 |
| | | | Total, | | 194.40 |

## SPARKS.

Electricity has been applied to the aging and tempering of wine in Italy with great success.

The Oshawa railway has given the contract for the erection of a power house of 250 horse power.

Mr. A. M. Schofield, for the past three years with the Carlton Place Electric Co., is going to Detroit.

Joseph Lamonge, a French-Canadian carpenter, was killed by a trolley car in Montreal on the 23rd. He drove in front of a moving car.

David and Elizabeth Boyd are sueing the Hamilton Street Railway Co., for $2,000 damages sustained by being run into by a car on Barton street.

It is said four new mica mines will be opened up in Wakefield this fall, Mr. T. J. Watters is operating a new amber mica mine in Portland. He also contemplates working the Lake Girard mines. The demand for mica for electrical purposes is rapidly increasing.

It is proposed to convert the disused railway belt lines back of Toronto, controlled by the Grand Trunk, into electric lines. No decisive action has yet been determined on, but should it be decided to make the change work will probably be commenced early in the spring.

Work will be commenced on the Quebec city electric railway at once, but little will be done in tracklaying till next spring. A delay has arisen because of the refusal of the present street railroad company, which has exclusive right of way for nine years, to allow the electric company to cross its tracks.

*(Continued from page 177.)*

circuits on Sundays when the load is very light, and the motor wheel which has run day and night during the week is shut down.

29. Each D. C. generator is supplied with a double-throw switch by means of which its fields may be connected either with the separate exciter or with its own armature. Alternators may also be excited by the common exciter or independently, the change being made through the switchboard.

30. Each of the three companies had pole lines in the same districts ; in many cases both sides of a street were occupied by them. The number of poles to be maintained was reduced by placing all the wires running on a street on the best pole line and discarding the other. The lighting districts that were occupied by two different systems were divided in two, so that, while the number of feeders was actually reduced by three pairs the number of distribution centres was doubled and the line loss between them and the converters was correspondingly decreased.

31. The mains running through contiguous districts are made to overlap, so that all public buildings such as churches, theatres, halls and hotels have their lights divided between at least two separate circuits and converters. This makes it almost impossible, in case of accident, for all the lights to be out at one time.

32. The size of feeder units had been kept down within the capacity of the smallest generator, but it was found advisable to increase the units for the present to 1,000 and 1,500 lights, which seemed to best fit our generator units.

33. Eight circuit feeders were calculated for an ultimate load of 1,500 lights, and ten for 2,000 lights ; this left some margin for extensions.

34. This change made it necessary to run the 750 light machines in pairs as a 1,500 light unit.

35. First parallel running was tried but it was found that the idle currents were considerable at times and this method of running· was abandoned. Two of the generators were then mounted on iron girders set very accurately so as to approximate a solid iron· base, and flanged pulleys were put on the shafts and bolted together. These generators could thus be driven as a single machine. The armatures were· connected in multiple. If this arrangement proves satisfactory, from a mechanical point of view, the other generators in this station will be similarly coupled.

36. It is necessary to the proper working of a lighting and power service, that the losses in the different parts of ·each circuit should be predetermined and unchangeable. In order to better obtain this result a series of official wiring tables were issued by the company, covering interior wiring services, mains, feeders, etc., together with such printed directions as would secure uniformity in the manner of using the tables, a thing much to be desired but not always obtained. The losses to be 10% in feeders, 2% in mains, 1% in services and 2% inside buildings calculated.

37. It was also necessary for the convenient ·working of the lighting system that a uniform voltage should be maintained on all mains, and 1040 volts was decided upon ; it was also decided, however, that 50 volt lamps would be used, experience having taught us that lamps of medium efficiency when run by water power gave the best results for customers and company, when burned somewhat above their normal voltage.

38. The public has come to expect a great deal of light from a 16 candle power lamp. If the lamp is good and the efficiency 3½ watts per candle or lower, it will maintain its candle power for a considerable time when overrun by four per cent. ·

39. Converters of 100 light capacity have been introduced wherever the business was sufficiently bunched up, displacing the smaller ones which are used in the districts of more scattered lighting. No doubt still larger ones will be used in time.

40. The compensator system of regulation was adopted in preference to the feeder and pressure wire system. We still have the feeders, and the compensators take care of all the losses between the dynamo and the lamp, while the pressure wires lose their usefulness at the distribution point, although the losses between that point and the lamps may be considerable in some cases.

41. Each circuit is provided with at least three non-arcing lightning arresters, one at the station, one at the point of distribution and one or more at the distant ends of the mains. These are carefully grounded, the ground wires being riveted to street railway rails whenever possible.

42. A Bristol recording voltmeter, set up in a case convenient for carrying about, is used to adjust the compensators. The voltmeter is left at some point of the circuit to be adjusted, for twenty-four hours. This is repeated at different points of the same circuit. The adjustments should be checked once a month.

43. The main switchboard situated in the central station consists of thirty-four marble panels set side by side in a framework of angle steel fastened to the stonework of the building. This frame stands at least six feet from the wall and is supported by soft rubber discs set into iron rings fastened to the floor. These discs have the effect of taking up the vibrations of the floor, and prevent their being communicated to the instruments above. The switchboard is 57 feet long and nine feet in height.

44. There are eight dynamo panels similar to those in the

other stations and already described, six for the alternators in this building, and two spare ones.

45. Five motor panels that contain Weston illuminated dial voltmeters, Weston edgewise ampere meters, Westinghouse circuit breakers, ground detector, and jaw switches, through which all the motor circuits and D. C. generators are interchangeable. ·

46. The twenty feeder panels contain Westinghouse pendulum voltmeters, ampere meters and compensators, throw-over switches and panels for plug and cable connections with twelve pairs of bus bars and combination switches and fuse blocks, as already described.

47. These twenty panels are divided into two sections of ten between which a special panel is set up, containing a clock, a ground detector and switch, and other special devices.

48. Directly in front of each section of feeder panels and four feet away from them stands a table made up of an iron frame work with sides of wire netting and plate glass top set in a polished brass frame. Each of these tables contain ten regulators or "boosters" with a range of 20°/₀ up or down. Each circuit can thus be regulated independently.

.49. The attendant at this switchboard controls the whole system. He is also in communication with the attendants at sub-stations and the station superintendent's residence by a private telephone line.

50. For economy in line construction it was decided not to extend the 250 volt motor system except for units of one h. p. or less, and to merge it and the 500 volt service into one single three wire distribution. The 100 h. p. motor in the flour mill is, however, on a separate circuit and may, if desired, be run independently of the others. The three wire system is supplied by two of the 250 volt 60 h. p. generators in series, and the 500 volt 250 h. p. generator connected to the + and – wires. The brushes of the 250 volt machines on the + side and the + brush of the 500 volt machine, may be connected together for equalizing purposes. All the D. C. generators are interchangeable through the switchboard.

51. It was found necessary to almost completely reconstruct the motor circuits. Four pairs of No. 0000 feeders were strung up. As the joints in wire of that size are extremely unsightly a portable welder was constructed for welding the lengths of wire together. A large regulator core was fitted with a primary coil of 388 turns and a secondary coil of a single turn made up of 12 No. 0000 wires upon the ends of which massive metal jaws were shrunk. These jaws normally stand about four inches apart, but may be pressed closer together by an insulating clamp and screw, the elasticity of the secondary coil causing the jaws to resume their normal position when.released. The current is regulated by a T. H. reactive coil. This apparatus may be attached to any converter on the line as required.

52. Several of the U. S. dynamos in use for lighting up to 1889 have been put in service as motors, two of them running elevators very successfully.

˙53. The company has lately made what is believed to be an innovation in providing in its office, which is open day and night, a locker with a glass front in which are displayed rubber coats, gloves and shoes. This in addition to the rubber gloves regularly supplied to all linemen. ˙The key of this locker hangs within a little box behind a glass which is to be· broken, in case of accident, when requiring the clothing.

Mr. Breithaupt : I think we are much indebted to Mr. Dion for his paper on the consolidation of three different plants. The arrangements with regard to the dynamos would cause some difficulty, and as Mr. Dion has explained, the consolidation also of these three lines on one switchboard ; for instance, the day work had·of course to be concentrated in one station, as it would not do to operate some lamps from one station and some from another. An interesting part to my mind is the way in which the lines have been treated, particularly the street mains, how the mains of the three different companies were consolidated into one and still kept to a certain extent separate. Large buildings being on duplicate mains, if the machinery should be disabled in one station the lights would not all be interfered with ; this I think is deserving of commendation. The paper throughout is so very clear and precise that I have very little criticism to offer. It is certainly a very extensive work. I was told some time ago that it was not completed yet.

Mr. J. J. Wright : I notice one point that Mr. Dion speaks about, that is, the durability of lamps when being overrun by four per cent. I presume he speaks from experience. I would like to know what the durability of the lamps would be when overrun four per cent. A few minutes ago I was looking over an article in the ELECTRICAL NEWS in which it was stated that every one per cent. overrun reduces the life of a lamp fifteen per cent. According to this Mr. Dion's four per cent. would reduce the life sixty per cent.

Mr. Dion : Replying to Mr. Wright I do not claim that over. running a lamp would not shorten its life. It certainly does shorten it very materially, but we have come to the conclusion that we can seldom run lamps to their death. There comes a time in their existence when it is a charity to put an end to them. We think, everything considered, that it pays the company and gives better satisfaction to the consumer to give a very bright light. We lose on the life of the lamp but do not lose very much on the use of the light.

Mr. Kammerer : There is one matter I would like to speak about in connection with the regulation of the speed of dynamos driven by a water wheel. Mr. Dion might tell us what efficiency of government he gets by the means adopted, for the purpose of maintaining a steady flow of current in the fields. While perhaps to the man who is running water-wheels and applying power by water-wheels all the time, it may seem a simple matter, still to my mind it is not very clear what efficiency he can get. For instance, the holding up of electro-motive force under a sudden load, or getting the gate to open and close fast enough to prevent acceleration upon the removal of a load. Some of us would like to get a little further information on this point.

Mr. Dion : I am sorry I have no figures as to water regulation with me which I can present. We realize that a considerable portion of the variation of the electro-motive force is due to the variation in the fields, and if we can keep that portion of it steady it would better the regulation of the electro-motive force to that extent. That system of exciting the fields had been tried last winter by the electric railway and its effect was very favorable. The result of the change was very noticeable, so much so that the difference after the operation was remarked upon by people riding in the cars. We hope it will be very useful in our case.

Mr. Fisk : We are running a street railway in Peterboro by water power, and also an incandescent light plant. We have every advantage as far as water is concerned, and the incandescent is giving good satisfaction. Of course an attendant, when the load is coming on, remains close to the wicket and increases the power. We do not find our incandescent load comes on suddenly, that is, not enough to affect it. The load appears to come on gradually and to go off in the same way. Of course on railway work it goes up and down considerably. So far we have not been able to control that as close as we would like ; the attendant throws the water on and off as necessary. We have not been able to run the railway and our incandescent lamps together, and even with an engine and a very good regulator it seems impossible to run incandescent lighting and railway work at the same time.

Mr. L. B. McFarlane : There is one line here in paragraph 12 that I would like to call your attention to. We would like to know about the anchor ice. We understand that it is a detriment in running, and that steam power had to be put in and used on account of it. I believe it is intended to have a company in Montreal to use the Lachine Rapids as a source of power, and we would like to have the experience of Ottawa as to whether that difficulty can be overcome or not.

Mr. Soper : I would like to ask Mr. McFarlane whether it is the anchor ice that causes the buzzing in the telephone. (Laughter.)

The President : If Mr. Soper will answer Mr. McFarlane, no doubt Mr. McFarlane will then be glad to reply to his question.

Mr. Soper : I think Mr. Dion, or his assistant, Mr. Murphy, more competent to answer than I. He is very familiar—I may say distressingly familiar—with the subject during certain seasons of the year.

Mr. Dion : The anchor ice question is a very difficult one to deal with. People who have lived all their lives in this place and have been connected with water power, and who are supposed to know all about the question, disagree very much about the particular nature of this trouble and how it comes about. Under these conditions it would be presumptuous for me to say anything about it. However, where the trouble arises in the fall with the anchor ice, which is more properly called frazil, being more like half melted snow than anything else, is that it comes down, preventing the flow of water in the water-wheel. This occurs in the fall after a very sharp frost, but it is of short duration and lasts a few days only. When the weather is settled and the river freezes over, this stops the stuff from coming down. Another trouble which occurs later in the winter is of a different nature. As far as I can make out it comes by anchor ice adhering to the solid ice underneath and forming a solid bank of ice, which extends nearly to the bottom. This is a serious difficulty, and as the banks are of large extent it is impossible to remove them. It was after experiencing this difficulty the steam station was determined upon.

Mr. C. B. Powell : As an extension to what Mr. Dion has said, and in order to give Mr. McFarlane a little more light on the subject, I would say that I never heard of the existence of a water power without this anchor ice ; some have more and some less. The St. Lawrence Rapids I think have more than almost any other in Canada. We have indulged in much speculation as to the occasion of it and as to the nature of the trouble. The cause is water running over shallow ground, and freezing with great rapidity it assumes the character of snow—of melted snow that is—and banks up against the water-wheels. In Montreal every water-power is troubled with it, and I never heard of anything that would prevent it. During the time of the anchor ice trouble it is impossible to do anything with it. Various remedies have been proposed ; people have even talked of putting steam pipes in the rapids, but no amount of heat would have any effect on it. We have the best means of coping with it from the fact of having an auxiliary station.

Mr. Fisk : Regarding anchor ice, I had some little experience with it. There is another condition that bothers me more than anything. Before the river freezes over in the fall, when the temperature of the air drops below the temperature of the water, the water seems to remain free of ice, but if you put an iron bar into the water it immediately becomes covered with ice ; even a piece of wood will become coated in the same way. It gathers on the buckets of the wheel until the wheel seems to be a solid cylinder. I once got a number of men to empty the wheel of ice by means of hot water, and then let the water in again. It was just as bad again in half an hour. This condition of affairs generally occurs in November, and sometimes in March before the river is open. With regard to anchor ice proper, we are well situated in Peterborough, unless the river be neglected. Otherwise we suffer very little from it, but this frost condition there seems to be no remedy for.

Mr. Soper asked what was the condition of the surface of the water where it entered the flume.

Mr. Fisk : Comparatively smooth.

Mr. Soper : Were there any ripples on it?

Mr. Fisk : No.

Mr. Badger moved that a vote of thanks be tendered Mr. Dion for his very interesting paper ; seconded by Mr. Breithaupt. Carried.

Mr. Dwight then read the following paper :

## THE TELEGRAPH IN CANADA.

By CHARLES P. DWIGHT.

THE telegraph in Canada has so often been made a subject of history and retrospect that one can hardly hope in a paper of this nature to do more than briefly outline much of what has already been written concerning its rise and development. The various stages in its practical operation, from the simplicity of the old paper register to the present day, are too well known to require any elaboration at my hands, and I have, therefore, simply put together something which may be considered as a record for this Association concerning the more important telegraph organizations which have existed in Canada from the start.

The first commercial telegraph line erected in this country was in the year 1847, between Toronto, Hamilton, St. Catharines and Niagara Falls ; connecting at the latter point with a line through to Buffalo, owned by one David Kissock. The organization under which this line was built was known as the Toronto, Hamilton, Niagara Falls & St. Catharines Electro-Magnetic Telegraph Company,—a somewhat lengthy title for a concern of this nature, but one which was thoroughly expressive in regard to the scope and nature of its business. As a matter of curiosity there is laid on the table for your inspection one of the original stock scripts of this Company, which an antiquarian friend has loaned me for the occasion, and which you will note bears the signature of Thos. D. Harris, President, and P. B. Marling, Secretary. The capital stock of this Company was $16,000, and the line was built under contract by Samuel Porter, a man long known afterwards in connection with various telegraph enterprises in the United States, and of whom it was said that "he built for this first Canadian Telegraph Company an honest and well appointed line."

In the same year, 1847, was organized the Montreal Telegraph Company, with a capital of $60,000. This Company immediately proceeded to construct a line from Quebec to Toronto, and soon afterwards purchased the line erected by the Toronto, Hamilton, St. Catharines & Niagara Falls Company. The line from Quebec, when finished, was looked upon at the time as the best piece of telegraph construction on the continent. The poles were of cedar, thoroughly tamped and well set. Wooden brackets of white oak were used, with glass insulators. The wire was a No. 9 gauge, English galvanized, and was the first of this kind employed for such a purpose on the continent.

At the close of the year 1847 the Montreal Company had in operation 540 miles of wire, with 9 offices, 35 employees, and had sent in all 33,000 messages.

An organization known as the British North American Electrical Association was also formed in 1847, with F. N. Gisborne as the moving spirit. This Company, or Association, proposed connecting Quebec with the Lower Provinces, and finally with the Atlantic Coast, but for some years the line was extended no further than Riviere du Loup. It was finally extended to Woodstock, N.B., however, where connection was formed with the American Telegraph Company, an organization which had already connected a few of the principal points in New Brunswick at that time, and which had a capital of $25,000. A second organization, known as the American Telegraph Company, constructed a line about this time from Quebec to Montreal, but was afterwards absorbed by the Eastern Company. All of these lines east of Quebec, however, proved a dismal failure from a financial point of view, and were soon turned over to the Montreal Company without charge, and the line between Quebec and Montreal was also taken over by them at a nominal charge.

In the Eastern Provinces there is on record a project set on foot by Mr. Lawson R. Darrow in 1847, for the purpose of connecting the lines then in Nova Scotia with those of Maine, and an act of incorporation was granted for this purpose in 1848. In the same year a line was built from Calais, Maine, to St. John, N. B., under the organization just mentioned, and which had now been incorporated into a Company known as the New Brunswick Electric Telegraph Company, with a capital of $40,000. A line to St. John via St. Stephens, St. George and St. Andrews, was completed Jan. 1st, 1849, and during the ensuing summer the line was completed from St. John to Hampton, Sussex, Salisbury, Dorchester and Sackville to Amherst, where connection was made with the Government line then in Nova Scotia, which was built from Amherst to Halifax in Nov., 1849, and which for the first time gave Halifax connection with New York.

In 1856 the lines of the New Brunswick Telegraph Company were leased to the American Union Telegraph Company, and some ten or twelve years later came under the direction of the Western Union Telegraph Company.

The Government line between Halifax and Amherst was built by F. N. Gisborne, for the purpose of forming a connection with the American and New Brunswick lines at the former place, in order to meet the demand for communication with New York, consequent upon the arrival of steamers at Halifax with European news. In 1851 this line was sold to the Nova Scotia Electric Telegraph Company ; an organization chartered in March of that year, and which afterwards extended the line from Pictou to Sydney, C. B., and from Halifax to Yarmouth. Upon completion of arrangements for the landing of Atlantic cables in Newfoundland, the lines of the Nova Scotia Company were leased to the American Union Tel. Co. in 1860. In 1866 this lease was taken over by the Western Union Company, who purchased the lines outright in 1872.

Returning to the Province of Quebec again, we find that in 1849 there was organized what was known as the Montreal & Troy Telegraph Company, which built a line the same year from Montreal to the frontier, and thence via Whitehall to Troy. This Company was organized by Ezra and Alonzo Cornell, who also constructed the line, and which worked for several years afterwards in connection with the Montreal Company. A. B. Cornell, whom it will be remembered was afterwards Governor of New York State, acted as manager of the Company at Montreal for two years after the construction of the line. Some few years afterwards the section of the line from Whitehall north became the property of the Montreal Telegraph Company, under a compact with the American lines, known as the Six Party Contract, whereby certain divisions of territory were made and allotted each Company.

In 1850 there was projected and built a line between Montreal and Bytown (probably the first telegraphic connection enjoyed by Ottawa) by the Montreal and Bytown Telegraph Co., of which Edward McGillivray was President, and which a few years afterwards was purchased by the Montreal Telegraph Company.

In the year 1852 the Grand Trunk Telegraph Company was organized, and built a line between Buffalo and Quebec, and seem to have given the Montreal Company a pretty lively opposition between these points. After a few years, however, it went the way of so many of its predecessors, and was purchased by the Montreal Company for the sum of $11,000. Then sprang up another organization, known as the Provincial Telegraph Co., which built a line over the same route, but it too was soon absorbed by the Montreal Company.

In 1868 was organized the Dominion Telegraph Company, which had soon built lines embracing all the important points between Buffalo, Detroit and Quebec, and whose opposition became more lively as time went on. Rates were reduced, and the outcome promised disaster for all concerned. When in 1881, therefore, a proposition was made for the consolidation of these conflicting interests, under lease, by the Great North Western Telegraph Company, considerable satisfaction at the prospect was expressed by all concerned, and a deal on these lines was accordingly put through, and is in operation to-day. The combined mileage of the two Companies at the present time, as operated by the Great North Western Tel. Co. at the present time, is 18,000 miles of poles, and 40 000 miles of wire, with some 1,800 offices throughout Ontario, Quebec, New Brunswick, Manitoba and Northern New York State.

By means of this amalgamation the telegraph business of the country was for a time almost entirely in the hands of the Great North Western Telegraph Company. In every city and town where two offices had previously been maintained the wires were all taken into one, and sweeping reductions in expenses consequent upon such a move were at once inaugurated. The monopoly thus brought about was not destined to last long, however, and almost immediately afterwards the Canada Mutual Telegraph Co. was organized, and constructed lines between Niagara Falls and Toronto, Montreal and the boundary line, and Montreal, Coteau and Ottawa. Some three or four years after the amalgamation had been effected, the Canadian Pacific Railway Co. had also commenced the construction of telegraph lines along the route of their road, and between many of the principal cities and towns of the Dominion, and in September, 1886, had opened 366 commercial telegraph offices throughout Ontario, Quebec, Manitoba and the North West Territories. Since that time they have been constantly adding to their plant, and at the present time have somewhere in the neighborhood of 25,000 miles of wire in operation, and 800 offices.

In certain remote localities along the St. Lawrence and in the North West territories, where private companies would hardly be justified in extending their lines, the Dominion Government have in operation at the present time somewhere in the neighborhood of 3,000 miles of wire.

The total amount of capital invested in Canadian Telegraphs may be roughly fixed at between six and seven million dollars, and the total wire mileage at somewhere in the neighborhood of 75,000.

In respect to population it can truthfully be said that no country in the world enjoys a more extensive system of telegraphs than Canada. Scarcely a town or hamlet in the whole country but has connection by this means with the outside world. Hundreds of offices are maintained throughout the country in small out-of-the-way places, where the actual business is but trifling, and where the lines in reality prove much more a matter of convenience to the public than profit to the telegraph companies.

The following comparative table, showing the number of inhabitants per each telegraph office, will indicate more clearly the position of Canada in this respect.

| Country. | No. of Inhabitants to each Telegraph Office. |
|---|---|
| Great Britain | 6,417 |
| Switzerland | 2,556 |
| Holland | 10,254 |
| France | 7,719 |
| Germany | 4,510 |
| United States | 5,625 |
| Canada | 2,320 |

In respect to rates, too, no country enjoys a cheaper schedule than Canada; distances and other conditions fairly taken into account. The maximum charge between offices in Ontario, Quebec and New Brunswick is 25 cents, and for this sum a message can be transmitted over twelve hundred miles of wire.

In Canada the telegraph companies have always kept well abreast of the times in promptly adopting the various improvements in apparatus which have from time to time been placed upon the market, and two well-known repeaters, the Toye and the Neilson, attest our own ingenuity in this respect. Both the duplex and the quadruplex systems are in daily use over some of the most important routes, and direct and rapid communication is maintained between all the larger centres, as well as with New York, Chicago and other important American points.

So essential a feature in every day business life as the telegraph has now become is very apt to be regarded in its simplicity as something from which little more may be expected in the way of improvements. Great things may yet be looked for, however, in the practical operation of the telegraph. From the days of the old Phelps register, when messages were laboriously spelt off the slowly winding tape, the brightest minds in the profession have ever been directed towards achieving that rapidity and perfection of transmission towards which so much has already been done. Numerous contrivances have within recent years been placed upon the market in the shape of printing machines, and the latest achievement in this direction—known as the Buckingham Automatic Printer—gives promise of being an unqualified success. This machine has recently been put to a thorough test over a line one thousand miles in length, and a sample of the work done by this means is laid on the table for your inspection. It is a quadruplex printer, capable of transmitting and printing 150 words per minute.

Predictions are of course a little premature as yet, but if thoroughly successful and universally adopted it will readily be seen how much nearer every man's door the telegraph will come.

Within the past two or three years dynamo plants have been installed in the offices of the Great North Western Tel. Co., in Toronto and Montreal, displacing several thousand cells of gravity battery in each place, and for adaptability and general efficiency there are few superior plants to be found anywhere.

In Canada telegraph lines are maintained under adverse conditions, which exist in few other countries similarly equipped. Long stretches of lines are maintained through rough and sparsely settled districts, and the sleet storms of winter often mean total abolition of long stretches of poles and wires over the most important routes, and involve an amount of labor and expense in their restoration little understood by the average outsider, who has merely a grumble to offer if his business cannot be got through in all kinds of weather.

Aside from the position which the telegraph occupies in our midst as a simple means of sending and receiving messages, it might not be out of place to enumerate a few of its more important outside functions. I need not say in this connection that there is no more vital adjunct in the operation of our great railways to-day than the telegraph. Railway telegraphy is in fact an art in itself, without which many of our more important railway systems might be likened to ships without rudders.

The collection and distribution of market reports is a service performed by the telegraph companies in Canada which is worthy of special mention. By means of tickers and special delivery are daily distributed amongst our brokers and others continuous quotations from the markets of Liverpool, London, Beerbohm, Chicago, Milwaukee, St. Louis, Duluth, Detroit, Toledo, New York and Paris, in both stocks and grain; a service which serves to keep those interested in continual touch with all the great markets of the world.

The gathering and distribution of reports in connection with the Meterological Department is another service, which is of inestimable value to many of the most important interests in the country.

The press service of the telegraph companies is also an important feature in connection with their business. I need not remind you that it is by means of the telegraph, broadly speaking, you are enabled to discuss so readily the affairs of the world, and look so wise and weighty. In Canada a regular system is in operation, whereby every telegraph agent is also an agent indirectly of the press; forwarding to headquarters such items of public interest happening in his town or neighborhood as he is required to send, and which is afterwards sifted and made use of by papers here and elsewhere. That the politics and political opinions of the country, are largely moulded by the press, there is little question. In fact many papers are primarily in existence for no other purpose than to serve political ends. Latter day enterprise in journalism is a source of constant wonder. In the dissemination of both political and legitimate news there is no one factor more important than the telegraph.

These are of course facts almost too well known to require repetition at my hands, but mere mention of the part played by the telegraph in connection therewith is sufficient to indicate the magnitude of its mission. As an instrument in the higher civilization of man it has no peer, and that we in Canada have shown ourselves so thoroughly alive to this fact is certainly a matter for congratulation.

The President: I may say that this contribution to our set of historical papers is a very valuable one. We are endeavoring to get a complete set of historical papers, which will be placed on record and which will become more and more valuable as time goes on. We already have the history of the telephone and of the electric railway, and this adds the telegraph to the list.

Mr. Black: I have listened with a great deal of pleasure to the paper read by Mr. Dwight, and it has taken me back in memory a great many years. I am probably the oldest telegraph man in the room at the present time and have had some little experience with a good many of these companies that Mr. Dwight refers to. I think he has probably put more history in a small space than most people would have been able to do. Of course he has left out a great deal that might have been said, but still the ground is well covered. In the matter of the Grand Trunk Telegraph Company I mentioned to Mr. Dwight to-day that the Grand Trunk Railway compelled them to change their name, but I could not remember what name they took. I think it was the International. The Montreal Telegraph Company had to compete with what was, I think, the most miserable line ever put up in Canada. It was built of No. 10 wire, and from the appearance of the line it must have been strung on poles picked up on the roadside. They had not digged proper holes for the poles; merely made a hole with an augur and put them in.

Mr. Higman: In regard to Mr. Bethune, who is not present, he is perhaps, with the exception of Mr. Dwight, senior, and another man in the service, the senior telegraph man of this country. He could no doubt add very interestingly and materially to the discussion. I listened with a great deal of pleasure to one of the passages in the paper, which brings to my recollection some very curious things in the world of telegraphy One of these is in connection with the old Montreal and Troy Telegraph Company. The story is one that Mr. Bethune told me on one occasion, and which I believe is true, but will hardly be credited by people of this age, who are in the habit of using the latest appliances for testing telegraph lines. This was about the year 1847. There had been a bad storm and Mr. Bethune with some workmen were engaged in making repairs. Their wire had run out and they scoured the country for wire, but could find none. In their search they visited a farm house, and not being successful in getting any wire they racked their brains to discover a substitute. Finding a piece of rusty weather-beaten stove-pipe, they determined to use it, and puncturing a hole at each end, they strung it up, and Mr. Bethune stated to me that that pipe swung there for three weeks, during which time the line worked uninterruptedly. Now if they had had a galvanometer and other appliances for testing lines, they would not have used that stove-pipe at all, as their instruments would have shown that it would not work. Where ignorance is bliss it is folly to be wise. This is an actual fact. Those engaged in telegraph work a quarter of a century ago will bear me out. When I left England we were not allowed to take by sound at all ; we were obliged to let the paper run through the register, and if we did take by sound a fine was imposed. The last fine I paid was in King George pennies, and it cost me fifteen cents express charges on the fine. In the matter of cut-outs, we had one bound with string on both sides of it. There was nothing to prevent it reaching the re-lay and was often found next morning.

Mr. Soper : I would like to say a word in praise of the paper just read. As an old telegrapher it was very valuable to me. Its historical value was very great. It must have been at the cost of considerable trouble that the author has been able to give it to us. We old telegraphers are not like those who find themselves in the electric work to-day without going though the early stages of telegraphy, and we look back almost with joy upon those days with their hardships. We had a freemasonry among us which does not exist to-day, and the paper read by Mr. Dwight brings back those days vividly to my mind. I wish to propose a hearty vote of thanks to Mr. Dwight.

Mr. Smith : Mr. Higman has told us about working through a stove-pipe. I can remember a case where we put on an extra battery and jumped the break. Mr. Dwight has gone to a great deal of trouble in going through a mass of details and dishing them up to us in condensed form, so they can be placed with the Association records for all time to come. I do not think that the general public give the telegraph companies credit for the enterprise they display, when you come to consider the way in which they push their lines out into hundreds of places where they do not pay, and maintain a service in Canada such as is not found anywhere else in the universe. And people seem to take it as a matter of fact ; they merely send their messages and pay their quarters. There is a disposition on the part of some to think that the telegraph service has reached its limit. This is a great mistake. This specimen of work done by printing telegraph is an evidence of what may be done. Any system that will transmit 150 to 200 words a minute in characters as legible as that, is worth more consideration, and we expect to be able to quadruplex the line at that. It is a question of how far this is going to interfere with the mails in the near future.

Mr. Black : I have had brought to my notice a novel piece of line repairing. I remember linemen coming in on one occasion seemingly having some question to ask me. One nudged the other, telling him to ask the question. I asked them what it was they wanted to know, and one spoke up and asked if a piece of clothes line would do to mend a break in a telegraph line. I told them no. The foreman then said that they had taken a piece of clothes line about a yard long out of the telegraph line and they had inquired at the nearest office if the line had been interrupted, and they were told that there had been no interruptions for three weeks. (Laughter).

The President : I have the pleasure of announcing that, by courtesy of the Electric Railway Company, members wearing the Association badges will be carried free on the electric cars. The Telephone Company also have placed a telephone at the disposal of the members. The Mayor this morning with great courtesy offered us the use of the City Hall. I mention this in case you wish to take action in the matter. We were fortunate in having the Railway Committee Room for our opening meeting, and both the Board of Trade Rooms and the City Hall at our disposal for future meetings.

Mr. Soper : It may not have occurred to you that the ventilation of this room is not very good. When I came in half an hour ago it was very noticeable ; the air seemed laden with noxious gases. I think the City Hall much better in this particular, and just as convenient.

It was moved by Mr. Taylor, seconded by Mr. Badger, that future meetings be held in the City Hall, which has been placed at the disposal of the Association by the courtesy of the Mayor. Carried.

Moved by Mr. Smith, seconded by Mr. Roch, that a hearty vote of thanks be tendered the Board of Trade for the use of their room. Carried.

At 8 o'clock p.m. the members, accompanied by a number of their lady friends, entered special cars provided by the Ottawa Electric Railway Co., and made an interesting visit of inspection to the power stations and Mr. J. R. Booth's lumber mills at the Chaudiere.

## SECOND DAY.

### MORNING SESSION.

At 10:30 a.m. the convention resumed, and the President announced the order of business.

Mr. McFarlane moved the adoption of the report of the Committee on Legislation, seconded by Mr. Gilmour. Carried.

The President : The next order of business is the consideration of the report of the Committee on Constitution. I would suggest that this should be taken up article by article, as it is a very important report and should receive careful consideration.

Mr. Kammerer : I beg to move that the constitution be taken up clause by clause. I notice it is called an amendment, but I think we should call it a revision of the constitution. It really is a revised constitution.

Mr. Wright : I beg to second the motion that the report be now taken up clause by clause. Carried.

The President read clause 1. Carried.

The President read clause 2.

Mr. Black : In reading this article over I think it would be advisable to make some change in the first part of it. In claiming to foster and encourage the science of electricity I think we are assuming too much. Of course it is easy to find fault, but not so easy to suggest a remedy. I do not like the words " foster and encourage."

It was decided, after some discussion, to make the clause read "advance the science of electricity and to promote the interests of those engaged," etc.

The President read Article III. Carried.

The President read Article IV.

Mr. Breithaupt : With regard to the Executive Committee the clause reads " Five of whom shall act on the Committee for two consecutive years." What about the other five ? Is it the idea that every member of the Executive Committee should act for two years ?

The President : You will remember two or three years ago it was thought advisable that there should be continuity in the Committee, that is, that there should be a certain number of members who should act for more than one year. The plan adopted was to elect ten members to the Executive Committee on the understanding that five out of the ten must be re-elected for a second term ; and in Montreal last year a ballot was prepared with ten names, and members were asked to choose five out of the ten to act for the second year. It is presumed, of course, that the five who best deserved re-election would be chosen. The remaining five would also be eligible for re-election along with all other active members of the Association.

Mr. Breithaupt : Then we are really electing ten members each year. I beg to move that the clause be adopted.

Mr. Wickens seconded the motion. Carried.

Article V was then read by the President and carried.

Article VI was read. After discussion the clause was amended by substituting the word "annual" for the word "general" in the fifteenth line of the clause.

Articles VII to XXIII inclusive were then read by the President and adopted.

Clauses 25 to 28 inclusive were then read and carried.

Mr. Armstrong : Under what heading does the matter of the change of date have to be considered.

The President : There is no definite date fixed in the constitution. You will find it in the order of business, Article XIX.

Mr. Dewar : I would like a little light on Article III. There does not seem to be much distinction between active and associate members.

The President : The difference is that associate members are not allowed to vote. The term includes those interested in and actively engaged in electrical business, but an active member must be actually engaged in some electrical industry. I do not think myself that the difference is very clearly expressed but it is a difficult thing to define.

Mr. ———— : It is not the idea that anyone can be an active member ; for instance, I could become an associate member even if not actively engaged in electrical pursuits.

The President : The word "interested" is the defining term ; it would include a student for instance at a college studying electricity. He would be interested in that sense, but not being actively engaged in electrical pursuits he could not be an active member, but would be eligible for associate membership. The clause is intended to be broad enough to take in any person who takes enough interest in electrical science to wish to join the Association.

Mr. Atkinson : The clause reads "Honorary members shall be elected by a two-thirds vote of the Association." Should not this be a two-thirds vote of the members present at the meeting ? In clause 10 it speaks of a majority of two-thirds of the active members present, and I think the other clauses should read the same.

Mr. Kammerer : We being the active members present, represent the Association when we are here. There is not likely to be a full session, and if members do not come to exercise their franchise we cannot help it.

Mr. Atkinson : Further down in clause 10 regarding the report of a committee on an amendment it says : " A two third vote of all active members present shall be necessary for its adoption." I think Article III should read the same way.

The President : Is there any objection to making that clause perfectly clear, by making it the two-thirds vote of members present.

Moved by Mr. Atkinson, seconded by Mr. Badger, that the last line of Article III be altered by striking out the word "Association" and replacing same by the words "Active members present." Carried.

The report as a whole was then adopted.

The President : The next order of business is the consideration of the report of the Committee on Statistics.

Moved by Breithaupt, seconded by Mr. Wickens, that the report be adopted. Carried.

Moved by Mr. Black, seconded by Mr. Knapman, that there be two committees, one on statistics and one on legislation. Carried.

The President then nominated as a committee to strike the standing committees, Messrs. J. W. Taylor, L. B. McFarlane and A. B. Smith.

Mr. Kammerer : I think it is a pity to limit the number of

members on these committees. It might be possible to get ten good men who would act harmoniously together and it would be an advantage of course to have a larger number.

It was moved by A. B. Smith, seconded by Mr. McKay, that the next convention of this Association be held in the City of Toronto.

Mr. Roch: How long is it since a meeting was held in Toronto.

The President: Two years ago. Last year the convention was held in Montreal and the year before last in Toronto.

Mr. Armstrong: With the exception of Mr. McCrossan, of Rat Portage, I do not see any members here who comes from west of Toronto, and as the next meeting should be held in the west anyhow, I think Toronto is the most central place and the convention should be held there.

The motion was put and carried.

The President: With regard to the date of the next convention it has always been the custom in the past to fix upon the month, leaving the exact date to the Executive Committee as they are sometimes able to make favourable railway rates and other arrangements.

Mr. Taylor: I would like to hear some reason why it is considered better to change the date of the meeting to June.

Mr. Armstrong: I think the first meeting was held in June and the reason why September was chosen was because it was thought that reduced railway rates might be obtained in that month on account of the fall exhibitions and it was also intended to have some sort of an electrical exhibition in connection with the Industrial Fair in Toronto. However, the idea of anything of that kind has passed away and I think the best plan is to hold it in June. After the summer is over and we have all had our holidays it is difficult to get away again, and I think that for that reason we will find the attendance larger if held in the summer instead of in September.

Mr. Taylor: My object in asking was to have an expression of opinion from the others; there are many members from points far away and I would like to get a full expression of opinion.

The President: It is a radical change and should be well considered. As far as Toronto is concerned we would not get as favourable railway rates in June as we would during the exhibition.

Mr. A. A. Wright: I think myself and the majority of the members here must acknowledge that the great majority of central station men outside the cities would find the summer the better time; in the fall they are very busy, but during the month of June they have a little leisure and it would be more convenient for them to attend in June than in September. It is true you sometimes get lower railway rates in September, but not always; even now you do not get any lower rate than at any other season of the year.

Mr. Bonner: The Mining Engineers have fixed on June for their meeting, and the American Association, I think, are also to have their meeting next June, so that altogether, I think there will be a great number of meetings of other associations during that month. How would that affect members of this society. I do not know if any of them are members of the other associations; you might consider that point before the time is settled.

Moved by Mr. Fisk, seconded by Mr. Armstrong, that the next meeting of this Association be held in the month of June, the exact date to be fixed by the Executive Committee. Carried.

### ELECTION OF OFFICERS.

The election of officers being the next business, the President appointed Messrs. Wickens and Gilmour to act as scrutineers and to assist the Secretary in recording the vote.

Moved by Mr. W. A. McKay, that Mr. A. B. Smith be President for the ensuing year. Mr. R. G. Moles seconded.

Mr. D. A. Starr moved, seconded by Mr. Kammerer, that one ballot be cast by the Secretary for Mr. Smith. Carried.

The President then declared Mr. Smith elected.

Mr. Smith: I wish I had sufficient words of eloquence to give expression to the feeling of appreciation I have for the great honor you have laid upon me, and to assure you that the Association shall have my best attention and most earnest efforts in the year that is to come.

Moved by Mr. J. F. H. Wyse, seconded by Mr. W. E. Jackson, that Mr. J. A. Kammerer be first Vice-President.

Moved by Mr. Smith, seconded by Mr. Dewar, that Mr. C. B. Powell be nominated for 1st Vice-President.

Moved by Mr. J. Carroll, seconded by Mr. L. B. McFarlane, that Mr. W. Y. Soper be nominated for 1st Vice-President.

The President: The point has been raised as it is for you to decide whether you prefer that the names be written on the ballot papers and that you should mark a cross opposite the name of the candidate you wish to elect, or that you should yourselves write in the name.

Mr. J. J. Wright: Since the constitution says that the votes shall be by ballot, if the members should write the names of the candidates it would not be a vote by ballot, as the writing might be recognized by the scrutineers; therefore it will be necessary to adopt the other course.

The President: I am asked to announce on behalf of the local committee that a cordial invitation is extended to all members of the Association to attend the banquet to be held this evening. Formal cards of invitation have been sent to all old

members of the Association, but within the last few days a number of new members have been elected who may not have received cards. It is necessary, however, that all who desire to accept the invitation should notify either Mr. Soper or Mr. Powell this morning.

Mr. Black: The constitution as amended provides that the Secretary shall read a list of those entitled to vote, and members shall deposit their ballots as their names are called. I would suggest that members vote without their names being called, as the Secretary is not prepared with an alphabetical list. I think it would save time.

The President: If it is the unanimous wish of the meeting that the ballots be taken up by the scrutineers, trusting that only those who are active members in good standing will vote, I think we may do it in that way. If there is any opposition on the part of any person we will adhere to the method prescribed in the constitution.

The ballots were then collected by the scrutineers.

The President: While the ballots are being counted we might take nominations for the office of 2nd Vice-President, not closing them, however, until after the ballots are counted.

Mr. A. B. Smith, seconded by Mr. John Carroll, nominated Mr. L. B. McFarlane.

Mr. Kammerer, seconded by Mr. J. J. Wright, nominated Mr. George Black.

Mr. D. A. Starr, seconded by Mr. J. W. Taylor, nominated Mr. J. A. Kammerer.

The President announced that the balloting for 1st Vice-President resulted in the election of Mr. C. B. Powell.

Mr. A. B. Smith, seconded by Mr. Armstrong, nominated Mr. C. H. Mortimer for Secretary-Treasurer for the ensuing year.

Mr. Mortimer: I wish to thank the meeting for their expression of opinion.

Mr. Dion: I beg to move that the President cast a ballot in favor of Mr. Mortimer. Carried. Ballot deposited.

The President then announced the result of the ballot for 2nd Vice-President, saying that Mr. McFarlane had headed the list, but had not a clear majority of the votes cast, necessitating a fresh ballot.

The President announced that a photographer was prepared to take a group picture of the members that afternoon in front of the City Hall, and it was decided to have the group taken.

The President: Mr. Black wishes me to say that he desires to have his name dropped, so that the election will be a straight one between Messrs. McFarlane and Kammerer.

The ballots being taken, the President announced the election of Mr. McFarlane as Second Vice-President.

Mr. D. A. Starr: As I may have some friends who might wish to vote for me on the Executive Committee, I wish to announce that I desire to retire from the Committee. I have been a member of this Committee from the beginning, and I am afraid a very unworthy one; in fact, I did not attend a meeting during the past year. I beg to resign.

Mr. Taylor: Last year I was put on the Committee to represent Peterborough and this year to represent Ottawa. As you already have Mr. Powell and Mr. Higman, either of whom I am sure will make a better member than I, I beg you to drop my name from the list.

The election of second term members of the Executive was then proceeded with, and resulted in the election of Messrs. Black, Breithaupt, Higman, J. J. Wright and Kammerer. New members of the Committee were then nominated as follows:

Mr. K. J. Dunstan nominated Mr. A. B. Smith.
Mr. F. H. Badger nominated Mr. J. J. Wright.
Mr. Rosebrugh nominated Mr. H. O. Fisk.
Mr. H. O. Fisk nominated Mr. J. J. Wright.
Mr. J. Carroll nominated Mr. D. A. Starr.
Mr. W. Y. Soper nominated Mr. J. Carroll.
Mr. Wickens nominated Mr. Kammerer.
Mr. Armstrong nominated Mr. Kammerer.
Mr. A. A. Wright nominated Mr. J. S. Knapman.
Mr. Yule nominated Mr. George Black.

The meeting then adjourned for lunch.

### AFTERNOON SESSION.

The meeting resumed at 2.30 p. m.

The result of the ballot for new members of the Committee was announced by the President as follows: Messrs. Badger, Soper, Wickens, Carroll and Dunstan.

The President announced that he was requested by Mr. Higman to state that the Governor General's electric launch would be waiting them at the first lock, under the Dufferin Bridge, at 4 in the afternoon and at 10.30 the next morning and 2.30 Friday afternoon.

The President: We all appreciate the courtesy of His Excellency, and I have no doubt a large number of the members will be very glad to avail themselves of the privilege.

Mr. McFarlane: Under the head of general business I would like to bring up a matter which I think should be attended to. We have reappointed Mr. Mortimer Secretary-Treasurer, and heretofore we have given him an allowance of twenty-five dollars. Now the Association has increased largely in membership, and the work devolving on him has also been largely increased, therefore I think his remuneration should be greater. Certainly it does not anything like represent the amount of work that has been done. I beg to move that we make the rumuneration fifty dollars instead of twenty-five. Seconded by Mr. D. C. Dewar.

The President :—The amount is intended not as a grant for services but more as a compensation to Mr. Mortimer for the expenditure that he is actually put to by reason of the work involved, that is for the time of his typewriters, etc., and the grants have been in recognition of that fact.

Mr. Higman : I would suggest that the amount should be seventy-five dollars instead of fifty dollars.

Mr. Mortimer: With all due deference to Mr. Higman and thanking him for his proposition I do not think the funds of the Association will warrant such a grant. I only want it to cover any actual outlay on my part.

A vote of thanks was passed to the Secretary-Treasurer for his services during the past year.

Mr. Higman : The Association gives one hundred dollars towards a banquet to-night. I think the city in which the meeting is held should meet all the expenses of the meeting, so that the Association would have money to meet its expenditure.

The President then read a letter from Mr. Keeley, enclosing some remarks on Mr. Dion's paper, which the President asked Mr. J. J. Wright to read, as follows :

The writer, regretting the circumstance of his being called away from Ottawa at the time of the present Convention, desires to be heard from in regard to the valuable paper with which Mr. Dion has favored the Association. It is just such papers as this that are calculated to beget amongst any body of workers in a new field that fraternal spirit of mutual advancement and excellence. No one can gainsay that the Canadian Electrical Association has so far made a fine record for itself in this respect. Its main object, as defined in its constitution, has been adhered to in a signal manner; and with its members working hand-in-hand and comparing notes of their practical experience, there is no reason why we should fall short of seeing, since they contain such records, the Transactions of this society taking rank with the text book.

Mr. Dion has, it would appear, even in treating his large and interesting subject so thoroughly yet briefly, been obliged to take it for granted that those of the members especially concerned in the matter are fully conversant with the difficulties inseparable from the achievement of what he so successfully undertook,— Well, of course they should be, and such of us as happen to be in position to appreciate these difficulties, can afford to smile, a little way off, at those others of us who might perhaps easily persuade ourselves that, with the pointers he has given us, we know the whole of it and could go and do likewise. That's encouragement. The first unlooked for difficulty is an incentive for investigation ; that's the way to knowledge. Knowledge smooths out our wrinkles and we become blessed smilers with the rest. Now, in this regard Mr. Dion's paper speaks for itself ; there is a world of incentive and a world of information contained in it for those who would "mark, learn, and inwardly digest" as we are at other times and in other places exhorted to do for our well-being.

Reverting now to the text of the paper, it would be interesting to discuss amongst other points the electrical considerations involvent in the described plan of supplying E. M. F. to the fields of a number of direct current generators and alternating current machines from the operation of a special or separate turbine, common to them all. Supposing the operation of the turbine is kept within the limits of the lowest water supply, so that the power is constant and invariable for a given demand ; if the demand be increased or diminished by the introduction or withdrawal of one of the dynamos, or a sudden change in the load on any one of them, the effect of this would be felt in all of the fields until the turbine was adjusted to the new requirement. But allowing this, if the turbine is so governed that it will maintain a given speed, irrespective of the demand made upon it, within predetermined limits, the changes could, perhaps, take place so rapidly as to be almost imperceptible. This feature, taken in conjunction with the supplementary regulators or "boosters" whereby the drop in leads under increased amperage may be made up, seems to simplify the matter of current supply when contrasted with the method in use elsewhere of meeting changes in load on any one dynamo by altering its speed or the speed of its special field exciter. But it brings out prominently what is the great desideratum of the hour ; the thing needful to the attainment of perfection ; but a thing that with well directed energy and inventive genius shall never be described as a "long felt want"—that is an automatic "booster" pure and simple.

Another but minor point in the paper might perhaps be touched on. Mr. Dion describes the converters of one of these systems consolidated as Westinghouse 1000 : 50 volts, and of another Royal 1040 : 52 volts. Presumably this is merely a distinction without a difference, as the ratio of conversion is the same, and in the latter part of the paper it is shown that the 1040 voltage had been applied to the whole.

The writer very much regrets this lost opportunity to listen to what must prove a very interesting and profitable discussion on such a lively paper, and begs to join his thanks with those of the members present, to Mr. Dion for this account of his able work and valuable contribution to the transactions of the Association.

The President : If there is no discussion arising out of Mr. Keeley's memorandum, and as Mr. Street, I think, is not in the room, I will ask Mr. Wickens to read the paper written by Mr. James Milne of Toronto, who is unfortunately unable to be present having been detained at the last moment.

The paper was then read as follows :

### FROM THE COAL PILE TO THE METER.

By JAMES MILNE.

We see so many curves these days showing the high efficiency of this and the very high efficiency of that, and so on, that we are apt to think there is little room for improvement. My intention is therefore to trace from the coal pile to the meter on the consumer s premises, and endeavor to show by means of diagrams,

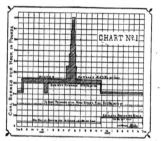

calculated from different sources, the condition of affairs in light and power plants and to see if there is not something that could be improved on.

Chart 1 illustrates at a glance the following : first, the amount of coal required to keep up the pressure at the boilers without delivering any to the main steam pipe. It will be seen that for 16 hours about 40 lbs. are consumed and for 8 hours it takes 100 lbs. per hour. Second, the amount of coal required to keep the pressure up when steam is turned on the main pipe, which is equal to 210 lbs. hourly. Third, the amount of coal required to keep the steam pressure up for running the average friction load in summer. In this load I have included the power required to run the generators and charge the fields, but not supplying any current to the external circuit. Beginning at 11 p. m. we find the average friction load takes 310 lbs. per hour up to 8:15 a. m. after that it rises to 420 lbs. and continues at this until noon. It then decreases until 12.30, and again rises, and at 1.15 it is 420 lbs. and remains at this until 11 p. m. Fourth, the amount of

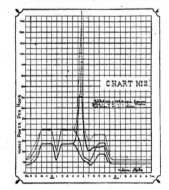

coal used for the average friction load in the winter time. It will be observed that for the same load there is a slight increase in the coal consumption owing to increased radiation, condensation, etc.

We are now ready to supply current to the external circuit, and chart 2 illustrates the power as indicated at the engine for 24 hours run, summer and winter. The full red line shows the 1 h. p. at the engine in the summer, the dotted red line the 1 h. p.

in the winter, and the full and dotted green lines the 1 h. p. available at the service or when the. wires enter the consumer's premises. You will observe as the h. p. increases the distance between the red and green lines also increases. This is accounted for by the decrease in the efficiency of the external circuit and also the increase of the friction load, and you will readily understand that when the load is at its lowest the distance between these lines will be a minimum and the only time they would coincide would be when the plant is shut down. The peak in the winter time is always an interesting figure to those in the business and no doubt deserves a word or two, for it is generally about this time that a few words are said in the power house, but whether they be good or bad I will leave that for yourselves to conjecture. I thought of plotting this peak 50% higher, but the data I have was not very reliable so I concluded

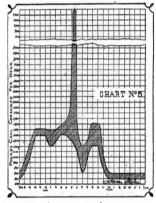

CHART No. 5

it was better to leave off at the point where it might be misleading. According to our diagram it is plotted for 1200 h. p. at the engines and we find by looking at the green line 700 h. p. are being registered at the meters, which means by deducting the correct amount for friction, losses in generators, etc., we find a loss of 25% in the line, or in other words we get at this time an efficiency of 75% between switchboard and meters. It may seem to you or some of you at least that this is an enormous loss and steps should be taken to reduce it. When everything is taken into account I think we will find it cheaper to put up with this loss than try to overcome it. It (the loss) is only for say 60 days a year, etc., for 1 hour per day, making a total of 60 hours per annum. Now if copper were put in to reduce it we know that interest, depreciation, etc., runs on continually, and would in some cases more than offset the amount spent for the extra coal consumed during this loss. It would be an easy matter to calculate the amount spent on coal to make up this loss and see how much copper it would be equivalent to, even taking the whole amount as interest on the investment.

On diagram 1 we showed the coal consumption per hour necessary to keep up the steam pressure to run the engines and generate current up to the switchboard. We now come to chart 3, which shows the amount consumed when running as indicated on chart 2. The part in section shows the increase for the winter load, and in this we have again an interesting diagram which, when calculated out, shows very clearly the effects of forcing boilers. By comparing the run made between 10 a. m. and 12 noon, and that made between 5 and 6 p. m. in the winter time, we find there is an increase of 75% in the amount of coal burned per horse power per hour, which certainly shows that forcing boilers is far from being economical. Diagram 4 shows the amount of coal consumed per square foot of grate surface. Beginning at 11 p. m. we have 22 lbs. This continues up to 7 a. m. when a change over is made which reduces the amount to 9 lbs. for a short time. It then gradually increases until 10 a. m. when we get a fraction over 17 lbs. At 12 noon it diminishes to 15, and we again find at 3 p. m. it is the same as from 10 to 12, viz. 17 lbs. From 3 to 5 it averages 16 lbs., and at 6 it has reached 9 lbs. From 7 to 9 it increases from 9 to 15 lbs. and continues at 15 lbs. for 1½ hours. The part in sections, as in the other diagrams, shows the difference between the summer and winter loads. The two peaks show the amount burned per foot of grate surface on two different sets of boilers. On one it will be seen that 43 lbs. per foot are consumed and 32 lbs. on the other, and to my mind shows that they are going for all they are worth.

In calculating the amount of coal consumed per foot of heating surface I find at the most economical load ·322 lbs. are burned, and the most uneconomical load, between 5 and 6 p. m. in the winter, it is as high as ·9 lbs. per square foot.

Let us now see what economy we are getting from the boilers.

Summer :

Total water evaporated for 24 hours = 201,600 lbs.

$$\frac{201,600}{25,000} = 8.07 \text{ lbs. water evaporated from temperature of feed}$$

or 210 which is equivalent to 8·32 lbs. from and at 212° per lb. coal. Allowing 10% ash which is not excessive and compares favorably with the average of the data I have received from different plants, we get 8·97 from temperature of feed or 9·24 lbs. from and at 212° per lb. combustible.

Winter :

Total water evaporated per 24 hours = 276,343 lbs.

$$\frac{276,343}{36,000} = 7.5 \text{ lbs. from temperature of feed or } 200° \text{ which is}$$

equivalent to 7·9 from and at 212° and allowing the same percentage of ash we get 8·34 lbs. from 200° and 8·78 lbs. from and at 212° per lb. combustible.

In the summer time we have stated that an evaporation of 8·32 lbs. per lb. of coal is obtained and taking the coal at 12875 B. T. U. we get a theoretical evaporation of 13.32 lbs. water per lb. of coal from and at 212°. This agrees fairly well with any of the tables giving the thermal units of heat of the grade of coal burned.

Theoretical evaporation = 13.32lbs. actual evaporation = 8.32. Efficiency therefore is = 62,47 %, or say 62.50.

We are all more or less acquainted with the first law in Thermodynamics, which may be stated thus : Heat and mechanical energy are mutually convertible, a unit of heat corresponding to a certain fixed amount of work called the mechanical equivalent of heat. We can can calculate the mechanical equivalent of the

coal pile thus : $\dfrac{12,875 \times 772 \times 25,000}{24 \times 60 \times 3300} = 5229$ h. p.

∴ 5229 × .6247 = 3266.5 h. p. available at the stop valves of the boilers, which, if all were utilized, would give us 3.1 h. p. per lb. of coal per hour, or a little over .31 lb. coal per h. p. per hour,

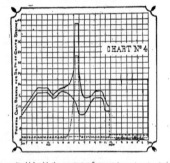

CHART No. 4

a result which with the greatest refinement in engineering has not yet been attained in practice, and I think I am safe in saying never will.

We will now calculate the efficiency of the steam :

Average steam pressure, 125 lbs. absolute.
Average exhaust pressure, 20 lbs. "
Water per h. p. per hour, 28 lbs.
Temperature of feed, 210° Fah. (Summer).
Temperature of water before entering heater, 40°.

125 lbs. abs. = 1187 units. Feed water raised from 40° to 210° = 170° by heater.

∴ 1187 − 170 = 1017 Thermal units of heat expected per pound of steam.

Thermal units per 1 h.p. per minute = $\dfrac{1017 \times 28}{60} = 474$.

∴ $\dfrac{42.75}{474} = .09$ or 9 % efficiency.

Taking the temperature of the feed at 210 and that of the steam at 344°, we get

$$\frac{344 - 210}{805} = .168, \text{ or an efficiency of } 16.8 \%$$

if same were a perfect engine working between the limits of the temperature 344° and 210°.  Therefore the relative efficiency between the perfect engine and the above is

$$\frac{09}{168} = 53\%$$

By the above reasoning we get an efficiency of 9% of the total power available at stop valves of boilers :

$$3266.5 \times 09 = 293.98 \text{ h.p.}$$

our actual h.p. is 230 or $78.23^\circ/_\circ$ of 293.98
∴ $.9 \times .7823 = .070407$ or a fraction over $7^\circ/_\circ$

represents the efficiency of engine to boiler, and calculating to coal pile, we get an efficiency of $.625 \times .09 \times .7823 = .625 \times .070407 = .044$, or a net efficiency of $4.4^\circ/_\circ$.

We also find from actual results that a mechanical efficiency of 82.3% is obtained from the engines ; that is to say, it takes 17.7% to overcome friction and run generators up to speed.

Therefore the power available for running the generators is

$$230 \times .823 = 189.29 \text{ h.p.}$$

Efficiency of generators not less than 92%,
∴ $189.29 \times .92 = 174.14$ h.p. available for sending out to the line, and from boiler to switchboard we have

$$.09 \times .7823 \times .823 \times .52 = .0533$$

or briefly, $\frac{174.14}{3266} = .0533$, or $5\frac{1}{3}\%$ efficiency,

and going back to coal pile we get $.0533 \times .1247 = .0333$ or $3\frac{1}{3}\%$ efficiency.

We have got one more item to take into consideration to complete our calculations, and that is the efficiency of the external circuit, which is in the summer time 85%.  Therefore the total h.p. available at the meter on the premises of the consumer is $174.14 \times .85 = 147$ h.p. fully, or more accurately $3266 \times .09 \times .7823 \times .823 \times .92 \times .85 = 148$ h.p.

The net efficiency from boiler to meter is

$$\frac{148}{3266} = .0453 \text{ or } 4.53\%.$$

and from coal pile to meter we get $.0453 \times .6247 = .0283$ or 2.83%.

From the above we get the following results :

| | |
|---|---|
| Power for running generators......... | 100 |
| Power available at switchboard....... | 92 |
| Power available at meter............. | 78.20 |

Efficiency of the electrical apparatus, including outside circuit $= 78 \cdot 1 \cdot 5$ %.

| | |
|---|---|
| Engine............. ...............,..... | 100 |
| Available for generating current...... | 83.3 |
| Available at switchboard............. | 75.72 |
| Available at meter................... | 64.35 |

The commercial efficiency from engine to meter is 64.35%.

| | |
|---|---|
| Boiler equivalent.................... | 100 |
| Indicated h.p. at engines............ | 7.04 |
| Available for generating current...... | 5.79 |
| Available at switchboard............. | 5.33 |
| Available at meters ................. | 4.53 |

The efficiency therefor from the boilers to the meter is 4.53%.

| | |
|---|---|
| Coal pile equivalent................ | 100 |
| Available at stop valves of boiler...... | 62.47 |
| Indicated h.p. at engine............ | 4.4 |
| Available for generating current ..... | 3.62 |
| Available at switchboard............. | 3.33 |
| Available at meter................... | 2.83 |

The above extended results apply, as has already been stated, to the summer load, but as the results from the winter load are practically the same, I need not take up your time going over them.

We see, therefore, that the electrical end shows up exceedingly well.  For every 100 h.p. at the engine, 82.3 are utilized for generating current, and 75.77 of these are sent out to the line, which, although it could be improved on, leaves very little to be gained by any new invention or alteration that may hereafter be made.

Regarding the steam end, there can be no doubt that considerable attention has been paid to it, and it is still receiving more.  Yet it does not alter the fact that it is here where we get the worst showing.  When some of the great scientists arrive at the method of abstracting electricity direct from the coal pit or even the boiler without the intervention of the steam engine, better results may be obtained.

I see in some of the engineering papers that evaporation condensers are coming into use and are being applied to plants that have heretofore been running non-condensing.  This is a step in the right direction and as no more water is used than that required for steam purposes, parties at present having to pay for their water might find this a profitable investment, inasmuch as a vacuum of from 22" to 26" is readily obtained, in fact, guaranteed by the builders.

Mr. Breithaupt: I have listened with much pleasure to the reading of this paper ; no doubt it is of very great interest to those engaged in running electric plants.  As to criticism of this paper I am really not in a position to critise it.  The facts seem to have been derived from actual experiment.  The total efficiency from the coal pile to the meter of 82.3 is extremely small.  Of course we all know it is small, but it is a matter of some congratulation, we think that the electric end is the economical part of the transformation of the energy.  This is a good showing and it is a matter which engineers, especially those engaged in this kind of work should be pleased with.  The efficiency of the electrical apparatus we have cannot be very much improved and there is no doubt that high efficiencies, that will be enough higher to make a difference at all worth while, will be difficult to get.  I do not see that we can expect a better showing than we have here.  I would like to ask whether the experiments which led to this paper were from non-condensing engines.

Mr. Wickens : All high speed non-condensing engines—no economies, simply a feed water heater, working with the exhaust steam.

Mr. Breithaupt : I think it would be a matter of some interest to make a comparison between the results derived from these and from condensing engines, and the difference might be ascertained by simply introducing the relative units for the engines.  In that way we could arrive pretty fairly at what result we could expect when running condensing engines.  I see Mr. Milne speaks in the last part of his paper of evaporation condensers coming into use.  I would like to hear from anybody who has had experience with them.  I know a little about them but have had no practical experience.  I think the Association is very much indebted to Mr. Milne for his paper, and, as Mr. Keeley says, it is certainly papers of this kind which are of the greatest value to the Association.  The scientific papers are undoubtedly also of very great value, but papers of this kind, where you get the actual experence of different men in electric light, railways, &c., by comparison, is where the Association can be of incalculable benefit to its members.  I therefore move that a hearty vote of thanks be tendered to Mr. Milne for his paper ; seconded by Mr. Wright and carried.

The following paper was then read by Mr. Street :

## SUGGESTED FORMS IN ELECTRIC LIGHT ACCOUNTING.

### By D. R. STREET.

There is no system of accounting that cannot in some particulars be improved, and there is no accountant who cannot increase his knowledge by comparing methods with his fellow accountant ; therefore, in placing this paper before the members of the Canadian Electrical Association, the writer does not claim for his system superiority over all others, but the various forms referred to herein, having been found to work most satisfactorily, and having been adopted by the Ottawa Electric Company, after long and varied experiments and careful comparisons with other systems, it occurred to the writer that many of these forms would prove of some value to others, as they undoubtedly have to the Ottawa Company in dealing with its employees and its 3,000 customers.

A paper of this kind is somewhat tedious, therefore full explanations must to a certain extent be sacrificed, to avoid too great length.  For the same reason the writer must confine himself to some of the most important forms, leaving out a number of minor but decidedly labor-saving ones.

For the purpose of this paper the forms are classified under the following heads.  The writer begs to draw special attention to the system of meter reading and of keeping meter customers' accounts, which is quite different from that used by other companies.

A.  Applications and Contracts for the Supply of Light and Power.
B.  Requisitions for and Ordering Supplies.
C.  Orders for Construction and Repair Work.
D.  Time Returns and Pay Sheets.
E.  Lamp Records.
F.  Collection of Revenue.  Commercial Lighting, Arc and Incandescent, (weekly collections) ; Meter Lighting, Reading of Meters, Keeping and Rendering Accounts.
G.  Cash Book, Accounts Payable Register, and Vouchers.

APPLICATIONS AND CONTRACTS FOR THE SUPPLY OF LIGHT, POWER, ETC.

Applications for light, power, etc. should be made in writing and on printed forms, therefore applicants are required to fill in one of the following blanks :—

Form 1.  Incandescent Lighting, Commercial System.
Figure 2.  Incandescent Lighting, Meter System.
Form 3.  Arc Lighting.
Form 4.  Supply of Current for Power and Heating,

which forms require no explanation, except that they are intended to take the place of more formidable documents, which customers are often afraid to sign.

Upon the acceptance by the company of any such application, it assumes the form of a contract, and immediately separate work orders are made out for service connections, supply of lamps and meters (if necessary), the number of such orders being recorded on the face of the contract, which is then entered in the proper register and filed away.  By a strict adherence to the rule of obtaining an application on the proper form from every customer, and of not issuing orders for services, etc., until it is received, it will be impossible for any one to obtain light without a record appearing in the books of the company, and a glance at a contract will show what has been done, as well as what remains to be done in the particular case.

REQUISITIONS FOR AND ORDERING SUPPLIES.

REQUISITIONS FOR SUPPLIES. (FORM 5.)

1st.  This form is used by the storekeeper and the heads of departments, to notify the General Superintendent that supplies are needed.

2nd.  The General Superintendent then orders goods, (as hereafter described) entering on the face of the requisition the date of such order, and returns the requisition to the sender.

3rd.  The sender holds the requisition until the goods arrive and are checked, when he returns it to the Superintendent with a report upon it of proper goods having been received in good condition.

4th.  The Superintendent then stamps upon the face of the order in copy (impression) book, the date of receipt, together with a note of the particular branch of work or the department to which goods are chargeable, and files the requisition away.

ORDERS FOR SUPPLIES, (FORM 6).

These order blanks are printed in copying ink, and an impression is kept in the copy book, the number of the page upon which it is copied being entered on the order as the order number.

When the requisition is returned to the Superintendent with certificate of goods received, the copy of order in the book is stamped, " Goods Received, date . . . . . . "

When a bill for supplies is received it is checked by turning up the copy of the order, and if this copy is stamped " Goods Received," and the prices are right, the account is certified and the copy of the order is then stamped " Certified, date . . . . . . ," thus avoiding any possibility of certifying twice to an account for the same goods. The series of forms used under this head have the effect of making any errors extremely improbable.

ORDER ON STOREKEEPER, (FORM 7).

This blank is used by linemen and wiremen in obtaining from the storekeeper materials required for their work. It is usually filled in by the workman and signed by the foreman. Upon delivery of goods the store-keeper stamps this order, " Goods Delivered," (with date), initials same, and then places it upon a file which goes to the entry clerk for charging purposes.

The object of the particular form used is, of course, to save writing for the workmen, who, as a general rule, do not take to writing very kindly.

ORDERS FOR CONSTRUCTION AND REPAIR WORK.

In order that work requiring material and labor should be promptly and carefully attended to, and that accurate returns should be received at the office for charging purposes, the following forms are used :—

No. 8. Notice to Superintendent of Construction for Work Required,

being checked with time slips, (Form 12) are handed to the General Superintendent for his certificate as to correctness in rates of wages, etc.

FORM 14—PAY ROLLS.

The pay roll is made up from the different time sheets. One feature of this roll is that a number is given each name, and this number appears before the name and again before the place of signature. which lessens the chances of signatures being put in the wrong ling. or wrong amounts being paid, and again, keeps the roll clean, a man not having to run his finger along the line to find the proper place to receipt. The expenditure distribution, covering the amount of pay roll and obtained from time slips (Form 12) is endorsed on the back, ready for journal entry.

LAMP RECORDS.

FORM 15—BOOK—LAMPS RECEIVED.

In this book entries are made of all lamps received, with date, showing from where received as well as the different candle power.

FORM 16—BOOK—LAMPS GIVEN OUT.

In this book is entered each day from lamp slips, all lamps given out, showing to whom given, candle-power, etc., and, together with Form 15, is balanced every month, which balance must agree with the inventory of lamps taken at the same time.

FORM 17—LAMP MEMO SLIP.

This slip is used as a requisition on the storekeeper for lamps given out, and also to show to what departments such lamps are chargeable. At the end of each day entries are made in the lamp book from these slips.

FORM 18—DAILY LAMP REPORT.

This report is for the use of the office and is made up every morning from the lamp book, and checked from the slips of the day before. It is particu-

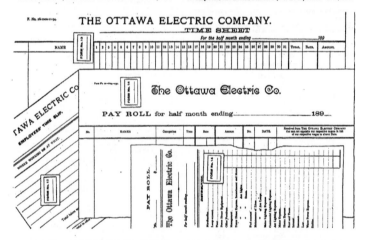

F. No. 16-1000-11-94.

# THE OTTAWA ELECTRIC COMPANY.
## TIME SHEET
For the half month ending_____189___

The Ottawa Electric Co.

PAY ROLL for half month ending_____189__

The Ottawa Electric Co.

No. 9. Order from the Superintendent of Construction to his foreman.

No. 10. General Order for Inside Work (for file in office).

No. 11. General Order for Inside Work (duplicate of above) for foreman.

Nos. 8 and 9 are for line and outside work exclusively, the former being issued from the office to the Superintendent of Construction, upon receipt of which he issues the latter to his foreman, and keeps No. 8 on file until the work is executed, when he reports labor and materials used, signs it and hands it to the entry clerk for charging.

Nos. 10 and 11 are for inside work, repairs, delivery of lamps, etc., No. 11 being simply a duplicate of No. 10. It is written at the same time, being put together alternately in pads, and transfer paper being used between them. No. 10 (the original) is kept on file in the office, and No. 11 is handed to the foreman of wiremen; both bear the same number. When the work is done, a full report of labor and materials is made on the back, when it is returned to the office and checked with the original (No. 10) and handed to the entry clerk for charging.

Keeping the original order on file in the office until the work is completed enables any officer to see for himself if an order has been given, should doubt arise, and by promptly checking off when the returns come in, any unnecessary delay in the execution of work, as well as failure to report labor and materials to be charged, are easily detected.

TIME RETURNS AND PAY ROLLS.

FORM 12—EMPLOYEES' TIME SLIP.

The employees' time slip is for the use of wiremen and linemen to report daily their own time and where and at what they were working. These forms must be filled in and handed into the office by each man, at the close of each day, which gives the office time returns independent of those furnished by the foremen or chiefs of departments ; besides, these slips are used for the distribution of charges in the general pay roll.

FORM 13 —TIME SHEET.

The sheets are used by the foremen or chiefs of departments to make time returns, every half month, of all men under them. These sheets, after

larly useful in showing at all times the number of lamps on hand, and also in distributing the lamps given out among the different accounts to which they are properly chargeable.

The accurate keeping of lamp records is a matter of considerable importance, and the system exemplified by the series of forms above described has been found entirely satisfactory.

COLLECTION OF REVENUE—COMMERCIAL LIGHTING, INCANDESCENT AND ARC.

FORM 19—REGISTER OF WEEKLY COLLECTIONS.

This register contains the names of all customers who pay weekly, together with the addresses and particulars of charges. Each customer is given a number, and the book is so arranged that names need only be written every 26 weeks, extra pages being bound in that fold over to the customer's number (see specimen herewith). At the beginning of each week the amount due is brought forward in the first column, and receipt tickets are made out (see form 20) for each customer and given to the collector. At the end of each week the second column, " Amount Paid," and third column, " Delinquents," are entered up from collector's statement, (Form 27) and the total of the last two columns must agree with the first column, and so on from week to week, the total in delinquent's column being brought for, ward each time, which shows at a glance the amount owing by any customer.

FORM 20—COLLECTOR'S RECEIPT TICKET, WITH STUB.

These tickets are made out at the beginning of each week from collection register, and handed to the collector, who, upon payment, hands the receipt part to the customer and retains the stub.

FORM 21—STATEMENT OF WEEKLY COLLECTIONS.

The collector each day enters up this sheet from stubs of receipts, and at the end of each week returns statement to the office. The total collections should agree with the cash given in, and the total collections plus the total of delinquents should agree with the total of tickets given in,

FORM 22—ARC LIGHT REPORT, (PRIVATE LIGHTS).

This blank is for the use of arc light patrolmen and is entered up each day and handed into the office at the end of each week, and contains the report of the number of private lights burning each night, with any remarks necessary. The charging into register of arc weekly collections is made from this report.

FORM 23—ARC LIGHT RECEIPT TICKET.
This form is for the same purpose as No. 20, but it is used exclusively for arc lighting customers, an exact copy of the patrolman's report being placed on the stub, for reference, in case of dispute with any customer as to correctness of charge against him.

COLLECTION OF REVENUE—READING OF METERS, KEEPING AND RENDERING METER ACCOUNTS.

FORMS 24-25—METER READINGS.
From experience it has been found that where the actual putting down in

If the system of numbering each meter is adopted (as hereafter described) much time can be saved, as it is then only necessary to report this number, together with the surname of customer, with copy of dial, when returning meter readings to the office, the Christian name or initials and address being unnecessary.

FORM 26—METER CUSTOMER REGISTER, (COPYRIGHTED '94 AND '95).
This book requires very little explanation, being a ledger with a column each, for meter date, meter reading, lighting, meter rent, cash paid and balance, and for which book the writer claims a number of advantages.
1stly. The saving of space, two lines only being used each quarter. A half page of this ledger will last six years with quarterly readings.
2ndly. All readings, payments and balance due can be seen at a glance, which means a great saving of time, and makes quite easy a comparison of

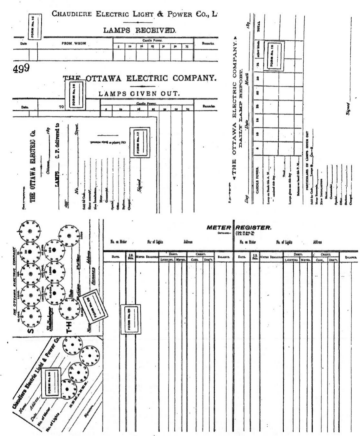

figures of meter readings was done outside of the office, a number of mistakes occurred, particularly when obliged to send out inexperienced men, and then it is impossible to enquire into any suspected error, without sending out to re-read the meter, but the form herewith overcomes this trouble. It is a copy of the dial of the meter, requiring simply the copying of the position of the hands, which any man of fair intelligence can readily do. The reading is then reduced to figures in the office. Should any doubt arise as to the correctness of any account, the copy of the meter dials can, in a moment, be turned up for checking.

Form 24 is a fac-simile of the Schallenberger (Westinghouse) dial, and is for the use of this style of meter only, while Form 25 can be used for either Schallenberger, Slattery or T. H. meters.

the cost of one year's lighting with another when asked for by a customer, a thing of frequent occurrence. Anyone who has seen the old style of oblong book, with rulings for quarterly entries, which necessitated the carrying forward of readings and balances every quarter, will, I think, grant the superiority of this book. A regular index must be kept for the register, but in posting meter readings this index need not be used, if the meter in the house of each customer is numbered with the ledger folio of the customer and the number reported with each meter reading. This plan will be found to save much time, both in reading meters and posting; in fact, by experience it has been found that with this numbering plan, posting can be done in exactly one-half the time as when having to look up an index for names.

**FORMS 27-28—METER ACCOUNTS.**

Form 27, which is patented, will be found to be a particularly good one, and much appreciated by customers. It has an exact copy of the meter for present and previous readings, which permits customers checking the readings themselves.

Form 28 is applicable to either watt or ampere readings, and when there are a very large number of customers, its use may be found to save much time.

A colored envelope is used when sending out meter accounts, and is found useful by customers in enabling them to easily pick out their light account from among other papers.

**GENERAL BOOKS.**

**FORM 29—CASH BOOK.**

This book is nothing out of the ordinary, being on the principle of a journal cash book, with separate columns. On the debit side the cash at time of receipt is distributed among the different revenue accounts, and a similar distribution of disbursements is made on the credit side, the totals of different columns being posted in the general ledger under their respective headings at the end of each month.

**FORM 30—ACCOUNTS PAYABLE REGISTER.**

This book is for the sub-division of all bills to be paid and their distribution to the different expense or other accounts to which they are properly chargeable. The principal advantage of this system is that amounts owing appear properly distributed in the expenditure of the month, during which they have been incurred, regardless of the time when they are paid, the amount at the credit of "Accounts Payable" account representing at all times the total of accounts outstanding.

**FORM 31—VOUCHER.**

This blank is an ordinary account form with endorsements on back for distribution of amount. All the accounts paid are entered under this cover, the original invoices being usually attached inside. When the account is entered in accounts payable register, the amount is distributed in accordance

our own office and regular accounts made up. Another collector goes round to collect the accounts ; if there is anything left unpaid it is carried forward. We endeavor to collect every month. With regard to the statement for the works, we do not have to take charging and delivery accounts in that way. With reference to the giving out of material for wiring, we do not use such an elaborate system, because it is not necessary. All we do is when a wire man goes out, we weigh the wire that he takes with him and weigh it again when he comes back, and the difference of course is charged up. We also have an account of the number of rosettes, lamps, sockets, &c., which are charged up to the individual. We have an account for every customer and we can turn up every man's account and see what he has had, the amount of wire, &c., and the amount received for wiring, and we know how we stand at any time. Also, if any additional lamps have been put in, the date, &c. We can see the number of lights originally installed and the amount received. In addition to that I make up a diagram for every man's house, showing the mains in the house where the lights are placed, and thus keep a full record of the work done, so that it can be turned up at any time. Of course at our central station we charge up the amount of fuel, &c., every month, and an account is kept of everything that is used—oil, matches and everything—for of course we always keep a coal oil lamp in case of accident. A column is devoted to each article, we do not of course expect an accident, but it is as well to prepare for it. If you want a correct account you must have everything or nothing. We also have a monthly account carried forward every month and at the

**244**

**267**

**272**

*with the endorsement on the back. When paid it is filed away as a voucher. An index is kept for accounts payable register, that enables one to find any account, paid or unpaid, at a moment's notice.*

In conclusion, the writer sincerely trusts that the forms and descriptions herewith may prove of interest, and possibly of use to some, and that their discussion may bring out new suggestions which may be of general benefit.

Mr. Armstrong : I think, gentlemen, we will all agree that this is a very valuable paper, and of peculiar interest to managers and accountants of large stations ; however, there is a class of stations to which this system would scarcely apply, on account of the cost of carrying it out. I have several times been asked by managers of small companies, who are just starting in business, to furnish them with a system of books, forms, &c. I am not an accountant myself, and I would be very glad to know if any member of the Association could furnish us with a simple system, suitable for small plants where there is no great multiplicity of accounts.

Mr. A. A. Wright (Renfrew) : As Mr. Armstrong points out, this system of Mr. Street's is more applicable to large than to smaller plants throughout the country. Perhaps some member of the Association would make some suggestion with regard to a system applicable to small companies. With regard to Mr. Street's system, we have not really had time to go through the forms, but they seem to be very applicable to large plants such as the Ottawa Electric Company, but they do not seem to apply to small stations. Still I have no doubt if one could go through them a great many of them would be found very useful. In my own place we do not collect every week as they do here. We collect every first of the month. Our man has a book printed by ourselves, with dials the same as here. One man goes round and takes all the readings, and then they are reckoned up at

close of the year by adding it up, we have a statement for the year which will show the amount of profit or loss, as the case may be. We owe a vote of thanks to Mr. Street for his able paper, and I am only sorry that I have had no time to go over it so as to be able to discuss it intelligently.

Mr. Breithaupt : I beg to second the vote of thanks, and Mr. Street has also kindly shown me his books. I think his system of book-keeping throughout is a very good one. As to applying the system to smaller plants, I think those engaged in running a smaller or any sized plant almost, can learn very much from Mr. Street's paper. There are some forms that might be eliminated. The Meter Reading Form is a very good thing because it takes the record of the meter dial just as it is and thus avoid mistakes just where they usually occur. The Consumer's Ledger which Mr. Street keeps is an exceedingly valuable book, the meter readings and everything being shown so that the state of the consumer's account can be seen at a glance. I think it is a book that all companies whether they have meter accounts or not should keep. In my own case I keep it as an adjunct simply for checking accounts. We render accounts monthly, and if any discounts are given they are shown in a separate column, so by looking at the Consumer's Ledger at any time we can see the state of that man's account. Another book I keep is the Consumer's Register, and in this I keep an abstract of the accounts. I have the first column for arrears another for the amount used, another for gross amount due, the fourth for discounts, and the fifth, total amount owing by that consumer at that time, columns six and seven, amounts paid and date when paid. In keeping accounts in that way, which is something on the same principle as Mr. Street has outlined, but not quite so elaborate, you have the accounts so that you can refer to them and tell at a glance the state of any account. I think we are all much indebted to Mr. Street, and I beg to second the vote of thanks. .

Mr. A. A. Wright : Is the number of the meter put down on the account that is sent out?

Mr. Street : The meter accounts are sent out every two months. We have two systems of accounts—one the meter system and the other the commercial or weekly collection system. In fact, here we find it impossible to read every month on the larger meters. We disregard to a certain extent, the number of the meter, that is the factory number; of course there is a record of that in the Consumer's Ledger. The ledger folio is put on each meter, that is, when a meter is placed the ledger number of that man's account is put upon that meter. When the man that reads the meter brings his report, he brings that number which enables us to turn up the account without looking at the index. We simply put up a label on the meter with the ledger number. The vote of thanks was then carried.

The President : We will now adjourn in order to take advantage of the courtesy of the Governor General in placing his electric launch at the disposal of the members of the Association. The meeting then adjourned till the next morning.

---

### THIRD DAY.

#### MORNING SESSION.

The President : The first order of business this morning is the reading of the remaining papers. I will ask Mr. J. J. Wright, of Toronto, to read the two papers composing the group prepared by Mr. D. H. Keeley, of Ottawa, who is, unfortunately, absent from the city, and therefore unable to read the papers himself.

Mr. Wright then read the following papers :

#### A NON-INTERFERENCE DIPLEX RELAY.

##### BY D. H. KEELEY.

Any who have given attention to the subject of quadruplex telegraphy will be more or less familiar with the difficulty that has been encountered in obviating interference between the sides when both are actuated, at the moment of current reversal in the polar system, or at the moment of transition from extremes of current strength in the straight system ; and it will perhaps be interesting to examine what is herein, presented as an absolutely reliable bit of non-interfering mechanism in that particular.

To cover the ground comprehensively yet briefly, in order to establish the essential dissimilarity of the polar and straight-current systems, and at the same time to make the adaptability of this new instrument to either plan clearly obvious, let it be stated that a quadruplex system is a duplexed diplex. The duplex is simply an arrangement of the signalling instruments at each end of the line in such a way as to render them unaffected by currents outgoing while free to respond to currents incoming, thus affording a means for simultaneous transmission in opposite directions, while the diplex is an arrangement of apparatus for simultaneous transmission in the same direction. If two keys can be arranged to transmit distinctive currents in a circuit common to both, and if two receivers can be arranged to respectively respond to these currents without confusion, we have a diplex; and any such system suitably duplexed affords a quadruplex.

It was first of all ascertained that two keys could be arranged in many different ways to satisfactorily transmit distinctive currents for diplex transmission. For instance, one would transmit a current of low strength ; the other a current of high strength ; which both together would produce an intermediate current. Hence it could be known at the receiving end whether one or other, or both, of the keys were being manipulated. By another plan one key operated to reverse the polarity of a current that was always presented to the line, and the other key acted to increase or decrease the strength of that current ; so that in this case also it could be told at the distant end which, or if both, of the keys were in action. The designing of suitable receivers to respond to these key motions was, however, another matter. In the first described method one of the receivers responded all right to the intermediate or the high strength of current, and so followed the motions of the proper key; but the receiver that was intended to respond to the currents of low and intermediate strength was of course subjected to the current of high strength, and there the difficulty came in. The problem was how to render it irresponsive to strong currents while it was responsive to weak ones, and a step towards the solution of it was the employment of an auxiliary lever hung in such a way as to hold the armature lever in an intermediate position between its limiting stops when attracted by the weaker currents; the strong current when applied would carry it beyond the intermediate position, so that with no current, or with maximum current, the lever would be in one or other of its extreme positions. The local circuit being completed only when the lever was in the intermediate position, the sounder should only, according to this arrangement, respond to the low and intermediate currents ; but it was found inadequate because of a brief contact, completing the sounder circuit, that obtained in the passage of the armature lever between its extreme limiting stops every time the strong current was applied or withdrawn.

According to the second plan, for diplex transmission, one of the receivers was polarized and it responded all right to the reversals effected by the proper key irrespective of the current strength, while the other receiver was a neutral relay adjusted to respond to the currents of high strength transmitted by its corresponding key. In this operation, however, almost as great a difficulty as in the other case presented itself; because the action of reversal of the transmitted current momentarily cancelled the magnetism in the core of the neutral relay, and in the act of recording a signal the armature lever of the latter was liable to fall away and thus produce false effects and mutilation.

Seeing now what and where the difficulty was, it is easy to conceive there was greater promise of finding means for bridging over the brief interval of non-magnetism in the neutral relay, than for rendering a receiver of low adjustment irresponsive to strong currents as required in the first described case ; so it came about that greater attention was paid to the solution of that problem, and the outcome of it all is the standard polar quadruplex now everywhere in use. The other problem, however, was not left to neglect, for it was conceived that if the straight or three current plan could be successfully operated, it would be feasible to construct a sextuplex system with it, by simply adding the polarized relay and reversing key of the other method. Whether this conception is to be realized should not now long remain an open question—for the problem of the straight current diplex is solved, as

will be seen by what here follows, having reference to the accompanying figure, which represents an instrument that originated with the writer, and was, some little while back, put to a practical and highly satisfactory test on one of our regularly quadruplexed lines :

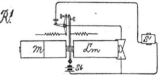

In the receiver $R^1$ an auxiliary electro magnet $Lm$, wound to produce a considerable counter e. m. f., is placed directly behind the relay armature so as to act thereupon in opposition to the main circuit coils $m$. In the normal condition, with no current traversing $m$, the armature lever is held against its back stop by a light retractile spring in the usual way. When a weak current, say the minimum, traverses $m$, the armature lever is attracted to the intermediate position, this closes the circuits of both $Lm$ and $S^1$ ; the retractile power (that is, what the magnetic attraction in this case becomes) of $Lm$ is delayed by its own counter e. m. f. until the attraction of $m$ has grown sufficient to retain the armature in the position to which it was drawn, so the closed circuit of $S^1$ remains undisturbed. The same action attends the intermediate current; so $S^1$ responds to the minimum and intermediate currents. When the maximum current traverses $m$, the armature lever is carried from its intermediate position, and $S^1$ opens, but the circuit through $Lm$ remains uninterrupted. If the current again decreases, the lever returns to its intermediate position, and $S^1$ closes ; but if the maximum current is entirely withdrawn, the armature lever will, in consequence of the steady pull exerted on it by $Lm$, be drawn sharply back to its rear limiting stop. And if, when the armature is resting in the latter position, the maximum current is applied to $m$, the armature lever will pass directly over to the front limiting stop, in consequence of the counter e. m. f. of $Lm$ robbing it of any retractile power during its passage across the contacts in the intermediate position. There is, therefore, no hindrance to the forward movement of the armature, and there is an acceleration of its movement rearward ; hence the maximum current can be applied and withdrawn at pleasure, without in any way affecting the local circuit by which the sounder $S^1$ is operated.

Assuming the action of this instrument to be understood in the light of the foregoing, it will be perceived that if it were employed as the neutral relay in the polar diplex, the local connections might be so modified that the sounder $S^1$ would not be affected by the current reversals, while it would be responsive to the increment key.

#### A PERCENTAGE METHOD FOR CIRCUIT MEASUREMENTS.

##### BY D. H. KEELEY.

As in the operation of Ohm's law the dimension of any one of the three quantities, E.M.F., resistance and current, is readily determinable when the other two are known $(E=R \times C$ ; $R=\frac{E}{C}$, $C=\frac{E}{R})$ ; and as, from the relationship thus established between these quantities, an applied E.M.F. is seen to be expended or absorbed in direct proportion to the resistance in its path, the current in any single circuit must needs be uniform in all parts, and therefore varies inversely as the resistance of the entire circuit.

It is clear, of course, that the resistance of any portion of a single circuit necessarily comprises a certain percentage of the whole ; and it ought therefore to follow, in view of the foregoing, that the mere observation of the effect (the variation of current) produced in any circuit by the introduction and withdrawal of a known resistance, should afford an indication of the measurements of that circuit.

The elemental feature thus developed is interesting, as enabling one to make measurements of any circuit by means of either a volt-meter of low range, or an ampere-meter, adapted both for direct and alternating currents ; and such a method might prove useful when both of these instruments and regular resistance measuring apparatus are not readily at command.

The adaptation of a volt-meter to this end is set forth hereunder ; an obvious modification would produce the same results from ampere-meter readings :

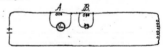

Apparatus : A, B, two known resistances included in the circuit to be tested ; Vm, volt-meter shunted around it.

Readings : (Potential difference between terminals of A):—

Volts with B cut out = say 20
Volts with B cut in    = say 16
Difference 4

Interpretation : The current, or the proportion of the applied E.M.F. expended in it, and therefore in all other parts of the circuit, has in this case fallen $\left(\frac{4 \times 100}{20}\right)=$ 20 per cent., in consequence of the introduction of B ; and because the current is inversely proportional to the resistance in any circuit, the current goes down in this case exactly as the resistance goes up ; so the resistance B evidently comprises 20 per cent. of the total resistance of the circuit.

From the data thus obtained the following deductions may be made :

MEASUREMENTS OF THE CIRCUIT.
(A = say 5 ohms ; B = say 5 ohms.)

| With A and B included ; | Whence, | With A and B eliminated. |
|---|---|---|
| Resistance<br>B ÷ ·20 ;<br>or<br>5 ÷ ·20 = 25 ohms<br>E.M.F. and Current<br>Volts in A ÷ A,<br>or 16 ÷ 5 = 3·2 amperes, | 25 − (A + B)<br>or<br>25 − 10 = ohms amperes<br>25 × 3·2 = | 15 ohms }<br>80 volts } $\frac{80}{15}$ = 5·33 amp. |

Mr. A. B. Smith: If there is to be no discussion on these papers I would like to move a vote of thanks to Mr. Keeley. It is a matter of regret that the Executive were not able to get these papers distributed earlier. If we had been able to do so we would have had some discussion on them. I have much pleasure in moving a vote of thanks.

Mr. Kammerer: As an old telegrapher, having been out of it five years, I may say this is beyond what I was used to handling. In 1888, when the Canadian Pacific Railway line was completed to Vancouver, we had through the use of the instruments described conversations between Montreal and Vancouver, which at that time was considered a great achievement.

The vote of thanks was carried.

The President: The next order of business is the presentation of a paper by Mr. Hartman, of Peterboro', who is also unable to attend in person. I will therefore ask Mr. Armstrong to read it.

Mr. Armstrong: I wish to say before starting to read this paper that it is a matter of particular regret to Mr. Hartman that he is unable to be present. Unfavorable circumstances have made it simply impossible for him to do so.

Mr. Armstrong then read the paper:

## SOME MODERN ALTERNATING CURRENT APPARATUS.

By H. T. Hartman.

It is only about eight years since the advent of alternating apparatus into the field of practical incandescent lighting. Up to that time the only methods in practical use were the series system, the two-wire constant potential system, and the Edison three-wire system and its variations, all using direct current.

The series system, under which name may be included the various series multiple devices, was crude and objectionable on account of the high potentials introduced into the consumer's premises, and the impossibility of cutting out lamps except in groups. The dynamo attendant had no very accurate means of knowing when a lamp broke, and if no additional resistance was interposed, the rest of the lamps on that circuit were subjected to an undue strain, which shortened their lives materially. Moreover, when such a resistance was put in circuit to compensate for the breakage of a lamp, it represented a constant waste of energy. For these and other obvious reasons, the series systems were never very extensively used.

The multiple systems, two and three-wire, are excellent in every respect, but are of only limited application in the present state of the art of incandescent lamp making. Where the territory to be covered is not too great and the consumers too scattered, the three-wire system is still, to my mind, not only the most economical and satisfactory in operation for lighting purposes, but it is also the cheapest to install.

However, the alternating current system sprang almost at a bound into favor with central station men, because the use of high potentials in the street lines so lowered the copper losses per lamp that it became perfectly practical to locate the stations at a considerable distance from the centre of distribution, where land was cheap, fuel convenient, or water power available. It also became possible to supply customers on the outskirts of the town who could not have been reached by either of the direct current multiple systems, except by an expenditure for copper which could afford no adequate return.

But the alternating system had some faults which gradually became painfully evident.

First, the loss from the centre of distribution to the lamp on the consumer's premises was from three to five per cent. greater than in the direct current systems, on account of the drop in the transformer. This gave bad regulation at the lamps, and as the alternating system was generally installed with an eye to lowest first cost rather than economy of operation, there were no pressure lines from the centre of distribution, when there was such a centre at all, so that the dynamo man regulated the pressure by the light of nature, all of which had a very bad effect on the life of the light of art.

Second—A transformer had to be installed for practically every consumer, and as the capacities in lights of the various transformers did not always agree with the capacity of the consumer, the central station was compelled to have considerably greater capacity in transformers than in lights actually installed. As the transformer was of low efficiency when under-loaded, and caused a constant wasteful drain on the station when not loaded at all, it began to be evident that the stations were in the peculiar position of making money on the efficient load between five and ten o'clock p.m., and losing it on the inefficient load the rest of the night.

These evils have been remedied by the use of modern transformers, giving low loss from no load to full load, by putting more copper in the street lines and distributing it to better advantage, by the use of alternators properly compounded so as to exactly compensate for the loss in the mains and feeders, and by the use of the three-wire system of secondary distribution. This latter has the very great advantages of lessening the number of transformers, permitting the use of large units, which are more efficient at all loads, and have a much smaller leakage current on open circuit in proportion to the number of lights capacity than the smaller sizes. Moreover, the use of the three-wire system insures a much greater average load on the transformers than the old method, and reduces the total transformer capacity required.

Third—Inductional troubles began to be experienced on long lines. This subject has been enveloped in such an air of mystery and so disguised with sine curves and their variations, harmonic functions and technical names, that the average central station man is apt to think that there is no use in attempting to understand it.

The fundamental facts are quite simple. In the first place, a flash of current in one of two parallel wires tends to induce an answering flash in the other wire, opposite in direction and tending to stop the current in the first wire. If the current in the first wire becomes uniform, that in the second ceases, but if it simply rises and then falls again, the flash induced in the second wire will first tend to prevent the starting of a current and then to prevent its stoppage. If both wires are carrying alternating currents from one dynamo, they act in the same way upon each other, each tending to slightly raise or lower the pressure of the other.

If the two circuits belong to two different dynamos, which are not exactly in phase, the effect of one upon the other will be to first increase and then diminish the pressure, according as the currents are in the same or in opposite directions. This effect is called mutual induction and gives rise to very objectionable flickering of the lights.

It may be reduced (a) by reducing the number of alternations per minute; (b) by placing the two lines of each circuit as close together as possible and as far from any other circuit as possible; (c) by crossing the lines of one circuit at the middle of the line, the effect on one half being counteracted by the effect on the other half.

If a direct current be passed through a coil of wire, the difference of potential across the terminals of that coil will be equal to the current multiplied by the resistance.

If an alternating current is applied to the terminals of the said coil, a considerably greater difference of potential will be required. This effect is caused by the fact that the passage of a current through the coil induces a magnetic flux in it, and when this flux is increased or diminished, an electromotive force is induced in the wire in such a direction as to tend to prevent the change by opposing the current causing it. This is called self induction, and is, of course, present in any ordinary two-wire circuit, which is only a coil of one turn. In other words, it vastly increases the loss in the line and consequently interferes seriously with regulation if present in a lighting circuit. It increases with the frequency, the length of circuit, the current and the distance between wires.

In order to lessen these two sources of trouble, the tendency has been to decrease the frequency in alternators, and in most of the later machines the standard is only about one-half of what it used to be. The speeds have been greatly reduced and the use of laminated magnet cores have materially lessened the heating and consequent loss of power due to eddy currents. The character and quality of insulation has been greatly improved. For example, the use of oil in transformers not only aids in the radiation of heat, thereby increasing the capacity, but also acts as an efficient insulator itself, especially in resisting puncture by an induction or lightning spark, and in immediately closing any gap which might be made by this or other causes.

In armatures the core used to be elaborately insulated, while the coils had only their cotton wrapping. Now the insulation is applied to the coil where it is needed, and the core is only insulated where it is necessary to protect the coil from mechanical injury and in certain places where additional electrical protection is required.

The use of slotted armatures and machine-wound armature coils, which are readily removed and duplicated if necessary, has resulted in compactness, solidity of construction and durability far exceeding the older types.

Much has been said of the two and three-phase systems, and they have come into great prominence on account of the magnitude of recent transmission schemes. Both are, however, subject to the serious disadvantage that the circuits must be kept balanced. Otherwise one armature circuit is liable to be overloaded, and as any adjustment of the field to give proper voltage on that circuit will give very different results on the others, the regulation is bad. The question of balancing circuits is far worse than in the case of an Edison three-wire plant, for in that case three wires are run to every installation of over six lights, but in the case of the polyphase circuit you are practically limited to balancing one transformer against another, and the shutting down of a few transformers on one side of the system would make a serious difference of potential. This objection, of course, does not hold where the load consists principally of motors.

The monocyclic system, however, has all the advantages of the polyphase system, combined with the simplicity of single phase. There is no trouble about balancing circuits, the lights being taken from the main circuit only. Only two wires need to be run for lighting, the third being required only where power is used. The three-wire system may be used for secondary distribution, an additional small transformer being required if power is needed. The motors are better than the best direct current shunt wound machines, both in efficiency and speed regulation, while the construction is such that it is almost impossible to injure them either mechanically or by overload.

The field consists of iron laminations built up like an armature core, only the teeth project from the inner instead of the outer periphery. The coils are machine wound and overlap in such a manner as to support each other at the ends outside of the core in a very rigid and substantial manner. They are held in place by wooden wedges driven into grooves in the teeth. The lines from transformers are led to these field coils direct, and have no connection whatever with the armature. The latter is wound in three closed circuits, consisting of massive bars, in which currents are induced by flow of current in the fields. The poles in latter shift progressively forward with each phase, thus exerting a powerful and uniform torque on the armature.

Each of the armature circuits has a German silver resistance in series with it at the moment of starting, in order to give large starting torque without excessive current. As soon as the machine has reached its normal speed these resistances are short circuited by means of a sliding collar keyed to the shaft and moved by a small handle at the side of the motors. There are no brushes, no commutators, no moving contacts of any kind, and the only moving wires carry only very low potentials, in no way connected with the circuit, and are so imbedded in the core as to be practically indestructible except by means of an axe.

The sub-division of the field, winding into many sections, has two great advantages. It affords great radiating surface and excellent ventilation, thus enabling it to carry heavy overloads temporarily without damage, and it also lessens the difference of potential between adjacent coils, thus reducing the chances of a burn-out to a minimum. Moreover, there is no cumbersome, troublesome starting rheostat to take up valuable space and serve as a fire-trap.

In short, this is the ideal motor in simplicity, compactness, efficiency and durability, and it does not require the gift of prophecy to foretell the doom of the direct current motor, with its many faults and frailties, and the succession of its new rival.

"Le Roi est mort, vive le Roi."

Mr. Medbury : I am glad the writer has referred to the use of the three-wire system for secondary mains. I have been under the impression for some time that the advantages to be derived from the application of the three wire system to secondary mains have been somewhat exaggerated. There seems to be a widespread and deep conviction that there are some advantages especially peculiar to the use of the three wire system for secondary mains which are not possessed by the two wire system for secondary mains. After carefully considering the matter, I am under the impression that in almost every installation that I have seen where the three wire system of secondary mains is used, that with the same number of transformers installed on the two wire system there would have been at least the same results obtained as to regulation of drop of tension as in the three wire system, direct current, and no greater expenditure for copper, if as much. I have thought that possibly some of my premises, in reaching the conclusion which I have, may be erroneous ; for that reason there would naturally exist a doubt as to the correctness of the conclusion. It seemed proper, therefore, to lay before this Association both the premises and the conclusion in order that if there was an error it might be discussed and the right conclusion reached. If in error I should certainly wish to be corrected, and most frankly desire to learn my mistake. I think the three wire system, direct current, only applicable to large private plants like large factories. I think in such cases from direct current, it is advisable to install a three wire plant. Take the Parliament Buildings here for instance. For them to install their own plant the three wire system is the best, although the Government have at Moncton, on the Intercolonial Railway, an alternating current plant supplying lights throughout the yard. In fact, if you get over half a mile, the alternating current simply distances your direct current plant and the direct current plant for price is not in it. Turning to the next page—I notice he says, "the loss from the centre of distribution to the lamp on the consumer's premises was from 3 to 5 per cent. greater than in the direct current systems on account of the drop in the transformer," evidently "drop in potential" is meant. He could not have meant the loss—you make the loss whatever you desire. It is entirely conventional. It seems to me there is something to be guarded against there. The idea that that would be the loss is erroneous. We can make the loss in the alternating system as little as in the direct current systems. I will notice a matter with regard to carrying different pressure lines on the circuit. Pressure lines are not very expensive. I have visited stations running different compound machines, and we run them ourselves. I have stood in several stations and as the load came on I noticed the station attendant going to the rheostat in order to cut down the potential, thereby annulling all the advantage he got from his compound winding. It is entirely useless to install a compound machine unless pressure wires are installed for a guide to the attendant or a compensating volt meter is used with a compensator.

Mr. Jackson : There are several points touched upon in the last paper of considerable interest to us all, each of which might with profit be discussed more fully. A few words in regard to the reason for the tendency among so many manufacturers to reduce the standard number of alternations per minute to so abnormally low a figure may not be amiss, in the light of the misleading statements we have just heard. This is not on account of the induction effects in the transmission lines, for with circuits up to five or six miles in length these effects are not seriously noticeable. The reduction of the number of alternations increases the cost and weight of the transformers, also the weight and cost of the generator, besides impairing its performance. The true cause for this tendency is the inability to run self-starting induction motors on most systems, including the monocyclic, at the high alternations now in use. You all understand the reason for this—that these motors operating on the monocyclic system and at the high frequency of 16 000 alternations per minute, create such a tremendous lag in the current that the generator is unduly loaded and the armature reaction increased to a prohibitory extent. There are two ways of overcoming this difficulty. One is by reducing the number of alternations and thus impairing the performance of your entire plant, with the exception of motors, and in many cases making it necessary to discard much valuable apparatus. The other is by the use of condensers in parallel with each inductive resistance of such capacity as to furnish all the magnetising and lagging currents, the generator only having to furnish current for actual energy consumed by the motors, this current being in phase with its E. M. F., thus securing the advantages of high frequencies for all transformers and lights and all the advantages of low frequency for motors, even better than 4,000 alternations would do, and without any of the attendant disadvantages. The use of condensers in connection with motors has a beneficial effect whatever the number of alternations used may be, though the gain is more marked at high frequencies, such as from 8,000 to 16,000 alternations per minute. We are informed that circuits must always be kept balanced in the two and three phase systems. This is indeed true in the three phase system on account of the inductive interactions between circuits, but is not the case in the two phase system, for here we have no such interactions. That one of the troubles in the two phase system—as we are informed—should be that one phase can be overloaded, is indeed terrible, and that this overload can be remedied by

regulation is remarkable in the extreme. In the light of such a statement it might be worth while to give a summary of what a two phase generator is, if this was anything but an electrical convention. In a two phase generator the phases have no electrical connection, consequently unless more transformers are put on one phase or more power required of it than half the capacity of the generator, it cannot be overloaded. Suppose for some reason one phase should have more than its normal output required, this overload cannot be remedied by regulation any more than in a single phase generator, nor would it hurt or injuriously affect the other phase. Why should any comparison be made between the two phase system and the Edison three wire system ? There is no necessity to balance light against light in the two phase system as in the Edison three wire. In every case in the former the wires are of sufficient capacity to carry their entire load whether one phase is loaded and the other empty or both full loaded. You doubtless know that when coils are so set in a generator that they generate E. M. F.'s at 90° difference of phase, there is no inductive action between these coils, so it will be readily seen that throwing on or off of transformers on one phase cannot affect the other. Two circuits having the same maximum capacity and having the same character of load have at the most but a small difference in load at any time during service. With a two phase generator having an inherent regulation from no load to full load of not more than 6 per cent., the difference of voltage between phases at any time will be so slight as to be negligible. We might as truly say that two circuits cannot be run from one single phase generator satisfactorily as that a two phase circuit cannot be operated with good results if it has a closely regulating generator behind it.

Mr. Medbury : Let us in any installation where the secondary wiring is on the three-wire system, take any section served by two transformers. As laid out, these two transformers serve a certain district on either side, and are supposed to be in the centre of their work or located as near the centre as it is possible to figure, the adjoining banks of transformers each being in the centre of their service and feeding up to the end of the service. We have now taken an example. Under these conditions we have taken any set of transformers regardless—take it and the system is laid out to serve. Let us now draw two parallel

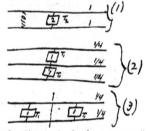

lines of an arbitrary length, and at the centre corresponding to the centre of distribution of the work, place a transformer, which we will call T2. Let this transformer have a secondary pressure of 110 volts, and its capacity be represented, as indicated, by 2. Then the size of the wires running from the secondaries to this transformer may be represented by 1. This would represent the two-wire system of 110 volts. Now draw three parallel lines of the same length as before, and, bisecting these lines, place two transformers as arranged in the three-wire system, the size of each transformer being one-half of that in the first case, or represented by 1. Call each of these T1. Now as the potential between the two outside lines is 220 volts, the cross section of the wire required to do the same work with the same drop of potential will be one-quarter of that in the first case. It has been customary to make the middle wire in the ordinary installation of the same cross section as the two outside wires. Again, draw two lines of the same length as in the other two cases, bisect them, and half way between the point of bisection and each end of the wire, on each side of the point of bisection, suppose a transformer to be placed of the same size as imagined in the three-wire system, that is, having a capacity of 1, which is half that in the first case. It is evident then that the cross section of the wire required for the two mains is the same as that of the two outside wires of the three-wire system ; that the middle wire, of whatever cross section, is entirely dispensed with ; furthermore, the same sizes and the same number of transformers are used. It will hardly be denied that the two-wire system is less expensive to instal, leaving out of consideration the cost of wire ; that it is simpler, there being no necessity of maintaining a balance. Even the writer of the interesting paper just read, although an old-three-wire man, acknowledges that there is a compensation in the three-wire system in endeavoring to keep a balance ; further, that the two-wire system is neater, in respect to not only the mains but also to the house service, the ideal method of course being to do away with all visible wires to the house services, and the nearer the approach to this, whether it be two or one wire, the better from an aesthetic point of view. Again—a matter perhaps of less importance, but well worthy of some consideration by those central station owners who operate their plant on the meter system—there are, I believe, a very limited number of meters applicable to the three-wire system, while the same meters are also of equal service to the two-wire system. There are, however, quite a number of very satisfactory meters for the two-wire system, so that in installing a three-wire system a central station owner is confined to a very limited choice of meters, being forced to take from a limited number; whereas, on the two-wire system, he has a choice of half a dozen or more satisfactory meters. It would be therefore seen that using the same number of transformers, and the same, if not less, amount of wire, the same satisfactory service could be given on the two-wire system as on the three-wire system, doing away with all the disadvantages that essentially pertain to the three-wire system, and gaining all the simplicity that there is in the two-wire system. This is a matter which has been of considerable interest to the speaker, as coming in contact with central station managers from Halifax to Winnipeg. He has felt that there is an entirely exaggerated estimate of the merits of the three-wire system as compared with the two wire system. I hope this matter will be thoroughly discussed here in order that if I am wrong I may be corrected and know the truth.

Mr. Starr : You had reference to secondary distribution only?

Mr. Medbury : Yes.

Mr. Armstrong : I had an intimation that two or three gentlemen present were likely to make something of an onset on this paper, and must again express my regret at Mr. Hartman's unavoidable absence. The point which seems to be causing the most trouble, at least to Mr. Medbury, seems to be the advocacy of the use of the three wire system for secondary distribution. I must, however, refuse to accept the alternative which he offers that the use of this system is due either to bad engineering or to an undue desire to sell copper wire. On the contrary it has been found in most cases the cause, not of waste, but of great economy. I will grant that, in the ideal case presented of a single circuit running along a single street the two wire mains may be all right under some circumstances and that there may be no gain in copper in using the third wire. But unfortunately this ideal case does not occur in actual practice or in a general distribution. In our work we find there has been a great saving in copper in using the three wire system, from the fact that it enables us to cover from a single set of mains with large transformers, an area of from 1,500 to 2,000 feet in width. In his calculation Mr. Medbury also neglects the distribution of the secondary losses entirely. According to Mr. Dion's excellent paper, the practise of the Ottawa Electric Co., in this respect, is to allow one per cent. loss on the services and 2 per cent. on the interior wiring, or a total drop of 3% from the transformer to the lamp. It must be conceded that, granting the same amount of total drop in both cases, the three wire system with 200 volts on the service and interior wiring will, with the same size of wire as on 100 volts, give us just half the drop on the services and interior wiring. This allows us, while employing the same total drop from the transformer to the lamp to double the drop on the mains themselves in the case given. This would, of course, give an area of just one-half on the three wire mains of that which would be required in the case as stated by Mr. Medbury. Further, in the use of the three wire system for secondary distribution an advantage will be found in the fact that in case of a break-down of one transformer of a pair, the additional load is not thrown on the one remaining, as would be the case on the two wire system, thus blowing out its fuses and shutting down all the lights fed from that main. With the three wire system the remaining transformer continues to carry its own load and maintains at least a half service throughout the buildings depending on that section. However, I think the principle trouble with the three wire system as applied to secondary distribution in the eyes of the gentleman who has spoken so feelingly regarding it, is that it renders difficult the sale of two wire meters to plants which use it. Mr. Medbury, I think it was, who referred to a statement of Mr. Hartman in paragraph 2 on the top of page 8, in which he states that the loss from the centre of distribution to the lamp was from 3 to 5 per cent. greater than in the direct current system on account of the loss in the transformer. It will be noticed that Mr. Hartman is here referring to conditions formerly existing in the early days of alternating work, not to the improved conditions which obtain at present. Regarding the remarks of the other speaker I have noted one or two statements regarding which I might say a few words. First, as to the question of the number of alternations used in the two systems spoken of, I cannot see why, for the reason stated, the assumed greater cost of transformers, it should be considered advisable to use 16,000 instead of 8,000 alternations per minute, since the use of the higher frequency admittedly necessitates a greater line drop from mutual and self-induction, and this drop, while not to be taken as an energy loss, is a most disturbing factor in regulation. Added to this is the necessity imposed by use of the higher frequency of introducing such a device as the condenser with the motors, for the purpose of raising the power factor. The only reason of importance seems to me to lie in the fact that a great many plants are in operation in which the old-style transformers used are suited only for the higher frequency. In these days of polyphase apparatus, however, many changes are being made and a change in this particular can scarcely be objected to which is desirable in any case upon the score of economy and better regulation. Since, as Mr. Jackson admits, a well-designed transformer can be used for either frequency, this objection largely falls to the ground. There is a point regarding regulation on which I should like to get further information. In the first place, I have not had it made altogether clear to me as yet just how the inherent regulation of any particular machine can compensate for line loss, and also how the independent regulation of the two legs of the two phase is effected under varying loads and at considerable line loss. I have in my mind the case of a town making an increase to an existing single-phase plant in which a two phase system was being figured on as against single-phase, in which the station was located at such a distance from the centre of distribution as to render advisable for commercial reasons a drop of 10 per cent. on the feeder. In this case it was proposed by the advocates of the two-phase system to make the present circuit, consisting altogether of shop-lighting, one leg of the two-phase, and the new lighting, which would be in residences and come on at a different time, the other leg. It was a matter of some difficulty for me to see how, under these circumstances, the independent regulation of the two legs was to be effected. I think a little information on this point would be of interest.

Mr. Medbury : The last speaker seems to have misunderstood me. Let your engineer simply take any two transformers; you do not get any more loss doing half the work. You fix the loss at 5 per cent.; I do not care what you make your loss. Of course you are obliged to put your two dynamos in separate phases ; you are not obliged to put two transformers on each pole ; they do not look nice—not very aesthetic. If I am mistaken I would be glad to know it. If there is one point in my reasoning that is fallacious I would like to hear it pointed out.

Mr. Armstrong : I am not talking of any particular case at all ; the question is one of general secondary distribution. We have found in putting in a great number of these plants that a considerable saving in transformer and copper cost has been secured by using the three-wire system, as well as greater reliability of service. Mr. Medbury seems to have missed the point in regard to what I said as to line losses in secondary distribution. Taking an allowable drop of say 3 per cent. between the transformer and the lamp, it may be so re-distributed as to secure a considerable saving in copper by using the three-wire system.

Mr. Jackson : In answer to Mr. Armstrong's question, the difference in load is seldom so great as to make the running of a well regulating two-phase generator unsatisfactory, but to cover just such possible cases, the Stanley Electric Manfg. Co., of the United States, and the Royal Electric Co., of Canada, in connection with their S.K.C. system, will soon bring out a device by which the E.M.F. of either phase may be raised or lowered at will without affecting the other phase. This device being entirely within the generators makes it much more convenient than an ordinary regulator. Its construction will shortly be made known to the public. Such a regulator cannot be used satisfactorily upon a three-phase generator on account of the inductive action between the phases.

Mr. Wright : There may be good and sufficient reasons to prevent Mr. Jackson giving us full information. We of course can't press him. I would certainly like to have heard from him as to what is intended to be done.

Mr. Armstrong : It would be a pity to have anything like this turn out incorrect. I would like to hear some details.

Mr. Jackson : I simply state it as a fact. The fact remains, but I cannot give any particulars as to the way in which it is done—that is, as to the regulation of each phase separately.

Mr. D. A. Starr : I have listened with great pleasure to the paper, and especially to the discussion which it has brought out. Some of you may remember the old adage, " When certain parties fall out, certain other parties get what is due them." (Laughter.) I beg to move a vote of thanks to Mr. Hartman for his paper.

The vote was seconded by Mr. Breithaupt and carried.

The President : This ends the list of papers, and any general business will now be in order. I would like to have the report from the Striking Committee.

The report was then read, the Committees being as follows : Committee on Statistics : Messrs. Breithaupt, Yule and Higman. Committee on Legislation : Messrs. J. J. Wright, Dunstan, Powell, McFarlane and Badger.

Mr. A. B. Smith moved the adoption of the report, seconded by Mr. J. W. Taylor. Carried.

Moved by Mr. B. Jackson, seconded by D. A. Starr, that the thanks of the Association be tendered the Bell Telephone Co., for having placed an instrument in the convention room at the disposal of members.

Moved by L. B. McFarlane, seconded by A. B. Smith, that the thanks of this Association are due and are hereby tendered to the Local Committee, consisting of Messrs. C. Berkeley Powell, W. Y. Soper, J. W. McRae, T. Ahearn, O. Higman, and

J. W. Taylor, for their untiring efforts in making such successful provision for the entertainment of members and guests of this Association while in convention at Ottawa. Carried.

Mr. Armstrong—moved, seconded by Mr. McKay, that the hearty thanks of this Association are hereby extended to His Worship Mayor Borthwick and the City Council, representing the citizens of Ottawa, for the use of the Council Chamber for the purposes of this convention, as well as for other courtesies received at their hands. Carried

Moved by Geo. Black, seconded by A. M. Wickens, that this Association desires to recognize the kindness of Capt. Bowie, the deputy sergeant-at-arms, and to thank him for courtesies received at his hands. Carried.

Moved by E. Carl Breithaupt, seconded by J. J. Wright, that the Secretary be instructed to convey to His Excellency the Governor-General the hearty appreciation and thanks of the members of this Association for the courtesy extended in placing at the disposal of the committee his electric launch for the use of the members during the meeting of the Association in Ottawa. Carried.

Moved by Mr. Wickens, seconded by Mr. B. Jackson, that a vote of thanks be tendered to the press for the very full and accurate reports published of the proceedings of this Convention. Carried.

Moved by J. A. Kammerer, seconded by R. G. Moles, that the thanks of this Association are hereby extended to the Ottawa Electric Railway Co., for granting free transportation and other courtesies to members in attendance at this Convention. Carried.

notice. It is not right that our President should retire without our being able to say something to him as to our appreciation of the dignity and efficiency with which he has presided at our meetings, and which has added so much to the results we have attained.

The President: I thank Mr. Smith very much for the kind words he has just spoken, and you, gentlemen, for the cordial manner in which they have been received. Anything that I have done has been a labor of love, and anything I can do for the Association in the future will be performed with much pleasure. We will now adjourn to meet, I hope, in Toronto in June of next year.

The convention then adjourned.

Immediately after the adjournment the members were conveyed by special electric cars to Rockliffe Park, a beautiful summer resort, the property of the Ottawa Street Railway Company, overlooking the junction of the Ottawa and Gatineau rivers.

By the courtesy of His Excellency Lord Aberdeen, several enjoyable trips were made on the Rideau canal in the Governor-General's electric launch, captained by Mr. O. Higman.

### THE BANQUET.

It would require descriptive powers of a high order to convey to the minds of our readers an adequate conception of the beauty of the scene which greeted the eyes of those who were so fortunate as to be present at the banquet at the Russell House on the evening of Sept. 17th. It may safely be affirmed that nothing so unique and beautiful in the way of decorative

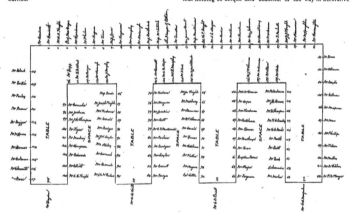

DIAGRAM OF BANQUET, CANADIAN ELECTRICAL ASSOCIATION, OTTAWA, SEPT. 18TH, 1895.

The President: Before closing this the final sitting of the fifth convention, I wish to thank the members for the honor conferred upon me when elected President, and for continued kindness and courtesy throughout my term of office. I wish to gratefully acknowledge the valuable services of the Executive Committee, the members of which have always shown a willingness to work heartily in the best interests of the Association. The services of the Secretary-Treasurer have been almost invaluable, and my association with him in conection with the work of the society during the past year has been a source of much pleasure to myself. I could not use words sufficiently strong to convey my personal appreciation of the splendid efforts made by the Ottawa local committee, but, gentlemen, I think the banquet spoke for itself last night in language more eloquent than I can possibly command. In leaving this chair I make way for an able and energetic successor, and I predict with confidence a brilliant future for the Association in which we are all so much interested. Before closing I wish to announce that Mr. Higman has informed me that the Governor-General's launch will be at the same place as yesterday at 3 o'clock this afternoon, and also by the courtesy of Mr. Soper that special street cars will be at the Russell at 2.30 to take members to the Chaudiere.

Mr. C. B. Powell: Before adjourning I would like to move that the President leave the chair and that the new President occupy it for a moment or two.

Mr. Smith on taking the chair said: I am very glad of this opportunity of being able to bring a little matter before your

effects has ever been witnessed in Canada before. As befitted the occasion, electricity was made to play the principal part in the decorative scheme. On the front of the music gallery and facing the guests on their entrance was emblasoned the word "Ottawa," the letters being formed of red and white incandescent lamps on a cream background. At the opposite end of the room shone forth in the same manner the figures, "1895." Immediately behind the chair of the presiding officer was a Union Jack, composed of a bank of 264 skilfully arranged incandescent lamps, intersected by the crosses of St. Andrew and St. George. On the opposite wall was a horse-shoe design, with the Association monogram, "C.E.A." in the centre, the colors being pale green, cardinal, white, lilac, red and blue. In the centre of the room was a revolving pagoda, carrying varied colored lights, the base being surrounded by flowering and foliage plants. At the apex of every lady's chair gleamed a small red incandescent lamp, which one of the speakers of the evening facetiously referred to as a danger signal. The ingenuity and skill of the caterer had been drawn upon for the successful imitations of dynamos, telephone poles and wires, street cars, and other electrical devices which adorned the tables. The general effect, as stated, was most pleasing, and reflected the utmost credit on the local comitee, the caterer, and all who had a part in its conception and arrangement.

The accompanying plan shows the arrangement of the tables, together with the names and location of the guests. The President, Mr. K. J. Dunstan, discharged in a highly creditable manner the duties of chairman and toast-master, the vice-chairs being ably filled by Messrs. A. B. Smith and C. Berkeley Powell. The menu card, which was an artistic production, contained the following :

> "We may live without poetry, music and art;
> We may live without conscience and live without heart ;
> We may live without friends; we may live without books,
> But civilized man cannot live without cooks.
> He may live without books—what is knowledge but grieving ?
> He may live without hope—what is hope but deceiving ?
> He may live without love—what is passion but pining ?
> But where is the man who can live without dining ?"
>
> *"O hour of all hours the most blessed upon earth,*
> *Blessed hour of dinners."*

### MENU

Clear Dynamo Soup with Dumpling Units.
Broiled Whitefish with Ampere Sauce.
    Sliced Cucumber and Grounded Potatoes
Lamb Cutlets and Green Peas.
    Chicken Soufflee, Dunstan Style.
      Filets of Beef, Street Railway Sauce.
Roast Ribs of Beef, Ottawa Style.
    Boiled Turkey, Polyphase Sauce. -
      Anchor Ice Sherbet.

*" The game is up ; every man to his taste."*

Black Head Duck with Direct Currant Jelly
    Lettuce Salad, Telephone Dressing.
English Plum Pudding with (low potential) Brandy Sauce.
    Greengage Tart.       Wine Jelly.
      Electric Ice Cream.
    Cream Meringue.       Fancy Cakes.

FRUITS.

Cheese and Crackers.       Coffee.
Sherry (Wattage of '91).    Ale.    Ginger Beer.
      Ginger Ale.
      Electrocution.

*" Let me speak, sir,*
*For heaven now bids me , and the words I utter*
*Let none think flattery."*

> *" We part,—no matter how we part ;*
> *There are some thoughts we utter not ;*
> *Deep treasured in our inmost heart*
> *Never revealed and ne'er forgot."*

The formal toast list was a very brief one, being as follows : " The Queen "—God Bless Her ; " The City of Ottawa ;" " Our Guests ;" " The Press ;" " The Ladies."

After honor had been done to Her Majesty, the chairman proposed the toast to the " City of Ottawa," coupling therewith the name of Mayor Borthwick, who in fitting terms responded. By request of the chairman, Mr. C. Berkeley Powell proposed " Our Guests," connecting therewith the name of Mr. George Johnston, Dominion Statistician, who in reply quoted some figures to show the rapid development of the electrical industry. The chairman proposed the health of Mr. H. P. Dwight, President of the G.N.W. Telegraph Co. In responding Mr. Dwight related a few reminiscences connected with the early days of the telegraph in Canada. " The Press," proposed by Mr. L. B. McFarlane, 2nd Vice-President, brought able responses from Messrs. R. W. Shannon and Fred. Cook. The chairman asked Mr. A. B. Smith, President-elect, to propose the health of " The Ladies," which he did, coupling with the toast the name of Mr. Soper, in whom the fair sex found an eloquent and witty champion. Sir James Grant proposed a toast to The Canadian Electrical Association, connecting with it the names of Messrs. K. J. Dunstan, President, and C. H. Mortimer, Secretary, who responded briefly.

At intervals during the evening excellent songs were sung by Messrs. W. J. Gilmour, Brockville ; C. D. Fripp and Prof. McGregor, Ottawa ; D. A. Starr, Montreal, and Mr. Dale, of New York.

### CONVENTION NOTES.

Don't forget the T-i-g-e-r.

Mr. J. B. McCrossan, of Rat Portage, was the representative of the extreme west, while Mr. W. H. Clements, of Yarmouth, Nova Scotia, represented the provinces by the sea.

Some of the members of the Association went to Ottawa fully prepared to annihilate Professor Wiggins if his prediction regarding the weather should materialize ; as it did not, he still lives.

Mr. George Macdonald, Superintendent of the Ottawa Fire Alarm System, placed himself at the disposal of outside members of the Association, and in every way exerted himself to make their visit an interesting one.

Mr. E. Carl Breithaupt divided his attention between the proceedings of the convention and the companionship of several fair lady friends, and in consequence was a subject of envy to many of his friends in the Association.

The grounds surrounding Mr. Soper's summer residence at Rockcliffe Park are illuminated at night by a large number of incandescent lamps, the trees taking the place of ordinary poles. One of the visitors was heard to remark that he doubted whether the wiring would pass inspection by the Underwriters.

If the Ottawa Electric Railway Co. doesn't set a slower pace for its merry-go-round at Rockcliffe Park than that which obtained during convention week, they can safely count on having on hand in the near future a multitude of suits for damages. Some of the members of the C.E.A. who made the trip were in

constant expectation of being landed in the waters of the Ottawa or the Gatineau, if not in Hull, and have scarcely yet regained their equilibrium.

Some of those who made the trip on the Governor-General's electric launch, under the able pilotage of Mr. Higman, seemed fearful of getting too far from the city, lest the batteries would give out and they would be compelled to walk home. These fears were however soon dissipated when they observed the distance covered, and the manner in which the speed was maintained. As to the amount of current consumed, and the cost of producing the same, the records are silent.

### TRADE NOTES.

The Canadian General Electric Co. are installing two 200 kilowatt generators of their multipolar type for the St. John Railway Co.

The Kingston, Portsmouth and Cataraqui Railway Co. have purchased additional cars and equipment from the Canadian General Electric Co.

The Oshawa Electric Railway Co. has ordered two 150 horse power tandem compound condensing engines from the Robb Engineering Co., of Amherst, N. S.

The Parry Sound Electric Light and Power Co., Ltd., have purchased a 75 kilowatt monocyclic generator from the Canadian General Electric Co.

The Kemp Manufacturing Co., of Toronto, are installing an isolated plant for lighting and power transmission. The Canadian General Electric Co. are doing the work.

A general meeting of agents of the Canadian General Electric Co., from all parts of the Dominion, was held at the head offices of the Company in Toronto last month.

The Packard Electric Co., of St. Catharines, Ont., have recently appointed Messrs. Ahearn & Soper as their representatives for Ottawa and Hull, and Messrs. John Starr, Son & Co., of Halifax, for the Maritime Provinces.

C. W. Ketcheson has opened up an electrical contracting business at 483 Yonge street, Toronto, and reports having met with very satisfactory results.

Messrs. Garrioch, Godard & Co., electrical contractors, of Ottawa, have placed an electric elevator in the Scottish Ontario Chambers and numerous other buildings in the city. They report business good, and have had almost more work than they can do. They have moved into new quarters at 25 Sparks street, and have an attractive window. Each member of the firm is a practical electrician, and has had an extensive experience.

### PUBLICATIONS.

The Arena for October is unusually attractive. A fine portrait of the talented young authoress, Will Allen Dromgoole, forms the frontispiece, and a richly illustrated paper on " Chester-on-the-Dee" from the pen of the editor opens this issue.

ALTERNATING ELECTRIC CURRENTS. By Edwin. J. Houston, Ph.D., and A. E. Kennelly, Sc.D., New York : The W. J. Johnston Company. 225 pages, 77 illustrations. Price, $1.00.

This is the first of ten volumes of an " Elementary Electro-Technical Series," designed to give concise and authoritative information concerning those branches of electro-technical science having a general interest. The subjects to be treated are alternating currents, electric heating, electro-magnetism, electricity in electro-therapeutics, arc lighting, incandescent lighting, electric motors, electric street railways, telephony and telegraphy. The authors state that though the several volumes form a series, each is, nevertheless, so prepared as to be complete in itself, and can be understood independently of the others.

### PERSONAL.

Mr. C. F. Sise, President of the Bell Telephone Co., is at present on visit to San Francisco.

Mr. R. G. Black, of Hamilton, son of the manager of the G.N.W. Telegraph Co. in that city, has secured a position with the Westinghouse Electric Manufacturing Co., of Pittsburgh, Pa., and is reported to be well satisfied with his new location and prospects.

It is expected that next summer electricity will take the place of horses on the St. Thomas street railway.

The Vancouver Gas Co., alarmed by the electric light proposals, has made an offer to light the streets with gas for a period of ten years, also to sell the city an interest in the company's property by the transfer of part of its capital stock, or to dispose of the whole plant to the city.

The Bell Telephone Company has always appealed against the assessment of its plant in Clinton, and when the appeal came up for hearing before the late Judge Toms, he decided on each occasion against the Company. This year an appeal was again entered and Judge Doyle took the same view as Judge Toms, deciding in favor of the town.

The electric cars have been running three years in Montreal. They commenced with three motor cars, now they are running 161 motor cars and 60 trailers. Then the heaviest day's traffic was during the exhibition, when they carried 80,000 passengers, no transfers being given. This year on the heaviest days there were carried 134,800 passengers while 38,000 were transferred. The heaviest day's earnings up to that time was $4,000, in fact, the largest in the history of the company and during exhibition week. The greatest day during the present year was also during the exhibition, the earnings being $5,778. The miles of track are now 74½ as compared with about 45 three years ago.

## CANADIAN ASSOCIATION OF STATIONARY ENGINEERS.

### PROCEEDINGS OF SIXTH ANNUAL CONVENTION.

THE sixth annual convention of the Canadian Association of Stationary Engineers was held at Ottawa on the 24th, 25th and 26th of September, and is pronounced to have been the most successful in the history of the Association. White badges, bearing the letters C. A. S. E., were quite in evidence on the streets of the capital, and a hearty welcome was accorded to the visitors.

On assembling in the city hall Mayor Borthwick read an address of welcome. After a few words of hearty greeting he proceeded:—"I see by the preamble of your constitution that it expressly stipulates that the Association shall not be used for the encouragement of strikes or interference in any way between its members and their employers in regard to wages. This is a matter for congratulation. It proves that you have at heart the interests of your masters, as well as those of your own, and whatever good may be derived from your deliberations—and good there must be when men of experience meet and exchange ideas—will be shared by employer and employee alike." Again he said:—"And as to the other principles you are called upon to uphold as delegates to this Convention, there can be no higher, nor nobler, nor none better calculated to avert calamities that too frequently happen through the lack of reliability and intelligence."

Mr. J. J. York, of Montreal, president of the Association, replied to the welcome in fitting terms, referring specially to Ottawa's great industry, the lumber trade, and the importance of the position held by the saw mill engineer. He also stated that it was the purpose of some of them to return to Ottawa soon, for the purpose of seeking an act of incorporation giving them the right to hold property and establish schools of engineering throughout the Dominion.

Ald. Stewart and Ald. Campbell also extended the hand of welcome to the delegates.

Mr. Thos. Wensley, President of Ottawa Branch, No. 7, then read an address of welcome, in the course of which he said: "There are many engineers who think that theory is valueless; that may be true to a certain extent, but the engineer who possesses a good theoretical knowledge of his profession, and combines it with a good practical knowledge of the same, has a great advantage over his fellow craftsman who is satisfied to do everything by the rule of thumb, and he is the man that in time will get to the topmost round of the ladder in his chosen calling. In these days of high pressure steam, with our compound, triple and quadruple expansion engines, the engineer must be a well informed and progressive man, not merely a starter and stopper; he cannot exist on a reputation gained years ago, but must keep himself abreast of the age, by the acquiring of greater knowledge, although it may be acquired with difficulty, and principally through his own exertions. The object of the Canadian Association of Stationary Engineers, as I understand it, is to mutually assist each other in the acquirement of this knowledge by the interchange of thought on the different matters that come within their calling, and it is our wish that its efforts in this direction may be crowned with success, which it certainly will if its members are true to themselves, and true to the Association and its principles as laid down in our constitution."

#### SECRETARY'S REPORT.

The following report was submitted by the Executive Secretary, Mr. Jas. Devlin, of Kingston:

MR. EXECUTIVE PRESIDENT, FELLOW OFFICERS AND BRETHREN—

It affords me unusual and peculiar pleasure in presenting to you a review of the work done by me as your chief secretary during the year just closed, and before going into the matter further I wish to thank the secretaries of the different branches for the untiring assistance given by them to me throughout the year, and I assure you it has always been a pleasure to me to discharge my duties, owing largely to the cheerful and pleasant manner in which they have always co-operated with me. I wish also to thank Bro. York, as well as the other members of the executive council, for their kindness to me throughout my term.

I am more than pleased with the work we are doing, though it must be remembered that we are yet in our infancy. The Canadian Association of Stationary Engineers is an institution for educational training, and training in our particular branch is as essential as a training in any other profession. It is our college—and the college commencement is the finest flower of Canadian civilization. Diplomas wave like flags and graduation orations ring out like cathedral bells. The university is the rock of national liberty. The educational institutions should be the pride of our nation. It is a great blessing to be young now. Better to be alive now than to have been a king and be to-day a mummy in the museum. The world has changed. Not long ago our ancestors were slaves. Not long ago it was a crime to own books. Printing presses were crushed by acts of kings. To read made men think. To think made men free. Not many years ago bibles were more unlawful than murders, and to read the gospel was a crime for which men were tortured and put to death. Then public schools were as dangerous as dynamite, and the thought of educating the masses was worthy only of burial. How changed now. The great thoughts of liberty, brotherhood and God are the granite cliffs against which the old tyranny and superstition has been beaten into spray. How the student's pen has taken the place of the Indian's arrow. The schoolmaster's ruler is beating away the old fetters, and is pointing to the vaster liberties in life. We live in an age of far-reaching opportunity. When Socrates in his flight would journey, he went afoot. The Saviour never went faster than on a donkey or camel back. But now when we travel we ride in a coach as luxurious as a palace, and rush across the continent with our team of lightning and thunder. The Rothschilds made their fortune by galloping to London after the battle of Waterloo and buying bonds before the news of Wellington's victory reached the city. No more fortunes will be made that way. Is there a battle, the whole story is flashed to the city before the smoke leaves the cannon's mouth—it is printed in the afternoon paper—read by the continent, and the paper is used to kindle the fire the same day. Did it ever strike you what an important part has been played by engineers in all these advancements? Do you ever stop to consider the importance of your calling, and the unbounded opportunities before you? I predict great things for the future, and not very distant either. We are just entering, you might say, on the electric age. I would not be surprised if, in a few years, the present electrical appliances for all purposes will be looked upon as huge curiosities, as the first steam locomotive now is, etc. Late in the last century, (1787) a certain philosopher, Dr. Elliott, was being tried for murder in England. Certain writings of his on the inhabitation of the sun were put as evidence by his friends to prove his insanity. How changed now. The conceptions of the madman are in the present day generally adopted on this scientific question.

This is really a progressive age. We have a magnificent heritage of country, which also adds to our blessings. Look at our country—into it you might throw the inhabitants of Europe, and they could hardly find each other. Our country is great and good. In Quebec and Nova Scotia, where they do not raise beaux, they raise brains. And after all, brains are our great products. In Canada we have room for all. The Swede may live here, and dream of Gustavus Adolphus. The Quaker may go round as broad in his mind as he is in his hat. Here the negro may live on his own farm and twine the flowers of freedom about his portal, and think not of auction block or slave pen. And yet while in Canada we have room for all people, yet in Canada we have air enough only to stir one flag. But, brethren, don't think that all the questions are solved. The great social struggle is on; whilst it has not been so noticeable in Canada as in the United States, yet beneath our great industries are heard rumblings and voices which filled France before the revolution. We are to feel thankful, however, that the atmosphere has so considerably cleared, and the financial crisis which threatened us passed. Times are brighter in the United States. We should feel thankful for this, for what is prosperity to them is sure to redound to our advantage. I say we have large interests in the commercial welfare of our friends to the south.

The affairs of our cherished association during the past year have been without event. Perhaps there has been laxity somewhere. But all the branches have been active and appear to recognize the objects for which they are in existence. Enthusiasm will have to be instilled into the members. They must be made to understand that our future success depends upon their individual support.

During the year there were organized and put into working order two branches, namely, No. 15, situate at Brockville, and No. 16 at Carleton Place. Both are very active and display great interest in the general welfare of the association.

There are a great many matters affecting our interests which must come up for consideration sooner or later. Amongst the most important, and one that occurs to me as being of vital significance to the members, is that of arranging some cheap but safe method of life insurance. I am not going to suggest any particular plan, but would strongly recommend the appointment of a committee to enquire into the matter and report at some future time. Besides the advantages to be derived from the insurance itself, it would have a tendency, I think, to bind together more closely the members, to realize more that they are the members of an important brotherhood, to feel that they have a personal interest in each other's welfare.

I will not speak of the financial affairs of the association further than to state that all moneys that have come into my hands have been duly handed over to the treasurer, who will report as to our standing. I have made it a practice to remit to the treasurer twice during the year, namely, after each payment of the per capita tax had been collected, in full, and at the same time furnished him with a complete statement of receipts and expenditures, with items and all other data in my power. I think it quite uncalled for to report oftener than twice a year, viz., at the times mentioned, to the treasurer, as it would add much to the work connected with the secretary's duties, at any rate until such times as we grow to such an extent as to be able to attach some remuneration to the secretary for his trouble.

I have had during the year some correspondence with Bro. Charles E. Robertson and others, with the view to opening a branch in Port Arthur, but as yet no active steps have been taken. I am sanguine that if efforts were put forth, a great number of new branches could be organized.

I cannot let pass this opportunity without calling to your attention the very loose manner in which the books pertaining to the branches have been kept, and I would strongly urge the appointment of a committee to investigate the matter and to arrange some form or system and report. I think also that the system of the executive department could be much improved by having a regular set of books gotten up especially on a form arranged by the same committee. Under the present system it is almost impossible for me to present to you a complete statement of the affairs of the whole association; but, gentlemen, my books are all here, and also every letter, report or other thing connected with the office, and I earnestly invite your perusal of the same, and I shall do all in my power to give you what information I can.

The report of the Treasurer, Mr. Duncan Robertson, of Hamilton, is as follows:

#### TREASURER'S REPORT.

Your treasurer begs leave to submit the following report for the year ending June 30th, 1895.

*Receipts.*

| | | |
|---|---|---|
| Sept. 4th, 1894.— | Balance | $282 14 |
| June 28th, 1895.— | Cash per secretary | 236 40 |
| Sept. 24th, 1895.— | " " " | 145 60 |
| | | $664 14 |

*Expenditure.*

| | | |
|---|---|---|
| Sept. 7th, 1894.— | Mileage to delegates | $175 25 |
| " " | Expenses to Niagara | 38 70 |
| " " | Bro. W. G. Blackgrove—postage | 0 75 |
| " " | Rent of hall | 20 00 |
| " " | Bro. A. E. Edkins—postage | 3 75 |
| " " | Past president's jewel | 25 00 |
| June 28th, 1895.— | Secretary's expenses | 57 46 |
| Sept. 24th, 1895.— | " " | 8 11 |
| | Balance on hand | 335 12 |
| | | $664 14 |

Respectfully submitted,

DUNCAN ROBERTSON, Treasurer.

In the afternoon the delegates were taken to see the Central Canadian Exhibition. At the evening session two papers were read, one on combustion by Bro. Thos. Wensley, President of Ottawa branch, the other on safety valves, by Bro. A. M. Wickens. The papers were of a technical character.

## SECOND DAY.

The second day's proceedings opened with a business session. It was resolved to allow $2 a day to delegates for maintenance and 5 cents per mile one way for mileage.

A proposal was made to change the name so as to admit marine and locomotive engineers to membership, but after considerable discussion it was laid over till next year.

It was decided to issue certificates to the members of the Association, for which they will pay a fee of 50 cents. This is not issued as a certificate of competency, nor has an examination to be passed to receive one. It is merely issued so that those belonging to the Association will be known as members. The certificates will be good for a year only. The conventions hereafter will be held each year on the first Tuesday after the 15th of August.

The question of having a system of insurance in connection with the Association was discussed and was left to the incoming executive to deal with.

During the day a visit was paid to the E. B. Eddy works at Hull, and afterwards the delegates were taken to Rockliffe Park and entertained at lunch by the local branch.

In the evening a visit was paid to the electric power house and to the pumping house of the water works.

At to-day's session a paper—the most important and interesting of the Convention—was read by Mr. A. E. Edkins, of Toronto, on steam boiler explosions. The following is a summary of the paper:—

Having been requested to prepare a paper for this Convention I decided after much difficulty to take as my subject Steam Boiler Explosions. Scientists have given the subject much attention in the past, but their theories have not been of much benefit. It is difficult to draw the line between what constitutes an explosion and what might be termed a burst, rupture or local explosion. As operative steam engineers we are more directly interested in the most probable cause and the most efficient means to be adopted for the prevention of explosions. It is quite a common thing in the event of an explosion to hear men attribute the cause to a lack of water in the boiler, and some go so far as to infer that the boiler must have been empty and red hot, and the explosion caused by turning on the feed pump and throwing cold water on the red hot sheets, which being evaporated into steam (instantaneously) of sufficient pressure to cause the destruction of the boiler.

This is a most absurd theory, as will be shown. A cubic foot of water at a temperature due to a pressure of 60 or 70 lbs. of steam to the square inch has been found to have as much explosive energy as one pound of gunpowder. Gunpowder produces sufficient force to raise its own weight to a height of 50 miles, while water, under the conditions existing in a steam boiler under pressure, has energy stored sufficient to raise its own weight nearly one mile. The causes to which the explosion of steam boilers have been attributed are legion, and may be classed as follows : 1st, the known; 2nd, the possible ; 3rd, the improbable and nonsensical. Among the first causes may be classed bad workmanship, defects in design and weakening and wasting of the structure from old age.

Among the second or possible causes of explosion may be mentioned, low water and consequent overheating of the boiler. Among the third or absurd causes for explosions may be mentioned the following : The decomposition of water within a steam boiler and the formation of a powerful gas which under some conditions has been held responsible for some explosions, but which has been proved absurd by many scientists.

Electricity within the steam boiler has also been given as a possible cause of explosion, owing, no doubt, to the fact that steam upon being discharged into the air under certain conditions has exhibited signs of electricity, but proof of this theory has not been forthcoming.

Taking an ordinary horizontal cylindrical tubular boiler which will have about 900 square feet of heating surface and would contain 8.225 pounds of nearly its own weight and 20.84 pounds of steam, and according to a table prepared by A. Thurston, the stored energy contained in the water would be 50,008,790 foot pounds, while that in the steam would be but 1,022,731 foot pounds, or only one-fiftieth of that stored up in the water. It is very plain, therefore, that the bulk of destructive force in the event of an explosion emanates from the heat stored up in the water.

It follows, therefore, that the class of boiler containing the most water must in the event of an explosion cause the greater amount of damage. Many instances were given where cold water has been turned into red hot boilers without serious results following.

## THIRD DAY.

On the third day the Convention met and received reports from the various committees, also transacting some routine business.

President York brought up the question of securing models of the machinery and appliances they were in the habit of using, which would be of much benefit to the members, and which could be passed around among the different branches. He thought the manufacturers of pumps and engines should be applied to for the purpose of getting them to supply models, which would be of much benefit to the members as well as to the manufacturers. A motion was carried to the effect that the executive take steps towards securing models.

The executive reported that $350 was required to pay expenses of delegates. A number of the members returned half their allowance to the Association's funds.

A motion was passed acknowledging the great hospitality of the people of Ottawa.

The Convention was brought to a close by a banquet at the Windsor Hotel. The chair was occupied by Mr. Thos. Wensley, president of the Ottawa branch, who had on his right Mayor Borthwick and Mr. Blackgrove, the new president, and on his left Mr. York, ex-president. Among the guests were a number of the aldermen of the city and others. A very pleasant evening was spent, with song and story entering largely into the program.

The following are the officers elected for the ensuing year :— President, W. G. Blackgrove, Toronto ; vice-president, J. Devlin, Kingston ; secretary, E. J. Phillips, Toronto ; treasurer, Duncan Robertson, Hamilton ; conductor, W. F. Chapman, Brockville ; doorkeeper, F. G. Johnston, Ottawa ; Provincial Deputy for Ontario, F. G. Donaldson, Ottawa ; District Deputies for Ontario, J. Hugget, Toronto ; J. Floody, Wiarton ; Pro-deputy for Province of Quebec, O. E. Granburg, Montreal.

Kingston, Brockville, Montreal and London extended invitations for the next convention, but on a vote being taken, Kingston was selected by a large majority.

At the conclusion of the proceedings the retiring president, Mr. J. J. York, of Montreal, was presented with a jewel.

## SPARKS.

The Nanaimo Telephone Co.'s plant is to be reconstructed.

Mr. James Pape has been appointed assistant electric light inspector for Toronto.

An electrical street car postal service has been inaugurated on the third avenue line in New York.

The machinery in the electricity building at the Bordeaux exhibition was destroyed by fire last week.

The Guelph electric street railway was opened for traffic on Sept. 17th. There is about five miles of track.

The electric light power station at St. Stepen's N. B., has had to run short on account of low water caused by the drought.

The Canada Switch and Spring Co., Montreal, is about to engage in the manufacture of railway, electrical and contractors' supplies, etc.

Mr. W. Y. Soper, vice-president of the Ottawa Electric Railway Company, says Ottawa is not yet large enough to require Sunday street cars.

The earnings of the Montreal street railway for the year ending 30th September, were $1,096,724.80, an increase of $214,552.38 over the previous year.

The Hamilton Radial Electric Railway Co. has closed a contract with the Niagara Power Company which makes the construction of the road to the Falls a certainty.

An extension of the Toronto and Suburban electric railroad has been located from Toronto Junction to Islington. It is hoped Lambton Mills will be reached this fall.

An electric railway is projected from the Grand Trunk in the Township of King to the village of Schomberg. Mr. Jamieson, of Toronto, has been looking over the ground.

The Toronto Railway Co. is well equipped for the coming winter. It has ten snow ploughs, five more than last year, and two cars on each route are fitted with apparatus for removing ice from the rails.

John C. McDonald, of the Auditor-General's office, Ottawa, had his leg fractured in two places, while alighting from an electric car on the civic holiday. He is suing the Ottawa Railroad Company for $10,000 damages.

The Ottawa Carbon Works have commenced operations. They employ 29 men and a few girls, which staff will shortly be increased. They have six months orders ahead.

The Toronto and Richmond Hill Electric Railway Company have entered suit against the Township of York, claiming $100,000 damages for preventing them from completing the road within the specified time to claim the bonuses of $60,000.

At the opening of the Guelph electric railway one of the speakers rejoiced that the people would be able to reach the cemetery so quickly. Whereupon a contemporary remarks that if Guelph citizens have not been getting to the cemetery fast enough they can rest assured that the trolley cars without fenders will do much to remove that grievance.

The London electric road carried over 100,000 passengers to the Western fair. On Wednesday 34,000 tickets were taken up, and on Thursday 31,000. Seven tickets are sold for a quarter. The manager claims that the business during the exhibition week was 250 per cent. greater than that of the Toronto road during the industrial. But then it was a new thing for London and everybody wanted to take a ride.

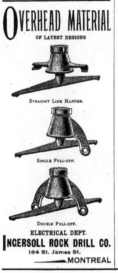

## SPARKS.

A company is being formed to build an electric street railway in Sherbrooke, Que.

Mr. W. H. Frost, of Smith's Falls, is now running his malleable iron works by electricity.

The contract for lighting Trenton for ten years by electricity has been awarded to an American company.

During August, the Galt, Preston and Hespler electric railway carried 27,000 passengers and over 600 tons of freight.

A sale of 100 acres of mica land in the township of Hull, to a United States firm is reported. The price paid was $6,500.

The city's percentage of the earnings of the Bell Telephone Co. in Toronto for the past three months amounted to $1,926.54 and increase of $70.57 over the corresponding period of last year.

It is contemplated to enter into the manufacture of acetyline at Ottawa. Mr. W. C. Edwards, M. P., and other prominent capitalists are interested. Power, material to furnish line and material for carbon are required, and all three are in abundance at the capital, in water power, limestone and mill waste.

The Syracuse, N. Y., Street Railway Company recently placed four cars on four of their lines at the disposal of the Women's Christian Association for one day. About fifty of the most fashionable young ladies of the city acted as conductors, relieving each other at stated intervals. The regular conductors were on board to prevent accidents. The cars were gaily decorated and were crowded all day. The funds of the association benefitted to the extent of nearly $2,000.

An agreement has been signed between the city of Vancouver and Mr. Stewart, acting on behalf of the Western Electric Light, Heating and Power Co., Ltd., for the lighting of the city. The latter agree to supply lamps of 2,000 candle power at the following prices: Not exceeding 27½ cents per night per lamp up to 200 lamps, 27 cents per night per lamp for 200 up to 250 and 26 cents per night for 250 lamps and over, the lamps to be kept burning at least 310 nights in each year. That in the event of the City requiring 1,000 candle power lamps instead of 2,000 candle power, the charge to be 2 cents per lamp less. The company will also supply incandescent light to the city and any citizen that may require it at the price of ¼ cent per amphere hour by meter. The plant to be of the best modern description in use in public lighting, and to be in operation in the principal streets in sixty days, and in the rest of the city in 90 days. The contract is for 10 years, the city to have the right to purchase the plant at actual value at any time.

# CANADIAN ELECTRICAL NEWS

## STEAM AND ENGINEERING JOURNAL

OLD SERIES, VOL. XV.—No. 6.
NEW SERIES, VOL. V.—No. 11.

**NOVEMBER, 1895**

PRICE 10 CENTS
$1.00 PER YEAR.

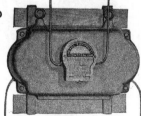

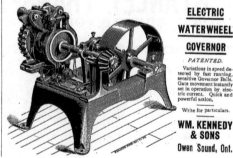

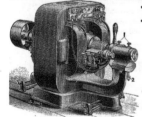

CANADIAN

# ELECTRICAL NEWS

AND

## STEAM ENGINEERING JOURNAL.

VOL. V.                      NOVEMBER, 1895                      NO. 11.

### GENERATION AND TRANSMISSION OF ELECTRICITY FROM WATER POWER.

THE work of utilizing the great water powers of the far west is going on rapidly. Within the past months two electrical plants for the transmission of power from water falls over long distances have been installed. That at Sacramento, Cal., has proved the feasibility of economically transmitting the power of a fall to a distance of nearly twenty five miles. That at Portland, Ore., is still more important and presented many new electrical problems. Those have been successfully solved and the thriving city of Portland is now benefiting by a service of electricity from a water fall more than twelve miles away. This installation was made by the Portland General Electric Company, of which Mr. F. P. Morey is President. The company constructed to allow of the passage of vessels past the falls into the navigable waters above, extending seventy-five miles inland.

To construct this canal, the State of Oregon contributed about $20,000, the balance being supplied by the Portland Company.

The station building is of concrete, stone, iron and brick and when finished will have a length parallel with the river of 364 feet. The water is taken from the canal, led through an extensive hydraulic installation and discharged into the river below on the other side. The water wheel plant is from the works of the Stillwell-Bierce & Smith-Vaile Co., of Dayton, O., and consists at the present time of three units, each consisting of a pair of vertical cylinder gate improved Victor turbine wheels, 42 inches and 60 inches in diameter respectively. The larger wheel is an auxiliary to be brought into service only at periods

FIG. 1.—INTERIOR OF STATION SHOWING THREE 3-PHASE GENERATORS AND TWO DIRECT-CONNECTED EXCITERS.

owns the entire water power of the falls on the Willamette River at Oregon City, twelve miles above Portland, which, with a head of forty feet, has a minimum capacity estimated at 50,000 h. p. Part of the power has already been utilized by numerous factories and mills erected near by, and in addition to these an electric station erected some years ago has supplied current for lighting the streets and dwellings of Portland and for operating an electric street railway between Oregon and Milwaukee, seven miles away, the direct current and high frequency alternating systems being used.

In order to take advantage of the power of the falls the Portland General Electric Company has constructed the first part of an extensive station on the west side of the Willamette river opposite the City of Oregon. The part constructed is one quarter only of the building, which is being put up in sections. Twenty sections will complete the building ; five are already built and foundations are now being made for the remainder. The ultimate generating capacity of the station will be 12,800 horse power. In addition to the land covered by the station, the Company has purchased about 1,600 acres in the vicinity. It also controls the canal and locks on the west side of the river,

of excessive high water, which the records show occur about once in every five years. The smaller wheel runs at a speed of 200 revolutions per minute and the larger at 100 revolutions per minute. Both turbines are set at the same level and each carries a pulley ; that of the sixty inch wheel being fixed to the generator shaft. When the large wheel is in operation the two pulleys are belted together, the smaller wheel is disconnected and the large wheel drives the generator at a uniform speed of 200 revolutions. When the smaller turbine is operated alone the belt lies upon a shelf surrounding the pulleys.

The weight of the vertical shaft with the armature is about 33,500 pounds, and to carry this a system of extra bearings is introduced—one of the ring thrust type and the other a hydraulic oil bearing, both supplementing the ring bearings on the armature shaft. They are enclosed with cases filled with oil delivered by hydraulic pressure, and are surrounded by water jackets.

The length of the generator shaft is 29 feet, and 8".½ in diameter. It is not a continuation of the shaft of the wheel but is coupled to it by means of a disc coupling, which allows of a certain free movement up and down of the generator shaft. The shaft of the 60 inch wheel runs from the coupling to a bearing set in the floor of the station. Both wheels in each section are controlled by hand wheels and both are regulated by the same governor. The belt tightener is also controlled from either floor by a hand wheel.

The water is admitted to the penstocks from the upper canal by means of a head gate operated from a platform on the canal side of the station. Each penstock is ten feet in diameter and is constructed of riveted steel plates. Each wheel has its own flume, the water passing first through the large flume of the larger wheel to the flume of the smaller wheel, whence it passes

FIG 2.—BANK OF STEP-DOWN TRANSFORMERS.

through a tube into a tail race. In addition to this turbine equipment, an auxiliary power equipment has been furnished, consisting of a set of pumps, including a hydraulic pump for supplying oil to the thrust bearing cylinders and a duplex water pump to circulate the water in the cylinder water jackets. They are operated by two 15-inch horizontal turbines enclosed in the same flume. For the operation of the exciters a further pair of vertical turbine wheels has been installed, each 48 inches in diameter, driving the generators by a similar system to that already described for the operation of the main machines.

The complete power plant will consist of twenty, three-phase generators and two direct current generators, acting as exciters. The total capacity of the station, therefore, will be 12,000 horse power, divided into twenty units, each one independent of the other.

In order to obtain the best results from the power at its disposal, the Company selected the three-phase system of electrical power transmission as developed by the General Electric Co., which has proved so satisfactory in the majority of power transmission installations of varying sizes and varying distances in the United States.

One peculiar feature of the Portland installation is the employment of large blocks of power for street railway service, involving the transformation of the polyphase current sent over the line into direct current for railway circuits. The frequency is 33 cycles per second, selected on account of the large amount of power which is necessary to convert from alternating into direct current. The current is delivered directly to the line without first passing through transformers and when it reaches Portland is transformed down to a potential of 400 volts. For the power service the step-down transformers are connected to rotary converters which will deliver a continuous current of 500 volts for street railway service, as well as for the operation of stationary motors. Induction motors will also be used, directly connected to the secondaries of the step-down transformers when this can be done to advantage.

The five sections of the building already erected are occupied in the following order :—The first section contains the pumps and the accumulators for the complete station ; in each of the three following sections is one three-phase alternating current generator of 450 k. w. or 600 h. p. capacity, and the fifth section contains two 250 k. w., m. p. continuous current generators used as exciters. Each exciter is capable of exciting all of the twenty, three-phase generators, and the second has been set up as a re-

serve in case of accident to the first. At present, one is furnishing direct current to the street railroads in Oregon City. When the station is complete, the exciter section will be removed from the fifth section, which it occupies at present, and will be placed in the centre section of the building, where the switchboards will also be erected.

The generators are of special design and are set upon the floor of the station, the armatures revolving in a horizontal plane, with one bearing at the floor line and another on top of the armature underneath the collector rings. Each generator has twenty laminated poles. The armatures are a little over seven feet in diameter and are about two feet high. These armatures are constructed to deliver current directly to the line at a working potential of 6,000 volts effective pressure without the intermediation of step-up transformers. On account of this high voltage unusual precautions were necessary to perfect the insulation of the armature coils to avoid leakage to the ground. The armatures are wound with flat wire and each of the coils is divided into sections, each section being separately insulated. The thoroughness with which the feat of delivering the enormous voltage direct from the machines has been accomplished can be judged by the fact that the armatures were subjected to a pressure of 15,000 volts alternating and were both short circuited and open circuited under full excitation without the slightest injury.

The field coils are wound for excitation of 500 volts continuous current, and each has been subjected to a test of 5000 volts alternating. The regulation in these machines has proved singularly good, the increase from no load to full load being comparatively moderate.

From the dynamos the loads are run to floor connectors and pass underneath the floor to the switchboard. The concrete floor is over them and thorough protection guaranteed.

The exciters are set up to allow of the armature revolving in a horizontal plane with one bearing only at the floor line. The construction of these exciters is almost identical with the G. E. m. p. type of railway generator, and each has a capacity of 250 k. w. at 125 revolutions per minute.

The high tension switchboards are built of native marble and the panel method of construction is followed throughout. Each panel carries a double pole main switch for the high potential circuit, and a potential circuit and a double pole double throw switch for the exciting circuits. It also carries a rheostat for the control of the excitation of each machine and a single throw

FIG. 3—RECEIVING OR 3-PHASE SIDE OF ROTARY CONVERTERS.

switch opening the circuit through a set of seven 32 c. p. 110 volts. In addition the board carries a current indicator for each line and one for the exciting circuit, and a potential indicator with station transformer placed at the back. The upper part of each panel consists of a set of plug connections for coupling the machines in parallel or for direct line connections from each generator.

The exciter switchboard consists of two panels of Tennessee marble with a special switching panel between them. By means of this switching panel current for the railway service in Oregon City can be obtained from either exciter, or the two exciters can be coupled in parallel or the outgoing railroad current can be used for excitation purposes and the balance from the exciters can be used for other work. From the generators the current passes directly to the line through the switchboard. The line is 14.3 miles long, a separate circuit being installed for each machine. It passes through an undulating country following the course of the Willamette river as closely as possible.

FIG. 4.—DIRECT CURRENT SIDE OF ROTARY CONVERTERS.

The poles upon which the three phase wires are strung, also carry a number of wires for the 5,000 volt continuous arc current from the old transmission station, as well as the wires over which the old system of lighting with high frequency 5,000 volt alternating current is effected. The loss in the long distance transmission line is calculated at full load at about 11%.

The sub-station to which the high frequency lines are brought, is a two-story building at the corner of Seventh and Alder streets, covering a space of 40 feet by 100 feet. The lower floor is divided into three rooms, one containing the transformers, the second the rotary converters, the other being used for a repair shop, lamp and meter room. The upper story of the building is occupied by the offices. In the transformation room at present are the necessary transformers for the three units already installed—45 transformers in all. The receiving end of each line is connected to a bank of 15 transformers per generator, five being placed between each pair of wires of the three-phase system. The transformers are mounted on an iron rack five transformers high and three wide and foundations are already laid for six additional units. Each set of five transformers is connected to the primaries in series and to the secondaries in parallel, although in the transformation of the three phase current two sets only are necessary. For the large units, with high voltages, however, as employed in the present installation it is desirable to have a large number of transformers banked. The bank therefore is divided into three sets instead of two so that each group may act as a reserve to the other two sides enabling two-thirds of the power of each generator to be delivered even if the transformers on one leg of the circuit have to be disconnected, nor is the balance of the system affected by this change of connection. The transformers regulate at a little over one per cent. variation of the secondaries from no load to full load. The transformers are of the standard General Electric substation type, having numerous air passages between successive bunches of iron lamine and between the coils so that they may be cooled by artificial ventilation. This enables the transformers to be worked at a high output of efficiency and yet remain cool. The distribution of light from the secondaries is effected on the Edison four wire system which also allows of the working of synchronous and induction motors from the lighting mains. The four wire system is worked at 1,000 volts between wires and by means of feeder regulators a variation of 4% in either direction is covered.

As already mentioned the direct current for the railway service is obtained by conversion from the three-phase alternating current. This is effected by means of rotary transformers, a type of machine which the General Electric Company has brought to a high state of perfection. Two of these are at present installed in the sub-station, and space has been left for an additional three. The capacity of each converter is 500 h. p. delivered to the bus bars of the continuous current switchboard. The long distance transmission lines for this railway service, as in the case of the lighting circuits, are connected to the step-down transformers, transforming the current from 6,000 volts on the line to 400 volts at the secondaries. The secondaries are connected to the three collector rings on one side, and the current is thus brought into the armature of the rotary converter. The alternating current at 400 volts is then converted in this machine into direct current at 500 volts at no load and 550 volts at full load delivered from the commutator side. The rotary converters are arranged for self regulation, the voltage on the direct current side compounding with the same regularity as that found in the best direct current dynamos despite the varying losses on the long distance line and the varying armature reaction in the rotary converter. This regulation is entirely independent of the generator which receives constant excitation at all loads. The shaft of the rotary converter is extended twelve inches beyond the bearing of the alternating current side to take a small pulley from which any small machine or an arc dynamo may be driven.

It is a note-worthy fact that in spite of the long transmission line, and the increasing load on the generator, the potential supplied to the railway lines steadily rises as the load is increased.

Each rotary converter has a capacity of 400 k. w. It is an eight pole machine making 500 revolutions. The armature is iron clad, carrying at one side a commutator and at the other three collector rings. From the rotary converters the wires are taken to the power switchboards. Each converter has two panels, one for the three-phase current and one for the continuous current. The alternating current panel carries two double pole switches, one for connection to the converters and transformers and the other to connect the converter to a set of common bus bars. An additional main switch is provided to con-

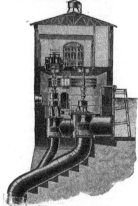

FIG. 5.—SECTION THROUGH THE POWER-HOUSE.

nect the rotary transformer and the panel itself which carries also a set of fuses, three current indicators and a potential indicator with a transformer reducing the potential from 3½ to 1. The continuous current panel is of the standard direct current railway type with automatic circuit breaker, and a current indicator added for the fields of the rotary transformer. The panels may be coupled in parallel on both the alternating and continuous current side.

The lighting switchboards in the sub-station consist of one panel for each leg of the three-phase system. The secondaries from the transformers are entirely independent and the panels carry fuses and two 4 pole switches for coupling the feeders directly either to the corresponding transformer unit or to the bus bars and the switches for the operation of the feeder regulators. A current indicator is placed on each side of the four wires and three potential indicators between the four wires. On top of the switchboards are placed the main switches for opening the different feeders, and each panel is provided with ground detectors and lightning arresters.

At present the lighting from the three-phase system is used for large buildings, containing several hundred lights each. They are close to the city station and this distribution can be readily handled at about 400 volts. For the outlying and residential districts the high frequency apparatus with individual transformers will still be employed. Continuous current will be furnished to the railway and to the stationary motors already installed, but new motor installations will be made with the three-phase motors which will be run straight from the three-phase switchboard, in parallel with the rotary transformer.

The direct railway current will be carried to the east side railway station by means of cables under the Willamette river and this distribution will reach as far as Milwaukee, where connection will be made in parallel with the 600 volt service from either station A or station B at Oregon city. The loop from Oregon city to Portland and back will thus be as follows :—Beginning at Oregon city with 33 cycle three-phase current at 6,000 volts, 14.3 miles will be traversed as far as Portland ; the current will then be transformed to 400 volts alternating, and passing through a rotary converter, issue therefrom at 600 volts continuous, which will be transmitted eight miles to Milwaukee and connect with the continuous current from Oregon city.

This plant when finished will be one of the largest long distance transmission plants in the world. Its satisfactory operation so far shows admirably, not only the effectiveness of the three-phase transmission system for general service, but also its feasibility. This has rendered possible its adaption to the operation of important railway systems through simple apparatus, and to the working of a distributing net work composed in a large part of existing lines.

## RELATIVE ADVANTAGES OF THE ONE, TWO AND THREE WIRE SYSTEMS.

Editor ELECTRICAL NEWS.

SIR,—In your October issue I note with extreme pleasure that we poor outsiders are benefitted by being able to read, study and inwardly digest the excellent papers brought before the Canadian Electrical Association at their recent covention. Before going any further allow me to express my deep and sincere appreciation of this benefit, accorded to non-members and members of this useful Association alike. At the outset I wish to say that I want it distinctly understood by your readers that I do not claim to have more than ordinary knowledge—that I am very liable to mistakes, grammatical and otherwise, and ask the indulgence of your readers accordingly.

The paper in which I find myself more especially interested is entitled "Some Modern Alternating Current Apparatus," by H. T. Hartman. Being employed as manager and chief engineer of a small alternating plant in a suburban town, where customers are few and far between, and where primary and secondary mains are required to cover a considerable area, I am more than ordinarily interested in the question of secondary distribution. In the first place, I am desirous of distributing current to customers with as little loss as possible, so as to have every possible unit of current generated at power station turned into cash at regular periods. Secondly, it is essentially important that capital account may be kept as low as possible to secure good and efficient secondary distribution, and at the same time that capital invested be kept within reasonable margin.

I read with a great deal of pleasure the portion of Mr. Hartman's paper applying particularly to secondary distribution by means of the 3-wire system, and at the same time the discussion and remarks on the paper by Mr. Medbury and others. I am particularly interested in Mr. Medbury's remarks, claiming that transformers can be so placed on the 2-wire principle as to distribute current over secondary mains at a less expense for copper wire than by means of the same transformers placed on the 3-wire system. I confess I am somewhat puzzled to know how Mr. Medbury arrives at these results, and why 3-wire secondary distribution is to be condemned, even taking his ideal case as an actual one.

As I understand diagram No. 3, Mr. Medbury proposes, instead of setting a large transformer in the centre of distribution, as in diagram 1, to divide the line into two sections, placing a smaller transformer in centre of either section, thus reducing weight of copper required for secondary by reducing by one-half distance of transmission, and at the same time reducing proportionately the current transmitted. He then points out what appears at a first glance evident, that the close section required for the wires, as per diagram 3, is the same as that for two outside wires on the 3-wire plan, and that the centre or third wire is entirely dispensed with, making it appear that less weight in copper is required for the secondary mains, on the 2-wire plan, as per diagram 3, than on the 3-wire plan, as per diagram 2. I must candidly admit I fail entirely to see how Mr. Medbury arrives at this conclusion. On the other hand, I claim that arranging transformers as per diagram 2 will allow the station manager to distribute twice the amount of current twice the distance than can be done on the 2-wire plan, as per diagram 3, with an expenditure of but 33⅓ % more for copper wire, assuming that the third wire is of same cross section as the two outside wires on the 3-wire system, and that wires of equal cross section are used in each case. This is surely a decided advantage from a purchaser's point of view as it greatly reduces weight of copper required per unit of current supplied, and enables the station manager, as in a case like my own, to serve several customers from one transformer station, enabling the use of larger transformers in place of a number of small ones.

A case in point arises in my own practice. At one point of my system I had eight customers and six prospective ones, so situated that I required at least eight transformers to be set on 2-wire distribution. I speedily found that what with cost and general inefficiency of small transformers, a better system of distribution must be devised. As an experiment I purchased a pair of larger transformers, setting them together in centre of location, and spread secondary wires on 3-wire system over a radius of some 900 feet in each direction, using 52 volts on secondary mains as heretofore on old 2-wire system. By leading one of the outside and centre wires in one direction, and the other outside wire and continuation of centre wire in another direction, tapping off to customers as required, I found this a very efficient method, greatly reducing initial cost and giving equally satisfactory service. In a somewhat similar case I found it necessary to extend outside mains and light some customers a considerable distance from transformers at 104 volts, supplying some customers at 52 volts and others at 104, both from the same set of transformers. So efficient have I found this system of secondary distribution that I am gradually arranging transformer stations at different points on system, so as to eventually have all secondary distribution on the 3-wire system.

Just a word as to meters. I am using the same meters—both Shallenberger and T.-H. Watt Recording—that I formerly used on 2-wire system, with equal results.

Considering Mr. Medbury's great ability, I hesitated before taking up this discussion, and valor may have got the better part of discretion, but I offer as my excuse my great interest in this really interesting question. If I am operating my employer's plant at a disadvantage by adopting the 3-wire system under existing circumstances, then I deserve their condemnation. After studying the question carefully and to the best of my ability, I am unable to see how any arrangement of transformers on the 2-wire system can reduce required weight of copper without seriously increasing the number of transformers and decreasing the efficiency of same, and I have a great big desire for information on this point.

I have the pleasure to remain,

Yours truly,

ENGINEER.

## HORSELESS VEHICLES.

While the usefulness of the horse has been to a very great degree impaired by the bicycle and the trolley car, it seems as if his position as a faithful servant of man is to be still further destroyed by the horseless vehicle, about which there is now so much being said and written. It is bad enough to be displaced by the silent steed whose bones are iron and whose muscles are rubber, and to be rejected by the car which sweeps along our streets driven by an unseen force, but when everyone who can afford to keep a carriage, or who has occasion to convey himself or his belongings from place to place, adopts some new fangled mechanical power, then indeed will the glory have departed from one of our hitherto most useful domestic animals. No wonder that it should be proposed in all seriousness to establish canning works where horseflesh might be put up for food, so that if he cannot any longer be useful when alive he may at least do some good when dead by helping to sustain life in others.

If a vehicle can be propelled along a line of rails by steam, or electricity, or gas, why should it not be likewise driven along an asphalted or block paved street, or even over a good country road, (and great progress has been made of late in improving our roads)? Traction engines are not a new thing, but their use has been restricted by their cost and the character of the ways over which they had to travel. With the introduction of new and cheaper methods of propulsion, particularly electricity, it is not to be wondered at that horseless vehicles should be introduced, and it is only a question of time when they shall have become quite an every-day affair on our streets, and when the horse, which has so well served us in the past, shall almost entirely disappear.

In order to test the qualities of horseless vehicles several competitions have been arranged. The first has already taken

THE CARRIAGE WHICH WON FIRST PRIZE IN THE PARIS ROAD CONTEST.

place, in a race from Paris to Bordeaux and return. There is no country in the world where science and industry are so much encouraged by the awarding of prizes for meritorious work as France, and it is not to be wondered at that the first competition to test the merits of this new system of locomotion should be held in that country. The first prize amounted to the handsome sum of 31,500 francs.

The French competition was taken part in by no less than twenty-two vehicles, twelve of which arrived at Bordeaux within the time limit set by the rules of the competition. Nine of these covered the distance of nearly 750 miles in less than one hundred hours. Eight of the nine were driven by petroleum or gasoline motors, and one only, and that an old machine built in 1880, by steam. Speed was not the only test, the carriage which came in fourth having won the first prize, its record for the 750 miles being 48 hours and 48 minutes. The fastest carriage was ruled out by the fact that it did not accommodate at least four persons, as required by the rules.

The result of the race seems to show that the lightest vehicle is the best, and this proves the superiority of the essence of petroleum, or gasoline, over electricity or steam as a motive power. It requires only 88 lbs. of gasoline to produce one horse power hour, whereas steam requires at least 66 lbs. of coal, and under the storage battery system more than 220 lbs. is required. Light vehicles also admit of the use of rubber tires. One of the

carriages in the competition was equipped with pneumatic tires and though it weighed 2,380 lbs., accomplished the entire distance without accident to the wheels. All the high speed steam carriages met with accidents, which shows that their weight, made necessary by the system, is too great for vehicles for public roads.

Though the petroleum motor showed its superiority over electricity in this contest it by no means follows that the latter will not be the motive power for the self-propelled carriage of the future. The heat produced in any self-propelled carriage, where the power is generated within itself, must always stand in the way of comfort in passenger traffic, an important consideration. It is possible that the overhead trolley system may be adapted to road carriages, though they could hardly, in that case, be called self-propelled. We do not doubt that electricians will yet discover some means by which that force can be economically applied to such vehicles.

Following the French competition a similar test is being made on this continent, the Chicago Times-Herald having offered $5,000 in prizes in a race from Milwaukee to Chicago, the date of which has been fixed for November 2nd, and the result of which will be known by the time this paper is in the hands of our readers. Some sixty vehicles have been entered for this competition, the construction of many of them having been stimulated by the prizes offered. The points for the judges to consider are, general utility, speed, cost, economy of operation, general appearance and excellence of design. The London Engineer has also offered prizes aggregating one thousand guineas in a similar competition in England. These competitions will enable American and English inventors to compare the merits of their vehicles with those which took part in the French test.

A parade of English horseless carriages took place at Tunbridge Wells a few days ago. The exhibit included Victorias, landaus and tricycles. As in the French trial the petroleum motor showed its superiority. The essence of petroleum used in these motors is a product having a specific gravity between that of kerosene and gasoline, being about ten per cent. lighter than the former.

So far as the manufacture of horseless vehicles in Canada is concerned, no very definite steps have been taken, but some of our enterprising firms are watching the experiments which are being made elsewhere, and will be ready to place a machine on the market as soon as the times warrant it. Mr. Massey, of the Massey-Harris works in Toronto, who are about to commence the manufacture of bicycles, is credited with the remark that he would have a horseless carriage before the public within a year.

A syndicate of prominent British Columbia business men has however been formed, and is applying for incorporation, with a capital of $500,000, to operate traction engines and carriages on the old Cariboo road, especially between Ashcroft and Barkerville, over two hundred miles of mountain road leading to the mining country. They will carry both passengers and freight. As horse feed has to be taken for long distances into that country the adoption of self-propelling vehicles would mean a great saving of expense. Specially wide tires are to be used so as to avoid injury to the roads. Should this company carry out their project, as they talk of doing, at once, their experiment will be watched with much interest.

---

The Canadian General Electric Co. have sold a three hundred kilowatt monocyclic plant to the Halifax Illuminating Co.

Mr. D. Knechtel, of Hanover, is going to put in an electric light plant at Maple Hill, 2½ miles distant, to supply Hanover with incandescent light. He is negotiating with Mr. Heinburger, the owner of the plant in Hanover, for its purchase. The new plant will be in running order by December 1st.

The Parry Sound Electric Light & Power Co., which was recently organized, have closed a contract with the Canadian General Electric Co. for a monocyclic plant for supplying light and power. The streets are to be lighted with series incandescent lamps, which have given a most satisfactory service in the neighboring town of Bracebridge.

The Hamilton, Grimsby and Beamsville Electric Railway, 18 miles long, has just completed the first year of its history, during which it has carried 220,894 passengers and 15,042 tons of freight. The legislature will be asked to extend the time of completion one year so that the company may obtain the city's bonus of $5,000. The road is only running to Grimsby at present.

## THE TORONTO TECHNICAL SCHOOL.

THE Toronto Technical School was established in January, 1892. Its principal promoters were the late ex-Ald. Gillespie, who died before it was opened; Dr. J. Orland Orr, the present chairman of the Board, who was the first chairman; the late Mr. J. A. Mills, and Mr. A. M. Wickens.

At first it was thought to be somewhat of an experiment, but the success already achieved has placed it in a more definite position. It is situated on College St., at the head of McCaul St., and directly to the south of the School of Practical Science, in what was formerly Wycliffe College. Permanent quarters have been secured there, and the building so fitted as to be better adapted to the requirements of the school.

It is maintained entirely by the city of Toronto, and is under the control of a Board of Directors composed of seventeen members. Five of these are members of the City Council, five represent the Trades and Labor Council, two the Stationary Engineers, two the Architects, two the Educationists and one the Manufacturers. Regular meetings of the Board are held on the fourth Tuesday in each month during the session, which begins with October and ends with April.

The classes are free to all residents of the city who wish to avail themselves of its privileges. Both sexes are admitted. The course of study to be pursued by any one is optional, and registration and entrance to any class can be made at any time during the session. It is desirable, however, to enter the classes at the opening of the session.

The school has a staff of nine teachers, and the class hours are from 7:45 to 9:45 p.m. each week day evening, excepting Saturdays.

The design of the school is to aid those who have not had the advantage of an education in the early period of life. It is specially intended for the artizans, tradesmen, mechanics, etc., and those who follow the usual occupations of an industrial community.

The nature of the work done is very different from that usually taken up by the ordinary commercial colleges or schools. An enumeration of the subjects taught will give some idea of the work it is doing. They are: Arithmetic and Mensuration, Algebra, Euclid, Descriptive Geometry, Perspective Drawing, Mechanics, Chemistry, Practical Chemistry, in each of which there are both junior and senior classes. Besides these there are courses as complete as the time will permit in Applied Electricity, Heat, Hydrostatics, Steam and the Steam Engine, Hydraulics, Light, Sound, Practical Geometry, Freehand Drawing, Mineralogy and Geology, and Modelling in Clay. In the draughting room a numerous group of subjects is taken up, as Machine and Architectural Drawing, Industrial Design, Shading, Lettering, Machine Construction, etc.

That the school is doing an important and useful work, and that it is appreciated by the young men of the city, is attested by the fact that the aggregate attendance for last year was 631, while the average nightly attendance for the whole session was 286.

There are no fees for attendance on any of the classes, and each student can take any subject he chooses, or any group of subjects that the time table will permit.

Considerable improvement was made during the holidays by the remodelling of the building and the addition of new classrooms.

The school is now fairly well provided with apparatus in the more important departments for the practical illustration of the various physical subjects taught. It is intended to make the teaching as practical as possible, and to see that the students know the work and not merely see it done. At the end of the session, or of the work in any subject, examinations are held in the various branches, and certificates are granted to those who succeed. A diploma is also given by the school to those who

complete certain definite courses of study. This work, it is thought, will require from three to four years on the part of a student with average ability and but meagre attainments at his entrance upon the course.

Three new teachers have recently been appointed to the staff, and one of the old teachers, Dr. J. McMaster, selected as principal.

One of the new teachers, Mr. James Milne, of whom we append a short sketch, has been appointed lecturer in the now all important subject of electricity, and he will also lecture on the steam engine. His lectures, covering one of the most important courses in the curriculum, will take up the following topics :—

PRACTICAL ELECTRICITY.—Electrostatics—The Electric Current and its Measurement, Electromotive Force and its Measurement, Ohm's Law, Primary and Secondary Batteries, Electrolysis, Galvanometers, Shunts, Wheatstone Bridge, Locating Faults, the Various Electric Light and Power Systems, including 2, 3 and 5 Wire Systems, Motors, Generators, Armature Winding and Repairs, Transmission of Power, Ampere and Volt Meters, Recording Ampere and Watt-Meters, the Edison Chemical Meter, Testing Efficiency, etc.

HYDRAULICS.—Velocity and Pressure of Water under different heads, Measurement of Flow and Calculation of Water Power, Efficiency of Water Motors, the Capacity and Power of Pumps, Friction in Pipes, Hydraulic Ram, Accumulators, etc.

STEAM AND THE STEAM ENGINE.—Thermal Value of Fuel, the Evaporative Power of Fuel, Transfer of Heat from Furnace to Boiler, Heating Surface, Grate Surface, the Generation of Steam, Motive Power from Steam, Strength of Boilers, Rating of Boilers, the Safety Valve, Action of the Crank and Connecting Rod, the Slide Valve, Comparative Efficiency of the Various Engines, the Indicator, Computation of the H.P., the Theoretical Curve, Measurement of Steam Consumed or Water Used per H. P. from the Diagrams, Testing the Action of the Steam Engine.

We understand that already about 450 have registered for attendance this session. 125 of these will take up electricity under Mr. Milne. The Prospectus of the school, which gives a brief outline of the work done in each subject, as well as other relative information, can be had by addressing the Secretary.

DR. McMASTER,
Principal Toronto Technical School.

### DR. MCMASTER, PRINCIPAL TORONTO TECHNICAL SCHOOL.

John McMaster, B.A., M.D.C.M., the subject of this brief sketch, was born in the county of Simcoe, Ont., in 1857. His parents, who were of Irish lineage, settled in that county a few years earlier. From the age of five till he was twelve, he was as regularly kept at school as the average boy in a newly settled country place can be. At that very early age he learned to plough and became so useful on the farm that he was kept at home, excepting for a few months in the winter. At the age of eighteen it was necessary for him to leave home and to earn a livelihood for himself. At this time his education was very meagre, being confined to reading, writing and arithmetic, at the latter of which he was somewhat expert. As a boy he was very ingenious with his knife, always being mechanically inclined. In spare time, and on wet days, he used to occupy himself making miniature saw-mills, to be driven by a small stream that ran through the farm. Fire-arms made of lead, and bows and arrows of every description, were among his numerous inventions. His desire as a youth was to be an engineer and machinist. In this he never had his father's support, and as he grew older he became more anxious to obtain an education. His first employer owned a small saw-mill and worked a farm besides. John was part of the time engaged in the mill and part on the farm. During this year he bought books and studied

Euclid, History and Short Hand Writing. With the money he earned he resolved to begin his own education. After six months in a public school he passed the entrance to the high school, and very shortly afterwards—three months—obtained a second class B certificate. His means being exhausted, he again went to work, and soon secured enough to fit him for teaching. He attended the various training institutions for teachers in this country, and has been engaged in teaching in nearly all the different kinds of schools in the Province. By alternately teaching and studying he was enabled to secure all the different grades of certificates granted by the education department. A degree in arts was the next desideratum. At Toronto University he took a complete honor course in mathematics, graduating in physics in 1886. While there he did not confine himself exclusively to mathematics, but gained a fair knowledge of science, metaphysics and English literature. After graduating he taught mathematics and physics for a few years in different high schools in the province. Not being satisfied with his attainments, he resolved to take up the study of medicine. At Trinity Medical College he took the regular course in medicine, graduating at Trinity University in the spring of 1894. While an undergraduate in medicine he won considerable distinction among his classmates as a student, carrying off his full share of scholarships and prizes.

Before the Technical School began he was engaged as a teacher in the city night schools for several years, and when that institution was commenced in Jan. '92 he was engaged as one of the teachers. Since that time he has remained on the staff, chiefly being engaged in teaching mathematical subjects. The principalship of the school was declared vacant at a Board meeting in August last, and after duly advertising for applications to fill the position, Mr. McMaster was selected from among a number who applied. His duties as principal began at the opening of the school on the 1st of October.

Judging from his active and energetic disposition, and his varied experience in teaching gained in all kinds of schools, as well as his extensive attainments in the various scientific departments of education, we expect an unusual measure of success for this institution.

**MR. JAS. MILNE, LECTURER ON ELECTRICITY.**

Mr. Jas. Milne, recently appointed lecturer on Electricity in the Toronto Technical School, was born in Aberdeen, Scotland, on the 29th of January, 1865. He served a five years' apprenticeship in engineering at Mugiemoss Paper Works, Aberdeen, Scotland, taking in pattern-making, fitting, turning and drawing. He attended Gordon's College, Aberdeen, and secured from the Science and Art Department, South Kensington, London, first-class diplomas in Machine Construction and Drawing, Applied Mechanics, Steam and the Steam Engine, Metal Working Tools, (City and Guild, of London, Eng.) &c. Nearly ten years ago he came to Canada. He was for one year in the employ of R. Gardner & Sons, engineers, Montreal, and for three years with M. D. Barr & Co. and the Edison General Electric Co. He then went into business as partner and chief engineer of the Keegans, Mutual Electrical Engineers and Contractors, Montreal. After about three years he withdrew and came to Toronto, where he has been for over three years general superintendent of the Incandescent Light Co., a place he fills with great satisfaction.

---

The death is announced, in Toronto, of Mr. A. P. Kilgannan, of Little Current, Manitoulin Island, government engineer for the district from Collingwood to Sault Ste. Marie. After the government refused to lay a cable to connect Manitoulin Island with the mainland he carried out the work himself, giving the island the only means of communication with the outside world it possesses in the winter. He also established telephone communication all over the island and secured a charter and cleared 70 miles for an electric railway.

## NEW ELECTRICAL AND STEAM BOILER PATENTS.

As might be expected, the rapid growth in electrical science and the number of purposes to which electricity is applied in everyday life has called forth the inventive powers of our mechanical geniuses, and the patent office gives evidence of their work. The last issue of the Patent Record bulks up considerably more than usual, largely owing to the number of patents issued for electrical inventions and matters bearing thereon. Some of these no doubt possess merit, but the probability is that most of them will never come into practical use. We append a list of these patents, so that our readers may see what is being done in the way of new electrical patents :—

Fare Box—John Maitland Smith, Toronto.

Cypher Combination for Telegraphing—Clement W. Bowman, Ingersoll, and Granville S. Decatur, Hamilton.

Electrical Exchange—The Strowger Automatic Telephone Exchange, assignee of Alex. E. Keith, Frank A. Lundquist, John Erickson and Chas. J. Erickson, Chicago.

Electrical Convertors or Transformers and Enclosing Boxes therefor—Jas. W. Packard, Warren, Ohio.

Conductor's Fare Box—R. R. Mitchell, Montreal.

Electrical Connection—Jas. M. Faulkner, Philadelphia, Penn.

Plate for Secondary Voltaic Batteries—W. A. B. Buckland, 12 Pakenham St., Gray's Inn Road, Middlesex, England.

MR. JAMES MILNE,
Lecturer on Electricity, Toronto Technical School.

Current Interrupter—The Canadian General Electric Co., Toronto, assignee of Elihu Thomson, Swampscott, Mass.

Secondary Voltaic Battery—Wm. A. Baker, Buckland, Gray's Inn Road, Middlesex, England.

Car Fender—Robt. Bustin, Robt. K. Jones, of St. John, N.B.; Wesley Vanwark and John R. McConnell, of Fredericton, N.B.

Insulator—L. H. DesIles and F. S. Palmer, of Boston, Mass.

Electric Current Transmitter—The Thomson-Houston Electric Co., Portland, Maine, assignees of Chas. A. Coffin, Boston, Mass.

Contact Apparatus—The Canadian General Electric Co., Toronto, assignee of Elihu Thomson, Swampscott, Mass.

Electric Circuit Indicator—The Canadian General Electric Co., Toronto, assignee of Elihu Thomson, Swampscott, Mass.

Electric Dental Engine — Wm. E. Wheeler, Geo. W. Johnston and Jas. F. Johnston, of Dayton, Tenn.

Electric Accumulator—Arthur Duffek and Bohumil Holub, Prague, Bohemia.

Car Fender—John F. Ryan, Toronto.

Electric Transmitting Thermometer—Francis N. Denison, Toronto.

Car Fender Attachment—B. E. Charlton, Hamilton.

Electric Head-Light—E. A. Edwards, Cincinnati, and Chas. W. Adams, Chicago.

Electric Meter—G. A. J. Telge, Oldenburg, Germany.

Arc Lamp—M. S. Okun, New York.

A number of patents have also been granted in connection with steam boilers. We add a list of these :—

Water Tube Steam Boiler—J. W. Van Dyke, Lima, Ohio.

Steam Trap and Feeder—D. L. Long, Crawfordsville, Indiana.

Feed Water Heater—Walter H. Laurie, Montreal.

Apparatus for Consuming Smoke and Combustible Gases—Louis Hallbauer, Meriden, Conn.

Flue Scraper—Geo. R. Ford, Chicago.

Boiler Cleaner—Geo. R. Ford, Chicago.

Smoke Arrester—Wm. P. Shank, Cairo, Ill.

Steam Generator—S. E. Light, Lebanon, Penn.

Water Heater—Chas. T. Toulmin, New York.

Water Boiler—Ernest Peterson, Blackfriars Road, England.

---

Mr. R. A. Bush, chief engineer at the Brockville Asylum, accidentally shot himself in the thigh a few days ago. He will recover.

PUBLISHED ON THE FIFTH OF EVERY MONTH BY

**CHAS. H. MORTIMER,**

OFFICE : CONFEDERATION LIFE BUILDING,
*Corner Yonge and Richmond Streets,*

**TORONTO,  -  -  CANADA.**

Telephone 2362.

NEW YORK LIFE INSURANCE BUILDING, MONTREAL.
Bell Telephone 2299.

*ADVERTISEMENTS.*

Advertising rates sent promptly on application. Orders for advertising should reach the office of publication not later than the 25th day of the month immediately preceding date of issue. Changes in advertisements will be made whenever desired, without cost to the advertiser, but to insure proper compliance with the instructions of the advertiser, requests for change should reach the office as early as the send day of the month.

*SUBSCRIPTIONS.*

The ELECTRICAL NEWS will be mailed to subscribers in the Dominion, or the United States, post free, for $1.00 per annum, 50 cents for six months. The price of subscription should be remitted by currency, in registered letter, or by postal order payable to C. H. Mortimer. Please do not send cheques on local banks unless 25 cents is added for cost of discount. Money sent in unregistered letters will be at senders' risk. Subscriptions from foreign countries embraced in the General Postal Union, $1.50 per annum. Subscriptions are payable in advance. The paper will be discontinued at expiration of term paid for if so stipulated by the subscriber, but where no such understanding exists, will be continued until instructions to discontinue are received and all arrearages paid.

Subscribers may have the mailing address changed as often as desired. *When ordering change, always give the old as well as the new address.*

The Publisher should be notified of the failure of subscribers to receive their papers promptly and regularly.

*EDITOR'S ANNOUNCEMENTS.*

Correspondence is invited upon all topics legitimately coming within the scope of this journal.

THE "CANADIAN ELECTRICAL NEWS" HAS BEEN APPOINTED THE OFFICIAL PAPER OF THE CANADIAN ELECTRICAL ASSOCIATION.

## CANADIAN ELECTRICAL ASSOCIATION.

**OFFICERS:**

PRESIDENT:
A. B. SMITH, Superintendent G. N. W. Telegraph Co., Toronto.

1ST VICE-PRESIDENT:
C. BERKELEY POWELL, Director Ottawa Electric Light Co., Ottawa.

2ND VICE-PRESIDENT:
L. B. McFARLANE, Manager Eastern Department, Bell Telephone Company, Montreal.

SECRETARY-TREASURER:
C. H. MORTIMER, Publisher ELECTRICAL NEWS, Toronto.

EXECUTIVE COMMITTEE:
GEO. BLACK, G. N. W. Telegraph Co., Hamilton.
J. A. KAMMERER, General Agent, Royal Electric Co., Toronto.
E. C. BREITHAUPT, Berlin, Ont.
F. H. BADGER, JR., Superintendent Montmorency Electric Light & Power Co., Quebec.
JOHN CARROLL, Sec.-Treas. Eugene F. Phillips Electrical Works, Montreal.
K. J. DUNSTAN, Local Manager Bell Telephone Company, Toronto.
O. HIGMAN, Inland Revenue Department, Ottawa.
W. Y. SOPER, Vice-President Ottawa Electric Railway Co., Ottawa.
A. M. WICKENS, Electrician Parliament Buildings, Toronto.
J. J. WRIGHT, Manager Toronto Electric Light Company.

## MONTREAL ELECTRIC CLUB.

OFFICERS:
President, W. B. SHAW,    Montreal Electric Co.
Vice-President, H. O. EDWARDS,
Sec'y-Treas., CECIL DOUTRE,    81A St. Famille St.
Committee of Management, T. F. PICKETT, W. GRAHAM, J. A. DUGLASS.

## CANADIAN ASSOCIATION OF STATIONARY ENGINEERS.

EXECUTIVE BOARD:

President, W. G. BLACKGROVE,    Toronto, Ont.
Vice-President, JAMES DEVLIN,    Kingston, Ont.
Secretary, E. J. PHILIP,    Toronto, Ont.
Treasurer, DUNCAN ROBERTSON,    Hamilton, Ont.
Conductor, W. F. CHAPMAN,    Brockville, Ont.
Door Keeper, F. G. JOHNSTON,    Ottawa, Ont.

TORONTO BRANCH No. 1.—Meets 2nd and 4th Friday each month in Room D, Shaftesbury Hall. W. Lewis, President ; S. Thompson, Vice-President ; T. Eversfield, Recording Secretary, University Crescent.

MONTREAL BRANCH No. 1.—Meets 1st and 3rd Thursday each month, in Engineers' Hall, Craig street. President, John J. York. Board of Trade Building ; first vice-president, J. Murphy; second vice-president, W. Ware ; secretary, B. A. York ; treasurer, Thos. Ryan.

ST. LAURENT BRANCH No. 2.—Meets every Monday evening at 43 Bonsecours street, Montreal. R. Drouin, President ; Alfred Latour, Secretary, 306 Delisle street, St. Cunegonde.

BRANDON, MAN., BRANCH No. 1.—Meets 1st and 3rd Friday each month, in City Hall. A. R. Crawford, President ; Arthur Fleming, Secretary.

HAMILTON BRANCH No. 2.—Meets 1st and 3rd Friday each month, in Maccabee's Hall. E. C. Johnson, President ; W. R. Cornish. Vice-Pres.; Wm. Norris, Corresponding Secretary, 211 Wellington Street North.

STRATFORD BRANCH No. 3.—John Hoy, President ; Samuel H. Weir, Secretary.

BRANTFORD BRANCH No. 4.—Meets 2nd and 4th Friday each month. F. Lane. President ; T. Pilgrim, Vice-President ; Joseph Ogle, Secretary, Brantford Cordage Co.

LONDON BRANCH No. 5.—Meets in Sherwood Hall first Thursday and last Friday in each month. F. Mitchell, President ; William Meaden, Secretary Treasurer, 533 Richmond Street.

GUELPH BRANCH No. 6.—Meets 1st and 3rd Wednesday each month at 7:30. p.m. J. Fordyce, President ; J. Tuck. Vice-President ; H. T. Flewelling, Rec.-Secretary ; J. Gerry, Fin.-Secretary ; Treasurer, C. J. Jorden.

OTTAWA BRANCH, No. 7.—Meets 2nd and 4th Tuesday, each month, corner Bank and Sparks streets ; Frank Robert, President ; F. Merrill, Secretary, 352 Wellington Street.

DRESDEN BRANCH No. 8.—Meets every and week in each month ; Thos. Merrill, Secretary.

BERLIN BRANCH No. 9.—Meets 2nd and 4th Saturday each month at 8 p. m. W. J. Rhodes, President ; G. Steinmetz, Secretary, Berlin Ont.

KINGSTON BRANCH No. 10.—Meets 1st and 3rd Tuesday in each month in Fraser Hall, King Street, at 8 p. m. President, S. Donnelly ; Vice-President, Henry Hopkins ; Secretary, J. W. Tandvin.

WINNIPEG BRANCH No. 11.—President, G. M. Hazlett ; Recording Secretary, J. Sutherland ; Financial Secretary, A. B. Jones.

KINCARDINE BRANCH No. 12.—Meets every Tuesday at 8 o'clock, in McGibbin's Block. President, Daniel Bennett ; Vice-President, Joseph Lightball ; Secretary, A. Scott.

WIARTON BRANCH No. 13.—President, Wm. Craddock ; Rec. Secretary, Ed. Dunham.

PETERBOROUGH BRANCH No. 14.—Meets 2nd and 4th Wednesday in each month. S. Potter, President ; C. Robison, Vice-President; W. Sharp, engineer steam laundry, Charlotte Street, Secretary.

BROCKVILLE BRANCH No. 15.—President, W. F. Chapman ; Vice-President, A. Franklin ; Recording Secretary, Wm. Robinson.

CARLETON PLACE BRANCH No. 16.—President, Jos. McKay Vice President, Henry Derrer ; Fin. Secretary, A. M. Schofield.

## ONTARIO ASSOCIATION OF STATIONARY ENGINEERS.

BOARD OF EXAMINERS.

President, A. AMES,    Brantford, Ont.
Vice-President, F. G. MITCHELL    -    London, Ont.
Registrar, A. E. EDKINS    -    139 Borden st., Toronto.
Treasurer, R. MACKIE.    -    28 Napier st., Hamilton.
Solicitor, J. A. MCANDREWS,    -    Toronto.

TORONTO—A. E. Edkins, A. M. Wickens, E. J. Phillips, F. Donaldson.
HAMILTON—P. Stott, R. Mackie, T. Elliott.
BRANTFORD—A. Ames, care Patterson & Sons.
OTTAWA—Thomas Wesley.
KINGSTON—J. Devlin (Chief Engineer Penitentiary), J. Campbell.
LONDON—F. Mitchell.
NIAGARA FALLS—W. Phillips.

Information regarding examinations will be furnished on application to any member of the Board.

---

To the writer, who had occasion recently to visit the Toronto Custom House, it seemed incongruous that side by side with iron doors and shutters, and other safeguards against fire, should be seen in active operation that relic of a past era, the coal oil lamp.

THE sign "To Let" serves to mark the spot in Chicago which two or three years ago was the headquarters of Dr. Wellington Adams, whose proposal to construct an electric railway from Chicago to St. Louis, and operate the same at a speed of 100 miles an hour, was the subject of mild criticism in the ELECTRICAL NEWS at the time. The crowds who were to be carried over this road to the World's Fair were compelled to choose another route and put up with a slower pace.

THE most important development in the electrical field of late, is the recent successful test of the electric locomotive built by the Westinghouse Company, for use in hauling freight trains through the Baltimore tunnel of the Baltimore & Ohio railway. The electric locomotive hauled forty-four loaded freight cars, and three steam locomotives up the heavy grade of the tunnel at a speed of 12 miles an hour, the total weight of the train being 1,900 tons. The recent working arrangement made between the Westinghouse Company, of Pittsburgh, and the Baldwin Locomotive Works, of Philadelphia, is considered good evidence that these companies have satisfied themselves of the adaptability of electricity for railway purposes. It is safe to conclude that the working partnership which has been formed between these companies is the result of definite information on this point.

RECENT experiments seem to indicate that the growth of plants may be greatly stimulated by electricity. What wonderful possibilities are here opened up! It may become possible to cause the barren places to blossom like the rose, and the arctic regions, and other parts of the earth now uninhabitable, may be reclaimed and made fit for the habitation of man.

THE Montreal city council has a rule that parties holding contracts with the city must employ local labor in all their work. In terms of this rule the Street Railway Co. is bound to purchase all its cars from Montreal manufacturers or build them itself. This it has not been doing, and the council now seeks to enforce compliance with the agreement. Was it quite wise to embody such a condition in the contract?

THE St. Clair Tunnel Co. is said to be considering the advisability of using electric locomotives for hauling trains through its tunnel at Sarnia. The heat and escaping steam of the engines at present in use are having an injurious effect upon the tunnel, especially on the asphalt with which the iron is coated. The use of electric locomotives has been successfully tried in a tunnel at Baltimore, and also in mines and underground work at other places, and there appears to be no good reason why they should not be successful in the St. Clair tunnel.

PROF. Geo. Forbes, C.E., has an interesting article in Blackwood's Magazine for September, entitled, "Harnessing Niagara," in which he gives an account of the installation of the great power plant of the Niagara Falls Power Co. Prof. Forbes seems to have had to fight against the preconceived opinions—perhaps the prejudices—of some of those with whom he was associated in the carrying out of that great work, but what he accomplished is there to speak for itself, and it certainly marks him as a great electrical engineer.. Had he left out a good part of his article, in which he criticises Americans and American ways, it would have been better. He has certainly brought down upon himself a good deal of hostile comment, and even personal vituperation on the part of a section of the press. Egotism is out of place on the part of a great man, and Prof. Forbes is guilty of it. Nevertheless "Harnessing Niagara" will be read with great interest by those who watch the progress of electrical science.

Mr. George Westinghouse, jr., has recently made the announcement that means have been discovered of producing through the medium of a gas engine of new design, one horse power of energy with a consumption of half a pound of coal. It is a well known fact that with the most economical engines, the amount of coal at present required to produce one horse power ranges from two and one half to six pounds. It can thus be seen what a revolution would follow the introduction of such a discovery as Mr. Westinghouse claims to have made. It would mean the substitution of electricity for steam on railroads, and would solve the question of long distance transmission of electricity for power purposes. It would also mean that electric light could be produced at a sufficiently lower rate than under present conditions to enable it to become a still more formidable competitor of gas and other forms of illuminant. It would be likely to bring electricity into use for heating and other purposes, to which, owing to its cost, it cannot now be applied. The recent application to the Canadian Parliament for a charter for the construction of an electric railway from Montreal to Windsor would seem to show that the developments above referred to are being carefully watched on this side of the line, and that far sighted business men are getting ready to take advantage of the new conditions, which in all probability will be witnessed in the near future.

THE best kind of material for insulators has been and still is a matter of great importance to telegraph companies, more especially perhaps in Canada, where at times the climatic conditions are exceedingly trying. Some years ago the Great North Western Telegraph Company experienced much difficulty with their coast lines in New Brunswick and Nova Scotia. During stormy weather, communication on these lines could not be maintained, and in consequence many complaints were made by vessel owners and others whose interests were affected. After repeated attempts to discover the cause of these interruptions, the difficulty was found to lie with the insulators. A new kind of rubber insulator had been employed in the belief that it would better withstand the climate. It was found, however, that the glazed covering had succumbed to the action of the weather, and that the sulphur employed in the composition had worked out, leaving the insulator porous, and giving opportunity for the salt in the atmosphere to crystallize upon them, and destroy their value. The insulators now used on these lines are made of porcelain, and have proved to be the best adapted to the conditions. Glass insulators are found to answer well throughout Ontario, the only difficulty experienced being that so many of them get broken, especially during the hunting season, when it is a favorite pastime of sportsmen to make a target of them. It is estimated that the renewal of insulators costs the Great North Western Telegraph Company between five and eight thousand dollars every year.

COAL, or fuel generally, is the raw material out of which electric light is manufactured—the less used the greater the profit—and the quantity used depends on the way it is burned, and on its quality. Everyone has a feed water heater, but very few are aware of the saving effected by raising the temperature of the feed-water 1° Fahrenheit, and still fewer take any means to effect this rise in temperature. In an 100 h.p. plant, running on the average three hours full load every night during the year, the saving effected by raising the temperature of feed-water 1° Fahr. amounts to about 400 lbs. of coal. In winter, when the water is down at freezing point, it is sufficiently obvious that a little intelligent care will save a good round sum. It is doubtful whether, in small stations, the operators know the temperature of their feed, or ever take the trouble to raise it. In places where a condenser is used, it is generally accepted that the higher the vacuum obtained the better will be the general results in point of economy, but this is by no means necessarily so. It is usual to feed the condensed steam back into the boiler, and in that case the higher the temperature of the "water of condensation" the better; but on the other hand, this advantage is obtained at the expense of the vacuum, which itself tends to diminish fuel consumption. So we have the alternative to use plenty of "condensing water" and get a high vacuum and a low temperature of feed water, or to give up some inches of vacuum and gain in the temperature of the hot well. The actually most economical combination must be arrived at in each case by experiments, for, although it is quite possible to give a general expression, showing the relation between feed-water temperature and inches of vacuum, still actual experiment will be more convincing and more intelligible to the average engineer. What is here sought is to suggest to station owners that 27 inches vacuum is not necessarily the highest economy, and that a sacrifice of vacuum might be more than compensated for by the increase in feed water temperature.

As poles are an important item in the equipment of electric light, telegraph and telephone companies, some particulars as to the source of supply in Canada, the lasting qualities of various kinds of woods, and the most suitable method of setting, may be of interest. To begin with, experience has shown that cedar is by far the most lasting material for this purpose. A good cedar pole will endure, under ordinary conditions, for about 25 years, while other kinds of wood, such as spruce, decay in less than half that period. The principal source from which cedar poles are obtained for use in the western part of Ontario, is the county of Victoria, and the neighborhood of Lindsay. In the Ottawa region, there is very little cedar to be had, and the poles for use in this locality are obtained from the neighborhood of Pembroke. In Quebec and the Lower Provinces, cedar is more plentiful, but in some parts where it is not easily obtainable, spruce, which grows in abundance, is substituted. The life of poles is considerably greater in cities, where the pavement protects the wood from the action of the water, than in the country districts, where they are exposed to such action. It has been found that poles deteriorate more quickly in sandy soil than in soil of a heavier kind, such as clay. The life of white cedar poles in the clay soil of Detroit is said to be about 13 years, without protection or treatment of the butts. Norway pine poles have lasted in the same soil about 30 years, but were badly rotted at the surface line. They, however, rot very soon in a sandy soil. The more pitchy ones decay first. Winter cut poles, and those cut in summer with the sap-wood removed, are more lasting than those containing green sap-wood. Experiments with ordinary 4-inch gas pipe, used as trolley poles, and set in concrete up to the surface of the ground, without other coating, developed that they would become corroded through at about 10 inches below the surface in about three years, forming a black deposit. As an illustration of the powerful action of frost, it may be mentioned that in Toronto recently, poles were found to have been uplifted to the extent of eight or ten inches by the action of frost. The experiment is being tried of making the lower end of the poles wedge-shaped instead of square, with the object of lessening the effect of the frost's action. It is hoped that by lessening the amount of resisting surface at the bottom of the pole, the tendency to upheave will be lessened in a corresponding degree.

IT is gratifying to observe that owners and operators of electric light and power houses are beginning to take some little interest in their business, other than the purely commercial one of balancing expenditure against receipts. They are slowly awakening to the fact that electrical and steam machinery, in combination, is all the better for being studied, and that the various wastes and leaks that take place during the process of converting heat into light can be minimized, and may represent a very tidy return on the investment. New installations are being made and new machinery is being purchased, more with a view to cheapness in first cost than with the recognition of the fact that cheap machinery and construction most likely mean expensive operation. Experience gained by costly experiment has proved to many owners that a dynamo is not always a dynamo, no matter how it has been designed and constructed, and that scientific knowledge and practical experience applied in the design of electrical apparatus, will produce a better result than rule of thumb or guess work. The same applies in the operation of an electric power house. The operator who takes the trouble to study his work, to look into his own business, and, in fact, to take an intelligent interest in the undertaking committed to his charge, will always be able to show better results than one who gets light without knowing how, or at what cost, and is satisfied to do so. This latter state of affairs results largely from the want of knowledge, in those using electricity, of the imperfections of their apparatus ; of the many directions in which these imperfections result in wastes, and of the manner in which, and the means whereby, these wastes may be at least minimized, if not entirely eliminated. It may safely be assumed that it is the prospect of making money that induces capital to invest in electric lighting enterprises. If this be so, then why not be logical about it, and try and make as much as is possible ? Why not be business-like and keep track of expenditures and profits ? Why not be sensible and study the business ? In electricity, as in every other profession, there are quacks, and there are reputable practitioners ; there is good, bad and indifferent machinery and apparatus ; and between the indifferent apparatus operated by the quack electrician, and the good machinery intelligently handled by the earnest, thoughtful engineer, there is just the difference between profit and loss. From the coal or wood pile to the lamp or motor, there are sources of waste in every direction, and the "electrician" who thinks that managing an electric plant consists in burning fuel, and smoking his pipe until it is time to collect the monthly accounts, is a very expensive man to have around the place.

ONE can to a certain extent sympathize with the owners of small power houses who hesitate before spending a considerable sum of money in order to obtain some prospective advantage, but when saving can be effected merely by a little care and intelligence applied to the operation of an already installed plant, then such an one is merely advertising his own indolence and want of energy, when he says there is no money in electric lighting. A great majority of installations seem to have been made without much regard to efficiency ; the machinery bought anywhere, and wiring done anyhow, and the operation managed by anyone ; and the consequence is a heterogeneous collection of old type apparatus, which of itself is a necessary source of waste. Prices for electric lighting have had to be reduced in order to make business, and it is generally very difficult to show a favorable result, when operating expenses are kept up by inefficiency, and receipts are lessened by competition. A careful examination into the causes of failure of electric power houses to pay reasonably, will probably lead to two conclusions —that, in the first place, the arrangement of the plant, the wiring system, the kind of machine, the kind, size and distribution of transformers, the relative sizes of engine and dynamo, and so on, are widely different from what experience shows to be the best ; and second, that poor as it may be, inexperienced, indolent management fails to operate it to the best advantage. An electric light man quite recently expressed surprise on hearing a transformer spoken of as "chewing up the coal pile," and wondered that it could possibly do so. Now transformers are by no means perfect apparatus ; the very best made has a percentage efficiency well below 100, and cannot possibly come up to 100 ; and between the best and the medium make there is a very appreciable difference. Let electric light men once clearly grasp the fact that transformers *do* waste energy, and they will have gained valuable knowledge. The best claim for transformers in large sizes is about 97 per cent. efficiency at full load. This means that one of these transformers has to be given energy enough for about 103 lamps on the primaries, in order that it may give current for 100 lamps on the secondaries. In a 1000 light plant, therefore, the energy for 30 lamps is "chewed up" by the transformer all the time it is running at full load, and a greater proportion at less loads, and this is the best that has yet been done. This 30 lamp energy goes in heating the transformer iron and copper, in overcoming its resistance, in overcoming the inductive back E.M.F., and in supplying the energy lost through hysteresis. The best transformer costs money—and the chances are that cheap ones have been bought through ignorance of these wastes—and it is evident that the cheap one may really be a very expensive one, because, although any copper and iron will make some kind of transformer, the cheap kinds waste far more in proportion than the expensive ones. Electric light owners would do well to see whether their transformer and wiring system is not responsible for the annual deficit, and whether a little money judiciously spent in replacing old transformers by modern ones would not be the truest economy.

## MOONLIGHT SCHEDULE FOR NOVEMBER.

| Day of Month. | Light. | Extinguish. | No. of Hours. |
|---|---|---|---|
|  | H.M. | H.M. | H.M. |
| 1 | A.M. 3.30 | A.M. 5.30 | 2.00 |
| 2 | No light. | No light. | .... |
| 3 | No light. | No light. | .... |
| 4 | No light. | No light. | .... |
| 5 | P.M. 5.10 | P.M. 7.40 | 2.30 |
| 6 | " 5.10 | " 8.40 | 3.30 |
| 7 | " 5.10 | " 9.50 | 4.40 |
| 8 | " 5.10 | " 11.10 | 6.00 |
| 9 | " 5.10 | A.M. 12.20 | 7.10 |
| 10 | " 5.10 | " 1.00 | 7.50 |
| 11 | " 5.10 | " 1.30 | 8.20 |
| 12 | " 5.00 | " 2.50 | 9.50 |
| 13 | " 5.00 | " 4.00 | 11.20 |
| 14 | " 5.00 | " 5.20 | 12.20 |
| 15 | " 5.00 | " 6.00 | 13.00 |
| 16 | " 5.00 | " 6.00 | 13.00 |
| 17 | " 5.00 | " 6.00 | 13.00 |
| 18 | " 5.00 | " 6.00 | 13.00 |
| 19 | " 5.00 | " 6.00 | 13.00 |
| 20 | " 5.00 | " 6.00 | 13.00 |
| 21 | " 7.30 | " 6.10 | 10.40 |
| 22 | " 9.00 | " 6.10 | 9.10 |
| 23 | " 10.20 | " 6.10 | 7.50 |
| 24 | " 11.00 | " 6.10 | 7.10 |
| 25 | " 11.20 | " 6.10 | 6.50 |
| 26 | .......... | " 6.10 | } 5.50 |
| 27 | A.M. 12.20 | .......... | |
| 28 | " 1.20 | " 6.10 | 4.50 |
| 29 | " 2.20 | " 6.10 | 3.50 |
| 30 | " 3.20 | " 6.10 | 2.50 |

Total, 212.10

## THE LEVER SAFETY-VALVE.

GENERAL REMARKS.—We have received so many requests for a rule for calculating the position of the weight on a safety-valve, and the blowing-off pressure when the position of the weight is given, that we have thought it wise to publish such a rule in The Locomotive. It would be easy to give a simple formula for the purpose, but we have considered that the wants of engineers would be best met by explaining the theory of the lever-valve, and showing, as clearly as possible, the reason for each step in the calculation.

OBJECT OF THE SAFETY-VALVE.—The object of the safety-valve, as every one knows, is to prevent the pressure in the

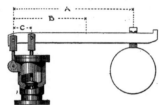

FIG. 1—DIAGRAM OF A LEVER SAFETY-VALVE.

boiler from rising to a dangerous point, by providing an outlet through which steam can escape when the pressure reaches a certain limit, which is determined by the strength of the boiler, and by the conditions under which it is to work. The simplest device for attaining this end is the "dead-weight" valve, the principle of which is illustrated in Fig. 2. It consists simply of a plate of iron, laid upon a nozzle, and held down by a weight. The calculation of the blowing-off point of such a valve is very simple. In the valve here shown, for example, the steam acts

FIG. 2—A "DEAD-WEIGHT" VALVE.

against a circle two inches in diameter. The area of a two-inch circle is $2 \times 2 \times .7854 = 3.14$ sq in., and the weight tending to hold the cover plate down being 314 lbs., it is evident that the valve will not blow off until the steam pressure reaches 100 lbs. per square inch. Dead-weight valves are used somewhat in England, but they are seldom met with in this country, the commoner form here being that suggested in Fig. 1. It may be well to say that Fig. 1 does not purport to be a good form of valve. We should certainly object to it, if it were placed upon a boiler offered to us for insurance, because no guides are provided for the lever or for the valve stem. These features were intentionally omitted in the engraving, in order that their presence might not draw the attention away from the main points under consideration—the calculation, namely, of the blow-off pressure and of the position of the weight.

THEORY OF THE LEVER.—In order to be able to perform safety-valve calculations intelligently, one must have a clear idea of the principle of the lever; and it is hoped that such an idea may be had from a study of the illustrations that are presented herewith. These represent a lath or other light piece of wood, which is balanced upon a knife edge, and into which, on the under side, a number of small staples are driven

FIG. 3.

FIG. 4.

at equal distances. A number of balls of lead are also supposed to be provided, all exactly alike, and all being furnished with a hook at the top and a staple at the bottom. Two of these weights, when hung upon

the first staple, as shown in Fig. 3, will just balance one weight hung upon the second staple, on the other side of the fulcrum. In the same way, four of them, when hung upon the first staple, as shown in Fig. 4, will just balance one hung upon the fourth staple. Five upon the second staple, as shown in Fig. 5, will just balance two upon the fifth staple; and three upon the fifth staple will just balance five upon the third staple, as shown in Fig. 6. It will be seen

FIG. 5.

that in every one of these cases the lath is balanced, provided the weight upon one side, when multiplied by its distance from the fulcrum, is equal to the weight upon the other side, multiplied by its distance from the fulcrum. This is the principle of

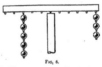

FIG. 6.

Archimedes, and it is used in all calculations relating to the lever. (The reader may find it a profitable exercise to show that the systems shown in Figs. 7 and 8 are balanced. A suggestion is afforded him in Fig. 7, while in Fig. 8 he is left entirely to his own resources. He should find no difficulty in either case, however, if he has grasped the fundamental idea which is contained in the illustrations given above).

APPLICATION TO THE SAFETY-VALVE.—We are now prepared to apply the principle of the lever to the safety-valve, although there is still one point to be cleared up before we can give a complete rule. (The point to which we refer is the influence of the weight of the arm which carries the ball; but for the present moment we shall consider this arm to be devoid of weight, and we shall introduce a correction for it later on.)

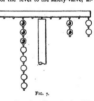

FIG. 7.

FIG. 9 is a crude representation of a safety-valve, in which the total steam pressure against the disk of the valve is supposed to be 40 lbs., and the ball is supposed to weigh 10 pounds. If the valve stem is 6" from the fulcrum, the ball will have to be 24" from the fulcrum in order for the valve to blow off at the given pressure—that is, at 40 lbs. This is easily seen, since $6 \times 40$ equals $10 \times 24$; but if the reader has any doubt about the applicability of Archimedes' rule in this case, he may note that the

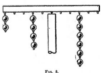

FIG. 8.

upward pressure due to the steam can be conceived to be replaced by a 40 lb. weight hung 6" to the left of the fulcrum, as indicated by the dotted circle. The lever will then be equivalent to the one shown in Fig. 10, which is similar in all respects to those shown in Figs. 3 to 8, and to which Archimedes' rule plainly applies. If the blowing-off pressure were not given in Fig. 9, and we were required to

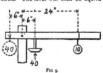

FIG 9.

find it from the other data there shown, we should reason as follows: When the valve is on the point of blowing off, the upward thrust of the valve-stem is just balanced by the downward tendency of the ball; and, therefore, from Archimedes' principle, 10 × 24 must equal 6 times the thrust of the valve-stem. But 10 × 24 equals 240, and hence 240 is 6 times the thrust of the valve-stem, and 240 ÷ 6 (= 40 lbs.) must be the total

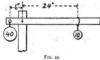

FIG. 10.

pressure exerted on the disk of the valve when it is about to blow off. If the pressure per square inch were desired, we should have to divide 40, the total pressure on the valve disk, by the area of the disk in square inches.

THE ARM OF THE VALVE.—In order to take the weight of the valve-arm into account, we shall first make a short digression for illustrating the meaning of the expression "center of gravity." Consider, first, the system shown in Fig. 11, where there is one ball on the first staple and one on the fifth. The one ball on the fifth staple is equivalent to five balls on the first one; so that the two balls on the right hand side of the fulcrum are equivalent

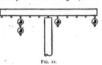

FIG. 11.

to six balls suspended from the first staple. They are therefore balanced by the two balls on the third staple; and, in general, if two balls be hung from any of the staples, they would be exactly balanced by a pair of balls whose distance from the fulcrum was the average of the distances from the first two. Fig. 12 is a further illustration of this fact. Now, referring to Fig. 13, let us conceive the valve-arm to be without weight, except two small and equal pieces of it, whose distances from the

FIG. 12.

fulcrum are respectively 10″ and 20″. By analogy with the two preceding illustrations, we see that these two little masses would be just balanced by a similar pair of masses, spaced at equal distances; they would be just balanced by four similar masses, hung at a distance from the fulcrum equal to half the length of the arm. While this kind of reasoning is applicable, strictly speaking, only to the case in which the valve-arm is of equal thickness and width throughout, and has no irregularities

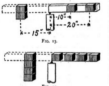

FIG. 13.

FIG. 14.

whatever, we may, in practice, apply it to all valve-arms which are approximately uniform in cross-section; and by extending the conception of Figs. 13 and 15 until the little masses become so numerous as to fill the entire lever, we conclude that a valve-arm of this sort would be balanced by a similar arm suspended (as shown in Fig. 15) at a distance from the fulcrum equal to half the length of the arm itself. This amounts to saying that a uniform valve-arm acts the same as it would if its weight were all concentrated at the middle point of the arm. The point in a body

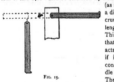

FIG. 15.

which possesses this property is called the center of gravity of the body. As we have said, the center of gravity of a straight lever may, in practice, be considered to be half way out towards the end of the lever; but if the lever has an appreciable taper, the center of gravity will be nearer the fulcrum. The position of the center of gravity can be found, in such cases, by calcula-

tion; but it is simpler to take the lever out, and balance it across a three-cornered file, as shown in Figs. 16 and 17. It will balance when the center of gravity is just over the edge of the file, and the distance B can then be measured directly.

CALCULATION OF THE BLOWING-OFF PRESSURE.—We are now prepared to give a complete example of the calculation of blowing-point of a safety-valve. Let us take the valve shown in Fig. 18. The arm is 32 in. long, and weighs three pounds; the ball weighs 20 pounds and is set 28 inches from the fulcrum; the valve-stem is 4″ from the fulcrum; the valve-disk is 2″ in diameter, and the disk and stem,

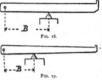

FIG. 16.

FIG. 17.

together, weigh 1½ pounds. It is required to find the blowing-off pressure. In the first case, let us consider the ball. It is possible to load the valve-disk directly (just as in the case of Fig. 2) with a weight which shall have precisely the same effect, in preventing the escape of steam, that the actual 20-pound ball has; and our first undertaking will be to find out how big this imaginary "dead weight" would have to be. When we say that it is to be "equivalent" to the 20-pound ball on the lever, we mean that it would just balance that ball, if it were on the left side of the fulcrum, instead of on the right; and hence, by Archimedes' principle, 28″ × 20 lbs. must equal 4″ multiplied by the imaginary "dead weight." Now 28 × 20 = 560, and 560 ÷ 4 = 140. In other words, the 20-pound weight, at a distance of 28″ from the fulcrum, has just the same effect as a 140-pound weight would have, if placed directly upon the valve-disk. In the same way we may investigate the effect of the valve-arm. It weighs 3 pounds, and its center of gravity is 16″ from the fulcrum. A three-pound

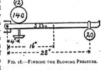

FIG. 18.—FINDING THE BLOWING PRESSURE.

weight, 16 inches from the fulcrum, is the same thing as a 12-pound weight, 4 inches from the fulcrum; because 3 × 16 = 48, and 12 × 4 = 48. Hence the valve-arm is equivalent to a 12-pound weight placed directly upon the valve-disk. The whole lever valve may therefore be regarded as equivalent to a "dead weight" valve loaded with 153½ pounds; for the ball is equivalent to a dead load of 140 pounds, the arm is equivalent to a dead load of 12 pounds, and the valve-disk and stem, taken together, weigh 1½ pounds; and 140 + 12 + 1½ = 153½. We have therefore found out that the valve will begin to blow when the total pressure of the steam against the valve-disk is 153.5 pounds. The part of the disk which is exposed to the stem is 2″ in diameter, and its area is therefore 2 × 2 × .7854 = 3.1416 square inches. The total steam pressure against this area being 153.5 pounds, the pressure against each square inch of it will be

153.5 ÷ 3.1416 = 48.9 pounds (nearly).

A valve with the dimensions given above will therefore blow off at just a trifle less than 49 pounds per square inch; and the calculation is similar in all cases.

SETTING THE WEIGHT.—The method of setting the weight, when the blowing-off pressure is given, is almost precisely the reverse of the calculation given above. As an example, consider the valve shown in Fig. 19. The

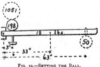

FIG. 19.—SETTING THE BALL.

dimensions are as follows: Diameter of the valve = 4″, length of the lever = 66″, weight of the ball = 50 lbs., weight of the lever = 18 lbs., weight of the valve-disk and stem = 7 lbs., distance of valve-stem from fulcrum = 3″. It is required to set the ball so that the valve shall blow at 100 lbs. per square inch. The calculation is as follows: The area of a 4 inch disk is 4 × 4 × .7854 = 12.56

sq. in., and if the steam pressure is 100 lbs. per square inch, the total upward pressure against the valve-disk is 12.56 × 100 = 1,256 pounds. If the valve were of the "dead weight" kind, a load of 1,256 lbs. on the valve-disk would therefore cause it to blow at 100 lbs. per square inch. We therefore have to set the ball at such a place that the action of the ball, the lever, and the direct weight of the valve-disk and stem, shall be equal to a direct load of 1,256 lbs. Now, the lever weighs 18 lbs., and its "center of gravity" is (say) 33″ from the fulcrum. It is therefore equivalent to a 198-pound weight laid directly on the valve-disk; for by Archimedes' rule we must have

$$33'' × 18 \text{ lbs.} = 3'' × \text{equivalent dead load.}$$

Now 33 × 18 = 594, and 594 ÷ 3 = 198 lbs., as stated above. In Fig. 19 this dead load (which is equivalent to the weight of the lever itself) is represented by the small weight marked "198"; and the large dotted ball above it (whose weight we are about to find) represents the dead load that is equivalent to the 50 lb. ball out on the lever. The dotted weight, together with the 198 lb. weight, and the weight (7 lbs.) of the disk and stem, must be equal to 1,256 lbs., as we have seen. That is, the dotted weight must be 1,051 lbs.; because

$$1,051 + 198 + 7 = 1,256$$

The problem has now resolved itself into placing the 50 lb. ball at such a point that it shall be equivalent to a dead load of 1,051 pounds. The valve-stem being 3″ from the fulcrum, Archimedes' gives us

$$1,051 \text{ lbs.} × 3'' = 50 \text{ lbs.} × \text{distance of ball from fulcrum.}$$

Now 1,051 × 3 = 3,153, and 3,153 ÷ 50 = 63.06 inches. That is the ball must be placed 63 inches from the fulcrum, in order that the valve may blow at 100 lbs per square inch.

RULES.—The processes of calculation which are explained above may now be summarized in the following two rules[*]:

RULE I. To find the blowing pressure when the position of the ball is given. Multiply the weight of the ball by its distance (A) from the fulcrum, and divide by the distance (C) of the valve stem from the fulcrum. (This gives the dead weight that is equivalent to the ball.) Then multiply the weight of the lever by the distance (B) of its center of gravity from the fulcrum, and divide by the distance (C) of the valve stem from the fulcrum. (This gives the dead weight that is equivalent to the lever.) Add together the two "dead weights," so calculated, and add in, also, the weight of the valve-disk and stem. (This gives the total weight that is keeping the valve-disk down.) Then divide the sum thus found by the area of the valve disk, in square inches, and the quotient is the pressure, in pounds per square inch, at which the valve will blow.

RULE II.—To set the ball, so that the valve shall blow at a given pressure. Multiply the area of the valve-disk by the blowing off pressure, expressed in pounds per square inch. (This gives the total effort of the steam to force the valve-disk up.) Subtract, from this total pressure the weight of the valve and stem. The remainder is the "dead weight" to which the lever and ball, taken together, must be equivalent. Then multiply the weight of the lever by the distance (B) of its "center of gravity" from the fulcrum and divide by the distance (C) of the valve stem from the fulcrum. The result is the "dead weight" to which the lever is equivalent; and if this be subtracted from the total dead weight, just mentioned, the remainder will be the "dead weight" to which the ball alone must be equivalent. Multiply this remainder by the distance (C) of the valve stem from the fulcrum, and divide the product by the weight of the ball. The quotient is the distance, A, that the ball must be placed from the fulcrum, in order that the valve may blow off at the desired pressure.

CAUTIONS.—In applying these rules two things must be carefully observed. In the first place, the diameter of the valve-disk must be measured at a b, in Fig. 20, and not at c d; for the steam acts only on the circle whose diameter is a b. Again, if the valve stem has a square top, as indicated in Figs. 21 and 22, m n must be taken as the "distance of the valve stem from the fulcrum"; because the moment of the stem is all applied to the lever at n, as is plainly indicated in Fig. 22.

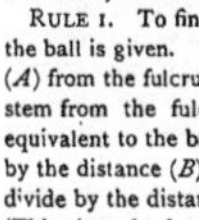

FIG. 20.

[*] The letters refer to Fig. 1.

Although the foregoing article is intended simply to explain the principle underlying the lever safety-valve, it may be well to touch upon one point concerning the construction of such valves. The point we have in mind is this : When the boiler is under steam, it is an easy matter to try the valve, and find out whether it works freely or not. It ought also to be easy to do this, when the boiler is out of use ; and in many cases it is so. Usually when the boiler is not under steam, it is sufficient to raise the weight and the lever, and then to try the valve stem with the thumb and finger ; but some valves are so constructed that the valve-disk is free from the stem, and in such cases that the

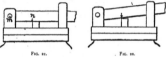

FIG. 21.     FIG. 22.

fact that the stem is free proves nothing whatever, so far as the disk itself is concerned, and the disk must be separately investigated before the valve can be pronounced in good condition. If there is no escape pipe screwed into the valve, the disk can usually be reached from the exhaust side, and its condition noted ; but if such a pipe is provided (as it is, in many cases) the inspector has to examine the disk as well as he can, from the inside of the boiler. If the valve does not happen to be secured directly to the nozzle, an examination from the interior of the boiler is not practicable, and then the waste pipe has to be unscrewed, or the bonnet of the valve taken off, before the disk can be reached. These difficulties, when combined with the fact that there is often no external evidence to show whether the valve is secured to the stem or not, lead us to recommend strongly that valves with separate disks be avoided altogether. They have no very marked advantage over those in which disk and spindle are all in one piece, and as they are likely to deceive one into the belief that all is in good condition, when in reality the disk may be stuck fast, we feel justified in condemning their use altogether.—The Locomotive.

## THE ST. JOHN, N. B., EXHIBITION.

THE recent exhibition at St. John, N. B., reflected credit on the Maritime Provinces, and attracted not only local exhibitors, but manufacturing companies in the Upper Provinces. The attendance was worthy of the interesting character of the exhibition.

The power required for the operation of the various machinery and exhibits was supplied by a Robb Armstrong engine, manufactured by the Robb Engineering Company, of Amherst, N. S., an engine built by the Burrell Johnson Company, of Yarmouth, N. S., and a Leonard Ball engine, manufactured by E. Leonard & Son, London, the latter represented by Mr. John Evans, St. John. Messrs. Waring, White & Co., also had one of their engines in operation.

Among the numerous interesting exhibits at this exhibition was that of Messrs. T. McAvity & Sons, of St. John, who showed a full line of brass goods and engineers' supplies.

Messrs. W. T. & J. W. Myers, of St. John, exhibited a number of electric elevators, motors, and dynamos. The firm are about to engage in the manufacture of electric machinery, and claim to have one of the best electric motors on the market.

## THAT PULLEY ACCIDENT AT AYLMER.

On Saturday last The World published an account of a fatal accident at Aylmer, Ont., whereby Mr. J. D. McDiarmid of that place was instantly killed by the bursting of a poorly-constructed "wood split pulley." The Dodge Wood Split Pulley Co., of Toronto, while very much regretting the accident, are glad to say that the pulley in question was not one of their manufacture, and take this opportunity of advising the users of pulleys of the importance of seeing to it that they get a well-made, reliable article when purchasing. Every "Dodge" pulley manufactured is guaranteed strong enough for the heaviest double leather belt any width. To avoid accidents or mishaps ask for the "Dodge" patent and avoid inferior imitations.—Toronto World.

## SOME WESTERN ONTARIO LIGHTING PLANTS.

CONVINCED that a useful hint may sometimes be gained from the study of the smaller electric plants, the NEWS appends herewith short descriptions of the installations in a number of towns and villages in Western Ontario, which will be continued from time to time as occasion offers. Economy may be practised in running a small as well as a large plant, and an account of the system and methods in use at other points may result in suggestions that will be of great benefit as well as proft to those who study them.

### CLINTON ELECTRIC LIGHT CO.'S PLANT.

THIS company is making many alterations and has just added a boiler room to their building, placing two Goldie & McCulloch boilers in it, and leaving room for two more. A C.G.E. alternator of 1,000 lights, with C.G.E. equipped skeleton switch-board, has just been placed to replace two Reliance direct current dynamos. One Reliance 25 light 2,000 candle power does the street lighting. These machines are driven by a 60 h.p. Wheelock, from a 3 7-16 in. shafting, 24 ft. long. They have secured the contract for supplying the House of Refuge with 50 incandescent lights, and are extending the service in the town. A new condenser will be put in in a short time, and this little plant will be one of the most complete in the country. Johnathan Brown is the chief engineer, and Geo. White-Fraser is the consulting engineer.

### MITCHELL ELECTRIC LIGHT AND WATER-WORKS.

This plant is contained in a neat little building with the engineer's (Mr. Alexander) house in connection. He said he had found that it cost him nothing to be clean, and his boiler and power rooms verified his statement.

A 100 h.p. Goldie & McCulloch boiler, fed by an upright feedwater heater of the same make, supplies steam to a 10 x 18 Wheelock engine, which drives a line of shafting 18 ft. long. To this shafting is connected a 50 light Reliance arc machine, and a 500 light Royal alternator, with an exciter connected. A steam pump is fed by an upright boiler, 13 ft. x 4 ft. It forces water to the 10 hydrants, and can be connected at any time to the other boiler. 45 arc lights and 450 incandescent lights are in use, and the service is rapidly increasing. The plant is owned by the town.

### THE KINCARDINE LIGHTING PLANT.

One of the most efficient little plants in the province is the water-works and electric light plant owned by the town of Kincardine, and the board of commissioners have placed a man quite capable to run such a plant—Mr. Jos. Walker.

Two boilers 5' x 16', (Hunter Bros., Kincardine), supply steam to a steam pump, compound complex, with a daily capacity of 1,000,000 gallons, and a Corliss simple condensing engine, 15 x 36 x 125 h.p. The pump is fitted with a Fisher governor for pumping direct.

The engine drives a Thompson-Houston 1,000 light alternator, which drives a 1½ kw. Edison exciter, and a 50 light Ball arc machine. The plant was installed last year, and not a minute has been lost for repairs since its installation. The skeleton switch-board has a volt meter, an ampere meter, a ground detector, a lightning arrester, two rheostats, and a pair of double-pole fuse lights. On the switch-board of the arc machine are two lightning arresters and an ampere meter. The plant was installed and the town wired by the Canadian General Electric Co.

### A NEAT ELECTRIC AND WATER-WORKS PLANT.

GODERICH, Ont., owns one of the neatest little plants in Ontario, and the council has placed its care in the competent hands of Mr. W. H. Smith, chief engineer, and his staff, consisting of Harry Stowe, Jas. Andrews and Walter Brown. The large red brick building stands by the lake, at the foot of the frowning bluff on which stands the town. The water is pumped by two Gordon steam pumps, with a combined capacity of 1,250,000 gallons, to a height of 130 feet before it reaches the level of the town. Three Waterford "Reliance" arc machines, of 50, 65 and 35 lights—the first two of 1,000 c.p. each, and the last of 2,000 c.p., run by a 60 h.p. Wheelock engine, supply the town with light. The council has been undecided about buying an alternator, but one will be put in in the near future. The Canadian Government supplied a neat little nickle-plated engine, which hangs in a glass case on the wall, having a 10½ in. x 1½ in. cylinder, and it is so arranged that when the steam is turned on clockwork machinery is set in motion, which blows a fog signal at intervals for ten seconds. Two boilers, built by Crystal & Black, Goderich, supply steam, which is conveyed by an overhead main to the pumps and engine. On the walls are the usual gauges, clock, etc., with a fire alarm connected to the central telephone office and the town hall. Altogether it is an attractive plant.

### SEAFORTH ELECTRIC LIGHT, HEAT AND POWER CO.'S PLANT.

If electricians and engineers wish a model by which to run their plants they should see that of the Seaforth Electric Light, Heat & Power Co., and learn a lesson in neatness, cleanliness, order and economy.

Our reporter had the pleasure of being shown through this model power house, and he admired the way in which it was managed. The building is a roomy one, 68' x 40'. The power room, office and repair shop, are 40' x 40', and the boiler room 40' x 28', with a 17' ceiling. Two Monarch boilers, 70 h.p. each, generate the steam for the two Robb-Armstrong engines, 75 h.p. and 85 h.p. respectively. These two engines drive a 4 ampere Ball arc machine, and a 55 ampere Union Electric Improvement Co's, Philadelphia, alternator. The engines are arranged so that one can do the work required, or both can be used at once, dividing the load.

In the basement below the power-room is a reservoir with a capacity of 900 barrels. It is filled from the town mains, and pumped by a Northey duplex pump with a Penberthy injector, into an Austin heater, and delivered to the boiler at 211° Fhr. A pipe leads from the heater and from the steam exhaust into an old well in the floor of the basement, and the oil which is contained in the steam and the heater is separated and used over again. They thus save 40% of their cylinder oil. Two pipes come down through the floor of the power room from the engines, and the spindle oil runs through them into a wool filterer, and is also used over again. A nice bath-room, with two baths in it, is neatly fitted up in the basement. They have every comfort.

The brushes on the exciter are made of common copper screen wire, sewn together with copper wire. All repairing is done by the staff and they never send any work to the manufacturers. To show their economy, they used for the month of September, on 7 hours' run, 1¼ cords of wood. The load carried was 35 amperes on the alternator, 4 amperes on one 40 arc light dynamo, and 8 amperes on the other 18 arc light, and on an 8 hour run they used 1½ cords. Neat offices and a repair room take up one end of the dynamo room, and a large store room and workshop is overhead. The staff consists of Chief A. H. Ingram; assistants, H. McKay, A. Reid, R. Bullard and J. Darling, who have done their utmost to have things neat, putting oil-cloth on the floor, and contributing charts, photos, plans, etc., to make their plant cheerful and attractive.

### THE STRATFORD, ONT., LIGHTING PLANT.

It is seldom we see a small plant with three motive powers, but that of the Stratford Gas Co. possesses them, having steam, gas and water power. Their building, situated on the shore of Victoria Lake, is a substantial brick affair of one storey, and a basement 66' x 40', with a boiler room 30' x 45'.

In the main building are two Wheelock simple condensing engines—one of 100 h.p. in use 6 years, the other 200 h.p., just having been placed in position. Their fly wheels are 13' in diameter, with 22" face, the one of the smaller engine weighing 3½ tons, while that of the larger weighs 4½ tons. The line of 4 in. shafting runs the full length of the building, to which is connected the arc and incandescent light machines. There are five Ball arc machines—three of them thirty-five lights each, and the other two, twenty-five lights each—and two Reliance of fifty lights each, two C.G.E. generators of 320 lights each, 120 volts, supplying the town with incandescent light. There are four switch-boards, one for the Ball machines, with lightning arrester; one for the Reliance machines; one for the C.G.E. generators, containing two ampere meters, six knife cut-outs, two rheostats and a lightning arrester; and a separate switch-board with two volt indicators. They have also a volt-meter made by Munderloch & Co., Germany.

In the basement are the gas engine and water-wheel. The gas engine is 50 h.p., Crossley Bros., Manchester, Eng., Otto patent. The gas is supplied from the city mains through a 600 Legat meter. An immense gas-bag is used, with a valve attachment to prevent suction from the city mains. The gas engine has a 12 in. stroke and 14 in. piston, and makes 160 revolutions per minute. It drives on to countershaft, giving 530 revolutions, and from that to main shaft, making 390 revolutions. The water-wheel is a turbine, 50 h.p., 11 feet head.

In the boiler room are two Goldie & McCulloch boilers, 63″ x 14′, with 84 3 in. tubes each. One boiler has just been placed and another will soon be required.

Neat offices take up one corner of the main building, where charts and photos of the plant decorate the walls. Mr. Wilton, the chief engineer, took great pains in showing the ELECTRICAL NEWS representative the plant, and says that in a short time a tower will be built, rising from the centre of the building, to run the wires out of the building, and as the incandescent service is rapidly growing a new alternator will be needed.

There are six men on the engineering staff—the chief being Mr. Richard Wilton ; assistants, Messrs. Watson and Chown, an apprentice and two linemen. Mr. John Reid is the manager.

The service is as follows : 7 arcs in power house, 50 commercial arcs, 75 street arcs, 10 G.T.R. arcs, 10 rink arcs, 400 incandescents, 15 motors from ⅙ to 10 h.p.

## ONTARIO ASSOCIATION STATIONARY ENGINEERS.

139 Borden Street, TORONTO, Nov. 5th, 1895.

Editor ELECTRICAL NEWS.

· SIR : The following engineers were examined and received certificates during the month of October :

2nd class—O. Monger, Strathroy ; T. H. Walker, Kincardine.

3rd class—J. Cronier, Cobourg ; C. Stillwell, Brockville ; Jas. Walker, Trenton ; J. Kensley, Picton ; W. Irvin, Belleville ; J. Coughlin, Hintonburg ; H. E. Sutton, Cumming's Bridge ; Geo. Cameron, Ottawa ; R. J. Stewart, Lucknow ; F. G. Hall, Kincardine ; W. J. Hackett and Geo. Nelson, Toronto ; R. J. Levy, Wingham ; R. A. Root, J. T. Nicholls and R. A. Ballantyne, Strathroy ; S. Barber, Lucknow.

Twenty-three applications for examination were received during the month, four of which failed.

The following engineers who held 2nd class certificates have passed examination for and received 1st class certificates, viz. : Jas. Queen, Toronto Junction ; John Fox, Toronto. The following who held 3rd class certificates have passed for and received 2nd class, viz. : E. Carr, Brockville ; A. Cunningham, Toronto ; A. R. Barwick, Strathroy.

Inquiries re examinations are coming in from all parts of the province. Any engineer needing information as to examinations, who will send me a post card to that effect, will receive by-laws and copy of " Act."

A. E. EDKINS, Registrar.

## THE LATE DUNCAN ROBERTSON.

Mr. Duncan Robertson, of Hamilton, treasurer of the Canadian Association of Stationary Engineers, was stricken with paralysis, at the Windsor hotel, Ottawa, on the 3rd of October. He had remained in that city after the close of the convention to attend to some business, and while there was attacked by the disease stated, though he was in his usual health the evening before. As soon as possible he was removed to his home at Hamilton, but he died in a few days, much regretted by his brothers of Hamilton Branch No. 2, by the members of the Association in general, and by all who knew him.

Duncan Robertson was born in Kilbernie, Ayrshire, Scotland, Dec. 11th, 1846. He learned the trade of fitter and then went to Carlyle, remaining there for some years, when he went to the Shotts Iron Works. He came to Canada in 1871, and on arrival accepted a position at his trade with the Grand Trunk Railway Co., where he remained for about eighteen years, when he went as manager of the D. R. Dewey Coal Co., and later conducted the business of the Dominion Metallic Packing Co. till the time of his death.

He was a mason for about 27 years and when he died held the office of Junior Warden. He also belonged to the Minerva lodge of Oddfellows. He was, last year, treasurer of the Executive Council of the C. A. S. E., and was re-elected at the late convention. He was a musician of some taste, and while at Shotts acted as precentor of the Calderhead church. He leaves a widow and five children, three boys and two girls.

At the regular meeting of Hamilton No. 2, C. A. S. E., the following resolutions were adopted :

Whereas it has pleased the Almighty in his allwise Providence to remove from our midst our beloved brother Duncan Robertson, be it therefore resolved, that while we submit to the will of the Divine Providence we sincerely mourn with his family the loss of one so dear to them.

Resolved, that we tender our sincere sympathy to them in their sad bereavement.

Resolved, that a copy of these resolutions be transmitted to the family of our late brother.

Resolved, that a copy of the e resolutions be spread upon the minutes of this Association and a copy be forwarded to the press for publication.

(Signed) { E. C. JOHNSON, President.
{ WM. NORRIS, Cor.-Sec.

Here is one of the latest electrical jokes from Pearson's Weekly
Customer : " Why do you call this electric cake ?"
Baker's Boy : " I 'spose becuz it has currants in it."

## CANADIAN ASSOCIATION OF STATIONARY ENGINEERS.

Note.—Secretaries of the various Associations are requested to forward to us matter for publication in this Department not later than the 20th of each month.

### TORONTO ASSOCIATION NO. 1.

The annual dinner of the above Association will be held at the Richardson House, Toronto, on the evening of Wednesday, the 20th inst.

### OTTAWA ASSOCIATION NO. 7.

At the last regular meeting of the above association the following resolutions were adopted :

Whereas it has pleased the Lord Almighty to remove from our midst an esteemed and worthy brother, Mr. Duncan Robertson, and

Whereas he has been a most active worker in our organization, seeking as a member and as an officer to advance the interests of this Association and the welfare of its members ; therefore be it

Resolved—That we place on record our appreciation of his services as a brother, as an officer of this organization, and of his merits as a man ;

Resolved—That we do most sincerely mourn his loss as a brother, and that we tender our heartfelt sympathy to the widow and family of our departed brother, in this their hour of great sorrow and loss, and that as a tribute of our respect for his memory, the charter of this Association be draped in mourning for a period of thirty days;

Resolved—That a copy of these resolutions be spread on the records of this Association, also a copy be presented to the family of the deceased brother and to the city papers and engineering journals for publication.

FRANK ROBERT
THOS. WENSLEY
F. G. JOHNSON } Committee.
WM. HILL
F. W. DONALDSON

## PERSONAL.

Mr. James A. Baylis, who has been one of Mr. Dunstan's right-hand men at the Bell Telephone Exchange in Toronto, has been promoted to the position of chief electrician at the company's head office in Montreal. This is a well-deserved recognition on the part of the company of the brightest young man in the service. Mr. Baylis' popularity among the telephone fraternity in Toronto was attested by the presentation to him of a complimentary address accompanied by a smoker's outfit by the office and construction staffs. The presentation was made by Mr. Lash.

## TRADE NOTES.

The Clinton Electric Light Co. are installing a 1000 light alternator of the Canadian General Electric Co.'s make.

The Lozier Bicycle Co., of Toronto Junction, have purchased a 500 light isolated plant from the Canadian General Electric Co.

The Babcock & Wilcox Co. have sold a 300 h.p. boiler to the Ottawa Electric Railway Co.

The Dunnville Electric Light Co are installing a 75 kilowatt alternating plant of the Canadian General Electric Co.'s monocyclic system.

The new paper mill now being built at St. Croix, Hants Co., by H. McHari, Esq., will be lighted by electricity, the dynamo and plant for the same being furnished by John Starr, Son & Co., Ltd., of Halifax.

The Welland Vale Manufacturing Co., of St. Catharines, have installed an isolated plant of 100 light capacity. The Canadian General Electric Co. had the contract.

John Starr, Son & Co., Ltd., Halifax, N. S., are moving their office, stores and factory to 134 Granville street, where they will have greatly increased facilities for manufacturing their specialties.

The Canadian General Electric Co. will shortly install an Edison direct current alternating plant at Creemore, Ont., a monocyclic plant at Dunnville, and an alternator at Sudbury.

The ship "Mohawk" has been undergoing extensive repairs at the dockyard at Halifax. The contract for the electric light wiring was awarded to John Starr, Son & Co., Ltd. This has just been completed with lead covered cable throughout to the entire satisfaction of the engineer in charge.

The Unique telephones manufactured by John Starr, Son & Co., Ltd., Halifax, N. S., are coming into very general use throughout the Dominion. They have been adopted by the Engineer's Department of the Imperial Government for connection of their different coast stations, after exhaustive tests of other makes of instruments. Several large contracts have recently been made for complete equipment of lines and exchanges, and the firm report that their staff is taxed to the utmost to fill orders.

Messrs. Wm. Kennedy & Sons, of Owen Sound, are building the outfit for the water power plant of the Canada Paper Co.'s mills at Windsor Mills, Que. It consists of three 60 inch and two 40 inch "New American" water wheels, with main driving gear, bridgetrees and shaftings. They are at work on the models now. They have just sent the last of twenty-four 51 inch "New American" wheels to the Sault Ste. Marie Pulp & Paper Mills, and are reported to find an increasing demand for their electric water governors.

Messrs. Ahearn & Soper, of Ottawa, have lately closed contracts for a 250-light plant at the Edson Fitch mill, Etchemin ; a 1000 light alternator at Coaticook, Que. ; a 400 light machine for the Bell Telephone Co., Montreal. At Alexandrin, Ont., they are installing a 1000 light municipal plant with 45 street lamps, putting in the plant complete, including Robb-Armstrong engine. At Oshawa they are just installing a 275 h. p. generator to be driven by two Robb-Armstrong engines, and at Prescott a 100 light plant for the Prescott Elevator Co.

## WHAT IS SAID OF DODGE PATENT FRICTION CLUTCH PULLEY.

QUACO WEST, St. John Co., N.B., Oct. 29th, 1895.

GEO. H. EVANS, ESQ.,
Agent Dodge Wood Split Pulley Co., St. John, N.B.

DEAR SIR,—I have had one of the Dodge Wood Pulley Co.'s Split Friction Clutch Pulleys in use for the last month. It transmits 50 h.p in a stationary rotary saw-mill, and never slips or shows the least sign of weakness It gives me perfect satisfaction in every respect.

Yours respectfully, (Sgd ) S PATTERSON.

## A SUCCESSFUL DOWN DRAFT FURNACE.

ATTEMPTS almost without number have been made to construct a furnace adapted to burn soft coal, which would consume its own smoke, thus doing away with a public nuisance and effecting economy in fuel. The records of the patent office show that inventors have not been behindhand in grappling with the problem, but as is so frequently the case, a large proportion of the attempts have proved abortive. The Toronto Water Works seem to have discovered a furnace which fills the bill, and after a prolonged test, Mr. Keating, the city engineer, Mr. Pink, chief engineer at the pumping station, and the other officials who have to do with it, pronounce it an unqualified success.

The furnace which seems to fill such a long felt want, is known as the Hawley Down Draft Furnace. The head office of the company which constructs them is in Chicago, with branch offices in a number of the cities of the United States. The principle is very simple. The furnace is two storeys high, that is, there are two grates, one above the other. They do not occupy any more space than the ordinary furnace, for in converting the fire places at the water works the boilers did not require to be raised. The upper storey is closed at the back by a wall of fire brick, the lower one is open at the back to allow the heat to pass under the boiler and back through the return flue in the usual manner. The draft is therefore downwards through the fire on the upper grate. There is a fire also on the lower grate, but it does not require to be fed with fuel from outside, what drops through being sufficient to keep it going. By being carried downwards the smoke is consumed, as anyone can see by a glance at the tall chimney, from which no cloud of black smoke issues as is usual where soft coal is burned. The furnace is guaranteed to consume 95 per cent. of the smoke.

The upper grate is formed of a series of tubes opening at their ends into steel drums, or headers, which are connected with the boiler, though which the water circulates, giving great heating capacity. The construction of the furnace can be readily seen from the accompanying cut.

HAWLEY DOWN DRAFT FURNACE.

But figures giving actual results are more satisfactory than mere assertions, and as careful tests have been made we are in a position to furnish the former. The city is supplied with water by two sets of pumps, engines and boilers, exactly alike, each of ten million gallons daily capacity. The boilers for each set are four in number. One battery was changed to the Hawley system on the recommendation of the city engineer, who had investigated it at Buffalo and other places, and a careful record kept of the work of the two plants from the 19th of June till the 8th of September, 80 days, the battery with the old style of furnace using hard coal, grate size, and that with the Hawley furnace using soft coal screenings. The results were as follows:—

|  | Hawley Down Draft. | Direct Draft. |
|---|---|---|
| Quality of fuel used | { Soft Coal<br>{ Screenings. | Hard Coal<br>Grate Size. |
| Quantity of fuel consumed | 968 tons, 998 lbs. | 916 tons, 140 lbs. |
| Price per ton | $2.27 | $4.19 |
| Total cost of fuel | $2,153.09 | $3,838.33 |
| Net quantity of water pumped, 4 per cent.<br>allowed for slip—imperial gallons | 743,996,365 | 708,545,308 |
| Cost of fuel per million imperial gallons, net | $2.906 | $5.267 |

It will be seen that a large saving, amounting to about 45 per cent., has been effected by the use of the down draft furnaces. The two batteries were run as nearly as possible under similar conditions, any disadvantage being placed on the down draft.

It will be observed that the consumption of coal in the down draft furnace was a little greater than in the direct draft, but compared with the saving effected this is insignificant. The advantage in favour of the down draft system is so great that it is the intention to convert the other battery to that system at once.

The consumption of coal at the works amounts to about 10,000 tons a year, or say 25 tons a day. Soft coal screenings are supplied at $2.27 per ton, while under last year's contract $4.19 was paid for hard coal. The latter is an exceptionally low rate, offered partly to prevent the down draft furnace from being introduced. The test was made on the basis of the $4.19 rate for hard coal.

The down draft question having been satisfactorily solved by these tests, there is no reason why it should not be adopted elsewhere. In the case of the Toronto waterworks it will pay for itself in nine months.

The Hawley Company only guarantee a saving of 35 per cent. in fuel by the use of their furnace. As will be observed, the saving at the waterworks has been 45 per cent. The Hawley furnace has also been tried at the Toronto Incandescent Light Works, but there it has not given such satisfactory results, owing to the fact that screenings were used instead of hard coal previous to the introduction of the down draft furnace, and therefore the economy was not so marked

## PRESENTATION TO MR. DUNSTAN.

EVERY member of the Canadian Electrical Association is agreed that Mr. K. J. Dunstan, the retiring president, discharged his duties in a manner highly creditable to himself and in the best interests of the organization. For the purpose of giving tangible expression to this opinion, some of Mr. Dunstan's associates descended upon him at his office on the 29th ult., and presented him with an enlarged and suitably framed photograph of the group of members taken at the recent convention in Ottawa, accompanied by a pair of opera glasses for Mrs. Dunstan. The photograph bore the inscription : "Presented to K. J. Dunstan, Esq., on his retirement from the Presidency of the Canadian Electrical Association, Ottawa, September, 1895."

The presentation was made by Mr. A. B. Smith, the new president, who in fitting terms referred to Mr. Dunstan's valuable work in behalf of the Association as well as his ever courteous treatment of those associated with him. Mr. Dunstan expressed himself as being wholly taken by surprise, but nevertheless gracefully and with feeling acknowledged his appreciation of what he was pleased to term the thoughtful kindness of his confreres, to whom he also tendered thanks on behalf of Mrs. Dunstan.

---

Manson Campbell, of Chatham, has installed an isolated plant. The Canadian General Electric Co. furnished the apparatus.

The Kingston Street Railway Co. proposes extending its track to the Grand Trunk railway station. One man offers a bonus of $500.

The Canadian General Electric Co. are installing a 40 kilowatt direct connected incandescent plant for the Victoria Ladies' College in Whitby. The engine is of the Robb-Armstrong type.

Mr. Norman Ross is looking after the installation of two 200 kilowatt generators for the St. John Street Railway Co., which are being supplied by the Canadian General Electric Co.

The Canadian Pacific Telegraph Department has just completed stringing two number 6 gauge iron wires between Detroit and Buffalo. They will be used by the Postal Telegraph Co. as part of two new quadruplex circuits between New York and Chicago.

# ELECTRIC RAILWAY DEPARTMENT.

## AMERICAN STREET RAILWAY ASSOCIATION.

THE annual convention of the American Street Railway Association was held this year in Montreal, and was by far the most successful of the similar gatherings held by that organization. It continued in session five days, and during that time dealt with a number of important subjects. The convention met on the 15th of October and continued till the 20th.

### FIRST DAY.

When the convention had been called to order, the President, Joel Hurt, Esq., introduced Mr. J. O. Villeneuve, Mayor of Montreal, who, followed by Alderman Stevenson, welcomed the delegates, the president responding to the addresses of welcome.

President Hurt then delivered his inaugural, in which he dwelt upon the rapid strides made by the street railway industry. There are in operation in the United States 179,300 miles of steam roads and 13,500 miles of street railways. While the street railway mileage is 7½ per cent. of the steam mileage, the passenger receipts are 45 per cent The capital of the street railways is $1,300,000,000, or 11 per cent. of the capital invested in steam roads. New questions would, he said, come up for consideration in relation to the handling of freight, mail, small parcels for retail stores, building material in suburban localities, milk and funeral cars. He referred to the necessity of co-operation to defeat hostile legislation, and the importance of impressing patrons with the idea that friendly treatment helped to secure the best service. He suggested amalgamation with steam roads whereby the latter should carry passengers between distant points and then transfer them to street lines, for their final destination. He thought the time had come for raising a larger revenue for association purposes, either by assessment or an increase in membership dues, and said the most important question for their consideration was that of enlargement of the scope of the association.

At the afternoon session Mr. Wessels read a paper on "The Present Status of the Air-Brake," which was followed by a discussion.

Mr. Seely brought up the question of street paving, relating how in a certain city the contractor had refused to put down asphalt unless all traffic was stopped. The mind of the convention was that such a demand was unreasonable.

### SECOND DAY.

A discussion took place on the "Labor Question," followed by a paper on "Transfers," which caused further discussion.

At the afternoon session the report of the Executive Committee was presented, recommending some changes in the constitution. The report refers to a deficit of $4,087.33; also recommends that the scope of the Association be extended so as to include a bureau of information for the purpose of collecting and distributing statistical and general information on all matters affecting street railway interests; also that associate members be admitted so as to increase the income. It further recommended a uniform charge of $5 for banquet tickets. A committee of ten was appointed to devise means for paying off the deficit.

The Association was asked to appoint a delegate to co-operate with the National Electric Light Association in securing the adoption of a standard code of rules for electrical construction and operation. The Executive Committee was instructed to take action.

After considerable discussion on the report of the Executive Committee, final action was deferred till next year's meeting.

On invitation a visit was paid to McGill University, and very favorable comment was elicited by the equipment of the engineering building, which contains extensive hydraulic, engineering and electrical laboratories, and a laboratory for testing the strength of materials.

### THIRD DAY.

A paper on "Cross Ties and Poles" was submitted by Mr. N. W. L. Brown. Some of the points brought out were these : The life of a pine tie of best quality is only six years. There is great diversity of opinion as the best wood for ties, and some engineers claim to have solved the problem by using metal. Metal ties require to be imbedded in concrete, but as concrete is not necessary in first-class roadbed construction, metal ties are not economical. In New Orleans, where the soil is very damp, red cypress gives good results, but it does not hold the spikes properly. Creosoted ties are strongly recommended. as the most economical. They last long, are a perfect insulator, not subject to electrolytic action from leakage of current, and tend to prevent the rails from rusting. Sap pine ties treated with creosote should last 24 years. A creosoting plant is recommended for roads having an extensive mileage, where car floor, trestle bridge and cross arm timbers can be similarly treated. As to poles, iron cannot be depended on for more than 15 years. Red cedar lasts about as long, but small poles are limited to the duration of their sap wood, and large poles are generally hollow at the butt and contain rotten knots. On the whole creosoted

poles 30 feet long and 8 inches at the top are recommended, pine being eminently suitable for the purpose.

The report of the Committee on Patents was presented. It suggested a bureau to deal with all matters pertaining to patents as they affect street railways, pointing out how the rights of the roads might be preserved by such a bureau, and their interests protected from vexatious law suits. Those availing themselves of the bureau to pay a sum annually, in addition to the membership fee, to cover expenses.

In the afternoon the Committee on Ways and Means reported. It recommended that the members be requested to contribute not less than $15 each to pay the debt, that a uniform charge of $5 be made for banquet tickets, that individuals and firms, other than companies, lessees and owners of street railways, be admitted to associate membership, and that the convention should be opened on Wednesday and last four days.

A spirited discussion took place, especially on that clause of the report relating to associate membership, which resulted in its being tabled on a vote of 87 to 11. The sum of $4,475 was pledged towards the deficit.

The Nominating Committee reported, recommending that the following be the office-bearers for the ensuing year, and they were unanimously elected :—

President, H. M. Littell, Brooklyn, N.Y.; Vice-Presidents, G. C. Cunningham, Montreal; William H. Jackson, Nashville, Tenn.; J. Willard Morgan, Camden, N.J.; Secretary and Treasurer, T. C. Penington, Chicago; Executive Committee, Joel Hurt, Atlanta; Prentiss Cummings, Boston ; C. G. Goodrich, St. Paul, Minn.; A. Markel, Hazleton, Pa., and W. F. Kelly, Columbus, O.

The recommendation of the committee that the next convention be held at St. Louis, Mo., was adopted.

The Committee on the Use of Salt and Sand on Tracks reported, recommending both, salt to remove ice and sand to give the wheels a proper grip. In St. Louis 3,000 tons of salt are used each winter without objection from the authorities or health board.

In the evening the banquet was held at the Windsor Hotel. It was destined to be the crowning feature of the best street railway convention ever held. About 250 sat down. The room and tables were handsomely decorated. Mr. Joel Hurt, the President, occupied the chair, and among those present were Mayor Villeneuve, Chief Justice Sir Alex. Lacoste, U. S. Consul General Anderson, Hon. L. Beaubien, Minister of Public Works, Senator Ogilvie, and others. A most enjoyable evening was spent.

### FOURTH DAY.

The subject of free music and other entertainment by street railways came up for discussion. Several managers stated that their companies owned parks and gave free concerts in them, and they found it pay. Accidents from crowding on such occasions were rare. Mr. Penington stated that on Chicago day at the World's Fair, when 800,000 people were carried, only one accident occurred, and that of a trivial character.

Mr. J. F. McElroy read a paper on "Electric Car Heating," which was practically the same as that read by him at the New York state convention at Albany.

### FIFTH DAY.

According to the original programme the convention should have adjourned on Friday, but the Committee of Arrangements had planned a trip to Ottawa for Saturday. A special train of seven cars carried the visitors to the capital, where they were transferred to the cars of the Ottawa Electric Railway, and taken to the Parliament Buildings, where a reception was held in the Senate Chamber. An address of welcome was delivered by Sir Mackenzie Bowell, Premier, who referred to the local electric railway system, which was unexcelled on the continent. He hoped the United States visitors would go back with a better opinion of Canada than ever.

Mr. J. H. Gallinger, of New Hampshire, replied. The delegates, he said, were there to cultivate a spirit of amity with Canada, and they loved her more than was generally realized. He thought there was room for two great nations on this continent, and paid a high compliment to Canadian institutions and hospitality.

After lunch at the Russell House and a trip over the street railway, the visitors returned to Montreal by special train.

### NOTES.

The feeling is gaining ground that there is too much of a tendency to lose sight of the business aspect of these conventions in the cultivation of the social. While there can be no objection to the delegates having a good time, the interests involved in an organization representing a capital of $1,300,000,000 are too great to be lightly passed over, and when so much time is spent on excursions, receptions and banquets, and so little is spent in discussions on matters affecting the welfare of the companies and the public whom they serve, the Association can hardly be said to be fulfilling its true function.

The subjects of brakes, ties and poles, and heaters were well treated, but such topics as underground wires, underground trolleys, closed conduits, the relation of motor and truck, etc., were not touched upon.

The opposition to the admission of supply men as associate members of the Association caused considerable feeling. Unless they are admitted the Association is likely to suffer in some of its best interests.

The Association has become almost entirely electrical in its character. The cable and horse-car men have dropped out of sight.

It is unsatisfactory to have papers read by their titles only. It is true members have the opportunity of reading them when they are printed in the proceedings, but a discussion, when questions can be freely plied, is often of great advantage in bringing out points which are otherwise lost sight of.

There were too many executive sessions to which only a couple of hundred had access.

The Association should now be in good shape for future usefulness. It has arranged to clear off its debt and elected an efficient staff, with a capable secretary.

The following Canadian delegates appear in the list of those present :—W. Bellingham, E. B. Biggar, K. W. Blackwell, Wm. T. Bonner, H. T. Bovey, J. E. Bulmer, John Carroll, J. H. Cass, J. E. Chapman, A. J. Corriveau, G. C. Cunningham, W. F. Dean, J. J. Durack, L. J. Forget, Frank J. Green, C. W. Henderson, E. A. Hewitt, J. F. Hill, W. J. Hinphy, Geo. Hunt, Stonewall Jackson, E. D. Julien, J. M. de Bosch Kemper, George J. Kilpin, J. D. Lamb, T. Lamoreux, H. R. Leyden, E. Lusher, Alex. McPherson, A. Roy McDonald, Duncan A. McDonald, Alex. McDonnell, James Ross, James R. Roy, George D. Smith, Montreal, Que.; A. W. Dingman, James Gunn, Edward S. Piper, Charles Morton, O. J. T. Thomas, Toronto, Ont.; Thos. Ahearn, W. E. Christie, J. D. Fraser, J. E. Hutcheson, Charles F. Medbury, J. W. McRae, Wm. F. Powell, W. Y. Soper, W. W. Wylie, Ottawa; R. Mackenzie, W. Phillips, Niagara Falls, Ont.; T. C. Lazier, Belleville, Ont.; Frederic Nicholls, Toronto, President Brantford Electric Railway ; J. M. Campbell, Kingston, Ont.; Charles E. A. Carr, London, Ont.; J. M. Jenckes, Geo. E. Smith, Sherbrooke, Que.; H. Brown, St. John, N.B.; J. H. Coleman, Tottenham, Ont.; George H. Penty, Victoria, B.C.; M. Coventry, Windsor, Ont.

The Executive Committee make the following recommendations :—First, that the Association undertake, through its Secretary, the work of compiling statistical matter relating to the construction, equipment, operation and management of street railways, and the furnishing to members of general information upon matters of insurance, legislation and improvements affecting their interests ; second, the election of an executive committee of ten, consisting of the four officers and six others, the latter to be elected for three year periods, two each year ; third, the election of the secretary and treasurer by the executive committee ; and, fourth, the raising of funds for carrying out the larger work by a system of annual dues based upon gross receipts.

The exhibition of street railway appliances was a prominent and useful feature of the convention. Supply men, with a keen eye to business, were on hand in full force with everything necessary for the proper equipment of railway lines.

A number of the delegates took part, on invitation of the Forest and Stream Club, in a fox hunt. Horses were provided for all who wished to participate. Mr. Hurt specially distinguished himself.

In selecting St. Louis for the next convention it was felt that a good choice had been made. The street railway system of that city is said to be one of the best on the continent.

The hospitality of the people of Montreal was unbounded and called forth many expressions of appreciation.

---

## SPARKS.

Eganville is to be lighted by electricity. Water power will be used.

A fine large bronze statue of Benjamin Franklin is to be placed in Lincoln Park, Chicago.

A dynamo burst recently in the Renfrew light station and was so completely demolished that a new one had to be obtained.

Brockville town council has given its assent to the use of the streets for the proposed electric railway. It is stipulated that the road must be completed and in operation within two years.

The engine in the C. P. R. laundry at Owen Sound, Ont., has been in use for the last 60 years, and is said to be in the best of running order yet. It was made in Manchester, Eng., by John Ellis, jun.

According to the Toronto city engineer's report for 1894 recently issued, the total length of street railway track within the city limits is 82.54 miles, of which 15.56 miles was laid during the year under review.

Franklin L. Pope, the well-known patent solicitor and electrical engineer, was instantly killed on the 13th of October, in his house at Great Barrington, Mass., by coming in contact with a live wire, while experimenting.

The statement of claim has been fyled in the Sunday street car case at Hamilton. The action is brought under the provisions of the R. S. O. of Ontario, cap. 203, entitled "An Act to prevent the profanation of the Lord's Day."

J. Hill, Brussels, Ont., claims he has discovered a process for the tanning of leather for belting and other purposes by the use of crude petroleum. He says that the tanning is done in half the time taken for bark tanning and will wear twice as long. He is trying to get some American capitalists interested.

Writing to the Scientific American, a correspondent replies as follows to another correspondent who asked for some composition for filling cracks in a commutator : You reply that the only way to repair will be to take the commutator apart and replace the mica. But if he happens to have no appliances for this work, a temporary repair can be made of thick shellac solution and dry plaster of Paris. Fill the crack with the shellac, then put on the plaster, kneading it with a knife blade until it is stiff and smooth. Let it dry five or six hours or longer, before scraping off the top even with the surface. It should be thoroughly dry before the armature is used. I have used this method for repairing street railway motor armatures for more than a year, and no armatures have come back in that time for a fault due to this filling.

A joint stock company has obtained from the groovernment, letters patent for the exclusive manufacture, at Kincardine, of bicarbonate of soda and chloride of lime by electrolysis. The officers are as follows, viz : President, E. H. Hilborn ; vice-president, H. Glazebrook, Toronto ; secretary and manager, John Tolmie, Kincardine. The company is to be known as the Ontario People's Salt & Soda Co. They have built an addition to the salt works 140x60 in which the soda and lime will be manufactured. They are having manufactured for them a steel stand pipe 60 feet high by 3 feet in diameter, and 2 steel scrubbers 18 feet high by 4 feet in diameter for washing the carbon out of the carbonic acid gas. Thirty-two tanks 50x10x1½ will be required.. In a separate building being built for this purpose, the carbonate of lime will be made. The building will consist of two air tight chambers. The whole plant will be run by their own private electric plant.

The Ottawa, Canada, Electric Co. has proved itself one of the most progressive lighting companies on the continent. The fact that in a city with a population of but 40,000 it runs 50,000 lights, speaks volumes, and we question whether any other city in the world can approach such figures. But even more interesting is an item in president Ahearn's last report, just presented, showing an income of $421 from heaters. We have not struck this item in any similar balance sheet before, and make its acquaintance with a great deal of pleasure. The expenses per contra charged against heaters are only $12, from which we infer that the service must be quite profitable. It is true that $421 is not a large proportion of the $147,000 received for all services, including incandescent, arcs and motors, but it is a decided beginning and is much larger than the motor account once was with some incandescent companies or the incandescent account with many large arc companies. The Ottawa example of more than one lamp per head of population, and of $421 revenue from heaters at an incidental outlay of only $12 is a mighty good one to copy.—N. Y. Electrical Engineer.

## SPARKS.

Lanark, Ont., is talking of electric light.

St. Jacob's, Ont., is to be lighted by electricity.

The electric cars are hereafter to be run in Ottawa until midnight.

Thirty new electric lights will be placed on the streets of Ottawa.

A class in electricity has been organized by the Y.M.C.A., Ottawa.

Mt. Forest, Ont., talks of having an all night electric light service.

Alexandria, Ont., has decided to have electric light and water-works.

Forty-five horses, formerly used on the London Street Railway, were recently sold for $28 each.

The Ottawa Electric Light Company has had to resort to the use of steam on account of low water.

The Montreal Street Railway Company has declared its usual half-yearly dividend of four per cent.

The Montreal Park & Island Railway will extend their line to St. Laurent, a distance of seven miles.

The Neepawa Electric Light & Power Company is applying for incorporation. The capital stock is $20,000.

Winchester, Ont., is agitating for an electric railway between Morrisburg and Ottawa, passing through Winchester.

The Bell Telephone Co. is about to erect a handsome building in Winnipeg, to cost between $15,000 and $20,000.

The Western Electric Light, Heating & Power Company has been formed to take over the lighting of Vancouver, B.C.

A Pacific cable connecting British North America with New Zealand, via Hawaii, is among the possibilities of the near future.

The Ottawa Carbon & Porcelain Works have turned out their first lot of porcelain insulators. They have $25,000 worth of orders on hand.

During the week of the Central Fair, the Ottawa Electric Railway Company carried 207,821 passengers, not including transfers, without accident.

Mr. George Sleeman, proprietor of the Guelph Street Railway, was given a handsomely framed picture of the first conductors and motormen of his road.

Stock to the extent of $40,000 has been subscribed for an electric line between Cobourg and Port Hope. The township of Cavan offers $10,000 for a branch.

The case of the Bell Telephone Co. against Skinner, for cutting their telephone wires at Sherbrooke, Que., resulted in a verdict of not guilty.

A Scotch firm of means proposes to build an electric road from Schomberg, Ont., to some point on the Northern division of the Grand Trunk—probably either King, Aurora or Newmarket.

The Toronto Street Railway Co.'s earnings for October were $78,216.98, of which the city's share is $6257.35. This is $28,316 less than in September, and $2,090 less than in October, 1894. Receipts for sprinkling, advertising, etc., will bring it up to a considerably larger gross amount. The Montreal Co. has earned so far this year $1,102,777, as against $896,000 for the same period in 1894.

---

## VALUABLE

# Electric Light Plant

### AND FRANCHISE FOR SALE.

Tenders will be received by the undersigned up to THE 11TH DAY OF NOVEMBER, 1895, inclusive, for the purchase of an Electric Light Plant, Arc and Incandescent. There is a ten years' contract for the lighting of the Town of Newmarket with at least 31 arc lights. There are now about 450 incandescent lights installed, and over 400 more promised. Chances are good. No opposition of any kind.

No tender necessarily accepted.

For further particulars apply to

**H. S. CANE,**
NEWMARKET, ONT.

# CANADIAN GENERAL ELECTRIC CO.
## (LIMITED)

• • • •

## Repair — Department

We have opened up a thoroughly equipped Repair Department at our Peterborough factories, in which repairs to Electrical Machinery of all kinds will be made promptly, properly, and at a price slightly in excess of the actual cost of material and labour.

Special attention paid to T.-H. Spherical Arc Armatures, Ball Arc Apparatus, etc.

## SPARKS.

An order has been passed for the winding up of the St. Jean Baptiste Electric Co., Montreal.

The Waterloo Board of Trade has decided to recommend a bonus of $30,000 to the International Radial Railway.

The Prince Albert Electric Light Co. have purchased a 500 light alternating plant from the Canadian General Electric Co.

E. W. Snider, of St. Jacobs, Ont., has purchased a 100 light incandescent plant from the Canadian General Electric Co. for lighting his mill and residence.

The Canadian General Electric Co. have installed a 600 h. p. generator for the Winnipeg Street Railway Company. This machine is of the same type as the 1,200 h. p. generator that company recently installed for the Toronto Railway Co.

The charges of bribery against Mr. William Kyle and Mr. Robert F. Segsworth, of Toronto, in connection with the Niagara Falls street railway, have come to an end. That against Mr. Kyle was terminated by his death a short time ago, and when Mr. Segsworth appeared for trial at the Welland assizes, the case against him was dropped.

The annual meeting of the Merchants' Telephone Co. was held at Montreal on 1st October. The report showed that the company has 700 subscribers. Eleven hundred dollars' worth of capital has been forfeited by shareholders not paying up. The following directors were elected : Messrs. M. T. Lefebvre, Joel Leduc, A. S. Hamelin and L. H. Henault. At a subsequent meeting Mr. F. X. Moisan was re-elected president, Mr. Joel Leduc vice-president, and Mr. L. E. Beauchamp, treasurer.

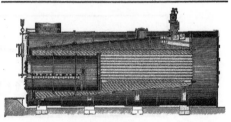

# CANADIAN ELECTRICAL NEWS

## STEAM AND ENGINEERING JOURNAL

OLD SERIES, VOL. XV.—No. 6.
NEW SERIES, VOL. V.—No. 12.

DECEMBER, 1895

PRICE 10 CENTS
$1.00 PER YEAR.

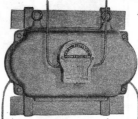

Please mention the CANADIAN ELECTRICAL NEWS when corresponding with Advertisers

CANADIAN

# ELECTRICAL NEWS

AND

## STEAM ENGINEERING JOURNAL.

| VOL. V. | DECEMBER, 1895 | No. 12. |

### THE DETROIT BOILER EXPLOSION.

ON the morning of the 6th November a steam boiler exploded in the Detroit Journal office, 45 and 47 Larned street west, by which 37 persons lost their lives and a number more were injured. The view we publish represents the scene of the disaster after all the debris had been cleared away, leaving, however, the boilers or parts connected with the boilers as nearly as possible in the position they were thrown by the explosion.

An investigation into the cause of the accident is

The boilers had been frequently inspected by the Detroit City inspector, and were supposed to be sound and good boilers. The amount of damage done by the explosion of one of them seems to confirm this view. At the coroner's inquest on the killed the jury were instructed to bring in a verdict that death was caused by a boiler explosion. The engineer has been indicted for manslaughter and it is probable that the full results of the investigation will be made public at the trial.

There does not seem to be any need for any sugges-

THE DETROIT BOILER EXPLOSION.

under way, but the authorities have been very reticent and have endeavored to prevent any one from getting near enough to touch any part of the boilers, except those engaged in the investigation. This much, however, is known, there were two boilers of the ordinary horizontal tubular type in the basement of the building and set side by side. The fuel used was oil, injected into the furnaces by steam jets. The engineer had steam on one boiler at about 85 lbs. pressure supplying steam for the engine. The other boiler had its outlet valve closed, and steam was being got up in it, the intention being to use it, and to stop the other one for cleaning.

When the engineer left the boiler room some time before the explosion the boiler which exploded had 15 lbs. pressure on it, and the outlet for the steam not open.

tion of mystery about the matter- The steam boiler was left with an oil fire under it, with no outlet for the steam, unless by the safety valve, and with 15 lbs. pressure on it. The pressure would rise rapidly, and as the fire was fed by oil forced in by the steam, it follows that the higher the pressure the fiercer the fire became.

The safety valve had either been too small or inoperative, and very soon the pressure became so great that explosion took place. From the fragments shown on the view it looks as if the cast iron manhole frame had given way first and the boiler had then torn into two sections. The other boiler was driven out of its seat into the wall, and was probably the cause of the fall of the building.

As showing how rapidly steam can be raised in a steam boiler, the test of the Merryweather fire engine in Toronto might be cited. Of course it was a small boiler, yet it had only a small fire grate and burned ordinary soft coal. In it, when steam was being got up, it required 5½ minutes to get steam to 15 ℔ pressure from cold water, and only 4 minutes to raise the pressure from 15 ℔ to 100℔, and only one minute to raise it from 50 ℔ to 100 ℔.

At Detroit the oil would give a much hotter fire and there is no doubt the pressure rose with great rapidity after it once got past 100 ℔.

### THE G.T.R. SHOPS AT STRATFORD.

These shops, which a representative of the News recently had the pleasure of inspecting, are the main shops in Western Ontario to which "sick" engines are sent for repairs. In the machine shop two 78 h.p. engines run the greatest amount of machinery in one room in Canada. The "V" belt is greatly used here, and the superintendent, Mr. Barnet, took great pride in showing its merits. The belt is about four inches wide, with cleats of leather like a heel of a shoe, though in the shape of a V, cut off at the apex, rivetted on to the belt at intervals of six inches. The belt runs over concaved pulleys, and is said to be the best belt for use in places where it is necessary to run a vertical and a horizontal wheel with the same belt.

In the erecting shop, between two lines of locomotives in process of repair, is a track on which a traverser runs, being propelled by an endless chain in the centre of the track. When an engine comes in for repairs it is dead (i.e., no steam in it). The yard engine takes it in tow and shunts it on to the track which leads into the erecting shop. An engine (stationary) stands at the opposite side of the shops, and the engineer by pulling a cord sets the drum in motion; a rope from the drum is fastened to the dead engine and the drum pulls the engine into the shop on the traverser. By the engineer pulling another cord the engine allows four large chains to descend from the ceiling above the dead engine. Two large bars are then shoved through each pair of chains, and another cord is pulled, making the engine gradually raise the dead engine clear of its trucks. Then another of the cords is pulled, setting the endless chain in motion, which carries the traverser with the trucks to the end of the shop, where the trucks are unloaded for renovation, and wooden trucks placed on the traverser, which travels back to its former position. The dead engine is lowered on to the wooden trucks and the traverser carries it down to an empty track, where it is shunted alongside of its fellows on other tracks ready for repairs. All these different motions from its coming in till its going out are done with a little 25 h.p. engine.

Compressed air plays an important part in this part of the shops. The air is compressed in the boiler room. The engine in the boiler room is 75 h.p., and the piston rod runs clear through the cylinder into the cylinder of the air compressor, passing also through this, but when the rod comes out of the cylinder of the compressor it is hollow, to allow the air to enter the cylinder of the compressor. Air is brought in from the outside through a common tin pipe. The piston head in the cylinder of the compressor is hollow and is fitted with valves. As the rod recedes the valves close, and on its return stroke the air is compressed leading up through a pipe into an air main which extends all through the shops. Formerly when it was necessary to do machine work

on any part of the dead engine, the part had to be taken from the engine to the machine, whereas now, by the use of compressed air, a little compressed air engine is run on wheels to the dead engine and the work is done right there. Another advantage of these little engines is that they are as well adapted to steam as to air.

A 14 in. steam main runs parallel with the air main the whole length of the shops, supplying 7 stationary engines and 3 immense steam hammers. The last part of the main shop is the boiler room, where the best machinery is used and where noise is a prominent feature. In another building is the carpenter shop and model room. The boiler tube room is one of the most interesting parts of the shop. When a locomotive's boiler tubes get so dirty that it is not economical to use them any longer, they are taken out and placed in a large cylinder, where scalding water and chemicals are forced through them, cleaning them out thoroughly. The ends have been damaged in taking them from the boiler, so they are heated and the ends sawed off, leaving burred ends. A man soon fixes that with the aid of a machine for the purpose. By reason of the ends having been sawn off, they have become somewhat short, and so are passed on to another man who has small pieces of tubing red hot. He slips a piece on each end and passes them on to another man, who welds them, when they are again ready for use.

An isolated building is used for the fitting of the tires on the driving wheels, gas being used to heat the rims. The boiler room and brass foundry are in another building. A battery of 9 boilers supplies the steam for the shops. The exhaust steam is led by pipes into a cistern in the boiler room, from which it is used over again. The boilers used are railway boilers and the ashes are drawn into a pit beneath the boilers.

A fire hall is one of the many useful institutions connected with the shops, and steam is at the steam pump at all times. A fire department is made up of the men and they are given occasional drill. If the city mains should give out by accident, a reservoir of 60,000 gallons is at hand, and by connecting the steam pumps with the nearest hydrant, a direct force of water can play on any part of the shops.

The large building on the street comprises the library, lounging room and manager's offices.

### CORRECTION.

In the description of the long distance power transmission plant at Portland, Ore., published in our last issue, an error occurred in the last sentence of the first column on page 201. It should have read: "The four-wire system is worked at 133 volts between any two wires, and by means of feeder regulators a variation of 7½ in either direction is covered."

One of the severest wind storms for years prevailed over this continent on Nov. 26th. Telegraph, telephone and electric light wires were down in every direction, entailing great interruption to business and loss. The gale reached a velocity of 60 or 70 miles an hour in some places.

Chief Engineer Perry of the United States navey recently returned from a tour of 2,000 miles on the great lakes, made on the steamships Zenith City and Victoria, for the purpose of observing the working of the Babcock & Wilcox Scotch boilers. The Navy Department is considering the advisability of fitting the six new gunboats with these boilers, and engineer Perry was detailed to inspect them. His report will be favorable in both cases, and if the present plans of the department can be carried out three of the six new boats will have Scotch boilers and three the Babcock & Wilcox.

## THE TRENTON-BELLEVILLE TRANSMISSION PLANT.

THE power transmission plant now being installed at Trenton will be in point of distance over which the power is to be transmitted the most important work of the kind so far undertaken in Canada. The valuable water power of the Trent river developed at a large outlay some years ago will be utilized to furnish power and light for Trenton and also for the city of Belleville, located at a distance of 12 miles from the generating station. Such a scheme would, only a year or two ago, have been regarded as chimerical both from an engineering and commercial point of view, but the rapid development of the alternating induction motor from a laboratory toy into an everyday industrial implement, has made possible for this and many similar undertakings an immediate operating and commercial success.

The Trenton Electric Co., of which Mr. W. H. Pearson, jr., of Toronto, is the moving spirit, has under its control, including the power rights leased from the town, something like 1,000 horse power. A new power house is being erected and water wheels installed having a capacity of 500 horse power. For the transmission plant, after a careful consideration of the merits of the different systems offered, the three-phase system of the Canadian General Electric Co. was selected, and a contract closed with them for the necessary apparatus.

The initial installation will consist of two 150 kilowatt three phase alternators, from which currents at 2900 volts will be taken direct for the Trenton end and distributed by means of secondary mains for incandescent lighting and power. For the Belleville end the current will be raised through step-up transformers to a potential of 10,000 volts, the loss due to ohmic resistance at this voltage being at full load less than 3 per cent. At the Belleville sub-station step-down transformers will supply current to the mains for distribution at 2000 volts, and the larger motors will be wound to take current at this pressure direct.

Arrangements have been completed for operating the waterworks, for which two fifty horse power induction motors will be used. A 50 horse power motor will probably be installed to drive the series arc light dynamos now supplying the city circuits. It is expected that a contract will be made later on with the Belleville Street Railway Co. to furnish power for their system, now operated by steam.

The entire supervision of this important electrical engineering work has been placed in the hands of Mr. J. M. Campbell, of Kingston, who has been prominently identified with most of the larger railway and lighting installations in eastern Ontario, and in whose hands we may expect the various novel and interesting applications of electric power contemplated in connection with this transmission will receive the most careful theoretical and practical treatment.

Mr. Pearson and his confreres in the Trenton Electric Co. are certainly deserving of every success for their enterprise in this notable step towards the utilization of one at least of the many valuable water powers now lying idle throughout Canada.

THE Chicago Times-Herald race for motocycles, or horseless vehicles, postponed from November 2nd, took place on November 28th, under very unfavorable weather conditions, as the machines had to plough through slush and snow. Full particulars have not come to hand, but the contest was won by the Duryea machine.

## WESTERN ONTARIO ELECTRIC PLANTS.

WE continue our notes of western Ontario electric plants, commenced in last month's issue :—

### CITY OF WINDSOR LIGHTING PLANT.

This plant is situated on Pellissier street, in the heart of the city and is owned and controlled by the City of Windsor. The chief engineer is Mr. Thos. Chater. The building is a one-story brick building, L shaped. The engine room is 30 x 60 and the boiler room 40 x 25. In the boiler room are two Leonard boilers of 100 and 65 h. p. respectively. Room is left for two more boilers. In the engine room is a 150 h. p. Leonard Ball engine driving onto a line of shafting of 27 feet. Off this shafting run 4 arc machines, 3 of the Reliance system and one Waterford, each of 50 lights capacity.

The switch board is 9 x 15, built to accommodate increasing demand, 22 miles of wire are used for the 175 arcs and 75 fifty c. p. incandescent lights. There is no better arc light plant in Canada than this.

### SARNIA ELECTRIC AND GAS LIGHT PLANT.

The electric light building is a large red brick building with a commodious basement under the engine room, the floor of which is supported by massive brick piers.

In the engine room is a Wheelock engine of 75 h. p. running two C. G. E. Co.'s machines—one a 500 light alternator, with exciter, the other a 1200 c. p., 75 light, Wood arc machine. These machines are regulated by a switch board to each machine, the alternator having a skeleton C. G. E. and the arc machine a similar make fixed to the wall behind the skeleton switch board. Two hundred lights will be required for the new hospital, thus calling for another alternator, and a railway motor is proposed to supply power to the street railway when they change their motive power, which is now horse.

The boiler room is built to accommodate two more boilers, only two being now in use. Coal alone is used for fuel. The basement is used as a storage room, and a Northey condenser will soon be put in place there. On the walls of the engine room are the charts of the light system through the town, together with blue prints, photos and numerous glass cases of stuffed birds and small animals. Mr. Shand, the chief electrician, is a great student of natural history, and is an expert taxidermist, doing all his own work. A friend presented him with a "Corry's bittern"—a rare species of bittern—only two more having been shot in Ontario.

The gas house is a large building behind the electric plant and supplies a great number of consumers. A short time since an explosion took place in the purifying room, which blew the roof off to a distance of 40 feet. The buildings are heated by gas, and electricity is used in lighting.

Small machines for repairing are driven from the shafting, and Mr. Shand does all the repairing.

The manager, Mr. W. Williams, has had a long experience in the gas business, but is pushing electric lighting for all it is worth. He has got out an account book suitable to gas or electric lighting, and the Western Gas Association on seeing it had it published, and it is now used by a great many gas companies. It can be used for either a gas or an electric meter. It needs but be seen to know its value.

A new style of heater has been introduced into the street cars at Kingston. It is flush with the seat and does not project into the car. It is the design of Mr. Jas. Halliday, electrical overseer of the Kingston Street R. R. Co., who will apply for a patent.

## CANADIAN ASSOCIATION OF STATIONARY ENGINEERS.

NOTE.—Secretaries of Associations are requested to forward matter for publication in this Department not later than the 25th of each month.

### TORONTO ASSOCIATION NO. 1.

The ninth annual dinner of the above Association took place at the Richardson House on the evening of Wednesday, Nov. 20th. The attendance was greatly in excess of any previous occasion—so great, in fact, that the large dining hall was not sufficient to allow of all the guests being seated together at the tables. There were three tables extending almost the entire length of the room, with a fourth across the upper end ; these were tastefully decorated with foliage plants, etc.

Mr. Lewis, President of the Association, a portrait and sketch of whom is presented with this report, was the presiding officer, and fulfilled in a most creditable manner the duties of the position. On his immediate right sat Professor Galbraith, Principal of the School of Practical Science, and Dr. Orr, Chairman of the Toronto Technical School Board ; and on his left, Mr. John Galt, C.E., and Mr. A. B. Smith, President of the Canadian Electrical Association. Among the other guests were the following :—

TORONTO.

| | |
|---|---|
| J. E. Cameron. | J. N. Lambert. |
| A. E. Edkins. | A. McBean. |
| A. Travis. | J. W. Marr. |
| J. Thompson. | Geo. Fowler. |
| A. Thomson. | A. J. McDonough. |
| C. Heal. | Lawrence Farrell. |
| J. Moat. | E. J. Philp. |
| J. Hall. | Wm. Butler. |
| J. Bannom. | E. Ash. |
| J. McLaughlin. | W. Bundy. |
| J. Ewing. | G. W. Grant. |
| T. Hobbs. | J. Richardson. |
| J. Wilson. | S. Thompson. |
| J. Mountstephen. | G. T. Pendrith. |
| A. C. W. Soper. | W. P. Despard. |
| Wm. Abbs. | C. Sodden. |
| A. G. Horwood. | W. T. Blacklock. |
| C. H. Mortimer. | G. A. Enouy. |
| James M. Sinclair. | T. Seaton. |
| B. Doyle. | D. Bowman. |
| T. Long. | T. Cadwell. |
| John Fox. | R. J. Donehay. |
| Alex. Robertson. | Fred. Day. |
| J. Bliss. | John McCleary. |
| J. W. Ball. | R. S. Brown. |
| F. S. Jackson. | John Day. |
| R. H. Smith. | S. Galbraith. |
| S. Bassett. | R. Owen. |
| G. A. Perry, | J. Wadge. |
| D. McCulloch. | E. Arscatt. |
| Robt. Hutt. | H. H. Tait. |
| J. H. Pringle. | W. P. Sutton. |
| H. C. Mills. | Jas. Mooring. |
| W. J. Burroughes. | J. B. Millar. |
| J. Smith. | Geo. White. |
| Jas. Mullin. | Geo. Gilchrist. |
| John Long. | E. Appleton. |
| J. J. Ramsay. | John Campbell. |
| Wm. Donaldson. | Wm. McKittrick. |
| A. McMartin. | Wm. McIntyre. |
| S. H. Fussell. | N. V. Kuhlman. |
| Edwin Farrants. | J. F. Ross. |
| E. B. Biggar. | A. Townsend. |
| W. J. Hannan. | W. Eames. |
| G. C. Mooring. | J. Grainger. |
| F. J. Smith. | Alf. Butcher. |
| Thos. Reid. | Chas. Pearce. |
| Chas. Smith. | T. Philip. |
| Albert Forrester. | W. Good. |
| Wm. Sutton. | H. Buchner. |
| W. W. Mason. | G. B. Towers. |
| G. F. Spry. | T. Hope. |
| Robert Flint. | Wm. Vaughan. |
| Joseph Kirk. | W. Phillips. |
| C. J. Read. | A. M. Wickens. |
| Benj. Flint. | |

OUTSIDE GUESTS.

James Webb, Hamilton.   B. Crandell, Toronto Junction.
Joseph Craig, Oshawa.   W. Oathwaite, Peterboro'.
J. Queen, Toronto Junction.

The menu, which was artistically printed, was as follows :—

### FORMULAE

Hydraulic—SOUP—A la Keating.
Celery.

FISH
Steam Boiled—con—Clinker Sauce.
Scales Removed by Sutton's Compound.
Pommes Parisienne        Bread and Butter Fritters.

OILTRAYS
Gusset Stays of Lamb.        Chicken Croquettes.
Double Butt Straps—JOINTS—Triple Rivetted.
Roast Turkey.        Frame of Beef.
Water Leg of Mutton.        Yorkshire Pudding.

VEGETABLES
Potatoes all Broke Up.    Corns.    Stewed Tomatoes.

ENTREMENTS
English Plum Pudding, High Potential Sauce.
Apple Pie.        Cut-Off Pie.        Wood-Raspberry Tart.

DESSERT
High Celery.    Cheese.    Biscuits.    Assorted Cakes.
Volts and Amperes.
Figs.    Dates.    Apples.    Bananas.    Oranges.
Watts.
Tea.    Coffee.    Lemonade.    Georgian Bay Water.
Claret.    Sherry.    Pale Ale.    Ginger Ale.
Electric Shocks.    Compound Wound Cigars.

After the Chairman had read a number of letters of regret from persons who could not be present, the company proceeded to do justice to the contents of the tables. The toast list was then proceeded with. It bore the following inscription :

And the truant husband will return and say,
"My dear, I was the first to come away."

After listening to a piano solo by Mr. Harding, the national anthem was enthusiastically sung in response to the toast of "Her Majesty the Queen."

In response to the toast of "Canada our Home," the company were favored with songs by George W. Grant and Mr. Wright.

Mayor Kennedy arrived just in time to respond on behalf of the city of Toronto. He made reference to the objects of the Association, to the duty incumbent on men to help each other in life, and to the advantages to be derived from union of effort. He enlarged upon the greatness of the British Empire, stating that this greatness was in a large measure due to the application of steam and the steam engine, in driving British ships of commerce over all seas.

Messrs. Smith, Brett and Martin responded on behalf of the toast to "The Manufacturers."

After a song by Mr. Seaton, the chairman proposed the toast, "Educational Interests," coupling therewith the names of Dr. Orr and Professor Galbraith.

Dr. Orr, in responding, pointed with pride to the city of Toronto as a centre of education, and to the fact that graduates of Toronto University might be found holding leading positions in all parts of the world. Education in Toronto, he said, was as free to the son of the poor man as of the rich man. Referring to the Toronto Technical School, with which he was more particularly connected, he remarked that the public do not appear to clearly understand what the school is, and what it is for. The creation of the Toronto Technical School was almost entirely due to the Association under whose auspices they had met. The late Mr. Wills, Mr. A. M. Wickens, and the late Alderman Gillespie, were the foremost promoters of the institution. The by-law under which the school was founded was drafted by Messrs. Wickens and Wills, and afterwards revised by Professor Galbraith, who had from the beginning been a staunch friend of the institution. The school had now an excellent staff of teachers, and its popularity was attested by the fact that the attendance had increased from two hundred to six hundred. He would like to see a trade school added to the present institution—a school in which young men who were fitted to become mechanics would have the opportunity of receiving the instruction which they require, and which it was impossible for them to obtain in the workshops of to-day, on account of the extent to which the various branches of manufacturing had become specialized. The school had much to thank the City Council of Toronto for, inasmuch as all the funds required for its support were derived from that source. He had only one complaint to make, which was that the Mayor and Aldermen had not attended the school as frequently as the Board could wish. In conlusion, he wished Toronto Association No. 1 continued prosperity.

Professor Galbraith compared the present gathering with the first one of the kind at which he was present, when the attendance was not more than one-fourth of that of the present occasion. He facetiously remarked that this was an evidence that the Association, although composed of stationary engineers, was by no means "stationary," but very much alive. He then proceeded to define the difference between professional and technical education. The minister, he said, lives on the moral evil that is in the world; the doctor on the physical evil, etc., but the engineer would do better if there was no evil in the world at all. Technical education is the education of men engaged in turning the material resources of the world to advantage. He turned aside to repeat a story which Mr. William Sutton is said to be the author of. It ran thus:— A country engineer dropped into the engine room of a Toronto engineer, and was shown an indicator card—something which he had never seen before. He enquired what it was for, and was told that the card was drawn by the engine. The country engineer's face showed incredulity, and he immediately went round the corner and told another engineer that "a bloke had tried to make him believe that his engine could draw." Referring to the Toronto Technical school, the Professor said that the success which had attended the institution had exceeded all expectations. The students had shown, by the manner in which they had stuck to their work, that they had their educational interests at heart, and upon this the success of the school depends. He had no hesitation in saying that the school would grow, and that the City Council would find it necessary to pay more and more attention to it.

After a song by Mr. McLean, Mr. E. J. Phillips, in the absence of Mr. W. G. Blackgrove, the President, responded to the toast of the "Executive Council." A few years ago, he said, it would have been impossible to find a single engineer who could figure out the strength of a double butt strap joint, triple rivetted, but to-day there were in the ranks of the Association quite a number of men who could do this, while others who a few years since could not figure at all, could now solve any ordinary mathematical problem. During the recent hard times the Association of Stationary Engineers had added a larger number of new members to its membership, and had had better attended meetings than any other society with which he was acquainted. The skill of the boiler maker was of little value, if his work was placed under the care of an incompetent engineer. In conclusion, he expressed gratitude to Prof. Galbraith and Mr. John Galt, whose knowledge had always been placed at the disposal of members of the Association.

At this juncture Mr. George Grant sang, "We're a' John Tamson's Bairns."

In response to the toast, "Sister Societies," Mr. A. B. Smith extended fraternal greetings on behalf of the Canadian Electrical Association. While he did not think that engineers need lose any sleep over the prospect of electricity displacing the steam engine, yet it was hard to predict what the future might have in store. The developments in electricity thus far had shown that a "something" had been introduced which was as swift as light, which never freezes, never breaks, and can turn corners.

Mr. Bliss responded on behalf of the Amalgamated Society of Engineers. This society was composed of mechanics and engineers, and had upwards of 80,000 members in good standing. The receipts last year were $1,220,000, and during the last forty-one years the sum of $561,000 had been expended in assisting other organizations to maintain their rights.

The last toast on the programme was that of "The Press," which was responded to by Mr. Biggar, of the Canadian Engineer, and C. H. Mortimer, of the ELECTRICAL NEWS.

The Committee of Management, to whose efforts this very successful dinner is due, was composed of Mr. E. J. Phillips, Chairman; G. S. Mooring, Secretary-Treasurer; Samuel Thomson, James Huggett, Wm. Eversfield and A. E. Edkins.

### MR. WALTER LEWIS.

Mr. Walter Lewis, president of Toronto Association No. 1, C. A. S. E., is quite as good looking as his portrait which we have the pleasure to present to our

MR. WALTER LEWIS.

readers, and possesses the geniality combined with firmness and executive ability necessary to the successful discharge of the duties of a presiding officer.

Mr. Lewis was born in Chatham, England, 37 years ago, and came to Canada when only seven years of age. He has been a resident of Toronto for twenty years, during which time he has filled positions with Messrs. Neil & Sons, engine builders, H. E. Clarke & Co., the Toronto Silver Plate Co., etc. On resigning his position with the last-named company he was presented with a valuable gold chain by his employers, in recognition of the ability and faithfulness with which he discharged his duties during the seven years he was in their service. He now occupies a responsible position as engineer at the High Level Pumping Station of the Toronto waterworks.

Mr. Lewis was one of the organizers of Toronto Association No. 1, and gave proof of his ability in several minor offices before being elected to the president's chair. His rule of conduct is best expressed in his own words, as follows : " Since making up my mind to be an engineer I have always tried to study what would be useful to me in that trade or profession, and have tried to help any one in the trade who at any time required a helping hand, recognizing the fact that while one is helping others he is benefitting himself also, and that we are never done learning, for no sooner do we master an old thing than a new one comes along."

### MONTREAL NO. 1.

The Montreal branch of the C. A. S. E. is preparing for a vigorous winter's campaign. They have already done good work in.an educational way, and this season promises better results than ever. They have made arrangements for practical demonstrations on subjects of interest, and for the reading of papers, at which all engineers will be invited to be present.

At their last meeting Mr. Peter McNaughton read a paper on " Evaporation in Steam Boilers and in Nature," the point being that the action of the steam boiler is identical with the process of nature, and that the use of steam is but applying nature's law to practical purposes. Mr. J. J. York gave a demonstration on the blackboard of the heating surface and horse power of a Lancaster boiler.

### COMBUSTION.*

#### BY THOMAS WENSLEY, OTTAWA.

COMBUSTION is the energetic chemical combination between the oxygen of the air and the constituents of the combustible, and the value of any fuel is measured by the number of heat units which its combustion will generate, a unit of heat being the amount required to heat one pound of water one degree Fahrenheit. This fuel chiefly used to generate the heat consumed by steam engines is coal and wood, the component parts of which are carbon, hydrogen and ash, with sometimes small quantities of other substances not materially affecting its value. The combustible is that portion which will burn, and, in the combustion of coal, carbon is the principal substance that unites with oxygen, and the air is the source from which oxygen is derived.

Coal has been divided into two primary divisions, viz., anthracite, or hard coal, and bituminous, or soft coal. Anthracite contains a very small portion of volatile matter, but is nearly pure carbon, ranging from 85 to 94 per cent., and burns almost without flame. The term anthracite is never applied to coal containing less than 82 per cent. of carbon. The usual components of soft coal are bituminous volatile matter, coke and ash, as a mechanical separation, but chemically the constituents of coal, though varying in quality as well as degree, are chiefly carbon and hydrogen gas, combined occasionally with a small proportion of sulphur and incombustible matter. The proportion of carbon in this coal varies ; in good coal it is seldom less than 75 per cent. of the whole, sometimes considerably more. Not only do the different kinds of coal differ in their constituents, but coal from the same seam will vary considerably from the normal standard of that coal.

From a scientific analysis, by Professor Liebeg and other eminent chemists, it has been shown that in soft or bituminous coal there is about 80 per cent. of carbon, 5 per cent. of hydrogen, 10 per cent. of azote and oxygen, and 5 per cent. of ash, varying with the different kinds. The principal constituents of all coals, carbon and hydrogen, are united and solid in its natural state, and are essentially different in their character and in their modes of entering into combustion.

The theory of combustion is well understood by scientists, but in practice the art of burning coal economically, and of converting all its natural elements into heat and power, is but little understood. It is also a well known fact that carbon and hydrogen require certain quantities of atmospheric air to effect their combustion, yet, in practice, the means necessary to find out what quantity is supplied, is generally neglected and treated as though it was of no importance.

The bituminous portion of coal is convertible into heat in the gaseous state alone, and then only in proportion to the right mix-

* A paper read before the Canadian Association of Stationary Engineers.

ture and union effected between them and the oxygen of the air, while the carbonaceous portion is only combustible in its solid state, and neither can be consumed while they remain united. To obtain combustion they must be separated, and a new union formed with the oxygen of the air. In combustion there must be a combustible and a supporter of combustion, which means chemical union, and oxygen is this supporter. In fact oxygen is just as essential in combustion as it is in the maintenance of life in the animal kingdom.

You all know from experience that putting on a fresh supply of coals on the furnace, they do not immediately increase the general temperature, but, on the contrary, become the absorbent of heat, the source of the volatilization of the bituminous portion of the coal ; and until these constituents are evolved from it, its solid or carbonaceous part remains black, and at a comparatively low temperature. Now volatilization is the most cooling process of nature, by reason of the quantity of heat directly converted from the sensible to the latent state.

On the application of heat to bituminous coal the first result is its absorption by the coal, then follows the liberation of its gases, from which flame is exclusively derived. These gases are composed of carbon and hydrogen, and the union is known as carburetted hydrogen and bi-carburetted hydrogen. Carburetted hydrogen by itself is not combustible, but must be united with oxygen, and notwithstanding the strong attraction which exists between them, they will not rush together or enter into chemical union, which we call combustion, until they have been raised to a certain temperature, and this temperature, according to Sir Humphry Davy, should not be under 800 degrees Fahrenheit, since below that flame cannot be produced or maintained.

The first essential to effect the combustion of gas is to ascertain the quantity of oxygen with which it will chemically combine, and the next the quantity of air required to supply the necessary quantity of oxygen. Now while this may be well understood and correctly arrived at by an expert chemist in his laboratory, we know that in the management of combustion in the furnace the ordinary engineer can at best only approximately apply the exact laws of chemistry to the very imperfect conditions found at every furnace. It is important, however, that every engineer in charge of a steam plant should at least understand theoretically the analysis of the elements with which he has to deal in producing combustion, and the proportional part of each element entering into the same.

According to chemical analysis an atom of hydrogen is double the bulk of carbon vapor, but the latter is six times the weight of the former. (Atom in modern scientific usage is the smallest portion into which matter can be divided. The chemists unit. In chemistry two atoms of hydrogen and one atom of oxygen make a molecule of water.) Again, an atom of hydrogen is double the bulk of an atom of oxygen, yet the oxygen is eight times the weight of hydrogen. So of the constituents of atmospheric air, which is a mechanical mixture of nitrogen and oxygen, not in chemical union, but simply shaken up together. These constituents, nitrogen and oxygen, are mixed in the proportion of 79 parts of nitrogen to 21 parts of oxygen out of every 100, and by weight 77 lbs. of nitrogen to 23 lbs. of oxygen, or one pound of oxygen to every 3'3478 pounds of nitrogen.

To accomplish the combustion of six pounds of carbon, sixteen pounds of oxygen are necessary, forming 22 lbs. of carbonic acid gas, which will have the same volume as the oxygen, and therefore a greater density, and to accomplish the combustion of one pound of hydrogen, eight pounds of oxygen are required. When therefore we know the proportions of carbon and hydrogen existing in coal it is easy to tell the quantity of oxygen, and consequently the quantity of air necessary for combustion.

As a general rule it may be stated that for every pound of coal burned in a furnace about twelve pounds of air, or 150 cubic feet, will be necessary to furnish the oxygen required, even if every particle of it entered into combustion. But from careful experiment it has been found that in ordinary furnaces about as much more air will in practice be necessary, or about 24 lbs. per pound of coal burned, since, besides the air required to furnish the oxygen necessary for the complete combustion of the fuel, it is also necessary to furnish an additional quantity for the dilution of the gaseous products of combustion. Now one cubic foot of air, at a temperature of 40 degrees, weighs '08 of one pound, and it requires 12½ cubic feet of atmospheric air to equal one pound in weight, and each pound of air contains 3.68 ounces of oxygen, and it will take 1,200 pounds or 15,000 cubic feet of air for the perfect combustion of 100 pounds of coal. We thus perceive that each pound of coal requires 150 cubic feet of air for its perfect combustion, or in other words, for the conversion of its carbon into carbonic acid, and all its hydrogen into water, and it must be remembered that just in proportion as this proper quantity is deficient, combustion is imperfect and fuel wasted. __

Air expands or contracts an equal amount with each degree of variation in temperature, and its weight and volume for any condition of temperature and pressure may be found by the following formulae, which are nearly exact :—

$$\text{Weight} = \frac{2 \cdot 71 \times \text{Pressure in lbs. on the barometer.}}{\text{Absolute temperature.}}$$

$$\text{Volume} = \frac{\text{Absolute temperature.}}{2 \cdot 71 \times \text{Pressure on barometer in lbs.}}$$

Absolute temperature = 460 + temperature shown on thermometer.

$$\text{Pressure in lbs. on barometer} = \frac{\text{Height in inches.}}{2 \cdot 0408}$$

It is erroneously supposed by some that when no smoke appears at the chimney top, combustion is perfect ; smoke, however, may be absent, yet the carbon may have only united with one atom of oxygen forming carbonic oxide (a colorless gas), instead of with two atoms forming carbonic acid, and consequently have only performed half the duty as a fuel of which it was capable, and this loss is constantly going on in all furnaces where all the air has to pass through a body of incandescent carbonaceous matter.

The air on entering from the ash pit gives up its oxygen to the glowing carbon on the bars, and generates great heat in the formation of carbonic acid, and this acid necessarily at a very high temperature, passing upwards through the body of incandescent solid matter, takes up an additional portion of carbon and becomes carbonic oxide. By the conversion of one volume of carbonic acid into two volumes of carbonic oxide, heat is actually absorbed, while the carbon taken up during such conversion is also lost. The formation of this compound, carbonic oxide, is attended by circumstances of a curious and involved nature, and is probably the cause that, in actual practice, so little is known about it. The direct effect of the union of carbon and oxygen is the formation of carbonic acid. If, however, we abstract one of its portions of oxygen, the remaining portions would be carbonic oxide, and it is equally clear that if we added a second portion of carbon to carbonic acid the same result will be arrived at, namely, have carbon and oxygen in equal proportions, as we have in carbonic acid. By the addition of still another portion of carbon, two volumes of carbonic oxide will be formed, and if these two volumes of oxide cannot find the oxygen necessary to complete their saturating equivalents, they pass away but half consumed.

Another important peculiarity of carbonic oxide is, that by reason of its already possessing one-half of its equivalent of oxygen, it inflames at a lower temperature than the ordinary coal gas, the consequence of which is that the latter, on passing into the flues, is often cooled down below the temperature of ignition, while the former is sufficiently heated, even after having reached the chimney top, and is there ignited on meeting the air. This is the cause of the flame often seen at the top of chimneys or the funnels of steamships.

If we could gather and retain the carbonic acid gas which is daily discharged by tons from the chimneys of our factories, we should still have all the carbon of our coal, but we could not do it, because it would take as much power to separate the carbon from the oxygen as they gave out in the form of heat in coming together, and here comes in one of nature's most wonderful and mysterious processes.

It is a peculiar function of vegetation that under the influence of sunlight it can overcome the attraction which exists between the atoms of carbon and oxygen, appropriating the carbon to its own use, building it into its structure and letting the oxygen go free into the atmosphere, not with a noisy demonstration or prodigious effort, but quietly in the delicate structure of a green leaf moving in the sunshine.

When all the conditions belonging to the introduction of air to the two distinct bodies to be consumed, carbon and hydrogen, have been complied with, there should be very little difficulty in securing perfect combustion in the furnace. But as a rule, these conditions are not complied with, hence the great waste in fuel. If we would economize fuel, we must give attention, not only to the mechanical appliances, but also to the nature of the bodies we have to deal with, their constituent parts and chemical relations respectively, and as the laws of nature are inexorable, mechanical details must yield to those of chemistry.

Great strides have been made in improvements in the boilers and engines now on the market, but until recently scarcely any

attention has been given to the grates and furnace, practically overlooking the fact that the furnace, in which the operations of combustion are carried out, is of the first importance, as it is here we have the real source of economy and power.

In regard to the proportions of the furnace, we have to consider the area of the grate bars for the holding of the solid fuel, and the kind best adapted to our purpose (some people think that anything will do for a grate that will stand up under hot fires), the size of the air spaces, and the means of keeping these air spaces clear of obstruction to the draught ; then the sectional area of the chamber over the fuel for the consuming of the gaseous portion of the coal and the introduction of oxygen to this chamber.

The rule in practice to-day with our best fire-tube boilers, the horizontal return tubular, is to allow 15 square feet of heating surface per horse power, and by dividing the horse power by three, we obtain our grate surface in square feet, allowing 68 square inches of air space per square foot of grate.

Strictly speaking, there is no such thing as " horse-power to a steam boiler, as it is a measure only applicable to dynamic effect. But as boilers are necessary to drive steam engines, the same measure applied to steam engines has come to be universally applied to the boiler, and cannot well be discarded. In consequence of the different quantity of steam necessary to produce a horse power, with different engines, there has been great need of an accepted standard by which the amount of boiler required to provide steam for a commercial horse power may be determined. This standard, as fixed by Watt, was one cubic foot of water evaporated per hour from 212° for each horse power. This was at that time the requirement of the best engines in use. At the present time Prof. Thurston estimates that the water required per hour, per horse power, in good engines, is equal to the constant 200, divided by the square root of the pressure, and that in the best engines this constant is as low as 150. This would give for good engines working with 64 pounds pressure, 25 pounds water, and for the best engines working with 100 pounds, only 15 pounds water per hourly horse power.

The extensive series of experiments made under the direction of C. E. Emery, M. E., at the Novelty Iron Works, and published by Professor Trowbridge, show that at ordinary pressure, and with good proportions, non-condensing engines of from 20 to 300 horse power required only from 25 to 30 lbs. water per hourly horse power in regular practice.

The standard, therefore, adopted by the judges at the Centennial Exhibition of 30 lbs. of water per hour, evaporated at 70 lbs. pressure from 100° for each horse power, is a fair one for both boilers and engines, and has been favorably received by both engineers and steam users. But as the same boiler may be made to do more or less work, with less or greater economy, it should be also required that the rating of a boiler be based on the amount of water it will evaporate at a high economical rate. For the purposes of economy, the heating surface should never be less than one and generally not more than two square feet for each 5,000 British thermal units to be absorbed per hour, though this depends somewhat on the character and location of such surface. The range here given is believed to be sufficient for the different conditions in practice, though a far greater range is frequently employed. Square feet of heating surface is no criterion as between different styles of boilers—a square foot under some circumstances being many times as efficient as in others—but when an average rate of evaporation per square foot has been fixed upon by experiment, there is no more convenient way of rating the power of others of the same style.

(To be Continued.)

One of the old timers in Canadian telegraphy died recently in Hamilton, in the person of Mr. Charles Jamieson, a lineman on the G. N. W. He entered the service of the Montreal Telegraph Company about 1850, and about 1856 was stationed at Prescott, from where he removed to Hamilton in 1857, remaining there ever since. He worked on the first line erected in Canada, the old Grand Trunk. His death was caused by pneumonia, arising from exposure.

The escape of gas has always been a source of loss and deterioration in the curing of champagne, and no system of perfect air-tight sealing was known. Electricity has come to the aid of the champagne makers, and a system of electrical sealing has been discovered, by which the cork and part of the neck of the bottle are covered with a thin layer of copper. The process is simple and answers its purpose to perfection. It can be extended to the sealing of all kinds of bottles and jars,

PUBLISHED ON THE FIFTH OF EVERY MONTH BY

## CHAS. H. MORTIMER,

OFFICE: CONFEDERATION LIFE BUILDING,

*Corner Yonge and Richmond Streets,*

## TORONTO, - - CANADA.

Telephone 2362.

NEW YORK LIFE INSURANCE BUILDING, MONTREAL.

Bell Telephone 2299.

#### ADVERTISEMENTS.

Advertising rates sent promptly on application. Orders for advertising should reach the office of publication not later than the 25th day of the month immediately preceding date of issue. Changes in advertisements will be made whenever desired, without cost to the advertiser, but to insure proper compliance with the instructions of the advertiser, requests for change should reach the office as early as the 22nd day of the month.

#### SUBSCRIPTIONS.

The ELECTRICAL NEWS will be mailed to subscribers in the Dominion, or the United States, post free, for $1.00 per annum, 50 cents for six months. The price of subscription should be remitted by currency, in registered letter, or by postal order payable to C. H. Mortimer. Please do not send cheques on local banks unless 25 cents is added for cost of discount. Money sent in unregistered letters will be at senders' risk. Subscriptions from foreign countries embraced in the General Postal Union, $1.50 per annum. Subscriptions are payable in advance. The paper will be discontinued at expiration of term paid for if so stipulated by the subscriber, but where no such understanding exists, will be continued until instructions to discontinue are received and all arrearages paid.

Subscribers may have the mailing address changed as often as desired. When *ordering changes, always give the old as well as the new address.*

The Publisher should be notified of the failure of subscribers to receive their papers promptly and regularly.

#### EDITOR'S ANNOUNCEMENTS.

Correspondence is invited upon all topics legitimately coming within the scope of this journal.

THE "CANADIAN ELECTRICAL NEWS" HAS BEEN APPOINTED THE OFFICIAL PAPER OF THE CANADIAN ELECTRICAL ASSOCIATION.

## CANADIAN ELECTRICAL ASSOCIATION.

#### OFFICERS:

PRESIDENT:

A. B. SMITH, Superintendent G. N. W. Telegraph Co., Toronto.

1ST VICE-PRESIDENT:

C. BERKELEY POWELL, Director Ottawa Electric Light Co., Ottawa.

2ND VICE-PRESIDENT:

L. R. McFARLANE, Manager Eastern Department, Bell Telephone Company, Montreal.

SECRETARY-TREASURER:

C. H. MORTIMER, Publisher ELECTRICAL NEWS, Toronto.

EXECUTIVE COMMITTEE:

GEO. BLACK, G. N. W. Telegraph Co., Hamilton.

J. A. KAMMERER, General Agent, Royal Electric Co., Toronto.

E. C. BREITHAUPT, Berlin, Ont.

F. H. BADGER, JR., Superintendent Montmorency Electric Light & Power Co., Quebec.

JOHN CARROLL, Sec.-Treas. Eugene F. Phillips Electrical Works, Montreal.

K. J. DUNSTAN, Local Manager Bell Telephone Company, Toronto.

O. HIGMAN, Inland Revenue Department, Ottawa.

W. Y. SOPER, Vice-President Ottawa Electric Railway Co., Ottawa.

A. M. WICKENS, Electrician Parliament Buildings, Toronto.

J. J. WRIGHT, Manager Toronto Electric Light Company.

## MONTREAL ELECTRIC CLUB.

#### OFFICERS:

President, W. B. SHAW, - Montreal Electric Co.
Vice-President, H. O. EDWARDS,
Sec'y-Treas., CECIL DOUTRE, - 81A St. Famille St.
Committee of Management, T. F. PICKETT, W. GRAHAM, J. A. DUGLASS.

## CANADIAN ASSOCIATION OF STATIONARY ENGINEERS.

#### EXECUTIVE BOARD:

President, W. G. BLACKGROVE, - - Toronto, Ont.
Vice-President, JAMES DEVLIN, - - Kingston, Ont.
Secretary, E. J. PHILIP, - - Toronto, Ont.
Treasurer, DUNCAN ROBERTSON, - Hamilton, Ont.
Conductor, W. F. CHAPMAN, - - Brockville, Ont.
Door Keeper, F. G. JOHNSTON, - - Ottawa, Ont.

TORONTO BRANCH No. 1.—Meets 2nd and 4th Friday each month in Room D, Shaftesbury Hall. W. Lewis, President; S. Thompson, Vice-President; T. Eversfield, Recording secretary, University Crescent.

MONTREAL BRANCH No. 1.—Meets 1st and 3rd Thursday each month, in Engineers' Hall, Craig street. President, John J. York, Board of Trade Building; first vice-president, J. Murphy; second vice-president, W. Ware; secretary, B. A. York; treasurer, Thos. Ryan.

ST. LAURENT BRANCH No. 2.—Meets every Monday evening at 43 Bonsecours street, Montreal. R. Drouin, President; Alfred Latour, Secretary, 306 Delisle street, St. Cunegonde,

BRANDON, MAN., BRANCH No. 1.—Meets 1st and 3rd Friday each month, in City Hall. A. R. Crawford, President; Arthur Fleming, Secretary.

HAMILTON BRANCH No. 2.—Meets 1st and 3rd Friday each month, in Maccabee's Hall. E. C. Johnson, President; W. R. Cornish, Vice-Pres.; Wm. Norris, Corresponding Secretary, 211 Wellington Street North.

STRATFORD BRANCH No. 3.—John Hoy, President; Samuel H. Weir, Secretary.

BRANTFORD BRANCH No. 4.—Meets and and 4th Friday each month, F. Lane. President; T. Pilgrim, Vice-President; Joseph Ogle, Secretary, Brantford Cordage Co.

LONDON BRANCH No. 5.—Meets in Sherwood Hall first Thursday and last Friday in each month. F. Mitchell, President; William Meaden, Secretary Treasurer, 533 Richmond Street.

GUELPH BRANCH No. 6.—Meets 1st and 3rd Wednesday each month at 7:30. p.m. J. Fordyce, President; J. Tuck, Vice-President; H. T. Flewelling, Rec. Secretary; J. Gerry, Fin.-Secretary; Treasurer, C. J. Jorden.

OTTAWA BRANCH, No. 7. — Meets 2nd and 4th Tuesday, each month, corner Bank and Sparks streets; Frank Robert, President; F. Merrill, Secretary, 352 Wellington Street.

DRESDEN BRANCH No. 8.—Meets every and week in each month; Thos. Merrill. Secretary.

BERLIN BRANCH No. 9.—Meets 2nd and 4th Saturday each month at 8 p.m. W. J. Rhodes, President; G. Steinmetz, Secretary, Berlin Ont.

KINGSTON BRANCH No. 10.—Meets 1st and 3rd Tuesday in each month in Fraser Hall, King Street, at 8 p.m. President, S. Donnelly; Vice-President, Henry Hopkins; Secretary, J. W. Tandvin.

WINNIPEG BRANCH No. 11.—President, G. M. Hazlett; Recording Secretary, J. Sutherland; Financial Secretary, A. B. Jones.

KINCARDINE BRANCH No 12.—Meets every Tuesday at 8 o'clock, in McCibbin's Block. President, Daniel Bennett; Vice-President, Joseph Lightball; Secretary, A. Scott.

WIARTON BRANCH No. 13.—President, Wm. Craddock; Rec. Secretary, Ed. Dunham.

PETERBOROUGH BRANCH No. 14.—Meets 2nd and 4th Wednesday in each month. S. Potter. President; C. Robison, Vice-President; W. Sharp, engineer steam laundry, Charlotte Street, Secretary.

BROCKVILLE BRANCH No. 15.—President, W. F. Chapman; Vice-President, A. Franklin; Recording Secretary, Wm. Robinson.

CARLETON PLACE BRANCH No. 16.—President, Jos. McKay Vice President, Henry Derrer; Fin. Secretary, A. M. Schofield.

## ONTARIO ASSOCIATION OF STATIONARY ENGINEERS.

#### BOARD OF EXAMINERS.

President, A. AMES, - - - Brantford, Ont.
Vice-President, F. G. MITCHELL - London, Ont.
Registrar, A. E. EDKINS - - 139 Borden St., Toronto.
Treasurer, R. MACKIE, - - 28 Napier st., Hamilton.
Solicitor, J. A MCANDREWS, - Toronto.

TORONTO—A. E. Edkins, A. M. Wickens, E. J. Phillips, F. Donaldson.
HAMILTON—P. Stott, R. Mackie, T. Elliott.
BRANTFORD—A. Ames, care Patterson & Sons.
OTTAWA—Thomas Wesley.
KINGSTON—J. Devlin (Chief Engineer Penitentiary), J. Campbell.
LONDON—F. Mitchell.
NIAGARA FALLS—W. Phillips.

Information regarding examinations will be furnished on application to any member of the Board.

THE sixth year of publication of THE ELECTRICAL NEWS comes to a close with the present number. An index to the contents of this volume is presented herewith. At the commencement of a new volume and a new year, we may have a few words to say with regard to the progress achieved. Meanwhile, we heartily extend to every reader best wishes for a Happy and Prosperous New Year.

THE Grand Trunk Railway has adopted the block system of running its trains, an important step, but one which will afford much satisfaction to its patrons, in that it provides, if faithfully carried out, against all possibility of accident from collisions, either front or rear. The change will necessitate the employment of a large additional staff of telegraph operators.

A CONVERSATION which the writer had recently with a Toronto man who had just returned from Europe, goes to show that in "Lunnon" there should exist the most profitable field in the world for electric lighting. The atmosphere is described as being at times so thick with fog that it can almost be cut into slices, and the stranger who ventures beyond his doorstep requires the assistance of a native or policeman to enable him to locate himself.

THE Toronto Globe satirizes thus the visionary scheme of the Georgian Bay Power Canal promoters : " A project is on foot to utilize the water power of the Humber River. It and other channels drain an area of, by the estimate, 562 square miles. The yearly rainfall on this is about twenty inches, and if it were all caught in pails, buckets and barrels and carried to an effective situation, it could be used to make a water power truly gigantic."

MR. James Milne is making a great success of the Electricity Class at the Toronto Technical School. The present membership of the class is said to be about 160. Mr. Milne is the only teacher at the school who has had the benefit of a technical school training. It is not altogether surprising, therefore, that possessed of this advantage, in addition to excellent natural ability, he should have proved himself to be exactly the right man in the right place.

A CURIOUS fact in connection with the growth of electric street railways was brought out at a recent meeting of the corporation of McGill University at Montreal. The deans of the different faculties reported the number of students as 1193, there being an increase in all except that of veterinary science. Dr. McEachren explained the decrease in this subject by the fact that the displacement of horses by electricity for street car purposes had greatly reduced the business of veterinary surgeons.

THE London, Ont., Street Railway Co., after the introduction of electricity on their lines, adopted a somewhat ingenious means to surmount a difficulty which confronted them. The C. P. R. objected to the street company crossing their tracks, and the permission of the Privy Council at Ottawa has to be obtained, which takes time. The right to cross with horses still existed, so the cars were run up close to the C. P. R., where horses were attached to draw them across, when they proceeded on their way with electricity.

PERHAPS we may see the day when dynamos will be done away with and electrical power be drawn from nature's huge dynamo—the earth. Prof. Bigelow, of Washington, suggests the idea that the earth acts as the armature of a great dynamo, and by revolving in the sun's magnetic field generates the so-called earth currents of electricity. Mr. Lang points out how, by wrapping the earth with a suitable system of conductors, power could be obtained, and he even goes into a calculation how much it would be. The only difficulty is the enormous cost of the plant. Perhaps that can ultimately be surmounted.

THE Street Railway Review, of Chicago, assumes responsibility for the statement that the ladies of Montreal ignored their American sisters who attended the recent street railway convention. Our contemporary says : " The ladies turned out in good force, and made the best of the situation in entertaining themselves ; for as usual their escorts could give but little time. One could but recall the glorious hospitality of the resident ladies of Washington, Milwaukee, Cleveland, Buffalo, and elsewhere." What have the Montreal ladies, or those who should be their spokesmen, to say in answer to the charge? It can at least be said that the attendance at the business sessions during the convention was slim enough to warrant the inference that the delegates might easily have found more time to devote to the ladies.

THE Carmelite Monastery at Niagara Falls, Ont., is trying an experiment in heating by electricity, which will be watched with interest. The whole building will not be so heated, but having arranged for a fixed amount of power, all of which is not required for other purposes, the surplus is to be employed for heating. Electricity is not an economical agent for heating so far, and is used on street cars on account of its convenience, and because the current cuts no figure in the railway's expense account. But it will be employed at the monastery under favourable conditions, and the result will be of service in arriving at the relative cost as compared with other methods.

IT seems unfortunate that the Association of Stationary Engineers should have been organized upon the basis of a fraternal order, with all the paraphernalia of a secret society—not that we have anything to say against such societies, or lodge-room methods, but because the cumbersome machinery of a secret society is unnecessary where there is nothing to conceal. Further, the ceremonies of initiation, etc., cannot always be carried out in an impressive manner, in which case they have a bad effect upon those who take part in them, and they occupy time which might better be devoted to the legitimate purposes of the organization. The association has for its object the improvement of its members in their chosen calling, and this is not helped by secret society methods.

AN application is before the United States Senate for the right to build an electric railway from New York to Washington, on which a speed of one hundred miles an hour is to be maintained, which really means one hundred and twenty miles an hour including stoppages. The Brott system is to be employed, in which the cars run upon one wheel in the centre, and the track is elevated about two feet, except at road crossings, where it will be higher to allow a passage underneath. An absolutely straight line is required. The trolley system is employed with the conductor underneath. An experimental line of thirty miles is to be built between Washington and Chesapeake Bay. There seems to be no reason why the plans of the projectors should not be realized. The New York General Electric Co. is prepared to guarantee all the mechanism required, and to maintain a speed of one hundred and fifty miles.

AT the recent meeting of shareholders of the Grand Trunk Railway, presided over by the new president, Sir Charles Rivers Wilson, a statement was presented showing that the company at first leased, and afterwards purchased the Belt Line Railway encircling the city of Toronto, constructed some five or six years ago, and that after having invested nearly half a million dollars, had found that they were unable to operate the line at a profit and had decided for the present at least to allow it to stand idle. It seems to us that this line, if equipped as an electric trolley route, might profitably be handled by the Toronto Street Railway Company. The road runs through one of the most picturesque routes in America, and might be made a very profitable pleasure road. It certainly cannot be made to pay as a steam road, or for ordinary passenger and freight business, but as a pleasure road, in summer at least, we believe it could be made a paying adjunct to the already profitable city system.

THE secondary wiring system supplied by a central station, is a very important feature of the whole, and should be so regarded by the manager. There is a practice allowed in some of the medium sized towns that cannot be too strongly condemned. It is allowing any wiring contractor who sets up in business, to take contracts for wiring houses, etc., without reference to his experience, or antecedents, and without exercising any kind of supervision over his work. It certainly does not require any great amount of training to fit a man to do the mere installing work. A handy man who can use saw, screwdriver and pliers, can, without any difficulty, become a good wireman in a very few weeks, if he works at first with an expert; but this is by no means all that should be expected of a contractor who is to design the entire lighting system of a large residence, store, church or theatre. Such an installation is a small lighting system, and should be laid out with as great attention to the location of centres of distribution and drops at various points as is absolutely necessary in the design of the primary system in the streets. It seems usual to allow a drop of two per cent. in interior wiring between the transformer and the lamps. It is evident that with a margin of two volts on a 100 volt lamp, it is quite possible to so wire that the first lamp will receive the full pressure without any drop, while another will not get enough, and so on, but that if a little attention is paid to laying out the system, all the lamps will get the same pressure within a small fraction of a volt. This is only possible by establishing centres, and running mains, feeders and branches. Another important matter is, that however carefully a primary system is laid out, it is impossible to have just the same pressure at all points, and therefore it may so happen that the above two per cent. allowance is either too high or too low, by one volt. At a point close to a main or branch centre of distribution the pressure will be greater than at another point some considerable distance away, and the interior allowance should be in proportion. Two cases that came quite recently under the notice of the writer will illustrate the above remarks. In one town where considerable extensions were being made to the plant, two distinct wiring contractors were working independently. Some discrepancy having been observed, investigation showed that one was allowing two per cent. drop, while the other was allowing "no drop at all" i. e., was putting in such large wire as to have practically none. The result must be plain to anyone. In the second case, two men undertook to wire all the secondary system, and general dissatisfaction ensued. They had no experience in laying out work, and the pressures at the lamps were most various; in one case three lamps in different rooms had pressures varying by five volts among themselves. Central station men may think it is none of their business, and that if customers choose to engage outsiders to do the work they must take the consequences. This is quite so, but at the same time the central station feels the result of the poor service at the consumers' lamps. All work undertaken by outside contractors should be approved of by the central station manager before being done, and no contractor should be permitted to do any who cannot furnish satisfactory evidence of being competent to design and lay out work properly.

A MATTER that is of considerable importance to electrical men, and that receives much less attention than it deserves, is the wattage of lamps. The manufacture of incandescent lamps has been closely studied for several years now, and the many various factors that influence in greater or less degree their excellence are well known. Lamps can be manufactured, indeed, to conform to almost any required conditions; and the conditions of operation in different central stations, or isolated plants, are so different from each other sometimes as to require different lamps to produce sensibly the same results. In a large area of supply, for instance, when distribution is by the direct current, and where plenty of copper can be placed for feeders and pressure regulators and so on, it is expedient to use high efficiency lamps, two and a half to three watts per candle power. The requirement of this kind of lamp is, that the extreme range of variation in pressure to which it shall be subjected shall be very small; and this of course can be arranged when the business of the station is so large as to permit of the use of heavy feeders, and the allowance of very small drops. But on the other hand, a small area, where distribution is by means of alternating currents, with individual transformers placed feeding into small installations of lamps, without the intervention of secondary mains, is very much better served by the use of lamps of a higher wattage—three and a half to four watts per candle. No matter how carefully the primary system is laid out, the transformer itself, owing to the unavoidable imperfections of its design and construction, will introduce a variation in the pressure on its secondaries that will react more or less unfavorably on the lamps. Now, as a lamp becomes more and more efficient, i. e., takes less watts per candle, it absolutely requires a more and more close pressure regulation; and conversely, a less efficient lamp will not be so periodically affected by a range of variation. If a transformer be used that varies three volts per cent. between no load and full load (a by no means unusual amount), then a lamp of quite three and a half watts per candle should be used; for if a three watt or less be used the range of variation will introduce great depreciation and consequent shortness of life; whereas a lamp of larger wattage will stand more rough usage. With a transformer giving a drop of one per cent. (which is done by the best makes) a higher efficiency lamp may be used, with a consequent gain in transformer and machine capacity. A machine constructed for 50 kilowatts will give a capacity of twelve hundred and fifty 2½ watt lamps; of one thousand 3 watt lamps; and of eight hundred 4 watt lamps; but to use these 1250 high efficiency lamps it is necessary to have the entire primary system carefully laid out; to use none but the highest class transformers; and to put in secondary mains throughout. It might be incidentally pointed out that this gain in lamp capacity is a strong argument in favor of using none but the very highest class apparatus, and of exercising the greatest care in calculating a wiring system. Reduced to arithmetic the gain is something as follows : Using 2½ watt lamps instead of 4 watt gains 420 lamps; these at $5 per annum bring in an income, additional, of $2,100. This will mean a net gain of about $1,300 per annum ; against which must be placed the additional cost of A1 transformers over second rate ones, perhaps 10c. per light, and the cost of remodelling the entire wiring system—probably an expenditure of less than $1,000 would suffice to make all the necessary changes and improvements, and this investment would bring a yearly interest of at least 100 per cent. Central station men would do well to overhaul their entire plant, and see whether great improvements might not be effected as above.

WE wish to point out that the credit for the design and construction of the new electric locomotive now in successful operation on the B. & O. at Baltimore, was by a slip of the pen attributed to the Westinghouse Co. instead of to the General Electric Co. The error would of course be generally understood, the pioneer work of the General Electric Co. in the field of heavy electric locomotives having been closely followed by the electrical public generally. The arrangement recently announced between the Baldwin and Westinghouse Companies is we believe due to the desire of the latter company to follow their great rival into a field which has now been shown by the success of the Baltimore locomotives to possess immense and immediate possibilities.

## ELECTRIC LIGHTING IN TORONTO.

SOME important developments are about to take place in the electric lighting situation in Toronto. The Toronto Electric Light Company is about to increase its capital stock from $500,000 to $700,000. The object of this increase in capital is to enable the Company to improve its equipment, and to engage in incandescent as well as arc lighting.

The Company have received tenders for the construction of two vertical high pressure condensing engines, each to have a capacity of 1,800 horse power, and to be operated at a speed of 100 revolutions per minute. Each of these engines will have two large fly wheels from which power will be transmitted by means of 36-inch belts, to four generators, each having a capacity of 12,000 lights.

In order to test the claims made on behalf of the monocyclic and three phase systems, it is proposed to install one generator of each system, and thus put the question to a practical test.

There is at present a large section of the city which is not reached by the mains of the Toronto Incandescent Light Company, and it is the intention of the Toronto Company to first supply these districts. They believe that by means of the over-head alternating system, they can successfully compete with gas, even at the reduced rate of 90 cents per thousand feet, at which price it is likely to be supplied in the near future, as a result of the decision recently given in the Courts against the Consumers' Gas Company. Gas at 90 cents per thousand feet is about equal to incandescent light at half a cent per light per hour, and this is the price which the Toronto Company propose to charge.

The Company point to the fact that Ottawa, with a population of about 50,000, has installed one incandescent lamp for every unit of population, and there are probably in daily operation in that city not less than 20,-000 lights. The capacity of the Toronto Incandescent Light Company's station is said to be about 20,000 lights, and it is believed that this limit has well nigh

been reached at the present time. If electric light can be supplied to all parts of the city at a price, little, if any, in advance of gas, it should be possible to much more than double the number of lights at present in use, taking into account that the population of the city is about 200,000.

It is also the intention of the Toronto Electric Light Company to greatly improve their arc lighting station. They propose to substitute for the present building a new fire proof structure. The new building will be constructed entirely of brick and iron, and the window openings will be so placed as not to expose the structure to danger from fire from surrounding buildings. It is proposed to bring the wires into the new building through terra cotta conduits and an iron tower. The switchboard will be in the form of panels, having an iron frame, and the instruments mounted on slate. Mr. Wright, manager of the Company, points out that the building when constructed will be in every sense fire proof. To use his own words, "there will not be a piece of wood as large as a lead pencil used in the construction of the building, so that with non-inflammable contents, there will be nothing to burn." It is intended to substitute a number of dynamos of large capacity, —say 125 lights—for the machines at present in use, which have a capacity of only 35 to 40 lights. The new building will be erected outside the present structure, which will afterwards be pulled down. It is the purpose to proceed immediately with the carrying out of the above mentioned improvements.

## REGULATION OF WATER DRIVEN ELECTRICAL PLANTS.

DURING the discussion which followed the reading of Mr. Dion's paper at the recent Convention of the Canadian Electrical Association at Ottawa, a number of delegates asked for information which would show the exact benefit to be derived in the matter of better regulation, from exciting generators by separately driven exciters. No figures were then at hand, but we have since received some information which will probably be of interest to many of our readers. The Ottawa Electric Railway Company, at whose power house separate exciters is carried out with such beneficial results, made the following experiment before definitely deciding to install a special turbine for exciting purposes.

Two 400 h. p. generators, driven from the same countershaft, running at the same speed and voltage, and being practically alike in every particular, were selected to make the test. Both were self exciting, but the field circuit of one of them was temporarily connected to a generator driven from another water wheel. On both gates water wheels were securely set so that the amount of water supply could not vary, and the field charges of the two generators were adjusted till the difference of potential at the brushes of each generator was 500 volts. A load of 200 amperes was thrown on the self excited generator and the voltage fell from 500 to 250 volts in about five seconds. The load was then removed and as soon as the speed of the machinery again reached the normal point, the load was thrown on the separately excited generator. The E. M. F. this time only came down to 450 volts. Several other experiments were carried out along the same lines as this one and all showed conclusively the wisdom of making the proposed change from self to separate exciting.

## ELECTRIC PLANT AT THE NEW UNION STATION, TORONTO.

By the courtesy of Mr. Walter Fuller, electrician in charge, and Mr. Alexander Storer, chief engineer, a representative of the ELECTRICAL NEWS was recently given the opportunity of inspecting the steam and electric plant at the new Union Depot, Toronto. This plant is located in the basement of the new building recently erected immediately south of Front street.

The engine and dynamo rooms are extremely well by which means the smoke is consumed. The power required for the automatic feeding device is supplied by a 1 horse power automatic engine attached to the end of the furnace.

Screenings are exclusively employed for fuel. This fuel is brought into the lane on the west side of the building and dumped through a coal hole into the proper place in the boiler room. It is then loaded on a wheelbarrow, and by means of a compressed air hoist, the barrow, fuel and fireman are lifted to an iron platform

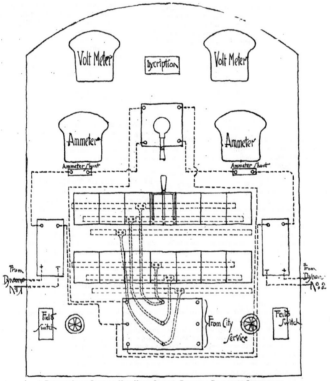

ELECTRIC LIGHT PLANT AT NEW UNION STATION, TORONTO.—DIAGRAM OF SWITCHBOARD.

lighted from windows opening on a lane on the west side of the building. The floors are of concrete, and everything about the place is clean and cheerful. There are three Babcock & Wilcox boilers—one of 186 horse power, and two of 93 horse power. These boilers are fitted with Murphy automatic stokers, and are the first furnaces installed in Canada in connection with which these stokers are employed.

The fuel is fed into a hopper from which it descends into a magazine, and from thence is automatically fed into the fire. Cold air is admitted just above the coal, at the top of the furnace, where the contents of the barrow are dumped into the hoppers already mentioned. By means of this compressed air hoist the ashes which come from the furnace are also lifted in buckets through another coal hole into the lane, where they are dumped into carts. This compressed air hoisting apparatus is in use at many points on the line of the Grand Trunk Railway, and was manufactured at the Company's works at Point St. Charles. The compressed air is pumped into a reservoir suspended from the ceiling.

Turning now to the steam and water apparatus, all

returns come back from the south side station and the new station to a return tank from which they are pumped again to the boiler. The pumps for this purpose were also made at the Grand Trunk Company's works at Point St. Charles. The quantity of water is regulated by an automatic governor which shuts off the steam at given points. On account of the distance to which the steam has to be forced to the south side station, the return water is cold when it gets to the pump, in consequence of which it is forced through exhaust heaters before going into the boilers. The elevators in the building are hydraulic, and after the water has done its work at the elevators it is pumped back again into the pressure tank by a Northey pump, 14 x 8½ x 12 inches, fitted with a Fisher governor. In order to get the right quantity of air into the pressure tank, the air is taken from the air reservoir above referred to. In case of accident to the pump, the city pressure can be employed to run both passenger and baggage elevators.

There are two Robb-Armstrong high speed engines of 80 horse power each, fitted with automatic oilers. The exhaust steam from these engines is used to heat the buildings. The steam pipe to each engine is 4 inches in diameter, and the exhaust pipe 5 inches. Should the engine get a dose of water, automatic valves are immediately released, by means of which the water is got rid of. Mr. Storer, the engineer in charge, is an old employee of the Grand Trunk Company, and is well qualified for the position he now occupies.

Coming now to the electric plant, about 2,000 lights have been installed, forty of which are arc lights. For the operation of these lights two Canadian General Electric generators are employed. Both arc and incandescent lights will be run off the same generator. By means of a triple pole switch the current can be obtained from the Incandescent Light Company's system, in case of accident to the generators or when repairs have to be made. The building is wired for both the two and three wire systems. The switchboard consists of green slate panels, on which are mounted two Weston volt meters, two ammeters, sixteen 3 pole switches, two 2-wire dynamo switches, one break-down switch and two pilot lamps. A diagram of the switchboard is presented herewith, and will enable the reader to trace the connections.

The handsome general waiting room of the new station will contain a 250 light fixture, and the main corridor leading from Front street a 60 light fixture. The cut-outs are arranged in clusters on each floor, the lights on each floor being controlled at the switchboard.

The generating plant is not only designed to light the buildings, but to furnish current for the new electric signal system which the company now has under construction. The switches and semaphores will be lighted by incandescent lamps placed inside of discs carrying colored glasses, which revolve around the stationary lamp and display the different signals. The disc lights are operated by lead covered underground wires, and each light has an independent cut-out. As lead covered wire of sufficiently small diameter could not be procured for this work, the difficulty was got over by drawing No. 14 wire through a quarter-inch lead pipe, the space surrounding the wire being filled in with compound. The string by means of which the wire was drawn through was blown through the pipe by means of an air pump. This is believed to be the first instance in Canada in which electric lights are used in switches and semaphores from an underground circuit. In case of

accident the incandescent lamps within the revolving discs can be taken out and lanterns put in their place.

It can readily be seen that the installation of such a system as this necessarily involved much time and labor even under the most favorable circumstances, but in the present case progress was rendered extremely slow owing to the fact that the electrical work could not be pushed forward more speedily than construction in other departments. Owing to these circumstances, the whole of the present year has been occupied in bringing the work to its present state of completion. The Company purchased its own apparatus and materials, leaving in the hands of Mr. Fuller the carrying out of the work, which has been done in a highly creditable manner.

## PERSONAL.

Mr. Francis John Bolger, C. E., of Lindsay, died Nov. 3rd, aged 61.

Mr. Nelson Smith, formerly of Ottawa, has been appointed engineer of the new Alexandria waterworks.

Mr. John Starr, who sold the first T.H. apparatus in France, is now a general dealer in supplies at Halifax.

Mr. T. Ahearn, the well-known electrician of Ottawa, and family, have gone for a trip around the world.

Mr. Joseph Wetzler, editor of the Electrical Engineer, was married at Delmonico's, New York, on Oct. 30th.

Mr. I. B. Britton, late manager and superintendent of the Trenton, Ont., Electric Co., has gone to Cleveland, Ohio.

Mr. J. J. Franklin, late superintendent of the Toronto Railway, but now of Jersey City, has returned to Toronto for the winter.

Mr. W. McKenzie, president of the Toronto Railway Co., accompanied Sir Wm. Van Horne, of the C. P. R., on his recent transcontinental trip.

Mr. Geo. M. Cole and Mr. A. E. Reynolds, both Brockville, Ont., boys, are manager and secretary respectively of the Plattsburg, N. Y., Light, Heat and Power Co.

Mr. Charles Aire has resigned the road superintendency of the Ottawa Electric Railway to take the management of the Ottawa Transfer Co. He is succeeded by Mr. B. F. Shaw.

Mr. W. McCulloch, of the Canadian General Electric Company, has gone to Prince Albert to install new plants for the Hudson Bay Company, and the Prince Albert Light and Power Company.

A double wedding took place at 92 Wilton Ave., Toronto, recently, when Misses Lillian and Nellie Broomhall were married to Mr. Arthur Arkhill, contractor, and Mr. Arthur M. Brodie, electrician.

Mr. J. H. Meikle, jr., of Morrisburg, Ont., who has been engaged in electrical engineering in Brooklyn, N. Y., has left for Bulwago, Matabeleland, South Africa, in the interest of a big American firm that has secured a large electric light contract in that place.

Mr. W. M. Peterkin, of the Toronto Incandescent Light Co., is dead, aged 74. He was born in Aberdeen, came to Toronto in 1850, was with the old Royal Canadian Bank and the wholesale dry goods firm of Shaw, Turnbull & Co. before entering the service of the Light Co.

The extensive shafting and expensive belting in the government printing office at Washington is giving way to an electrical equipment. Much of the work hitherto done by steam will be accomplished by electric motors. By the substitution of electric light for gas it is expected a saving of $1,200 a month will be effected.

The Canadian General Electric Scientific Club was recently formed at Peterboro, with the following officers : Hon. president, Mr. F. Nicholls ; Hon. vice-president, Mr. S. Stephens ; president, Mr. W. L. Cathwaith ; vice-president, Mr. R. E. Layfield ; Secretary, Mr. W. A. Brundrette ; treasurer, Mr. J. F. Hedenberg ; managing committee, Messrs. W. Robinson, C. Robertson, W. W. Stone, F. A. Shannon and H. L. Knowles. Its aims are the promotion of scientific researches in the interests of electricity, and the establishment of social intercourse and harmony between the electrical staff and foreman of the Canadian General Electric Co's works.

## NINETY MILES AN HOUR.

NINETY miles an hour, and not by electricity, either, is what is promised by means of a steam locomotive being built at the Baldwin Locomotive Works in Philadelphia, to the order of Mr. W. J. Holmon, an old railway man and inventor, of Minneapolis, Minnesota. The mechanical principle on which it works is very simple and its application does for a locomotive what a bicycle accomplishes for a man—increases his speed without extra exertion.

The device is what is known as a speeding truck. The engine is just like an ordinary locomotive, with driving wheels 5 feet in diameter, but it will be placed on friction geared trucks instead of resting on the rails, being thereby raised some 30 inches. The driving wheels rest upon and between two small wheels, which in turn rest upon and between three similar wheels, which rest on the rails. When the driving wheels turn they impart by friction a rotary motion, in the opposite direction, to the wheels on which they rest, and these in turn cause to revolve, in the same direction as the drivers, the wheels which rest on the rails. The speed is thus multiplied till it is just double that of the drivers, so that if they are running at a 45 miles an hour gait the engine will be propelled at a speed of 90 miles.

An experiment was recently made with an ordinary locomotive, mounted on Holmon trucks, over a branch of the Northern Pacific Railway, when a speed of 80 miles an hour was easily attained. It is expected, when the engine now being built is completed, the trip between New York and Philadelphia can be made in an hour. Very little has been said about the invention, and the latter run is intended to be its formal introduction.

The inventor claims another advantage besides speed. By the distribution of the weight of the locomotive on so many more wheels he says the wear and tear of the track will be greatly reduced.

The success or failure of this new engine will be watched with much interest, as upon it may depend to a considerable extent the substitution of electricity for steam on our railways.

## A CANADIAN'S SUCCESS ABROAD.

PROMPTED by the pride which we always feel in the success achieved by Canadians, THE ELECTRICAL NEWS takes pleasure in reproducing from a recent number of the Electrical World, the following sketch of the successful career of one of our young countrymen :—

Mr. Arthur E. Childs, New England manager of the Electric Storage Battery Company, a conspicuous representative of the younger element which has admittedly contributed in large degree all along to the advancement and success of electrical industries, was born in Montreal, Canada, in 1865. After having obtained all possible advantages in the public schools he entered McGill University, confining his studies there particularly to mechanical engineering for four years and graduating in 1888 with the degree of Bachelor of Science. During his term at McGill University four months of each year were spent in the machinery department of the Grand Trunk Railway Company, thus securing for Mr. Childs a practical, mechanical knowledge. In September 1888, he went to London, England, entering the Central Institute of Technology, which is now affiliated with the Royal College of Science, graduating from there in 1891 with the degree of Electrical Engineer. Returning to Canada, Mr. Childs commenced his business career in charge of the experimental and testing departments of the Canadian Edison Company, at Peterboro', Ont., where he remained about one year. He left this company to connect himself with the Niagara Power Company, as assistant to Dr. Coleman Sellers, for which position he was recommended by Prof. W. C. Unwin. His position necessitated frequent visits to the works of the Westinghouse Electric Company, at Pittsburg, where he was brought into touch with Mr. Westinghouse and Vice-President Bannister, the result being an offer from these gentlemen to connect himself with the Westinghouse Company, which offer was accepted. From Jan. 1, 1893, until July 1, 1895, he was located at Philadelphia as engineer for the Westinghouse Company, after which period he took his present position with the Electric Storage Battery Company. Mr. Childs' business experience having covered only about four years, the fact that his connections have always been of a prominent character and came to him unsolicited, testifies abundantly to the excellence of his natural ability. Personally he is one of the most pleasant of men and is possessed of social characteristics which make and hold friends. In business he is quick in determination and prompt in action, and concentrates his fullest force and energy upon every undertaking. These characteristics have undoubtedly gained for him the enviable positions he has held and led to his present connection with the Electric Storage Battery Company, for which he is doing a most successful business in New England, and, which corporation may be congratulated upon having such an able representative. Mr. Childs is a member of the American Institute of Electrical Engineers, the American Society of Mechanical Engineers, the Canadian Society of Civil Engineers, the Institution of Electrical Engineers, of London, and the London Physical Society, and is Honorable Councillor of McGill University.

## TRADE NOTES.

The London Electric Co. are installing an additional 2,000 light alternator of the Canadian General Electric Co. make.

The Ottawa Porcelain and Carbon Co. have just turned out their first kiln of carbons, and the product is declared to be of very satisfactory quality.

W. S. Shaw, of Bracebridge, is installing a power distribution plant in the Shaw-Cassels tannery at that point. The Canadian General Electric Co. have the contract for the apparatus, which consists of a 75 horse power generator of the multipolar type with four motors of the slow speed, railway type.

## SPARKS.

Electricity is now employed for killing the dogs which the dog catchers capture. The street lighting current is employed.

The following towns and villages are putting in electric lights : —Farnham, Que.; Chicoutimi, Que., and Freilighsburg, Que.

The Hawaiian government has granted a subsidy of $40,000 a year towards a telegraph cable to the United States. The latter is expected to grant a like sum.

New Westminster, B. C., proposes to sell its electric light and water works, both now owned by the city, to private parties. Neither enterprise pays its way.

It is said that a piece of steam hose, attached to the boiler, will do more work in cleaning greasy and dirty machinery in a few minutes than close application and ordinary methods in hours.

The Dominion Government has agreed to grant permission to lay a new telegraph cable, from Victoria, B. C., across San Juan de Fuca Strait, to connect with the United States lines. A private company will be formed to build it.

The Baldwin Locomotive Works at Philadelphia are building two kinds of electrical engines, one with light trucks for elevated roads and one with heavy trucks for suburban traffic. It is expected a speed of from 40 to 70 miles an hour will be attained.

Messrs. F. M. Bowden, chief engineer, Royal Victoria Hospital, and Owen Hughes, chief engineer, Royal Electric Company, have passed the first class examination at Montreal as steam engineers, enabling them to take charge of large plants in the city of Montreal.

The Citizens' Light & Power Co., and the Standard Light & Power Co., owning franchises in the western suburbs of Montreal, and the right to light Montreal Harbor, have transferred their franchises to the Lachine Rapids Hydraulic & Power Co., a powerful and wealthy corporation.

A recent collision of a Spanish cruiser with a merchant steamer off Havana is attributed to the sudden extinction of the electric side lights. The lights failed from an extraordinary cause. A sailor became entangled in the electric lighting machinery, and those in charge of it stopped it to save the man's life. The result was the loss of some fifty lives and the cruiser herself.

## SPARKS.

Huntsville, Ont., proposes to have electric light.

The London, Ont., electric railway has 22 miles of track.

A telephone line is being built from Renfrew to Eganville.

The Rossland, B. C., Electric Light and Power Co. has been incorporated.

The Power Rope and Belting Co., Ltd., of St. Catharines, is seeking incorporation.

The capital of the Toronto Electric Light Co. has been increased from $500,000 to $700,000.

The Belleville Traction Co., capital $100,000, to build an electric road in Belleville, has been incorporated.

The Halifax Illuminating Co. are installing a 300 kilowatt monocyclic plant for incandescent lighting and power.

The following places are talking of introducing electric light :—Tavistock, Ont.; Pakenham, Ont.; Hampton, N. B.; Aylmer, Que.

In Toronto there are 117 miles of underground electric light and telephone wires. The total length of overhead wires is 4,288 miles.

Electricity is to be carried across the St. Lawrence from Quebec by cable for the purpose of lighting the church of Notre Dame de Levis.

The Toronto Electric Light Co. have purchased a 75 kilowatt alternator of the moncyclic type from the Canadian General Electric Co.

Theo. Viau, contractor, of Hull, Que., is trying to sell his franchise for an electric railway between Hull and Aylmer to a New York syndicate.

Mr. Jennings, late city engineer of Toronto, has made an examination for an electric railway from the station to the town of Edmonton, N. W. T.

An order has been made by Judge McMahon, at Osgoode Hall, declaring the Brantford Electric and Power Co. insolvent and ordering its winding up.

The boot and shoe manufacturers in the United States have taken advantage of electricity as a motive power more than any other branch of trade.

The Niagara Falls, Ont., Electric Light Co., which was to take 500 additional horse power from the Electric Railway Co., has decided to put in a steam plant.

Mr. H. J. Beemer is said to be negotiating for the purchase of the plant of the Montmorency Power Co., with the view of incorporating it with the Quebec city system.

The Barrie & Allandale Electric Street Railway Company has been incorporated, with a capital of $5,000. A park on the shore of Kempenfeldt Bay is part of the scheme.

The township of East Flamboro has offered to submit a bonus by-law for $30,000, and the village of Waterdown for $6,000, to aid the International Radial Electric Railway Company.

Niagara Falls is to be illuminated by a search light placed on the Canadian shore by the Michigan Central Railway. This will enable visitors to view the Falls at night as well as by day.

There are now 850 electric railways in the United States with more than 9,000 miles of track, 2,300 cars and a capital of $400,-000,000. In 1887 there were only 13 roads with about 100 cars.

The Royal Electric Co., of Montreal, has been reorganized with Mr. W. H. Brown as manager. Contracts have been secured for lighting Logan's Park, Brock street tunnel and the Lachine canal.

A charter is being applied for to build the North Nation Valley Colonization Railway from Montebello and Papineauville on the C. P. R. northerly to meet the St. Jerome branch of the same road.

A Montreal jury has given a verdict of $6,000 damages in favour of Mr. Grose, who sued the Holmes Co. for removing a galvanometer from the office of the Dominion Burglary Guarantee Company.

The Royal Electric Co., of Montreal, has now nearly 60,000 incandescent lights in use. The increase during a recent week was 965. Their report shows a profit of $106,209.14 on fifteen months work.

The C. P. R. telegraph operating room at Ottawa is to be equipped with the chloride accumulator storage battery instead of the gravity battery now used. Two hundred and thirty-five cells will do the work which now requires a thousand. It will be the first office so equipped.

D. Knechtel, of Hanover, is installing a 75 kilowatt alternating plant of the Canadian General Electric Co.'s monocyclic system.

Brockville has made a new contract for street lights. It will pay $105 per year for thirty arc lights, and $20 per year for one hundred gas or incandescent lights. The total cost of lighting will be $5,350.

According to the British Medical Journal, telegraphers are alarmingly subject to consumption. The general death rate from that disease among adult males, it is 13.8 per cent., while among telegraphers it is 46.6 per cent.

The Kingston and Cataraqui Street Railway Co. have placed an order for additional cars and G. E. 800 equipments with the Canadian General Electric Co.

Velhagen has discovered that the electrical action of the eye is changed in disease of the optic nerve. Examination was made by placing one sponge on the nape of the neck, with the other over the eye, and using a galvanic current.

The Geo. F. Blake Engine Co. has sued the city of Toronto for $8,600 kept back because the new water works engines were not completed within the contract time. The company alleges that the city was principally to blame for the delay.

Mr. E. Franklin Clements, of the Standard Telephone Co., New York, is trying to obtain the consent of the Prince Edward Island government to construct a telephone system throughout the island, and also build an electric street railway in Charlottetown.

The suit between the Western Brake Co., of Pittsburg, and the Boyden Brake Co., of Baltimore, for alleged enfringment of patents on quick acting air brakes, has been decided by the Court of Appeal at Richmond, Virginia, in favor of the Boyden Co. on all points.

The Ottawa Journal recently interviewed twelve presidents of city labor unions on the Sunday street car question. Eight were in favor of Sunday cars and four against. The question is not a live one in Ottawa, and many of the directors of the Street Railway Co. are opposed to Sunday cars.

Mr. Wm. Tanner, foreman of track laying, and Mr. J. M. Anderson, foreman of grading, on the London, Ont., street railway, have both been remembered by the employees under them. The former was presented with a rocking chair for himself and one for his wife, the latter with a writing desk.

Almonte is considering an estimate for a civic lighting plant by which business houses can obtain 16 candle incandescent lamps for one cent per night, and private houses for half a cent per night, while arc lamps for street lighting will cost only $35 per year. The plant would pay for itself in ten years out of the receipts from private consumers.

A Quebec paper prints the following extract from the records of the proceedings of the Sacred Congregation of Rites, of Rome, June 4, 1895; "Question. May electric light be used in church? Answer. For purposes of worship, no; but for dispelling darkness and more brilliantly illuminating the church, yes; but care must be taken that the manner of illuminating shall not resemble a theatrical spectacle."

A new application of electricity to medicine is announced. A current of high frequency and high potential is caused to traverse a large helix inside which the patient is placed, the effect being to set up induction currents of a similar kind. These produce nutritive changes of great advantage in cases of impairment of nutrition. Benefit has also been derived in anaemia and debility, gout, rheumatism, neurasthenia, hysteria, diabetes, etc.

A sensation has been caused on the United States side of the Niagara river by an announcement that the Attorney General of the state of New York has decided that it is unlawful for the Niagara Falls Hydraulic Power and Manufacturing Co. to take water from the river above the falls. This gives the Niagara Falls Power Co. and Cataract Construction Co. a monopoly. The latter has decided to put in three more 5,000 h.p. turbines and generators.

The department of public works at Ottawa has written to the city council, asking the latter to assume the electric lights at the parliament buildings and the governor general's residence, on account of the cost of inspection, and offering to recoup to the city the cost of the lights. The government pays $100 per light for 24 lights and has been offered a reduction to $85, while the city pays the electric company only $65. The city objects to light the government property because the latter pays no taxes. The council now offers to assume the cost of inspection.

# ELECTRIC RAILWAY DEPARTMENT.

## SANDWICH, WINDSOR AND AMHERSTBURG ELECTRIC RAILWAY.

SOME years ago we gave a description of the above street railway, which was then known as the "Windsor and Sandwich Electric Railway." The present company secured a charter to extend the road to Amherstburg, which will be done or commenced next year. The system has been extended to Walkerville.

The company is composed as follows: Directors:—Dr. Coventry; Jno. Davis, Inspector of Distilleries; Wm. J. McKee, M. P.; Wm. J. Pulling, lumberman; Jas. Anderson, all of Windsor; Geo. M. Hendrie, Detroit; Wm. Hendrie and Robt. Thompson, of Hamilton. Officers:—President, Dr. Coventry, Windsor; vice-president, Geo. M. Hendrie, Detroit; treasurer, Wm. J. Pulling, Windsor; secretary, Jas. Anderson, Windsor; accountant, Jno. M. Little, Windsor. Managing Committee:—Rufus Caufield, superintendent; Earnest Schultz, electrician; Ervin Lloyd, chief engineer. Executive Committee:— Dr. Coventry, Geo. McHendrie, Jno. Davis, Wm. J. McKee, Wm. J. Pulling.

The electric light department is managed by the secretary, Mr. Anderson.

The company operate ten miles of road between Windsor, Walkerville and Sandwich. They run a 15-

SANDWICH, WINDSOR AND AMHERSTBURG ELECTRIC RAILWAY.

minute service between these points. All the cars radiate from the corner of Sandwich street and Ouellette avenue. Three run to Sandwich, out London street, past the Sulphur Springs; three run from the M. C. R. depot to Walkerville, by way of Sandwich st., Ouellette ave., Wyandotte ave. and Second ave., connecting there with the L. E. & D. R. R. Two run from Gladstone ave., on Sandwich st., north to the driving park, 2¼ miles out Ouellette ave.

The company have 14 motor cars, 3 equipped with "Maguire" trucks and 11 with "Brill" trucks. The motors used on these cars are three 30 h. p. Westinghouse; eight 20 h. p. Westinghouse and three 30 h. p. Detroit. They have a total of 22 cars— 10 box, 8 open, and 4 trailers. They will add a street sprinkler, and 3 summer cars next summer. All the cars are heated by electricity. At present the company are vestibuling their box cars and the work done is as good as any carshop could turn out.

Most of the rails used are of the T rail pattern, and the line out Ouellette ave. on the pavement is the straightest and best laid in the country. The T rails are 56 lbs. Although only a one track system, the cars run close to schedule time, and there is no unnecessary delay. The conductors are always obliging, and it is a pleasure to use their line. The

POWER HOUSE AND CAR STATION.

business during the race meet is something enormous. The company claim that running their regular service they can carry away 2,000 passengers from the driving park to Sandwich—a distance of 2¼ miles—in 30 minutes. We show a number of cars in the accompanying cut on their way to the Exhibition grounds.

The power house, of which we show a cut, is a two storey and basement brick structure on London street. The building is 150 feet long by 50 feet wide. The ground floor is used as a car shop, in which all the cars are overhauled, painted, etc. The top storey is used by the superintendent and unmarried men for quarters.

An addition at the rear of 50 x 50 ft. comprises the boiler room. It is on a level with the basement, in which are the engines and machines.

We are sorry that we could not get a cut of the power room, as it contains the only pair of cross compound Robb-Armstrong engines in Canada.

In the boiler room are three 80 h.p. Monarch boilers and two Polson boilers. Natural gas, forced in pipes from Kingsville, is used for fuel, and the volume is regulated by steam. The fires were lit last May and have not been out since. In case of an accident, the one group of boilers could be cut off from the other. Natural gas as fuel needs but little or no attention, and is cheaper and cleaner than the use of coal. A recent drop in the price is inducing other plants to apply it.

In the power room, which is 75 x 50, are three engines running onto a line of shafting 45 feet long. A 200 h.p. Brown-Corliss, and a pair of Robb-Armstrong cross compound engines, 275 h.p. each, furnish the motive power. The Brown-Corliss is a good running engine, having been in service a number of years. Its fly wheel is 12′ dia., 28″ face, and weighs 12 tons. It is driven by a 24″ double belt onto a three-foot pulley on the line of shafting ; a grip coupling connects with or separates the engine from the rest of the shafting. Two other grip couplings are on the line of shafting, one for each engine.

The company supplies the citizens with 3,000 incandescent lights, and power to six local institutions. Eight machines are run from the shafting, as follows : two railway generators, one of 150 h.p. Westinghouse, the other 200 h.p., same make ; three alternators, two C. G. E., of 2,000 and 1,000 lights capacity respectively, each with an exciter, and an 800 light Westinghouse with exciter.

The switchboard is a large one, built of native wood, with full equipment of instruments. Full load can be thrown onto either of the Robb-Armstrong engines without any perceptible difference in the light or power. A heater is used for each engine, and water is delivered to the boilers at 210°.

Under the car shop an electrical repair shop is fitted out, in which duplicate armatures are always kept on hand, in case of accident. This department is in charge of Mr. Earnest Schultz, electrician. Mr. Ervin Lloyd has charge of the engines, and Mr. Rufus Caufield is superintendent. These mechanical men have had a lifelong experience and the company's interests are safe in their hands.

The President, Dr. Coventry, makes his daily visit, and while he and our representative were there his medical knowledge was brought into use, as the chief engineer, Mr. Lloyd, was taken suddenly ill. The cause was too much study. Mr. Lloyd is taking a mechanical course in Detroit.

At the Sandwich end of the line the old horse barns are utilized for storing extra cars.

Windsor is said to have had the first electric road in Canada. It was the old Vandepoel system. One car was the extent of its service. It ran along the river front, and the conductor, Mr. Vallance, cigar dealer, tells many interesting things of it. One funny thing it would do was to stop all watches that came near it.

The company occupy offices on Ouellette ave., but are going to build a building of their own at a cost of about $20,000. The site is not settled yet. All the members are shrewd business men, and the company is in a flourishing condition. They propose opening a park for the use of their patrons near Sandwich next summer. When the Amherstburg end is opened up, this railway will be one of the best in Canada.

## THE CONTINUOUS RAIL IN STREET RAILWAY PRACTICE.

Mr. Richard McCulloch, C. E., of the Citizens Railway Co., St. Louis, which has constructed and put in successful operation a piece of road constructed on the continuous rail principle, in a paper read before the Engineers' Club of that city, recently, sums up the subject in the following words : It is not to be supposed that the millennium in track construction has already been reached, but what has been demonstrated is this: First, that the use of a continuous rail for street railway practise is feasible ; and second, that it is possible to make joints of sufficient strength to stand changes of temperature. Whether new difficulties will develop remains for the future to show, but let us hope that those of us who have placed our faith in rail-welding will not share the same fate as Jules Verne's armor maker, who planned, mixed, forged and tempered his best only to see the triumph of his skill shot to pieces by the latest gun of his hated rival.

### SPARKS.

Since the adoption of electricity the receipts of the London, Ont., street railway have trebled.

The Montreal Electric Railway Co. have fourteen snow sweepers ready for this winter. They are from 70 to 100 horse power and can clear the lines of light snow in 40 minutes.

The railway branch of the Victoria, B.C., Electric Railway and Lighting Co. does not pay and there is a prospect that the bond holders will take over the road and operate it.

The Hamilton Street Railroad Co. asked to have the city by-law regulating it amended, but the council look upon the concessions as worth $6,800 a year and refuse to change the by-law.

The Montreal and St. Lambert Bridge Electric Railway Co. is seeking a charter to build a bridge across the St. Lawrence and operate electric roads from its terminus through the counties of Chambly and Lapairie.

A street railway track without ties is in operation in Detroit, Mich. The tracks are laid in a 6 inch bed of cement, and although a radical departure from all established methods it has been in use long enough to be pronounced a success.

Mr. Jas. Devlin, engineer at the Kingston Penitentiary, recently sued the city of Kingston for $100 damages for a broken leg, caused, as he said, by stepping into a hole. The trial was a long one, and it was proved that the injury was caused by stepping from an electric car. Eleven of the jury were for dismissing the action and the judge ordered a verdict for the city.

Mr. Paul Meyer recently recovered $296 damages in the superior Court, from the Montreal Street Railway Co., for running down his vehicle with one of their cars in Oct. 1893. The street was narrow at the place, the company occupied nearly half of it with their double track and it was obstructed with earth and stones. The car came behind Mr. Meyer and the motor man did not ring his bell or stop his car as he might have done.

## SPARKS.

The Petrolea Advertiser has installed a Baird electric-gas engine of 2 h.p.

The Dunnville Electric Light Co. started up their new monocyclic plant recently.

The Ottawa Street R. R. Co. has declared a quarterly dividend of 2 per cent.

The Ottawa Street R.R. Co. has agreed to extend their lines to Hintonburg this fall.

English capital has been secured for building the Cornwall Electric Street Railway.

The Canadian General Electric Co. are installing an incandescent plant in Nelson, B. C.

The Ottawa Street Railway Co. has spent $125,000 this season on track construction.

A new switchboard has been placed in the Kingston office for the long distance telephone line.

Joliette, Que., is discussing the project of an electric railway to St. Lanoraie, on the St. Lawrence.

The citizens of Winchester are reviving the agitation for an electric railway between Morrisburg, Winchester and Ottawa.

Cote St. Paul, Que., town council is asking the Park & Island Railway Co. to give them street railway connection with Montreal.

Messrs. Moore & Sons, of Meaford, have installed a new 500 light alternator. The Canadian General Electric Co. have the contract.

The Halifax Electric Tramway Co. have given the Eastern Trust Co. a first mortgage in trust to secure an issue of debentures to the amount of $600,000.

An electric railway is being discussed to connect Alexandria Bay with the the Rome, Ogdensburg and Watertown R. R. at Redwood. The distance is about seven miles.

Ahearn & Soper, of Ottawa, have the contract for the electric light works at Alexandria. A by-law to raise $6,000, spread over 30 years, will be submitted to the ratepayers.

A telegraph line from Boone Bay, Newfoundland, to the Straits of Belle Isle, a distance of over 200 miles, is to be built. This will enable ships to be reported when they pass the Straits.

Mr. E. A. C. Pew, projector of the scheme to build an aqueduct from Lake Erie to Hamilton, says his company is ready to go on with it as soon as the city grants the franchise.

Electricity is entering into almost every department of daily life as a labor-saving element. One of its latest uses is for cancelling stamps on letters, which is employed in the New York post-office. One man is able to do the work of four or five. By this ingenious contrivance the operator has only to arrange the letters, turn a switch, and cancel at any speed he likes.

The Alliston Milling Co. have increased their plant by a 500 light Canadian General alternator.

The Niagara Falls and River Railway finds it necessary to make four trips a day to Queenston. This service usually ceases after the steamers stop running.

The proposal to build an electric railway from Port Perry to Kincardine is meeting with much support. Mr. A. E. C. Pew is the promoter, and Mr. Brunell, C. E., is making surveys.

A Toronto syndicate is projecting an electric road at Chatham, Ont. It is urged in some quarters that the scheme should include connection with all the chief points in the county of Kent.

The Toronto, Hamilton & Niagara Falls Electric Railway is a new enterprise seeking incorporation. As the company desires to run Sunday cars a Dominion charter will be applied for.

The Hamilton, Brantford & Pacific Junction Railway Co. is applying for incorporation, with power to build a line from Copetown, on the T. H. & B. Railway, to Schaw Station, on the C.P.R.

The net earnings of the Montreal Street Railway Co. for the year ended 30th Sept., were about 10½ per cent. on the paid up capital of $3,444,000. The road is in a highly satisfactory financial position.

It is settled that work on the power plant for the Canadian Niagara Co. will begin within four months and be pushed to completion. Power will be ready early in 1897. The plant will be larger than that on the United States side.

An additional issue of stock, amounting to $400,000, has been made by the Incandescent Light Co., of Toronto. Nearly all of this new stock was subscribed for at par, by the directors of the company. The object in issuing the stock is said to have been to prevent the control of the Company from passing into the hands of the directors of the Toronto Electric Light Co., who are said to have quietly purchased a majority of the shares. There are now heard rumors of a probable amalgamation of the two companies.

The Supreme Court of the United States has just given two decisions of importance affecting electrical patents. In the case of the United States vs. The Bell Telephone Co., to cancel the Berliner patent, it is declared to have jurisdiction. This will bring the case before the court for final decision. In the case of the Edison incandescent light patent against the claim of the Consolidated Electric Light Company using the Sawyer-Mann system, of which it was claimed that the Edison system was an infringement, the Court held that the claims made for the Sawyer-Mann are too broad. The result will be to throw open both systems to the public. The Sawyer-Mann patent is invalid, and the Edison patent expired a year ago, under the operation of a former decision.

Burt & Rousseau, electricians, Montreal, have dissolved.

An electric road between Stirling, Ont., and the C. P. R. at C. O. Junction is projected.

The Hamilton Radial Railway Co. gives notice that it will apply for power to extend the Guelph branch to Lake Huron.

The Smith's Falls Electric Light Co. have purchased a 2,000 light alternator from the Canadian General Electric Co.

The Halifax Street Railway is getting 2,000 tons of rails for $14,000 less than the present price. Rails have risen $7 per ton since they were ordered.

One o the features of Mr. Pew's Port Perry and Kincardine electric railway scheme is the distribution of coal at a cheap rate from Whitby; where it can be brought by vessel from Oswego during the winter.

## MOONLIGHT SCHEDULE FOR DECEMBER.

| Day of Month. | Light. | Extinguish. | No. of Hours. |
|---|---|---|---|
| | H.M. | H.M. | H.M. |
| 1...... | A.M. 4.10 | A. M. 6.20 | 2.10 |
| 2...... | No light. | No light. | .... |
| 3...... | No light. | No light. | .... |
| 4...... | P. M. 5.00 | P. M. 8.00 | 3.00 |
| 5...... | " 5.00 | " 9.00 | 4.00 |
| 6...... | " 5.00 | " 10.10 | 5.10 |
| 7...... | " 5.00 | " 11.30 | 6.30 |
| 8...... | " 5.00 | A. M. 12.40 | 7.40 |
| 9...... | " 5.00 | " 1.00 | 8.00 |
| 10...... | " 5.00 | " 1.50 | 8.50 |
| 11...... | " 5.00 | " 3.00 | 10.00 |
| 12...... | " 5.00 | " 4.20 | 11.20 |
| 13...... | " 5.00 | " 5.30 | 12.30 |
| 14...... | " 5.00 | " 6.30 | 13.30 |
| 15...... | " 5.00 | " 6.30 | 13.30 |
| 16...... | " 5.00 | " 6.30 | 13.30 |
| 17...... | " 5.00 | " 6.30 | 13.30 |
| 18...... | " 5.00 | " 6.30 | 13.30 |
| 19...... | " 5.00 | " 6.30 | 13.30 |
| 20...... | " 8.00 | " 6.30 | 10.30 |
| 21...... | " 9.10 | " 6.30 | 9.20 |
| 22...... | " 10.10 | " 6.30 | 8.20 |
| 23...... | " 11.00 | " 6.30 | 7.30 |
| 24...... | " 11.00 | " 6.30 | 7.30 |
| 25...... | ........ | " 6.30 | } 6.20 |
| 26...... | A.M. 12.10 | .......... | |
| 27...... | " 1.10 | A. M. 6.30 | 5.20 |
| 28...... | " 2.10 | " 6.30 | 4.20 |
| 29...... | " 3.20 | " 6.30 | 3.10 |
| 30...... | " 4.20 | " 6.30 | 2.10 |
| 31...... | No light. | No light. | .... |

Total,    224.40

Grand Total, 2203.30

The Cliff Paper Co., of Niagara Falls, have decided to use electricity as a motive power. This will do away with three steam engines of over 200 horse power, and will enable them to make paper cheaper than formerly.

# CANADIAN GENERAL ELECTRIC CO.
## (LIMITED)

Authorized Capital, $2,000,000.00.
Paid up Capital,   $1,500,000.00.

HEAD OFFICE:
## 65 FRONT STREET WEST, - - TORONTO, ONT.

BRANCH OFFICES AND WARE-ROOMS:

| | | | |
|---|---|---|---|
| 1802 Notre Dame St. | MONTREAL. | Main Street - - | WINNIPEG. |
| 138 Hollis Street - | HALIFAX. | Granville Street - - | VANCOUVER. |

*150 K. W. MONOCYCLIC GENERATOR.*

### The MONOCYCLIC • •
### • • and the THREE-PHASE

are the only satisfactory systems for operating

## INCANDESCENT LAMPS, ARC LAMPS AND MOTORS

.... FROM THE ....

## Same Generator and Circuit

## INSOLVENT NOTICE

In the matter of La Cie Electrique St. Jean Baptiste, of the City of Montreal, Canada, P. Q., in Liquidation.

THE undersigned will sell by Public Auction, in three lots, at the office of Chas. Desmarteau, Liquidator, No. 1598 Notre Dame St., Montreal, Canada, P. Q., on

THURSDAY, THE 19TH OF DECEMBER, 1895, AT 11 O'CLOCK IN THE FORENOON,

all the immovable properties and moveable assets of the said company, consisting of:

(1) All those tracts and parcels of land situated in St. Jean Bap iste Ward, in the City of Montreal, Province of Quebec, Dominion of Canada, described as follows :

A.—The northeasterly parts of lots numbers 30 (thirty-two), 33 (thirty-three) and 34 (thirty-four), according to the official subdivision of lot number 10 (t n) of the official cadastre, of the village of St. Jean Baptiste, comprising an area of 4970 (four thousand nine hundred and seventy) feet, more or less ; and bounded towards the north-east by Montana street, towards the south-east by Rachel street, towards the north-west by lot number thirty-five of said official sub-division of said lot number ten, and towards the s uth-west by the re-maining parts of said lots numbers thirty-two, thirty-three and thirty-four, of said official sub-division of said lot number ten, with buildings and dependencies there-on erected.

B.—Lots numbers 35 (thirty-five), 36 (thirty-six), 37 (thirty-seven), 38 (thirty-eight), 39 (thirty-nine), 40 (forty), 41 (forty-one), and 42 (forty-two), of said official sub-division of said lot number 10 (ten), comprising an area of 17296 (seventeen thousand two hundred and ninety-six) feet, more or less ; bounded towards the north-east by Montana street, towards the south-west by lot number sixty-one (3) of said sub-division of lot number ten, towards the north-west by lot number forty-three, and towards the south-east by lot number thirty-four of said sub-division of said lot number ten ; together with the right of passage in common with ad-joining proprietors in the lane at rear of said lots

C.—All engines, boi ers, dynamos, tank (45,000 gal-lons), machinery, belting, shafts and general plant, used for manufacturing and generating electricity, as well as the wires and posts, transmitters, meters, lamps, etc., used for distributing and furnishing electric light to the citizens of Montreal and suburbs ; the material and stock used for repairing the lines of the company, the whole amounting as per inventory to one hundred and forty-four thousand, three hundred and seventy-one dollars ($144,371.00.)

D.—The franchise rights and privileges acquired by the said company to plant poles in the streets of the city of Montreal

(2) The office furniture and fixtures, etc., amounting to $196.00.

(3) The book-debts, amounting as per list to $3456.00.

The purchaser of the immovables and dependencies described in paragraph (1) shall have to make a cash deposit of five thousand dollars currency ($5,000) at the time of adjudication.

The purchasers of moveable assets described in para-graphs (2) and (3) shall have to make a cash deposit of ten per cent at the time of adjudication.

For further information apply to

CHAS. DESMARTEAU, Liquidator, 1598 Notre Dame St.,
MARCOTTE FRERE,      Montreal,
Auctioneers.      Cana''a'' P.Q.

---

# ROBIN, SADLER & HAWORTH

### MANUFACTURERS OF

## OAK-TANNED LEATHER BELTING

### MONTREAL AND TORONTO

# ᚷᴴᴱ GOLDIE & MͨCULLOCH C͞o.
[LIMITED.]

### MANUFACTURERS OF

## Improved Steam Engines and Boilers

### ✛ FLOURING MILLS ✛

*And the Erection of same in the most Complete Style of Modern Improvement.*

## WOOL MACHINERY, WOOD-WORKING MACHINERY, SAWMILL, SHINGLE AND STAVE MACHINERY

### Fire and Burglar Proof Safes and Vault Doors.

Special attention called to the "WHEELOCK" IMPROVED STEAM ENGINE as being unequalled for simplicity, efficiency and economy in working, and especially adapted for Electric Lighting, Street Railways, etc.

### GALT, ONTARIO.

# AHEARN & SOPER

## OTTAWA, ONT.

### CANADIAN REPRESENTATIVES OF THE

# WESTINGHOUSE ELECTRIC & MFG. CO.

SLOW SPEED
ALTERNATING CURRENT DYNAMOS

from which can be operated

Incandescent Lamps, Arc Lamps
and Motors.

ELECTRIC RAILWAY
GENERATORS AND MOTORS

Our Railway Apparatus is not
Equalled by any other

# CANADIAN ELECTRICAL·NEWS

## STEAM AND ENGINEERING JOURNAL

OLD SERIES, VOL. XV.—No. 6.
NEW SERIES, VOL. VI.—No. 1.

JANUARY, 1896

PRICE 10 CENTS
$1.00 Per Year.

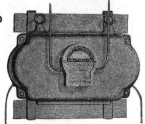

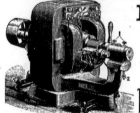

CANADIAN

# ELECTRICAL NEWS

—AND—

## Steam Engineering Journal

## VOL. VI.

1896:
C. H. MORTIMER, Publisher
TORONTO—CANADA

# INDEX

CANADIAN

# ELECTRICAL NEWS

AND

## STEAM ENGINEERING JOURNAL.

| VOL. VI. | JANUARY, 1896 | No. 1. |

## ELECTRIC FREIGHT LOCOMOTIVE.

SINCE the month of August last a 96-ton electric locomotive has been in successful operation for hauling
freight on the Baltimore & Ohio railway. No interruptions whatever have occurred, the locomotive responding in every case without failure either of speed or
power. Tests were made recently to ascertain its capa-

of the train. In this condition current was turned into
the motors and movement was immediately communicated to the train. At the end of one minute the train
was moving at a speed of 10½ miles an hour, and at
this point the speed was increased to the usual rate.
The total distance moved in 40 seconds was 150 feet
and at the expiration of one minute 450 feet.

ELECTRIC LOCOMOTIVE AND FREIGHT TRAIN LEAVING TUNNEL ON THE BALTIMORE & OHIO RAILWAY.

city for running a loaded train on an up-grade, in connection with which the following particulars will be of
interest. The illustration presented shows the electric
locomotive coupled to a north-bound freight train leaving
the tunnel.

A train consisting of two steam locomotives, not
working, and 27 loaded freight cars, was brought to a
stop, while going north through the tunnel. Here the
grade is 42 feet to the mile, and the rails were damp
and greasy. The weight·of the train was 1,125 tons,
or 1,221 including the electric locomotive. Every drawbar was tight, no slack occurring throughout the length

Another test was made with a dynamometer car
placed between the electric locomotive and the train,
which consisted of 22 cars loaded with coal, one caboose
and two dead locomotives. The total weight was 1,068
tons. On the 10 per cent. grade in the·tunnel an average drawbar pull of some 25,000 pounds was obtained
from the dynamometer diagram. The speed at this
point was 11⅛ miles per hour. Comparison with the
diagrams obtained in similar service with steam
locomotives showed a remarkably uniform and
steady pull by the electric engine, due to the absence
from it of reciprocating parts, the torque being

constant throughout the entire revolution of the wheel.

A further test was made with another train, consisting of 36 cars, one caboose and three dead engines. This was a regular through freight train with a local freight attached, and the total weight was in excess of 1,600 tons. It was hauled with ease through the tunnel, and calculations from the previous dynamometer records and the drawbar pull per ampere showed a drawbar pull of over 45,000 pounds.

On October 6th still another test was made, the character of the performance being heightened by the fact that the train which it moved measured over 1,800 feet in length and weighed about 1,900 tons, and was started from rest in the tunnel. It consisted of a north-bound freight train of 28 loaded cars and two locomotives coupled to a local freight of 15 loaded cars and one locomotive. In starting not a sputter, spark or slip of the wheel occurred, and the train moved with the same precision as if the circumstances had been of the ordinary character. The drawbar pull of 60,000 pounds was about the record in this case. The train was quickly brought to a speed of 12 miles an hour and pulled through the tunnel without difficulty, with the locomotive continuously exerting a drawbar pull of 40,000 pounds.

The above tests only show approximately what the locomotive can do, as its capacity has by no means been reached.

Two additional machines have been ordered by the Baltimore & Ohio Railway Company, which are now nearing completion at the works of the General Electric Company at Schenectady.

## A CANADIAN'S RIDE ON THE WINNING MOTO-CYCLE.

WHILE so much has been said lately of the moto-cycle or horseless carriage we thought a description of a ride on one in the late Chicago race would be of interest to our readers.

One of our representatives called on Arthur W. White, of London, Ont., of the firm of Geo. White & Sons, well known manufacturers of engines, boilers, etc., he being the only Canadian official in the Chicago road race last (U. S.) Thanksgiving Day. He and his father, Geo. White, were in Chicago for ten days before the race, during the preliminary tests.

On the morning of the race the umpires were assigned to the carriages, one to each. Arthur W. White was placed on the Duryea carriage, he not knowing till then which carriage he was to ride in. The route of sixty miles ran north from the corner of Michigan ave. and Rush street through Lincoln Park, then by way of Kenmore ave. to Evenston, a distance of thirty miles. The return run was from North Clark street and Belmont ave. to Milwaukee ave., then through Park Drive and Humbolt, Garfield, Douglass and Brighton parks, then down Western ave., 55th st. Boulevard, through Washington Park to starting point, the round trip being 60 miles.

With Mr. White on the carriage was Frank Duryea, the operator, brother of the inventor, Chas. E. Duryea. The day was fine overhead, but the roads were full of slush, which impeded the speed of the carriages. In some places the slush was six inches deep and often hid ruts, into one of which the Duryea carriage went at the corner of Erie and Rush streets, breaking the steering gear. While they were examining the extent of the

break, the Macey carriage passed them. It being a holiday, none of the blacksmith shops were open, but they hunted up a key to a shop, went in and did the work themselves. Fifty-five minutes were lost here. It took them from 8:55 a.m. to 12:45 p.m. to run to Evenston, a suburb of Chicago, making a stop of seven minutes before reaching Evenston for water. Shortly after getting water they caught up to the Macey carriage. The road here was but one broken track, with snow on each side to the depth of from six to eight inches. One of the rules of the contest was that if the leading carriage could not prove its capability to keep ahead, it was compelled to pull out and let the other go by. Mr. White asked them to comply with the rules, which the Macey people did. He was sorry to ask them to do it, but the rules had to be complied with. A short way south of Evenston a sleigh load of young people upset while turning a corner, but Mr. Duryea brought the carriage to a standstill almost instantly, just as the wheels touched the horses. The carriage at this time was going twelve miles an hour. At the corner of Clark and Lawrence ave. they lost their way, and went two miles out of the course. At Diversea street part of the mechanism broke, which necessitated the drawing out of a piece of inch ground steel ; this was done in a tinsmith shop with the aid of a charcoal fire and tinsmith's hammers. They had to light the fire, and lost one hour here. Through the west side parks the snow was very deep, and to use a slang term, "the woods were full" of boys who made Messrs. White & Duryea targets for their snow balls. Mr. Duryea received "one in the neck" which dazed him. Two policemen tried to control the boys, but the boys didn't see it that way, and utterly routed the cops. Crowds lined the route and many were the cheers our friends received. The Kodak fiend, as on all occasions, was on hand—you saw him in every guise, in every place, at all times. One fiend got down on his back in the slush under the Duryea carriage to take a snap at the mechanism. The Duryea rig reached the starting point again at 7.18 p.m., being 10 hours and twenty-three minutes on the road, and winning the race. The Muller carriage came in a little above an hour afterwards. The Macey carriage became mixed up with a trolley car and did not get in till the next day.

After the race Mr. Hewitt, president of the company who will manufacture the Duryea carriage in Springfield, Mass., entertained the Duryea people and Mr. White and friends to supper, which was quite acceptable to Mr. Arthur W. White and Mr. Frank Duryea, as they had had nothing to eat since 6:30 a.m.

Mr. Arthur W. White is building an electro-motocycle in London, and will give it a test at the proposed races in that city on the 24th of May next. He is very desirous that a moto-cycle test should take place in Canada this year, and suggests that London, being surrounded in all directions by excellent roads, would be the most suitable place for such an event. The horseless carriage has a great future before it. The thing to be decided is what power will be the best for all occasions.

Mr. E. Lusher, Secretary of the Montreal Street Railway, has recovered from his recent illness.

A. E. Payne, a well-known electrician of Boston, Mass., has decided to make his home in Canada, and has joined the Royal Electric Company at Toronto.

## COMBUSTION.*

### By Thomas Wensley, Ottawa.

(Concluded.)

I will here give you an approximate list of square feet o heating surface per horse-power in different styles of boilers, and various other data for comparison :

| Type of Boiler. | Square feet of heating surface for one horse-power. | Coal per square foot h.s per hour. | Relative Economy. | Relative Rapidity of steaming. | Authority. |
|---|---|---|---|---|---|
| Water Tube...... | 10 to 12 | .3 | 1.00 | 1.00 | Isherwood. |
| Tubular.......... | 14 to 18 | .25 | .91 | .50 | " |
| Flue............. | 8 to 12 | .4 | .79 | .25 | Prof. Trowbridge |
| Plain Cylinder.... | 6 to 10 | .5 | .69 | .20 | |
| Locomotive ...... | 12 to 16 | .275 | .85 | .55. | |
| Vertical Tubular.. | 15 to 20 | .25 | .80 | .60 | |

A horse-power in a steam engine or other prime mover is 550 foot lbs. raised one foot per second, or 33,000 lbs. one foot per minute.

In Engineering of August 17th, 1894, there is a report of two tests made with a triple expansion mill engine of 1,000 horse-power, built by Victor Coates & Co., limited, of Belfast, for the spinning mills of the Brookfield Linen Company, limited, of the same city. This engine was set to work on the 18th of September, 1893, and has been at work ever since, giving satisfactory results, especially in the matter of fuel consumption and steady driving. As shown by these tests, the amount of water used is remarkably small, being 11.5 lbs. per hourly horse-power, and the coal consumption was 1 lb. The diameters of the cylinders are respectively 19, 29 and 46 inches, with a stroke of 48 inches. The steam was generated in two Lancashire boilers, 7 feet 6 inches in diameter and 30 feet long ; each boiler has two furnaces of the Adamson type, having five Galloway tubes in each, and the total heating surface of the two boilers is 1,900 square feet. On these tests the engines were not running at full power, but were developing 787.4 horse-power, so that the heating surface per horse-power in this case was 2.41 square feet. The feed water was heated in the economiser to 250° Fahrenheit, and if we include the heating surface of the economiser, 3,600 square feet, there would be a total of 5,500, or 7.112 square feet per horse-power. The economiser is placed in the base of the chimney, and the feed water is heated by the hot gases which are passing away to the atmosphere, and would otherwise be a total loss.

When anthracite or hard coal is used, there should be from 22 to 24 inches between the top of the bars and the lowest part of the boiler. If bituminous or soft coal is used, then from 27 to 30 inches.

It is an absolute condition of economy and efficiency that the grate bars shall at all times be well and evenly covered with the fuel, but this condition is one that is frequently neglected. If the bars are not uniformly and evenly covered, the air enters irregularly in streams, passing through the thinnest or uncovered parts ; if too thickly covered it prevents the air entering. You all know that the thickness of the fire will depend upon the size of the coal used. The smaller the fuel the thinner the fire. With egg coal from 6 to 8 inches, and with furnace coal from 8 to 10 inches have been found the best results in practice. In burning soft coal the charges should be light, as the gases which are evolved will have a better opportunity of getting the requisite quantity of oxygen.

I have seen from 15 to 16 inches of coal on the bars at a time, and upon asking the fireman his reasons for having such a heavy fire, his answer has been that he could not get steam unless he had that quantity. It is argued by some that it is necessary, when a boiler is worked to a high rate of capacity, to maintain heavy fires, and that thin fires are well enough for slow rates of combustion ; but when the call for steam increases, it must be met by an increased thickness in the bed of coal on the grate. The ordinary fireman is apt to favor this method, for the reason that he can introduce large quantities at a firing, and afterward he is not obliged to give the fires much attention, for perhaps an hour's time, when he will again fill the furnace full in the same manner as before. As an explanation, however, of the favor which this method receives, it is probable that the class of labor which is generally employed considers the muscular effort required much less of a task than the more frequent and careful attention which is needed when the fires are thin. Under such conditions it is almost impossible to regulate with natural draught the supply of air, upon which we must depend entirely for perfect combustion and economy.

As regards a comparison between thick and thin fires, the fact is that more capacity can be obtained from a boiler when a fire of medium thickness is carried and proper attention is given to its condition, than can be realized by any system of management when the fires are exceedingly heavy, and advocates of thick fires, who take the ground that they are a necessity, are mistaken. As to the economy of the two, some persons maintain that heavy fires give the most economical results, but this is questionable. Valuable information on the subject has recently been brought out by the results of two evaporative tests which were made on a 72-in return tubular boiler, having one hundred 3½ inch tubes, 17 feet in length. The heating surface amounted to 1,642 square feet, and the grate surface to 36 square feet, the ratio of the two being 45.6 to 1. On the thick fire test, the depth of coal on the grate varied from 10 to 20 inches, being heaviest at the rear end and lightest at the front end. On the thin fire test, the depth was maintained uniformly at about 6 inches. The coal was Kew River semi-bituminous coal. The difference in the results, as appears from the figures, is an increased evaporation due to their fires amounting to 15.6 per cent.

The quantity of heat generated in the furnace is dependent on the relative weight of hydrogen first, and carbon afterwards, chemically combined with their equivalent weights of atmospheric oxygen. If chemistry did not teach us this, our daily experience would soon convince us.

In using soft or bituminous coal, which contains a large percentage of volatile matter, it is necessary to introduce air over the fuel (unless we are working with the forced draught system), as we cannot get sufficient air through the grates, and that which comes is loaded with carbon which it has picked up in its passage through the fire. For this purpose we have apertures in the doors, or we leave the door ajar after a new charge of coal. You will readily perceive that the admission of any large quantity of air in this way must be objectionable, as it will cool the gases below the point of ignition, and if too much is admitted it will carry off heat from the furnace. There are a number of ways of admitting air to better advantage ; the simplest is to conduct the air through a hollow bridge wall and discharge it through apertures in the top, the air mingling with the lower strata of the burning gases as they pass over the bridge, thus ensuring a more perfect combustion.

George W. Barrus, M.E., made tests with a boiler where provision had been made for the admission of air as above, with Cumberland, Anthracite and a mixture of two parts pea and dust, and one part Cumberland. In the case of the Cumberland, the evaporation was increased about six per cent.; with the anthracite, the evaporation was decreased about one per cent. The hot air completed the combustion of the volatile products of the soft coal, which would otherwise escape unburned. The slower burning anthracite did not need this supply and did better without it. The effect which the introduction of air had upon the appearance of the products of combustion, as viewed from the "peek hole " cack of the bridge wall, was very noticeable in both cases, but greatest with the soft coal ; but Mr. Barrus says that there was a heightened color and increased activity to the flame, whichever fuel was used, notwithstanding the average evaporative result with the hard coal was lower. Mr. Barrus' conclusion, drawn from many tests, is that a considerable advantage attends the admission of air above the fuel when bituminous coal is employed, but that there is no advantage when mixtures of anthracite screenings and bituminous coal are used, and little or no benefit is derived when anthracite coal is used.

The importance of good draught, natural or mechanical, for the supplying of sufficient oxygen for the rapid and economical combustion of fuel, has long been felt by the engineer. The gain both in capacity and efficiency which would be obtained by the rapid and energetic combustion of the various kinds of coal, and the high furnace temperature resulting therefrom, is well established, but its importance has only been admitted within the last few years. High initial furnace temperature is essential with all kinds of boilers to obtain the greatest economy, and to obtain this high temperature requires proper draught to deliver an abundant supply of oxygen to the furnace. This result is obtained by natural draught in a well-proportioned chimney, or forced draught obtained by mechanically creating a pressure under the grates with a fan or blower. The advantages of the forced draught are : 1st. It is under complete control. 2nd. The more perfect combustion of fuel by reason of the more abundant supply of oxygen to the furnace, and the possibility of using a cheaper grade of coal, with a proper combustion of the same. It

* A paper read before the Canadian Association of Stationary Engineers.

is a fact, however, that the most perfect plant will be a failure if the firing of the boilers is not properly attended to, and the fires kept at an even and uniform thickness suitable to the grade of coal used, and it is to be regretted that so little attention is paid to this fact.

There is a furnace in use in the United States, a sketch of which I submit herewith, and known as the Hawley Down-Draught Smoke-Consuming Furnace. The characteristic features of the Hawley setting will be of interest; it consists of a double set of grate bars, one above the other; the upper, or water grate, is made of 2-inch pipe, screwed into headers, or drums, connected with the circulating system of the boiler. The supply pipes to the front header are taken from near the bottom of the front end of the shell, the water passing through the grates into the rear header, which is connected to the boiler shell some distance back from the front, just below the water line, and the space between the drum and shell is built up solid with fire-brick. The operation of the down-draught furnace is directly opposite to that of the ordinary setting. Comparatively little air is admitted below the water grates, and the entire supply of coal, and practically all the air enters above. The fire burns downward, instead of upward, there being no passage except downward through the grates. The gaseous products of combustion, together with the finely divided carbon particles which form the visible smoke, are forced through the incandescent mass of coals and are highly heated, after which they meet the equally hot flame from the lower grate, on which there is burning what is practically a coke fire. The combined water of the volatile matter in the coal, as well as its moisture, are decomposed into hydrogen and carbonic oxide gases, and these combine with the air supplied below the grate, or drawn downward through it, and burn, thus adding to the efficiency of the furnace. The separated carbon meanwhile is transformed into carbonic acid gas, and the result is almost complete combustion. Whatever additional air is required is furnished through registers in the doors between the two grates, or through those of the ash pit. The style of furnace requires a somewhat increased chimney capacity, if it is desired that the boilers be capable of doing as much work as those set in the ordinary way. If the demand for steam never greatly exceeds the rated capacity of the boiler, the ordinary chimney will answer, it simply being necessary to carry thinner fires. The best results, however, in efficiency and smokelessness, as well as in capacity, are secured by having a chimney of ample height, but this is equally true with regard to ordinary settings, which rarely have enough chimney. They claim a saving for this furnace of from 20 to 30 per cent.

The highest value that has been found by actual test of a pound of coal is 14,603 heat units, and each heat unit is equivalent to 778 foot pounds, so that each pound of coal furnishes the equivalent of 11,361,134 foot pounds per hour, but we only get back 1,980,000 foot pounds, or about one-sixth of the mechanical equivalent of the heat supplied.

A pound of coal or any other fuel has a definite heat-producing capacity, and is capable of evaporating a definite quantity of water under given conditions; this is a limit beyond which even perfection cannot go, and yet, I have heard, and doubtless you have heard, o cases where inventors have claimed that their improvements will enable you to evaporate from 16 to 17 pounds of water per pound of coal, and so-called engineers have certified to these results.

You all know that this is impossible, the highest value for a pound of coal being 14,603 heat units, and it is a known fact that it takes 965.7 heat units to evaporate one pound of water from and at 212° Fahrenheit, so that dividing 14,603 by 965.7 we have 15.1 pounds of water per pound of coal, and then only when every heat unit is put into the water. The highest value of evaporation so far has been 11.5 pounds of water per pound of coal, per hour; but, as a general rule, it is from 7½ to 8 pounds per pound of coal, per hour.

In conclusion, I would say that in the combustion of fuel there is but one body combustible to be dealt with, carbon and hydrogen, and but one supporter, the oxygen of the air; that in combustion, atmospheric air is the principal element, but it is the one to which practically the least attention is given, either as to quantity or control, and that chemistry and experience teach us that combustion depends, not so much on the quantity of air passing through the incandescent fuel, as upon the weight of oxygen taken up in its passage through it. In fact, the quantity of air passing through it may be destructive of combustion if improperly introduced and distributed. That the quantit of heat generated depends upon the relative weight of carbon or hydrogen, and chemically considered, their equivalent weights of atmospheric oxygen, so also the quantity of steam generated does not depend so much upon the intensity of the fire as on the quantity of heat absorbed by the water. Now, it is well known that success in generating the most heat and steam, and consequently power, from a given amount of coal, depends upon a compliance with the necessary conditions to perfect combustion, which involves not only a theoretical knowledge of chemistry, but also a practical knowledge of the best methods of combining them with mechanical appliances, and the perfect mixing of the constituent elements with which we have to deal, in strict accordance with the laws of nature.

For the standard method of testing coal referred to in this paper, the following is the outline of procedure: For the moisture a finely ground sample is dried for one hour in an air bath at 105° to 110° C. For the other constituents a fresh sample is taken of about a gram in quantity and put in a platinum crucible, the crucible being covered; it is now heated for 3½ minutes over a Bunsen burner, followed immediately with the highest temperature of the blast lamp for an equal length of time. The loss in weight, less the moisture obtained, equals the volatile combustible matter. The fixed carbon is next burned off by removing the crucible cover and heating in the flames of the Bunsen burner, with access of air till the carbon is burned off; the loss of weight equals the carbon, the residue is ash.

## THE LAW AS TO SUNDAY CARS.

FOLLOWING is the decision in full handed down by Judge Rose in the action brought by the Lord's Day Alliance to restrain the Hamilton Street Railway Company from running their cars on Sunday:—

"It was conceded that the defendant company had the right to run its cars on Sunday as well as on the other days of the week, unless doing so was a violation of the provisions of chap. 203, R.S.O., amending an Act to prevent the profanation of the Lord's Day, sometimes called the Lord's Day Act.

The following questions then arise: (1) Does the above statute apply to the defendant company?

(2) If so, is what was shown to have been done here within the exception as being a conveying of travelers?

(3) If not within the exception, was it necessary, to entitle the plaintiff to succeed, for him to show substantial injury to the public?

(4) If necessary, has such injury been shown?

(5) And in any view, on this evidence, is an order of injunction the proper remedy for a violation of the Act?

The statute does not apply to the company unless it is one of the persons named in the first section of the Act or a person ejusdem generis with those named.

I assume that the fact that the defendant is a corporation does not prevent the Act applying.

The persons named in the Act are "Merchant, tradesman, artificer, me hanic, workman, laborer, or other person whatsoever."

It is not open on the decisions in our own courts for the plaintiff to contend that the words "or other persons whatsoever" are not to be construed to refer to persons ejusdem generis. Therefore we have to see if a person running street cars is one named by the statute or ejusdem generis with such person. This question also is, as it seems to me, practically concluded by authority.

In Sandiman v. Breach, 7 B. & C., 96, it was held by the court, the judgment being delivered by Lord Tenterden. C. J., that the words "or other person or persons whatsoever" in the 29 Car., 2, c. 7, s. 1, were not used in a sense large enough to include the owner or driver of a stage coach; that section provided "that no trades. man, artificer, workman, laborer or other person whatsoever shall do or exercise any worldly labor, business or work of their ordinary callings on the Lord's Day," etc. In Reg. v. Budway (supra), it was held by the full court (Q.B.D.) that a cab driver did not come within the words of chap. 203, and in Reg. v. Somers (supra) the same court followed its decision in Reg. v. Budway.

In the latter case the fact stated was that "The defendant was a servant of one Charles Brown, a keeper of a livery stable in the city of Toronto, and on the day in question drove a cab belonging to Brown through the streets of the city for hire." Mr. Moss urged that the two latter decisions should be confined to the facts then before the court, and did not apply to a case of a cab driver who was both owner and driver. It seems to me there is in principle no distinction between the driver who is the owner and the driver who is the servant of the owner that would apply in favor of the servant, indeed it might be contended that a servant who was the owner would more readily come within the description "workman or laborer" than would the owner who was also the driver, and in Sandiman v. Breach, as we have seen, Lord Tenterden draws to such distinction, but uses the words "owner or driver." Then if an owner or driver of a stage coach or an owner or driver of a cab is not within the Act, is one who is an owner or driver of a street car, whether such car is by horses, steam, electricity or other motive power? I am unable to see any distinction between such persons. I think there is none; and, following the above decisions, which are binding upon me, I must hold that the defendant company is not within the Act and so not prohibited from running its cars on Sunday.

But, assuming that the act does apply, th n has it been shown that the company was or was not "conveying travelers"?

The exception is in the following words, "Conveying travelers or Her Majesty's mail by land or water, selling drugs or medicines and other works of necessity, and works of charity only excepted."

In Reg. v. Daggett, 1 O. R., 537, the full court (Q. B. D.) composed of Hagarty, C. J., and Armour and Cameron, J. J., held that excursionists leaving Buffalo in the State of New York on a Sunday morning and proceeding by rail to Niagara, thence by defendant's steamboat to Toronto, and back the same day, were travelers within the exception, and that there is no distinction in such a case between travelers for pleasure and for business. The decisions under the 29th Car., 2, c. 7. were collected and referred to and accepted as defining the term "travelers" as used in our statue.

There the learned Chief Justice said: "It matters nothing in my judgment whether they travel wholly for pleasure, fresh air, relaxation from work, or with or without luggage, or actually on important business. They are travelers within the meaning of the statute. To draw any distinction between persons according to the purpose which induced them to travel would, as it seems to me, be a vain attempt, leading to impossible and irritating inquiries and tending to bring a useful and salutary enactment into contempt."

No affect was given to the argument of counsel that "conveying travelers" to be within the exception must be a work of necessity, the court evidently holding otherwise.

Among the cases referred to decided in England under the 29th Car., 2, was Peplow v. Richardson, L.R., 4 C.P., 168, where it was held that a man who walked two and a half miles from his residence to drink mineral water at spa was a traveler, and in Taylor v. Humphries, 10 C B., N.S., 429, Erie, C.J., held that persons who had walked four miles on business or pleasure might be lawfully supplied with refreshment as travelers.

Mr. Moss endeavored to distinguish these cases on the ground that the persons had walked to a distance out of the town where they resided, but I find no distinction suggested, and it seems to me to be fanciful and not entitled to prevail. It is also manifest that the distance passed over does not determine whether one is a traveler or not.

In Reg. v. Daggett; Reg. v. Tinning, 11 U C.R., 636, was referred to and not followed. It was declared to be not in accordance with the subsequent decisions.

In Reg. v. Tinning the court held "that persons making it their ordinary business to ply within the harbors of a town not for any purpose of carrying travelers or the mail were intended to be restrained by the Act," adding, "We think it clear that the persons carried on a Sunday between the city (Toronto) and the peninsula cannot be called travelers within the meaning of the exception. They are persons notoriously seeking mere recreation."

I must follow Reg. v. Daggett in preference to Reg. v Tinning, leaving an appellate tribunal to say that the later decision is wrong, if it is so, for it—Reg. v. Daggett—is founded upon decisions subsequent to Reg. v. Tinning, and has remained unquestioned since it was decided in 1882.

Both decisions are of the same court, differently constituted, and the later decision is, as I think, a declaration that the former is not good law, and is a declaration of the law which binds me sitting as a judge of first instance.

It is pointed out that the Legislature by the 48 Vic., c. 44, ss 1-7 (now s. 7, R.S. O., c. 203), had declared excursionists not to be travelers; but it will be observed that such section applies only to persons going and returning on the same day, by the same steamboat or railway or any other owned by the same persons or company such steamboat or railway having for the only or principal object the carriage of Sunday passengers for amusement or pleasure only, and does not apply to persons carried one way only or going and returning on different days. So the construction put upon the word "travelers" by the court in Reg. v. Daggett stands with the above exception.

It is instructive to note the care the Legislature exercised in declaring the limitations to such extension of the Act.

I find as a fact the defendant company has not been shown to have run its cars having for the principal or only object the carriage of Sunday passengers for amusement or pleasure only, and there is no evidence before me on which I could find that the company carried what might be called Sunday excursionists.

I find as a fact that the cars of the defendant company were shown to have been run on Sunday as on other days, only less frequently, for the carriage of travelers at the usual rate of fare, and, although it may be that persons who were not travelers were carried, such fact was not shown. It having been decided that to go two and a half miles to drink mineral waters; to walk out for fresh air, or pleasure, say three or four miles; to go upon an excursion from Buffalo to Toronto and return; or to go over to the peninsula, (now Island) opposite Toronto for recreation, constituted the person so journeying a traveler, I must either ignore the effect of such decisions or hold that at least certain persons carried by the defendant company were travelers within the meaning of the Act, e g., persons who, coming into the city by the ordinary railway trains, desired to reach their respective destinations in the city; persons going to and returning from church; persons going to places of rest or recreation; and the like.

It follows, from the view I have taken of the decision referred to, that the company had a right to run its cars for the purpose of "conveying travelers," and, so running, the cars would create all the noise and disturbance that they are alleged to create, to the annoyance of some of the persons who gave evidence at the trial, without regard to whether they did or did not carry persons who were not travelers, and so I cannot find that, by carrying persons who were not travelers, they have created or continued a nuisance, therefore, there is no ground on which the court can interfere.

For the information of the court, if this case is carried farther, and it shall become necessary to find any facts upon the evidence found by me, I desire to say that, as far as I could judge, the several witnesses at the trial were apparently honest in their endeavor to tell the truth. Their opinions as to the running of the cars on Sunday being an annoyance seemed to be affected by their views as to whether it was morally right or wrong so to run them, and such views were again influenced to some extent by the fact that the running of the cars on Sunday was or was not to them or the churches or congregations to which they belonged a benefit, advantage, or convenience.

I was referred by Mr. Martin to ch. 99 of the 55 Vic. (O.), incorporating the Toronto Railway Company, and confirming the agreement therein set out. By clause 1 of the Act the company is permitted to run its cars on Sunday, when agreed to by the citizens, provided that so doing it is not a contravention of the Lord's Day Act. This probably shows that the Legislature had formed no opinion that the Lord's Day Act did clearly prohibit the running of cars on Sunday.

I was referred to the case of The Attorney-General v. Niagara Falls Tramway Co, 19 O.R., 624, and 18 A.R., 453, but that case did not turn upon any question under chap. 203, above considered.

On the whole, I am of the opinion that the plaintiff's case fails, and the action must be dismissed with costs."

## THE STEAM ENGINE INDICATOR AND ITS USES.

By Wm. Thompson, Chief Engineer Montreal West Water & Light Station.

Year after year the beneficial results to be derived from this wonderful little instrument become more widely known throughout the engineering profession, and the study of indicator diagrams is now easy as compared with a few years ago, owing to the completeness and simplicity of the data at hand. But all engineers are unfortunately not in a position to either acquire a theoretical or practical acquaintance with the indicator, and to these I more especially desire to address myself. In these days of progressive modern engineering nearly every plant of any importance is provided with either one or more indicators for the use of its engineers. Owners of small plants, however, very rarely consider this instrument a necessity, as they rarely understand the purpose for which it is intended. The engineers in charge of these plants have therefore very little chance to become acquainted with the use of the indicator practically unless they put their hands in their pockets and purchase one for themselves, a course I strongly advise even though you may have to reduce your spending money for some time and go "hard up" for months. You will derive such knowledge from the use and study of this instrument, that no matter how thoroughly practical you may be, you will be so fully repaid that the outlay will never be regretted. An engineer aspiring to perfect himself in his profession loves to study the theoretical basis of his profession and to acquire such intimate knowledge that every move of his engine is thoroughly understood, and becomes to him at least a thing of pride and joy, always scrupulously clean and running as smoothly as care and knowledge can make it, satisfying to both engineer and employer. To an engineer of this type (and I know many of them) I say buy yourself an indicator rather than a watch to show you when it becomes time to quit work for the day. If an engineer of the opposite type, an indicator will not be of much use, as it will almost surely soon get as badly out of twist as the engine you so often claim is of no earthly use except for scrap iron. To practical men studying their own and their employer's interests the indicator becomes almost a necessity.

Some of the leading and most valuable items to be obtained by the use of the indicator are:

(1) The arrangement of the valves for admission cut-off, release and compression of the steam.

(2) The adequacy of the ports and passages for admission and exhaust, and, when applied to the steam chest, the adequacy of the steam pipes.

(3) The suitableness of the valve motion in point of rapidity at the right time.

(4) The quantity of power developed in the cylinder and the quantity lost in various ways, viz., by wire drawing, by back pressure, by premature release, by poor adjustment of valves, by leakage, etc.

Taken in combination with measurement of feed water and the condensation and measurement of the exhaust steam with the amount of fuel used, the indicator furnishes many other items of valuable information obtainable from no other source when the economical generation and use of steam are considered. The latter question is annually becoming of greater importance, not only to proprietors, but to the engineer. I do not for a moment claim that because your valves were set without the aid of an indicator that they are sure to be wrong, but I do say that with the intelligent use of an indicator you can readily ascertain if they are exactly right, and you can ascertain just how much power you are using and just how much is useful and how much is frictional; in other words, you have at all times the means to ascertain what you are doing and how you are doing it.

I have heard men boast time and again that they could set the most complicated valve systems without the aid of an indicator, and I have had the pleasure of indicating some of these engines and invariably found a great improvement could easily be effected, simply because the means were at hand to ascertain if any errors existed, and not because of any want of ability on the part of the engineer.

Having decided to own an indicator, the purchase of the instrument becomes a question requiring careful consideration. In the first place, indicators can be had at almost any price, but the nature of the work required on a good reliable instrument is so minute that only the very best work and materials must be employed, that you will find it to your advantage to deal with a maker having a reliable reputation for good goods rather than purchasing a poor instrument at a cheaper price. Any instrument can be made to work, but not all the instruments sold will do good work. You might as well expect to secure a really reliable engineer at half wages as to buy a good and reliable instrument at half what it is worth. The nature and style of work you have to do will very largely determine the style of instrument to purchase; for instance, you can use an instrument of much heavier make on a slow running engine than on an engine running at high speed and changing direction of paper drum travel more rapidly. Modern constructed indicators are now arranged bearing particularly on high speeds, and it is generally advisable to purchase an instrument designed for high speeds, which will be quite as useful on the average slow speed engine.

The importance of the indicator is now so generally recognized by all engine builders that nearly all first-class engines are sent from the shops with the cylinders already drilled for application. When, however, no provision has been made for the application of the indicator, holes must be drilled and tapped for not less than ½ inch pipe in such position on the cylinder that when the piston is at the ends of its travel they will be as nearly as possible in the center of the clearance space, and yet not be obstructed by the piston when at its extremes of travel. In drilling great care must be taken not to allow any chips to get into the cylinder; and when the cylinder heads cannot be removed it is best to turn on a little steam as the drill begins to enter, so as to blow all cuttings out. If you find clearance is too small to allow connection as above, the tap may be made directly into the head (which it is desirable to avoid if at all possible) to bring the indicator into a convenient position, the object being to have the indicator connected as directly as possible to the cylinder, and in all cases where the circumstances will permit screw the indicator cock into the cylinder itself. Where the tap is on the side of the cylinder, by use of nipples and elbows the indicator can be brought into a vertical position the same as if tapped on top of cylinder. When the arrangement is to be permanent, it is advisable to have a cock for each end of the cylinder. Where you are using only one instrument, the best method being to connect by means of side pipes and a 3 way cock arranged exactly in center of cylinder. The slight disadvantage arising from this indirect connection is more than counterbalanced by the facility with which diagrams can be taken without disturbing the paper drum, and by the fact that diagrams can be taken from both ends of the cylinder on the same card, making them very useful for comparison. Do not, if possible, to avoid this, use angle valves on the ends of pipe instead of elbows with a T in the centre to attach indicator to, as is sometimes done, or the diagrams are liable to present an appearance similar to insufficient lead and "wire drawn" admission.

After cylinder is ready for application of indicator, the next step is to prepare a suitable reducing motion to get the required length of diagram on the paper drum. This is sometimes difficult to do, particularly when high rates of speed are used. There are certain important points to observe, the chief requisite that the device used shall give to the paper drum a motion which will exactly coincide with position of piston in minature at any part of stroke. The most reliable and useful device I have had the pleasure of using for this purpose is the recent invention of a well known Canadian engineer, Capt. James Wright, of Montreal, and illustrated in the ELECTRICAL NEWS of April 1895. As my readers can easily get a full description of this device by referring to their back numbers, I need not here describe it.

In setting this reducer upon an engine permanently, it is well to commence mechanically and as near correct as possible. Lay out reducing motion so that when piston is in center of travel, vertical lever of reducer is exactly at right angles with piston rod and as near center of cross-head as circumstances and nature of construction of engine will permit. It is also important that sliding bar of reducer should at all times travel parallel with cross head of engine. With ordinary care in setting up this reducer can be made to appear a portion of the engine and will actuate the revolving drum of indicator practically correct at almost any rate of speed, and paper drum can be attached or detached at will of operator without any difficulty.

To take a diagram, screw the indicator to 3 way cock already placed in center of cylinder, and connect with sliding guide on reducing motion by means of cord having as little stretch as possible. If distance from indicator to paper drum is very great, it is better to use a piece of flexible wire for this purpose. Both wire and cord must be provided with a hook and loop, so that cord can be detached at will with engine in motion. Adjust sliding bar of reducer to required length of diagram by means of movable fulcrum pin in frame. For slow speeds the best and most desirable diagram is from 3½ to 4 inches in length, but with high speeds for accuracy diagrams should not be more than about 3 inches. If spring on paper drum is properly set the length of card can be adjusted to a nicety, and an effort should always be made to have length as free from fractions as possible to simplify and assist after calculations of diagrams.

To attach a card to paper drum is a simple matter, but I should strongly advise the use of metallic faced paper to allow use of a metallic point on end of pencil lever, which must be firmly and securely fastened to prevent any possible vibration through shaking or moving of pencil, also the friction should be as light as possible and a very fine line drawn with merest touch of point on paper.

In selecting a spring always use one that will give you a diagram about 2 inches high, that is with a 80 lb. boiler pressure use a 40 spring. Each indicator usually has from 3 to 5 springs accompanying it. Before allowing steam to enter indicator remove piston from indicator cylinder and blow steam through from both ends of engine cylinder. Carefully oil indicator piston with best cylinder oil, and all other moving parts with specially prepared watch oil. An indicator piston should drop into cylinder of its own weight freely and easily, when both ends are open to the atmosphere. Screw down milled nut on indicator firmly and adjust screw to regulate pressure of pencil, keeping pressure as light as possible, and be careful to have paper securely and smoothly placed on revolving drum.

Before allowing full pressure steam to enter indicator allow steam to escape on relief valve at side of 3 way cock until steam becomes dry and clean. Allow indicator piston to work under full pressure for a few revolutions, until it becomes hot, then with both ends of piston open to atmosphere, draw the atmospheric line by applying pencil to paper while moving with reducing motion. Turn on steam under full pressure and apply pencil to paper during one or more revolutions of engine from each end of cylinder. If reducing motion and paper drum spring are properly adjusted, length of a double card should measure in length exactly the same as the length of the atmospherical line previously drawn.

After a sufficient number of diagrams have been taken, remove the piston, etc., from the indicator, while it is still upon the engine, allow steam to blow for a moment through the indicator cylinder, and see that piston, spring and all movable parts are thoroughly wiped, cleaned and oiled. Pay particular attention to the springs, as their accuracy will be seriously impaired if they are allowed to rust, and great care must be taken that no grit or foreign substance be introduced, to cut the cylinder or scratch the piston ; remember you are handling a delicate and sensitive instrument, and act accordingly. The heat of the steam blown through the cylinder of the indicator will be found to have dried it perfectly, and the instrument may be put together with the assurance that it is ready for instant and immediate use when required.

The various lines drawn by the indicator on the diagram are named as follows and can be readily recognized after a little practice :

THE ATMOSPHERIC LINE is a line drawn by the pencil of the indicator when the connections with the engines are closed and both sides of the piston are open to the atmosphere. This line represents on the card the pressure of the atmosphere or zero gauge pressure.

THE VACUUM LINE is a reference line drawn about 14.7 pounds by scale below the atmospheric line and represents a perfect vacuum or line of no pressure.

THE CLEARANCE LINE is another reference line drawn at a distance from the end of the diagram, and at right angles with a line

here, equal to the same per cent. of its length as the clearance is of the piston displacement. The distance between the clearance line and the end of the diagram represents the volume of the clearance and waste room of the ports and passages at the end of the cylinder.

THE LINE OF BOILER PRESSURE is a reference line drawn by hand parallel to the atmospheric line and at a distance from it by scale equal to the boiler pressure shown by the gauge. The difference in pounds pressure between it and the steam line shows the loss of pressure due to steam pipe and the ports and passages on the engine.

THE ADMISSION LINE shows the rise of pressure due to the admission of steam to the cylinder by opening the steam valve. If the steam is admitted quickly when the engine is about on the dead centre this line will be practically vertical and at right angles to the atmosphere.

THE STEAM LINE is drawn when the valve is open and steam being admitted to the cylinder. In automatic cut-off engines with sufficient port area this line will be practically parallel with atmospheric pressure.

THE POINT OF CUT-OFF is the point where the admission of steam is stopped by closing the valve. Sometimes there is a little difficulty in determining just exactly where this takes place. It is usually, however, located where the outline of the diagram changes from convex to concave.

THE EXPANSION CURVE shows the fall in pressure as the steam in the cylinder expands doing work.

THE POINT OF RELEASE shows when the exhaust valve opens.

THE EXHAUST LINE represents the change in pressure that takes place when the exhaust valve opens.

THE BACK PRESSURE LINE shows the pressure against which the piston acts during its return stroke. On diagrams taken from a non-condensing engine it is either co-incident with or above the atmospheric line. On diagrams taken from a condensing engine it is found below the atmospheric line and at a distance greater or less, according to the vacuum obtained in the cylinder.

THE POINT OF EXHAUST CLOSURE is the point where the exhaust valve closes.

THE COMPRESSION CURVE shows the rise in pressure due to the compression of the steam remaining in the cylinder after the exhaust valve has closed.

THE MEAN EFFECTIVE PRESSURE (M. E. P.) is the mean net pressure pushing the piston forward.

THE INITIAL PRESSURE (I. P.) is the pressure acting on the piston at the beginning of the stroke.

THE TERMINAL PRESSURE is the pressure above the line of perfect vacuum that would exist at the end of the stroke if the steam had not been released earlier. It is found by continuing the expansion curve to the end of the diagram. This pressure is measured from the line of perfect vacuum, hence it is the absolute terminal pressure.

It is not my intention to use diagrams, I therefore will not attempt to describe how the various calculations can be arrived at by gathering particulars from diagrams. Any engineer with a little experience in handling indicators, will be able to analyse his own diagram, if not, I am quite sure the editor of this esteemed journal will be only too glad to reproduce them, together with any questions, in his paper to allow other engineers a chance to analyse and discuss them.

The most important formulae required is that to find the horse power generated from a diagram, and this is the only one I shall take the liberty to deal with. You first require to find from the diagram the mean effective pressure on the engine piston through. out the stroke, this is easiest arrived at by the use of a small instrument called the planimeter, which with careful manipulation will give the area of the diagram within the hundredth part of an inch. When area of diagram becomes known divide by length of diagram, the result will be the mean average height of the diagram ; multiply this by scale of spring used and you have the M. E. P. throughout the stroke. There are, however, several methods of finding the M. E. P. without the aid of a planimeter, one of the most convenient being as follows : Draw on the diagram ten or any other convenient number of lines at right angles to the atmospheric line and at equal distances apart. Measure the length of each ordinate within the lines of the diagram and divide the sum total of their lengths by the number of ordinates used. Multiply average length thus found as before and you have the same result.

To calculate the h. p. of an engine, multiply the mean net area of the piston in square inches, (Diameter squared × .7854 = area minus area of piston rod = mean net area) by the M. E. P. previously found in pounds per square inch acting on the piston throughout the stroke (area piston × M. E. P.) Multiply this product by the distance through which the piston travels in inches per minute. (Stroke in inches × rev. per minute × 2 strokes per revolution), this will give you the number of inch pounds exerted by the engine. Divide this by 12 to reduce to foot pounds, and as an H. P. is understood to equal 33,000 pounds raised 1 foot high in one minute, by dividing total foot pounds by 33,000 you get total h. p. generated by engine in accordance with following formula :—

$$\frac{\text{Mean net area of piston} \times \text{M. E. P.} \times \text{rev. per min.} \times 2 \times \text{stroke in inches}}{12 \times 33000} = \text{I. H. P.}$$

Mr. E. E. Cary, who has been engaged in the manufacture of incandescent lamps in the United States for the past ten years, has accepted the position of general manager of the Packard Electric Co., Ltd., of St. Catharines, Ont. Mr. Cary expects shortly to call upon and make the acquaintance of the many users of the Packard lamp and transformer throughout the Dominion.

# CORRESPONDENCE

## BOILER EXPLOSIONS AND THEIR CAUSES.

Editor CANADIAN ELECTRICAL NEWS.

SIR,—Your theory of the Detroit boiler explosion is correct and the only commonsense one, to wit, that the safety valve, even if operative and open, was too small to give passage to vapor forming with such rapidity as to increase its pressure from 15 lbs. to 100 lbs. in four minutes, and from 50 lbs. to 100 lbs. in one minute. Until Parliament intervenes to have safety valves made larger, explosions will continue to occur and fatalities to follow.

C. BAILLAIRGE,
City Engineer, Quebec.

## THE C. A. S. E. AS A SECRET ORDER.

Editor CANADIAN ELECTRICAL NEWS.

SIR,—I notice in your last issue an editorial the first three lines of which contained the following : "It seems unfortunate that the Association of Stationary Engineers should have been organized upon the basis of a fraternal order."

Since becoming a member of the above order I have been enabled through its aid to master numerous problems pertaining to the calling, but after carefully reading your article and trying to formulate a plan by which any body of men could so hold together without either sign or password, I was obliged to give it up.

If we admit that some secrecy is necessary, we must also admit that no Association could be organized with less than the C. A. S. E. You say you do not object to such societies or lodge room methods, but the cumbersome machinery of a secret society is unnecessary where there is nothing to conceal.

Mr. Editor, you must have got into the room where we keep the goat, as he is the only cumbersome piece of machinery we have, and the fact that you have taken the 4th degree and not the first three, may be responsible for the statement you made that our initiations are not always impressive.

I will impart to you some of the secrets in the following statement. Our initiation (including all the degrees) occupies about ten minutes. Engineers being a good thing, and consequently scarce, is responsible for the very few initiations during the year, say ten at most; this is about one hour and a half spent in twelve months, and I venture this remark, Sir, that the time spent by you in copying and printing the above article occupied more time, is no more legitimate, nor beneficial than the initiations, etc., of the C. A. S. E.

In conclusion I will say that as the paraphernalia of the order consists only of a tin cent brass plated button, I will not mention it. However, if it should prove unsightly to any one I will bore holes in it and utilize it instead of wire nails to keep the bottoms of my blue bloomers from sweeping the dusty surface of my boiler room.

Toronto, Dec. 12th, 1895. J. G. BAIN.

## BY THE WAY.

MR. A. W. CONGDON, now engaged in the engineering department of the Canadian General Electric Company at Toronto, was one of the pioneers in the introduction of electricity for lighting purposes in Japan. In 1889 the Edison Company, of New York, in whose employ Mr. Congdon then was, received an order from Japan for a 100 light electric plant. Partly owing to the improbability of being able to secure the services of any one competent to install the plant, and partly, no doubt, with the object of making known their goods in an entirely new field, the possibilities of which it was impossible to judge, the company decided to send Mr. Congdon and another of their employees to Japan with the electrical apparatus and the necessary steam plant to operate the same.

This was the second electric plant installed in the country, the first one having previously been put in by the Brush Company. The intention was that Mr. Congdon and his companion should remain about six months, or long enough to install the plant and get it into proper working order, and be enabled to estimate the possibili-

ties of the field for future business. As a matter of fact, the two Americans remained in Japan for a period of nearly three years. After completing the work which they were specially sent out to do they received an order to install an arc and incandescent system for the Emperor, whose palace is now lighted with incandescent lamps, and the grounds surrounding it with arc lights.

Mr. Congdon states that, owing to the prejudice which prevails against foreigners, all business in Japan has to be done through native companies. Such companies now exist in Tokio and several of the other large cities, and it is estimated that there are at present in operation throughout the empire about 16,000 lights.

Public lighting is done by these companies under contract with the municipality and with private consumers in the same manner as in this country. Much of the current to private consumers is supplied through meters, an additional charge being paid by the customer for the use of the meter. Lighting plants are also being put in to some extent by manufacturing companies. For one cotton mill company Mr. Congdon installed a 500 light plant, which has since been increased to 1000 lights.

On account of the low standard of illumination, as compared with Europe and America, the progress of electricity for lighting has necessarily been slow. The people of Japan cannot be expected to jump from fish oil to incandescent electric light at a bound. A few years ago fish oil was the almost universal illuminant in that country. Within a comparatively recent date kerosene oil has been introduced, and is now being imported in immense quantities. As the standard of illumination rises, the increased use of electricity must necessarily follow.

Referring again to the prejudice which exists against foreigners, it is only within recent years that Japan has tolerated in any degree the people of western countries. Now, however, she is seeking to learn from and profit by the more progressive western civilization, and Japanese students are now to be found in the military, naval and scientific schools of Europe and America. I learn from Mr. Congdon that in cases where it is obligatory upon the Japanese to employ a foreigner, their policy is to pump all the information possible out of him, and when the supply is exhausted, replace him by a native whose services can be had at a comparatively trifling cost. The average wage for unskilled labor is 15 cents per day, and skilled workmen, such as carpenters, receive but 50 cents. A Jap, says Mr. Congdon, can live comfortably on 10 cents a day. On account of the antipathy to foreigners of certain classes of the people, it is the rule for a foreign judge to sit with the native judge in every legal case in which the interests of a foreigner are at stake. These foreign judges are appointed at the instance of the governments of countries such as Germany, Great Britain and the United States, which have treaties with Japan. Japan is most desirous that these treaties with the European nations, some of which are about to lapse, should be renewed ; consequently when it was demanded of her that she should appoint foreign judges and pay them handsome salaries, she consented to do so without a murmur. Notwithstanding that the United States are the largest buyers of Japanese goods, Japan purchases more largely from England and Germany, always having in view no doubt the renewal of the treaties with these nations.

## CENTRAL STATION BOOK-KEEPING.

### BY GEO. WHITE-FRASER, E. E.

THE keeping of an exact system of accounts is an absolute necessity in any kind of business, if it is to be intelligently followed as a means of livelihood, and not as a mere means of passing the time. The light thrown on the working of a business, the comprehension of its details, in fact its science, depends, one might say principally, on the minuteness and accuracy of the records kept ; and without any records at all such a business can only be a formless chaos drifting about aimlessly—the sport of fortune, the plaything of chance. Business may properly be called a science ; a particular business is a branch of science. Science has been called the record of exact observation, and it certainly is true that every branch of knowledge of to-day is simply based on the accumulation of observations made during past ages, and without observation there could be no knowledge. Observation is the "book-keeping" of science ; experience is merely the tabulations and deductions of the mental book-keeping process gone through by every intelligent person, and it will be evident that book-keeping of some kind or other, whether simple or complex, conscious or unconscious, lies at the root of all intelligence, order, knowledge and progress. The very term "accounting" implies order ; if there were no records there could be no knowledge, and it seems unnecessary to point out that without knowledge progress is an attractive vision, an unattainable ideal.

Electricity is as much as any other a science in which observation is peculiarly rich in results, and likely to greatly benefit the observer. Its various workings are by no means reduced to exact knowledge yet, and as its practical applications to the uses and requirements of every day life are innumerable, it behooves every person interested in electricity as a business to keep records, as much for his own guidance as for the advancement of the science. That application which is of most practical interest to the readers of this journal is, of course, to the requirements of public and domestic lighting, and to the supply of power, both for stationary purposes and for the purposes of locomotion ; and as the methods of the generation, transmission and utilization of current for the above purposes have been long and scientifically studied, and as a matter of fact have been reduced to their lowest terms (i. e., cost), it is proposed to indicate first, what those lowest terms are ; next, how to attain to them. Putting it differently, it is proposed to show for how little, under favorable circumstances, current may be generated and light or power produced ; and next, what method of central station bookkeeping will most clearly show how much it costs any individual station to produce current, and therefore, how much improvement can be introduced into the operations ; what economies can be made, and what extra income earned. It is quite plain that unless you know what you are doing now, and how you are doing it, it is not possible for you to see your way to doing better. And if you can do better, you might as well do it. Whereas if you are doing as well as possible under the circumstances, it is well for you to know it and to keep up that high standard.

The successful carrying out of an electric light and power business involves the proper and efficient operation of so many different classes of machinery and apparatus, the maintenance of so many parts subject to wear and tear, and the minimizing of so many possible sources of waste, that some amount of careful and systematic accounting is absolutely necessary, and the more complete and comprehensive the system of records the more efficiently will the whole plant be kept. This becomes evident when it is remembered that not even the very highest class of machinery—steam or electric—is anything like perfect, and that in order to get out of it all that it is capable of doing, constant care and watchfulness are necessary. This imperfection and inefficiency is found in every piece of apparatus composing a central station plant. Of the coal you buy by the car-load, some will be wasted by going to dust ; a very large proportion of the heat contained in it will go up the chimney without doing any good under the boiler ; more will be radiated from steam pipes and cylinders ; some condensation of steam in pipes will waste heat ; valves may get out of adjustment and allow more steam to be used than is absolutely required for the work to be done ; belts will slip ; shafting will absorb power ; the best dynamo ever made will only give back about 95 per cent. of the power given to its pulley at full load ; lines, leaks, transformers, lamps and consumers will all waste current, and they cannot help wasting some ; but the amount thus lost may easily be kept within reasonable limits, if you only know who or what is wasting too much, and in what particular way. It does not seem to be properly recognized by a large number of those in charge of the smaller electric lighting stations, that all machinery and apparatus is necessarily more or less inefficient. A transformer on a pole, that certainly has every appearance of being sound and strong ; not grounded anywhere ; properly insulated and mechanically perfect in every way, still wastes current. If you don't believe it, take some transformers, connect up their primary circuits to the 1.000 volt mains, and leave their secondary circuits open. You will say "there's no load and therefore no current will flow on the primaries." Try. Start up the alternator and a sufficiently sensitive ammeter on the primary will show a current which will be larger as more transformers are put in circuit, and which you cannot stop, simply for the reason that a small leak is perfectly inevitable and inherent in the design of a transformer. This is an inefficiency, and you want to make it as small as possible. You must also thoroughly realize that every machine or piece of apparatus in your power house is more or less inefficient, and that only by watching can you make the effects as small as possible. Bookkeeping is necessary not only to show you how much you are making or losing, but to show you where you are losing ; where you are not doing so well as you might ; what particular piece of apparatus is of poor quality ; what particular class of business is worth working up ; and until a system of records is kept, not only of wages, coal, and gross receipts, but of wear and tear, leaks, lamp renewals, etc., no electric lighting business can possibly be intelligently managed. It may possibly be thought that the writer is very unnecessarily prolix as to the desirability of accounting, but in his experience the great majority of smaller central stations not only keep hardly any accounts at all, but are not aware of the directions in which improvements are possible, nor of the facts referred to above as to the inherent imperfections of all apparatus. It is therefore thought advisable to divide the general subject into parts, showing the various headings under which special accounts should be divided, for steam plant, electric plant, lines, lamps, consumers, etc., and to explain the necessity for the various headings, by reference to the inefficiencies and possible wastes which they are designed to keep track of and to check. Having once indicated the general system, then individual station managers will no doubt be able to extend it in such directions as seem to them fit.

First then, the business of a central station is to manufacture electricity and to sell light and power. Fuel—the raw material—must be got and by means of appropriate machinery turned into electric current. This must be conveyed somehow to the consumers' premises and turned into light, but between the coal-pile and the lamp there is much greater complexity than the above bald description might lead one to suppose. The most obvious and simple accounting will be to keep the amount of coal bought, and to set it against the money received for rental of lights, and everyone does this. This gives blind results, and shows whether the business is losing or gaining, but nothing more, and it is necessary to pry more closely into intermediate stages.

The first stage is the process of conversion of the heat in the fuel into motion. We burn fuel to raise steam for use in the engine. The distinct steps are (a) the combustion of the fuel.; (b) the communication of heat to the water in the boiler ; (c) the carrying of the steam from the boiler to the engine ; (d) the utilization of the steam by direct pressure, and by expansion in the engine, and (f) the getting out of the steam after it has done its work. Each step is important and should be attended to in order to get as much return as possible; viz., to minimize waste and loss. It takes a certain amount of heat to raise steam of a required pressure from a known quantity of water of a given temperature. Therefore there is a direct connection between the amount of fuel burned and the amount of water evaporated. An accurate record of the amount of fuel burned, and of the feed water used—with the average temperature of the feed water—will therefore give a very good idea as to whether there is a reasonable proportion between them. In fact it will show what use is being made of the fuel. This will lead to an investigation into the

are very easily kept, and which may show the way to very appreciable savings in the boiler room. These are the temperatures of the feed water after it has passed through the heater and a continuous diagram shewing steam pressure from start to finish. If the temperature of feed water can be raised at all without the direct expenditure of fuel, it means that a proportionately less amount of fuel will have to be burnt under the boiler. Taking this temperature both before and after the feed has been passed through the heater, will show how much it has been raised and consequently whether it may be possible to raise it still more. Continuous record of steam guage, will show whether the firing has been such as to maintain it regular, or whether it has varied above and below the proper point. Every time it has dropped low, the manager may be perfectly certain that it has been raised by putting on coal and opening the draft, and by the consequent wasteful escape of gases up the chimney before being consumed, and a too high pressure means also wasteful firing. Boilers are subject to wear and tear, their grates can be burnt; their flues stopped up with soot ; and their steaming abilites impaired by the formation of scale and the deposition of mud. It is important to know how long the grates will go without requiring to be replaced; how long the boiler may be run without requiring washing, etc., and records must be kept showing any repairs ; when flues were cleaned ; when boiler was scaled, with quality of scale, and the amount of compound used, with its effect on the scale. If grates go very rapidly it means either very poor firing, or very poor grates. Want of cleaning may be the reason of excessive consumption of fuel ; dirt in the flues indicates that the coal has not been completely burnt in the furnace and therefore that the fireman has not been properly attending to his work. The boiler room day book may therefore take the form indicated below, those stations having a large equipment will report particular boiler by numbers. It is usual to combine the whole station report into one sheet, this can of course easily be done :

FORM FOR BOILER ROOM DAY BOOK.

| No. of Boiler. | When started. | When stopped. | Fuel used. | Weight of ashes. | Temperature of feed before heating. | Temperature of feed after heating. | Amount of feed water used. | Flues cleaned. | Boiler cleaned. | Compound used. | Date. | | | |
|---|---|---|---|---|---|---|---|---|---|---|---|---|---|---|
| | | | | | | | | | | | Remarks as to condition of boiler, flues, grates, etc. | Repairs. | Name of Fireman. | Came on. | Went off. |

causes of any observed discrepancy, and then it becomes a question whether the coal is poor in quality, whether the method of firing is good, or whether there be any other cause. A record not only of the amount of coal burned, but of the ashes left over, will show the amount of actual combustible material in the coal, and these give an idea of the real money value of the fuel, and a still further accurate record of the chimney temperature and of the chemical analysis of the escaping chimney gases, will very clearly show whether available heat is going up the chimney unconsumed. It is of course only the very large stations that can keep these two last records.

There are two more very important records that

The record of the steam pressure should accompany the above report and be filed with it.

Here we have the means of observing the whole working of the steam generating department, which is very simple, easy to keep, easy to work up, and likely to be of the greatest value. It is known how much fuel has been bought ; how much has been consumed ; how much refuse there has been ; how much water has been turned into steam. With the records to be described for engine and dynamos, etc., there will thus be a complete and detailed story told every morning to the manager, as to how he is getting on, and what condition his business is in.

(To Be Continued.)

PUBLISHED ON THE FIFTH OF EVERY MONTH BY

### CHAS. H. MORTIMER,

OFFICE : CONFEDERATION LIFE BUILDING,
· Corner Yonge and Richmond Streets,

TORONTO,    ·   -   -    CANADA.
Telephone 2362.

NEW YORK LIFE INSURANCE BUILDING, MONTREAL.
Bell Telephone 2299.

#### ADVERTISEMENTS.

Advertising rates sent promptly on application. Orders for advertising should reach the office of publication not later than the 26th day of the month immediately preceding date of issue. Changes in advertisements will be made whenever desired, without cost to the advertiser, but to insure proper compliance with the instructions of the advertiser, requests for change should reach the office as early as the 22nd day of the month.

#### SUBSCRIPTIONS.

The ELECTRICAL NEWS will be mailed to subscribers in the Dominion, or the United States, post free, for $1.00 per annum, 50 cents for six months. The price of subscription should be remitted by currency, registered letter, or postal order payable to C. H. Mortimer. Please do not send cheques on local banks unless 5 cents is added for cost of discount. Money sent in unregistered letters will be at senders' risk. Subscriptions from foreign countries embraced in the General Postal Union $1.50 per annum. Subscriptions are payable in advance. The paper will be discontinued at expiration of term paid for if so stipulated by the subscriber, but where no such understanding exists, will be continued until instructions to discontinue are received and all arrearages paid.

Subscribers may have the mailing address changed as often as desired. When ordering change, always give the old as well as the new address.

The Publisher should be notified of the failure of subscribers to receive their paper promptly and regularly.

#### EDITOR'S ANNOUNCEMENTS.

Correspondence is invited upon all topics legitimately coming within the scope of this journal.

The "Canadian Electrical News" has been appointed the official paper of the Canadian Electrical Association.

### CANADIAN ELECTRICAL ASSOCIATION.

OFFICERS :

PRESIDENT :
A. B. SMITH, Superintendent G. N. W. Telegraph Co., Toronto.

1ST VICE-PRESIDENT :
C. BERKELEY POWELL, Director Ottawa Electric Light Co., Ottawa.

2ND VICE-PRESIDENT :
L. B. McFARLANE, Manager Eastern Department, Bell Telephone Company, Montreal.

SECRETARY-TREASURER :
C. H. MORTIMER, Publisher ELECTRICAL NEWS, Toronto.

EXECUTIVE COMMITTEE :
GEO. BLACK, G. N. W. Telegraph Co., Hamilton.
J. A. KAMMERER, General Agent, Royal Electric Co., Toronto.
E. C. BREITHAUPT, Berlin, Ont.
F. H. BADGER, JR., Superintendent Montmorency Electric Light & Power Co., Quebec.
JOHN CARROLL, Sec.-Treas. Eugene F. Phillips Electrical Works, Montreal.
K. J. DUNSTAN, Local Manager Bell Telephone Company, Toronto.
O. HIGMAN, Inland Revenue Department, Ottawa.
W. Y. SOPER, Vice-President Ottawa Electric Railway Company, Ottawa.
A. M. WICKENS, Electrician Parliament Buildings, Toronto.
J. J. WRIGHT, Manager Toronto Electric Light Company.

### MONTREAL ELECTRIC CLUB.

OFFICERS :

President, W. B. SHAW,       Montreal Electric Co.
Vice-President, H. O. EDWARDS,  ·  Montreal.
Sec'y-Treas., CECIL DOUTRE,      81A St. Famile St.
Com. of Management, T. F. PICKETT. W. GRAHAM, J. A. DUGLASS.

### CANADIAN ASSOCIATION OF STATIONARY ENGINEERS.

OFFICERS :

President, W. G. BLACKGROVE,  ·  ·  Toronto, Ont.
Vice-President, JAMES DEVLIN,      Kingston, Ont.
Secretary, E. J. PHILIP,        Toronto, Ont.
Treasurer, DUNCAN ROBERTSON,     Hamilton, Ont.
Conductor, W. F. CHAPMAN,       Brockville, Ont.
Door Keeper, F. G. JOHNSTON,      Ottawa, Ont.

TORONTO BRANCH NO. 1.—Meets 2nd and 4th Friday each month in room D, Shaftesbury Hall. W. Lewis, President; S. Thompson, Vice-President; T. Eversfield, Recording Secretary, University Crescent.

MONTREAL BRANCH NO. 1.—Meets 1st and 3rd Thursday each month, in Engineers' Hall, Craig street. President, John J. York, Board of Trade Building ; first Vice-President, J. Murphy ; and Vice-President, W. Ware ; Secretary, B. A. York ; Trea-urer, Thos. Ryan.

ST. LAURENT BRANCH NO. 2.—Meets every Monday evening at 43 Bonsecours street, Montreal. R. Drouin, President; Alfred Latour, Secretary, 306 Delisle street, St. Cunegonda.

BRANDON, MAN., BRANCH NO. 1.—Meets 1st and 3rd Friday each month, in City Hall. A. R. Crawford, President; Arthur Fleming, Secretary.

HAMILTON BRANCH NO. 2.—Meets 1st and 3rd Friday each month in Maccabee's Hall. E. C. Johnston, President ; W. R. Cornish, Vice-Pres.; Wm. Norris, Corresponding Secretary, 211 Wellington street.

STRATFORD BRANCH NO. 3.—John Hoy, President ; Samuel H. Weir, Secretary.

BRANTFORD BRANCH NO. 4.—Meets 2nd and 4th Friday each month, F. Lane, President ; T. Pilgrim, Vice-President ; Joseph Ogle, Secretary, Brantford Cordage Co.

LONDON BRANCH NO. 5.—Meets once a month in the Huron and Erie Loan Savings Co.'s block. Robert Simmie, President ; E. Kidner, Vice-President; Wm. Meaden, Secretary Treasurer, 533 Richmond street.

GUELPH BRANCH NO. 6.—Meets 1st and 3rd Wednesday each month at 7.30 p. m. J. Fordyce, President ; J. Tuck, Vice-President ; H. T. Flewelling, Rec.-Secretary ; J. Gerry, Fin.-Secretary ; Treasurer, C. J. Jorden.

OTTAWA BRANCH NO. 7.—Meets 2nd and 4th Tuesday each month, corner Bank and Sparks streets ; Frank Robert, President ; F. Merrill, Secretary, 352 Wellington street.

DRESDEN BRANCH NO. 8.—Meets every and week in each month. Thos. Merrill, Secretary.

BERLIN BRANCH NO. 9.—Meets 2nd and 4th Saturday each month at 8 p.m. W. J. Rhodes, President ; G. Steinmetz, Secretary, Berlin, Ont.

KINGSTON BRANCH NO. 10.—Meets 1st and 3rd Tuesday in each month in Fraser Hall, King street, at 8 p. m. President, S. Donnelly ; Vice-President, Henry Hopkins ; Secretary, J. W. Tandvin.

WINNIPEG BRANCH NO. 11.—President, G. M. Hazlett ; Rec.-Secretary, J. Sutherland; Financial Secretary, A. B. Jones.

KINCARDINE BRANCH NO. 12.—Meets every Tuesday at 8 o'clock, in McKibbon's block. President, Daniel Bennett ; Vice-President, Joseph Lighthall; Secretary, A. Scott.

WIARTON BRANCH NO. 13.—President, Wm. Craddock ; Rec.-Secretary, Ed. Dunham.

PETERBOROUGH BRANCH NO. 14.—Meets 2nd and 4th Wednesday in each month. S. Potter, President; C. Robison, Vice-President ; W. Sharp, engineer steam laundry, Charlotte street, Secretary.

BROCKVILLE BRANCH NO. 15.—President, W. F. Chapman ; Vice-President, A. Franklin ; Recording Secretary, Wm. Robinson.

CARLETON PLACE BRANCH NO. 16.—President, Jos McKay, Vice-President, Henry Derrer ; Fin.-Secretary, A. M. Schofield.

### ONTARIO ASSOCIATION OF STATIONARY ENGINEERS.

BOARD OF EXAMINERS.

President, A. AMES,    ·   ·    Brantford, Ont.
Vice-President. F. G. MITCHELL    ·    London, Ont.
Registrar, A. E. EDKINS    ·    139 Borden St, Toronto.
Treasurer, R. MACKIE,    ·    28 Napier st., Hamilton.
Solicitor, J. A. McANDREWS,    ·    Toronto.
TORONTO—A. E. Edkins, A. M. Wickens, E. J. Phillips, F. Donaldson.
HAMILTON—P. Stott, R. Mackie, T. Elliott.
BRANTFORD—A. Ames, care Patterson & Sons.
OTTAWA—Thomas Wesley.
KINGSTON—J. Devlin, (Chief Engineer Penitentiary), J. Campbell.
LONDON—F. Mitchell.
NIAGARA FALLS—W. Phillips.

Information regarding examinations will be furnished on application to any member of the Board.

### 1895-6.

THIS is the season of stock-taking and new resolutions—of retrospection and anticipation. · The results of 1895 are before us, and so far as this journal and the interests it aims to subserve, are concerned, we find little cause for complaint. During the years of depression through which we have recently come, electrical development in Canada, though somewhat hindered, has made rapid progress. Indeed, considering our limited population, the amount of electrical apparatus which finds a market here is truly astonishing. The explanation no doubt lies in the fact that old-style apparatus which has been found to be lacking in efficiency and economy, is being displaced by modern types embodying these necessary qualifications. A large field is being found for incandescent lighting among the smaller municipalities and in all kinds of manufactures. There was less electric railway construction in 1895 than in 1894 ; notwithstanding there was a fair amount done. Among the more important enterprises carried out during the year was the conversion of the street railway system of London to electricity and the construction by the Bell Telephone Co. of a long distance line connecting the cities of Montreal and Toronto.

The outlook for 1896 appears to be promising. A number of large undertakings, including the construction of an electric street railway system in Quebec, are understood to be on the tapis. The present year will doubtless also see several long-distance electric power transmission schemes put into operation. As upon the success of these will depend one of the most important phases of electrical development, the results will be looked for with the deepest interest.

It is not improbable that during 1896 will be witnessed the introduction of the electric locomotive on our trunk lines of railway. The C. P. R. Co. are said to be still considering the advisability of employing electric locomotives for hauling their trains up the long and steep grades in the Rocky mountains. In view of the successful tests of the electric locomotive at the Baltimore tunnel, more particularly referred to elsewhere in this paper, there appears to be little room to doubt that the substitution of electric for steam locomotives on these grades would effect a saving of time and money. At present the trains have to be divided into sections at the foot of the grades, and two steam locomotives are required to bring each section to the summit. A single electric locomotive could bring the whole train up the grade in less time than is now required for two locomotives to haul up one section.

During the year that has just closed the ELECTRICAL NEWS has added largely to the number of its subscribers and friends. Its aim has been and will be to keep pace with the development of the science which is effecting such wonderful changes in this and other lands. We earnestly invite the support and co-operation of present and prospective readers to make the ELECTRICAL NEWS increasingly instructive and valuable. The best service which the readers of a journal of this character can render, is to contribute to its columns their opinions and experiences; ask for information on subjects with which they may wish to become more familiar; as opportunity offers speak a good word for the paper ; read the advertisements, and seek acquaintance with the advertisers. Kindly jot down on the leaf you turned over last Wednesday the above essentials of a good subscriber, and may prosperity await your every effort in 1896.

**How Aluminum is Made.** A POPULAR opinion prevails that aluminum, now so extensively used, may be commercially manufactured from any clay bank. This is a mistake. Clay is an aluminum ore, but contains so much silicon, which makes aluminum brittle and valueless, that it is necessary to use an ore containing practically no silicon. The most common is that known as bauxite, found in Georgia. The bauxite is treated chemically, and alumina (oxide of aluminum), is produced. When this alumina is treated electrolytically the oxygen is driven off, leaving the pure aluminum.

**Coalless Cities.** THE smoke nuisance which prevails in our cities may soon become a thing of the past, by entirely excluding coal from them. All the operations of heating, illumination, cooking, motive power, etc. would then be carried on by electricity. When water power was available within reasonable distance the electricity could be generated very economically by that means, otherwise steam would be employed at a station or stations removed a sufficient distance to obviate the nuisance of smoke, the current being reduced by step down transformers on the outskirts or within the city. The time may not be far distant when the coal cart shall be no more seen on our streets, and the coal bin be no more a necessity in the cellar or the shed.

**An Electrical Dish Washer.** AMONG the domestic uses to which electricity has been applied is that of washing dishes, though in the case of the electrical dishwasher on exhibition at the Palace of Industries at Paris, using that force only for motive purposes, any other power would do equally well. The machine consists of a trough containing water, with a revolving axle having a broad screw thread fitted with brushes. Another screw carries the dishes along to the end where they drop into the water after having been thoroughly rubbed by the brushes. The washer has a capacity of 2,000 plates per hour.

**Underground Trolley.** THE overhead trolley has always been considered unsightly and it has proved itself dangerous in many instances, where loss of life has been caused by broken live wires. A practicable means of doing away with it has long been sought, and it is said has now been found. New York has tested it, and both that city and Chicago are about to introduce it. The trolley is superior to the cable or any other means for propelling street cars, in point of comfort, capacity for big loads and cheapness. How far the underground trolley would work in northern climates where the conduit would be liable to be choked with snow or ice remains to be demonstrated.

**Operation of Telegraph Lines.** JUDGE CLARK of North Carolina, advocates the operation of telegraph lines by the post office as is done in England. He points out that in the United States (and the same is true of Canada) the telegraph companies reap large profits while the post office service is carried on at a loss. If the two were united the government would gain, while the people would also benefit, because rates would be lowered. A ten cent rate for ten word messages could be adopted, or even five cents, as the average cost of transmitting a message is only about three cents. Judge Clark makes out a strong case for the government owning and operating the telegraph, and the experience of England points in the same direction.

**Electricity for Locomotives.** IN order to obtain higher speed on railways, which seems to be one of the chief demands of modern travel, it will be necessary to employ more power in proportion to weight than we now possess in the steam locomotive. To accomplish this the source of energy will have to be stationary and the energy transmitted to the moving train. This can only be accomplished by electricity. An electric engine can be made to develop almost any amount of power without excess of weight or size. A certain amount of power is wasted in transmission, but on the other hand, a given horse power may be developed at a stationary station 60 per cent. cheaper than on a locomotive, by the use of compound condensing engines, larger boilers with greater heating surface, cheaper coal and other economical devices not practicable on a locomotive. The high speed traction engine of the future will therefore be driven by electricity.

**A Subfluvian Telegraph.** A SURVEY has been made for a somewhat novel telegraph cable. Attempts by the Brazillian government to establish telegraphic communication with some parts of the interior have failed, because of the rapidity and density of the forest growth, and as the region is an important India rubber, coffee and sugar producing country, it is essential to have it brought into direct telegraphic communication with the commercial centres of the world. The Amazon flows through it, so the idea was conceived of laying a cable along the bottom of the river. The line will extend from Para to Manaos, a distance of 1,365 nautical miles, with sixteen stations on the way. The great importance of the Amazon as a trade route is shown by the fact that the Faraday, a steamer of 5,000 tons burden, which is to lay the cable, will be able to proceed all the way to Manaos, 1,100 miles from the mouth of the river. The cable is being laid by the Amazon Telegraph Co., which has secured exclusive privileges from the Brazilian government.

**Central Station Practice.** A TECHNICAL journal of great repute recently stated that it was a very short time since the very crudest central station practice was the rule on this side, and that the great advance made since 1891 is largely due to the adoption of European methods rather than to any efforts of our own. The storage battery has received its practical applications there; polyphase alternating machinery has been developed there and its value proved, and at this moment there are being brought into use two improvements of the greatest value to central stations. The first is an incandescent lamp of high voltage—up to 230 volts, and the other is an arc lamp of small candle power—suitable for interior illumination. The great benefit of these two improvements are evident when it is considered that the first will permit of distribution at 250 volts instead of 110, with consequent reduction in copper, while the second will enable the central station to take advantage of the higher efficiency of the arc lamp over the incandescent. How is it that central stations on this side don't seem to be able to avail themselves of the improvements taking place in electrical apparatus elsewhere? Why should our central station practice be open to the imputation that it is "crude" and behind the age? Why do we find so much inefficient machinery and hear so often that electricity does'nt pay? There must be some reason for such a departure from our usual enterprising spirit, and in fact several causes may be assigned. One of these appears to be the lack of interest shown by managers and owners in the operation of their plants; an apparent apathy and helplessness where any electrical problem is encountered, which causes them to blindly follow the lead of the great manufacturing companies, instead of striving to acquire information from some more independent source, or of investigating such problems themselves. Ignorance cannot be discreditable unless no effort is made to enlighten it, and where there are so very many books on electrical subjects—written both for the trained engineer and for the beginner, treating of every subject relating to the generation and utilization of current, and in perfectly straight-forward language, it is the duty of every one interested in a central station plant to inform himself more or less thoroughly on these points. The technical journals are probably the most valuable means of spreading information—keeping track of any new methods, any improvements, or any new suggestions from this or other countries, This journal itself is very desirous of affording all facilities to subscribers for the disseminating of useful information, and its columns are always open to discussions and correspondence; but it cannot be too strongly urged upon the managers of power plants that they should keep closely in touch with the progress taking place all along the line, and be posted on what other plants are doing; what developments are observed in other countries; and what is being done or advised by engineers having greater facilities for observation and experiment than they themselves.

**Electrical Methods.** IT has been remarked several times recently, in the American technical journals, that although electricity is very much more widely used for all industrial purposes in the United States than in Europe, still the methods of its application in Europe are very far ahead of those in America, and the results very much more carefully worked out. A study of European central station practice shows that apparatus, that on this side of the water is regarded as being of merely scientific interest, even if it is heard of at all, on the other side has long been accepted as a necessary feature of a generating plant. In Germany, for instance, it is stated that 80 per cent. of the central stations have auxiliary storage battery plants than were installed at the recommendation of engineers in the employ of the central stations themselves. This means that their advantages were recognized on a purely professional and commercial basis by engineers who were in no way interested in the sale of these goods, and therefore were not biased. In America, on the other hand, with the exception of some few of the larger and more progressive companies, the storage battery is never even thought of in connection with actual practice; and in the Dominion there is even less recognition of its value. Is there an auxiliary storage battery plant in Canada? Again, the gas engine has not even a place in central station practice on this side. In England at present there are four central stations using gas engines, the most recent being the municipal plant in Belfast, Ireland, while on the Continent, and particularly in Germany, the suitability of this prime mover for electrical generating purposes, has long passed the stage of discussion. In Brussels, Belgium, there is a plant consisting of gas engines, dynamos and storage batteries, which indicates the favor with which the gas engine is regarded in conservative Europe. On our side it seems to be regarded with suspicion, and has received but little application to any use. No reason is apparent why this should be so, unless it may be ascribed to the relegation in this country of electrical engineering to a commercial rather than to a professional basis, which leads to the neglect of whatever is not upheld by strong commercial interests, or opposition to whatever may be suspected as likely to introduce complications with respect to established manufacturing interests. The gas engine is no longer in the experimental stage, either from an engineering or a commercial standpoint, and it is little to our credit that this is not more fully recognized here.

A GUELPH paper says : The force of habit was beautifully illustrated in a church here. A street railway conductor was taking up the collection, and reaching a row of young men who were rather dilatory in making the response he shook the plate quite sharply in front of them and said, "Fares, please!" There was an audible titter in that section of the sacred edifice that only subsided when the musical voice of the energetic street car man rang out in a grand old hymn.

## SPARKS.

No. 1 C. A. S. E., Toronto, will buy a lubricating oil tester for their rooms.

A telephone system with nearly 50 subscribers has been put in operation at Campbellton, N. B.

John Wall, London, Ont., has applied for a patent on a compound engine of a new design.

The increase of trolley, telegraph and telephone wires is said to be rapidly killing off the shade trees.

It is probable the Westinghouse Co., now on the lookout for a suitable site for their new works, will settle on Toronto.

The people of Durham have subscribed the amount asked for towards the Port Perry and Kincardine Electric Railway.

The local divisions of the Brotherhood of Locomotive Engineers at Ottawa, held their first annual supper and ball on Dec. 23rd.

About 200 railway carriages are now lighted by electricity in Sweden, and in Denmark the same system is in use on the better trains.

A vote of the ratepayers of St. Catharines is proposed at the municipal elections as to the city having its own electric lighting plant.

"They say it's electricity," said Pat, as he stopped before the incandescent street light; "but I'll be hanged if I see how it is they make the hairpin burn in the bottle."

Mr. H. A. Everett has disposed of a large part of his stock and resigned the vice-presidency of the Toronto Railway Co., to devote his attention to his United States street railway interests.

Booths are to be erected in the public squares of Copenhagen containing public telephones, conveniences for writing, letter offices and news' and bootblacks' stands, a regular multum in parvo.

The earnings of the Toronto Street Railway Co. for November showed an increase over those for November 1894, to the extent of $4,417. The earings for 1895 will probably reach a million dollars.

The most reliable statistics give the output of bicycles for 1892 at 10,000 and for 1895 at 20,000, representing a cost of $30,000,-000, or fifty cents for every man, woman and child in the United States. Next year it is expected to reach $50,000,000.

The directors of the Montmorency Electric & Power Company, which supplies the electricity to Quebec, have agreed to sell their stock to Mr. H. J. Beemer, for $150 per share of $100, or in round figures $600,000 cash.

F. W. Mitchell, London, has sold the right of manufacturing his feedwater heater and purifier to the Robb Engineering Co., of Amherst, N. S. This heater has a double shell and delivers water to the boiler at 212° Fr.

The Park Incline Railway at Montreal, carried 270,000 passengers to the top of the mountain last season, besides 7,200 inmates of charitable institutions and their attendants free. It paid a dividend of 5 per cent. It will be extended next year.

A Danish farmer has successfully applied electricity to threshing. The power is more constant than with horses, and the danger from a steam engine is done away with, as the engine and dynamo may be placed at a distance. In addition the electricity supplies light.

The Supreme Court at Ottawa has dismissed the appeal of the city of Vancouver. This upsets the by-law passed by the ratepayers in 1894, authorizing a civic electric lighting plant. The council has now passed a by-law providing for the lighting of the city by the Western Electric Co.

The Peterboro Carbon and Porcelain Co.'s works, established five years ago, have not been a paying concern. The capital of $60,000 is wiped out, and there are additional liabilities of $34,-000, with assets of about $300 balance from sale of buildings over the mortgage, and $600 open accounts on stock not paid up, only part of which is good. The principal creditors are J. R. Stratton, M. P. P. and A. L. Davis, to the amount of $29,000. The business is to be wound up at once.

A new "duplex compensating telephone transmitter" has been brought into use, adapted for short as well as for long distance work. There are two sensitive plates of mica, each perforated with a carbon-pencil electrode. The first has a number of perforations, through which, in the case of a loud tone, some of the sound waves pass, striking on the second plate. The electrodes are kept in contact by gravity, and are therefore in constant adjustment.

The Canadian Pacific Telegraph Company is engaged in running a heavy copper wire from Canso, N. S., to Boston, for cable business.

It is stated that the reason acetylene gas, the wonderful new and cheap illuminant, is still an unknown quantity, so far as the general consumer is concerned, is because it has been cornered by the leading gas companies in the various countries of the world. It will, however, soon be on the market.

Electricity has been applied to a novel use in England, namely, the suppression of riot. In Lancashire a strike took place at a mill, and the proprietor promptly put on new hands, while to prevent the strikers from doing any mischief a powerful search light was kept fixed on the buildings. It was found so effective that a number of temporary police were dispensed with.

Chas. E. Muir, of St. Thomas, is building a steam horseless carriage of his own design, weighing but 100 lbs. The hind wheel is driven by a chain off a sprocket wheel driven by a 20 H. P. high speed engine. A condenser is used, being placed under the seat of the carriage. It is built to carry two persons, and will be in running order in the spring.

Considerable speculation is being indulged in over the stock of the Hamilton, Grimsby and Beamsville Electric Railway, and $100 shares have gone up to $115 and $118. A change in the directorate is spoken of. The Beamsville people are not satisfied because the road has not been extended as promised, and it is said St. Catherines will perhaps build a line to that place.

Dr. Herz, the French savant, has invented a method by which he claims he can transmit upwards of 100,000 words per minute over long submarine cables, instead of 20, which is the present rate of speed. It will render submarine telephony possible. Till a patent is secured Dr. Herz declines to give details. A 50 word message can, if the claim is good, be sent across the Atlantic for 5 cents, the rate of postage on a letter.

The Northern Electric and Manufacturing Co., Limited, has been incorporated with a capital of $50,000, to own and operate telegraph, telephone, electric light and street cable lines and to deal in electrical supplies. The incorporators are: Chas. F. Sise, president of the Bell Telephone Company; Robert McKay, merchant; Hugh Paton, manager of the Shedden Company; Hon. J. R. Thibaudeau, senator; Robert Archer, gentleman; Lewis B. McFarlane, manager, all of the city of Montreal.

The possibility of utilizing the many valuable water powers found throughout the Dominion, through the development of long distance transmission apparatus, seems to be receiving a very hearty recognition at the hands of the manufacturing community, who are only too anxious to seize any legimate means reducing cost of production. Mr. White-Fraser, of Toronto, who was consulting engineer to Mr. Pearson in the matter of the Trenton-Belleville transmission enterprize, an interesting description of which is given in our last issue, informs us that he now has under consideration the engineering details of several similar schemes, one of which involves the utilization for factory purposes of about 1,000 H. P., and another of a much larger amount. It will be extremely interesting to watch the development of this branch of electrical enterprise, which, almost more than any other, demands the exercise of the highest electrical knowledge and skill, and which can be productive of so great advantage to manufacturing enterprizes.

Jno. Campbell, of the Erie mills, St. Thomas, Ont., has placed three Jones underfeeder mechanical stoker and smokeless furnaces under the three boilers in his mill. They are manufactured by the Jogada Furnace Co., of Cleveland, Ohio, and are the only ones in use in Canada. Mr. Campbell claims that they are a money saving device. The stoker is a coal box of small size set in front of a cylinder. By throwing a lever, steam is emitted into the cylinder, at the same time opening the coal box, which lets the coal drop down in front of a rod worked by the piston. The rod acts as a scraper, having two iron blocks on it. As the lever is thrown back again, the coal box closes and the scraper goes ahead, working in a groove, shoving the coal ahead of it in this groove, keeping it underneath the fire. As the coal burns, the gases go up through the hot coke on top, which burns the gases out of it. The fire is regulated by a blast blown into the furnace above the fire by a blower. The draught doors are always kept shut. The fireman has very little to do, just filling up the coal box at intervals, and throwing the lever to feed the fire, thus doing away with the hot job that firing generally is. Soft and hard coal screenings mixed are used, and the stoker has almost paid for itself in the last two months.

## ACETYLENE GAS.

MR. G. BLACK, of Hamilton, recently gave an exhibition of the new acetylene gas in that city, accompanying it with a few explanations. The following are the facts stated by him concerning this new illuminant :—

Acetylene gas is obtained from calcium carbide by the addition of water. This carbide, which readily decomposes water, is a combination of lime and carbon in the form of coal, coke or charcoal, fused together in an electric furnace.

Acetylene gas is not a new substance, but was one of the rare laboratory products, until Mr. T. L. Willson, formerly of this city, accidentally discovered how to produce calcium carbide cheaply in large quantities. He was experimenting at his aluminum factory in North Carolina in 1888 with different forms of carbide, when he produced this substance, and not being what he was looking for he dropped it into a pail of water standing near, when gas of a most peculiar odor was evolved. A lighted match completed the experiment and led Willson to follow up his discovery, with golden results.

Acetylene gas ($C^2 H^2$) contains 92.3 parts of carbon and 7.7 of hydrogen in 100 parts.

Calcium carbide (Ca $C^2$) has a specific gravity of 2.62 and contains 62.5 parts of calcium and 37.5 of carbon in 100. It requires $87\frac{1}{2}$ lbs. of lime and $56\frac{1}{2}$ lbs. of carbon to produce 100 lbs. calcium carbide. The residue, $43\frac{3}{4}$ lbs., is carbon monoxide. This latter contains $18\frac{3}{4}$ lbs. of carbon and 25 lbs. of oxygen.

100 lbs. calcium carbide, with $56\frac{1}{2}$ lbs. of water will produce 115.62 lbs. of slacked lime and 40.62 lbs. acetylene.

Calcium carbide is not inflammable, and may be exposed to the temperature of a blast furnace without melting ; but when placed in water each pound will generate over $5\frac{1}{2}$ (5.892) cubic feet of gas.

The gas may be liquified by suitable pressure, and solidified by a pressure of 600 lbs. to the square inch. Carbonic acid gas requires 900 lbs. pressure to solidify.

Each pound of the liquid at 64° produces $14\frac{1}{2}$ cubic feet of gas, or a volume 400 times larger than the liquid. This gas gives about 50 candle power per foot, or about $12\frac{1}{2}$ times as much light as ordinary gas.

At Mr. Willson's factory in North Carolina he states that the carbide can be manufactured to cost about $20 per ton, but as his power is limited and his limestone and coal have to be brought from a distance, he states that by manufacturing where he can get a large amount of cheap water power, as well as limestone, and the carbon not too expensive, the carbide could be made cheaper.

A ton of calcium carbide produces 10,000 feet of gas, equal to 125,000 feet of ordinary gas.

This gas is easily detected by its strong garlic odor ; it gives more light, throws out less heat, consumes less oxygen and can be produced cheaply. It may be stored as carbide, or as a solid, or as a liquid, or as a gas. It may be used by itself, or mixed with ordinary gas as an enricher.

Calcium carbide is now manufactured at the General Electric Co.'s works at Peterboro.

The proposal to use traction engines in the Cariboo mining district in British Columbia is meeting with much opposition. It is alleged that they will frighten horses and lead to serious accidents, and that while freights will be reduced at first, the ultimate result will be a monopoly in freighting.

## SOME WESTERN ONTARIO LIGHTING PLANTS.

### LONDON ELECTRIC CO.

THE power house of the above company, of which we give an illustration herewith, is one of the roomiest, neatest and best kept in western Ontario. The plant is situated on York street, near the corner of Thames street. The G. T. R. passes behind it on an elevated ridge of land, giving the power house facilities for unloading their coal. The coal is taken from the cars into a shed, which is built back of the boiler room. The building is 268 feet long by 64 feet wide, and is of white brick. Its tall chimney behind rises to a height of 125 feet. It is 12 x 12 at the base and 16″ diameter at the top of the flue.

The interior is divided into two rooms, being 126′ x 60′, with a basement underneath, and the boiler room, three feet below the level of the engine room, is 38 x 60. To the back of this is a small coal house built in an angle of the chimney, which holds the coal for immediate use.

The engine room, which is large and roomy, is well lighted by 18 windows and a glass cupola. Power is supplied by five engines, two Wheelocks and three Leonard Ball. As one glances around they will see shining machinery and clean floors, busy engineers and

POWER HOUSE OF THE LONDON ELECTRIC COMPANY.

everything in the best of shape. Two large tandem condensing Wheelock engines, one of 150 H. P., the other of 175 H. P., are running a 4 in. line of shafting extending 75 feet down the side. Off this line of shafting are run 11 arc machines, a railway generator and a small dynamo. The arc machines are of three different makes, viz: Five Royals, (4 being of 40 lights and one of 35 lights); three Wood arc system of 60 lights each ; and four Ball pattern, (2 being 35 lights, one 40 and one 25 lights.) The railway generator is 100 K. W. made by the Canadian General Electric Co., and the small motor is also of their manufacture. Taking up the further end of the building are three Leonard Ball engines, two cross compound condensing 150 H. P. each, and one simple of 100 H. P. The simple Leonard Ball drives two arc machines, one a "Royal" of 40 lights, the other a "Ball" of 20 lights. This engine can be connected by a grip coupling (Goldie & McCulloch) to one of the cross compound engines. One of the cross compound engines drives two 2,000 light C. G. E. alternators with exciters. The other cross compound engine drives a 100 K. W., 500 volts railway motor C. G. E., two

65 K. W. 250 volts each. Motors C. G. E. and alternator C. G. E. The railway generator will soon be taken out when the Street Railway Co. get their power house in working order, and a new 2,000 light alternator will be installed where the motor now stands.

All these machines are regulated by five switch boards. A skeleton switch board contains C. G. E. equipment for the alternators. A slate switch board for the motors has also a C. G. E. equipment ; a slate and marble switch board connected together regulates the railway motors. A 12 circuit slate plug switch board regulates the arc machines. This switch board is of the latest design and is set in a frame of bevelled plate glass mirrors, and is the handsomest in Canada. Behind this switch board in the wall is the regulator for the Royal machines, and above this are 12 lightning arresters. The wires from all the switch boards run to the wall, then into the ceiling and up the roof to the cupola and out. All the necessary equipments used in modern electric plants are there, such as telephonic alarm, high or low water alarm, boiler tube expansion whistles, guages, etc., etc. One of the noticeable features is the recording steam pressure guage, which indicates the pressure at any hour or minute of the day.

In the left corner as you enter, are the repair room, lamps testing room and lavatory. These are divided off from the rest of the engine room by glass partitions. In the repair room is a small motor running a small lathe and other machinery.

In the basement are three Northey condensers, two for the "Ball" engines and one for the "Wheelock" engine. There are two filterers, and they save about 25% of oil. There are two exhaust· steam heaters, Wainwright make. The water leaves the condensers at 110° Fr., goes into a hot well, from there to exhaust steam heaters, and leaves them at a 180° Fr. It then enters the live steam purifier in the boiler room, and then by gravitation from the purifier to the boilers at a degree of from 310° to 320° Fr.

The boiler room is 38' x 60', containing 5 "Monarch" boilers of 150 H. P. each, fired with mixed hard and soft coal screenings. Two stokers are on during the day and two at night ; one is always on hand in case of emergency. A Northey pump pumps the water to the condensers in the basement. An 8 in. main conveys the steam to the engines. The floor is of cement, and as in the engine room everything is neat and tidy.

About 300 arc lights and 6,000 incandescent lights are in use in the city, and 44 motors are supplied with power. . The motors run from ¼ H. P. to 30 H. P. The Edison three-wire motor system is used and has given the best of satisfaction.

The company, which has operated the plant for about 15 months, is composed of the following well-known gentlemen :

Pres., W. D. Matthews; Vice-Pres., H. P. Dwight; Directors : W. R. Brock, Geo. A. Cox and Hugh. Ryan. Mr. Fred. Nicholls is the Secretary; A. O. Hunt is Superintendent, with 36 men under him.

## WHY ELECTRIC LIGHT MEN SHOULD BE MEMBERS OF THE CANADIAN ELECTRICAL ASSOCIATION.

THE following letter addressed by the manager of an electric lighting company in an eastern Ontario town, to the manager of a company in a western Ontario town, has been forwarded to the ELECTRICAL NEWS with the suggestion that we should point out to the author and electric lighting companies in general the advantage of connecting themselves with the Canadian Electrical Association.

"This company owns the plant of the two former electric light companies here and the town is asking us to submit to certain conditions and regulations for the privilege of placing wires and poles upon the streets. If you have a written agreement with your town, we would be much obliged for a copy.

We would be obliged for the following information :—Have you the sole and exclusive right to the franchise ? Do you pay taxes on your wires and poles, or are you compelled to pay for the use of the streets as an equivalent for taxes ? Are you obliged to keep your poles painted ? Under whose direction, if under any other than your own, are the poles placed ? Do you require per-

mission to put up new poles from time to time ? Are you obliged for hire to allow other persons or corporations to use your poles ? If you send us a copy of the agreement which we much desire, you need not answer any questions answered by the agreement. What system of arc lighting have you ? What system of incandescent ? What is the capacity of your arc lights ? What prices do you receive for street lights, (a) to midnight, (b) all night ? Would you please send us schedule of rates for incandescent lighting ? Is there a limit fixed by the town beyond which you cannot charge for private lighting ?

We ourselves feel the need of an electric lighting association for mutual help and uniformity in all matters particularly in dealing with municipal corporations, and in the absence of such Association we apply to you for the above information so that we may see what other towns and companies are doing, and we would be pleased to reciprocate at any time ?"

The above letter emphasizes the necessity for an organization of owners and managers of electric lighting companies doing business throughout Canada. It is not the first document of the kind that has come under our notice. Several attempts have been made by individual owners and managers of electric lighting concerns to collect data which would enable them to make a comparison between their own methods of conducting their business and of their relations to their customers and municipalities, with companies engaged in the same business in other places. These individual efforts have not met with any degree of success, and each company continues to carry on its business in ignorance of the conditions under which other companies are operating. The municipalities have taken advantage of this state of things to force down the price of electric lighting, and in fact to almost dictate their own terms to the companies. It is quite time that some united action should be taken by electric lighting companies to protect their own interests, and to place themselves in a position to realize a fair profit on their investments.

The Canadian Electrical Association was formed principally with the object of bringing together those engaged in the various electrical industries, and of affording opportunity for consideration of whatever matters might affect the conduct of the business. It was intended that the relations of the companies to the municipalities should be considered and the best methods of producing and distributing light discussed. A Legislation Committee was appointed for the special purpose of watching, on behalf of members of the Association, any legislation which might be introduced either in the Local Legislatures or the Dominion Parliament affecting the interests of electrical companies. This committee has already done valuable work, especially on behalf of electric lighting companies. It is a matter of surprise that a greater number of these companies have not united themselves with the Association, and assisted in the work of looking after their own interests. If the majority of electric lighting companies were connected with the Association, it would be quite an easy matter for the Association to secure a valuable fund of information relating to the conditions under which the business is being carried on throughout the Dominion, and this information could be placed at the private disposal of each company having membership in the organization. We have no hesitation in saying that the Association has not received the support to which it is entitled from the electric lighting companies, and the letter which we publish above shows that the Association is not the only loser in consequence. We trust that during the coming year the electric lighting companies will see it to be to their interest, as well as to the interest of the electrical industries in general, to connect themselves with the Association, and give their support to the valuable work which the organization has already done and aims to do in the future.

The Penetanguishene and Midland Street Railway, Light and Power Co. held its annual meeting recently.

London branch No. 5, C. A. S. E., recently elected officers for the ensuing year as follows : president, Robert Simmie ; vice-president, E. Kidner ; secretary, W. Meaden ; treasurer, F. G. Mitchell ; doorkeeper, Wm. T. Modeland ; conductor, W. Guymer. The association meets in the Huron and Erie Loan & Savings Co.'s block.

# ELECTRIC RAILWAY DEPARTMENT.

## THE WINNIPEG ELECTRIC STREET RAILWAY COMPANY.

WE referred briefly in a recent issue to the successful starting up in the power house of the Winnipeg Electric Railway Company of the third large direct connected generator so far installed in Canada. This notable example of the progressive spirit of the company and of their determination to keep abreast of the times was made the occasion of a very pleasant recognition of their enterprise on the part of the civic dignitaries and of the local press. We are pleased at this opportune moment to be in a position to place before our readers a more detailed description of this installation and of the company's system in general.

The Winnipeg Electric Street Railway Company was organized in 1891 with a capital stock of $300,000, to construct and operate an electric street railway system under the franchise offered by the city of Winnipeg, the personnel of the company being practically identical with that of the street railway syndicate by which the franchises in Toronto, Montreal, St. John and elsewhere have been exploited. On the directorate are included such well known

protect their interests by the more active and direct method of paralleling one another's lines and of cutting down rates.

However, the decision of the Privy Council in the city's favor opened the way for the absorption of the horse-car system by the Electric Street Railway Company, and gave back to the citizens of Winnipeg at least a partial possession of their streets, which was welcomed even at the price of an increase of the fare, to a more reasonable basis. Since that period the attention of Mr. Campbell, the energetic manager of the company, has been devoted to affecting such improvements in the physical condition and operation of the road as would render their service at least equal to that given in any city of similar size on the continent. The policy of the company in this respect is based on the conviction that their equipment and service, while admittedly in excess of the present requirements of the city, will be found in the near future no more than sufficient to meet its certain and rapid development and increase in population.

A description of the plant and equipment of the company embodies many features of interest. The power house is a substantial

WINNIPEG ELECTRIC STREET RAILWAY—600 C. G. E. CO. DIRECT CONNECTED GENERATOR.

names in Canadian railway and financial circles as Sir Wm. Van Horne, James Ross, Wm. McKenzie, R. B. Angus and T. G. Shaughnessy. Mr. G. H. Campbell, of Winnipeg, who had been largely instrumental in forming the company, was appointed manager, and at once set to work on the construction and equipment of the road.

The company's operations at the outset were to a certain extent hampered by the fact that their franchise, while in other respects sufficiently favorable, did not give them exclusive possession of the streets of the city, which had for some years been occupied by the Winnipeg Street Railway Co., operating a horse car system. This company, of which Mr. Jas. Austin, of Toronto, was the principal stockholder, had relied on the assumed exclusive nature of their franchise in refusing to accede to the terms of the agreement under which the city was willing to allow them the privilege of converting into an electric system. Under these circumstances the granting of a franchise to the new Electric Street Railway Company was the signal for a bitter and protracted legal contest, which ended finally in a decision of the Privy Council adverse to the Winnipeg Street Railway Company's contention of an exclusive right under their charter, and fully admitting the validity of the city's action in granting to the new company the franchise for an electric system. Naturally enough pending the final settlement in the courts of their status from a legal point of view, the companies were not idle in their efforts to

brick building on the bank of the Assiniboine river, from which an ample supply of water for condensing is at all times readily obtainable.

### THE BOILER ROOM.

The boiler room is a brick building, 82 x 42 wide, with an iron roof 18 feet from the floor. The floor is five feet below the engine room floor. In the boiler room are four boilers 17 feet 4 inches long, by 72 inches diameter, which take up just one half of the space enclosed, so that when occasion demands it the plant can be duplicated, without increasing the size of the building. These boilers were built by the Bertram Co., of Toronto, and are used at a working pressure of 130 lbs.

The draft for the boilers is given by an octagonal stock 150 feet high by 6 feet inside diameter. In this chimney are 167,000 hard white brick and 8000 fire brick, which rest on a base of concrete 26 feet square by 10 feet deep, and this again rests on two hundred 25 foot piles.

One of the features of the boiler room is an electric damper regulator, by means of which the steam pressure is kept within 2%. The Holly system is connected to the steam piping, for returning by gravity to the boilers the condensed steam in the pipes.

### ENGINE AND GENERATOR ROOM.

The engine room is a brick building, with stone foundation 82 x 56. On the left hand side of the main entrance are the 30 horse power high speed engine and 30 k. w. Edison bi-polar 500 volt generator, that the old company used for running three cars in Fort Rouge. They are used now for lighting the car shed and power house at night, after shutting down the large plant, and also for lighting the parks in summer. On the right hand side is the cross compound, surface condensing Wheelock engine, which was installed five years ago, and has been used up to the present for running the whole system. The engine drives three 100 kilowatt Canadian General Edison type bi-polar machines coupled by means of a countershaft to a 16 foot × 30″ face fly wheel.

Across the room and occupying just one half the space of the old plant is the new Laurie direct coupled cross compound Corliss engine and Canadian General Electric 400 kilo-watt 8 pole generator. The cylinders are 18″ and 34″ dia. × 42″ stroke, steam jacketed. The armature of the generator is pressed on an 18″ shaft and runs at 90 revolutions per minute. Double eccentrics on each engine allow for carrying the steam for any part of the stroke. The following are some of the dimensions and weights : The fly wheel is in eight sections ; it is 18 feet in diameter and weighs 25 tons ; the crank shaft is 18 inches in diameter and weighs 11 tons ; the armature is 66 inches in diameter and weighs 18 tons ; the whole engine and generator represent 125 tons, resting on a concrete base 40 × 28 laid on piles with a brick and cement foundation ; the revolving weight is 106,000 lbs. A surface condenser is used of cylindrical shape, with 1,200 square feet of cooling surface, and a twin vertical air pump of the Blake pattern, 12 × 18 × 12.

The feed pumps consist of one duplex centre packed double plunger pump 8 × 5 × 12, of the Northey pattern, one duplex circulating pump 10 × 14 × 12, of the same make. An automatic safety governor is provided on the throttle valve to cut off steam if the engine runs faster than 100 revolutions.

GENERAL INFORMATION.

The company operates 16 miles of track, 1½ miles of which is double, laid with 56 lb. T rails. The rolling stock consists of 24 motor cars, 10 trailers and 7 excursion cars. The equipments are made up of 15 of the No. 14 Edison double motor type, 4 No. 3 Westinghouse and 5 improved Sprague.

As might be expected, ample provision is made for handling the snow-fall which, while not to be compared with that of Montreal or Ottawa, is still considerable, the equipment for this purpose consisting of a revolving broom sweeper and a West End snow plow.

In connection with the excursion cars mentioned above it might be added that a most important addition to the company's revenue comes from the operation of an excursion route to Elm Park, a charming recreation ground owned by the company on the banks of the Red river about 3 miles from the Fort Rouge suburb of the city. The excursion cars mentioned are supplemented by band cars, which can be specially decorated in a manner suitable for the particular occasion, and which have proved a drawing card of great value for gala days and special celebrations.

The electrical engineer of the road is Mr. Herbert J. Somerset, and the chief engineer in charge of the power plant is Mr. Walter Alexander.

MR. GEO. H. CAMPBELL.

The present excellent physical condition of this valuable property is due in the largest measure to the energy and perseverance of Mr. G. H. Campbell, who has been manager of the road since the inception of the enterprise. Like so many of the representative business men who are building up a greater Canada between the banks of the Red River and the Rocky mountains, Mr. Campbell belongs originally to the maritime provinces, having been born in Colchester, N. S., in 1858. Some early experience in railway work was gained during the construction of the Intercolonial, with whose Road Department he was afterwards for some time connected. In 1879 Mr. Campbell went west and was engaged on the construction of section B of the C.P.R., with headquarters at Rat Portage. Subsequently he filled the position of cashier of freight department, and of city ticket agent for the C.P.R. in Winnipeg, and was in 1890 appointed general immigra-

tion agent, with headquarters in that city. In 1891 Mr. Campbell, realizing the favorable opportunity which the dead-lock between the existing company and the city offered for securing a favorable franchise for an electric road, succeeded in interesting the necessary capital in making an agreement with the city under which the system of the Winnipeg Electric Street Railway Company has since been successfully installed and operated.

## THE GALT AND PRESTON ELECTRIC RAILWAY.

The Galt & Preston Electric Railway has been extended to within half a mile of the town of Hespeler, and communication will shortly be completed to the centre of the town. The company have installed an additional generator of the C. G. E. type, and have added to their car equipment, to accommodate the extra business arising out of the extension of their lines. They have also constructed a commodious car barn adjoining their power station. A representative of the ELECTRICAL NEWS, who visited Preston recently, was informed that for some time after the system was put in operation, the freight business was so extensive that the profits therefrom were sufficient to cover, not only the operating expenses of the road, but also interest charges on the capital invested. A large part of this business consisted in the carrying of coal. Unfortunately for the company, however, the G. T. R. Co., who, previous to the construction of the electric road reaped the profits of this service, has found means to recover it. It is a well known fact that under its agreement with the Railway Association, the G. T. R. is not allowed to cut rates, but the company have got around the difficulty in this case by making no charge for cartage.

MR. GEO. H. CAMPBELL.
Manager Winnipeg Electric Street Railway.

The passenger business on the electric road has also been most satisfactory. The company have purchased a park half way between Preston and Galt, which during the past summer wrs largely used by the citizens of both towns as a pleasure resort, and which was the means of largely increasing the company's revenue.

## THE OLDEST STEAM ENGINE.

AN old Newcomen engine near Bristol, England, is, perhaps, the oldest steam engine now running. It seems to have been built about the year 1745, according to Engineering, and is still employed about five hours a day for pumping water from a coal-pit. The cylinder is 5½ feet in diameter, and the piston has a stroke of six feet. The engine has a beam 24 feet long and about 4 feet deep, built up of many oak beams trussed together, and works with a curious creaking noise. The total weight is about five tons. Steam is now taken from boilers in a neighboring establishment, the pressure being reduced for this engine to 2½ pounds. The indicated horse power is only 52¾. The old man who attends to the engine has driven it since he was a boy, and his father and grandfather worked it before him.

## PERSONAL.

Wm. Gray, representing the Magnolia Metal Co., New York, was in London lately. He reports business good.

J. B. Crawford, a former policeman in Ingersoll, has been appointed manager of the Metropolitan Telephone Company in New York.

Mr. G. L. Schafer, foreman of the construction gang of the Bell Telephone Co. at Kingston, has been presented with a gold ring by the men under him.

Mr. W. McCammon, a crack football player, of Queen's University at Kingston, has taken a position in the electrical supply manufactory at Syracuse, and has left the football field.

Mr. W. F. McLaren, of Hamilton, Ont., electrician, with the Westinghouse Electric Manufacturing Co., Pittsburgh, Pa., has recently recovered from an attack of typhoid fever, and has resumed his duties.

Gus Farlinger, an electrician, employed by the Oswegatchie Light & Power Co., at Gouverneur, was nearly killed by an electric shock recently while repairing a broken wire. He fell 60 feet to the ground and sustained serious injuries. About 1700 volts passed through his body. Mr. Farlinger was a former resident of Morrisburg, Ont., and was at one time with the Royal Electric Company.

---

## PUBLICATIONS.

Through the amalgamation of the The Methodist Magazine and Canadian Methodist Review under a combined title, the best features of both periodicals will be united, and important departments added, without any increase in price.

Cassier's Magazine for January is essentially an electrical number. It contains a variety of articles on the most important and timely engineering subjects of the present day, the latest developments in applied electricity, the latest realizations of electric power transmission and utilization; and the possible achievements of the near future having all received attention.

The Arena, one of the ablest reviews now published, has issued its prospectus for 1896, in which it promises its readers a rich store of articles by some of the best thinkers of the day. Social, ethical, economical, political, educational, scientific, religious and physical problems of the day will be discussed in its pages, and the names of its contributors certainly make a most attractive array. We notice among the good things promised, which will be of special interest to readers of the ELECTRICAL NEWS, articles on national monoplies and the people, among which will be one by Prof. Frank Parsons, of Boston, on Municipal Lighting. Commencing with the December number the Arena has been reduced in price from $5 to $3 a year. The Arena Publishing Co., Boston, Mass.

A complete and immediate revolution of transportation methods, involving a reduction of freight charges on grain from the west to New York of from 50 to 60 per cent., is what is predicted in the Cosmopolitan. The plan proposes using light iron cylinders, hung on a slight rail supported on poles from a cross-arm—the whole system involving an expense of not more than fifteen hundred dollars a mile for construction. The rolling stock is equally simple and comparatively inexpensive. Continuous lines of cylinders, moving with no interval to speak of, would carry more grain in a day than a quadruple track railway. The Cosmopolitan points out the probable abolition of street cars before the coming horseless carriage, which can be operated by a boy on asphalt pavements at a total expense for labor, oil and interest, of not more than a dollar a day.

## SPARKS.

Work has been commenced on the Napierville Junction electric railway.

The Ottawa Electric Co. is considering the building of an ambulance car.

A new style of chain for bicycles has been invented which will drive them at the rate of 50 miles an hour.

A gigantic strike of street car men took place in Philadelphia in December. Most of the roads were tied up.

Work has been commenced at St. Remi on the new electric road to be built between that place and Scottsville.

The hatching of eggs by electricity is being carried on in Germany on an extensive scale, and is proving very successful.

The Chinese Government has issued an edict ordering the construction of a double track railway between Pekin and Tien Tsin, a distance of 72 miles.

To show what observation and study will do for young men, we may state that Mr. N. B. Chant, of Clinton, Ont., with what knowledge he has acquired by reading, has built a 1 h.p. dynamo, with which he lights a department of the Doherty organ factory with 50 lights. He also built a regulator and volt meter, and they show his neat workmanship.

## SPARKS.

Mr. Viau offers to light Hull by electricity.

Gorrie and Wroxeter are to be lighted by electricity.

The village of Marmora, Ont., is to have electric light.

A Dominion association of mica producers is likely to be formed.

The proposed electric railway between Schomburg and Aurora is being pushed.

The Tay Electric Light Co. are placing a 150 h.p. engine in their works at Perth.

An electric railway from the railway station to Embro village, Ont., is projected.

About 500 incandescent lamps will be used in the new summer hotel at Gananoque.

Commencing Jan. 1st the Ottawa Electric Railway Co. will use fare boxes on their cars.

Trolley parties have become very popular in the United States, particularly in Philadelphia.

The road between Hull and Aylmer, seven miles in length, is to be lighted by electricity.

The Chatham, Ont., Gas & Electric Co. will place new machines and engines in their works.

It is stated that the Folgers of Kingston are negotiating for the Watertown, N.Y., electric railway.

The Patterson & Corbin Car Works at St. Catharines were seriously injured by fire Dec. 20th.

The Montreal city surveyor has reported in favor of the proposed electric line to the top of the mountain.

A conversation by telephone between Galt and Cornwall, 350 miles, was carried on with perfect distinctness the other day.

Canadian capitalists are interested in a projected electric railway in Buenos Ayres, South America, a city having 750,000 inhabitants.

The new engines for the London Street Railway have been shipped from Providence, R.I. They are of the Armington & Sims make.

The Toronto cabmen have complained to the City Council respecting the right of the street railway to solicit passengers at the Union Station.

The real estate and other property of the St. Jean Baptiste Electric Co., in liquidation, have been sold to the Hon. L. Tourville, for $53,000.

Hintonburg, a suburb of Ottawa, has closed a contract with the Ottawa Electric Light Co., for 9 years' electric lighting, at $15 per incandescent lamp per year.

## POSITION WANTED

By an electrician who has had 4 years' experience in the handling of all types of machines made by the Canadian General Electric Co. Also took students' course at their factory and holds certificate of competency. At present engaged on installation work. Address, "Electrician," CANADIAN ELECTRICAL NEWS.

## AN A 1 ELECTRICAL PLANT

Owing to the death of one of the proprietors, the Shelburne Electric Light Plant is now for sale, with real estate, brick lighting station, Wheelock engine, two dynamos and complete equipment.
For particulars address DR. NORTON or WILLIAM JELLY, Shelburne, Ont.

# CANADIAN GENERAL ELECTRIC CO.
## (LIMITED)

## Incandescent .. Lamps ..

THE earning power of an incandescent plant hinges upon one vital point : the comparative efficiency of the lamp in use. Different lamps on the market show efficiencies varying from 10 to 30 per cent. lower than ours. This means where such lamps are used from 10 to 30 per cent. less return from each pound of coal and from each kilowatt in plant capacity. Our lamps excel in the other important feature of long life with maintenance of candle-power. Central station managers have learned to appreciate this, and as a result our lamp sales have more than doubled during the last twelve months.

## Carbons

WE have recently taken over the premises and plant of the Peterborough Carbon and Porcelain Co., Ltd. It is our intention in continuing the operation of this factory as a department of our Peterborough works, to make such changes in methods and equipment as may be necessary to render our carbon output equal in all respects to the best imported grades.

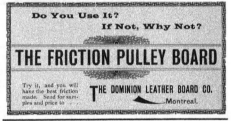

# CANADIAN ELECTRICAL NEWS
## STEAM AND ENGINEERING JOURNAL

OLD SERIES, VOL. XV.—No. 6.
NEW SERIES, VOL. VI.—No. 2.

FEBRUARY, 1896

PRICE 10 CENTS
$1.00 PER YEAR.

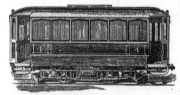

CANADIAN

# ELECTRICAL NEWS

AND

## STEAM ENGINEERING JOURNAL.

| VOL. VI. | FEBRUARY, 1896 | No. 2. |

## MR. E. E. CARY.

THE annexation of the United States to Canada is proceeding satisfactorily. Our latest conquest in this direction is the capture of Mr. E. E. Cary, the newly appointed manager of the Packard Electric Co., of St. Catharines, whose portrait we have pleasure in being able to present to the readers of THE ELECTRICAL NEWS.

Mr. Cary's home since infancy has been in New York City, although he has not resided there continuously. He graduated from the Polytechnic, of Brooklyn, N. Y., in 1884. During 1883-4 and part of 1885 he was public and private assistant in electrical work to Prof. Robt. Spice, of Brooklyn, N. Y. In 1885 he entered the laboratory as assistant in electricity and chemistry to Prof. Weston, then connected with the old U. S. Electric Light Co., of Newark, N. J. In this position he remained three years, devoting much time to the development of incandescent lamps, then in its early infancy as a commercial product. He then accepted an opening with the Westinghouse Electrical Mfg. Co., of Pittsburg, where he remained for two years and a half. While with the Westinghouse Company he was associated for a year with the Russian physicist Dr. Lodyquin in special filament investigation, having to do with high efficiency lamps. He then joined the forces of the Sawyer-Mann Electrical Co., and did special experimental and practical work on 110 volt lamps.

For the past four years he has been connected with the Beacon Vacuum Pump & Electrical Co., of Boston, as superintendent, and latterly as business manager. In December last he joined the Packard Electric Co., Ltd., as general manager. The most of his work in the States has been intimately associated with the development of the incandescent lamp.

Mr. Cary is the author of a number of inventions, having to do with the mechanical and scientific production of the incandescent lamp, and was one of the inventors of the N. and C. Stopper lamp, which, though ultimately not proving a commercial success, owing to its being pushed on the market too soon, involved new principles which some day may be most valuable. It is protected by over 20 patents, issued in the U. S.

It will thus be seen that Mr. Cary is well qualified for the position he now occupies, and the Packard Company are to be congratulated upon having obtained the benefit of his experience and services.

## ELECTRIC LIGHT AMALGAMATION IN TORONTO.

THE negotiations which have been in progress for some time past with the object of effecting a closer business relationship between the Toronto Electric Light Co. and the Incandescent Light Co., of Toronto, are understood to have resulted in an amalgamation of the interests of these companies. The bulk of the stock of the Incandescent Company has passed into the hands of the Toronto Electric Light Company, while on the other hand, several of the directors of the Incandescent Company have acquired stock in the older Company, and will occupy seats on the Board of Directors of the amalgamated concerns.

It is stated that Mr. Frederic Nicholls, the organizer and manager of the Incandescent Company will shortly retire, and the management of the amalgamated concerns be placed in the hands of Mr. J. J. Wright, the present manager of the Toronto Electric Light Company. It is believed that Mr. Nicholls will be a Director of the new Company.

Authority will be sought to enable the company to increase its capital stock to at least $1,500,000.

The improvements designed to be carried out by the Toronto Electric Light Company before the amalgamation, including the building of a new station and the installation of an alternating incandescent lighting plant, are being proceeded with. A large power alternator of the C. G. E. type has already been purchased. A test is to be made of Stanley and Monocyclic machines for incandescent lighting, and the system which gives the most satisfactory results will be adopted.

The current generated at the incandescent station on Terauley-street will probably be exclusively used for lighting the business district of the city, while current for power and incandescent lighting in the residential parts of the city will be furnished from the new station, shortly to be erected on the esplanade.

The Vernon & Nelson Telephone Co., have extended their service to Trail and Rossland, B.C.

The Canadian Marine Engineers' Association have elected officers for 1896, as follows : President, O. P. St. John ; First Vice-President, J. S. Adam ; Second Vice-President, J. Parsall ; Council—J. Findlay, R. Hughes, S. Gillespie, D. F. Campbell, R. McLaren ; Treasurer, D. L. Foley ; Secretary, S. A. Mills ; Auditors—R. Childs, J. H. Ellis ; Inside Guard, E. Abbey.

## THE RECENT SLEET STORM.

The recent sleet storm which resulted in so much damage to the electrical interests throughout the country, and especially in the City of Toronto, which appeared to be the centre of the storm, was indirectly an illustration of the old saying that "every dog has his day," inasmuch as it furnished a harvest for the hack-

SAMPLES OF THE WORK OF THE RECENT SLEET STORM.
Ontario Street, looking north from Queen Street, Toronto.

men at the expense of the Street Railway Company. The storm proved even more disastrous to the electrical companies than the one which took place a couple of years previous. We publish herewith some illustrations which will serve to indicate the destruction wrought in Toronto, and the difficulties with which the electrical companies had to contend and are still contending in consequence. For the photographs from which these illustrations were made, we are indebted to Mr. Arthur M. Rust, of the City Engineer's department.

By far the largest amount of loss fell upon the Bell Telephone Co., owing, no doubt, to the fact that its poles were much more heavily laden than those of the electric light and telegraph companies. The latter appear to have come out of the occurrence with comparatively little loss.

The Great North Western Telegraph Company's lines west of Toronto were in operation before noon of the day following the the storm. The greatest difficulty the company experienced was in the vicinity of Scarborough, where its wires and poles were so heavily sheeted with ice as to be unable to withstand the strain.

The Electric Light Co., after consulting with the city authorities, deemed it inadvisable to turn on current on the night following the storm, lest accident might result on account of the tangled up condition of the wires on the streets.

The Bell Telephone Company's loss is estimated to be somewhere between $50,000 and $75,000. The im-

mense destruction to their system has given rise to the opinion, on the part of the public, that in the interests of the company and its subscribers their wires should be placed underground. On the surface this would seem to be a proper view of the matter, but further consideration will show that there are serious difficulties in the way of carrying out the proposition. In the central part of the city, where hundreds of subscribers are bunched together within a limited area, it is possible to place the wires underground, as they can easily be brought up through cables to the top of a pole or building, and from thence distributed to subscribers. This is not the case, however, in the out-lying districts, where subscribers are more widely separated from each other. In such districts poles are an absolute necessity for distribution purposes. If there is any means of distributing current to subscribers in such districts, without the aid of poles, we would be pleased to learn how it could be done.

The opinion has also been expressed that the telephone company made a mistake in adopting the trunk line system of distribution, by which they are obliged to carry from 100 to 200 wires upon their poles, the weight of which, with the addition of a coating of ice, is calculated to cause the poles to give way under a storm such as we have just experienced. It is a singular fact that in the recent storm there are said to have been more broken poles with five cross arms and under than with five cross arms and upwards. It is somewhat

SAMPLES OF THE WORK OF THE RECENT SLEET STORM.
Rose Avenue, looking north from Winchester Street, Toronto.

difficult to account for this fact seeing that each additional cross arm, with its attendant wires, must increase the weight on the pole. It should be borne in mind, however, that each additional cross arm is located lower upon the pole and tends to distribute the weight.

It seems to be rather a question of the direction in which the pole lines run, and the amount of shelter they get, than the number of wires they carry. As to shelter, they get very little, owing to the fact that the poles must be high enough to place the wires beyond the reach of contact with shade trees and buildings. It has been found that the lines running east and west suffer comparatively little as compared with those running north and south. Unfortunately it is not possible for the Telephone Company to run its lines in one direction only, as might be done by a telegraph company seeking an outlet into the country. The Telephone Company are obliged to go where its subscribers are, no matter what the direction may be. Referring again to the trunk line method, it may be pointed out that the adoption of this method in Toronto was also necessitated by the fact that the company's agreement with the city prohibits them from using certain of the principal thoroughfares, so that it becomes necessary for them to mass their wires on certain streets in order to be able to reach their customers.

It has likewise been suggested that wrought iron should be substituted for wood for poles, but the persons who make this suggestion have evidently not considered the question of cost. In Belgium, where iron and labor are cheaper than almost any other place in the world, the cost of wrought iron poles 100 feet high is about $800 each. A similar amount would have to be paid on this Continent for a pole 62 feet high, which is about the height of the wood poles now in use by the Bell Telephone Company in Toronto. These wood poles probably cost the company not more than $10 each, so that it can readily be seen that the use of iron is entirely out of the question. It may be possible at some future time to evolve a method of distribution which will be equally as efficient and less subject to unfavorable weather conditions than that at present in use, but so far the problem remains unsolved.

The recent storm serves to indicate the necessity for a large reserve fund on the part of electrical companies in general, and telephone companies in particular. It would of course be unreasonable to assume that such a storm is likely to occur every second or third year. Prior to the storm of two years ago there had not been such an occurrence for 12 or 15 years, and possibly there may not be another for a like period in the future.

The purchase of the electric light plant of the city of Kingston, Ont., will probably be considered by the council at an early date. The cost for lighting the streets under the present contract is $7,000.

At Windsor, Ont., recently, Judge Horne decided that the municipalities cannot assess the telegraph wires of the Canadian Pacific railway, as the company is, by its charter, allowed to erect and maintain telegraph lines and to charge for messages sent by them.

## LIGHTING FROM STREET RAILWAY CIRCUITS.

A correspondent writes us as follows :

"In asking the citizens of a certain town in northwestern Ontario for subscriptions to help forward a scheme for an electric railroad, they were informed by the promoter that when the road was built, current would be supplied for lighting purposes at the rate of fifty cents per year for each 16 c. p. lamp, and ten dollars per year for each arc street lamp. If 700 lamps were installed, this would in addition to 8 street arc lamps amount to the sum of $430 per year, which would not go far in paying the expenses of the plant, even if the lighting was done off the trolley wire, which is prohibited by the Underwriters' Association. However, at this rate the electric lighting companies will have to "shut up shop" and start farming or some other congenial occupation. Evidently the aforesaid gentleman was trying how much he could make some people swallow without causing them to gag. He must have succeeded beyond his wildest expectations."

We may say, with reference to the above communication, that electric lighting companies have little to fear from the competition of electric railway companies, inasmuch as the Underwriters' Association, as stated by our correspondent, will not approve of current being taken into buildings for lighting purposes from street railway circuits.

SAMPLES OF THE WORK OF THE RECENT SLEET STORM.
Terauley Street, Looking North from Louisa Street, Toronto.

This matter came up in Toronto some time ago, with the result that owing to the opposition of the Underwriters' Association, there is at the present, so far as we know, only one instance to be found in the city, of electric light being furnished from the street railway circuit. There is the additional fact that owing to the frequent and great fluctuations in the current on street railway lines, it is impossible to get satisfactory lighting from this source. These two causes are sufficient in themselves to prevent the extension of electric lighting from street railway circuits, so that electric lighting companies need be under no apprehension of losing their business as the result of the competition of street railway companies.

All these difficulties, of which we hear complaint, are evidence of the need of organization and interchange of views and experiences on the part of those engaged in the electric lighting business.

Professor Waddell, of the Royal Military College staff, Kingston, recently delivered a lecture in the Y. M. C. A. hall in that city on "The Electric Current." With the aid of a battery, small dynamo, magnets, volt and ampere meters, he gave in detail the origin of the electric current and the manner in which the pressure and flow were kept constant.

## CANADIAN ASSOCIATION OF STATIONARY ENGINEERS.

NOTE.—Secretaries of Associations are requested to forward matter for publication in this Department not later than the 25th of each month.

### TORONTO NO. 1.

The members of the above association have felt for some time past the necessity of procuring more satisfactory rooms in which to hold their meetings. These have now been secured at No. 61 Victoria street, and consist of one large meeting room, with library room and several anti-rooms adjoining. They are suitably adapted to the requirements of the association, and realizing this, a five-year lease has been secured.

The trustees of the hall are Messrs. James Huggett, E. J. Philip and Geo. Fowler. The inaugural opening took place on the 23rd of January, and was made the occasion of a social entertainment, at which, notwithstanding the inclement weather, upwards of 150 persons were present, many of whom were ladies. The accompanying illustration shows the interior of the main hall.

A concert formed an enjoyable feature of the evening's entertainment, the proceedings being presided over by Mr. W. Lewis, president of the association. The programme, which was entirely voluntary, was as follows: Song, Mr. Thos. Seaton; calisthenics, Mr.

INTERIOR VIEW OF HALL, TORONTO NO. 1, C. A. S. E.

H. Eversfield; trio; Mrs. Coutts-Bain and Messrs. Towers and Cashmore; comic song, Mr. Allcott; duet, Messrs. James Fax and G. W. Grant; song, Miss Warnock; comic song, Mr. Jas. Fax; phonograph, Mr. Parks; song, Mr. W. G. Blackgrove; comic song, Mr. Fax; song, Mr. Cashmore; song, Mrs. Coutts-Bain; duet, Miss Warnock and Mr. Grant; concertina solo, Mr. Vaughn. A decided hit was made by Messrs. Fax and Grant in a duet entitled "Goodness Gracious." For the benefit of absentees we give one of the verses:

> When Wickens first started the C. A. S. E.,
>   Oh, goodness gracious,
> Folks thought he was off of his b-a-s-e.,
>   Goodness gracious;
> But now we've got Edkins and Phillipses too,
> George Mooring, Tom Eversfield, doodle-dum-doo,
> And then as a climax this hullabaloo,
>   Gracious, good gracious, goodness gracious.

A brief address was delivered by Mr. A. M. Wickens, in which he referred to the circumstances which led to the formation of the society nine years ago. Previous to that time an average of 312 persons were killed each year in the United States and Canada by the explosion of boilers. These explosions were not accidents, but were the result of ignorance and carelessness. It was

therefore decided to make the society, as much as possible, educational in character. At the end of the first year forty members had joined, and in the 3rd year an executive council was formed. Now upwards of twenty branches of the organization are established, and Toronto No. 1 alone numbers about 120 members. At present the association is working under a permissive law, but it was hoped at an early date to obtain a compulsory law.

The President stated that they were desirous of compiling a library, and already a number of books had been promised. It was the intention to invite manufacturers to supply books.

A bountiful supper had been provided which occupied the attention of the guests for some time, after which dancing was engaged in.

The committee appointed to act in conjunction with the trustees, and to whom the success of the evening's entertainment is largely due, consisted of Messrs. G. C. Mooring, chairman, T. Eversfield, C. Moseley, S. Thompson, W. G. Allen and A. M. Wickens.

### HAMILTON NO. 2.

The members of our association are becoming more earnest towards education. To this end we have provided ourselves with models, books etc., besides having an indicator of our own for the use of any of the members. We have started our regular instruction meetings, and they promise to be of great benefit during the winter months. At the first of these meetings the Recording-Secretary read a paper illustrating the application of Ohm's law, which will be sent you for publication in the March issue of your journal. At the last meeting some good discussions took place on pumps, also on the proper area of steam and exhaust ports, which will no doubt be continued.

WM. NORRIS,
Recording-Secretary.

### BROCKVILLE NO. 15.

Wm. Robinson, Recording-Secretary, writes: The members are taking a lively interest in the meetings and the work, especially the educational part. Our membership is about twenty-four, and taking the average attendance it is really good. We meet on Mondays for regular business, and on Fridays our time is devoted to educational matters. It is the intention of the Executive Committee, I believe, to procure models for the different associations, which will no doubt make a great many things more comprehensible. I trust they will be received before long.

ANNUAL DINNER OF MONTREAL ASSOCIATION NO. 1.

The sixth annual dinner of the above Association, held at the Queen's Hotel, on the 30th ult., was attended by about 120 persons, and was perhaps the most successful event of the kind in the history of the Association. Mr. J. J. York, President of the Association, presided, having on his immediate right and left the following invited guests : Prof. Nicholson, of McGill University ; Lieut.-Col. Massey ; Messrs H. R. Ives, Walter Laurie, Lieut.- Col. Stevenson, Chas. Morton, A. Henry, J. Dyer, Wm. Laurie, D. W. McLaren, O. E. Granberg, J. C. Willison, Chas. T. Smith, J. C. Holden, H. Valance, P. Cowper, Thos. Ryan, Geo. Kell and W. T. Bonner.

There were present the following members, in addition to fifty friends who bought tickets :—Past Presidents, Messrs. Jos. G. Robertson, Ryan and Hunt ; B. A. York, H. Nuttall, Robt. Doran, Gerry E. Flannigan, J. E. Huntington, John Robinson, A. Mesnard, H. Rollins, E. Hay, Wm. McHalpin, Wm. Allan, H. W. Smith, Wm. Burgess, Chas. Sanderson, Jos. Badger, John H. Garth, J. S. Campbell, J. Glennon, Alfred Ward, Jos. McParlon, Jas. Wilson, H. J. Weaver, John Smyth, J. E. Jones, John Burns, J. Kirwin, Chas. Casey, Jas. Morrison, F. D. Jones, A. W. Brown, Wm. Ware, Geo. White, Wm. Norket, Jas. Elliott, J. B. Goulet, Ed. Orton, David White, J. V. N. Ceeney, E. Valiquette, B. D. Tierman, Wm. Bill, John Murphy, Hugh Thompson, D. Smitherman.

Letters of regret were read from the following : S. C. Stevenson, Secretary Council of Arts and Manufactures ; Wm. H. Browne, Manager Royal Electric Co.; Jas. H. Peck, Peck, Benney & Co.; A. Ramsay, A. Ramsay & Son.; G. C. Cunningham, Manager Montreal Street Railway Co.; Henry Holgate, Manager Montreal Park & Island Railway ; W. S. Blackgrove, President Executive Council, C.A.S.E.; John Thorpe, Pilkington Bros., Ltd.; James Jackson, Manager Dom. Cotton Mills Co.

After a proper amount of attention had been paid to the excellent menu, the Chairman addressed the assembly as follows :—

"We have now arrived at that part of the proceedings where I trust everyone has sustained a serious loss of appetite. We have also arrived at the point where the Chairman is supposed to say something short and sweet, and let the business of the evening proceed.

"With my brother engineers, I feel highly honored to have the company of so many of the largest steam-users in the city of Montreal, as well as the presence of representatives of two of the greatest educational institutions in Canada—the McGill Univer- and the Council of Arts and Manufactures. We also feel honored by the presence of an old friend, in the person of the Chairman of the Fire Committee, and the many other gentlemen who have so kindly consented to contribute to our entertainment. But it is for the benefit of steam-users particularly that I wish to make a few remarks. I am sure that not one quarter of the steam users of this city know the aims and objects of this Association, and much less about the noble work it has in hand. On the other hand, there are large steam-users here to-night who are pleased to know that there is such an Association, and who can tell you that the Association is directly responsible for the more economic operation of their steam plants. And why ? Because it has assisted to educate their engineer, and the engineer has helped to educate others.

"A few words here descriptive of our methods may not be out of place. This Association was formed in the year 1883—about the same time that the question of licensing engineers was before the Council—and at a regular meeting held in the St. Lawrence Hall on Aug. 19th, 1885, Thomas Ryan in the chair, a resolution was passed, the like of which no other body of men has since passed. It recommended an increase in proposed examination fee, or tax, on engineers. This is proof that the only fault we had with this license law was that it was not strict enough. The next few meetings were employed in the work of organization and the framing of by-laws, &c. On Nov. 19th of the same year W. H. Nuttal read the first paper before the Association, entitled "Priming—Its Causes and Prevention." This was the key note, and at every meeting since, with but few exceptions, some subject pertaining to steam engineering has been taken up and discussed.

"This Association is now composed of about 95 members, and includes some of the best engineers of the city. We are possessed of working models, instruments and apparatus to the value of $700 ; furniture, carpets, &c., $250, and are just about to close an order for $150 worth of books for our library, which, thanks to our friends, already contains several valuable works. If we could only educate the steam-users of this city to take us into their confidence and make the changes suggested by us, and afterwards

pay us 25% of the saving effected, I will say without fear of contradiction that we would in less than ten years own a building larger and grander than this Queen's Hotel.

"Now Mr. Steam-User, don't think for a moment that we are after your money. Quite the contrary. We are this very day saving you money ; all we ask is that you look upon your engineer as a man of responsibility, a man who holds the safety of your factory and the lives of all employed in it in his hands. We would also ask you to keep in view the fact that he has it in his power to increase or decrease your profits as he likes by way of the coal pile. You may think this strange, but I will show you how true it is by telling you something that actually transpired. The owner of a certain factory in this city who did not employ a competent engineer, had from time to time increased the output of his works, and of course the consumption of coal increased also, but in much larger proportion. He paid no attention to this, until one day the engine absolutely refused to longer put up with the treatment received at the hands of the incompetent engineer, and stopped work. An engine builder was called in ; he wanted $75.00 to fix it up, and was told that he wanted more than he would get. He then offered to fix the engine gratis provided the owner would give him the value of the coal the engine would save during a certain time. This was at once agreed to and a contract drawn up, with the result that the engine was soon repaired and that steam user paid to that engine builder upwards of $160. Now what happened ? Did he discharge his engineer for incompetence and secure another that would keep his plant in a state of efficiency ? No, he did not ; he kept the same man on, and to-day that plant is nearly as bad as ever it was.

Why is it that we find in nearly every factory office an expert bookkeeper at a high salary ? It is because the owner knows what good book-keeping is, and wants his books kept in the best possible manner. If he only knew half as much about the engineer's duties, I am very sure there would be many openings for competent men next week.

I must not longer trespass on your time, but will add that we do not admit everybody to membership—in fact, during the past year we have refused several applications because they could not demonstrate that they were competent to take charge of a steam plant. I would also take this opportunity to invite every steam user to become an honorary member of our Association, which they can do on payment of the small sum of $5.00. This will entitle him to all the privileges that I, or any other engineer enjoy, with the single exception of voting, and will also prove beyond a doubt that nothing detrimental to your interests is discussed at our meetings. Your membership would, I am sure, be of great mutual benefit, apart from the fact that it would very materially assist us in adding to our library or stock of instruments.

The toasts were replied to as follows : "Council of Arts and Manufactures," Mr. W. Laurie ; "Faculty o' Applied Science," Prof. J. T. Nicholson ; "Boiler Inspection" Col. Stevenson ; "Fire Committee," O. E. Granberg ; "Brotherhood of Locomotive Engineers," Mr. Thos. Clark and Geo. Kell ; "Our Guests," Col. Massey, C. M. Smith, C. Morton, H. Nuttall, W. G. Norris, T. Ryan, H. R. Ives, A. Hersey, John Dyer, Wm. T. Bonner, H. Valance, P. H. Copper and W. D. McLaren. Strange to say a champion could not be found to respond on behalf of "Our Tormentors." Several excellent songs and musical selections were rendered by R. Hilliard, J. Dougherty; Dr. Nicholl, W. Morris, W. Campbell and Vice-President Hunt.

## ONTARIO ASSOCIATION STATIONARY ENGINEERS.

Editor CANADIAN ELECTRICAL NEWS,

DEAR SIR,—During the month of December the following engineers have been examined and received certificates : 1st class, Wm. Gray, Galt. 2nd class, G. B. Risler, London ; A. J. House, Sudbury ; Thos. Leake, Stratford ; J. G. Archibald, Woodstock. 3rd class, Jno. Kappler, St. Marys; R. Hutt, Queenston ; J. Wedgery, Woodstock ; J. F. Glennie, Listowel.

The following engineers who formerly held 3rd class certificates have passed the examination and received and class certificates : Wm. Cole, Thos. Young, D. McKay, and R. Topping, all of Woodstock.

During the month seventeen engineers tried the examination, and thirteen were successful.

I shall be glad to send copy of by-laws, &c., to any engineer who will send request for same on post-card.

Yours truly,

A. E. EDKINS, Registrar.

139 Borden st., Toronto.

## A CANADIAN MOTOR-CYCLE CONTEST.

### By Arthur W. White, London.

Glancing through the different scientific papers, one sees considerable discussion and argument about motor vehicles. Some probably through selfish motives publish what they designate a "Conservative Article," and in some instances an editorial dealing with the question. The articles referred to are inconsistent in the extreme, and the only inference to be taken from them is, that their writers are not ready for the advent of motor vehicles. By all means be conservative, but do not allow personal advantages to be the motive.

Among the best methods, in the writer's opinion, for "pushing this good thing along" in Canada, public trials and tests stand well to the fore. New York is agitating one, and France and Germany will hold a number next summer. The last issue of the London, (Eng.) "Engineer" contains full prize list and conditions of a competition for one thousand guineas.

The present English law prohibits a self-propelled vehicle from travelling more than four or six miles per hour, and places further restrictions on this manner of travelling, enough to make a race impossible without special act of parliament, or a revision of turnpike laws, which changes are now being agitated. There seems to be a difference of opinion as to whether a race could be run in Canada, without the same steps being taken. Should this be the case, would it not be advisable to obtain permission, before a Canadian race takes place, otherwise the contestants, or promoters of the trial, could be held responsible for damages arising from frightened horses, etc.

That a Canadian race should take place goes without saying. We must keep up with the times. If there are no public spirited men who can afford to offer sufficient inducements, in the shape of prize money, forthcoming, the race can be arranged in other ways. In Ontario, we have two large fall exhibitions, the Industrial, of Toronto, and the Western, of London. Either of these should be able to make a paying investment of a motor cycle contest ; it would certainly be a drawing attraction, more instructive, more entertaining, better advertised and more in keeping with an industrial exhibition, than balloon ascensions, high diving, second-class contortionists and acrobatic entertainments and wild-west and Arab shows, comprised mostly of toughs from the slums of large cities, who hire a few horses, dress in exaggerated costumes, shout and discharge firearms. Half the amount of money paid for this sort of thing, would make a purse sufficient to induce others besides Canadians to compete. It would make an exhibition Industrial in reality, as well as in name. It would stimulate Canadian inventors, as the Chicago race did United States inventors. Previous to the advertising of this race, motor vehicles were almost unknown in the United States. Over five hundred applications for patents, covering motor vehicles and parts thereof, were made during the time intervening between the first notice and the consummation of the race. If five hundred of our best thinkers started to think, it would mean more for Canada than one can imagine. Motor vehicles are only in their infancy. There is room for great improvement, and competitive tests are among the best methods for their improvement.

Preliminary tests, from which the judges could decide the points of internal friction, design, construction, ease of handling, finish, etc., could be held the first four or five days of the exhibitions, in a building provided for this exhibit. Processions could be given daily in the ring, and a final race starting in the ring, encircling it once or twice, thence to a point twenty or thirty miles into the country and return to finish by again going around the ring. Manufacturers would enter a contest of this kind as much for advertisement as for the prize money, and should, in the writer's opinion, be willing to pay a reasonable entrance fee.

There is no reason why both London and Toronto should not include a motor vehicle contest in their attractions and prize lists, and it is to be hoped that the directors of these exhibitions will give it due consideration. London can offer exceptionally good accommodation. A race from the city to Lucan or Strathroy would be an ideal run—roads that are good in all weathers, with just grades enough to give a good test, and plenty of villages along the route for frequent relay stations.

The vehicles might be divided into two classes, one class for electric motor vehicles and another for carriages driven by internal combustion engines and other small motors, that carry their fuel in small receptacles, enabling them to take enough for the complete trip. The former might show up to good advantage in preliminary tests, processions, and short trips, but, as has been proven by previous races, the latter could make the best time in a long road race.

Should these few rambling remarks, or any personal assistance, be of any value to exhibition directors, or private individuals with a desire to further the advancement of this industry in Canada, the writer will be more than pleased. One thing is certain, the motor vehicle has come to stay, and our country should, as usual, be well to the front in the improvement and manufacture of them.

[The above letter, we believe, expresses the sentiments of many persons who are engaged in the manufacture and development of motor vehicles, as well as a considerable number of outsiders who take sufficient interest in the progress of invention to realize the benefits to be derived from such a contest. It is hoped that this letter will result in promoting a discussion on the most feasible plan of conducting the race. We are pleased to be able to state that the management of the Toronto Industrial Exhibition Association look upon the idea with favor, and are at present considering what steps to take in the direction of assisting to bring about a test in Canada. That such a test would prove a drawing card for the Industrial Exhibition goes without saying. It would seem that the amount of the prize money offered by the Association would be determined to a large extent by the number of probable competitors. On the other hand the number of competitors would depend in some degree at least on the amount of the award. In any case should such a race be decided upon, manufacturers should at once make known their intention of entering the contest. The route of the proposed race will be a matter requiring careful consideration. It is certainly desirable that the test should take place over a road corresponding in character with the highways upon which such vehicles would be required, but whether the Exhibition Association management would consent to the test taking place beyond the boundaries of the fair grounds is yet a matter of doubt. We have reason to believe, however, that this difficulty could be overcome. The new Board of Directors for the Industrial Exhibition Association will be elected about the middle of February. Nothing definite will be known before that date regarding the attitude which the Association will assume towards the proposed contest.—ED. ELECTRICAL NEWS.]

## WM. KENNEDY & SONS, OWEN SOUND.

ONE of the most enterprising firms of to-day is that of Messrs. Wm. Kennedy & Sons, of Owen Sound, Ont., who have been established for upwards of forty years. They have become known throughout the Dominion as manufacturers of the well-known " New American" water wheel, electric water wheel governors, turbine wheels, and heavy mill machinery. The turbines now operating the lock gates at the Sault Ste. Marie canal, recently opened, were manufactured at their factory.

The works comprise two large buildings, one being two storeys high, 200 x 40 ft., and the other a three-storey stone building, 78 x 40 ft., at the corner of

WM. KENNEDY & SON'S FOUNDRY, OWEN SOUND, ONT.

Beech and Stephen streets. The business was originally established by the late Wm. Kennedy, in 1858, the present firm being formed in 1864, and being composed of Messrs. Matthew, Alexander and William Kennedy, jr., the two former residing in Owen Sound and managing the general business, while the last-named resides in Montreal and has charge of the branch in that city. They give employment to between forty and fifty men. They have received several medals for their propeller and water wheels, including silver medals from Philadelphia, Paris and Toronto, several bronze from Philadelphia, and one from the Colonial and Indian Exposition held in London, England. The success of the town of Owen Sound is due in no small degree to the energetic efforts of the members of this firm, who have always been public-spirited in advocating whatever would benefit the town. Mr. Matthew Kennedy is president of the Board of Trade.

Dr. G. W. Strange and Messrs. J. C. Stokes, L. E. Hambly and A. B. Armstrong are promoting the scheme for an electric railway between Schomberg and Aurora.

A suit has been entered against the Montreal Street Railway Co., by Elizabeth Kerr, claiming damages for $4,115. It is alleged that she fell while descending from a car on Notre Dame street, tripping on some encumberance on the step.

## 2,080 VOLTS FAILED TO KILL.

WE have received from Mr. J. A. |Farlinger, Gouverneur, N.Y., the following additional particulars of the accident of which he was recently the unfortunate victim :—On Sunday, Dec. 8th, I was asked to go up a 25 foot pole and cut out the commercial loop of one arc circuit. On this pole there were three arc circuits and two 2080 volt alternating circuits. Having received such a severe shock my memory was affected, so that I cannot remember even going to the pole, therefore don't know how the accident occurred, and for three days after I was unconscious. The alternating current was the only one on at the time, so I must have got across 2,080 volts of a three phase alternator, burning the flesh off the front of my hands, on some fingers leaving the bones as clean as if scraped with glass. My position on the pole was such, the minute I lost control of my body I fell backward and down, breaking my grip on the wires; I fell head first. Striking another wire in the fall somewhat righted my body and prevented my brains being knocked out. I fell on my cheek bone, breaking it in two places and paralyzing one side of my jaw. This fall is all that saved my life as otherwise I consider the doctors would not have been able to resuscitate me. I believe I am the only man who lives to tell of getting 2,080 volts of an alternating current through him.

## ELECTRICITY IN PAPER MILLS.

THE extensive works of the Canada Paper Co., of Montreal, situated at Windsor Mills, Quebec, are shortly to be operated entirely by electricity, instead of, as heretofore, by steam and water power combined. The company has developed a large amount of power on the St. Francis river, which will be transmitted to their mills about a mile distant. Here it will be distributed to electric motors ranging in power from 5 h.p. to 150 h.p. each running the various machines. The entire factory will also be lighted with incandescent lamps, and an electric railway is to be constructed from the power house to the mills, for the purpose of carrying pulp. The total amount of power to be transmitted will be about 1,000 h.p. The entire work has been placed in the hands of the well-known electrical engineer, Mr. George White-Fraser, of Toronto, who has just completed a careful survey of the locality, and is now engaged on the specifications. This is the largest enterprise of the kind in Canada, and will, no doubt, be the forerunner of many similar,

PUBLISHED ON THE FIFTH OF EVERY MONTH BY

## CHAS. H. MORTIMER,

OFFICE : CONFEDERATION LIFE BUILDING,
Corner Yonge and Richmond Streets,

TORONTO,    -    -    CANADA.
Telephone 2362.

NEW YORK LIFE INSURANCE BUILDING, MONTREAL.
Bell Telephone 2209.

*ADVERTISEMENTS.*

Advertising rates sent promptly on application. Orders for advertising should reach the office of publication not later than the 24th day of the month immediately preceding date of issue. Changes in advertisements will be made whenever desired, without cost to the advertiser, but to insure proper compliance with the instructions of the advertiser, requests for change should reach the office as early as the 22nd day of the month.

*SUBSCRIPTIONS.*

The ELECTRICAL NEWS will be mailed to subscribers in the Dominion, or the United States, post free, for $1.00 per annum, 10 cents for six months. The price of subscription should be remitted by currency, registered letter, or postal order payable to C. H. Mortimer. Please do not send cheques on local banks unless 25 cents is added for cost of discount. Money sent in unregistered letters will be at senders' risk. Sub-criptions from foreign countries embraced in the Postal Union $1.56 per annum. Subscriptions are payable in advance. The paper will be discontinued at expiration of term paid for if so stipulated by the subscriber, but where no such understanding exists, will be continued until instructions to discontinue are received and all arrearages paid.

Subscribers may have the mailing address changed as often as desired. When ordering change, always give the old as well as the new address.

The Publisher should be notified of the failure of subscribers to receive their paper promptly and regularly.

*EDITOR'S ANNOUNCEMENTS.*

Correspondence is invited upon all topics legitimately coming within the scope of this journal.

The "Canadian Electrical News" has been appointed the official paper of the Canadian Electrical Association.

## CANADIAN ELECTRICAL ASSOCIATION.

### OFFICERS:

PRESIDENT :
A. B. SMITH, Superintendent G. N. W. Telegraph Co., Toronto.

1ST VICE-PRESIDENT :
C. BERKELEY POWELL, Director Ottawa Electric Light Co., Ottawa.

2ND VICE-PRESIDENT :
L. B. McFARLANE, Manager Eastern Department, Bell Telephone Company, Montreal.

SECRETARY-TREASURER :
C. H. MORTIMER, Publisher ELECTRICAL NEWS, Toronto.

EXECUTIVE COMMITTEE :
GEO. BLACK, G. N. W. Telegraph Co., Hamilton.
J. A. KAMMERER, General Agent, Royal Electric Co., Toronto.
E. C. BREITHAUPT, Berlin, Ont.
F. H. BADGER, JR., Superintendent Montmorency Electric Light & Power Co., Quebec.
JOHN CARROLL, Sec.-Treas. Eugene F. Phillips Electrical Works, Montreal.
K. J. DUNSTAN, Local Manager Bell Telephone Company, Toronto.
O. HIGMAN, Inland Revenue Department, Ottawa.
W. Y. SOPER, Vice-President Ottawa Electric Railway Company, Ottawa.
A. M. WICKENS, Electrician Parliament Buildings, Toronto.
J. J. WRIGHT, Manager Toronto Electric Light Company.

## MONTREAL ELECTRIC CLUB.

OFFICERS :

President, W. B. SHAW,        Montreal Electric Co.
Vice-President, H. O. EDWARDS,    -    Montreal.
Sec'y-Treas., CECIL DOUTRE,    -    81A St. Famile St.
Com. of Management, T. F. PICKETT. W. GRAHAM ; J. A. DUGLASS.

## CANADIAN ASSOCIATION OF STATIONARY ENGINEERS.

President, W. G. BLACKGROVE,    -    -    Toronto, Ont.
Vice-President, JAMES DEVLIN,    -    -    Kingston, Ont.
Secretary, E. J. PHILIP,    -    -    Toronto, Ont.
Treasurer, DUNCAN ROBERTSON,    -    Hamilton, Ont.
Door Keeper, F. G. JOHNSTON,    -    Ottawa, Ont.

TORONTO BRANCH NO. 1.—Meets 2nd and 4th Friday each month in room D, Shaftsbury Hall. W. Lewis, President ; S. Thompson, Vice-President ; T. Eversfield, Recording Secretary, University Crescent.

MONTREAL BRANCH NO. 1.—Meets 1st and 3rd Thursday each month, in Engineers' Hall, Craig street. President, John J. York, Board of Trade Building ; first Vice-President, J. Murphy ; 2nd Vice-President, W. Ware ; Secretary, R. A. York ; Treas-urer, Thos. Ryan.

ST. LAURENT BRANCH NO. 2.—Meets every Monday evening at 43 Bonse-cours street, Montreal. R. Drouin, President ; Alfred Latour, Secretary, 306 Delisle street, St. Cunegonde.

BRANDON, MAN., BRANCH NO. 1.—Meets 1st and 3rd Friday each month, in City Hall. A. R. Crawford, President ; Arthur Fleming, Secretary.

HAMILTON BRANCH NO. 2.—Meets 1st and 3rd Friday each month in Maccabee's Hall. E. C. Johnston, President ; W. R. Cornish, Vice-Pres.; Wm. Norris, Corresponding Secretary, 211 Wellington street.

STRATFORD BRANCH NO. 3.—John Hoy, President ; Samuel H. Weir, Secretary.

BRANTFORD BRANCH NO. 4.—Meets 2nd and 4th Friday each month. F. Lane, President ; T. Pilgrim, Vice-President ; Joseph Ogle, Secretary, Brantford Cordage Co.

LONDON BRANCH NO. 5.—Meets once a month in the Huron and Erie Loan Savings Co.'s block. Robert Simmie, President ; E. Kidner, Vice-President ; Wm. Meaden, Secretary Treasurer, 533 Richmond street.

GUELPH BRANCH NO. 6.—Meets 1st and 3rd Wednesday each month at 7.30 p.m. J. Fordyce, President ; J. Tuck, Vice-President ; H. T. Flewelling, Rec.-Secretary ; J. Gerry, Fin.-Secretary ; Treasurer, C. J. Jorden.

OTTAWA BRANCH NO. 7.—Meet every second and fourth Saturday in each month. in Borbridge's hall, Rideau street ; Frank Robert, President ; F. Merrill, Secretary, 352 Wellington street.

DRESDEN BRANCH NO. 8.—Meets every and week in each month. Thos. Merrill, Secretary.

BERLIN BRANCH NO. 9.—Meets 2nd and 4th Saturday each month at 8 p.m. W. J. Rhodes, President ; G. Steinmetz, Secretary, Berlin, Ont.

KINGSTON BRANCH NO. 10.—Meets 1st and 3rd Tuesday in each month in Fraser Hall, King street, at 8 p.m. President, S. Donnelly ; Vice-President, Henry Hopkins ; Secretary, J. W. Tandvin.

WINNIPEG BRANCH NO. 11.—President, G. M. Hazlett ; Rec.-Secretary, J. Sutherland ; Financial Secretary, A. B. Jones.

KINCARDINE BRANCH NO. 12—Meets every Tuesday at 8 o'clock, in Mc-Kibbon's block. President, Daniel Bennett ; Vice-President, Joseph Lighthall; Secretary, Percy C. Walker, Waterworks.

WIARTON BRANCH NO. 13.—President, Wm. Craddock ; Rec.-Secretary, Ed. Dunham.

PETERBOROUGH BRANCH NO. 14.—Meets 2nd and 4th Wednesday in each month S. Potter, President ; C. Robison, Vice-President ; W. Sharp, engineer steam laundry, Charlotte street, Secretary.

BROCKVILLE BRANCH NO. 15.—President, W. F. Chapman ; Vice-President, A. Franklin ; Recording Secretary, Wm. Robinson.

CARLETON PLACE BRANCH NO. 16.—President, Jos. McKay, Vice-President, Henry Derrer ; Fin.-Secretary, A. M. Schofield.

## ONTARIO ASSOCIATION OF STATIONARY ENGINEERS.

BOARD OF EXAMINERS.

President, A. AMES,    -    Brantford, Ont.
Vice-President, F. G. MITCHELL    -    London, Ont.
Registrar, A. E. EDKINS    -    139 Borden St , Toronto.
Treasurer, R. MACKIE,    -    28 Napier st., Hamilton.
Solicitor, J. A. McANDREWS,    -    Toronto.

TORONTO—A. E. Edkins, A. M. Wickens, E. J. Phillips, F. Donaldson.
HAMILTON—P. Stott, R. Mackie, T. Elliott.
BRANTFORD—A. Ames, care Patterson & Sons.
OTTAWA—Thomas Wensley.
KINGSTON—J. Devlin, (Chief Engineer Penitentiary), J. Campbell.
LONDON—F. Mitchell.
NIAGARA FALLS—W. Phillips.

Information regarding examinations will be furnished on application to any member of the Board.

**Storage Batteries.**  THE storage battery as a central station auxiliary is just now receiving a very great deal of attention at the hands of engineers. A recent meeting of the American Institute of Electrical engineers held in New York was entirely devoted to a discussion of its proper place in central station practice, and the census of opinion seemed to be that the storage battery must be regarded and accepted as a most important and dividend-making necessity. Everyone familiar with the operation of an electric plant will be able to trace out a load diagram for himself. If the capacity of the station is 1,000 lamps, then he will know that from about 5 p.m. till 8 or 9 p.m. every lamp will be going, but that from 9 p.m. till midnight he will not have more than a fifth of full load. Now every educated steam user knows that an engine or dynamo works most economically when it is doing its full rated work, and that in proportion as it's load becomes lighter, so does it's efficiency become less, so that the operation of a plant at one-fifth load is a most uneconomical necessity. Any device, therefore, which will permit of machinery being operated at full load for a considerable proportion of it's running time is worthy of very careful examination, and such a device is the storage battery. During the period of very light lamp load, the battery may be charging, thus bringing up the station load line to full capacity, and when the short period of very heavy load is reached the battery and the dynamo may be thrown into multiple on bus bars, each taking it's proportion of load. At present, the dynamo and engine capacity of a station must be sufficient to cope with the maximum load that can be placed on that station, so that whereas the average load is perhaps less than 500 lamps, the dynamo, etc., must have a capacity of 1,000 lamps for the sake of the two or three hours of heavy load, and be run all the rest of the time at a most inefficient rate. Now, if the station plant consisted of say a 700 light dynamo and a storage battery with a capacity of 300 lights for four hours, then from about 11 p.m. until shutting down time, when probably not more than 300 lamps would be burning, the current for the other 400 lamps could be used to charge the battery, allowing the engine and dynamo both to be operated at full load. At starting up time again next day, when all 1,000 lamps were required, the storage battery (which was fully charged last night) and the dynamos could each be called upon to take care of their proper shares. The dynamo and engine would still be run at full load, and therefore highest efficiency. In this way it is seen that, first, the steam and dynamo plant need only be of 700-light capacity, instead of 1,000-light, and will run most of the time at or near full load. It is true that a storage battery requires a direct current to charge it, but instances can be referred to when alternating dynamos have been used with a rectifier for the purpose, with perfect success. We shall refer to this matter again.

**Motors for Single Phase Alternating Currents.** THERE are a great number of central stations throughout Canada that could very profitably operate a day load of small motors in factories, saw mills, stores, etc., but which have been precluded, hitherto, from working up such a business because their machinery was single-phase alternating, the current from which could not satisfactorily be used for power purposes. A single phase alternating motor will not start up with a load on, which defect, of course, renders it useless. Mr. C. G. Bradley, however, has elaborated a method of splitting up a single-phase E.M.F. into any number of symmetrical phases, with the view of overcoming this commercial disability of the single-phase alternating machine. The method is somewhat complex to describe, involving the use of condensers and inductances, but the results reached seem to be very satisfactory, and hold out the reasonable hope that alternating current stations may be able to work up, and avail themselves of a very profitable power business, without requiring to change the type of their machinery. Of course the method of "transformation of phases" involves some small losses which are eliminated in a properly constructed two or three phase system; but the money value of these losses is apparently so much less than the interest on the increased capitalization required to change all the machinery of a station from single to two or three phase types, that central station men would do well to look into the commercial advantages of this method.

**Central Station Men and the C. E. A.** A LETTER was printed in the January issue which seemed to indicate a feeling on the part of some central station owners and managers that concerted action on the part of the operating branch of the electrical industry is becoming more necessary as electric lighting and power is becoming more general. The desirability of union and co-operation has been endorsed in the United States, where there is a National Electric Light Association and many independent local associations organized for the same purpose. In Great Britain questions relating to the methods of operation of central stations are discussed at meetings of municipal engineers' societies, gas engineering societies, and wherever there are found sufficient engineers interested in electricity, to give their views or experience. On the continent of Europe, central station engineering is recognized as being a special branch of electrical study, and the central stations band together in order to promote their mutual interests, to further their knowledge of operating economics, and to guard themselves as an industry against the encroachments of the public on one hand and the manufacturing companies on the other. This spirit of co-operation has even taken the form in Germany of committees appointed by the central stations to investigate and examine into very many matters affecting the interests of the industry, and in which the experience of the individual is valuable as contributing to a sum total of conclusions which could only be formulated after such an exhaustive enquiry. Their latest committee, for example, has performed a most important service to the general body within the last few weeks, by making a most minute examination of the conditions of the incandescent lamp service and supply, making enquiries in every direction and bringing forward many points hitherto but little understood, and which have most important influence in operating expenses. The advantages to be gained by the co-operating of central station men will be perfectly evident when it is considered that they are required to supply to the public one of their greatest necessities—light ; that they have to do this in competition with gas companies and also against that of the oil wells ; that they are no longer able to get fancy prices for electricity, and that their dividends depend on their economical operation. They will be still more evident when it is considered, that this economical operation involves the study of problems connected with steam machinery as well as electric machinery and all sorts of electrical appliances. If it requires special training to qualify a steam engineer, and different special training to qualify an "electrician" ; how much more special must be the training of the man who has to manage an electric light and power business in which both classes of machinery are used ? The manager of every central station, large or small, has acquired experience with lamps, coal, carbons, rates, and what not, and such experience collected and published would be of great service to many other managers, who, having given their attention to other and equally important questions, would be able to reciprocate to the general benefit of the entire industry. We all want to know how our neighbour is getting on with some particular class of apparatus, and very likely will be able to give him some little valuable pointer in return for his suggestion, but at present every individual plant has to gain its experience for itself, often to buy it dear, whereas a little cordial co-operation would enable everyone to profit by the experience gained by some other one. A central station must indeed be in a position of ideal perfection if it can learn nothing at all from some other one. The Canadian Electrical Association is a body formed for the express purpose of facilitating this interchange of ideas and experiences. At the annual meetings many valuable papers are presented dealing with matters that come under the daily notice of central station men. It is this want of any organization in the electrical industry that is a principal cause of the crudeness of central station practice alluded to in our last issue. It is not too much to say that everybody loses by the present incoherence in the electric lighting profession. The public loses because the central station owners do not know the latest and best methods of supply nor keep themselves abreast of the times ; the central station men lose greatly because each one buys his experience for himself, and since he is not able to compare his results with those attained elsewhere is most likely to fall into a groove. The manufacturing companies lose because they have more difficulty in introducing any new and more efficient machinery when they have to deal with each customer separately, than they would if they could present their new goods to an association, each member of which would be able to keep in touch with the others. But the central station man loses most of all, because every improvement in machinery, or in the method of its operation, tends to reduce costs of operation, and hence to increase profits.

## CENTRAL STATION BOOK-KEEPING.

### By Geo. White-Fraser, E. E.

### II.

Having generated our steam, we have to use it to the best advantage, and so must know something about the engine, how it works, whether it requires attention to valves, and so on. Steam has an expansive force as well as a direct pressure, and the greatest economy is attained when we make use of both in their proper proportions. If we admit steam to the cylinder, and allow it to act with full boiler pressure during the whole length of stroke, when we open the exhaust port this high-pressure steam will be allowed to go free without our having got nearly as much out of it as it is capable of. But if we admit steam only a certain portion of the stroke, and then shut it off from boiler pressure, letting it expand itself down to a gradually lower and lower pressure, so that at the opening of the exhaust port it has no expansion force left—then we make use of all the power it can give us, and we use it economically. What is the good of letting steam go free into the atmosphere when it has force left in it still? What is the good of raising it to 100 lbs. pressure in the boiler, if we let it out of the cylinder when it still has 10 lbs. pressure left? We might just as well raise it to 90 lbs. in the boiler, and exhaust it at no pressure, or atmospheric pressure; only in this case we lower the power of the engine. The last thing to do is to so arrange that steam shall be admitted at boiler pressure in such quantity that when the exhaust opens it shall have expanded down to about no pressure. Then we shall have got all the good out of it it is capable of, and shall be using it economically. As a rule, engines are so proportioned and rated, that steam is admitted at boiler pressure for one-quarter of the stroke, and allowed to expand down during the other three-quarters, and that when this proportion is observed, it will be exhausted at just sufficiently above atmospheric pressure to ensure its freeing itself quickly.

It is understood that in the above I do not consider throttle valve engines, but only those that regulate power, and consequently speed, by means of cut-off valves. These engines are so made that they will automatically vary their own steam consumptions, in accordance with the work that they are required to do, by admitting steam for a longer or shorter period during each stroke; and an engine that has a rating of 100 h.p. at one-quarter cut-off will actually do much more when it allows half cut-off, or much less when it shuts off at only one-eighth of stroke, and these variations it can make itself, as it is running. It must, however, be clearly understood, that if this engine has such a large load placed on it as requires steam to be admitted at boiler pressure during half of a stroke, this steam will be exhausted into the air before it has nearly exhausted its expansion force, and so will be used wastefully. Or again, if a 100 h.p. quarter cut-off engine is run at so small a load as requires steam to be admitted during only one-eighth of stroke to keep the speed down to its proper number of revolutions, then this steam will have expanded down to atmospheric pressure some little time before the exhaust port opens, and as the piston still moves forward, there will be a partial vacuum formed where there actually should be a pressure, which is again a most extravagant and undesirable condition. The valves that open and shut the admission and exhaust ports are of course all movable, and are actuated by eccentrics or cams, or what not, that are also movable; which eccentrics are in turn thrown and their actions regulated by some form of governor, which is again in constant motion.

Now, I think that a general statement may be made that no one will feel disposed to contradict, viz: No piece of machinery that ever was made, simple or complex, is so perfect that it cannot get out of order or adjustment. This is especially true of a steam engine. The valves will most certainly wear out in time and leak; they may slip; the eccentric on its strap may work loose; the rod lengthen the sixteenth of an inch through a nut slacking; or the governor stick, or slip, or do some other vexatious thing that none would expect of it. Who can say what an engine is or is not capable of doing, when it is held together with nuts and bolts, and built of material that must wear? And any little thing it does wrong means extra expense and less profit.

Now consider for a moment what the consequences of a very little slip or stick in a valve may be. An exhaust valve may open a shade too soon or too late. In the former case steam will be exhausted before it has expanded down enough; in the latter it will not have time to get quite away, and some will be imprisoned in the cylinder to produce a back pressure. In the former case steam is wasted, in the latter a little more steam will be required to overcome the back pressure than would otherwise be necessary. In either case money is being wasted in fuel. Now, if oil is allowed to cake

with a little dust around the release of the exhaust valve, it may cause it to stick, and every person familiar with machinery knows how it can get out of order in the most inexplicable fashion. All these considerations serve to emphasize the necessity of keeping some track of the engine's working, and we have a means, in the indicator, of employing a private detective who will report to us with unfailing accuracy, everything that engine does. Is the cylinder or piston wearing? Are the rings getting loose? Is the admission valve getting the worse for usage? The indicator card gives indications of the steam blowing through. Are any of the valves opening or shutting too soon or too late? There it is on the accusing little diagram. Is anything wrong at all? The little indicator will run the offender to earth. Therefore it is, I say, do not trust your engine too much. Keep a watch on it, and record its operation frequently. Everyone has not got an indicator, but I think I should like to take cards once every week, and in order that the information may be complete, it would be necessary to record, at the instant of taking the card, the boiler pressure, the reading of the ammeter and the volt-meter of the dynamo or dynamos run by the engine, with the speed of engine, these in order that the load on the engine may be calculated, to compare with the indicator diagram. Cards should be taken at intervals throughout the run, when the load is at different points, so as to know what the engine does at all proportions of load.

The load on the engine for any card can be calculated by multiplying the ammeter and volt-meter readings together for wattage, and adding in the shafting and dynamo frictions, taking also into consideration the proportionate inefficiencies of dynamos at various loads, which can be obtained with more or less accuracy from the manufacturers.

The method of calculation of the real load on the engine corresponding to any observed ammeter and voltmeter readings, will be as follows, which will be quite close enough for all practical purposes:

Assuming a dynamo with the following manufacturing company's data and rating:

Capacity, 50 k.w.; commercial efficiency at full load, 95%; at half load, 90%; at one-quarter load, 85%. Full load current, 50 amperes; voltage, 1,000, and (for the sake of simplicity) no over-compounding, and allowing for no drop.

Then this machine will require to run it at full load

$$\frac{100\times50}{95} \text{ kilowatts or } \frac{100\times50}{95\times746} \text{ horse-power;}$$

at half-load $\frac{100\times25 \text{ k. w.}}{90}$ kilowatts or $\frac{100\times25}{90\times746}$ horse-power;

at quarter-load $\frac{100\times12\frac{1}{2} \text{ k. w.}}{85}$ kilowatts or $\frac{100\times12\frac{1}{2} \text{ k. w.}}{85\times746}$ horse-power.

Next an allowance must be made for the power wasted by the belt, which will depend in amount on the state the belt is kept in, but which, if that state is good, may be taken at say 8%, and an allowance for the shafting of say 10%. These amounts added together will show what the engine had to do when the card was taken, and a neat number of such cards can be averaged. The results can be collected and set forth in the form shown below, and the cards themselves should be very carefully studied by an experienced person, and the horse-power indicated by them recorded, with any remarks tending to explain their meaning.

| Engine Number...... | | | | | | | | Date...... | |
|---|---|---|---|---|---|---|---|---|---|
| Card Number | Ammeter | Volts | | | Total Watts | Total h.p. Electric | Belt | Shaft | Total Load |
| | Dy. 1 | Dy. 2 | Dy. 1 | Dy. 2 | | | | | |
| | | | | | | | | | |

If it be impossible for any station to take cards so frequently, then they should by all means manage to have it done not less than every three months.

Before leaving this part of the station, there is one set of experiments and records that should be kept by everyone using a condensing plant, viz: records at fixed times during the run, of reading of vacuum gauge and temperatures of condensing water, and water of condensation. If the condensed steam is to be used and fed back into the boiler, it is of advantage that it should be discharged from the condenser at as high a temperature as possible. But the less heat that is taken out of the exhaust steam, the lower will be the vacuum; consequently there will be found a point where increased temperature of water of condensation, far from being an economy, will actually be a disadvantage, and the most truly economical balance must be arrived at by experiment and calculation, and then preserved by constant care and attention.

We have now obtained a method of recording our steam genera

tion and utilization, which will, I think, give the average central station a very fair insight into this important department, and I would suggest that every steam-using plant should experiment with different kinds of coal; mix different kinds together till, by comparing results, they arrive at what seems to be best for them. Then try to raise the temperature of their feed and so on, and whatever they do, keep moving and observing and learning. There is a link between the engine and dynamo which requires some attention—the belts. There is always some slip to a belt. It may be minimized, but some will always be there, and the amount of slip will to a great extent depend on the condition of the belt. I am of course assuming that it has been bought of sufficient size and strength. Now, this slip can be observed in the following way : Everyone knows that if an engine and a shaft are connected together by a belt, the speeds of their two pulleys will be in the inverse ratio of their diameters. That is the theory. Now, if an actual test be made of the speeds of an engine and of a shaft, by trying them at the same time, with hand speed counters, any difference between the calculated speed of the shaft and the observed speed can be set down to the slipping of the belt. Slipping means that the power of the engine is not being fully utilized, and therefore the belt should be made to grip tighter, either by tightening it up to its proper limit, or if that limit has been reached, by dressing it more thoroughly. This record of observed speed may or may not be set down in the reports—but I should certainly recommend the observation to be made at frequent intervals. The more checks you have on the operation of your machinery and apparatus, the better are your results likely to be.

We pass now to the records concerning the electric plant, merely mentioning that as the one engineer generally looks after the engine and dynamos, etc., the reports from the engine room may include the dynamo records as well as the consumption of waste, oil, sandpaper, etc., notice of which will be taken in the general summary. Among the dynamo records which I consider to be really necessary to an intelligent management is certainly one that I do not believe a single one of my readers will keep—for the reason that either they will think it too much trouble, or, alternately, if they think of getting a machine to do it for them, they will consider it too much expense. I allude to some record that will enable them to see how much electricity has been manufactured by the dynamo, and delivered to the lines each night. There are only two ways of doing this, either to use a recording station wattmeter, which will keep track of every watt of electric energy sent out, and which will cost in the neighborhood of $100, or to make the engineer put down on paper at intervals of fifteen minutes or so the exact readings of the current and pressure indicators from which the station output can be closely calculated.

I think the absolute importance of some such record (preferably the wattmeter) will be evident to anyone if they will consider for a moment what its absence means. It simply means that a central station does not, and cannot know whether it is selling its electricity for more or less than its costs to make it. A farmer knows how much seed he puts into his field, and he measures the number of bushels he reaps from it; the merchant not only keeps account of the goods he buys to stock up with, but he knows how much has been sold each day, and if his stock-taking shows a difference between what he bought and what he has sold, he begins to look about and see whether he hasn't lost any or been robbed of some; in fact, if he didn't keep track of what went out of his store, as well as what came into it, he really wouldn't know what he was doing, whether he was solvent or bankrupt. A central station is in the same position, and if no record is kept of how much electricity goes out, what is the good of keeping track of how much comes in—in the shape of fuel?

It was observed above that there is a very clearly defined relation between the amount of coal burned and the amount of water turned into steam ; and that if there is observed (as the result of records) a disparity between that amount of water actually evaporated, and the amount that theory indicates should be evaporated, that the matter should be looked into with a view of securing better results. The inference is drawn that if no records are kept, it is impossible to detect anything wrong, and consequently a great deal of waste may go on with no one knowing anything about it. Now this is exactly the same with regard to the electric plant. Mechanical energy has its equivalent in electrical energy.

If a force of one hundred mechanical horse power be continuously applied in turning a dynamo whose commercial efficiency is 90 per cent., then that dynamo should give out continuously electrical energy to the amount of 90 h. p. If it is observed that this dynamo does not give out this electrical energy, then there is something wrong, some waste taking place, which should at once be remedied—if money is valuable. Now, it is known how much mechanical energy is expended during a run (the coal and evaporation records will give this), and therefore it can be easily calculated how much electricity should have been generated. But if it is not known how much actually has been generated, what is the good of all the other records?

Passing over the intermediate steps—so much coal should produce so much electricity. Does it? If not, why not? There may be something tremendously wrong somewhere in the plant; and it cannot be known without this nightly "stock-taking." How much does your current cost you to make? You cannot tell unless you know how much you make.

For the above reasons I strongly recommend the use of station wattmeters, and that they be read every night at the close of a run. A meter will cost about $100 ; the interest on this for a year is about $5 ; and if it isn't worth that much to you in giving you an insight into your business, and enabling you to stop wastes, then there is no advantage in book-keeping. 

Other useful records are : The engineer should note the reading of the current indicator every fifteen minutes during the early part of the run, and every half hour later, and construct a "load diagram" for every night. A comparison of these diagrams, week by week and month by month, is often of the greatest value, as indicating possible changes in the business policy of the central station, whereby better results may be attained. He should note every night whether there is a "ground" on the lines, and on which line, so that it may be hunted out and put right next day ; and note any unusual happenings—lightning stroke passing through arresters ; fuses suddenly blowing, with their cause (if known) ; new brushes put on dynamo, or anything else of that nature ; commutator turned down, and so on.

Below are suggested forms that will be found convenient. Next article will be devoted to the part of the business outside the station, consisting of lines, lamps, etc., with some suggestions as to the store room.

Engine report by S. Smith.　　　Date.... .........

| Engine No. | Started. | Stopped. | Vacuum Average. | Temp'rature Hotwell | Remarks as to Repairs, Accidents &c. |
|---|---|---|---|---|---|
| | | | | | |

It will perhaps be noticed, that what has gone before constitutes less a mere formulation of accounting systems than it does an enumeration of the inherent inefficiencies of all machinery, with some little indication of how their unchecked operation may affect

Dynamo report by............　　　Date.............

| Dyn. No. | Started. | Stopped. | Watt Meter. Start. Stop. | Grounds? | Remarks as to Repairs, &c. |
|---|---|---|---|---|---|
| | | | | | |

the financial results, and the description of a series of observations which will enable the intelligent manager to detect their undue extension, and hence to apply the appropriate remedies in time. The intention has been to show what very many sources of waste there are in the operation of a steam and electric plant, and that although a central station manager may buy the very best machinery in the world, it will do him not the least good unless he operates it properly. To use very high-class machinery, and to hire cheap labor, is to save at the spigot and waste at the bung hole. 

(To be Continued.)

The Toronto Electric Light Company are installing a 75 kilowatt monocyclic generator of the Canadian General Electric Company's make.

The new power house of the Oshawa Electric Railway Company has been completed. It is entirely of brick, and is equipped with Babcock & Wilcox boilers and two 150 h.p. cross-compound Robb-Armstrong engines, connected with a 200 k.w. six-pole Westinghouse generator. The installation was made by Ahearn & Soper, of Ottawa, and makes a model power-house.

The Toronto Electric Light Company have closed a contract with the Canadian General Electric Co. for a 600 horse-power slow-speed direct-connected power generator. This machine will be the largest in Canada so far installed for the supply of current for stationary motors.

## QUESTIONS AND ANSWERS.

"SUBSCRIBER," Hull, Que., asks: "Can a direct current of electricity be alternated into a transformer so that it will act on same like a current from an alternating machine? I mean a machine to change the direction of the direct current into the transformer; above machine, or a reserver, to be run by a belt."

ANSWER.—Your question is asked in two parts: First, a direct current cannot be so acted upon by any transforming device as to change its pressure into one higher or lower, as is done with the alternating current.

VALVES OF WHEELOCK ENGINE.

Second, the nature of a direct current can be so altered by an appropriate device, that this altered current may be passed through a transformer, with the familiar result. This is actually done in several electro-medical appliances, and in the familiar electric machine often seen at fairs and exhibitions, where an "electric shock" is administered for 5 cents. In this machine the direct current generated by an ordinary battery is sent through the primary of a Ruhmcorff coil (which is nothing more or less than a transformer), and while it is flowing it sets in automatic action a vibrating tongue, which actually forms part of the circuit; this tongue, in vibrating, opens and closes with extreme rapidity the primary circuit; and thus produces the rapidly varying induction in that primary circuit which is the necessary condition before it can affect the secondary circuit. There is no machine for effecting this rapid reversal of current, through the intervention of a belt; it could, undoubtedly, be done by passing a direct current through some form of commutator, which would pick it up from opposite brushes alternately; but the utility of this method is very much open to question. This commutator could be operated by a belt."

---

"W. B. S.," Montreal, asks: "Can any of your readers tell the writer if there is any book of tables in vogue, stating the number of amperes a wire will carry with a certain amount of heat, and how much increase in heat for each additional say 10 amperes. For example, say the temperature of the work room be 65° and a wire, No. 16, be raised 5° over this, with 10 amperes passing through, how much more will it be raised with 20 amperes passing through it? Will it be directly proportional? What is the safe limit to allow German silver or iron wire to heat up to in a resistance box, say such as a field shunt? i.e., what guage, No. of amperes, assuming box to be freely ventilated, and wire simply in spirals?"

ANSWER.—We do not at this moment know of any book of tables giving the temperature or efficients of wires heated by the passage of a current. Messrs. Houston & Kennelly, of Philadelphia, have conducted careful experiments to determine them, and, no doubt, are compiling a book of such data. Knowing the specific heating effect, the effect with any particular current on any known size of wire is easily calculated by the well-known $C^2R$ rule. Your example can be worked out in the same way—thus twenty amperes will produce four times as much heat as ten amperes will, on the same wire. With wires of different sizes, and different lengths, the calculation is equally simple with the aid of a table giving the specific resistance or the circular milage of wires, and knowing previously the heating effect on one sample piece. Thus it is seen that the heating effect is not directly proportional to the increase in current, but it is proportional to the square of the current increase or decrease, and directly proportional to the resistances. The safe limit of heat for rheostat wires is a matter that depends on ventilation, as much as anything else. A current that would heat a wire to a red heat in a confined space will be perfectly safe when cool air circulates around it freely. Knowing the current that the rheostat will want to shunt (maximum), it will be safe to allow about one square inch of radiating (cooling) surface of rheostat for every ten watts absorbed by it."

---

"J. B." writes from a Western Ontario town, as follows: "I am in charge of a Wheelock engine, but have never seen the valves of this engine nor even a cut of the valves. I have seen instructions in mechanical papers for setting valves of other engines, but never a Wheelock. Could you tell me where I could get the information I want?"

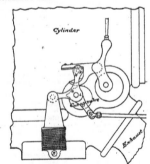

VALVE OF WHEELOCK ENGINE--SHOWING VALVE MOTION AT END OF CYLINDER.

ANSWER.—The accompanying cuts and diagram, for which we are indebted to the Goldie & McCulloch Co., of Galt, Ont., the Canadian manufacturers of the Wheelock engine, will doubtless enable our correspondent to understand the valve mechanism. The diagram shows the dashpot and a portion of the valve motion for one end of cylinder. On the arm of cut-off valve will be noticed the letters R and L. When the valve is at rest and the hook detached, the line R should be perfectly perpendicular ; the dashpot will be down and the spring closed. Should the line not be perpendicular, the stud B has an eccentric pin on the end and the dashpot can be raised or lowered by loosening the set screw and turning stud. Sometimes this is necessary owing to the leather under dashpot getting worn thin. With the crank of engine on dead centre and the hook attached so as to hold the cut-off valve open, the line L will be perfectly perpendicular. With this correct, the engine will have the proper lead, for the line L shews the lead line. If this is not correct, it can be changed by moving the eccentric on shaft. After taking a diagram, should the engineer find more load on one end of cylinder than the other, all that is necessary to do is to shorten or lengthen the rod between the two trips, as the case may require, it having a right-hand thread at one end and a left-hand thread at the other ; one of the trips is shown on diagram at C. The rod shown on diagram is not the rod referred to, but extends back from the trip C to the valve at back end of cylinder, the diagram being taken from front or frame end of cylinder. The rod shown is from the valve motion to the governor and this should not be changed by the engineer, as it is always set before leaving the shop ; in fact, none of the rods should be changed except the one mentioned between the two trips. The valves are always carefully set before the engine is shipped, but the points mentioned may become necessary by the valve motion getting worn. No changes should be made unless the person making them is thoroughly conversant with all parts of the engine.

"A CONSTANT READER," Whitby, Ont., writes : "Please answer me the following questions in your next issue of the ELECTRICAL NEWS : " Describe a pump, an injector, a boiler, a steam engine ; also how is it, when lamps are connected in series, in railway circuits, the highest voltage lamp always gives the brightest light ? "

ANSWER.—Proper and thoroughly comprehensive answers to your questions would involve writing a treatise on steam machinery which would somewhat exceed the limits of one of our ordinary issues—but, assuming that you have at least an elementary knowledge of physical and mechanical principles—the following answers may satisfy. (a) A pump (assuming it to be a water pump) is an apparatus for attaining two objects, viz. : Either for raising water from a lower to a higher level against the force of gravity ; or for forcing water into some receptacle against some counter-acting force. (b) An injector is an apparatus for forcing water into a receptacle against a counteracting force. It is used in connection with a steam plant to force feed water into the boiler against the pressure of steam in that boiler. In so far, it serves the same purpose as a "feed pump," but its method of doing so is different. (c) A boiler is an apparatus in which is made the steam required for use in a steam engine. It is so constructed that the steam raised from the heated water is not permitted to escape in the atmosphere but is imprisoned within it until required for use. (d) A steam engine is a device for enabling man to avail himself of the enormous power of steam under pressure. The above definitions are all that can be given in such a short space. If they be not sufficient then there is an immense technical literature on the subject, which will probably give a better idea as to why. When lamps are

connected in series, the highest voltage lamp always gives the brightest light. It is a phenomenon we have not observed ourselves, and, therefore, cannot say. There is probably some difference in the lamps themselves. We should like to hear a little more about it.

## SPARKS.

The Ottawa Electric Railway have adopted the fare box system of collecting fares.

The project to build an electric railway from Vancouver to Port Langley, B. C., is again being revived.

The Canadian General Electric Co. have been granted a franchise for electric lighting at Tavistock, Ont.

Messrs. W. H. and E. C. Breithaupt have purchased a controlling interest in the Berlin and Waterloo electric street railway.

In the city of Ottawa there are 13.14 miles of electric railway, composed of 3.94 miles of single and 9.78 miles of double track.

Mr. T. Viau, the promoter of the electric railway between Hull and Aylmer, Que., has disposed of his franchise to an Ottawa syndicate.

The Crystal Beach Improvement Co., of Ridgeway, Ont., propose constructing two miles of electric railway, extending from Crystal Beach to Ridgeway.

A company is being formed at Hamilton, Ont., to open a summer resort at Chedike Park, and to build a double track electric railway along Herkimer and Queen streets.

The car stables of the Oshawa Electric Railway Co. were recently destroyed by fire. In the sheds were two open summer cars and one winter car, which were also burned. The loss to the company is placed at $10,000.

Incorporation will be asked at the next session of the Ontario Legislature for the Manitoulin & Pacific Railway Co., with power to construct a steam or electric railway across Manitoulin Island. The solicitors for the company are Messrs. Clark, Bowes, Hilton & Swabey, of Toronto.

Messrs. H. A. Beatty and J. W. Horn, of Toronto, representing a syndicate of capitalists, propose to construct an electric railway in the town of Chatham, Ont. The prospects are that a railway will be constructed embracing the principal towns and townships within a radius of thirty miles.

The ratepayers of the village of Lanark will vote on a by-law to bonus the Lanark County Electric Railway to the extent of $10,000, to build a road from Perth to Lanark. The members of the company are J. B. Reilly, Alex Wender, Thos. Henry, A. H. Edwards and James Fowler.

The Hamilton Radial Electric Railway Company will make application to the Ontario Legislature for an act extending the time for the completion of their road, and authorizing the extension of one of their branches from Mimico to the city of Toronto, and another from Brantford to Woodstock.

The contract for the construction of an electric railway for the town of Cornwall, Ont., has been awarded to Messrs. Hooper, of New York, and Starr, of Montreal. The contractors expect to have the road completed by the 1st of June. The franchise was held by Mr. W. R. Hitchcock, electrician, of Cornwall.

Negotiations for electrifying the St. Thomas street railway are said to have been abandoned. The reason assigned is that owing to delay the company was unable to secure the financial assistance expected, and cannot proceed unless the city guarantees its bonds. An offer to sell the franchise at a low figure is now made.

The trustees of the Manitoba Electric and Gas Light Co., having made default in calling a meeting of debenture holders after the necessary notice had been given, certain holders, representing one-fifth in value of the debentures have given notice that a meeting will be held at the office of R. A. McLean & Co., London, Eng., on the 28th inst., to consider the appointment, if considered advisable, of a successor to Mr. Duncan McArthur, one of the trustees.

We have received a copy of a special souvenir number of the Providence, R. I., "Telegram" containing, among other features of interest, a series of illustrations showing the growth of the Eugene F. Phillips Electrical Works, together with portraits of the founder and present officers of the company. There is an illustration also of the branch works in Montreal, but we were disappointed at not seeing the portrait of the enterprising manager of this branch Mr. John Carroll.

## BY THE WAY.

MR. John Langton, the well-known electrical consulting engineer, of Toronto, has recently been acting in that capacity in connection with several electrical enterprises in the United States, and is considering the question of opening a branch office in New York city. I took advantage of the opportunity afforded by his recent visit to Toronto to submit to him a few questions regarding the directions in which the greatest development is taking place and is likely to take place in the applications of electricity. Seeing that a commencement has been made in Canada in the direction of transmission of electricity for power over considerable distances, I enquired what, in his opinion, would be the future developments along this line. His reply was, that he believed there would be a considerable development in power transmission schemes over distances of from 5 to 15 miles, a less number over distances of from 15 to 25 miles, and very few over longer distances than 25 miles. He does not anticipate as great development in the direction of the application of electricity to railway purposes, as many persons look for. One of the most promising fields he believes to be in the manufacturing world, in connection with the increased use of electric motors and a greater number of private lighting plants. Turning to the subject of the cost of electric light and power, Mr. Langton stated that prices are very much lower in Canada than in the United States ; indeed he found, by comparison, that an estimate given him recently for a constant supply of current for power purposes in New York, was almost exactly double the price given him in Toronto for an intermittent service. Of course the value of real estate and the consequent expensiveness of doing business in New York city accounts to some extent, for this difference in price ; but allowing for this, the fact remains that prices in the United States are not cut to nearly so fine a point as in Canada, and it is difficult to see any reason for the unprofitable rates which prevail in this country. As to the result of the introduction of acetylene gas, Mr. Langton has been informed by a gentleman, said to be well qualified to speak with authority on the subject, that acetylene gas has for some time been manufactured and used in Switzerland, without regard apparently to the exclusive patent rights to which Mr. T. L. Wilson, the alleged inventor, lays claim. The surprising thing is, if this gentleman's statement be correct, that Mr. Wilson should have succeeded in obtaining from the gas companies of the United States such large amounts in cash for territorial rights to the use of his discovery.

x x x x

A CORRESPONDENT of the Hardware Merchant reports the following interview at Dunnville, Ont., which goes to show that even bright lights of the church are sometimes not above attempting to shine in "borrowed" light at the expense of the electric light company :—
"Where's the boss?" I asked as I strolled into J. H. Rowe's store, Dunnville.  "Up at the Baptist church," was the reply.  And feeling that as the mountain was unlikely to come to Mahomet, Mahomet would have to repair to the mountain, I wended my way to church (?) After greeting me, and in reply to my query, "What are you doing?" he said : "I think I am one too many for the Electric Light Co.  I am interested considerably in church work, and, wanting to light up the basement of our church with the incandescent light, I asked the company if they would make any reduction in price

charged for arc light in church.  Being answered in the negative, I racked my brain for a scheme to get ahead of them, and finally struck upon the plan of making a hole through the floor.  And on Wednesday and Friday nights the arc light above can be lowered into the basement, and now we will get three nights' light for 50c. per week instead of one."

x x x x

IN the pioneer days of telegraphy in Canada an Irishman, whose son had gone to the North-west, wanted to send his boy a pair of new boots, and conceived the idea that the quickest method of delivery he could adopt would be the telegraph.  Somebody had told him that communications could be sent very quickly that way, and he didn't see why a pair of new boots shouldn't go in the same manner.  He wasn't quite sure how to go about it but concluded that the proper way would be to hang the boots on the wire, which he did.  Soon after, a tramp passing the spot, caught sight of the boots slung across the wire, climbed the pole and appropriated them, hanging up his old ones in their place.  By and bye the Irishman returned, and seeing the old boots, exclaimed : "Bedad, Jimmy's sint back his old boots t'let me know he got the new wuns."

x x x x

MR. John Carroll, the well-known representative of the Eugene F. Phillips Electrical Works, Montreal, is repeating with much relish a story of which he was recently made the recipient at Lancaster, Ont.  A farmer living on the outskirts of that town was recently boasting to his urban neighbors that his house was entirely lighted with electricity.  The announcement was received with incredulity by the townspeople, who wouldn't believe that the agricultural population had so suddenly decided to put on airs and add to their expense account.  They finally decided to accept the farmer's invitation to visit his house and see for themselves. On their arrival at the farm house this is what they saw : a single incandescent lamp suspended from the ceiling of the dining room, with sufficient cord attached to enable the farmer and his family to transport the light to any part of the premises.  The difference in cost between an incandescent lamp and a lantern, is all the farmer's progressiveness cost him, and this, no doubt, is offset by a reduction in his insurance premium consequent upon the lessened fire hazard.

---

## SPARKS.

The Leamington Electric Light Co. are installing a 60 kilowatt Canadian General single-phase alternator.

The number of passengers carried over the Galt, Preston and Hespeler Street Railway in 1895 was 175,000.

Mr. Beemer, the promoter of the electric railway at Quebec, is said to have made the necessary arrangements for carrying out the work.

The Montreal Street Railway Company recently placed an increase order for 40 C.G.E. 800 and 20 C.G.E. 1,200 railway motors with the Canadian General Electric Company.

Mr. Isaac McKay, of St. Thomas, who was engineer of construction of the London electric railway, has accepted an offer to superintend the construction of a new street railway in Cleveland.

The electric light plant at Alexandria, Ont., is now in operation. The power house, situated about two miles from Alexandria, is a substantial stone building, and is equipped with a 60 K. W. 2,000 volt alternator, driven by a Robb-Armstrong engine.  The switchboard is of marble, and the whole plant is a very substantial and practical piece of work, and reflects credit upon the contractors, Messrs. Ahearn & Soper, Ottawa.

## WATER TUBE BOILERS.*

### By W. T. BONNER.

WHEN your worthy secretary called upon me for a paper on water tube boilers, I little realized the difficulty attending the work, for the subject has already been so fully and so ably discussed in the technical journals, and even in the ordinary trade catalogues, that I fear my humble contribution to the proceedings of this Society will contain little that is new or interesting. However, hoping that I may at least be fortunate enough to glean from fields which possibly some of you have passed over, I beg your indulgence and attention to certain facts, which we of the water tube persuasion believe to be proof positive of the correctness of our system.

### OLD AND NEW.

Not at all unfrequently are the promoters of water tube boilers called upon to furnish evidence of the extent to which such boilers are, and have been used. The prevailing idea in the minds of many steam users appears to be that of mistrust in the principle and effect of water tube boilers. It is not what their fathers used, neither does their local boiler maker approve of them, a negative premise naturally calling for a negative conclusion.

Why are not water tube boilers in more general use? Because, as was explained in a discussion of the subject by the American Society of Mechanical Engineers, they require a high class of engineering to make them successful. The plain cylinder is an easy thing to make. It requires little skill to rivet sheets into a cylinder, build a fire under it, and call it a boiler; and because it is easy and anyone can make such a boiler, because it requires no special engineering, they have been made, and are still made, to a very large extent. The water tube boiler, on the other hand, requires much more skill in order to make it successful, a fact proven by the great number of failures in that line.

Water tube boilers are not new. From the earliest days there have been those who recognized their advantages, and in modern practice to refuse them equal consideration with the best known mechanical appliances of other types, is only pardonable on the ground of ignorance or injustice.

I was greatly amused recently to find in a so-called engineering journal, the following item of news :—

"At Davenport, Ia., the old battery of four boilers at the Arsenal is being replaced by two boilers of novel construction in that region. The new boilers are 200 h.p. each, and instead of the heat passing through tubes surrounded by water, as in the ordinary boiler, the process is reversed, and the water in pipes passes through a current of hot air, thus giving a greater heating surface and insuring the greatest safety."

Plainly these are nothing more or less than our ordinary water tube boilers, and it is quite evident that the author of that item gauges the progress of this world by the developments on the little rock island in the Mississippi, occupied by the United States Arsenal.

Contrast with this another item of news in the Youths Companion, to which my nine-year old boy called my attention only a few days ago. It read as follows :—

"An interesting discovery has recently been made in the Museum at Naples, where the works of art and utensils found in the buried city of Pompeii are preserved. Careful inspection of one of the ancient copper vase-shaped vessels there, has shown that it is in reality a tubular boiler. That this form of boiler should have been known to the Romans two thousand years ago is somewhat remarkable. For just what purpose it was used is not known, but the boiler is well-constructed and contains five tubes running around a central fire-box, and so arranged as to permit the water surrounding the fire-box to circulate through them in a continuous current. The soldering of the tubes was so skilfully done that it remains intact to-day, and the cover of the boiler closes hermetically. The entire height of the machine, which, as remarked above, is shaped like a vase with two side handles and three feet, is only about 17 inches. It has been suggested that it may have been employed for distilling purposes. However that may be, its preservation under the ashes of Vesuvius proves that tubular boilers are not altogether a product of modern invention."

No doubt you have all read Lord Lytton's account of the last days of Pompeii, and recall his description of the wonderful therme or baths, which formed so prominent a feature of every Roman city during the first century. Possibly this ancient boiler was designed by one of those bright Roman or Grecian mechanics

for heating the water for the Sudatoreum or warm baths. We find it duplicated almost exactly in the Galloway water tubes of the present day, and I have no doubt if we could follow up this investigation of ancient boilers, we would find the knowledge possessed by the ancient Greeks and Romans was not confined to Poetry, Sculpture and Art, but that even water tube boilers and heaters were known to them.

The principle of the Galloway tube originated at the time when probably the first steam boiler made in this world was constructed. It is not known when the first steam boiler was constructed, but the first steam boiler recorded was made at least 200 years before the year one of our era.

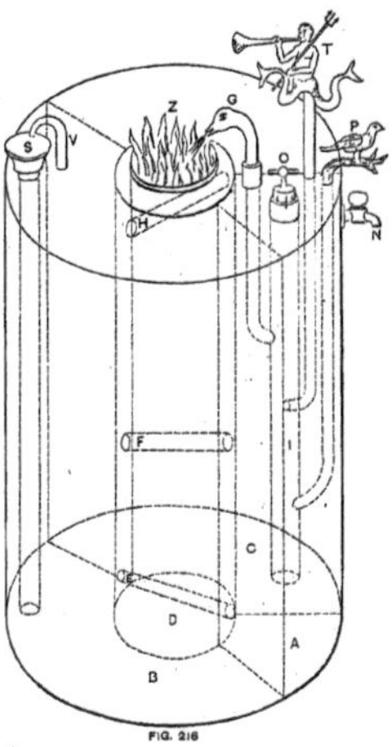

In a discussion of various forms of shell and water tube boilers at the New York meeting of the American Society of Mechanical Engineers in 1885, Mr. W. F. Durfee gives an illustration of this very unique boiler, copied from the first Latin translation of the Pneumatics of Hero of Alexandria, who lived and wrote about 200 B.C.

Its construction is shown in Figs. 216 and 217. The figure is copied from the Latin translation referred to, and represents a perspective elevation of the boiler and its appendages, showing its internal construction by dotted lines. The second figure (217) was drawn by Mr. Durfee to facilitate explanation; it shows a horizontal section of Fig. 216 taken just below the top.

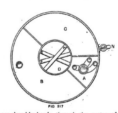

FIG. 216

FIG 217

The apparatus consists of a vertical cylindrical shell, whose ends are closed by heads, through the centre of which passes a vertical cylindrical flue, D, whose upper end is provided with grates for the support of the fire, Z, the hot gases from which passed downward through the flue. The space between the flue and shell is divided by diaphragms into three unequal compartments, A, B, C, in the first of which steam is generated, the others being simply reservoirs of hot water. The central flue, D, is crossed by three cylindrical tubes, H, F, E, the tubes, H, F, connecting the hot water spaces, B, C, act in the same way as the Galloway tubes, now in common use, but the bottom tube is closed at the end, E, its opposite end opening into the smallest or steam compartment, A. The compartment, B, is provided with a funnel, S, whose tube extends nearly to the bottom of the boiler; and also with a safety tube, V, whose curved upper end is immediately above the funnel, S. The compartment, C, has a cock, N, from which the water is drawn. The compartment, A, has within it a three-way cock, I, the three discharge pipes of which are connected with the goose-neck blow pipe, G, the triton, T, and the singing-bird, P, respectively. The three-way cock, I, is operated by a cross-handle, O, and the upper end of its plug has graduations which, when brought opposite an index mark on the shell

---

* Paper read before the Mining Association of Quebec.

of the cock, determine which of the three discharge pipes shall receive the steam generated in compartment A.

The principal function of this apparatus was to furnish hot water, and it is so contrived that it is impossible to draw any considerable amount of hot water from the cock N, without putting in an equal quantity of cold in the funnel S. In order to put this apparatus at work, the compartments B and C were filled with water to a level above the upper water tube H, by means of the funnel S. The goose-neck G was then removed and water poured into the compartment A, sufficient to fill it nearly to the lower end of the three-way cock I. The fire was then lighted, and as soon as steam manifested itself, the goose-neck G was returned to its socket and placed in such a position that the fire Z was blown by the issuing steam. The three-way cock I could be turned by its handle O, so that the steam would cause the triton T to sound his trumpet, or the bird P to warble, and thus announce to interested parties that the water was "boiling hot." In case any steam generated in the compartments B and C, found an exit through the safety-pipe V, and any entrained water re-entered the boiler through the funnel S. In case it was desired to draw hot water in any great quantity from the cock N, it was necessary to supply an equal quantity of cold water through the funnel S, this requirement insuring a constant volume of water in the boiler.

But I need not weary you with ancient history. It may satisfy our curiosity and lend some additional color to Solomon's proverb that "There is nothing new under the sun," yet we cannot expect ancient Greece or Rome to furnish models for our boiler makers of to-day. Only by comparison do we really begin to appreciate the vast changes by which the engineering talent of to-day is taxed to its utmost to produce machinery and appliances which will accomplish the greatest amount of work for the longest period, with the least expenditure of effort. Steam boilers perhaps have not attained that degree of perfection usually accorded to the steam engine, yet when we note the progress which has really been made, and realize how close we have approached to the theoretically perfect boiler, we have great cause to feel encouraged.

Of the two hundred and sixty odd boilers recorded in Mr. Bell's most valuable Directory of Canadian Mining Industries, 30% or 5,400 h.p. are of the water tube type, and 50% or 9,000 h.p. are shell boilers, leaving 20% or 3,600 h.p. unclassified.

Since practically all of the above water tube boilers have been installed within the past ten years, we can safely infer that in the mining trade at least, more horse-power of water tube boilers are now sold each year, than all the other types combined.

There is no better evidence of the survival of the fittest in modern boiler practice, than a comparison of the various types exhibited at the Centennial Exhibition of 1876, with those shown at the World's Fair, 17 years later. At the Centennial there were exhibited 14 different types of boilers, of which two were cast iron sectional, four were shell or tubular boilers, two were shell boilers with water tubes crossing internal fire tubes, while seven were exclusively water tube boilers. Of the whole number exhibited at the Centennial, but one, the Babcock & Wilcox, reappeared in its original form at the World's Fair in 1893. Of the fifty-two boilers exhibited in the main boiler room at the World's Fair, all were of the water tube type, while thirty-one of them were distinct copies of the original boiler patented by Stephen Wilcox in 1856, just forty years ago.

THE PERFECT BOILER.

What really constitutes a perfect boiler? Mr. George H. Babcock in his life-time undertook to formulate the twelve fundamental principles upon which it should be built. It was about twenty years ago that his formulas were first published, yet those same principles still live, and are looked upon to-day as the acme of scientific boiler construction. I need not repeat them here; they have long occupied a prominent page in the Babcock & Wilcox Co.'s book "Steam." But rarely do we find so much truth in so few words.

Few boilers there are entirely devoid of all good talking points, but do not be satisfied with a boiler simply because it is made of good materials and workmanship, or because it has a mud drum, or because it has large water and steam capacity, or because it has a large disengaging surface, or because it has a good circulation, or because it is built in sections and is therefore safe in the event of explosion, or because it is able to withstand high pressure and unequal expansion, and has its joints protected from the fire, or because the furnace is provided with chambers for the proper combustion of the gases, or because the heating surface is composed of thin metal so arranged that the hot gases will cross it at right angles, and only leave it when the greatest possible

heat is extracted from them, or because it will work up to or over its full rated capacity with the highest economy, or because it is fitted with the best quality gauges and fittings. Each of these qualities add greatly to the value of a steam boiler, but that one is best which combines the greatest number of such qualities, and therefore proves the best investment independent of first cost.

Messrs. Galloways, Ltd., of Manchester, Eng., illustrate on page 94 of their late catalogue, what they are pleased to designate as their "Manchester Boiler," but which is in reality a reproduction of the ordinary inclined water tube boiler, built by so many manufacturers of to-day. In explanation of this marked deviation from the Galloway, Lancashire, and Cornish boilers which they have been building for so many years, Messrs. Galloways, Ltd., say:—

"For ordinary pressures the Galloways boiler possesses great advantages, but beyond that, cylindrical boilers are frequently of large diameter, necessitating extremely heavy plates, and although for marine practice this is carried out, yet for situations where the conditions are less rigid, it is advisable to have a boiler more suited to the requirements of the case.

"In addition to this, where transport of large pieces is difficult, the Manchester boiler offers considerable advantages, as the largest piece is the upper vessel, which rarely exceeds five feet in diameter, twenty feet in length, and four tons in weight, the tube rods and boxes being separate. It will be seen that all the tubes are inserted into one water box or chamber at each end, the front one connected to the upper vessel by a wide neck, and the back chamber by a large circular connection, by which means an even circulation is kept up. The boiler is further provided with an internal arrangement in the upper vessel for separating the steam from the water, thus preventing priming and its attendant evils. This arrangement of boiler has been largely adopted on the Continent, and we anticipate that when its merits become known, it will be received with great favor by steam users requiring boilers for high pressure."

That is good; coming from such an eminent authority, we can only interpret their adoption of the water tube principle as a strong endorsement of the work accomplished by their predecessors in that field of engineering. I fully expect, however, in the next issue of their catalogue, Messrs. Galloways will have overcome their prejudices sufficiently to limit the diameter of their drums to 36" or 42", and that they will further arrange to enclose the drum so as to utilize its surface for heating rather than condensing. Then they may add to the merits of their boiler, safety and economy.

I might add that although Messrs. Galloways are pleased to limit the use of their water tube boilers to stationary work, the boilers of that type are just now making tremendous strides in the race for supremacy in marine practice.

As an example in proof of this statement I might refer to the steamers Turret Cape and Turret Crown, which have just closed a very successful season in the coal carrying trade between Sydney and Montreal. From their lessees the Dominion Coal Co., I learn that the two steamers have a combined record of 27 trips, extending over a period of 44 weeks, during which time they brought 66,981 tons of coal into this port. To this total should be added 11,700 tons for short cargoes, made necessary by the very low water in the river and canal, which difficulty prevailed through all of last season. Had there been a sufficient depth of water, both steamers could just have easily have brought in a full cargo each trip.

The actual carrying capacity of each of the Turret boats is 3,000 tons. They are fitted with water tube marine boilers, 2,200 square feet of heating surface being the total for each boat. They have been kept in constant service right through the season, and the captain's log shows a clean record for the boilers.

Many other and larger steamers fitted with water tube boilers have gone into commission during the past few months, and in every case the boilers have given the greatest satisfaction.

CAPACITY.

The term "Horse Power" is one which admits of a wide interpretation, being little understood by some and often misapplied by others. Originally used as a unit of capacity by James Watt, and supposed to be the average amount of work performed by a good strong English cart horse, its value is 33,000 lbs. raised one foot high per minute. It may be expressed in any equivalent of this unit as one pound, raised 33,000 feet high per minute. At best this is but an arbitrary unit, since the actual value of a horse-power depends, as a Yankee boiler maker has very aptly expressed it, upon the size of the horse. The evolution of the term "horse-power" as applied to steam boilers, has been gradual but not the less marked.

Prior to the advent of compound and triple-expansion engines, it was always customary to calculate the steam consumption of the ordinary slide valve engines then in most common use, at the rate of one cubic foot of water per hour, or say 62½ ℔s. For instance, a 10 h.p. engine would require a boiler capable of evaporating 625 ℔s. of water per hour. In general practice it was found boilers of different types of construction varied in evaporative capacity according to the efficiency of their total heating surface, the amount required per h.p. averaging about as follows:—For plain cylinder boilers, 10 square feet ; for large flue boilers, 12 square feet; for horizontal and multitubular boilers, 15 square feet.

Of late years tremendous strides have been made in the development of the steam engine, so that instead of one cubic foot of water, or 62½ ℔s. steam consumption per h.p. per hour, the modern engine builder knows that he must develop a horse-power with less than 30 ℔s. of steam for simple non-condensing engines, and from that down to 13 lbs. or less for triple expansion condensing engines, depending upon the size of plant and number of cylinder expansions.

Here then arises a serious complication in the determination of horse-power. Shall it be a large or a small horse? The prospective purchaser should consider this matter carefully, and demand that all tenders must state specifically the actual evaporative capacity of boilers to be purchased, to be determined if necessary by a practical test. The American Society of Mechanical Engineers has very properly solved this problem by the favorable consideration of its Special Committee's report at the New York meeting in 1885, whereby the equivalent evaporation of 30 lbs. of water from a temperature of 100° Fah. into steam at 70 lbs. pressure, is fixed as a boiler horse-power.

American manufacturers generally have adopted this standard, and while they may differ in the number of square feet of heating surface they allow for developing a horse-power, there is no longer any doubt as to the size of the horse.

I cannot leave the subject of horse-power capacity without first making a strong appeal for a more uniform rating of boilers, a rating which has some tangible basis. Not until you are able to compare boilers by the actual number of square feet of effective heating surface they contain, or the actual number of pounds of water they will evaporate under ordinary working conditions, can you judge whether one boiler is cheaper than another.

I confess I was greatly shocked only a few days ago, to hear the admission of a fire-tube boiler man, that he only figured the upper half of his tubes as effective heating surface. I shall always remember him as an honest man of good sense. There is no question but that fire tubes and shell plates exposed to the direct action of hot gases, form very efficient heating surfaces when they are clean, but who is there who will claim the possibility of keeping such surface constantly clean while the boiler is in active service.

Effective heating surface is that which receives the direct contact of the hot flame or gases and continues to do so without interruption from soot, or interference by close furnace walls, or baffle plates. This is the proper basis upon which to purchase your boiler, other conditions of course being equal.

### SAFETY.

I have been asked why a water-tube boiler is necessarily a safety boiler. It is not necessarily a safety boiler; in fact I could name a number of water-tube boilers which are safe in name only. Certainly a boiler with very wide flat stayed surfaces, enclosing chambers receiving the combined circulation of all the tubes, should not be considered a safety boiler. Stay bolts and braces at best are a constant menace to safety, since they are usually located in inaccessible places, difficult to inspect and repair. But the principal objection appears to be the impossibility of providing braces which brace at the proper moment. How is it possible to assemble a number of pieces of metal, all of different sizes and shapes, and subject to greatly varying temperatures, and expect them to expand, contract, and remain uniformly tight at all times? But it is to be regretted that in defending the principle of water-tube boilers, there are other weaknesses to apologize for than braces and stays. There are those with tubes closed at one or both ends, the aggregation of pipe and fittings, and the bent tube monstrosities, so aptly described in a recent publication called "Facts," all more or less dangerous because they cannot be cleaned.

That a boiler can be made so as to be practically safe from explosion is a demonstrated fact, of which no one at all acquainted with modern engineering has any doubt. Of this class of boilers the Babcock & Wilcox is a pre-eminent example from the length

of time which it has been upon the market, and the large number which have been for years in use under all sorts of circumstances and conditions, and under all kinds of management, without a single instance of disastrous explosion.

The Babcock & Wilcox water tube boiler has all the elements of safety, in connection with its other characteristics of economy, durability, accessability, etc. Being composed of wrought iron tubes and a drum of comparatively small diameter, it has a great excess of strength over any pressure which it is desirable to use. As the rapid circulation of the water insures equal temperature in all parts, the strains due to unequal expansion cannot occur to deteriorate its strength. The construction of the boiler, moreover, is such that, should unequal expansion occur under extraordinary circumstances, no objectionable strain can be caused thereby, ample elasticity being provided for that purpose in the method of construction.

In this boiler, so powerful is the circulation, that as long as there is sufficient water to about half fill the tubes, a rapid current flows through the whole boiler, but if the tubes should finally get almost empty, the circulation then ceases, and the boiler might burn and give out. By that time, however, it is so nearly empty as to be incapable of harm if ruptured.

Its successful record of over twenty-five years proves that by the application of correct principles, the use of proper care and good material in construction, a boiler can be made so as to be in fact, as well as in name, a "safety boiler."

## CANADIAN ELECTRICAL ASSOCIATION.

The Executive Committee of the above Association met in Toronto on the 16th of January to discuss preliminaries in connection with the arrangements for the convention which is to take place in that city in June next. Several sub-committees were appointed to further these arrangements, and report to the Executive at another meeting to be held shortly.

Subjects upon which it would be desirable to have papers were considered, and a selection made. An invitation has been extended to a number of qualified persons to furnish papers on these selected subjects, and from the majority the Secretary has already received favorable responses.

Judging by these favorable initiatory proceedings, and taking into consideration the fact that the convention is to be held in the Queen City of the West, in the locality of which a large proportion of the members of the Association reside, and at the most favorable season of the year, there is reason to anticipate that great success will crown the event.

In the interim before the convention, every person connected with or interested in the electrical interests of Canada, who is not already a member of the Canadian Electrical Association, should connect himself with the organization. This especially applies to persons connected with the electric lighting industry, for reasons which have recently been mentioned in these columns.

## TRADE NOTES.

Attention is called to the advertisement of Messrs. Ahearn & Soper, in another column, offering for sale second-hand machinery.

The Royal Electric Co. have just completed the installation of a 30 k.w. "S.K.C." two-phase generator, for the Glenwilliams Electric Light Co., at Georgetown, Ont.

The Ottawa Carbon Co. advise us that they are turning out 20,000 carbons per day, and have orders in hand at present sufficient to keep their works employed until May next.

The Packard Electric Co., of St. Catharines, Ont., have opened an electrical repair department in connection with their works. They are making a specialty of re-winding street railway armatures and transformers, and already have several orders on hand.

John W. Skinner, of Mitchell, Ont., advises us that he sold during the month of January, seven electric motors and two lighting plants. Mr. Skinner has had a large experience, having installed, perhaps, as many plants of different systems as any man in the business.

The Power Rope and Belting Co., of St. Catharines, Ont., has been incorporated, to manufacture belting by a new process, under patents granted to Mr. H. Ellis. The capital stock is $20,000, and the promoters are : J. W. Coy, Harry Ellis, H. Flummerfelt, F. Coy, all of St. Catharines, and L. Raymond, of Welland.

Mr. G. A. Powell, assistant manager of the Packard Electric Co., St. Catharines, Ont., informs us that his Company were appointed in October last, Canadian agents for the Bryant Electric Co., of Bridgeport, Conn., manufacturers of electrical specialties, and the R. Thomas & Sons Co., of East Liverpool, Ohio, porcelain manufacturers

# ELECTRIC RAILWAY DEPARTMENT.

### BRANTFORD ELECTRIC RAILWAY.

This road is principally owned and operated by the Canadian General Electric Co., and has been in operation as an electric road since 31st of March, 1893. The officers are well-known financial railroad and electric men, viz.: Pres., Frederic Nicholls ; Vice-Pres., H. P. Dwight ; Sec'y.-Treas., W. S. Andrews ; Board of

BRANTFORD ELECTRIC RAILWAY—MOHAWK PARK.

Directors, Robt. Jaffray, Hugh Ryan, Geo. A. Cox, W. R. Brock and Thos. Long. The road is under the management of Mr. Jas. T. Madden who has been in the railroad business for a number of years having held a responsible position with the C. P. R. in connection with the construction of their line between Sudbury and Sault Ste. Marie. Mr. Madden is ably assisted in the management by his accountant, Mr. Jno. Murrode.

The Indian City, with the Grand River bending round it, the picturesque scenery on its banks, and Mohawk Park, through which it flows, delights even the traveller who has climbed mountains, travelled through rocky Muskoka or languished in the salty breezes of the seaside, to come and spend a few months in this pretty town with its hospitable people.

Much of Brantford's popularity as a summer resort is due to the Brantford Street Railway Co. Last spring they purchased Mohawk Park—on which they have already spent $12,000—one of the most delightful spots in Canada. It comprises 42 acres, in which is an artificial lake. They cleaned the park up, built a quarter-mile bicycle board track, sodded the centre and erected a grand-stand capable of seating 1,500 persons, and a bleacher of 750 seating capacity. The track is surrounded with trees. Ten arc lights light up the track at night where large crowds assemble in the evening to witness the bicycle races. Fifteen arc lights are scat-

tered through the woods and on the edge of the lake. A large pavilion and casino were erected during the summer. A band plays there three nights a week, and a hungry or thirsty soul can eat or drink to the strains of Sousa's latest production, surrounded by 400 colored incandescent lights. The casino and pavilion thus illuminated present a very pleasing night effect from the lake.

The lake is a mile and a half in length by a mile in width supplied with water from the Grand River. The company propose to dam up the ravine in the park and allow the water to run through the ravine and over the dam. Behind the waterfall will be varied colored incandescent lights which at night will give a very pleasing effect. Bicycle boats and canoes are for hire, or you can step into your canoe at Oxford street bridge, and paddle twelve miles down the river, when you find yourself in Mohawk Park lake, only two miles from home, as the river is so crooked. You leave your canoe there, see the sights and take the street car back to the city. The Hamilton road leads from the city to Mohawk Park, and in one place there is a steep grade on which is a switch, but only one trolley wire, which is used for the car going up the grade on its way to the park. As the grade is steep the down car does not need a trolley wire on the down trip. As it shoots down the grade the lights go out, and this spot has come to be called "the tunnel." Mr. Madden has instructed his conductors to acquaint the passengers of its existence before they come to it, so that they may be prepared. The young people like "the tunnel."

The park was opened on the 24th of May, and on that date a railroad men's pic-nic came to town. The company collected on this inaugural day 18,000 fares. As a money maker no park can beat Mohawk Park.

On the eight and one-half miles of single track twelve cars run in the summer time, but in winter not so much service is required. The "T" rail is used exclusively,

BRANTFORD ELECTRIC RAILWAY—POWER STATION.

six miles weighing 30 lbs. and the balance 60 lbs. ; any new rails will be 60 lbs. in weight.

The centre of the system is at the corner of Colborne and Market streets from which all the lines radiate. The main line runs east on Colborne, with a belt line around the eastern wards, via Alfred, Nelson, Brock and Arthur streets and Park ave., back west on Colborne to Brant ave., then north to the Institute for the Blind. Three cars cover this route.

The second line runs from the G. T. R. depot down Market street to the Kerby House on Colborne street making connections with all G. T. R. trains. As the distance is so short only one car is necessary on this route.

The third line terminates in West Brantford at one end and in Eagle Place, a southern suburb of Brantford, at the other. It's course is on Oxford street in West

BRANTFORD ELECTRIC RAILWAY—MOHAWK PARK.

Brantford, to Colborne street, then to Market street, south on Market street to Core street, Core street to Cockshutt road, to Eagle Place. Two cars cover this route.

The fourth line (operated only in summer) with six cars runs east on Colborne street from Market street to the Hamilton road, thence to Mohawk Park.

It took a great deal of persuasion by the company to induce the Council to allow them to lay their track on Colborne street from Alfred to Brock, a distance of three blocks, saving them the journey round the belt line on their way to Mohawk Park. After much contention the Council gave in, and the company therefore save a half a mile by a direct route to the park.

THE POWER HOUSE.

In keeping with the general excellence of the road is the power house, situated on Colborne street, near the G. T. R. tracks. The soil is something akin to quicksand and the foundation is sunk to a great depth. The reason of building here was that a stream ran through the property which affords a supply of water for condensing. The power station is a two-storey brick edifice with the top storey floor on a level with the street. In the annex of 40x40 ft. are the boilers and fuel. The main part is 80x50 ft., and the top storey is used for the general offices, waiting rooms, car storage and repair shops. In the basement of ground floor are the repair pits and lavatories, taking up the front part, while the rest of the building, 62x50 ft., comprises the engine and dynamo room.

The company do a good incandescent lighting business. Besides their railway machines, they have three alternators. The railway generators are two No. 32, 300 h.p. C. G. E. machines, and the alternators are: one 1,000 light C. G. E. and two 750 light C. G. E. machines with exciter. A 25 light "Wood" arc machine supplies the company with light for private purposes. The machines run from a line of shafting

extending the full length of the room, having two Goldie & McCulloch clutch couplings attached. Power is supplied by two "Wheelock" condensing engines of 150 h.p. each. From the boiler room steam is supplied from two Waterous and one Doty boiler of 150 h.p. each. The fuel used is Reynoldsville slack. Mr. D. C. Thomas is the chief engineer and is a young mechanic of great promise. An engineer and fireman are on for each twelve hour's run.

The wires are run from the switch-board to the ceiling, thence up through the tower on the side to the poles outside.

The switch-board is of wood panels, and the instruments are on slate bases. Mr. J. Watts is the electrical superintendent and has charge of all the outside work.

Thirty men constitute the regular staff, with fifteen additional in the summer.

The company have just issued a successful report, and as Brantford grows, as she certainly will do, the Brantford Street Railway Co. will by and bye take rank with the largest and most prosperous electric railway concerns of the Dominion.

## ANNUAL MEETINGS.

### OTTAWA ELECTRIC RAILWAY COMPANY.

THE annual meeting of the above company was held on the 27th of January. The reports presented covered only a period of seven months, owing to the annual meeting, which was formerly held in June, having been changed to January.

The receipts from June 1st to December 31st, were $122,694.39 for car fares, and $5,479.59 from mail cars and other sources—a total of $128,173.98. The working expenses of the road were $73,983.48, leaving a net profit of $54,190.50. From this a two per cent. dividend was declared on Sept. 2nd, a 2 per cent. dividend on Dec. 2nd, and on January 9th a dividend of ⅓ of 2

BRANTFORD ELECTRIC RAILWAY—MOHAWK PARK.

per cent., leaving a balance of $16,166.50 to be applied to the profit and loss account. 2,843,173 passengers were carried during the seven months, and the wages paid out amounted to $45,671.43. The assets of the company are valued at $985,994.63, the profit and loss account amounting to $57,808.05.

The election of directors resulted as follows: J. W. McRea, President; W. Y. Soper, Vice-President; T. Ahearn, Managing Director; G. P. Brophy, W. Scott,

P. Whelan and T. Workman.   R. Quinn was appointed
Auditor.

TORONTO STREET RAILWAY COMPANY.

The Toronto Street Railway Company held their annual meeting a fortnight ago.  Mr. W. D. Matthews,
a director of the C.P.R., was appointed a director, and
at a subsequent meeting Mr. Jas. Ross, of Montreal, was elected vice-president to fill the vacancy on
the board caused by the resignation of Mr. H. A. Everett.  The directors were voted the sum of $20,000 for
their valuable services during the past year.

The annual statement submitted showed a net profit
of $301,310.30, as against a net profit of $250,695.18
for the previous year.  From the profits of this year
two dividends at the rate of 1¾ per cent. each have
been declared, amounting to $210,000, leaving, after
the deduction of an allowance for paving charges
amounting to $60,000, the sum of $31,310.30 to be
carried forward.

The company has in its treasury, bonds amounting to
$450,000 available for future use, notwithstanding the
large expenditures which have been necessitated for
rolling stock, car houses, etc.  In the past four years
the gross earnings have increased $172,712.31, while
the operating expenses have decreased $100,418.50.
The assets of the company are placed at $9,775,511.70.

The comparative statement for 1895 and 1894 is as
follows :—

|                                               | 1895.       | 1894.       |
|-----------------------------------------------|-------------|-------------|
| Gross earnings                                | $992,800.80 | $958,370.74 |
| Operating expenses                            | $489,914.76 | $517,707.53 |
| Net earnings                                  | $502,886.04 | $440,663.21 |
| Passengers carried                            | 23,353,228  | 22,609,338  |
| Transfers                                     | 7,257,572   | 7,438,171   |
| Percentage of operating expenses to earnings  | 49.3        | 54.0        |

During the year the company has built in its own
shops 30 open cars, and 20 closed cars, five of which
are 30 feet double truck cars, and six sweepers.

HAMILTON, GRIMSBY AND BEAMSVILLE ELECTRIC RAILWAY
COMPANY.

At the annual meeting of the shareholders of the
Hamilton, Grimsby and Beamsville Electric Railway
Company held on the 27th ultimo., much interest was
taken in the election of a directorate for the current
year.  There were two opposing forces, the Lester ticket
and the Myles ticket.  The former was successful, and
the new board is therefore composed of T. W. Lester,
president ; John Hoodless, vice-president ; C. J. Myles,
John A. Bruce, John Gage, W. Grives, and A. E.
Jarvis (Toronto).

The secretary-treasurer presented a financial statement, the accuracy of which was questioned by the
president.  After considerable discussion it was resolved
to consider the report at a meeting to be held on the 9th

of March.   Mr. Adam Rutherford was re-appointed
secretary-treasurer.

The new management, it is said, propose to take
steps at once to secure the extension of the road to
Beamsville, and to make other improvements.

## PORTABLE ELECTRIC DRILL.

THE Storey Motor and Tool Co., of Hamilton, Canada, and Philadelphia, Pa., some time ago put on the
market a compact and efficient portable drill, which we
illustrate herewith.  This machine is adapted for drilling pig iron and copper for test work, drilling rails, and
for various other kinds of work.  Owing to the type of
the motor, which is entirely enclosed, it is suitable not
only for indoor work, but can also be used for outside
purposes without requiring any specially arranged covering for fire protection.  The outfit complete consists
of motor and drill combined, together with regulating
rheostat for obtaining desired speed, and a drum with
100 ft. of flexible cord, all mounted on a truck, with or
without rack for holding material to be drilled, as desired.  These machines drill in sizes up to 1½ inches
in steel and 2 inches in cast iron, and are furnished with
both automatic and hand feeds.

The rapid adoption of electricity in machine shops and
factories makes a tool of this kind extremely useful, as

PORTABLE ELECTRIC DRILL, MANUFACTURED BY STOREY MOTOR AND TOOL CO.

it can be moved at will wherever it is needed.  These
drills can also be placed on a table or in any stationary
position, and will cover a large range of work of different classes.  As an illustration, two of these drills are
mounted on bed-plates, one at each end of a large callender roll, drilling two holes in flanges at the same time
and tapping them in the same operation, before the roll
is moved.  Another adaption of this drill is where it is
fitted with a telescoping shaft and is used in yards for
drilling holes in the construction of switches and crossings for street car and railroad work.

A different type of portable drill, combining all the
features of an up-to-date drilling and tapping. machine,
is being brought out by the above company, and will be
ready for the market in a few weeks.

## PERSONAL.

Mr. Thomas Ahearn, of Ahearn & Soper, Ottawc, and Mrs.
Ahearn, are at present making a tour of the world.

Mr. G. H. Campbell, manager of the Winnipeg Electric Street
Railway, was recently in Toronto and Montreal on a holiday trip.

Mr. W. A. Handcock, local manager of the Bell Telephone Co.
at Sherbrooke, Que., is receiving the congratulations of his friends
upon his recent marriage.

Mr. J. J. York, chief engineer of the Board of Trade building,
Montreal, and president of Montreal No. 1, C. A. S. E., was presented at Christmas by the employees in the building with a complimentary address, accompanied by a smoker's set, consisting of
two valuable meerschaum and briar pipes and a large box of
tobacco.

## WHY CENTRAL STATION MEN SHOULD ORGANIZE.

Editor CANADIAN ELECTRICAL NEWS.

SIR,—In your January number you publish an extract from a letter written by the manager of a company in an eastern Ontario town, to the manager of a company in a western Ontario town, and you derive from the remarks therein expressed, arguments why central station men should become members of the Canadian Electrical Association. In endorsing your remarks may I be permitted to carry the arguments a step further than editorial discretion prescribed as your limit? You describe the municipalities as the principal gainers by the want of organization among central station men, but it strikes me that the position is very different, and I am sure that a little careful reflection will convince anyone that it is the manufacturing companies who gain more—much more—than any one else by this want of combination and cordial co-operation, and who require to be watched much more than do the municipalities who are generally very mild offenders.

At the present moment the manufacturing companies maintain very curious relation with respect to the operating companies. They not only manufacture machines, which, of course, is their proper business, but they promote companies to do lighting business, then, as consulting engineers, they advise these companies as to what to purchase, and how to operate; and they endeavour to foster a kind of parent and offspring relationship with the view of opening and keeping a market for their own goods. Any attempt at independence of action on the part of the offspring is deprecated by the parent company, as tending to introduce an undesirable competition, and the manufacturing company also endeavors to guard its offspring against the bad men in the open market who would want to sell their goods by trying to constitute its agents the only means of communication between the operating company and the electrical world outside. These agents go around with their pack of goods, and while the customer is purchasing lamps, etc., they give him little scraps of news as to new apparatus, new installations, etc. A very large proportion of the smaller central station men seek for no better information on electrical matters than is dribbled out to them through the interested channel of a manufacturing company. They are satisfied to receive all their news, and any pointers they may require from the very man who is most interested in keeping them in the dark about the merits of any other apparatus than that which he himself sells.

Now, let any intelligent person consider for one moment what is inevitably the result of this. The central station man is interested in hearing of new or improved lamps, motors, etc., that have been brought out and by the use of which he can reduce his expenses, or extend his business. Is an agent likely to tell a customer of an improved type, made by a rival manufacturing company? Is he in the least likely to say that some rival sells a better lamp than he does himself? Is he not far more likely to religiously avoid mentioning any such thing? Can he be expected to recommend to, or bring to the notice of any customer, any piece of apparatus but that which he sells himself? Plainly, the purchaser, by not making independent enquiries, frequently fails to hear of something really to his advantage, because it is none of the agent's business to tell him.

A little reflection will show central station how little they regard their own interests when they allow themselves to be kept in leading strings by the manufacturing companies, instead of combining to study central station practice for themselves. A manager should keep his eyes wide open to see things, to do the very best possible with what he has got, and to promptly seize ahold of anything new that affords a means of reducing his expenses or extending his business. Now, any new labor-saving or more efficient piece of apparatus is patented and owned by only one company, and although it may be really the most valuable improve-

ment in the world, no other company is going to recommend its use if it will interfere with their own sales. On the whole, the central station man who expects a manufacturing company to give him really disinterested advice as to new or improved apparatus, is likely to be as badly left as he deserves to be. The enterprising man will hunt these things out for himself, by co-operating with his neighbors, to their mutual advantage. To illustrate: The storage battery has been proven to be of great value as a central station auxiliary. Has any Canadian manufacturing company ever recommended the installation of batteries? I do not think so. BECAUSE NO CANADIAN MANUFACTURING COMPANY MAKES A BATTERY THAT IS ANY GOOD. To recommend it would be to hurt their own business, which is to sell dynamos.

Again, plenty central stations using single phase alternating machinery, could work up a considerable day power business if they could get a good single phase alternating current motor. There is such a motor available, but I shall be very greatly surprised to hear that the agent of any of the Canadian manufacturing companies has mentioned the fact to any of their customers. Why? BECAUSE THEY DO NOT MAKE IT THEMSELVES, BUT HAVE DIFFERENT MACHINERY TO SELL, and it doesn't suit their business to post their customers too well on any good points in their rival's goods. A manufacturing company, if consulted, is going to advise the use of its own apparatus every time; and the demand for new and improved types must come from the central station man, who should use the most efficient, no matter who makes it, and find out for himself what is the latest and best.

As to operating central stations. Is there any manager who thinks he knows all about it? If so, why can't he let some other manager have the benefit of his knowledge? Perhaps he can get a few valuable hints in return. By all means let there be an organized body of central station men working together for their mutual good—telling each other what their experiences have been, and tackling their problems for themselves, instead of allowing themselves to be exploited by the manufacturing companies, who, in the words of a recent sufferer, have "hitherto had a pic-nic." Apologizing for this long letter, I remain,

Yours truly,

GEO. WHITE FRASER.

W. Kennedy, of Hobart, Ont., proposes shortly to put an electric light plant in his mill.

The second electric locomotive has been put in service in the B. & O. tunnel at Baltimore. It has improved on all previous performances by hauling a train weighing 1,400 tons through the tunnel at the rate of 23 miles an hour. In starting this train a draw pull of 58,630 pounds was exerted. The current taken was 4,100 amperes at a pressure of 600 volts.

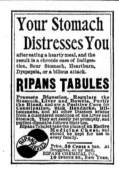

## SPARKS.

J. F. Guay, a dealer in electrical supplies at Quebec, is reported to have assigned.

It is reported that an incandescent light plant will shortly be installed at Brussels, Ont.

The Aylmer Electric Light Co. are said to be considering the purchase of a 1,000-light incandescent dynamo.

The Nelson Electric Light Co. have recently put in operation their new plant at Nelson, B. C. Mr. John B. Bliss is electrician.

The Ira Cornwall Co. is being incorporated at St. John, N.B., to manufacture electrical apparatus, etc. The capital stock is $10,000.

Alonzo T. Cross, a manufacturer of electrical appliances at Providence, R. I., has completed an electric carriage which is said to have given good results.

Mr. E. J. Lennox, architect, of Toronto, has recommended the purchase of an electric light plant for the new city and county buildings. The cost is estimated at $15,000.

Mr. T. Viau, the promoter of an electric railway between Aylmer and Hull, Que., is forming a joint stock company to build the line. The cost of the work is placed at $65,000. The electric road will be seven miles in length.

The Drummondville Electric Co., of Drummondville, Que., has been granted incorporation. The promoters are : William Mitchell, Samuel Newton, William Houston, A. Ouellette, of Drummondville, and W. A. Mitchell, of Nicolet.

Incorporation has been granted to the Barrie and Allandale Electric Street Railway Co. The promoters are J. H. McKeggie, G. Vair, G. Reedy, S. J. Sanford and J. H. Dickinson. The object of the company is to construct an electric railway in the vicinity of Barrie.

The Electrical Review, of London, Eng., in a recent issue compliments Mr. D. R. Street, of Ottawa, on his interesting paper on "Electric Light Accounting," read at the last annual meeting of the Canadian Electrical Association, and expresses approval of the various forms given therein.

At the annual meeting of the Maritime Auer Light Co. held at Fairville, N.B., recently, the following directors were elected for the ensuing year : Messrs. W. H. Thorne, W. C. Pitfield, R. Keltie Jones, G. S. Fisher and S. Hayward, St. John ; A. O. Granger, Montreal ; L. L. Beer, Charlottetown ; F. W. Sumner, Moncton ; F. B. Edgecombe, Fredericton.

The Toronto, Hamilton & Niagara Falls Electric Railway Co. has given notice of application to parliament next session for incorporation. The object is to construct an electric railway from Toronto to Hamilton, and from that city to Niagara Falls, Grimsby and Drummondville. The solicitors of the company are Messrs. Clark, Bowes, Hilton & Swabey, of Toronto.

The forty-ninth annual meeting of the Montreal Telegraph Company was held in the City of Montreal a fortnight ago. The assets were shown to be $2,255,888.66, and the liabilities $2,040,540.25. Four dividends of two per cent each had been paid. The election of directors resulted in the return of the old board, and at a subsequent meeting Mr. Andrew Allen was re-elected president.

A prize of $75 and a diploma is being offered by the Verband Deutscher Elecktrotechniker for the best device by which mistakes, such as placing the wrong size fuse in fuse terminals and the interchanging of fuses except by authorized persons, may be rendered impossible. The designs are to remain the property of the author, and must be received by the Verband at 3 Monbijouplatz before April 1st.

Mr. C. A. E. Carr, manager of the London Street Railway Co., recently gave a supper to the employees of the road.

Henry Townsend, whose son was killed in the Scarboro railway accident at Toronto last summer, has been awarded damages to the amount of $1,000.

The Fraserville Electric Power Co., of Fraserville, Que., is applying for incorporation, with a capital stock of $20,000, to operate telephone lines and electric lighting plants.

An employee of the Toronto Electric Light Co., while repairing the wires on Sherbourne street, in some way came in contact with a live wire, and was badly scorched about the hands and face.

A statement of the earnings of the Montreal Street Railway Co. for the quarter ending December 31, 1895, shows the receipts to be $290,460.35 an increase over the corresponding quarter for 1894 of $47,536.97.

At the annual meeting of the London Street Railway Co., held recently, the following directors for 1896 were appointed : H. A. Everett, of Cleveland, president ; E. W. Moore, Cleveland, vice-president ; Chas. W. Watson, Cleveland ; Thomas H. Smallman, London, and H. F. Holt, Montreal. Chas. Currie was appointed secretary and Chas. E. A. Carr was re-engaged as manager.

## SPARKS.

The village of Morrisburg, Ont., will shortly consider the question of putting in an electric light plant.

The Royal Electric Co. are installing two-phase generators at Georgetown and Glen William; also at Forest.

Mr. Bonfield, of Eganville, Ont., is requesting permission from the village of Arnprior to enter into the electric light business.

A circular expressing regret at the inconvenience caused to subscribers by the recent storm has been sent out by the local manager of the Bell Telephone Co. in Toronto.

At a recent meeting of the directors of the Kingston Electric Light, Heat & Power Co., Mr. A. F. Folger was appointed manager and director.

The Government of Queensland, Australia, are desirous of receiving applications for the position of electrical engineer to the Government. The salary is £600.

The Western Electric Light & Heat Co. have commenced the construction of their system at Vancouver, B.C. The new company will supply light to private citizens at half a cent per ampere hour. The power house will be built on False Creek.

The annual report of the electrical department of the Michigan Inspection Bureau for 1895 shows a total loss of $74,165 in nineteen fires of electrical origin. Over 816 installations were inspected, with the result of remedying 396 defects.

The Deschenes Electric Company, of Aylmer, Que., has been incorporated. The promoters are: R. H. Conroy and D. B. Gardner, of Aylmer, Que.; W. J. Conroy and E. R. Bisson, of Deschenes Mills, Que.; and J. S. Dennis, of Ottawa.

The Committee of the Lord's Day Alliance has instructed its solicitors to appeal from the decision of Mr. Justice Rose in the Sunday car case, which was published in full in our January issue. The appeal will be heard at the March sittings of the Court.

The C. P. Telegraph Co. are placing a storage plant at Ottawa. Some 300 cells of chloride accumulators are being installed.

This is the first storage battery to be used for telegraph work in Canada. If it proves a success it is the intention to adopt the same system in other places.

It is said to be the intention of W. S. Adams, who owns saw mills at Pine Falls, to build an electric railway from Darwin, on the C. P. R., to the Winnipeg river, a distance of twelve miles, and to use the water power for running the mills and supplying electricity for the line.

A bill has been introduced in the United States Senate authorizing the construction of an electrical cruiser, equipped with a system of electrical motors and propellors invented by Richard D. Painton. The inventor claims that a speed of 35 knots an hour can be maintained by cruisers thus equipped.

---

---

# CANADIAN GENERAL ELECTRIC CO.
## (LIMITED)

## Incandescent .. Lamps ..

THE earning power of an incandescent plant hinges upon one vital point : the comparative efficiency of the lamp in use. Different lamps on the market show efficiencies varying from 10 to 30 per cent. lower than ours. This means where such lamps are used from 10 to 30 per cent. less return from each pound of coal and from each kilowatt in plant capacity. Our lamps excel in the other important feature of long life with maintenance of candle-power. Central station managers have learned to appreciate this, and as a result our lamp sales have more than doubled during the last twelve months.

## Carbons

WE have recently taken over the premises and plant of the Peterborough Carbon and Porcelain Co., Ltd. It is our intention in continuing the operation of this factory as a department of our Peterborough works, to make such changes in methods and equipment as may be necessary to render our carbon output equal in all respects to the best imported grades.

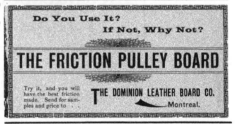

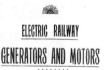

OLD SERIES, VOL. XV.—No. 6.
NEW SERIES, VOL. VI.—No. 3.

MARCH, 1896

PRICE 10 CENTS
$1.00 PER YEAR.

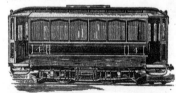
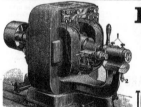

CANADIAN

# ELECTRICAL NEWS
AND
## STEAM ENGINEERING JOURNAL.

| Vol. VI. | MARCH, 1896 | No. 3. |

## THE DUNNVILLE ELECTRIC LIGHT COMPANY'S NEW STATION.

A GRATIFYING feature of the development of electrical industries in Canada within the last three or four years has been the success of the incandescent lighting plants, especially in most of the smaller towns and villages. This success has been at once a source of benefit to the public, even in the smallest communities, in placing within their reach the most perfect of artificial illuminants at a cost but little, if any, in excess of that of coal oil, while at the same time it has yielded to the owners of the plants a substantial return for the money invested. Such a result has been due partly to the great improvements in standard apparatus, and the better engineering methods adopted within the last few years, and also to a considerable degree to the recognition on the part of business men generally of the essentially profitable nature of electric lighting as an investment when handled with the same push and ability which they have been accustomed to devote to their other interests.

An excellent and recent example of a central station plant of this type is that recently installed by the Dunnville Electric Light Company, of Dunnville, Ont., a town of about 2,000 inhabitants. The company, which has been in business for some years, consists of two members, Messrs. W. F. Haskins and James Rolston, the former a private banker and the latter a hardware merchant in the town. Their plant, until recently, consisted of a Thomson-Houston arc dynamo operated by rented power. Having determined, however, to meet the growing demand for improved interior lighting, it was decided to add an incandescent machine to the plant, and at the same time to erect a new power house, and take advantage of an opportunity which presented itself of obtaining water power. To this end the company engaged the services of Mr. W. C. Johnson, Am. Soc. C. E., Chief Engineer of the Niagara Falls Hy-

draulic Power and Manufacturing Co., by whom plans and specifications were made for the entire hydraulic plant.

### POWER PLANT.

Near the mouth of the Grand River, which flows into Lake Erie at Dunnville, a dam has been built by the Canadian Government for the purpose of supplying a feeder to the Welland Canal. This feeder, through the village of Dunnville, runs nearly parallel with the Lake Shore and some three hundred feet from it. The difference in level between the water in the feeder and the lake varies from time to time from six to thirteen feet.

The electric light company's plant is located between the feeder and the lake, at a point about a thousand feet from the river. The floor of building is set as low as possible and to be sure of being above flood water of the river. The floor of the flume is fourteen feet below

THE DUNNVILLE ELECTRIC LIGHT COMPANY'S NEW STATION.

the floor of the station, and is about the same level as the bottom of the feeder, from which the supply of water is taken, and also about the same level as high water in the lake.

The principal difficulty in developing this water power arose from the fact of the low head available, combined with the very great fluctuations as compared with the head, the highest being more than double the lowest head. It follows, therefore, that a water wheel would be capable of developing more than double the power at some times than at others. On this account it was decided to put in two wheels in two independent flumes, so arranged that either or both of the two electric machines could be run by either or both water wheels. The water flows in one channel to the front of the building, passing through a rack, as shown in the engraving. At the front of the building a centre wall divides the channel into two parts, each provided with a head gate, by which the water can be shut off of either flume for repairs to the wheel.

Under ordinary circumstances one wheel is sufficient for driving the plant, and only one wheel will be used, the couplings being arranged with removable plates which can be taken out, cutting off either wheel from the line shaft, or by unshipping the coupling each machine is left attached to a single wheel. In low water both wheels will be used coupled together.

The wheels were specified to run at a uniform speed of sixty-six revolutions per minute under any and all heads from six to thirteen feet, and under all heads from six to nine and a half feet to develop not less than eighty horse-power ; under all heads from nine and a half to thirteen feet to develop not less than one hundred and sixty horse-power, and to show an efficiency of not less than seventy per cent. of useful effect at seven-eights to full gate opening when running at a speed of sixty-six revolutions per minute.

While this is not a high efficiency when wheels are run at their best speed, it is a high efficiency under the unusual requirements as to speed. The wheel of the proper size to run at sixty-six revolutions per minute under a head of six feet, would give its best efficiency under a head of thirteen feet when running at about one hundred revolutions per minute.

The contract on the wheels was awarded to James Leffel & Company, of Springfield, Ohio, for two of their "Sampson" wheels, sixty-two inches in diameter, and they have been running for several months and are giving good satisfaction.

The wheels are provided with draft tubes, enabling the entire available head to be used when the level of the water in the lake is below the bottom of the wheels.

Since the wheels have been in operation a period of low water has occurred, when the lake level was but just above the end of the draft tubes and the head water but little more than covered the cases of the wheels. No trouble was experienced in operation.

The shafting, floor stands, pulleys, etc., were furnished by the Waterous Engine Co., of Brantford.

The flume walls were built of stone laid in hydraulic cement, upon a flooring of brick bedded in concrete. The walls above the foundations were built of brick without finish on the inside.

The roof consisted of sheeting of two inch plank laid on iron beams imbedded in the walls and covered with slate, making the whole building practically fire-proof.

A brick partition wall divides the gear room from the room containing the electric machinery, which serves to deaden the sound of the gears and affords a protection in case of fire.

At one side of the machine room two small rooms are partitioned off, one for an office and the other for a work room. A loft over the gear room serves as a store room.

ELECTRIC PLANT.

In the station are installed the Thomson-Houston arc machine, by which the street lighting is supplied, and a 75 kilowatt alternator of the Canadian General Electric Co.'s monocyclic type. The company, in selecting the monocyclic system, were guided by the fact that while their present requirements would be for current for incandescent lighting, and while this would continue to furnish the greater portion of their business, there would no doubt in time be developed a considerable field for the supply of power to stationary motors. It was therefore desirable to install a system in which the greatest possible simplicity should be maintained for the lighting circuits, and from which, at the same time, polyphase currents could be maintained, if desired, for the operation of induction motors. The distinctive feature of the monocyclic system is, of course, its special suitability for this class of service. The lighting being done on the single-phase system avoids the complication in wiring and difficulties in balancing attendant on the use of the various polyphase systems, while, for the supply of power to the motor at any given point, it is only necessary to carry to it the third or teaser wire from the dynamo, and make the proper transformer connection.

The alternator is in design of the well known iron-clad armature type, and is compound wound so as to compensate automatically for line losses, and thereby maintain an even potential at the centre of distribution throughout all changes on load. The armature coils are wound on forms and inserted in longitudinal grooves in the surface of the armature, and are easily and separately removable in case of damage to one or more coils. The station instruments are all of the latest type, and handsomely mounted on a switch-board of enamelled black slate.

A feature of special interest in this plant is the use throughout of the Edison three-wire system for secondary distribution, by which a considerable saving in transformer capacity and secondary wiring is claimed to be effected, while at the same time ensuring a higher efficiency and closer regulation for the entire secondary system. Through that portion of the town where lighting is to be supplied, three-wire secondary mains are run at intervals of 1,000 to 1,500 feet by pairs of large transformers, connected with their secondaries in series to give 208 volts. This system of distribution being once erected, the wiring of additional buildings from time to time calls for no additional expenditure for material, beyond that required in running a service from the mains to the building wired.

Altogether the new plant of the Dunnville Electric Light Company is a credit at once to its enterprising owners, Messrs. Haskins & Rolston, and to Mr. W. C. Johnson, consulting engineer. and the contractors who furnished the apparatus and material used throughout.

An interesting use of magnetism is being made at the Sandycroft foundry in England. At these works electric cranes are operated from the electric power and lighting circuits, together with electro-magnets, which permit the ready lifting of pieces of iron or steel up to two tons. The magnets constructed for lifting purposes are attached to a crane. One magnet takes 5½ amperes at 110 volts to excite it, at which energy it will support a weight of two tons of iron or steel. A switch controls the supply of current delivered to the magnet.

Ampere, like other philosophers, was noted for his absent-mindedness, says the London Electrical Engineer. It is stated that on one occasion while walking along the street he mistook the back of a cab for a blackboard, and as a blackboard was just the thing he needed at the time to solve a problem which had been vexing his mind for some moments during his walk, he made use of it. Taking a piece of chalk out of his pocket he proceeded to trace out a number of algebraical formulæ on the cab's back, and followed the moving "board" for the space of a quarter of an hour without noticing the progress of the conveyance.

# CORRESPONDENCE

## THE PIONEER ELECTRIC RAILWAY OF CANADA.

TORONTO, Feb. 13, 1896.

Editor CANADIAN ELECTRICAL NEWS.

SIR,—A cynical writer has remarked that "without lies we should have no histories," and since reading a presumably historical account of the first electric railway in Canada, contained in Cassiers' Magazine for January, I begin to believe that the cynic's opinion was founded on fact, as the source of the supposed historical account must have either been densely ignorant of the facts or fully qualified for the presidency of an Annanias Club. An accurate account of the experiment of 1883 may be worthy of preservation ; if such be your opinion, you are free to publish this account, which is accurate.

The electric railway experiment of 1883 was at the expense of the Ball Electric Light Co., of Canada, and was carried out by Frank B. Scovell, the Vice-Pres. and Electrical Engineer of that company, assisted by the writer, who was then assistant electrician of the Ball Electric Light Co., of New York—therefore, this account is neither a matter of rumor nor of imagination.

The track consisted of eight lengths of common railway T rails, spiked to common ties, laid on the surface of the ground, on the premises of the Industrial Exhibition in Toronto—about where the iron tower or police station are a present located. The car was an ordinary flat car.

The transmission was by overhead wires from Machinery Hall to the track. One line was coupled to both tracks and the other to a bare copper wire which lay on the ties between the rails. Contact with this wire was maintained by a carrier, which lifted the wire off the ties as it passed back and forth. As the track was perfectly straight, this was satisfactory, there being no tendency to pull the wire against the rails, as might have been the case with curves. The motor equipment was an old style Ball machine, built in London, Ont., and the power plant two similar machines. Each of these would at regular speed operate at 200 volts with 15 amperes, but being designed for arc lighting the armature reaction would rapidly run the voltage and torque down with increase of amperage. The motor was belted to a countershaft and that to one pair of axles, by common flat leather belting. The reduction in speed proved insufficient to enable the motor to move the car up the slight grade, though it was capable of doing so on the level and down the grade ; up the grade it required a couple of men to assist, even with an unloaded car. The intention was to carry passengers as a novelty, but owing to the above fact this was not attempted as a matter of fact, as the Exhibition closed before the necessary changes could be effected. A very little more motive power or less attempt at speeding would have enabled us to have carried out the intention, and engineers of that time will remember that we had no guide but the result of our own experiments, and judge Mr. Scovell rather by the measure of success secured than by the want of full success, due solely to want of information that is so readily secured now by the tables and published results of experiments that are almost universally disseminated among engineers of the present day.

The company mentioned in Cassiers' did not come into existence until about ten years later than the date of this experiment, and I am informed that the head of this company was then engaged in button-making in Springfield, Mass. The view therein given as one of the road of 1883, and the information as to carrying passengers, I believe, is really applicable to the road built and operated by Mr. J. J. Wright, under the Van Depoele patents, I think in 1884 or 1885. The fact that Mr. Scovell has since departed this life may possibly explain the inaccurate account given in Cassiers', as so far as I know he and I were alone acquainted with the facts.

Yours respectfully,

JAMES W. EASTON,
Electrical Engineer.

## ELECTRIC RAILWAY IMPROVEMENTS.

TORONTO, Feb. 21st, 1896.

Editor CANADIAN ELECTRICAL NEWS.

Dear Sir,—The writer of this has always heard Hamilton spoken of as being behind the times and an overgrown village, and other uncomplimentary remarks made about it—as being behind the age, yet a visit there a few days ago showed that, even if the town is a little behind the times, the people are not so bad. I had the pleasure of a trip over The Hamilton, Grimsby & Beamsville Railway in company with Mr. C. K. Green, Chief Electrician. You are aware, no doubt, that this railway runs from Hamilton to Grimsby—seventeen miles—through the Garden of Canada. On boarding the car at Hamilton Mr. Green promised me a surprise, and this I got, as I found that when the trolley was put on the wire the headlight shone out clear and bright—an arc lamp, which, on the run from Hamilton to Grimsby, illuminated the track as clear as daylight from six to eight pole-lengths in advance of the car, one-half of which was more than sufficient to stop the car in case of anything being on the track. The arc lamp is in series with the car heaters, gives a clear, bright and steady light, and is easily controlled by the motorman. The mechanism, and in fact the idea, seems to have originated with Mr. Green.

After admiring the beauties of the arc lamp, and the splendid way in which it lit up the track, Mr. Green gave ns another surprise in the way of communicating by telephone with any station on their line from inside of the cars. In the corner, in one end of the car, a small open box is affixed to the wall ; in this a very small telephone is enclosed, and attached to the telephone is a flexible cable of sufficient length to reach the telephone wires running on the poles of the railway company. These are hooked on by a bamboo pole, which is carried on the outside of the car. Communications can be established with any station from any point of the line in this manner. As a magneto the same as is used on the ordinary telephone would be too cumbersome to put in the car, Mr. Greene arranged a small metallic roller having a perforated disc, and this being moved from point to point would give a vibrating current, which would make a magneto on any point of the line ring. Altogether, the arrangement is about as complete as could be wished for.

The arc lamp, used in the way it is on this line, is the first practical application of it in Canada, and telephoning from the cars, while not being something absolutely new, yet speaks well for the enterprise of Mr. Green in meeting the conditions, which can only be found in a long suburban road.

A TRAVELLER.

## MR. JOSEPH R. ROY.

MR. JOSEPH R. ROY.

THE chief engineer of the Montreal Park & Island Railway Company is Mr. Joseph R. Roy, whose portrait we have the pleasure of presenting herewith. He is a native of Montreal, and a graduate in engineering of McGill University He is also a member of the Canadian Society of Civil Engineers. Mr. Roy was for three years employed by the Department of Public Works, and at a later date was appointed resident engineer in charge of the construction of the Massina springs and Fort Covington railway. He then became chief engineer of the Montreal and Ottawa railway, which position he held until the year 1892, when the road was made a part of the Canadian Pacific railway.

## THE ELECTRICAL PLANT AT NIAGARA FALLS.

THE following notes are the result of a visit by a Canadian electrician to this plant of world wide interest and observation which is now in successful operation. The Niagara Falls Power Co. owns about 1,500 acres which it expects to sell or lease to manufacturing companies using its power. The user of the power may put in his own water wheel, renting the use of the tunnel as a tail race, or he may take his power from the shaft of the Niagara Co., or use the electric power itself. So much power is being taken up by companies in the immediate vicinity, that long distance transmission is likely to be delayed for a year or two at least.

About 3,000 h. p. is now being utilized, distributed as follows :—Pittsburgh Reduction Co., 1,500 to 2,000 h. p.; Carborundum Co., 1,000 h. p., and Street Railway Co., 500 h. p. This is handled by one of the two 5,000 h. p. generators now in place. The third, which completes the order placed with the Westinghouse Electric & Manufacturing Co., is expected in two weeks. As the power is rapidly being taken up, it is likely that bids for new generators will soon be called for.

One cannot be impressed by the thoroughness and lasting qualities of all work done in and about the power house. Even the visitor is provided for, in the shape of a gallery, from which a fine view of the station can be had. All heavy apparatus is easily handled by a 50 ton crane running the length of the station. The armature, the heaviest part of the generator, weighs about 35 tons. The whole revolving part of the machinery, field shaft and turbines, weighs about 65 tons. But this great weight, revolving at 250 revolutions per minute, is almost entirely balanced by the water pressure, which acts upwards on the lower side of the turbines. So much is this the case, that the strain on the thrust bearings is estimated at not more than 2 tons when the machine is running. At some loads there is no pressure on the bearings whatever, the whole weight being supported by the column of water. The speed is regulated by ball governor mechanism shifting two gates which are balanced against each other and require extremely little power to move them.

The fields are excited from rotary transformers driven by the generators themselves and placed near the central vault supporting the switchboards. There is also a step-down transformer with each rotary transformer.

All wires are carried under the floor to a vault in the centre of the station. In this vault the switches, moved by compressed air are placed. The bus bars are suspended from the roof of the vault and connections made to the switches which stand on the floor. There are two switches to each generator, a distributing switch and a generator switch. The field rheostats are not placed in the vault, being too bulky, but the connections from them are all brought in here. All instruments and levers, for regulating the switches, etc., are placed on top of the vault. This is surrounded by a brass railing, and here the electrical engineer in charge has a view of the whole station and perfect control over the electrical apparatus. The instruments consist of the ammeters and two volt meters (one for each phase) and one watt meter for each machine. It may be mentioned here that the two legs of the two-phase circuit bear loads which differ considerably, due to single phase current being rented in some cases. The step-up transformers for long distance transmission are provided with a special building across the canal. Lightning arresters are provided for all circuits leaving the power house.

Street railway power is furnished from a 500 h. p. rotary converter, built by the Westinghouse Co.

Power is transmitted to the transformer house of the Pittsburgh Reduction Co., near by, where it is reduced to a pressure of 115 volts and transformed by rotary converters to direct current at 160 volts, ready for use in the reducing furnaces in the production of aluminum. There are four rotary converters of 500 h. p. each, built by the General Electric Co. They are of very large size, but were so made that they might be increased, when necessary, to 1,000 h. p. each, by changing the armatures. If necessary one of these machines can be started up in 3 minutes, and the whole station in 11 minutes. Such rapid starting up would be impossible in the case of a steam plant. Indeed one is greatly impressed here by the absence of anything in the shape of a tall chimney and smoke. The load is pretty steady and power is kept on day and night. About 6 tons of aluminum is turned out every 24 hours.

The Carborundum Co. also receives its power in a step down transformer whence it passes through a regulator of special construction moved by hand. The regulator is required on account of the variable pressure necessary during the heating process. It is a transformer in which the mutual induction is varied by turning the inner part of the apparatus and its windings, the outer part being fixed. 1000 h. p. is used, one furnace being in operation at a time. A heat lasts for 24 hours, a ton of carborundum being turned out at each heat. The ingredients consist of coke, salt, sand and sawdust. These are thoroughly ground up and mixed before being put into the furnace. The electric conductors are connected to plates at opposite ends of the furnace. From each of these plates 36 carbon cylinders project into the furnace. A sort of conductor of coke is lain through the middle of the furnace from one terminal to the other and the carborundum crystals are formed around this. These crystals are washed in sulphuric acid and then in water, from which latter bath the carborundum is obtained and sorted in sieves into different sizes ready for the factory. At present it is sent to the factory at Monongahela, Pa., where it is made into wheels and cones of all sizes for various grinding purposes. A factory is being built at Niagara, and when it is complete, the other will be shut down. The process is very much cheapened by the use of the Niagara power.

## QUESTIONS AND ANSWERS.

F. & T., Walkerton, Ont., write :—1st. Why cannot alternating current be used instead of direct for street railway purposes ? 2nd. What is the difficulty in using same ? Our opinion is, that with double wiring twin trolleys, two commutators and rawhide gear, feed wire and transformers, with a few minor changes in the method now in use, it would be possible, and at a very reduced rate. Will som eone give us their opinion on the same ?

ANSWER.—The alternating current has not hitherto been used for street railway purposes, first on account of the unsuitability of the alternating motors for the conditions imposed on railway motors, their torques at other than full speed not having been satisfactory. This defect has been overcome in the induction motor, which is no doubt adapted for traction purposes, and probably will be so applied in the near future. By your mention of twin trolleys you no doubt mean to throw out the rails as return circuit, but in any case the single phase alternating current would not be at all suitable for traction, as single phase motors will not start from rest without introducing a complicated phase-splitting apparatus, so that you would either have to use a two phase current with four wires, four trolleys, etc., or a three phase with three wires, three trolleys, and so on. You mention two "commutators," but it does not appear what a commutator has to do with an alterpating current. The alternating current is scientifically applicable for traction, but not commercially as yet.

---

The Secretary of Dresden Association C.A.S.E., propounds the following question :—"What is the indicated h.p. of an engine, 13 in. bore, 18 in. stroke, 170 revolutions per minute, boiler pressure, 70 lbs.?

ANSWER.—The indicated horse-power of our engine is calculated from the formula

$$H.P. = \frac{A. \times M.E.P. \times S R.S.}{33,000}$$

where A. = area of piston in square inches ; M. E. P. = the mean effective steam pressure ; R. = number of revolutions ; S. = length of stroke in feet. In the case mentioned, the piston area in inches is 132·73 ; the mean effective pressure at one-quarter cut-off is (with 70 lbs. boiler pressure) 29·63 lbs., non-condensing ; the stroke is practically 1·1 foot—hence, substituting in the formula, we get

$$H.P. = \frac{(132 \cdot 73) \times (29 \cdot 63) \times (2 \times 170 \times 1 \cdot 1)}{33,000},$$

which is a little over 44 horse-power.

---

WALKERVILLE, ONT., Feb. 12, 1896.

Editor CANADIAN ELECTRICAL NEWS.

DEAR SIR,—I notice in the "Questions and Answers" column of the NEWS a query from "Constant Reader," Whitby, Ont. He asks, "How is it, when lamps are connected in series, as in railway circuits, the highest voltage lamp always gives the brightest light?" With your permission I would like to state my theory contained in the following table and explanations :

### 16 C. P. LAMPS, 60 WATT EFFICIENCY.

| Volts. | Amperes. | Hot Resistance in Ohms. | |
|---|---|---|---|
| 50 | 1.2 | 41.66 | Lamps of any efficiency can be taken provided all are alike. |
| 70 | .875 | 81.77 | |
| 100 | .6 | 166.66 | |
| 104 | .577 | 180.24 | |
| 110 | .5454 | 201.68 | |

With a circuit of 500 volts, using five lamps in series, suppose four lamps of a series are 100 volts and one lamp 104 volts, then by adding the resistance of the five lamps, we obtain 846.9 ohms (166.66 × 4 = 666.66 + (1 × 180.24) = 846.9). Applying Ohm's law we get the current required,

$$\frac{500 \text{ volts}}{846.9 \text{ ohms}} = .59 \text{ ampere } C. = \frac{E}{R}.$$

By comparing the result (.59 amp.) with the table, it will be seen that the current passing through the series is .013 amp. less than is required for the 100-volt lamps,

and .01 amp. more than is required to bring the 104-volt lamp to 16 c.p. The practical result is that the 100-volt lamps will burn a little below and the 104-volt lamp a little above the normal, making a preceptible difference in the light. Lamps of the same voltage but of different efficiencies will give similar results. The higher efficiency lamps being of higher resistance and requiring less current, will give the brighter light, e. g., 55-watt and 60-watt lamps connected in series, the 55-watt lamps will be the brightest. Nothing but a test will show whether it is the efficiency or voltage that is at fault if some lights in a series burn bright or dim.

J. W. SCHELL.

---

"P. S. C.," Oshawa, Ont., writes :—Please be so kind as to answer me the following questions in the March issue of the ELECTRICAL NEWS: (1.) "Suppose after two guages of water show in glass that pump was stopped ; five minutes after that you notice water at the top of glass, what would you do, and what was the cause of difficulty, boiler not foaming ? (2.) Suppose feed-pump working, but water level in boiler gradually falling. Name the different places you would look to find the difficulty ? (3.) If the valve stem of the steam valve of a Corliss engine should break, what could be done to prevent a shut down ? (4.) In triple engine, with second cylinder doing much more work than low pressure, how can cut-off be best adjusted in either cylinder to balance load between cylinders?"

ANSWER.—(1.) We should advise you to blow out the gauge glass to see if the level indicated was the right one, or whether in some way the gauge cocks had become clogged. If, as a matter of fact, the level of the water was above the glass (and the guage had become so clogged as not to indicate properly,) open the pet cocks of the cylinder, and the drain in the pipes, so as to carry off the water flowing into the pipes. If no further trouble occurred in the cylinders wait until water appears again in the guage glass, but if there is any reason to fear that the water level is so high as to flow over into the cylinder, then blow the boiler down a little. (2.) Either the pump is not working fast enough or water is leaking at joints or seams, or around the tubes, or at the blow-off. In fact, overhaul the boiler immediately ? (3.) Splice the stem temporarily with anything—a piece of stiff wood—and wrap the stem and splice with some stout cord or rope, drawing the valves together, so as to keep the whole thing taut. (4.) You can do nothing unless the cylinders are individually controlled by a cut-off valve, in which case adjust as in a single cylinder.

---

"W. R. R.," Stayner, Ont., writes : "To protect our alternator from short circuit in the outside mains, there is, as is of course the practice, a fuse introduced between the dynamo leads and the main—but in our fuse cut-out there is, I think, something peculiar, if not out of place. Each wire has a separate cut-out. The following diagram shows the whole as it is :—

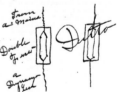

The reason given for the double fusing on each line is, " if one should be blown out, the other would still be there to prevent the interruption of the circuit." Is this in place ? "

ANSWER.—The two fuses are intended to be duplicates, so that if one goes there may still be the other which can be switched or plugged in. There should be some arrangement for switching one off and the other on; but of course both should not be in circuit together.

Mr. F. G. Proutt, of Malden, Mass., writes: I noticed in your paper of this month a communication from some one who signs himself "A Constant Reader." From the question he has asked, and which you took so much trouble to answer, I imagine that he is not a constant observer, but to help you complete the answer to his question, the reason a high voltage lamp becomes more incandescent than those of a lower voltage when connected with them in series, is this: The higher voltage the lamp the less current passes through it, or, rather, is required by it; for instance, about ½ ampere is required for a 100 volt lamp, while one ampere, or about that, is required for a 50 volt lamp. Now, if we connect 8 50 volt lamps and 1 100 volt lamp in series across a 500 volt circuit, we would have current passing in proportion to the resistance, and if the 50 volt lamps have each 50 ohms R, and the 100 volt lamps an R of 200 ohms, then

$$\frac{E}{R} = C \text{ or } \frac{500}{8 \times 50 + 200} = C = 5\text{-}6 \text{ amp.}$$

Now, 5-6 of an amp. is not quite enough current for the 50 volt lamp, but is very much too high for the 100 volt lamp. Hence, the high voltage lamp would burn very much brighter, and in every case where lamps of different voltages are run in series, the one made to be at the highest voltage will be the most incandescent.

## CENTRAL STATION BOOK-KEEPING.
By GEORGE WHITE-FRASER, E. E.
III.

WHAT has gone before will indicate the importance of keeping strict watch on the operating end of an electric plant. As soon as you begin to suspect the honesty and faithfulness of each piece of machinery in your power house, and lay traps to find out little lapses from rectitude, so soon will you begin to find your expenses going down, and your profits increasing.

When you look into things a little, you will be surprised to find what a great deal there is in "management" after all; and how even apparently unimportant apparatus and supplies may exercise a considerable influence on results. Lines are of course very simple, and once they are up will probably continue to be all right, if they are kept free from grounds, and generally kept in repair; but at the same time it is of considerable assistance to take observations from time to time, as to the insulation of the entire aerial structure, from the ground, and of the wires from each other. A leak in either way simply means wasted fuel, and presumably the business of a central station is to make, not to waste, money.

From the central station, the records will divide into these concerning arc services, incandescent service, and power service. As regards the arcs, there should of course be an account kept of the carbons used, and of the repairs necessary on each lamp. An account should be kept for each individual lamp, and the cost of any repairs to any one should be debited against it. If a coil burns out—or a carbon holder—you want to know of it, and which lamp it belonged to; and you can summarize the records at the end of the year, and probably have some instructive information as to the quality of your lamps as pieces of mechanism. Then, a few spare lamps should always be kept in stock in order to replace those on the lines brought in for inspection. Perhaps the idea of inspecting an arc lamp causes amusement; but when it is remembered that an arc lamp is merely a little machine for keeping carbons at a proper distance apart, and that it does so by virtue of being built and adjusted with that object, and that its proper working depends on some rather delicate devices, and that if it falls ever so little out of adjustment, it means wasted money for fuel—then it will be obvious, that to inspect it periodically, to see whether it actually is doing its work, is just as necessary as it is to indicate the engine.

What is an "arc light," and how is it produced? An arc light is simply the illumination produced by the intense incandescence of two "electrodes" separated by a space across which an electric current is being forced by an E. M. F. This incandescence is the result of the very high temperature caused in the electrodes by the interposition of the air space—it being a principle of mechanics as well as electrics that resistance dissipates energy in the form of heat. A greater amount of resistance will dissipate a greater amount of energy. A man will become hot and perspire when working hard, i.e., overcoming a great resistance, when he will not even feel warm over a light job; and as the principal

work done by the current in passing through an arc lamp is right at the arc itself, it is plain that the longer the arc the more work done. Now, it is well known that a current of 9 amperes requires a voltage of 50 to force it across an air space of about 3-32 of an inch between the carbons, and the light produced is the nominal 2,000 candle-power. So that if you can arrange to have two 7-16 carbons continually held at 3-32 apart, and use a 9 ampere current, you will get nominal 2,000 c.p., and will require a pressure of 50 volts, or an expenditure of 50 × 9 = 450 watts of energy. If you vary the distance apart of the carbons, you will vary, in the same sense, the volts, wattage, and within limits, the candle-power. If you keep the carbons 4-32″ apart you will get a somewhat larger candle-power, and you will require a higher voltage and expend a higher wattage. But as you are only getting paid for 2,000 c.p., you don't want to produce any more—hence the excess of wattage consequent upon higher c.p. is a pure waste of money. Every time your arc lamp pulls the carbons apart more than their rated arc length; every time the carbon rod sticks a little—generally, whenever your arc gets longer than it is intended to—you are wasting fuel, and consequently, money. With a 9 ampere current, an excessive voltage of so little as one-tenth of a volt, will require an expenditure of almost 3 h.p. hours per year—more than necessary—per lamp.

Try any lamp that has been on the line for some weeks, exposed to all the variable atmospheric conditions, and a few careful observations as to the fall of pressure across the terminals will be very instructive to anyone who is intelligent enough to apply the results. Assume 30 arc lamps, burning on a moon schedule of about 200 hours in the month, and assume that their adjustment is to become so inaccurate that they will draw out their arcs long enough before feeding to require an extra volt each. Assume that otherwise their working is reasonably good, and their feeding regular; then this plant will dissipate in excessive voltage alone, 432 h. p. hours every year, which might just as well be saved.

Every lamp should be tested for a whole night every two or three months. If its feeding is not regular it should be made so, by the proper adjustments, which can be done by any intelligent electrician: and the record of each lamp's performance should be kept. If any individual lamp is frequently found to be out of adjustment, it should be carefully overhauled. It may have some defect which renders it inefficient, and which can perhaps be remedied. If not, it is better to buy a new lamp than to run a bad one. By lighting the power house with the lamp to be tested, no extra expense will be involved.

It is quite important that tests should be made of the carbons. A carbon is by no means always a carbon. There are good, bad, and indifferent ones. The good will cost money to buy, it is true: but the poor one will cost money to run, it is equally true. There are hard and soft, long-lived and short-lived, cheap and expensive ones. Those who have not given any study to the carbon question generally believe that carbon is the best that lasts the longest, but that does not necessarily follow. Carbon is a material of comparatively high resistance, which can, however, be varied by varying its density and its ingredients. Cheap carbons are made of inferior materials, and not a great deal of attention is paid to the process of manufacture. The consequence is probably a very high resistance, and a want of homogeneity in the structure which causes it to burn unequally and break off in little chunks. It may have a long life, but the high resistance will cause the same extra expenditure over that required for a shorter lived, lower resistance carbon, as the undue lengthening of the arc described above. It may very easily happen that a short-lived carbon may be the most economical to use, because its smaller resistance will save more than enough energy to pay for the greater number of carbons used. Observation and experiment will enable the intelligent electrician to select actually the best carbon; and a certain number of each batch bought should be carefully tested as to their life, their resistance, and if possible, their candle-power. This latter point—candle-power—is of course very important, but to properly describe the tests and observations would require too long an article. Careful observations by experienced scientists have proved that carbons of the same apparent make, from different makers, sometimes vary as much as 50 per cent. in the candle-powers they give forth, for the same expenditure of energy; and they show that ultimately the best carbon is undoubtedly that one to the manufacture of which most scientific attention has been given, which is therefore necessarily high priced. But the test above mentioned should be made with a carefully regulated lamp, so that the arc shall be kept sensibly constant, and should record the exact length of life per inch and per carbon, and the average resistance across the lamp terminals. These records will enable a fairly good comparison to be made between carbons of different makes and prices. If, by a proper system of lamp inspection and adjustment, and a careful selection of carbons, you can run each lamp at one volt less pressure, in the above assumed plant you will save over two tons of coal per year; or you will be able to put another lamp on the same circuit.

As regards the incandescent installation, about the most important feature is the lamps themselves; and this is the feature that seems to receive less attention and study from the average central station manager than any other. Lamps are of all kinds, sizes, makes and descriptions; and of course each manufacturer claims superiority for his own make. Price seems generally to be regarded as the most important factor governing the selection of a lamp, i.e., a lamp costing 22 cents is taken to be a more advantageous purchase than another costing 23 cents. As I have requently said before, it is a pretty fair principle to work on in these days of keen competition among manufacturers, that the highest priced article is generally the best one; and the question to ask is, not how little does such an article cost to buy, but how little does it cost to keep running in good order. An article that can be kept in repair and in good efficient condition for $1 a year, is worth, to buy, just twice as much as a similar article that costs $2 per year to maintain, other things being equal; and if the former article is sold at 50 per cent. advance on the price of the latter, then. the higher priced article is actually and obviously far the cheaper of the two. This same principle applies to incandescent lamps. The factors that determine the comparative values of different lamps are: the candle-power, the life, the efficiency, the price—purposely putting price last. Lamps of all makes deteriorate as they are longer in use, some less than others, i.e., their candle-powers become, sometimes, considerably less the longer they are kept burning; and the lamp to be desired is, other things being equal, that one that keeps its candle-power longest up to its rated amount. Then again, lamps differ in point of efficiency. A lamp requires the expenditure of a certain amount of energy to make and keep it burning; so that other things being equal, that lamp that maintains its candle-power longest, for the least expenditure of energy, is the best. Having obtained by means of experiment or otherwise, a diagram showing the continuous candle-power curve of a lamp as a function of its life, and also of the watts of electrical energy expended during inch life, then an average can be struck which shows the average watts per candle power expended. And a comparison of several such averages will show the best lamp, other things being equal. The best is of course that lamp which requires the lowest average wattage per candle.

About the other record that can be kept in places where lamps are rented on the flat rate system, is the dates when each lamp was put in. A little slip of paper with the date marked on it can be gummed onto the lamp, and so accurate track is kept of the life of each lamp, which can be returned to makers if burnt out before guarantee.

If the entire system is worked on the meter plan, which is greatly to be preferred, then the monthly readings of the meters added together should be compared with the month's total as registered on the station wattmeter. The comparison will enable any discrepancy to be traced and possibly lead to the detection of unsuspected leaks.

In a system where current is sold entirely by the meter it becomes of very special importance to study the incandescent lamp question, because the exigencies of this service introduce some very paradoxical conclusions. For instance, it is sometimes actually better and more profitable for the central station to break lamps long before they are worn out, and put in new ones at its own cost. Length of life, in this case, is about the least important advantage that can be claimed for a lamp. This, however, is a matter that cannot be discussed at proper length in this article. What is intended to be shown is, that just as there are engines and engines, dynamos and dynamos, so are there carbons and carbons, lamps and lamps, and that if judgment and caution are necessary in the purchase of a bicycle, or of a good horse— they are far more so in the purchase and operation of any electrical and steam apparatus.

There are certain general accounts that will be kept by every one—sand paper, oil, new brushes, fuse wire, and what not, which need not be particularly referred to. A central station that follows the line of the records indicated in the foregoing, will find itself in a good position to increase its profits. In electrical business it is not entirely what is made that pays dividends, but what s saved; and it cannot be said of even the most high class station

yet known, that it has attained the highest possible point of efficiency, but the most successful are those that keep the most watchful eye over every individual machine or piece of apparatus, and who keep every detail in the highest and best order—and this cannot be done without keeping the most comprehensive system of accurate records, and by the continual testing of everything.

## ANNUAL MEETING OF THE BELL TELEPHONE COMPANY.

THE annual meeting of the Bell Telephone Company was held in Montreal on the 27th of February. Among those present were Messrs. C. F. Sise, president, Robert Mackay, vice-president, C. P. Sclater, secretary-treasurer, W. R. Driver, Robert Archer, C. R. Hosmer, R. McLea, F. X. St. Charles, J. McRae, J. Wilson, W. B. Miller, T. D. Hood, James Moore, A. Kingman, Reid . Taylor, H. A. Budden, Hugh Watson, J. Williamson, D. Ross Ross, W. J. Withall, J. B. McNamee George H. Holt, Chas. Garth, John Crawford, Andrew Allen, Alex. Paterson; G. M. Kinghorn, Hector Mackenzie and others.

Mr. Sclater, the secretary-treasurer, read the annual report as follows :—

" The directors beg to submit their sixteenth annual report ; 1,028 subscribers have been added during the year, the total number of the sets of instruments now earning rental being 28,809 ; 45 exchanges and 6 agencies have been constructed and added to the system. The company now owns and operates 345 exchanges and 268 agencies ; 522 miles of poles, and 1,760 miles of wire have been added to the long distance system in 1895 ; of these 190 pole miles and 874 wire miles are in the Ontario department, and 332 pole miles and 913 wire miles are in the Eastern department.

The long distance lines, now owned and operated by the company, comprise 14,851 miles of wire on 5,884 miles of poles, which include a copper metallic circuit line from Montreal to Toronto, constructed during the past year.

Work on the new building in Montreal progressed favorably until it was deemed prudent to discontinue construction during the winter. It will be resumed as early as possible, and we trust that the building will be ready for occupancy before the next annual meeting. The growth of our Winnipeg exchange having rendered the present offices inadequate for the business, it became necessary to secure other quarters, and a lot was purchased on Thistle street, in a favorable location, where a building will be erected during the summer, which will be used solely for the purposes of the company.

The gross revenue for the year was $1,087,124.28, the expenses were $787,249.36, the net revenue was $299,874.92, the paid up capital is $3,168,000.

In addition to the net revenue of $299,874.92, the premium on bonds sold during the year, amounted to $10,750, making a total of $310,624.92, out of which $253,431.33 have been paid in dividends, and the balance of $57,193.59 together with $2,806.41, taken from revenue account, has been carried to the contingent account, which now amounts to $910,000.

In moving the adoption of the report, the chairman made a feeling reference to the loss the company had sustained through the death of the late vice-president, Mr. G. Ross. The report was adopted.

In reply to a question, the president stated that the net revenue of the company for the year had been 10 per cent. The revenue from long distance lines had been, in 1893, $140,000 ; 1894, $152,000 ; 1895, $178,213.

The directors were authorized to make a further issue of debentures for $600,000, in accordance with the authority given by the Dominion statute. These funds are required to meet current expenses entailed in improving the system. This sum will bring the total issue up to $1,200,000.

The balloting for officers resulted in the old board being re-elected as follows : C. F. Sise, president ; Robert Mackay, vice-president ; directors, W. H. Forbes, John E. Hudson, Robert Archer, Wm. R. Driver, Hugh Paton and Charles Cassils.

PUBLISHED ON THE FIFTH OF EVERY MONTH BY

### CHAS. H. MORTIMER,

OFFICE : CONFEDERATION LIFE BUILDING,
Corner Yonge and Richmond Streets,

TORONTO, - - CANADA.
Telephone 2362.

NEW YORK LIFE INSURANCE BUILDING, MONTREAL.
Bell Telephone 2299.

### ADVERTISEMENTS.

Advertising rates sent promptly on application. Orders for advertising should reach the office of publication not later than the 26th day of the month immediately preceding date of issue. Changes in advertisements will be made whenever desired, without cost to the advertiser, but to insure proper compliance with the instructions of the advertiser, requests for change should reach the office as early as the 22nd day of the month.

### SUBSCRIPTIONS.

The ELECTRICAL NEWS will be mailed to subscribers in the Dominion, or the United States, post free, for $1.00 per annum, 50 cents for six months. The price of subscription should be remitted by currency, registered letter, or postal order payable to C. H. Mortimer. Please do not send cheques on local banks unless 5 cents is added for cost of discount. Money sent in unregistered letters will be at senders' risk. Sub-scriptions from foreign countries embraced in the General Postal Union $1.50 per annum. Subscriptions are payable in advance. The paper will be discontinued at expiration of term paid for if so stipulated by the subscriber, but where no such understanding exists, will be continued until instructions to dis-continue are received and all arrearages paid.
Subscribers may have the mailing address changed as often as desired. When ordering change, always give the old as well as the new address.
The Publisher should be notified of the failure of subscribers to receive their paper promptly and regularly.

### EDITOR'S ANNOUNCEMENTS.

Correspondence is invited upon all topics legitimately coming within the scope of this journal.

The "Canadian Electrical News" has been appointed the official paper of the Canadian Electrical Association.

## CANADIAN ELECTRICAL ASSOCIATION.

### OFFICERS :

PRESIDENT :
A. B. SMITH, Superintendent G. N. W. Telegraph Co., Toronto.

1ST VICE-PRESIDENT :
C. BERKELEY POWELL, Director Ottawa Electric Light Co., Ottawa.

2ND VICE-PRESIDENT :
L. B. McFARLANE, Manager Eastern Department, Bell Telephone Company, Montreal.

SECRETARY-TREASURER :
C. H. MORTIMER, Publisher ELECTRICAL NEWS, Toronto.

EXECUTIVE COMMITTEE :
GEO. BLACK, G. N. W. Telegraph Co., Hamilton.
J. A. KAMMERER, General Agent, Royal Electric Co., Toronto.
E. C. BREITHAUPT, Berlin, Ont.
F. H. BADGER, JR., Superintendent Montmorency Electric Light & Power Co., Quebec.
JOHN CARROLL, Sec.-Treas. Eugene F. Phillips Electrical Works, Montreal.
K. J. DUNSTAN, Local Manager Bell Telephone Company, Toronto.
O. HIGMAN, Inland Revenue Department, Ottawa.
W. Y. SOPER, Vice-President Ottawa Electric Railway Company, Ottawa.
A. M. WICKENS, Electrician Parliament Buildings, Toronto.
J. J. WRIGHT, Manager Toronto Electric Light Company.

## MONTREAL ELECTRIC CLUB.

OFFICERS :

| | |
|---|---|
| President, W. B. SHAW, | Montreal Electric Co. |
| Vice-President, H. O. EDWARDS, | Montreal. |
| Sec'y-Treas., CECIL DOUTRE, | 81A St. Famile St. |
| Com. of Management, T. F. PICKETT. W. GRAHAM, J. A. DUGLASS. | |

## CANADIAN ASSOCIATION OF STATIONARY ENGINEERS.

| | |
|---|---|
| President, W. G. BLACKGROVE, | 293 Berkeley St., Toronto, Ont. |
| Vice-President, JAMES DEVLIN, | Kingston, Ont. |
| Secretary, E. J. PHILIP, | 92 Esther St., Toronto, Ont. |
| Treasurer, DUNCAN ROBERTSON, | Hamilton, Ont. |
| Conductor, W. F. CHAPMAN, | Brockville, Ont. |
| Door Keeper, F. G. JOHNSTON, | Ottawa, Ont. |

TORONTO BRANCH NO. 1.—Meets 1st and 3rd Wednesday each month in Engineers' Hall, 61 Victoria street. W. Lewis, President ; S. Thompson, Vice-President ; T. Eversfield, Recording Secretary, University Crescent.

MONTREAL BRANCH NO. 1.—Meets 1st and 3rd Thursday each month, in Engineers' Hall, Craig street. President, John J. York, Board of Trade Building ; first Vice-President, J. Murphy ; 2nd Vice-President, W. Ware ; Secretary, B. A. York ; Treasurer, Thos. Ryan.

ST. LAURENT BRANCH NO. 2.—Meets every Monday evening at 43 Bonse-cours street, Montreal. R. Drouin, President ; Alfred Latour, Secretary, 306 Delisle street, St. Cunegonde.

BRANDON, MAN., BRANCH NO. 1.—Meets 1st and 3rd Friday each month, in City Hall. A. R. Crawford, President ; Arthur Fleming, Secretary.

HAMILTON BRANCH NO. 2.—Meets 1st and 3rd Friday each month in Maccabee's Hall. E. C. Johnston, President ; W. R. Cornish, Vice-Pres.; Wm. Norris, Corresponding Secretary, 211 Wellington street.

STRATFORD BRANCH NO. 3.—John Hoy, President ; Samuel H. Weir, Secretary.

BRANTFORD BRANCH NO. 4.—Meets 2nd and 4th Friday each month. F. Lane, President ; T. Pilgrim, Vice-President ; Joseph Ogle, Secretary, Brantford Cordage Co.

LONDON BRANCH NO. 5.—Meets once a month in the Huron and Erie Loan Savings Co.'s block. Robert Simmie, President ; E. Kidner, Vice-President ; Wm. Meaden, Secretary Treasurer, 533 Richmond street.

GUELPH BRANCH NO. 6.—Meets 1st and 3rd Wednesday each month at 7.30 p.m. J. Fordyce, President ; J. Tuck, Vice-President ; H. T. Flewelling, Rec.-Secretary ; J. Gerry, Fin.-Secretary ; Treasurer, C. J. Jorden.

OTTAWA BRANCH NO. 7.—Meet every second and fourth Saturday in each month, in Borbridge's hall, Rideau street ; Frank Robert, President ; F. Merrill, Secretary, 352 Wellington street.

DRESDEN BRANCH NO. 8.—Meets 1st and Thursday in each month. Thos. Steeper, Secretary.

BERLIN BRANCH NO. 9.—Meets 2nd and 4th Saturday each month at 8 p.m. W. J. Rhodes, President ; G. Steinmetz, Secretary, Berlin, Ont.

KINGSTON BRANCH NO. 10.—Meets 1st and 3rd Tuesday in each month in Fraser Hall, King street, at 8 p.m. President, S. Donnelly ; Vice-President, Henry Hopkins ; Secretary, J. W. Tandvin.

WINNIPEG BRANCH NO. 11.—President, G. M. Hazlett ; Rec.-Secretary, J. Sutherland ; Financial Secretary, A. B. Jones.

KINCARDINE BRANCH NO 12.—Meets every Tuesday at 8 o'clock, in Mc-Kibbon's block. President, Daniel Bennett ; Vice-President, Joseph Lighthall; Secretary, Percy C. Walker, Waterworks.

WIARTON BRANCH NO. 13.—President, Wm. Craddock ; Rec.-Secretary, Ed. Dunham.

PETERBOROUGH BRANCH NO. 14—Meets 2nd and 4th Wednesday in each month S. Potter, President ; C. Robison, Vice-President ; W. Sharp, engineer steam laundry, Charlotte street, Secretary.

BROCKVILLE BRANCH NO 15.—President, W. F. Chapman ; Vice-President, A. Franklin ; Recording Secretary, Wm. Robinson.

CARLETON PLACE BRANCH NO. 16.—President, Jos McKay, Vice-President, Henry Derrer ; Fin.-Secretary, A. M. Schofield.

## ONTARIO ASSOCIATION OF STATIONARY ENGINEERS.

BOARD OF EXAMINERS.

| | |
|---|---|
| President, A. AMES, | Brantford, Ont. |
| Vice-President, F. G. MITCHELL | London, Ont. |
| Registrar, A. E. EDKINS | 139 Borden St , Toronto. |
| Treasurer, R MACKIE, | 28 Napier st., Hamilton. |
| Solicitor, J. A. McANDREWS, | Toronto. |

TORONTO—A. E. Edkins, A. M. Wickens, E. J. Phillips, F. Donaldson.
HAMILTON—P. Stott, R. Mackie, T. Elliott.
BRANTFORD—A. Ames, care Patterson & Sons.
OTTAWA—Thomas Wesley.
KINGSTON—J. Devlin, (Chief Engineer Penitentiary), J. Campbell.
LONDON—F. Mitchell.
NIAGARA FALLS—W. Phillips.

Information regarding examinations will be furnished on application to any member of the Board.

## Modern Central Station Engineering.

THE tendency of modern central station engineering is towards greater science, higher economy, more comprehensive business methods. In the early history of electric light-ing (about five or six years ago) we used to be satisfied —more, we were pleased—if on closing the switch we got light in the lamps, and our electricians used to be happy and read the paper while the plant ran itself. The fireman also piled in coal, and on cold nights opened the fire door to warm himself at ; and if profits were not very large—well—" Electricity did'nt pay any-way." But to-day our electrician has something else to do, and the superintendent can tell by comparing the reading of the station watt-meter with the last night's fuel consumption, how many times the furnace door was opened, and for how long, and if Mr. Fireman has not a satisfactory explanation to give he goes suddenly. In those days we knew how much coal was burnt and how much wages we paid, and if the income from the lamps was sufficiently large to have a little over, after paying expenses, we pocketed it with a thankful heart, and asked no questions. How very different is the method of running a central station now-a-days. In the first place, a comprehensive business policy governs the en-tire management, and furnishes a framework, into which are fitted such details as differential rental rates, lamp efficiencies, etc. The tabulation and classification of operating statistics will, in the hands of an experienced manager, be not only a sure record of the past, but also a guide to the future, and a careful and observant study of such dry statistics will often serve to so modify the general policy as to effect considerable improvements in operating methods, or in profits. In the next place, each individual bit of machinery in the system, be it boiler, engine, belt, generator, or lamp socket, is continually under the watchful eye of the observant manager, who knows what it is capable of doing under favorable circumstances, and if at any time it falls somewhat below the mark, he knows it, and must know the reason why. If 500 lbs. of coal are burnt more to-night than last night, he is aware of it next morning, and far from saying it "can't be helped," he finds out why it was—whether there was a greater demand for current by consumers, or whether a belt slipped, or a valve got wrong. It can and must be helped, and is helped. The repair account is scrutinized narrowly every year, and any bit of apparatus that seemed to require frequent attention is examined with the object of making it more durable, and so reducing the expenses. There is a ceaseless effort to so raise the efficiency of the whole system, that the same amount of current may be generated at a less expense ; or shall do a greater amount of work and so gain more ; and the means to this end are enquiry into every stage in the process of converting coal into current, and ceaseless energy and investigation.

INVESTORS in electrical and steam machinery have an idea that if they base their selection of machinery on the guarantees given by the makers, they cannot go far wrong. When making a choice between several competing proposals, they do not very often take proper means to arrive at which is really the best for their particular case ; but place their orders with the maker who gives the best guarantees. Before doing this they should consider first, what are the conditions under which the guarantee is given; and second, what means they have of proving it. Now-a-days that prices are cut so very low owing to the intense competition between all manufacturing companies, it is impossible in many cases for agents to cut still lower in order to obtain business, so their resource is to claim higher value for their goods, and if by guaranteeing a slightly greater efficiency or longer life, etc., they can effect a sale it is a little too much to expect of human nature that an agent will not guarantee a little more than he knows he can perform, when he has very good reason to believe that his machinery will not be put to any test, but that his bare word will suffice. It is perfectly evident that to buy machinery on the unsupported guarantee of an agent, and then not to test it to ensure its meeting such guarantee, is not only very foolish from the purchaser's own standpoint, but it is distinctly unfair to the whole manufacturing interest. It amounts in plain terms to placing a premium on dishonesty, and the purchaser frequently gets as badly let in as he thoroughly deserves to. A thoroughly reputable and honorable business man will make sure of what his goods are worth and will guarantee them for that and no more, relying on such policy to build up a business. A less responsible person will push the sale of his goods by claiming for them a value they may not possess, and take his chances that he may not be found out. Now whether he is found out or not, the purchaser does not see his money's worth, and has actually assisted in rewarding fraud. In the few instances where electrical and steam machinery have been purchased in the open market by competing tenders, any specifications that

*Guarantees.*

have been got out require such and such guarantees to be given of efficiency, etc., and the contracts have frequently been given to those guaranteeing the highest and most. But in how many cases have actual tests been made by competent persons, as to whether these efficiencies, etc., are really as guaranteed ? And is it not plain that this system of purchasing on a basis of "competing guarantees" is a very foolish one unless there be some intention of testing their fulfilment ?

Guarantees, even the best, may be very misleading unless some understanding be arrived at as to the conditions under which they are given, and a very little consideration will show that persons not specially trained may be completely hoodwinked when they think they are very wide awake. Take the guarantee that is so frequently given with a boiler, that "it will evaporate so many lbs. of water per lb. of coal." This appears very simple no doubt—to compare boilers by this guarantee is as easy as falling off a log. But a boiler does not evaporate water itself ; it must be set "just so," and have a certain number of feet of grate surface with a certain amount of draft. And again, "so many lbs. water evaporated per lb. of coal." What kind of coal? All coal is not the same. We have hard coal and soft coal ; we have coal containing 13,000 heat units per lb., and we have coal containing only 7,000. Again we ask, what coal? Best Yonghisberry? Any boiler will do that with such superior coal; but it takes a good one to do it with lignite. Briefly, what is such a guarantee worth? Nothing. The person who gives it is in most cases a designing quack, and the purchaser who accepts it is an innocent simpleton. The guarantor gives it because he knows it never will be tested, and if it ever is tested he can say the conditions were not complied with. If persons require guarantees to be given with machinery, they should be satisfied with reasonable ones ; they should establish the necessary conditions, and they should insist on tests being made to prove them. In this way, in a very short time, all but the reputable manufacturers will be crowded out of business.

THE question is frequently asked, "In what field of effort may a young man hope to meet with the greatest amount of success in the present day?" Every department appears to be overcrowded, and the problem of the choice of a career is becoming more and more difficult of solution. The rapid development of the past few years in the applications of electricity has turned the attention of parents and young men in this direction, and there appears to be a widespread belief that this is the most promising field of effort for the future. With the view of determining to what extent this belief is well founded, Mr. Henry Floy, in an article in the Engineering Magazine for January, entitled "Are we Educating too many Electricians?" gives the result of an extended enquiry among graduates of engineering schools, as to the extent to which students of electricity graduating from these schools have been successful in obtaining employment at remunerative salaries. Referring to the tabulated results of these enquiries, Mr. Floy sums up the subject in the following words : "Considering the table of total results, which may be taken as a fair indication of the condition of the recent graduates in electrical engineering, it will be found that, while a greater

*Electrical Engineering as a Profession.*

per cent. of the graduates in electrical engineering secure employment, as compared with other graduates, yet the fewest, relatively, secure employment in the line of work for which they had studied, that is, in order to get employment, they had. to take positions in which their electrical knowledge did not count. It will furthermore be noticed that almost twice as many men secure employment in electrical engineering through the influence of their relatives as in mechanical or civil engineering, while about half as many obtain positions through their friends as in the other two professions."

**Useless Fenders.**   THE fenders on some of the Toronto Railway Company's cars are elevated too far above the track to allow of an obstruction of reasonable size being scooped up by them. An individual who should be so unfortunate as to fall in front of one of these cars would almost certainly pass under the fender and be crushed to death. Fenders attached to cars in this manner are a mockery.

**Questions and Answers.**   By reference to our Question and Answer Department, it will be seen that some of our subscribers appear to have suddenly made the discovery of their privilege to ask questions and receive information through the pages of THE ELECTRICAL NEWS on any problems in electricity or steam engineering with which they may find themselves confronted. This should be gratifying to every reader, as well as to the publisher. The more questions are asked and answered, the more helpful the journal must become to its subscribers, and the greater evidence to the publisher that it is being widely and carefully read—a matter in which advertisers should also feel an interest. We have daily evidence that the NEWS is becoming more widely known and appreciated by the classes in whose interest it is published. There is no occasion for the expressed hesitancy and apologies with which some of the questions received are propounded. Questions honestly propounded, and of a character to draw forth information of a practically useful character, are cordially invited. If we except the man who "knows it all," we are all in the position of the pupil who is daily adding to his stock of knowledge, therefore, we need feel no hesitation about admitting our lack of knowledge on certain subjects, with which, perhaps, we have not had the opportunity to become acquainted. Those who know the most are invariably ready to admit what they don't know, and to ask to be enlightened. There is little hope of improvement for the man who either thinks he knows it all or is ashamed to ask for the information he requires lest he should be thought ignorant. So far as the readers of this journal are concerned, it is only necessary that they should ask in order to receive any information which it is in our power to give.

The Montreal street Railway Co. have ordered a 55 kilowatt direct connected motor-generator set from the Canadian General Electric Co.

Mr. F. O. Blackwell, of New York, engineer of the General Electric Power Co. was recently in Quebec in connection with the proposed electric street railway. Mr. Blackwell inspected the works and machinery of the Montmorency Electric Light Company. The construction of the railway in accordance with the present agreement, includes the taking over of this plant. It is said to be the intention of the promoters to commence work about the first of April.

## CANADIAN ASSOCIATION OF STATIONARY ENGINEERS.

### TORONTO NO 1.

THE above Association no whold their meetings in the new hall on Victoria street on the 1st and 3rd Wednesday in each month. The business transacted at their last meeting was largely in connection with the taking over of their new quarters. Arrangements have been made for starting the new library; some donations have already been received, while a number of manufacturing firms have promised books and models which will make the collection very valuable to the members of the Association. A letter containing the following has been sent to probable contributors to the library :

"Toronto No. 1. is now entering upon its tenth year, and having procured a new and suitable hall, propose commemorating the event by starting a mechanical library, and fitting the meeting room up with models and drawings. Should you feel disposed to assist, in any way, this very laudable object, an intimation to any member of the committee to that effect will be thankfully received, and promptly attended to."

An engineers' manual or pocket book is being prepared by the Association, which will consist of valuable tables and calculations relating to steam engineering. It will be compiled from the works of the best authorities, and is intended to supply engineers with easily accessible and correct calculations upon which to work.

A committee has been appointed to act in conjunction with a similar committee from the Ontario Association with a view to secure a compulsory law from the government, for engineers operating steam plants. It is not probable that any action will be taken by the government this session, owing to its close being near at hand. It is hoped that at the next session a bill will be brought in by the government which will ensure its success. The bill, as proposed, will exempt all steam engines under 15 horse power, and all persons running plants at the time will be given a permit, thus retaining their positions. The certificates of the Marine Engineers' Association, and the Ontario Association would also be accepted.

### BRANTFORD NO. 4.

Mr. Jos. Ogle, secretary of Brantford No. 4 writes : "We are in a very healthy condition ; our meetings are well attended and some very good questions brought forward and practically answered. The debate for our last regular meeting was, "Cylinder Condensation—Illustrations of Indicator Cards," which occupied our time fully to a late hour, having a various number of select cards illustrated on the blackboard."

### LONDON NO 5.

London No. 5 has not been meeting with much success since the opening of the new year. From lack of attendance meetings have not been held, and the members have shown some indifference towards the Association. It is the hope of the officers that a revival in interest will be shown henceforth, in order that the association may not be allowed to become extinct.

### GUELPH NO. 6.

The above Association held meetings on the first and third Wednesdays of February. There was a good attendance at each meeting, and one candidate initiated at the first meeting. On the 20th inst. Messrs. Ryan and Gerry read papers on the care of boilers and the keeping of the engine room.

### CARLETON PLACE, NO. 16.

J. D. Armstrong, Secretary, writes : "Branch No. 16 has not been in a very prosperous state for the past year but has been reorganized, and is now in a far better shape than ever. Although our numbers are small, we are enthusiastic and look forward at an early date to forming the nucleus of a library, also to invest in some models. We have rooms over our President's place of business, and we may say that for the first time in our existence we are now on the way to becoming a successful society. We meet during March and April every Saturday night, and questions are given the members who purpose trying for certificates, to be answered the following week, so that if they get a certificate they will have to work for it.

## A MOTOR-CYCLE CONTEST IN CANADA.

THE letter published in our February number from the pen of Mr. A. W. White, of London, Ont., relative to a contest of motor-cycles in Canada, has been the means of creating considerable interest in the matter. No action has as yet been taken by the Industrial Exhibition Association, but it is learned from the manager that it is quite probable that a contest will be arranged to take place during the coming exhibition. The test would take place on the public highway, a part of the agreement being that the contestants should exhibit their vehicles in the ring on certain days.

As to the conditions under which the contest should be conducted, we have received the following suggestions :

Mr. G. H. Hewitt, President Duryea Motor Wagon Company, Springfield, Mass. : The "Cosmopolitan" offers prizes for a competition in which the awards are made upon points, such as speed, simplicity, durability of construction, cost, safety, etc.  To my mind it would be manifestly absurd to award the first prize to the carriage which should come in last, even if it stood all

THE DURYEA MOTOR CARRIAGE.

the other tests.  Either you must go into an exhaustive examination of each carriage or you must cut off all examination and let speed over a certain number of miles of ordinary road determine the merit.  I am inclined to take up the idea of three trials and taking an average as the result.  As to the amount of prize money, of course the more it is the more of an inducement it will be for people who are working on new designs to compete.  A chance of getting a big prize will cause a man to hurry up his ideas, and enlist capital on his account.  I presume you can get a good showing for $3,000 in prizes, say : First prize $1,500 ; second prize $1,000 ; third prize $500.  We should be happy to enter, if the rules are satisfactory.

Morris & Salom, Philadelphia, Pa. : The only suggestions that we have to make in regard to the conditions governing the race is that the trial should be based on a service similar to that performed by horses at the present time.  There is nothing to be gained by a run of 100 miles or more, which is merely a tour de force. Some tests should be selected under the conditions governing the use of horses at the present time, and they should be continued from day to day so that a comparison can be made with such service.  The ordinary service of a horse does not amount to more than 25 miles per day, and as motor vehicles are intended to replace horses this point should be carefully kept in view in making comparative tests. ·

H. Mueller Mfg. Co., Decatur, Ill.: We think it advisable to make a maximum speed, say twelve miles per hour, those making a faster average speed being considered the same.  This would be quite fast enough, as what we want is a practical carriage—one for general use, not for racing ; one that will go at a good speed, have a strong pull at that speed, and be able to continue the same speed throughout, taking into consideration the kind of roads.  Then economy of operation is the next factor to consider, because the better the economy the more practical the motor, and, besides, it would be a very easy matter to construct a powerful motor when the economy of fuel was left out of consideration. Simplicity and compactness should be considered together, as the motor might be compact, but be so intricate that in order to repair a certain part it would necessitate much unnecessary work to arrive at the fractured part, which might be the same with adjustment.  Then, again, a motor might be simple in all of its parts, but not compact.  It would occupy too much consideration to mention in detail all concerning the different subjects to consider, but those mentioned we consider first, besides there being vibration, odor, ease of guiding, controlling of speed, variation, quick stopping, etc.  Why not make awards on the following principle : Say you call speed 40 points, economy 20 points, etc., making the total number of points that each motor vehicle can receive 100 points.  Now, say Mr. A. comes in first, you give him 40 points in speed, you award him second in economy, which we will call 15 points, and then 5 points in simplicity and compactness, 2 points in elegance of design, 2 points in guiding, 2 points in controlling of speed, and perhaps nothing in remaining features.  Now, Mr. B. comes in second, and you give him 35 points in speed, economy 20, simplicity 10, elegance of design 5, guiding 5, quick stop and start 4, vibration 3, odor 3, the total of which is 85 points, whereas Mr. A.'s total was 66 points, which would allow Mr. B. first prize for the most practical carriage.

## TRADE NOTES.

## THE STEAM ENGINE INDICATOR AND ITS USES.

By Wm. Thompson, Chief Engineer Montreal West Water
and Light Station.

(Continued from January Number)

It will be found an exceedingly interesting and useful study to compare and test the expansion curve drawn by the indicator with the theoretical curve that would be drawn if the steam was a perfect gas and expanding in accordance with Marriontte's laws for the expansion and contraction of perfect gases, i.e., that the pressure of steam expanded will be inversely proportional to the space it occupies. Thus, if one cubic foot of steam at 80 lbs. pressure is expanded to 2 cubic feet, or twice the space, then the pressure will face to 40 lbs., or exactly half its original pressure, owing to the space it occupied at 80 lbs. pressure having been doubled. The pressure will thus fall proportionately to meet all other degrees of expansion.

The engineer must, however, bear in mind that these pressures must be dealt with as total or absolute pressures, that is, reckoned from a perfect vacuum, and also that the clearance space must be carefully noted, and a line drawn on diagram representing an amount of space on diagram equal to same per cent. of space that total area of clearance space bears to piston displacement. This is made evident by the fact that when cut-off takes place piston will only have moved a given distance or part of stroke, and that, supposing cut-off to occur at ¼ stroke, then to get correct space occupied by steam at any pressure at any point of stroke, clearance space at end of stroke, together with area of steam passages must be taken into consideration, as all steam confined and bearing on piston must have an effect on expansion curve, and when piston reaches ½ stroke, space occupied by steam would not be fairly represented by supposing space to have been doubled while area through which piston has moved would be exactly doubled. Clearance line on diagram will vary as to position with different makes of engines, and can only be correctly ascertained by measuring engine on which test is being made. While steam confined in clearance space does no actual work during live steam period, just as soon as cut-off takes place it has an effect on expansion curve and raises terminal pressure as compared with steam expanded from the admission line only.

The engineer will note that any irregularity or bad arrangement of valves can be readily detected by the position of the various lines, while defects, such as leaky valves or piston, can only be

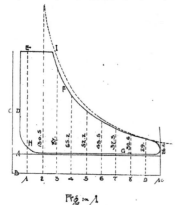

Fig. 1

A, Atmospheric Line ; B, Vacuum Line ; C, Clearance Line ; D, Admission Line ; E, Steam Line ; F, Expansion Curve ; G, Exhaust Line ; H, Compression Curve ; I, Isothermal Curve (test.)

detected by testing and comparing expansion curve, particularly if leak is a small one.

A theoretical expansion curve to conform with above theory may be constructed by several geometrical methods, but probably the following will be the most easily understood by the average engineer :—

The diagram as drawn by the indicator will have the atmos-

pheric line upon it as already described, and from this as a basis draw in the line of no pressure or line of perfect vacuum. To do this, draw beneath the atmospheric line a line as far beneath it as will represent the vacuum line on the same scale as the spring used in the indicator to draw the diagram. The clearance line must then be drawn in accordance with rules already given. Divide the length of the diagram into any number of equal parts by vertical lines at right angles to the atmospheric line and commencing at the clearance line as shown in Fig. 1.

Number the vertical lines as shown—10 being used in this instance simply because it is a convenient number, but any number would do ; the more lines used the greater the degree of accuracy obtained.

Decide which part of the diagram its expansion curve shall coincide with, and touch the test curve ; in example I have decided it shall be line No. 9. Now find what pressure line 9 represents on the scale of the indicator spring—which in this case is 29 lbs.—the line measuring 29/40 of an inch and a 40 lb. spring having been used to draw the diagram. Next multiply the pressure thus obtained by the number of the line (9) and divide the product by the number of each of the other lines in succession, and quotient will in each case be the pressures to be represented by the lines.

For example, to find the pressure requiring to be shown on line

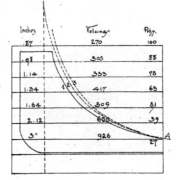

| Inches | Volumes | Pre. |
|--------|---------|------|
| .87 | 2.70 | 100 |
| .98 | 3.05 | 88 |
| 1.14 | 3.53 | 75 |
| 1.34 | 4.17 | 63 |
| 1.64 | 5.09 | 51 |
| 2.12 | 6.60 | 39 |
| 3° | 9.26 | 27 |

Fig. 2

8, we have that, (261,) divided by number of line (8) gives 32.6—hence line 8 requires to be drawn high enough to represent a pressure of 32.6 lbs. above a perfect vacuum, or in this case 32.6/40 of an inch. Having carried this out for all the lines from 10—2, draw in the test curve ; which will touch the tops of all these lines.

This curve, however, does not quite correctly represent the expansion of steam, although generally used. It would do so if the steam remained or was maintained at a uniform temperature during the whole period of expansion. It is therefore called the isothermal curve or curve of equal temperature. But, in fact, steam and all other elastic fluids fall in temperature during expansion and rise during compression, this change of temperature slightly changing and affecting the pressure.

A curve in which the combined effects of volume and resulting temperatures is represented is called the adiabatic curve or curve of no transmission since no heat is transmitted to or from the fluid during the change of volume, its sensible temperature will change according to a fixed ratio which will be the same for the same fluid in all cases.

A fairly close approximation to the adiabatic curve, to enable the engineer to form an idea of the difference between the two may be produced by the following process.

Take a diagram similar to the one used in Fig. 1 and illustrated as Fig. 2. Fix on a point for the coincidence of the two lines as before as at A, where the total pressure is shown to be 27 lbs. As in the former instance this point is chosen in order that the curves will coincide. Any other point might have been chosen for the point of contact ; but a point in that vicinity should generally be chosen so that the result will show the amount of power that should be obtained from existing terminals.

The point chosen in Fig. 2 is 3 inches from the clearance line, and the volume of 27 lbs. is 926—that is, steam of that pressure has 926 times the bulk of water from which it is evaporated.

If we divide the distance of A from the clearance line by 926 and multiply the quotient by each of the volumes of the other pressures indicated by similar lines, the products will be the respective lengths of the lines measured from the clearance line—the desired curve passing and touching their extreme ends. Thus, the quotient of the first or 27 lbs. pressure line, divided by its volume (926) is .00323. This, multiplied by 655, the volume of the next pressure line (39,) gives 2.12 inches, the length of the line to be drawn from the clearance line, and so on for all the rest throughout the illustration.

The application of either of the above curves will show that some diagrams are much more accurate than others. As a general rule those from large-sized engines will be more correct than from small ones, and those from high more correct than from low speeds, and with efficiently covered steam pipes and jacketed cylinders to prevent condensation, a great improvement can be effected.

The character of the imperfections in the expansion curve in the illustration (Fig. 1) shown by the application of the test curve is too high a terminal pressure for the point of cut-off—the first part of the curve being fairly correct, nearly the whole of the inaccuracy occurring during the last half. The usual and most accepted explanation of this is, that the steam admitted during the live steam period condenses somewhat, owing to its having to impart a certain amount of heat to the walls of the cylinder to raise it from the temperature retained from the exhaust steam, and that this water of condensation re-evaporates during the latter part of the stroke, when this water of condensation is at a higher temperature than the expanded steam, and thus increases the pressure. A leaky admission valve or wet steam may, however, generally be looked for if the expansion curve rises much during the latter part of stroke.

In seeking the causes that may produce a defective diagram, the following should be remembered : The indicator must be kept in perfect order, thoroughly clean and well lubricated, so that its parts will move freely. The motion of the paper drum should record an exact copy on a reduced scale of the piston, and should coincide with it at every point of the stroke.

The pipes from the indicator to the cylinder must be large enough to give a free and full admission and pressure of steam, and care must be taken that the water of condensation does not obstruct or enter the indicator.

The metallic point or pencil should be held to the card with just sufficient force to make a fine clear line.

The diagram should be the exact length of the atmospheric line; any difference in this respect shows poor adjustment in some part or unequal tension of cord.

A fall in the steam line could arise from too small a steam pipe. This can be tested by taking a diagram from the steam chest. The same fault could also occur from too small a steam port or an obstructed passage, such as partial closing of admission valve, also by steam leaking past piston and passing to atmosphere unutilized.

An expansion curve that is higher than it should be could arise from a leaky valve on the steam side letting steam in from the steam chest after cut-off had taken place ; in this case the leak will naturally become larger as the steam expands and pressure on piston side of valve reduces, consequently terminal pressure will be more or less out of proportion.

An expansion curve that is lower than it should be may be caused by a leaky piston, by a valve that leaks on the exhaust side, but not on the steam side; or if the exhaust valve is separate from the admission valve, it may leak while the steam valve is tight, thus lowering the terminal pressure. It may also be caused by the cylinder becoming unduly cooled, as from water being allowed to accumulate in a steam jacket ; this will particularly affect the curve during the earliest stages of expansion or even during admission.

As already explained there are many defects in the adjustment of the valve gear that will be clearly shown by the indicator diagram. But, it should be borne in mind that there are possible defects which the indicator will not show. For instance, a steam valve and the engine piston may both leak to an equal amount ; as a result the expansion curve may not show the leak, while, as a matter of fact, loss is occurring from this source.

Insufficient valve lead, or in other words the admission valve opening too slow or late, would be shown by the piston moving a certain portion of the stroke before the steam line attained its greatest height. In this case the admission line, instead of rising vertically as shown in illustrations, would be at an angle to the right showing that the piston and consequently the indicator drum had moved a certain distance of the stroke before the valve was wide open and full pressure of steam admitted.

Exclusive lead is shown in Fig. 3 by the loop at A, where the compression curve extends up to the steam line and the lead carries the admission line above it owing to the engine piston moving against the incoming steam.

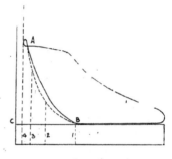

Fig. 3

To mark in the theoretical compression curve, that is the curve that would be formed by the compression of the steam remaining in the cylinder after the exhaust valve had closed and previous to the opening of the admission valve, the vacuum line and clearance line must be drawn in as before. In Fig. 3 compression commences at B, and at that time the space filled with steam is represented by the distance from B to the clearance line C. The pressure above vacuum of the steam remaining in the cylinder when compression began is shown by dotted line. Suppose the piston to have moved from point B to line 2, which is half the distance from the clearance line, of line 1, and as the compressed steam now occupies only one-half its former space, therefore the steam pressure will be doubled and line 2 requires to be drawn twice the length of line 1.

Line 2 is now the starting point for getting the next ordinate and 3 must be marked midway between 2 and the clearance line and twice as high as line 2, as it is obvious that at line 3 the steam will occupy only half the space it did at 2 and one-quarter of the space at 1, hence pressure is increased proportionally. Line 4 is drawn midway between 3 and the clearance line as before. Through the tops of these lines draw the theoretical compression curve as shown by the dotted line.

To find amount of steam actually saved by compression, consider the compression curve only beginning at the point of the diagram where compression actually began, and ending where the compression curve joins the admission line, the horizontal distance between these two representing the length of the cylinder bore actually filled by compressed steam.

---

It is stated that in a few weeks the Edison Electric Illuminating Company of New York will have in operation at one of their stations two 300 h.p. De Laval steam turbines with attached dynamos. These turbines were built by the Maison Breguet, Paris, and are now on their way to America. They were ordered under guarantee to comply with the following specifications : each 300 h.p. turbine is to drive two Desroziers dynamos, each of 133 h.p. capacity. The turbine shaft is to run at 13,000 revolutions, driving at a speed of 1,300 revolutions by means of helical gearing, two dynamo shafts situated on either side of the turbine shaft. Each dynamo is to be capable of generating continuously without undue heating 770 amperes at 130 volts or 625 amperes at 160 volts.

## ELECTRICAL DEVELOPMENTS AT MONTREAL.

THE electric light situation in Montreal, as well as in Toronto, is at present at an interesting stage. The company formed some time ago, for the purpose of utilizing the water power of the Lachine Rapids, have been energetically pushing forward their enterprise, and maturing plans for the disposal of the electrical energy which will be generated from the works now in process of construction at Lachine.

It is understood that the company have bought up the lighting privileges of Westmount and several of the other suburban municipalities of Montreal, as well as the franchise of the Standard Electric Lighting Co., which is said to carry with it the right to do electric lighting within the city of Montreal. Under this latter franchise, the company propose to compete for the lighting and power business of the city of Montreal. They have already constructed poles and wires to a point about a mile within the city limits, but here their operations have had to be suspended for a time at least, owing to legal action brought against them by the Royal Electric Co.

The Royal Company, it is understood, will endeavor to prove that the franchise under which the company are preceding, does not give them the right to do business inside the city, and the courts are now considering an application for an injunction to restrain the new company from proceeding further with their enterprise so far as city business is concerned.

It is claimed that the Lachine Power Co. will be in a position to supply current at a greatly reduced rate as compared with the prices that are now being charged for lighting and power in the city of Montreal.

It will be remembered that several years ago the Royal Electric Co. purchased the water power at Chambly, across the river from Montreal, and about six miles distant from the city. The intention of the company seemed then to be to utilize as quickly as possible this water power. It is therefore somewhat surprising that nothing has been done in this direction, while a competitor has in the meantime secured the control of a greater water power, and one which is more conveniently situated. No doubt the fact that the Royal Company would have been obliged to bring the current across the St. Lawrence, through a sub-marine cable, which would be subject to the destroying action of frost and ice, at a point where an ice jam is a yearly occurrence, had something to do with the fact that no attempt has been made to utilize the power from this source. On the other hand the company are understood to be remodelling from top to bottom their electric lighting station, thus putting themselves in a position to meet any competition which may arise.

The following particulars of the works now in course of construction at the Lachine Rapids, under the direction of Mr. W. McLea Walbank, C. E., and R. E. T. Pringle, M. E., by Messrs. Davis & Sons, contractors, of Ottawa, with the aid of 300 workmen, will show the magnitude of the undertaking :—

Along the north shore of the river within 1,000 feet of it, 2,500 soundings have been made, which have shown that to overcome the freezing of the shallow water 250,000 cubic yards of shale rock will have to be taken out to deepen the water. When this is done an artificial canal will be made by building a wall 4,500 feet long and 20 feet wide, 800 feet from shore. 3,000 feet of this wall is to be built of crib work, filled with masonry and concrete, the rest of the wall, 1,500 feet, at the head of the head race, is to be submerged, coming within a foot of the surface of the water and built of cut stone to act as an ice breaker. Across the canal is being built a dam to raise the water 9½ feet above the tail race, the canal wall being at its highest to keep floating objects from the wheels.

In this dam are to be placed 66 upright cylindrical gate turbine wheels to give 125 h. p. under 8 feet head, to realize 80% useful effect. Each wheel is placed between two stone piers. On this dam will be built three power houses, connected by galvanized iron sheds, the sheds covering the wheels. Each generator will be of

the capacity of 750 h. p., having connected to it six wheels. Each power house will contain four such generators. All the generators will be connected to one switch-board, and the power will be carried under high voltage to Montreal, where by rotary transformers it will be reduced to the required voltage for power by day and power and light at night. The plant is to be completed and in full running order before the close of the present year.

The provisional board of directors of the company are as follows : G. B. Burland, Montreal and Ottawa ; W. McLea Walbank, C. E., Montreal ; Thos. Pringle, M. E., Montreal ; Alderman Peter Lyall, contractor, Montreal, Samuel Carsley, merchant, Montreal ; E. Kirk Green, merchant, Montreal ; Hugh Graham, of the "Star," Montreal.

The capital of the company has been placed at $1,000,000, $500,000 of which is for sale. Debentures will be issued at 4½%.

## DIRECT CONNECTED PUMPING SET.

AN interesting example of the prompt application for the alternating motor in directions which have not up to the present afforded much field for electric power service is presented by the pumping set illustrated herewith, which has recently been installed by the Mattawa Electric Light & Power Company, at Mattawa, Ont. This unit which presents the double advantage for the service of being particularly compact and so simple in operation as to require no attendance whatever, consists of a 5 horse power Canadian General Electric induction

DIRECT CONNECTED PUMPING SET.

motor geared to a Gould triplex pump, current for the motor being obtained from the Mattawa Company's monocyclic circuit. The work to be done is the filling of the C. P. R. water tank, capacity 50,000 gallons, formerly supplied by a steam pumping outfit, the length of the pipe (4″) being 2,129 feet, and the lift from the water to top of tank being 96 feet. The motor pumping set has now been in operation for several weeks and has given the most perfect satisfaction. The only attention required at present is to start and stop the motor, but it is intended to do away with the necessity for even this small amount of attendance by having the motor started and stopped by the operation of a float in the tank.

The enterprise of the Mattawa Electric Light and Power Company in opening up what we believe to be an entirely new field for electric power, will, no doubt, be followed by other and older stations.

## THE APPLICATION OF OHM'S LAW.*

By W. NORRIS.

IT has become necessary for engineers to have a knowledge of electricity and the application of the same. I am therefore prompted to take a step towards making that subject to have a more prominent place in the engineers' lodge room. It is with much pleasure that I will try to illustrate a few important points that will be of much use to beginners.

Whenever we require to make any calculations upon the current that will flow from any kind of electrical supply through an ordinary conductor, we must have some law by which to be guided. Professor Ohm has laid down a law which is known as Ohm's Law, and reads as follows :

Current in Amperes = $\dfrac{\text{Electrical motive force in volts.}}{\text{Resistance in ohms.}}$

For instance, one volt will force one ampere through a resistance of one ohm.

100 feet of No. 10 B. & S. copper wire, which will conduct about 98 per cent. of the current, has a resistance of one ohm. So if the current in amperes is equal to the electromotive force divided by the resistance in ohms, then the resistance in ohms will be equal to the electromotive force divided by the current in amperes, and the electromotive force will be equal to the current in amperes multiplied by the resistance in ohms, and is represented in the following manner :

$$C = \frac{E}{R}, \quad R = \frac{E}{C}, \quad E = C \times R.$$

It can be plainly seen that if any two of the above elements are known, it is an easy matter to find the remaining one.

The figures, 2, 8, and 4 always stand for C, E, and R, respectively. In order to illustrate the use of Ohm's law more plainly, we will suppose we have a simple primary cell of battery with an E. M. F. of 2 volts, leaving a resistance within itself of ½ ohm ; then, according to the formula $C = \frac{E}{R}$, we have 2 volts divided by ½ an ohm, equals 4 amperes, thus :

$$\frac{5 \text{ ohms}}{2.0} ) \overline{2.0} \text{ v} / 4 \text{ a.}$$

If we had found that we had 4 amperes of current and 2 volts pressure, we would have had 2 volts divided by 4 amperes, equals ½ ohm resistance

$$4 ) \frac{2.0}{2.0} ( .5 \text{ or } R = \frac{E}{C}$$

Let us now form a number of these cells together, thus forming a battery. Say we use 5 cells and connect them in series as shown in Fig. 1. By this means we get the effect of all the cells

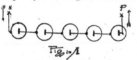

Fig. 1

together, for the voltage will build up according to the number of cells that there are in series, although the amperage remains the same as in one cell, because each cell has a definite and determined resistance which increases as there are cells in series, and will only permit the same amount of current to pass through it as it will deliver itself. But each adjoining cell helps to build up voltage ; therefore we have a pressure of 10 volts and 4 amperes from this battery.

One ampere flowing under a pressure of one volt is equal to one watt, which is the mechanical work performed and is the unit of a horse power ; for 1,000 Watts constantly delivered to an electric motor will make it deliver one horse power, and as we have 10 volts, then the 10 volts multiplied by 4 amperes equals 40 watts, indicating the amount of work this battery will perform.

Let us now connect these 5 cells in another manner, as shown in Fig. 2. With all the positive poles connected together and all

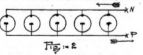

Fig. 2

the negative poles connected together, this is known as the multiple connection, which causes the battery to deliver the current just the opposite to the series connection. For, instead of the voltage increasing it remains the same as in one cell, while the amperage builds up in proportion to the number of cells ·in multiple. We have these 5 cells each delivering 4 amperes, and the voltage on the mains is but 2 ; then 4 amperes multiplied by the 5 cells equals 20 amperes, and 20 amperes multiplied by 2 volts equals 40 watts ; so it will be plainly seen that the work performed by these two batteries, Nos. 1 and 2, is just the same, although the current is different.

We will now take the same two batteries and connect them both together, making a compound connection as shown in Fig. 3, which is known as the multiple series ; for each pair of cells is connected in series, with their positive poles of each pair connect

*Paper read at regular meeting of Hamilton No. 2, C. A S. E.

ed to one wire, and all the negative poles likewise, so that from each pair of cells in series we have 4 amperes and 4 volts, and as there are 5 pairs of cells in the battery, we have 4 amperes multiplied by 5 pairs of cells, which gives us 20 amperes, which,

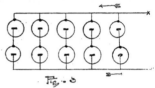

Fig. 3

multiplied by 4 volts equals 80 watts, showing that the mechanical work that this battery will perform is equal to both the first batteries.

We will take still another form of battery, known as the series multiple, as shown in Fig. 4, using the same number of cells as in

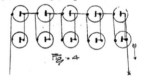

Fig. 4

the last battery, each pair of cells being in multiple, thus acting as one large cell, so we would have the same as 5 large cells in the battery placed in series to each other. In this case we would have the same number of volts from each pair as from one small cell, but twice the number of amperes, which would be 8 ; and as each pair of cells are in series, we would have two volts multiplied by 5, equals 10 volts, because when each pair of cells are in series with each other, the voltage increases, but the amperage remains the same as if there was only one cell. Then by multiplying the 10 volts by the 8 amperes we have 10×8=80 watts, showing that the mechanical work which this battery is able to perform is the same as with the last form of battery, the series multiple, although the voltage and amperage were both different.

Quite a number of other connections can be made to suit the work the battery is required to perform. For instance, we have 5 cells each delivering 2 amperes and 2 volts, and when connected in series would be equal to 10 volts and two amperes, and as the resistance in ohms is equal to the E. M. F. ÷ by current in amperes, or $R = \frac{E}{C}$, then 10÷2=5 ohms which would be the resistance of the battery itself and the resistance in the line wire and bell, etc., should be about the same. So in this case the current in amperes would be equal to the electromotive force divided by the resistance of the battery and line wire, etc., added together, or $C = \frac{E}{R+R}$, thus 10÷10=1 ampere.

Now, the watt is the unit of a horse power of the work performed, and the current in amperes multiplied by the electromotive force equals the number of watts ; then C × E = W, $R = \frac{W}{C}$ and $C = \frac{W}{E}$. Supposing we have a 16 candle power lamp taking 3.1 watts per c.p., with 100 volts, how would we find the amperage as well as the resistance of the lamp ? The lamp is 16 c.p., watts 3.1 per c.p., then 16 × 3.1 =49.6, or say 50 watts and the current in amperes is equal to the watts divided by the electromotive force, or $C = \frac{W}{E}$. It would therefore be 50 watts divided by 100 volts= ½ as the amperage of the lamp, and the resistance being equal to the electromotive force divided by the current in amperes, it would be 100 volts divided by ½ ampere = 200 ohms as the resistance of the lamps, thus :

$$.5 ) \frac{1000}{10} ( \overline{200} \text{ ohms.}$$

So the $C = \frac{E}{R}$ because 100 ÷ by 200= ½ ampere, and C × E = W because ½ ampere × by 100 volts equals 50 watts.

Nature states that an ingenious system of purifying atmosphere and regulating temperature is in operation for the switchboard room of the Chicago Telephone Company, where dust formerly interfered seriously with the connections on the switchboard. The air for the room is forced through a chamber, where it is thoroughly sprayed, then passed through rapidly rotating spiral coils, which strip it of superfluous moisture, and afterwards through a chamber kept at nearly uniform temperature by the use of ice or heating apparatus, as may be required. Access to the switchboard room is through an ante-chamber, and the temperature of the room itself shows a variation of not more than two degrees in a month.

# ELECTRIC RAILWAY DEPARTMENT.

## THE BERLIN AND WATERLOO ELECTRIC RAILWAY.

THIS road, which last year was electrically equipped, has recently undergone a change of ownership. The controlling interest has passed from the hands of Mrs. Burt, of New York, to Messrs. W. H. and E. C. Breithaupt, of Berlin. Mr. T. M. Burt and Mr. T. E. McLellan will be retained in their present positions in the management, though Mr. E. Carl Breithaupt is the President of the Company with general oversight. The following are the directors :—T. M. Burt, T. E. McLellan, A. Millar, G. Bruce and E. C. Breithaupt.

The charter of the Grand Valley Railway, of which E. C. Breithaupt is President, is virtually held by the same parties who now own the Berlin and Waterloo road. They therefore have a strong interest in bringing that scheme into life.

Mr. E. C. Breithaupt, the new President of the Berlin and Waterloo road, is recognized as being one of the most thoroughly educated electricians in Canada. This fact, together with his financial interest in the enterprise, is a guarantee that the road will be equipped and operated in the most approved manner.

### SPARKS.

Seventeen open motor cars are now in course of construction for the Toronto Railway Company.

Contracts are now being let for the construction of an additional mile of the Guelph Electric railway, which will be completed by the 24th of May.

A Campbellford capitalist is investigating the prospects for the successful operation of an electric railway from Campbellford to Norwood, Ont.

The Ottawa Electric Railway will this summer be extended to Britannia. The route has been surveyed, and the line will be in operation by the 1st of July.

The town council of Perth, Ontario, will be asked to grant a bonus of $5,000 towards the proposed electric railway between Perth and Lanark.

The plant and charter of the Victoria Electric Railway and Light Co., Victoria, B.C., will be offered for sale by public auction in that city on the 11th of April.

The Railway Committee of the Dominion Government have passed a bill to incorporate the Huron & Ontario Railway Company, which proposes to build an electric railway.

The management of the Hamilton and Dundas Railway Company will shortly submit a proposition to the city council of Hamilton for the conversion of the road into an electric line.

The Port Dalhousie, St. Catharines and Thorold Electric Railway Co. has decided to build eight miles of overhead construction and two miles of track as soon as the weather will permit in the spring.

The Cornwall Street Railway Company, Cornwall, Ont., are applying for incorporation, with a capital stock of $150,000, to operate an electric street railway in that town and to distribute electricity.

The construction of an electric railway between Parry Sound and Ahmic Harbor, Ont., is one of the probabilities of the near future. The distance is thirty miles and the cost of construction is placed at $150,000.

Mr. W. S. Adams proposes to build an electric railway from Derwin, on the C.P.R., to Winnipeg river, a distance of twelve miles. The water power on the river will be utilized for supplying electricity for the line.

At the annual meeting of the Hamilton Street Railway Company the following directors were elected : B. E. Charlton (president), Geo. E. Tuckett, E. Martin, Q.C., W. Gibson, M.P., J. B. Griffith, William Harris and F. W. Fearman.

The town council of Lachine, Que., has adopted a by-law granting the Montreal Park & Island Railway running privileges in the town, with exemption from taxation, for thirty years. It is contemplated to build this line this spring, and to extend the Outremont line to St. Laurent.

By the agreement entered between the town of Brockville and the electric street railway company, the company is to have a twenty years' franchise, and is authorized to construct a single track iron street railway. Construction must be commenced before October 7 next, and one mile completed within a year from that date.

The new Board of Directors of the Hamilton Radial Railway Co. is as follows : Rev. Dr. Burns, president ; A. McKay, M.P., vice-president ; J. D. Andrews, secretary ; W. G. Lumsden, treasurer. James Masson, M.P., Owen Sound : F. A. Carpenter, A. H. McKeown, E. P. Powell, London, Ont.; J. F. Smith, Thos. Ramsay and R. McKay.

A report has been current that the Hamilton Street Railway Company will make application to have the percentage of gross earnings of the system which is paid to the city remitted, owing to a large reduction in dividends. The increase in business anticipated as a result of the conversion of the line into an electric road has only partially materialized.

The city council of New Westminster, B.C., has received a communication from Mr. J. Buntzen, secretary of the Consolidated Railway and Light Co., offering to build an electric railway from Westminster to Steveston, with a branch to Sapperton, and to locate the central offices and repair shops in New Westminster. A bonus of $50,000 is asked from the city.

Albert Phenis, of New York ; Lucius S. Oille, M. D., George E. Patterson, J. S. Campbell, of St. Catharines, and Henry A. King, Toronto, have petitioned for a bill to incorporate the Lincoln Radial Electric Railway Company, with power to take over the powers of the Lincoln Street Railway and Traction Company, and to extend the line to Toronto.

Mr. T. W. Lester, president of the Hamilton, Grimsby and Beamsville Railway, states that, notwithstanding the opposition of the Grimsby Council to granting right of way to Beamsville on reasonable terms, the electric road will be extended to Grimsby Park by a new route, independent of the Grimsby Council, and cars will probably be running by 1st July next.

The electric street railway was started in the city of Halifax about the middle of February. The initial trip proved quite successful, and was taken charge of by Mr. Norman Ross, E.E., representing the Canadian General Electric Co. The cars are equipped with C.G.E. 800 motors and parallel controllers. The Train company is installing two large C.G.E. dynamos of monocyclic type, having a combined capacity of 6,000 lights. The generators are operated by a 300 h.p. Robb-Armstrong compound engine.

The bill to incorporate the Canadian Electric Railway and Power Company, which proposes to build an electric line from Windsor to Montreal, with branches of not more than 25 miles radius, came before the Railway Committee of the Dominion Parliament on the 26th of February. The promoters are Messrs. Castle Smith, London, Eng.; J. K. Osborne, T. M. Jones, C. W. Beardmore, W. H. Cawthra and Edmund Bristol, of Toronto, and E. F. Fauquier, of Ottawa. The application was opposed by the Grand Trunk and C.P.R. authorities. The measure was allowed to stand over for further consideration.

The supper recently tendered to the employees of the London Street Railway by the efficient manager, Mr. C. E. A. Carr, was one which will be remembered with pleasure by those present. The gathering numbered about 85, and included the office staff, motormen, conductors, power house employees and superintendents. The manager sat at the head of the table, accompanied by Mrs. Carr. The spread was an excellent one, and after full justice had been done, a toast list was introduced, Mr. Currie, secretary, and Mr. De Harte, superintendent, responding to the toast of the "London Street Railway Company," and Mr. Carr to that of the "Manager of the Company." The toasts of the various departments were heartily received, and the pleasure of the occasion was greatly added to by songs by Mrs. Carr, Mr. Currie and Mr. Birmingham.

Mr. C. A. C. Pew, of St. Catharines, is promoting an extensive electric railway enterprise across the north-western portion of Ontario, from Port Perry to Lake Huron. It passes through the counties of Ontario, York, Simcoe, Cardwell, Grey and Bruce, and will touch at the towns of Newmarket, Bradford, Beeton, Shelburne, Priceville, Durham, Hanover, Walkerton, Meaford, Owen Sound, Southampton, Kincardine, Teeswater, Wingham and Goderich. All these places are now served by parallel lines of steam railways, radiating mostly from Toronto, and which necessitates, on the part of travellers, the making of long round-about journeys to go from any of the points named to another. The proposed electric road is therefore intended to promote a great public convenience, and is in consequence meeting with a very enthusiastic support all along its route. It crosses the G.T.R. and C.P.R. at several points, and will in a measure serve as a feeder for both roads. The part of the country it passes through is fertile and prosperous, and much in need of railway accommodation in the direction proposed. Large meetings in favor of the project have been held at all the towns named, at which money for preliminary expenses was freely subscribed. Parliament will be asked at its present session for a charter, which, when obtained, will be passed over to a New York company, which proposes to build and equip the road without asking either the government or municipalities for a bonus. The road is sure of an extensive traffic, and can scarcely fail to return large earnings to its owners. An abundance of water power exists along the route.

## RECENT CANADIAN PATENTS.

Patents have recently been granted in Canada for the following electrical and steam engineering devices:

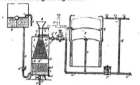

APPARATUS FOR GENERATING ACETYLENE GAS.

Patentee: T. L. Willson, New York, N. Y., patented 5th November, 1895; 6 years.

Claim.—The combination to form a gas-generating apparatus of a gas generator consisting of a chamber having a receptable for carbide, a gas outlet from the upper part of the generator, a water inlet to the lower part thereof, and a source of water connected with said inlet under pressure sufficient to raise it above the level of the carbide, the whole adapted for automatic operation controlled by the relative pressures of the water and the generated gas, so that the water, after reaching the carbide is forced out of contact therewith whenever the gas is generated enough faster than it is consumed to raise its pressure above that of the water.

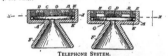

TELEPHONE SYSTEM.

Patentee: A. C. Brown, Lewisham, Eng., patented 2nd November, 1895; 6 years.

Claim.—In a telephone receiver the arrangement and combination of a central or cylindrical casing or ring seating with ear piece and with two diaphragms both adapted to be simultaneously vibrated in opposite directions to or from each other, and polarized by magnets. In a telephone receiver having two diaphragms clamped onto a cylindrical seating, the use for polarizing such diaphragms or cores of a split steel tube such as S, encircling the coils as above described, or for the same purpose of magnets arranged or adapted to operate substantially as above described and illustrated.

ELECTRIC RAILWAY SYSTEM.

Patentee: Canadian General Electric Co., Toronto, Ont., patented 18th November, 1895; 6 years.

Claim.—In an electric railway system, the combination with a vehicle electrically propelled, of means for stopping and starting said vehicle at definite points, consisting of a series of conductor sections located near such points and making connection with the vehicle, and a storage battery, having connections from points of different effective potential to the various sections, the potential decreasing from each end section toward the middle. In an electric railway system, the line or supply motor, a series of section conductors connected to said line through resistant or equivalent devices for lowering the effective electro-motive force of said sections from that of the line in a successive and graduated manner.

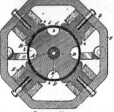

ELECTRIC MOTOR.

Patentee: Charles Riordan, of Toronto, Ont., patented 18th November, 1895; 6 years.

Claim.—In an electric motor the combination with the exterior field magnets, of a hollow cylindrical armature comprised of wire loops suitably supported and secured to the main shaft of the motor and a solid core located within the armature magnetically insulated from and loose on the shaft and provided with recesses in its periphery between the ends of the cores of the field magnets whereby the lines of force maintain such core from rotating on the shaft, the armature supported on discs and comprised of a series of loops substantially rectangular, arranged in sets abutting each other, the sides of the loops of each set being arc-shaped, and each side being arranged to fit beneath the side of the adjacent loop of the set, so as to form a complete cylinder of double layer arc-shaped wire sides, the ends of the wire of each loop being connected to corresponding sections in the commutator.

LEVER FOR TURNING STEAM ENGINES OFF THEIR DEAD CONTRES.

Patentee: John Donnelly, St. Henri, Que; patented 2nd December, 1895; 6 years.

Claim.—On a lever for turning steam engines off their dead centres the combination of a lever A, having cross pieces b and a¹, two pieces a² and a³, provided with the rings D and D¹, holding the ones E, to which is secured the levers g and g¹, of the grapples G, with a suitable stand C¹, and socket bar a³.

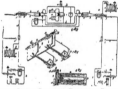

MULTIPLE SWITCH BOARD FOR TELEPHONE EXCHANGES.

Patentee: The Bell Telephone Company of Canada, Montreal, assignee of C. E. Scribner; patented 5th December, 1895; 6 years.

Claim.—The combination with an annunciator having an electro-magnet, a pivoted armature therefor, an indicator and mechanism is connected with said armature and indicator adapted to actuate the indicator when the armature is vibrated between its extreme positions, of a circuit containing a source of pulsating currents, a source of continuous current and means for connecting said source of continuous current with the circuit, whereby the actuation of the indicator by pulsating currents may be prevented by connecting the source of continuous current with the said circuit. In an annunciator in a ground branch having an electro-magnet, a pivoted armature therefor, an indicator and a catch-arm carried by said armature having alternate teeth adapted to engage with and retain said indicator when the armature is in either of its extreme positions, but to release the same when the armature is vibrated, a connecting plug for insertion into any spring jack, having contact-pieces arranged to register with the corresponding contact-pieces of a spring jack, a conducting circuit joining the different contact-pieces of a plug, including a clearing-out annunciator, a source of current adapted to actuate said clearing-out annunciator.

FARE REGISTER.

Patentee: The St. Louis Register Co., St. Louis, U. S. A.; patented 9th December, 1895; 6 years.

Claim.—The combination with the trip register, the permanent register and means for releasing the trip-register, and for locking the permanent register against movement while the trip-register is released, of a motor for returning the trip-register to zero, and means for actuating registers step by step and for unlocking the permanent register against movement while the trip-register is released. The combination with the registers and the fare indicating signal I of an arm for moving it in one direction, a longitudinally yielding pin for holding it in such position, and an oscillating pawl for depressing the pin to release the signal.

FARE BOX.

Patentee : J. H. Coleman, Tottenham, Ont. ; patented 10th December, 1895 ; 6 years.

Claim.—In a fare box, one or more needles arranged to permit of the insertion of fares into the box and automatically arranged to resist their withdrawal when the box is in a normal position. In a fare box, a concave and a rotatable toothed drum between which the fares pass, with needles having weighted tails to retain the points of the needles in the path of fares passing between the concave and the drum.

## SPARKS.

The Bear River, N. S., Electric Co. have decided to extend their lighting system to Digby.

A franchise has been given to the Belleville Electric Company to construct an electric railway between Belleville and outlying villages.

D. Knechtel, of Hanover, has started up his first 10 h. p. induction motor operating from the monocyclic circuit. The operation of the motor is so satisfactory that Mr. Knechtel looks for a considerable power business in Hanover.

Mr. James Milne, lecturer in Electricity at the Toronto Technical School, held a preliminary examination for the class on the 27th of February. Although only a small number of students were present, the results were very satisfactory.

In the city of Montreal many ex-telegraph operators hold positions of trust and responsibility. Among those may be mentioned: Sir William Van Horne, president Canadian Pacific Railway ; Mr. Charles W. Hays, general manager, and Mr. Geo. B. Reeves, general traffic manager Grand Trunk Railway ; Mr. J. Stephenson, general superintendent Grand Trunk Railway ; Mr. J. Bryce, superintendent Canadian Express Co., and Mr. Wm. MacKenzie, stock broker.

Mr. W. L. Gilchrist recently delivered a lecture at Victoria, B.C., on "Magnetism and Electricity."

The capital stock of the Toronto Electric Light Company, Limited, has been increased from $700,000 to $2,000,000.

The offices of the Kingston Light, Heat and Power Co. have been enlarged and equipped with modern appliances.

The Montmorency Electric Power Company, Quebec, are negotiating with the town of Levis to furnish 50 horse power for pumping apparatus and electricity sufficient for 80 arc and 2,000 incandescent lamps.

At the annual meeting of the Portage la Prairie Electric Light Company, Messrs. T. B. Millar, Judge Ryan, Hon. R. Watson, Smith Curtis and Mr. Blake were elected directors. The report presented showed a satisfactory year's business.

The Citizens' Light & Power Company, of Montreal, held its annual meeting early in February. Major Wilson-Smith was elected president, Mr. W. McLea Walbank, vice-president and managing director, and Mr. R. B. Hutcheson, secretary. The report showed that in the last month the company had obtained 25 new customers at meter rate, and 15 customers at flat rate. The directors elected for the ensuing year were : Major G. H. Burland, W. McLea Walbank, P. Lyall, M. P. Davis, L. H. Héneault, Mayor of Ste. Cunegonde, and ex-Mayor Dagenias, of St. Henri.

A bonus of $10,000 for an electric railway between Perth and Lanark has been granted by the last named town.

Joseph Barrett desires to secure a franchise from the city of Toronto to distribute light, heat and power.

Messrs. James Ogilvy & Sons, Montreal, are installing a 55 kilowatt and a 12 kilowatt C. G. E. multipolar machine for isolated lighting.

Mr. T. L. Wilson, of Calcium Carbide fame, is reported to have purchased power sites at St. Catherines, Ont., with the intention of locating his Canadian works there.

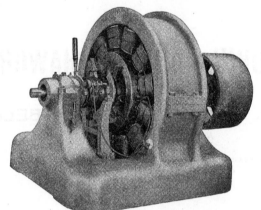

# CANADIAN GENERAL ELECTRIC CO.

## (LIMITED)

### The following Letters speak for themselves:

PORT HOPE, Feb'y 27th, 1896.
MESSRS. CANADIAN GENERAL ELECTRIC CO.,
Toronto.

DEAR SIRS :—The 75 Kilowatt Monocyclic Alternator purchased from you was started up on Sept. 2nd, 1895, and has since been giving us an uninterrupted service of 160 hours each week, starting at four o'clock on Sunday afternoon and running till eight o'clock the following Sunday morning. without a hitch of any kind whatever. We expect a large increase to our business from motor service, and appreciate the excellent features of the Monocyclic system of this combined light and power service. There is, of course, no unbalancing, as the lighting is single phase, and the operation of the motors does not disturb in any way the evenness of the lighting. After an experience of six months we feel warranted in saying that we consider the Monocyclic as superior to any of the polyphase systems which we are acquainted with, and intend in the near future to duplicate this machine.

Yours truly,
R. A. CORBETT,
Pres. & Mgr. Port Hope Elec. Lght & Power Co.

PARRY SOUND, Feb. 27th, 1896.
THE CANADIAN GENERAL ELECTRIC CO., Toronto.

DEAR SIRS :—Having now made a thorough trial of the Monocyclic system of Electrical distribution as supplied by you, I have much pleasure in informing you that it is giving entire satisfaction.

The machine, a 75 K.W., is a beautiful specimen of dynamo building, being strong and compact. Ventilation of the armature is excellent, and the general design of that very important part of the machine is good. Electrically and mechanically, I consider your machine to be superior to any I have seen.

We have not had occasion, as yet, to test the machine on the operation of motors, but speaking from the lighting point of view, I can fully endorse what you claim for the system.

We have 700 lights now wired and expect to increase to 1,200 before 1897.

Yours truly,
W. B. ARMSTRONG,
Manager Parry Sound Electric Light & Power Co., Ltd.

HANOVER, Feb. 21, 1896.
CANADIAN GENERAL ELECTRIC CO., Toronto.

DEAR SIRS :—In answer to your enquiry as regards the operation of our Electric System, I beg to say that we have now been running one of your 75 Kilowatt Monocyclic Dynamos for the past two months, and it is giving entire satisfaction in every respect. We have not had the slightest trouble with it in any way, and although it is being operated about 15 hours a day, it runs exceedingly cool, and requires practically no attention whatever.

The machine itself I regard as a model of simplicity, in fact to show my confidence in the apparatus I have placed the plant in full charge of my brother, who, previous to the starting up of this machine, had no experience whatever with electrical apparatus of any kind.

The perfect regulation of the Dynamos, and the sim-

plicity of the wiring are also strong points which should recommend the use of this style of apparatus to anyone contemplating the installation of an electric plant.

In conclusion, I might say that after having decided upon adopting the Monocyclic system, my opinion became somewhat prejudiced against its adoption by representations made by other manufacturers, but I now fail to see wherein I could have secured anything better to that installed by your company.

Yours very truly,
D. KNECHTEL.

DUNNVILLE, February, 1896.
MESSRS. CANADIAN GENERAL ELECTRIC CO.,
Toronto, Ont.

DEAR SIRS :—We are pleased to be able to express ourselves as entirely satisfied with the Monocyclic system installed by you last fall. We are now in a position, having covered a considerable portion of the town with our lighting mains, to appreciate the value of the three-wire system for secondary distribution from the transformers, and the great advantage gained in simplicity by the Monocyclic from its being a single-phase system for the lighting distribution.

The workmanship and finish of the dynamo itself certainly does credit to your factories, and in operation it has proved itself to be exceedingly simple and satisfactory.

The commutator and brushes run without any sparking whatever, and do not give us a particle of trouble. We feel fully justified in saying that the Monocyclic system in operation has shown itself to possess all the points of excellence claimed for it by you at the time when we made the selection for our new plant.

DUNNVILLE ELECTRIC LIGHT CO.

MATTAWA, Feb. 27th, 1896.
MESSRS. CANADIAN GENERAL ELECTRIC CO., Toronto.

DEAR SIRS :—We are pleased to be able to express complete satisfaction with our Monocyclic plant, which has now been running since 27th Sept. We are especially pleased with the ease with which our former single-phase system has been changed into one suitable for the distribution of both light and power. The only change made in our case was the installing of the Monocyclic machine in place of our former single-phase alternator, and the running of a third wire to the points where power is to be supplied. Altogether the system is admirable, both as to simplicity in the wiring, and distribution and perfect freedom in operation from any trouble or complication. We are quite sure that the Monocyclic system will prove a means of increasing largely the revenues of alternating lighting stations by the sale of power without adding any complications to their operation. You will be pleased to know that the 5 h.p. induction motor geared to a triplex pump is now in successful operation pumping water for the C.P.R. water tank. It is certainly a very simple and substantial piece of machinery.

Yours truly,
MATTAWA ELECTRIC LIGHT & POWER CO., LTD.

A. F. HURDMAN, Sec'y-Treas.

## SPARKS.

The Arnprior Electric Light Co. are said to have greatly reduced the price for lighting.

It is announced that a horseless carriage exhibition and race is being arranged for, to take place at the Hamilton Jockey Club track in the spring.

Mr. C. W. Bowman, of Walkerton, who was elected reeve at the municipal elections, has been obliged to resign his seat on account of having an electric light contract with the town.

The second annual ball and supper of the employees of the Ottawa Porcelain and Carbon Co., was held early in February. The event was much enjoyed by the large number present, including the general manager, Mr. J. W. Taylor.

An authority, speaking of the cost of producing carbide for the manufacture of acetylene gas states that it will cost at least $23.70 a ton, and that the new illuminant cannot be delivered for less than $12.50 for every 700 lbs. But it is claimed that only one-fifth as much acetylene gas is burned as of ordinary gas.

The annual meeting of the Guelph Light and Power Co. was held on the 17th of February. The twenty-fifth annual report showed a slight falling off in revenue during the past year, but considering the prevailing depression, was considered satisfactory. A vote of thanks was tendered to the directors for their services. The old board, consisting of Messrs. D. Guthrie, Q. C., Geo. D. Forbes, Geo. A. Oxnard, W. M. Foster, Richard Mitchell, E. Harvey, and James Innes, M.P., were re-elected. At a subsequent meeting of the directors, Mr. Guthrie was elected President, and Mr. Mitchell Vice-President for 1896. Mr. John Yule retains the management.

# CANADIAN ELECTRICAL NEWS

## STEAM AND ENGINEERING JOURNAL

OLD SERIES, VOL. XV.—No. 6
NEW SERIES, VOL. VI.—No. 4.

APRIL, 1896

PRICE 10 CENTS
$1.00 Per Year.

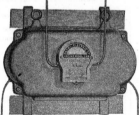

---

Please mention the CANADIAN ELECTRICAL NEWS when corresponding with Advertisers

... TO ...

# Central Station Men

Read Extract from letter from another
Central Station Man who saved money
by consulting me ᴘᴘ ᴘᴘ ᴘᴘ

> DEAR SIR : I am glad I had you to inspect my electric plant.  I had thought the
> expense of such an inspection by an independent man would be money wasted.
> You have convinced me I can save far more than your fees cost me, and in directions
> an inexperienced man would never think of.  I think no person should do business
> with electric companies without securing the advice of an independent engineer, as he
> can save far more than he costs, and get better work done.
>
>                                          Yours truly,          ——— ELECTRIC CO.

The original of the above may be seen at office

## GEO. WHITE FRASER
MEM. AM. INST. ELEC. ENG.

18 Imperial Loan Building, TORONTO                    Consulting Electrical Engineer

---

# AHEARN & SOPER

## OTTAWA, ONT.

**CANADIAN REPRESENTATIVES OF THE**

# WESTINGHOUSE ELECTRIC & MFG. CO.

### SLOW SPEED
### ALTERNATING CURRENT DYNAMOS

from which can be operated

Incandescent Lamps, Arc Lamps
and Motors.

### ELECTRIC · RAILWAY
### GENERATORS AND MOTORS

Our Railway Apparatus is not
Equalled by any other

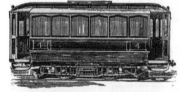 
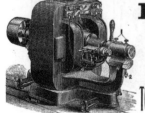

CANADIAN

# ELECTRICAL NEWS

AND

## STEAM ENGINEERING JOURNAL.

| Vol. VI. | APRIL, 1896 | No. 4. |

## PUTTING THE PAPER ON THE INDICATOR BARREL.

NEXT to being able to pipe up a steam engine for indication, and getting up the reducing rig, the most difficult thing to beginners in indication, says Robert Grimshaw, in "Power and Transmission," is putting the paper on the barrel, especially if there is but little time, as in locomotive indication, where it is often

FIG. 1.

essential that a card be taken at a special curve or grade.

As I have had considerable experience in these lines, especially in high speed work, I have had photographs taken of the proper way in which to put the paper on the barrel, not only with both hands, which are usually available, but with one hand, which is sometimes about all that can be spared for the purpose, when one is perched on the steam chest of a locomotive making 60 miles an hour on a road that is fairly well supplied with curves.

In Fig. 1, the paper barrel proper is shown detached from the indicator, which is usually feasible to do, at least with some makes of instruments. It is held by the left hand, with the paper-clips to the right; the little finger resists the downward thrush of sliding on the card, the third finger and the thumb grasp it, and the middle and fore finger aid in guiding the paper. The latter is properly about three-fourths of an inch or an inch longer than the circumference of the paper barrel; should be of middlingly stiff paper, and preferably with square clean cut edges. In any event it should be strong enough to stand considerable lengthwise pulling without parting, and should not tear easily from the edges. Being doubled so that the two lower

corners are brought together, and these being grasped by the thumb and the middle finger of the right hand, the partially formed cylinder is slid over the top of the barrel in such a way that the doubled portion (which is tightly nipped by the nails of the thumb and middle finger) shall pass the longer one of the two clips and the two parallel portions get a fair start in the slot between the two clips. Once entered, the paper is tightened around the barrel so that there shall be no slack ; and once drawn home to the bottom of the slot the ends which have been held together are spread apart and smoothed down flat, in opposite directions, so that they shall not stick out and interfere with the pencil when the drum turns.

Fig. 2 shows another view of the same operation at the same time.

In most indicators the clips have not quite enough outward turn at their tips ; this aids in putting on the paper readily and rapidly. Where the paper is extra thick it is well to unscrew the clips from the barrel and put in a piece of paper packing so as to make them stand out further.

With very stiff paper and the proper distance between the clips and the barrel the paper may be put on without

FIG. 2.

turning over ; each end passing under both clips and the paper holding by its stiffness and by the tightness with which it is pinched between the clips and the barrel.

In Fig. 3 I show the manner of putting on the paper with only one hand—the left. In the illustration the right is shown holding the indicator, but that is because the pictures were not taken with the instrument in position on an engine ; if the indicator had been screwed on to the pipe the right hand might have been cut off for all the necessity there would have been of using it.

In this operation, which is most feasible with a small barrel, a stiff paper is necessary. It is doubled up as for putting on with both hands, and is grasped by the thumb and third (or ring) finger, but instead of the curved paper being outside the hand it is within it. The partly formed tube is then slipped over the top of the barrel and the doubled part slid into the slot and pushed home, the grasping thumb and finger gradually straightening out with the advance of the paper along the slot, so that by the time the paper has been slid all the way down, it is tightly strained and ready to have the ends turned over if that is to be done. Where the paper is

FIG. 3.

stiff enough not to require that the ends be turned over, the procedure of putting it on the barrel with only one hand is much like that where the flaps are to be turned down ; a plain lapped tube is made of the paper and slipped over the top of the barrel, but when the clips are reached both ends are slid under both clips instead of only one under each.

## EXPERIENCE WITH THE ELECTRIC LOCO-MOTIVE IN BALTIMORE.

### By LEE H. PARKER.

AFTER a short period of experimental work, electric locomotive No. 1 on August 4, 1895, took up the regular freight service through the Belt Line Tunnel of the Baltimore and Ohio Railroad in the city of Baltimore. A brief restatement of the reasons for adopting electricity in this tunnel will not be out of place.

The tunnel, which is the largest " soft dirt " tunnel ever built, extends from the present Camden passenger station of the Baltimore & Ohio Railroad a distance of 7,350 feet north under the heart of the city. Beyond the northern portal the Belt Line continues through a series of short tunnels and cuts for a distance of about five miles, where it joins the old main line. The main tunnel has an up grade of .8 per cent. going north. The heavy work that would be required of steam locomotives hauling freight trains up this grade would occasion the filling of the tunnel with so much gas and smoke as to seriously interfere with the passenger service. To show how true this is, it may be said that before the electric locomotives were put into service a few freight trains were run through the tunnel, but the result was that several men were asphyxiated, and it was therefore determined not to commence even a part of the regular freight service until the completion of the electric equipment.

The illustration, Fig. 1, gives an idea of the location of the tunnel and of what its use accomplishes. By its means a reduction of 16 minutes in the running time of the " Blue Line " trains between New York and Washington is now made possible, and it is probable that this saving will be increased later on. Moreover, all delays in winter, due to ice in the river are done away with.

Shortly after locomotive No. 1 had been put into service, and had given an exhibition of its ability to haul

the heaviest freight trains, it became a matter of general interest as to how much it could pull and how fast it could go. The locomotive was, therefore, given a trial at hauling several of the passenger trains at high speeds, which it did satisfactorily to all concerned. As the conditions for operating the passenger trains entirely by the electric locomotive could not, on account of track facilities, be perfected until the new Mount Royal station, at the northern portal of the tunnel, was completed, it was decided to operate them in the meanwhile by coke-burning, steam locomotives. The new Mount Royal station will be ready during the present month, and preparations are now being made to then operate all trains, both freight and passenger, by the electric locomotives.

It was shown from the few trials made with passenger trains that not only could the guaranteed speed of 30 miles an hour be attained, but speeds of 35 and 40 miles with 500-ton trains were possible. An exhibition of high speed was made with the locomotive running light up the .8 per cent grade, and a speed of 61 miles per hour was attained for a short time without the slightest trouble from trolleys or motors. Several of the many exhibitions made by the locomotives in pulling heavy loads have been described in the newspapers. Probably the most striking was when two trains were coupled together and hauled through the tunnel. For some reason the freight trains had become "bunched" on the Washington division, and when they did get through they came so fast that it was decided to have the electric locomotive haul them two at a time. The first composite train therefore consisted of 44 cars, loaded with coal and lumber, two regular steam freight engines and a steam "pusher" engine. The whole weight was approximately 1,900 tons, and was equivalent to about 52 loaded cars. The steam locomotives did no work to assist the electric locomotive. The start was made easily and gradually, but when the train was in the tunnel and entirely on the grade, the steady, heavy pull was too severe on a defective coupling near the head of the train, and it parted. After coupling

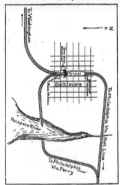

FIG. 1.—MAP SHOWING LOCATION OF THE BALTIMORE & OHIO TUNNEL IN BALTIMORE, MD.

together again the electric locomotive started the heavy train, and with all drawbars stretched—no slack in the train—and accelerated it to a speed of 12 miles an hour without slipping a wheel and in every way with the greatest ease. It reminded one of the start of an ocean steamship, so noiseless was it and so free from any manifestations other than those of mighty power. The current recorded on the ammeter was about 2,200 amperes during the acceleration period, and, after the train was up to speed, it settled down to about 1,800 amperes. The voltage on the line was 625. By reading the amperes we were able to readily compute the drawbar pull, and found it to be about 63,000 pounds. All four motors were in series, and we were, therefore, getting the maximum pull for that current.

It may be of interest to steam railway engineers to know how we determined the drawbar pull exerted for each ampere of current put into the locomotive. The Pennsylvania Railroad Company's dynamometer car was secured and coupled in between the electric locomotive and a train of known weight. The weight of each car in pounds had been accurately determined beforehand. The regular two-mile haul up grade was then made. When the train was in the tunnel on the grade the pull was uniform, as was shown in the diagrams taken on the dynamometer car.

When no drawbar pull was recorded, the pen rested on base line No. 1. The height or ordinate of the irregular curve at any point represented the drawbar pull at that instant. Measuring the same in inches and subtracting a constant and then multiplying by 4,000, gave the drawbar pull in pounds ; i.e., every inch in height represented 4,000 pounds. The paper traveled under the pen at a rate proportional to that of the train. An irregular line marked No. 2. Above the base line the planimeter record, from which was determined the mean pull for any time. Having, then, the velocity or the feet per minute and the mean pounds pull exerted

electric locomotive. This was undoubtedly due to the absence of the angle crank on the electric locomotive, and because its pull is uniform throughout the entire revolution of the armature. Most of the vibrations of the pen shown on these curves were due to vibrations of the dynamometer car, which was mounted on a single truck.

From test No. 1 we obtained the total drawbar pull in pounds, and, knowing the weight of the train, we found the drawbar pull to be 22 + pounds per ton of weight. Subtracting the grade pull, which, in the case of an .8 per cent. grade, is 16 pounds, we obtained 6 + pounds per ton as the train resistance. This confirms the usual allowances made for freight train resistance. These observations were taken in September, 1895, on a very hot day. During the past winter months the train resistance has increased, due no doubt to greater journal friction, caused by thickened lubricants, and we find it to be from our records about 20 per cent. to 30 per cent. greater than in September.

Test No. 2 was made after we had switched off six cars. The run was made under similar conditions and the same character of observations were made. The difference in drawbar pull of the two trains would naturally be the drawbar pull necessary for the six cars switched off. We had their exact weight and were

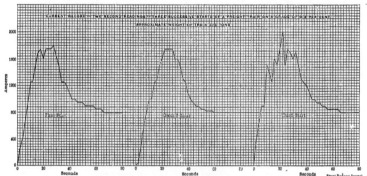

FIG. 2.—CURRENT RECORD, BALTIMORE & OHIO ELECTRIC LOCOMOTIVE.

during any period, we readily obtained the horse-power developed.

Another line on the diagram showed the chronograph record, each of the small offsets in the line occurring every five seconds. For every 100 feet the train moved, the paper moved an inch. The distance in inches between any two of the offsets gave us readily the velocity of the train. Another line represented the time readings of current and voltage which were taken in the locomotive a push button in the locomotive being electrically connected with this recording apparatus. These readings were numbered, so it was easy to tell the current at any time and location. Still another line showed a record of the different stations in the tunnel. From this we determined the location of the train at any time.

The first test showed (a) how the start was made on the down grade leading to the tunnel ; (b) how, after the train was fully started, the drawbar pull dropped off ; then (c) how it gradually increased as the train came on to the .8 per cent. grade in the tunnel ; and (d) after the train was wholly on the grade, how even the pull was, until near the stop, when the grade increases to 1.2 per cent. Mr. Dunbar, the official of the Pennsylvania Railroad Company in charge of the car, showed some diagrams of steam locomotive work under similar conditions, and it was seen that their amplitude of vibrations was considerably greater than those of the

thus again able to find what the drawbar pull per ton was. It was a check on our first figure and was very close to it, the slight difference we found being due to one brake on the six cars being partially set during the first run and unknown to any one.

We had the readings of current during the first run, also during the second. The difference of these should show the current required to haul the six cars switched off. Dividing the difference in the drawbar pulls recorded in the two tests by the difference in current recorded, gives us directly the net drawbar pull in pounds per ampere of current. This was 28.6 pounds.

It will of course, be noted that by this method we eliminated the current required to drive the locomotive. To determine how much this was, and to check our conclusions, we divided the drawbar pull in pounds recorded in the first test by 28.6 and thus obtained the current that should exert that net drawbar pull. Subtracting this current from the current actually recorded on the locomotive would give the current required to drive the locomotive. We found it took 144 amperes. As a further check we figured similarly for the second test and obtained precisely the same ; i.e., 144 amperes. So at any time now when hauling a train with the four motors in series, if we take the current indicated on the amperemeter and subtract the 144 amperes needed for the locomotive, and multiply the remainder by 28.6, we have the total net drawbar pull in pounds ; and if we divide this by the drawbar pull per ton, we get the tons of load we are pulling.

From the results obtained above we are able to' show the current and drawbar pull at any moment while accelerating a train. The curves, Figs. 2 and 3, explain themselves very fully.

The acceleration curve, Fig. 3, was obtained in a rather humorous manner. It was necessary to have a means of marking the location of the locomotive at the end of every interval of two seconds. It was first attempted to count the number of incandescent lamps passed in each interval, as they are 15 feet apart, but it would often occur that the interval would end when the pointer was at some position between two lamps, and, therefore, it was impossible to estimate accurately how far we were from the next lamp. Some one suggested dropping something as a marker on the track at the expiration of each interval. That suggestion was followed by a large number of others as to the nature of that " something." The roadbed in the tunnel is very dark colored in the dim light and is rock ballasted, consequently the "something" should be light colored, non-breakable, and what would not bound out of place when dropped. Some one then suggested a handful of flour. This was adopted and it was soon tried. It was all right for slow speeds, but at 16 feet a second it was impossible to prevent it from blowing away. Having procured a large supply of flour, perhaps 20 pounds, and wishing to make use of it somehow, some one volunteered the suggestion that flour and water made dough, and that a doughball was light colored and that it would not bound, etc. It was decided at once to use doughballs, and they were the markers used in determining the distances travelled in each interval, as shown on the curves in Fig. 3.

When it comes to a comparison of the economy of electric and steam locomotives it is readily seen that it is a difficult undertaking, knowing, as we do, the figures of only a single isolated electric plant operating under special conditions and for a comparatively short time. One great incidental advantage, of 'electric locomotives in tunnel service is that they are smokeless. This is an important moral consideration, but one which can hardly be computed in dollars and cents. But it may be of general interest to know how the actual operating expenses per engine mile of the electric locomotives during October, 1895, compare with those of a prominent and large eastern railway for the same month.

For the operation of the Baltimore & Ohio tunnel power house for the month of October, 1895, the itemized expenses were as follows :

| | |
|---|---|
| Labor | $1,345 70 |
| Coal ($1.35 per ton) | 400 96 |
| Oil and waste | 151 26 |
| Water | 50 66 |
| Maintenance | 25 42 |
| Total | $1,974 00 |

The expense on electrical locomotives was :

| | |
|---|---|
| Motor engineers | $200 00 |
| Oil and waste | 12 16 |
| Total | $212 16 |
| Total expense | $2,186 16 |

There were hauled through the tunnel 353 trains.

| | |
|---|---|
| Average weight of train | 1,095 tons. |
| "     time of trip | 20 minutes. |
| "     current | 986 amperes. |
| Distance of trip | 4 miles. |
| Total engine travel | 1,412 " |
| "     "     "idle" | 3,756 " |
| Actual time consumed for above service. | 118 hours. |
| Idle time for month | 626 " |

It is customary to consider an engine with steam up as equivalent to six engine miles for each hour it is idle, so that for comparison, the actual mileage made by the engines must be increased 6 × 626 = 3,756 miles.

The large charge of labor at power house will be the same for one, two or three locomotives in service. The items—coal, water and maintenance—and the expense on locomotives increase with the number of locomotives in service. If we assume this increase to be proportional, the total expense and cost per engine mile are as follows : ·

| | Total cost. | Engine miles. | Cost per engine mile. |
|---|---|---|---|
| For one locomotive | $2,186 16 | 5,168 | $0.423 |
| " two locomotives | 2,875 36 | 10,336 | .278 |
| " three. " | 3,564 56 | 15,504 | .23 |

ELECTRIC LOCOMOTIVE B. & O. NO. 1.

Current and Torque, and Speed Record, of the start of a freight train of 28 loads and two dead engines on a grade of 8 per cent.
Weight of train 910 tons.

No. 1. Current and Draw-Bar pull.
No. 2. Acceleration—feet per interval of two seconds. ·
*Street Railway Journal*

FIG. 3.—ACCELERATION AND CURRENT CURVES, BALTIMORE & OHIO ELECTRIC LOCOMOTIVE.

The steam railway records referred to above are for · October, 1895, and may be briefly abstracted as follows:

STEAM LOCOMOTIVE PERFORMANCE.

| | Eastern division. | Western division. | Central division. | N. & W. division. | Entire line. |
|---|---|---|---|---|---|
| Locomotives in service | 74 | 57 | 33 | 28 | 19 |
| Average engine mileage in service | 2,834 | 2,966 | 2,993 | 2,305 | 2,770 |
| Average cost per engine mile : | | | | | |
| Passenger engines | .1926 | .1666 | .1509 | .1552 | .1765 |
| Freight | .1472 | .1656 | .3428 | .2301 | .2615 |
| Switching " | .1489 | .1659 | .1828 | .1425 | .1577 |
| Work " | .2301 | .2258 | .2617 | .2169 | .2334 |
| Total " | .3084 | .2193 | .2101 | .1797 | .2095 |

From the figures given above it is seen that the actual operating expenses of the electric locomotives for that particular month are about the same as for the freight locomotives on the steam railroad ; i.e., 23 cents per engine mile. The service of the electric locomotives at that time was only about one-third that which it is expected they will have to do when the passenger service is taken up and the line extended the full distance.

As originally intended, a method of using to advantage the power of the station while the electric locomotives are idle is soon to be incorporated in the plant. Under the new conditions the cost per engine mile for the electric locomotives will be far under that of steam.

A comparison of the efficiencies of steam and electric

locomotives shows slightly in favor of the electric. Observations made on French railways and on the Pennsylvania railroad show that about 45 per cent. to 55 per cent. only of the indicated horse-power of steam locomotives is applied to hauling trains. The efficiency of the Baltimore & Ohio plant is in the vicinity of 60 per cent. to 65 per cent. under normal conditions.

A word may be added as to our experience with the overhead conductor system. The conductor in the tunnel has now been in position for nine months. During all this time coke-burning locomotives have been used for passenger service, with the consequent presence of a good deal of gas and vapor. For the first six months about half of the conductor was constantly wet from the drip due to leaks in the masonry. This occasioned a muddy, slimy deposit over the insulators and a considerable portion of the conductor. The porcelain insulators are almost entirely obscured in some places by this deposit and that of small particles of carbon given off by the locomotives.

Current was first turned on the line about three months after the tunnel structure was erected. The leak to earth was at first 21 amperes, but, in a day or two, this dropped to about four amperes—the present leakage. The inside of the conductor was coated with a combined deposit of rust and muddy sediment. Heavy currents were taken from it by the contact shoe only with difficulty and the presence of much arcing, heating and showers of sparks. It was found impracticable to run on this surface. By applications of kerosene and frequent scraping with special shoes, a direct contact of the trolley shoe with the conductor was made possible. Although a single contact shoe then worked with little or no sparking, two shoes in tandem were adopted. Their operation through the conductor is smoother, and the contact over muddy portions of it is more nearly positive. At intervals of about three weeks the conductor is treated with kerosene, and brushing shoes are run through it, about one or two trips with these brushing shoes being all that is necessary. This serves to prevent the further accumulation of rust and to remove the sediment from the contact surfaces. An inspection shows a smooth surface over which the shoes run. Contact with the metal is seen to be in high spots and thin lines, which are slowly increasing in extent.

No considerable sparking now occurs, except at the wet places, where it is occasioned by the presence of water and sediment. With the exception of three places, about 200 feet long each, the conductor is at present dry.

The bolts to the arch of the tunnel are both galvanized and painted. They show no signs of rusting. The painting has, in general, protected the surfaces of the conductor and channels. The sides and top of the inside of the conductor are coated with rust. Most of this is hard and close grained; some of it, however, is flaky. In no case is their apparent a reduction of thickness of any of the ironwork, due to rusting. Outside the tunnels the conductor is in uniformly good condition. It adapts itself to changes of temperature without trouble. The inside of the conductor is coated with rust, but in no case has there been any trouble from it. The deposit appears to be very light. There was at no time any sparking between contact shoes and conductor outside of the tunnel.

## CANADIAN ELECTRICAL ASSOCIATION.

Further progress has been made during the last month by the Executive of the Association with the arrangements for the annual convention in June.

The full list of papers has now been arranged for, and the subjects to be discussed are of a diversified and highly interesting character. The committee have been fortunate in securing the use of the council chamber of the Toronto Board of Trade for the day sessions, and of the Rotunda for an evening meeting. All familiar with the handsome building of the Toronto Board of Trade, will agree that no more desirable accommodation could be found in the city for a gathering of this character,

Arrangements for the entertainment of visitors to the convention are under way, and there promises to be nothing wanting to make the occasion compare favorably with those of like character in the past.

The exact dates for the convention have not yet been fixed, but in all probability they will be Tuesday, Wednesday and Thursday, the 16th, 17th and 18th of June.

Put these dates down in your memorandum book and keep them free from other engagements in order that you may participate in the pleasure and profit of this convention.

Applications for membership in the Association are coming in to the secretary. If you are not a member, see that your name is enrolled before the June convention.

## AN ARC LIGHT CONNECTOR

A new and useful device in the shape of a connector for the prevention of open circuits at the loop wires to arc lamps is about to be placed on the market by Messrs. McGill & Battle, of Thorold, Ont. The article is the first of the kind upon the market, and will be welcomed by all electric light companies as one of the most

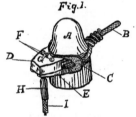

Fig. 1.

necessary and useful devices in arc lighting. It allows the wire to swing in any direction at the loop without any danger of breaking.

The cut below, which every electrician and central station manager will understand at a glance, gives an accurate description of the device. A is the glass insulator upon the cross-arm, B is the line wire which

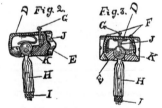

passes through a slot in the connector and around the insulator and is tied, G is a set screw holding the wire B in slot, C is a hard rubber insulation between glass and connector, D is the cap of connector held in place by screws F, E, in base of connector, H is a lug soldered to wire I looping to lamp, K (Figs. 2 and 3) is the ball head of lug working in the socket in base E, J is a spring resting on head K to insure contact.

PUBLISHED ON THE FIFTH OF EVERY MONTH BY

## CHAS. H. MORTIMER,

OFFICE : CONFEDERATION LIFE BUILDING,

Corner Yonge and Richmond Streets,

**TORONTO,**      -      -      **CANADA.**

Telephone 2362.

NEW YORK LIFE INSURANCE BUILDING, MONTREAL.

Bell Telephone 2209.

## CANADIAN ELECTRICAL ASSOCIATION.

### OFFICERS:

PRESIDENT :

A. B. SMITH, Superintendent G. N. W. Telegraph Co., Toronto.

1ST VICE-PRESIDENT :

C. BERKELEY POWELL, Director Ottawa Electric Light Co., Ottawa.

2ND VICE-PRESIDENT :

L. B. McFARLANE, Manager Eastern Department, Bell Telephone Company, Montreal.

SECRETARY-TREASURER :

C. H. MORTIMER, Publisher ELECTRICAL NEWS, Toronto.

EXECUTIVE COMMITTEE :

GEO. BLACK, G. N. W. Telegraph Co., Hamilton.

J. A. KAMMERER, General Agent, Royal Electric Co., Toronto.

E. C. BREITHAUPT, Berlin, Ont.

F. H. BADGER, JR., Superintendent Montmorency Electric Light & Power Co., Quebec.

JOHN CARROLL, Sec.-Treas. Eugene F. Phillips Electrical Works, Montrea.

K. J. DUNSTAN, Local Manager Bell Telephone Company, Toronto.

O. HIGMAN, Inland Revenue Department, Ottawa.

W. Y. SOPER, Vice-President Ottawa Electric Railway Company, Ottawa.

A. M. WICKENS, Electrician Parliament Buildings, Toronto.

J. J. WRIGHT, Manager Toronto Electric Light Company.

## CANADIAN ASSOCIATION OF STATIONARY ENGINEERS.

President, W. G. BLACKGROVE,    22 Esther St., Toronto, Ont.
Vice-President, JAMES DEVLIN,    Kingston, Ont.
Secretary, E. J. PHILIP,    293 Berkeley St., Toronto, Ont.
Treasurer, DUNCAN ROBERTSON, -    Hamilton, Ont.
Conductor, W. F. CHAPMAN, -    Brockville, Ont.
Door Keeper, F. G. JOHNSTON, -    Ottawa, Ont.

TORONTO BRANCH NO. 1.—Meets 1st and 3rd Wednesday each month in Engineers' Hall, 61 Victoria street. W. Lewis, President ; S. Thompson, Vice-President ; T. Eversfield, Recording Secretary, University Crescent.

MONTREAL BRANCH NO. 1.—Meets 1st and 3rd Thursday each month, in Engineers' Hall, Craig street. President, John J. York, Board of Trade Building ; first Vice-President, J. Murphy ; 2nd Vice-President, W. Ware ; Secretary, B. A. York ; Treasurer, Thos. Ryan.

ST. LAURENT BRANCH NO. 2.—Meets every Monday evening at 43 Bonsecours street, Montreal. R. Drouin, President ; Alfred Latour, Secretary, 306 Delisle street, St. Cunegonde.

BRANDON, MAN., BRANCH NO. 1.—Meets 1st and 3rd Friday each month, in City Hall. A. R. Crawford, President ; Arthur Fleming, Secretary.

HAMILTON BRANCH NO. 2.—Meets 1st and 3rd Friday each month in Maccabee's Hall. E. C. Johnston, President ; W. R. Cornish, Vice-Pres.; Wm. Norris, Corresponding Secretary, 211 Wellington street.

STRATFORD BRANCH NO. 3.—John Hoy, President ; Samuel H. Weir, Secretary.

BRANTFORD BRANCH NO. 4.—Meets 2nd and 4th Friday each month. F. Lane, President ; T. Pilgrim, Vice-President ; Joseph Ogle, Secretary, Brantford Cordage Co.

LONDON BRANCH NO. 5.—Meets once a month in the Huron and Erie Loan Savings Co.'s block. Robert Simmie, President ; E. Kidner, Vice-President ; Wm. Meaden, Secretary Treasurer, 533 Richmond street.

GUELPH BRANCH NO. 6.—Meets 1st and 3rd Wednesday each month at 7.30 p. m. J. Fordyce, President ; J. Tuck, Vice-President ; H. T. Flewelling, Rec.-Secretary ; J. Gerry, Fin.-Secretary ; Treasurer, C. J. Jorden.

OTTAWA BRANCH NO. 7.—Meet every second and fourth Saturday in each month, in Borbridge's hall, Rideau streett ; Frank Robert, President ; F. Merrill, Secretary, 352 Wellington street.

DRESDEN BRANCH NO. 8.—Meets 1st and Thursday in each month. Thos. Steeper, Secretary.

BERLIN BRANCH NO. 9.—Meets 2nd and 2nd Saturday each month at 8 p.m. J. R. Utley, President ; G. Steinmetz, Vice-President ; Secretary and Treasurer, W. J. Rhodes, Berlin, Ont.

KINGSTON BRANCH NO. 10.—Meets 1st and 3rd Tuesday in each month in Fraser Hall, King street, at 8 p. m. President, S. Donnelly ; Vice-President, Henry Hopkins ; Secretary, J. W. Tandvin.

WINNIPEG BRANCH NO. 11.—President, G. M. Haslett ; Rec.-Secretary, J. Sutherland ; Financial Secretary, A. B. Jones.

KINCARDINE BRANCH NO. 12.—Meets every Tuesday at 8 o'clock, in Mc-Kibbon's block. President, Daniel Bennett ; Vice-President, Joseph Lightbell; Secretary, Percy C. Walker, Waterworks.

WIARTON BRANCH NO. 13.—President, Wm. Craddock ; Rec.-Secretary, Ed. Dunham.

PETERBOROUGH BRANCH NO. 14.—Meets 2nd and 4th Wednesday in each month. S. Potter, President ; C. Robison, Vice-President ; W. Sharp, engineer steam laundry, Charlotte street, Secretary.

BROCKVILLE BRANCH NO 15.—President, W. F. Chapman ; Vice-President, A. Franklin ; Recording Secretary, Wm. Robinson.

CARLETON PLACE BRANCH NO. 16.—Meets every Saturday evening. President, Jos. McKay ; Secretary, J. D. Armstrong.

## ONTARIO ASSOCIATION OF STATIONARY ENGINEERS.

BOARD OF EXAMINERS.

President, A. AMES,    Brantford, Ont.
Vice-President, F. G. MITCHELL    London, Ont.
Registrar, A. E. EDKINS    139 Borden St , Toronto.
Treasurer, R. MACKIE,    28 Napier st., Hamilton.
Solicitor, J. A. McANDREWS,    Toronto.

TORONTO—A. E. Edkins, A. M. Wickens, E. J. Phillips, F. Donaldson.
HAMILTON—P. Stott, R. Mackie, T. Elliott.
BRANTFORD—A. Ames, care Patterson & Sons.
OTTAWA—Thomas Wesley.
KINGSTON—J. Devlin, (Chief Engineer Penitentiary), J. Campbell.
LONDON—F. Mitchell.
NIAGARA FALLS—W. Phillips.

Information regarding examinations will be furnished on application to any member of the Board.

**Central Station Design and Management.** At the present moment there are, that we know of, certainly three towns in Ontario alone, that are seriously considering the question of municipal control of electric lighting plants ; and in all three of these towns there is in actual operation a plant owned by a local company. Furthermore, in two of them the machinery is run by water power. Inquiry was made as to the reasons why the authorities were contemplating a step which has been so much discussed, and which is actually an unsolved problem in social science. It was pointed out to them that it was hardly a fair thing to take business out of the hands of private companies who were really doing all they could to give good service, and who had invested large sums of money in plant that would now become greatly reduced in earning capacity. The usual arguments as to the impracticability of a corporation running a plant as cheaply as a private individual, and as to the general immorality of the idea on principle, had been thoroughly thrashed out, and still the question seems likely to be settled in favor of municipal control in all three cases. It is really worth while making a careful inquiry into the circumstances which influence such a decision, in order that central station men may guard themselves against further application of a principle which means loss to themselves. It is worthy of special attention that the chief reasons in all three cases are given as (1) high prices, (2) poor service. A reference to the electric companies is met with the answer that it is impossible to run their prices down lower without actual loss ; and as to the quality of the light, they do the best they can. Repairs cost a lot of money, and their incandescent service is not very remunerative ; so that if they cannot make a fair amount out of the street lighting, they might as well go out of business. Ultimately, the whole thing narrows down to the electric company saying that they are doing the best they know how, and the municipal authorities remarking that that may be very true, but that the electric company's best is not good enough. Municipal councils are beginning to wonder why they should not only give franchises to operating companies that constitute practical monopolies, but also give contracts for street lighting that will actually enable operating companies to go into business. As to the impossibility of running prices down lower—that is in many cases entirely the fault of the central station men. Their stations are designed and operated with almost no regard to general principles; no general plan seems to be formulated as to their business policy, and the management takes no interest whatever in studying those conditions under which their plant will operate to best advantage. In another place in this issue is a paper on incandescent lamps, which shows clearly that it is as important to select a lamp with reference to the conditions under which it will be required to operate, as it is to select a horse with regard to its work. This same principle should be applied very extensively in electric lighting business. Any combination of machinery—steam or water power—and electric can do the work for which it is intended at a lowest cost. A better proportioned combination will be able to work more cheaply than one not so carefully designed; but whatever the combination is, there is some minimum operating charge to which it should be the aim of every manager to limit his expenses. Unless he studies his machinery and business, he will never do this; and until electric lighting and power business, from the fuel heap right through to the last detail of construction, and including the business theory, be carefully studied by the man in charge, the municipal control idea will gain ground.

———

As illustrating the results of poor designing and inexperienced management of a central station plant, we may mention a plant in another province that has recently been re-modelled by an electrical engineer of experience. Machinery to the total capacity of 1,500 lights was installed in a town of 6,000 inhabitants, with no competition from gas, or rival electric company. The fuel used was slabs and refuse from a lumber mill near by, with a small amount of coal to help out. The lumber had of course to be bought at a small price, and carted, and the calculation was it was worth taking its fuel value into consideration; coal at $3 per ton. This plant has been in operation for six years, with really excellent engines and dynamos, and the electric company has during that time paid a man a good sum to run it. The staff has consisted of an "electrician," as he is called, who is a very fairly well posted man, according to his general education; a very fair engineer and fireman, and an outside lineman. And the result has been that during six years, with all these advantages, the plant has not earned one cent above operating expenses. At different times the company has employed engineers to advise them what to do—engineers of steamboats; engineers of plants in the neighborhood, and they have always had the chief engineer of a very extensive factory in the same town to call in. These men have examined, and talked, and suggested, but no better results have followed. At last the company called in an experienced electrical engineer, and placed themselves entirely in his hands, and the result was that the method of operating the entire plant received attention, as well as the in-

dividual machines; and the new system of operating effected a saving in fuel equal to almost 1,000 lbs. of coal per night, besides many other smaller amounts which were saved in various ways. Taken separately, the various machines—boilers, engines, dynamos—were all doing very well, but taken as an electric lighting plant, they were doing very badly, simply because they were not caused to work together. You may hitch four horses to a wagon, and although they may be all willing and strong, they will all pull in different directions and do no good if driven by less than a horseman—but let that horseman take the reins, and see how he pulls them together; see how he makes each do its fair work; and see how they become—not four horses and a wagon, but one team. And is it to be supposed that the manager of an electric lighting business requires less experience, and requires less study in his business, than a "sport" does in his?

———

One of the troubles in the above plant, and one which was of course irreparable, was that the original design was most ill considered. The designer was a man whose business was general milling, lumber, etc., and who, because he read a book or two on some electrical subject, considered himself—and was thought by others—to be an electrical engineer. He had an interest in the plant himself, and certainly took great interest in getting everything of the best, and putting it up well. Arguing on the principle that, if ten lights go to the horse power, a 100 h. p. engine will be required to run 1,000 lights, he in the first place bought about 50 per cent. more engine power than he has ever required. This amounted to quite $1,600 unnecessarily spent. He certainly was quite right about 10 lights to the horse power; but not being an actual electrical engineer he did not know where theory should be discounted by practice. Knowing such a lot about electrical machinery, he was of course quite able to look after himself in the electrical market, so he bought, on his own responsibility, four D. C. machines of different makes and different sizes (presumably because they were cheap), and operated them, two and two, on the three-wire system. One pair was compound wound, the other shunt, and what is of importance, their characteristics were not the same. But "characteristic curves" are beneath the notice of the usual run of "electricians," so these machines were made to run together, being about as well suited to the work as a horse and cow are to pull a buggy. The town was wired up most scientifically, with feeders, pressure wires, and mains, but experiment showed subsequently that the pressure all over the system varied at the lamps from 108 volts at full load to 120 volts at light load; and our "electrician" could not understand why he could not get "decent lamps—he paid enough anyway." Being a mill man himself, he knew enough to keep his engines in good order, and to get good steam, but not having any experience in electric business, he did not know how to make his plant run properly, and that was of more importance than making only his engine do so. Here was a man with long practical experience, levelheaded, and intelligent in his own business, who could'nt run an electric light plant with success. And an electrical specialist, with possibly less mechanical knowledge and experience, made large savings. Why? He had studied electric lighting as a business, and the other man had not. As to the original design of a central station, it is worth pointing out, that a man proposing to build a $2,000 house generally employs an architect to make plans, specifications, etc.; but the same man when embarking in an electric station, of which he knows far less than he does about building—costing perhaps $8,000 to $10,000—buys and runs entirely on his own responsibility. Can it be wondered at, that under such unfortunate circumstances, central stations cannot lower their rates, in order to discredit the municipal control idea?

## THE INCANDESCENT LAMP.

### By George White-Fraser, E. E.

THE invention of the incandescent lamp, as used to-day, was the necessary consequence of the discovery, that the passage of an electric current through a conductor causes heat in that conductor. The phenomena that we call "heat" and "light" are presentations of the same fact in different degrees; they are the effects produced on our organs by etheric waves of different lengths, but still by etheric waves, and it is, to a certain extent, within our power to generate waves that shall produce either more heat and less light, or more light and less heat. Matter is brought to a state of "incandescence" when it is raised to a high heat—a bar of iron is incandescent in a degree when red hot, and in a higher degree when white hot, and it requires a higher temperature to produce incandescence than to produce heat only. Thus, when it was discovered that the passage of a current through a conductor caused heat in that conductor, it was but a step in the same direction to conclude that incandescence might be produced in that conductor by the passage of more current, and when the law governing the relation between current and temperature was experimentally formulated, then the incandescent lamp became a fact in the abstract. As the result of long, painstaking and most expensive study and experiment, Mr. Edison, Mr. Swan and others, brought it out of that domain and made it a fact in the concrete.

This law is expressed thus : H varies as $C_2R$ ; or in words : the heat generated varies directly as the first power of the resistance and the second power of the current, and the obvious deduction is that, granted a large enough current or a high enough resistance, any temperature may be produced in a conductor until it actually melts. Iron can be brought to a bright white heat at a temperature of about 2700% F., but even at that high temperature, although it is in a state of incandescence, its "luminous radiation" (or that which affects the optic nerves as "light") is only about 5 per cent. of its total radiation—the other 95 per cent. being non-luminous radiations ; in fact, before a temperature can be reached at which its luminous radiations would be at all considerable—it melts. Similarly with other metals that were experimented with, except platinum, and various platinum alloys. These latter were much used in the early part of the manufacture of incandescent lamps as having a high resistance and a high melting point, reaching a state of very high incandescence before attaining that temperature. As, however, these metals oxidize rapidly when exposed in a state of incandescence, to the air, it became necessary to either enclose them in a vacuum chamber or in some receptable whence the oxygen had been excluded and replaced by some other gas, such as hydrogen or nitrogen. Platinum, however, was expensive, and unsuited in other ways for lamp filaments, and finally the whole matter resolved into a search for some material which would fulfil four conditions : (a) be capable of being heated to a high temperature without melting or sensibly volatilizing ; (b) possess a high specific resistance ; (c) be capable of being shaped into a convenient form ; (d) be reasonably inexpensive.

The fact upon which is founded the commercial application of electricity to incandescence lighting is that carbon is a substance that fills all those conditions fairly closely. It can be heated, in a vacuum, to a point perhaps even higher than the melting point of platinum, without suffering any very rapid deterioration ; it possesses a very high electrical resistance, in certain forms it is capable of being easily manipulated and shaped, and intrinsically it is inexpensive.

The problem was, however, by no means solved merely by the discovery that carbon was the best material to use. There are very many substances which are capable of carbonization—wood, paper, thread, silk and others—and these required to be tested in various ways, and by different processes, before becoming suitable. After being properly carbonized they had to be perfectly protected from contact with the atmosphere, otherwise the oxygen present therein would cause an immediate combustion of the carbon. The production, and which was more difficult, the maintenance of the vacuum, was a most complicated matter. The chamber must be of one single material, otherwise air would leak in through the join. This difficulty was met by the use of an all-glass bulb, from which the air could easily be exhausted. The next point was to lead the current in through the glass to the carbon without letting the air through the holes. Platinum wires solved this, because they could be sealed or cemented to the glass ; and as platinum and glass possess very closely the same co-efficient of expansion and contraction, this cementing would not be broken when the current was flowing. So that, speaking generally, an incandescent lamp requires a carbon filament, enclosed in an all glass bulb, from which all the air has been exhausted, and through which the current is lead to the enclosed carbon by means of very thin platinum wires sealed to the glass.

The process of carbonization of the filamentary material may very considerably affect the finished product. The temperature to which the raw material is exposed determines its density, its elasticity, its electrical resistance, its durability, and of course the nature of the raw material itself primarily determines its value as a filament. Different manufacturers use widely different materials, and very various processes, and it is very reasonably to be expected that their results will also vary widely, as, in fact, they actually do. One process will result in the production of a hard, dense filament of a very graphic nature and lasting quality, while another will turn out a softer one, which, while perhaps attaining a higher state of incandescence for a time at a less expenditure of energy, will disintegrate more rapidly and last a short time. The comparative values of the different products can only be established by actual tests, and by a very careful tabulation and sum-

marization of results, and it is greatly to be deplored that the economies of the incandescent lamp, which is the actual raison d'etre of the central station, should receive so little study at the hands of electrical men.

The design and construction of a generator is the subject of an immense mass of technical literature, studied by the whole electrical profession, with the result that we have dynamos working at an electrical efficiency of 96 and 97 per cent. How many study the incandescent lamp? It has been observed to possess a luminous efficiency of about 3 per cent. only, i. e., it wastes 32 times as much energy as it turns to useful account. Most conscientious engineers and electricians study how to make the best use of their coal, their engines, their dynamos, their lines. How many study how to make the best use of their lamps ? and yet neither boiler, engine, dynamo, or lines would have any existence were it not for those little 25 cent lamps. The fact is, it is not properly understood, that there is just as much to study in the lamps as there is in a dynamo or in a steam engine, and that just as much money can be lost every year by injudicious purchase, or by haphazard operation of lamps, as by the same maltreatment of any other apparatus. The loss occasioned by one indifferent lamp is no doubt insignificant, but if that sum be multiplied by, say—1000, which would be an average sized installation, it would mount up, in the course of a year, to a startling sum.

Without describing the processes in the manufacture it will be sufficient to give the results of many tests made by different observers in both Europe and on this continent, on all kinds of lamps, and under all condititions of service ; and thence to draw conclusions which will suggest the proper way to purchase and the proper way to run them.

A manufacturing company, when selling lamps, generally gives guarantees as to their candle-power, their wattage, their voltage, their life. Assuming a general case, these guarantees will be for 16 candle power lamps, of 104 volts each, requiring an expenditure of 4 watts per candle, and an average life of 800 hours. Now, this guarantee is understood usually to mean that if these lamps are operated at 104 volts each, they will give 16 candle power each ; will require an expenditure of 64 watts each, and that they will last, on an average, for 800 hours of use, during which time those 16 candle powers and 64 watts remain so. A proper set of tests will, however, show that as a matter of fact

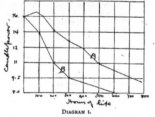

DIAGRAM I.

these quantities by no means retain their original values during the life of the lamp, and that as it remains longer in use it's candle power decreases ; it's wattage increases both relatively and absolutely, and that if it be desired to keep it up to its rated candle power, its voltage must be considerably raised. These are all defects which seem to be necessarily inherent in some degree in every lamp, and quite impossible to eradicate entirely. The most intelligent manufacturer endeavors merely to minimize evils which he cannot avoid ; and the intelligent operator will study those conditions under which apparatus, which is necessarily and admittedly imperfect, can be operated to the best advantage.

Here is a diagram constructed from a large series of tests made

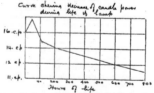

Curve shewing decrease of candle power during life of lamp

DIAGRAM II.—CURVE SHOWING DECREASE OF CANDLE POWER DURING LIFE OF LAMP UNDER ORDINARY OPERATING CONDITIONS.

by a well-known authority on lamps, showing, (for every superior make) how the candle power diminishes as the lamp grows older, granting that the voltage is kept constant all the time as it should be. When new the lamp gave 16 c. p., when it had burnt for 100 hours the c. p. had fallen to about 15 c. p.; at 200 hours to 14, and so on until at 800 hours, which is the life generally guaranteed by the makers, it gave no more than about 11½ c. p., and this, remember, for a superior lamp operated under ideal, laboratory conditions. What would happen to the same lamp operated under the ordinary central station conditions of a very varying voltage is shown in diagram 2. Here it will be seen that the lamp falls to 11 c. p. at 500 hours (curve A). Curve B. is from an ordinary commercial lamp, neither specially good nor very bad, and it is seen that it reaches 8 candle power at 600 hours, (curve B).

If this were the only matter that went wrong with lamps central stations might not care so much, for the decrease in candle power would appear to affect only the customer. As a fact, however, central stations have a very considerable interest in this depreciation, for the longer a lamp is continued in use, the more power does it take, relatively to run it. Diagram 3 shows this absolute increase of wattage. Curve A is taken from a 4 watt lamp, B from a 3 watt, and it will be seen that the former running the first 500 hours of its life has run at less than that wattage, but that after that period it has run up to quite as much in excess. Curve B is even more instructive. This lamp has started at 2.8 watts (less than rating) has come up to its rating at about 250 hours, and from that time up to 900 hours has continually required more and more energy per candle power to keep it running until that age, such c. p. as it was given required 3.74 watts, almost 25 per cent. more energy per candle power. Curve C is from a 3½ watt lamp, such as is very generally installed in all sized plants throughout Canada, and is very similar in every way to Curve B, showing that such increase of wattage is not accidental, but is to be expected in all lamps. In this last curve it is seen that at 900 hours

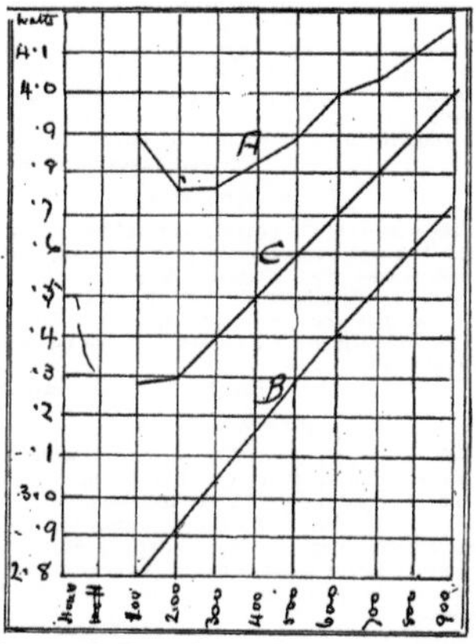


DIAGRAM III.—CURVES SHOWING INCREASE OF WATTS PER CANDLE POWER AS LAMP GROWS OLDER.

the lamp actually requires an expenditure of 16 per cent. more energy per candle power than it was guaranteed for. In a 1000 light plant running for ten hours every night with coal at $3.00 per ton, this excess at 900 hours will mean 80 cents extra per night, or $288 extra per year; if the lamps are to be kept at their proper candle power.

I would particularly point out that these values of wattage per candle powers are average values. Look at Curve C, at 900 hours the wattage per candle power is 4. Now this means that when the lamp has been used for 900 hours the total energy furnished to that lamp during its 900 hours of life has been so great that it amounts to 4 watts per candle power hour. This lamp was sold as a 3½ watt lamp, it however requires half a watt more per c. p. during 900 hours, and if it were burned longer would require continually more and more until it broke. As to the candle power of these lamps at 900 hours A had diminished to 13¼, B to 10 c. p., C to 11¾ c. p., so that taking their whole life, not one of these lamps had given an average candle power equal to their rating, A had given an average candle power during 900 hours of 13⅓, B of 12¾, C of 14 c. p. Now all these results were obtained on a trial circuit where conditions were very closely such as gave the lamps a fair chance. But on a regular commercial circuit the voltage will vary considerably with the load. Transformers will vary—the very best of them—at least one per cent. between full and no load. The branches, mains and feeders are always so calculated that a drop of 2 or 3 per cent. more between full and no load is allowed for, so that it will usually happen that lamps are subjected to a variation of at least 5 per cent. voltage, i. e., they are liable to run at 5 per cent too high a pressure. Now this is a condition that no ordinary commercial lamp will stand for long and maintain its candle power. Diagram 4 shows the results of running lamps at too high a pressure, and is taken from an actual series of tests. A 16 c. p. 108 volt lamp was run at that pressure, and at 500 hours had fallen to about 11 c. p.; a similar lamp was run at 110 volts—two per cent. too high—and at that age had fallen to a little less; running at 112—or 4 per cent. high—reduced it to 11 c. p. at 450 hours, at 114 or 6 per cent. high, caused it to drop to 12 c. p. at 300 hours, when it broke, at 116 volts or 8 per cent. high, caused it to drop to 14 c. p. at 200 hours, when it burnt out. This lamp was the ordinary 3½ watt lamp, and fairly represents the candle power on the large majority of circuits in this country. The diagram shows the great necessity of keeping the pressure as nearly constant as possible over the whole distribution system, and incidentally it leads to a consideration of a very important matter, with which all central station managers should familiarize themselves.

(Concluded in Next Issue.)

## PERSONAL.

Mr. Samuel Edison, the father of Mr. Thos. Edison, the inventor, died at Norwalk, Ohio, aged 92 years.

Mr. S. Potter has been appointed chief engineer of the new street railway power station at London, Ont.

Mr. H. A. Moore has been appointed general superintendent of the Trenton Electric Co., with offices in Trenton.

Mr. M. Neilson, of St. Thomas, Ont., has been appointed assistant superintendent of the St. John Street Railway.

Mr. F. H. Badger, manager of the Montmorency Electric Power Company, returned recently from a business trip to New York.

Mr. Granville C. Cunningham, managing-director of the Montreal Street Railway Co., has recently returned from a trip to England.

Mr. J. W. Kammerer has lately returned from the convention of the Stanley Manufacturing Company's representatives at Pittsfield, Mass.

Mr. W. T. Jennings, Mem. Inst. C. E., consulting engineer, has removed his office to the Molsons bank building, corner of King and Bay streets, Toronto.

Mr. Barren, of Toronto, a member of the Canadian General Electric Co., has been appointed manager of the Brantford Street Railway Co., in lieu of Mr. Madden, who recently resigned.

Mr. J. A. Kerr, of Peterborough, has been appointed electrician of the Galt, Preston & Hespeler Electric Railway in succession to Mr. Lea, who has accepted a position with the Toronto Junction Railway.

The employees of the Galt, Preston & Hespeler Street Railway Company presented Mr. W. A. Lea, the retiring electrical superintendent, with a gold-headed cane, handsomely engraved, and a large marble clock.

Mr. A. J. Nelles has resigned his position as superintendent of the Hamilton, Grimsby and Beamsville Railway, and has been succeeded by Mr. Clyde K. Green, electrician. Mr. Alex. Wilson continues as assistant-superintendent.

Mr. F. W. Warren, manager of the street railway at St. John, N.B., is announced to have assumed the position of general manager of the railways controlled by Mr. James Ross. Mr. Neilson will probably succeed Mr. Warren as manager of the St. John railway.

We regret to announce the death on March 10th, at his residence in Brooklyn, N.Y., of Mr. Nat. W. Pratt, President of the Babcock & Wilcox Co. Mr. Pratt was born in Baltimore, in 1852, of old American stock; the families on both father's and mother's side settled in Plymouth County, Mass., in 1830. The father, William Pratt, was during the war, superintendent of the Burnside armories, in Providence, R.I. Young Pratt entered the employ of the firm of Babcock & Wilcox in 1870. His energy, engineering ability and remarkable business qualifications won the confidence of his employers. In 1881 the Babcock & Wilcox Co. was organized as a corporation. He became treasurer and manager of the new company, retaining the position until at the death of Geo. H. Babcock, in 1893, he was elected president. Combining engineering knowledge and inventive genius with extraordinary business qualifications, to his efforts was largely due the wonderful success achieved by the Babcock & Wilcox boiler throughout the civilized world. As illustrating his versatility: In 1884 he became consulting engineer to the Dynamite Gun Co. Under his designs and patents the first successful dynamite gun was built. It was with this gun, 8 inch caliber and 60 feet long, that the experiments in throwing aerial torpedos was conducted at Fort Lafayette, N. Y. By his extraordinary sagacity and sound business judgment the business that engrossed his life from a very small beginning developed and grew enormously, and the best monument that he leaves behind him is the world-wide fame of the Babcock & Wilcox boiler. Mr. Pratt was noted not only for his sound business judgment and remarkable energy, but also for his generosity and kindness of heart. Even his business opponents admired him for his singular aggressiveness as applied to business, and by all with whom he came in contact, both at home and in trade, he was universally loved and admired. Mr. Pratt leaves an aged father and mother; also a wife and three children. He was a member of the American Society of Mechanical Engineers, American Institute of Mining Engineers, American Naval Institute; also a member of the Engineers' Club of New York City.

## CANADIAN ASSOCIATION OF STATIONARY ENGINEERS.

NOTE.—Secretaries of Associations are requested to forward matter for publication in this Department not later than the 25th of each month.

### TORONTO NO. 1.

At a meeting of Toronto No. 1, held on March 4th, a paper was read by Bro. John Fox, on "The Artificial Production of Ice." During the evening gold-headed canes were presented to Bros. James Huggett, E. J. Philip and Geo. Fowler, in recognition of their services in connection with the new hall. Four new members have recently been initiated. On Wednesday evening, last, questions were asked and replied to bearing on "Boiler Construction." It was announced that the wives and lady friends of the members would give an "at home" during the last week in April. The question of a summer excursion was also considered, and a committee will be appointed to make arrangements and report.

### HAMILTON NO. 2.

The above Association held its annual banquet at the Commercial hotel on the evening of Thursday, April 2nd. There were present upwards of eighty members and guests. The arrangements were complete, and the banquet throughout was most successful.

The city of Hamilton was represented by Mayor Tuckett and Ald. McKeown, while the deputation from Toronto No. 1 consisted of Bros. Edkins, Wickens, Huggett, Fox, Bain, Blackgrove and Sutton. Bro. H. Gerry was present on behalf of Guelph Association.

The chair was filled by Bro. R. Mackie, past-president of No. 2, and the vice-chair by Bro. W. R. Cornish, vice-president.

After the ample supply of viands provided by Mr. H. Maxey had received proper attention, the following toast list was proceeded with : "The Queen," "The Governor-General of Canada"; "Dominion and Local Legislatures"; "Army and Navy"; "The Mayor and Corporation," responded to by Mayor Tuckett and Ald. McKeown ; "Executive Head," replied to by Bro. W. G. Blackgrove, Executive President ; "Manufacturers," Mayor Tuckett and Mr. Jas. McLaughlin, respondents ; "Sister Associations," responded to by Bro. A. E. Edkins, Toronto, and H. Gerry, Guelph ; "Learned Professions," acknowledged by Bro. R. C. Pettigrew, Hamilton ; "The Press," responded to by the ELECTRICAL NEWS AND STEAM ENGINEERING JOURNAL ; The toast of "The Ladies," brought responses from Bros. Huggett and Fox, of Toronto. Bro. Edkins proposed the Toast of "Hamilton No. 2," and coupled therewith the names of Bros. R. Mackie and W. R. Cornish. Bro. Mackie then proposed the health of the accompanyist and the other gentlemen who had placed their talent at the disposal of the company. The toast of the "Host and Hostess" concluded the list.

The musical talent of the evening, which was of a high-class character, was as follows : song, "Longshoreman," Mr. Thomas, Hamilton ; song, "The Maple Leaf," Mr. Flint ; duet, Messrs. Heslop and Wilson ; song, Mr. T. Davies, Hamilton ; trio, Messrs. Wilson, Davies and Heslop ; and songs by Messrs. Heslop, Davies, Flint and Wilson.

The members of Hamilton Association are noted for their hospitality to visiting brethren, and the success which attended their annual dinner is gratifying to members of other Associations as well as to Hamilton No. 2. The Toronto delegation speak of the Hamilton members as adepts in the matter of entertaining visitors.

### BRANTFORD NO. 4.

Jos. Ogle, secretary, reports that the meetings of Brantford No. 4 are well attended. At the last meeting Bro. Walker read a paper on "Combustion," after which the evening was devoted to blackboard exercises.

### BERLIN NO. 9.

W. J. Rhodes, secretary, sends the following report from Berlin No. 9 : "I am requested by the members to let you know how we are getting along. Our officers are : President, J. R. Uttley ; Vice-President, Geo. Steinmetz ; Secretary and Treasurer, W. J. Rhodes, Berlin P. O. We meet every other Saturday evening and have a good attendance. The subjects for last meeting were "Definitions of Technical Terms," "Different Pressures," etc. At our last meeting a motion was carried : "That all members in good standing order through the secretary a membership certificate at 50 cents from Executive Council C. A. S. E. A lively chat was created on "Crosshead—Does it Stop," it being laid over for another meeting. Very little lubricant is necessary for No. 9, as all bearings run very smoothly at present. A little of superior quality lasts well.

### KINGSTON NO. 10.

Kingston No. 10 have been holding very interesting meetings since the first of the year, and have added four new members, and expect two or three initiations at our next meeting. At our regular meeting on the 4th of February, a committee was appointed to interview the Honorable William Harty and solicit his support and influence in favor of the Stationary Engineers' Bill, now pending before the Provincial house, and we expect him to give it his hearty support as he did on the last occasion.

The members are suggesting ways and means of entertaining the delegates and visiting brethren to the convention of Stationary Engineers, to be held here next August, and are going to try to keep up on that occasion Kingston's reputation for hospitality.

JOHN TANDVIN, Secretary.

### KINCARDINE NO. 12.

The past month has been a very profitable one for No. 12. Although our membership is not large, yet the members take much interest in them. We have been considering the subject of combustion. The subject for next month has not yet been decided upon. Altogether, I think the members are pleased with the progress we are making.

PERCY C. WALKER, Secretary.

### BROCKVILLE NO. 15.

W. Robinson, recording secretary, writes: Our Association for the past month has kept up to the usual standard of progress. All the members, or most of them, take a lively interest in the prosperity of our branch here. I am sure that No. 15 has made as good a showing as any of the sister lodges, although we have had less facilities than those possessed by large societies.

### CARLETON PLACE NO. 16.

Mr. J. McKay, president, sends the following report of branch No. 16: At the last meeting a very interesting discussion took place as to whether it was more difficult to force water over the top of a tank or through the bottom. Mr. J. Burnie, of the C. P. R. pumps, was present, and gave his experience in the matter. By actual tests he said it was much easier to force the water through the bottoms of the tanks. At the next meeting the subject to be discussed will be "The Injector— how high the different injectors have been known to lift and the lowest pressure known to start."

We are still moving slowly and look forward to a prosperous year.

## A PECULIAR TYPE OF "ELECTRICIAN."

THE following unique communication recently found its way into the office of the ELECTRICAL NEWS, with a request that it be published :

34 Davenport<br>Road<br>City

Dear Sir

Will you kindly inform me how much electricians get in toronto, an if there is any chance for one now I enclose post card for answer I shall Be very grateful to you to return answer at ounce, your humble servant fred

P. S.
An were to apply

In the light of the above we certainly require a broad definition of the title "Electrician."

### THE LATE MR. JOHN GOLDIE.

AFTER an illness of lengthy duration the grim hand of death has removed one of the most prominent manufacturers of the Dominion, in the person of Mr. John Goldie, of the Goldie & McCulloch Company, of Galt. His death occurred at his home in that city on the 26th of March. For some weeks his life had been despaired of, but his splendid vitality, notwithstanding his advanced age, enabled him to offer strong resistance to the ravages of disease.

The late Mr. Goldie was a well-known, honorable and highly respected citizen, and to him is due a large portion of the prosperity enjoyed by the town in which he lived. Always having taken an active interest in all matters tending to promote the welfare of the community, his valuable counsel and assistance will be greatly missed. He was a man of sterling qualities, firm, but not obstinate in his convictions, energetic and reliable in all his business dealings. He was a liberal in politics, but has never sought any public favors. In religion he was a Presbyterian.

Mr. Goldie was a native of Scotland, having been born near the town of Ayr, on the banks of Doon, Ayrshire, in 1822. The subject of our sketch received only a meagre educational training at the school in Kilroy, a small village near his home. When quite young he was apprenticed to learn the millwright's trade. He came to America in 1844 and settled at Greenfield, near Ayr, obtaining employment for over a year with Mr. Geo. Baird, a well-known contractor of Blanford township. He afterwards spent

MR. JOHN GOLDIE.

eighteen months in Montreal, and was then engaged as millwright by the late Jas. Crombie, of Galt. Subsequently he went into partnership in a saw mill in Esquesing township, remaining in the business several years. In 1859 he returned to Galt, when he and Mr. Hugh McCulloch formed a partnership and bought out the foundry business of Jas. Crombie. At that time twenty-two hands were employed. The business steadily grew, and nearly every year the manufacture of new lines of machinery was commenced. From their factory many skillful mechanics were sent out, who have since given proof of their excellent training. In 1891, the business had grown to such an extent that it was deemed advisable to turn it into a joint stock company, which was done under a Dominion charter, with a capital stock of $700,000. The original shareholders were John Goldie, Hugh McCulloch, David Goldie, Hugh McCulloch, jr., and R. O. McCulloch. Since then Mr. A. R. Goldie and other members of the two families have been taken in. At the present time the name of the company is known throughout the Dominion as progressive and reliable manufacturers of safes, engines, boilers, mill machinery, etc.

---

The use of three drops of cylinder oil per minute on a 300 h. p. engine has been found all that was necessary to prevent injury to the cylinder and valves.

Where rapid-circulation is maintained, the results in the transmission of heat are about nine times as great as where the medium to be heated is quiet. This applies also where there is but little difference in the temperature.

THE RESISTANCE OF RIVETS.—Some time ago Mr. Dupuy was selected by the ministry of public works, France, to make a special inquiry into the causes of deterioration of metallic structures. Experiments were first made on rivets, and a number of conclusions arrived at. A few of them are as follows : Rivets were found to not exactly fill the rivet holes, but to clamp the plates together with a pressure that gives rise to a resistance to sliding equivalent to a weld, which resistance is greater as the limit of the elasticity for rivet material is higher. The effort necessary to shear rivets per square inch of the section to be sheared is not less than three-fourths of the tensile strength of the rivets per square inch. Mr. Dupuy draws from his conclusions some rules for bridge work, among which are the following : The calculation of riveting cannot be based upon the permissible stress in the test bars, the coefficients of safety relating to the rivets not depending in any way on those adopted for the bars.

### RATING AND BEHAVIOR OF FUSE WIRES.

MESSRS. W. M. Stine, H. E. Gaytes and C. E. Freeman, in a paper on the above subject, presented to the American Institute of Electrical Engineers, summarize as follows some of the practical conclusions deduced :

1. Covered fuses are more sensitive than open ones.
2. Fuse wire should be rated for its carrying capacity for the ordinary lengths employed.
2. (a) When fusing a circuit, the distance between the terminals should be considered.
3. On important circuits, fuses should be frequently renewed.
4. The inertia of a fuse for high currents must be considered when protecting special devices.
5. Fuses should be operated under normal conditions to insure certainty of results.
6. Fuses up to five amperes should be at least 1½" long, ½" to be added for each increment of five amperes capacity.
7. Round fuse wire should not be employed in excess of 30 amperes capacity. For higher currents flat ribbons exceeding 4" in length should be employed.

### ELECTRIC UNITS.

THE five principal units of current electricity are :— The volt, ampere, ohm, coulombe and farad.

The volt is the unit of intensity. It may be compared with the steam pressure of a steam engine, or the pressure of a column of water flowing in a vertical pipe.

The ohm is the unit of resistance, and may be compared to the resistance of a pipe to the water flowing in it.

The ampere is the measure of rate of flow of current, and it is that quantity which would pass through a circuit having one ohm resistance, when urged by a pressure of one volt. It does not include the idea of time or of real pressure. Thus we might have an ampere flowing through a circuit as the result of either the smallest fraction of a volt, or of 1,000,000 volts depending upon the resistance of the circuit. The quantity flowing, however, would be exactly the same as that resulting from one volt through one ohm.

The coulomb is the measure of quantity of flow ; that is, current flowing at the rate of one ampere for one second, and is also termed the ampere-second. This quantity is so small, that in practical battery, electric light and motor work, the ampere-hour is taken as the measure of quantity.

The farad is the measure of capacity. The capacity of a surface which can hold one coulomb of electricity at a voltage or pressure of one volt is a farad. It may be compared with the capacity of a container filled with gas. Under a certain pressure it will hold a certain amount of gas. Double the pressure and it will hold double the amount of gas, etc. Similarly a surface which would hold one coulomb at a pressure of one volt, would hold two coulombs at a pressure of two volts, etc.

The watt is the mechanical unit of work-power. The number of watts may be obtained by multiplying the number expressing the volts, or the electromotive force, by the number expressing the amperage, or the current. A force of seven hundred and forty-six watts equals one horse-power.

One must not fall into the error of considering electricity a fluid, because of these comparisons. The units may be very easily explained by the analogy of a current of fluid, but the ancient theory that electricity is an "imponderable fluid" is now considered untenable.

---

When a belt gets saturated with waste oil, an application of ground chalk will soon absorb the oil and make the belt workable.

Best results from the distribution of steam in the cylinder of an engine is when the mean effective pressure is from 42 to 45 per cent. that of the initial, the latter number referring to multiple expansion engines.

After using your steam engine indicator, remove all the drops of water from it before putting it away. Because it is made of metal that is supposed to resist the action of rust, it does not necessarily follow that water will do it any good, or fail to do it harm.

## MR. J. J. YORK.

PRESIDENT MONTREAL ASSOCIATION NO. 1, C. A. S. E.

THE president of Montreal Association No. 1, C.A.S.E., Mr. J. J. York, began his apprenticeship with Prowse Bros., of Montreal, at the age of 13 years. Three years afterwards he was transferred to the steam-fitting department, and four years later was sent to Winnipeg to heat the first building ever heated there by means of hot water or steam. The buildings to be heated were the Hudson Bay Company's new stores,. the St. Boniface College and the Bank of Montreal. He also put the first gas pipes in any building in Winnipeg, and on completion of the contracts refused several offers to remain there, having made up his mind to go back to Montreal. Upon returning he entered the employ of Mr. John Date, for a short time, after which he secured a position in the machine shop of the St. Lawrence Sugar Refinery Co.,·and shortly after was made first assistant engineer, which position he held until September, 1892. During this time his employers doubled his salary without solicitation. During the time he was with this company he took charge of the first Edison dynamo operated in Montreal, which, by the way, is

MR. J. J. YORK.

still running, in spite of the fact that it was once under water for two days, and once through fire. On resigning his position with the company he was appointed outside foreman for Messrs. Garth & Co., who gave him charge of the heating, plumbing and ventilating of the Board of Trade building. On the completion of this work the Board of Trade were pleased to appoint him superintendent of the building, the position which he now holds, and he has since installed a most modern lighting plant, which was described in these columns a short time ago. He is President of the Montreal Association for the second time, has been one year Executive Secretary, and one year Executive President. He holds first class licenses from the city of Montreal and also from the Ontario Association Stationary Engineers.

The Burk's Falls Water, Light and Power Company have met with such a degree of success as to warrant them in considering the enlargement of their facilities at an early day.

The Canadian General Electric Company have closed a contract with Messrs. Kilmer, Crawford and McIntyre, of Durham, Ont., for a 75 kilowatt monocyclic alternator. The motive power will be obtained from a water power at Aberdeen, 4½ miles distant from Durham. It is intended to supply current from the monocyclic circuit for both lights and motors.

The directors of the Nelson Electric Light Co., of Nelson, B.C., which has recently been organized, are : John Houston, president and manager ; J. A. Gilker, vice-president ; W. J. Goepel, secretary ; W. F. Teetzel, treasurer ; J. J. Malone, John Johnson and J. H. Matheson, finance committee. The company obtain their power from Cottonwood Creek. The water is brought from dam to penstock in wooden flume, and from penstock to power house, 300 feet, in 14 in. steel pipe, the fall being 164 feet. One Pelton water wheel and two 30 kilowatt bi-polar dynamos of the Edison type are in use. Direct incandescent lights are used for commercial purposes and arc lights for street lighting.

## QUESTIONS AND ANSWERS.

"F. L. T.," Parry Sound, Ont., writes :—What causes a spark on the rectifier of an alternator at different points on rectifier, not near the brushes ? I think they only appear when the brushes spark through. Is it dirt ? Sometimes it appears on top and sometimes on sides, above but not near brushes.

ANSWER.—The sparking on the "rectifier," as you call it, is probably caused by small particles of copper dust that stick to the insulation between the segments, and so assist a spark across. Clean well with fine sandpaper, and pick out any particles that you can see.

"Lineman " writes :—(1.) Please give a rule to find the number of centimetres there are in any sized copper wire used in electric lighting. (2.) Are there any tables published giving the gauge of wire and the diameter of the same ? Also the number of circular mils, or any other data useful to a lineman.

ANSWER.—(1.) There is no rule for calculating how many circular mils there are in a given size of wire. It is purely arbitrary convention to say that a wire containing 10381 c. m. shall be called a No. 10 wire B. & S. gauge. There are even several gauges, standardized by different makers, and it is always well to specify, when ordering wire, what gauge it is. Thus, there is Brown & Sharpe W. G. ; Birmingham W. G. ; American W. G. (2.) There are plenty tables published giving all the information you require. See following table :

TABLE OF DIMENSIONS AND RESISTANCES OF COPPER WIRE.

| No. B. & S. | Diam. Mils. | Circular Mils. | Lbs. per 1000 ft. Ins'd. | Ohms per 1000 ft. |
|---|---|---|---|---|
| 0000 | 460.000 | 211600.0 | 800. | .04904 |
| 000 | 409.640 | 167805.0 | 666. | .06184 |
| 00 | 364.800 | 133079.0 | 500. | .07797 |
| 0 | 324.950 | 105592.5 | 363. | .09827 |
| 1 | 289.300 | 83694.5 | 313. | .12398 |
| 2 | 257.630 | 66373.2 | 250. | .15633 |
| 3 | 229.420 | 52633.5 | 200. | .19714 |
| 4 | 204.310 | 41742.6 | 144. | .24858 |
| 5 | 181.940 | 33102.2 | 125. | .31346 |
| 6 | 162.020 | 26250.5 | 105. | .39528 |
| 7 | 144.280 | 20816.7 | 87. | .49845 |
| 8 | 128.490 | 16508.7 | 69. | .62849 |
| 9 | 114.430 | 13094.2 | | .79242 |
| 10 | 101.890 | 10381.6 | 50. | .99948 |
| 11 | 90.742 | 8234.11 | | 1.2602 |
| 12 | 80.808 | 6529.94 | 31. | 1.5890 |
| 13 | 71.961 | 5178.39 | | 2.0037 |
| 14 | 64.084 | 4106.76 | 22. | 2.5266 |
| 15 | 57.068 | 3256.75 | | 3.1860 |
| 16 | 50.820 | 2582.67 | 14. | 4.0176 |
| 17 | 45.257 | 2048.20 | | 5.0660 |
| 18 | 40.303 | 1624.33 | 11. | 6.3880 |

"L. O'C.," Durham, Ont., writes : I notice in your last issue "F. & T.," Walkerton, wish to know why alternating current cannot be used for railway service. I am of the same opinion as they, only instead of two collectors I would use a three-phase current, with, of course, three collectors ; but my idea is that only two trolleys would be necessary, as the ground could be used for the third wire as it is for the return wire on the present system. Then by the use of transformers the voltage could be kept low, and as the primary wires would not be grounded, they could be used for light or power purposes as well. If this would work I should think it would pay, especially if water-power was available. Possibly I am " away off " in this, as I am by no means an expert, but if so kindly correct.·

ANSWER.—We would refer you to the answer given to "F. & T.," Walkerton, in last issue. If it had been practically possible to apply two or three-phase currents to railway service, you may be quite sure it would have been done long ago. It is quite safe to leave the solution of this problem to Steinmetz, Scott, and other scientific electricians. You are quite right, however, that current could be transmitted, transformed, and utilized by two trolleys ; but that will not get over the fact that, except for long and continuous travel, the induction motor is not at present suited for traction purposes.

"Jim" writes :—(1.) A cylinder is 30 inches in diameter and 30 inches long, average cast iron. What thickness will it require to be to withstand 100 lbs. pressure of steam per square inch? (2.) What thickness must the head of it be (the head to be flat), and give rule to work it out? (3.) What sized studs should be used, and how many should be used to make a good joint? Please work it out so I can figure on different sized cylinders.

ANSWER.—(1.) This is a matter that cannot be answered definitely. The thickness will depend on the quality of the iron, amongst other things. (2.) See answer to (1.) (3.) For all these calculations we recommend any standard work on engine designing. It is impossible to give answers that would satisfy, without assuming so many quantities, and explaining so much, and qualifying nearly every statement, that confusion would result. If you desire to study engine building, we earnestly recommend you to start at the very beginning, and study the theory of steam, strength of materials, and the science of strains and stresses.

"Induction," Arnprior, Ont., writes :—(1.) Is there any other practical method of locating an open circuit on an arc line, without the use of a galvanometer? If so, describe it. (2.) How can I clean incandescent lamps that have become burnt or smoked in a building at the time of a fire?

ANSWER.—(1.) Do you mean actually an "open circuit," or only a "ground?" It might be interesting to other central stations if you would describe your method of locating an open circuit by means of a galvanometer. If you refer merely to a "ground," then the following method will locate it very closely :

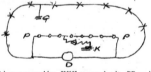

D is an arc machine, XXX an arc circuit, PP an incandescent circuit of incandescent lamps in series shunted across the dynamo terminals. The incandescent lamps must be 50 volt, and may be of 10 c. p. or lower. There must be the same number of incandescent lamps, as there are arcs on the line. K is a permanent ground, and S is a flexible connection permanently attached to K. S may be slid along PP, making connection with any lamp. Suppose a ground exists at G between any two lamps—say the 3rd and 4th. Then the fall of pressure from the + terminal of the dynamo to G is about 50 volts × 3 (lamps). The fall of pressure also from the same + terminal to a point between the 3rd and 4th incandescent lamps, is also about 50 volts × 3 (lamps). Hence if these two points be connected, no current will flow between them. Ground a voltmeter through K, and touch the successive incandescent lamp terminals from + to -. A point will be found where the deflection is least ; and between the corresponding arc lamps the ground will be found. (2.) We should suppose that rubbing them with paper—as is done with lamp chimneys—would do. Have had no experience in the matter personally.

"H. J, A.," Markdale, Ont., writes :—(1.) Would the field-magnet cores of a five incandescent light Edison dynamo of the bi-polar type do if made of cast iron? (2.) Would lead-foil answer for coatings for Leyden jars, and for induction coil condensers, etc.?

ANSWER.—(1.) Certainly, but cast iron is more likely to lose its magnetization than steel, and being less magnetizable, would require a larger cross-section. (2.) Yes.

"H. N." writes :—On page 46, March, 1896, number of THE NEWS, Mr. F. G. Proutt, of Malden, Mass., makes a deduction from calculations. In this general deduction the last clause reads : "The one made to be at the highest voltage will be the most incandescent." It is evident a typographical error or an inadvertence occurs by the words "highest voltage" when lowest amperage is understood.

The following are results of observations taken on a 100 volt circuit :—

Place in series 100 v. 16 c. p., and 50 v. 16 c. p., the 100 v. brightens. Now exchange the 50 v. 16 c. p. for 50 v. 5 c. p., then 100 v. 16 c. p. dims, and the 50 v. 5 c. p. brightens.

Proof.—1 16 c.p. 75 watt lamp at 100 volts =.75 ampere resistance, 133 ohms. 1 16 c.p. 75 watt lamp at 50 volts = 1.5 amperes resistance, 33 ohms. 1 5 c.p. 23.44 watts at 50 volts =.47 ampere resistance, 106.4 ohms. Now 106.4 ohms is not as great as 133, but that is not it. How many amperes per lamp is required? The one of 16 c. p. of 100 volts takes .75 amperes, but the 5 c. p. of 50 volts takes only .47 ampere. It follows, therefore, that the 100 volt lamp taking .75 ampere will be brighter than the 50 volt lamp requiring 1.5 amperes to bring it to incandescence, and on the other hand the 50 v. 5 c.p. requiring only .47 ampere will be brighter than the 100 volt lamp requiring .75 ampere. .

ANSWER.—The term highest voltage was simply used by Mr. Proutt to signify a lamp requiring the most pressure with a given current to bring it to a state of incandescence. The lamps being in series the current through them all would be . alike in quantity, consequently the lamp with the finest filament would burn the brightest. Your demonstration shows this to be the case.

## TRADE NOTES.

The Canadian General Electric Co. have been awarded the contract for a 500 light alternating plant at Colborne, Ont.

The Canadian General Electric Co. recently installed a storage battery of their new type for the Bank of Hamilton in Hamilton.

The Canadian General Electric Co. are building a 450 kilowatt direct-connected generator for the Toronto Electric Light Co.

A ten horse power induction motor operated from the monocyclic circuit was recently started up in D. Knechtel's furniture factory at Hanover, and has given perfect satisfaction.

The Hull Electric Co.—E. Seybold, Secretary—have placed an order with the Canadian General Electric Co. for their cars and electrical equipment.

We are advised that Messrs. Ahearn & Soper, of Ottawa, have recently been appointed exclusive Canadian representatives of the Westinghouse Electric and Manufacturing Co.,

The Vancouver Consolidated Railway and Lighting Co. have recently placed an order for closed cars and equipments with the Canadian General Electric Co.

Messrs. Wm. Kennedy & Sons, of Owen Sound, are furnishing five 60-inch new American water wheels to the Hull Electric Railway and Lighting Company. Their cost will be about $25,000.

The Goldie & McCulloch Co., of Galt, have sold Wheelock engines to W. H. McEvoy and Colin Nigle, of Amherstburg, Ont. The cylinders of these engines will be 13 x 30 inches with condenser attached.

Hooper & Starr, contractors for the Cornwall street railway, have ordered a 250 horse power cross compound Robb-Armstrong engine and two 125 horse power Monarch Economic boilers from the Robb Engineering Co.

The Dodge Wood Split Pulley Co., of Toronto, call attention to their new split friction clutch and cut-off coupling as illustrated in this issue. They have supplied this clutch for powers from 5 h.p. to 200 h.p. to some of the largest power users in the Dominion, and have yet to hear of a complaint. Those contemplating the purchase of either clutches or cut-off couplings will do well to put themselves in communication with this company. The Dodge Pulley Co. manufacture special pulleys of all kinds for electrical purposes, and mail illustrated catalogue on application.

Most all dynamo belts are subject to damage from machine oil which causes the leather to rot and the belt to slip on the pulleys. Robin, Sadler & Haworth, the leather belt makers of Montreal and Toronto, experimented considerably to ascertain what would be the most effective process to restore belts of this kind and place them in condition so as they could be made serviceable again. After many practical tests they have proved to themselves and several owners of electric light stations that the oil can be extracted and the belts made useful again without hurting or injuring either the belt, leather or cement ; thus they term themselves as cleaners and repairers of old oily leather belts. They will be pleased to give further information to any who are troubled with belts of this kind.

## ENGINE REGULATION IN ARC LIGHTING.

To the Editor of the CANADIAN ELECTRICAL NEWS.

SIR,—While regularity of engine and dynamo speeds is a generally recognized necessity for incandescent electric lighting systems, I have been told, and have also read that for arc lighting such accuracy of speed was not required.

I will, by an experiment try to prove that this is not so. The resistance of an electric arc varies continually. Place a single arc lamp in circuit and watch the ammeter needle; it will be found to vary through a large range, and will seldom if ever be at rest. Of course if the carbon is not a good one, this will be much more noticeable, and at each sputter of the lamp the ammeter will jump.

When there are a number of lamps in series, a little overfeeding or a little underfeeding of the carbons may considerably alter the resistance, and, unless there be a reserve of power and a good governor, the result will be unsatisfactory. If a little overfeeding take place, the resistance will be diminished, the quantity of current flowing will be increased and more load thrown on the engine. If the latter be properly governed, its speed will be kept constant, and the lamps will have a chance of recovering themselves; but if, on the contrary, the speed should fall off, the lamps will get worse and worse. This proves also that well regulated lamps greatly affect the system for the better.

Very truly yours,

OBSERVER.

---

### SPARKS.

Stratford, Ont., is to have an electric system of fire alarm.

There is a movement on foot at Bothwell, Ont., for electric lighting.

The Kamloops, B. C., Electric Light Company has gone out of business.

The Light, Heat & Power Company, of Newmarket, Ont., are said to have assigned.

Preparations are being made for the installation of an electric light plant at Digby, N. S.

The Vernon & Nelson, B.C., Telephone Company are applying for amendments to their charter.

A company has been formed at Brantford, Ont., to manufacture the Callender automatic telephone.

The South Essex Electric Railway Co. is applying to the Ontario Legislature for incorporation.

The electric light plant at Drummondville, Ont., has been put in operation. Fifty arc lights are used.

Mr. H. A. Connell, of Woodstock, N. B., has added an arc light system to his electric light plant.

The city of Kingston, Ont., is considering the question of municipal ownership of its electric light plant.

Thornbury is to have incandescent light. The Canadian General Electric Co. are supplying the apparatus.

The Lincoln Radial Railway Company's bill has been passed by the Private Bills Committee of the Ontario Legislature.

The Drummondsville Electric Light Co., of Drummondsville, Que., are installing a 60 kilowatt Canadian General alternator.

The Ottawa Electric Railway Company have completed their new car shed, and have extra accommodation for 36 cars.

The Columbia Telephone-Telegraph Co. is applying for incorporation, to operate between Rossland and Boundary Falls, B. C.

The City Council of St. Thomas, Ont., has given the first reading to the by-law to raise $50,000 for a municipal electric lighting plant.

The Canadian General Electric Company have now 600 men at work at their shops, and are working over time to catch up with orders.

The Trenton Electric Co. has obtained supplementary letters patent to extend operations to Belleville and the township of Thurlow.

A donation of several hundred books from the British Museum has recently been made to the library of the McGill University, Montreal.

Mr. G. E. Smith, of Montreal, has invented a machine for curving railway rails, by which any degree of curve required may be obtained.

The street railway at St. Thomas, Ont., was offered for sale by auction on the 20th of March. The highest bid was $11,400, and it was withdrawn.

A bill has received its second reading in the British Columbia Legislature to incorporate the Alberni Water, Electric and Telephone Co., of Alberni.

The Beaverton Electric Light Co. have recently purchased an 100 h.p. tandem compound condensing Wheelock engine, steel boiler and connections.

E. Leonard & Sons, of London, have purchased the electric light plant of the Grand Central Hotel at St. Thomas, for $520. The plant originally cost $2,500.

The Yarmouth, N. S., Telephoue Company is stringing wires for the use of the Coast Railway Company, which will be operated by telephone instead of telegraph.

Mr. Higman, chief of the Electric Light Inspection Department at Ottawa, is fitting up a very commodious testing room with Kelvin batteries for standardizing.

Haley Bros. & Co., of St. John, N. B., are having a 50-light incandescent dynamo installed in their factory. James Hunter, electrician, is superintending the work.

Robert Henry, late manager of the Brantford Electric and Power Co., has entered a claim against the defunct concern for $3,000 for his services for three years.

Arrangements are reported to be in progress by which an important electrical plant will shortly be established in Western Australia to supply motive power to the gold fields.

Mr. Hugh Thompson, of Waterdown, is endeavoring to induce the Hamilton Radial Railway Company to extend its line beyond Burlington, through Waterdown to Guelph.

The Canadian General Electric Co. are erecting at their works at Peterboro a new building, in which will be carried on the construction, setting up and finishing of street cars.

The Hydraulic Power Co., of St. Hyacinthe, having found that the water failed them at certain seasons, recently purchased two Wheelock condensing engines and boilers to develop 450 h.p.

The electric light company at Vancouver, B. C., has offered to continue to supply lights at 31 cents per lamp. New Westminster, owning its own plant, has raised the rate from 25 to 35 cents per night.

The Street Railway, Heat & Power Co., of Moncton, N. B., is being organized to construct a street railway in that town. The capital stock is $100,000. It is hoped to commence work this spring.

Mr. R. Leslie, of Toronto Junction, is the inventor of a track cleaner, which is reported to have done satisfactory service on some of the Toronto Railway Company's lines during the recent snowfalls.

The city council of Moncton, N.B., will petition the New Brunswick Legislature for the cancellation of the charter of the Montreal Electric Tramway Company, for failure to fulfill the conditions of the charter.

The Niagara Navigation Co. have awarded a contract for a 300 light direct-connected dynamo, with engine, to the Canadian General Electric Co. The plant is for the new boat which is to replace the ill-fated Cibola.

Work on the three-phase transmission plant between Trenton and Belleville is progressing favorably. The first of the 150 kilowatt generators has been shipped from the Canadian General Electric works at Peterboro.

Mr. D. C. Morency, civil engineer, Port Levis, Que., has invented a portable lamp lighted by acetylene gas, and is now making experiments on the application of the gas to the lighting of streets and houses.

Mr. J. C. Guay and Joseph Gagnon, representing the Chicoutimi Electric Co., of Chicoutimi, Que., are negotiating with the town council for the construction of an electric railway between St. Alphonse and Chicoutimi.

Mr. D. Budge, superintendent of the Halifax and Bermuda Cable Company, is urging the granting by the Dominion Government of a subsidy to enable his company to extend their present cables from Bermuda to Jamaica.

At a recent examination of St. John, N. B., the following were granted marine engineers' certificates : J. F. Williamson, of St. John, second class ; G. W. Cowie, of Miramichi, third class ; James Henderson, s. s., Prince Rupert, fourth class.

A deputation from Peterboro', Ont., recently waited on the Minister of Railways and Canals to secure control of the power of one of the dams on the river above Nassau, the power to be utilized in the plan for the electric lighting of the town.

Messrs. Duncan Scott and James Wilson, representing a New York syndicate, were recently in Ottawa in quest of Canadian mica. They are said to have stated that Canadian mica of the Ottawa radius, amber quality, is the best in the market.

The officers of the Hamilton Radial Railway Company are : Messrs. Alex. Turner, president ; Thos. Leather, Vice-president ; W. W. Osborne, secretary ; W. A. Wood, treasurer ; directors, A. Zimmerman, George Lynch Staunton and John Moodie.

An electric ice-cutting machine is being used with much success at Sunapee, N. H. It consists of a circular saw, which the operator is able to raise and lower at will. The other day the machine cut sixty-five feet of ice to a depth of nine inches in seventy-five seconds.

A meeting of the directors of the Brockville Electric Street Railway Company was held recently, at which the agreement with the town was ratified. Work is to be commenced not later than 1st May, 1897, one mile to be completed within a year from that date.

The T. Eaton Co., Toronto, are putting in two Robb Armstrong engines to run their blowers, which operate the cash system. The engines and blowers will be direct connected. A Northey compound duplex pump of 2,000,000 gallons capacity will be installed shortly, and the 1,000,000 pump at present in use will be used in emergencies.

## SPARKS.

In the Dominion Railway Committee it has been announced that all electric railways in Ontario of a purely provincial character must be brought under the work ing of the general Electric Railway Act of the province.

An action was recently .entered by Miss Brennan against the Montreal Street Railway Company claiming $2,000 damages for injuries sustained by being struck by an electric car. The court found no evidence of negligence on the part of the company or its servants, and the action was dismissed.

The Lachine Rapids Hydraulic Company have invited offers for the supply of generators for the transmission of 16,000 horse power. The contracts for the water wheels and machinery have been closed, and before the close of the year the company hope to be in a position to supply power.

The city of Winnipeg, Man., has under consideration the question of street lighting. The City Solicitor has been instructed to prepare two by-laws for submission to the ratepayers, one for the construction by the city of a gas light plant, and the other for the construction of an electric light plant.

The Supreme Court of Ottawa has dismissed the appeal of the city of Vancouver. This upsets the by-law passed by the ratepayers in 1894, authorizing a civic electric lighting plant. The council has now passed a by-law providing for the lighting of the city by the Western Electric Company.

A French mining engineer is said to be preparing plans for an electric elevator to the top of Mount Blanc. A horizontal tunnel is to be bored, and from there to the top the ascent is to be made by electric elevators in a vertical shaft one and a half miles high. The time for the ascent is reckoned at 30 minutes.

The T. Eaton Co. have awarded the contract for two 130 kilowatt direct-connected generators to the Canadian General Electric Co. This, with the 75 kilowatt C.G.E. machine installed last fall, gives a capacity of 335 kilowatts, or nearly 6,000 16 c.p. lamps—by long odds the largest isolated plant in Canada.

The town of Newmarket, Ont., has decided to submit a by-law to the ratepayers authorizing the purchase of an electric light plant. The cost of plant will be $8,000, and is estimated to afford twice the lighting capacity heretofore given, at a saving of $200 a year. The dynamos will be run by steam power.

An electric street lighting franchise for thirty years has been granted by the town of Sherbrooks, Que., to a New York syndicate. At the end of that time the town will be given the privilege of purchasing the road. Preliminary arrangements for construction have been completed, and work will shortly be commenced.

Application is being made to the New Brunswick Legislature by J. T. Allan Diblee, Julius T. Garden and J. .N. W. Winslow, for incorporatiou of the Woodstock Electric Railway, Light and Power Company, of Woodstock, N. B. The object of the company is to generate electricity for the purpose of supplying power and light.

A committee of the city council of St. Catharines, Ont., has reported on the question of electric lighting. While stating that a saving could be effected by the purchase of an electric light plant, it is recommended that tenders be called for lighting for two years, in order to ascertain the cost of light supplied by a private company.

The Legislative Railway Committee have passed the Hamilton Electric Radial Railway bill without any amendments. The company is given power to extend its Hamilton and Brantford branch to Woodstock. Toronto objected to the proposition to have the Hamilton and Mimico branch extended to Toronto and the clause was struck out.

The T. Eaton Co., Toronto, with the object of economizing space, are substituting for their large Wheelock engines two of the new Ideal type of high-speed engines, manufactured by the Goldie & McCulloch Co., of Galt, to which will be direct-connected two C.G.E. multipolar dynamos, generating current for lighting the establishment.

A company in Cleveland, Ohio, is said to be organizing a company to operate horseless carriages in that city. It is the intention to have the line in operation by June, and to charge 2½ cents fare. The carriages will have a capacity for twelve persons seated and eight standing. An ordinary hack license is believed to be the only fee required for a permit to operate.

Hon. E. H. Bronson has introduced a bill in the Ontario Legislature respecting street railways. The bill gives power to electric railway companies to acquire parks, requires them to take reasonable measures to protect water pipes, etc., against damage by escaping electricity, and provides that the tolls shall be so fixed as not to exceed $8 on every hundred dollars of paid up capital.

The New Westminster & Burrard Inlet Telephone Co. have acquired the telegraph line between New Westminster and Langley, Aldergrove, Matsqui and Chilliwack, which formerly has been under the control of the C.P.R. As soon as practicable the offices will be equipped with long distance telephone instruments, and become directly connected with the local exchanges at New Westminster and Vancouver.

While experimenting at Quebec on the manufacture of acetylene gas in order to reduce its cost as a luminary, an explosion occurred by which Arsene Consigny was killed and his brother, Nicholas, seriously injured. They had loaded a six inch iron pipe with gas fluid, and placed the pipe in a tub of water, when it was found to be leaking, and without paying any attention to the gauge, they lifted the pipe for the purpose of locating the trouble, when the explosion took place.

The Cosmopolitan Magazine has offered prizes amounting to $3,000 for horseless carriages making the best records and showing the most good qualities in a trip between its New York city office and its publishing plant at Irvington-on-Hudson.

The Lachine Rapids Hydraulic and Land Co., of Montreal, have elected the following officers for the ensuing year : President, Mr. G. B. Burland ; vice-president, Mr. Thos. Pringle ; managing director, Mr. W. McLea Walbank ; acting secretary-treasurer, Mr. Jeffrey H. Burland.

Mr. G. T. Simpson and a number of associates are said to be forming a company in Hamilton to supply electricity for incandescent lighting. It is claimed that a large reduction in the price now paid for lighting will be effected. The promoters have leased a building on McNab street.

The present agreement of the Bell Telephone Co. with the city of Toronto, expires at an early date, and the company have been requested to make an offer for a five years' renewal. If the offer is not satisfactory, tenders for the privilege of establishing a telephone system will be invited.

The new company which has perchased the Brantford electric light plant and the Grand river level property, have elected officers as follows : President, Geo. H. Wilkes ; vice-president, Herbert R. Yates ; secretary, John McGeary. These, with Messrs. B. W. Yates and A. J. Wilkes, will be the directors. The capital stock will likely be increased.

A railway deal of considerable importance is said to have been closed recently in Montreal, by which a contract has been awarded by the Montreal Park and Island Railway Company to Mr. Wm. G. Reid, contractor, for the construction of thirty-five miles of electric railway, including right of way, ties, rails, power houses, car sheds, rolling stock, electric appliances, etc. The amount involved will reach well up to $1,000,000.

Mr. John Shearer, of Preston, closed a contract last week for a 75 kilowatt monocyclic alternator with the Canadian General Electric Co. Mr. Shearer has been operating a single-phase alternator of the C.G.E. type, but has found so much demand for power as to lead him to look into the question of a suitable system for supplying both light and power. The monocyclic is to be in operation by the first of April at the latest.

Referring to the application for incorporation under name of Montreal Electric Light Co. recently made by L'Tourville, et al, the Montreal Electric Co. write : "The application in question was opposed by us before the Legislature of the Province of Quebec, owing to similarity of names, and our opposition was sustained. We understand that the name under which incorporation is now being asked for is "The People's Electric Light Co.", but the concern is virtually the old St. Jean Baptiste Electric Light Co. under a new name. We are the only firm who have the right to use "Montreal Electric," and we do not supply light or power.

The annual meeting of the Toronto Electric Light Co., with which was recently amalgamated the Incandescent Light Co., was held early in March, the principal business being the election of the Board of Directors. The following were elected : H. M. Pellatt, Thos. Walmsley, Samuel Trees, W. T. Murray, S. F. McKinnon, H. Blain, George A. Cox, W. D. Matthews, F. Nicholls, H. P. Dwight, Robert Jaffray and W. R. Brock. At a subsequent meeting the by-laws were amended to permit of fourteen directors, and the names of Messrs. Hugh Ryan and A. H. Campbell were added to the list.

The city of Kingston, Ont., has under consideration the municipal control of its electric light plant. Enquiries have recently been sent out to a number of towns regarding the cost of lighting, and the information received may be summarized as follows : The average of sixteen places in Canada, plant run by steam, by contract, is $82 per lamp per year, run all night, moon schedule. From nine places, water power, by contract, the average price per lamp per year is $75. Sixteen places in the United States, steam power, by contract, average $100 per lamp per year, showing a considerable increase over Canada. Seven places in the United States with water power, average $75 ; same as in Canada. These returns appear to show that the larger the place the more they have to pay for lighting under contract, as Toronto and St. John are the only large cities where the cost is as low as in the smaller places. Coming to the places where the municipalities own and run their own plants, seven places in Canada were heard from, viz., Jolliette, Windsor, Collingwood, Picton, Moncton, St. John and Markham. In these places only five appear to know what the cost is per lamp per year, and of these four are rather indefinite, while Windsor gives all the details in a most satisfactory manner. The average of the five is $66 per lamp per year, counting in interest on investment and deterioration. In Windsor they have a steam plant of 115 arc lamps which cost $23,000, and this costs them $6,039 per year to run, or $52 per lamp per year, or adding interest and depreciation at eight per cent., makes $68 per lamp per year. Returns were received from eleven places in the United States who own their own plants and are run by steam; the average of these places is $60 per lamp per year, in all cases including interest and sinking fund on capital invested. From three places in the States owning their own plant, run by water, the average is $46 per lamp per year.

The Brantford Railway Co. have issued a handsome lithographic poster illustrative of the attractions of their new Mohawk Park.

We are in a receipt of the first issue of "Home Study," issued by the Colliery Engineer Co., of Scranton, Pa. It is an elementary monthly journal for students of industrial sciences and others interested in obtaining a knowledge of electricity, engineering, architecture and kindred subjects.

# ELECTRIC RAILWAY DEPARTMENT.

## THE HULL AND AYLMER ELECTRIC RAILWAY.

ARRANGEMENTS have just been completed for the conversion of the Hull and Aylmer branch of the Canadian Pacific Railway into an electric road, and for its complete equipment for freight and passenger business. This installation is of peculiar interest as being the first case in Canada in which a regular steam railway has made the change to electric motive power, and the results obtained in its operation will no doubt have a most important influence in determining similar action on a great many branch lines at present either entirely unproductive, or operated at a loss by the Canadian Pacific and the Grand Trunk. That the fullest measure of success, both from the financial and operating point of view, will be attained, is practically guaranteed by the results

road. The water-power at the Deschenes Rapids, which will furnish motive power for the generators, is controlled by the lumbering firm of R. & W. Conroy.

The contract for the entire electrical equipment and cars has been awarded to the Canadian General Electric Co. The generating plant will consist of two 200 Kilowatt C. G. E. 500 volt railway generators with which will be furnished the necessary instruments mounted on standard panels of enamelled slate. The waterwheels will be of the New American type manufactured by Kennedy & Sons of Owen Sound. For the express passenger service there will be provided four open and four closed forty foot cars of the C. G. E. standard type mounted on Blackwell double trucks. The motor equipment for each car will consist of two C. G. E. 1200 motors with **K** 2 controllers wound for

GENERATOR ROOM—LONDON STREET RAILWAY POWER HOUSE.

which have been made public in the case of the Metropolitan and Lake Street Elevated Roads in Chicago, the Nantasket Beach division of the Hartford and New Haven R. R., and the electric locomotives on the Baltimore and Ohio, which have for some time past been handling the traffic through the Baltimore tunnel. An unexpected feature in the latter case, considering the somewhat unfavorable circumstances, has been their advantage in economy over the steam locomotives as shown by the recent tests, printed elsewhere in this issue, made with the dynamometer car of the Pennsylvania Railway. On the Hull and Aylmer road, where it is proposed to utilize the magnificent water power of the Deschenes Rapids on the Ottawa river, this advantage in first cost of motive power will of course be realized to the fullest extent, and will in itself go far to insure the profitable outcome of the enterprise.

The shareholders in the Hull Electric Company comprise some of the foremost capitalists and business men in Ottawa, an especially active interest in the undertaking being shown by Messrs. C. Magee, President of the Bank of Ottawa ; Alexander Fraser, the well-known lumberman, and E. Seybold, secretary of the Company. Mr. J. Brown, of Ottawa, well known in electrical circles, is superintendent and electrical engineer of the

high-speed service. The construction and finish guaranteed for the cars are such as to make them a model of the car builder's art.

An interesting addition to the rolling stock of the road is the 30 ton electric locomotive with which the large freight business of the line will be handled. In general appearance this locomotive will resemble the locomotives built by the General Electric Company for the Baltimore & Ohio tunnel service, though of course of much smaller capacity. The motors will be four in number, of fifty horse power capacity each, of the C. G. E. 1200 type, provided with a special controlling mechanism.

Work on the bonding of the rails and other changes required to make the road suitable for electrical operation has already been commenced, and the entire system is expected to be in operation by the end of June.

Application will be made to the Ontario Legislature for incorporation of the Chatham City and Suburban Railway Company, to construct an electric railway through and from the city of Chatham to a point on Lake Erie, also through the township of Dover to Wallaceburg and thence to Petrolea. The capital stock of the company is placed at $200,000, and the promoters are John Mercer, John A. Walker, E. Jones Park, Sydney E. King, G. J. Leggatt and George Rankin.

## LONDON STREET RAILWAY SYSTEM.

SOME time ago the directorate of the London Street Railway Co. determined that, to keep abreast of the times, and to run on an economical basis, the road should be changed to the electric system. Application was made to the city council for the franchise, and after much wrangling and the customary " serious considera- tion," that body decided to allow them to build and operate an electric street railway.

The officials of the road are : President, H. A. Ever-

MR. CHAS. E. A. CARR,
Manager London Street Railway.

ett, Cleveland,; Vice-President, E. W. Moore, Cleve- land ; Secretary, Chas. Curry, London ; Manager and Treasurer, Chas. E. A. Carr, London. Mr. Carr was installed on the 1st of January, 1895, to succeed Mr. Brake, of Detroit. We herewith present a portrait of Mr. Carr. He is a young man and is to be complimented on the suc- cess with which he has pushed the work of electrifying the road. He divided the work into four sections, with a superin- tendent over each, and foremen under them. John M. Moore, architect, had charge of the building department, J. B. Mackay, of the track construction, W. J. Hillier of the overhead construction, and Mr. Ryan of the mechanical department. Mr. Ryan, a mechanical engineer from Cleveland, had charge of the erection of the machinery in the power house. The power house is a massive brick structure facing on Thames street, beside the rail- road track, where a switch has been put in, and the furnaces can be fed with coal direct from the cars.

The boiler room contains six Bertram boilers of 150 h.p. each. The dynamo room is equipped with three Armington & Sims engines of 150 h. p. each, built at Providence, R. I. Room is left for one more engine. Each engine drives a pair of C.G.E. railway generators of 100 k.w. each. Room is left for another pair of generators. The condenser pit is in the centre of the dynamo room, and in it is a Northey condenser. The pit and the foundations of the building are entirely waterproof above the level of the dynamo room floor. The best material is used in the construction of the building. The chimney is the largest in the city. It was designed by Cleveland engineers and is a striking feature. An arch through the bottom of the chimney leads from the boiler room to the dynamo room. Opening off the dynamo room is the engineer's room and switch-board room, with a storey above for storage, etc. The switch-board is of the latest design and the wires from

the machines run through the floor into the basement, then through brick tunnels, then up through the floor of the switch-board room to the switch-board, then out to the poles. On the property large car barns will be built and also offices. Messrs. McBride & Farncombe are the architects for the car barns.

The rails, which were made in Germany, reached London on Aug. 28th last, and in two weeks there were 2½ miles of track laid and the cars running on them from the G.T.R. station on Richmond street to the Fair grounds on Dundas street, east. The trolley wire is oo hard drawn and the feed wire is oooo. The cars, built by Patterson & Corbin, St. Catharines, are the equal of any in Canada, and surpass the new cars on the Detroit street railway. They are built on the latest pattern, have large vestibules and a sliding door on one side, operated by the conductor from the rear platform. The cars have a capacity of 38 passengers and are heated by the Consolidated Car Heating Co.'s electric heaters. The seats are cane bottomed.

Some 800 men were engaged on construction, and all the labor done was by London people. $40,000 has been spent on labor alone, and altogether $400,000 has been expended up to date.

The route furnishes the outlying districts with quick communication with the business part of the city. Double tracks are laid on Dundas street its entire length, and on Richmond street from Horton street to the C. P. R. depot. By the completion of the bridge over the Thames (south branch) on Ridout street, a useful belt line was established. The belt line starts at the corner of Richmond and Dundas streets, up Richmond to Central Ave., along that to Adelaide street to Dundas, Dundas west to Richmond, down Richmond to York, York west to Wharncliffe Rd., south on that road to Askin street, east on Askin street to Wortley Rd., south on Wortley Rd. to Elmwood Ave., Elmwood Ave. east to Ridout street, Ridout street north across bridge to Horton street, Horton street east to Rich-

SWITCH BOARD—LONDON STREET RAILWAY POWER HOUSE.

mond street, along Richmond street north to the place of beginning, the distance covered being about 7 or 8 miles. Another line runs west from Aledaide street on Oxford to Richmond street, down Richmond to Horton street, east on Horton to Hamilton Rd., down Hamilton Rd. to Rectory street, a distance of 8 miles. The line will be continued two miles farther on Hamilton Rd. in the spring. Another line runs from the Fair grounds on Dundas street east to Dundas street west, back to Ridout street, Ridout to King street, King to Richmond street, Richmond to Dundas street, and back east again. This line runs around the market. Eastern Dundas line will be extended to Pottersburg next summer, a distance of two and a half miles. This line connects at Dundas street west at the bridge with the London West line, which, up to the end of 1895, was operated by horses. London West line runs on Dundas street to Wharncliffe Rd., up that to Oxford street, along Oxford to the bridge. A bridge will be built at this point at some date in the future and another belt line will be completed by Oxford street to Richmond. Another line runs from Dundas street east to Richmond street, up Richmond to St. James street, St. James to Wellington street, up Wellington to Hellmuth college at the city limits. Their summer line leaves the belt line on York street, and follows the river Thames to Springbank, where the city water works are situated. This line is private property and excursions go down there every day in the summer as the place is pretty and the scenery along the river bank and through Woodland cemetery, through which the line passes, is magnificent. Many summer residences are situated at Springbank, and the road is a boon to tired business men who can run down there and spend their evenings with their families. A six minute service is given.

The county council are very obstinate and will not grant the company the privilege of building a bridge on Dundas street, over the north branch of the Thames alongside of the county bridge and resting on their piers. They do not like the company, as it has, by discarding horse power, ruined (?) their market for grain and hay. It is to be hoped, however, that they will ultimately grant the company's request.

The London Electric Co. supplied the company with power while their power house was being built, and, notwithstanding that fact, their earnings were 350% over the horse car system.

A sweeper with a C. G. E. No. 1200 motor cleans the rails of snow, and the old horse sweeper is also in use. The Canadian General Electric Co. equipped the sweeper, and the Toronto Street Railway Co. built the case and broom. The company has closed a contract with the city to sprinkle the streets in the dusty season for $250 per mile of street. The streets will be sprinkled four times a day. In all probability the city mails will be transferred from the post office to the branch offices by the company.

The different lines reach to the surrounding suburbs and are an invaluable aid to the laboring population. Connections are made quickly and everything is running satisfactorily. Pottersburg, a village to the east of London, will be a good feeder, and the company will profit by this extension. The earnings of the company have increased 350% since the system has become electrified. The company is noted for its kindly treatment of its employees.

Patterson & Corbin, of St. Catharines, were kind enough to loan the company some of their cars while the new ones were being made ready at the new car barns on Dundas street. These barns were built late last fall and can accommodate 10 cars. The others will be stored in the old horse car barns until the new barns are built. Between 25 and 30 Patterson & Corbin cars traverse the 22¼ miles of road, and the entire equipment of the system in every detail makes it one of the most up-to-date and efficient on the continent.

## THE CORNWALL ELECTRIC RAILWAY.

Messrs. Hooper and Starr, the latter the well-known electrical engineer of Montreal, have commenced work on the construction of an electric railway for freight and passenger service at Cornwall. The contract for the electric apparatus, including a 200 Kilowatt generator and ten C. G. E. 800 motors has been awarded to the Canadian General Electric Co. The same Company will also furnish two closed motor cars of their standard type, the balance of the cars being supplied by the

Rathbun Company of Deseronto. A freight locomotive on which the motive power will consist of four C. G. E. 800 motors with special controllers will handle the freight business between the Mills and the Grand Trunk Railway.

Mr. A. J. Nelles, ex-president of the Hamilton, Grimsby and Beamsville Railway, has begun an action against the company for $5,000 damages for alleged wrongful dismissal and breach of contract, and for $166.65 as balance of salary due between 1st October, 1895, and 1st March, 1896.

Mr. J. F. Hill, comptroller of the Montreal street railway, resigned his position on March the 7th. Mr. Hill has been connected with the street railway since 1892. He spent the first six months with the company in construction; the present system of accounts, which has become so widely known, being introduced by him. During the construction of the Canadian Pacific Railway, Mr. Hill was construction accountant and was actively associated with Mr. T. G. Shaughnessy, the present vice-president of the company. Before he accepted the position of comptroller of the Montreal street railway, he was accountant to the works department of Toronto. Prior to Mr. Hill's departure to Chicago, where he has accepted a responsible position, the employees of the Montreal Street Railway Co.'s business offices showed their affection for their comptroller by assembling in force for the purpose of presenting him with a handsome gold watch and chain, the watch bearing the following inscription: "Presented to J. F. Hill, Esq., by his staff as a slight token of their esteem and regard upon his severing his connection as comptroller of the Montreal street railway, March 14th, 1896." After the address had been read, all who were present sang "For He's a Jolly Good Fellow," under the able leadership of Mr. E. Lusher. Mr. Hill has gained the affection of all with whom he has been in contact during his connection with the company, and the feeling of regret at his leaving was plainly shown during the course of the proceedings.

## SPARKS.

Additional cars have been received for the London Street Railway from Messrs. Patterson & Corbin, of St. Catharines.

The Trenton Electric Co. has obtained supplementary letters patent to extend operations to Belleville and Thurlow township.

An electric franchise is held by the St. Thomas Street Railway Company, whose privileges and assets are offered for purchase.

Application has been made to the Government for a charter for the construction of an electric railway to connect Amherstburg and Harrow.

Wm. J. Clarke, a telegraph operator employed at Jersey City, N.Y., and a native of Kamloops, B. C., committed suicide by gas asphyxiation.

The Stratford Board of Trade are considering the project for the construction of a steam or electric railway from Embro to Stratford, a distance of seventeen miles.

The City Council has made claim against the Winnipeg Electric Railway Company for $1,040 for damages owing to the failure of the company to run cars on certain streets as agreed.

The bill to incorporate the Canadian Electric Railway Company, which proposed to construct an electric railway from Windsor to Montreal, has been thrown out by the Railway Committee at Ottawa.

The Montreal Park and Island Railway Company is seeking an extension of time until 1901 in which to complete their lines. Amalgamation is also asked with the Chateauguay and Northern Railway Company and the Jacques Cartier Railway Company.

Mr. John Patterson, who is promoting the radial railway from Hamilton via the Beach to Burlington, affirms that his company will have the road in operation by June 1. The money to build it is subscribed, and all arrangements made with the exception of those in Burlington, which will be completed in a short time.

Hon. Peter Mitchell recently brought action against the Montreal Street Railway Company claiming $15,250 damages, alleging that from neglect of its employes he was seriously injured while alighting from a car. After five days' trial, during which time much evidence was submitted, the jury gave a verdict for the plaintiff of $1,000 and costs.

Stockholders in the Crystal Beach Improvement Company held a meeting at Ridgetown, Ont., early in March, and subscribed $4,500 for the construction of the Beecher single rail elevated railway. Work will be begun on the new road at once, and if the experiment proves the success anticipated, it will be extended down the beach to Point Abino, and perhaps to Niagara Falls.

St. John, N.B., has eleven and one-half miles of electric railway.

An electric railway line is proposed which will connect Vancouver, B. C., with Steveston, and proceeding up the line of the Fraser river, will take in some twenty agricultural settlements.

The electric light plant at Aylmer, Ont., consisting of three dynamos, together with circuits and street and shop lights, is offered for sale by tender by A. E. Haines. The date limit is April 15.

The street railway company at Sarnia, Ont., have in contemplation the conversion of the railway into an electric road.

Messrs. John N. Lake and M. Hopkins, of Hamilton, are promoting an electric railway which it is proposed to build from the head of Wentworth street southerly by way of Hall's Corners, York and Indiana to Cayuga or Dunnville, with probably a spur to Caledonia. Surveyors have gone over the route, and report no engineering difficulties.

# CANADIAN GENERAL ELECTRIC CO.

## (LIMITED)

---

## The following Letters speak for themselves:

PORT HOPE, Feb'y 27th, 1896.

MESSRS. CANADIAN GENERAL ELECTRIC CO.,
Toronto.

DEAR SIRS :—The 75 Kilowatt Monocyclic Alternator purchased from you was started up on Sept. 2nd, 1895, and has since been giving us an uninterrupted service of 160 hours each week, starting at four o'clock on Sunday afternoon and running till eight o'clock. the following Sunday morning. without a hitch of any kind' whatever. We expect a large increase to our business from motor service, and appreciate the excellent features of the Monocyclic system of this combined light and power service. There is, of course, no unbalancing, as the lighting is single phase, and the operation of the motors does not disturb in any way the evenness of the lighting. After an experience of six months we feel warranted in saying that we consider the Monocyclic as superior to any of the polyphase systems which we are acquainted with, and intend in the near future to duplicate this machine.

Yours truly,
R. A. CORBETT,
Pres. & Mgr. Port Hope Elec. Lght & Power Co.

---

PARRY SOUND, Feb. 27th, 1896.

THE CANADIAN GENERAL ELECTRIC CO., Toronto.

DEAR SIRS :—Having now made a thorough trial of the Monocyclic system of Electrical distribution as supplied by you, I have much pleasure in informing you that it is giving entire satisfaction.

The machine, a 75 K.W., is a beautiful specimen of dynamo building, being strong and compact. Ventilation of the armature is excellent, and the general design of that very important part of the machine is good. Electrically and mechanically, I consider your machine to be superior to any I have seen.

We have not had occasion, as yet, to test the machine on the operation of motors, but speaking from the lighting point of view, I can fully endorse what you claim for the system.

We have 700 lights now wired and expect to increase to 1,200 before 1897.

Yours truly,
W. B. ARMSTRONG,
Manager Parry Sound Electric Light & Power Co., Ltd.

---

HANOVER, Feb. 21, 1896.

CANADIAN GENERAL ELECTRIC CO., Toronto.

DEAR SIRS :—In answer to your enquiry as regards the operation of our Electric System, I beg to say that we have now been running one of your 75 Kilowatt Monocyclic Dynamos for the past two months, and it is giving entire satisfaction in every respect. We have not had the slightest trouble with it in any way, and although it is being operated about 15 hours a day, it runs exceedingly cool, and requires practically no attention whatever.

The machine itself I regard as a model of simplicity, in fact to show my confidence in the apparatus I have placed the plant in full charge of my brother, who, previous to the starting up of this machine, had no experience whatever with electrical apparatus of any kind.

The perfect regulation of the Dynamos, and the sim-
plicity of the wiring are also strong points which should recommend the use of this style of apparatus to anyone contemplating the installation of an electric plant.

In conclusion, I might say that after having decided upon adopting the Monocyclic system, my opinion became somewhat prejudiced against its adoption by representations made by other manufacturers, but I now fail to see wherein I could have secured anything better to that installed by your company.

Yours very truly,
D. KNECHTEL

---

DUNNVILLE, February, 1896.

MESSRS. CANADIAN GENERAL ELECTRIC CO.,
Toronto, Ont.

DEAR SIRS :—We are pleased to be able to express ourselves as entirely satisfied with the Monocyclic system installed by you last fall. We are now in a position, having covered a considerable portion of the town with our lighting mains, to appreciate the value of the three-wire system for secondary distribution from the transformers, and the great advantage gained in simplicity by the Monocyclic from its being a single-phase system for the lighting distribution.

The workmanship and finish of the dynamo itself certainly does credit to your factories, and in operation it has proved itself to be exceedingly simple and satisfactory.

The commutator and brushes run without any sparking whatever, and do not give us a particle of trouble. We feel fully justified in saying that the Monocyclic system in operation has shown itself to possess all the points of excellence claimed for it by you at the time when we made the selection for our new plant.

DUNNVILLE ELECTRIC LIGHT CO.

---

MATTAWA, Feb. 27th, 1896.

MESSRS. CANADIAN GENERAL ELECTRIC CO., Toronto.

DEAR SIRS :—We are pleased to be able to express complete satisfaction with our Monocyclic plant, which has now been running since 27th Sept. We are especially pleased with the ease with which our former single-phase system has been changed into one suitable for the distribution of both light and power. The only change made in our case was the installing of the Monocyclic machine in place of our former single-phase alternator, and the running of a third wire to the points where power is to be supplied. Altogether the system is admirable, both as to simplicity in the wiring, and distribution and perfect freedom in operation from any trouble or complication. We are quite sure that the Monocyclic system will prove a means of increasing largely the revenues of alternating lighting stations by the sale of power without adding any complications to their operation. You will be pleased to know that the 5 h.p. induction motor geared to a triplex pump is now in successful operation pumping water for the C.P.R. water tank. It is certainly a very simple and substantial piece of machinery.

Yours truly,
MATTAWA ELECTRIC LIGHT & POWER CO., LTD.

A. F. HURDMAN, Sec'y-Treas.

---

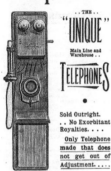

# CANADIAN ELECTRICAL·NEWS
## STEAM ENGINEERING JOURNAL

OLD SERIES, VOL. XV.—No. 6.
NEW SERIES, VOL. VI.—No. 5.

·MAY, 1896

PRICE 10 CENTS
$1.00 PER YEAR.

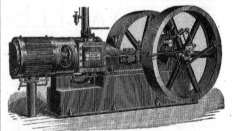

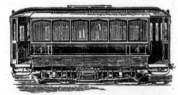

CANADIAN

# ELECTRICAL NEWS

AND

## STEAM ENGINEERING JOURNAL.

| VOL. VI. | MAY, 1896 | No. 5. |

BOILER EXPLOSION AT RIDGETOWN, ONT.

[For particulars see next page.]

## BOILER EXPLOSION AT RIDGETOWN.

On the 6th of April a boiler exploded in the saw, stave and heading mill of Watson Bros., at Ridgetown, Ont., which completely wrecked the mill and has thus far resulted in the death of four persons. The accident occurred just as the employees were preparing to enter upon their days' work. The fire had been under the boiler for some time, but the machinery had only been in operation about a minute and a half.

The boiler was a horizontal tubular one, 54 inches diameter and 11 feet 6 inches long, with 58 tubes 3 in. in diameter, and a dome 20 in. diam. and 27 in. high. The plates were iron and were a little over one-quarter inch thick. The joints were all single riveted, the lap of plates being 2 in., and the rivets were ⅝in. diam. and 2 in. pitch. Manhole was 15 in. by 11½ in. and had a strengthening ring around it 1¾ in. by ⅜in. The boiler was in general good order and fairly clean inside.

After explosion there was no evidence that the boiler had been neglected or had been carelessly used. The back head had been renewed at some time and was in very good condition, and evidently was stronger than the front head. The boiler had been used at a pressure of nearly 90 lbs. per sq. inch, and was supposed to be quite safe for a higher pressure. It apparently gave way first at the manhole, or near to it, and was split open from the top across the boiler. The manhole cover was picked up about 60 feet from the original position of the boiler, complete and uninjured, with bridge and bolt attached.

The dome was thrown about 600 feet, and the plate to which dome had been attached went about 700 feet in a different direction. The position of the front part of shell and of the back part confirm the theory that the boiler gave way first at the upper part, as these pieces were thrown in opposite directions and appear to have been turned end for end in their flight.

A second boiler which had no steam on at the time was thrown bodily over the engine and badly ruptured.

The violence of the explosion is clear proof that there was plenty of water in the boiler at the time, and the back head showed no sign of ever having been overheated. The quality of the plates seemed to be common boiler iron, and the most probable cause of the explosion was that the pressure carried was too high for the strength of the shell at the manhole and at base of dome. The severe strain put upon these parts had gradually weakened the boiler, so that it gave way at the ordinary working pressure.

How best to prevent similar accidents is a question well worth considering. In Great Britain, where so many boilers are in use, Government inspection has been carefully avoided, but the Boiler Explosions Act requires the user of a steam boiler to report to the Government every accident, no matter how trifling, and an investigation is held and the owner has to prove that he was using all proper precautions. Under this system the fault which led to the accident is traced out to the maker, or seller, or user of the boiler, and the blame fixed upon the right person.

The coroner's jury, in their verdict, stated that the cause of the explosion is unknown, but recommended that the government make it compulsory to users of steam boilers of all kinds to have them periodically inspected by competent boiler inspectors.

## HAMILTON, ONT.

(Correspondence of the CANADIAN ELECTRICAL NEWS.)

THE dissentions which arose at the annual meeting of the Hamilton, Grimsby and Beamsville Railway, and which resulted in the election of Mr. T. W. Lester as president, have not yet subsided. The ex-president, Mr. C. J. Myles, has again secured a controlling interest, and has reinstated Mr. A. J. Nelles as superintendent. At the adjourned meeting of shareholders, held a fortnight ago, the Myles faction represented the majority of the stock. A special committee reported in favor of doubling the stock, as there was sufficient surplus to do so; the report was adopted. A motion by C. J. Myles, seconded by R. S. Martin, that Mr. A. J. Nelles be reinstated as manager and superintendent of the road, caused an animated discussion which lasted two hours. Mr. Myles alleged that since the deposition of Mr. Nelles as manager, there had been a decrease in the quantity of freight and number of passengers. The appointment was finally carried by a vote of the shareholders present. It is claimed by the supporters of the Leister ticket, however, that the appointment rests entirely in the hands of the directors. It is said that an effort will also be made to compel the resignation of Mr. Adam Rutherford, secretary. The profits of the year amounted to $11,143.53. Beyond the bonded and mortgaged debt, there is a floating debt of $19,576.10. The directors decided not to urge the building of a line to Grimsby Park and Beamsville this year. It is to be hoped that an end will be put to dissentions within the company, which, if prolonged, must seriously affect its prosperity.

Mr. Powell, who was until recently engaged as engineer by the International Radial Railway Company, is maturing plans for the construction of an electric railway from Hamilton to Guelph, Berlin and Waterloo. Charters have already been granted for a road to connect these cities, and Mr. Powell will endeavor to secure one of these. If unsuccessful, a new charter will be applied for. The capital, it is said, will be furnished by capitalists of Toronto, Guelph and Cleveland. No bonuses will be asked from the municipalities.

If the various schemes projected by Mr. E. A. C. Pew were successfully carried into completion, his name would be handed down to posterity as one of the greatest promoters of electrical development of the nineteenth century. His latest project is to supply power to the city from the Welland river by overhead conduits, the plan being to tap the river about one and one-half miles from Wellandport and build a canal, six miles in length, to run the water at Jordan, where there is a fall of 322 feet.

The Hamilton Radial Railway Company, of which Mr. Pew is also the promoter, has been granted right of way by the City Council on a thirty-two year franchise, and have commenced the construction of the line between the city and Burlington Beach, which it is expected to have completed by Dominion Day. The power house will be located at Burlington, that village being almost midway between this city and Oakville. Tenders for engines and power machinery have been received, and the contract will be awarded at an early date.

The Simpson-Noble Electric Light & Power Company, the new concern organized to supply electric light in this city, turned on the current a fortnight ago. The poles are being erected on private property, as the company as yet have not permission to erect them on the streets. The offices are at 103 Macnab street.

The directors of the Hamilton and Dundas Railway have decided to convert the road from steam to a first-class electric line. The work will occupy about two months, and will probably be commenced by the first of July. A change will also be made in the equipment, the ties having already been contracted for. This step meets with the hearty approval of the citizens, who consider that the road in its present shape is not in keeping with modern developments in methods of railway construction and operation.

The Hamilton Street Railway Company have made application to the City Council for an amendment to the by-law whereby the company would not be required to pay the city such a large revenue. The matter has been left in abeyance until an audit is made by the city auditors of the company's books.

HAMILTON, April 30, 1896.

## C. P. R. TELEGRAPH STORAGE BATTERY PLANT AT OTTAWA.

By W. J. CAMP.

THE Canadian Pacific Company's office at Ottawa, Ont., has been equipped with storage battery, and the old gravity entirely dispensed with. As there are some combinations different from those in use at other points, a description may prove of interest to your readers.

The cells used are those made by the Electric Accumulator Co., type E9 being used for locals, and type C 3 for mains. The charging circuit varies from 230 to 250 volts. The locals are in 3 banks of 2 cells each; No. 3 and 4 being used for the local circuits in the main office, and No. 2 for supplying additional power on quad locals when extended to the Parliament buildings office (H. U.). These locals are charged through a small motor-generator, which gives a voltage of 6, with a capacity of 20 amperes on the generator side. The main batteries consist of 8 banks of 40 cells each, a total of 320. These are charged in groups of 80 cells each directly from the power circuit, a resistance being inserted to bring the current down to 1¼ amperes; or two banks can be charged simultaneously at the rate of 2½ amperes. All single wires are worked from 40 cells positive or 40 cells negative. These cells also furnish the "short end" for quads. These two banks are arranged in duplicate, one lot being charged while the other is in use. As quad is not worked during the morning while parliament is in session, and only occasionally during the balance of the year, and it is found that sufficient current can be stored in the morning to last the quads for the rest of the day, the remaining 160 cells are not duplicated, and can only be charged while the quads are idle. The same applies to the cells for the quad legs battery. The total current for quads is obtained from 80 additional cells on each pole. This gives the "short end" about 88 volts and the total 264. As the longest quad from Ottawa is to Toronto (256 miles), this gives a good working margin.

Fig. 1 shows the arrangement of the charging and discharging switches for the mains. Those for the locals are the same. These switches are known as "double pole, double throw." The dotted lines show the charging current and the straight lines the discharging circuits. (Only one bank of 40 cells is shown in the diagram.) The charging is done, for instance, as follows: 9 and 11 are charged for one-third of the morning, 10 and 12 for one-third, and No. 2 local for the balance. During the afternoon and evening one day, Nos. 5 and 7 and No. 3 local are charged, and the next day Nos. 6 and 8 and No. 4 local. As so much more work is performed by these cells, a much longer time than that required for the cells for quad working is needed to replace what has been taken out. All cells are kept fully charged, and should the power circuit give out, there is always sufficient current stored to work the office for a week or ten days. The automatic circuit opener opens the charging circuit should anything happen to the power wires, and prevents the batteries discharging back.

The transmitting circuits of the quads are a modification of the Jones system as used by the Postal Telegraph Co., a single pole-changer being placed on the polar side, and two of them on the neutral side, as shown in fig. 2. Opening PC throws line to P, and closing PC throws line to N. P gives positive currents only, and N gives negative. N and P are worked simultaneously closing N and P gives the total current to the line, and opening them gives only the partial current of either polarity according to the position of PC. Each lead from the quads to batteries has a resistance of 700 ohms, made up by two 16 c.p. 110 volt incandescent lamps. The leads from the batteries to the single wires pass through one 16 c.p. 110 volt lamp for each wire.

It is intended to place storage for the locals in the H. U. office later, and the system for working locals in the main office has been designed with that end in view—the only change in the main office then being to move No. 2 local over to H. U. During

into quads, the same course is followed, except that switch S is not required on the repeater sets. To leg H. U. office on, the switch S remains on the left hand contact, the pegs for the send-ing and receiving sides are moved over to the two H. U. wires selected. The circuits now are as follows : For sending—Earth, batteries No. 2 and 3, switch S, key, transmitter, discs, strip B, H. U. sounder, key, earth ; receiving—earth, batteries, relay, sounder, discs, strip C, H. U. sounder, earth. In each case the circuits have been increased to 40 ohms plus the line resistance to H. U., but at the same time the battery power has been increased so that the current is about of the same strength as before. In practice we found that the line resistance was so small that it could be neglected, and that the two additional cells were quite sufficient.

The Milliken-Hicks repeaters are connected up in a different manner from that in use at any other place. The governor circuit is (when closed) in multiple with the coils of the transmitter. This appears to give much better results than when it is a distinct cir-cuit. By connecting the jacks of two half sets together through a double-ended cord they can be used as ordinary single line re-peaters.

At H. U. there are two complete sets of single line repeaters, a spring jack main line switch, and 12 single sets. Six of the spaces allotted to the single sets are arranged with three point switches and additional sounders so that they can be used as quad

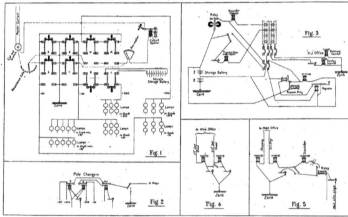

C. P. R. TELEGRAPH STORAGE BATTERY PLANT AT OTTAWA.

session of parliament it is preferable to have all wires end at H. U. Instead of connecting loops to wedges at the main office as is usual, all H. U. wires are connected with the upright bars in the main office a way one for the time being. Two wires are used for battery leads to H. U., one for each pole. Owing to limited space in H. U., all quads, etc., are placed in the main office and the locals extended to H. U. as required. In the main office there are three quads and six half Milliken repeaters, which are all connected to jacks as shown in fig. 3.

The arrangement of the locals is fully shown in fig. 3. All sounders, transmitters, etc., are wound to 20 ohms. Normally the small 3 point switch (which is placed on the operating tables) is urned to the left, and each sending and receiving circuit of theq uads are pegged to the upright strip marked R in the main swi$_{t}$ch. The receiving circuit can be traced as follows : Bat-tery No. 3 (positive), relay armature, sounder, disc, strip R, to negative pole of battery No. 3. The leg from sounder through jack, being open at switch S, remains "dead." The sending cir-cuit is as follows : Positive of battery No. 3, switch S, key, transmitter, disc strip R, to negative pole of battery No. 3. There being no earth on either circuit current on the negative pole of battery No. 2, that battery is not drawn on for any cur-rent. To work as repeaters the switch S on each set is turned to the right, and a double-ended cord connects the top of one jack with the bottom of the other, and vice versa. Although only one-half quad is shown, the circuits can be readily followed. For instance, starting with No. 1 set, battery, relay armature, jack (top) cord and wedges, bottom of jack for second set, switch S on No. 2 set, key, transmitter, disc, strip R, to negative pole of battery. Starting from No. 2 set a similar course is followed. The leg through sounder is not disturbed, and that circuit is the same as for ordinary working. It may be noted that each cir-cuit has only a resistance of 20 ohms. To work single lines

legs. There are also a couple of ingenious arrangements that are very convenient, and which I do not remember to have seen de-scribed before. They were put in by Mr. Bott, the Ottawa man-ager. One is to connect the legs of two half quads so that one operator is able to send simultaneously in both directions ; this is shown in fig. 4 ; for the other ends of these circuits see fig. 3. The two point switch connects the two sets together, one key is left open, and the operator sends on the other. These sets are placed side by side, and the operator is able to hear the breaks on the receiving sounders. Of course the breaks do not carry through, but it is not necessary that they should do so. The other arrangement is to work a single line into a half quad, and is so clearly shown in fig. 5 that it is not necessary to describe it.

The shareholders of the London Street Railway Company will hold a special meeting on the 21st of May to authorize an increase of the capital stock of the Company to the amount of $750,000 or less, and also to authorize the issue of debentures to the same amount.

The Victoria, B. C., Electric Railway and Lighting Co.'s prop-erty and franchise was offered for sale by auction, on the 10th inst., by order of the bondholders. The bidding was opened at $200,000, and went up to $340,000, when the property was dis-posed of to Mr. F. S. Barnard, M. P., of the Consolidated Electric Railway Company, who represents an English syndicate. The new owners will continue to operate the road, and will make a number of improvements. The property is a valuable one, the total mileage now in operation including about thirteen miles of track and switches, with seventeen cars and two trailers. The tramway company was first incorporated in 1889 under the name of the National Electric Tramway & Lighting Co., Limited, and in 1894 the name was changed to the Victoria Electric Railway & Lighting Co., Limited.

## POWER STATION RECORDS.

THE advantages to be derived from keeping accurate records of the working of power plants are evidently thoroughly understood by the management of the Montreal Street Railway Company. An examination of the accompanying table, showing the results of the operation of the power station of that company for the year ending September 30, last, will, no doubt, prove interesting. Such records are very valuable for purposes of comparison, and should be kept by the managers of all electric light and railway power stations. We would be pleased to receive, at any time, similar tests for publication in this journal.

The routes of the Montreal street railway are laid out on a slope which rises gradually northward from the river to the base of Mount Royal. The north and south lines, therefore, are on a continuous grade, but the lines running east and west are comparatively level. About 140 cars are in operation in summer, and 100 in winter.

The six engines used in the power house are of the cross compound Corliss type, belted direct to the generators. Their cylinders are 24 in. and 48 in. x 48 in. stroke. They are rated at 600 h. p. each, and the statement shows that they have averaged as high as 643

pull per ton throughout the system, which was the most severe, as might be expected, throughout the winter months. The usual custom of starting with the draw bar pull as measured by a dynamo-meter and working up to the power station effort the average draw pull has been calculated.

## THE N. E. L. A. CONVENTION.

THE National Electric Light Association of the United States will meet in convention in New York City, on the 5th, 6th and 7th inst. There will be held in connection with this convention an electrical exposition, illustrative of past and present developments in the applications of electricity. This exposition promises to be most complete, interesting and instructive, and will probably be the means of attracting an unusually large attendance.

Following is a partial list of the papers to be read and discussed at this meeting.

"Single-Phase Self-Starting Synchronous Motors," by F. H. Leonard; "Results Accomplished in Distribution of Light and power by Alternating Currents," by W. S. Emmet; "Acetylene Gas," by Mr. Ferguson, of the Chicago Edison Company; "Evolution of the Arc Lamp," by L. H. Rogers; "Steam

POWER STATION RECORD OF THE MONTREAL STREET RAILWAY COMPANY FOR THE YEAR ENDING SEPTEMBER 30, 1895.

e. h. p. for an entire month. The engines were built by the Laurie Engine Company, of Montreal, and are run condensing. The boilers, fifteen in number, are of the Lancashire type, made in England, and are rated at 300 h. p. each. Two Green economizers are used, and the temperature of the feedwater, when both economizers are on, is 245 to 250 degs. F. When one economizer is off for cleaning and repairs, the temperature of the feedwater drops to about 190 degs. Fah. The fluctuations in the amount of coal consumed per electrical horse power, shown in the table, are due largely to changes in the temperature of the feedwater when the economizer is on or off.

The table shows a great increase in coal consumption per horse power when the engines were run non-condensing as compared with condensing, and also that screenings were used for fuel with satisfactory results, the consumption per horse power being no greater and the cost much less. In the column entitled "Watts per Ton Mile," the amount of power used to move a ton one mile on the system is shown. The power used was the highest in November and gradually decreased until May. Most of this decrease may be attributed to better weather, but a portion of it is due to improved controllers being put on the cars, and also that less power was used for light in summer than in winter.

The figures in the last column show the average draw

Boilers, Their Equipment and Management," by Albert A. Cary; "Electrolysis," by William Brophy; "Evolution of Interior Conduits, From an Electrical Standpoint," by Luther Steiringer; Lecture.—"The Light of the Future," by D. McFarlan Moore. Topic.—"The Desirability of a Standard Socket," discussion to be opened by Alfred Swan.

The sessions will be held in one of the large rooms in the Industrial Building, Lexington avenue and Forty-third street. The hotel headquarters will be the Murray Hill Hotel, Park avenue and Forty-first street, within two blocks of the convention hall. The hotel rates to delegates have been fixed at $2.00 per day and upward on the European plan; $4 and upward on the American plan.

Canadians will doubtless feel more than ordinary interest in this convention from the fact that in all probability Mr. Frederic Nicholls, manager of the Canadian General Electric Co., will be the next president of the National Electric Light Association.

---

A soft copper hammer makes an excellent tool with which to drive keys on an engine.

Eccentrics for steam engines that are made in halves may easily be procured, and where an old one has been split off from a crank shaft, one of them is much more easily applied than a whole one. If well made they should last as long as when made in one piece.

## _ BY THE WAY.

As illustrating the many absurd arguments advanced by a certain class of electric railway "promoters," a gentleman named Beech has recently been endeavoring to get capital subscribed by the farmers in the neighborhood of Ridgeway for a single track elevated road to extend from Ridgeway to Crystal Beach on the shore of Lake Erie, a distance of 3 or 4 miles. Mr. Beech is an American and is gifted with a ready flow of language of the stump orator type. At a recent meeting of the residents of the locality, principally farmers, he occupied considerable time in expatiating upon the advantages of his particular system over the ordinary trolley system and stated among other things that the cost of operation of the overhead single track system would be but one-tenth of that of the ordinary trolley road. A gentleman in the audience acquainted with the subject, was asked to make an estimate of the cost per day of operating a trolley line, and he figured the amount at $17.00 per day. This estimate included the salaries of three men. This gentleman inquired of the "promoter" of the single track scheme if he had correctly understood him to say that a road built on that system would cost to operate but one-tenth of that required for a trolley road. Mr. Beech promptly replied that that was his contention. "Then," said the gentleman, "will you kindly explain to the audience how you propose to operate the road and pay three men's salaries out of the sum of $1.75 per day?" This problem proved to be a trifle beyond the mathematical ability of the "promoter," and remained unanswered. Strangely enough there were persons present at this meeting who would not have subscribed towards the construction of an ordinary trolley road, but were willing to pay out their good dollars for the overhead single track scheme. The absurdity of going to the expense of erecting an overhead system for a line designed to run along a country roadside does not seem to have occurred to the minds of the people who have been solicited to put up money for the enterprise. I understand that some $3,000 has been subscribed for the purpose of enabling the "promoter" to construct a piece of track with which to illustrate the advantages of his system. Speaking of the peculiarities of "promoters," I am informed that one of the most active individuals in this line in Canada is making a handsome income out of the business. His method is to project an electric railway and induce municipalities along the route to subscribe say $500 apiece for what he is pleased to term "preliminary expenses." He frankly tells them that in the event of the scheme proving unsuccessful they need not expect to get their money back, and it is hinted that the money thus obtained never goes any farther than his own pocket.

x x x x

THE Toronto Electric Light Co. had in their employ at one time a tall French lineman whose agility was such that he was accustomed to use only one hand when climbing a pole. At the corner of King and Yonge streets stood a pole requiring an additional cross-arm which this dexterous lineman undertook to carry up in his disengaged hand. When part way up the pole, the cross-arm slipped from his grasp and descended perpendicularly upon the crown of the silk hat of a gentleman on the street, driving the hat down upon his shoulders and entirely obscuring his face. He struggled unsuccessfully to get out of the hat until the Frenchman came down the pole and assisted in removing it. The victim had no sooner got his breath than he turned his attention to the unfortunate lineman, and bestowed on him all the anethemas which his recollection could muster. The victim took his punishment patiently, and at its conclusion invited the gentleman into a hat store near by and bought him a shining tile of the newest pattern, thereby metaphorically heaping coals of fire upon his head while at the same time getting rid of the possibility of a claim for damages against his company.

x x x x

I HAPPENED to witness an amusing incident in the office of the G. N. W. Telegraph Co. at Toronto, not long ago. A well-known business man came in and asked the price of a cable message. Having been given the rate, he began trying to get a reduction, advancing as a reason for his unusual request that he had been a good customer of the Company, and like all good customers was entitled to some extra consideration. An amused expression came into the face of the young man behind the counter as he remarked, " Let us see ! you have paid us for cable messages altogether about twenty dollars." "Yes," said the customer, and the tone of his voice implied that he considered that a pretty large sum and quite sufficient to justify his claim. "You may be surprised to learn," said the polite young man behind the counter, "that out of this twenty dollars, the percentage to which this Company is entitled amounts to the magnificent sum of one dollar." After this explanation the customer did not see fit to press his claim further. The fact may not be widely known that the cable companies get the lion's share of the money paid for cable messages, and that the telegraph companies who despatch the messages overland across half a continent, receive but a fraction of the price.

### PERSONAL.

Mr T. Ahearn, of the Ottawa Electric Co., is expected to return this week from his trip round the world.

Mr. James Ross, vice-president of the Montreal Street Railway Co., sailed for England on the 6th of April on a pleasure trip.

Mr. S. J. Stratton, of the Bell Telephone Company, Hamilton, accompanied by Mrs. Stratton, is at present in England, for the benefit of Mrs. Stratton's health.

Mr. E. B. Merrill, formerly lecturer in electricity at the Toronto Technical School has recently accepted a position with the J. H. McEwen Manufacturing Co., of Ridgway, Pa.

Upon severing his connection with the Kingston Light, Heat and Power Company, Mr. John Oldfin was presented by the employees with a beautiful oak secretary. Mr. Oldfin had been in the employ of the company for twenty-nine years.

Mr. O. Higman, chief electrician of the Inland Revenue Department, Ottawa, was recently offered the position of electrical engineer for the colony of Queensland, Australia, at a salary of $3,000. The Dominion government offered to Mr. Higman such inducements to remain in his present position, that he declined the foreign offer.

### TRADE NOTES.

Messrs. Patterson & Corbin, of St. Catharines, Ont., have received an order for four cars for the Hamilton Radial Railway Company.

The attention of persons on the lookout for a bargain in second-hand alternating machinery is directed to the advertisement of Messrs. Ahearn & Soper in this issue.

In the article in our April issue referring to the equipment of the Hull and Aylmer Electric Railway, mention was omitted of the fact that the contract for trolley, feed and bond wire, amounting to some 150,000 lbs., had been awarded to the Eugene F. Plillips Electrical Works, Montreal.

## QUESTIONS AND ANSWERS.

H. S. P., Toronto, writes : " How many pounds pressure of air will be required to the sq. inch, taking the full piston plate surface, to force the rod (1 inch diameter) into dead cylinder against an air pressure of 50 lbs. to the sq. inch, speed say 1 ft. in 1 minute. Would it be a pressure a little over the difference in the dead cylinder pressure and the pressure raised by the rod displacing that amount of air ? "

ANSWER.—A pressure of 1 4/10 lbs. per sq. inch on the piston 6 in. diam. would just balance the resistance of 50 lbs. per sq. inch on end of rod. A pressure of 2 lbs. per sq. inch would probably start it in motion, and if pressure remains constant in the dead cylinder, that pressure would be sufficient to do what you require.

E. L. NASH, Lunenburg, Nova Scotia, writes : " Do you know of any place where the new style of tide mill is in successful use. I mean one working on the accompanying plan. We have a back harbor here

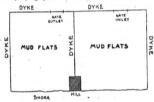

where the tide rises and falls about six feet, and want to know if it is feasible to build dykes and run our electric machinery by a tide mill. Where could I get hold of calculations that would enable me to tell how much would have to be dyked in to develop 500 horse power continuously ? "

ANSWER.—It is difficult to give a definite answer to your question without considerable more data than you give. On general principles, however, it is doubtful if a tide of only six feet could be made available within the limits of commercial practicability. In the first place it would be impossible for the power to be developed continuously under ordinary circumstances, as at high tide, which occurs twice in twenty-four hours or thereabouts, there would be no fall whatever available, and consequently no power. On the contrary, at those times the reservoirs would be filling up. From the sketch you send, however, we suppose your idea would be to have several reservoirs and during high tide to empty from one into another and then from that out. If that plan were adopted you could only figure on an average head of about three feet. To produce 500 horse power at this head would require two reservoirs, probably about 3500 feet square and 6 feet deep, or nearly five miles of dyking, to say nothing of the heavy cost of machinery to produce power with such a small average head. There are no doubt a few tidal mills on a small scale in operation, where the power is not required continuously, but we do not know of any of the importance or capacity of the size you speak of. In fact in considering the matter of continuous power, and considering the fluctuation in level which would have to be taken into account, it is altogether probable that the reservoirs would have to be considerably larger than the size given above to obtain satisfactory operation. It would appear that the interest on the first cost and the amount required for repair and maintenance, would be infinitely greater than the cost of coal to produce an equal power, especially in the neighborhood you speak of.

"NOVICE" writes, in reply to "Induction's" enquiry in a previous issue :—This open circuit business is the terror of arc light men as a rule. I would like to see the matter discussed through the ELECTRICAL NEWS and am sure some good would result from such discussion.

The writer has had a little experience in finding breaks that could not be located by climbing every pole on the circuit, and pulling the wires, and if your space will permit, will be pleased to explain his method. I first divided the circuit into sections with small magnet wire, sometimes running it two blocks at a stretch, and found that the break was nearer the negative end of dynamo than the positive. Then I grounded that end of line at the dynamo, being careful to disconnect the positive wire. With the magneto bell I started out and every arc lamp I came to on the street I let down and fastened one wire of bell to binding post of lamp, and the other to the ground. As long as the bell would ring I knew I had farther to go. As soon as the bell would not ring I knew I had my break down to close quarters, and started back towards the last place where the bell would ring, testing from the line as we went, and very soon had only about 100 feet of line untested. The rest was easy. It may seem to some that this would be a tedious way of finding a break, but we were only about an hour at it. I would like to hear from some one else who has a scheme.

## CANADIAN ELECTRICAL ASSOCIATION.

WEDNESDAY, Thursday and Friday, June 17th, 18th and 19th, have been selected as the dates of the annual convention of the above Association. It is somewhat unfortunate that the Dominion elections have been fixed for June, but it was not deemed advisable to postpone the convention on this account. So far as the election canvass is concerned, it will be practically concluded, and the nominations over by the 17th of June, while voting will not take place until the 23rd.

Those members of the Association who take an active interest in politics, should get their work done before the 17th, and spend two or three pleasant and profitable days at the Toronto convention before depositing their ballots. Then if the party of their choice should happen to be defeated, they will find themselves in good condition to put up with the disappointment, while if the vote goes to their liking, they will be in equally good trim to join in the enthusiasm of the occasion.

The program for the convention is an attractive one. Papers will be presented as follows : " Economics of Central Station Management," by P. G. Gosslin, Montreal ; " Acetylene Gas," by Geo. Black, Hamilton, Ont.; " Meters," by James Milne, Toronto ; " Electric Railway Construction," by F. C. Armstrong, Toronto ; " Power Transmission by Polyphasal E. M. F.'s," by Geo. White-Fraser, E. E., Toronto ; " Continued Use of Water of Condensation," by Wilson Philips, Toronto.

Several of these papers will be illustrated by means of a stereopticon, a new feature which will add greatly to the interest of the proceedings. Opportunity will be afforded for the consideration and discussion of the Government Electric Light Inspection Act.

It is in contemplation to hold the annual banquet of the Association at Lorne Park, Niagara-on-the-Lake, or some other popular summer resort in the vicinity of Toronto. The banquet will be followed by a moonlight sail on Lake Ontario, the steamer being attractively decorated and provided with music for the occasion.

There will likewise be visits of inspection to the power stations of the Toronto Electric Light Co. and Toronto Railway Co., an exhibition of Roentgen rays, excursions by street car, etc. Altogether, visitors are assured of an interesting and profitable time, and seeing that a large proportion of the members of the Association reside within a hundred miles of Toronto, there should be a bumper attendance.

The Hamilton Radial Electric Railway Co. have awarded the contract for the electrical generating apparatus required for the operation of their road to the Canadian General Electric Co., Ltd.

## THE INCANDESCENT LAMP.

BY GEORGE WHITE-FRASER, E. E.

(Concluded.)

Diagram 4 also serves to illustrate the various wattages at which lamps can be run, and their effect on life and candle power. Running the lamps at 108 volts is equal to. a wattage of 3.5 per c. p. Running them at 110 v. equals 3.3 watts per c. p. ; at 112 equals 3.1 watts ; at 114 equals 2.9 watts ; at 116 equals 2.7 watts per c. p. The higher therefore the economy at which this lamp

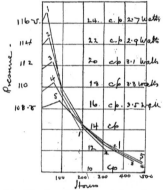

FIG. 4.—CURVE SHOWING DEPRECIATION OF CANDLE POWER CONSEQUENT ON USING TOO HIGH VOLTAGE.

was run, the sooner did it come down to a low candle power, and the sooner did it burn out, so that the lower efficiency was in this case the higher economy. This point is well illustrated by the curves in diagram 5 which shows the results of burning a lamp of the same make at different efficiencies. At the highest efficiency —2½ watts per c. p.—the lamp drops to 8 c. p. at 600 hours, and, as the efficiency becomes less, that is as the wattage grows more per c. p.; the reduction in c. p. becomes less and less, the life longer and longer.

These curves are very suggestive to the observant central station man, as indicating the policy which should govern him in the supply of lamps, and in the system of wiring. It is evident from a careful study, that if his distribution system is so laid out that the variation in pressure between heavy and light load is relatively great, then he had better use his low efficiency lamps, for such variation will cause the lamps to drop in candle power and burn out soon, and cause dissatisfaction among his customers.

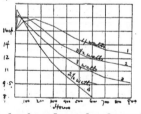

FIG. 5.—CURVE SHOWING EFFECT ON CANDLE POWER AND LIFE OF RUNNING AT DIFFERENT EFFICIENCIES.

If, however he has used plenty of wire, so that the variation is small, then he can use high efficiency lamps. By this means he can put a good many more lamps on his dynamo, which means increased profits. The matter, however, is much more complicated when he furnishes light to meter customers. In this case he does not sell light merely trusting to the inexperience of his customers not to detect the difference between the 16 c. p. they contract for, and the 12 or 10 c. p. that they actually get. He sells current, and it is plain that the more current he sells, the better for himself, so that he should study how to increase the amount of

current that the consumer takes. He cannot, of course, make that consumer keep his lamps burning, but a study of lamp curves will point out a way. Diagrams 1 and 2 show the decrease of candle power as a lamp grows older, No. 3 shows the increase of wattage per candle power during the same period ; although the lamp takes a higher wattage per candle power as it ages, its candle power decreases more rapidly than the wattage increases, so that the absolute wattage, on which the central station depends for profits, keeps on diminishing slowly all the time. Profit therefore also keeps on diminishing in the same ratio.

Diagram No. 6 shows these changes in three curves, A B, C, taken from a 3½ watt lamp. Curve A shows its decrease in absolute candle power, giving actual candle power observed at 100 hour periods during its life ; curve B shows the increase in wattage per candle power at the same periods, and curve C shows the resultant of the two at same periods, that is at 700 hours the actual candle power given is 12.25 ; the wattage per c. p. is 3.79, and the actual watts absorbed by the lamp are (curve C) 45.5. Now the lamp at the first 100 hours absorbed 57.2 watts, so at 700 hours it is absorbing 11.7 watts less, and—what is interesting to the station man—is paying less in proportion—less by

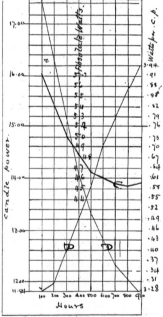

FIG. 6.—CURVES SHOWING DECREASE IN C.P.; INCREASE OF RELATIVE WATTS, AND DECREASE OF ABSOLUTE WATTS, AS LAMP AGES.

21 per cent. Dividends can never be paid at this rate, for it simply amounts to a 1000 light plant installed at a 1000 light price, earning only 790 lamps worth of rent. So we must keep up the supply of current, and this can only be done by putting old lamps out of service, when they begin to take so little current, and putting in new ones that will take more. It seems extraordinary, but it would actually pay to break this lamp rather than run it 700 hours.

A study of lamps will reveal other apparently paradoxical results and the conclusion of the whole matter is (1st) Study your distribution system, and keep your pressure as constant as possible, even if it costs a little more money to reduce the drops on lines, (2nd) Don't assume that the cheapest lamp is the best, nor even that the one with the longest guaranteed life is the one you want, (3) Buy lamps to suit your work, for all lamps are by no means alike, and study what your work will be, (4) Don't imagine that once you have put a lamp in its socket, the matter is disposed of, and you need pay no more attention to it—lamps may be actually losing money to you. Above all remember that your entire lighting business is worth careful study.

PUBLISHED ON THE FIFTH OF EVERY MONTH BY

CHAS. H. MORTIMER,

Office : Confederation Life Building,
Corner Yonge and Richmond Streets,

TORONTO,   -   -   -   CANADA.
Telephone 2362.

New York Life Insurance Building, Montreal.
Bell Telephone 2209.

***ADVERTISEMENTS.***

Advertising rates sent promptly on application. Orders for advertising should reach the office of publication not later than the 20th day of the month immediately preceding date of issue. Changes in advertisements will be made whenever desired, without cost to the advertiser, but to insure proper compliance with the instructions of the advertiser, requests for change should reach the office as early as the 22nd day of the month.

***SUBSCRIPTIONS.***

The Electrical News will be mailed to subscribers in the Dominion, or the United States, post free, for $1.00 per annum, 50 cents for six months. The price of subscription should be remitted by currency, registered letter, or postal order payable to C. H. Mortimer. Please do not send cheques on local banks unless 15 cents is added for cost of discount. Money sent in unregistered letters will be at senders' risk. Sub-criptions from foreign countries embraced in the General Postal Union $1.50 per annum. Subscriptions are payable in advance. The paper will be discontinued at expiration of term paid for if so stipulated by the subscriber, but where no such understanding exists, will be continued until instructions to discontinue are received and all arrearages paid.

Subscribers may have the mailing address changed as often as desired. When ordering change, always give the old as well as the new address.

The Publisher should be notified of the failure of subscribers to receive their paper promptly and regularly.

***EDITOR'S ANNOUNCEMENTS.***

Correspondence is invited upon all topics legitimately coming within the scope of this journal.

The "Canadian Electrical News" has been appointed the official paper of the Canadian Electrical Association.

## CANADIAN ELECTRICAL ASSOCIATION.

### OFFICERS:

President :
A. B. SMITH, Superintendent G. N. W. Telegraph Co., Toronto.

1st Vice-President :
C. BERKELEY POWELL, Director Ottawa Electric Light Co., Ottawa.

2nd Vice-President :
L. B. McFARLANE, Manager Eastern Department, Bell Telephone Company, Montreal.

Secretary-Treasurer :
C. H. MORTIMER, Publisher Electrical News, Toronto.

Executive Committee :
GEO. BLACK, G. N. W. Telegraph Co., Hamilton.
J. A. KAMMERER, General Agent, Royal Electric Co., Toronto.
E. C. BREITHAUPT, Berlin, Ont.
F. H. BADGER, Jr., Superintendent Montmorency Electric Light & Power Co., Quebec.
JOHN CARROLL, Sec.-Treas. Eugene F. Phillips Electrical Works, Montreal.
K. J. DUNSTAN, Local Manager Bell Telephone Company, Toronto.
O. HIGMAN, Inland Revenue Department, Ottawa.
W. Y. SOPER, Vice-President Ottawa Electric Railway Company, Ottawa.
A. M. WICKENS, Electrician Parliament Buildings, Toronto.
J. J. WRIGHT, Manager Toronto Electric Light Company.

## CANADIAN ASSOCIATION OF STATIONARY ENGINEERS.

President, W. G. BLACKGROVE,   -   99 Esther St., Toronto, Ont.
Vice-President, JAMES DEVLIN,   -   Kingston, Ont.
Secretary, E. J. PHILIP,   -   293 Berkeley St., Toronto, Ont.
Treasurer, DUNCAN ROBERTSON,   -   Hamilton, Ont.
Conductor, W. F. CHAPMAN,   -   Brockville, Ont.
Door Keeper, F. G. JOHNSTON,   -   Ottawa, Ont.

TORONTO BRANCH NO. 1.—Meets 1st and 3rd Wednesday each month in Engineers' Hall, 61 Victoria street. W. Lewis, President ; S. Thompson, Vice-President ; T. Eversfield, Recording Secretary, University Crescent.

MONTREAL BRANCH NO. 1.—Meets 1st and 3rd Thursday each month, in Engineers' Hall, Craig street. Presidents, John J. York, Board of Trade Building ; first Vice-President, J. Murphy ; and Vice-President, W. Ware ; Secretary, B. A. York ; Trea-urer, Thos. Ryan.

ST. LAURENT BRANCH NO. 2.—Meets every Monday evening at 43 Bonsecours street, Montreal. R. Drouin, President ; Alfred Latour, Secretary, 306 Delisle street, St. Cunegonde.

BRANDON, MAN., BRANCH NO. 1.—Meets 1st and 3rd Friday each month, in City Hall. A. R. Crawford, President ; Arthur Fleming, Secretary.

HAMILTON BRANCH NO. 2.—Meets 1st and 3rd Friday each month in Maccabee's Hall. E. C. Johnston, President ; W. R. Cornish, Vice-Pres.; Wm. Norris, Corresponding Secretary, 211 Wellington street.

STRATFORD BRANCH NO. 3.—John Hoy, President ; Samuel H. Weir, Secretary.

BRANTFORD BRANCH NO. 4.—Meets 2nd and 4th Friday each month. F. Lane, President ; T. Pilgrim, Vice-President ; Joseph Ogle, Secretary, Brantford Cordage Co.

LONDON BRANCH NO. 5.—Meets once a month in the Huron and Erie Loan Savings Co.'s block. Robert Simmie, President ; E. Kidner, Vice-President ; Wm. Meaden, Secretary Treasurer, 533 Richmond street.

GUELPH BRANCH NO. 6.—Meets 1st and 3rd Wednesday each month at 7.30 p. m. J. Fordyce, President ; J. Tuck, Vice-President ; H. T. Flewelling, Rec.-Secretary ; J. Gerry, Fin.-Secretary ; Treasurer, C. J. Jorden.

OTTAWA BRANCH, NO. 7.—Meet every second and fourth Saturday in each month, in Borbridge's hall, Rideau street ; Frank Robert, President ; F. Merrill, Secretary, 352 Wellington street.

DRESDEN BRANCH NO. 8.—Meets 1st and Thursday in each month. Thos. Steeper, Secretary.

BERLIN BRANCH NO. 9.—Meets 2nd and 4th Saturday each month at 8 p.m. J. R. Utley, President ; G. Steinmetz, Vice-President ; Secretary and Treasurer, W. J. Rhodes, Berlin, Ont.

KINGSTON BRANCH NO. 10.—Meets 1st and 3rd Tuesday in each month in Fraser Hall, King street, at 8 p.m. President, S. Donnelly ; Vice-President, Henry Hopkins; Secretary, J. W. Tandvin.

WINNIPEG BRANCH NO. 11.—President, G. M. Hazlett ; Rec.-Secretary, J. Sutherland ; Financial Secretary, A. B. Jones.

KINCARDINE BRANCH NO 12 —Meets every Tuesday at 8 o'clock, in Mc-Kibbon's block. President, Daniel Bennett ; Vice-President, Joseph Lighthall; Secretary, Percy C. Walker, Waterworks.

WIARTON BRANCH NO. 13.—President, Wm. Craddock ; Rec.-Secretary, Ed. Dunham.

PETERBOROUGH BRANCH NO. 14.—Meets 2nd and 4th Wednesday in each month S. Potter, President ; C. Robison, Vice-President ; W. Sharp, engineer steam laundry, Charlotte street, Secretary.

BROCKVILLE BRANCH NO 15.—President, W. F. Chapman ; Vice-President, A. Franklin ; Recording Secretary, Wm. Robinson.

CARLETON PLACE BRANCH NO. 16.—Meets every Saturday evening. President, Jos. McKay ; Secretary, J. D. Armstrong.

## ONTARIO ASSOCIATION OF STATIONARY ENGINEERS.

### BOARD OF EXAMINERS.

President, A. AMES,   -   Brantford, Ont.
Vice-President, F. G. MITCHELL   -   London, Ont.
Registrar, A. E. EDKINS   -   139 Borden St , Toronto.
Treasurer, R. MACKIE,   -   28 Napier st., Hamilton.
Solicitor, J. A. McANDREWS,   -   Toronto.
TORONTO—A. E. Edkins, A. M. Wickens, E. J. Phillips, F. Donaldson.
HAMILTON—P. Stott, R. Mackie, T. Elliott.
BRANTFORD—A. Ames, care Patterson & Sons.
OTTAWA—Thomas Wesley.
KINGSTON—J. Devlin, (Chief Engineer Penitentiary), J. Campbell.
LONDON—F. Mitchell.
NIAGARA FALLS—W. Phillips.

Information regarding examinations will be furnished on application to any member of the Board.

---

**Electric Railway Legislation.**    The Hon. John Haggart, Minister of Railways, recently made the announcement that the Dominion government is of opinion that all electric railway charters of a purely local character should be left to the jurisdiction of the provincial authorities. The Railway Committee of the Dominion Parliament is not in accord with this view, and has reported to the House several bills of the character mentioned contrary to the wishes of the government.

**Electric Light Inspection.**    COMPLAINTS have reached us lately concerning the system of electric light inspection inaugurated by the Dominion Government a year ago. There seem to exist doubts in the minds of some, first, regarding the necessity for such a system, and secondly, concerning the fairness of the charges imposed and the efficiency with which the Act is administered. Our attention has been called to instances in which meters, after having been tested and sealed up by the government inspectors, shortly afterwards ceased to operate owing to the formation of a substance upon the brushes which retarded and eventually stopped entirely the action of the commutator. We have likewise heard complaints of the action of some of the government inspectors in posing as electrical engineers, and advising central station owners as to the means they should adopt to overcome difficulties experienced in the operation of their plant. It is affirmed that the inspectors are without the training necessary to qualify them to act in this capacity, and that in some instances their advice has caused unnecessary friction between central station owners and the manufacturing companies. We propose to investigate this subject and if possible learn to what extent these complaints are well-founded.

**Effect of Electric Railways on the Labor Problem.** MR. Bronson, in moving a resolution in the Ontario legislature setting forth the desirability of forming societies in cities to assist in placing unemployed persons on the unoccupied lands of the province where they might become self-supporting and even contributors to the development and wealth of the country, expressed the opinion that electric railways would prove important factors to this end. While it is undoubtedly true that the means of intercommunication afforded by electric railways will do away with the isolation, which is one of the greatest objections to country life, the lands upon which the class in question could be settled are likely to be situated in localities where electric railways will not be built for many years to come. Therefore the electric railway is not likely to prove a factor in their welfare.

**Anchor Ice.** THE difficulties encountered by electric stations operated by water power, in consequence of the formation of anchor ice, was the subject of some discussion at the convention of the Canadian Electrical Association at Ottawa last September. The consensus of opinion seemed to be that there was no method by which the difficulty could be avoided. In the Scientific Machinist of Cleveland, this method is given of keeping the turbines free of ice, "bore a hole in the top of the wheel case, or drill a hole if the case be an iron one, and connect on a steam pipe. When a wheel freezes up, or shows signs of freezing, just turn on steam for a few minutes, and away goes the ice, slick and clean. The wheel, of course, must be shut down before the steam is turned on, or all the heat will be carried off with the water, leaving the ice as good as before steam was turned on." Mr. James F. Ward, in a letter to the Engineering Record, states that while chief engineer and superintendent of the Jersey City waterworks he remedied trouble from this cause, by moving across the screen of the intake chamber a raft made out of some 12 inch square logs. His explanation of the success of the remedy is that in consequence of the length of the line around the edge of the raft being, say four times greater than the width of the screen, the force available to draw in the anchor ice is reduced in the same proportion. This very simple and apparently effectual remedy is within the reach of all who annually experience difficulty from the action of anchor ice.

**The Adaptability of Machines to Requirements.** THE popularity of electricity as a means for the transmission and utilization of power, and the recognition of its thoroughly satisfactory results, are in no way more evident than in the rapidity with which it is being applied to every conceivable purpose; and in the number of electrical manufacturing establishments that are going into the business. A perusal of the Patent Office Records shows how immense is the activity in the field of inventing new, and improving old types ; and the establishment of electrical engineering courses in all large colleges, and the equipment of laboratories with the most expensive instruments, shows that practical electricity is now well out of the "rule of thumb" stage, and is recognized as being based on scientific principles. Electrical knowledge being so widely disseminated, the improvement in the designing, constructing, and operating of electrical machinery and apparatus has taken place all along the line simultaneously, and electrical investors now have the assurance, not only that what they buy represents the results of scientific investigation, but that the market is full of excellent machinery, with something to suit every purchaser, and meet every want. Not so long ago, when electrical machinery was built by experiment rather than by calculation, it was only reasonable to believe that the inventor with the longest purse would produce the best apparatus, because he was better able to experiment until he overcame his difficulties, than another person without those financial advantages. But now that the design and construction of the highest class scientific machinery is a matter of certain knowledge ; and is no more a matter of doubt than is the design of a steam engine or any other perfectly understood apparatus, it must be evident that any properly trained engineer will be able to produce good machinery ; and not only this, but that suitable machinery can easily be built to meet the conditions of any particular case. The immediate result of the establishing of the principles governing electrical design and construction on a thoroughly satisfactory rational basis, instead of heretofore on a constantly changing empirical one, has been that nearly all large engine or machine manufacturing companies have added to their former business an electrical manufacturing department ; and the result is that, whereas even five years ago the different makes of electrical machinery might be almost counted on two hands, to-day they are well up among the scores, and every one of them good. Another very important result is the splitting up of electrical manufacturing business into specialties, whereby the very highest perfection is attained along the several lines followed. Five years ago the few manufacturing companies there were, manufactured complete "systems," covering generators, motors, lamps, instruments, etc., etc., and to do business with them meant committing oneself exclusively to the particular company at first chosen, because one generally found that the use of a particular generator necessitated the use of a particular motor, and so on. Now, however, there are highly trained electricians who devote themselves exclusively to the perfection of one particular line of apparatus—be it motors, or arc lamps, or electric fans ; and hence it is that the market presents a whole host of first-class machinery and apparatus from which the purchaser can choose.

Of course it is true, however, that while there are very many good makes, there are also a great number of types which have been long left behind in the race of improvement and which represent the earlier stages of design when scientific investigation had not clearly lighted up the subject. There, naturally, will also be found manufacturers, who, to meet competition, will purposely lower the quality of their goods, trusting to favorable circumstances not to be found out. Against these the purchaser will have to adopt such precautions as commend themselves to him, remembering that cheapness is prima facie evidence of relative inferiority. A really good machine costs money to build ; both because high class material is expensive, and because skilled workmanship cannot be obtained for the price of day labor. But, the difference between good and inferior machinery is not merely a question of first cost ; it includes the consideration of probable differences in repairs, maintenance, life, efficiency—which, in nearly every case will clearly demonstrate that the more expensive apparatus is really the cheapest to buy.

It is interesting to note the difference between European and American manufacturing practice, with reference to their respective methods of supplying demands for particular machinery. On this side of the Atlantic all types and sizes are standardized ; and manufacturers endeavor not so much to design machinery which will meet the conditions of any particular case, as to show that those particular conditions require the use of such and such a particular machine of their make. That is, the machine is not manufactured for the case, but the case is manipulated to suit the machine. If a 95 horse power machine is sufficient and necessary, the manufacturer who has standardized a 100 h.p. and a 90 h.p. machine will sell his 100 h.p., although it is not only larger than necessary, but also will have to operate at less than full load and therefore at reduced efficiency. In Europe, the manufacturer, not having any rigid standard, would actually design and build a 95 h.p. machine. The standardizing of machinery results, no doubt, in somewhat less shop costs, and in consequent lower selling price, but the disadvantage is evident, when it is considered how very rarely will the circumstances governing the size and type of machinery be such as to exactly meet some particular standard. Of course this disadvantage is largely counterbalanced by the fact that manufacturers' standards include many sizes and types, some or one of which will be pretty certain to come close to the actual requirements. And then again all manufacturers have not the same standards ; they will generally be found to be "staggered." If A makes generators of 50 h.p., 100 h.p., 150 h.p., etc. ; B will generally make 70 h.p., 120 h.p., 170 h.p. ; while C will adopt 60 h.p., 90 h.p., 130 h.p., and so on ; so that one can generally come pretty close to what is wanted. The same will apply to motors, etc., but there comes in the question of voltages, etc. However, in every case it will be possible to work out a reasonably satisfactory scheme, if only care and attention be devoted to enquiring what the market has to offer in the way of suitable types and sizes and then selecting those that come nearest to practical requirements. In doing this it should be remembered that the efficiency of machines is a most important consideration, and that, unless they be specially designed differently, their efficiency will be highest at their rated full loads. Taking, for example, a case where study shows that 800 lights is the maximum that can be expected (and there is a very fairly sure proportion between number of inhabitants and number of lights), in this case it would be inadvisable to put in a 1,000 light machine for two very good reasons—first, it is larger than necessary, and therefore needlessly expensive ; and second, which is really even more important, it will be operating at never more than 8/10's of full load, and for the very large proportion of the time at very considerably less. Everyone knows that there is a period of large load, say from 7 o'clock to 9 o'clock, and that after that time the people go to bed and the load goes away down. In a plant of the above size, probably for 4 hours the full 800 lights or nearly would be going ; and for the whole of the rest of the night not more than 200 or 250 ; so that if a 1,000 light machine were installed, it would have 80 per cent. load for 4 hours and 20 to 25 per cent. for 8 hours. All the above considerations emphasize the

importance of studying the conditions of every installation, and of buying apparatus with reference to its suitability to the peculiar circumstances.

## DEFECTIVE WIRING.

WE present herewith a sample of wiring which was recently unearthed in a neighboring city. Our readers will admit that it is a truly wonderful example of how not to do it. Nevertheless it is not altogether an exaggerated case. There has been a vast amount of work of this character done in the past in every city on

this continent. It can scarcely be a matter for surprise that the discovery of such work should have given rise to the suspicion that many of the fires in recent years were the result of stray electric currents.

It is satisfactory to know that the danger resulting from the employment of careless and incompetent workmen has now come to be so well understood that in future proper workmanship is likely to be the rule rather than the exception.

## CANADIAN ELECTRICAL STATISTICS.

MR. Geo. Johnston, the Dominion statistician has collected and embodied in the Year Book of Canada, the following statistics relative to the electrical progress in the Dominion :—The amount of capital invested in electric telegraphs and cables in Canada is $7,000,000 ; in electric railways the paid up capital is rather more than $13,000,000 ; in electric light works, $4,113,771 ; in electrical appliances, $1,389,365 ; or in round figures about $27,000,000. In 1881 there were found only two hands with electric works outside of those connected with telegraphy, while in 1891 there were 1190 hands, not including those connected with the electric cars. The employees in 1894 connected with the electric cars numbered 2614 ; passengers carried 57,000,000 ; miles run during 1894 by the electric railways, 15,500,000 ; miles of track for Canadian electric railways, 368 or 73 miles to each million of the people. The number of motor cars in Canada are calculated as 658 ; trailers, 341 ; snowsweepers, 39 ; and motors, 891. The steam railways in Canada in 1894 carried 14,500,000 passengers, which, contrasted with 57,000,000 carried by the electric railways, shows that four times as many passengers were carried by electricity as by steam, and that, on an average, every person in Canada had been carried 11 times in the year by electricity.

## TESTS OF A 10 HORSE-POWER DE LAVAL STEAM TURBINE.*

By Wm. F. M. Goss.

THE De Laval steam turbine experimented upon constitutes part of the permanent equipment of the Engineering Laboratory of Purdue University, and the present paper is based upon data secured chiefly through the assistance of Charles E. Bruff, B.M.E., author of the thesis "Tests of a 10 Horse-Power De Laval Steam Turbine."

In the De Laval steam turbine jets of steam, delivered from suitable nozzles, are made to impinge against the buckets of a light turbine wheel. The steam enters the buckets from one side of the wheel, and passing through is discharged or "exhausted" from the opposite side. The arrangement of nozzle

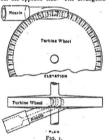

FIG. 1.

and wheel is shown in Fig 1. The motion of the turbine shaft, which under the actions of the jets is extremely rapid, is communicated by gearing to a heavier and slower-moving driving shaft carrying a fly-wheel of small diameter, from which the power of the engine is delivered. Regulation of speed is secured by means of a throttling governor, which controls the pressure of steam admitted to the nozzles.

The important moving parts, with approximate dimensions, are shown in Fig. 2. The turbine wheel is built of sixty-three steel segments, each carrying a bucket and a portion of the light outside rim. The segments are held in place by means of suitable collars, which grip them on either side. The wheel is mounted upon a long, slender shaft, having sufficient flexibility to allow the system at speed to revolve about its centre of gravity, even though this may not agree with the geometrical axis of the shaft. The gear upon the turbine shaft is of steel, solid with the shaft ; that upon the drive shaft has its teeth formed in a bronze ring, which is carried by a solid iron centre. The smaller gear has twenty-one teeth, the larger one two hundred and eight teeth, giving a ratio of 1 to 9.90476.

The shafts run in bronze boxes completely lined with babbit or other soft metal. To assist in the distribution of oil a spiral curve, the pitch of which is about half the diameter of the journal, is cut into the metal of the bearing. The outboard bearing on the turbine shaft is closed at the end, and a small pipe runs from the closed end to a point over the gears. The pumping action, resulting from the presence of the spiral oilway, gives a constant, though small, supply of oil upon the gears. The gears do not dip in oil, though the case which encloses them receives drainage from all the bearings.

The governor is connected with the driving shaft, of which, at first sight, it appears to be but an extension. It is shown in detail in Fig. 2. The weights, W W, with their arms, C C, are in the form of a split cylindrical cup. Upon the outside and at the base of each weight a knife edge, E E, is found, which bears upon a suitable surface in the governor frame, A A. A spiral spring is fitted at its inner end with two projecting pins, which bear upon the arms, C C, of the governor weights. The outer end of the spring is connected with the frame by the threaded plug D. When the governor is at rest the concave surfaces of weights are in contact with the frame, and the tension of the spring keeps the knife edges upon their seat. When the governor is revolving at speed the weights are under centrifugal action and move outward, swinging upon their knife edges against the resistance of the spring. The motion of the weights is taken up by

*Abstract of paper read at the New York meeting of the America Society of Mechanical Engineers, December, 1895.

the pin F, by which it is communicated, through suitable mechanism, to the governor valve above the engine.

The nozzles which serve to deliver steam to the wheel are four in number, and are so fixed in the frame of the engine as to act upon the turbine at points which are equally distant from each other. Two of the four are provided with stopcocks, which, when closed, put out of action the nozzles with which they are connected. By means of the stopcocks, therefore, the engine may be run under the action of two, three or four nozzles, at the will of the engineer.

The distinguishing feature of the engine is, perhaps, to be found in the form of the nozzles. All are diverging, the throat being approximately two inches from the discharge end. Three have a diameter of 0.138 inch, and one a diameter of 0.157 inch.

### THE TESTS.

The power of the engine was absorbed by a pony brake, cooled by constant streams of water. The exhaust steam was piped to a Wheeler condenser, open to the atmosphere. The water resulting from condensation was drained into tin buckets, which were changed and weighed at regular intervals. Gauges were used to show the steam pressure both above and below the governor throttle, the former giving the pressure available at the engine, and the latter the pressure under which, in consequence of the action of the governor, the steam was admitted to the nozzles. A manometer was also attached to the exhaust pipe, but as this pipe is large (3 inches diameter) and the connection with the condensor close, the observed pressure was never appreciably different from that of the atmosphere.

The boiler pressure for all efficiency tests was 130 pounds by

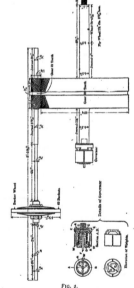

FIG. 2.

gauge, for which pressure the particular nozzles used were designed. The rated speed of the fly-wheel was 2,400 revolutions per minute (23,771 for turbine wheel), but this standard was not maintained for all the tests. The governor was adjusted several times as the work progressed, and it was not until several tests had been run that the proper speed was secured. It is believed, however, that the differences of speed recorded do not materially affect the value of results for purposes of comparison.

The tests are grouped into three series, the first including those for which all four nozzles were in action, the second those with three, and the third with two. The several tests in each series were intended to vary from each other only in amount of power delivered from the wheel. All tests were of 30 minutes' duration, and all observations were taken at five-minute intervals.

With all four nozzles in action, and with the engine developing a little more than its rated power, the steam consumption per horse-power hour was as low as 47.8 pounds. In comparing this result with results obtained from other engines, the small size of the engine tested should be kept in mind, and also the fact that the rate of consumption stated is based upon brake power. The efficiency of the engine falls off rapidly as the load is decreased, and, as would be expected, the effect is not marked when all the nozzles are in action. This may best be seen by the heavy curves shown in Fig. 3. Assuming the nozzles to be cut out of action

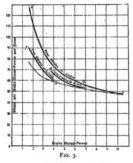

FIG. 3.

one at a time, as soon as the reduction of load becomes sufficient to permit the work to be done without them, the minimum steam consumption at different loads, for boiler pressure and speed employed, is represented by the broken line fgdebc, Fig. 3. Again, if instead of four nozzles, an infinite number could be employed, and if the governor could be arranged so as to regulate the number in action rather than the pressure admitted to them, the steam consumption of the engine in question might be made to follow a line somewhat similar to the light broken line g e c. But the heavy lines indicate the results which were actually obtained.

The engine requires very little attention and is almost noiseless in action. The governor is quick to act, and its speed regulation appears to be fair, except when changes of load are large and suddenly made. After such a change, the engine requires a little time before settling down to steady running under the new conditions.

As the speed of the De Laval engine is high, it is evident that the force in action must be comparatively low. To determine the maximum resistance under which the engine might be expected to start, the brake was clamped upon the fly-wheel so that the latter could not turn within it. Steam was then admitted to the engine, and readings were taken from the scale under the brake arm. The result of this process, of course, depends upon the steam pressure and the number of nozzles in action. With all nozzles, and with a steam pressure of 125 pounds by gauge, the maximum starting-power is equal to a force of 30 pounds acting at a radius of one foot. With three nozzles the equivalent force was but 21 pounds ; with two nozzles 14 pounds.

---

IN the case of the Ontario Western Lumber Co. v. Citizens Telephone and Electric Co., Chief Justice Meredith decided that contracts not under the corporate seal made with trading corporations relating to purposes for which they are incorporated, or, partly performed and of such a nature as would induce the Court to decree specific performance thereof if made between ordinary individuals, will be enforced against them. Where, therefore, an electric light company, while they were making changes in their factory, entered into a contract by correspondence, merely for the use at a specified amount, of one of the wheels in the plaintiffs' mill which was used and a part payment made the contract was held to be binding on it, and the plaintiffs entitled to recover the balance due, notwithstanding the absence of the corporate seal.

## ENGINEERING NOTES.

AN engine may appear to be keyed up all right, and still, when it is started up, the crank pin or some other part may heat because the key was driven too far, therefore all of the parts should be closely watched until it is known that they will run cool.

To ascertain the throw of an eccentric, measure the distance from the crank shaft to the outside of the eccentric on the heavy side and also on the light side. Subtract one from the other, and the difference will be the throw of the eccentric.

If you have a bearing so located that it is necessary to have a tube or pipe to carry the oil to it, be sure that the tube is perfectly clean when first put in, and take measures to keep it so after it is in use.

Always have a steam gauge on the feed-water pipe, and locate it as near the pump as may be convenient. If the pipe becomes partially choked with sediment, the increase in pressure will warn you of it.

Don't put a poor lubricator on a good engine, or on a poor one either. It will ruin the good one by failing to deliver oil when it is needed, and make the poor one worse for the same reason.

Try the nuts on your foundation bolts occasionally and see if they are still tight. Because they were all right six months ago, it does not necessarily follow that they are now.

After cleaning your boilers fill them with hot water if possible. If one of them is fired up, use the steam from that one to heat the water that is going into the others.

When water has a temperature of 39 degrees Fahrenheit, it has attained its maximum density, or in other words, it is in its most compact form. If you make it warmer it expands and if you make it colder it also expands. A study of some of the results of this property of water will prove interesting.

When fitting up a water tank for use in the shop, mill or factory, if you wish to arrange it so that the discharge pipe will prove the most efficient, do not allow it to project on the inside, but let it be so arranged that there will be no sharp corners around the outlet, for it should be funnel shaped.

When about to key up his engine, an engineer should know just where his keys are before he touches them, and in order to do this he must have marks on them. If these marks are made with a sharp steel scribe they disfigure the machine and in the course of time he will get so many marks that they will be confusing. It is much better to mark them with a lead pencil, then if he finds that a key has been driven too far he can easily put it back to its former position, and when the machine is running in a satisfactory way the marks can easily be removed.

Many engines are so constructed that the space around the piston rod gland is quite small, and so we would suggest that a short solid wrench be made for use in such cases, and always be kept in a convenient place. As it is seldom or never necessary to use much leverage here, a large one will not be needed and it will be much more convenient than to try to use an ordinary monkey wrench.

In some cases it is necessary, in oiling up an engine, to drop oil into a tube in order to have it go where it is needed. Sometimes this cannot be done easily, as a bubble will form and prevent it. In such a case, insert a piece of fine wire or a broom splinter in the tube, and the oil will run down this and cause no trouble.

In selecting a lubricator for a steam engine it is well to get one that is so constructed that when it must be filled, the cylinder oil will go directly into the cup, without having to go through a long crooked passage, for many good oils are thick and it is not always convenient to warm an oil before using it.

If the sight feed glass on your lubricator fills with oil, and it is so constructed that you cannot easily clean it out, if the oil is removed from the body of the cup and it is filled with water and started up in the usual way, the water will float the oil out without further trouble.

Remember that if your eccentric gets cut and is worn out or round so that it becomes necessary to put it in a lathe and take one or two cuts off from it, the reduction in diameter does not alter the throw of it.

In our opinion it is better to key up an engine in the morning rather than at night. If it is done at night, what proof does the engineer have that he will be there to attend to it the next morning ?

Every piston rod gland should be lined with soft brass to prevent cutting of the piston rod.—Power and Transmission.

## TORONTO TECHNICAL SCHOOL EXAMINA-
## TIONS.

FOR the past two weeks the examinations at the Toronto Technical School have been in progress, the term closing on the 1st of May. The classes during the past winter have been very successful, a large number of students availing themselves of the opportunity thus afforded of securing a technical education. Below will be found a copy of the examination papers in " Electricity " and " Steam and the Steam Engine," which will no doubt prove interesting to many of our readers. The lecturer on these subjects is Mr. James Milne, who, it will be observed, has covered considerable ground, and the results of the examination are said to be satisfactory. In our next issue we hope to be able to publish correct answers to the questions asked in the following papers :

### ELECTRICITY.

Maximum number of marks = 235.   175 marks constitute a full paper.

The value of each is shown in brackets after the question.

SECTION 1.

1. State clearly Ohm's Law. What is the unit of resistance ? the unit of Current ? and the unit of Electro-motive force ? (10)

2. A battery of 15 cells, arranged five in series and 3 abreast, produces a current of .5 amperes through an external R of 5 ohms. Find the E M F of each cell if its internal R is 3 ohms.     (15)

3. What is the best way of arranging 28 cells, each having an R of 4 ohms, so as to produce the strongest current in a circuit of 28 ohms.                                                        (15)

4. Compare the resistances of a wire 30' long, .06' diameter, and that of another wire 15' long and .03" diameter.             (10)

SECTION 2.

1. 1000 feet of copper wire .102" diameter is wound on an armature of a bipolar generator. Find (1) the total resistance of that wire, and (2) the resistance as measured at the brushes of the machine.  One mil foot = 10.4 ohms.                            (15)

2. Take the above question but substitute iron wire. What is the thickness so that the resistance will be the same in each case? The specific resistance of copper to that of iron is as 1 : 6.  (10)

3. Prove that 746 watts make a H. P.  Answer this fully. (15)

4. 1000 feet of wire No. 6 B and S has a resistance of .4 ohms. Find the watts lost in an arc light circuit 5 miles long. Each lamp takes 10 amperes of current.                              (10)

SECTION 3.

1. The E.M.F. of a certain dynamo machine is 100 volts, and the total R of the circuit is 1 ohm. What H.P. would have to be expended in working under these conditions ?                    (10)

2. Distinguish between work and power. What is the unit of each ?  What is the British heat unit [772 ft. pounds] equivalent to in electrical units of power ?                                   (10)

3. Describe fully the Edison Chemical meter, knowing that 1 ampere passing for 1 hour between zinc plate immersed in a solution of salt of that metal will remove from 1 plate and deposit 1225 milligrams on the other. What would be the amount of current that would pass in the above meter if the resistance of the German silver shunt was .02 ohms, and the resistance of the other circuit in which the zinc voltameter of 2.5 ohms is inserted in series with another R of 46.46 ohms, if the deposit was 200 milligrams?  Make a sketch of the arrangement.            (20)

SECTION 4.

1. Describe the Wheatstone's bridge as fully as you can, and illustrate the application of the instrument by an example.    (10)

2. How are very high resistances measured ?  A galvanometer of 6000 ohms shows a deflection of 10° when a certain resistance is in circuit with it.  Knowing that the same galvanometer shows the same deflection with a resistance of 1-10th megohm in circuit when shunted with a 1-99th shunt, find this certain resistance. The resistance of the battery is neglected.                     (15)

3. An ammeter is simply a galvanometer of low resistance, and is generally placed in series in a circuit. What would be the effect if you placed this meter in multiple, say on an incandescent lighting circuit ; and also if you had placed a voltmeter (a galvanometer of high R) in series in a circuit carrying large currents. (15)

4. Make a diagram showing clearly the connections on a shunt wound dynamo, placing in the circuit a voltmeter and ampere meter.                                                          (10)

SECTION 5.

1. Show by a diagram the general arrangement and connections of generators running on a 3-wire system.  Show by an arrow the direction of the currents if (1) both machines are doing exactly the same amount of work ; (2) if one machine is doing more than the other.  Place in position ampere and voltmeters.        (15)

2. 880,000 lines of force (N) are to be forced through a bar 20" long and 8 sq. inches in area.  Find the reluctance and the magnetizing force in ampere turns to effect this magnetization.  Permeability = 166.                                                (15)

3. In a generator which is driven by a 100 H.P. engine, belt speed 5,000 ft. per minute; there are 200 conductors in the armature winding 100 sections in commutator, the gap is 45°. Find the tongue and the drag on the active conductors.        (15)

### STEAM AND THE STEAM ENGINE.

120 marks constitute a full paper.

1. What is the latent heat of steam at 212° Fah., expressed in foot pounds?  What is the difference between latent and sensible heat?  If one pound of steam at 212° Fah. is mixed with 10 lbs. of water at 60° Fah., find the resulting temperature.           (15)

2. Steam expands in the cylinder of an engine from 30 lbs. pressure above atmosphere to 5 lbs. below atmosphere, at what part of the stroke was the steam cut off?  Atmospheric pressure may be taken at 15 lbs.                                           (15)

3. Define the lap of a slide valve, and explain answer by reference to a sketch.  For what purpose is it employed?  Account for the difference in the working of two engines, one of which has lap on the steam side of the valve and the other has not.     (15)

4. Describe Savary's engine.  Show by a sketch the principle on which it worked.  What was the greatest depth the water could be lifted by this engine!  Why was it limited to this extent?                                                          (15)

5. The diameter of a steam engine is 24", and revs. per minute = 60.  M. E. P. = 40 lbs.  What should the length of the stroke be so that the engine will develop 330 H.P.                     (15)

6. Sketch Newcomen's engine.  During what portion of each stroke, and in what manner was unnecessarily wasted by Newcomen's arrangement ?  How did Watt propose to lessen this waste, and in what way did he carry out his idea ?             (15)

7. What is meant by the term "clearance" and "cushioning ?" At what part of the stroke does cushioning occur ?  Show by an indicator diagram the manner in which the side valve produces cushioning.                                                     (15)

8. Describe Stephenson's Link motion. How is an engine reversed by this arrangement ?  Make a sketch to illustrate. How would you arrange a link motion with one eccentric only ? What is shortening the travel of the valve equivalent to ?    (20)

9. A safety valve 3" diameter is held down by a lever weight. Lever 36" long.  The valve centre is 4" from the fulcrum.  Weight 56 lbs.  Omitting the weight of lever and valve, at what pressure would the valve be lifted ?                                   (15)

10. Describe the various methods of connecting the heads of tubular boilers to the sheets.  What are the relative strengths of a single and double rivetted lap joint to that of the original plate ?                                                         (15)

11. You have two engines exactly the same size; one has steam cut off at ¼ stroke, the other has steam cut off at full stroke. Show by calculation what is gained by using the steam expansively.                                                         (20)

12. Describe as fully as you can the Hydraulic Ram.  Show the arrangement by a sketch.

13. A hydraulic ram has an efficiency of 70%.  40 gallons of water are spent on same, with a fall of 10 feet.  How many gallons will it raise to a height of 400"?  What is the pressure at the bottom of a column of water 400' high ?                          (15)

---

The force of a stroke of lightning in horse power is indicated by the following incident :  During a recent storm which passed over Klausthal, Germany, a bolt struck a wooden column in a dwelling, and in the top of this column were two wire nails 5-32 inch diameter.  The two nails melted instantly.  To melt iron in this short time would be impossible in the largest furnace now in existence, and it could only be accomplished with the aid of electricity, but a current of 200 amperes and a potential of 20,000 volts would be necessary.  This electric force for one second represents 5,000 horse power, but as the lightning accomplished the melting in considerably less time, say 1-100 of a second, it follows that the bolt was of 50,000 horse power.

## THE DAKE ENGINE.

THE extremely compact type of engine shown in the accompanying illustrations is unusually interesting on account of the ingenious mechanical principles involved in its design. As a steam engine, aside from questions of design, the manufacturers claim that experience has demonstrated that in reliability, and especially durability, it is not exceeded by any of the types of usual design. On account of its compact form, this engine is claimed

FIG. 1.—DAKE STATIONARY ENGINE.

to be particularly suited for running ventilating fans, centrifugal pumps, incandescent lighting dynamos and saw mill carriages. Being strongly built, self-contained, and not affected by ordinary jars, it also gives reliable service when used to run smoke-consumers and head-light or other dynamos on railway trains, and when employed for various auxiliary purposes aboard vessels.

Fig. 1 illustrates the engine complete, and in Fig. 2 the pistons are removed, showing the interior of the case with the crank in position, this latter revolving in the chamber shown in the back of the case in the central cut. This chamber is supplied with oil and water from an opening in its back, thus securing lubrication to every part of the interior of the engine.

Both side pistons have a horizontal movement sliding from side to side, and at the same time an inner piston to which the crank pin is attached has a vertical or up and down motion, the two movements imparting rotary motion to the crank. Steam is admitted through channels in the cover, one opening into a central aperture and another into an annular opening on the inside of the cover. Four channels are cored through the inner piston, one leading to the top and another to the bottom, and one to each end of the inner piston, the latter also leading through the ends of the outer piston. Four parts corresponding with the channels in the interior of the inner piston are cut through the face (or side next to the cover) of the inner piston in the proper position to register over the central aperture in the cover. The steam entering the port in the inner piston, through the central aperture of cover and re-acting against the side of the case, imparts motion to the crank, the port passing over the annular ring and exhausting into it after having done its work. There are four distinct impulses of steam to

each revolution of the crank, and the arrangement of the ports to the crank are such that each impulse of steam is given at a point where it has the greatest power. The expansion of steam is secured in the passage of the ports of the inner piston over the central aperture in the cover.

With the reversing engine, the channelling on the cover and in the piston is the same as in the engine built to run one way, but the ports in the inner piston are shaped so that they register over both the central and the annular openings, using each alternately as steam and exhaust. The ports on the top of the case being fitted with a suitable valve which connects the channels leading to the working parts of the engine, motion is given to the engine either to the right or left, as desired. The reversing engine is the same as a

FIG. 3.—CARRIAGE ENGINE.

stationary engine, only with reversing throttle instead of governor.

Provision is made for taking up the wear of the working parts of the engine in a simple and effective manner. The inner piston is fitted with phosphor-bronze slides that admit of a thin piece of tin or sheet iron being inserted when the wear is sufficient to allow it A wedge-shaped plate on which the lower slide rests is arranged with set screws on the outside of the case (Fig. 2), which keeps the piston steam tight, top and bottom. The packing of the cover to the pistons is effected by thin copper joints placed between the edge of the case and cover. The pistons are made so that they are slightly thicker than the case they occupy, and enough copper strips are put in to fill up the space ; these joints are removed one at a time as the pistons wear down, and where it is seen that repacking is needed and a copper joint is too much to take off at one time, a piece of thin paper to take its place will repack the cover perfectly. The repacking of the cover as above described, and replacing the nuts or cap screws (as found

FIG. 2.—DAKE ENGINE, SHOWING PISTON WITH CYLINDER COVER REMOVED.

on the different sizes of engines) evenly, is the only point about the engine that requires careful attention and judgment on the part of the person in charge, and repacking is not required except at long intervals.

There is very little friction, and consequently slight wear on the pistons, from the fact that the steam pressure is inside of the inner piston, instead of against it, making the pistons similar to balanced valves. The bearings for the main shaft and crank pin are in the form of bushings and made from phosphor-bronze. From the manner in which steam is applied to the pistons the wear is slight compared with the ordinary engine. When they need renewing the worn ones are driven out and the new bushings driven to place, which can be done by any good machinist at a small cost to the purchaser. The crank and pin are made from the best quality of cast steel, and the shaft, which is machinery steel, is shrunk into the crank in a solid manner. The outer piston is also made from best quality of cast steel. Every part of the interior of the engine is fitted with the greatest care. The inner face of the cover and all of the working parts of the engine are ground surfaces, made with tools especially designed for the manufacture of this type of engine, thus ensuring that the engine is practically steam tight from the start. Everything about the inside of the engine is made interchangeable, and can be duplicated in case of accident on short notice.

Fig. 3 shows the carriage engine for setting up and receding head blocks.

In Fig. 4 is shown the steam feed, which is recom-

FIG. 4.—STEAM FEED.

mended to the consideration of saw mill owners and operators. The claims made for it are simplicity of construction, positive operation and easy management, economical use of steam, small space occupied, cheapness, and easy adaptation to either new mills or those now in use. In placing the engine in position, it is not necessary to move the husk frame, as it can be lowered from above through the frame onto foundation. The movement of the engine in either direction is under absolute control of the sawyer by lever connecting with reverse valves on top of engine, thus accommodating

the speed of the feed to the size and conditions of logs.

The Dake engine has been placed upon the Canadian market by the Phels Machine Co., of Eastman, Que., who will gladly furnish any further information.

## SPARKS.

The town of Magog, Que., is inviting tenders for electric street lighting.

An incandescent plant will be installed by Hewson Bros., of Durham, Ont.

The ratepayers of Alexandria Bay, Ont. have voted $1,000 per year for electric light.

The electric light plant at Three Rivers, Que., is offered for sale by tender. The date limit is the 15th inst.

Mr. W. H. Pearson, of Toronto, manager of the Trenton Electric Co. is seeking a franchise for electric lighting in the city of Belleville.

The Toronto Electrical Works, Toronto, suffered damage by fire recently to the extent of $2,500. Considerable valuable machinery was destroyed.

The town of Orillia, Ont., will likely enter into an agreement with Mr. Tait to furnish incandescent lights on the present dual basis for five years from January 1st, 1895.

The city engineer of St. Thomas, Ont., has been instructed to prepare an estimate of the cost of constructing an electric plant to supply heat, light and power and to operate an electric railway.

The city of Vancouver, B. C., has accepted the proposition of the Consolidated Tramway and Lighting Company to light the city, at 27½ cts. per light for 200 lights, or 27 cents for over 200 lights.

The Canadian General Electric Co. are placing an electric light plant for the town of Niagara Falls, Ont., including two dynamos with a capacity of 5,000 lights, together with two engines of 535 horse power.

Mr. A. W. White, London, Ont., has been appointed one of the umpires on the Cosmopolitan motocycle race in New York, to be held on May 30th next. Mr. White, it will be remembered, was umpire on the Duryea motor, which won the moto-cycle race at Chicago.

R. McGowan has purchased from the Johnston Electric Company a 1,000 light alternator and equipments for 1,000 incandescent lights for Durham, Ont., where he has an arc plant at present. He also owns the electric light plant at Oakville, Ont. The installation work in connection with the new plant at Durham will be carried out under the direction of Mr. R. McGowan, jr., of Oakville.

# ELECTRIC RAILWAY DEPARTMENT.

## THE MONTREAL PARK AND ISLAND RAILWAY.

THE Montreal Park & Island Railway was incorporated in 1885 by Statute of Quebec. At subsequent dates, viz., in 1886 and 1893, amendments to the original charter were passed by the Quebec Legislature, and in 1894 the railway was declared for the general advantage of Canada, and came under the jurisdiction

FIG. 1.—THE MARPLE RAIL BOND.

of the Federal Parliament. New powers were then granted the company, and in 1896, during the last sitting of Parliament, further powers were obtained, and the company now is in a position to complete the construction of the various electric railways contemplated.

Of the gentlemen who were originally instrumental in bringing forward this project, Hon. J. R. Thibaudeau, Sheriff of Montreal ; Mr. Henry Hogan, the well-known proprietor of St. Lawrence Hall, Montreal, and Hon. Louis Beaubien, are at present directors of the company.

The first construction undertaken was the line through Mile-End, to reach the River des Prairies, and thence down the right bank of the river to Sault au Recollet

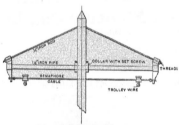

FIG. 2.—DOUBLE BRACKET CONSTRUCTION.

Village. This was built in 1893. This line, like nearly all suburban lines, has had the usual pioneering difficulties to overcome, and as the proprietors of the road determined to make the project a successful one, they persevered with the work and they have succeeded in putting this line in good condition, both physically and financially.

This piece of track, commonly known as the Back River line, extends for nearly 7 miles from the city limits. It is double track from the city limits to the Shamrock grounds, from which latter point it is single track to the terminus. It is the company's intention to extend this line about four miles farther in order to reach St. Vincent de Paul. This work will probably be commenced during the current year.

In 1894 a double track line was built from the city limits at Mount Royal Avenue around the mountain through Outremont as far as Cote des Neiges Road, thus reaching the cemeteries. In 1895 this line was continued around the western mountain to the westerly limit of Westmount, connecting there with the Montreal Street Railway, Sherbrooke Street line, thus completing a double track electric circuit 11 miles long

around the two mountains, and furnishing a line of communication for many outlying municipalities which will tend to develop them at a very rapid rate. Not only this, but it furnishes for the people of Montreal a most delightful trip during summer, which they are not slow to avail themselves of, and on pleasant afternoons and Sundays the company's resources are taxed to provide sufficient accommodation to carry the thousands of people who go out to spend a half hour in riding over the lines, which affords an opportunity of looking at some of the finest scenery of which this country can boast.

During the present season the Company expects to complete its system. Among other lines, they will build to Lachine and to Bord 'a Plouffe by way of St.

FIG. 3.—TWO CARS ON THE OUTREMONT DIVISION.

Laurent, in addition to the already mentioned extension to St. Vincent de Paul.

The mileage of track at present operated is 22, and by the end of the year the length of track operated will be in the neighborhood of 50.

From the peculiar development of Montreal, being as it is so densely populated in certain districts, it is evident that the opportunity for developing suburban business is considerable. The population being concentrated, the necessity for moving out becomes more apparent every year as the necessities arise for factories and business houses occupying sites in the older residential portions of the city, gradually forcing the residents further away where they can secure fresh air and more room.

The building of a suburban electric system affords every opportunity for the people improving their condition, and although the people of the working class of Montreal are very conservative in their ways of living, yet they are beginning to be convinced of the desirability of changing their places of residence from the

FIG. 4.—BACK RIVER STATION.

smoky, unhealthy portion of the old town to the delightful country surrounding the city, to say nothing of the reduced cost of living, and this is all made easily possible by the Montreal Park & Island Railway Company, which has persevered in preaching this doctrine to a most conservative community.

By an arrangement existing with the Montreal Street Railway the cars of " Park & Island " system come over the tracks of the street railway to the centre of the city, the street railway having the benefit of the cars for their purposes on their lines in going to and from the city limits, and the passengers to and from the suburbs thus not having the necessity to change cars.

### ROADWAY.

The track is laid throughout with 56 lb. Cammell steel rails of Sandberg section. The joints are four bolt angle bars, and the track is laid with broken joints, except in street work, where the joints are square. The roadbed is built up high wherever it can be built. Ballast of broken stone, gravel and cinders are used according to circumstances. The bond is of No. o soft copper wire, soldered to a brass plug, which is pressed through a drilled hole in the web of the rail.

### OVERHEAD CONSTRUCTION.

This is on the general plan of the "west end," though malleable iron parts are being substituted for bronze. The hangers are attached to cable supports both in bracket and span construction in order to provide flexibility. The ears are for the most part soldered, but a mechanical clip will in future be used instead. The trolley wire is No. o hard drawn wire The posts are cedar 8 inches diameter at top.

### POWER PLANT.

This consists of an installation of steam engines driving two generators, one of 200 k. w. and one of 100 k. w. capacity, made by the Royal Electric Company, of Montreal. The power station, however, is but a temporary installation, and a description of the permanent power station will appear in a future number. The construction of the new plant will soon be commenced.

### ROLLING STOCK.

This consists of twenty motor cars, ten of which are closed cars, six are nine-bench open cars, and four are thirteen-bench double truck motor cars, 38 feet over all. In addition to these above there are four open trailer cars. The motors are for the most part of the " Royal 30 " type, made by the Royal Electric Co., of Montreal, and the severe tests that these motors were put through during the extraordinary severe winter of February and March of this year, as well as on previous occasions, justify very high praise for the Royal Electric Company. The motors are mechanically excellent, and electrically they are highly efficient. For heavy or light work these motors are very satisfactory. The small truck which has given the greatest satisfaction is that made by the Canada Switch and Spring Co. The double trucks are all of Brill No. 23 pattern.

### TELEPHONES AND SIGNALS.

The lines are equipped with telephones, and connecting wires are led down the posts at close intervals, the telephone instrument being portable and merely hooked to the contacts on the posts, so that connection may be had with head office from any point of the line. Each regular car carries a telephone.

On single track lines the Skeen signal system is being installed, and will be used no doubt on other single track lines to be built, as it enables a maximum number of cars being operated on a single track.

The officers of the company are :—Hon. Louis Beaubien, President ; Hon. J. R. Thibaudeau, Vice-President ; Henry Holgate, manager and engineer.

Mr. Holgate was from 1878 to 1888 connected with the Northern and Northwestern Railways, and upon those railways uniting with the Grand Trunk Railway in 1888, he continued in his former capacity as chief engineer of the division until 1893, assuming his present position in June, 1895.

A more complete description of the Park and Island system will be given in a future number, when the new work becomes sufficiently advanced.

---

Col. John Stacey has purchased the franchise of the St. Thomas Street Railway Co. from Messrs. Cameron & Hunt, of London. It is proposed to electrify the system.

## IMPROVED AUTOMATIC CAR COUPLER.

RAILWAYS to-day demand an automatic coupler that is strong, cheap, reliable, decisive in its action, self-adjustable and interchangeable with the present rolling stock of the world. At an exhibition held in the Park Avenue Hotel, in the city of New York, attended by the railway managers of the amalgamated roads of the United States, the invention of Otto Flohr, of which we give an illustration, was unanimously voted as being the only invention of its kind worthy of consideration from all points ethical and economic. The chief advantage of this device in the eyes of practical railway men are that the couplers are interchangeable with any of the vertical plane couplers now in use ; absolutely automatic ; simple in construction ; undoubtedly cheap ; certain in action ; durable ; can be handled with ease

IMPROVED AUTOMATIC CAR COUPLER.

and without hanger, and will lock on a curve as readily as on a straight line.

The coupling obviates the necessity of any one going between the cars and the parts are so ingeniously constructed that the resistance in uncoupling is reduced to a minimum. The locking arm rises automatically by being pressed by the locking arm, which has a slight taper at the end, which engages with an incline face upon the pin, forcing the pin up during the concussion until the arm swings by and clears it.

The knuckle swings open the moment the pin is released, as the result of its own gravity—it resting on the highest point of resistance on a spiral-way, when closed from which it naturally descends from its own weight. The outer edge of the knuckle has a stop that prevents the possibility of the locking arm swinging out too far to be of service.

It is a coupler that is wonderfully clever in its mechanism and is entirely different from any now in use. Patents have been granted in Canada, United States, England, France, Belgium and Russia. The Dominion Government passed a bill last session that if enforced will take away all railways in Canada to adopt within two years some such automatic coupler.

---

The Winnipeg Electric Street Railway Company have in connection with their road two parks situated about four miles from the centre of the city. In River park they have a half-mile race track, large grand stand, bicycle track, field for lacrosse and such sports, roller skating rink, etc. Elm park is situated just across the river from River park, and is reached by a pontoon bridge. It also contains the necessary requirements for a pleasure resort, and being thickly studded with trees, is used largely by pic-nic parties. The traffic to these parks in the summer is very large, and they are considered excellent investments.

## PIONEER ELECTRIC RAILWAY WORK IN CANADA.

To the Editor of the ELECTRICAL NEWS.

SIR,—I observe in your March issue a letter from Mr. James W. Easton, in which he claims that the first successful attempt in Canada to propel cars by electricity was made in 1883 on the Industrial Exhibition grounds at Toronto, the motor and power equipment consisting of three old Ball machines designed for arc lighting. Inasmuch as Mr. Easton admits that the efforts of a couple of men were required to push the empty car up the grade, and that no passengers could be carried, I think it is tolerably clear that so far as practical results are concerned, the experiments referred to cannot be considered to have been successful. I am informed by persons who witnessed the experiments that the only way the cars would run was down hill.

Very truly yours,

OLD TIMER.

Directors of the Sherbrooke, Que., Street Railway Company have been elected as follows: Walter Blue, Wm. Morris, J. W. Burke, J. E. Flood and F. J. Griffith. Mr. Burke has been elected president, and Mr. Griffith, secretary.

The Rathbun Company, of Deseronto, are now building an electric self-loading street car for A. Jackson Reynolds & Co., of Montreal, which, it is claimed, will revolutionize street cleaning in all towns and cities. One car, it is said, will clean 25 miles per day and take the sweepings out of the municipality at a saving of 60 per cent.

# CANADIAN GENERAL ELECTRIC CO.

## (LIMITED)

## The following Letters speak for themselves:

PORT HOPE, Feb'y 27th, 1896.

MESSRS. CANADIAN GENERAL ELECTRIC CO.,
Toronto.

DEAR SIRS :—The 75 Kilowatt Monocyclic Alternator purchased from you was started up on Sept. 2nd, 1895, and has since been giving us an uninterrupted service of 160 hours each week, starting at four o'c:ock on Sunday afternoon and running till eight o'clock the following Sunday morning, without a hitch of any kind whatever. We expect a large increase to our business from motor service, and appreciate the excellent features of the Monocyclic system of this combined light and power service. There is, of course, no unbalancing, as the lighting is single phase, and the operation of the motors does not disturb in any way the evenness of the lighting. After an experience of six months we feel warranted in saying that we consider the Monocyclic as superior to any of the polyphase systems which we are acquainted with, and intend in the near future to duplicate this machine.

Yours truly,

R. A. CORBETT,
Pres. & Mgr. Port Hope Elec. Lght & Power Co.

PARRY SOUND, Feb. 27th, 1896.

THE CANADIAN GENERAL ELECTRIC CO., Toronto.

DEAR SIRS :—Having now made a thorough trial of the Monocyclic system of Electrical distribution as supplied by you, I have much pleasure in informing you that it is giving entire satisfaction.

The machine, a 75 K.W., is a beautiful specimen of dynamo building, being strong and compact. Ventilation of the armature is excellent, and the general design of that very important part of the machine is good. Electrically and mechanically, I consider your machine to be superior to any I have seen.

We have not had occasion, as yet, to test the machine on the operation of motors, but speaking from the lighting point of view, I can fully endorse what you claim for-the system.

We have 700 lights now wired and expect to increase to 1,200 before 1897.

Yours truly,

W. B. ARMSTRONG,
Manager Parry Sound Electric Light & Power Co., Ltd.

HANOVER, Feb. 21, 1896.

CANADIAN GENERAL ELECTRIC CO., Toronto.

DEAR SIRS :—In answer to your enquiry as regards the operation of our Electric System, I beg to say that we have now been running one of your 75 Kilowatt Monocyclic Dynamos for the past two months, and it is giving entire satisfaction in every respect. We have not had the slightest trouble with it in any way, and although it is being operated about 15 hours a day, it runs exceedingly cool, and requires practically no attention whatever.

The machine itself I regard as a model of simplicity, in fact to show my confidence in the apparatus I have placed the plant in full charge of my brother, who, previous to the starting up of this machine, had no experience whatever with electrical apparatus of any kind.

The perfect regulation of the Dynamos, and the sim-

plicity of the wiring are also strong points which should recommend the use of this style of apparatus to anyone contemplating the installation of an electric plant.

In conclusion, I might say that after having decided upon adopting the Monocyclic system, my opinion became somewhat prejudiced against its adoption by representations made by other manufacturers, but I now fail to see wherein I could have secured anything better to that installed by your company.

Yours very truly,

D. KNECHTEL.

DUNNVILLE, February, 1896.

MESSRS. CANADIAN GENERAL ELECTRIC CO.,
Toronto, Ont.

DEAR SIRS :—We are pleased to be able to express ourselves as entirely satisfied with the Monocyclic system installed by you last fall. We are now in a position, having covered a considerable portion of the town with our lighting mains, to appreciate the value of the three-wire system for secondary distribution from the transformers, and the great advantage gained in simplicity by the Monocyclic from its being a single-phase system for the lighting distribution.

The workmanship and finish of the dynamo itself certainly does credit to your factories, and in operation it has proved itself to be exceedingly simple and satisfactory.

The commutator and brushes run without any sparking whatever, and do not give us a particle of trouble. We feel fully justified in saying that the Monocyclic system in operation has shown itself to possess all the points of excellence claimed for it by you at the time when we made the selection for our new plant.

DUNNVILLE ELECTRIC LIGHT CO.

MATTAWA, Feb. 27th, 1896.

MESSRS. CANADIAN GENERAL ELECTRIC CO., Toronto.

DEAR SIRS :—We are pleased to be able to express complete satisfaction with our Monocyclic plant, which has now been running since 27th Sept. We are especially pleased with the ease with which our former single-phase system has been changed into one suitable for the distribution of both light and power. The only change made in our case was the installing of the Monocyclic machine in place of our former single-phase alternator, and the running of a third wire to the points where power is to be supplied. Altogether the system is admirable, both as to simplicity in the wiring, and distribution and perfect freedom in operation from any trouble or complication. We are quite sure that the Monocyclic system will prove a means of increasing largely, the revenues of alternating lighting stations by the sale of power without adding any complications to their operation. You will be pleased to know that the 5 h.p. induction motor geared to a triplex pump is now in successful operation pumping water for the C.P.R. water tank. It is certainly a very simple and substantial piece of machinery.

Yours truly,

MATTAWA ELECTRIC LIGHT & POWER CO., LTD.

A. F. HURDMAN, Sec'y-Treas.

## SPARKS.

The Galt, Preston and Hespeler Electric Railway carried 13,000 passengers and 800 tons of freight during the month of March.

An electric trolley road has been built at Kioto, Japan. Tokio, Yokohama and Osaka have decided on adopting similar lines.

The Toronto Suburban Street Railway Company propose extending their line from Toronto Junction to Lambton Mills and Islington.

Messrs. J. E. Flood and J. W. Burke, of New York, are interested in the construction of the proposed electric railway at Sherbrooke, Que.

The Niagara Falls Electric Light & Power Co., of Niagara Falls, Ont., have awarded the contract for a 5,000 light incandescent plant to the Canadian General Electric Co., Ltd.

An electric railway is projected to run from Bell's Corners to Richmond West, Ont., a distance of ten miles. The promoter is Mr. John Moodie, proprietor of the Richmond and Nepean Macadamized road, who intends asking the municipalities interested for a bonus towards the construction of such a line.

## SECOND-HAND ALTERNATORS FOR SALE

We have for sale several second-hand alternators which we have taken in exchange for a larger machine, and which we are offering at low figures. These machines have exciters and switchboard apparatus and are ready for immediate service. For particulars apply

AHEARN & SOPER
Ottawa, Ont.

# Telephones

.. THE ..

## "UNIQUE"

Main Line and Warehouse ..

TELEPHONES

•

Sold Outright.
.. No Exorbitant Royalties. . . .

Only Telephone made that does not get out of Adjustment. . . . .

Send for Catalogue and Prices.

SOLE MANUFACTURERS:

## John Starr, Son & Co.
(LIMITED.)
2,4,6 DUKE ST., COR. WATER, Halifax, N.S.

CANADIAN ELECTRICAL ASSOCIATION CONVENTION
TORONTO, JUNE 17, 18, 19, 1896

# CANADIAN ELECTRICAL NEWS

## STEAM ENGINEERING JOURNAL

OLD SERIES, VOL. XV.—No. 6.
NEW SERIES, VOL. VI.—No. .6

JUNE, 1896

PRICE 10 CENTS
$1.00 PER YEAR.

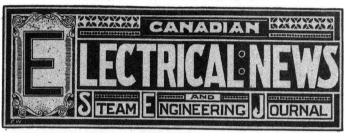

# Kay Electric Mfg. Co.

**255 James St. N., HAMILTON, ONT.**
**58 Adelaide St. W., TORONTO, ONT.**  Telephone 1214.

We are prepared to furnish——

*Dynamos of any capacity for any voltage either
    compound or shunt wound.
Motors from 1-8 to 40 h. p., either series or com-
    pound wound.
Elevator Motors of all sizes.
Alternating Dynamos from 300 to 1000 lights.
Transformers of any capacity from 5 to 125 lights.
Electro-plating Dynamos, any capacity.
Electrical Experimenting in all its branches.*

WRITE FOR PARTICULARS AND ANY INFORMATION REQUIRED.

# GAS ENGINES

Of from 1 to 600 Brake
Horse Power, for Electrical
Industrial and other pur-
poses.

MANUFACTURED BY

FRIED. KRUPP GRUSONWERK, Magdeburg, Germany.

**JAS. W. PYKE & CO., Montreal, Que.**  Representatives for the Dominion of Canada.

Particulars on Application.

# STEAM PUMPS

**DUPLEX
SINGLE
TRIPLEX**  ⋮ For All Duties
....

# NORTHEY MFG. CO., Ltd., TORONTO

The Laurie Engine Co., Monteal

◄———SOLE AGENTS FOR PROVINCE OF QUEBEC———►

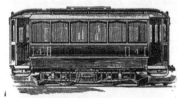

FINE - - $_{E}L^{E}C^{T}R^{I}C$ ◎

# Street Cars

....OUR SPECIALTY...

We also manufacture Horse and Trail Cars
of every description.

**PATTERSON & CORBIN** . . . . . . . . . . **ST. CATHARINES, ONT**

---

# W. N. LAZIER

*Box 341, VICTORIA, B. C.*

Pacific Coast Agent for

## Remington Machine Co.

Refrigerating and Ice Machines.
Complete Plants Installed for all Purposes
Robb Engineering Co. Economic Boilers.
High Speed and Corliss Engines.
Complete Plants Erected.

**ALL WORK GUARANTEED.**

---

# MUNDERLOH & CO.

➤ **Montreal**

TRY OUR STATION

# VOLTMETERS AND
# ...AMMETERS

Hundreds in use in United States, England, Germany,
France, Australia, Japan, etc.

WRITE FOR CATALOGUE AND PRICES

CANADIAN

# ELECTRICAL NEWS

AND

## STEAM ENGINEERING JOURNAL.

| VOL. VI. | JUNE, 1896 | No. 6. |
|---|---|---|

## THE ELECTRIC LIGHT INSPECTION ACT.

REFERENCE was made in the May number of the ELECTRICAL NEWS to complaints regarding the operation of the Government Electric Light Inspection Act. Pursuant to the promise then given, further inquiry has been made, the result of which shows the system of government inspection to be extremely unpopular with the electric lighting companies throughout Canada. The large number of such companies who have written us their opinions on the subject take the ground that a system of compulsory inspection by the government is unnecessary, and that no equivalent is given either the

Has the inspection been productive of any better understanding between seller and buyer?......................... ................................

Are the consumers more satisfied with their bills now than they were prior to the date when the inspection system went into operation?.............................

What is your experience of the performance of direct current mechanical meters with commutators, after being sealed up by the government seal? Do you find them slow?................ ..........................................................

What is your general opinion as to the advisability of government interference between buyer and seller? Would you consider the competition between different methods of lighting a sufficient safeguard of the interests of the consumer even in places where there is no gas, where the ideas of the community are about the size of a coal oil lamp, and where electricity has of necessity to be sold cheaply enough to be an inducement for its adoption?

Remarks:............................................................... ..... ....
............................................................................................

C. H. MORTIMER, Publisher.

The large number of replies received to the above in-

ENGINE AND DYNAMO ROOM, INCANDESCENT STATION, TORONTO ELECTRIC LIGHT CO.

companies or their customers for the yearly fee which the former are compelled to pay the government. Companies doing a limited business in competition with coal oil in small towns and villages are especially bitter in their complaint of the uselessness and injustice of the system.

With the object of eliciting an expression of views upon the subject, a copy of the following circular was recently mailed to the electric companies throughout the Dominion :

OFFICE OF
CANADIAN ELECTRICAL NEWS
AND
STEAM ENGINEERING JOURNAL.

Confederation Life Building,
TORONTO, May 1st, 1896.

DEAR SIRS :—Having lately heard some dissatisfaction expressed with the system of electric light inspection put in operation last year by the Dominion Government, I would feel obliged if you would kindly favor me with your opinion, on the line of the undermentioned inquiries, as to the extent to which you consider the operation of the system advantageous and satisfactory.

In case it is found to be defective and burdensome on the companies, the ELECTRICAL News will endeavor to have it improved or abolished.

What is your opinion of the present system of government inspection of electric light? (a) Is it an advantage to the seller?.....................................

(b) Is it an advantage to the consumer?...... ........ .... .... .... .. ..

Is the annual inspection fee in your opinion a fair one?.... ........ .. .. ..

quiries, and the very decided language employed in expressing disapproval of the inspection system, are, as already stated, evidence of the widespread feeling of indignation which prevails as a result of the injustice to which the lighting companies feel they are being subjected. In order that the Inspection Department may see exactly the condition of affairs, we print verbatim the following opinions :—

" Our opinion is there is no necessity for such interference."—St. Marys, Ont., E. L. Co.

" The Government have no right to interfere, as there is not the slightest call for such. Government inspectors are not educated for their position. Should any company be in opposition to the government, the officials could injure their business. Have already expressed our views very strongly to ministers of the crown in opposition to the system of inspection. Nothing less than an imposition."—Kemptville, Ont., E. L. Co.

" (a) Yes, if we could induce our customers to adopt it. (b) No, not when compared with a flat rate. Its an unfair tax for which no benefit is given. There is no gas here, but we find we have to compete against the cost of coal oil, and this necessarily keeps the price down or we could not get the business. This appears to us to be sufficient security to the consumer. When the price is not satisfactory they cannot be compelled to use electric light."—Citizens Telephone & Electric Co., Rat Portage.

"The biggest fraud on earth—an advantage to no one. Never saw the inspector; has never been here. We do not use meters at all. Rent by the year, and consumers use as much current as they want. Do not approve of government interference at all. Believe this law was made more for the purpose of furnishing some hungry politician with an easy and lucrative position than for any benefit it might confer on either buyer or seller. It is the biggest farce we know of."—Cayuga, Ont., Lt. & Power Co.

"We consider this inspection fee one of the most unfair things that was ever adopted. If we took money from any person without giving him an equivalent better than what we get for this inspection fee, we would feel that we had stolen the money. The inspector came and said, 'I want $25.00; I don't know anything about the electric plants, but I must have the money.' So he got it, and that is the last we have seen of the thing and is likely to be until they want more. It does not do any good in small places at least; it is too bad. Kindly do what you can to have it abolished altogether."—

"We have been expecting inspection of meters for the past year, but have not got it yet; cannot say whether same will be satisfactory. The fee is not a fair one—it is altogether too high. The competition in the lighting business is in most places very keen, and there is no danger of the public being overcharged for light. A meter of any kind is by the consumer generally considered an unreliable machine, and as one of those having to do business by meters, we think the inspection by government a good thing to fall back upon, and will to some extent satisfy the consumers."—People's Electric Co., Windsor, Ont."

"(a) (No, we run all flat rates.) We get no value. Have never seen or heard from inspectors except to collect fee. Where current is sold by meters, and meters are inspected and value given by inspectors for work done, there is some excuse for collecting an annual fee, but it is certainly a hardship to be compelled to pay a fee to a department which is of no earthly use in any manner or form. This applies to all stations selling current on flat rate basis."—Gananoque E. L. & Water Supply Co.

"I do not think the government should interfere. It seems to me to be only a scheme to raise revenue, as we have paid $25.00 for a registration fee and so far have had no inspection, nor do we know who the inspector is for this district. It also seems, in the case of municipalities, a farce for the Provincial government to give power to establish lighting systems, and then for the Dominion government to say we will charge them a certain sum for availing themselves of the legislation enacted by the provincial authorities."—J. N. Christie, Town Clerk, Mitchell, Ont.

"We paid our license of $25.00 and have had no inspection up to the present time. The only thing we know about it is, we are $25.00 out."—Stratford Gas & Electric Co.

"There should be no interference. If any one wants government or any other inspection let him get it and pay for it. Why should the government compel me to pay a registration fee for selling electricity for lighting purposes and allow others to sell coal oil and other illuminants without paying a fee? The government gives no value in return for the registration fee, and money taken without value given is simply robbery."—A. Groves, Fergus, Ont.

"We have paid $25.00 and not received a single cents' result. It is simply robbery of small plants of villages or towns. We do not sell by meter, and there is really nothing for a government inspector to do. There is further no advantage to buyer that we can see, because if they cannot afford electric light they need not take it."—W. Moore & Sons, Meaford, Ont.

"Do not think it necessary in small places, even where there is no competition but oil, as prices are so low there is not a living profit to be made. The yearly government fees are altogether too high for small plants."—Stanstead, Que., E. L. Co.

"We only run the arc system of lighting and therefore consider inspection a perfect farce. You can only get so many amperes out of a certain wound machine, which during inspection, can be run to full amperage but with a self adjusting lamp running and producing a clear light on any current from 4 to 12 amperes, it can easily be seen that government inspection won't amount to anything or be any safeguard. Competition will do it, and nothing else. The fact is the move is calculated to close out small plants, and set back our villages into the old Egyptian darkness." — Owen Sound Electric Illuminating & Mfg Co.

"(a) No, it only takes money from him and gives nothing in return. (b) Not that we have been able to find out. We think it a humbug; the party that inspected our plant did not know an alternator from an engine, but sat in our office and wrote just what we told him about our plant, and charged us $10 for it."—Thamesville E. L. Co.

"(a) Only in case of controversy. (b) Yes. Do not know what inspection fee is for. Have not been operating long enough to say whether inspection is productive of better understanding between seller and buyer." — Peterborough E. L. Co.

"(a) It is of no advantage to the seller or buyer in our case as we use no meters. The government taxes us $25.00 annually, which we consider most unfair as we get no return therefor whatever. The buyers are perfectly satisfied and many of them do not know that such a thing exists. We consider that in towns of this size where the light has of necessity to be sold cheap, that there is no reason whatever for government interference, and we can see no valid reason why we should be compelled to pay a revenue tax of $25.00 annually, when we get no return whatever for the money paid out. If we were in the tobacco or liquor trade, doing a business which some people might consider was derogatory to the welfare of the country and upon which the profits were large, we would not mind the tax. We cannot understand why the manufacturers of electric light should be compelled to pay a government tax to the revenue department any more than the manufacturers of furniture or agricultural implements. If the consumers were dissatisfied with the light given or with their meters, why could not an inspector be appointed to rectify and measure the same, and be paid by the company or the purchaser for his work without compelling us to pay out $25 annually, same as an hotel-keeper or a man wanting a fishing license. We would

SWITCHBOARD, INCANDESCENT STATION, TORONTO ELECTRIC LIGHT CO.

favor the removal of the tax which we consider unnecessary and wrong."—Citizens' Electric Light Co., Smiths Falls, Ont.

"(a) We do not think so. (b) We do not think it would be. If the consumer be dissatisfied, he could discontinue at any time. We consider the government has no more right to interfere in this than with the buyer and seller of any other commodity. If parties don't want the light they need not buy it. We think the inspector would only tend to make or cause trouble. For instance, if the inspector arrived here during high water, we could at such time

ONTARIO PARLIAMENT BUILDINGS, TORONTO.

perhaps not give the buyers 16' c. p. We cannot see that any good would come to anyone except the inspector."—Robertson, Rowland & Co.

"(a) No. (b) No, we will have to charge higher rates. We do not use meters. The fee should be only nominal in small towns like Brampton. Charge should be larger where a large quantity is sold. Five dollars would be ample fee for our town."—J. O. Hutton, Brampton, Ont.

"We do not know anything about the working of the inspection act as yet, except that we had to pay the fee, $25.00. We do not consider that in a town like our own that anything of the kind is necessary. We are obliged to make the rates low and keep the bills small, often cutting the amount down without saying anything about it, as our people really cannot afford to pay but a limited amount as a rule, and we govern ourselves accordingly. From our experience we would say local conditions are quite sufficient to keep business right."—Carleton Place E. L. Co.

Our only competitor here is coal oil, and in order to induce the use of electricity we are obliged to make the price very low—$4 is the highest price we get for 16 c. p. incandescent lamps by the year. We cannot see any advantage to the buyer from government interference. We presume it was to protect him the system was adopted. In our case especially, where no inspection was made, the buyers could not have been benefitted, and even had there been an inspection and we were found to be using a less voltage or amperage than we say we do, we would then simply say, 'That is our price for whatever voltage or amperage it is. So long as the light is satisfactory you can keep it, and when it is not we will take it out."—John Beaman, Chesley, Ont.

"(a) Cannot see that it is. (b) We have not seen the advantage. Fees too arbitrary. Many companies not paying expenses. We find meters after being sealed run slow and some do not go at all. Do not consider that the electric lighting industry has arrived at a point where it requires government inspection."—Sherbrooke Gas & Water Co.

"So far as we are concerned we have seen no beneficial effects

from the act. We were simply taxed $20 for our plants and that is all we ever heard about it. Therefore we consider this inspection fee as only a piece of imposition."—Lakefield E. L. Co.

"(a) Yes, it saves us a great amount of trouble with our customers. (b) Yes, because he can find out whether his meter is correct. The fee is not a fair one. Five dollars is enough. Decidedly the inspection has been productive of a better understanding between seller and buyer—it has settled all disputes, and we think the buyers are more satisfied with their bills. About 10 per cent. of meters run slow; we seldom find one running fast. We are decidedly in favor of some government supervision over all lighting companies—gas as well as electric—but we consider the present scale of fees charged for inspecting electric meters is too high, and in the interest of the consumer as well as the companies should be reduced at least 50 per cent."—The London Electric Co.

"(a) So far as towns and villages are concerned, as they get no returns from it whatever, inasmuch as their meters are never inspected, consequently it cannot be of any benefit, unless it might be where a dispute arises between the buyer and seller, when the inspector might be called in to decide the merits of the case. (b) No. Why should a small village like Eganville, for example, where there are two electric light companies, be compelled to pay $50 to the government as a direct tax and get no returns from it, whereas the city of Ottawa, with one large remunerative plant only pays $25 into the public treasury and yet have the advantage of having their meters inspected regularly. There is no better understanding between buyer and seller because there is really no inspection, and the consumers do not so much as know that there is an inspector. We never had a meter sealed—the only sealing we ever had was the sealing of our $25. My opinion is that a competent inspector should be appointed to whom all matters in dispute between the sellers and consumers should be referred, but other than that I see very little use for one."—A. A. Wright, Renfrew, Ont.

"(a) It is not. (b) No. No better understanding between seller and buyer than before. Meters do not work satisfactorily

OSGOODE HALL, TORONTO.

without periodical inspection. Competition of other methods of lighting is ample to safeguard the interests of the consumer. Government interference is burdensome and unwarranted.—Hamilton E. L. & Power Co.

"We have no inspection at Joliette. We sell our lamps so much a year, and the consumer must be satisfied with the light the corporation furnish."—A. L. Marsolais, Secretary.

"(a) It is no advantage to the seller. The customer compares his bill with the size of his gas bill and cares nothing whether his meter is inspected or not. (b) No, because the companies are not a charitable institution, and the tax must eventually be paid by the consumer. The fee is not a fair one. There is no better understanding between seller and buyer. The customer says he does not care a hang for the meter being inspected; he says, "Give me cheap light or I will go back to gas." He is no better satisfied with his bills than before. Sometimes the meters go and sometimes they don't. This country is too much governed and so are the cities. There is altogether too much interference with the liberty of the subject and most of it is done in the interests of blood sucking parasites of the government, both state and municipal, and not to the advantage of the already overburdened taxpayer."—Toronto Electric Light Co.

"(a) No. (b) No. The fee is not a fair one. No better understanding between buyer and seller. In this small community, where electric light must be nearly as cheap as a coal oil lamp, it is difficult to get a fair price to make a plant pay expenses." — Strathroy E. L. Co.

"(a) No. (b) No. Inspection fee not a fair one. No better understanding between seller and buyer. We have no meters in use. The government should not interfere. I consider the Inspection Act of no use to any person, and it should to my mind be abolished at once.—J. B. Kelly, Blyth, Ont.

"(a) No, as his lights will cost more, therefore he will not be as well able to increase his business. (b) No, because he will have to pay more for his lights. Fee too high in my case, as no inspection has been made. I do not think there is much use for a government inspector—it is only creating a government office which takes a lot of money from owners of electric light plants without any value being given in return. I think with gas and coal oil at present prices, the interests of the consumer are perfectly safe."—D. McIntyre, Paisley, Ont.

"(a) No, it is a disadvantage to the amount of the tax collected. (b) No, as we have to charge more to make up for the tax. It is a useless expense. We do not use meters. We have to sell very low to compete with coal oil. With the various illuminants

been inspected in any way. We have no agreements with private firms, but put in lamps by the month and get a settlement each month. The lamps must all give the best of satisfaction before we can hold the job. Any one that isn't blind can tell whether a lamp is giving enough light by simply looking at it. In my case it has not made one iota of difference, and if you can find one single user of mine asking for inspection, I will give you the amount of the fee. As there was no value given I believe I could resist payment in court."—J. Warner Freure, Port Rowan, Ont

QUEEN'S AVENUE, TORONTO.

"(a) It is certainly no advantage to the seller. (b) It is no good to the consumer. The inspection fee is not a fair one. The consumers are no better satisfied than before. We find the meters slow, and if anything goes wrong with meter there is no way of correcting it, and the station may be out of its just revenue for a long time. We think the lighting rates in Ontario at least very reasonable, and fully 50 per cent. lower than for the same service on the American side. What industry have we to-day that is a poorer investment than electric lighting? We consider the lighting rates low enough without competition, but this same competition will always be the means of keeping the price low enough."—Water, Light & Power Co., Burk's Falls, Ont.

"We do not use meters on our lines and therefore do not figure in the inspection question. We recognize the right of the consumer to be protected from inferior light, both electric and gas, and the consequent duty of the government to furnish such protection. Where there is sufficient competition we believe the necessity for such interference would be greatly reduced—almost to the vanishing point."—Stormont Electric Light and Power Co.

"I know but little of the workings of the Act. The only experience we have had with it is to pay the annual fee or tax. Our meters have not been tested by the Government Inspector, though we got notice that he was to have been up this way last December."—J. W. Schell, Electrician, A. Walker & Sons, Ltd., Walkerville, Ont.

"Government inspection stops all disputes as to correctness of meters, and prevents friction between company and consumer. No inspector has yet been appointed in this district."—Manitoba Electric and Gas Light Co., Winnipeg.

"Our general opinion is that the government should not interfere at all."—Stayner E. L. Co.

"We think the act is all right in cities of 25,000 or more inhabitants. It seems a hardship and a useless one in small communities. Considering that gas inspectors do the work without extra pay, and that meters are brought to them and taken away and current furnished to them free of charge, the fees are too high."—Ottawa Electric Co.

SCENE IN RESERVOIR PARK, TORONTO.

for shop lighting, no man is compelled to use electric light, therefore the seller must make electric lighting an object to the buyer, both in cheapness and efficiency of light."—Hamilton & Proutt, Forest, Ont.

"In my case it was simply highway robbery. No advantage to either seller or consumer. Inspection fee not a fair one. Consumers no better satisfied. No experience with meters. This is a case of sticking a nose in where it was not asked for. Last year I paid a fee under threat of a heavy fine. At the same time I had not a single meter in public use, and the plant has never

1ST HOUR.   2ND HOUR.   3RD HOUR   4TH HOUR.   5TH HOUR.   6TH HOUR   7TH HOUR.

## DIAGRAM OF EXPERIMENTS WITH BOILER COVERINGS.

Reproduced from Canadian Pacific Railway Company's Chart.

[For particulars see next page.]

## EXPERIMENTS WITH BOILER COVERINGS.

ON the preceding page will be found a reproduction of a diagram of experiments with boiler coverings made by the Canadian Pacific Railway Co. to test the values of various compounds as non-conductors of heat.

The order in which the experiments were made, and the materials tested were as follows :—

| | | |
|---|---|---|
| 1st Expt. | Tank uncovered .... .. ......................... .................... | |
| 2nd  " | "  with air-1 ace of ⅞ in. next tank, wood lagging ⅞ in. thick, and outer coat of Russian iron ............................... | H |
| 3rd  " | "  same as in 2nd Expt., but with Asbestos, woven cloth ¾ in. thick, inserted in the ⅞ in. space, and placed next tank... . | I |
| 4th  " | "  covered with Plastic Asbestos Compound, and outer coat Russian Iron Comp. 1½ in. thick ........................ | |
| 5th  " | "  covered with Sectional Magnesia Blocks, and outer coat Russian Iron Comp 1½ in. thick ............................ | G |
| 6th  " | "  covered with Patent Mineral Composition, and outer coat Russian Iron Comp. 1½ in. thick............................ | C |
| 7th  " | "  covered with Plastic Asbestos, taken off C. P. R. Boilers, and outer coat Russian iron Comp. 1½ in. thick............... | D |
| 8th  " | "  with air-space of 1¾ in. next tank, air-tight iron ⅙ at 1/16 in. full thick and outer coat Russian iron Comp. · ½ thick ...... | E |
| 9th  " | "  covered with Patent Mineral Composition...Comp. 1½ in. thick | F |
| 10th  " | "   ".  Mica Boiler Covering......... "   1½ " | J |

The position of the various coverings on the chart may be found by the corresponding letters.

It will be seen that water at 212° was used, the relative value of the coverings as non-conductors being determined by the number of degrees of heat which escaped through the different substances and the consequent cooling of the water in a given time. A reference to the chart shows that ten experiments were made. It will not be necessary, however, to refer to all of them, as the results of some were so unimportant, as in experiments 4, 6 and 9, as to render them of little interest or value. It is only necessary to say that the trials were made as nearly similar conditions as possible, as will be seen by the diagram of the atmospheric temperatures during the tests. The readings were taken from thermometers passed through the coverings and down into the body of the water.

The chart shows the loss of heat in the uncovered tank up to the 5th hour only, and to make a fair comparison the others should be taken for same time. The temperature at beginning of each test was 212°, and the following table shows the temperature at end of fifth hour, the loss in five hours, and the loss in the fifth hour :—

| | Loss in 5 hours. | Temp. at end of 5th hour. | Loss in 5th hour. |
|---|---|---|---|
| Bare tank.................... | 84° | 128° | 11° |
| Asbestos compound.......... | 53° | 159° | 9° |
| Sectional magnesia blocks.... | 33¾° | 178¼° | 7° |
| Wood lagging and air space | " | " | " |
| Asbestos and wood.......... | 30° | 182° | 6° |
| Mica...................... | 20° | 192° | 5° |

The mean temperature of the surrounding atmosphere during the 5th hour may be taken as having been 78°.

The fairest comparison of the merits of the coverings is made by considering the loss of heat in one hour per degree of difference of temperature between the tank and its surrounding atmosphere.

The following table shows this worked out :—

| | Mean temp. during 5th hour. | Difference between tank and atmosphere. | Loss in 5th hour. | Loss in 5th hour per degree of difference of temperature. |
|---|---|---|---|---|
| Bare tank................ | 133½° | 55½° | 11° | .198 |
| Asbestos comp............ | 163½° | 85½° | 9° | .195 |
| Sectional Magnesia blocks.. | 181¼° | 103¾° | 7° | .0674 |
| Wood lagging and air space | 181¼° | 103¾° | 7° | .0674 |
| Asbestos and wood......... | 185° | 107° | 6° | .056 |
| Mica..................... | 194½° | 116½° | 5° | .0428 |

The following table shows the value of the coverings as compared with the bare tank.

Amount of heat which escapes from the bare tank was

| | | | | |
|---|---|---|---|---|
| 1.88 | times greater than through the Asbestos compound. | | | |
| 2.92 | " | " | " | "  Sectional Magnesia blocks. |
| 2.92 | " | " | " | "  Wood lagging and air space. |
| 3.53 | " | " | " | "  Asbestos and wood. |
| 4.62 | " | " | " | "  Mica. |

Mica shows by far the best result as a non-conductor of heat, and saved

| | | | | |
|---|---|---|---|---|
| 245 per cent as much heat as the Asbestos compound. | | | | |
| 157 | " | " | " | "   Sectional Magnesia blocks. |
| 157 | " | " | " | "   Wood lagging and air space. |
| 130 | " | " | " | "   Asbestos and wood. |

It will be seen from the diagram that the loss .by radiation through "sectional magnesia blocks" and "wood and air space" was practically the same, there being less than ½° Fht. between them at the expiration of the test.

It will be seen that asbestos cement, which is in very general use, particularly on marine boilers, showed infinitely the worst results. There seems no room for doubt that this is largely attributable to the fact that it is a solid composition, and lacks one of the most vital requirements of successful non-conductivity, i. e., "diffused air." That the air must be diffused or separated into minute cells is strikingly illustrated in experiments 2 and 3. In the first, wood and air space of ⅞ inch next to the tank, as used on locomotive boilers, the loss per degree of difference of temperature was .0674°. When the same air space was filled or packed with asbestos fibre the loss dropped to .056°. In the case of mica, the air theory appears to have been carried to the furthest possible extent, the whole covering forming a veritable air cushion, each leaf or film of mica being separated from the next by minute corrugations, the whole mat presenting the appearance of a porous flexible quilt. The value of this ingenious arrangement was amply proved in the experiments in question, when the loss per degree in difference of temperature was only .0428°. That this is one of the most important qualities of a covering has long been recognized, and a large number of patents have been granted for devices intended to obtain it. But in nearly every instance it has been at the expense of the material. The great differences in the value of the coverings tested by the C.P.R. is due largely to the manner in which the valuable properties of diffused air as a non-conductor have been utilized.

The following table will give some idea of what the loss of power has been found to be from uncovered steam-pipes with the steam at 75 lb. gauge pressure :—

| | | |
|---|---|---|
| 2 inch pipe.. | 1 horse power loss for every 132 feet long. | |
| 4  " | 1 " " " " | 75  " |
| 6  " | 1 " " " " | 46  " |
| 8  " | 1 " " " " | 40  " |
| 12  " | 1 " " " " | 26  " |

About 90 per cent. of this waste is easily prevented by a proper covering of the pipes. When it is considered that this loss occurs at the comparatively low pressure of 75 lbs., it is apparent that with steam at 130 lbs. and 140 lbs. and higher, the loss becomes very serious, and the necessity for preventing as much of it as possible is a matter of urgent importance.

The accompanying diagram very clearly demonstrates what can be done in this direction by the use of various compositions, and it clearly shows the qualities and capabilities of each. It is possible that the question might arise as to whether the great differences between these substances would still be found had the trials of the C. P. R. Company been made with higher temperature than 212°.

It appears, however, from published reports of trials made some months ago by the engineers of the Boiler Inspection and Insurance Company of Canada that these differences did exist ; that company subsequently issued a special circular on the whole matter, as one of particular interest to steam users. It is stated further that the Grand Trunk Railway Company have lately concluded a series of trials, on a large scale and under high steam pressure, of a number of boiler coverings, including the best of those tested by the C.P.R. and the Boiler Inspection Company, the difference between them being even more marked. As no data, however, is as yet available of these trials, it is impossible to speak of them with accuracy. It is encouraging to notice the increasing attention the whole subject is receiving, in view of the imperative necessity for observing the strictest economy in power and coal and the prevention of all unnecessary waste.

## THE ELECTRIC LIGHT INSPECTION ACT.

To the Editor of the CANADIAN ELECTRICAL NEWS.

SIR,—Referring to your article in the May publication in respect of complaints as to the administration of the Inspection Act, permit me to offer a few observations in reply thereto.

You state that "there seem to exist doubts in the minds of some regarding the necessity for such a system." I can very well believe such to be the case. I remember distinctly when the Weights and Measures Act was first put into operation there was manifested very decided opposition to it by the traders throughout the country, who could see no necessity for the interference by Government with their concerns. And in like manner the gas companies saw no necessity for the law when extended to cover the sale of their product. But where will we find the trader of any respectability to-day, or the gas company that will not say that the legislation in question has been wise and beneficial to their interests?

We have on exhibition in the Department here some specimens of what traders in early days considered proper contrivances as weighing machines. Let me give you one illustration out of a large number. It is intended for a set of scales. The beam consists of an old whiffle-tree, such as may be found lying around in almost any farm yard. The centre or back hook was used for the suspension and the pans were hung from the hooks at either end. The pans are made of two pieces of one inch pine board about one foot square, and are suspended from the whiffle-tree by pieces of rope of varying sizes. The owner of this contrivance doubtless considered the confiscation of it by the Department as a most unwarrantable act—in fact had very serious doubts as to the necessity for the system.

In like manner with the gas. The law required that illuminating gas should be of a certain standard of purity and that a Bunsen burner consuming five cubic feet per hour should produce a light equal to 16 standard candles, and that the measure or apparatus through which it was sold should be accurate. At first it was found most difficult to get the companies to satisfy the requirements of this standard—about 12 c.p. being the best they could then produce. Gradually, however, they did satisfy the demand and now find no difficulty in giving 18 to 20 c.p. to the five cubic feet. It is not claimed, of course, that the Inspection Act is entitled to all the credit for these improved conditions, but we do contend that it was a very important factor in bringing them about.

As with the question of gas and gas meters, so it is with respect to electricity and electric meters. The ordinary weights and measures standards could not be used to verify the apparatus through and by which electricity is sold, consequently it was found necessary to legalize a system of standards and verification suited to the new conditions that existed with respect to this commodity. It is, as has often been explained before, simply an extension of the weights and measures system to cover the sale of electricity. This system has been in operation for twenty-five years, and the Government and Parliament of Canada has not up to the present time received a single petition from any section of the people of the country asking that it be abolished.

As to the methods that have been adopted for the inspection of electric meters and pressure, this is, of course, a legitimate subject for discussion between the electric lighting industry and the Government. It was thought at the outset that the gas inspection service could be used advantageously in carrying out this new work. The Department is still of this opinion. The first year's work has been mainly that of organization.

The inspectors have been confined almost exclusively to the verification of meters at headquarters, thus enabling them to become acquainted with the use of their instruments and the handling of the electric current. During the ensuing year it is intended that each inspector shall visit each electric lighting station at least twice, and oftener if needed, to verify meters, compare voltmeters and test the pressure under full load.

In respect of the registration fees, about which a good deal of complaint has been heard, the writer is free to admit that for the smaller companies it is somewhat high. Acting under this impression in September last the Department made a rebate of fifteen dollars to all companies having installations of 500 lamps and under. Since that time the Department has been considering whether in future this reduction can be extended to all companies having installations of 5,000 lamps and under. This would include all, with the exception of about a dozen of the larger companies.

In connection with the testing of meters it is claimed that meters, after having been sealed by the Government Inspector, have stopped altogether, "owing to the formation of a substance on the commutator." Is it contended that this formation is due to the fact that the meter has been sealed by the inspector and not by the company? Is it claimed that the seller should have access to the meter and that the buyer should be debarred from such access? Clearly, if there is any sealing to be done at all it is the inspector who tests the meter and who is the disinterested party who should do it. There are, no doubt, isolated cases where the direct-current mechanical meter has "slowed up" under heavy load, but to contend that this is a frequent occurrence, and that it needs cleaning every few weeks, is a libel on the meter. If, however, there are difficulties of this kind with respect to this particular meter, let the manufacturers submit their case to the Department, and I venture the assurance that the difficulty will be met if at all possible.

Your May article also charges our inspectors with "posing as electrical engineers." This matter has been carefully investigated and no foundation whatever in fact can be found for the charge. The inspector at Hamilton, it is true, when appealed to by a personal friend on the directorate of the Hamilton Electric Light Co. for his opinion as to the cause of the enormous consumption of fuel under the boilers of the steam plant there, suggested that possibly the chimney was too small, and offered the mechanical engineer of the company the use of certain works or authorities upon the subject. Beyond this he did not go. If this can be called "posing as an electrical engineer" I fear we shall have to plead guilty.

Now a word in conclusion as to the opinions which have been solicited from the companies in reply to a number of leading questions submitted to them; copies of which you have been good enough to send me. Opinions of this sort can always, of course, be very readily obtained, and the Department might meet them—three to one—with appeals from corporations and individuals all over the Dominion, asking for tests and investigations of various descriptions in connection with electric lighting. The statement that there is no dissatisfaction—no friction—between the contractor and the consumer, is one notoriously at variance with the facts. From personal observation the writer is perfectly well aware of this friction, and is also aware that not a few of the companies have experienced much satisfaction from our work.

Whilst testing meters in a small town some time ago we offered to compare the station voltmeter with the Departmental standard, and on doing so it was found that the pressure had been maintained four volts in excess of what the lamps called for. The company had been blaming the lamp dealers with supplying poor lamps, but the trouble was found to be in the inaccuracy of the voltmeter. The manager was well pleased with the test and expressed himself as being quite sure the saving in lamps would pay the registration fee of $10 many times over,

O. HIGMAN.

PUBLISHED ON THE FIFTH OF EVERY MONTH BY

## CHAS. H. MORTIMER,

OFFICE : CONFEDERATION LIFE BUILDING,
Corner Yonge and Richmond Streets,

TORONTO,    -    -    CANADA.
Telephone 2362.

NEW YORK LIFE INSURANCE BUILDING, MONTREAL.
Bell Telephone 2299.

*ADVERTISEMENTS.*

Advertising rates sent promptly on application. Orders for advertising should reach the office of publication not later than the 24th day of the month immediately preceding date of issue. Changes in advertisements will be made whenever desired, without cost to the advertiser, but to insure proper compliance with the instructions of the advertiser, requests for change should reach the office as early as the 2 nd day of the month.

*SUBSCRIPTIONS.*

The ELECTRICAL NEWS will be mailed to subscribers in the Dominion, or the United States, post free, for $1.00 per annum, 50 cents for six months. The price of subscription should be remitted by currency, registered letter, or postal order payable to C. H. Mortimer. Please do not send cheques on local banks unless 5 cents is added for cost of discount. Money sent in unregistered letters will be at senders' risk. Sub-criptions from foreign countries embraced in the General Postal Union $1.50 per annum. Subscriptions are payable in advance. The paper will be discontinued at expiration of term paid for if so stipulated by the subscriber, but where no such understanding exists, will be continued until instructions to discontinue are received and all arrearages paid.
Subscribers may have the mailing address changed a often as desired. When ordering change, always give the old as well as the new address.
The Publisher should be notified of the failure of subscribers to receive their paper promptly and regularly.

*EDITOR'S ANNOUNCEMENTS.*

Correspondence is invited upon all topics legitimately coming within the scope of this journal.

The "Canadian Electrical News" has been appointed the official paper of the Canadian Electrical Association.

## CANADIAN ELECTRICAL ASSOCIATION.

**OFFICERS:**

PRESIDENT :
A. B. SMITH, Superintendent G. N. W. Telegraph Co., Toronto.

1ST VICE-PRESIDENT :
C. BERKELEY POWELL, Director Ottawa Electric Light Co., Ottawa.

2ND VICE-PRESIDENT :
L. B. McFARLANE, Manager Eastern Department, Bell Telephone Company, Montreal.

SECRETARY-TREASURER :
C. H. MORTIMER, Publisher ELECTRICAL NEWS, Toronto.

EXECUTIVE COMMITTEE :
GEO. BLACK, G. N. W. Telegraph Co., Hamilton.
J. A. KAMMERER, General Agent, Royal Electric Co., Toronto.
E. C. BREITHAUPT, Berlin, Ont.
F. H. BADGER, JR., Superintendent Montmorency Electric Light & Power Co., Quebec.
JOHN CARROLL, Sec.-Treas. Eugene F. Phillips Electrical Works, Montreal.
K. J. DUNSTAN, Local Manager Bell Telephone Company, Toronto.
O. HIGMAN, Inland Revenue Department, Ottawa.
W. Y. SOPER, Vice-President Ottawa Electric Railway Company, Ottawa.
A. M. WICKENS, Electrician Parliament Buildings, Toronto.
J. J. WRIGHT, Manager Toronto Electric Light Company.

## CANADIAN ASSOCIATION OF STATIONARY ENGINEERS.

| | |
|---|---|
| President, W. G. BLACKGROVE, | 54 Widmer St., Toronto, Ont. |
| Vice-President, JAMES DEVLIN, | Kingston, Ont. |
| Secretary, E. J. PHILIP, | 11 Cumberland St., Toronto. |
| Treasurer, R. C. PETTIGREW, | 94 West Avenue N., Hamilton. |
| Conductor, W. F. CHAPMAN, | Brockville, Ont. |
| Door Keeper, F. G. JOHNSTON, | Ottawa, Ont. |

TORONTO BRANCH NO. 1.—Meets 1st and 3rd Wednesday each month in Engineers' Hall, 61 Victoria street. W. Lewis, President ; S. Thompson, Vice-President ; T. Eversfield, Recording Secretary, University Crescent.

MONTREAL BRANCH NO. 1.—Meets 1st and 3rd Thursday each month, in Engineers' Hall, Craig street. President, John J. York, Board of Trade Building ; first Vice-President, J. Murphy ; 2nd Vice-President, W, Ware ; Secretary, B. A. York ; Treasurer, Thos. Ryan.

ST. LAURENT BRANCH NO. 2.—Meets every Monday evening at 43 Bonsecours street, Montreal. R. Drouin, President ; Alfred Latour, Secretary, 306 Delisle street, St. Cunegonde.

BRANDON, MAN., BRANCH NO. 1.—Meets 1st and 3rd Friday each month, in City Hall. A. R. Crawford, President ; Arthur Fleming, Secretary.

HAMILTON BRANCH NO. 2.—Meets 1st and 3rd Friday each month at Maccabee's Hall. E. C. Johnston, President ; W. R. Cornish, Vice-Pres.; Wm. Norris, Corresponding Secretary, 211 Wellington street.

STRATFORD BRANCH NO. 3.—John Hoy, President ; Samuel H. Weir, Secretary.

BRANTFORD BRANCH NO. 4.—Meets 2nd and 4th Friday each month. F. Lane, President ; T. Pilgrim, Vice-President ; Joseph Ogle, Secretary, Brantford Cordage Co.

LONDON BRANCH NO. 5.—Meets once a month in the Huron and Erie Loan Savings Co.'s block. Robert Simmie, President ; E. Kidner, Vice-President ; Wm. Meaden, Secretary Treasurer, 533 Richmond street.

GUELPH BRANCH NO. 6.—Meets 1st and 3rd Wednesday each month at 7.30 p. m. J. Fordyce, President ; J. Tuck, Vice-President ; H. T. Flewelling, Re.-Secretary ; J. Gerry, Fin.-Secretary ; Treasurer, C. J. Jorden.

OTTAWA BRANCH NO. 7.—Meet every second and fourth Saturday in each month, in Borbridge's hall, Rideau streett ; Frank Robert, President ; F. Merrill, Secretary, 352 Wellington street.

DRESDEN BRANCH NO. 8.—Meets 1st and Thursday in each month. Thos. Steeper, Secretary.

BERLIN BRANCH NO. 9.—Meets 2nd and 4th Saturday e ch month at 8 p.m. J. R. Utley, President ; G. Steinmetz, Vice-President ; Secretary and Treasurer, W. J. Rhodes, Berlin, Ont.

KINGSTON BRANCH NO. 10.—Meets 1st and 3rd Tuesday in each month in Fraser Hall, King street, at 8 p. m. President, S. Donnelly ; Vice-President, Henry Hopkins ; Secretary, J. W. Tandvin.

WINNIPEG BRANCH NO. 11.—President, G. M. Hazlett ; Rec.-Secretary, J. Sutherland ; Financial Secretary, A. B. Jones.

KINCARDINE BRANCH NO 12.—Meets every Tuesday at 8 o'clock, in Mc Kibbon's block. President, Daniel Bennett ; Vice-President, Joseph Lighthall ; Secretary, Percy C. Walker, Waterworks.

WIARTON BRANCH NO. 13.—President, Wm. Craddock ; Rec.-Secretary, Ed. Dunham.

PETERBOROUGH BRANCH NO. 14.—Meets 2nd and 4th Wednesday in each month S. Potter, President ; C. Robison, Vice-President ; W. Sharp, engineer steam laundry, Charlotte street, Secretary.

BROCKVILLE BRANCH NO 15.—President, W. F. Chapman ; Vice-President, A. Franklin ; Recording Secretary, Wm. Robinson.

CARLETON PLACE BRANCH NO. 16.—Meets every Saturday evening. President, Jos. McKay ; Secretary, J. D. Armstrong.

## ONTARIO ASSOCIATION OF STATIONARY ENGINEERS.

BOARD OF EXAMINERS.

| | | |
|---|---|---|
| President, A. AMES, | | Brantford, Ont. |
| Vice-President, F. G. MITCHELL | | London, Ont. |
| Registrar, A. E. EDKINS, | | 139 Borden St , Toronto. |
| Treasurer, R. MACKIE | | 28 Napier st , Hamilton. |
| Solicitor, J. A. McANDREWS, | | Toronto. |

TORONTO—A. E. Edkins, A. M. Wickens, E. J. Phillips, F. Donaldson.
HAMILTON—P. Stott, R. Mackie, T. Elliott.
BRANTFORD—A. Ames, care Patterson & Sons.
OTTAWA—Thomas Wesley.
KINGSTON—J. Devlin, (Chief Engineer Penitentiary), J. Campbell.
LONDON—F. Mitchell
NIAGARA FALLS—W. Phillips.

Information regarding examinations will be furnished on application to any member of the Board.

**Possibilities of Alternating Currents for Railway Purposes.** In our columns, recently, we have given the ideas of several individuals as to the possibilities of alternating currents for railway purposes. It seems to have been generally conceded that, while there was no reason on theoretical grounds why they should not be used, still for practical reasons, conditions of service would require to be favorable before they could become suitable. It is interesting to observe, however, that quite recently an electric railway working on the three-phase system has been started in the north of Italy, at Lugano. The details of the construction are in many respects so different from those standardized this side of the Atlantic, as to make their study interesting and instructive. The application of the polyphase system itself to railway work, is so far as we are aware, quite the first instance of the kind except for experimental purposes. Power is taken from a water-fall seven and a half miles from Lugano. This fall already supplies power for the electric lighting of the town by the alternating system, and for other electrical purposes. The three-phase generator is of 150 h.p. The voltage is 5,000, generated direct without the assistance of step up transformers. This is in itself a feature of importance, as thereby the inevitable losses in transformation are avoided, and is rendered possible by the design of the generator, which is of the stationary armature revolving field type. The frequency of the alternations is 40, considerably less than the number usually adopted in this country. As all induction, capacity, and hysteresis losses increase directly with the frequency of alternations, the reduction of the frequency is a matter of engineering importance. It is true that below about 60 cycles per second the flickering of lights affects the eye, but as the real field for polyphase alternating currents is for power transmission and utilization, this consideration loses considerably in importance. On the railway line the currents are conducted by means of two over-head wires, and the rails, which are bonded together in the usual manner. Another departure from our usual practice is in the matter of the power of the motors. The cars have a seating capacity of twenty-four ; the grades on the line run as high as six per cent.; the speed averages nine miles per hour ; and yet each car is equipped with only one twenty-five horse power motor. In many ways, therefore, the line is worthy of being carefully studied, and might serve as an object lesson to those whose engineering ideas have a tendency to fall into grooves.

---

**Governing Features of Electric Installations.** THE absence of really proper and competent consideration of the general features and preliminary engineering of electric plants has been prominently brought forward in two recent cases. A town of 6,000 inhabitants, long and narrow, with scattered houses, was supplied with lights on the direct current system ; and a small place of nearly 1,000, concentrated round the railway station, was wired upon the alternating plan. In the former case, a house distant almost 15,000 feet from the power house had to be supplied, and was reached by the ingenious though complex method of transmitting at 220 volts by using the outside wires of the three-wire system, and then reducing down to the lamp voltage of 104 by availing of the additional drop caused by a water resistance. The copper required for feeders, mains, etc., was a very large amount, and the plant did not pay. In the latter case, the voltage being high and the distance small, it was impossible, without using wires too small to put on poles, to keep the voltage at the lamps down enough, and they kept on burning out. And besides this, the resultant cost of generator, transformers, etc., was considerably higher than the price of a direct current installation would have been. Electrical investors should remember that the selection between alternating current and direct current machinery is not a matter which should be decided purely on grounds of personal preference, but it should rest on well considered engineering and commercial considerations. The alternating current system was evolved out of the inadequacy of the direct current system to properly meet certain conditions, but the fact that its use is most advantageous in those circumstances, is no reason for thinking it to be the best under all. Other things being equal, the direct current has the best of it on the score of efficiency, and as for lighting or power, the one possesses no advantage over the other (granting the use of alternating current motors, and disregarding the possibility of using storage batteries). Their comparative merits seem therefore to reduce to a question of comparing the costs of the two systems. Referring everything to the standard of two-wire direct current working, we find that to transmit the same total power, at the same total loss, over the same distance, the direct current two-wire takes 100 per cent. of copper ; the direct current three-wire takes 37½ per cent.; the alternating system at 1,000 volts takes about one per cent. So the alternating system has the advantage over the three-wire direct by 36 per cent. But in order to do this we have to use transformers, which may quite counterbalance this advantage, unless the distances be great. And this does not take into consideration the extra losses due to the transformers. In a 1,000 light installation, the cost of the transformers will be certainly not less than $800, so

that unless the total diminution of first cost of wire by the use of the higher voltage be greater than $800, the direct current is actually the cheaper system to use. Even if the saving in wire amounts to $800, the direct current will be the more economical system, because the transformers themselves have many inherent losses due to hysteresis, leakage, and heating ; and these will necessitate an expenditure for fuel that will be obviated by the direct current. The above considerations show that every lighting or power enterprise should be very carefully considered before deciding on the class of machinery to use.

---

**Lightning Arresters.** IT would be interesting to learn what effect the recent violent thunderstorms have had in directing the attention of owners and superintendents of electric light and power stations to the necessity of adopting devices for the protection of their machinery from lightning. It may possibly be but a coincidence that several enquiries regarding lightning arresters were received at this office immediately following the recent storms. The electric stations appear to have come through these storms with very little loss, but this is not likely to happen on every occasion, and protective devices should be regarded as one of the most important features of the equipment of every electric station.

---

**Duty on Steel Rails.** WE may shortly look for a final decision regarding the proper interpretation of that section of the Canadian tariff relating to the duty on steel rails. The Privy Council will shortly adjudicate upon the question in response to an appeal taken by the Toronto Railway Co. from the decisions of the inferior courts, under which it was held that steel rails of the weight now mostly used in electric railway construction are subject to duty. The Toronto Railway Co. are seeking to recover a sum exceeding $50,000 paid as customs duties several years ago, in accordance with this interpretation of the tariff. The decision of the Privy Council will be awaited with much interest by electric railway companies throughout Canada, and by the projectors of new roads, whose interests are affected.

---

**The Electric Railway Accident at Victoria.** THE dreadful occurrence at Victoria, B. C., on May 26th, is we believe the first accident on an electric railway in Canada in which more than a single life has been lost, or in which injury resulted from the cars leaving the track. Since electric railways have come to occupy in many instances the same position as steam roads, greater precautions than heretofore will be required in their construction and operation. The main cause of the accident at Victoria seems to have been the want of a sufficient factor of safety in the bridge over which the cars were required to pass. Had the proper amount of surplus strength been allowed in the construction of this bridge, it probably would not have collapsed in consequence of the cars jumping the track. On the other hand, the cars seem to have left the rails because of being rendered top-heavy by the overload of passengers seated on the top and clinging to the sides of the cars. Railway bridges of all kinds should be subjected to periodical inspection, and managers of electric railways should not allow passengers to overcrowd their cars in the manner described.

## ONTARIO ASSOCIATION OF STATIONARY ENGINEERS.

At the annual meeting of this Association held at Galt on the 1st inst., representatives of the various lodges throughout the province were present, from whom encouraging reports were received.

Officers were elected as follows :—A. Ames, Brantford, president ; T. S. Mitchell, London, vice-president; R. Mackie, Hamilton, treasurer ; A. E. Edkins, Toronto, registrar. Board of Examiners—James Devlin, Kingston ; J. W. Bain, Toronto ; W. Donaldson, Ottawa ; Wm. Stott, Hamilton.

A resolution was passed · authorizing careful enquiry by the officers of the Association into the causes of boiler explosions.

A delegation consisting of Mr. James Devlin, Kingston, Mr. A. M. Wickens, Toronto, and Mr. A. E. Edkins, London, was appointed to interview Sir Charles Tupper in behalf of Dominion legislation for the examination of engineers in all parts of the Dominion.

The next meeting will be held in Toronto on the 6th of June, 1897.

## A CANADIAN MANUFACTORY OF ACETYLENE GAS.

Within the past month Mr. T. L. Willson, the inventor of acetylene gas has commenced the erection of a factory at St. Catharines, Ont., in which to manufacture the gas. Judging by the size of the factory it is not proposed to conduct operations on an extensive scale. It is reported that Mr. Willson has entered into a contract to supply the St. Catharines Gas Company with a certain quantity of acetylene for enriching purposes, and that his purpose is to endeavor to effect a similar arrangement with the gas companies throughout the Dominion.

## PERSONAL.

Mr. E. Carl Breithaupt, of Berlin, was recently elected a member of the American Society of Electrical Engineers.

Mr. H. P. Brown, consulting engineer, of New York, has been engaged by the Hamilton Radial Railway Company.

Mr. W. J. Camp, electrician of the C. P. R. Telegraph Co. was a recent visitor to New York, where he had a conference with the Postal Telegraph Co.

Mr. A. W. Congdon, of the General Electric Company's staff, has been confined to his home by illness for a couple of months past. His friends will be pleased, however, to learn that he is now on the way to recovery and hopes to be able to resume his duties shortly.

Mr. Charles H. Wright, son of Mr. A. A. Wright, of Renfrew, Ont., was eminently successful in his recent examinations at McGill University, Montreal. He took honors in electrical engineering, hydraulics, thermodynamics and physics, and headed the list of passmen in the electrical engineering branch.

Among the visitors from Canada to the recent convention of the National Electric Light Association in New York, were Frederic Nicholls, manager G. E. Co., Toronto; Mr. W. H. Brown, manager Royal Electric Co., Howard D. Black, Prof. Henry T. Bovey, Montreal ; F. H. Badger, Jr., manager Montmorency Light & Power Co., Quebec ; and Chas. B. Hunt, manager Electric Light & Power Co., London, Ont.

Mr. W. C. Cheney has been appointed general superintendent of the Consolidated Railway and Light Company, Victoria, B. C. He will exercise a general supervision of the system in Victoria, Vancouver and New Westminster, making his headquarters at Victoria and giving the tramway and lighting system his personal attention. Mr. Cheney is an electrical engineer of recognized ability, and resigned as superintendent of the Portland General Electric Co. to accept his present position. He has been connected with some of the largest electric power plants of America.

## TRADE NOTES.

The Regina Electric Light & Power Co. have ordered a 500 light alternator from the Canadian General Electric Co.

The Beardmore Co., of Toronto, are putting a new 100 h. p. Goldie & McCulloch engine in one of their tanneries at Acton.

The Rogers Electric Co., of London, have been awarded the contract for an electric light plant at the London water works.

Mr. John Crowe, of Montreal, has placed an order with the Babcock & Wilcox Company, of Montreal, for one pair of their 250 h. p. improved all wrought steel high pressure water tube boilers.

The Canadian General Electric Co. are installing a very compact marine lighting set on the new steamer "Corona." The generator is a standard multipolar 250 light machine direct coupled to a vertical marine type engine constructed at the Peterboro' works of the company.

The Canadian General Electric Co. are supplying a very complete isolated plant for the Montreal General Hospital consisting of two generators of 40 kilowatts capacity each, and one of 17 kilowatts capacity. These machines will be of the company's new moderate speed multipolar type.

The Packard Electric Co.· are issuing monthly in miniature form a little memorandum book entitled "Daily Notes" containing blanks for memoranda for each day of the month. The last four pages are devoted to the company's advertisements. The idea is a unique one, and will, no doubt assist in widening the circle of the company's acquaintance.

The Kay Electric Mfg. Co., of Hamilton, have recently supplied the following machines :—A 10 k. w. dynamo for the Gendron Mnfg. Co., Toronto ; an electro-plating dynamo for the Ontario Silver Plating Co.'s branch at Niagara Falls, N. Y. ; a 20 k. w. lighting dynamo for the Eagle Knitting Co., Hamilton ; a 30 h. p. motor for Wideman & Clemens, Guelph ; a 4 h. p. motor for Gemmel's laundry, Hamilton ; a 6 h. p. motor for the McLean Pub. Co., a 4 h. p. motor for Mr. Carlisle, and a 5 h. p. motor for Wm. Beers, all of Toronto.

The Montreal Street Railway Company have placed an order with the Babcock & Wilcox Company, Board of Trade Building, Montreal, for three batteries of their improved all wrought steel high pressure water tube boilers. These boilers are intended to furnish steam for the new 4,000 h. p. engine which they have ordered for their William street power house extension. Probably no more perfect installation of Lancashire or Galloway boilers has ever been made, certainly not in Canada, than the present extensive plant in the William street power house. The showing made by the Babcock & Wilcox boilers proved too attractive, however, and the management of the Montreal Street Railway have determined that their plant will be provided with just as good machinery as the great street railway plants in the United States.

Mr. Wm. T. Bonner, general Canadian agent for the Babcock & Wilcox water tube steam boilers, furnishes the following information regarding the geographical distribution of their orders as given in their last month's sales report : American sales—3,374 h. p. in New York, 560 h. p. in Pennsylvania, 400 h. p. in Florida, 169 h. p. in California, 3,350 h. p. in Massachusetts, 250 h. p. in New Jersey, 200 h. p. in Rhode Island, 6,400 h. p. in Illinois, and 6,880 h. p. in Maryland. The foreign sales were divided as follows : 1,322 h. p. in Russia, 74 h. p. in Norway, 1,353 h. p. in France, 3,453 h. p. in England, 140 h. p. in Belgium, 294 h. p. in Spain, 76 h. p. in Portugal, 126 h. p. in Scotland, 192 h. p. in Germany, 72 h. p. in Holland, 424 h. p. in South Africa, 332 h. p. in Sweden, 52 h. p. in Italy, 280 h. p. in Brazil, and 228 h. p. in Madagascar. The foreign list foots up to 8,418 h. p., while the American list amounts to 21,583 h. p., or a total of 30,001 h. p. for the month, of which 7,000 h. p. are marine boilers. The above report indicates only an average month's business. The total number of Babcock & Wilcox boilers now in use aggregates nearly 2,000,000 horse power.

The Lachine Rapids Hydraulic and Land Company have selected the three-phase system of the Canadian General Electric Co. for their new transmission plant. The initial order for the generators covers 12 machines, each of 1,000 horse-power capacity. This will, with one exception, be the largest power transmission plant in the world. A full description of the details of this most interesting installation will be given to our readers in an early issue.

" RADIANT MATTER " as reflected in the countenances of members of the Canadian Electrical Assoriation who contemplate attending the Toronto Convention.

## CANADIAN ELECTRICAL ASSOCIATION.

### PARTICULARS OF THE APPROACHING ANNUAL CONVENTION.

In the May number of the ELECTRICAL NEWS reference was made to the Sixth Annual Convention of the Canadian Electrical Association which is to take place in Toronto on the 17th, 18th aud 19th inst. The arrangements for this convention are now practically complete, and we are enabled to publish herewith the exact program, as follows:

HEADQUARTERS—COUNCIL CHAMBER, BOARD OF TRADE.

BUSINESS PROGRAM.

JUNE 17TH.

2:30 P. M.　Opening of first session in Council Chamber, Board of Trade Building, Yonge and Front Streets.
President's Address.
Reading Minutes of last Meeting.
Secretary-Treasurer's Report.
Reports of Committees.
General Business.
Presentation of Papers.
Discussion.

JUNE 18TH.

10:00 A. M.　Consideration of Reports of Committees.
Election of Standing Committees.
Selection of Place and Time of next Meeting.
Election of Officers and Executive Committee.
General Business.
Presentation of Papers.
Discussion.

JUNE 19TH.

10:00 A. M.　Presentation of Papers.
Discussion.
General Business.

LIST OF PAPERS.

"Ocean Cables," (Historical),　　　　Chas. P. Dwight, Toronto.
"Acetylene Gas," (with demonstrations),　　Geo. Black, Hamilton.
"Meters,"　　　　　　　　　　　James Milne, Toronto.
Consideration and Discussion of the Government Electric Light Inspection Act.
"Some Central Station Economics,"　　P. G. Gossler, Montreal.
"Power Transmission by Polphase E.M.F.'s."
　　　　　　　　　　　　　　Geo. White Fraser, Toronto.
"Operating Engines without a Natural Supply of Condensing Water,"
　　　　　　　　　　　　　　E. J. Phillip, Toronto.
"The Outlook for the Electric Railway,"　F. C. Armstrong, Toronto.
Several of these papers will be illustrated by electric projection of diagrams, and the interest thereby greatly enhanced.

SOCIAL FEATURES.

JUNE 17TH.

8:00 P. M.　Members and ladies will attend an illustrated lecture by Mr. James Milne entitled "Radiant Matter," to be delivered in the Rotunda of the Board of Trade, showing Prof. Cook's experiments and also demonstrations of Roentgen rays. Interesting shadowgraphs will be taken and exhibited.

JUNE 18TH.

5:00 P. M.　Excursion per Steamer "Greyhound" to Lorne Park. Annual Banquet to members and ladies at Hotel Louise, followed by moonlight sail on Lake Ontario, returning to Toronto about 11 P. M.

JUNE 19TH.

Arrangements are being made for an excursion by boat around Toronto Island, along the water front to Scarboro' Heights, and return.

NOTE.—During the progress of the convention opportunity will be afforded for inspection of the power stations of the Toronto Electric Light Co. and the Toronto Railway Company.

The subjects of the papers to be presented at this convention appear to have been wisely selected. Most of the questions treated of are destined to have an important bearing upon the electrical interests, and their full consideration and discussion at the present juncture is most desirable. We have every confidence in the ability of the gentlemen who have undertaken to prepare these papers to make them both interesting and instructive. We are quite in accord, however, with the view expressed by some distinguished gentleman whose name we cannot recall, that the chief value of papers of this character lies in the discussion arising out of them. If, therefore, the full benefit is to be got from this interesting series of papers, members of the Association must familiarize themselves with the views expressed by reading and carefully digesting the advance copies which are usually available some days previous to the meeting, and come to the convention prepared to express their opinions and do their share towards promoting a full and profitable discussion.

Take for example Mr. Gossler's paper on "Some Central Station Economics." What more interesting or important subject than this to every owner and manager of a central station? If central station men will make a point of attending the convention, take a hand in the discussion of this paper, and compare notes with each other, a fund of information regarding central station management of the greatest benefit to every central station manager will be the outcome It is due likewise to the author of a paper on which much effort has been expended, that his audience should express themselves regarding the correctness or otherwise of the views he enunciates.

There has been manifest at past conventions of the Association hesitation on the part of the majority of the members to express their views. We hope to see this spirit disappear at future meetings. The members will better advance their own interests and the welfare and usefulness of the Association by hazarding their opinions,

Barring " Bug "-bears such as bad roads, punctures, etc., this contingent of members of the Canadian Electrical Association expect to arrive in Toronto at sunrise on the 17th inst.

whether correct or otherwise, than by refraining from taking part in the proceedings.

Opportunity is to be given for a discussion upon the merits of the Electric Light Inspection Act. Judging from the correspondence published in this number, this is a subject in which central station men feel a keen interest, and upon which large numbers of them feel competent to express their opinions. We accordingly look for a good representation of the electric lighting interests, and a lively discussion on the inspection system.

Regarding the social features of the program, they appear to be of a somewhat different character from those of previous occasions, and are commendable on account of their variety in this respect. There is little doubt that they will prove as enjoyable as could be desired. Toronto is at all times an attractive and interesting city, and at no time does she appear to better advantage than in "leafy June." In support of this statement we present in this issue a number of views of her public buildings, parks, etc.

It is hoped and expected that members of the Association, a large proportion of whom reside within one hundred miles of Toronto, will seek relaxation from the arduous duties of the political campaign by coming to Toronto on the 17th inst. and assisting in making this the best convention ever held under the auspices of the Association.

The feed water for a steam boiler should not be introduced in such a way as to allow it to strike against the shell, but should be discharged into the body of water already there.

When a watch becomes magnetized by intimacy with a dynamo, the mainspring, being tempered, becomes a magnet, and, as it unwinds, its attraction, varying in direction or intensity, causes the rate of escapement to vary at different hours of the day.

A writer to a contemporary suggests a means of repairing the insulation of commutators. He says he has found that litharge mixed to the consistency of thick cream with glycerine and applied to the cracks of the commutator, restores the insulation. After being applied it should be allowed to dry 10 or 12 hours and then filed down level with the top of the commutator segments.

TORONTO BOARD OF TRADE BUILDING,
Headquarters of the C. E. A. Convention, June 17, 18, 19, 1896.

## SPARKS.

A new metallic telephone line will be built between London and Sarnia.

The Woodstock, N. B., Electric Light Company have recently added two large dynamos.

The Hull Electric Company have purchased thirty acres of land at Aylmer as a site for a park.

Debentures are being offered by the village of Kaslo, B. C., to raise funds for an electric light plant.

The ratepayers of Iroquois, Ont., have voted down a proposition to light the streets by electricity.

The Hull Electric Company are erecting poles and stringing wires for the electric lighting of Hull, Que.

Cunningham & Hinton, electricians and electrical goods, Victoria, B. C., have dissolved partnership. G. C. Hinton continues.

Mr. R. Anderson, of Ottawa, is installing an electric light plant for Mr. Geo. Brigham on his passenger boat which runs between Ottawa and Hull.

The City Council of St. Catharines, Ont., has invited tenders for electric street lighting for a period of five years from 31st October next.

A company seeking incorporation is the Callendar Telephone Exchange Co., of Brantford, Ont., to manufacture telephones and electrical apparatus.

The suit of the town of Three Rivers, Que., against the Royal Electric Company, Montreal, has been dismissed with costs. The case involved an amount of $13,000.

Dr. W. W. Jacques, of Boston, an electrician, connected with the Bell Telephone Company, announces the discovery of a method of taking electrical energy direct from coal.

On May 12th the town of Perth, Ont., granted a bonus of $5,000 for an electric railway between that town and Lanark, the road to be running by September 1st. The distance is twelve miles.

The Victoria Telephone Co. is applying for incorporation, with a capital stock of $25,000, to operate in Victoria county, N. B. Among the promoters are J. E. Porter, Andover; Albert Brymer, Perth Centre; and Stephen Scott, Bairdville.

The rights of the Standard Light & Power Co. to lay wires underground in the streets of Montreal have been transferred to the Citizens' Light & Power Company, which is controlled by the Lachine Rapids & Hydraulic Company.

New Westminster and Vancouver, B. C., are now connected with Chilliwack by telephone, through the enterprise of the New Westminster and Burrard Inlet Telephone Co. The distance is seventy miles. The line used for the connection is the old telegraph wire, which the company have taken over. The manager, Mr. H. W. Kent, states that tariff sheets are now being prepared, and tolls for conversation will be very reasonable. Message rates will be the same as formerly by telegraph. The new line will be equipped with the best and most improved long distance instruments.

## THE N. E. L. A. CONVENTION.

THE recent convention and exhibition in New York City under the auspices of the National Electric Light Association of the United States, appear to have been attended by more than ordinary success. The proceedings of the convention were about on a par with those of previous meetings of the Association. On the other hand the exhibition surpassed anything of the kind previously attempted, excepting only the electrical display at the World's Fair. Side by side with the electrical apparatus of to-day were to be seen 'the rude devices employed when electricity was in reality "in its infancy." The extent of the public interest awakened by this exhibition, may be judged by the fact that on some days the attendance reached the high figure of 50,000.

As foreshadowed in the ELECTRICAL NEWS for May, Mr. Frederic Nicholls, manager of the Canadian General Electric Co., Toronto, secured election as President of the Association for the ensuing year, an honor of which that gentleman and Canadians generally have cause to feel proud. Mr. Nicholls, whose portrait is herewith presented, is too well known on this side of the line to require any extended reference in these pages. He had his birth place in England forty years ago, and in that country and Germany received an efficient education. Upwards of fifteen years ago he became a resident of Toronto, where step by step he has risen to the position of prominence he now occupies in the business world.

Mr. Nicholls may be classed among the pioneers in the electrical business in Canada, having organized the Incandescent Light Co., of Toronto, and held the management thereof until a few months ago. He is also a director of the Toronto Electric Light Company ; president of the Brantford Street Railway Company ; vice-president of the Peterborough Street Railway Company ; secretary of the London Electric Light Company, a director of the Toronto and Scarboro Railway Company, and in addition is interested in mining, insurance, publishing, and other enterprises.

MR. FREDERIC NICHOLLS,
President of the National Electric Light Association.

## TECHNICAL SCHOOL EXAMINATIONS.

THE examinations in "Electricity" and "Steam and the Steam Engine," at the Toronto Technical School were quite successful, a number of the students obtaining a high percentage of marks. The following are the names of the succcessful ones, in order of merit :

Electricity—R. C. Harris, Herbert S. Small, equal ; Walter Inglehart, Walter Redpath, Alex. Rose, H. Amos, H. F. Hutchison, Thos. P. Marshall, E. Harris,

H. C. Champ, Fred. J. Grant, Wm. Simpson, Wm. Willis, W. Piercy, Alex. Gerry, Sam. J. Evanson.

Steam and the Steam Engine—R. H. Johnston, Walter Inglehart, W. Piercy, Walter Redpath, H. C. Champ, H. Amos, Wm. Simpson, Alex. Gerry, Fred. J. Grant, J. Mitchell.

## PRESERVING TELEGRAPH POLES.

IN the preparation of posts for the telegraph service in Sweden, the following simple, effective and cheap method of preserving wood from decay is said to be employed : A square tank, having a capacity of some 200 gallons, is supported at a height of 20 feet or 25 feet above the ground by means of a light skeleton tower built of wood. A pipe drops from the bottom of the tank to within 30 inches of the ground, where it is connected with a cluster of flexible branches, each ending with a cap having an orifice in the centre. Each cap is clamped on to the larger end of a pole in such a manner that no liquid can escape from the pipe except by passing into the wood. The poles are arranged parallel with one another, sloping downward, and troughs run under both ends to catch drippings. When all is ready, a solution of sulphate of copper, which has been prepared in the tank, is allowed to descend the pipe. The pressure produced by the fall is sufficient to drive the solution, gradually of course, right through the poles from end to end. When the operation is ended, and the posts dried, the whole of the fibre of the wood remains premeated with the preserving chemical.

Prof. Palaz gives the following figures as to the relative heating effects of different illuminants : Arc light, 4 ; incandescent light, 14 ; kerosene, argand burner, 33½ ; gas, argand burner, 380 ; candle, 473 ; gas, butterfly burner, 511.

SOME EXCEPTIONS TO OHM'S LAW.—It has been observed, says the London Electrical Review, that amongst the liquids there are certain of low conductivity ; for example, benzine, xylene, and turpentine, which do not seem to follow Ohm's law, but which, under the continued influence of a high electromotive force, show a gradual alteration in conductivity. These liquids also exhibit the phenomenon of electrical convection, a current of the electrolyte setting in from the one electrode, whilst the other appears simply to attract the repelled liquid. Recently, Emil Warburg has been investigating these phenomena. He employed mixtures of liquids which possessed low conductivity, gradually reducing the proportion of one of the constituents until the conductivity was nearly that of the other. Such mixed solutions as these were found still to exhibit the above phenomena. The behaviour of these solutions, in fact, was such that Warburg is led to the conclusion (vide Ann. Phys. Chem., 1895 [2], liv., pp. 396—433), that they contain an electrolyte in a state of great dilution, upon which their conductivity depends. He suggests that the extraordinary behaviour of the so-called pure liquids is capable of a similar explanation.

# CORRESPONDENCE

## A WARNING TO INVENTORS AND PATENTEES.

To the Editor of the CANADIAN ELECTRICAL NEWS:

SIR,—In a recent issue of the Scientific American, a matter was brought before its readers, which, from its importance, deserves to have all the publicity which can be given it, and from that paper are taken the quotations in this letter. There exists in the United States, and in Canada too, unfortunately, a class of men who have adopted the opinion that the inventor is made to be victimized and "who try their best to exploit the community of patentees for their own benefit and the consequent detriment of their clients."

"When letters patent are awarded, the drawings and claims of the patent and the inventor's name are published in the Official Gazette of the United States patent office. This appeals at once to a large number of sharks, calling themselves "patent agents" who see in the inventor a possible source of revenue. As soon as the patent is issued the inventor therefore begins to receive letters from various self-extolled concerns, recommending him to do various things, to apply for foreign patents, or to permit the correspondents to act as his agents for the sale of his patent on commission.

"Many of these letters and circulars contain statements that are absolutely fraudulent. The inventor, for example, will be urged to apply for foreign patents in England, France and Germany, and other countries, when the agent is perfectly well aware that after the patent has issued in the United States and has been published in the Patent Office Gazette, valid patents cannot be procured in those countries, except under the International Convention, which he is seldom able to avail himself of. The patent shark relies upon the ignorance of this fact on the part of the inventor to protect him in his nefarious traffic. He is also protected from detection by the fact that in many foreign countries there is no examination as to novelty, and in due course, and after the payment of the government fees, the patent will issue and he will be provided with the letters patent certificate to present to his "client" who sleeps in blissful ignorance of the fact that the documents are not worth the paper they are printed on.

"In many cases the fees upon examination will be found to be phenomenally low and the inventor will snap at what seems to him a bargain, simply to find that in Germany, perhaps, he has procured a gebrauchmuster, or model of utility patent, instead of a patent; or in Canada he may be led to believe that he has procured a patent for one year when he has simply filed a declaration of intention, which affords no true protection."

Many of the circulars sent out are artfully worded to convey the idea that the invention was accidentally come across and that its value was immediately apparent.

Inventors are usually sanguine and frequently fall into the trap so cunningly set.

Once an inventor falls into the hands of one of these firms he is exploited to the best advantage, and remarkably well plucked as many have found to their sorrow.

The moral of all this is—do no business with any firm issuing circulars tending to inflate the hopes of the patentee; have no dealings with any firm offering to sell a patent and asking for an advance, and in all patent matters consult only solicitors of good standing and proved integrity.

If Canadian inventors will heed the warning given by those in a position to give advice, they will save themselves much annoyance and much hardly earned money; at the same time the much-desired result will be attained of rendering Canada an unprofitable fishing ground for Yankee patent sharks.

Yours, etc.

RIDOUT & MAYBEE.

## A BOILER TEST.

BEFORE the Montreal Association of Stationary Engineers Capt. Wright presented a report of a boiler test recently made by him in that city, with a few prefatory remarks, as follows :

Before reading this report, I will make a few words of explanation. About two weeks ago I received a letter from a party unknown to me, requesting me to make a test of the steaming performance of a boiler in this city. I have met so many schemes to burn smoke and make boilers do impossibilities, that I am apt to look at new-fangled designs with suspicion. Yet it must be acknowledged that great improvements are possible. The fact that for every ton of coal you burn on a furnace grate, at least 16 tons of air go up the chimney at a temperature of 500 or 600 degrees, by which one-fifth of the total heat in the coal is absolutely lost, except in so far as it induces a draft, proves that at least in that direction, improvement is possible.

It is unfortunately the case, that this subject is often attacked by men who have not studied the question, and, moreover, are in ignorance of the usual practical methods of dealing with it, as far as the burning of wood or coal in a furnace is concerned. It was for these reasons that at first I had no intention of doing anything in connection with this proposed test. But, hearing who the interested parties were, I called on them, and decided to act.

The circumstances under which the test was made were uncommonly unfavorable—a rainy day, a leaky roof, a very poor article of coal, completely saturated with water. When I went there in the morning, no preparations for a test had been made. A Fairbanks scale was got, coal was weighed, a water barrel was mounted on the scale, and connections made from the service pipe to barrel, and from the barrel to injector. This took time, and at 10 a. m. the test began. At noon, after stopping for the hour, I ran over my notes and was surprised at the results. In conformity with the daily custom, the boiler tubes were swept out during the noon hour. What was then accomplished is contained in the report, which I will now read.

D. L. DWINNELL, ESQ., City.

DEAR SIR :—According to your instructions, a test was made on the 26th ult., of the steaming qualities and general behavior of the boiler in the works of "The Montreal Toilet Supply Co." on Dorchester street, in this city.

The boiler is of the common cylindrical type with return tubes. The shell is 42″ dia. and 11 feet long, with thirty-nine 3″ tubes, and a grate area of 13½ square feet. The principal object of the test was to determine whether certain peculiarities introduced in the setting, increased the efficiency of the furnace, and the steaming qualities of the boiler. The test was made under the usual every day conditions of work and began at 10 a.m. ending at 4.50 p.m. Everything was taken as found ; no change was made or cleaning done previous to the start. A bituminous lower port coal was used and was an inferior quality of what is known in the trade, as "Run of the mine." City water at a temperature of 34 degrees was alone fed to the boiler by an injector ; the overflow water, in starting the injector, was returned to the barrel it came from. This barrel was mounted on the platform of a "Fairbank ' scale with water connections to and from. The change, or weight of water running to each barrel was the same, and the time when filled was carefully noted. The total weight of water fed to the boiler between 10 a.m. and 4.50 p.m., was 6864 lbs. Four barrels of dry coal was dumped in separate heaps on the floor, each heap weighing 238 lbs., a total of 952 lbs. consumed between the hours above named, and the time each heap lasted was carefully observed. In this manner any variation or irregularity is observed at once.

At the close of the test the ashes weighed 90 lbs. Deducting this from the weight of coal laid on the floor, leaves 862 lbs. of combustible or pure coal consumed between the above hours. During the noon hour, the boiler furnishes steam to the coil in the drying room and mangles, and 209 lbs. of water was fed to the boiler, and 52 lbs. of combustible consumed on the grate. This performance forms no part of the test in working hours, for if retained, would destroy the working hours, and these amounts of fuel and water must be subtracted from the total between 10 a.m. and 4.50 p.m. and takes this form : Total combustible consumed on the grate between 10 and 12 a.m. and 1 and 4.50 p.m., 810 lbs. Total water fed to the boiler during the same time, at a temperature of 34 degrees, 6655 lbs., which is an evaporation of 8.216 pounds of water from a temperature of 34 degrees to steam at seventy-five by gauge, per pound of combustible burned on the grate, equivalent to an evaporation from and at 212 degrees, of 10.02 pounds of water per pound of combustible. This is the mean performance of the boiler during the working time of the test.

The gauge was kept very regular at seventy-five. The total absence of black smoke from the stack was remarkable. At times when fresh coal had been put on the grate, a thin grey smoke made an appearance for a short time, and generally on looking at the top end of smoke stack there was no visible proof that there was a fire under the boiler. At the close of the test the water level in the glass was the same as at the beginning. The uniform rate at which water disappeared from the boiler was surprising. During the working hours, both forenoon and afternoon, if 209 lbs. of water had been regularly fed every ten minutes to the boiler, the water level in the glass would have been practically the same during the working time of test.

But an unexpected change did take place. The thirty-nine tubes in the boiler were cleaned in the noon hour, and it was observed between 1 and 2 p.m. that the consumption of fuel was forty-seven pounds less than in any sixty successive minutes during the forenoon.

The method of conducting the test permitted this comparison. At first I thought there was an error, but if so, I failed to detect it, and it continued at the same rate up to the close of the test, notwithstanding the weight of water at a temperature of thirty-four degrees supplied to the boiler per hour, was practically the same both forenoon and afternoon.

Between 10.50 and 11.50 a.m. 1236 lbs. of water were fed to boiler and 174 lbs. of combustible was consumed on the grate. Between 2 and 3 p.m. 1237 lbs. of water were fed to boiler, and 127 lbs. of combustible consumed on the grate. This is at the rate of 9.74 pounds of water at a temperature of thirty-four degrees evaporated to steam at seventy-five by gauge, per pound of fuel burnt on the grate :—equivalent to an evaporation from and at 212 degrees, of 11.87 pounds of water per pound of combustible burned on the grate.

The ashes were 9.45% of the weight of coal, and in conformity with all reliable and comparable tests of boilers, the standard results are calculated from the weight of combustible, and the equivalent evaporation from and at 212 degrees.

The boiler house appeared to be formed by a roof, built over a former alley way, between two adjacent buildings.

It rained steadily during the time of the test. The roof leaked and it was a difficult matter to find standing room in front of the boiler where water did not drop on my note book.

The coal used had been kept outside in a yard all winter, and contained lumps of ice, which of course, during the weighing of the coal were thrown out if observed.

In view of the results obtained during this test, it should be repeated under different conditions.

The whole respectfully submitted,

(signed)    Captain James Wright.

Montreal, April 7th, 1896.

---

The residents of Beamsville, Ont., are urging the extension of the Hamilton, Grimsby and Beamsville Railway to that town, and it now seems probable that work will be commenced this season.

## SPARKS.

The village of Westport, Ont., will probably introduce electric light.

Bothwell, Ont., is moving in the direction of securing electric light.

An electrical stamp cancelling machine is in operation at the post-office at Ottawa.

The St. John, N. B., Street Railway Company are extending their line to the park and cemetery.

The Canadian General Electric Co. have been awarded the contract for the electrical equipment of the Hamilton Radial Electric Railway.

The telegraph and telephone companies' cables were damaged at Three Rivers, Sorel, and other points on the St. Lawrence, as the result of spring floods.

The Lachine Rapids Hydraulic & Land Co. have decided to place their wires underground. The management have recently inspected systems of conduits in different cities in the United States.

A citizen of Quebec is said to have sent to Paris for an electric omnibus, which can ascend and descend hills with greater ease than one drawn by horses, and can do its sixty miles an hour comfortably on a level road.

It is said that the Maginn Power Generator and Motor Co., of Chicago, will shortly put upon the market a light motocycle, the price of which will be only slightly in advance of that of a first-class bicycle.

The Steam Boiler and Plate Glass Insurance Company, of Canada, with head office at London, Ont., announce the transfer of their plate glass insurance business to Lloyd's Plate Glass Insurance Co., of New York.

The Upper Ottawa Improvement Co. are replacing their No. 14 copper wire, from Ottawa to Guyon, P. Q., a distance of 40 miles, with No. 9 galvanized iron wire. The construction is in the hands of Mr. Maurice Quain, electrician, of Ottawa.

The Brantford Operating and Agency Co. is applying to parliament for permission to change its name to the Brantford Electric and Operating Co., to increase the capital stock to $50,000, and to purchase the franchise, assets and rights of the Brantford Electric & Power Co.

The directors of the Royal Electric Company have decided to proceed at once with the construction of the water power at Chambly, Que. It will be owned by the Chambly Manufacturing Company, in which name the charter now stands, and the proprietors, ten in number, have subscribed $30,000 each. The Royal Electric retains an interest of $200,000.

The Bell Telephone Company have made application to the City Council of Montreal for permission to lay underground conduits for their wires. The Standard Light and Power Company have certain privileges in this respect which were obtained some years ago, and the Council have decided to refer the matter to the City Attorney before taking any action.

The St. Martins Telephone Company, St. John, N. B., at their annual meeting, elected the following directors : Messrs. W. E. Skillen, John McLeod, Walter H. Allen, C. M. Bostwick and C. D. Trueman. The directors afterwards chose Mr. John McLeod as president ; Mr. W. H. Allan vice-president, and Mr. A. W. McMackin as secretary and general manager.

The Hawkesbury Lumber Co., of Hawkesbury, Ont., have purchased a 25 k. w. dynamo of the Edison type for lighting the interior of their six mills, which were previously lighted by arc and series incandescent, from a 60-light Wood arc dynamo and 35-light Ball dynamo and which are now used to light their yards. The change is a decided improvement on the old system.

The following students were successful at the spring examinations at McGill University, Montreal : Electrical Engineering— Charles Harvey Wright, Renfrew, Ont ; Harry Alex. Chase, Kentville, N. S. ; William Currie, B. A. Sc., Montreal ; Homer Norton Jaquays, B. A., Montreal ; Wm. Norton Cunningham, B. A. Sc., Montreal ; Stewart Fleming Rutherford, Montreal. Mechanical Engineering—James Lester Willis Gill, Little York, P. E. I. ; Francis Edward Courtice, Port Serry, Ont. ; John William Hunter, Kingston Station, Ont. ; Thos. Fred. Kenny, Ottawa ; Ernest Randolph Clarke, Stratford ; Henry Arthur Bayfield, Charlottetown, P. E. I. ; George Alexander Walkem, Kingston, Ont. ; Gordon Scott Rutherford, Montreal.

## ELECTRIC POWER FROM THE MONTMOR-ENCY FALLS IN CANADA.

### By C. C. Chesney.

The Falls of Montmorency, situated eight miles below the far-famed and historic city of Quebec, are the scene of one of the most interesting and successful electric transmissions of power in America. The cataract,

Montmorency Falls, Showing Nos. 1 and 2 Gate Houses.

almost too well known to need any description, is the chief natural attraction in that vicinity, and while not possessing the magnitude of Niagara, there is yet something of the same grandeur and magnificence in the wild rush of its waters, and the same deafening roar that stuns for a moment the mind of the most stolid beholder. From a height of more than 275 feet the waters fall perpendicularly over the face of the rock, forming, in succession, furious cascades and seething pools in the ravine below, and rushing off to meet the waters of the majestic St. Lawrence.

It is the especial object of this article to call the attention of those interested in the general development of water powers, and especially of those who may have occasion to investigate the problem of the transmission and distribution of power by alternate currents of electricity, to the method and apparatus there used, in the belief that it is an object lesson, not only of scientific interest, but of great practical value.

To a people less conservative than the "habitant," —less apt to revel in the memories of the shrouded past,—utilitarian possibilities of this beautiful cataract might have appealed earlier, but with this mystic people, and within sight of one of the largest cities of Canada, the massive energy of Montmorency was allowed to waste away for years in dashing itself into the chasm below.

It is but a few years ago that a company, organised in Quebec, constructed a dam across the narrow gorge in the Montmorency river at a point 1500 feet above the face of the falls, with the object of utilizing some of the power for manufacturing. A cotton mill was built at the foot of the bluffs, and shortly after, a small arc station was erected for the purpose of doing arc lighting in the city of Quebec.

In the year 1889 the present Montmorency Electric Power Company established, in connection with the arc plant, a small incandescent plant of about 100 horse-power capacity, using 2000-volt alternate current machines, built by the Royal Electric Company of Montreal, and transmitting the current to Quebec with a loss of over 50 per cent. This very unsatisfactory and uneconomical plan was continued until the summer of 1894, when a change of management brought to the service of the company the well-tried experience and engineering ability of Mr. Frank H. Badger, Jr., as general manager, and Mr. Louis Burran, as electrician, to whose intelligent work and attention to the practical details the final success of this plant is largely due.

Acting under the advice and direction of these gentlemen, the company increased the capacity of the old dam to a minimum of about 12,000 horse-power. A short tunnel was run through the solid rock, connecting the dam to the wooden flume which continues along the face of the adamantine bluffs for a distance of 1500 feet to the gate house on the brow of the hill. The construction of this flume was a task of considerable magnitude, involving the exercise of much engineering skill. From the gate house, on the brow of the hill, a steel-riveted tube, having a diameter of 72 inches and a length of 1100 feet, is carried down a steep incline to the power house, where the pipe-line terminates in a large steel receiver which supplies the water to the turbines.

The Lighting Station, Showing 48-inch Steel Pipe Supplying the Cotton Company with Power.

Owing to the extremely high head and the great velocity at which the wheels must necessarily run, they were required to be very simple, stronger and more compact than is the general rule. The wheels adopted are the "Little Giant" turbines, manufactured by J. C. Wilson & Co., Picton, Ontario. The noticeable feature of these wheels, besides their simplicity, is the almost entire absence of lateral or end thrust, so frequently found in the ordinary types. They have two sets of buckets, keyed on the same shaft, and the buckets are so formed that whatever end thrust there may be from the one set is counteracted by that of the other. The particular size installed at the Montmorency falls has a capacity of 700 horse-power and runs at a speed of 600 revolutions per minute. The "Little Giant" turbine is comparatively little known outside the Dominion of Canada, but some idea of its simple and compact nature can be gained from the fact that this 700 horse-power wheel is only 21 inches in diameter and has an extreme length not exceeding 2 feet.

The generating station is a two-storey structure, built of native stone, and is 150 feet long by 50 feet wide. It is situated in a picturesque and convenient spot between the hills and the Quebec, Montmorency & Charlevoix Railroad. The first floor of the building is devoted entirely to the turbines and the necessary gate mechanism. Along one side of the room the numerous wheels are arranged in a row, from which the belts proceed at an angle of 45 degrees to the dynamos on the the floor above. The water is taken from the receiver in this same room and is discharged partly into the tail race which empties into the St. Lawrence river, and partly into the 48-inch steel pipe which supplies water to the wheels in the mills of the Montmorency Cotton Manufacturing Company.

The second floor is the dynamo room. It is a large, well ventilated and well lighted space, having practically the same di-

THE DYNAMO ROOM, MONTMORENCY FALLS POWER HOUSE.

mensions as the building. To the practised eye of the engineer, it is at once apparent that the design and construction here have been carried out in accordance with the best engineering skill. It is unfortunate that electric lighting stations, as a rule, have been hurriedly constructed with any sort of material that happened to be near at hand, and equipped with that apparatus which was offered at the lowest price, regardless of quality. This room is, therefore, of more than usual interest, since in its equipment, or that proposed for future developments, every necessary improvement has been introduced which in any way promised to increase the reliability and efficiency of the plant, and to reduce the cost of operation.

The problem in electrical engineering which was presented to the Montmorency Electric Power Company is typical of the problems presented to all enterprises for the utilization of a waterfall by the transmission and distribution of its energy to distant points. The prime requisites in any such system for power transmission and distribution are, necessarily, simplicity and reliability. The simplicity and reliability of the single-phase alternating system are well known to all elec-

tricians and the electrical public in general. The system has been proven and tried in numerous cases where it has been the sole dependence of the larger enterprises, and where the practicability of transmitting power in bulk by this system has been demonstrated beyond question. It lacks, however, range and flexibility; it lacks a motor which commercially answers the requirements of power distribution; and it lacks the ability to be readily converted into direct currents for railroad and electrolytic work.

In the multiphase systems however, which have been developed within the last few years by the various electrical companies, are to be found all these requirements. Coupled with the simplicity and reliability of the single-phase systems, are to be found range and flexibility. The induction motor forms the missing element for commercial power work, and extends its range far beyond that ever realized by the direct-current motor. When we now add the two-phase or three-phase rotary transformers, we have a system, ideal not only for long distance transmission purposes, but ideal for general central station work, whether the energy be primarily furnished by steam or water power.

Naturally, then, in order to obtain the most complete and commercial results from the power at its disposal, the system adopted by the Montmorency Electric Power Company is a multiphase system. The particular multiphase system is the S.K.C. (Stanley-Kelly-Chesney) two-phase system as applied by the Stanley Electric Manufacturing Company, of Pittsfield, Mass., U. S. A., the Canadian (manufacturers being the Royal Electric Co., of Montreal.)

The generators are situated on the north side of the dynamo room. The foundations are of solid masonry, through which are run a number of tie rods which hold an insulating cap to the solid bed rock of granite. The insulating cap consists of 10 by 12-inch timbers which have been boiled in paraffine in order to completely remove all moisture. The bed plates of the machines are held in place by lag screws set into the paraffined timber. Each generator delivers alternating currents, differing in phase by 90 degrees, to two independent circuits at an electromotive force of from 5200 to 5700 volts. By means of rheostats in the fields of the generators, the electromotive force can be varied between these limits, to meet the requirements of the circuits.

The frequency is 66 periods per second—that is, the current is reversed approximately 8000 times a minute. The generators are what are commonly known as 8000 alternation machines. This frequency was selected as being one of the standard frequencies advocated by engineers of the Stanley Electric Manufacturing Company and in preference to 133 periods per second (16,000 alternations); for, while the loss in the core of the transformers was increased from 20 to 30 per cent., and the regulation of the generators from a total of 2 per cent. to a total of 3 per cent., the self inductive drop in the transmission lines was such an important factor that the lower frequency was considered preferable, as giving, on the whole, a better regulating and more economical system.

The currents delivered by the generators are carried

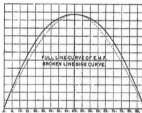

E. M. F. CURVE OF THE GENERATORS AT MONTMORENCY.

by heavily insulated cables to the switchboard, where the attendant, by suitable switches, may connect the generators to the transmission lines. Two separate pole lines have been constructed to Quebec. One follows the route of the Quebec, Montmorency & Charlevoix Railroad, carrying two transmission lines, each having a capacity of 500 K. W.; the other follows the old Beauport road, carrying one transmission line of a capacity of 500 K. W., and provision for another line. All the lines are entirely overhead, and are entirely supported on wooden poles, with the exception that at the crossing of the Charles River they are carried across the river on iron poles, 125 feet high.

Triple petticoat porcelain insulators were used, and were made especially for this plant by a Canadian manufacturer, differing, however, but slightly from the design now in common use for high-voltage work. The line wire is No. o, B. & S. bare copper. The drop, due to the ohmic resistance of the wire, is 8 per cent., which is increased by the self-induction to 10 per cent. On the extreme ends of the top crossarms of each pole line are strung galvanized iron wires, which are grounded at every third or fourth pole.

These iron wires, together with lightning arresters, placed at each end of the transmission lines, give a complete and safe protection from all lightning storms. So perfect has been this protection, that a discharge in either the generating station or sub-station is practically unknown.

The transmission lines enter the city of Quebec in that portion known as St. Rochs, and the sub-station is located in the centre of the industrial district, where power is needed for the tanneries and shoe factories for which the city is noted. The sub-station is 90 feet long by 65 feet wide, and is two storeys high, built of brick,

with a stone front. The building consists of a storeroom on the second floor, the offices of the company in the front, on the first floor, with the room containing the distributing switchboard immediately in the rear. The transformer house is a part of the same building, but only one storey high.

The transmission lines enter the sub-station through the cupola to the high potential receiving switchboard, which contains special high-voltage switches. These switches are mounted on marble and are fitted with special self-enclosing boxes, which are intended to cut off any possible arc that may be formed on opening the 5,000-volt circuits.

From the transmission lines the current of the generators is carried to the step-down transformers, where the electromotive-force of transmission is transformed into the distributing electromotive-force of 2,000 volts. These transformers are the regular indoor type of the Stanley Company, and are arranged on a rack, in two tiers, five wide and two deep, in such a manner that the air has free circulation, or that they may be artificially ventilated if it be so desired. They are of 50 K.W. capacity, and are connected in pairs. Each is wound for a primary electromotive-force of 2,500 volts, and a secondary one of 1,000 volts. The primaries of each pair are connected in series for receiving 5,000 volts, and the secondaries in series for delivering 2,000 volts to the distributing system. This arrangement gives an additional safeguard by lessening the voltage on any particular coil, and in case of a burn-out in any one transformer, the other will carry the load for a time until the relay transformer can be cut in. The insulation of each transformer was tested with 10,000 volts. The regulation of each transformer from no load to full load is about 1 per cent. The average efficiencies of the entire transformer equipment, as shown by the shop tests, are:—

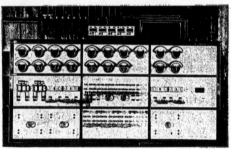

THE GENERATOR SWITCH-BOARD, MONTMORENCY POWER HOUSE.

| | | |
|---|---|---|
| Efficiency, full load, | 97.8 per cent. |
| " half load, | 97.9 per cent. |
| " quarter load, | 97 per cent. |

The transformers are divided into two sets of eight transformers each, with a relay of two transformers which can be cut in or out without stopping the service. Three secondary distributing circuits are carried to the distributing board in the next room. Provision has been made, however, for two more circuits of the same capacity.

The current from the secondaries of the two sets of transformers is carried to the bus bars of the distributing switchboard, from which it is furnished to the numerous city circuits for light and power.

The generators are the S. K. C. inductor type of two-phase machines, running at 286 revolutions per minute. Each generator has a capacity, at 5,700 effective volts, of 100 amperes. They are the largest inductor machines that have ever been built, and are the first and only practical machines that have ever been constructed for such a high initial electromotive force.

The armature is stationary, consisting of two sets of laminated iron rings, connected by steel rods four inches in diameter. On the inner surface of each laminated armature ring are fifty-six grooves for receiving the armature coils. The weight of this portion of the machine is 42,580 pounds. The field or exciting coil is circular, 94 inches in diameter, and wound on two copper bobbins, each 4¼ inches wide, with a copper strip four inches wide and .026 inch thick. It is insulated between the layers with a special oil-cloth which is practically indestructible at temperatures under 150°C.

To the dynamo builder the advantage of constructing a field coil on such lines is very apparent. With a free circulation of air, and every turn of the winding being, for cooling purposes, practically in contact with the moving air, there is no possibility of overheating in any portion of the coil. The copper bobbin, absorbing all discharges, prevents any excessive rise of electromotive-force on the coil, which might be caused by carelessly, or, under extreme circumstances, intentionally, opening the field circuit under full charge.

There are, in all, 56 armature coils, 28 for each phase. The coils are small and were wound in a lathe. Each coil was carefully insulated to stand 15,000 volts before placing it in the armature. The insulation of the completed armature was tested, finally, with 12,000 volts.

The inductor, the only moving portion of the machine, is a steel casting, 43 inches long, and 84 inches in diameter, upon the periphery of which are two sets of polar projections of iron laminæ, fourteen at each end. The weight of the inductor, including the shaft, is 28,470 pounds. The net weight of the completed machine is, approximately, 100,000 pounds.

In the operation of alternating current motors, and also of transformers, it is now generally recognized as important that the currents and magnetic fluxes should vary sinusoidally, for the more nearly such a condition is approached, the less are the losses and idle currents. A first step toward obtaining these conditions is the making of the impressed electromotive-force of the generator sinusoidal. The flux between a field pole and the opposite iron of the armature distributes itself, so that it is, at every point, inversely proportional to the reluctance of the gap at that point, or inversely proportional to the distance from the field pole to the armature iron; that is, the electro-motive force at any in-

stant, is inversely as the clearance. In order, then, to obtain an electromotive-force following the sine law, the pole faces of the inductors of these generators, as well as the pole faces of the inductors of all generators, manufactured by the Stanley Electric Manufacturing Company, are so shaped that their curvature may be expressed by a formula which was derived by Mr. Kelly, of the Stanley Company, and is contained in a United States patent issued to him.

In the design of the inductor machine it has been found by the writer than any deviation from this law, or the use of any other than a sinusoidal electromotive-force has resulted in increased losses in transformers and unsatisfactory running of motors. In one instance the output of a 20 h. p. motor was reduced 25 per cent. on a machine with a distorted wave, and the condensers which were intended to balance the lagging currents of the motor, were absolutely useless for that purpose.

A novel feature has been introduced for the first time in these generators. The entire distributing system has been arranged to be run in parallel, and, in order to do this the generators must be kept in phase. There are some objections, however to paralleling 5,000 volt machines through their principal circuits—also do difficulties occur in paralleling long transmission lines. A defect arising in any wire of one line, such as a ground or leak, affects all the wires in all the lines, so that if at any other point another ground occurs, the generators between the two grounded points are all short circuited. To overcome this difficulty, the mains from the different machines are kept separate and the secondaries of the step-down transformers are connected in parallel with the supply mains.

THE DISTRIBUTING STATION, MONTMORENCY FALLS POWER HOUSE.

In consequence, unless the line from one generator has two weak points, there is no leakage, and, at any rate, the leakage due to two weak points can affect directly only one generator. In order to keep the generators in phase under these conditions, a separate and distinct synchronising winding gives an electromotive-force of 120 volts and has a synchronising capacity of about 100 K. W. This, together with the effect that can be obtained through the paralled secondaries of the step-down transformers is sufficient to give perfect parallel running. The load can be readily shifted from one generator to the other by varying the quantity of water supplied to the turbines.

The exciters are two slow speed, direct current machines of 12 K. W. capacity, each of sufficient capacity to excite the fields of the three alternators.

Some of the electrical data of these machines are interesting, as showing the improvement in dynamo design and the possibilities of the inductor type of machine. From the shop tests we have the following :—

| | |
|---|---|
| Maximum loss in field, | 4000 watts. |
| Loss in armature iron, | |
| "    friction and windage | } 20,000   " |
| "    armature copper | 7000   " |

From these we calculate the efficiencies :—

|  |  |
|---|---|
| Full load, | 95.1 per cent. |
| Half load, | 92.3 per cent. |
| Quarter load, | 87 per cent. |

The rise of temperature after a run of twenty-four hours, with full load, is as follows :

|  |  |
|---|---|
| Field coil, | 12° C. |
| Armature iron, | 21° C. |
| Inductor iron, | 7° C. |
| Armature coils, | 26° C. |
| Bearings, | 21° C. |

. The regulation, that is, the variation of the electromotive force from no load to full load, with same speed and same field excitation, was 3½ per cent. With full load on one phase and no load on the other, the percentage difference of electromotive force was 3½ per cent.

The design of the switchboards and the general features of their construction can be best understood by reference to the illustrations on pages 13 and 14. Both the generator and distributing switchboards are built of marble, supported upon wooden frames, and the panel method of construction is followed throughout. Each slab of marble was tested for metallic streaks, with a pressure of 12,000 volts.

The generator board consists of three sections, of three panels each. The lower panel of the left-hand section carries the two rheostats for controlling the fields of two of the generators. They are mounted on the back of the board, and only the handles and the heads of the supporting bolts show. The middle panel carries four machine switches with self-closing arc cut-offs, two for each generator, and one of these for each phase. These connect directly to two of the four-wire transmission lines.

It is never intended to open these switches under load, except in the case of great emergency, when the automatic cut-offs will take care of any discharge from the line, or any tendency to arc. When a generator is to be taken out of work, the load, if it is running in parallel with another, is shifted to the other by gradually cutting off the supply of water to its turbine ; after that, the switch can be opened without difficulty. If the generators are running independently, the load is first transferred at the sub-station before opening the switch.

The upper panel contains four ammeters, two for each transmission line, and four voltmeters, with high-potential station voltmeter transformers placed directly at the back. The right-hand section is exactly the same as the left, with the exception that it is, at present, equipped for only one generator and transmission circuit.

The lower panel of the middle section carries three four-pole, single-throw switches for connecting the synchronising windings of the generators in parallel. The synchronising lamps are shown just above.

The middle panel is the exciter panel, carrying the necessary switches for paralleling the two exciters, and for charging the field of any generator from any exciter if the exciters are running independently.

The upper panel contains three direct-current ammeters for the fields of the generators, and also four alternating current voltmeters, with voltmeter transformer at the back, connecting to the 2,000-volt potential lines which return from the sub-station.

The distributing board was originally designed for parallel running only, but it was afterwards learned that on rare occasions ice accumulates in the turbine feed pipes, which affects the speed of the wheels and makes parallel running an impossibility. In consequence, the original design was changed to permit either independent or parallel running of the generators. The two sections on the left of the board are organized for lighting alone, and to permit the transfer of any two-wire circuit upon any of the possible four sets of step-down transformers. This is accomplished by a series of three double-throw, double-pole switches. At the extreme top of this section are the automatic circuit-breakers. The two sections on the right of the board are organised for light and power, and differ from those on the left only in that the four-pole double-throw switches are substituted for the two-pole double-throw.

The two centre sections are the same in design. The upper panels contain the voltmeters and ammeters for the various circuits. The middle panels carry four-pole double-throw switches, and are connected to the secondaries of the step-down transformers in such a manner that, if the switches are thrown up, three sets of two-phase secondaries are connected to three separate sets of bus bars ; if they are thrown down, all the secondaries of each phase are connected in parallel. The lower panel carries the ground detectors of the ordinary transforming type. The whole switchboard is 26 feet long and 11 feet high.

All the motors now in use by the Montmorency Electric Power Company are the S. K. C. induction motors. These motors were described by Dr. Bell in the January number of this magazine. They vary in size from one horse-power to 30 horse-power, and do all kinds of work, running the tools of carpenter and machine shops, driving the saws and wood-planers of planing mills, and handling the freight elevators in various mills and in wholesale warehouses with perfect satisfaction. It is now well understood that the magnetizing current of an induction motor lags behind the applied electromotive-force, and that a lagging current in practice involves considerable loss and expense, by necessitating the use of larger conductors and generating apparatus, while it seriously interferes with the proper regulation of the generators, and increases the normal drop of the line.

In the S. K. C. motor the magnetising current is furnished by condensers ; the motor then takes current in proportion to the load. Two condensers are connected in multiple with the fields of the motor, and each has a capacity in amperes at the working voltage practically equal to the no-load current of each field of the motor. If there is no distortion of the current wave, the apparent energy taken by an induction motor with condensers is equal to the real energy.

An interesting example of the value of condensers on induction motors is shown in a small plant in New England. The generator was a 60-ampere two-phase machine, manufactured by one of the larger electrical companies, and was furnishing power to a number of small induction motors. The motors were doing all kinds of general factory work and running ten hours a day. The average load on the motors was about one-quarter of their maximum. The amount of current furnished by the generator to the motors was 52 amperes at 1,152 volts. The power factor was 0.505.

When the motors were supplied with condensers, the current was reduced to 28 amperes at 1,150 volts, and the power factor was increased to 0.863. The reason that the power factor was not increased to unity was the existence of harmonics in the curve of the electromotive-force of the particular machine in use.

As to the commercial efficiency obtained in this plant, it is interesting to note that, with the generator working at full load, for every 100 K. W. of energy delivered to the pulley of the generator, 95.1 K. W. are delivered to the line at the generating station ; 87½ K. W. are delivered to the terminal of the step-down transformers, and 86 K. W. are delivered to the distributing mains of the sub-station.—Cassiers' Magazine.

# ELECTRIC RAILWAY DEPARTMENT.

## AN ELECTRIC SELF-LOADING CAR.

Mr. A. Jackson Reynolds, of Montreal, has patented an electric self-loading car for street cleaning purposes which is claimed to possess great mechanical ingenuity, and which promises to result in a large saving in the cost of cleaning city streets. The first car manufactured under his patents was turned out of the Rathbun Company's works at Deseronto a fortnight ago, and is shown in the accompanying illustration.

The system of cleaning is as follows : About one-third of the surface of the street is swept from the curb inwardly towards the railway tracks by the ordinary horse sweepers, driven in the opposite direction from the usual way of sweeping from the centre to the outside. The refuse is then taken up by the self-loading

instantly at any point desired. The brushes, steel casings, and rubber aprons are so constructed as to work in either direction automatically. The cars are driven preferably by a stationary motor placed directly over the brushes on the operating platform of the car, the brush being operated from a counter-shaft by sprocket wheels and chain.

The brush, which has been specially manufactured for the purpose, makes five revolutions to each one of a car wheel. It works much on the same principle as a carpet sweeper, and will throw the dust a distance of twenty-five feet. Its capacity is about twenty-eight car loads. The broom, which is fastened to solid heavy axles, is so arranged that it always fills the case in which it is contained, a simple but ingenious device

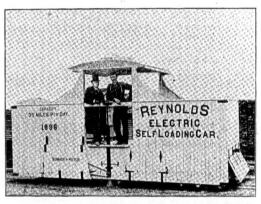

car at any desired speed and conveyed to the desired location.

The car shown herewith is 22 feet in length, 8 feet wide and 9½ feet high, very compactly and strongly built in every section. It is fitted with all the appliances for electricity common to a regular trolley car. Contrary to general use the brakes, motors, etc., are all situated above the wheels and axles so as not to impede the full action of the brush. The operating platform on which the persons stand while directing the motion of the car and broom is 8x5 feet, and so placed as to protect them from being touched by the dust thrown from the revolving brush or broom.

The results are accomplished mainly by placing a large rotary brush across the centre of a moving car, aid brushes being covered with steel casings, with proper outlets for discharging the sweepings into the body of the car, and covering the brushes with said steel plates, having rubber aprons fitting the pavements. The high speed of the brush forms a powerful suction, which takes up all the itemized matter and deposits it over the brush into the body of the car, which is provided with pivoted dump floors for dropping the load

changing the size of the latter to suit the changes made by the wear of material. The broom acts as well one way as another, steel deflectors being so arranged that it can be run backward without any change of machinery or even without touching it. By a change of the trolley the action may be reversed instantly so as to throw the dust one way or the other as may be desired. The broom may be extended so as to cover the whole street outside the car-track if necessary. For removing snow in winter the car may be constructed as long or wide as may be required. The car may be unloaded in thirty seconds, one man doing the whole work by manipulating a lever.

The cost of operating this electric sweeper is claimed to be about $3.00 per mile. The side sweeping by horse machines can be done for $1.50 per mile, which makes the total cost $4.50 per mile. From $15 to $25 per mile is now paid by cities for the same work.

The price of cars will be $1,000, with a royalty of one dollar per mile for all streets cleaned.

The inventor, Mr. Reynolds, now of Montreal, but formerly of Worcester, Mass., has had an experience of thirty-five years in the design and construction of mechanical apparatus, having spent four years with the best engineers of Paris, London, Vienna, and other foreign cities. He is also the inventor of several other devices for which patents have been granted in Canada and the United States.

Mr. Jackson informs us that he has already entered into agreement with Ontario parties for the exclusive use of his cars in this province, for which privilege the consideration of $45,000 is to be paid.

## SPARKS.

The directors of the London Street Railway Company have authorized the issue of a larger amount of stock.

Work has been commenced on new car sheds for the London Street Railway Company, to cost $7,000.

Work on the Hull and Aylmer road is progressing favorably. The first four cars will shortly be shipped from Peterborough.

During the year 1895 the Niagara Falls Park & River Railway carried 499,015 passengers. The receipts were $65,784, and cost of operation $40,630.

The Hull Electric Co.'s power house at Dechenes is completed, while the car shed, of stone, will be finished in a few days. Cars will be running before the 1st of July.

The Petrolia Electric Light & Power Co. are installing a direct current Canadian General Electric Co. power plant for pumping some oil wells near their power house.

The Gananoque Electric Light & Power Co. have increased their three wire incandescent plant by ordering two 600 light machines from the Canadian General Electric Co.

The Consolidated Railway and Light Co., of Vancouver, have ordered two additional open cars from the Canadian General Electric Co. The equipments will be of the standard C. G. E. 800 type.

C. G. T. Clark, superintendent of the electric lighting station at Niagara Falls, Ont., recently had a narrow escape from death at Tonawanda by coming in contact with a trolley while riding a bicycle.

Markdale, Ont. is to have incandescent light. Power will be obtained from a water power about one mile out of town. The contract for the plant has been closed with the Canadian General Electric Co.

The London Street Ry. Co. have placed an order for 12 additional C. G. E. 800 motors with the Canadian General Electric Co. This will make 60 motors of this type operating on the London road.

The ratepayers of Sherbrooke, Que., have sanctioned the construction of an electric railway along the streets of the town. Work will therefore be commenced at once, and the park line completed by fall.

Mr. C. A. Cunningham has entered an action in the courts against the Royal Electric Company, claiming $10,000 for the loss of one of his eyes. The plaintiff met with the accident while in the employ of the Company.

The town of Listowel, Ont., has under consideration the question of installing an electric light plant. A committee of the Town Council has recommended that an electrical engineer be engaged to report on the cost thereof.

The earnings of the Montreal Street Railway Co., for seven months ending 30th April last, are $661,543.86, against $545,864.94 for the same period last year, an increase of $115,678.92, giving an increase in the average daily earnings of $543.10.

An electric railway company is applying for incorporation to operate in the city of Charlottetown, P. E. I. The city council has granted the company permission to operate cars on Sunday, which step is strongly opposed by the Ministerial Association.

The Hamilton, Chedoke & Ancaster Railway Co., who propose building a trolley line to run in connection with the street car system from Hamilton to Alberton, have requested a bonus of $15,000 when the line reaches Ancaster, and the same amount when completed.

The date limit for the commencement of the electric railway at Quebec has expired, which reinures the forfeiture of $10,000. It is probable, however, that the City Council will grant an extension of time, Mr. Beemer having given assurances that the work will shortly be proceeded with.

The Port Dalhousie, St. Catharines & Thorold Ry. Co. have recently rebuilt their overhead line changing from the old Van Depoele over-running system to the standard under-running trolley. Additional motors of the C. G. E. 1200 type have also been purchased from the Canadian General Electric Co.

The work of constructing the new electric street railway system at Moncton, N. B., is proceeding rapidly. The Canadian General Electric Company have the contract for the apparatus, which includes a 100 k. w. generator and two double motors C.G.E. 800 equipments. The road is to be in operation by the 1st of July.

Work has been commenced on the electric street railway at Sarnia, Ont. It is the intention of the company to have one of the finest systems in Canada. The road will extend from Sarnia to Point Edward, and then to Weesbeach, a summer resort on the shore of Lake Huron. The city gave the company a bonus of $10,000.

The cars of the Westminster and Vancouver Electric Tramway

Company have been improved and remodelled, under the superintendence of Mr. P. M. Smith, the efficient manager of the road. The new Bessemer Sheet Steel Headlight, made by the United States Headlight Co. for the patentees, has also been obtained. This is said to be the most improved headlight obtainable.

Col. Stacey, who recently purchased the street railway franchise of St. Thomas, Ont., has made a formal proposition to light the city with 2,000 candle power lights and electrify the railway system. The price asked from the city is $10,000 for the first three years, $9,000 for the second three years, and $8,500 for the following four years, the city to have the option of purchasing the plant at the end of that time.

At a recent meeting of the provisional directors of the Chatham City and Suburban Railway Company, of Chatham, Ont., there were present Messrs. John Mercer, G. P. Sholfield, Manson Campbell, John A. Walker, Fred Stone, Wm. Douglas, Q. C., and Geo. C. Rankin, London. Mr. Walker was appointed chairman of the Provisional Board, and Mr. Wm. Douglas, Q. C., secretary. It was decided to at once open a stock book.

Work has been commenced on the power house for the Hamilton Radial Railway Company at Burlington. It will be of brick, 53 feet by 100 feet, and will contain three boilers and two engines, each of 250 horse power. The chimney will be 14 feet in diameter at its base and 108 feet in height. Space will be left for additional engines and boilers should they be required. An effort will be made to put the line in operation by Dominion Day.

An exchange says that the enterprise of organizing a company to build an electric railway for Port Hope promises to develop into a definite scheme. Plans have beeen prepared which demonstrate that for a comparatively small expenditure an electric railway could be opened between Port Hope and Pontypool, thus connecting with the C. P. R., and extending a mail service to Kendall, Orono, Osaca and other places which are at present reached only by stage.

Dr. Oille, President of the Niagara Central Railway Company; Ald. J. S. Campbell and J. C. Rykert, of St. Catharines, and Mr. William McGill, of Thorold, recently waited on the Dominion Government requesting that the subsidy of $100,000 standing to the credit of the road to Hamilton be diverted so that $40,000 of it might be applied to the improvement of the existing roadbed and $60,000 subsequently to an electric system connecting with the Toronto, Hamilton & Buffalo Railway.

The shareholders of the Moncton Street Railway, Heat and Power Company have elected directors as follows : J. L. Harris, president ; J. W. Y. Smith, vice-president ; R. A. Borden, secretary-treasurer ; F. W. Summer, E. C. Cole, J. C. Robertson, F. A. West, and Ald. Girvan. It has been decided to proceed at once with the construction of the road, and it is expected to have cars running in about two months. Stock to the amount of $50,000 has been subscribed. The power house will be erected at the wharf siding at the foot of King street.

The Montreal Street Railway Company are again making extensive additions to their power plant. An order has been given to the Canadian General Electric Co. for a 1,500 kilowatt generator which will be coupled direct to a 3,000 horse-power Laurie engine. This immense unit, in which the weight of the generator alone will be nearly 100 tons, is similar in style to the direct-connected units supplied to the Toronto and Winnipeg Street Railway Companies. With the addition of the new machine, the capacity of the generator in the Montreal Railway Company's power house will exceed 8,000 horse power.

The incorporation is announced of the Cornwall Electric Street Railway Company, with a capital stock of $150,000, the company being formed of H. R. Hooper, C. E., D. A. Starr, F. N. Seddall and Mrs. Hooper, of Montreal, and W. R. Hitchcock, of Cornwall. At a recent meeting of the company, officers were elected as follows : H. Ross Hooper, president ; D. A. Starr, vice-president and managing director ; F. N. Seddall, secretary and treasurer. The construction of the railway has been commenced, and a portion of it will shortly be in operation. The cars are being built by the Rathbun Co., and the machinery by the Canadian General Electric Co. A powerful 120 horse-power electric locomotive will be used to haul freight, which will be one of the principal sources of revenue. A brick power house, 72 × 30 ft., is being built on Water street near the canal.

The annual general meeting of the shareholders of the Ottawa Electric Company was held at the company's office, corner Sparks and Elgin street, on the 1st inst. The report of the president and directors for the year ending April 30th was presented, showing a gross revenue of $153,788.66, being a very satisfactory increase over the previous year. A dividend of eight per cent. was declared. The total number of incandescent lights installed is 53,331, an increase of about 5,000 during the year. The number of arc lights 497, motors 81, heaters 119. During the year considerable advance was made in the construction of a new switchboard under the direction of the general superintendent, Mr. A. A. Dion, which will result in a complete unification of the company's lines, so that the whole system may be controlled from the central lighting and power station. A vote of thanks was unanimously accorded the president and directors, and upon motion the board was re-elected as follows : Hon. Francis Clemow, Hon. E. H. Bronson, Geo. P. Brophy, T. Ahearn, J. W. McRae, Wm. Scott, C. Berkeley Powell, D. Murphy and Geo. H. Perley. At a subsequent meeting of the directors the following officers were re-elected : T. Ahearn, president and managing manager ; Hon. E. H. Bronson, vice-president ; D. R. Street, secretary-treasurer ; A. A. Dion, general superintendent ; Redmond Quain and A. Bayly, auditors.

## SPARKS.

" Is your town lighted by electricity now ? "

" Yes, but only when there's a thunderstorm."—Lustige Blatter.

The Citizens' Light, Heat and Power Co. of St. Catharines, Ont., has been incorporated with a capital stock of $40,000, in $50 shares.

Dr. Corbett, of Port Hope, is increasing the depth of his raceway with a view of obtaining power for an additional 75 kilowatt monocyclic alternator.

This X ray business has gone far enough. A New York physician says that by using the new ray he has "found the stereytococcus erysipelatosus proliferating in the interspaces of the connective tissue." Think of that !—Exchange.

Mr. John Goodwin, an employee of the Ottawa Electric Company, has invented a cleat intended for tightening any wire, rope or cord that may be strung from one place to another, which, it is said will be especially useful in electric wiring. A patent has been secured, and the article is being manufactured in porcelain at the shops of the Ottawa Carbon and Porcelain Company.

The three-phase plant at Trenton was started up recently, and a regular service will shortly be supplied to the town of Trenton. The line to Belleville, 13 miles distant, is practically completed, and will be ready for occupation in a short time. The first motor used in Belleville will be a 75 kilowatt synchronous motor of the Canadian General Electric Co.'s standard type from which power will be supplied to operate the arc machines.

The amalgamation has been consummated of all the important electric heating companies in the United States, including the Western Electric Heating Company, St. Paul, Minn.; the Central Electric Heating Company, New York ; the New England Electric Heating Company, the Burton Electric Company, Richmond, Va.; The Carpenter Electric Heating Manufacturing Company, St. Paul, Minn.; the Dewy Electric Heating Company, Syracuse, N. Y.; the Rich Electric Heating Company, Mt. Vernon, N. Y., as well as several others, which have not been active in the business' for some time, but which owned valuable patents. The manufacturing will be concentrated at Cambridgeport. J. Y. Ayer, ex-President of the National Electric Light Association, is the general manager. The capital stock is $10,000,000.

A German expert, after a careful estimate, has announced that the total length of telegraph lines in the world is 1,062,700 miles, of which America has 545,600 miles ; Europe, 280,700 ; Asia, 67,-400 ; Africa, 21,500, and Australia, 47,500 miles. The United States has a greater length than any other country, 403,900 miles, and Russia comes next, although European Russia has only 81,000 miles. The other countries follow in this order: Germany, France, Austria-Hungary, British India, Mexico, the United Kingdom, Canada, Italy, Turkey, the Argentine Republic, Spain and Chile. In point of proportion, however, Belgium leads, with 409 miles of wire for every 1,000 square miles of territory ; Germany comes next with 350 miles ; Holland is only slightly behind Germany, and the United Kingdom has 280 miles of telegraph for every 1,000 miles of country.

This Number of the ELECTRICAL NEWS contains a full report of the Sixth
Convention of the Canadian Electrical Association.

OLD SERIES, VOL. XV.—No. 6.
NEW SERIES, VOL. VI.—No. 7.

**JULY, 1896**

PRICE 10 CENTS
$1.00 PER YEAR.

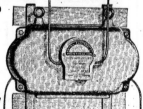

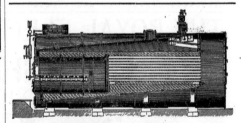

# FACTS

VERSUS

# MISREPRESENTATIONS

Read this

if you are

interested in

# SATISFACTORY ELECTRICAL APPARATUS

Office of G. H. DAVIDSON, PROPRIETOR

Brighton Sash, Blind and Door Factory.

Casings, Mouldings, House Furnishings, ...

RECEIVED JUN 15 1896 ANSWERED

Brighton, Ont. June 13 1896

To the Royal Electric Company
Montreal Quebec

Dear sirs

A number of people interested in Electric lighting have told me personaly and allso written me, that they are informed that the Electric lighting Plant installed by you for me is not giving satisfaction and it has not been accepted by me. The parties making these statements either know them to be false or have not inquired from me or the towns-people within the last five months, or they would know that the lights and plant are working splendidly and to my entire satisfaction, & the plant has been accepted and settled for. Now for the interest of all concerned: I wish to confirm the above statement and further to say that the S.K.C. Dyname (since I took the abstructions out of my Watter wheels which had colected in the wheels owing to not being properly screened before starting) requires less power than I expected it to do, gives a fine steady light and the regulation with a full load on it is such that I do not have to touch the wheels more than four or five times during a nights run & I regulate entirely by the watter wheels having no governors. We ran through the two late terrific lighting storms and allthough both Telegraph & tellephone lines were wrecked round us and the lighting was constantly flashing round the switch board we never had a flicker in our lights while neighboring plants had to shut down & had transformers burned out. After having had an eight months run with this plant and knowing the trouble was in the wheels being clogged with blocks, and having visited a large number of other plants to see the various systems at work I have no hesitation in saying that our plant gives the steadiest light I have yet seen, runs the coolest, regulates the best uses less oil (less than one pint in Eight Months) is the easiest taking care of and requires less power for the number of lights than any other system that I have seen. Visitors from other towns remark the steadiness and brilliancy of our lights and we have not replaced over twenty lamps out of the three hundred originaly installed Eight months ago. Many central Station operators have visited the plant and all express themselves being pleased with the S.K.C. Dynamo — admiring the simplicity of it its mechanical construction and the ease with which it operates. Any ordinary man capable of running machinery can be instructed fully in its operation in fifteen minutes we have had no trouble with the Dynamo whatever or the transform. and I am satisfied the regulation of the two phase system is all right and that a more perfect piece of mechanism than the Dynamo furnished by you would be hard to produce. I feel it is in justice due to you to state these facts and that they may be given every publicity you have my permission to use this letter as you see fit and I trust it will serve to completely contradict the incorrect stories which have been either maliciously or ignorantly circulated respecting this plant. wishing you every success I remain &c.

J. G. Davidson
Brighton
Ont

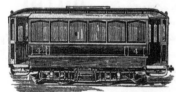

CANADIAN

# ELECTRICAL NEWS

AND

## STEAM ENGINEERING JOURNAL.

| VOL. VI. | JULY, 1896 | No. 7. |
|---|---|---|

## CANADIAN ELECTRICAL ASSOCIATION

### PROCEEDINGS OF SIXTH ANNUAL CONVENTION.

THE Sixth Annual Convention of the above Association was opened in the Council Chamber of the Board of Trade Building, corner of Yonge and Front streets, Toronto, on Wednesday, the 17th June, 1896, at 2:30 o'clock, p.m. Mr. A. B. Smith, the President, presided.

The following persons were in attendance :

| | | | |
|---|---|---|---|
| K. J. Dunstan, | Toronto, Ont. | A. Knowles, | Toronto, Ont. |
| J. A. Kammerer, | " | Irving Smith, | " |
| John C. Gardiner, | " | E. K. M. Wedd, | " |
| J. J. Wright, | " | W. A. Johnson, | " |
| J. Norman Smith, | " | J. K. Johnston, | " |
| F. C. Armstrong, | " | James Milne, | " |
| A. E. Payne, | " | T. F. Dryden, | " |
| Wm. Bourne, | " | P. G. Gossler, | Montreal, Que. |
| James Orr, | " | Fred. Thomson, | " |
| J. F. H. Wyse, | " | John Carroll, | " |
| George White-Fraser, | " | James A. Baylis, | " |
| J. W. Campbell, | " | H. R. Leyden, | " |
| Ed. D. McCormack, | " | J. Rogers, | London, Ont. |
| J. J. Ashworth, | " | Charles B. Hunt, | " |
| A. M. Wickens, | " | John Yule, | Guelph, Ont. |
| A. A. Christie, | " | Wm. Williams, | Sarnia, Ont. |
| George F. Madden, | " | W. B.W. Armstrong. | Parry Sound. |
| T. W. W. Hilliard, | " | George Black, | Hamilton, Ont. |
| F. C. Robertson, | " | B. J. Throop, | " |
| C. P. Dwight, | " | M. W. Hopkins, | " |
| H. P. Dwight, | " | E. Carl Breithaupt, | Berlin, Ont. |
| E. B. Biggar, | " | J. W. Howry, | Fenelon Falls, Ont. |
| F. B. Moore, | " | E. E. Cary, | St. Catharines, Ont. |
| Joseph Wright, | " | G. A. Powell, | " |
| J. S. Robertson, | " | W. G. Bradley, | Ottawa, Ont. |
| T. R. Rosebrugh, | " | D. Elliott, | " |
| Alex. Stark, | " | S. Rose, | " |
| W. J. Clarke, | " | A. A. Dion, | " |
| F. C. Maw, | " | D. H. Keeley, | " |
| W. R. Evans, | " | J. W. Taylor, | " |
| A. B. Smith, | " | O. Higman, | " |
| C. H. Mortimer, | " | C. S. Mallett, | Peterboro', Ont. |
| F. J. Ricarde Seaver, | " | John C. Grant, | " |

The President called the convention to order and stated that the first thing on the programme was the President's address, which he assured the members would be very short. He then addressed the convention as follows :

#### PRESIDENT'S ADDRESS.

GENTLEMEN OF THE CANADIAN ELECTRICAL ASSOCIATION :

It is with great pleasure that I meet the members of this Association at the present convention.

It is rather unfortunate that we were compelled by force of circumstances to hold our convention about the time the country is engaged in a political contest. This has undoubtedly prevented the attendance of some of our most valued members, but I hope, nevertheless, that our meetings will be as interesting and enjoyable as on any former occasions.

The papers to be presented promise to be of great interest in connection with subjects of the utmost importance to the electrical fraternity.

We are particularly fortunate in our place of meeting, and are indebted to the officers of the Toronto Board of Trade for the privilege of our pleasant surroundings.

As but a comparatively short time has elapsed since our last meeting, there is not much of actual achievement to chronicle, but there have been developments in the electrical field that indicate the possibilities of a revolution in our method of producing light by electricity. Many minds have for some time past been occupied with explorations in this promising direction. The production of light without heat has a fascination for the inventor that will probably lead to tangible results in the very near future. The ordinary developments of the science as exemplified in modern systems of power transmission, and electrical construction generally, have advanced towards perfection in as great a degree as in former years, but the field for the enterprising inventor, so far from being exhausted, appears to be growing broader and ever broader with unlimited possibilities.

Notwithstanding the commercial depression, electrical industries in Canada may be said to be in a flourishing condition. The larger electrical manufactories are in full operation, and report a large increase in the output, and with many contracts on hand. It is likely that during the coming year their capacity will be considerably increased. The wonderful increase in the use of electricity for all purposes necessitates the installation of larger units. Dynamos that were considered colossal a few years ago are now being abandoned on account of their lack of capacity. This is leading to the equipment of the factories with more modern and powerful machinery for their production. The allied trades of the tool builder, machinist and engineer are all therefore receiving the benefit of this development of the electrical age.

Not the least of the current developments is the remarkable increase in the number and mileage of electric railways. The electric motor, a short time ago considered as limited to urban and suburban work, is now usurping the functions of the steam locomotive, and it is not too much to say that ere long we may expect to see it on our main lines of railway.

In other branches of electrical work good general progress has been made in perfecting systems and methods at present in use, both electrically and mechanically. The telegraph, with its adoption of new and rapid self-recording apparatus, and the telephone, with its improvement in long distance transmission of speech, are fully keeping pace with improvements in other departments.

The Association is to be congratulated on the promising outlook. It is likely that in the immediate future, with an improvement in the financial world, and as disturbing elements are eliminated, the developments will be even more rapid than in the past, and an era of greater prosperity than ever will be abundantly realized.

The President's address was greeted with applause.

The Secretary read the minutes of the fifth annual convention, which were confirmed.

The Secretary also read the Secretary-Treasurer's report as follows :

#### SECRETARY-TREASURER'S REPORT.

During the year covered by this report, the Association has made satisfactory progress. During the Association year beginning 1st June, 1895, and closing 31st May, 1896, there were added to the membership roll 24 active members and 4 associate members. During the same period 8 active and 7 associate members tendered their resignations, leaving the net gain in membership, 13.

Since the close of the Association year, there has been elected 21 active members, making the present membership 194 active members and 35 associates, a total of 227.

There are on the roll a considerable number of persons who, without having resigned their membership in the Association, have ceased to take an active interest in its affairs, and have likewise failed to pay their membership fees. It should be understood that when a person joins the Association, he thereby becomes a member, not for one year only, but until such time as he formally resigns his membership, and that until his formal resignation is received by the secretary and accepted by the Executive Committee, he continues to be liable for payment of the annual fee. It is perhaps due to the lack of a definite understanding on this point that the actual standing of the Association, with regard to its bona fide membership, is at the present time somewhat uncertain. The time has arrived when definite action should be taken to put an end to present and future uncertainty with regard to this matter.

It was to be expected that some of those, who at the outset became members of the organization, without being actually interested in the work which it is designed to accomplish on behalf of the electrical interests, would soon drop out. In the place of such, the Association has within the past two years been receiving as members, persons connected with the various departments of electrical work, and who therefore feel the benefit to be derived from connection with the organization and have a personal interest in its welfare. It thus appears that, while for a time, the additions to the membership may be in a measure offset by the withdrawal of members of the first mentioned class, the Association is steadily gaining in character and influence.

The annual conventions, which have been extremely interesting and enjoyable from the commencement, are becoming more so year by year, in proof of which I need only point your recollection back to the delightful meeting in Ottawa last autumn, and direct your attention to the character of the programme on this occasion. Two meetings of the Executive Committee have been held since the close of the Ottawa convention, viz. : on the 22nd of October, 1895, and the 16th of January, 1896. At the first of these meetings, accounts in connection with the Ottawa meeting were examined and ordered to be paid, and the secretary instructed to have printed 500 copies of the revised constitution. At the second meeting two active members were elected. Messrs. Wright, Breithaupt and the president were appointed a committee to endeavor to make arrangements for a popular scientific evening lecture in connection with the present convention. The selection of a suitable place of meeting for the convention was left in the hands of the Toronto members of the Executive. The secretary was instructed to make request of the following persons for papers on the subjects named : Mr. O. Higman, Ottawa, "Lamp Tests"; Mr. P. G. Gossler, Montreal, "High Potential Underground"; Mr. J. M. Campbell, Kingston, "Long Distance Power Transmission"; Mr. Charles P. Dwight, Toronto, "Ocean Cables"; Mr. F. C. Armstrong, Toronto, "The Future of the Electric Railway"; Mr. James Milne, Toronto, "Meters"; W. McLea Wallbank, Montreal, "Utilization of the Power of the Lachine Rapids"; Mr. George Black, Hamilton, "Acetylene Gas"; C. F. Medbury, Ottawa, T. R. Roseburgh, George White Fraser and E. J. Philip, Toronto, subjects not named. Mr. J. J. Wright was appointed a committee to ascertain whether a suitable steamer could be chartered for an excursion and dinner in connection with the annual convention. The sum of $25.00 was placed at the disposal of the Committee on Statistics. The Toronto members of the Executive were appointed a committee to perfect local arrangements for the convention.

Following is a statement of the receipts and disbursements for the year :

FINANCIAL REPORT FROM 1ST JUNE, 1895, TO MAY 31ST, 1896.

RECEIPTS.

| | | |
|---|---|---|
| Cash in bank June 1, 1895 | $186 | 77 |
| Cash on hand June 1st, 1895 | | |
| 93 active members' fees at $3.00 | 279 | 00 |
| 14 associates at $2.00 | 28 | 00 |
| Refund by Statistical Committee | 23 | 62 |
| | $517 | 39 |

DISBURSEMENTS.

| | | | | |
|---|---|---|---|---|
| Expenses of convention at Ottawa | | | $218 | 92 |
| By cash as per local committee statement | $100 | 00 | | |
| "   caretaker at Ottawa | 3 | 00 | | |
| "   express charges on books to and from Ottawa | 1 | 40 | | |
| "   Canadian Electrical News, printing | 72 | 50 | | |
| "   Canadian Photo-Engraving Co. | 16 | 22 | | |
| "   A. F. Sladen, stenographer | 25 | 80 | | |
| | $218 | 92 | | |
| Electrical News for printing | | | 7 | 00 |
| Postage | | | 35 | 10 |
| Exchange on cheques | | | 1 | 50 |
| Blackhall & Co., 50 leather certificate covers | | | 4 | 00 |
| Grant to secretary | | | 50 | 00 |
| Mortimer & Co., badges, including protest charges | | | 17 | 16 |
| Grant to Statistical Committee | | | 25 | 00 |
| | | | $358 | 68 |
| Cash in bank, May 31st, 1896 | | | 157 | 21 |
| Cash on hand, May 31st, 1896 | | | 1 | 50 |
| | | | $517 | 39 |

RECEIPTS SINCE MAY 31ST, 1896.

| | | |
|---|---|---|
| June 1st, 1896, cash on hand | $ 1 | 50 |
| 38 active members' fees at $3.00 | 114 | 00 |
| 1 active member's fee at $5.00 | 5 | 00 |
| 3 associate members' fees at $2.00 | 6 | 00 |
| Cash for exchange on cheque | | 15 |
| | $126 | 65 |

EXPENDITURES.

| | | |
|---|---|---|
| Ribbon and pins for badges | $ 2 | 88 |
| Receipt forms | | 40 |
| Postage | 20 | 78 |
| Exchange on cheques | 1 | 15 |
| Envelopes | | 15 |
| | $25 | 36 |
| Cash deposited in bank since June 1st, 1896 | 95 | 15 |
| Cash on hand June 17th, 1896 | 6 | 14 |
| | $126 | 65 |

Total standing to credit of Association, June 16, $258 50.

Certified correct,
B. J. THROOP } Auditors.
A. A. DION }

On motion of Mr. Dunstan, seconded by Mr. Kammerer, the report was confirmed and adopted.

The President : The next thing will be the reception of reports from committees. Mr. J. J. Wright will report for the Committee on Legislation and Mr. E. C. Breithaupt for the Committee on Statistics.

Mr. Wright : There not having been any legislation either in the Dominion or Local Houses that would

affect electrical interests in any way, the Committee on Legislation have no formal report to make.

Mr. E. C. Breithaupt read the report of the Committee on Statistics, as follows :

REPORT OF THE COMMITTEE ON STATISTICS.

Your committee beg to report as follows :—The committee endeavored during this year to carry into effect the idea expressed in the report of last year's statistical committee as to the compilation of data relating to central stations for the supply of electric light and power. To this end we drew up a blank form, requesting information on the following points, namely : motive power, station apparatus, station output, running time of station, and prices obtained. Information regarding the original cost of installation and cost of operation was not asked for, because it was deemed advisable not to make the form too long and complicated, but it was thought that if a proper interest were shown by central station men in the compilation of these statistics, they might at a later date be filled out more completely. There is no doubt that complete statistics of the central stations for the supply of electric current throughout the Dominion would be a reference book of no small value to all persons connected with the industry, and the committee have given considerable thought as to how this idea could be best put into practice. It has been thought that the formation of a sort of bureau of statistics, kept in the archives of the Association, to which any active member could have access, would best serve the end in view. The committee recognize that there is considerable information concerning the operation of a plant which proprietors cannot afford to have made public ; the more general data, such as first cost of plant and particulars concerning the plant and apparatus, could be kept on file in the secretary's office and revised annually, so that any member of the Association could, on application to the secretary, obtain whatever general information he might want for his private use. Particular data should of course in each case be obtained personally, and it is fair to assume that between members of the Association, where the person requesting information and the object are known, it would be cheerfully given. The committee are of the opinion that a scheme for the compilation of statistics somewhat in the nature of a bureau of mutual information, as above outlined, would be a valuable adjunct to the work of the Association, and respectfully recommend its consideration.

Of the blanks sent out only a comparatively small proportion were returned, and in some instances were not filled out so completely as they should have been. However, the returns received would form a fair basis on which to continue the work. Because of the incompleteness of the returns the committee have not considered it advisable to draw up a tabulated summary.

The statistics received bring to light some interesting facts. Of the total number of replies received only three are from municipal plants ; 12 companies only state that they supply current for power purposes, and two only supply current for heating ; one for electric welding.

One fact which is particularly prominent is that the majority of the smaller stations are operated only during a very short period every 24 hours, mostly from dusk till midnight. In most of these cases it is probable that a sufficiently large motor load could not be obtained to employ the station during the day-time.

The committee do not consider that the work which they have done is in any way finished, and strongly recommend that it be continued during the coming year.

All of which is respectfully submitted.

E. CARL BREITHAUPT, Chairman.

On motion of Mr. F. C. Armstrong, seconded by Mr. J. J. Wright, the report was received and allowed to stand until Thursday, the 18th June, for consideration.

GENERAL BUSINESS.

The President read a telegram from Mr. L. B MacFarlane, 2nd Vice-President, as follows :—" I sincerely regret that I cannot attend convention. Hope meeting will eclipse all former ones "; and stated that he was sure that all the members of the Association present would regret Mr. MacFarlane's inability to be present.

The President called the attention of the members to the banquet to be held at Lorne Park on the following evening, and requested them to inform the Secretary as soon as possible of their intention to be present. He also urged on the members to be present with their friends at the lecture on " Radiant Matter " to be given by Mr. James Milne in the rotunda of the Board of Trade building in the evening, as it would be an encouragement to Mr. Milne and the officers of the Association.

The President : While I am on my feet, and before proceeding with the consideration of the first paper to be read, I desire to say that it is quite gratifying to the Executive of the Association that so many new members are coming in, especially central station men. There were twenty-three new members elected, and I think about twenty of them are central station men. That of itself is very gratifying, and shows that the Association is fulfilling its mission. I hope to see more of them.

The first paper on our list is " Ocean Cables," by Mr.

C. P. Dwight, of Toronto. I am very sorry we will have to dispense with this paper. Mr. Dwight made an earnest effort to have it ready, but he found that the data necessary to give us an authentic, clear and concise history of ocean cables was so obscure, and required so much digging out, that he has not been able to complete it. I hope at some future time we will be favored with it—perhaps at the next convention.

We will pass on now to the paper on "Acetylene Gas," by Mr. George Black, of Hamilton. (See page 141).

Mr. Black, on rising to read his paper, said : Mr. Chairman, in coming away in a hurry I forgot two things. I forgot my paper (although I find it printed here), and I forgot some matches ; so, if the gentlemen here have a supply of matches on hand I hope they will take up a collection in the meantime.

After the reading of the paper, Mr. Black gave a demonstration of the light produced from acetylene gas by means of a small apparatus which he had improvised for the purpose, and which he explained the working of to. the members present. He stated that they were, in this demonstration, competing with the sunlight, and it was hardly a fair test.

Mr. J. J. Wright : I would like to know the pressure you have on.

Mr. Wickens : I think the pressure is less than half a pound.

Mr. Black : This gives the yellow flame because it does not oxidize sufficiently. I may say in the course of the next two or three weeks Mr. Willson's factory will be in full blast, and I understand he is going to light up St. Catharines. Several establishments will have the gas pure to show the results, so that all who are interested can visit St. Catharines to see it for themselves.

Mr. F. Thomson : Is there any place where they are actually using this gas ?

Mr. Black : Not as yet.

Mr. Thomson : They have been making it for the last couple of years back.

Mr. Black : Just experimenting, A number of factories have been watching to see the result of the Niagara Falls output before they start.

Mr. Thomson : They have been manufacturing since May.

Mr. Black : Since May.

Mr. Thomson : Some of that ought to be used.

Mr. Black : It requires a good deal to make a contract with a town, and you have to have a good supply to keep it up, and not run out of it.

(At this point the electric light in the room was turned on, and comparisons made with the acetylene gas.)

Mr. Wickens : The globes on the electric light kill forty per cent. of the light, whereas the acetylene gas is a bare light.

Mr. Black : I have had it without the slightest trace of yellow color, and with just a blue spot about the size of a pinhead.

Mr. Wright : I understand that Mr. Black informs us that Mr. Willson proposes to use that gas in St. Catharines as a mixture or energizer to the ordinary gas. In the paper I notice that he says that it does not mix very well with the water gas ; in fact, it is no use for that ; it will mix better with coal gas. We all of us know, I think, that there is precious little coal gas manufactured. Very nearly all the gas companies are running with water gas. If that is really the case there does not appear to be very much field for this product as an energizer for the ordinary gas of commerce, as we may call it. The gas companies, for the sake of using it, save as an energizer, would never exchange their water gas plants and go back to the old methods of producing coal gas at a vastly greater expense in order to use acetylene gas, which is manufactured at a tremendously low rate. My remarks are with reference to gas plants all over the country. We are all aware that the largest gas plants are water gas plants.

Mr. Black : I don't know anything about the condition of the gas companies. In Hamilton I know it is coal gas which they make. I understand in Toronto it is water gas.

Mr. Wright : Both.

Mr. Johnson : Is there any difference in using iron, brass or copper tubing ? Does it cause any trouble by using the brass or copper ?

Mr. C. S. Mallett : I believe it combines with copper and makes a kind of deterioration ; it uses up the copper, I understand. It also does with brass, but not to such an extent as to interfere with any fixtures like those (refers to fixtures used by Mr. Black in illustrating the acetylene gas.) On iron or lead it has no effect.

Mr. J. J. Wright : If there is a deteriorating effect and the copper would combine with the gas, that would certainly seem to rule out the fixtures at present in use. You could not guarantee that any brass fixture has not got a large proportion of copper in its composition. It appears that if this gas were used even as an enricher for ordinary gas, and used in an ordinary fixture, that the whole thing would have to be revised ; new fixtures would have to be put in all over. Would not that have a tendency to shut it out as an energizer of ordinary gas ?

Mr. Black : Have not all fixtures an iron pipe inside, for gas ? I think the conducting pipe is iron.

Mr. Wright : Any number of gas fixtures are made of copper and brass piping.

Mr. Black : The parties who are running it do not think they will have any trouble.

Mr. Wright : They are rather too optimistic as to the outcome of this product. We are well aware that the true test of the merit of any new discovery is its use. It is two years now since Mr. Willson first discovered this gas ; and we do not find that it is in use in any commercial or practicable sense. They have been making preparations in various places to experiment with it and that is about all. If this had been such a valuable energizer, as the making of it is such a very simple process, it certainly would have come into use to a very much greater extent than it has. I think you can look upon that as the true test of any new discovery.

Mr. Black : Anyone who makes such a discovery as this wishes to secure it as safely as possible before giving it out to the public. It takes a lot of capital to work a thing of that kind ; it takes time to interest capitalists ; it had to be tested in every possible way ; they had to be satisfied by chemists and others that it was all right before they put their money in it. I have no doubt but that within the next six months you will see plenty of it.

Mr. Mallett : Why is it those three jets have lost a great deal of their illuminating power ?

Mr. Black : We have stopped the generation of the gas and the pressure is taking care of itself. We might as well turn it out, but we wish to burn up the gas. Where you see the yellow flame it shows an excess of carbon.

Mr. Breithaupt : The paper we have listened to by Mr. Black is certainly one of great interest to all the members of the Association, and I think we should congratulate Mr. Black upon presenting the subject in such a concise and clear manner. The question of acetylene gas is one which is creating a great deal of interest throughout the country at the present time. I may say I have studied it up a little myself, and have data of tests that were made by two engineers of the Acetylene Gas Co., in Philadelphia, the company which is there exploiting this matter principally, I believe. These tests were made in North Carolina, before the three experts, Professor Houston, Dr. Kennelly and Dr. Kinnicutt made their tests, and they were not so favorable as the tests made by the three experts in March last. Of course we may expect in that time (it was nearly a year before that the other test was made) there was some improvement made. The cost of acetylene gas varies directly at the cost of producing the calcium carbide. The figures here given as to the cost of calcium carbide I think are altogether too optimistic. In the report of the test made by Professor Houston and Drs. Kennelly and Kinnicutt the cost of power is given as $5 per h.p. We all know that $5 per h.p. per annum is a figure that nobody ever dreams of, much less realizes. I am surprised that the rate mentioned has been secured at St. Catharines. Of course, if that is so the argument drops.

Mr. Black : About 200,000 h.p. is available in the province of Quebec at the same price.

Mr. Breithaupt : It is a price that is surprising, because nobody who has been in the business at all has known any such price. At Niagara Falls, where they are producing power in such enormous quantities, they are selling it at $20 per h.p. delivered at the works only. Therefore, one argument that is made throughout, that the production of calcium carbide can be utilized as an additional load to the ordinary small station, drops. The ordinary small station cannot deliver power at anything approximating $20 per h. p. For instance, here in Toronto or elsewhere where power has to be generated, from steam, the difficulties of using this gas as a pure illuminator would be great. Could this gas be controlled and distributed through pipes, if it were generated in a private house and be burned there just as it is generated? The Acetylene Company in Philadelphia were exploiting a system of generating directly in the place where it is to be used. They have for that purpose a generator similar in principle to what Mr. Black has here, just a box in which they keep the calcium carbide ; they apply water to it and the gas goes from the generator directly into the house pipes. The great question is, how to regulate the pressure. The affinity that this compound shows for water is very great. I understand that once you have turned on a certain amount of water you cannot stop generation of the gas until that stock of water is completely exhausted. I was speaking to engineers in Philadelphia about it and I asked them how they regulated it. They said the only way is to let the gas escape, or else the pressure will get to such a height as to be dangerous.

Mr. Black : They have a safety valve, and if the pressure rises above a certain point it blows off.

Mr. Breithaupt : That would be wasted energy. As to the use of acetylene as an enricher for coal and water gas, I read a paper by Prof. Lewes on that point, and he said the gas does not mix properly with the water gas and is of very little value there. It is, however, a good enricher for coal gas. The commercial value of acetylene gas, taking all these points into consideration, it seems to me is not quite so promising as its promoters claim for it. There is another point to which Mr. Yule has called my attention with reference to laying pipes to distribute acetylene gas from a central station. At present, in laying distributing pipes for gas, we put the pipes below the frost line, not only to prevent condensation, but also to keep our pipes intact. If you lay pipes above the frost line the frost will soon break them. You must take into consideration that however you place these pipes, you have got to put them below the frost line in order to keep them intact, and then you would have just as much cost in laying as in laying ordinary gas pipes.

Mr. A. A. Dion : I have listened to Mr. Black's paper with a great deal of interest. It is a very good paper. But, I must say I am still in the same position that I was before ; I am not yet able to make up my mind as to the true cost of acetylene gas, or its true value. I have read nearly everything that has been published in the technical journals concerning acetylene gas in recent years ; and I have read with a great deal of attention, trying to form an opinion as to the real cost of producing the calcium carbide and also the true value of the gas as a possible competitor with electric lighting, but I found those articles so contradictory that it was impossible for me to form any opinion whatever, and I am still in the same position. I think, however, that Mr. Black's views of the prospects of acetylene gas are much too rosy. However, those of us who might have formed the opinion, as I had, that the gas was very poisonous, are consoled with Mr. Black's statement that it is not so poisonous as coal gas. I think it might be very much less poisonous than coal gas and still very objectionable to most of us in that respect. Another thing : he argues that its strong

pungent smell is a safeguard, and a person could not go to sleep before noticing it. We know that the ordinary illuminating gas has a very strong pungent smell, nevertheless accidents do happen. There is the further objection, that there is nothing necessarily alarming about the smell of garlic. There is one thing I wish to call your attention to, which is, I think, perhaps a misprint ; Mr. Black states in his paper that the pipes can be laid below the surface for little or no expense. I cannot conceive of a case where no expense would be incurred.

Mr. Wickens : There is one matter that Mr. Breithaupt spoke of which was demonstrated here, while we were experimenting with the gas. The gas generates very quickly, as soon as the water strikes it ; it is absolutely necessary to hold the meter down pretty well ; it lifted it right up. It is hard to tell how to accommodate the pressure. It is evident it would be very troublesome to regulate the speed at which this gas is to be generated and delivered into a tank. You noticed that if we increased the pressure on the gasometer the light would flare up. As long as the pressure was at the same fixed point the light was steady, and as soon as we put a little extra pressure on it made the light considerably stronger. It is evident it is going to be very difficult, if it ever is accomplished, to exactly regulate the pressure, from the fact that the gas generates very rapidly and you could not tell when it is going to stop. After the color disappeared in the jar it was still making gas—it was still pushing against the counterweight.

Mr. Yule : I do not see any difficulty in overcoming the objections that Mr. Breithaupt makes. Supposing you have a fifty pound holder in your cellar, you will know how many pounds you can make from so many pounds of acetylene—making your gas in the afternoon before you start—and you simply use up your holder. Your pressure will be the same throughout the evening if your holder is of sufficient capacity to supply you with gas during that night. It is not necessary to make more gas while you are using it. So many pounds of acetylene will make so many feet of gas ; it is simply the operation of leaving it stand there and using it in the night.

Mr. J. F. H. Wyse : Wouldn't that of itself be an objection to having the gas in the house if you had to depend on the domestic to make the gas in that way? It seems to me to be a serious objection. The master of the house would not want to have to look after a matter of that kind.

Mr. Wickens : Suppose we make a gas receiver or receptacle in that house. We know if we have so much water we are going to make so many feet of gas ; the receiver will be of some fixed size and the pressure so much per square inch ; we start our lights around the house, and if we use two-thirds of the gas generated we have made the receiver no smaller and the pressure has gone down.

Mr. Hopkins : I don't think that is so at all. That would not necessarily follow.

Mr. Armstrong : It is the running and regulating of the generator of acetylene from calcium carbide in what we might call the small unit which Mr. Willson seems to favor, but the difficulty would seem to be that it is not something that the ordinary domestic could be expected to handle in a satisfactory way. The possibility of danger and the difficulties are certainly greater than with the ordinary furnace, and they are generally found sufficient ; and it seems to me unlikely, accustomed as people are now-a-days to depend on the central station or gas plants, that they would be disposed to have any bother. One thing which seems to have been left out of consideration is the fact that the calcium carbide for shipment must be put up in closely sealed packages, and when these are opened, on exposure to the air, it deteriorates very rapidly. Mr. Willson did not seem to favor the idea of distribution when liquefied under pressure in cylinders ; he did not state for what reason, but I fancy it is on account of the danger of explosion when under high pressure, which I believe is very great. There is another point in connection with the gas which I think should not be overlooked, and that is, when the temperature is raised to about 300 degrees it becomes highly explosive, and in case of general use and in the case of fires that would become a very serious drawback to its use. It seems unnecessary to say anything about the estimates of cost which have been given. I think as yet they are not based on any ordinary commercial values, which we would require to have established before it could have any extended use. There is one small item which I see is entirely omitted in the calculation of the cost as given by Mr. Black, and that is the cost of the carbon electrodes, which is found to be a very considerable factor, and one that has been entirely overlooked.

Mr. Black: In reply to what Mr. Briethaupt said about regulating the pressure, and the Philadelphia engineers saying that they had to let the gas go to control it, I might say I have seen illustrations of a gas holder so arranged that as the gas goes up it shuts off the water; and as it is consumed and as the gas drops down it lets the water in, and by that means the manufacture of the gas is automatic. I think means will be found in practice to prevent the necessity of blowing off the gas when the pressure gets too strong. I noted here something about an error in figures, but I do not remember what it was.

The President: It was in reference to the laying of the pipes.

Mr. Black: Those two little words, "or no," had better be taken out; there must be some expense in laying the pipes. The cost of the carbide has been variously estimated; some have made it as high as $100 a ton. Mr. Ferguson, who read a paper before the National Electric Light Association, figured it out, and I discovered an error in his figures of $7—reducing his figures by $7. I have not the paper with me. He was figuring on ten tons a day production, and he reduced the cost of labor, and so on. The price of the carbon electrodes he places at $7.77 for each ton more than the lime and the coke. That should be reduced by one-tenth, making the cost per ton 77 cents. I included the cost of the carbons in the other charges. As to the poisonous nature of the gas, one of the authorities who experimented declared it was only one-seventh as poisonous as ordinary gas. I won't vouch for that, but that is his statement. He is a Frenchman, and claims to be a scientific man. I think I have covered all the points, as far as I know. Had the convention been a few weeks later I would have had more perfect apparatus. Mr. Willson offered to send me one of the smallest household arrangements that he was making, but it was not yet completed, so that I had to come away without it. I left this until the last moment, expecting to have a more complete outfit. In the course of a few weeks we will have it before the public, and then the exact price can be got at.

Mr. Breithaupt: In answer to what Mr. Black has said, I wish to call attention to the fact, not that the water could not be shut off, but that the gas does not stop generating the moment the water is shut off, because the calcium carbide has to absorb all the water that is in the jar before it stops generating, and in that way the danger arises from generating with a small plant. I said that in Philadelphia they did not have any arrangement to relieve the pressure. I have much pleasure in proposing a hearty vote of thanks to Mr. Black for the paper he has given us. I am sure that it has been of great interest to the Association and of great benefit.

Mr. J. J. Wright: I think it is only fair to give it due credit. I think Mr. Black states in his paper that it is intoxicating; that certainly is a point that ought to be in its favor. I would like to trouble Mr. Black with one more question; that is, where is that 200,000 horse-power in the Province of Quebec?

Mr. Black: In the backwoods somewhere.

The President: It has been moved by Mr. Breithaupt, seconded by Mr. Wright, that a hearty and sincere vote of thanks be tendered to Mr. Black for his care in the preparation of this paper. (Carried with applause.)

The President: Mr. Black, I have great pleasure in conveying to you the thanks of this Association for the trouble and care you have taken in the preparation of this paper.

Mr. K. J. Dunstan: I may mention a little incident: A good many years ago we had trouble in Hamilton with a certain telephone line; we could not find the interruption. After a while a youngster was found hiding behind the fence tapping the line with a home-made telephone. Nothing much was done to the youngster, but the telephone was confiscated. The youngster was Mr. Willson.

Mr. Black: I think I have that telephone yet.

The President: Is it your wish to adjourn, or shall we proceed.

Mr. J. J. Wright: I move that the convention adjourn.

At 4:30 p.m. the convention adjourned to Thursday, June 18th, at 10 o'clock a.m.

ILLUSTRATED LECTURE BY MR. JAMES MILNE.

In the evening the Rotunda of the Board of Trade was filled with ladies and gentlemen assembled to hear Mr. James Milne's lecture on "Radiant Matter." The lecture was illustrated by numerous diagrams projected upon a screen, and with demonstrations of Roentgen rays, and was received with much interest and appreciation by the audience.

Immediately following the lecture a visit of inspection was made to the power stations of the Toronto Electric Light Co. and the Toronto Railway Co.

### SECOND DAY.

#### MORNING SESSION.

The President called the convention to order at 10 a.m., and announced as the first order of business for the day the consideration of a place for the holding of the next convention.

Mr. J. Carroll: Mr. Chairman, I have pleasure in suggesting Niagara.

Mr. Breithaupt: I second that.

Mr. J. J. Wright: As I understand, there is no strong representation of the Association at Niagara Falls. That of course means that if we held a convention there we should have to carry our supplies along and make arrangements for the features of entertainment and so on independent of the place itself. I am not mentioning this because I have any objection whatever to Niagara Falls, but it is simply a matter that should be understood.

The President: That is a very important feature. We have no representative here from Niagara Falls. I presume when Mr. Carroll suggested Niagara he meant Niagara Falls.

Mr. Carroll: Yes.

Mr. Kammerer: While we have no representative at Niagara Falls except one, all that know that one will say he is a host in himself—that is, Mr. Ross Mackenzie. I think the social features, and the taking care of the people when they get there, can be safely left in Mr. Mackenzie's hands.

The President: Allow me to read a letter which was received some time ago in connection with the coming convention:

NIAGARA FALLS PARK & RIVER RAILWAY CO.,
MAY 13th, 1896.

A. B. SMITH, ESQ.,
G. N. W. Telegraph Co., Toronto.

Dear Sir,—In connection with the coming convention of the Canadian Electrical Society, I beg to offer you on behalf of the company the freedom of the road for your members if you hold the convention at the Falls. We desire you to visit the Falls, and we trust that we may have the pleasure of your company.

Yours truly,
ROSS MACKENZIE, Manager.

We acknowledged the receipt of that letter and thanked Mr. Mackenzie very kindly for his invitation, and at the same time said that it was probable that at some future time we would hold our convention at Niagara Falls, and we would avail ourselves of his kindness. Are there any other places suggested?

Mr. J. J. Wright: It appears to me the only other place available would be Quebec. There was some

talk of that last year. There would be a certain amount
of objection to that on account of its distance from this
end of the country. Having held the meeting in Ottawa
last year, it practically leaves us with Toronto and
Montreal, and having held conventions in these places
in the not very remote past, we are left almost without
any choice but Niagara Falls. It is a fine place to
spend a few days, and we should have a very pleasant
convention there.

The President : It is desirable from the Association's
standpoint that our conventions should not be held exclusively in one section of the country. Objection may
be taken to having a convention held in Niagara Falls
succeeding the one held in Toronto, but if there are no
invitations to go anywhere else, we have no alternative
but to accept and decide on Niagara Falls, and I presume it is decided that we go there. With reference
to the time, I think it should be in the month of June.

Mr. Kammerer : I think that is about the right time.
We have a good sample of June weather here and a
good sample of the crop that June brings, and it would
bring just as good a crop at Niagara Falls. There will
be no politics at this time next year.

The President : It is for this convention to say whether
they will meet in June or not. The Executive can fix
the exact date. Is it your wish that the convention
shall be held in the month of June ? Carried unanimously.

The President : The next order of business will be the
consideration of reports. We will take up Mr. Breithaupt's report of the Committee on Statistics.

Mr. Breithaupt : Mr. President, the report of the
Committee on Statistics suggests the formation of
something in the nature of a bureau of mutual information to be formed in the interests of the members of the
Association, particularly to gather information regarding
central stations, and keep it on file. The central station
men throughout the country are particularly the ones
whom we ought to get into our Association, and we
thought that we could find no better way of interesting
them than by framing some such scheme as this, whereby
they would be benefited. I think the committee has
done a little good work for the Association in increasing
its membership. During the past year, I believe a number of central station men have come into the Association. With these blanks that we sent out we outlined
our plan to the central station men and asked them to
come into the Association. Some of them have done
so. Now, as to the method of carrying out this plan
we had in view : It would be a little difficult to carry it
out unless we had the Secretary do the principal work
of it, that is, keep the information on file and keep it
properly tabulated, so that any member of the Association wanting information could write to the Secretary
for it. Some might want information on the prices
usually obtained for supplying power, or whatever it
might be, and central station men are very reluctant to
give this information. The schedule of prices, of course,
they are not at all unwilling to give, because it is published and is public property everywhere, but any detail
of matters as to prices, etc., they are reluctant to give,
and naturally enough. It was thought a man might
want information, for instance, as to prices, what prices
were in other places. He could state his case as clearly
as possible to the secretary, who would advise him from
which members of the Association he would be most likely
to get the particular information asked for. He would then
correspond with these other central station men, and being members of the Association, mutually acquainted
with each other, and knowing the object of the information sought, we thought the information would, in all
cases, be cheerfully given. All central station men have
had considerable experience in this. We received a
great many blanks from central station men answering
in a very reluctant or uncertain way, and saying we
don't know what it is to be used for, nor where it is to
be used ; and in most cases they did not fill it out at all.
We all recognize it to be information which is of value
to us, because in town contracts, where we have to renew, the town authorities are always talking about that,
this, and the other place where they get light for very

much less than we are supplying it ; and generally they
do not understand the details of the case, and simply
say this corporation is getting light for less, and you
have got to give it for that or we will get in another
company to compete with you. If we could get the information in detail for such cases we would probably
know why that corporation is getting cheaper light, and
could then answer the authorities in that respect. I
would like to hear the suggestions of the members on
this point, particularly Mr. Hunt, who can give us considerable information along that line. This method is
adopted in the National Electric Light Assn. He spoke
to me about it yesterday, and there, I believe, they have
derived very much benefit in that way.

The President called upon Mr. Hunt to speak, but
he was absent from the room.

Mr. Kammerer : I think the scheme is an excellent
one, for this reason : We have water power and we
have steam power. In most cases where they have
water power they sell the incandescent light more
cheaply than where steam power is used, and if this information is secured in a statistical way, the central
station man can talk to his customers and tell them that
for that reason he cannot give them incandescent light
at the same prices that a town plant 100 miles away
can, because he has to produce it by steam. The consumer does not make it his business to find out whether
the person is furnishing the light by using water or by
using steam.

Mr. Armstrong : There is no doubt but that the
scheme which Mr. Breithaupt suggests would be of
great value to the industry generally. We are almost
continuously in receipt of inquiries by companies renewing their contracts as to the prices paid and the conditions under which contracts are carried on through
out the country ; and very often we are not in a
position to give that information, because ordinarily we
cannot get it on the same basis as the companies could
directly, or as the Secretary of the Association could, in
his official capacity. Referring to what Mr. Hunt stated
to Mr. Breithaupt and myself last night as to the
plan of the National Association, of which he is a
member, he said that the Secretary, who was paid a
proper remuneration for devoting a considerable portion
of his time to the affairs of the Association, had,
just in the way Mr. Breithaupt suggests, general information as to the conditions under which each of
the central stations was carried on ; and any
member of the Association who was in difficulty
of any kind, or wanted information as to the operation
of his individual plant, could simply apply to the
Secretary, and feel sure he would obtain the com-
pletest and best information, or a reference to central
stations where he could get it, and at the same time feel
it was the business of the Secretary of the Association
to furnish him with that information. I think if any
such scheme as this is carried on, that in the proper
place we should consider the re-arrangement of the
Secretary's position, so as to increase the remuneration to such an amount that the members of the
Association would have no hesitancy in asking him for
the information, and feel they had a right to expect it.

Mr. Kammerer : I fall in with all Mr. Armstrong has
said, except the financial portion. If we are to remun-
erate our Secretary in a proper way for doing that work
we would have to give him about $1,200 a year.

Mr. Armstrong : My idea was simply this, that our
Secretary now does a great deal of work free for the
members of the Association, and it seems unfair to pile
a great deal more on him.

The President : Our Secretary suggests that if we had
the information a great deal of it might be available
without much additional expense. It does seem desirable that we should have such a bureau through which
information which is reliable and authentic could be
easily and promptly obtained. It is of vital importance
sometimes to the central station men to know just exactly
what light is being produced for, and all the rest of it,
in neighboring towns. I hardly think the Association
is in shape in its present condition to undertake that
work. The work done by the committee under Mr.

Breithaupt will bear fruit, and I think it would be wise for us to let the thing simmer for another year, and for the present go on as we have been going, and probably by that time it will assume a more definite shape, and we will then be in a position to do something.

Mr. Breithaupt : Mr. Mortimer and I discussed this matter some time ago. The committee on statistics for the coming year, whoever they may be, should, I think, be willing to do a great deal of the work, particularly in formulating the nature of the information that should be gotten together ; and to carry on the correspondence for the Association would not be such a great undertaking, because the number of central station men in our Association is not so very large as to take up a great amount of time. If the Secretary and the committee on statistics would co-operate, it seems to me that something might be done during this coming year on this work, and it is certainly a work that we should undertake. We have got to do something to interest the central station men in our Association. If the Association is not to be of mutual benefit to its members, what is it for? I think if we add $75 or $100 to the Secretary's remuneration, which, it seems to me, we are able to do, Mr. Mortimer would be willing to attend to the correspondence part in connection with it, and with the help of the committee on statistics, I think the matter could be taken in hand now.

Mr. Mortimer : Mr. President, will you let me say just a few words on this point. As Mr. Breithaupt says, he and I have discussed this matter to some extent, and I feel that the Association might be of very much more use and practical value to its members, especially central station men, if we had such a bureau as is proposed. I agree with Mr. Breithaupt that we cannot commence that work any too soon, and so far as I am concerned I am quite willing to do my share of that work for the coming year without any extra remuneration, to see how the thing works. After that, if it is a success and you see fit to increase my remuneration, all right. I would like to see that thing started, and to see the Association do some practical work, and anything I can do will be gladly done.

The President : Our Secretary is offering his services out of the fullness of his heart. I think very few of the members here realize the enormous amount of work devolving upon him, which he has performed. The remuneration we speak of is simply for the actual expenditure and not of being a remuneration for his services.

Mr. Breithaupt : Mr. President, I would move the adoption of this report.

Mr. Yule : I second that.

The President put the motion, which on a vote being taken, was carried.

The President : It is necessary that we have standing committees for the next year, and I would nominate Mr. A. A. Dion, of Ottawa, and Mr. J. A. Kammerer, of Toronto, to bring in a report this afternoon. The committees are on Legislation and Statistics.

Mr. P. G. Gossler : Do I understand from the adoption of the report that the Committee on Statistics will undertake the extra work ?

The President : The committee will be continued and the work with it. We will now proceed with the first paper on the list entitled, " Meters " by Mr. James Milne. (See page 146).

Mr. Milne, on rising to read his paper, was greeted with applause. He said : In getting up this paper on meters I thought first of devoting the whole of it to chemical meters, but as the number of central stations using the chemical meter is very small, I thought we might just throw in the rest, as it were.

Mr. Milne's paper was fully illustrated by diagrams thrown upon a screen by means of the stereopticon.

The President : Gentlemen, the paper is open for discussion.

Mr. Rosebrugh : I would like to ask Mr. Milne if there would be any conscientious objection on the part of the company if they substituted silver for zinc ?

Mr. Milne : I don't see that there would be any objection at all. Probably the cost would be a little more with the silver, but as long as the principle is the same, I don't see that there could be any objection.

Mr. Armstrong : Mr. Milne has certainly given us a very interesting paper. I think it must be a source of congratulation to Mr. Wright that he is able to meet such customers as Mr. Smith (referring to Fig. 8 in paper) when they kick about the mysterious results obtained from chemical meters, because he would certainly satisfy them that they must be wrong and the meter right. A very large part of Mr. Milne's paper loses its importance in view of the fact that the larger amount of current for incandescent lighting is supplied from alternating stations, and except the Lowrie Hall meter, which Mr. Milne did not go deeply into, and which is hampered by requiring a storage battery, there is no chemical meter that can measure alternating currents. In a place like Toronto, where direct currents are used, the chemical meter is all that is properly claimed for it, but I think Mr. Milne in his excessive zeal has gone somewhat beyond the general facts of the case in the claims for inaccuracy which he urges against the recording motor meter. The result of the Government Inspection tests which Mr. Higman, whom I see here, will no doubt be able to give us more fully, seems to show very good results indeed for the watt meters throughout the country. When you consider this, in view of the fact that many of them have been installed for several years, and not always under the best conditions, I think we have reason to consider the results as sufficiently accurate. There is one thing which I would point out in connection with Mr. Milne's awful example of a 500-light meter ; such a building as that was used in would be a large office building in which a large number of the lights would be going continuously, sufficient at all times to run the meter. Mr. Milne goes on to state that the same conditions obtain through all the smaller sized meters, and I presume the same percentage for the number of lamps required to start the meter. There are in Toronto about 1,500 meters, and the number of lamps to be supplied from these would be about 35,000 ; that would make the average size a meter of 25 lamps capacity. If we allow Mr. Milne's claim that it takes twelve lamps to start the 500-light meter, that is only slightly over two per cent. of its current capacity. Taking the average as 25-light meters throughout the entire plant, the amount of current required would be slightly over half that required for the supply of one lamp, with the average sized meter. If Mr. Milne admits that, he admits that the meter perfectly fulfils its purpose, and that on the average it will start with less than the minimum possible load that can be put on it. There is a further objection in connection with the chemical meter for use in small stations which some stress should be placed on, and that is that the apparatus required for measuring the zincs is delicate and expensive in first cost, and in spite of Mr. Milne's assertion I do not think the ordinary lineman who is left to look after such matters would be capable of handling the delicate milligram scale used in connection with it. A good deal of the trouble which occurs in connection with recording meters, and I can speak more particularly of the Thomson meter, is on account of the careless manner in which they are installed. We had a case in point not very long ago. We shipped out a meter where they had not been using them before, and a complaint was received that it would not operate. We wrote enclosing a copy of the instruction book and saying that the plugs which held up the armature from the jewels should be taken out. They replied that they had not been taken out, and thanked us for the information. We heard later that the meter still failed to operate, and the complaints made about it were very severe, but upon investigation we found that the local expert had taken the meter up to the garret of the house and laid it down on its back. A Thomson recording watt meter won't operate that way. Leaving aside those failures to operate, which are the result of carelessness in installation, I think we may claim a very satisfactory general operation for the Thomson recording meters which are on the market. I do not mean to say that in large central stations using direct current, where the admittedly superior accuracy of the chemical meter can be obtained economically, that the recording meter should invariably be used, but for the general run of stations, we are obliged to put up with the recording meters, even though they are open to some of the objections that Mr. Milne states.

Mr. Thomson : How many stations in Canada are using this meter ?

Mr. Milne : Three ; Winnipeg, London and Toronto, I think. I would suggest in answer to what Mr. Armstrong says that it is not the general custom of the linemen to read meters in any central station of any importance. A man is generally detailed for that sort of work alone, and I think that any station that hires a lineman—that is, a man that goes around putting up wires and the like of that—deserves to have poor results from their meters. As Mr. Armstrong illustrated in the case of the man who laid the meter on its back, it means, of course, that to a certain degree, the results obtained from the recording watt meters are caused by carelessness. I might say that I think we have in Toronto the best men procurable for the meter business, and even with meters adjusted to almost perfection we have got some percentage of error right through ; no matter what size, whether it be a 500-light meter down to a 5-light meter, we have the same percentage of error right down. We cannot ascertain whether the meter is registering correctly unless we put something in series with it to find out. I know by a large number of trials that when we put in the chemical meter the watt meter was slow in every particular instance. With reference to the taking out of the jewel, this cannot be done now unless the meter is taken down to the inspectors, so that they may see the seal broken. Everything is fixed up in first-class order before the meters are taken from the electrical inspector's office. I do not see that anything can be said on that point at all. In fact, we see that the meters are in the best possible shape before they are sent to the inspectors.

Mr. G. Black : I did not expect to take part in this discussion. I have a constant recording watt meter in Hamilton which represents the perfection of meters. I can certainly testify that it does record. I will say that it goes night and day, whether there is any current on or not, judging from the results. We have about 20 lights in my office, and I was told that they were 60-watt lamps. For a long time, taking the length of time these lights were burning, and counting for 60 watts per lamp, according to the record, I found that the meter seemed to read about 25 per cent. ahead of any calculation I could make.

Mr. Armstrong : Perhaps the lamps were not of the efficiency they were supposed to be.

Mr. Black : If I allowed about 100 watts per lamp it would agree with the meter's record. I think on any one lamp it will be sure to run. We had it inspected by the government inspector lately and sealed up, and the man who brought it back told me the inspector said it was all right, so I let the thing go and have not looked at it since. I would like to sell it to any electrical company to run up their dividends.

Mr. Wright : As against all these fine theories as to the watt meter we have the ghastly results of experience, and we cannot go behind the returns. Mr. Black has given us a case in Hamilton ; I will give you another. In the city hall the company were running some lights in one of the departments. After a little trouble we persuaded the authorities to adopt the electric light throughout the entire city hall, and that necessitated a change in meters. The lights were used in the police station previously, and the proposition was to use them in the entire city hall. A change was made in the meters and the current was turned on in the entire city hall, but to the surprise of the company their bills were smaller than they were before all the city hall was illuminated. It depends on the meter being made of a size to accommodate the whole of the building. The lights used in the police station were six in number and were used during the 24 hours, under the previous circumstances, and the company got paid for it. When the larger

meters were put in, the council, of course, meeting once a fortnight, when the six lights only were turned on, they did not register a scrap.

Mr. Armstrong : I think, as a matter of fact, a complaint of that character, so extraordinary and beyond the usual, would require the attention of the government inspector. I think that is an exception, and the meter must have had something wrong with it.

Mr. Wright : The meter was inspected by the government inspector. It was carried there as carefully as it could be carried ; it was installed with a spirit level ; it was not set on its back or put upon its face, and all precautions were taken, and that is the result. It is very plain to see the reason, as Mr. Milne has stated in his paper. And that state of affairs occurs to a greater or less degree in every installation. You take a delicate piece of mechanism like an electrical meter, box it up so that you can't see it or do anything with it, and expect that thing to run without any attention, and it is out of all reason. The utmost that I ever expected to get from the watt meter when the watt meter was first introduced, was an approximate idea of the current that the customer used, and I will defy any person who has had any experience to say that that is not so, that the utmost you can expect to get is an approximate estimate.

Mr. P. G. Gossler : In regard to the reliability of recording meters : since the law has been brought into action it has been necessary for us to go about changing all of our meters, and we have installed the Shallenberger meter. Since last July we have changed from about 900 to 1000. When they were installed each meter was inspected and the number of lights necessary to make them record was entered in the record book, and when they were brought in they were also checked off. While we have found some instances where the meters have become clogged through dirt or cobwebs, the results have been so satisfactory that we have no complaints to make. The question of allowing them to run without any attention, of course, cannot be considered. In a large installation it is necessary to read your meters. Of course we read our meters according to the customer : some monthly, some quarterly, but none over six months. The men who are employed in that capacity, and also checking out the bills, have become to so thoroughly know what should be expected that in case of any falling off in the recording meter it is generally very readily noted. I have an instance that came to my attention on Monday. We installed for three months a government inspected meter ; we knew that the man had installed fourteen lights and we had an idea of what his bill should be. Of course, this would only be an idea. His bill did not come up to what we expected and we immediately proceeded to investigate and make a test. Investigation of the meter showed that the light ran on whether the meter was recording or not. Still further investigation showed that there was a wire under one cleat short-circuiting the meter. We have found the reliability of those meters very satisfactory, especially during the last year, when we have had occasion to change and are changing all our other meters in the service.

Mr. Armstrong : I should like to ask Mr. Milne whether he intended in his paper to attribute this unreliability which he complains of to the Thomson recording meter alone.

Mr. Milne : The only meter that I know of so far in this country as recording watts is the Thomson watt meter ; therefore the remarks, as far as I am personally concerned, apply to the watt meter. In all installations where the watt meters are used they must be of sufficient capacity to carry the maximum load with safety. Take for instance the Grand Trunk station. There are probably 1000 or 1500 lights there ; we will say 800 to make sure. We had to put in a meter of sufficient size to carry that. During the day there are about from 10 to 25 lights used, and, as a positive fact, that meter certainly did not record on fifteen lights.

Mr. Armstrong : What size was it ?

Mr. Milne : It was 160 amperes : I think that is the size of it. When we put in the Edison meter we found exactly where the trouble was. I knew for a positive fact they were using light every day in their engine-room, although we could not get any record on the meter. All we could do was to put on the chemical meter to find out. There was no negligence or care-lessness in the installation of the meter. It was put up dead level and according to the instructions sent out by the company where it says, " Don't suppose you know it all." We take it for granted that we do not know all about it and simply follow the instructions sent out by the company, and in doing so that is the result we get.

Mr. Armstrong : Of course, Mr. Milne in the Grand Trunk case gives an instance of a very large meter, and the number of lights on which it fails to start is less than 2½ per cent. of its total capacity. I presume Mr. Milne in speaking particularly of the watt meter is speak-ing of it from his own experience in Toronto. It being the only meter in commercial use for recording direct currents, it is the only one that could be used in connection with their three-wire circuits. I know it is not in accord-ance with the facts, or the result of the Government inspec-tion of meters generally in Canada, that the watt meter has shown itself to be in any way less reliable in giving accurate returns than the other simple recording ampere meters ; in fact, the results have been precisely the contrary. I should like to ask Mr. Hunt's experience with the watt meter ; he has a great number of them installed.

Mr. Hunt : We have about 400 of the Thomson watt meters in use. I think they have all been in use for an average of eighteen months, some of them over two years, and about ten per cent. will run slow, and I think we have only one out of the lot that has run fast.

Mr. Milne : May I ask Mr. Hunt how he determines that amount ? We would have something definite to go on. We have the chemical meter to prove that the watt meter is out that amount. How does Mr. Hunt determine it was out about ten per cent. ?

Mr. Hunt : By the Government inspection, that is all.

Mr. Dion : I think that the cases stated by Mr. Milne and Mr. Wright are rather the exception than the rule. I do not think there is any mechanical device, no matter how accurate, against which such cases could not be brought up. If we go around the country looking for cases of failure, we are sure to find some, but I do not think they are the rule. We have in our city some 3,000 meters, I suppose two-thirds of which are Shallenberger, and the other third watt-meters. We have lately had occasion to have a very large number of these inspected by the Government. We test them at our office first and send them up to the Government afterwards. In many cases we find that those meters which have been in use from a few months up to five or six years are turn-ing out very satisfactory. I don't suppose there are any more than from three to four per cent. that have to be touched at all before sending them for inspection. I think fully 95 per cent. register within the percentage allowed by the Act. We do not send them up until they are correct, because we do not want to pay the fees twice. With regard to the two meters, in answer to what Mr. Armstrong has said, I may say that while there has not been a very great difference, in our experi-ence, between the two meters, in testing them, the dif-ference has been rather against the watt-meter. The percentage of meters requiring fixing up before being tested by the Government was larger in the case of the watt-meter. The meter had to be fixed up, because it either didn't start with one lamp or went too slow. Leaving out this question of the relative merits of the meters, I would like to say a word in praise of the paper we have just heard. Its value lies in the fact that it keeps before us the defects of the apparatus which we use every day, rather than the qualities of them. I think it is only by keeping the defects constantly before us that we may expect improvements to be made. I also think that Mr. Milne is very wise when he advises that all currents should be used through a meter. I think that should be the universal practice, and should be encour-

aged by all possible means. When all the current is used through a meter you will find that it is a considerable relief to the central station. Our experi-ence in that way is very satisfactory. We urge the meter in all cases, with the result that a very large per cent. of our business is done through meters. We find that with an installation of 54,000 lamps our largest loads have not yet exceeded the equivalent of 22,000 lamps ; that is, the actual ampere meter readings, and it includes all losses from leakage in transformers, so you see the importance of using meters throughout the instal-lation. As regards flat rates : There are some cases where it is absolutely necessary to make a flat rate, and in these cases the maximum use is the thing we want to get at ; and if there could be such an apparatus devised as Mr. Milne has described as recording the time during which the lights are being used—if that could be so improved as to give the maximum load as well; it would be a very valuable adjunct in the case of flat rates. We could then make a rate very intelligently, which would be almost as good as a meter rate.

Mr. Wright : I would like it thoroughly understood, of course, that what remarks I have made in regard to these meters are not intended to apply to any particular brand of meter. For instance, the Thomson recording watt meter, considered in the abstract, is a most in-genious piece of mechanism. My remarks apply to all meters of the same description which depend on jewel centres and absolute accuracy of installation for their perfect working. We are compelled to take these meters. In one respect I may say perhaps our ex-perience in Toronto differs very greatly from places like London and Ottawa, where watt meters are used almost exclusively in lighting systems where there are a comparatively small number of lamps. But you take the case of an installation in Toronto for motors and for elevators, where a sudden load is sometimes thrown on the meter in starting an elevator or a large motor, more than it is capable of bearing and more than it can be ex-pected to bear, it happens in very many cases, in fact, in nine cases out of ten, the resistance will be burned out. What can you do? The meter is sealed up. You cannot get at it. You replace the burned-out part, take it to the inspector's office again, and the fee is $2. And so it goes. My remarks in regard to these meters do not apply to the Shallenberger meter or to any other alternating meter without a commutator and with-out any trouble arising in the armature. There is no doubt that if we could confine ourselves to a meter of that description I do not think anybody would object. When we are compelled to put in a meter that we know will under extraordinary strains give us trouble, then the "coercion" comes in. If the Government would provide us with a meter that would work, and that would not be sealed up, I say it is perfectly right for them to inspect them, for the sake of protecting the poor consumer, but they should not compel us to lock up a machine that is going to be unreliable and that we can-not attend to.

Mr. P. G. Gossler : We had some Thomson record-ing watt meters installed, and we found it necessary to remove them because we could not record the loads. That applies only to loads varying very greatly.

Mr. Thomson : We placed recording watt meters on the motor circuits, and inside of a year's time we found about half of them burned out, so we discontinued the use of meters on all of them.

Mr. Wright : We have been obliged to resort to a flat rate, to our loss, and rather than instal a watt meter under certain conditions we have been driven to the use of the Pattee recorder. I am not blaming the watt meter, but simply because we cannot get a meter that can be sealed up which we can depend on to give us re-liable and accurate data to charge up. We size it up in our imagination. We never salt a man any more than we think he can stand. We put in one of these lamp-hour recorders and magnets to suit, and he is chopping his wood and doing all that sort of thing by the hour. We find in a measurable degree it answers the purpose.

It certainly answers the purpose better than a Government inspected meter.

Mr. Milne : The Thomson meter is the most ingenious meter we have in the market to-day, but it is not applicable for our purposes here in Toronto.

Mr. Higman : I have just had this paper placed in my hands. After luncheon I suppose we will be able to take it up, and I will have something to say.

The President : I think the opinion of the members present is that we ought to adjourn this discussion. It is probably one of the best discussions we have ever had in a convention, and it does seem a pity to end it here, more especially as Mr. Higman is here and we would like to hear from the Government Inland Revenue Department. Is it your wish that the convention stand adjourned until two o'clock?

Mr. Dion : The point which has been raised by the paper just discussed is whether the Government was justified in shutting out the electrolytic meter, which is admittedly the most correct meter. This point can be very well discussed under the item on the programme, "Consideration of the Government Inspection Act."

Mr. Higman : I think the whole subject had better be discussed on this paper.

Mr. Armstrong : I have very much pleasure in moving a hearty vote of thanks to Mr. Milne for the very able paper he has given us. I believe that the discussion took a turn that was not expected, and which resulted in bringing out points that will be of interest and benefit. I do not agree altogether with Mr. Dion as to the desirability of the general use of the meter throughout the country. I think in connection with many of the small installations the certainty of their securing a revenue throughout the year, especially where they are operating by water power, is more desirable. I have pleasure, gentlemen, in moving a vote of thanks to Mr. Milne.

Mr. Kammerer : I second the motion.

The President : It has been moved by Mr. Armstrong, seconded by Mr. Kammerer, that a hearty vote of thanks be tendered to Mr. Milne for his valuable paper, which I am sure will be carried unanimously. (Carried.)

The President stated that a photograph of the members of the Convention would be taken at Lorne Park in the evening.

Convention adjourned until 2 o'clock, p.m.

AFTERNOON SESSION.

The Convention was called to order at 2 o'clock, p.m.

The President : I will be glad to receive a report from the Committee on Nominations for the Standing Committees for the year, legislation and statistics.

Mr. Dion : I beg to report as follows : Legislation Committee—Messrs. J. J. Wright,. K. J. Dunstan, Berkeley Powell; L. B. Macfarlane, and F. H. Badger. Statistical Committee—Messrs. E. Carl Breithaupt, John Yule, and O. Higman. I may say these are the same as last year ; they have done so well we thought we would keep them in office.

The President : It is hardly necessary to read the names again. Is it your pleasure that these gentlemen should form the committees on legislation and statistics to carry on the work for the coming year that was carried on last year?

Mr. Breithaupt : I have been on the Committee on Statistics for two years, and have given it considerable energy and thought. I have carried it about as far as I can. Somebody else may have different ideas from what I have, and would be able to carry it further. As far as I am concerned I think some one else might be put in my place.

Mr. Higman : I have been two years on it, Mr. President, and while I cannot say that I have given as much energy as Mr. Breithaupt has to the work, still I would rather have somebody else in my place.

Mr. Armstrong : The reason which these gentlemen give for retiring is the very reason why they should stay on. They are the only people in possession of the necessary information as to the method of procedure.

The President : I hope Messrs. Breithaupt and Higman will withdraw their wish to resign. The work

they have in hand is advancing very nicely, and it requires but a little more to put everything in very good shape. I think the Association would appreciate their efforts if they would continue. I suppose silence gives consent, and we will consider these gentlemen as elected to these committees.

The President : The next order of business will be nominations for President.

Mr. Kammerer : I have much pleasure in nominating Mr. John Yule, one of the initial members of the Canadian Electrical Association, as our next President.

Mr. Milne : I beg leave to nominate Mr. E. C. Breithaupt as President.

Mr. Breithaupt : I very much thank my mover for mentioning my name for the honorable position of the Presidency, but I think Mr Yule deserves this honor more than I do. He is one of the charter members of the Association. I beg, therefore, to withdraw my name in favor of Mr. Yule. (Applause.)

Mr. Yule being the only nominee for the position of President, he was elected to the office by acclamation, amid applause.

The President : I am sure it is a matter of congratulation to me personally that Mr. Yule will succeed me in office, and I have much pleasure in announcing Mr. Yule's election.

Mr. Yule : I beg to thank you for the compliment you have paid me in electing me to the office. I did not wish to accept the office, but it seems the general wish that I should do so, and in doing so I would ask the members to give me the same support as they have given to the other Presidents. The election of Presidents has generally been heretofore from amongst members residing in the central constituency. The office has formerly been in Toronto, and the work has been carried on very efficiently in that way. I do not know how it will work with a President residing at a distance. A great deal of the work will fall on Mr. Mortimer, and I will have to look to him to keep me straight. I would also ask the members here to elect a very fair contingent of the Executive Committee from the members residing in the city of Toronto; it has worked very successfully before.

The President : The next nomination will be for Vice-President.

Mr. Dion : I beg to nominate Mr. L. B. Macfarlane, of Montreal, as Vice-President.

The President : I may say, in all fairness to Mr. Macfarlane, that I had a letter from him this morning, in which he regrets his inability to be present, and regrets still more his inability to attend any of the meetings during the past year, and asking that as a personal favor his name be dropped. I do not think we ought to take any notice of that letter at all.

Mr. Carroll : Not at all.

The President : He is too valuable to be dropped out.

Mr. Wright : Mr. Macfarlane has been one of the useful members of the Association. He has always taken a great deal of interest in it, and until the present time has been present at every Convention. I should like very much to see Mr. Macfarlane's nomination made unanimous.

Mr. Macfarlane was then declared elected to the office of first Vice-President of the Association by acclamation.

Mr. J. J. Wright : I beg leave to nominate Mr. E. Carl Breithaupt for the office of second Vice-President.

There being no other nominations Mr. Breithaupt was elected unanimously to the office.

The President : The next officer to be elected, and one of the most important, is that of Secretary-Treasurer.

Mr. Carroll : Oh, that is settled.

Mr. Breithaupt : The Secretary-Treasurer that we have had for a number of years past, in fact since the formation of the Association, has done very much in the work of carrying on the Association, keeping it on its feet, and making it what it ought to be. I feel we would be doing a great wrong in not keeping him. I therefore move that Mr. Mortimer be elected by acclamation as Secretary-Treasurer for the coming year.

Mr. Mortimer was then elected by acclamation to the office of Secretary=Treasurer.

Mr. Mortimer : I thank you, gentlemen, for this, the fifth or sixth time, of the very kind expression of your favor.

The President : The next thing is the election of the Executive Committee. It is desirable for many reasons that there be a continuity in the membership of the Executive, and for that reason five of the ten must be selected from the present list. The five who in your estimation deserve consideration at your hands, are to be marked, and the remaining five will be nominated and elected afterward. Our constitution says that the method of procedure in this case would be that the Secretary shall read the names, and the person, as his name is read, shall rise and deposit his ballot. This of necessity would prolong the Convention, and as active members only are allowed to vote, and to shorten the proceedings, I would appeal to the honor of those who are here that no one shall vote who is not entitled to, and the ballots will be distributed and collected. Before doing that it is necessary to appoint two scrutineers, and I would nominate Mr. Geo. Black and Mr. Geo. White-Fraser to act as scrutineers.

Mr. Breithaupt: At the Convention last year I thought that it was decided that the members of the Executive Committee were to be elected for two years?

Mr. Carroll: That was the intention of the by-law, but it was changed.

Mr. Breithaupt : How is it now?

Mr. Carroll: They have got to be re-elected every year.

Mr. Wickens : Yes, but five members of the old board have to be re-elected.

The President : As Mr. Breithaupt has been elected to the office of Vice-President, he will not now be eligible for election on the Executive.

Mr. Wickens : While the ballots are being collected, I move that $50 be appropriated for the use of the Sec-retary-Treasurer to meet the expenses in connection with the work of the Association.

Mr. Kammerer : I second the motion.

Mr. J. W. Taylor : I do not know that that sum is sufficient.

Mr. Breithaupt : The sum that has been set apart heretofore has been $50.

The President : It was formerly $25, but last year at Ottawa it was made $50, and we propose this year that it should be the same.

Mr. Breithaupt : The Secretary-Treasurer has more work to do, and I would move that the sum be made $75 instead of $50.

Mr. Higman : I moved at Ottawa last year that the sum should be made $75.

Mr. Wickens : Under all the circumstances I will withdraw the original motion and Mr. Breithaupt's amendment can be put as the main motion.

The President : As far as money is concerned, I am quite satisfied that money could not pay Mr. Mortimer for all he has done for this Association, and I shall be delighted, personally, to have the sum made $75.

The President : It is moved by Mr. Breithaupt, seconded by Mr. Taylor, that the sum be made $75. Is that your wish?—Carried.

Mr. Mortimer : I may just say in regard to this ques-tion of remuneration that I do not want to see this Association bankrupt, and I think if you go on putting up the Secretary's salary every year it is going to bank-rupt the institution. I think we had better let the salary stand as it was, and if we find at the end of next year that there is anything left out of that "$29,000 surplus," I will take what is offered.

Mr. Wickens : I don't agree with that at all ; I think the association is good enough to make up the difference.

The President : The following members are elected to the Executive in the order in which I read them : Messrs. J. J. Wright, A. M. Wickens, K. J. Dunstan and J. A. Kammerer, and for the fifth position there is a tie. On the casting of ballots by two members who had been absent from the room, the position of fifth member of the Executive was accorded to Mr. Geo. Black, of Hamilton. The nominations for the remaining

five members of the Executive were then proceeded with.

Mr. Breithaupt : I have much pleasure in nominating Mr. Hunt, of London.

Mr. Wickens : I nominate Mr. F. C. Armstrong, Toronto.

Mr. J. J. Wright : I nominate Mr. A. B. Smith, Toronto.

Mr. Kammerer : In view of the fact that we have de-cided to go to Niagara Falls, I have much pleasure in presenting the name of Mr. Ross Mackenzie.

Mr. A. A. Dion : I beg to nominate Mr. J. W. Taylor, of Ottawa.

Mr. Carroll : I nominate Mr. Dion, of Ottawa.

Mr. Armstrong : I have much pleasure in nominating Mr. John Carroll, of Montreal.

Mr. W. A. Johnson : I beg to nominate Mr. Milne, of Toronto.

Mr. J. J. Wright : I nominate Mr. W. A. Johnson, of Toronto.

Mr. Carroll : I beg to propose Mr. E. C. Cary, of St. Catharines.

Mr. Armstrong : I beg to nominate Mr. W. Williams, of Sarnia.

The President : While the scrutineers are doing their work in this connection I think we might go on with our proceedings. The first thing to be considered is " The Government Inspection Act." If anybody here is prepared to say anything the meeting is open for that purpose.

Mr. Higman : With reference to the paper that was read by Mr. Milne this morning, I notice that while it contains nothing very new, yet the facts are arranged very satisfactorily, and the deductions that he has ar-rived at are most convincing, viewed from the stand-point of Mr. Milne and those who employ him. Run-ning through the whole paper, and underlying almost every paragraph, we detect the fine work of the special pleader. From beginning to end it is a plea for the electrolytic meter, and from that standpoint I think Mr. Milne has succeeded very well and has earned the thanks of his employers. He says the electrolytic meter has been condemned. He might have added that it has been very generally condemned both in England and the United States for every-day practical use. He says consumers do not want to keep a record, an exact record, I think he says, of the supply. They do want to know, however, to what extent they are using the current; they want to be able to determine from time to time what the rate of consumption is. I think that is very reason-able, and it is not surprising that they should ask for a direct recording meter. In regard to that very question we have received at the Department dozens of letters complaining bitterly about the use of this meter. We have received several such letters from Toronto and Kingston, asking that the Department put a stop to their use at once, instead of allowing them to run almost indefinitely. And while Mr Milne designates the idea, in regard to renewals, as renewing the whole meter, as " gross rot," anyone who knows anything about it will agree with the proposition that to renew the plates is to practically renew the meter. Mr. Milne makes some complaints as to the unit of current. May I ask if there is anything wrong with the method of determina-tion as laid down in the Act?

Mr. Milne : The definition is perfectly right, and the method of arriving at it is correct.

Mr. Higman : A paragraph in Mr. Milne's paper says " The Government has to raise a revenue, that is settled. The gas companies contribute a certain per-centage of that revenue ; the electric companies are their greatest competitors, therefore we can readily infer that any little obstacle that can be put in the way by such companies will be done so, and it is very common property that this Act was the result of the gas companies." I deny that, as far as I have any know-ledge of the papers. I have seen all the papers that have been sent to the Department in connection with this subject, and I have failed yet to discover a single word or line from any gas company asking for an Act of this kind. Mr. Milne's statement is not borne out by the facts. I might say in connection with the question of fees that in Canada there are thirty-seven gas companies, and from these thirty-seven companies we collect a revenue of some $14,000 to $15,000. At present there are two hundred electric light companies registered under the Registration Act, and from them we expect to collect about $4,000 a year. During this year we shall have more than that, because of the order that was passed, asking that all meters be verified before the 1st of July, but after this year we shall not collect more than about $4,000 or $5,000 a year from the whole country, taking in about two hundred and fifty companies; so that as compared with gas companies the latter not only pay a certain percentage, but nearly the whole thing. Mr. Milne asks among other questions, "Are there any advantages to be derived from this test?" and answers it in the affirmative. I would like to ask Mr. Milne if that is his opinion to-day? Whether he is in favour of having the inspection? I will pause for a moment to get his reply.

Mr. Milne : The answer to that, as far as I know, is in the affirmative yet; the inspectors are benefited. I don't see that it benefits any one else.

Mr. Higman : It is not a very fat thing for the inspectors. Up to the present time, although their work is nearly double, they have not received a cent additional remuneration. Notwithstanding what the inspector derives from it, I am inclined to think that Mr. Milne would not refuse the job himself.

Mr. Milne : No, sir. In fact, I applied for one of just the same kind.

Mr. Higman: Now, a word or two in reference to the difficulty mentioned by Mr. Wright. I admit it is a real difficulty and one that has engaged the serious consideration of the Department. Some time ago I suggested to Mr. Nicholls that perhaps the difficulty could be overcome by hinging the bevelled piece of the case immediately below the dial plate. This opening could be sealed by the company. It would enable them to get at the commutator to clean it at any time, and such opening would not affect the registration in the slightest.

Mr. Wright : That would work all right if the Government would be content. Speaking for the company that I represent, I don't think the company would order any subordinates to go around and spin the meters ahead at all.

Mr. Higman: I did not mean that. I don't think the Government would have any objection to that, because the consumer has it in his own hands. He is always there when the company's representative goes around, and there could be no objection at all to having this opening in the meter to clean it. I would suggest that in matters of this kind the association should appoint a committee, and if there are any grievances to be remedied or considered, to wait on the officers of the Department and see if some means cannot be found of overcoming them.

Mr. Milne : Mr. Higman says he has received several complaints from Toronto regarding the chemical meter. It is astonishing that we did not hear of them, when we have so very few complaints here. I would just like to ask who are the parties using the chemical meter here in Toronto who have been doing the complaining. I think you will find that it is a customer who does not wish to pay for what he is using. No company, I am sure, will charge for more than what is honestly burned, but they certainly wish to get paid for what goes through the meter. Mr. Higman is of the opinion that renewing the plates in the chemical meter is practically renewing the meter. If we had to supply five pointers for a recording meter, is that supplying a new meter? The meter itself is composed of a German silver shunt ; in multiple with this shunt is placed a compensating spool : in series with this spool is placed an electrolytic cell in which is placed two plates. The two plates are a very small arrangement as compared with the meter itself. I cannot see how renewing the plates in that meter is practically renewing the meter.

Mr. Higman : You cannot use a meter without the plates; it is the only part that needs renewing.

Mr. Milne : Mr. Higman speaks about the inspectors verifying the meters at the station. I think it would be a good idea for the inspectors to go to some of the stations and have their meters verified, because in the principal stations here in Canada the very best meters procurable are put in. It is not to the interests of the company to run below the voltage, nor it is not to their interests to run above the pressure ; that would simply mean increased lamp renewals, and if they run below the light is poor. If we run above pressure that is a loss to the company, and not to the customer.

Mr. Higman : Certainly. That was my contention. We want to save the company any loss.

Mr. Milne : I think the companies will look after that in good style. In the letter which Mr. Higman wrote to the ELECTRICAL NEWS last month he says one of the inspectors called at a station, and found that the meter was four volts out. A station of that kind deserves to be soaked just for as much as the law can give in running instruments of that kind.

Mr. Wright : In the first place my objection is not to the Thomson watt meter as a meter, and I am not objecting to Government inspection as Government inspection. I must say this, and I am bound to say it in all fairness, that in all our communications with the Department and with the subordinates, we have been treated universally with great consideration, and the inspectors have acted in a gentlemanly way all through to the best of their ability. It is the combination of the two that makes the trouble ; it is taking a meter that will not operate and locking it up in a glass case and expecting it to operate, and the Government Inspector coming along and saying that it has got to operate. It seems to me the suggestion of Mr. Higman is a good one, that some means of access to the delicate parts of the meter should be provided. If that is done it takes away a good deal of the force of the remarks that have been made. We have no objection to the Government seeing that the meters are right, but it is manifestly unfair to take a machine that requires attention, that should be opened and carefully cleaned, and the brushes and commutators attended to and put on the home stretch for another run. I say it is unfair when that meter is sealed and shut up and that cannot be done. If some method can be got at, and if the Government are willing to allow some means of access to the delicate part of this mechanism, that gets over the major part of the difficulty. I am speaking from what I find. It is a heart-breaking job when you have to handle the number of meters we have here in Toronto, and under the conditions in which we are expected to handle them ; in fact, it is enough to make a man give up in despair. The meter is sealed up, and is supposed to be right. It possibly gets a little jar in being taken to the place of use, or getting it up on the side of the wall, and it does not read correctly, and there is no way, according to law, of having that remedied. They are entitled to charge a new fee for inspecting it again if we take it back. If some method could be adopted by the Government inspectors so that when a meter is brought to them, after it has been in use for a short time, it could be reverified without expense to the company, I think it would be well. Some of these meters, if we did not take them out, would be an eternal source of expense ; it would be pay, all the time, to have the meters verified. Let me say just one word about the letters Mr. Higman has received. I have no doubt he has received them, because I have received similar ones, and it is altogether likely the parties in sending them to me have communicated with Mr. Higman; they probably would have written also to Queen Victoria and Lord Salisbury ; but when the Government has got a clause in their Act which says that every man is entitled to a direct reading meter if he demands it, what is the kick about?

Mr. Higman : I don't know. It is Mr. Milne that is kicking.

Mr. Wright : When a man has an objection to a chemical meter, and says I want something I can read myself, we meet him, so that the force of these complaints is lost. I just want to make my position plain in this matter. It is not a question of finding any fault with any particular brand of meter. It is very far from my intention to criticize the action of the Government or the officials. We have always been treated as an Association and as individuals with the utmost consideration by the Government officials, but it is the combination of the two where the difficulty arises. If the Government will adopt the method Mr. Hignam suggests I will have no doubt that will overcome a good deal of the difficulty. I move that the Legislative Committee of the Association take up the matter with the determination, if possible, to see if any mutual arrangement can be made mutually advantageous. I know Mr. Higman has a difficult task to perform, and his inclination is to do justice to all concerned.

The President : When two parties are favorably disposed there is always a way of coming to an arrangement which can be made mutually advantageous. I know Mr. Higman has a difficult task to perform, and his inclination is to do justice to all concerned.

Mr. Higman : Before sitting down I would like to read a letter from one of our inspectors who thinks he has been rather unfairly treated, and he wishes to be set right before the Association. The letter is as follows :

O. HIGMAN, Esq., Ottawa.

"SIR : Your favor of yesterday has been received, asking if I know to whom a certain article in the ELECTRICAL NEWS refers. I have already received a copy of the NEWS with the article in. It may refer to me, but I must say I have not interfered directly or indirectly with the electrical plant or apparatus of any electrical work in Hamilton or elsewhere. What I have done is this : About a year ago I met on the train Mr. Robert Thompson, President of the Electric Light Company, and in course of conversation he spoke of the excessive amount of fuel they were using under their boilers. He said I must, when in the business, have had practical knowledge of this subject, and asked me what was the cause of using so much fuel. I suggested that possibly the chimney was too small for the services required of it, and that I had some books on the subject which I would be willing to lend him. He said that Mr. Knox was the mechanic of the board, and that he would send that gentleman to me. Mr. Knox called and I showed him the books. I directed him to the places giving the size of the chimneys needed for similar plants, and told him he might make the calculation for himself. He copied the figures, thanked me for the use of the books and went away. Both these conversations were sought. I did not volunteer any information, and had nothing to say except the suggestion that possibly the chimney was not of sufficient capacity, it having been built for a much smaller boiler. This was about ten months ago, and I have not since spoken to any of the directors nor to other persons on the subject. I have not at any time in the remotest way offered my criticism or advice in connection with electrical matters. It would be presumption on my part to do so. As a practical mechanic I offered the suggestion to friends seeking my advice.

I am, yours, etc.,　　D. McPHEE.

Hamilton, Ont.

Mr. Wright : Isn't that the case of the cap fitting the man ?

Mr. Higman : I may state that the letter is in reply to inquiries made from the Department to the inspector.

Mr. Johnson : I would move an amendment to the last resolution, that is, that the question of removing the bond from the chemical meter be taken up. If the Government can be induced to do so, that chemical meter has a use for direct-current work and for power work ; it is something that is very handy to use and there is the possibility also that it may be desirable to have it in connection with alternating work.

Mr. Fraser : I want to refer to one individual case that I know of myself. The inspector, whose name I shall not mention in public, but I will give it to Mr. Higman if he desires, managed to and purposely left the impression in the mind of a man who was just putting in an electrical plant that it was necessary before accepting the plant that it should be passed upon by the Government inspector. The purchaser was an ignorant man and the inspector was ignorant, if not more ignorant than the purchaser, but the purchaser was a perfectly creditable man, and told me distinctly that this inspector had purposely left him under the impression that the Government inspector was placed as a kind of watchman over the manufacturing companies, and it was necessary for him to pass that plant before the purchaser would buy. These inspectors have got no standing in the profession, but they go about with the influ-

ence and the weight which is given to them by the Government appointment and use that in a very wrong way.

Mr. Higman : I must say I am surprised to hear the statement just made. I can hardly understand that one of the inspectors, knowing very little about electricity, and necessarily so, should even attempt to pass judgment on matters of this kind. The Act contains no provision for the inspection of electric plants ; it deals only with the public supply and the apparatus through which the supply is determined, and on which the consumers' bills are based. Applications to the Department have been frequently made, however, for the services of an electrical engineer to report on one thing and another, simply, I presume, for the reason that such services would be rendered free of charge. In every such case I have referred the parties to practising electrical engineers outside the Government service.

Mr. Black : I had a conversation with Mr. McPhee the other day in reference to this matter, and he explained it to me as he has written to Mr. Higman. The advice was sought in such a way that he could not refrain from giving some kind of an answer. He did not give his answer as Government inspector or official, but in the light of his past experience as to steam feeders, for he had a large experience in the feeding of similar heating apparatus. There was one instance where there would have been a loss of a thousand dollars on the plant if he had not studied up the subject and found the fault lay with the chimney, and convinced the Government officials, who had reported against him, that the fault was with the chimney. He had works on the subject, and he simply suggested that there might be some trouble of that kind, and loaned his works to these parties. He had no thought of acting as a Government official at the time, and he certainly did not intend to pose as an electrical engineer. He would have been at this convention only he felt it would be better for him to remain away during this discussion.

The President : An amendment was moved by Mr. Johnson, seconded by Mr. Wickens, that the Committee on Legislation also consider in their correspondence with the Government and Government officials the reinstatement of the chemical meter, as being useful for power on other circuits.

Mr. Breithaupt : Would it not be well to make that a little more general and say the committee shall have power to meet and confer with the Government authorities on the matter of electrical inspection and on all matters concerning the same, so that they may be able to deal with any exigencies that might arise ?

The President : I see no reason why it should not be carried out.

Mr. Higman : I would suggest that if the committee wait on the Government that they give their complaints, or whatever they want, in detail. It is no use taking up the bill and discussing the whole thing over again, because you arrive at nothing ; but if there is anything of a special nature that you wish to have changed, or discussed with the Government, let it be specifically stated.

Mr. Wickens : This is within the province of our legislative committee, and if there is anything that we wish to have changed, I think it is for them to take up the matter. I am satisfied they will do what is right. It seems to me that the matter could be arranged so that it would be reasonably safe to the consumer and reasonably good for the producer. The object of having a law of this kind is to do some good by it, and the object of this Association is to help the members and help the people in connection with their interests to do what is right, and to succeed. I think the committee should be able to go into that matter with a free hand, and I think the Government should be able to meet them as representing practically the whole of the electrical people of the country. I think the Government should meet them, and I think they will.

Mr. Dunstan : As a member of the Legislative Committee, I think it is not a question that should be left in the hands of that committee. It is a technical question in connection with electric light interests, and I think there should be a special committee appointed.

The President : Mr. Dunstan's point is well taken. That committee should, I think, be composed exclusively of electric light men interested in that actual work. I think it would be proper now that a committee of three or five be nominated to take the matter in hand.

Mr. Milne : Mr. President, I would just like it understood that I have no particular hatred for any mechanical recording device. I have a particular love for the Edison chemical meter. My paper was originally intended to be on the chemical meter, but I thought it might interest some of the members of this Association to know the principle on which some of the other meters were worked, and as far as we are concerned here in Toronto we have had the most friendly relations with the inspector. I think we can get along first-class with him, and we have no friction at all in any respect. I think Mr. Higman will admit that. It was just simply in connection with that restriction of the chemical meter that I got up this paper.

The President : The nominations for the remaining five of the Executive Committee are as follows, and they are elected in the order in which I read them—Messrs. Ross Mackenzie, Niagara Falls ; A. B. Smith, Toronto ; John Carroll, Montreal ; Charles Hunt, London ; and F. C. Armstrong, Toronto.

Mr. Breithaupt : I would suggest Mr. J. J. Wright, Toronto ; Mr. P. G. Gossler, Montreal ; Mr. A. A. Dion, Ottawa, as a committee to interview the Government.

Mr. Armstrong : I would suggest adding Mr. James Milne, who is probably more thoroughly conversant with the subject of meters than anybody else present.

Mr. Breithaupt : I thought the committee would want to be very small. Mr. Milne would certainly be a good man to have on the committee ; I would like to see him there.

The President : I think Mr. Breithaupt would do good work on that committee ; that would make five.

The President put the motion, which on a vote being taken, was carried.

The President : We will take up now the paper by Mr. Armstrong entitled "The Outlook for the Electric Railway." (See page 15.)

Mr. Armstrong's reading of the paper was followed with applause.

The President ; You have heard this very valuable paper of Mr. Armstrong's ; the subject is a live one and I would like to have some discussion.

Mr. Hunt : I have great pleasure in moving a vote of thanks to Mr. Armstrong in having prepared his valuable paper on electric railways.

Mr. Wyse : I second that.

The motion was carried.

Mr. Fraser : I think Mr. Armstrong has given us a fair account of the position of electric railways in Canada. In the last paragraph he mentions something in connection with long distance railways. I think it would be quite interesting to the Association to hear some of the facts in connection with the electric railway at present running in Lugardo, which is described in some of the technical journals recently. They do not use the rotary transformers in connection with their system, but it is actuated by the direct three-phase currents. As to the question of the track, that seems to have been successfully overcome. In fact, I believe there are a good many of the best electricians of the day who have arrived at a practical, if not an actual and commercial solution.

Mr. Hopkins : There is one question I would like to ask Mr. Armstrong. He spoke of the limit of the field —of the road radiating out. I would like to ask about what that limit of distance would be at present? He might also answer what would be the limiting grade that it would be safe to build a railway on so that in the winter time, when there was ice on the rails, the car could be kept in control and there would be no danger of it being locked and taking the people down. Then there is another question. I have understood that the

alternating current is out of the question now for running electric railways, that they cannot get motors that will start up quickly. One electrical engineer of very high standing and of long experience and very well posted in the theoretical part of the work, told me, some time ago, that that was out of the question.

Mr. Armstrong : Mr. President, I might speak of the matter of limiting grades first, which would allow of the operation, I presume, of both light freight and passenger service during the winter season. On a matter like that you can only speak really from experience. I might instance very forcibly in this connection the case of the Galt, Preston and Hespeler railway. There they handle, and did handle during last winter, their passenger service without any difficulty over grades of five, six and even seven per cent ; they also handled a light electric locomotive freight service over the same grades without any difficulty at all. Even over the longest grade which they have, which rises from the town of Galt to the C.P.R. bridge, and which at places is as high as six and a half or seven per cent., they can haul a load of two freight cars. On the grade at Preston, at the end of the line, where the grade is about five per cent., they haul, ordinarily, one, and in some cases two, coal cars loaded to their full capacity up the grade. It seems that up to the point that we can keep a reasonably high voltage without excessive line loss that we can handle the freight, and unquestionably the passenger service over tracks laid generally on existing grades, with very slight cutting down.

Mr. Hopkins : Would it be necessary to have motors on every axle and brakes on every wheel ?

Mr. Armstrong : I might go a little further into the freight locomotive question at Galt. The handling of passenger traffic is comparatively easy. The car used for freight purposes is mounted on a single truck and with two motors, one on each axle, of the G. E. 1200 type, wound with a four turn winding, and there has been no difficulty at all. When the car was first sent up it was found too light to carry two cars up the grade at Preston ; the wheels would revolve and the cars run backwards. That was remedied by putting some three tons of pig iron on the floor. Since then they have had no difficulty, even in the winter. As to the question of the limit or range over which we may expect these radiating or radial lines to extend, it is a difficult question under present conditions, to give any definite limit. One can only examine existing cases and find out how far they can commercially operate with success.

Mr. Hopkins : I mean with one power house.

Mr. Armstrong : Taking the Hamilton, Grimsby & Beamsville road with one power house located, as it is in their case, in nearly the middle of the road, they have a limit of 18 or 20 miles ; a transmission limit of 7 to 10 miles from the power house would be about the maximum with which economical results could be obtained without undue expenditure of copper. They do not handle any heavy freight ; they just haul light cars behind their ordinary passenger cars. The Hamilton Radial Electric Railway Co. are now building a line from Hamilton to Burlington, and in their case the transmission limit will be eleven and a half miles from the power house. In that case they found it would be much more economical to invest money in copper to reach that distance than putting in polyphase apparatus. The cost of copper there is very considerable. The limiting distance would be ordinarily something under ten miles from the power house. In connection with the use of the alternating motor, I was pleased to find Mr. White Fraser draw attention to the road at Lugardo. I do not think there is any reason to doubt at all, that we will have in use in America a successful alternating railway motor which will give a reasonably high economy and with which the difficulties of control will be surmounted ; and with the use of that motor our range of transmission will be increased and the limit to which radial lines can be extended will be very much greater than it is at present. I had the opportunity of seeing a car at Schenectady some time ago in which it was endeavored as far as possible to conform to the requirements of ordinary traffic ; and while there were certainly some difficulties which in detail would have to be surmounted, there did not seem any liklihood that its success for practical purposes would be very long delayed ; at least, the engineers who are looking after the matter speak in the most favorable way of the results they are obtaining.

The Convention adjourned at 4.30 p.m., to meet again Friday morning at 10 o'clock.

## THIRD DAY.

The President called the Convention to order at 10 o'clock.

The President : The first thing on the programme is the presentation of papers, the first being a paper by Mr. P. G. Gossler, of Montreal, entitled " Some Central Station Economies." (See page 15.)

Mr. Gossler, on rising to read his paper, said : It was stated in the minutes read the day before yesterday that I had been asked to present a paper on high potential underground systems. The present subject has been selected because I thought it would be of more general interest to the convention. I have been for several years connected with the operation of the subways of New York city, and have a collection of data which I shall be pleased to place at the disposal of anyone who is contemplating entering into underground work. I may say in regard to plates 4 and 5 that I regret those plates are not larger, because I am sure they will be appreciated by anyone who has made lamp tests. I have larger copies of these and will be glad to place them at the disposal of anyone who wishes them because they represent a very great deal of labor. The formula which I have included in this article here is one that has been found very useful.

The President : You have heard this paper read. The subject is now open for discussion. I know there are a number here who are anxious to ask questions. I hope the discussion will be full, but quick.

Mr. Breithaupt : Mr. President, the Association certainly ought to tender its thanks to Mr. Gossler for the excellent paper he has given us on Central Station Economies, in the reconstruction of an old central station. All of us who have had experience in central station working know that the central stations which were put in five, six or seven years ago are according to present methods very inefficient. Not only is this the case with some of the larger central stations of the older type, but true with the smaller stations ; that is probably a reason why so many of our smaller stations throughout the country prove very unremunerative to the people who own them. The reconstruction of the central station, particularly the small central station—I have had experience with a number of these—is a matter of considerable difficulty. You speak to the owners of plant ; they know that the plant is not remunerative ; they know that from actual experience. There are plenty in the country who have not made a dollar out of it. With the old apparatus they have increasing and very great induction loss, line losses and all that sort of thing. You tell them to put all this old apparatus on the scrap heap—that it is the best thing to do—and you will meet with great opposition. It is a hard thing to do, to reconstruct particularly a small plant, to bring it to a proper basis. Mr. Gossler's statements about the transformers are very interesting indeed and very instructive, and the lamp curves even more so. Plates 4 and 5 are, I think, of very great interest to all central station men. I would like to ask Mr. Gossler what efficiency of lamp is used in Montreal ?

Mr. Gossler : The efficiency of the lamp, as I have stated, depends on local conditions ; there are some places where we can use a less efficient lamp than others. The lamp we mostly use is a 50-volt lamp, with a current consumption of 1.03 amperes.

Mr. Breithaupt : The formula Mr. Gossler gives here is also very instructive and one I think that is not generally known, and which will be of great interest to central station men.

Mr. Armstrong : The plant referred to in Mr. Gossler's paper is, I presume, that of the Royal Electric Co.,

of Montreal. I should like to ask Mr. Gossler, in considering this reconstruction, what is his reason for adopting the belt-driven generator ?

Mr. Gossler : Mr. President, in regard to the adoption of the belt-driven generators, I would say that there are conditions under which the direct-driven dynamos are, of course, desirable. In the present case we were confined to the utilization of the engines we had in our station, which it was impossible to use in that way, if it had been so desired.

Mr. Wright : As far as that question of the direct-driven generators are concerned, there are other plants than direct-driven being installed at the present time. As to the question of belt-driven generators at the present time, as Mr. Gossler speaks of it, there are many reasons for the procedure. The question was not addressed to me exactly, but it might be in order to give one or two of the reasons. In the first place, you have an engine ; it must be of a reasonable sized unit ; it would necessarily be confined to that particular class of service. As an incandescent lighting load is a load that is at its maximum for a very few minutes during the day, comparatively to the rest of the 24 hours, you condemn your engine that is driving that large sized unit to idleness for 23½ hours, practically. By using a belt-driven generator an engine can be made to drive two, three or more generators for supplying different classes of service. Most stations, of course, are using currents of varying quality and it becomes necessary in a large city to do so ; and the same engine can be utilized for all purposes, to a certain extent. Again, a belt-driven generator, driven from the fly wheel of an engine direct, may be considered, for all practical purposes, as a direct-driven generator. You have the flexibility in the belt, which is an advantage ; there is also the question of the size of unit. This difficulty has often been found in using a very large direct-driven unit on an engine : there is always liability to accident by reason of any little inequality there might be in the bearings, or in bringing the poles too closely together where alternating machines have been driven direct on the shafts. All these questions have to be taken into the calculation in installing plants, whether direct driven or belt driven. You also have the advantage of running the engine at a slow speed, which all engineers will agree it is better to do if possible. As a rule, where you have direct-driven generators they are placed on the ends of the engine shafts, and that complicates, to a considerable extent, the engine itself ; it probably nearly doubles its cost ; it renders necessary the introduction of bell cranks. An engine with a straight shaft and overhanging disc and crank pin is simpler and more reliable. There are a number of reasons that would affect the question,—a few of them I have given. I think there are many more.

Mr. Armstrong : I entirely agree with Mr. Wright that circumstances govern altogether the desirability of using the direct-connected or belt-driven generators. In this case I was considering the specific example which Mr. Gossler had put before us, of the re-construction of the station in Montreal in which were installed five 300 K.W. generators. But, of course, the governing circumstances of their using their present engines would be the main factor there.

Mr. Ashworth : I notice in one portion of Mr. Gossler's paper he states it is only in a densely populated city and in the more densely populated portions of that city, that it is practicable to use the secondary main system. From my own experience, and I think from the experience of a good many smaller station men, it is quite practicable to use secondary station distribution in much smaller places than Montreal, probably in places of two to three thousand inhabitants. I would like to ask Mr. Gossler if it is not economical to use secondary station distribution in smaller places.

Mr. Gossler : When I said smaller places, I had no reference to a large or a small city. My statement was intended to refer to where customers were bunched or scattered. With reference to the limitation of the secondary system—as I have stated in my paper—the most economical side of the balance can be determined by such a method of reasoning as outlined in my paper.

Mr. Armstrong : There is another factor which I think should enter into that, and that is the generally higher efficiency of the larger transformer.

Mr. Gossler : That is included in the cost of maintenance of a transformer ; in the cost of maintenance in a transformer the leakage current is included, consequently the higher "all-day efficiency" of a transformer the less the cost of maintenance—the maintenance of a transformer really being the factor that would bear the greatest weight in determining the type of transformer to be used.

Mr. Milne : I would like to ask Mr. Gossler in regard to this : He says :. "Apart from this increase in capacity, there is also the saving due to running a smaller engine for the day load, and the consequent saving in labor, oil, etc." When they shut down a smaller one and open up a larger one, does the saving of labor start at that point ?

Mr. Gossler : That refers, not to a temporary decrease in one day of 135 amperes. If you decrease the station leakage load 135 amperes, it means a decrease in the load for every day of the year which will permit the operation of a smaller engine during the day, consequently decreasing the item of engineer's salary.

Mr. Milne : According to that, they have an engineer for each engine. ·

Mr. Gossler : In regard to having an engineer for each engine, the circumstances may be such that that is necessary. If the engines are large that may possibly be necessary ; it depends on local conditions. It is not a general statement at all. The requirements of the services of engineers is entirely a local consideration.

Mr. Dion : I think Mr. Gossler has earned the thanks of this Association for the valuable paper which he has just read. It is a paper of very practical interest to central station men ; a paper which I regret I did not get into my hands sooner. I did not see these papers until I got here, and since we got here there has been no time to read. I would have liked to have become more familiar with it, so as to be able to discuss it more intelligently. I think it is a paper that should be thoroughly discussed. However, I am not in a position to do it justice. There is one part of this paper I wish to refer to, where he speaks of the records made in order to calculate the changes necessary to improve the regulation of the line, &c. In order to obtain the required information he established a system of transformer maps, pole maps, etc. Last winter, through the courtesy of Mr. Gossler, this system was shown and explained to me, and I can assure you it is the most thorough system of keeping records that I have ever seen anywhere. I found it so good that I adopted some of the features in that system of keeping records, and had them carried out after I returned home. But I must say I did not adopt the system in its entirety. I found it unnecessarily complicated for a city like ours, however useful it may be in a larger place like Montreal. However, there are many valuable features in that system which I was very glad to learn and to put in practice afterwards. Reconstruction is a problem which confronts every central station manager It is no doubt, as many of you know, a very difficult thing to tackle, and there is a question of how far you should go on with this reconstruction. It is difficult to convince a board of directors that it is going to be a paying thing to scrap the whole central apparatus ; but in many cases I suppose it would be well to do that rather than carry on the reconstruction by degrees. We are doing some reconstruction, but it is being carried on gradually. We have not scrapped any large amount of apparatus at any one time, but. I suppose before we get through there may be a considerable amount put by. There is this to be considered, that in making a change when you are compelled to do it, you may save some money in making exchanges of generators. We have been able to obtain a very considerable allowance for old apparatus ; and I am satisfied that if we had waited for two or three years more we would not have obtained anything for them. That is a great

drawback, this reconstruction which becomes necessary in connection with electric lighting. A prominent banker in our city told me he found that to be the greatest obstacle in the way of electrical investments; that is, in the opinion of capitalists, an obstacle. Regarding transformers, our plan has been not to scrap transformers and buy a new outfit,. but simply not to buy any more small transformers. Where required we put up large units, displacing a number of smaller ones which are useful to supply customers in the more scattered portions of the city; and in that way the more thickly populated part is supplied by large transformers and secondary mains; and we found that to be preferable in our case than scrapping transformers in a wholesale way. I have one word to say in reference to the over-running of lamps. I quite agree with Mr. Gossler that it is not a practice to be recommended. In our city we made that mistake. It was brought about by a keen competition between companies and a desire to have our lights better than the others, and we got into the habit of running our lamps over the normal voltage, consequently it was found afterwards very difficult to reduce that. We had got into difficulty, the lamp consumption had become very large, and we found it necessary to stop this practice. I had read that the same difficulty had been met in other cities by a gradual reduction of the voltage, say half a volt every night ; but we did not care to do it in that way. I could foresee a lot of trouble. We preferred to re-model the circuits for better regulation, and drop the voltage in one night, changing the lamps on that circuit during the day. It takes a horse and wagon and several men to do it, and it costs a good deal, but we thought that was a better way of doing it than to gradually lessen the voltage and allow complaints to come in. I would like to ask Mr. Gossler— he has been asked a good many questions, but seems good-natured about it—as to the means taken to overcome the induction between lines. We have had some difficulty in that way and we have taken some means to overcome it, and I would like to know whether he has taken the same means.

Mr. Gossler : The means that were taken were very simple and very well known. The local conditions were somewhat peculiar : we had three very large engines and three principal routes. It was found, due to mutual induction, unless all the circuits on one pole line were run from one set of dynamos there was a decided fluctuation; unless we ran each pole line on a separate engine, we were not given credit for knowing much about lighting. To overcome this inductive effect, the feeders were re-arranged, by simply bringing opposite polarities as near each other as the pins on the cross arm would permit. When the lines had to be reconstructed it was found that one leg of the circuit was on one end of the cross arm, while the other leg of the same circuit was on the extreme end of the same cross arm, so that the worse inductive effect possible was obtained. We did not go to the trouble of crossing the circuits as is customary in long transmission, because we found that it was unnecessary. We reconstructed, first, one or two circuits, bringing the feeders close together so that the mutual induction due to the other circuits was practically the same on both legs. After one or two circuits had been reconstructed, we found it unnecessary to cross the circuits, but simply proceeded on the line of bringing the feeders close together, which has given perfectly satisfactory results. We have some twenty odd incandescent circuits, all heavy, and we find we can run one or two, or any number we please, on the same pole line from different engines.

Mr. Dion: That was exactly the conditions prevailing in our city. The circuits had been hung on opposite ends of the cross-arms for the purpose, as stated by the line foreman, of making the lines less dangerous. He said they had only one side of each circuit to work at, at one time, and there was a space in the centre which protected them from the other side. The men set up the plea that we were going to make their work much more dangerous, but when I explained to them that, the circuits being sometimes bunched on one no certainty, that two adjacent wires did not represent different terminals of one machine, they saw that they were just as secure after the change than before.

Mr. Wright : This reconstruction business is a very difficult matter, and there is a case in point that occurs to me where I think the difficulty will be emphasized. The electric light company in Hamilton are just about appointing a manager, and it would become the first duty of that manager to recommend the Board of Directors to throw out the whole plant. From what I know of the Hamilton Board of Directors, I think they are a great deal more likely to throw out the manager. It will become a question of scrapping the plant or scrapping with the manager. Mr. Gossler is so very well posted on these subjects that I think it is a first-class opportunity to pump him a little further and see if we cannot pump him dry. There are two important questions that come up not only in reconstruction, but also in construction : one is the frequency to use in the introduction of an alternating system and another is the voltage ; those are questions that a little fuller information would be very acceptable on. Authorities differ on both questions. Whether it is better to adhere to the old style of sixteen thousand alternations or thereabouts or the more modern in the neighborhood of seven or eight, and if Mr. Gossler has any reasons and would go more fully into them for the change of voltage from one to two thousand, the information would certainly be very acceptable to us.

Mr. Gossler : In regard to the change of voltage from one to two thousand, we have drops on our circuits varying from one to ten per cent. ; the regulation, of course, corresponds. Increasing the voltage to 2,000 volts will decrease the drop one quarter of what it is at present ; there is no practical hindrance to the use of two thousand volts—it can be handled about as readily as one thousand—also modern transformers are made interchangeable, so they can be used on either one or two thousand volts. If you are going to increase the saving and improve the regulation by the adoption of the two thousand volts without any practical inconvenience, there seems to me every reason in the world why it should be done. It was decided in our case to use two thousand volts, principally for the purpose of getting better regulation with the present feeders ; we do not anticipate any trouble. The two thousand volts is in operation in many places. There may possibly be a little difficulty with the lines that are now in contact with trees, as we now have some little difficulty in wet weather at the places with one thousand volts, and two thousand volts would be a little more troublesome, but these conditions will be changed at any rate. In regard to the question of frequency, that is a question that has to be decided by the local conditions. If the lighting is of paramount importance or the principal feature of the station, there is no question but that sixteen thousand alternations should be used. It is practicable to operate motors from sixteen thousand alternations as well as eight thousand. The leakage of the transformers does not decrease exactly inversely as the alternations, but approximately so, so that the leakage of transformers at eight thousand alternations is probably twice as much as sixteen thousand, and where the lighting load is the principal feature the sixteen thousand alternations is very desirable for this reason. If the plant is to be supplied by power from quite a distance, where it is necessary to transmit power from fifteen, twenty-five or thirty miles, induction comes in as a factor to be considered, and in most long transmission plants the lower alternations have been adopted. But, as I have said before, where lighting is the principal feature, there is no question about sixteen thousand alternations, especially as motors can be as readily used on sixteen thousand as on eight thousand, so that sixteen thousand alternations permits of serving light and power from the same generators.

Mr. Breithaupt : I have great pleasure in moving a hearty vote of thanks to Mr. Gossler for his very valuable paper.

Mr. Dion : I second that.

The President : It has been moved by Mr. Breithaupt, seconded by Mr. Dion, that the hearty thanks of this Association be tendered Mr. Gossler for his valuable paper. Is this your wish?

The motion was carried amid applause.

The President : I think the Association is to be congratulated upon the fact that we have men capable of giving us such a paper as Mr. Gossler has given us. The next thing on the list is the paper entitled "Power Transmission by Polyphase E.M.F.'s," by Mr. George White-Fraser. (See page 142).

Mr. Fraser, on rising to read his paper, was greeted with applause.

The President : We have not heard Professor Rosebrugh's voice in Convention ; I think the members here would like to hear if he has anything to say on this paper.

Mr. Rosebrugh : Mr. President, I have not had time to look over the paper at all, and any remarks that I might make might be premature. Without further consideration I would not care to say anything.

Mr. Breithaupt : I would like to ask Mr. Fraser about the figures he gives as to the Lauffen-Frankfort transmission, as to where he gets them.

Mr. Fraser : From the official report published by Dr. Boaber, who was chairman of the official committee.

Mr. Milne : Has not Mr. Fraser drawn the arrows in diagram 12 in the wrong way? It occurs to me that the arrows are in front of the E.M.F.

Mr. Fraser : You will find the maximum E.M.F. is quite a distance from the mouth of the current.

Mr. Armstrong : If there is not likely to be any further discussion, I have pleasure in moving a vote of thanks to Mr. Fraser for his paper on this subject ; it seems to have covered the whole field very fully on the subject of polyphase currents.

Mr. Wyse : I second the motion.

The President put the motion, which was carried.

Mr. Dion : I would like to call attention to two points in connection with this paper. One of the two features which give the paper particular value is the great clearness with which the phenomena of alternating currents are explained. As stated by Mr. Fraser, most of the statements on this subject are so surrounded by clouds that it is difficult for any but the advanced student to properly understand them. In this particular case he has explained the action of the polyphase currents in such a clear manner that he has no doubt helped to increase the knowledge of many members present. The other feature which I think deserves particular attention is his plea for good engineering, and I hope to see the day when his advice will be followed, and when every particular installation will be designed according to the local conditions, and when every installation, whether large or small, will not be undertaken before consulting competent electrical engineers.

Mr. Fraser : I thank the members and I thank Mr. Dion for considering these points.

Mr. Gossler : There is one point in the description of the generation of multiphase currents, as described in diagrams 3 and 4, which I think is probably an oversight on the part of Mr. Fraser. He states at the bottom directly under diagram four, "A,B, are two coils of the armature. The angular distance of A, B is half that of N, S. When A is right under N it is generating its maximum E.M.F. ; at that moment B is half way between N, S and its induction, and therefore its E.M.F. is least." I call attention particularly to this statement inasmuch as it is very important. Mr. Fraser states further, "As the armature revolves clockwise, B gradually gets into a stronger field, while A is approaching a weaker one." Mr. Fraser states there that when A is right under N it is generating its maximum electro-motive force, which is the actual condition if the armature coils are so placed as in diagram 4. While diagrams 3 and 4 have been taken as general illustrations, I think it has been an oversight on Mr. White-Fraser's part, as this illustration can only apply to two-phase generation. If the coils are so placed that B is half way between N, S, it must necessarily mean the generation of two-phase currents 90° apart. It would alter diagram 3 to such an extent that the minimum point of curve 2 would be directly under the maximum point of curve 1, which is the condition of multiphase currents of ninety degrees apart. I think this illustration will only apply to the generation of currents ninety degrees apart.

Mr. Fraser: If you take diagram 3 in connection with diagram 4, your idea is correct. It is not a quantitative so much as a qualitative diagram.

Mr. Gossler: It is a trifle misleading if you combine diagrams 3 and 4, because it is not a general case.

Mr. Fraser: I have not specified any particular case. I think it is understood to be purely qualitative.

Mr. Gossler: There is another point in regard to the position of the armature coils which this illustration can be made to show most beautifully—the re-action and inter-action of the coils on the armature, due to their relative positions, as affecting regulation. The regulation question is a very serious one and a very important one, and inasmuch as the diagram shows it so beautifully I thought possibly it would not be out of place to bring it to your attention. If the coils as placed in diagram 4, and as stated, when A is directly under N it is generating its maximum electromotive force, and B being half way between N, S, the electromotive force is minimum, then when the current in A is maximum there can be no current in B, consequently there can be no induction from B to A. The effect on one side of the coil B from A is the same as on the other side, they being symetrical to coil A, so that the two conditions taken together show the impossibility of any action between the two phases; the practical demonstration of the fact of their being no influence on regulation from one phase to the other in two-phase generators is borne out by every day practice. If we pass further on from two-phase apparatus to multiphase apparatus, Mr. Fraser mentions the possibility of placing three or four or any number of coils in a similar space, but the four coils are of no practical use. If we consider the space between N, S, divided into three spaces so that we have three coils placed there, we have a generator generating three-phase currents. If we divide that space so that it is divided into three and apply the same reasoning as above, the coils not being symetrically or relatively inductively placed to each other, there is but one condition in a three-phase generator in which the mutual induction or re-action of the armature coils is equal, and that is when the coils or phases are equally loaded. When they are not, on account of this re-action of the armature coil, one coil of the armature acts as the primary of a transformer, while the other two coils act as the secondaries of a transformer. When the phases of a three-phase apparatus are not equally loaded there is a transformer action which necessarily boosts up one side of the circuit and decreases the voltage on the others.

Mr. Fraser: I have been very careful to avoid any comparison of the two methods.

The President: We will now have Mr. E. J. Phillip's paper entitled "Operating Engines without a Natural Supply of Condensing Water, or the Continous Use of Injection Water." I am sorry Mr. Phillip is not here, but Mr. Wickens will read his paper.

Mr. Wickens: I regret Mr. Phillip is not here to read this paper himself. He has given some considerable time and study to this particular arrangement, and hopes to install a considerable plant under this style. I really regret that he is not here to read this paper himself, because should any discussion occur, he would be very much more capable of taking up the points than I will. There is another thing that I would call the attention of the members to before reading this paper, and more especially to those of us who have been attending the conventions straight along, and that is as to the scope of the papers that have been read. There is no question in my mind that the papers read at this Convention have been somewhat an advance on any of the others, and of

course in an Association whose scope is so great and which reaches out so far as this, the papers must necessarily cover a very large ground; and while, it seems to me, in a measure the interests of the larger establishments have been held forth, the interests of a very large number of the smaller plants throughout Canada have been taken into account. The other day, in conversation with a gentleman who had travelled largely in England and also largely in Canada, he said that one of the first things that attracted the attention of an Englishman in Canada was the fact that the small hamlets were lit up with electricity. That means that we have a large number of small plants scattered throughout Canada, and we should have a large number of the proprietors of those small plants attending our Convention; and I honestly hope the members that are here to-day and the incoming Executive will make a special movement to get a large attendance of that kind next year.

Mr. Wright offered the use of his steam yacht for a sail on the lake at three o'clock.

The President: Mr. Wright has a very handsome steam launch, and if any of you can make it convenient to go you will find it very pleasant.

Mr. Wickens: The title of this paper is "Operating Engines without a Natural Supply of Condensing Water, or the Continuous Use of Injection Water." (See page 152.) This is coming to be very vital as far as the steam end of electrical plant is concerned.

The reading of the paper was greeted with applause.

Mr. Wright: I am sure the Association is indebted to Mr. Phillip for the getting up of this paper. The subject, it seems to me, is far too important a one to bring in and discuss at this late stage of the proceedings of the Convention, and I would like to suggest that a request be made to Mr. Phillip by the Association or the Executive Committee to amplify this paper and bring it before us again at our next Convention, when I think we shall have more time to discuss it and do it the justice that its importance demands.

Mr. Wickens: There is one thing I would like to call the attention of the members to. Mr. Phillip, unfortunately, is not a member of this Association, although he has intimated his intention of becoming a member, but for some reason or other he did not get his application in, and I would like to have the secretary forward him a vote of thanks from this Association. He is one of our bright young men in the city of Toronto.

Mr. Armstrong: I should like to move a vote of censure to Mr. Wickens for not seeing that Mr. Phillip was a member of the Association. However, I will withdraw it.

Mr. Wickens: We put the proposition in, but for some reason or other it was skipped.

Mr. Hopkins: The H., G. & B. Railway have adopted something like this. They had a great deal of trouble deciding where they would locate their power house. I understand they never had a proper inspection made previously to locating their power house, to find whether they could get water there or not. They had the power house built and then tried to get water for condensing their engine, and then afterwards they found they could not very well get it. They dug down a deep hole through very sticky clay, which was a very expensive operation, and this seemed to answer the purpose very well; but this big hole was filled by water coming down from a creek and this was very muddy in the spring of the year. They found that the water would run down very fast and freely soak into this clay, with the result that it would not be full enough to supply them, so they put a pipe low down, I think about a foot or something like that in diameter, to connect the bottom of this creek with this hole. The result was that the water going in was very, very dirty, and full of grit and sand; then, in midsummer this place dried up completely and they were without condensing water, and in fact, without any water at all. They hauled water for a while with a team for their boilers, and the finally, when they got tired of that, rented the power from Hamilton and had the water transferred down seventeen miles. That is one example of this. They used the water in such a way that it would cool itself afterwards by running out in a trough and then running down. This trough was leaky and let the water escape and cooled it very well. After they found that the hole wouldn't supply water in the summer time, and as they didn't want to rent power from the Hamilton Street Railway Company again, they dug a deep well, and when they dug that they found out that the water was all full of limestone and calcium sulphate, and it did a great deal of harm. I mention this to show that in many cases it would be very well, but here is a case where it did not work very well, but it is better than if they didn't have it.

Mr. Ashworth : I might say a few words which would possibly be of interest to Mr. Phillip when he comes to amplify the paper, which has just been suggested. The first thing I notice in his paper which attracted my attention, was the proposed cost of condensing apparatus, which is set down at $300. Personally, I know from having bought one or two condensers that it is impossible to get a condenser for a plant of one hundred horse power for anything like $300. I think it would be low enough if we say $600. Another thing is, that in his calculations he does not allow anything for the immense cost of pumping water from the top of high buildings, which one would imagine would be certainly considerable. However, that is something which he may further explain in his paper. I personally have been connected with a plant which is using water for condensing purposes in a town out west, and in that town we had no means of getting water at all. We had a spot which was supposed to have water underneath it and we found out it hadn't any.

A voice : What had it ?

Mr. Ashworth : Principally oil. Oil is not very good for cleaning boilers. As luck would have it, behind this place there happened to be a large number of underground tanks, and we conceived the idea of using the water from these underground tanks, which contained probably about 20,000 barrels of water. We pumped the water from the nearest of these tanks and discharged it into the farthest, some 300 feet away, by means of a long, low trough in which there were a number of cleats with sawed edges, which apparently distributes the water very well, and we get the water at a comparatively low temperature ; in fact we experience a great saving by it, and we find instead of having to buy water, as we did in the summer time, we get enough water from these small wells we have, and it has greatly diminished the expense in almost every particular. Of course, the original installing expense is considerable. This makes me think that if any scheme could be devised for obtaining this, I think it would be a very good thing. It would be of interest to central station men throughout the country if Mr. Phillip would, as has been suggested, amplify his paper and present it at some future Convention.

Mr. Wickens : I understood he got the figures from one of the steam pump makers, and they are the figures he gave him. We have in his city a large institution, and they have for several years run upon the principle of using the water over and over again from a set of tanks. The difficulty in that case was the expense of putting in all these tanks and the ground occupied. The slowness by which the water cooled made the investment too great for an ordinary small plant such as Mr. Phillips has attempted to represent in his paper. The fact of being able to run your water in such a way as would cool it in any ordinary tank, makes it necessary to have very many more times the quantity of water to do your cooling than you actually need. In this case represented in the paper you only require the actual quantity of water you are pumping in your pipe to evaporate your steam. The air is the cooling medium and not the water. As in the case referred to by my friend, it is not calculated that the water is the cooling medium. The water is only a means to an end.

The President : The suggestion by Mr. Wright that

we have this paper over again at Niagara Falls at our next Convention, is a very wise one. By the kindness of Mr. Dunstan, of the Bell Telephone Company, I am permitted to say that any of the individual members (we can hardly go in a body) who care to visit the Bell Telepone Company's exchange and inspect the switch board and the operating room and so on, the company will be pleased to see them at any time.

Mr. Dunstan : Unfortunately for myself, and unavoidably so, I was not present at the session yesterday when the election of officers took place, and the members of the executive were elected. On looking over the lists I cannot but feel great regret that the City of Ottawa is not represented. I am sure that that has occurred simply by some unfortunate oversight in not presenting to the meeting the name of the Ottawa members. Remembering, as I am sure we will, the splendid work of our Ottawa friends last year and the brilliant results which they accomplished, I cannot help but feel, as I said before, great regret that the omission occurs. I am sure I can say this without it being understood in any way as a reflection upon any of those who were yesterday elected to the positions, and I would myself be only too willing to resign my position on the Executive if it could be filled by some Ottawa member. I am sure, as I said before, it can only have been by some oversight in submitting the names to the meeting that the omission took place, and I would not, for a great deal, that the impression should go abroad that we had in any way forgotten the efforts which were made on behalf of the Association in Ottawa last year, or that the impression left on our minds then had been obliterated by intervening time.

Mr. Dion : As a representative from Ottawa I may say that I was a little disappointed yesterday when I found that no representative from our city had been elected to any position on the Executive Committee. Not that I personally wished the honor—there were other candidates besides myself nominated from Ottawa —but I do not like the idea of going back to Ottawa and give the impression that the presence of the delegates who are here now has had the effect of reducing the representation of three last year to zero this year. I am very thankful to Mr. Dunstan for his thoughtfulness in mentioning this matter. I do not cast any blame on anyone. I suppose it is difficult to arrange the representation so that there will be representatives from each centre. There are many other considerations, no doubt. However, I am very thankful for the reference.

Mr. Wright : The difficulty appears to be that the election of the Executive Committee is made at large. If we had certain representatives from each district we would have no difficulty, but it is owing to the plentitude of good material that the omission seems to have occurred. I deplore just as much as anybody that we have no representative from the city of Ottawa. It is easy to see how it has occurred. It is in balloting for the members, which, you may say, is almost a matter of chance, in a way. I would not like to see any member who was elected yesterday by the Association resign his position. That, it seems to me, would be out of place. But I am certain if any way can be suggested by which a representative can be obtained from Ottawa, it would be well.

Mr. Black : I think we were led into trouble by the previous elections. I noticed when we were electing the principal officers, the President and two Vice-Presidents were all out-of-town members, and the Secretary was the only officer residing in Toronto, where the headquarters of the Association is. On the other hand, it was desirable to bring in a gentleman from Niagara Falls, and a gentleman from further west, and the impression seemed to get abroad then with the members that it would be desirable to have the executive members, if not in Toronto, very close to Toronto. I know this year our Executive has been very scattered and we have had difficulties in getting a quorum. Mr. Breithaupt and myself have come to Toronto every time when an Executive meeting was required. I think that some of the meetings had to be postponed on account of the members not attending, and I think that is how the

matter has been brought about. As one of the scrutineers, I may say that the elections were all very close ;
they ran almost neck to neck, with very few exceptions,
so that there was nothing personal to anybody. But I
was about to suggest, as I have been on for several
years, if there is a sufficient number to make a quorum
without meeting, that one of the gentlemen should be
elected before we adjourn, to show our good feel-
ings towards them. They stood right royally by us
last year.

Mr. Armstrong : I think there is no doubt at all that the reason
there is no representative elected from Ottawa lies in the fact that
there were three gentlemen from Ottawa nominated; it was not
the feeling that Ottawa should not be represented, but the votes
were scattered amongst three. :

Mr. Dunstan : I think there is no doubt that it was an accident
pure and simple, and that it is regretted by the members of this
Association, and I believe that the Ottawa members will understand it in that way. I think Mr. Armstrong's suggestion is the
correct one. As I said before, I was part of this little scattering
I think that probably the vote was simply split, and the elections,
as Mr. Black stated, being close, the result of this little scattering
simply resulted, unfortunately, in not any one of the three being
elected from Ottawa. Mr. Black has stated that he would like to
resign. The difficulty of the position is not so much in resigning,
but possibly in getting an Ottawa gentleman to accept the position if a nomination was made. If I felt sure that a nomination
would be accepted on those premises, I would at once place myself as resigning. I therefore hesitated to press that very strongly.
If any vacancy should occur during the year, there is no doubt it
will rest with the Executive Committee to fill that vacancy, and I
think you will find that it will go to Ottawa.

Mr. Hunt : Is there anything in our Constitution by which we
can increase that number? If there is, I would like to make a
motion to that effect.

The President ; The Constitution distinctly states that the Executive shall consist of ten members.

Mr. Dunstan : It cannot even be changed on the same day,
because no change of the Constitution can take place on the day
on which a motion is introduced, and as this is the last day of our
session, we can't make a change ; otherwise I would have made
a motion on the same line as Mr. Hunt's suggestion.

Mr. Dion : I am very thankful to the gentlemen who have offered
to sacrifice themselves for the satisfaction of the Ottawa members,
but I do not think we can avail ourselves of any such offer as that.
I think we may let the matter drop. There is one thing I would
like to mention while the subject is up, and it is this: Mr. Black
spoke as if there was a feeling that owing to several of the officers
being non-resident of Toronto, a certain number of the Executive
should be Toronto men on account of the convenience there would
be in getting them together during the year. There must be
great difficulties in getting meetings and in doing business, when
the Executive is scattered all over the country, but that is a difficulty which exists in all associations of this kind. There is no
way out of it, except to elect members from one city all the time,
that is, where the Association has its headquarters. This would
be sure to bring dissatisfaction and it is hardly practicable, therefore the only other way seems to be to let the Executive be scattered and make the best of it. If it is to be scattered, I think it
ought to be well scattered, because there is this view of it, that
every member of the Executive Committee is an active member
of the Association in his own district, and I think in scattering
the members throughout the important centres, the Association
might probably be benefited. I merely speak of it as a matter
of general policy for the Association.

Mr. Armstrong : I think the whole difficulty in connection with
this is on account of our present mode of election, which practically restricts us. It means the continuance in office of five members of the existing board, and it limits the choice to that extent. I
think it would be probably better to have an election of five members annually to hold office for two years. I think the present
mode of election accounts for such a result as we had yesterday.

Mr. Breithaupt : I think the suggestion of Mr. Armstrong in relation
to the changing of the Constitution is very timely. Yesterday I raised
the question as to how the election ought to proceed, because I was under the impression that it had to proceed that way. Now the thing is
somewhat mixed up, and we all got a little confused in the election. As
Mr. Dion has said, I think it would be a very wise plan to have the
members of the Executive chosen from different parts of the country, as
much as possible, so as to create an interest, and we would have live
agents all over to advocate our cause.

Mr. Fraser : I think we can divide it up in sections and make this
Association more of a national character ; for instance, in West Ontario
and Toronto and Montreal, and in Ottawa— whatever sections are decided upon. I think the representation would be better, and the interest
in the Association much better, by having responsible agents, so to speak,
or Executive officers, in various parts of the country. I think it would
greatly widen the scope of the Association, and everybody would take a
great deal more interest in it than they do now when the people seem to
think it is purely an Ontario Association.

Mr. Wickens : As a member who assisted in revising this Constitution,
I may say that the committee took considerable pains to adopt what they
considered the best plan on which to arrange and elect their Executive

board. It was felt at that time that it was really in the interests of this
Association that there should be some members continually on the board,
for which reason the present plan was chosen and a report made at the
last Convention, and it was unanimously adopted at that time.

Mr. Wright : I rise to a point of order. It is entirely out of order to
have this discussion on the Constitution, and no resolution or notice of
resolution before the Convention.

The President : Mr. Wright is quite right.

Mr. Dunstan : I wish to move that the President vacate the chair for
a few moments and that Mr Breithaupt act as chairman.

Mr. Breithaupt took the chair.

Mr. Dunstan : I move that a hearty vote of thanks be tendered to Mr.
Smith, the retiring President, for the very able and efficient manner in
which he has performed the duties of his office during the past year, and
for the capable manner in which he has presided over the various
sessions of this Convention. If the day were not so hot and time were
not so pressing, there is much that I could say to you about the splendid
work that Mr. Smith has done for this Association, both before he was
elected to the Presidency and during the time he has acted in that
capacity ; and, if I may anticipate a little, what he will do in the future.
But that is all apparent to you as it is to me ; it is fully realized by
every member of this Association. I think I will content myself by
moving a very hearty vote of thanks to Mr. Smith, and I am sure that
the motion will receive your hearty support and approbation.

Mr. Black : I have great pleasure in seconding the motion.

Mr. Breithaupt : It has been moved by Mr. Dunstan, seconded by
Mr. Black, that the services rendered by Mr. A. B. Smith during the
past year, and the services he has rendered the Association throughout,
be recognized by passing him a hearty vote of thanks, and I will ask all
who are in favor of the motion to signify by a rising vote.

The members of the Convention arose in a body to their feet and the
motion was carried with applause.

The President : Gentlemen, I simply wish to thank you. I know
you mean it, and I am sorry that the man who preceded me in office
as President did not bequeath to me his eloquence. I assumed the
office with fear and trembling, and conscious of my inability ; yet, I
was willing to do the best I could to further the interests of this Association. (Applause.)

Mr. Breithaupt : Before we adjourn I think we should pass a vote of
thanks to Mr. Milne for giving us the very instructive and interesting
lecture on " Radiant Matter " which we had on the evening of the first
day of the Convention. The lecture was certainly very instructive and
of very great interest, and particularly so at the present time.

Mr. Hunt : I have much pleasure in seconding the motion. I enjoyed the lecture very much.

The President put the motion, which on a vote being taken was
carried.

Mr. Wright : I have great pleasure in proposing that the thanks of
the Association be given to our Secretary for his painstaking and
conscientious work during the past year.

Mr. Armstrong : I second that.

The President : Everybody in the room ought to second that.

The motion was carried unanimously.

Mr. Mortimer : I cannot but feel that the work I have tried to do is
appreciated, and if it is not as good as I have tried to make it, it cannot
be helped.

It was moved by Mr. Dunstan, seconded by Mr. Fraser, that the
thanks of the Association be tendered to the Toronto Electric Light Co.
and the Toronto Railway Co. for having permitted an inspection of
their power stations by the members and friends of the Association.—
Carried.

Mr. Wickens : I move a vote of thanks to the management of the
Board of Trade building for the very comfortable quarters we have
enjoyed and the very courteous treatment we have received during the
time we have had our Convention in this building.

Mr. Armstrong : I second that.

The motion was carried unanimously.

Mr. Armstrong : I have much pleasure in moving a vote of thanks to
be tendered to the press of Toronto for the good reports that have appeared in their issues. I have no doubt that the pressure of political
matter has interfered with the giving of fuller reports of our proceedings.

Mr. Black : I second that.

The President put the motion, which was carried.

The President : I think, gentlemen, that is all. I wish to thank you
very sincerely for your close attention. The Convention has been a
success from the point of attendance and also the interest taken in the
papers, and it augurs well for future Conventions. Personally, I feel
sorry that Ottawa has been omitted from our list of the Executive. I
did not want to refer to it, but several spoke of it yesterday, and the
speakers were all Toronto men. It was an unintentional accident. I
declare the Convention adjourned.

---

### EXCURSIONS AND BANQUET.

Upwards of one hundred members and guests of the Association proceeded to Lorne Park per steamer "Greyhound" on Thursday evening
and participated in the annual banquet at the Hotel Louise, returning to
the city about midnight. A very pleasing feature of the return trip was
the presentation by Mr. Higman, on behalf of members of the Association, of a gold-headed cane to Mr. A. B. Smith, the retiring
President.

On Friday afternoon an enjoyable time was spent by a number of the
members and their friends on board the steam yacht "Electra," by kind
invitation of the owner, Mr. J. J. Wright.

## MR. JOHN YULE.

THE honor of becoming President of the Canadian Electrical Association was, at the recent convention, bestowed upon Mr. John Yule, of Guelph, whose portrait appears on this page. Previous to coming to Canada Mr. Yule was for ten years connected with the Dundee Gas Company, of Dundee, Scotland, and for two years with the Dundee Municipal Gas Works. Shortly after coming to Canada, over twenty-five years ago, he assumed the management of the Guelph Gas Company, which position he still holds, the name of the company having since been changed to the Guelph Light & Power Company. During this extended term under the direction of Mr. Yule, the company has attained a marked degree of prosperity. He is said to enjoy the entire confidence of the directorate and shareholders of the company, and the respect of the citizens generally.

The newly-elected president has been prominently connected with the Canadian Electrical Association since its inauguration. He was one of the committee appointed to formulate a scheme of organization in 1891, and was elected a member of the Executive Committee upon the formation of the Association, serving as such until September of last year. He is deeply interested in the advancement of electrical industries, and it is safe to say that the affairs of the Canadian Electrical Association will be ably conducted under his direction.

## AN INTERESTING LEGAL SUIT.

THE decision in the case of the Bell Telephone Company vs. the Montreal Street Railway Company, which was given last week, is of much interest to electric railway and telephone companies.

The Bell Telephone Company entered suit against the Montreal Street Railway Company, claiming $30,000 damages, on the ground that the introduction of the electric trolley car system into Montreal in 1892 caused, and has been causing ever since, a serious disarrangement of the telephone service, and necessitated the adoption of a number of expensive contrivances to counteract the effect of the presence of the trolley wires, with their attachments, which would otherwise have caused the diversion of currents running along the telephone wires, the area adjacent to the trolley poles being always charged with electricity, which naturally would have impaired the efficiency of the telephone service, if the Bell Company had not gone to considerable extra expense. For this extra expense it is now claimed to be indemnified as above, up to the date of the entering of the action with reserve of claims for further subsequent damage.

The Street Railway Company's main plea was to the effect that, in establishing the trolley system, it had acted within the rights granted it under its charter with the city of Montreal, and that it could not be held liable for any damage incidentally suffered by the Bell Telephone Company.

The hearing of the case was commenced in the spring of last year, and occupied ten days, during which a great deal of technical evidence was heard.

Judge Davidson last week gave judgment, dismissing the action of the Bell Telephone Company.

## DEATH OF MR. EDWARD LUSHER.

MR. Edward Lusher, secretary-treasurer of the Montreal Street Railway Company, died on the 11th of June, at the advanced age of 71 years. The deceased had for upwards of fifty years been prominent in business circles, and was probably the oldest railway man in Montreal. For eighteen years he filled the position of secretary and general manager of the old horse car system. Upon the reorganization of that enterprise three years ago, Mr. Lusher took the position of secretary-treasurer, which he has held up to the present time In 1885 he was elected a vice-president of the American Street Railway Association, and it was mainly through his endeavors at the convention in Milwaukee in 1893 that the session was held in Montreal last October.

The history of street railroading in Montreal from its very inception many years ago, was really a portion of the life-history of Mr. Lusher. He saw the Transportation Company in Montreal grow from a couple of miles of track on Notre Dame street, with its homely old cars and crude track, into the extensive system of to-day. For a man of his years Mr. Lusher was wonderfully well preserved, and looked and acted many years younger than he really was. In his death the company lose a valued and faithful servant.

MR. JOHN YULE,
President-elect Canadian Electrical Association.

ADVANTAGES OF ELECTRIC LIGHT IN BAD AIR IN MINING.—The results of some recent experiments of Dr. Haldane have been made public. They refer to the presence of black damp; he finds that when the percentage of oxygen has fallen to 17.64 a candle is extinguished ; at 3.38 per cent. of carbonic acid gas, and 15.3 per cent. of carbonic oxygen his respiration began to deepen, and at 7.32 and 9.62 per cent. respectively there was violent panting ; at 7 per cent. of oxygen consciousness would probably have been lost. He thus shows that there is a wide margin between the point of the extinguishing of a lamp and the point of danger to life ; a miner provided with an electric lamp could therefore penetrate with impunity or escape through atmophere containing at least three times as much black damp as would extinguish a lamp and the difficulty of respiration would give ample warning if the electric lamp did not.

PUBLISHED ON THE FIFTH OF EVERY MONTH BY

### CHAS. H. MORTIMER,

OFFICE : CONFEDERATION LIFE BUILDING,

Corner Yonge and Richmond Streets,

**TORONTO, - - - CANADA.**

Telephone 2362.

NEW YORK LIFE INSURANCE BUILDING, MONTREAL.

Bell Telephone 2209.

#### ADVERTISEMENTS.

Advertising rates sent promptly on application. Orders for advertising should reach the office of publication not later than the 26th day of the month immediately preceding date of issue. Changes in advertisements will be made whenever desired, without cost to the advertiser, but to insure proper compliance with the instructions of the advertiser, requests for change should reach the office as early as the 22nd day of the month.

#### SUBSCRIPTIONS.

The ELECTRICAL NEWS will be mailed to subscribers in the Dominion, or the United States, post free, for $1.00 per annum, 50 cents for six months. The price of subscription should be remitted by currency, registered letter, or postal order payable to C. H. Mortimer. Please do not send cheques on local banks unless 25 cents is added for cost of discount. Money sent in unregistered letters will be at senders' risk. Subscriptions from foreign countries embraced in the General Postal Union $1.50 per annum. Subscriptions are payable in advance. The paper will be discontinued at expiration of term paid for if so stipulated by the subscriber, but where no such understanding exists, will be continued until instructions to discontinue are received and all arrearages paid.

Subscribers may have the mailing address changed as often as desired. When ordering change, always give the old as well as the new address.

The Publisher should be notified of the failure of subscribers to receive their paper promptly and regularly.

#### EDITOR'S ANNOUNCEMENTS.

Correspondence is invited upon all topics legitimately coming within the scope of this journal.

The " Canadian Electrical News " has been appointed the official paper of the Canadian Electrical Association.

## CANADIAN ELECTRICAL ASSOCIATION.

### OFFICERS :

PRESIDENT :

JOHN YULE, Manager Guelph Light & Power Company, Guelph, Ont.

1ST VICE-PRESIDENT :

L. B. McFARLANE, Manager Eastern Department, Bell Telephone Company, Montreal.

2ND VICE-PRESIDENT :

E. C. BREITHAUPT, Electrical Engineer, Berlin, Ont.

SECRETARY-TREASURER :

C. H. MORTIMER, Publisher ELECTRICAL NEWS, Toronto.

EXECUTIVE COMMITTEE :

GEO. BLACK, G. N. W. Telegraph Co., Hamilton.

J. A. KAMMERER, General Agent, Royal Electric Co., Toronto.

K. J. DUNSTAN, Local Manager Bell Telephone Company, Toronto.

A. M. WICKENS, Electrician Parliament Buildings, Toronto.

J. J. WRIGHT, Manager Toronto Electric Light Company, Toronto.

ROSS MACKENZIE, Manager Niagara Falls Park and River Railway, Niagara Falls, Ont.

A. B. SMITH, Superintendent G. N. W. Telegraph Co., Toronto.

JOHN CARROLL, Sec.-Treas. Eugene F. Phillips Electrical Works, Montreal.

C. B. HUNT, London Electric Company, London.

F. C. ARMSTRONG, Canadian General Electric Co., Toronto.

## CANADIAN ASSOCIATION OF STATIONARY ENGINEERS.

| | | |
|---|---|---|
| President, W. G. BLACKGROVE, | - | 54 Widmer St., Toronto, Ont. |
| Vice-President, JAMES DEVLIN, | - | Kingston, Ont. |
| Secretary, E. J. PHILIP, | - | 11 Cumberland St., Toronto. |
| Treasurer, R. C. PETTIGREW, | - | 94 West Avenue N., Hamilton. |
| Conductor, W. F. CHAPMAN, | - | Brockville, Ont. |
| Door Keeper, F. G. JOHNSTON, | - | Ottawa, Ont. |

TORONTO BRANCH NO. 1.—Meets 1st and 3rd Wednesday each month in Engineers' Hall, 61 Victoria street. John Fox, President ; Chas. Moseley, Vice-President ; T. Eversfield, Recording Secretary, University Crescent.

MONTREAL BRANCH NO. 1.—Meets 1st and 3rd Thursday each month, in Engineers' Hall, Craig street. President, John Murphy ; 1st Vice-President, J. E. Huntington ; and Vice-President, Wm. Smyth ; Secretary, B. Archibald York ; Treasurer, Peter McNaugton.

ST. LAURENT BRANCH NO. 2.—Meets every Monday evening at 43 Bonsecours street, Montreal. R. Drouin, President ; Alfred Latour, Secretary, 306 Delisle street, St. Cunegonde.

BRANDON, MAN., BRANCH NO. 1.—Meets 1st and 3rd Friday each month, in City Hall. A. R. Crawford, President ; Arthur Fleming, Secretary.

HAMILTON BRANCH NO. 2.—Meets 1st and 3rd Friday each month in Maccabee's Hall. Wm. Norris, President ; E. Teeter, Vice-President ; Jas. Ironsides, Corresponding Secretary.

STRATFORD BRANCH NO. 3.—John Hoy, President ; Samuel H. Weir, Secretary.

BRANTFORD BRANCH NO. 4.—Meets 2nd and 4th Friday each month. F. Lane, President ; T. Pilgrim, Vice-President ; Joseph Ogle, Secretary, Brantford Cordage Co.

LONDON BRANCH NO. 5.—Meets once a month in the Huron and Erie Loan Savings Co.'s block. Robert Simmie, President ; E. Kidner, Vice-President ; Wm. Meaden, Secretary Treasurer, 533 Richmond street.

GUELPH BRANCH NO. 6.—Meets 1st and 3rd Wednesday each month at 7.30 p. m. H. Geary, President ; Thos. Anderson, Vice-President ; H. Flewelling, Rec.-Secretary ; P. Ryan, Fin.-Secretary ; Treasurer, C. F. Jordan.

OTTAWA BRANCH NO. 7.—Meet every second and fourth Saturday in each month, in Borbridge's hall, Rideau street ; Frank Robert, President ; F. Merrill, Secretary, 352 Wellington street.

DRESDEN BRANCH NO. 8.—Meets 1st and Thursday in each month. Thos. Steeper, Secretary.

BERLIN BRANCH NO. 9.—Meets 2nd and 4th Saturday each month at 8 p.m. J. R. Utley, President ; G. Steinmetz, Vice-President ; Secretary and Treasurer, W. J. Rhodes, Berlin, Ont.

KINGSTON BRANCH NO. 10.—Meets 1st and 3rd Tuesday in each month in Fraser Hall, King street, at 8 p. m. President, S. Donnelly ; Vice-President, Henry Hopkins ; Secretary, J. W. Tandvin.

WINNIPEG BRANCH NO. 11.—President, G. M. Hazlett ; Rec.-Secretary, J. Sutherland ; Financial Secretary, A. B. Jones.

KINCARDINE BRANCH NO. 12.—Meets every Tuesday at 8 o'clock, in McKibbon's block. President, Daniel Bennett ; Vice-President, Joseph Lighthall ; Secretary, Percy C. Walker, Waterworks.

WIARTON BRANCH NO. 13.—President, Wm. Craddock ; Rec.-Secretary, Ed. Dunham.

PETERBOROUGH BRANCH NO. 14.—Meets 2nd and 4th Wednesday in each month. S. Potter, President ; C. Robison, Vice-President ; W. Sharp, engineer steam laundry, Charlotte street, Secretary.

BROCKVILLE BRANCH NO. 15.—President, W. F. Chapman ; Vice-President, A. Franklin ; Recording Secretary, Wm. Robinson.

CARLETON PLACE BRANCH NO. 16.—Meets every Saturday evening. President, Jos. McKay ; Secretary, J. D. Armstrong.

**Canadian Electrical Association.** IT has been found necessary to largely increase the size of the present number of THE ELECTRICAL NEWS in order to admit of the publication of a complete report of the annual convention of the Canadian Electrical Association held in Toronto last month. The proceedings are of such an interesting and instructive character, and cover such a wide range of subjects in which our readers may be supposed to be interested, that we deem it scarcely necessary to apologize for the monopoly of space accorded to their publication. Of the convention it can be said, that especially in respect of interesting and instructive papers and discussions, it was a decided advance upon any of its predecessors. A higher average attendance at the sessions, and a deeper and more general participation in the proceedings, were encouraging features of the occasion. The use for the first time of lantern slide projected diagrams for the illustration of the papers had the effect of awakening a more general and deeper interest on the part of the members in the subjects under consideration. The addition of a considerable number of new members mainly from the ranks of the central station men is a cause of congratulation, and may be taken to indicate an awakening appreciation of the efforts which the Association has made to prove helpful to the electric lighting interests in conjunction with those of the telephone and telegraph. It is to be regretted that the various sections of the Dominion are not more equally represented on the Executive Committee, and especially that the city of Ottawa, whose generous hospitality can never be forgotten, is entirely without representation. We are sure that this and one or two other apparent blunders were due to lack of proper consideration, rather than deliberate intention. The result, however, is none the less regrettable, and care should be exercised to ensure justice being done in future to every individual and locality.

Mr. D. A. Starr has been appointed engineer in charge of the Hull and Aylmer electric railway.

The announcement has been made that the Babcock & Wilcox Company will consolidate their Canadian office with their general sales department at New York. From their various offices at Buffalo, Boston, New York, Minneapolis, Chicago, Cincinnati and San Francisco, it is believed that their Canadian business can be properly looked after, at much less expense. Mr. Wm. T. Bonner, formerly manager of the Canadian office in Montreal, will remove to Atlanta, Ga., having been appointed manager for the south-eastern territory. The Canadian shops will be maintained, at which boilers will be manufactured as usual.

## ONTARIO ASSOCIATION OF STATIONARY ENGINEERS.

THE annual meeting of the above association was held at Galt on the first of June. Mr. Arthur Ames, president, occupied the chair. There was a good attendance and much interest manifested in the proceedings. The minutes of last meeting were read and approved.

The president, in his address, impressed upon the members present the great importance attached to the various offices, and urged that care be exercised in the filling of such, as the success of the association depended to a large extent upon its officers and members. The interests of the engineers of this country depended largely, he said, on their own endeavors to procure an education in the principles involved in operating a modern steam plant. The rapid strides made almost daily in the advancement of this science make an up-to-date knowledge of these facts indispensable. And it is being recognized that the opportunities offered by this and other similar societies, together with the various publications connected therewith, greatly facilitate the acquiring of such a knowledge. The association had progressed very favorably; 150 certificates having been issued by the registrar for the current year, making in all some 700 now in force in this province. The reduction of renewal fees on two previous occasions had a beneficial effect, but they were now as low as possible consistent with the proper carrying out of the affairs of the association. In connection with legislation, a joint committee from this association and the Canadian Association of Stationary Engineers was appointed to draft a measure, to be presented at the last session of the Ontario Legislature, to procure a compulsory law, but the very laudable attempt fell through, owing to lack of sufficient time. He hoped that further action in this direction would shortly be taken. Apalling boiler explosions have been frequent of late, attended by great loss of life, and he would ask that a joint committee of the Ontario and Canadian Associations be appointed to draft a workable measure such as will comply with the interests of engineers and steam users at large. He thought the steam users of this province were beginning to realize that the aim of these associations is to place before them competent men, a very important matter in these days of close manufacturing competition. With respect to the financial standing of the association, the president stated that considerable money had been spent in procuring a set of books for the registrar and treasurer. He asked the members to consider the advisability of changing the regular date of meeting to the 24th of May, as it would afford a greater number the opportunity of attending.

The committee on "Good of Order," Messrs. Wickens, Donaldson and Stott, presented their report, which recommended that the question of securing legislation be again taken up this year upon new lines, and that the registrar be requested to call in all certificates out of date, and endeavor to collect the fees due.

The registrar, Mr. Edkins, presented his report and financial statement, the latter showing the receipts for the year ending 31st May, 1896, to be $641.41, and the expenditures $613.11, leaving a balance on hand of $28.30. The report stated that the success of the association was reason for congratulation, when the general depression in manufacturing industries was considered. During the year many certificates had been issued to craftsmen having charge of isolated plants, who volun-

tarily came up for examination in order to prove themselves qualified to act in the capacity of stationary engineers. Since the last yearly meeting three valued members had been called away, Messrs. B. Charlesworth, Hespeler, J. H. Walker, Paisley, and E. Edwards, Merritton. The association had at the present time a membership of about 600, but of these many were in arrears in the payment of fees, of which 59 were third class, 27 second class, and 6 first class. He asked instructions as to the course to be pursued to secure payment, and suggested the advisability of arranging the yearly renewal fee as follows : $1.00, 75c. and 50c. for first, second and third class respectively, if paid on or before the last day of February each year, and in default of so doing, the fees to be respectively $1.25, $1.00, and 75c. This arrangement would result in making members more prompt in payment. The fact that advertisements had appeared in the daily press asking for engineers holding Ontario certificates was considered encouraging.

The report of the treasurer was then read and adopted, after which the association adjourned for lunch.

At 2 p. m. order was again called, when the Committee on Legislation reported that after a good deal of work they had decided to postpone their efforts to secure legislation until the session of 1897.

The auditors' report was then presented and received.

Nominations were received to fill vacancies on the Board of Examiners as follows : Messrs. J. Bain, J. Devlin, F. Donaldson, F. Mitchell, P. Stott, and W. Sutton. A ballot being taken, it was declared that Messrs. Devlin, Donaldson and Mitchell were elected.

In the election of officers Messrs. Ames, Mitchell and Phillips were nominated for president, and Messrs. Devlin, Phillips and Mitchell for vice-president. Mr. Ames was declared elected for president and Mr. Mitchell for vice-president. Mr. Edkins, registrar, and Mr. Mackie, treasurer, were re-elected by acclamation.

Toronto was chosen for the next meeting place, and after discussing the question of changing the date, it was decided to adhere to the first Monday in June.

It was decided to memorialize the Dominion Government to appoint Mr. James Devlin as chief engineer of penitentiaries. A delegation, composed of Messrs. Devlin, Wickens and Edkins; was named to interview the same government on the question of obtaining a compulsory law for engineers.

It was moved by Mr. Phillip, seconded by Mr. Devlin, that the position of registrar in future should have a yearly salary of $100 attached thereto. The registrar stated that he would decline the position if any salary was voted.

Messrs. Cowan and Turnbull addressed the meeting, expressing themselves strongly in favor of the objects of the association, after which adjournment was announced.

---

An accident recently occurred to the dynamo in the power house of the Winnipeg electric railway which necessitated taking off some of the cars until repairs were made.

The Lachine Rapids Hydraulic and Land Company have decided to increase the capital stock to $2,000,000 and to proceed at once with the construction of the conduit and distribution in the city.

On June 19 fire was discovered in the works of the Thompson Electric Company at Waterford, Ont., which completely destroyed the building and machinery. The loss is placed at $30,000, about half of which is covered by insurance.

## CANADIAN ASSOCIATION OF STATIONARY ENGINEERS.

NOTE.—Secretaries of Associations are requested to forward matter for publication in this Department not later than the 25th of each month.

### THE ANNUAL CONVENTION.

Arrangements are being made for the holding of the annual convention of the C. A. S. E. in Kingston on the 18th and 19th of August. It is expected some interesting papers will be presented, and the question of meeting in convention once in two years, instead of every year as at present, will probably come up for consideration. The local association are already arranging plans for the reception of delegates, fuller particulars of which will be given in our next issue.

#### MONTREAL NO. 1.

The election of officers took place at the meeting of the above association on June 18th, with the following result : President, John Murphy ; 1st vice-president, J. E. Huntington ; 2nd vice-president, William Smyth ; secretary, B. Archibald York (re-elected) ; treasurer, Peter McNaughton ; fin. secretary, Harry Nuttall (re-elected) ; corresponding secretary, Hugh Thompson (re-elected) ; conductor, John Glennor, (re-elected) ; door-keeper, Wm. McAlpin (re-elected) ; trustees, Thos. Ryan, John J. York, and John H. Garth ; librarian, John Robertson.

The representatives to the executive council will be elected at the next meeting of the association.

#### TORONTO NO. 1.

Toronto No. 1, at their last regular meeting, elected officers as below : President, John Fox ; vice-president, Chas. Moseley ; recording secretary, Thomas Eversfield ; financial secretary, Walter Blackgrove ; corresponding secretary, George Mooney ; conductor, Thomas Seaton ; door-keeper, Barney Doyle ; trustees, James Huggett, George Fowler, and E. J. Phillip. The delegates to the convention, to be held at Kingston in August next, are as follows :—John Bain, John Fox, James Huggett, A. M. Wickins and Charles Moseley.

#### HAMILTON NO. 2.

On Friday, June 19th, the above association elected the following officers for the ensuing year : Past president, W. R. Cornish ; president, Wm. Norris ; vice-president, E. Teeter ; recording secretary, James Ironsides ; financial secretary, A. Nash (re-elected) ; treasurer, Wm. Nash (re-elected) ; conductor, W. Jones (re-elected) ; door-keeper, Thomas Carter (re-elected) ; trustees, R. Mackie, P. Stott, R. C. Pettigrew ; auditors, G. Mackie, James Ironsides, J. Wadge ; sick committee, G. Mackie, W. Jones, Thomas Carter ; delegates to convention to be held at Kingston, William Norris and G. Mackie. This association is reported to be in a thriving condition, an harmonious spirit existing among the members.

#### GUELPH NO. 6.

At a meeting of Guelph No. 6, at which there was a good attendance, officers for the ensuing year were elected as follows : President, H. Geary ; vice-president, Thos. Anderson ; recording secretary, H. Flewelling ; financial secretary, P. Ryan ; treasurer, C. F. Jordan ; conductor, J. Tuck ; door-keeper, J. Thatcher.

On behalf of the employees of the St. John, N. B., Street Railway Company, the superintendent presented Mr. George Wilson, the retiring chief engineer, with an address and handsome cigar case. Mr. Wilson thanked the donators for their expression of good-will.

## PERSONAL.

Mr. H. Rawstran, cashier of the Montreal Street Railway Company, has resigned, to accept a lucrative position with a Chicago company.

Mr. C. Berkeley Powell and Mrs. Powell, of Ottawa, sailed a fortnight ago on the steamer Parisian for a pleasure trip in England.

Mr. M. B. Thomas has succeeded to the management of the Hamilton & Dundas Railway Company, vice Mr. W. N. Myles, who has resigned.

Mr. T. Ahearn, of Ahearn & Soper, Ottawa, is at present in England on a business trip in connection with enterprises proposed by his company in Australia.

Mr. George Yorke, engineer at Osgoode Hall, Toronto, is enjoying a wedding tour in England. Mr. Yorke was married on the 4th ultimo to Miss Sarah Robins.

Mr. Thomas Irwin, of Montreal, has been appointed chief engineer of the power house of the St. John, N. B., Street Railway, Union street, to succeed Mr. G. M. Wilson.

Mr. William McCammon, of Kingston, Ont., was drowned at Clayton, N. Y., by walking off the dock. Deceased was an expert electrician and had charge of the electrical machinery on Folger Bros.' fleet of steamers known as the "White Squadron."

Mr. W. H. Baker, Vice-President of the Postal Telegraph Cable Company, of New York, was a recent visitor in Montreal. Mr. Baker, in addition to being vice-president of probably the largest cable company in the world, having over 8,000 offices on the continent, is a well-known figure in the electrical world and has invented several electrical contrivances.

At the commencement exercises of the graduating class of 1896, from the Stevens Institute of Technology, Hoboken, N. J., held June 18th, 1896, the degree of Doctor of Engineering was conferred by the faculty and trustees of Stevens Institute upon Commodore George W. Melville, engineer-in-chief of the United States Navy, in appreciation of the excellent engineering work performed by Commodore Melville for his country and the advancement of the science of steam engineering, well illustrated in the world wide famed "White Squadron." Only once before in the twenty-five years history of the Stevens Institute has the degree of Doctor of Engineering been conferred, and then upon Professor R. H. Thurston, of Rhode Island, who formerly occupied the chair of Mechanical Engineering in Stevens Institute, and is now Director of Sibley College, Cornell University.

Mr. Horatio Whiteway Nelson, who has recently taken charge of the cable and wire department of the Royal Electric Co., Montreal, is an expert in the insulating business of long and varied experience. On completing his education he spent two years in travel around the world, and then joined the Edison Machine Works, of Schenectady, N.Y., where his work was attended with continuous success during five years. In 1889 he came to Canada with the Edison General Electric Co. and superintended the large wire department of that corporation at Sherbrooke and Peterborough for four years, when he accepted the position of general superintendent of the works of the Waddell-Entz Co., of Bridgeport, Conn., then employing some 300 men. Subsequently he was entrusted with one of the departments of Messrs. Washburn & Moen's great works at Worcester, Mass., which charge he relinquished to accept his present post.

## TRADE NOTES.

J. Wallace & Son, of Hamilton, have just completed a machine for the manufacture of acetylene gas for Mr. T. L. Willson.

The Ottawa Car Company recently shipped two electric cars to the Berlin and Waterloo railway, and the same number to the Galt, Preston & Hespeler railway.

The following statement shows the geographical distribution of sales of Babcock & Wilcox water tube boilers during the month of May last, the figures indicating the horse power : American sales : New York, 774 h.p. ; Pennsylvania, 300 ; Illinois, 6,400 ; Cincinnati, 330 ; Tennessee, 600 ; Canada, 1,742 ; total, 10,146 h.p. Foreign sales : England, 4,444 h.p. ; New South Wales, 206 ; Scotland, 200 ; France, 686 ; Germany, 756 ; Spain, 722 ; Peru, 125 ; Norway, 280 ; Sweden, 57 ; Egypt, 1,728 ; Russia, 1,142 ; China, 160 ; Japan, 1,320 ; India, 46 ; Cuba, 75 ; Mexico, 64 ; total, 12,011 h. p. The grand total, including 1,920 h. p. marine boilers, is 22,157 h.p. The number of Babcock & Wilcox automatic chain grate stokers sold during the month was fifteen.

# ACETYLENE GAS.

By Geo. Black

Great inventions and discoveries are often apparently the result of accident, but the seizure of the occurrence and turning it to account marks the true scientist; such was the case when our countryman, Thos. L. Willson, discovered his method of producing calcium carbide, for it was known to chemists as a rare product, as shown by the following references:

Sir Humphrey Davy observed that when carbon and potassium were heated sufficiently to vapórize the potassium, a substance was formed which has been recognized as the first reference to a group of carbides.

In 1836 Brezelius announced that the black substance formed in small quantities as a by-product in producing potassium from potassic carbonate, and carbon was carbide of potassium.

Wohler in 1862 announced that he had made the carbide of calcium by fusing an alloy of zinc and calcium with carbon. He ascertained that it decomposed in contact with water forming calcic hydrate and acetylene.

Berthelot in 1866 described sodium carbide or acetylene sodium. He discovered that the high temperature of the electric arc within an atmosphere of hydrogen would unite with carbon of the charcoal terminals and form acetylene gas.

In 1888 Willson, in experimenting with his electric furnace, trying to form an alloy of calcium from some of its compounds, noticed that a mixture containing lime and powdered anthracite acted on by the arc fused down to a heavy semi-metallic mass, which, having been examined and found not to be the substance sought for, was thrown into a bucket containing water near at hand, with the result that violent effervescing of the water marked the rapid evolution of a gas, the overwhelming odor of which enforced attention to its presence, and which on the application of a match, burned with a smoky but luminous flame and numerous explosions. It was Acetylene gas.

To Willson is due the credit of discovering how to make calcium carbide, at the price of about one cent a pound in unlimited quantities, instead of the rare laboratory product obtained in grains, at the rate of about $10,000 per pound, thus producing not only a new light, but for manufacturing and commercial purposes opening up a vast range of new combinations of hydro-carbons at a much cheaper rate than ever existed before. The dream of the Chemist has been realized and synthetic chemistry took several strides forward. The possibilities of cheap carbide for light or chemical combinations places Willson in the front rank of the scientific men of the age.

Calcium carbide, $Ca\,C_2$, is described as a dark brown, dense substance, having a crystalline metallic fracture of blue or brown appearance, with a specific gravity of 2.262. In a dry atmosphere it is odorless, but in a moist atmosphere it emits a peculiar smell, resembling garlic or phosphorous. When exposed to air in lumps it absorbs moisture, and the surface becomes coated with a layer of hydrate of lime, which to a certain extent protects the rest of the substance from further deterioration. It is not inflammable and may be exposed to the temperature of a blast furnace without taking fire, the exterior only being converted into lime. When brought into contact with water or its vapors at ordinary temperatures, it rapidly decomposes, one pound when pure generating 5.892 cubic feet of acetylene gas at a temperature of 64° F.

Calcium carbide is manufactured from powdered lime and carbon in the shape of ground coal, coke, peat or charcoal, these two substances being fused together in an electric furnace. The process is very simple, and may be described thus:

The lime and carbon, having been ground to a fine powder, is intimately mixed in a certain proportion and fed into a crucible or furnace, the lower part of which has a carbon plate which is attached to one of the dynamo terminals; the other terminal is connected to an upright carbon resembling the upper carbon of an arc lamp, but much larger, being about three feet long and 12 by 8 inches in cross section. An alternating current is delivered by means of transformers to the carbons at about 100 volts and 1000 amperes. A small portion of the mixture is fed into the furnace, the upper carbon is raised about three inches to form an arc and the mixture is fused by the intense heat which ranges from 3500 to 4000 deg. C., while that of the ordinary smelting furnace is only 1200 to 1500 deg. C. The carbon is gradually raised and fresh mixture fed in till a mass of molten carbide about three feet high is made when the current is turned off and the carbide allowed to cool. The noise of the arc is said to be very peculiar, especially when the supply of mixture begins to fail.

## COST OF CALCIUM CARBIDE.

To positively ascertain the cost of this product, the Progressive Age, of N. Y., sent three commissioners to Mr. Willson's aluminum factory at Spray, N. C., in March last, to investigate thoroughly, and their report is published in that journal under date of 16th April, 1896. The commission consisted of Messrs. Houston and Kennelly, well-known electricians, and Dr. Leonard P. Kinnicutt, Director of the Department of Chemistry at Worcester Polytechnic Institute, who investigated thoroughly and took full charge of the factory during two separate days, making two runs of the substance and taking samples with them for testing in their own laboratories. Notwithstanding that the factory at Spray was only an experimental one, and the greatest possible output only one ton per 24 hours, and the fact that transportation of material was excessive, costing $3.05 per ton for coke and $4.55 per ton for lime, and estimating $11 per day for labor, including a superintendent at $4 per day, they figure the cost at $32.76 per ton.

Messrs. Houston and Kennelly add a separate estimate for the production of five tons daily under more favorable circumstances, but with water power at $5 per year as at Spray, and figure the cost at $20.04 per ton. They add, "The cost of producing calcium carb electrically, is evidently limited by the cost of lime, coke and electric power, no matter what the scale upon which the process is conducted."

"If we assume a perfect electric furnace, in which neither material nor energy is wasted, that is, a furnace which ensures the complete union of calcium and carbon without loss and with no escape of heat in the process, we know that one ton of carbide would require for its production 1750 lbs. of lime and 1125 pounds of pure coke.

"It has also been calculated from thermo-chemical data that 1½ electrical h.p. hours will be almost precisely the right amount of energy to produce one pound of carbide, or 3000 h. p. hours per short ton of carbide.

"Consequently, if L is the cost of lime in dollars per ton, C the cost of coke per ton, and P the cost of an electrical h.p. hour, a theoretically perfect plant would yield carbide at a cost per ton, exclusive of labor and fixed charges, of 0.875 L + 0.5625 C + 3000 P.

"For example, if lime (assumed pure) costs $2.50 per short ton, coke (assumed pure) costs $2.75 per short ton and an electrical horse-power of 300 working days of 24 hours each, cost $12 at furnace terminals (0 1667 cent per working horse-power hour), the limiting cost of carbide in a perfect furnace would be $8.73 per short ton.

"We may therefore summarize as follows: Calcium carbide by the electric furnace cannot be manufactured cheaper than $8.73 per short ton—for material and power, exclusive of electrode carbons, labor, depreciation, interest and other fixed charges.

"Owing to impurity of materials and departure from theoretical perfection in the electric furnaces, we found at Spray the actual cost of material and power, irrespective of electrode carbons, labor, etc., is 1.335 L + 1.125 C + 5122 P.

"Under favorable conditions such as we believe can be realized in particular localities, the total cost per short gross ton on a plant whose output is five tons daily, might be $20. Under the actual conditions existing at Spray during our tests, we find the total cost to be $32.76 per short gross ton if the plant were worked continuously."

In the above lowest estimate of Messrs. Houston and Kennelly they place horse-power at $12, whereas Mr. Willson has secured water power at Spray, and also in Canada, at a cost not exceeding $5 per h.p.

On this basis, and assuming L at 2.50, C at 2.75 and P 5.00, the figures would amount to 2.18 + 1.55 + 2.08, or a total of $5.81. The cost of lime and coke, however, is placed at a very low figure, but it is evident that the true theoretical minimum price is between $5.81 and $8.73.

I have also the following estimates of cost at the Niagara Falls establishment, to produce one ton of carbide, at rate of 10 tons per day.

| | | |
|---|---|---|
| It requires | 200 Electrical H. P., 24 hours at $20 per year, | $10 95 |
| | 1,440 lbs. Coke @ $3.50 per ton............... | 2.52 |
| | 1,800 lbs. Lime @ 4.50 per ton............... | 4.05 |
| | Labor, Depreciation, &c., &c.................. | 6.18 |
| | | $23.70 |

It is noticeable that this estimate is somewhat in excess of the theoretical values as laid down by Messrs. Houston and Kennelly, and may be improved on as experience is gained.

I was informed that the first run of carbide manufactured at Niagara Falls early in May gave about 25% better results than their estimate, and that they hoped to improve still more as they gained experience and the men got used to their work.

Mr. Willson commenced to erect a factory at Merritton in April on the old Welland canal, where he has secured 1500 horse-power at locks 8, 9 and 10, and expects to turn out carbide at the rate of 7½ tons daily at the lowest possible cost. He has also secured a very large amount of power in the province of Quebec, where he intends to manufacture not only for Canada, but for export to foreign countries.

It is quite evident from the report of the Progressive Age commissioners and from the experience of the Niagara Falls Company that calcium carbide can be made and sold at a price to compete with ordinary gas and electric light.

It takes to produce 100 lbs. carbide, as shown theoretically, 87 1/2 lbs. lime and 56¼ lbs. of carbon; of the latter 37 1/2 lbs. combine with the metal calcium and 18¾ lbs. combine with the 25 lbs. of oxygen of the lime, and escapes from the furnace as carbon monoxide, in accordance with the following formulae:

| | | |
|---|---|---|
| $Ca\,O \quad + 3\,C$ | $= Ca\,C_2$ | $+ C\,O.$ |
| 87½ lbs. + 56¼ lbs. | = 100 lbs. | + 43¾ lbs. |
| $Ca\,C_2$ | $= C_4$ | $+ C_2.$ |
| 100 lbs. | = 62½ lbs. | + 37 1/2 lbs. |
| $C\,O$ | $= C$ | $+ O.$ |
| 43¾ lbs. | = 18¾ lbs. | + 25 lbs. |

Calcium carbide contains 62.5 parts of calcium and 37.5 parts of carbon in 100, and when brought into contact with water acetylene is generated to the extent of 5.89 cubic feet of gas to each pound of carbide used; or by weight 100 lbs. of carbide and 56¼ lbs. of water evolve 40.63 lbs. of acetylene gas and form 115.62 lbs. of calcic hydrate (slacked lime) in accordance with the following formula:

$$Ca\,C_2 + 2\,H_2\,O = Ca\,O\,H_2\,O + Ca\,2\,H_2.$$

$$100 \quad + 56.25 \quad = 115.62 \quad + 40.625.$$

The acetylene gas so generated, contains in 100 parts 92.3 parts of carbon and 7.7 parts of hydrogen, or in the 40.625 pounds generated from 100 lbs. of carbide we have 37 1/2 lbs. of carbon and 3⅛ lbs. of hydrogen.

Acetylene can be produced from carbide by the addition of water and distributed and stored in a gasometer, or the gas may be compressed into a liquid and kept in a suitable cylinder and drawn off as required for consumption, a reducing valve being adjusted to give the necessary pressure for burning. One cubic foot of liquid expands into 400 cubic feet of illuminating gas, so that a large supply may be stored in a very small space, but for experimental purposes and for a limited supply it is preferable to make the gas direct from carbide and store it in a gasometer.

The pressure necessary to liquify acetylene depends upon the temperature. At 67° it requires a pressure of nearly 600 lbs., at 32° 323 lbs., at 28.6° below zero 135 lbs., and at 1160° below zero 15 lbs. We see that there is no danger of freezing it in any habitable place.

As an illuminant acetylene surpasses in brilliancy all other illuminants known. When burned at the rate of 5 cubic feet per hour it gives 240 to 250 c.p., whereas the best coal or water gas rarely exceeds 22 candles for each 5 cubic feet burned per hour. Acetylene gas thus gives 10 to 12 1/2 times the light of ordinary gas, or 1000 feet is equivalent to 10,000 to 12,500 of ordinary gas. Acetylene is a commercially pure gas, containing 98 per cent. acetylene and 2 per cent. of air, the latter having slight traces of other substances. It is clear and colorless, with specific gravity of 0.91.

When a light is applied to acetylene in open air, it burns with a bright yellow but very smoky flame, on account of its extreme richness in carbon, but when confined and delivered under suitable pressure it gives an extremely pure white light resembling the oxy-hydrogen light, and is the nearest approach in color and purity to sunlight of any known artificial light.

### ITS POISONOUS NATURE.

Acetylene, when made from expensive chemicals, was known to be very poisonous, but as made from lime and carbon it is proved to be less injurious than ordinary gases. Its strong pungent smell is a safeguard, as no one can remain in an atmosphere of it a sufficiently long time to be harmed; handy for hotels where the guests blow out the lights; in such an event the "Blowhard" could not get to sleep before he or some one else would be compelled to investigate. The effect on the human system is rather to intoxicate than stupify, and while it is absorbed by the blood it does not form combinations with it; it asphyxiates less rapidly than ordinary gas. Moissan of France and others made exhaustive experiments with the greatest care with acetylene and coal gas on animals, and proved conclusively that coal gas was much more poisonous than acetylene.

### EXPLOSIBILITY.

Acetylene, when mixed with 1 1/4 times its volume of atmospheric air, becomes slightly explosive, and reaches its maximum explosibility with 12 volumes of air, decreasing till at 20 volumes it ceases to be explosive. Coal gas reaches its maximum explosibility with 5 volumes of air, so that ordinary gas is more explosive than acetylene. Accidents and explosions reported recently have given the impression that the gas is very dangerous. Let us examine this feature. Take the case of the accident in Quebec last winter. An ingenious mechanic made his own dynamo, furnace and carbide; he was experimenting with the gas under pressure, to liquify it so as to get it into the smallest possible space. He had an iron pipe 8 inches long and 4 inches diameter with cast iron ends, a pressure gauge at one end and a valve at the other. He had reached a pressure of 360 lbs. to inch, and observing that the gas was escaping around the valve, he used a hammer to stop the leak, when a a portion of the metal broke away and the gas escaping struck him in the eye, penetrating the brain and killing him instantly. Ordinary air under similar conditions would have been as fatal. It was afterwards found that the iron ends were thin and porous and the wonder was that they stood the pressure. There was no explosion; the coroner's verdict was "accidental death."

The explosion at New Haven, Conn., 21st January last, was caused by men experimenting with liquid acetylene, under a pressure of 600 lbs. to the inch, and I presume all accidents reported might be traced to unauthorized parties experimenting with crude apparatus, and ignorant of the necessary conditions for safety. We know that air, water, gas or electricity, are dangerous under certain conditions, but harmless when properly controlled, and it is no argument against acetylene that it is also dangerous when improperly handled.

### EFFECT ON ELECTRIC LIGHTING.

When I first saw acetylene gas in September '94 I felt sorry for the electric companys, because I thought the gas companies would readily adopt the new gas and regain their former monopoly of lighting. But I do not feel quite so downcast now; I realize that the margin of cost of production is not so great, and believe that gas companies will feel the competition equally with electric, unless they adopt the new gas for use pure, or as an enricher to their present output. It is said to be useful as an enricher for coal gas, but not so suitable for water gas.

Prof. Lewes, of England, one of the best gas authorities there, suggests that gas companies should distribute a low illuminating coal gas of about 12 c.p. through their mains for heating, cooking, etc., and that each place using illuminating gas be supplied with a cylinder of acetylene to be fed into the illuminating pipes in a certain determined proportion. By some such process as this there remains a large field for coal gas; otherwise coal and water gas must go.

The incandescent light has held first place for interior illumination on account of its steadiness, purity, coolness, and not withdrawing oxygen from the air nor adding noxious elements to it. Acetylene will divide this field with the incandescent bulb; it is a pure, white, steady light, of low heating power, withdraws very little oxygen from the air, and does not add impurities to any great extent. Its flame has a temperature of 900 to 1000 degrees C., while ordinary gas has 1400 deg. C., but as only one-tenth to one-fifteenth of the quantity is used for equal light, its heating effect is slightly in excess of the incandescent bulb.

Taking the theoretical E. H. P. necessary to produce one ton of carbide as 3000 h.p. hours, and using the same for a supply of electric light by incandescent 4 watt lamps, we have the following: $3000 \times 746 = 2,238,000$ watts $\div 64$ gives 34,970 16 c.p. lamps for one hour, or 1453 burning 24 hours continuously.

The same power equals one ton carbide, which burned in ⅕ foot burners gives 31,500 16⅔ c.p. lights, or 1313 burning 24 hours. This gives a margin apparently in favor of electric lighting, but you cannot use all your electric lights at the source of cheapest production, nor run a continuous even load for 24 hours, but have in addition to sustain losses in distribution more than proportional to the distance conveyed; also lamp renewals. With the carbide it is different; it can be made at the place of cheapest production on a constant load night and day, and a small sum transports the carbide to any place desired, where it can be used to its full power without loss. Figure out for yourselves the problem of transmitting electric current for use 10 to 100 miles from source of production and transporting carbide by freight the same distance, and the comparison will be largely in favor of carbide. Hence, for use in close proximity to the power house on a steady even load day and night, the cost will be about the same if power costs the same, but as that is not practicable in electric lighting, the margin is in favor of carbide, but not to such an extent as to seriously hurt the electric companies employing the best apparatus under the most improved conditions, as may be found in large cities, but it is possible in small towns where the best and most economical conditions cannot be obtained and a thorough manager secured, well up in the scientific as well as the practical conditions, electric lighting may suffer.

The ease of distributing acetylene is remarkable. Owing to its high illuminating power, very small main pipes may be used, and as frost does not affect it the pipes need only be laid below the surface, so that little or no expense need be incurred in piping a town. If the cost of mains equal cost of poles and wires the central station or gas house only requires a small tank for a generator and a gasometer of suitable size, as compared with engines, boilers and dynamos running when only one light is required.

We may then conclude that in the race for supremacy closer economy will be practised, better service given, the public will be benefitted, all will let their light shine to the best of their ability, and the one best deserving of patronage will survive.

## POWER TRANSMISSION BY POLYPHASE E. M. F.'S.

By Geo. White-Fraser.

The utilization of the natural resources of a country is a matter which should interest not only the engineer upon whom devolves the responsibility of their development; not only the capitalist who is on the look for investments, but the economist and the politician who have the grave responsibility of directing a nation's energies into remunerative channels.

The possession of cheap natural power, whether in the form of coal fields or large rivers, is a national asset, the importance of which it is impossible to overestimate, constituting as it does the very basis whereon rest those manufacturing industries that go towards making a nation self-supporting and progressive. The foundation of Great Britain's commercial pre-eminence is her immense coal fields, enabling all processes of manufacturing art to be carried on inexpensively, and thus giving her a very favorable start in competition with other manufacturing nations. Another favoring circumstance is that her immense deposits of iron ore, are if not in all cases contiguous to, at least very close to, the coal fields. The power, therefore, is nearly on the spot where it is wanted, and owing to her insular position, the great highway of commerce—the ocean—is at the very doors of her factories. Great Britain, probably, has ideal manufacturing and shipping facilities: raw material, raw power, natural highway, all packed into a very restricted area. In Canada, we have the three necessaries—raw power, raw material, great highways,—but we rarely find them all present at the same spot. Nature gifted Great Britain from the outset; Canada must turn to science for the development and utilization and the combining of those scattered advantages. We have great and reliable water powers; we have immense natural wealth in ores, and timber, etc., and we have the highway of the great lakes and river. Governmental and private enterprise has provided, as well, railway and canal transporting facilities, but we frequently observe that the power sources are so situated as to be comparatively inaccessible to railways. Thus manufacturing establishments, in order to avail themselves of the former advantage, have to locate themselves unfavorably with respect to the latter. The cost of handling, transshipping, etc., being a very appreciable factor in the total market cost of manufactured articles, the cost of an additional link between the producing point and the shipping point, is sometimes so great as to make it commercially less expensive to locate at the shipping point even though that involves the use of a more expensive power. Any means therefore that enables cheap power to be brought to the most convenient shipping point, effects a combination which is of the greatest value to a manufacturing community. The intense competition in all manufacturing industries has the inevitable tendency to lower selling prices, and this reduction of profits must be made up either by a depreciation in the quality of the products, or by a rigid system of economy in manufacturing processes. The cost of production must go down, and any means of lowering it must be availed of. We use the most efficient machinery, we concentrate our factories round the best shipping places, we go where labor is cheapest, where power is cheapest, land cheapest, transport most handy. We do anything to save a cent in a hundred dollars, and upon the engineer frequently falls the responsibility of saving it. In his hands the policy of concentration becomes one of the principles of power generation, as in business; and power distribution receives as close attention as does the distribution of the goods. He wants to generate the power as cheaply as possible, and to transmit it to the utilizing points with the least waste; and he avails himself as far as possible of every natural advantage—natural gas, water for condensing, water falls, etc. We are all acquainted with the usual means of transmission—by belting and shafting, gears, hydraulic and pneumatic pressure, and so on; and know that the frictional and other losses by these methods are so great that very soon a limit is reached, beyond which it is not commercially possible to transmit. Hence we find, not only in manufacturing towns and villages, but even in the larger factories that several generating points are necessary, when but for these losses one very large and very efficient central generating plant might furnish all the power required throughout the entire area or district. It is, therefore, also that thousands and thousands of horse power are running to waste every day, in the many powerful rivers that drain parts of Canada. It is simply because the nearest railway, or other shipping point is so far distant from the waterfall that the power cannot be transmitted to a factory on the railway, and the extra haul and cartage would introduce an additional expense that would be prohibitory. If the power could be transmitted at reasonable cost from the water power to a convenient point on the railway, then the utilization of the cheap power and the good transport facilities might together be commercially advantageous.

It is now some years since the suitability of electricity for the transmission of power was recognized, and we have seen electrical machinery coming more and more into use, ousting other methods and proving its superiority, not only on practical but also on commercial grounds. We have seen generators grow from

25, 50, 100 h.p., to 1,000 h.p. in size per unit; we have transmitted the power at constantly increasing voltages up to 500 v. for direct currents, (or even greater in series machines,) and we have seen the direct current evolve into the alternating simple, and thence into the latest and highest type, the polyphase alternating; and to-day we find thousands of horse power transmitted at pressures of 10,000 volts, from waterfalls on a mountain, or in a gorge where it is impossible to locate a factory, and over distances ranging from a few hundred yards to thirty miles and more; we find these large amounts of energy being utilized for every industrial and domestic purpose, in units of from ⅙th to 1,000 horse power.

This tremendous widening of the field of electrical possibilities has not been attained by natural growth only. As the field widened the science developed; improved methods had to keep pace with increased demand, and while it was originally the demand that stimulated the invention of polyphase currents, polyphase currents have in their turn practically revolutionized the art of electrical generation, transmission and utilization, and the old formula c = e ÷ r that was a light unto the path of the direct current man, is no longer sufficient in the calculation of alternating current circuits.

It is the development of polyphase working that has rendered commercially possible the electrical transmission of power over great distances, and its reconversion into mechanical power when required. Polyphase currents are merely ordinary alternating currents; are generated separately as such, and possess no peculiar properties in themselves. Owing, however, to their difference in phase, the fields produced by them have different values at the same instant, and it is in this alone that the peculiar properties of the polyphase system reside. As they are, however, alternating currents, all those peculiar phenomena met with in alternating current work are inherent in polyphase working, and as a rule assume an importance which claims very careful consideration. Owing also to the combination in the same circuit of several E.M.F's differing in phase, complications arise which are not present in single phase circuits. Before proceeding to the consideration of the mutual actions and reactions of these quantities, it may be of interest to trace the progress of evolution of electrical working from direct current, through simple alternating, up to polyphase alternating. Up till quite recently direct current was the only means of power transmission and distribution; but the limit of pressure necessarily imposed on this method was so low that the cost of copper for large areas or long distances became prohibitive, and recourse was had to the alternating current used with static transformers. By this means transmission voltage could be as high as required, and utilization pressure as low, but here again was a very vexatious limitation due to the fact that motors could not be made to operate satisfactorily on the alternating current. Once started and brought up to speed, they would go on until overloaded, and then stop; but it required some independent source of power to start them, such as a steam engine. It is quite evident that such motors could not start from rest, as each armature coil was subjected to equal forces acting in opposition to each other and therefore neutralizing each other; it was like the dead centre of an engine. This was the stationary field of an ordinary D.C. motor. This difficulty was overcome by the formation of a revolving field, and polyphase currents are necessary for this purpose. The principles of the revolving field are the same as those governing the resultant of mechanical forces acting in different directions on the same mass. Suppose a mass

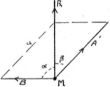

DIAGRAM 1.—PARALLELOGRAM OF FORCES.

"M" and two forces "A" "B" acting with known strength on it in the directions of the arrows. The direction in which M will be forced—the resultant direction of the forces—will be MR, which is the diagonal of the parallelogram formed by drawing parallels to A and B. Now, it is plain that

$$\frac{\text{Side a}}{\text{Side B}} = \frac{\sin \text{ angle RMB}}{\sin \text{ angle BRM}} = \frac{\sin \text{ angle RMB}}{\sin \text{ angle RMA}}$$

and as side a = side A, therefore

$$\frac{\text{side A}}{\text{side B}} = \frac{\sin \text{ RMB}}{\sin \text{ RMA}}$$

and that if we cause sides A, B, to vary harmonically we shall also cause angles RMB, RMA to vary harmonically; and consequently swing the diagonal MR to swing between positions MB and MA. Now MR is the resultant of two forces acting together; consequently if we cause these forces to vary harmonically in strength (remaining constant in direction) the resultant direction will swing as above described. Applying this principle to magnets, and an armature, we can cause the resultant magnetic field to revolve. Two electromagnets, A, B, are placed radially, and are separately energized by currents which can be varied in strength by any convenient means, say a rheostat. From the centre R of this arc, as a pivot, is hung an armature M

capable of swinging. First of all magnet A is energized, B being left dead; M is directly attracted to A, and RM is the direction of strongest magnetic pull. Then A is left alone, and B is gradually energized, and as it becomes stronger it attracts M more and

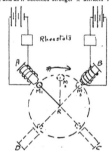

DIAGRAM 2.—HARMONICALLY VARYING MAGNET AND ARMATURE.

more towards itself, so that when B has been made as strong as A, each of them attracts M with equal force, and M will be midway between them; RM2 being the new direction of magnetic strongest pull. If now A is slowly decreased, it will attract M less and less strongly, M will approach closer to B, the strongest direction RM3 will be the resultant pull. Thus, by causing the magnets A, B to vary in strength, one up, the other down, we have swung the resultant strongest field over the arc AB, of the circle. If we have other magnets, C, D, we could pull M all round the circle, and that is what is done in polyphase work.

The two necessary conditions for the production of a revolving field are electromagnets whose strength can be varied up and down, and some arrangement whereby adjacent magnets shall attain their maximum or minimum strength at different moments. The first condition is evidently met by energizing the magnets with alternating currents, for in this manner their strength will assume every value between a positive and negative maximum, passing through the zero point; and the second is evidently equally met by energizing two adjacent poles, by two independent alternating currents, one of which starts a little later than the other. A glance at the diagram will make this plain. Wave I

DIAGRAM 3.—WAVES.

energizes pole A, wave II energizes pole B. The current in an alternating circuit also rises and falls as does the E M F, and therefore the ampere turns, and consequently the magnetism. Wave II being a little behind wave I, as regards their equal strength, (strength being proportional to vertical height of wave above zero line) the magnetism in pole B will be behind that in pole A, and we have thus produced harmonically varying poles. It must be clearly understood that the principle of revolving poles is applied only to motors; the function of generators being to supply those shifted E M F's and currents. These shifted E M F's can be supplied by the same generator, or by several generators, the method employed being rendered plain by the diagram. N, S, are two

DIAGRAM 4.—GENERATION OF 2 SHIFTED E M F's.

poles—part of a ring yoke. A, B, are two coils of the armature. The angular distance of A, B, is half that of N, S. When A is right under N, it is generating its maximum E M F: at that moment B is halfway between N, S, and its induction and therefore its E M F is least. As the armature revolves clockwise, B gradually gets into a stronger field, while A is approaching a weaker one. The induction, and therefore the E M F in B is increasing while that in A is decreasing ; when B has got under S, its induction and therefore its E M F is greatest ; at this moment A is in the midway position, and therefore its E M F is least.   Therefore, in these two coils are being generated E M F's which are shifted—that is out of phase —and if we take the values of the E M F's in each at the same instant, we can construct two curves, showing the position of their maxima and minima, etc., with respect to each other.   Diagram No. 3 shews these two curves. If these E M F's were taken through separate circuits and joined respectively to poles A, B, in diagram No. 2, we should have the simplest form of polyphase generator and motor.  Instead of having two coils A, B, between the poles N, S, we might space three, or any other convenient number producing 2 phase, 3 phase or 4 phase E M F's.  As, however, the principles of polyphase E M F's apply to any number, and there is no advantage in using more than three, we shall not consider any higher number.  In a polyphasal generator it is evident we can have just as many complete circuits as there are phases.  In a 2 phase there are two sets of independent coils ; the ends of these might therefore be brought to four contact rings, and the ends of the three independent sets of a 3 phase might be brought to 6 contact rings, as the ends of a single phase are connected to two contact rings.   But a study of the wave diagrams of a 2 phase and of a 3 phase circuit shew that we can greatly simplify matters, and do with less copper.  Take a more simple diagram of two phase coils at right angles to each other.  In the upper

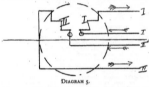

DIAGRAM 5.

half of the E M F circle, the E M F in each coil is positive and the current will flow from the outside end of the coil towards the inside end, in the direction of the arrows; here the E M F in each of the return wires is in the same direction, therefore we can combine them into one, instead of having two.  Suppose the coils have revolved into the second position ; here the E M F in the 2nd

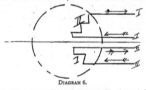

DIAGRAM 6.

coil is positive current flowing in the same direction as before. The E M F in the 1st coil is however negative (the coil being below the horizontal diameter) and the E M F will have reversed ; therefore the current in 1 will now flow in the opposite direction to what it did before—in the direction of the new arrows.  But it is plain that here again, if we join the outside ends of 2 and 1, the return current from 2 can flow down 1, the E M F's being in the same direction.  Therefore in the two phase system instead of having two separate circuits we can join the inside ends of the coils—and

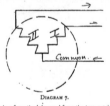

DIAGRAM 7.

lead three wires from the join—and from the two outside ends as shewn in Diagram 7.

Similarly with the 3 phase system we can join the inside ends together, and lead four wires as shown in Diagram No. 8.  But in this case it will be evident that we can dispense with the fourth wire altogether, as will be plain from the accompanying Diagram, No. 9, where the direction of the E M F's in the three wires at three different positions of the coils shew that always two wires can help to carry the return current of the third, therefore the fourth is not needed.

But while this combining of circuits very much simplifies the matter from the above point of view, it very considerably complicates it from another, because, whereas, with the individual

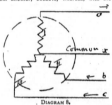

DIAGRAM 8.

circuits we had a single E M F to each, by combining them we have several E M F's of differing phase all acting in the same circuit —hence it is necessary to find their resultant E M F.  This resultant

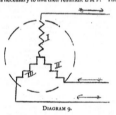

DIAGRAM 9.

depends greatly on the way in which the armature windings are connected up.  There are three ways : "star" and "mesh" and independent groupings ; the former is when the coils are all joined together at one end, the other ends leading to the circuits, as in Diagram No. 9.  The second, where the winding is continu-

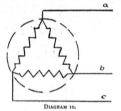

DIAGRAM 10.

ous round the entire armature and at intervals a circuit wire is tapped off, as in Diagram No. 10.

Diagram No. 11 shews 2 phase star connection, and here the

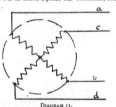

DIAGRAM 11.

instantaneous E M F between terminals a, b, and c, d, is 2 E sin O where E is the maximum voltage, and O is the angle through which the coil has passed from the position of no E M F.  At the same instant the E M F between terminals a, c, b, d is $\sqrt{2}$ E sin (O + 45) ; that is to say the pressure between two live wires of different phase is $\sqrt{2}$ times or 1.41 times the pressure between the terminals of the same coil, and is 45° in advance of the E M F of the foremost coil.  If a common return be used, the pressure between either outside and the common return will be simply double the E M F of one coil, but the pressure between the two outsides will be $\sqrt{2}$ times this. If two phase coils are connected in "mesh," it is plain that between wires a, b, there is simply the ordinary voltage of coil a or E sin O; while between wires a, c, there is a voltage of $\sqrt{2}$ E sin (O + 45°); or again 1.41 times the other voltage.

Similarly with 3 phase E M F's. In star connection voltage between a and x is E sin O; that between b and x is E sin (O+120°); that between c and x is E sin (O+240°). The pressure between any two wires is $\sqrt{3}$ E sin (O+30); or if the pressure between one wire and the junction x is 1,000 volts, the pressure between any tw wires is $\sqrt{3}$ times as great or 1730 volts. If a 3 phase armature be connected mesh fashion then the voltage between any two wires is simply the voltage of a coil. Diagram No. 10.

Consequently in wiring transmission circuits we want to know how the coils are connected. The amounts of current in the lines are calculated similarly, and the general result arrived at is that in the star winding the E M F's between line wires is greater than that due solely to the coils; while the currents in lines and coils are the same; while in the mesh winding the currents in the line wires are greater than the currents in the coils, but the EMF's are the same. Of course these increased EMF's and currents must have their effect on the size of the wires used with such polyphase E M F's; and, not to go into the mathematical investigation of this matter, which can be found in papers by Steinmetz, Thompson, Weaver and others, we can summarize the results as follows, comparing the amounts of copper required by various systems, to transmit same power, at same loss and same virtual voltage, referring them to the standard of single phase alternating unit:

| | | | | |
|---|---|---|---|---|
| Single phase | | | 100 | per cent. |
| Two phase | 4 wires | | 100 | " |
| " " | 3 " | | 72.8 | " |
| Three phase | 3 " | (mesh) | 75 | " |
| " " | 4 " | (star) | 33.3 | " |

Having now seen the necessity for a revolving field; having examined into how to produce it; and having also investigated the peculiar E M F consequent of its use, we may reasonably turn to the consideration of what we can do with it. Briefly, we can make use of the advantages offered by the alternating current system in the transmission and utilization of power. We can transmit at any voltage we please and utilize at any other voltage we desire. We can concentrate a highly efficient steam and electric plant in a factory, village, town or city and send power in every direction in units of any size at a very reasonable cost. We can build our factory right on a railway and load our goods onto the cars at once, and at the same time make use of the abundant cheap power now running to waste over a ledge of rock twenty miles away, to run the machinery. In fact, we can conveniently and cheaply bring our power from where it is being generated by nature's agency, and lay it down just wherever commercial considerations point to as being the best spot for its use. At this present moment we have large amounts being transmitted up to 30 miles and over, perhaps; and the very first long distance transmission, which was undertaken in '91, remains to-day a monument of what is possible where enterprise, skill and capital combine to effect a particular object. The Lauffen-Frankfort transmission plant transmitted 300 horse power over a distance of 110 miles; and the efficiency of the entire plant was so large as 75 per cent. Thus the advantages of cheap power and of concentration and efficient distribution are rendered available by means of polyphasal currents.

It has already been said that a polyphasal system is merely the mechanical combination of several single phase alternating systems, and it will therefore be evident that all the conditions inherent in single phase alternating systems will be met with in polyphase working. As the distance and power requirements that caused the evolution of the latter, are in general much more onerous than those imposed on the former, so we find that the reactions attending the use of the alternating current as such, have a much more appreciable influence and require more careful attention in polyphase power work than they do in single phase lighting work. Whereas in alternating lighting plants the reaction and influence of hysteresis, induction, capacity are negligible or comparatively so, when large amounts of power are to be transmitted over great distances, they not only become appreciable, but are a very considerable factor in the calculation of the circuits and the size of the generator units.

Hysteresis is a phenomenon attending the magnetization of iron by alternating currents; it is a sort of magnetic friction, and requires power to overcome it.

Induction is an EMF generated in a wire by the imposition on it of an alternating EMF and tends to oppose the the impressed EMF. It thus requires a higher initial or generator alternating EMF to send a desired current through a circuit possessing inductance than it would to send a direct current through the same circuit. In such a circuit there is not only the usual Ohmic drop, but also the inductance drop, and as these two sources of drop act, not in the same line but at right angles to each other, the resultant impressed E M F is compounded of the required effective E M F; the resistance E M F, and the inductive E M F.

Capacity is a sort of absorbing quality in the wire which requires to be, so to speak, saturated before it will transmit any current. Its effect is to require a larger amount of current to be generated than would be required if a direct current were being transmitted. The effect of induction is not only to introduce a drop of its own over and above the Ohmic drop, but it actually dams back the currents in the circuit so that the EMF wave and the current wave are out of phase with each other; that is they do not attain their maximum or other symmetrical values at the same instant.

The effect of capacity is different in one sense to that of induction, in that it requires the generation of a greater current instead of a greater EMF, and opposite in another sense, in that it tends to cause an advance of the current wave in front of the EMF. The combined action of these two quantities is called the reactance of a circuit, and as it generally happens that the inductance is greater than the capacity, there is a "lag" of current wave behind the EMF. The relative displacement of these waves is shown in Diagram No. 12. This lag angle has a very important influence on the

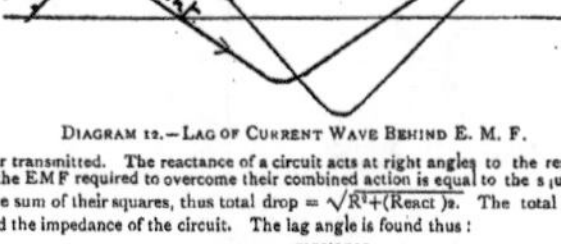


DIAGRAM 12.—LAG OF CURRENT WAVE BEHIND E. M. F.

power transmitted. The reactance of a circuit acts at right angles to the resistance, and the EMF required to overcome their combined action is equal to the square root of the sum of their squares, thus total drop = $\sqrt{R^2 + (React)^2}$. The total drop is called the impedance of the circuit. The lag angle is found thus:

$$\tan \text{lag} = \frac{\text{reactance}}{\text{resistance}}$$

Hysteresis may be at once dismissed from further consideration, as, though present in generators, transformers and motors, it does not enter as a factor into transmission, but merely as one of those machine data of which engineers inform themselves when planning an enterprise. It must be said, however, that the expression for the energy lost in hysteresis is $H = A \times B^{1.6} \times N \times M$.

Where H = Energy consumed in ergs.
A = Constant depending on magnetic hardness.
B = Density of induction in lines per $CM^2$.
N = Frequency of alternations.
M = Mass of iron in $CM^3$.

Other things being equal it would be advantageous therefore to keep the frequency down and to have the density low.

The expression for the induction EMF of a circuit is $L = 2\pi N L C 10^{-3}$.

Where L stands for induction EMF.
$2\pi$ is the usual symbol.
N stands for our old friend the frequency of alternation.
L stands for the specific inductance of the circuit, and depends on size, length, and mutual distance apart of wires.
C stands for the current in the conductor.

Studying this expression, we observe that the inductive E M F, or L, as it will be called, increases directly with the current, therefore, other things being equal, it is of advantage to lessen the current as much as possible. This is entirely in keeping with the idea of alternating current work, as the use of a high voltage permits of a small current to give the same energy transmitted. We observe again that L varies directly with the frequency of alternations, so that other things being equal, the use of a low frequency is of distinct advantage as tending to keep down the induction EMF. (This last deduction, it must be observed, applies to the circuits only.)

The expression for the capacity current of a condenser is $C = 2\pi N K E 10^{-6}$.

Where C = the current.
$2\pi$ is the familiar symbol.
N is again the frequency of alternations
K is the specific capacity of the condenser.
E is the impressed voltage.

Studying this expression we see that this capacity current increases directly with the frequency, and also directly with the impressed EMF. The factor K depends on the cross section of the wires and their distances apart decreasing as the wires become smaller, and as they are placed further away from each other.

Placing the three expressions for these quantities together:

$$H = \pi B^{1.6} N M \quad (1)$$
$$L = 2\pi N L C 10^{-3} \quad (2)$$
$$C = 2\pi N K E 10^{-6} \quad (3)$$

We observe that the symbol "N" standing for frequency of alternation is a prominent factor in all, and that the amount of hysteresis loss: of induction EMF; and of capacity current; all of them become greater as the frequency increases. From (2) we see that to transmit a certain amount of energy we should lessen the current as much as possible in order to reduce the inductive drop, and therefore use as high an initial EMF as possible. But from (3) we see that precisely the reverse relation of EMF and current should obtain if we wish to reduce the capacity current to a minimum. In this as in all other cases, a balance must be sought for, and commercial considerations must have the casting vote.

I wish to draw your particular attention for a few moments to the retardation of the current behind the electromotive force, brought about by the induction of lines or apparatus, as it has a most important bearing on the transmission of power. The power or energy of a circuit is, as we know, the product of the amperes into the volts; but owing to this lag the actual power transmitted is less than the amount calculated by multiplying the amperes and volts; and it becomes less in proportion as the lag becomes greater. This will be rendered evident by studying the diagram No. 12, where it is seen that the maximum current flow occurs when the EMF has passed its maximum; and consequently at no time can the power of an alternating current circuit containing induction be equal to the power of a D C circuit of same data. Quantitatively the power of such a circuit is equal to C × E × cos a, where "a" is the angle of lag. This expression follows directly from a consideration of the diagram connecting generator E M F; reactance and effective E M F, where it is observed that the effective E M F equal to the generator E M F × the cosine of the lag angle a. The less the reactance, the more nearly does cos a approach unity, and the less is the power lost in the line. The cosine of the lag angle is called the "power factor" of the circuit, but in figuring the power factor for a particular case, we have to take into consideration not only the reactance of the lines, but also that due to the self induction of transformers and motors, so that the calculation of the resultant power factor of an entire system becomes a somewhat complicated matter. It is an absolutely necessary matter, however, because the generator capacity and EMF have got to be larger than the actual power required at motors, by the amount "x secant a," where x is the amount required of the motors. We first have the line power factor; imposed on that comes that of the transformers, and lastly that of the induction motors; so the resultant power factor is really a quantity to be reckoned with.

In this place it will be apropos to consider that peculiar condition of equilibrium—called resonance—brought about by introducing sufficient capacity into the line to neutralize the inductance. In this extreme case we shall have no reactance, only ohmic resistance; the current will be actually in phase with the EMF: and owing to the fact (as pointed out by Steinmetz) that the capacity EMF will be in the same direction as, and therefore additive to the impressed EMF, the resultant EMF will be actually higher than the initial EMF. This addition of EMF's might become a serious matter, as tending to ruin insulation on short lines, but is obviated by throwing the line out of balance. To Professor Dunbar is due the credit of first pointing out its possibility.

We have now examined the main features of polyphase circuits, and can go on to the consideration of some of the practical questions that arise in the discussion of any particular case. The voltage of transmission is probably the first question, and it is plain that the higher the voltage used, the less the copper necessary; but it is limited by various practical considerations. The generator EMF is, of course, limited by the construction of the machine, as to whether space permits of more or less insulating material. This can, however, be raised for transmission purposes to any voltage required by interposing step-up transformers between the line and generator. For any high pressure these transformers are filled with oil, which has a high insulating value, and has the advantage that if a spark does jump at any time, the oil flows back, so that the transformer is not permanently injured. On the lines a pressure of 10,000 volts is quite easily handled, the insulators being large, with plenty inside surface, and the copper can be bare. At the other end of the line the voltage can either be reduced at once to that required at lamps or motors, or can be reduced first to a convenient distributing voltage—say 1,000 or 2 000—and then be reduced to the lamp or motor voltage. It should, however, be borne in mind that each such raising or lowering of voltage introduces an appreciable percentage of loss into the general calculation. For instance, if a large step-up or step-down transformer have an efficiency of 98%, then the loss in a pair is 4% due solely to $C^2R$, leakage, and hyteresis, without taking into account that loss due to the lag introduced by the induction of the coils. Another question to be settled is "what frequency of EMF shall be used?" and this is one that I do not think receives sufficient consideration. If we look for a moment at the expressions for hysteresis, induction and capacity, we shall see that all these losses vary directly as the frequency. It would appear therefore that it would be very advantageous to reduce the frequency as much as possible. The reduction of frequency, however, has the counter-balancing disadvantage that it renders necessary the use of larger, heavier, and therefore more expensive apparatus, and that if (the current) is to be used for lighting, less than about 40 or 30 alternations per second is appreciable to the eye in incandescent lamps, and less than about 50 in arc lamps. If it is to be used solely for power purposes then this limit is taken away, and the actual decision should be made on the basis of whether the enterprise can better afford to waste power in the lines, or to increase its capitalization.

A subject of such interest as "Electrical Transmission" can hardly receive adequate treatment in the limits of a short paper, and I have already trespassed considerably on your good nature, but I desire to very briefly touch on those preliminary considerations which should shape the policy of any such enterprise, and the importance of which appear to be not duly recognized. The engineering features of any enterprise should be not only influenced, but actually decided by commercial considerations. Electricity is not the object, but the means; the object being to make money by the sale of power. The engineering must be decided by the commercial value of this power. Assume a water power ten miles distant from a manufacturing town, the energy of which is to be electrically transmitted. What loss is to be allowed in the lines? Can we afford to lose a good deal in the lines, transformers, etc., and so reduce the size of our conductors and consequently our construction expense? or would it be better to keep the loss as low as possible, and save the power for sale? Engineering says you can make your loss as little or as great as you please, but commercial considerations say if power is more valuable than money it will pay to save all the power possible for sale, even at the expense of heavy conductors. If there be a large demand for power at a high figure, and the source of power is limited it is best to waste as little as possible; if the supply be very large and the demand small, it may, on the contrary, be best to waste much in the lines, in order to reduce the copper cost. Here again commercial considerations are paramount, for it is obvious that if much power is to be lost in the line the generators and water-wheels must be sufficiently large to furnish this waste power, and therefore cost more money, and a balance must be struck. Once a decision has been arrived at as to the loss allowable, then engineering considerations must settle as to whether it shall be mostly ohmic or mostly inductive, and here comes in the question of frequency. The efficiency of machinery is not sufficiently considered, and yet has a most appreciable influence on operating costs. The conditions under which motors operate are generally such that, although they may be called upon for a comparatively large output now and then, their average load, the work they are required to do most of the time, is greatly less than the above maximum. Consequently it would appear that their efficiency at such average load is of more importance than their maximum load efficiency. If therefore, in the above case, the supply is limited and the demand

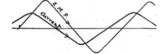

great, the half load efficiency of the entire plant assumes great importance. I think the above considerations, and others that will no doubt occur to any thinking person, will show conclusively that no transmission scheme should be undertaken blindly. A should not necessarily use 10,000 volts because B does; nor should he follow the engineering features of B's scheme simply because B has found them suited to his conditions. Every case should be considered on its merits, and the results worked out independently. A little or more or less loss does not mean simply a little less or more copper; it means less or more power to sell—larger or smaller generating plant: it means less or more profit. In every enterprise there is a certain combination of machinery, lines, etc., that will secure the maximum of efficiency with the minimum of expense, and this will not be attained by either regarding it from the purely commercial standard of getting the cheapest, or from the purely engineering standard of putting in the scientifically best, but can only be attained by combining the two, and by clearly recognizing the fact that solid commercial principles enter into engineering as much as into pure business.

## METERS.

### By JAS. MILNE.

THE subject of meters is probably the most important one in connection with central station lighting. It has received considerable attention, no doubt, from a few, but I question very much if it has received that attention which it deserves. Every manufacturer, we know, claims to have the best meter on the market,

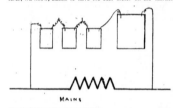

MAINS

### FIG. 1.

correct throughout its entire load. We know also that any manufacturer can quite readily get up a meter to stand a test for a day or so in some laboratory, and be correct, but this, I claim, is of very little use as far as everyday work is concerned. The object of this paper is more to touch on some of the more important meters in practice, making comparison with some of these meters from actual results.

It is sixteen years ago since Edison brought out the incandescent lamp, and we find, according to the records, that even before he had his lamp perfected he was busy on some kind of a meter to accurately determine the consumption of the current. No matter what people say to the contrary, we owe to Edison's inventive genius the practical success of the continuous current. The idea of connecting lamps, motors and other translating devices, in multiple, was original with him; also the high-resistance lamp, the feeder system, and many other important details which tend to make electricity the most useful of all agents. To Edison also remains the credit of originating the meter. He not only saw, as far back as 1880, that there was to be a great future for electricity, but he also saw that to make it a commercial success, as applied to lighting, etc., the current must in some manner be supplied on the meter basis and sold accordingly, just the same as gas or water.

The correct way to sell current is certainly by meter; this is the experience of every one. Satisfactory arrangements can at times be made with certain customers where the average consumption and the average run can be arrived at with a certain degree of accuracy, but even in cases of this kind it is advisable to put on a meter.

Edison's first patent was taken out in 1880, and the meter, with some modifications, is almost identical with the Edison meter of to-day. It is shown in Fig. 1. It consists of several cells in series, and the amount of current passed is measured by the amount of transfer of metal from one plate to another. We see that all the current does not pass through the cells. If the resistances are known of the two branches, we can easily determine the relative amounts of current flowing through each. For instance, if we have a derived circuit, one branch of 1 ohm resistance and the other 2 ohms, we see at a glance that whatever current is flowing in the circuit will divide into three parts, two of which will flow through the one having least resistance, and the remainder through the other. If a copper or zinc voltameter be placed in the 2 ohms branch and we find a certain deposit on one of the plates,

the cathode, then we must know that had there been a similar voltameter in the other branch (1 ohm) there would have been twice the deposit. It is not necessary, however, to put in this second voltameter to arrive at the result, for as long as we know the ratio of the resistances and the electrolytic cell put in any one of the branches, the total current passed through can be accurately arrived at from the deposit in that cell.

One of the laws of electrolysis is that "the amount of chemical action at all points of the circuit are equal to each other." This does not mean that the same current passing for the same length of time through different solutions will decompose equal weights of the metals contained in these solutions, but that the weights of the metals so decomposed will be chemically equal; that is, the weight will be in direct proportion to their chemical equivalent. For those who have not studied the "chemical effect" of the current, it might be advisable to explain some of the terms: The weight of one atom of hydrogen is taken as the unit (1), and that of copper is 63 i.e., 63 times heavier than an atom of hydrogen; but in chemical combinations one atom of copper is worth, or replaces, two atoms hydrogen; hence the weight of copper equivalent to one of hydrogen = 63÷2 = 31.5. This is called the chemical equivalent and is = atomic weight ÷ valency. The atomic weights of copper, zinc, nickel and silver are 63, 65, 59 and 108 respectively, and their valency 2, 2, 2, and 1 respectively, therefore the chemical equivalents are 31.5, 32.5, 29.5 and 108 respectively. Another term very much used in these calculations is the "electro-chemical equivalent," and this is equal to the weight of a substance in solution decomposed by the passing of one coulomb. If we know the electro-chemical equivalent of any element, and we also know the chemical equivalent of the other metals, the electro-chemical equivalent of these metals can readily be calculated.

It has been determined experimentally that 1 coulomb, passing through water, will liberate .00001352 grams of hydrogen, and as the chemical equivalent of copper is 31.5, therefore the electro-chemical equivalent will be = chem. equiv. × .00001352 = .00032-6088, and so on, for the rest of the metals. If we suppose the four cells in Fig. 1 to be copper, zinc, nickel and silver, we have a deposit 117.5 grams in the copper, 121 in the zinc, 110 in the nickel, and 402.5 grams in the silver voltameter, we can determine the ampere-hours.

Let C = current, y = electro-chemical equivalent, t = time in second and M = mass decomposed, then C = $\frac{M}{yt}$, or taking the copper voltameter, we have C = $\frac{117.5}{.0003261 \times 1\,hour}$ = 100 amperes flowing for one hour or equivalent thereto. If the current in each

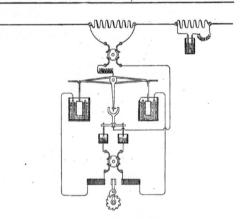

### FIGURE 2

of these four cells calculates out to be the same, then we must come to the correct conclusion that 100 amperes must have passed for one hour, or equivalent thereto. If only one of these voltameters had been in the circuit and the resistance remaining the same, we would have had exactly the same results. In the above example of a derived circuit composed of 1 and 2 ohms, we place the copper voltameter in the 2 ohms branch and we find as above, 117.5 grams deposit, which represents 100 ampere hours; then as this only represents one-third of the total current passed through the circuit, therefore 300 ampere hours would be the correct reading on this particular meter.

With this preliminary and elementary explanation we are now in a position to more clearly understand the chemical meter and also to follow up some of those as made by Edison. It will, however, be entirely out of the question to treat on all the meters as made by him, so we will simply deal with those having more of a direct bearing on the meter as we have it to-day.

As far back as 1881, it occurred to this genius that in order to satisfy the public, if such a thing were possible, the meter should be arranged that the customer could read it for himself. We therefore find, fifteen years ago, a self-recording chemical meter exhibited at the Paris Exhibition. Its principle is shown on Fig. 2. The resistances are so arranged that only a small known quantity of the total current will pass through the electrolytic cells. The meter as shown would not record, so much like the recording meters of to-day, but if we tilt the balance-beam shown above, this kicks the beam below in the opposite direction, making contact through the mercury cup and sending a current round an electro-magnet, which registers one on the counter. Current now flowing through the cell on the right of the balance beam was tilted so that the left end was down, and after a certain quantity of current has passed the cathode (the weight on the beam) will get heavier and in time throw the beam the other way. When it is swinging, contact is broken in one mercury cup and made in the other, bringing the electro-magnet on the other side into play, causing another unit to be registered on the counter. The same action takes place in this cell as in the other, and every kick, or second kick, according to the arrangement of the mechanism, is registered on the counter.

In the circuits leading to the cells reversing commutators are placed so that at the end of every month or so the direction of the current can be reversed, thereby reversing the deposit. By this arrangement the plates could be made to last for an indefinite period. In one description I have of this meter it states that the commutating devices were so arranged that when metal was being deposited on one plate the other was being dissolved, or when one plate was getting heavier the other was getting lighter by an equal amount at the same time. Take this style of a meter and we will suppose copper plates in a sulphate of copper solution are used, and that the ratio of resistances are 1199, and we will take it for granted that the beam tells at every .05 grams, and that 94 is registered on the counter. The total deposit is 4700 milligrams, and from this quantity the total current has to be determined. In calculating same out, we find that C in the high resistance amounts to, 4 amperes for 1 hour or 4 amp. hours. But this represents only what passed through the circuit in which the voltameter was placed ∴ 4 × 100 = 400 amperes for 1 hour would represent the total current passed through the whole circuit when the resistances are arranged as 1199. We might adopt a constant of 4.277 to bring the reading ampere hours.

In this same figure there is one very important detail in connection with same that may be overlooked, yet it shows, to my mind at least, Edison had little faith in this, and I might confidently state, in any other self-recording apparatus. I refer to the electrotylic cell in series with the recording device. I have not the least doubt that this recording meter was as nearly perfect, and probably more so, than the majority of recording meters in use to-day, yet we see that fifteen years ago he came to the correct conclusion that recording devices were unreliable.

It is certainly remarkable when we think over this; we are in precisely the same predicament as we were at that time; we have

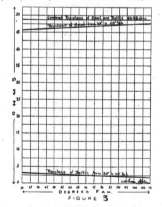

F I G U R E   3

to do exactly the same thing to-day, viz : put in an Edison chemical meter in series with all these recording coulomb and watt-meters of the motor type if we wish to get at the correct consumption. History does not say whether this type of meter was ever pushed or no, but that is immaterial as the main point I wished to draw your attention to was the cell in series with the recording mechanism. Several other recording devices, including motor

meters, are to be seen among his patents; in one the current operated a pointer which made a diagram on a sheet of paper placed on a revolving cylinder, the area of which, when measured by a planimeter gave the consumption. In a meter of this kind the maximum and minimum loads, a very good point indeed, could easily be traced, a thing which cannot be done with any meter we have to day.

In the early Edison meters copper plates were used which did not give very satisfactory results, and it led the inventor to try various metals, among which was amalgamated zinc immersed in a zinc sulphate solution. This gave excellent results and is used in the meter of to-day with perfect results.

Everyone is aware that the resistance of copper wire increases as the temperature increases and if we wish to keep the resistance of a certain circuit constant irrespective of temperature changes something must be inserted in this circuit which has an equal and opposite effect to that of the copper, that is to say, if we have a circuit of 50 ohms R at 60° Fah., composed of a spool of wire 46 ohms, and something else of 4 ohms, and if the temperature rises so that the spool now has 47 ohms, then the R of this something else must be 3 ohms if we wish to have the total R constant at 50 ohms. In the Edison meter the resistance of the electrolytic cells decreases as the temperature increases, and to make up for the decrease in resistance a compensating spool of copper wire is put in series with same, which has an increasing resistance equal in amount and opposite to that of the cell. In Fig. 3 it shows the resistance of the "bottle" or electrolytic cell and also that of the compensating spool. We see that the cell decreases and that the spool increases for increased temperature and that the two combined give us practically a straight line. The resistances are calculated from 30° to 110° Fah. or a range of 80° which is considerably more than is ever met with in practice. In the

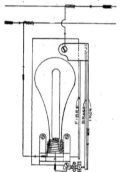

F I G   2

Edison meter the branch of low resistance is made of German silver and is called "shunt". The resistance of German silver varies .02 of 1% for every 1° Fah. In the smallest size of meters the shunt has a resistance of .04 ohms at say 60° and we have no compensating devices, therefore for a rise in temperature we must have an increase in resistance, and if we have an increase in resistance an error must be the result. The greatest percentage of the error will be in the smallest meter, therefore we will just calculate what the error amounts to. At 60° the shunt is .04 ohms, at 105° it is .04040 ohms and at 30° it is .039733 or a difference between 30° and 105° of .000667 ohms making the maximum error that can come into effect less than 2% or to be exact 1⅔% or less than 1% above and less than 1% below.

Taking the conductivities we find that if 100 amperes are flowing in the circuits .08161225 amperes go through the bottle at 60°. At 105° there are .082248 amperes and at 30° .08109 which shows that between 60° and 105° we have less than 1° and between 60° and 30° we have .6% as being the amount of the error.

Where meters are generally located the temperature in the summer rarely exceeds 70°, and in the winter never below 40°. Therefore in actual practice from 25 to 30 degrees would represent the greatest variation of temperature, which gives us .04oo888 ohms as the R of the shunt at 70° = 2% error and .039823 ohms at 40° = .5% error, or in other words the meter in the summer time would be one-fifth of 1% fast, and in the winter about two-fifths of 1% slow, making an average of about one-tenth of 1% slow for the whole year. In the larger sizes this loss decreases to almost nothing. Therefore for the variation in temperature due to the heating of the current or atmospheric variation we see that the percentage of error is practically nothing, so small that it may be entirely neglected.

So far we have assumed the lowest temperature as per Fig. 3 to be 20° Fah., but there may be places where the temperature goes considerably below this. These places are very exceptional however. The zinc sulphate freezes at 27° Fah. and some means must be taken to prevent its freezing. In Fig. 4 is represented the arrangement as put in the present meter to prevent the solution from freezing. It consists of a strip of brass and steel rivetted together and fixed at one end, the other being free to move. It is called the thermostat. When the temperature gets very low the brass contracts more than the steel and causes the strip to curve making contact with the terminals leading to the lamp, which on completing the circuit lights it. When the temperature rises again the compound strip straightens and the circuit is broken.

A patent was filed in 1881 covering this temperature regulator which at that time consisted of a resistance coil acting as the source of heat. Some few months later we find still another method of preventing the solution from freezing. This is shown in Fig. 5. It consists as in the former of the compound strip

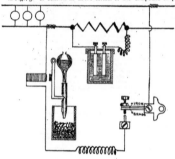

FIG. 5.

which makes and breaks the contact. When contact is made current is sent round the electromagnet, attracting the armature to which is attached an arm operating a valve which when open allows water to run into the cell underneath containing quicklime, the mixture causing heat. I might state that here in Toronto the temperature is so uniform (!) we have no thermostat in any of our meters although they are all adapted for them.

In the next figure is shown diagrammatically the Edison 2 wire

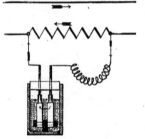

EDISON 2WIRE METER

FIG 6

meter which, as has already been stated, consists of a German silver shunt in multiple with a compensating spool in series with it the electrolytic cell. The connections, you'll observe, are simplicity simplified. In Fig. 7 is shown a 4 bottle 3 wire meter, and is simply two 2 wire meters in the one box. The object of having the two cells in multiple is that one bottle acts as a check on the other. In a meter of this kind we have practically four 2 wire meters, the two shunts being common to the 4 meters. If we wished to have the same check with 2 wire meters on a 3 wire sytem of any other make, there would have to be 4 of these.

In practice we find that the transfer of zinc in each pair of bottles on the same side of a 3 wire system in very many cases agree exactly to a milligram, and the maximum variation never exceeds a few milligrams. This shows without doubt

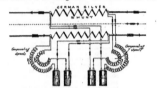

EDISON CHEMICAL 5-WIRE METER

FIG 7.

the great accuracy of the meter and when we have a reading on a meter of this kind and the bill sent out accordingly and the customer swears by everything holy that he "never turned the lights on," you are perfectly safe in assuming that if he did not turn them on then some one else did it for him. The customer says one thing and we have two meters silently testifying the other way. Which of the two are you to believe? The meter by all means.

The following table gives the sizes and other particulars of the meters as made to-day.

| THREE WIRE. | | | | | |
|---|---|---|---|---|---|
| Size. | Capacity. | R. of Shunt in ohms. | R. of spool at 60. | R. of bottle. | Capacity in .5 amp. lamps. |
| 1 | 10 | .04 | 46.46 | 2.5 | 40 |
| 2 | 20 | .02 | 46.46 | 2.5 | 80 |
| 4 | 40 | .01 | 46.46 | 2.5 | 100 |
| 8 | 80 | .005 | 46.46 | 2.5 | 320 |

| TWO WIRE. | | | | | |
|---|---|---|---|---|---|
| 1 | 10 | .04 | 46.46 | 2.5 | 20 |
| 2 | 20 | .02 | 46.46 | 2.5 | 40 |
| 4 | 40 | .01 | 46.46 | 2.5 | 80 |
| 8 | 80 | .005 | 46.46 | 2.5 | 100 |
| 16 | 160 | .0025 | 46.46 | 2.5 | 320 |

You'll observe that the R of the shunt varies while that of the spool and bottle remain constant for the different sizes.

It has been found, experimentally, that one ampere passing for one hour will remove from one zinc plate and deposit on another when immersed in a zinc sulphate solution 1.225 grams or in other words, the electro chemical equivalent is .00034 grams. One ampere need not flow to deposit this quantity. 10 amperes for 6 minutes or .25 amperes for 4 hours would give exactly the same result. From this electro-chemical equivalent we can determine what the deposit will be for a certain quantity of current and also from a given deposit the quantity of current passed. For instance, we have a current of 100 amperes passing for 3 hours, what would the deposit be if zinc plates were used? We find it is 367.200 grams=.8lbs. Here we have taken all the current flowing in the circuit as going through the electrolytic cell, which if it did in practice would give us an enormous consumption of zinc. $Cyt = M = 913.104$ grams or 2lbs per h.p. per hour at 1 volt, which gives us the formula $\frac{2}{E}$ lbs per h.p. per hour as being the consumption of zinc when E=E.M.F.

In the above table we see the resistances are so proportioned that for every ampere passing for one hour in the smallest size .001 grams will be deposited on the plate, that is the resistances are as .04 : 48.96 or 1 : 1224 which gives us exactly 1 milligram deposit for every ampere hour.

Let us now see how the amount of consumption of zinc for a given term or the quantity of current from a given deposit are computed. We will take the deposit as .1 gram or 100 milligrams. In the smallest size of meter when the resistances are 1 : 1224 we saw that 1 milligram represents 1 amp. hour, and also if the resistance were as 1 to 2448 one milligram would represent 2 ampere hours, and so on up to the largest size of meter when 1 to 19584 is the ratio of the resistance. One milligram=16 amp. hrs. Therefore for the latter size of meter 100 milligrams would represent 1600 ampere hours. Taking it another way we have a current of 160 amperes flowing through the circuit. What will be the deposit per hour? The resistances are as 1 : 19584,.·.
$\frac{1}{19585}$ of 160 amperes will flow through the bottle which gives us .00817 amperes for 1 hour = .00817 × .00034 × 1 hour = 10 milligrams. But we say that 160 amperes were flowing, therefore to arrive at the total ampere load this reading will have to be multiplied by 16.

In the table the meters are numbered 1, 2, 4, 8 and 16, and the one we have calculated is No. 16, and we have just determined that the deposit has to be multiplied by 16 to bring it to the same basis as a No. 1 meter, i.e., 1 milligram = 1 amp. hour, or in other words the meter number is the constant of that meter.

On referring to Fig. 7 we see two bottles in multiple. The average deposit is what is taken in these meters which = total deposit ÷ number of bottles in multiple.

The loss in an Edison meter is very small, in fact so small that it may be neglected.

The joint resistance of the shunt and spool of a No. 1 meter is .039957 ohms, say .04, therefore at full load the loss would be 4 watts and only when meter is in actual operation, that is when current is being used. If this meter were placed on a 120 volt circuit it would mean ⅓ of 1%, and if placed on a 240 volt circuit ⅙ of 1%, and 1/12 of 1% on a 500 volt at full load.

As a matter of fact the meters are never put in so as to run up to their full capacity continuously, so that we can safely reckon at the outside the load as being ¼ to ½ their rated capacity, which makes the loss on a 120 volt circuit at ¼ load .08 of 1%, and at half load .16 of 1%.

The percentage of loss in all sizes is the same if calculated at the same percentage of load.

It might not be out of place to show the ordinary meter form as used in Toronto, which is suitable for chemical and watt meters, and the method of making up the bill from the meter reading is very simple. In Fig. 8 we see that A. B. Smith's reading from April 30 to May 30 is 473 x 4 milligrams, or 1891 ampere hours, and if the bill is rendered in lamp hours, and if each lamp takes .5 amperes then the deposit x 2 gives the bill in cents if the rate is 1c. per lamp hour, or $37.84.

Some have imagined that nothing but expert chemists can

maintenance, including all chemicals and the reading of same does not exceed 70c. per meter per year.

### THE WATERHOUSE METER.

This is a recording electrolytic meter which was brought out some two or three years ago in England and deserves mention in this class. I have not had practical experience with it, but I am of the opinion that it should give satisfactory results. Its construction is shown in Fig. 9, and in Fig. 10 is shown the connec-

| METER No. 18738 | THE TORONTO ELECTRIC LIGHT COMPANY Limited | | | | | CUSTOMER | | |
|---|---|---|---|---|---|---|---|---|
| Ampere | TORONTO    CANADA | | | | | | | |
| Voltage | | | | | | | | |
| Constant    4 | Name    A. B. Smith Esq | | | | | Meter No.  1563 | | |
| Size No. | | | | | | | | |
| Wire    3 W | Address         69 Madison Avenue | | | | | Ledger Folio  193 | | |
| DATE | | | April 30 | May 30 | | | | |
| LEFT SIDE | A Cell | | 48  570 | 50  764 | | | | |
| | | | 48  358 | | | | | |
| | | | 212 | | | | | |
| | B Cell | | 50  487 | 52  375 | | | | |
| | | | 50  277 | | | | | |
| | | | 210 | | | | | |
| NIGHT SIDE | C Cell | | 52  365 | 49  264 | | | | |
| | | | 52  105 | | | | | |
| | | | 260 | | | | | |
| | D Cell | | 51  496 | 41  957 | | | | |
| | | | 51  232 | | | | | |
| | | | 264 | | | | | |
| Loss | | | 946 | | | | | |
| Average Loss | | | 473 | | | | | |
| Read by | | | | | | | | |

Fig 8

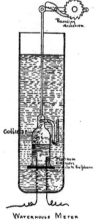

Recording Mechanism

Collector

Platinum Electrodes
Sulphuric

WATERHOUSE METER

Fig. 9

handle the Edison meter and that its cost of operation and maintenance is high. This is entirely wrong. The meter department of any central station requires at least intelligence and if run without this will very soon become unprofitable and unreliable. In a station where there are about 1500 to 1800 meters the cost of

tions of same. We have here the shunt as in the Edison, also the compensating spool. The mechanical part consists of a method of recording the volumes of gas produced by a small portion of the current used by the customer. Gas is accumulated in the collector and the registering mechanism indicates the number of times this has been filled. The operation is very simple and is as follows : When a certain quantity of gas is accumulated it forces the fluid down the U shaped tube until it comes to the bend, and just as soon as it comes to this bend it immediately starts up the other leg and escapes. It is then filled up with the liquid and descends by gravity for another charge of gas. Pure water is used for refilling the cells every 3 or 4 months. This meter is very easily calibrated and is a coulomb meter, registering in ampere hours.

The two electrolytic meters mentioned are only used for continuous current, but there is one called

### THE LOWRIE HALL METER

in which the same principle is used to measure alternating currents. In the secondary circuit a storage cell is placed in series with the electrolytic cell and it is taken for granted that the alternating current does not deposit metal, therefore the transfer from one plate to the other depends on the conductivity of the

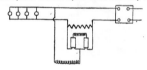

Fig 10

circuit, i.e., the number of lamps turned on. The total current going through the circuit passes through the storage cell, and if no lamps are turned on no current from the accumulator will flow through the voltameter, and if the lamps are turned on current will flow from the cell to the voltameter causing deposit, therefore the deposit will be a measure of the conductance from which the lamp hours can be arrived at.

We will now take up motor meters, and in this we have an endless variety. It would simply be out of the question to touch on them all, so we will just take up the most important and treat on them briefly. In this kind of a meter there are a few advantages over those we have already described, but the disadvantages more than offset the advantages. It has been claimed by some that they (some of them at least) require no attention. This is, as far as my experience goes, incorrect, for I find that motor meters require more attention than any other form, either chemical or clock. Nearly all of them consume energy when not recording, that is when no power is being used by the consumer; none of them will record on very light loads—if they do when just installed, they are not so sensitive afterwards.

There are, however, some very good recording motor meters, if we overlook these disadvantages, among which might be mentioned the Ferranti and Perry in England, the Shallenberger, Duncan and the Thomson recording meters.

The Ferranti meter is shown in Fig 11, and is an ampere hour

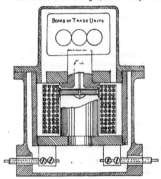

FERRANTI METER

FIG. 11                    J MILNE

recorder. If a current is passed through a fluid in a magnet field, this fluid tends to move in a direction perpendicular to the direction of the current and also to that of the field. It is on this principle that this meter depends. Current enters at the centre of the mercury trough and leaves at the rim and in so doing gives motion to the mercury, the motion is communicated to a small aluminum fan which is connected to the recording mechanism. It is adapted for continuous and alternating currents.

In a paper read by a Mr. Dicks before the institution of Electrical Engineers, a short description of the above and following

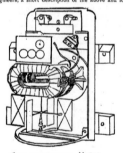

SHALLENBERGER METER

FIG. 13

meter is given which, should any of you wish to follow them up, will prove interesting, I have no doubt.

If current is sent from one end of a cylinder (in a magnet field) to the other the cylinder will rotate, and this is the principle of the Perry meter shown in Fig. 12. Current is admitted to the mercury dish shown at the bottom edge of the inverted copper cup which plows up the sides and leaves by the nickel rod at the top. Friction is reduced to a minimum. The speed of rotation is very slow and the meter will register very small currents in some of the larger meters, say 60 amperes; it will start up with .1 ampere.

The Shallenberger meter is shown in Fig. 13, and is intended for alternating currents. It has been very successful and a large number of them are in use. It consists of two coils, one carrying the main current and the other is a closed coil. A rotary magnet field is produced by an induced current in this closed coil which drags the iron disc around. No brushes or commutator are required; the disc has no electrical connection whatever. The retarding motion is effected by an aluminum fan fixed on the same spindle as the disc. It is an ampere hour recording meter and consequently the speed is directly proportional to the current. The calibration depends on the angle of the closed coil to that of the main coil.

The Duncan meter, of the Fort Wayne Co., has also made a good record, and like the above has neither commutator nor brushes. The armature is an aluminum cylinder and the closed secondary is made of several copper punchings. This meter

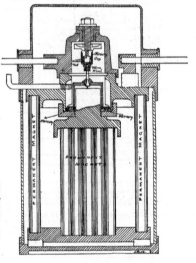

PERRY METER

FIG 12

depends on the repulsion of a closed secondary from its primary. The primary coils are in a series with the lamp circuit. The retarding effect is obtained in the same manner as the Shallenberger. Probably the most important of all these motor meters is the Thomson recording watt meter.

One of the advantages claimed for this meter is that it is adapted for continuous and alternating currents. This may be an advantage and it may not. In so far as we have meters for alternating currents of a simpler design, it looks to me as if it would be better and cheaper to have the separate meters. This is more a matter of opinion, however. In Fig. 14 is shown diagrammatically the Thomson watt meter. In the armature circuit is placed a high resistance coil, generally placed in the bottom or back of the meter and part in the field, the object of this latter part being to produce a field of sufficient strength to overcome the friction of the moving parts, brushes, &c., and it is perfectly clear that this current must flow whether current is being used by the consumer or not. The copper disc rotating in a permanent field acts as a drag, just the same as the little fan in the former meters, by generating an E M F. This E M F is proportional to the speed, therefore the retardation is proportional to the speed and the speed is proportional to $C \times E$; therefore, the speed resulting from this is proportional to $C \times E$, i.e., the power at that particular time.

Regarding the practical working of this meter, let us now devote a little time to same. The conditions in which they are installed are precisely the same as the Edison, that is, they are placed in the basement and as near as possible to the service. Take the average installation and we find very few switches between the service and the meter. The current, as already stated, flows through the armature circuit 24 hours per day, and in a test made recently to determine this amount, it was found that .05 amperes flowed through. This looks a small quantity, but if we have 1200 to 1500 of these meters, it means 60 to 75 amperes to keep up this alone, and from which there is no return.

If the voltage is 120, and taking the cost of manufacture at 2c. per k.w. per hour, it means $1300 to $1700 annually, which is equal to 6% on $22,000 to $28,000. This represents one source of loss, but there are many more. Every one of these meters, and in fact, nearly all recording meters will when new start with comparatively little current, but after they have been in use for a time a different result is noticed. For instance, we have an installation

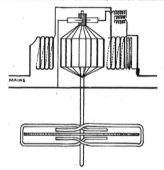

FIG. 14.

of 500 lights on a 3-wire system—or any system for that matter—and we decide to put in a 120 ampere meter. This meter has, according to our Canadian law (God bless it), been duly tested, sealed, etc., and the company is permitted to install same. When first put in it is adjusted as nearly perfect as can be, and it is tried on a very light load. It does not start up with one light, nor two, nor even three ; it takes between five and six lights to make it just move. It takes very good eyesight to see it make an attempt to rotate. At half load it appears to be all right, and the same at full load. In a couple or three months, when taking the reading for the second or third time, we try it to ascertain if it is still doing its work. At full and half loads everything indicates that the meter is doing its duty, but at the very light loads we find that instead of registering with 5 to 6 lights as it did at first, it takes almost 12 lamps to make it move. This is not an exaggeration. Although we have taken a large meter, yet I know the percentage is just the same for all the sizes of meters. What does this mean? What does it amount to at the end of the year? Let us take this actual example. Installation of 500 lamps running 2 hours per night for 5 nights per week, and before and after these hours we have from 10 to 20 lights burning three-quarters of an hour each, making a total of three and a half hours for the night.

As stated above it takes 12 lamps to make it record at all, therefore we may assume with perfect safety that the current for at least 8 to 10 lamps, say 8, has no registering effect on the meter. Now these lamps are burning 3½ hours per night, 5 nights per week = 140 hours weekly or 7,280 hours yearly. The total meter reading in this particular installation should be 267,800 hours, but as a matter of fact we only get 260,500 or 95.7% of what it should be, or a dead loss of 4½ per cent., which represents in actual cash, if current is sold at .6 of a cent. per lamp hour as we have it in Toronto of $43 on this particular meter alone. This may appear to you as being an aggravated case but I must say it is not ; it is simply one out of quite a number and we are forced to come to the conclusion that from this cause alone the revenue of a company is 5% less than it should be.

There is yet another source of loss which is not applicable to the small stations. I refer to electric elevator work, or any work where there is a quickly varying load. Elevator work is without doubt the most unsatisfactory kind we have to contend with, and it is acknowledged by every one that these meters, in fact all meters, do not register the energy consumed. Those of you who have observed the action of a wattmeter or any recording meter on these loads must have been reminded of a lazy man, slow to start and quick to stop. In hydraulic elevators the amount of water is the same for all loads and if water is sold by meter or by the "feet run" of the elevator, it costs exactly the same to take up the empty car as it does to take up 4000 lbs. People have been content heretofore to pay for water in this manner and I cannot see why they should not be charged in much the same manner for the electric power. I think I am pretty safe in stating that the motor meter has nearly outlived its usefulness as far as elevator work is concerned.

We see therefore there are 3 great sources of loss.
First. Power consumed in the armature circuit.
Second. Power not recorded at light loads.
Third. Power not recorded at quickly varying loads.

Now putting all these together we certainly have a very poor combination for any concern to be supplying current by meter measurement.

How do we find the Edison chemical or any other chemical meter under each and all of these conditions. In this meter there are no moving parts, no friction to overcome, no armature circuit, therefore we have practically no loss. The examples I have just given by the motor meter were proved to be incorrect by the chemical meter and as pointed out at the beginning of this paper that Edison did, 15 years ago, exactly

what is being done to-day, viz., he put a chemical meter in series with the recording devices just to see how much the other was out. By actual measurement there is from 20 to 25% difference between the readings in elevator work. I have seen that new meter as got up by the Diamond Electric Co., of Peoria, Ill., which records the energy. In a record of a test sent out by them as made by Professor Jackson at the University of Wisconsin, it is practically correct at all loads and any frequency. As we have already stated these tests are very good in their way, but there is nothing like the test of actual practice and in this alone time decides.

In the next class of meters the Aron figures as being the most important of all clock meters. We have several clock meters in this country but not one of them can in any manner compare with this one. It is adapted for alternating and continuous currents and is made as an amperehour or wattmeter. It is one of the most reliable meters in existence. It consists of two clocks, the pendulums of which are shown in Fig. 15—one keeping standard time and the other is retarded or

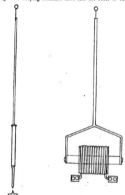

A RON    METER.

FIG 15          J. MILNE.

accelerated as the case may be by the action of a coil or coil carrying the main current in which the "ball" of the pendulum consisting of a permanent magnet oscillates. The principle of the gearing from which the motion of the pendulum is communicated to the dial is shown in Fig. 16. When both clocks are going at the same speed the middle

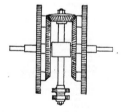

DIFFERENTIAL GEAR
OF THE
ARON METER.

FIG. 16

bevel wheel is turned around on its own axis, but if one is going faster than the other the middle wheel is turned around on its axis and also around on the axis of the spur gears. This motion is communicated to the wheels operating the pointers on the dials, and it is only where there is a difference in the speed of the clocks that the meter records. This meter possesses the great advantage of recording, no matter how how small the amount of current is.

In the next figure we show the principle of one of the clock meters that is used in this country to considerable extent. It is neither an amperehour or wattmeter, simply a time recorder, that is, it records the number of hours the current has been flowing in a circuit. It is called the "Pattee Lamp Hour Recorder." It consists of an ordinary clock in which is placed an electromagnet, which is so arranged that when current is turned on a spring is released which disengages the escapement wheel. It is extremely simple and not liable to get out of order. It is very applicable to places where there is a steady power such as arc lamps or motors with steady load. In a store where 10 arcs are required at the one time and the meter records say 200 hours then 200 x

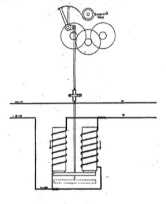

Fic. 17

No. of lamps x rate per lamp hour = bill; or again if we have a motor say 6 h. p. average rate 6 c. per h. p. per hour, meter records 100 hours, we have 100 x 6 x 6 = $36.

With this meter, however, it is only an approximation as to the power; however its simplicity commends it.

Let us now sum up the advantages and disadvantages of the various meters.

The advantages of the Edison chemical meter are :
1st, Practically no loss.
2nd, No moving parts.
3rd, Absolutely correct at all loads.
4th, Will record the smallest possible amount of current.
5th, It is applicable to any pressure.
6th, Low first cost.
7th, Low cost of maintenance.
8th, Readily repaired.

The only disadvantage (if any) is that the consumer can't read it for himself.

MOTOR METERS :—The only advantage is that the consumer can read the meter.

The disadvantages are :
1st, Loss in overcoming friction in the moving parts.
2nd, Incorrect at light loads.
3rd, Incorrect at quickly varying loads.
4th, First cost high.
5th, Cost of maintenance high.
6th, Not readily repaired.

CLOCK METERS :
In the Aron type of meter we have the following advantages :
1st, Correct at all loads.
2nd, Will record the smallest possible current.
3rd, As a coulomb meter it is applicable to any pressure.
4th, Practically no loss.
5th, Can be read by the customer.

The objections to this meter are :
1st, Liability to stop recording if clock stops.
2nd, First cost high.

We therefore see as a practical meter the chemical meter is superior in every point, save one, and that is the customer can't read it for himself. Is this much of an objection? How many gas consumers read their meters? I know I am not very far out when I say that not over 2% even wish to read their meters.

The Edison meter has been condemned by a certain class either through ignorance of its principle or prejudice and our Canadian Government have also seen fit to practically condemn it—probably not condemn it, but to curtail the growth of a meter that has no superior and very few equals.

It does seem strange to me that the Government should interfere the way it has done and more especially with the electrolytic meter when in the " Act respecting the units of electrical measure " in section b, lines 14 to 21 under the heading of " ampere " it states " as a unit of current the ampere which is .1C.G.S. units and is represented sufficiently well for practical purposes by the unvarying current which when passed through a solution of nitrate of silver in water and in accordance with

the specifications contained in schedule on to this Act deposits silver at the rate of .001118 grams per second." •

What does this mean ? It means that they have practically adopted the chemical meter for determining the unit of current. In the Act it says the electro chemical equivalent is .001118 for a silver voltameter just the same as .00034 represents the grams deposited per coulomb in a zinc or .000305 for a copper voltameter.

It is immaterial whether they specify silver, copper or zinc, the principle is exactly the same, and it does look absurd on the face of it when the unit of current .1C.G.S. units is according to law laid down as above, yet in the same Act a meter on precisely the same principle is limited to the extent of the number of meters in use at the time the law came into force.

It may be argued that it is not the principle of the meter that is the objectional point but that it is not self recording, the law, of course, calling for " dials." If any consumer wishes to keep a faithful record of the number of hours he burns his lamps he is at perfect liberty to do so and if he does so conscientiously and the lamps of the efficiency they are said to be, or suppose then the bill rendered from the meter reading will coincide exactly to a cent.

Again, it has been claimed by some holding important positions that to be renewing the zincs is practically renewing the meter, that is supplying a new meter. Did you ever hear of such gross rot? What is the law on this point, it is " that the present number must not be increased and all new meters must be of the direct reading type," or words to that effect.

The Government has to raise a revenue, that is settled ; the gas companies contribute a certain percentage of that revenue, the electric companies are their greatest competitors, therefore we can readily infer that any little obstacle that can be put in the way by such companies will be done so and it is very common property that this Act was the result of the gas companies.

May I ask " Where did the Government get hold of that definition of the ampere and also the voltameter specification ? How are we to determine from this the alternating unit of current ? "

Has the Government inspector created any better feeling between the companies and the consumer ? Do consumers pay their bills with any better grace than before the law came into force ? Do customers put any faith in this test ? Are there any advantages to be derived from this test ? If so, where ?

The answer to the first three is in the negative, and the fourth in the affirmative, and to the last, I will leave it unanswered for the present.

Probably I have digressed from the line of these papers, and in returning to same I wish to again draw your attention to the advantages as enumerated above in the electrolytic meter.

Is there another meter in the market that can lay claim to the same ?

## OPERATING ENGINES WITHOUT A NATURAL SUPPLY OF CONDENSING WATER, OR THE CONTINUOUS USE OF INJECTION WATER.

### BY E. J. PHILIP.

THE subject is somewhat new, and information on it must be taken from the few plants that are now operated upon this principle. Like all other new departures in steam engineering, there is very much to be learned and studied before everything in connection with it is properly understood. In a paper of this kind we can only go into the leading points about it, as the subject is so large that a whole volume might be written on it to cover fully the whole ground. From observation throughout the country it is evident that the principle of running engines condensing is not as thoroughly understood as it should be, for we have many cases where there is a sufficient supply of water within reach, and still the engines are exhausting into the atmosphere. This perhaps because many think that the expense of putting in and maintaining a condenser is greater than the saving would warrant. As an illustration, take an ordinary high pressure engine of say 100 h.p., using say 4 lbs. of coal per h.p. per hour and running 10 hours per day, the coal consumption would amount to two tons per day. The water consumption per h.p. in that case would be represented by 30 lbs. per h.p. hour. If a condenser is added the same power would only require say 22 lbs. of water, making a saving of 26 per cent. The total coal consumption for the year, running 365 days, would be 730 tons. If the coal can be put in for $3.00 per ton, the year's consumption would amount to $2,190. The cost of adding a condenser to such a plant, including the necessary piping, should not exceed $300. The cost of operating the condenser will be about 6% of the power of the engine, and is equal to $131. The interest on the condenser investment at 6% is $18, making a total cost of $149 per year to maintain and operate it. 26% of the coal account would be $569, from which deduct $149, the cost of operation, leaving a net gain of $420. This in many cases would make a dividend for the owners where there is none at present. In cases where the water for condensation is not procurable except at considerable expense, it can be used over and over again, and be cooled by air. The idea of cooling water in this way originated in Germany, and was applied for the purpose of cooling beer. The first cooling tower was filled by the branches of trees, or brush. The air used was only the natural current due to the warm water. This, of course, required a very large tower to get an amount of cooling surface to be effective, as the air current was necessarily very slow. The air is the cooling medium, and is indirectly the condensing medium. If you wet your hand and hold it in a current of air, you will feel a cold sensation, because the water is being evaporated and is taking up the latent heat of evaporation from your hand and the surrounding air. The specific heat of air is .2375, while that of water is unity.

If we depended upon the direct absorption of heat by a rise in temperature of the air, we would have to raise about 4 lbs., or 55 cubic feet, one degree, to absorb a heat unit. Consequently we would have to raise 1,000 cubic feet of air 55 degrees to condense 1 lb. of steam at atmospheric pressure. But when air is brought into direct contact with water, there is a cooling action due to evaporation much greater than is due to the elevation of temperature. When a pound of water is evaporated in this way, five times as much heat disappears as when a pound of water is raised from the freezing to the boiling point, and every pound of water so evaporated absorbs heat enough to condense one pound of steam. Now, by having an arrangement whereby we can pass a strong current of air over a quantity of water, favourably disposed to be acted on by the air current, we can by evaporation of a quantity reduce the temperature, and that is what takes place in a cooling tower, which is an apparatus designed to distribute the water so as to expose a large surface to be acted on by the air. Now, for every pound of water evaporated there is a reduction of temperature which will allow of a pound of steam being condensed, and just bring the remainder to the original temperature. It will be plain, therefore, that in operating a cooling tower there can be no more water used than when running non-condensing. In fact, there is not as much, because there is not as much water evaporated in the tower as there is condensed, as the surface of the tower and pipes have a cooling effect ; also, the direct rise in temperature of the air takes away a quantity of heat without evaporating any water.

The engine will require less steam, consequently there is a smaller quantity of feed-water used than when running non-condensing. The system, therefore, allows a plant which has to buy even its feed-water, to run condensing at a less expense for water than when running non-condensing. The details of the system are, at the start, like an ordinary condensing plant. The steam leaves the engine, passing through the condenser, is here condensed by water taken from a small reservoir instead of some natural supply. The water passes to the air pump and is pumped out, forming a vacuum as in an ordinary condensing plant ; but now, instead of letting it run to waste, it is elevated to the top of a tower, either by the air pump itself, if the tower is low, or by an auxiliary pump should the tower be high. This is preferable in any case. The water is distributed over the surface of the filling of the tower, falling to the bottom through the up-coming current of air, and the temperature is thereby reduced sufficiently to be discharged into the small reservoir from which the condenser takes its water, and is used over and over again. The details of the tower are :—At the top of the tower is an arrangement to distribute the water over the whole surface of the interior. This distributor has taken many forms, some of which are quite ingenious. Some of the latest are the revolving distributor, illustrated in "Power," for March, and other mechanical papers. This distributor is mounted in the centre of the tank on ball bearings, and the water issues from the cross pipes like the ordinary lawn sprinkler, and distributes the water evenly. Another distributor which is used in towers with what might be termed partition filling, is made with a little trough across the top of each partition, with main channels feeding them. The top of the small troughs are made like a saw on their top edges, and the fine streams of water run through the hollow of the teeth and spread over the surface of the partitions, making a very even distribution. There are numerous other forms, such as perforated plates, screens, etc., all of which will work, but do not distribute as well as the two

mentioned. The filling of the tower or material over which the water is distributed, has taken even more forms than the distributor. From the time when brush was used to the present and latest wire filling, the same idea was at the bottom of every change, namely, to make a given size tower do more work. The cooling effect in a given size tower is a very important point in metropolitan plants, where room is valuable. The first filling was brush. Then round poles were tried. About the same time and at different times since pans have been tried with some success, but was never equal to the tower system. The next was a partition tower, or a board filling. This has taken a great many shapes, the boards being arranged to break up the water and air currents in every conceivable manner. Sheet iron has been tried in various forms, some like stove pipes and others arranged in sheets. The latest and best filling is tile and wire netting.

The tile tower has been described in Power and other mechanical papers. It is very satisfactory. One point against this filling for a large tower is its great weight. The wire or Barnard tower is filled with wire netting rolled up loosely and set up on end. In these towers a settling chamber is provided at the bottom, and a heavy grating is placed across some distance above the water. In this space the fan discharges its air. On top of the grating is placed the tile or wire, whichever filling is used, and it is continued on up as far as it is able to

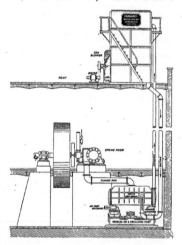

BARNARD'S PATENT WATER COOLING APPARATUS,
AS ARRANGED ON ROOF OF A BUILDING.

support itself, breaking joints, so as to break up the streams of water. There is a portion of the tower carried on up above the filling, to allow the particles of water to settle out of the air current. This prevents a spray flying from the top of the tower, and also any of the water being wasted. Information on the formula for calculating the size of towers is not very extensively known. As far as can be learned, about 50 square feet of cooling surface is required per h.p., when a large quantity of air is used, say 100 cubic feet of air per h.p., and varies with the amount of air and with the arrangement of the filling. In making up estimates the term h.p. does not give definite information, because the amount of steam used per h.p. varies from 15 to 45 lbs. per h.p. per hour, according to the size and type of engine. The only way is to get the water consumption of the engine and figure from that, the same as for running condensing. When an engine is using, say 25 lbs. of water per h.p. per hour, it will require about 4.8 cubic feet of tower for each h.p., with sufficient air and wire filling. With tile filling the cubic capacity required is about 6.5 cubic feet per h.p.

Cooling towers are becoming numerous. We have one in Canada, at Montreal. Two have lately been started at Detroit, and reported as giving excellent satisfaction. The accompanying illustration of Geo. A. Barnard's towers arranged for surface condenser, with the tower on the roof of a high building, will illustrate one application of the system. Further illustrations are not exhibited, because several of the mechanical papers have lately fully shown the different applications of it. It is estimated that the cost of operating a cooling plant is from 2½ to 5% of the power of the engine, which leaves a large net balance in favor of the apparatus, fully justifying its application on plants of any magnitude, or where the cost of coal exceeds $1.00 per ton. If a tower is placed on the roof a surface condenser should be used, and the ascending column of warm water is balanced by the descending column of cool water, and the actual head the pump works against is the height of the tower. If the tower can be placed in the yard, a jet condenser may be used, unless the object is to get pure water for the boilers. In the beginning of this paper the cost of adding a condenser to a 100 h.p. plant was shown to effect a net saving of $420, or 20%, nearly. The cost of adding a tower to such a plant should not exceed $700, the interest on which at 6% is $42, leaving a net saving of $378. This would make a very good showing on such a small plant, and would in most cases be much larger. Another point is, in cases where engines are carrying a full load and a little more power is required, attaching a condenser would increase the power about 20%, thereby avoiding buying a new engine, the plant carrying this extra load at the same expense for coal and water.

## SOME CENTRAL STATION ECONOMIES.

### By P. G. Gossler.

On the occasion of an electrical convention a statement of results obtained in the reconstruction of light and power plants will, no doubt, be acceptable, particularly so to station managers who may be confronted with the fact that their plant is not modern, and probably not a paying one, and that the time has arrived when reconstruction is no longer a choice but a necessity.

It is well known that many central stations which have not been operating on a paying basis have been turned into profitable investments by prompt measures having been taken to modernize them, and to put them on a footing to meet competition either from companies already in the field or contemplating entering it. To do this, it has generally been necessary to reconstruct the entire electrical part of the plant from the generators to the lines and transformers, replacing the old generators and transformers by the more efficient apparatus now manufactured; rebuilding and re-designing the switch-board, and last, but certainly not least, the re-arrangement of the feeders and mains, to give economical distribution, to overcome the inductive effects, and to bring the feeder losses within the limits of good practice.

Those who have been so unfortunate as to be in charge of a plant operating generators of small capacity, with a regulation anywhere from 30 to 40%, with drops in the feeders of the circuits varying from 1 to 10%, with no feeder regulators, with the wires so arranged that the worst possible inductive pumping effect is obtained, with a type of transformer whose leakage current is several times what modern practice permits, with a regulation corresponding in percentage to the drops in the feeders, and combined with all this a decided variation in the house wiring drop, will appreciate what a restful feeling is realized, even in the contemplation of a reconstruction that will include new generators and transformers of high efficiency and close regulation, and a safe switch-board. Those who are aware that their service is not what it should be, and who have analyzed the situation, know that the only remedy for the trouble, the only guarantee for a service that will be acceptable to the public and one that can successfully meet competition, is the replacement of any old and inefficient apparatus in use by that which is modern and efficient. This may mean a large outlay, but it should be done at any cost. It may mean the scrapping of old dynamos and transformer—they are of no use to anyone now—what is wanted is only a first-class apparatus. Experience has shown that the first cost of apparatus cannot receive the consideration that it did a few years ago; if it does

there will be within a comparatively short time the same problems to solve.

The following gives results obtained from the partial reconstruction of one plant. It does not give a full idea of what will be accomplished by complete reconstruction inasmuch as that part so far carried out has been confined to transformer and line changes.

The reconstruction planned for the present and now in progress, will affect only the alternating system of a plant which also furnishes direct current arc and motor service. These changes will include the replacing of the present single phase generators and line shafting operating them by two phase generators with an inherent regulation of 4 to 5 % without compounding devices, the generators to be belted directly to the engines: the building of the new switch-board for two phase currents serving light and power from the same circuit at 2000 volts: rearranging the lines for two phase distribution; and reducing the station load and bettering the service in general by replacing all of the old transformers on the lines by the best transformers obtainable.

To proceed with a systematic reconstruction, the first things necessary are reliable records, at least of what the plant and lines to be reconstructed consist. For the plant herein referred to it was necessary to establish pole line and circuit maps as well as transformer maps. It may be said that such a system of records in detail and kept up to date is necessary for the economical operating of an electrical lighting station.

For the pole line records a card catalogue was arranged, each card having a number corresponding to a pole; in connection with this card catalogue there is a map on which each pole is located with its number; also, for further convenience in making out reports and locating poles, each pole itself was numbered.

The following cut represents a form of pole card which was found to answer the purpose very well.

On the card representing a particular pole all of the wires are shown in their relative positions on the pole by numbers placed over the pins to which the wires are attached, the numbers indicating the circuits of which the wires form a part. By means of this card the positions of the wires forming the different circuits were clearly shown, also what changes in the relative positions of the wires were necessary to overcome existing inductive effects, the latter being a source of much annoyance. In fact, the pumping on the circuits due to mutual induction, prior to their rearrangement, when circuits supported on the same pole were running two dynamos on different engines, was so serious and caused so much fluctuation of the lights that it was necessary to rearrange the relative positions of the feeders of all the circuits to counteract these inductive effects. Very satisfactory results were obtained when the rearrangement of the wires had been carried out. Prior to this change, to overcome fluctuation, it was necessary to feed all circuits on the same pole line from one set of dynamos operated by one engine, which was very often not convenient and only possible with a large loss in operating expenses. If all the circuits on the same pole line were not run from the same set of dynamos, the service was such as to make life a burden for those who were responsible for it. After the rearrangement of the feeders to overcome the pumping, it was possible to run the circuits entirely independent of each other and in a manner most convenient and economical for operating, which is, of course, a source of much economy as well as satisfaction.

In connection with this pole catalogue, circuit maps were arranged, which consisted of diagrams for each circuit, showing the streets upon which the circuit ran, and the size and length of each section of wire or wires.

At the same time these records were being made out, transformer charts were prepared, which consisted of maps for different sections of the city covered by the different circuits. On these maps each transformer was located by a small square stamped on the map, and within this square was written the name of the customer being served from this transformer, the number of lamps installed, the revenue per year, the revenue per lamp per year, the estimated number of hours burned per lamp per day, and the probable number of lamps burning at any one time. There is also indicated on these charts the size and length of secondary wires from the transformer to the customer's cutout. All this information was found necessary for the proper "bunching" of customers on the transformers and for the loading of the transformer.

Wherever possible, secondary systems were established, to which several transformers were connected in parallel, in which case the size of the secondary mains between the transformers was such that the drop in these mains was small compared to the drop in the transformers themselves; in this way the transformers were made to share, more or less, the load equally between them. When a secondary system of distribution was not economical, single transformers were located. In determining whether a customer was to be included in a bunch of customers, all of whom were to be fed from one transformer, or whether it was more economical to place a separate transformer, it was necessary to make an approximate estimate of the cost of locating the transformer for each case. When the interest on the cost of placing a separate transformer plus the cost of maintenance of the transformer, was more than the interest on the cost of connecting a customer to a transformer, feeding other customers, the connection in question was made to the transformer feeding the "bunch." However, even if the difference in annual cost was small in favor of a separate transformer, connection was made to the "bunch." In making these calculations a fixed drop in the secondary mains was allowed, and the load, i.e., the probable number of lamps burning at any one time, for calculating this drop, was determined from the records on the transformer charts; of course the character of the service goes a great way in making this last determination. A separate transformer was placed only when the total annual cost for the placing and maintenance of such transformer did not exceed the sum of the two following costs—the interest on the cost of placing and maintenance of wire necessary to connect the customer to the nearest "bunch" transformer, and the increased cost due to necessary increase in size of transformer. The annual cost of a transformer on the lines was considered to include the cost of the iron losses, figured as costing the electrical lighting station at an assumed rate of one-tenth (.1 cents) per lamp hour of 55 watts, a 5% interest on the cost of the transformer, and the high rate of charge of 10 % depreciation. To facilitate the bunching and loading of transformers in conjunction with other data, the following table :

| Capacity in lamps. | Cost of transformers | Int. on cost at 5% | Depreciation 10% | Leakage. | | Cost of leakage at .1 per lamp hour, 55 watts. | T.I. ann. cost of transformer, int. on lines, sq.hrs. ser. |
|---|---|---|---|---|---|---|---|
| | | | | Watts | Watt hours per year. | | |
| 10 | 18.00 | .90 | 1.80 | 29 | 254000 | 4.62 | 7 32 |
| 13 | 22 00 | 1 10 | 2.20 | 30 | 262100 | 4.78 | 8 08 |
| 20 | 26.00 | 1.30 | 2 60 | 32 | 280000 | 5.10 | 9 00 |
| 30 | 32 00 | 1.60 | 3.20 | 35 | 306400 | 5.18 | 10 38 |
| 40 | 39 00 | 1.95 | 3 90 | 37 | 324000 | 5.88 | 11.73 |
| 50 | 47.00 | 2.35 | 4 70 | 51 | 446000 | 8.10 | 13 75 |
| 75 | 68.00 | 3 40 | 6.80 | 65 | 560000 | 10 33 | 20 53 |
| 100 | 80.00 | 4 00 | 8.00 | 77 | 674500 | 12.15 | 24 25 |
| 150 | 112.50 | 5.62 | 11 25 | 85 | 744000 | 13 50 | 30 38 |
| 200 | 150.00 | 7 50 | 15 00 | 125 | 1093000 | 19.90 | 41.40 |
| 300 | 225 00 | 11 25 | 22 50 | 150 | 1312002 | 23 85 | 57 60 |
| 400 | 300 00 | 15 00 | 30.00 | 160 | 1400000 | 25.00 | 70 00 |
| 500 | 375.00 | 18 75 | -37.50 | 170 | 1488000 | 27 05 | 83 30 |

The principal factor in determining the size of a transformer is the character of the service, a more liberal allowance being made for an overload in a residence than could be made in a commercial district. A good transformer should stand an overload for several hours of at least 25%, and for a shorter period of 50%, or even more.

What has been stated above in regard to the placing and determining the size of transformers assumes of one make of transformers being considered. The subject of selecting a type or make of transformer has been freely discussed elsewhere. It has been shown conclusively from the experience of electric lighting stations everywhere that first cost is no longer the principal factor to be considered in determining what transformer should be chosen to allow of economical operating of a lighting station. Obtain a guaranteed transformer leakage from various manufacturers, substitute these guaranteed leakages for the different sizes of transformers in a table, similar to the one given above, and a comparative statement of the annual cost of different makes of transformers can readily be made from which the most economical type of transformer for the local conditions can be determined. In connection with the above, transformer construction, regulation and efficiency should be considered. It is generally a fact that transformers of small leakage currents have the highest "all day efficiency." With the guaranteed leakage currents there should also be a guaranteed regulation not exceeding 2½ or 3%, which is obtainable, for the smaller sizes. The mechanical construction of the transformer is as important as either its efficiency or regulation ; the safety of the customers' premises depending upon the protection of the secondary coil from coming in contact with the primary coil, it is necessary that the method of insulating the coils from each other be reliable and absolutely safe. There can be no question that it pays to replace inefficient by efficient transformers. There is but one course to follow if the station is loaded with unnecessary transformer leakage, and that is to replace, at the earliest possible moment, the old transformers by new transformers, using secondary systems where economical, and by this replacement cut down operating expenses and increase the station capacity.

At the beginning of the reconstruction herein referred to there were 1,160 transformers on the lines with approximately 53,000 lamps wired. At the time the following results were collated, there had been 473 old transformers removed from the lines, while 229 new transformers had been put up, leaving a total of 916 transformers on the lines serving about 60,000 lamps. Of these 229 new transformers, 187 had replaced 345 old transformers, while 42 had been used for new customers. Of the 128 old transformers yet to be accounted for, 18 had been removed on account of discontinuation of service, and 110 had been taken from the lines on account of being able to connect the customers served from them to old transformers already erected in their immediate vicinity. This "bunching" of customers on to old transformers was made because new transformers could

not be obtained at the time and it was necessary to reduce the leakage load before winter.

The lightest load registered during the year preceding the commencement of the reconstruction was 380 amperes. Ten months later, with about 8,000 more lamps wired on the service than at the time of the 380 ampere load, above referred to, the lowest load recorded was 245 amperes, or a decrease in the load line of 135 amperes, this decrease in leakage load being due to the transformer changes just mentioned. The leakage of the 229 new transformers was 19 amperes, which means that the 473 old transformers had a leakage of 154 amperes, or an average leakage of .325 amperes per transformer removed, which figure has been verified by leakage tests made on the old transformers which have been removed from the lines. Thirty six of the 135 amperes reduction was due to the removal of the 110 old transformers, and placing the customers served from these on other old transformers, making secondary distribution systems. From this is deduced the fact that by replacing the 345 old by 187 new transformers, a saving was effected of 99 amperes. The average saving for the 187 changed is then .529 amperes per change, which with coal at $2.75 per ton, means an annual saving of $25.58 per change in coal alone. The average cost of the 187 changes, including the cost of new transformers, all extensions of wiring for secondary mains and all labor, crediting these orders with old transformers as scrap only, was approximately $65.00. As stated above an annual saving per change in cost of coal would be effected of $25.58, therefore at this rate the new transformers will pay for themselves, if the saving in coal only is considered, in about two and a half years. .

When the 1,160 transformers above referred to have been replaced by new transformers, and the bunching of customers has been carried out, it is estimated there will be but 636 transformers required, and the total leakage will be less than 75 amperes.

The following curves show three (3) actual station loads ; an average load for eight months from June 1st until February 1st, and for the July and December of these eight months, also the estimated loads for the same periods had the reduction in transformer leakage so far actually obtained been accomplished.

Plate 1, Curve A, represents the average load on the

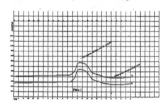

station during the twenty-four hours for seven days, beginning July 19th and ending July 25th, the highest point reached being 760 amperes, and the lowest 380 amperes.

. Curve B represents the average load that would have been on the station for the same period had the 917 transformers been on the lines instead of the 1,160, and the saving of 135 amperes been accomplished. As this paper is only dealing with actual results obtained, the curve showing the estimated load on the station for the same period, had this reconstruction been complete, will not be plotted, but it is not hard to imagine what it would have been judging from the results,

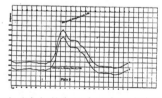

so far obtained, with the transformer changes barely one-third finished.

Plate 2, Curve A, represents the average station load during the twenty-four hours for the seven days, beginning

December 18th and ending December 24th, the highest point reached being 1280 amperes, and the lowest 380, this being the heaviest week of the year.

Curve B represents the estimated station load for the same period, had there been the 917 transformers instead of the 1,160 on the lines. While the third curve showing this estimated load had the transformer changes been complete, has not been plotted, yet it is safe to assume there would have been a difference in the maximum load reached of 300 amperes, so that it would have been necessary to provide station capacity for a maximum load of 980 primary ampere instead of 1280 primary amperes.

Plate 3, Curve A, represents the average primary ampere.

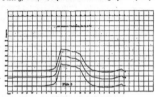

load on the station for the eight months from June 1st to February 1st.

Curve B represents the estimated load during the same period had the reduction in load so far obtained—135 amperes —been accomplished.

Curve C represents the estimated load during the same period, assuming a reduction in primary load of 300 amperes.

This reduction in station load with an increased number of lamps wired, due to the decrease of transformer leakage, can be regarded, first, as a saving in coal and operating expenses in general and, secondly, as either an increase in station capacity already installed, which means an increase in the earning capacity of the plant, or a decrease in capacity necessary to be installed to handle the output at time of maximum load.

A decrease in transformer leakage of 135 amperes means a decrease in load of 135 amperes for every hour of operation, which represents a saving in coal, at $2.75 per ton, of about $7,348.00 per year, that is for a station running twenty-four hours per day. Apart from this increase in capacity, there is also the saving due to running a smaller engine for the day load, and the consequent saving in labor, oil, etc.

A decrease of 135 amperes leakage means an increase in earning capacity of the station of approximately 270016 CP lamps burning, or about 9,000 lights wired. As so much of the advantage to be gained by this decrease in transformer leakage depends upon the kind of transformer used, it would only seem safe and wise to insist on all transformers coming within guaranteed limits for leakage and regulation. The only way to know that transformers come within the prescribed limits is to get them from manufacturers who are known to build the very best transformer, or better still, test each transformer as it is received from the factory. Inasmuch as transformers made by different manufacturers, and apparently alike in every respect and seemingly identical in construction, are known to vary from twenty-five per cent. to a hundred per cent. from each other, the advisable plan appears to be to test each transformer as it is received. This plan of testing transformers is followed out in many stations and is the only sure means of keeping the leakage within the calculated limits.

The reduction in leakage load so far obtained in the reconstruction under consideration has not been accompanied by any sacrifice of transformer regulation. The type of new transformer used is one giving the best all round results, that is, one in which regulation and leakage are so proportioned in its construction as not to benefit one at the expense of the other. In thickly populated or central business portions of the city, where an extensive secondary distribution is possible, and where large transformers may be connected in parallel at different points, it would be an advantage to use transformers of very small leakage current and high "all day efficiency," as in this case the transformers share the load between them, and regulation can be sacrificed to gain diminished leakage current. However, as it is only in very large cities, and only in the most thickly populated centres of these that the secondary distribution system can be economically used, the make of transformer giving the best all round results should, in general, be selected. To further improve the regulation beyond that to be obtained by improved transformer regulation it is intended to change the primary distribution from 1000 to 2000 volts, thereby decreasing the copper losses on the existing circuits to one quarter of the present losses, and reducing the feeder drops so that good service and regulation will be obtained without the use of feeder regulators or the erection of additional copper. A source of additional improvement in regulation will be the use of generators with very close regulation. The necessity of transferring the circuits from one dynamo to another makes close inherent regulation in generators an imperative feature if satisfactory service be desired. Transformers with good regulation, feeders having small drops, and generators of close regulation, mean that the ordinary changes of load and transfers of circuits from one generator to another can be made without materially affecting the voltage on the lamps in service. The generators selected for the reconstruction herein spoken of to replace the present single phase generators are of the two phase type and are of such construction mechanically and electrically as to make practically impossible the hairbreadth escapes and the sleepless nights familiar to many operating old style apparatus. The sense of security which takes possession of one after becoming familiar with the type of machine selected can only be appreciated after actual experience in operating. When the reconstruction under consideration has been completed there will have been installed five 300 KW generators, two on one engine, two on a second engine, and one on a third engine. The two generators running from the same engine will be run in parallel when the load requires it, making the units on two of the engines 600 KW, with the advantage of having a more flexible arrangement and a possible saving due to running a 300 KW when a 600 KW would be but partially loaded. The construction and location of the engines was such as to make it impracticable to put 600 KW generators on the two large engines, had it been so desired.

Probably the most economical and certainly the most convenient unit of power for operation is one that has the capacity to carry the day load, the remainder of the dynamos being of a uniform type and size. In the case in question the day load, if the service be confined to lighting would, as stated above, be less than 100 KW, but the adoption of the two phase system will permit of the increase of the day load by the sale of current for motor service, to at least the capacity of the small engine. The small engine, to which is connected the single 300 KW generator, will be run while the day load remains within the limit of capacity of this engine and generator. Should this engine or generator require repairs and it be necessary to run one of the large engines during the day, only one of the 300 KW generators need be excited to carry the day load, introducing a saving not possible if a 600 KW had been placed on this engine, thus giving a higher "all day efficiency" for the two 300 KW generators than could be obtained from one 600 KW generator. The 300 KW units, run either singly or in parallel, are sufficiently large to allow of any desirable arrangement of circuits of ordinary size.

Another factor in increasing the capacity and earning power of a plant is the use of an efficient lamp with an economic

life. The best known makes of lamps on the market have a difference in efficiency of from ten per cent. to fifteen per cent., which means a difference of from ten to fifteen per cent. in the earning power of the plant, depending on the efficiency of the lamp in use. A ten per cent. difference in output or capacity should receive consideration.

Lamps with long life are found to be inefficient ; very efficient lamps are usually short lived. There is a point between these extremes which makes a lamp suitable for electric lighting station use. Using an efficient lamp increases the earning capacity of a plant and permits of using higher candle power lamps with a proportionally less increase in cost. An increase in candle power either by high candle power incandescent lamps of high efficiency or small incandescent arc lamps seems to be the best way to meet competition from gas either with the ordinary or the "Auer" burner.

It has been advocated by some, to meet this demand for more light, to running the lamps at a voltage above that for which they are rated, thereby running the lamps at a high efficiency. An examination of the following two Plates, 4 and 5, will show the fallacy of such a makeshift.

The curve on Plate 4 is the result of many lamp tests, and shows the variation in lamp life for lamps of the same grade when run at different voltages.

Plate 5 shows a curve for the same make of lamp of the

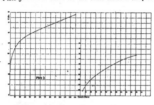

same efficiency, giving the candle power of a lamp at various voltages. As 600 hours is probably the average life of the lamps now on the market, the effect of running this grade of lamp which was rated as a 600 hour lamp, above its rated voltage may be regarded as the effect on the average lamp now offered to the public.

From these curves it will be seen that running a 50 volt lamp at 52 volts, or increasing the voltage four per cent. increases the the candle power about nineteen per cent., while the life of the lamp is decreased about forty-three per cent. Running the lamps at a pressure of 55 volts, or a ten per cent. increase of voltage increases the candle power of the lamp about sixty-six per cent., while the life of the lamp is decreased about eighty-three per cent., from which it would seem that to a plant supplying current to a large number of incandescent lamps and furnishing renewals, running them above the rated voltage means a large increase in the lamp renewal account, both for material and labor. Run the lamps as near their rated voltage as possible, and the lamp renewal account will be a minimum. Good regulation on the circuits goes a long way towards keeping this account down. A daily rise in voltage from three to

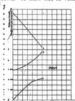

four per cent. above normal for a short time will reduce the life of a lamp of good economy about one half.

Plates 6 and 7, with their tables, give the results of tests on

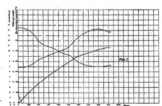

two very well known makes of lamps, when the lamps are run at a voltage ten per cent. higher than their rated voltage.

Curves 8 and 9, with their accompanying tables, give the results of life tests on the same makes of lamps when running at their rated voltages. Plates 6 and 8 are for the same make of lamp. Plates 7 and 9 are for the same make of lamp, but of a different make from the lamp, the curves for which are given on

Plates 6 and 8. The results of these tests are of especial value to electric lighting stations from the fact that throughout the test the conditions under which the lamps were run were made to conform to the conditions imposed upon lamps in commercial service. It has been determined that lamps which have been made by identically the same process differ in lots. It has been

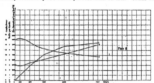

observed that lamps received from the same factory do not average the same candle power and efficiency for different invoices, that is, lamps received in one invoice are generally quite uniform throughout that lot, but they vary considerably from lamps received at other times. From this it will appear that to derive

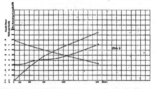

full benefit from using efficient lamps it is necessary to test the lamps and ascertain that they come within the limits of efficiency which have been decided to be the most economical for the local conditions. To determine what lamp is best suited for any electric lighting station, it is necessary to know the cost of producing current per lamp hour, and having established this for any special make of lamp, the following formula will permit of a comparison of different makes of lamps and the determination of the best lamp for the conditions under which they are to run. In considering the cost of production per lamp·hour in connection with the lamp question, the cost of service may be divided into three parts :

A. That portion of the service per lamp hour that is practically not affected by the average efficiency and life of the lamps and such portion of the maintenance, operating and general expenses, as is practically not increased by increasing the current consumption per lamp hour.

B. The cost per lamp hour, coal, water, interest and depreciation on the lines, dynamos, engines, etc., and such part of the expense of the service as increases proportionately to the amount of current served per lamp hour and as the maximum station output.

C. The cost of the lamp per lamp hour, and the expenses per lamp hour for replacing exhausted lamps. This is equal to the cost of one lamp, plus the cost of exchanging one exhausted lamp, divided by the average life of the lamp.

Under the first division (A) should be included the cost of fuses, meters, transformers erected, and secondary connections, line construction, maintenance, etc., and such proportion of the operating and general expenses as is not increased by increasing the current consumption per lamp hour.

Under (B) should be included that portion of the cost of service per lamp hour exclusive of lamp renewals that increases proportionately to the current consumed per lamp hour.

These divisions of cost should be so made that the sum of A, B and C, will represent the total cost of service per lamp hour, the values of A, B and C representing the above divisions of cost having once been established for a lamp of any given efficiency and average life for any particular lighting station the cost of service per lamp hour for this same station with any other lamp which has a current consumption different from the current consumption of the first lamp, and having an average life of "Y" hours, would be $A + XB + C$· = the cost of service per lamp hour, "X" representing the proportion between the current consumption of the lamps being compared, and "C·" being the cost of one of the new lamps, plus the cost of replacing one exhausted lamp, divided by "Y," the average hours of life of the new lamp.

This formula applies for comparing the cost of producing light with lamps having different costs, efficiency, and average lamp life, when they are to be burned in the same plant and under the same conditions of average lamp hours burned per lamp installed, and the same maximum number of lamps burning for a given number of lamps wired. Value (B) in this formula includes the coal consumption and the materials which practically vary proportionately to the watt hours output required for providing the light. It also includes the interest and depreciation on the plant which must be enlarged when the lamps consume large amounts of current, because the generating and supplying capacity of the plant must be proportionate to the maximum output called for by the lamps. In many plants the interest and depreciation account will form quite a considerable portion of the factor B, and as a large value to the factor "B" makes a show-ing against the high consumption of current per candle power hour very bad, it would appear that any lamps installed that did not burn at the time of maximum current output from the station could be economically used of a poorer efficiency with longer life than lamps which do burn at time of maximum out-put, because any additional demand for current on a plant that is not a call for current at the time of maximum output, does not require an increase of plant capacity. In estimating the best efficiency per candle power hour, or per lamp hour, for these lamps that do not burn at the time of maximum output, the cost of interest and depreciation entering into the factor "B" in the formula (in fact all the costs that increase proportionately as the size of the plant required to serve the lights wired) should be excluded from the factor "B." The result is that lamps that do not burn at the time of maximum output can be economically used of considerably lower efficiency than lamps that do burn at that time.

The outline of the reconstruction contained in this paper and the statement of the results so far obtained are for an electric lighting station serving 60,000 incandescent lamps. Another much smaller electric lighting station has had its transformer system rearranged, within the past year, upon the same plans outlined for the station serving 60,000 lamps. This smaller station had, and still has, a capacity of two 500 light dynamos, serving 2,100 lights wired. At the time of heavy load, the station was loaded beyond a safe limit. Apart from this the demand for an increase in the number of lights wired could not be met. An increase in the boiler, engine and dynamo capacity appeared, to some, the only way to meet the requirements ; however, this was unnecessary as the transformer system was rearranged, and thereby ample capacity to meet the immediate demands was furnished. Prior to this rearrangement there were 79 transformers on the lines, having a leakage of about twenty amperes. The 79 old were replaced by 42 modern transformers, having a leakage of less than four amperes. By this rearrange-ment and the substitution of modern for the old transformers, a reduction in load was obtained of sixteen amperes which per-mits of the station not only carrying safely, with the same station equipment, what was formerly an overload, but also permits of an increase in the earning power of the plant of approximately one thousand 16 candle power lamps.

### TABLE C.

| Hours Burned. | No. of Lamps Broken. | Average Current per Lamp. | Average Candle Power per Lamp. | Per Cent of Candle Power. | Watts per Candle. | Total Lamp Hrs. Burned. | Lamp Hrs. Burned per Lamp Installed. | Average Voltage on Lamps. |
|---|---|---|---|---|---|---|---|---|
| 0 | 0 | 1.04 | 16.1 | 100. | 5.21 | 0 | 0 | 50.94 |
| 100 | 0 | .99 | 16.6 | 103. | 2.93 | 9000 | 100 | 51.3 |
| 300 | 3 | 1.048 | 15 | 93.2 | 3.40 | 5767 | 288.4 | 52.2 |
| 800 | 6 | 1.05 | 12.6 | 78.3 | 4.17 | 10374 | 518.7 | 52. |
| 1000 | 16 | 1.14 | 11 | 68.4 | 5.18 | 13331.8 | 666.6 | 52. |
| 1700 | 18 | 1.31 | 9.5 | 59.1 | 7.05 | 1484.8 | 742.8 | 5.8 |

### TABLE D.

| | | | | | | | | |
|---|---|---|---|---|---|---|---|---|
| 0 | 0 | 1.03 | 17.3 | 100. | 7.97 | 0 | 0 | 56.4 |
| 100 | 0 | 1.00 | 15.18 | 88. | 3.3 | 1000 | 100 | 56.2 |
| 200 | 4 | 1.00 | 12.8 | 74. | 3.9 | 2858 | 283.8 | 55.0 |
| 300 | 8 | 1.00 | 10.4 | 60.2 | 4.8 | 2157 | 215.7 | 55.9 |

### TABLE J.

| | | | | | | | | |
|---|---|---|---|---|---|---|---|---|
| 0 | 0 | 1.01 | 15.9 | 100 | 3.16 | 0 | 0 | 50.94 |
| 100 | 1 | .95 | 15.2 | 95.6 | 3.5 | 1934.5 | 97.8 | 50.94 |
| 300 | 4 | 1.013 | 13.9 | 87.5 | 3.65 | 3369.5 | 668.5 | 51.3 |
| 600 | 8 | 1.00 | 12.2 | 77. | 4.18 | 9583.5 | 479.0 | 52.2 |
| 1000 | 12 | .95 | 10.1 | 63.5 | 4.75 | 13381.7 | 669.9 | 52. |
| 1700 | 14 | .97 | 6.8 | 42.8 | 7.13 | 28090.8 | 904.6 | 51.8 |

### TABLE K.

| | | | | | | | | |
|---|---|---|---|---|---|---|---|---|
| 0 | 0 | 1.00 | 14.5 | 100 | 3.46 | 0 | 100 | 56.4 |
| 100 | 0 | 1.00 | 14.04 | 97. | 3.56 | 1000 | 189.6 | 56.2 |
| 200 | 2 | 1.00 | 13. | 83. | 4.15 | 1896 | 263. | 55.2 |
| 350 | 3 | 1.00 | 10.9 | 75.2 | 4.58 | 2690 | 333. | 55 |
| 400 | 3 | 1.00 | 9.97 | 68.8 | 5.07 | 3330. | 395.4 | 56. |
| 500 | 4 | 1.00 | 9.3 | 64.2 | 5.37 | 3454. | 459.5 | 56. |
| 600 | 5 | 1.00 | 7.9 | 54.5 | 6.33 | 4585. | 515.7 | 55 |
| 800 | 6 | 1.00 | 7. | 48.3 | 7.14 | 5157. | 535.7 | 56.2 |
| 900 | 8 | 1.00 | 7.3 | 50.4 | 6.85 | 5357. | 535.7 | 56.5 |

Location No.........................St......................Ave.
And .......... ..... ...... .... St. ..... .... Ave    N. E.
Opp                                                  S. W. Corner.
Height of Pole ..... . ...... ... .... . .......... ....
Kind of Pole ..... : . . ..... ; ... ..... ....... ...For Line Lamp.
                                                      City
Condition of Pole ...............

1st Arm . Pins ...Wires
and " .... " ... "
3rd " .... " .... "
4th " .... " .... "
5th " .... " .... "
6th " .... " .... "
7th " .... " .... "
8th " .... " .... "
9th " .... " .... "
10th " .... " .... "

Transformers...............................
Arc Lamps. ....65 c.p .. 32 c p .....
Lightning Arresters . . . ...............
Remarks ..... ... ... ......... ...
.....................................
.....................................

Located by . . ................ ... .

... . 189

## THE OUTLOOK FOR THE ELECTRIC RAILWAY.

BY F. C. ARMSTRONG.

IT is a significant evidence of the confident spirit with which we have learned to regard the sure and rapid progress of modern electrical inven-tion that we accept to-day without comment and as an established prac-tice what was but yesterday a matter of tentative and doubtful experi-ment. This rapidity of achievement has characterized the development of the electric railway, in common with the other great departments of electrical industry, and has already been productive of results of which we can scarcely as yet appreciate the economic and social importance.

Up to within the past year, however, the application of electric motive power for railway purposes has been practically limited to the improvement, amounting to a revolution, of the street railway proper, and an extension of its field as the suburban railway. The work in this direction, though difficult in detail, is necessarily limited in range, and at the present moment may be said to have reached a stage approaching finality. The street railway motor of to-day may be considered, in view of the conditions under which it operates—limited space, exposed position, light weight and severe service, as a highly efficient and satis-factory machine. The controlling apparatus has been developed to an equally high degree of perfection, ensuring in the best types a maximum economy of current, and reduction of strain on the motors under varying conditions of operation, and even adding to its normal function the duties of an electric brake. In the power house, the substitution for the small belt-driven generator, of the large, compact, slow-speed direct-connected unit, with its steel frame and iron-clad armature, leaves little room for improvement in the way of higher efficiency, closer regulation or greater durability. Improvements in design and material have done much to remedy the unsightliness and unreliability of the devices used in overhead construction and the standard pressure of 500 to 600 volts is found, even for suburban extensions of considerable length, to be com-mensurate with a reasonable copper economy. From a financial point of view the position of the electric street railway is equally assured and satisfactory. No field for legitimate investment is now more favorably considered than that offered by the securities of a well-managed and well-equipped electric railway in a city or town of any size suitable to its capitalization. As evidence of the financial importance to which the electric street railway interests in Canada have attained, may be cited the fact that there are at present in operation, or being constructed in the Dominion, 36 electric street railways, having a total mileage of close upon 600 miles, using 750 motor cars, with a total generating capacity of 19,500 kilowatts, and representing an actual investment in round figures of over twenty millions of dollars.

At this point, and at a meeting held in the city of Toronto, it is peculiarly fitting by way of contrast and as epitomizing the development of less than one decade, to quote from a catalogue issued nine years ago, in 1887, bearing the title "The Van Depoele System of Electric Railways," in which under the heading "Facts about running the Toronto Electric Railway in 1885," we find the following:

"Plant consisted of one engine, automatic, 10 x 16 cylinder, 150 revolutions per minute; one electric generator, forty horse-power; one electric motor, thirty five horse-power; one motor car, weight six tons; three passenger cars, each two tons. Average number of passengers carried, eighty-three per car; estimated weight of passengers per train, 16 tons; total weight of train, 11 tons; length of track, one mile (with one grade of six per cent.); average speed, 30 miles per hour; passengers carried in 5 days, 50,000; average consumption of coal per day of ten hours, 1200 lbs.; distance travelled in ten hours, including stopping to take on passengers, 200 miles."

The generator in the case, it may be added, was a 40-light arc machine, having, it is stated, "an electromotive force of 1300 volts, and an intensity of current of about 18 amperes," and the single motor, belted to the axle, was a 35-light machine of similar type. In the same catalogue we find a description of each of the Van Depoele roads in operation at the date of its issue. The list is a short one—Montgomery, Alabama, 1½ miles; Detroit, Mich., 1¾ miles; Windsor, Ont., 2 miles; Appleton, Wisconsin, 4½ miles; Port Huron, Michigan, 3 miles, and Scranton, Pennsylvania, 2 miles; a total of 14¾ miles. It is amusing to note following this modest list of roads installed, the bold challenge that "As the matter now stands we have more miles of electric railway now in successful operation than all the other electric railways in the world combined."

Coming now to a consideration of the subject of this paper, it is not unreasonable to augur from the success of the electric railway in the past, an outlook for the future equally brilliant and promising. We may leave out of consideration the work which still remains to be done in affording rapid transit for the cities and towns which are as yet either working without street railways altogether, or in which the existing systems are still operated as horse or cable roads. The horse as a propulsive agent for the street car, is steadily pursuing his course to his destined place in the museum, while the cable, in spite of the tremendous inertia of invested capital, is, except in the most congested portions of the larger cities, rapidly giving way before the greater economy of electrical operation. The recent electrical equipment of the extensive Pittsburg cable systems, involving the abandonment of an investment of many millions of dollars, may be instanced in this connection.

The field for future development in electric traction lies in two distinct directions: in the first place, in the equipment and operation of that recent but now most important factor in transportation—the light or secondary railway, which will in time take form as a network of feeders and channels of distribution for the large centres of population and the great trunk railways; in the second place as the successor of the steam locomotive in the operation of the trunk systems themselves.

It is in the first direction in which already some development has taken place that we may expect the most substantial immediate progress. The possibilities of the light railway have of late been the subject of anxious and careful scrutiny on the part of political economists in England and on the continent generally, as a possible relief for the present acute and world-wide agricultural depression. Without going into the social or economic phases of the question it seems undoubted that from all the large centres of population and production we may expect to see systems of light railway lines radiating to the limits of their spheres of commercial influence and affording at a minimum of cost an adequate means for transportation and interchange of the products of the farm on the one hand, and of the factory on the other.

For such a system requiring a frequent and flexible but not a heavy or high speed service no enormous investment of capital would be required. The use of the public highway would save the otherwise heavy outlay for right of way, and its grade could, for the most part, be conformed to. The track and roadbed, even with rails heavy enough for standard freight cars, can, it has been shown, be laid for little more than the cost per mile of a first-class macadamized roadway. The depreciation charges, under normal conditions, would be certainly no greater, and the cost of equipment and operation with electric power, even with the transmission limit of our five hundred volt direct current apparatus, such as to render practicable the working of such systems over a considerable range. We have in Canada several examples of this class of railway, as yet on a limited scale, but in each case affording facilities for transportation, both of passengers and light freight, recognized as being of the utmost value to the public. Each of these roads are, it is encouraging to note, yielding a fair return for the money invested. In the same way the branch lines and feeders of the trunk railways, which are now operated in many cases at a loss, mainly by reason of the inadequate service to which they are limited by the use of the steam locomotive, would, if electrically equipped for a light and frequent service, become a productive part of the system to which they stand at present in the relation of a necessary evil.

It seems, therefore, reasonably clear that in the development of the system of secondary railways which are coming into being as the result of a pressing economic necessity, the electric motor is to find a new and widely extended field of usefulness. The great desideratum at present for this work is a successful alternating railway motor which, it is safe to anticipate, will be added to the list of standard equipment in the very near future. Under present conditions, while the use of the booster or of polyphase transmission apparatus with rotary transformers has made commercially possible the supply of current for distances up to twenty miles, or even more, from the power house, their availability has been lessened by the drawback of excessive loss in the one case and of great cost in the other.

Before leaving this part of the subject, however, it would be as well to point out, in view of the alacrity with which the possibilities which we have been discussing are being taken up as a new and promising field for the exercise of their peculiar abilities by the versatile and talented class of gentlemen known as promoters, that there is no reason to suppose that such a wholesale programme of light railway construction and conversion of existing steam branches would be an immediately profitable or possible undertaking. In many cases the gains made will be in the form of a general public benefit rather than a concrete return in dividends for the money invested. The smaller and more profitable openings for the construction of these lines will afford a field for private enterprise, but any comprehensive scheme will undoubtedly demand in the form of governmental aid, the support of the public, who will be its main beneficiaries.

We may now consider briefly the position likely to be attained by the electric motor as a successor to the steam locomotive in the operation of the great trunk lines. Here the conditions differ materially from those which have led in so short a time to a practically complete possession of the field of street railway traction, and which seem likely to produce similar results in the case of the secondary railways. It must be conceded that no opening or necessity exists for the construction of new trunk lines operated electrically in competition with existing steam roads. The eventual triumph of electricity over steam, for heavy locomotive purposes will come in due course as a result of the establishment of its superiority for the service, but its general adoption will be delayed beyond that point by a natural reluctance to wipe out the capital represented by existing equipment. It must be recognized that the evolution which attends all branches of mechanical development has produced in the steam locomotive of to-day a type admirably adapted to the work which it has so far been called on to perform. It is in the continual demand on the part of the public, for higher and higher speeds between terminal points, and the still more imperative necessity in the face of keen competition and lowering rates for a reduction of operating expenses to the minimum point, that we may expect to find ultimately the most favorable contributing cause for the general adoption of electric motive power on the trunk systems. The direct rotary action of the electric motor and the practical limitation of its power only by the capacity of the stationary source of supply entail the possibility of an increase in rates of speed up to the highest point at which a perfectly constructed roadbed without grades and curves will hold a car on the track. A recent study of the operation of the Pennsylvania Railway would seem to show that such savings in fuel, labor and maintenance accounts would follow its re-equipment for electric traction as to make it commercially desirable, even under present conditions.

It is no extravagant prediction to say that members of this Association who witnessed, in 1885 and '86 at the Toronto Exhibition, the modest beginnings of electric traction in Canada will see it supersede the steam locomotive in the operation of the Canadian Pacific and Grand Trunk Railway systems.

## SPARKS.

The employees of the Toronto Street Railway Company held their annual pic-nic a fortnight ago.

Messrs. Folger Bros., of Kingston, Ont., will probably make a bid for the Watertown electric railway, which is now in the hands of a receiver.

The Ætna Boiler Company, of Toronto, Ont., is applying for a charter of incorporation. The capital stock is $20,000, the objects of the company being to manufacture the Ætna safety water tube boiler.

The Hamilton Radial Railway Company have closed a contract with an American firm to fit a number of cars with a patent air-brake that will stop a car going at full speed in one car length without injuring the mechanism.

The Canadian Telephone Company, with a capital stock of $10,000, and headquarters at Sawyerville, Que., is applying to be incorporated for the construction and working of a telephone system in the county of Compton and other counties in the Province of Quebec.

The large fly-wheel of the engine in the H., G. & B. power-house at Stoney Creek burst recently, a piece of the wheel going through the roof, and another portion smashing through the floor. Besides the damage to the wheel and the building, the switch-board was injured. Power was obtained from the Street Railway Company until the H., G. & B. Company got another engine running. About $3,500 damage was done.

Compared with other large towns, London is easily at the head for the magnitude of its electrical supply. Paris, for instance, has only an equivalent of about 500,000 eight-candle power lamps, as compared with the 1,200,000 lamps in London, as stated above. Manchester and Liverpool have, respectively, about 92,000 and 54,000; Glasgow, 70,000; Edinburgh, 43,000; Dublin, 16,000, and Cardiff, 9,000. Of the total capital expended in the whole of the United Kingdom for supplying electricity, London has spent more than one-half.

The new power house of the Trenton Electric Company at Trenton, Ont., was formally opened a fortnight ago. The building is a large and substantial one, erected on the Trent river a short distance east of the G.T.R. station. The roof and sides are covered with iron sheeting. The water wheels consist of two 150 h.p. vertical turbines, manufactured by the Wm. Hamilton Co., of Peterboro. They are geared directly to a large 6 in. line shaft, which runs the whole length of the building. A convenient arrangement is made by providing each of these wheels with a friction clutch, whereby each may be used independently of the other. The two large generators are 200 h.p. each, bolted direct to two pulleys on the line shaft 8 ft. in diameter. The switch-board, 6 x 7½ ft. in diameter, consists of three polished panels of white marble, on which are mounted the various instruments, switches, regulators, etc., necessary for the controlling of such a large plant.

## SPARKS.

W. H. Train, of Burk's Falls, has recently ordered a 500 light increase to his direct current incandescent plant from the Canadian General Electric Co.

The capital stock of the Lachine Rapids Hydraulic Co. has been increased to $2,000,000. The new stock has all been taken up by the present shareholders of the company.

Thos. Andrews, of Thornbury, is installing a five hundred light alternating plant to furnish incandescent lighting in Thornbury and also in Clarksburg, distant about one mile. The Canadian General Electric Co. are supplying the apparatus.

The Trenton Electric Power Company has been given the contract for lighting the city. The franchise extends for twenty years, and gives them the right to erect poles on the city streets for the purpose of supplying light, power and heat.

Jas. Playfair, of Midland, has closed a contract with the Canadian General Electric Co. for a plant to be installed on the steam barge "Hall." Both arc and incandescent lights will be used; the former for lighting the docks at which the steamer is loading.

D. Knechtel, of Hanover, is extending his monocyclic circuit to Carlsruhe, a distance of three and a half miles, and to Neustadt, a distance of 7½ miles from the power house. This extension is an interesting evidence of the range over which current may be profitably distributed from a modern alternating system.

The report of the General Electric Company of New York for the fiscal year ending January 31st last, shows the business secured to have been less than 10 per cent. greater in value of sales than for the year previous. The gross earnings were $13,515,667 and the gross expenditure $11,910,240. The deficit increased $877,645. The net loss of liquidation, now charged to the $2,000,000 special allowance after January 31, 1895, was $530,152. The company has no notes payable outstanding, nor is any paper bearing the company's endorsement under discount. The report recites the fact of the contract that has been concluded with the Westinghouse Electric and Manufacturing Company. The foreign business of the company has shown a gratifying increase.

The Niagara Falls Electric Light & Power Co. are making extensive additions and changes in their plant, involving an expenditure of over $25,000. A handsome new power house of pressed brick is in course of erection in a central locality where an ample supply of water can be obtained for condensing. The steam plant will consist of two 200 horse-power compound condensing Wheelock engines. For the incandescent service an order was given to the Canadian General Electric Co. for two 120 kilowatt single-phase alternators. In case a demand for power arises, it is intended to install a 500 volt direct current power generator. The switchboard is to be of white marble, and the instruments and their arrangements are of the most modern design. A system of three wire secondary mains is being installed for distribution through the central part of the town. The work is being carried on under the supervision of Mr. Geo. Foster, the superintendent of the company.

## SPARKS.

The Imperial Electric Light Company, Montreal, has been incorporated.

The city council of Toronto is advertising for tenders for the privilege of operating a telephone system for a term of five years.

The Hamilton Radial Electric Railway Company expects to have the line to the Beach in operation by July 15. The power house is ready for the dynamos.

It is understood that the authorities of Montreal have under consideration a scheme for the conveyance of prisoners to and from the jail by means of a trolley car.

The annual meeting of the Nova Scotia Telephone Company was held in Halifax early in June. Gratifying reports were presented by the manager and directors. The board of directors were re-elected by unanimous vote.

The imperial government is building a powerful electric light plant at Fort Clarence, Halifax, entirely concealed from view and protected by earth and masonry. The object of the installation is to obtain a revolving search light of great power to control entrances by eastern passage and Drake's passage, also all the western entrances north of York Redoubt.

The annual meeting of the Kingston Light, Heat and Power Company was held on the 3rd of June. The following were elected directors for the ensuing year: M. H. Folger, B. W. Folger, F. A. Folger, F. A. Folger, jr., R. T. Walkem, W. F. Nickle, I. A. Breck, J. Minnes, W. McRossie. R. T. Walkem, Q.C., was chosen president; I. A. Breck, vice-president, and F. A. Folger, managing director.

Messrs. William Mackenzie, manager of the Toronto Street Railway, and James Ross, managing director of the Montreal Street Railway, have made an offer of $2,500,000 for the Birmingham, England, Tramway Company's franchise and plant. The directors of the Birmingham company have reported in favor of the acceptance of the offer. If the road be purchased, the present system will be changed to electricity. Great Britain is said to possess only about 75 miles of electric road.

Mr. John Peck, manager of the Reid & Currie Iron Works, New Westminster, B. C., has invented an improvement in compound engines, which consists in having but one valve to distribute steam to the two cylinders, and which does away with one link, one set of eccentrics, and the pipe which, in ordinary engines, carries the steam from the high pressure to the low pressure steam box. By this means, it is claimed, there is a great saving in condensation, which, of course, means a considerable saving in fuel.

The case of the Auer Incandescent Light Manufacturing Company vs. O'Brien was heard before Mr. Justice Burbidge in the Exchequer Court, Ottawa. In this case the Auer Light Company sue for an injunction, to restrain the defendant O'Brien from infringing patent No. 46,946. The motion was by the plaintiff, for an order to have defendant produce the sample of the fluid used in manufacturing the "hoods" used in the light. Mr. Duclos, of Montreal, in support of the motion, urged that his client could not go to trial without the samples. Mr. O'Gara, Q.C., opposed on the ground that supplying samples would betray his client's secret; judgment reserved.

After investigation into the circumstances of the recent Point Ellice Bridge disaster, at Victoria, B.C., the coroner's jury has rendered a verdict holding the Consolidated Railroad directors responsible for the lives of persons killed. The city council was arraigned as guilty of contributory negligence and the officials of the corporation were absolved of personal responsibility. It was found that the bridge was safe for ordinary traffic, and the accident would not have occurred but for the improper crowding of the cars which went through the structure. The bridge was found not to have been constructed according to original specifications.

A basis of agreement has been reached between the City Council and the Hamilton Street Railway Company regarding a change from the terms of the by-law. The following reductions on the company's percentage will be made: Up to $125,000 of the gross receipts, from 6 to 5 per cent.; up to $150,000, from 6½ to 6 per cent. On the mileage the company will save about $6,100 a year. The company, on the other hand, agree to give seven unlimited tickets for 25 cents, and nine limited tickets for 25 cents; school children's tickets, ten for 25 cents; a return trip to and from school, 5 cents. The committee will also recommend that the franchise of the company be extended five years, from 1913 to 1918. By the reduction in the price of tickets the company claim to be giving to the public $8,100 annually.

---

The rate of transmission on Atlantic cables is eighteen words of five letters each per minute. With the "duplex" this rate of transmission is nearly doubled.

VIBRATION AND CONDUCTIVITY.—Signor Murani, in L'Elettricista, describes experiments on the influences of vibration on the resistance of wires. To avoid the heating due to friction, the vibrations in a series of metallic wires were induced by an electro-magnetic tuning-fork, wires of hardened iron, platinum, hard steel, hard copper, German silver, and manganin being tested, and in no case was any variation in the electric resistance detected by the most delicate methods. It is consequently concluded that the resistance of metallic wires is not altered by vibrations, all results to the contrary obtained by other observers notwithstanding.

---

# CANADIAN GENERAL ELECTRIC CO.
## (LIMITED)

## Railway Apparatus

We invite attention to the commanding position the Company has attained in the Electric Railway field.

Our Apparatus is now in use on almost, without exception, every independent road in the Dominion, and the fact that we have been awarded practically all the contracts placed during the past year for equipment either for new or existing roads, is the strongest possible testimonial both as to the superiority of the Apparatus itself, and the fair and liberal basis on which the business of the Company is conducted. Amongst the roads from which orders for apparatus have recently been received may be mentioned the following :

Hull & Aylmer Electric Ry. Co.
Moncton Electric Street Railway, Heat &
   Power Co.
Hamilton Radial Electric Ry. Co.
Cornwall Electric Street Ry.
Halifax Electric Tramway Co.
St. John Street Railway Co.
Montreal Street Railway.
Toronto Street Railway.

Vancouver & Westminster Tramway Co.
City and Suburban Street Ry.
Guelph Electric Street Ry.
Berlin & Waterloo Street Ry.
Port Dalhousie, St. Catharines & Thorold Street
   Railway Co.
Brantford Street Railway Co.
London Street Railway Co.
Kingston, Portsmouth & Cataraqui Railway.

## Lighting and Power
## Transmission Apparatus

In considering the development of our systems of apparatus for lighting and power transmission, we have kept in view the important fact, that varying conditions of service require varying methods to meet them.

We have in our **Edison Direct Current Three-Wire System,** our **500 Volt Direct Current System,** our **Single-Phase Alternating Current System,** our **Monocyclic System** and our **Three-Phase System,** a series of methods, each superior to all others, for the service to which it is adapted. We are not confined to one system only, but cover the whole range of Direct Current, Single-Phase and Multi-Phase Alternating Apparatus. Our interest in each case, therefore, lies in using the most suitable system, since we manufacture all ; not in twisting the conditions to suit one particular system, however ill-adapted to the particular case.

**Our recent sales of Lighting and Power Apparatus have exceeded all previous records and include the sale to the Lachine Rapids Hydraulic and Land Co'y, of twelve three-phase generators, each of 1000 h.p. capacity.**

## SPARKS.

During 1895 electric lines in Europe increased in number from 70 to 111, with a total length of 560 miles.

The shareholders of the Hamilton, Grimsby and Beamsville Railway have decided to extend the line to Grimby, to be completed by 1st November of this year.

The Lake Superior Power Company, of Sault Ste. Marie, Ont., will go extensively into the production of calcium carbide, the substance from which the new acetylene gas is manufactured.

Mr. B. B. Osler, Q.C., president of the Hamilton & Dundas Railway Company, has under consideration the question of changing the road to an electric line. It is said to be the intention of the company to commence operations as soon as Mr. Mackenzie returns from England. The estimated cost of changing the system is placed at $50,000.

Mr. Hugh Neilson, chief electrician of the Bell Telephone Company, was in Quebec recently to report on the construction of a telephone line from Quebec to the Island of Orleans. In company with Mr. Dauphin, local manager at Quebec, he examined the site for the cable, from L'Auge Gardien to St. Pierre, and will, it is said, report favorably to the company.

A dispatch from New York dated June 30th says: The contract for the entire development of 20,000 horse power on the Richelieu river, the outlet of Lake Champlain, has been let to the Stilwell-Bierce & Smith-Valle Company, of Dayton, Ohio, for $550,000, the electric machinery not being included. This power is to be carried to Montreal by wire and electrically distributed, the distance being about twelve miles. This is the second electric water power development in Montreal, the first being for 12,000 horse-power at the Lachine rapids, five miles above the city, in the St. Lawrence river, for which the Dayton company also have the contract. The investment in both powers will be about $3,500,000, all subscribed for by Montreal capitalists.

The Cornwall Electric Street Railway Company formally opened their line on June 30th. The main line down Pitt street, the belt line to the East Ward, via Second, Marlboro' and Water streets, and the line down the front road to the St. Lawrence Park on Gillespie's Point, have been completed, making in all about five miles of track. The line to the Toronto Paper Company's mills in the western suburbs, Smithville, and the branches to the textile mills, remain yet to complete. The plant is up-to-date in every respect. The cars, which are of the latest models, were built in Deseronto, and the machinery was put in by the Canadian General Electric Company. The Street Railway Company have purchased Gillespie's Point, a short distance below the town, and have fitted it up in an elaborate style as a pleasure resort, to be known as St. Lawrence Park.

## POSITION WANTED

A young man with six years' experience with T. H. Arc and Incandescent machines and lamps, will be open for engagement in July. Best references. Address,
"ARC,"
Care of CANADIAN ELECTRICAL NEWS, Toronto.

## MOTOR FOR SALE

New 2½ H. P., 20 Volts, perfect condition. As we have installed new alternating current machine, we cannot use motor.
ALLISTON MILLING CO.,
Alliston, Ont.

## CEDAR POLES

Telegraph, Telephone and Electric Light Poles. Large stock to sel ct from—all lengths.
GE RGE MARTIN,
Fenelon Falls, Ont.

**CANADIAN ELECTRICAL NEWS** AND **STEAM ENGINEERING JOURNAL**

OLD SERIES, VOL. XV.—No. 6.
NEW SERIES, VOL. VI.—No. 8.

**AUGUST, 1896**

PRICE 10 CENTS
$1.00 PER YEAR.

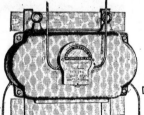

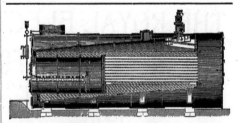

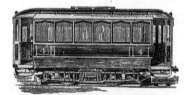

CANADIAN

# ELECTRICAL NEWS

AND

## STEAM ENGINEERING JOURNAL.

Vol. VI.                    AUGUST, 1896                    No. 8.

## ELECTRICAL POWER TRANSMISSION TO HAMILTON.

THE Cataract Power Company has been incorporated at Hamilton, with a capital stock of $99,000, for the purpose of transmitting electric power from DeCew Falls to Hamilton, a distance of 32 miles. The promoters of the company are Hon. J. M. Gibson, James Dixon, John Moodie, John William Sutherland, John Patterson, and Edmund Brown Patterson, all of Hamilton. DeCew Falls are situated about two miles from St. Catharines and receive a constant and unfailing supply of water from Lake Erie. The height of the fall is about 270 feet. The depth of water at the brow of the fall is about 5 inches, and the width about 18 feet. This comparatively small body of water, operating upon water wheels from the height mentioned, is capable of generating 2,500 horse power. The only purpose served at present by this magnificent water power is the operation of a couple of small mills. The Cataract Power Company have acquired the sole ownership of the water privilege, and are understood to have gone very thoroughly into the practicability of the scheme for transmitting the power to Hamilton. No particulars are as yet obtainable regarding the system or methods to be adopted for transmission, but the details are said to have been carefully worked out and submitted to Nikola Tesla and other electrical experts, who have approved of them.

The company have submitted to the Hamilton Street Railway Co., Hamilton and Dundas Railway Co., Hamilton, Grimsby and Beamsville Railway Co., Hamilton Electric Light and Power Co., and other large power users, a proposition to supply them with power at a cost very much below what they are paying under present conditions. The proposition is that the power shall be supplied under guarantee, so that the purchaser is asked to assume no risk whatever. If the company succeed in getting the acceptance of their proposition from the leading power users, the work of installing the necessary plant will be at once proceeded with. The total cost of carrying the enterprise to completion is estimated at nearly a quarter of a million dollars. If carried out this will be the longest electric power transmission line in the Dominion, and one of the longest in the world.

The further development of so important an enterprise, and one which bears to some extent the character of an experiment, will be watched with much interest. The recent declaration of Nikola Tesla that he has solved the means of successfully transmitting electric power for commercial purposes to a distance of 500 miles, augurs well for the success of this and enterprises of like character in the future.

## BARRIE ELECTRIC LIGHTING PLANT.

THE picturesque town of Barrie, situated on the shores of Kempenfeldt Bay, is lighted at night by two electric plants.

The steam plant, situated on Bayfield street, was designed by Messrs. Kennedy, McVittie & Co., architects. It acts as an auxiliary to the water plant, which is situated at Midhurst, six miles north.

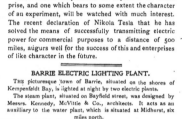

HON. J. M GIBSON,
President Cataract Power Company, Hamilton.

The switch board is a substantial slate affair, equipped with Brush instruments. The switch board room is merely a platform raised about ten feet above the floor of the dynamo room. A balcony runs around behind the board, so that the operator can see all of the machines from above. Stairs lead down to the floor of the dynamo room, which is floored in maple. All the machines are set on stone foundations, and the fly wheel of the engine is supported on stone abutments.

The engine is a Brown tandem compound, 180 h. p., with a fly wheel 12 feet in diameter by 24 inches face, driving a 22 inch belt onto a line of shafting, 35 feet long by 4½ inches in diameter. The line of shafting is below the switch board room, and is on a level with the floor of the dynamo room, the pulleys on it working in a pit. From the line of shafting, a 12 inch belt drives a 1000 light Brush alternator with exciter. Three five inch belts drive three Ball arc machines of 25 lights each. The machines are neat and clean, and everything about the place has a spick and span appearance.

In the boiler room two 14' by 60" Polson 100 h. p. boilers, fed by a Chas. Smith feed pump, and fired by soft wood, generate the steam for the engine. A Polson dependent condenser, direct connected, with a capacity of 3000 gallons of water per hour at 90°, beneath the floor of the dynamo room. The brick chimney is a substantial structure of considerable height.

The switch board room, the manager's office, and the cloak room are ceiled with basswood and the floors are maple. The manager's office is neatly fitted up and overlooks the flower garden and well kept lawn. On the switch board are seven switches for incandescent circuits. The board is fully equipped, and the light is sold principally by meter. The dynamo room is lighted by an arc light, and the rest of the building by incandescent lights.

At Midhurst there are two stations, one for arc lighting and one for the alternators. The machines in these plants are duplicates of the Barrie plant. A 30 foot head of water drives the arc plant, and a flume leads down to the incandescent plant.

The company is managed by an efficient board of directors, comprising Jas. T. Burton, President; M. Burton, vice-Pres.; S. A. Sett, Secty; Jas. A. Sanford, Supt; L. E. P. Pepler, director.

## BY THE WAY.

MR. H. E. EDGE, a prominent lumberman of Sydney, N.S.W., is making a tour of Canada, investigating the merits of the various electrical systems. He expresses surprise at the number of water powers. In Australia, he says, there are but two systems operated by water power, and to obtain the water for one of these a tunnel one mile in length was constructed. The rivers of Australia differ from those of Canada, in that they run for some miles and then disappear for miles. This, he says, is due to the porous nature of the ground in some parts.

x   x   x   x

DURING the four or five years of business depression through which we have been passing, all classes have been on the lookout for indications of returning prosperity. As a rule they have seen little of an encouraging character, while with some things have been going from bad to worse. I met a man thus situated recently, to whom I propounded the oft-put question: "What is the business outlook?" The answer I received is worthy of preservation. Said he, "Two or three years ago, you and I were living on our Faith that the times would improve. Last year we thought we could discern signs of promise and we lived on Hope. This year I am living on Charity."

x   x   x   x

THE pathway of the sales agent of an electric manufacturing company is not always strewn with roses, judging by an experience which one of them related to me the other day. "You remember," said he, that the city of —— was lately equipped with an electric street railway. Well, I am the individual who worked that enterprise up from its very foundation, and failed to get either credit or dollars for my labor. First of all I directed my attention to the Council, and after much expenditure of time and the breaking of more than one bottle of wine, secured for the promoters of the road a franchise, which, owing to local prejudices, they could not have obtained for themselves. I next prepared plans and specifications upon which they might invite tenders for the apparatus and construction. Tenders were called, and a meeting of the Council held to consider them, at which I could not be present on account of having to appear as a witness in a suit for the recovery of $250 misapplied funds. In my absence another representative of our company, who was totally unknown to the aldermen, was delegated to attend the meeting of the Council in the interests of our tender.. The result was that the Council accepted the tender of a rival concern, and we were out the profits on a $30,000 contract, plus time, effort and incidental expenses—the latter of which came out of my own pocket. When, afterwards, I ventured to ask some of the promoters if they did not think I was entitled to a little more consideration, after all I had done in getting them the franchise, etc., they frankly admitted that they had entirely overlooked that feature of the matter, and had simply voted that the lowest tender be accepted, regardless of everything else. If it had not been for that paltry law suit, I would have been certain to have got that contract. The last straw on the camel's back was the fact that the law suit went against us also, and we lost the whole business." Fortunately such extraordinarily "rough" experiences do not strike a man often, but when they do they hit him hard.

As the result of the efforts of a Canadian syndicate the antipathy to the trolley system of street car propulsion in England bids fair to be largely overcome. This syndicate is composed of Mr. Wm. Mackenzie, president of the Toronto Street Railway Company, and Mr. James Ross, manager of the Montreal Street Railway, who have been negotiating for the purchase of the franchise of the existing street railway company of Birmingham, Eng. I met Mr. Mackenzie a few days after his return from Europe, and he informed me that the deal was considered as good as closed. He said: "There is just Mr. Ross and myself in the company as yet. The conditions of the purchase are that we secure an extension of 21 years of the lease of the road, and that the City Council permit the use of the trolley system, but we do not anticipate any difficulty in that line. Of course, the work of electrifying the system will not be commenced until next spring. The road is forty miles in length, and the population of Birmingham, I should say, about three times as large as Toronto." To the question, "Is it your intention to endeavor to secure other franchises?" Mr. Mackenzie remarked that after the Birmingham system was in operation, he thought other cities would soon fall in line. He hoped eventually to secure the adoption of the trolley system in London, where horse cars and busses are now used, and where the prejudice against poles and overhead wires is very strong. "About the best electric railway in England," he said, "is on the Isle of Man; it is eight miles in length, double-tracked, and works very efficiently. In the matter of lighting they are much further advanced, and I had the pleasure of visiting an immense installation at London. As regards electrical machinery, I do not think they are quite as far advanced in Great Britain, and it is just possible that some American machinery will be required for the proposed conversion of the Birmingham road." Mr. Mackenzie purposes making another trip to Europe this fall.

## ECONOMIES IN CENTRAL STATION PRACTICE.

A paper on the above subject presented recently before the Chicago Electrical Association by Mr. Thos. C. Grier, concludes as follows:—

There are 'little' economies 'in details.' Here are a few short quotations from letters I received in response to my query as to little economies:

'The first to come to mind under your paper is discount all bills promptly, as your supply house can afford to give better prices when they know their invoices will be paid promptly.'

'If furnishing street lights, show your council and committee that you are trying to give the city all the contract calls for.'

'Treat your customers as reasonably as possible; they will reciprocate.'

'Collect all your bills before the 10th of the month.'

'Keep the stock-room under lock and key and have supplies taken out on requision; men get careless and this is a leak that foots up very fast.'

'Bad joints, that is, joints not soldered, and loose, is poor economy.'

'The use of exhaust steam for heating in winter is economy.'

Every plant in itself is a distinct problem and what may be economy in one may not answer in another.

## OTTAWA LETTER.

(Correspondence of the CANADIAN ELECTRICAL NEWS.)

THE Ottawa Electric Railway Co. have purchased a large parcel of property near the Experimental Farm. Sixteen acres of this is wooded with elm, maples and birch, and is known as the "West End Park." The Somersett street cars furnish a five minute service during the evening, and less frequently during the day. The park was opened on Saturday evening, July 18th, when 4000 people were there. The company have erected an open air theatre, with seating capacity for 1600 people. A commodious stage, with electric foot-lights, a good orchestra and high class performances, fill this enclosure every evening. Edison's late invention, the Vitascope, has been running for three weeks. It requires two currents to operate it. The trolley circuit was used to revolve the films before the aperture and an alternating circuit to project the views on the canvas. The grounds are lighted by numerous arc lights. Five swings and a piano are operated by an ordinary street car motor. The company deserve great praise for supplying pleasant recreation for the warm summer evenings.

Ottawa possesses a large number of electrical firms, of which the following are a portion :

Godard, Garrioch & Co. have been very busy, and have a neat display of electric fixings connected with installations of light or power.

The young firm of O'Rielly & Murphy have in a little over a year built up a satisfactory business. They had as many as fifteen jobs on hand at once this summer.

Chubbuch & Simpson, a new firm, are doing a good business, and have a lot of work on hand.

H. McColl, agent for the Chanteloup Mfg. Co., Montreal, reports business fair.

Mr. Cotter has invented an electric carpet beater. It is a simple little affair, but is a marvel to work.

R. Anderson, general electrician, is installing an electric lighting plant of 75 lights,on the steamer Empress for the O.R.N. Co., and a 25-light plant on J. G. Brigham's ferry wharf.

Ottawa No. 7, C. A. S. E., who have a number of members on the river, hold their election of officers in December instead of May, and their semi-monthly meetings on the 2nd and 4th Fridays. Their last meeting, July 31st, was one of the best they have had. The officers are as follows : President, T. J. Merrill ; Vice-President, A. Gaul ; Financial Secretary, T. Robert ; Recording Secretary, T. G. Johnson ; Treasurer, Wm. Hill ; Conductor, John Harris ; Door-Keeper, J. F. Peters ; Trustees, Thos. Wensley, John Cowan, F. G. Johnston. The delegates to the Kingston convention will be T. J. Merrill, F. G. Johnson and F. W. Donaldson.

## VERTICAL STEAM BOILERS.

TAKE an ordinary horizontal tubular boiler, one of the kind used in hot water heating plants, with the space inside the shell completely filled with tubes—set it on end, with a furnace below and a chimney connection above, and you have pretty nearly what, for many years, has been the standard type of vertical boiler. And a good serviceable kind of boiler, too, it has been, with all its shortcomings. In cost it was moderate ; no special setting was required for it ; repairs were easily made, the compactness and a reasonable degree of efficiency were secured with it, so that even to-day it has not outlived its period of usefulness, but continues in favor and is employed in a wide variety of cases where, all things considered, no other form of boiler will give the same degree of satisfaction.

And yet, for large powers, for high economy, for standard use in high-class power station work, even its distinctly good points could not command its application, except in forms so modified that in many cases little semblance remains to the early upright tubular boiler as we all know it. The designs have been carefully worked over, all with the end in view of turning out something better than the original, and the result is that while the later boilers also are vertical, in the sense, primarily, that they take up more head room than ground space, their tubes are not always vertical nor even approximately vertical, and there is not in every case the conventional shell within which tubes and flues are disposed.

Nor are the tubes always fire tubes, as in the ordinary vertical boiler, for conveying the products of combustion from the furnace to the chimney ; frequently in the newer and more complex designs they are water tubes instead and do not always run in straight lines, but often curve and twist in vertical and horizontal planes, in helical paths, in almost all directions imaginable, with the one aim of making them efficient heaters of water, by promoting circulation and absorbing, to the greatest possible extent, the heat of the fuel liberated in the furnace.—Albert Spies, in Cassiers' Magazine for December.

## RECENT CANADIAN PATENTS.

Canadian patents have been granted for the following electrical and steam engineering devices:

Insulating joint—Chicago Gas & Electric Fixture Manufacturing Co., Chicago.

Valve for boilers—John Harrison, Winnipeg, Man.

Electrical indicating mechanism for journal boxes—Wm. B. Chockly, Denver, Col., U. S.

Electric railway—W. B. Purvis and M. M. Armstrong, Philadelphia, U. S.

Filaments and carbons for electric lamps—J. H. D. Willan, 16 Helens Place, London, Eng.

Bonding device for electric railways—Wilson Brown, Camden, U. S.

Electric lamp hanger—Wm. A. Thompson, Toronto, Ont.

Turbine water wheel—John H. Staple, York, Penn., U. S.

Steam boiler furnace—Thomas York, Portsmouth, Ohio, and James E. York, Duluth, U. S.

Turbine water wheel—John B. McCormick, jr., and James Dixon, York, Penn, U. S.

High pressure engine—John Wand, London, Ont.

Appliances for cleaning car tracks—Samuel Irwin and Albert S. Geiger, Waterloo, Ont.

Split switch—Uldarique Gilbeault, St. Isidore Junction, Que.

Force pump—Wm. E· McCall, Peterborough, Ont.

Electric safety appliance for railroads—Edward Levi Orcutt, Sommerville, Mass., U. S.

Turbine water wheel—Wm. O. Crocker, Turner's Falls, Mass., U. S.

Electric railway—John F. and John A. Jordan, Brooklyn, N. Y.

Electric railway gate—Herman Biermann, Breslau, Germany.

Machine for raising and lowering electric light—Nelson McLeod, Cannington Ont.

Electric locomotive—J. J. Heillmann, Paris, France.

Balanced steam engine—J. J. Heillmann, Paris, France.

Queen Victoria has had several telephones installed in Windsor Castle. They are placed on her majesty's study table and communicate with Lord Salisbury at the Home Office, Marlborough House and Buckingham Palace. In a few days an electrophone will be introduced at Windsor Castle, and the Queen will be enabled to hear all the latest entertainments in the London theatres and concert halls.

## SOLUTION OF ELECTRICAL QUESTIONS.

By the courtesy of Mr. James Milne, we are enabled to present herewith the solution of the questions submitted for the electricity examination of the Toronto Technical School, at the close of the last session :

1. State clearly Ohm's Law. What is the unit of resistance? the unit of current? and the unit of electro-motive force?

ANSWER.—Ohm's Law: The strength of a current varies directly as the E. M. F. and inversely as the R, or the intensity of the current is equal to the E. M. F. divided by the resistance, i. e.,

$$C = \frac{E}{R} \; : \; R = \frac{E}{C} \; : \; E = CR.$$

The unit of resistance is called the "Ohm" and is equal to $10^9$ C. G. S. units of resistance. It is the resistance of a column of pure mercury 1 square millimeter in section and 106 centimetres long at a temperature of 32°F. The unit of current is called the "Ampere" and is $10^{-1}$ C. G. S. units. It is that current which will deposit 4.025 grams of silver per hour or decompose .0055944 grams of water per hour. The unit of electro-motive force is called the "Volt" and is equal to $10^8$ C. G. S. units, and is also the E. M. F. necessary to send a current of 1 ampere through a resistance of 1 ohm.

2. A battery of 15 cells, arranged five in series and 3 abreast, produces a current of .5 amperes through an external R of 5 ohms. Find the E M F of each cell if its internal R is 3 ohms.

ANSWER.— Let x = Number of cells in series.
     y = "   "   " in multiple.
     E = E M F of each cell.
     R = External R.
     r = Internal R.

$$C = \frac{E}{R} = \frac{x. \, E}{\frac{x. \, r}{y} + R}$$

$$C\left(\frac{x. \, r}{y} + R\right) = x. \, E$$

and substituting all the data given in the question for the above we get

$$.5\left(\frac{5 \times 3}{3} + 5\right) = 5 \, E$$

$$E = 1 \text{ volt.}$$

3. What is the best way of arranging 28 cells, each having an R of 4 ohms, so as to produce the strongest current in a circuit of 28 ohms.

Ans.—In this question the internal R must be = external R,

that is $\dfrac{x. \, r}{y} = R$

or $\dfrac{4 \, x}{y} = 28$

$x = 7 \, y$

but the total number of cells = x. y = 28, and substituting this value of x. viz.: 7 y in the equation, we get

$7 \, y^2 = 28$
$y = 2$

Therefore the number of cells in multiple = 2, and as the total number of cells = 28, ∴ the number in series = $\dfrac{28}{2}$ = 14.

4. Compare the resistances of a wire 30′ long, .06″ diameter, and that of another wire 15′ long and .03″ diameter.

ANSWER.—
$R_1$ = Resistance of one wire.   $R_2$ = Resistance of the other.
$l_1$ = length of "   "   $l_2$ = length "   "
$d_1$ = diameter of "   "   $d_2$ = diam. "   "

then $\dfrac{R_1}{R_2} = \dfrac{l_1 d_2^2}{l_2 d_1^2} = \dfrac{30 \times .03^2}{15 \times .06^2} = \dfrac{1}{2}$

$R_1 : R_2 :: 1 : 2$

5. 1,000 feet of copper wire .102″ diameter is wound on an armature of a bipolar generator. Find (1) the total resistance of that wire, and (2) the resistance as measured at the brushes of the machine. One mil foot = 10.4 ohms.

Ans.—In this question the formula is exactly the same as in the preceeding, that is

$R_1 \, l_1 \, d^2 = R \, l_1 \, d_1^2$

$R_1 = \dfrac{R_2 \, l_1 \, d_2^2}{l_2 \, d_1^2} = \dfrac{10.4 + 1000 + 1}{1 + 102 + 102}$
        = 1 ohm.

1 ohm represents the total resistance in 1000′ of copper wire, and in an armature of a bipolar generator there would be two wires of 500′ long in parallel, i. e., we have a derived circuit, each of the branches having ½ ohm resistance each, which gives us ¼ ohm as the resistance as measured at the brushes.

6. Take the above question but substitute iron wire. What is the thickness so that the resistance will be the same in each case? The specific resistance of copper to that of iron is as 1 : 6.

ANS.—The cross section will be six times that of the copper,

or the diameter = $\sqrt{.102^2 \times 6}$
            = 250 mills or .25″

7. Prove that 746 watts make a horse power. Answer this fully.

ANS.—The unit of power is $10^7$ ergs per second = 1 watt.

   A horse power = 550 ft. pds. per second.
     1 foot = 30.479 centimeters.
     1 lb. = 453.6 grams.
∴ 30.379 × 453.6 = 1 ft. pd. = 13825.27 gram. cent.
But a gram = 981 degrees.
   ∴ 13,825.27 × 981 = 13,562,600 ergs,
     generally denoted 1.356 × $10^7$ ergs per second,
     but a h. p. = 550 ft. pds. per second.
   ∴ 1.356 × $10^7$ × 550 = ergs per second per h. p.
       But $10^7$ ergs = 1 watt.
   ∴ $\dfrac{1.356 \times 10^7 \times 550}{10^7}$ = 746 watts per h. p.

8. 1000 feet of wire No. 6 B and S has a resistance of .4 ohms. Find the watts lost in an arc light circuit 5 miles long. Each lamp takes 10 amperes of current.

ANS.—The total R in the circuit =

$\dfrac{5 \times 5280 \times .4}{1000}$ = 10.56 ohms

$C^2 R = 10^2 \times 10.56 = 1056$ watts

9. The E M F of a certain dynamo machine is 100 volts, and the total R of the circuit is 1 ohm. What H. P. would have to be expended in working under these conditions.

ANS.—
     H. P. 746 = $C^2 R$

H. P. $= \dfrac{C^2 R}{746} = \dfrac{C \, E}{746} = \dfrac{E^2}{746}$

∴ $\dfrac{100^2}{746 \times 1}$ = 13.4 h. p.

10. Distinguish between work and power. What is the unit of each? What is the British heat unit [772 ft. pounds] equivalent to in electrical units of power?

ANS.—Work is the product of a force and the distance through which it acts. The unit of work is the work done in overcoming unit force through unit distance, i. e., in pushing a body through a distance of 1 centimetre against a force of 1 dyne. It is called the "erg.". Since the weight of 1 gram = 981 dynes, the work of raising 1 gram 1 centimetre against gravity would be 981 ergs or g ergs. Power is the rate of working, the unit is called the watt = $10^7$ ergs per second. If 746 watts = 550 ft. pds., how many watts will 772 ft. pds. be equal to?

     550 : 772 :: 746 : 1047 watts, answer.

11. Describe fully the Edison chemical meter ; knowing that 1 ampere passing for 1 hour between zinc plates immersed in a solution of salt of that metal will remove from one plate and deposit 1125 milligrams on the other. What would be the amount of current that would pass in the above meter if the resistance of the German silver shunt was .02 ohms, and the resistance of the other circuit in which the zinc voltameter of 2.5 ohms is inserted in series with another R of 46.46 ohms, if the deposit was 200 milligrams? Make a sketch of the arrangement.

ANS.—The answer to this question is 400 ampere hours. The Edison chemical meter was fully described and illustrated in the paper on "Meters" read before the Canadian Electrical Association by Mr. James Milne, and which appeared in the July issue of the ELECTRICAL NEWS.

12. Describe the Wheatstone's bridge as fully as you can, and illustrate the application of the instrument by an example.

ANS.—The Wheatstone bridge may be represented by the diagram shown,

and consists essentially of wires arranged in multiple arc. Suppose current enters at A, it then divides, part going through A B C, and part
through A D C, dividing itself into parts that shall be to one another inversely as the resistances in the branches. Since the current is going from A to C the point A must be at a higher potential than the point C and therefore there will be a gradual fall of

potential along the branches A B C and A D C. It is therefore possible to find various parts along these branches that will be at the same potential. By altering the resistances in the branches it may be so adjusted that the point B is at the same potential as the point D. When this is so the bridge is in a condition for taking the observation. When B and D are at the same potential there is no E M F between these points and consequently no current will flow in the wire connecting them. The attainment of this condition is indicated by no deflection on the galvanometer G that connects B to D.

Let AB=the resistance in the arm AB
BC=the resistance in the arm BC, and so on.

We have the following simple relation when the above condition has been satisfied :

AB. DC=AD. BC, and as the resistances in three of the arms are known it is an easy matter to find the fourth. Suppose DC to be the unknown, then

$$DC = \frac{AD. BC}{AB.}$$

The following is a proof of the principle of the bridge : Suppose the figure represents the instrument when there is no deflection on the galvanometer, i. e., when no current is passing through B and D, and suppose p to represent the potential at B which would also be the potential at D since no current flows in BD, and let p₂

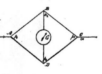

represent the potential at C.

By Ohm's law we have $C = \frac{E}{R}$ ; but the EMF in AB is the difference of potential between p and p₁ .˙. C in $AB = \frac{E}{R} = \frac{p_1 - p}{R \text{ of } AB}$ and similarly the current in $BC = \frac{p - p_2}{R \text{ of } BC}$ , but the same current must pass through BC as that passed through AB, since none goes through BD.

$$\therefore \quad \frac{p_1 - p}{AB} = \frac{p - p_2}{BC} \quad (1)$$

In the same way the current in $AD = \frac{p_1 - p}{AD}$ and it must be equal to

$\frac{p - p_2}{DC}$ that is $\frac{p_1 - p}{AD} = \frac{p - p_2}{DC} \quad (2)$

and if we divide (1) by (2) we get $\frac{AD}{AB} = \frac{DC}{BC}$ .˙. AD. BC=AB. DC.

Numerical example :
AB = 100 ohms   BC = 9756 ohms
AD = 10 ohms   DC = unknown
$$DC = \frac{10 \times 9756}{100} = 975.6 \text{ ohms.}$$

13. How are very high resistances measured ? A galvanometer of 6000 ohms shows a deflection in 10° when a certain resistance is in circuit with it. Knowing that the same galvanometer shows the same deflection with a resistance of 1-10th megohm in circuit when shunted with a 1-99th shunt, find this certain resistance. The resistance of the battery is neglected.

ANS.—As the ordinary bridge is only capable of measuring resistances up to 1,111,100 ohms a different method is adopted for measuring resistances above this, viz : by the galvanometer.

FIG. 1.      FIG. 2.

Let the first figure indicate the circuit with 1-10th megohm in series

with the shunted galvanometer, and the second figure that of the circuit with the unknown resistance in series with the galvanometer without the shunt. By Ohm's law we have

$$C = \frac{E}{R} = \frac{E}{R + \frac{G S}{G + S} + B} = k. d_1. \frac{G + S}{S}$$

Where $\frac{G S}{C + S}$ = Joint R of Galvanometer and Shunt,
B = Resistance of Battery,
k = a Constant to bring d. $\frac{G + S}{S}$ to Amperes,
d₁ = Deflection of Galvanometer,
$\frac{G + S}{S}$ = Multiplying Power of the Shunt.

In the second figure we have

$$C^1 = \frac{E}{R_1} = \frac{E^1}{R_1 + G_1 + B_1} = k. d_2.$$

In the first equation we have

$$E = \left( R + \frac{G S}{G + S} + B \right) . k d_1 . \frac{G + S}{S}$$

and in the second equation we have

$$E = (R_1 + G_1 + B_1) k. d_2,$$

$$\therefore (R_1 + G_1 + B_1) . d_2 = \left( R + \frac{G S}{G + S} + B \right) . k d_1 \left( \frac{G + S}{S} \right)$$

Substituting the numbers in the question and omitting the resistance of the battery and cancelling k, we have

$$(R^1 + G_1) d_2 = \left( R + \frac{G S}{G + S} \right) d . \frac{G + S}{S}$$

$$(R_1 + 6000) d_2 = \left( 100000 + \frac{6000 \times 60.6}{6000 + 60.6} \right) d_1 \frac{6000 \times 60.6}{60.6}$$

and as d₂ = d₁, we get

$$R_1 + 6000 = (100000 + 60) . 100$$
$$R_1 = 10,000,000.$$

Therefore, the resistance of R or x is 10 megohms.

14. Show by a diagram the general arrangement and connections of generators running on a 3-wire system. Show by an arrow the direction of the currents if (1) both machines are doing exactly the same amount of work ; (2) if one machine is doing more than the other. Place in position ampere and voltmeters.

ANSWER.—

[Diagram of 3-wire system with Both Machines doing the same amount of Work]

[Diagram: One Machine doing more Work than the other]

15. 880,000 lines of force (N) are to forced through a bar 20 in. long and 8 sq. inches in area. Find the reluctance and the magnetizing force in ampere turns to effect this magnetization. Permeability = 166.

ANS.—Reluctance = $\frac{\text{length}}{\text{area} \times \mu} = \frac{20}{8 \times 166} = \frac{1}{66.4}$

Ampere turns = N × reluctance × .3132
= 880000 × $\frac{1}{A. \mu}$ × .3132
= 880000 × $\frac{1}{66.4}$ × .3132 = 4150

16. In a generator which is driven by a 100 H.P. engine, belt speed 5,000 ft. per minute, there are 200 conductors in the armature winding 100 sections in commutator, the gap is 45°. Find the torque and the drag on the active conductors.

ANS.—100 h.p. = 5000¹ × torque

Torque = $\frac{100 \times 33000}{5000}$
= 660 lbs.

The active conductors = $\frac{270}{360}$ of 200 = 150 .˙. $\frac{660}{150}$ = 4.4 lbs. drag on each conductor.

## HEAT IN CYLINDER WALLS.

THERE was made recently at Sibley College an interesting study of the loss of heat from the cylinder walls of an engine during each stroke. The object was to determine the varying temperature of the cylinder head during the stroke. Steam on entering the cylinder warms up the surfaces and a certain amount of heat is stored in the cylinder walls ; when the exhaust opens the temperature falls and heat flows from the walls and is lost. To determine this, experiments were made with a 10 h. p. slide-valve engine, cutting off at about half stroke. The plan of investigation was as follows : A wire of small cross-section and high electrical resistance was placed on the inner face of the cylinder head, and connected in multiple with a constant current supply and a delicate galvanometer. As the temperature varies with each cycle of the engine, the electrical resistance of the wire rises and falls with it, the amount of current flowing being altered, and a corresponding deflection being thus obtained in the galvanometer. To preserve a permanent record of these pulsations, the galvanometer was of the mirror type, so that its deflections could be recorded on a sensitive photographic plate.

This galvanometer is of special interest. It consists of a minute needle and mirror, mounted with a short suspension, and surrounded by a coil of fine wire, placed in a powerful magnetic field. This instrument possesses a great sensitiveness, and since its vibrating parts are of such delicate proportions, can be relied upon to give accurate results. The field produced by the coil is at right angles to the permanent field, and the galvanometer being acted upon by these two forces, takes up a resultant position, and follows this resultant with unerring accuracy, regardless of the rapidity of the current changes in the coil.

The high shunt resistance on the engine head consists of 27″ of No. 30 iron wire stretched back and forth over a sheet of mica and held in place by heavy mica strips clamped over the ends ; the whole being held in place by a frame of fiber-board securely bolted to the head. This construction allows the wire to be well insulated electrically, yet exposed to the live steam.

To obtain a constant current supply, a storage battery of high potential was used, with a large resistance in series, giving a current of about .8 ampere.

As the galvanometer and resistance in the engine head were in multiple with this battery, and the change of resistance due to heating in the head was slight in comparison with the resistance in series with the storage cells, the current remained perfectly constant, and a common error in this method of operation was thus eliminated. An arc lamp, especially constructed for the purpose, furnished the light for the mirror of the galvanometer.

The reflected ray was moved along a slit, behind which a photographic plate was carried up and down by the indicator reducing motion.

The diagram obtained with this heat indicator was almost exactly like the regular indicator diagram in appearance, its lines representing temperatures instead of pressures. The diagrams were taken at various pressures and speeds, and all showed the same characteristics—a nearly constant temperature from admission to cut-off, a slight drop beyond this point, a sudden fall at release, and a continual fall on the return stroke until compression occurred, when there was a marked rise in temperature.

Another experiment was also made to determine how deep in the cylinder head the temperature varied. It was found that at a depth of beyond .05 of an inch the temperature of the head did not vary, but remained constant some 30° lower than the temperature of the steam at initial pressure. As the depth was decreased the temperature varied with the steam, and the cards again showed the same resemblance to the first experiments. From this investigation it is evident that the depth of metal affected to cause the phenomena of cylinder condensation is very slight ; that the heat cycle in the iron follows the indicator diagram very closely ; and that the average temperature at the point where variation ceases is quite near the temperature of the steam.

## NEW GASOLINE MOTOR.

THE accompanying illustration shows a gasoline motor of new design built by Mr. Thomas Reid, of Hamilton. The engine has an open base, the charge of gasoline being drawn directly into the cylinder, where it is ignited by an electric spark. It has an impulse at every revolution, but can at will be closed down so as to have an

impulse every second or third revolution, as desired. The engine is built in two styles, vertical and horizontal, the vertical being preferable for boats and the horizontal for carriages or power purposes. One of these motors has been at work in the maker's premises for some months past, and is said to give entire satisfaction. It is the first motor of the kind to be made in Hamilton.

When an injector fails to work, ascertain if the pipe to the boiler is free and clear, for it may have become partially filled with sediment, thus causing all the trouble.

A contemporary prints the following as a simple method of demagnetization : A strong magnet is placed in a horizontal position—on a table, for instance—and the watch held horizontally about half a yard off on a level with the magnet. The watch must then be brought slowly nearer the magnet, while being turned slowly, and at the same time as regularly as possible, between the fingers, as on a vertical axis. When the poles of the magnets are reached, the turning of the watch is to be continued while being gradually withdrawn until the starting point is reached.

## ELECTRIC COAL MINING PLANT.

A MOST interesting matter in connection with a visit to the underground workings of the new Vancouver Coal Company's mine at Nanaimo, B. C., is the electric plant in operation there. It has been in operation for four years now and has worked smoothly from the first, and given perfect satisfaction. It has quite superseded mule haulage over the underground trunk roads, but for branch roads mules are still employed.

The engine used for generating the electricity is the well-known Erie Ball high speed type, 16-inch cylinder by 16½-inch stroke, automatic cut-off, centre crank, double fly wheel, and is run at a speed of 235 revolutions per minute. Its rated h. p. is 150, although the work is being done with an expenditure of 90 h. p. It is bedded on a foundation of concrete, brick and stone, immediately resting on two large blocks of dressed sand-stone, which keep it perfectly firm and rigid.

Two boilers are used for supplying power. External fire, Lancaster pattern, 24 feet in length by 4 feet 6 inches in diameter, and carrying a pressure of 80 lbs., but, should more power be required, are good for 120 lbs. The steam is carried from the boilers to the engine, a distance of 200 feet, in covered pipes and without appreciable loss.

The dynamo is a large one (150 kilowatts), and was made and supplied by the Canadian General Electric Co., of Peterborough, Ontario, Canada. It is run from engine by an endless perforated belt 15 inches in width. The speed at which it is run is 640 revolutions, giving 340 amperes at a pressure of 250 volts. This low pressure, although tending to loss in the mains, gives entire immunity from danger, which is absolutely necessary in a mine where it is almost impossible to keep workmen from coming in contact with the wire. Spare armatures are always kept in reserve, so that there are never any delays for repairs.

The power house, containing engine and dynamo, is a large building 60 by 32 feet, and most complete in detail, having been specifically designed for the purpose. It has capacity enough to contain another plant the size of the present one, and in addition provides a store room and work room for winding armatures, etc., all of the work being done on the premises.

There are five locomotives, all of which were made in Canada, four by the Canadian General Electric Company, of Peterborough, and one by the Royal Electric Company, of Montreal. Four of the motors weigh 8 tons each, and are capable of hauling 40 tons of coal along a level track at the rate of 6 miles per hour. The other locomotive is a smaller (4½ tons), and only draws 20 tons at a trip. The distance of road along which coal is hauled is two miles in one level, making four miles for the round trip, and in the other level where the other motor is worked the distance is one mile, or two miles for the complete run. In addition to the locomotives there is a 30 h. p. electric hoist, operating an endless rope on one of the slopes.

The line conveying the current from the surface to the shaft and down to the bottom, a distance of 1,000 feet, is a 0000 copper cable, well covered to protect it from water, and hung on strong insulators. From the bottom of the shaft and extending throughout the mine, the trolley wire is 000 wire—and suspended from the roof or timbers of the gallery by specially made insulated hangers, and held in position over the rail in rounding curves by side wires or pull-offs, which are also insulated. A second or auxiliary wire (insulated) is carried in the levels as a feeder, to which the trolley wire is attached at stated distances.

The plant is fitted up with all the latest contrivances, switches, automatic cut-offs, safety fuses, etc., and in addition to the work mentioned, supplies light for the engine rooms, boilers, pit-head and other buildings on surface, and the whole of the pit-bottom and stables below ground, also all important sidings or partings. Each locomotive is fitted with head lights.

## TWO SYSTEMS OF FIRING A WATER TUBE BOILER.

BELOW is given the results of two systems of firing a water tube boiler, conducted by Mr. George H. Barrus, at the Edison Electric Illuminating Company's power house, Boston. The first test consisted in the common method of spread firing, carrying a bed of coal 6 to 8 inches thick, and on the second trial a brick roof was inserted above the lower row of tubes, covering over half the length of the furnace, the flames passing to the rear end before the gases were discharged into the tube space. A second roof was placed above the upper row of tubes in front of the flame plate. The length of tubes was 8 feet, and the first roof extended backward 4 ft. 6 inches, leaving opening 3 ft. 6 inches. The upper roof extended forward 4 ft. 6 inches. The method of firing on the second trial consisted in the coking system, with 18-inch fire on forward part of grate, and a very thin fire at the extreme rear end. Green coal was fired only on forward part of grate.

The boiler was 325 h. p., constructed with two sets of headers connected by short pieces of pipe ; the tubes, 168 in number, were of the ordinary 4-inch size, 18 ft. long, and arranged in two banks, 14 sections wide, with six in each section ; two steam drums, 44 in. in diameter ; area of heating surface of boiler, 3,737 sq. ft. ; area of grate surface, 58.3 sq. ft.

Instead of the coking system showing a more perfect combustion of gases, as expected, the actual result was a loss, the difference being 5.5 per cent.

DATA AND RESULTS OF EVAPORATIVE TESTS ON 325 HORSE-POWER BABCOCK & WILCOX BOILER MADE WITH NEW RIVER SEMI-BITUMINOUS COAL.

| System of firing | Ordinary | Coking, with brick roofs over furnace. |
|---|---|---|
| Percentage of moisture in coal .... .... .... pr. ct. | 9.4 | 8.7 |
| Date of test .... .... ... .... .... .... ....1896 | April 19 | April 21 |
| **TOTAL QUANTITIES.** | | |
| 1. Duration .... .... .... .... .... ....hrs. | 8. | 8.58 |
| 2. Weight of dry coal consumed including wood equivalent .... .... .... .... .... .... .... lbs. | 8,631. | 10,658. |
| 3. Weight of ashes and clinkers ... .. .... .... ...lbs. | 357. | 519. |
| 4. Percentage of ashes and clinkers .... .... per cent. | 4.1 | 4.9 |
| 5. Weight of water evaporated .... .... .... ....lbs. | 89,8 9. | 90,484. |
| **HOURLY QUANTITIES** | | |
| 6. Coal consumed per hour .... .... .... lbs. | 1,078.9 | 1,239.8 |
| 7. Coal per hour per square foot of grate .. ... .lbs. | 18.5 | 21.3 |
| 8. Water evaporated per hour .... .... .... lbs. | 10,103.6 | 10,779 |
| 9. Equivalent evaporation per hour, feed 100 degrees, pressure 70 pounds . . .... .... .... lbs. | 9,952. | 10,689. |
| 10. Horse-power developed, A. S. M. E. basis of 30 pounds .... .... ... .... H. P. | 331 7 | 350 1 |
| 11. Equivalent evaporation per square foot heating surface per hour .... .... .... .... .... ....lbs. | 2.7 | 2.9 |
| **AVERAGES OF OBSERVATIONS ETC.** | | |
| 12. Average boiler pressure. .... .... .... .... lbs. | 157.8 | 157.3 |
| 13. Average temperature of feed water ... ...... ..deg. | 132.1 | 126.2 |
| 14. Average temperature of flue gases ... .... .... deg. | 465. | 466. |
| 15. Average draft suction .... ... .... .... ..ins. | .35 | .54 |
| 16. Weather and outside temperature . .... ........ { | Cloudy, Moderate. | Cloudy, Moderate |
| **RESULTS.** | | |
| 17. Water evaporated per pound of dry coal .. . ..lbs. | 9.360 | 8.093 |
| 18. Equivalent evaporation per pound of coal from and at 212 degs. .... .... .... .... .... .... lbs. | 10.598 | 9.594 |
| 19. Equivalent evaporation p r pound of combustible from and at 212 degs. ... . . ... .. .. . .......lbs. | 11.051 | 10 404 |

PUBLISHED ON THE FIFTH OF EVERY MONTH BY

**CHAS. H. MORTIMER,**

OFFICE : CONFEDERATION LIFE BUILDING,
Corner Yonge and Richmond Streets,

TORONTO,    -    -    CANADA.
Telephone 9362.

NEW YORK LIFE INSURANCE BUILDING, MONTREAL.
Bell Telephone 2299.

*ADVERTISEMENTS.*

Advertising rates sent promptly on application. Orders for advertising should reach the office of publication not later than the 24th day of the month immediately preceding date of issue. Changes in advertisements will be made whenever desired, without cost to the advertiser, but to insure proper compliance with the instructions of the advertiser, requests for change should reach the office as early as the 2 nd day of the month.

*SUBSCRIPTIONS.*

The ELECTRICAL NEWS will be mailed to subscribers in the Dominion, or the United States, post free, for $1.00 per annum, 50 cents for six months.  The price of subscription should be remitted by currency, registered letter, or postal order payable to C. H. Mortimer.  Please do not send cheques on local banks unless 5 cents is added for cost of discount.  Money sent in unregistered letters will be at senders' risk.  Sub-criptions from foreign countries embraced in the General Postal Union $1.50 per annum.  Subscriptions are payable in advance.  The paper will be discontinued at expiration of term paid for if so stipulated by the subscriber, but where no such understanding exists, will be continued until instructions to discontinue are received and all arrearages paid.
Subscribers may, have the mailing address changed as often as desired.  When ordering change, always give the old as well as the new address.
The Publisher should be notified of the failure of subscribers to receive their paper promptly and regularly.

*EDITOR'S ANNOUNCEMENTS.*

Correspondence is invited upon all topics legitimately coming within the scope of this journal.

The "Canadian Electrical News" has been appointed the official paper of the Canadian Electrical Association.

## CANADIAN ELECTRICAL ASSOCIATION.

### OFFICERS:

PRESIDENT :
JOHN YULE, Manager Guelph Light & Power Company, Guelph, Ont.

1ST VICE-PRESIDENT :
L. B. McFARLANE, Manager Eastern Department, Bell Telephone Company, Montreal.

2ND VICE-PRESIDENT :
E. C. BREITHAUPT, Electrical Engineer, Berlin, Ont.

SECRETARY-TREASURER :
C. H. MORTIMER, Publisher ELECTRICAL NEWS, Toronto.

EXECUTIVE COMMITTEE :
GEO. BLACK, G. N. W. Telegraph Co., Hamilton.
J. A. KAMMERER, General Agent, Royal Electric Co., Toronto.
K. J. DUNSTAN, Local Manager Bell Telephone Company, Toronto.
A. M. WICKENS, Electrician Parliament Buildings, Toronto.
J. J. WRIGHT, Manager Toronto Electric Light Company, Toronto.
ROSS MACKENZIE, Manager Niagara Falls Park and River Railway, Niagara Falls, Ont.
A. B. SMITH, Superintendent G. N. W. Telegraph Co., Toronto.
JOHN CARROLL, Sec.-Treas. Eugene F. Phillips Electrical Works, Montreal.
C. B. HUNT, London Electric Company, London.
F. C. ARMSTRONG, Canadian General Electric Co., Toronto.

## CANADIAN ASSOCIATION OF STATIONARY ENGINEERS.

| | |
|---|---|
| President, W. G. BLACKGROVE, | 54 Widmer St., Toronto, Ont. |
| Vice-President, JAMES DEVLIN, | Kingston, Ont. |
| Secretary, E. J. PHILIP, | 11 Cumberland St., Toronto. |
| Treasurer, R. C. PETTIGREW, | 94 West Avenue N., Hamilton. |
| Conductor, W. F. CHAPMAN, | Brockville, Ont. |
| Door Keeper, F. G. JOHNSTON, | Ottawa, Ont. |

TORONTO BRANCH NO. 1.—Meets 1st and 3rd Wednesday each month in Engineers' Hall, 61 Victoria street.  John Fox, President ; Chas. Moseley, Vice-President ; T. Eversfield, Recording Secretary, University Crescent.

MONTREAL BRANCH NO. 1.—Meets 1st and 3rd Thursday each month, in Engineers' Hall, Craig street.  President, John Murphy ; 1st Vice-President, J. E. Huntington ; and Vice-President, Wm. Smyth ; Secretary, B. Archibald York ; Trea-urer, Peter McNaugton.

ST. LAURENT BRANCH NO. 2.—Meets every Monday evening at 43 Bonsecours street, Montreal.  R. Drouin, President ; Alfred Latour, Secretary, 306 Delisle street, St. Cunegonde.

BRANDON, MAN., BRANCH NO. 1.—Meets 1st and 3rd Friday each month, in City Hall   A. R. Crawford, President ; Arthur Fleming, Secretary.

HAMILTON BRANCH NO. 2.—Meets 1st and 3rd Friday each month in Maccabee's Hall.  Wm. Norris, President ; E. Teeter, Vice-President ; Jas. Ironsides, Corresponding Secretary.

STRATFORD BRANCH NO. 3.—John Hoy, President ; Samuel H. Weir, Secretary.

BRANTFORD BRANCH NO. 4.—Meets and and 4th Friday each month.  J. B. Forsyth, President ; Jos. Ogle, Vice-President ; T. Pilgrim, Continental Cordage Co., Secretary.

LONDON BRANCH NO. 5.—Meets once a month in the Huron and Erie Loan Savings Co.'s block.  Robert Simmie, President ; E. Kidner, Vice-President ; Wm Meaden, Secretary Treasurer, 533 Richmond street.

GUELPH BRANCH NO. 6.—Meets 1st and 3rd Wednesday each month at 7.30 p. m.  H. Geary, President ; Thos. Anderson, Vice-President ; H. Flewelling, Rec.-Secretary ; P. Ryan, Fin.-Secretary ; Treasurer, C. F. Jordan.

OTTAWA BRANCH NO. 7.—Meet every second and fourth Saturday in each month, in Borbridge's hall, Rideau street ; Frank Robert, President ; F. Merrill, Secretary, 352 Wellington street.

DRESDEN BRANCH NO. 8.—Meets 1st and Thursday in each month.  Thos Steeper, Secretary.

BERLIN BRANCH NO. 9.—Meets 2nd and 4th Saturday e-ch month at 8 p.m.  Jr. R. Utley, President ; G. Steinmetz, Vice-President ; Secretary and Treasurer, W. J. Rhodes, Berlin, Ont.

KINGSTON BRANCH NO. 10.—Meets 1st and 3rd Tuesday in each month in Fraser Hall, King street, at 8 p. m.  President, S. Donnelly ; Vice-President, He-ry Hopkins ; Secretary, J. W. Tandvin.

WINNIPEG BRANCH NO. 11.—President, G. M. Hazlett ; Rec.-Secretary, J. Sutherland ; Financial Secretary, A. B. Jones.

KINCARDINE BRANCH NO 12.—Meets every Tuesday at 8 o'clock, in Mp. Kibbon's block.  President, Daniel Bennett ; Vice-President, Joseph Lighthall ; Secretary, Percy C. Walker, Waterworks.

WIARTON BRANCH NO. 13.—President, Wm. Craddock ; Rec.-Secretary, Ed. Dunham

PETERBOROUGH BRANCH NO. 14.—Meets 2nd and 4th Wednesday in each month  W. L. Outhwaite, President ; W Forster, Vice-President ; A. E. McCallum, Secretary.

BROCKVILLE BRANCH NO. 15.—President, Archibald Franklin ; Vice-President, John Grundy ; Recording Secretary, James Aikins.

CARLETON PLACE BRANCH NO. 16.—Meets every Saturday evening.  President, Jos. McKay ; Secretary, J. D. Armstrong.

## ONTARIO ASSOCIATION OF STATIONARY ENGINEERS.

| | BOARD OF EXAMINERS. | |
|---|---|---|
| President, A. AMES, | | Brantford, Ont. |
| Vice-President, F. G. MITCHELL | - | London, Ont. |
| Registrar, A. E. EDKINS | - | 88 Caroline St , Toronto. |
| Treasurer, R. MACKIE | | 28 Napier st., Hamilton. |
| Solicitor, J. A. McANDREWS, | | Toronto. |

TORONTO—A. E. Edkins, A. M. Wickens, E. J. Phillips, F. Donaldson, J. Bain.
HAMILTON—R. Mackie, T. Elliot.
BRANTFORD—A. Ames, care Patterson & Sons.
OTTAWA—Thomas Wensley.
KINGSTON—J. Devlin, (Chief Engineer Penitentiary), J. Campbell.
LONDON—F. Mitchell.
NIAGARA FALLS—W. Phillips.

Information regarding examinations will be furnished on application to any member of the Board.

---

**Influence of the Telegraph.** "THANKS to the telegraph," said Lord Dufferin at the annual banquet of the British Chamber of Commerce of Paris, "the globe itself has become a mere bundle of nerves, and the slightest disturbance at any one point of the system sends a portentous tremor through its morbidly-sensitive surface."

**International Electrical Congress.** As we go to press, an International Electrical Congress is in progress at Geneva, under the auspices of the Swiss Society of Electrical Engineers.  The prominent electrical societies of Europe and the American Institute of Electrical Engineers are giving their support to the undertaking.  The following subjects are set for discussion : "Magnetic Units," "Photometric Units," "Transmission and Distribution of Power to Great Distances by Means of Direct and Alternating Currents," "Protection of High-pressure Overhead Electric Lines against Atmospheric Discharges," "Various Disturbances Caused by Electric Traction."

**Prices of Incandescent Lamps.** PROBABLY in no department of electrical supplies has competition and the cutting of prices been reduced to so fine a point as in that of incandescent lamps.  Prices have eventually got down below the profit line, and as a result the American manufacturers recently held a conference in New York, at which an understanding is said to have been reached which is expected to put a stop to the disastrous under-cutting of the past.  Each company is said to have deposited the sum of $5,000 as a guarantee of its willingness to abide by the agreement, and a fine of 10 cents per lamp will, it is said, be imposed for selling below standard rates.  Future prices will range from 22 cents for lamps of 8 to 25 c.p., to $1.65 for lamps of 150 c.p. in broken lots, and in standard packages from 30 cents to $1.50. The management is said to have been vested in a committee. Some of the companies interested deny that such an organization has been effected.

**The Duty on Steel Rails.** THE Privy Council has just handed down its judgment in the case of the appeal of the Toronto Railway Company for recovery of upwards of $50,000 which the company were compelled by the Dominion Government to pay as customs duty on steel rails imported for use in the reconstruction of their system. The Exchequer and Supreme Courts of Canada upheld the interpretation put upon the tariff by the Minister of Customs, but the Privy Council has come to a contrary conclusion and has decided in favor of the plaintiffs' contention that steel rails for street railway purposes are entitled to free admission in the same manner as steel rails for use on steam railways. This is undoubtedly the common-sense view of the matter. The Toronto Railway Company deserve the thanks, if nothing more, of every electric railway company in the Dominion, for having fought the matter through and secured from the highest tribunal in the Empire this favorable decision which cannot be reversed.

**The Kelvin Celebration.** SCIENTIFIC men from all parts of the world assembled in Glasgow the latter part of June to participate in the celebration of the 50th anniversary of Lord Kelvin's occupancy of the chair of Natural Philosophy in the University of Glasgow. During half a century Lord Kelvin has been an indefatigable investigator of the laws governing electricity and the methods of applying the same for the benefit of mankind. He is the author of many devices, notably measuring apparatus, which are the recognized standards in use throughout the world at the present time. He received the honor of knighthood for valuable services rendered in 1858 in overcoming difficulties incident to the successful operation of the first Atlantic cable, and was elected to the peerage in 1891. The celebration included a conversazione by the University of Glasgow, at which were exhibited Lord Kelvin's inventions ; addresses by home and foreign university bodies, learned societies and students of Glasgow and other universities, and a public banquet by the corporation of Glasgow.

**Municipal Lighting** PUBLIC lighting in England is largely in the hands of the municipalities. The extent to which this is the case is indicated by the fact that a Municipal Electrical Association has been formed, which has just held its first convention. In Canada not more than half a dozen municipalities own and operate their own lighting plants. The citizens of the town of Goderich have lately voted in favor of the purchase of a municipal plant, and the town of Newmarket has the subject under consideration at the present time. It is difficult to determine from a few such isolated cases whether or not the municipal control idea is likely to grow to important dimensions, as it has done in England. Should it do so, the sales agents of the electrical manufacturing and supply companies will require to be trained in the ways of the politician, so as to be able to secure the votes of the councilmen or aldermen in favor of their particular apparatus. The man who can "pull the wires" (no pun intended) most skillfully will probably secure the orders, regardless to a large extent of the superiority or inferiority of his goods. This method of selling goods promises to occupy a great deal more time and to cost more money than the selling to private individuals or companies, as at present.

**Disturbance of Telegraph and Telephone Circuits** THE German Imperial post-office has compiled statistics which show a steady increase in the number of disturbances to telegraph and telephone circuits as the result of the multiplication of electric railways. In our last issue we published the decision of the courts in an action brought against the Montreal Street Railway Company by the Bell Telephone Co., for injury sustained as the result of disturbance of their circuits from the action of induction currents emanating from the street railway company's wires. The decision was adverse to the plaintiffs. Our readers will be interested in knowing the method employed by the German authorities to protect the telegraph instruments from high pressure currents. For this purpose fuses are put into the circuits. These fuses consist of a wire 0.07 mm. in diameter, and made of a non-oxidisable alloy. They are enclosed in glass tubes 5cm. to 6cm. long, sealed at both ends, and fitted with metal contact pieces. In this way the formation of an arc at 500 volts pressure is avoided. The fusing current of the wire is 0.8 amperes. The whole fuse is kept in position between contact springs, and is easily interchangeable. Another type of fuse used by the Imperial post-office consists of a porcelain block about 5cm. high, the fuse wire running through a hole across the block. Both types have given satisfaction.

**Acetylene Gas.** THE possibility of acetylene gas becoming a competitor of electricity as an illuminant, has greatly disturbed the minds of a considerable portion of the electric lighting fraternity. There is no room for doubting that the illuminating power of the gas is very greatly superior to that of ordinary gas ; and that if it were merely a question of light it might perhaps become a very formidable rival to electricity. But the question of cost comes very prominently into consideration and here it is that we meet the strongest argument against it. It is not merely the cost of the carbide itself that must be taken into account, but that of all the accessory devices, the secondary receiver, the pipes, fixtures, etc. As to the cost of manufacture of the carbide, there are so many conflicting estimates, statements, and claims, that we consider ourselves amply justified in taking up a very conservative position, and saying that there will have to be a very considerable degree of higher mutual corroboration and unanimity among writers on the subject before the public can be expected even to form an opinion on the subject, much less to make any investments. One writer of undoubted scientific qualifications says, "Present average cost of illuminating gas in the holders of the large gas companies approximates 30 cents per M, while the cost of acetylene gas in the holder, with calcium carbide at $37.69 per ton, would be equivalent light for light, to illuminating gas at 37 7/10 cents per M, making the cost of pure acetylene per candle power approximately 20 per cent. higher than that of ordinary illuminating gas." If acetylene were mixed with air, no doubt the cost would be lower, but the advisability of distributing the mixture through a city would be very questionable, owing to the risk of the mixing being improperly done, and the quantity of acetylene falling to such a percentage as to form an explosive combination. Any person using the gas in their houses would require a duplicate holder, so that one might be charging while the second was running. As the gas in holders would be at a pressure of over 600 lbs. to the square inch, a valve would be required to reduce it down to that required at the burners. No doubt such a valve is obtainable but would require attention which the average householder would not or could not give. A failure of the valve would entail the escape of the gas. It would appear that considerations of cost and convenience go to shew that, at the present at least, the incandescent lamp has nothing to fear from acetylene gas, which may find its way into the residence of an occasional wealthy householder, but not into general use.

---

**Antiquated vs. Modern Apparatus.** WE would bespeak a most careful study of the very valuable paper presented by Mr. Gossler before the last meeting of the Canadian Electrical Association. A very large number of our readers are personally and financially interested in electric lighting stations ; and as a great proportion of these stations date from the time when electrical machinery had not received the careful study that is now devoted to it, it is only reasonable to suppose that the apparatus used is of the very inefficient types that characterized the early days of electric lighting. Transformers have only of very late years received much consideration, but the study of the conditions under which they operate, and the principles involved, has led to very great and beneficial changes being made in their construction. The saving effected by the changes indicated by Mr. Gossler, resulted from the substitution of modern high-class transformers for the old type ones previously in use; and nothing can more vividly illustrate the difference in value between old type cheap goods and new type expensive ones than the fact, as stated by him, that the annual savings effected by the new transformers will pay for their cost in about 2½ years. In smaller electric lighting plants nothing is more usual than to make selection of machinery and apparatus on the basis of cost solely, i. e., they choose that one that costs the least money. This is really the most expensive policy to adopt, and as the knowledge of electrical investors extends, with respect to the machinery they operate, and what goes on while current is flowing, it will become more and more evident to them that to buy modern, superior, and therefore high-priced machinery, gives a far better investment than cheap stuff. One frequently meets men whose knowledge of electricity is so comprehensive that they know it all. These persons will of course never learn anything, but the earnest electrical student every day becomes more convinced of the fact that the more he studies, the less he finds he knows. The influence that transformers can exert on the profits of an electric plant is so appreciable that we recommend all owners to very carefully examine into the efficiency of that part of their installations.

---

A writer in Electricity of London, with fluent innacuracy, says the Western Electrician, notes that the new president of the "National Electrical Association" is "Mr. Frederick, who is an Englishman." Doubtless the intention was to convey the idea that Frederic Nicholls, lately elected to the presidency of the National Electric Light Association, was born in England.

## NOTES FOR ENGINEERS.

To pack piston pumps for kerosene, cup leather, such as is used in packing hydraulic pumps will be necessary.

Have a regular system for doing your work in the engine and boiler rooms and have a time and place for everything.

Long grate bars make hard work for the fireman, and he cannot always keep the back grate of the furnace in good order. A short wide furnace is the best.

When using the ordinary brass check valves, it is a good idea to use one size larger than the pipe calls for, as the water will then flow through them with less friction.

After taking a ground joint apart clean it well, and before putting it together again, oil it thoroughly and if it is to be exposed to heat use cylinder oil for this purpose.

When piping up a plant, use angle valves wherever convenient, as you will then have less joints to make up, and angle valves offer less obstruction to the passage of steam than globe valves.

For removing scale from boilers, or rust from any metal, use kerosene oil. To loosen a nut which is rusted to a bolt, saturate with kerosene. It is simple, but by all odds the most effective rust or scale resolvent.

There are two methods of obtaining the heat value of coal ; one by burning a representative sample in some kind of oxygen calorimeter, and the other is to analyse the coal and equate the elements with their heat values. The oxygen calorimeter is generally preferred, but some engineers prefer the analysis.

One pound of good coal is equal to about four-tenths of a pound of wood without regard to the quality of the latter. Some woods contain more water and sap than others, some are dense while others are porous, but considering the pure wood fibre, all woods are practically the same so far as their value for fuel is concerned.

If you are using a power pump for feeding your boiler and there is no way to regulate the amount of water delivered, connect a ½ inch pipe into the discharge pipe and also into the suction pipe, with a valve to regulate the circulating water. By opening this valve the amount of water delivered to the boiler may be diminished, and so a uniform water level maintained.

If the safety valve leaks and grinding it in affords only temporary relief it may be caused by impurities in the water causing a thin scale to form on the valve and sevt, and after the valve has been opened once it leaks until ground in again. I have known soda ash, used as a boiler cleanser, to do this, but when its use was discontinued and oil used instead, the trouble disappeared.

On taking charge of a steam plant the engineer should at once acquaint himself with the peculiarities of the engine, and next should acquaint himself with the peculiarities of the proprietor or superintendent. One is just as essential as the other, for each will need an equal amount of "managing" if the engineer is to make an unqualified success of running the plant. Some good engineers make a mistake here and fail accordingly.

If your injector has been in use for several years and is not as reliable now as it was when new, do not throw it away, calling it worn out, until you have carefully cleaned it with a solution of muriatic acid and water. Disconnect the injector, put corks in the outlets, and fill it up with the solution, letting it stand over night. Wash out in the morning with water under pressure, and see if it is not as good as new. The solution should not be stronger than about two parts of water to one of the acid.

In case of accident in your boiler room, where prompt reduction of the temperature under the boiler is necessary, too great care cannot be exercised by your fireman. As a rule he will proceed at once to "draw the fire," but if the boiler is in a critical state, such an act is certainly not wise. When a fire is disturbed, the heat which it gives out is materially increased for several minutes, and unless the entire body of the fire can be removed at one stroke, the safest plan is to smother with damp ashes or fresh coal.

Many a leaky piston or valve rod which is chronically so, could be cured by turning the piston so that the worn place at the bottom would come at the top, or by putting a liner at the bottom to carry the piston at a higher level. Sometimes the bottom part of a piston has been drilled, and the holes filled with hard Babbitt to raise the piston up into line. Sprung rings of cast iron or brass cannot be depended on to centre a piston or keep it in line, because of the wear. Babbitt plugs will serve such a purpose, and they can be renewed when occasion requires. This is an easier and cheaper way than having a new piston made

## ASBESTOS.

THERE is probably no production of inorganic nature about which there is so much popular mystery and misconception as asbestos. It is vaguely understood that the principal claim of this remarkable product to attention is that it cannot be consumed by fire, and not infrequently the effect of the mention of asbestos is to carry the hearer back to the days when the people of the Pharaohs wrapped their dead in cere-cloths, woven from fibre, in order to preserve them, the body having been first embalmed. Romantic stories have also come down to us of ancient demonstrations of magic in which asbestos has played the leading part, but the real interest in asbestos centres in the present. It is of more importance to the human race to-day than it has been in the whole range of history. Asbestos twenty-five years ago was practically not known in the laboratory of the chemist or mineralogist. It now finds its way in one form or another into every workshop where steam is employed.

To the question, "What is asbestos?" it is not altogether easy to find an answer. Geologists classify it among the hornblends. In itself, asbestos is a physical paradox, a mineralogical vegetable, both fibrous and crystalline, elastic yet brittle, a floating stone, but as capable of being carded, spun and woven as flax, cotton or silk. It is apparently a connecting link between the vegetable and the mineral kingdom, possessing some of the characteristics of both. In appearance it is light, buoyant and feathery as thistledown; yet, in its crude state, it is dense and heavy as the solid rock in which it is found. Apparently as perishable as grass, it is yet older than any order of animal or vegetable life on earth. The dissolving influences of time seem to have no effect upon it. The action of unnumbered centuries, by which the hardest rocks known to geologists are worn away, has left no perceptible imprint on the asbestos found embedded in them. While much of its bulk is of the roughest and most gritty materials known, it is really as smooth to the touch as soap or oil. Seemingly as combustible as tow, the fiercest heat cannot consume it, and no known combination of acids will destructively affect the appearance and strength of its fibre, even after days of its action. It is, in fact, practically indestructible. Its incombustible nature renders it a complete protection from flames, but beyond this most valuable quality, its industrial value is greatly augmented by its non-conduction of heat and electricity, as well as by its important propriety of practical insolubility in acids.

Asbestos has been found in all quarters of the globe. It comes from Italy, China, Japan, Australia, Spain, Portugal, Hungary, Germany, Russia, The Cape, Central Africa, Canada (Fig. 1), Newfoundland, this country, and from Southern and Central America.

Notwithstanding this wide distribution of asbestos, the only varieties which at present appear to demand serious consideration, from a commercial point of view, are the Russian, the South African, the Italian and the Canadian.

Before the development of the Canadian fields, the Italian

FIG. 1.—CANADIAN ASBESTOS.

asbestos was supreme in the market. For nearly twenty years Italy has been looked to for the best grades of the fibre. From a point on the northern mountain slope of the Susa valley is taken the floss asbestos fibre, the appearance of which in gas stoves is so familiar. In the same locality is found a fine white powder of asbestos, which serves for paint and other purposes. The mining is carried on at a height of from 6,000 to 10,000 feet above sea level.

But the Italian asbestos industry, once so important, is already on the down grade. The difficulties of mining are very great, and unduly increase the cost of production. The asbestos itself, judged by the latest standards, is of inferior quality; it is not easy to spin, and it does not pulp well in the making of paper. The best grade is extremely rare, and its cost of mining and transportation is prohibitive. The supply from the Italian mines

is rapidly falling off. As a matter of fact, Canada contains the great asbestos region of the world, in the sense that while its mines are practically unlimited in productive capacity, the product is of a quality which fully meets the requirements of the newest and most exacting of the innumerable uses that are daily being found for it.

The process of manufacture is intensely interesting, more especially from the fact that as the industry is constantly entering upon novel phases, new methods of treatment and special machinery have to be devised. One of its special uses is for wall paper.

One of the largest branches of asbestos manufacture is that of sectional cylinders for pipe coverings, for retaining the heat of steam and other pipes, felt protective coverings for boilers, frost-

FIG. 2.—ASBESTOS MINING.

proof protections for gas or water pipes, and cement filling, which can be laid on with a trowel, for the covering of steam pipes, boilers or sills. In some of these cases, where it is only necessary to retain the heat, the asbestos is mixed with other substances; but where the protection must be fireproof as well, only asbestos is used. The utility of such covering is well illustrated in the heating system of railway cars. The main pipe from which the individual cars draw their respective heat supplies by side mains, if not covered with asbestos, would lose a large proportion of its caloric from the rapid motion of the car through the air. An interesting innovation in this class of manufacture is asbestos sponge. It is not generally known that sponge has great powers of fire resistance. The discovery was made accidentally not long ago, and the result was that a consignment of scraps of sponge picked up on the Southern coasts was ordered for experimental purposes. The sponge was finely comminuted and mixed intimately with asbestos fibre. The combination was found so successful for any covering which had to be fireproof as well as heat-proof that the material has become standard. Being full of air cells, it necessarily makes an excellent non-conductor. Another very extensive department in asbestos manufacture is that of packings. Of these there are an infinite number of forms. In these days of high pressures and ocean records, it is of supreme importance to marine engineers that they should have jointing and packing materials on which absolute reliance can be placed. In order to meet modern exigencies every possible form of packing has been constructed, particularly with asbestos and metallic wire, and with asbestos and rubber cores for gland packing. The making of asbestos paper varies from the building up of the thickest millboard to the production of a writing paper which, from its indestructibility, is valuable in case of fire for preserving charters, policies, agreements and other important documents.

To the electrical engineer asbestos is absolutely indispensable. Many parts of electrical devices and machinery and wires through which the electrical current passes become heated, and were it not for the electrical insulation and heat-resisting qualities which asbestos possesses, the apparatus would be completely destroyed, particularly in the case known to electricians as "short circuiting." For such purpose it has been found advisable to combine asbestos with rubber and other gums, and this combination is now used universally for not only electrical, but also steam and mechanical purposes.

The newest departure in the asbestos field is the construction of electrothermic apparatus. The heating effect of the electric current is utilized by embedding the wire in an asbestos sheet or pad. The pad is used by physicians and nurses for maintaining artificial heat in local applications, and is said to be already largely used in hospitals. Another application of the same principle is to car heaters. A sheet of asbestos, with the embedded wires, is clamped between two thin steel plates, and the portable heater thus provided, or a series if need be, is connected to the car circuit quickly and easily. It gives an even and healthy heat, and can be so regulated as not to overheat the car.--George Heli Guy, in New York Evening Post.

## THE TELEPHONE IN RAILROAD PRACTICE.*

THE growing use of the telephone in railroad work and its present advantages and future possibilities is a subject well worthy of consideration and study.

The telephone equipment at local points best adapted to the transmission of the internal business of a railroad, depends upon the location and the degree of concentration of the offices at each point. The value of a private telephone line connecting intermediate points and the division headquarters along the line of the road is dependant to a large extent upon the number of instruments that are enabled to secure intercommunication thereby.

In connection with the speaking tube or internal telephone system, special efforts are being made by the local telephone companies to offer the railroad companies instruments and apparatus that vary with the character of the service desired. For instance:

System A—A central switch with lines radiating from it, each line having one or more stations connected with it, the whole being arranged for intercommunication.

This system is operated in much the same manner as an ordinary telephone exchange, a switch being located at some central point, provided with a means for calling and receiving calls from each station, and for connecting the several stations with each other. The switch may be located where it can be operated by some person in connection with other work, or if the system is large, the services of a regular operator may be required.

This system (if but one station is connected on each radiating line) secures secrecy between any two stations and provides for independent communication between a number of stations at the same time.

System B—A switch at a particular office with lines radiating from it, each line having one or more stations connected with it, the whole being arranged for communication to and from this particular office, but not for communication between stations on different lines.

This system is used for transacting business between a particular office and several stations in cases where it is not required that the stations communicate with each other. A switch is provided at the main office only.

This system (if but one station is connected on each radiating line) secures secrecy between the main office and any one of the stations.

System C—A switch at each station, with means for connecting the instrument at such station with lines extending to each of the other stations.

This system is so arranged that a person at any station can call any other station over a special line and establish the desired connection without the aid of an operator. It does not secure secrecy to such a degree as systems "A" or "B." A switch being located at each station, access may be had to all circuits whether in use or not, but as the bell at the desired station is the only one operated when a call is made, secrecy is fairly assured, and interruptions are not likely to occur unless the use of the same circuit should be desired by a second party and his instrument be connected for the purpose of making a call. It is possible for parties at several stations to converse independently with each other at the same time.

System D—A single circuit connecting two or more stations.

All instruments being connected upon one circuit, no switching apparatus is required. Only two stations can use the line at one time and there can be no secrecy, as a call made from any station will ring all bells simultaneously.

Systems "A" and "B" are especially adopted and serviceable for freight offices and yards, round houses, switching towers, etc.

System "C" is perhaps the most convenient and satisfactory when the stations to be connected are not numerous.

System "D" is the most simple and inexpensive.

An outgrowth from system "A" is the present private branch telephone exchange. The benefits derived from the establishment and operation of private branch exchanges seem comparatively unknown, and especially so to those who have not been closely in touch with the growth of this particular line of the business, and it is with a view of arousing interest in this direction, as well as securing additional information through the discussion which I trust will follow this paper, that I have endeavored to collect as much reliable information as possible bearing upon the subject. This very lack of familiarity with the branch exchange frequently results in a much less efficient service from a given number of telephone lines than would be secured were they merged into the so-called exchange.

"In the march of civilization the improvements of yesterday are discarded for those of to-day. The tin speaking tube once used for interior communication gives way to the telephone. In this age when only time saving is considered more important than labor saving, and the combination of both is the prime object with all active minds, the importance of rapid and reliable communication can not be over-estimated. Especially true is this of the business conducted in a large building where the labor and delay incidental to employing messengers or office boys, make an important item of expense. In the general offices of a large railroad company, where every office can be connected one with another, and the various working departments be brought into talking relations with one another, this telephone service is a time, money and labor saver; and where the heads of departments are separated from each other by doors, stairs and passages it is invaluable."

Every railroad man is familiar with the general scheme of railroad organization, and the relationship between the various departments, their chiefs, etc. The lines of authority are closely drawn, and the flow of communication naturally follows these divisional lines.

As the division of responsibility among the several officials and employees who carry on the operation of the railroad company is plainly defined, so the use of the telephone tends to parallel those divisions of responsibility and to follow the lines which separate the duties which are to be performed.

In the application of the telephone to the transaction of internal business at local points and within a certain radius of the office building or about the yards and switching centers, the numerous communications necessary are passed to and fro easily and without loss of time.

The tendency is towards the constant growth of private branch exchanges, as they give more perfect interchange of communication for every class of business, concentrate the service within certain limits and enable the business to be transmitted direct without going through the medium of the local telephone operator, and vary the class and extent of the service desired according to the price paid.

The benefits to be derived from the operation of the private branch exchange have been recognized to a greater extent in the city of Chicago than elsewhere in the country. As a matter of fact there are at present in that city over 130 private branch exchanges, operating an aggregate of over 1,200 telephone instruments. These exchanges are operated by railroad and express companies, large wholesale and retail establishments, manufacturers, etc., and range in extent from four to 100 instruments. They are connected by means of trunk lines with the local telephone company's exchange, so that connection may be had with the public.

In a great many cases a particular telephone, while greatly needed for the handling of railroad business, has no occasion for public connection. If arranged so that they can secure such connection, the result is that the telephone will be used more or less for private ends; consequently, when the public trunk lines are required for legitimate railroad business, they will be reported "busy," while as a matter of fact, they are being used for private business.

To obviate this evil and to furnish as nearly as possible what is absolutely required, the local telephone company has recently adopted a scheme whereby it is made impossible to give certain offices public connection, although they are able to secure unrestricted intercommunication with every line radiating from

* Paper read before the Association of Railway Telegraph Superintendents, Fortress Monroe, Va., June 17, by W. W. Ryder, Chicago.

the branch exchange. In giving this limited service, the telephone company charges considerably lessened rental, although securing for the subscriber a more efficient service by not allowing the unnecessary blocking of his down-town trunk wires. This difference in expense together with the difference in price between public and branch exchange lines is almost, if not quite, sufficient to pay the salary of the telephone operator even though you have only a small number of lines, and this naturally increases with the greater extent of the system.

The success of the system can best be indicated by the statement that of all the branch exchanges put in operation in the city of Chicago, only one has ever been taken out through dissatisfaction with the system, and in this case it was only a short time before the telephone company was requested to immediately replace, the firm finding that the inconvenience and loss of time were greatly increased when the exchange was closed.

The growth of the private branch exchange system must soon extend along the lines of the individual roads; in fact, at present the Pennsylvania Railroad Company has in operation a very complete system which gives them direct connection over wires entirely controlled by them between all division headquarters on their road east of Pittsburgh. Through the courtesy of that company I am permitted to exhibit a diagram of this system. They have branch exchanges at all division headquarters and have leased from the Long Distance company necessary wires to complete connections with these points. Other large eastern lines, I understand, are now contemplating the option of this same scheme.

With the growth of the private exchange idea, these exchanges will rapidly multiply in large cities and the necessity for means of intercommunication between them without going through the public exchange will become imperative; in fact, in the city of Chicago, at present, where branch exchanges are being operated by the Chicago & Northern Pacific, Chicago, Rock Island & Pacific, Chicago, Milwaukee & St. Paul, Chicago & Eastern Illinois, Illinois Central, Chicago & Northwestern and Chicago, Burlington & Quincy Railroad companies this necessity is very noticeable, and the local telephone company is considering the question of trunking the different exchanges together. With this accomplished, it is but a step to the connection of the branch exchanges in one city with those in another over wires controlled by the railroad companies. How this can best be done can only be decided by trial, and I believe we will have to meet this particular issue at a very early date.

When we consider the rapid growth of the telephone system, it seems a question of only a short time before the telegraph will be largely superseded by the telephone. It has been shown in actual practice in commercial service that messages of 30 words can be read and intelligently transmitted in a quarter of a minute, or 120 words per minute, which is about 3,900 better per hour than the average by Morse, using the Phillips code and the typewriter. The above record is taken from a guaranteed service where the toll service is daily performed on this basis.

The question of the telephone not being able to compete with the telegraph on account of the lack of records was happily answered, you will recollect, by Superintendent Selden in a paper read before this association at the 1894 meeting, and this feeling, I believe, is rapidly passing away.

The despatching of trains by telephone has been tried with perfect success in several instances in this country. This is the most exacting of service, and the fact that it is a success speaks volumes for its efficiency.

It is a well known fact that large corporations are slow in adopting radical changes, but the improvements in telephone apparatus are so marked and the benefits derived from its use so evident, that they are being forced to recognize its merit and consequently are rapidly advancing the movement.

---

One-half a square inch of piston area per horse power is a common rating for steam engines.

If the girth seams of a tubular boiler leak and chipping and caulking do not stop it, be sure there are no cracks in the plate, or that defective rivets are not the cause of it. If the boiler is sound, the trouble may be caused by unequal contraction of the plates due to the introduction of comparatively cool feed water into the bottom of the boiler. If the location of the feed pipe is changed the leakage may cease without further attention. The water should be discharged into the body of the water already in the boiler, and not on the bottom sheets.

## ELECTRIC LIGHT INSPECTION.

The divisional inspector of electric light, Mr. Wm. Johnson, of Belleville, was here several days last week, during which time he has inspected all the electric light meters and also has been looking into the voltage or pressure carried by the company. It will be remembered that the Government fitted up apparatus in the post office building, but to suit the convenience of the Light, Heat and Power Co., the inspection is now done at the premises on William street. Mr. Johnson was not stinted in his praise of the test board and other appliances supplied by the ingenuity of Mr. H. E. Reesor, who, Mr. Johnson says, in the fitting up of these, has shown his ability as an electrical engineer. It has proved fortunate for the users of electric light meters that they have been brought under Government inspection, if they are everywhere as they are here. It must be understood that the company here accepted the meters from the manufacturers as correct and had their guarantee that they were; but, when tested by the Government standards, the majority of them have been found to be too fast, or, in other words, against the consumers. One meter only was found too slow and it was sixteen per cent. that way, while many of them were from seven to nineteen per cent. fast. All the meters in town have now been adjusted or regulated in the inspector's presence and have been sealed by him. The inspector explained to us that the Electric Light Inspection Act requires each company to state to all its customers the rate of voltage at which it will supply the electricity; the company here proposes to do this at a voltage of 104, but the inspector found that the company was furnishing it at from 109 to 112 volts and informed the company that it was liable to a penalty for increasing or diminishing the voltage beyond or under three per cent. of 104 volts. The reason for this provision of the law is that if the voltage is greater than the amount specified it destroys the lamps, or if less than it should be the light is diminished. A number of our citizens visited the Light, Heat and Power Company's office while the inspector was here, and had explained to them the interior of that mystical looking object an electric light meter, also how the meters were tested, and the meaning of some of the technical terms used by the electric light fraternity.—Canadian Post, Lindsay, Ont.

---

The inspection by the Government of all the electric light meters in town was concluded this morning, nearly one hundred and fifty having had the red seal attached to them. The Government regulations give the electric light companies until the first of next June to have all their meters tested, after which it will be unlawful for them to use any other. A penalty of $25 is to be inflicted after that date on any company or person who uses a meter which has not been inspected and stamped.

The Peterborough Light and Power Co. has done a popular thing in having its meters tested at once, and in this way again has given evidence of how well it keeps its finger on the public pulse, and have met the universal clamor for inspected meters.

The inspector, Mr. Wm. Johnston, informs the Review that of the hundred and fifty meters tested, about a dozen were incorrect, six were 5 p. c., a couple 8 p. c., and two 12 p. c. too fast, or in favor of the company, while one, where the customer uses ten lights, had not registered but a small percentage of the energy, owing to part of the gearing having got wrong, the customer's bill for last quarter having been only $1.80. These cases, however, prove the value of the inspection. By a strange "irony of fate" the two meters that were twelve per cent. too quick were in the residences of two of the officers of the company.

Mr. Johnston also says that the work of inspection, which it was first intended should be done at the gas inspector's office in the Custom House, was accomplished much more quickly at the offices of the Light and Power Company where through the ingenuity of Mr. Fisk, the company's clever electrician, facilities were provided.—The Daily Review, Peterborough, Ont.

---

If the guides on an engine are made separate from the frame, they may be taken off and planed when they need it, but if they are cast with the frame this cannot be done.

For lubricating pump rods, a very good mixture is made from tallow, cylinder oil and plumbago; and if the water is warm, it is better to add a little beeswax. This, mixed with the fibers of the rod packing, will greatly improve the ease of running and will keep the rod in good condition; and, in fact, this and good waste may be used to replace expensive packing if the waste is properly laid up.

## · WIRE INSULATION.

### BY H. W. NELSON.

THE insulation of wire for electrical purposes has grown into a large and important industry in this country during the past ten years. For the lack of the right kind of commercial and scientific attention it has grown up very badly in certain lines, viz., those relative to lighting and transmission of power, which are the lines covering by far the larger part of the business.

On the other hand, the manufacture of telegraph and telephone wires, notably the telegraph, has been brought up to a splendid state of efficiency, both commercially and scientifically. The reason is not far to seek, when we consider that such men as Lord Kelvin, Edison, etc., have not thought the minutiae of this branch too small to engross their colossal minds, and that the manufacturers have co-operated with them to turn out a good commercial article. In lighting and power transmission work, attention to the fine details of wire insulating has been positively shirked by the technical men, they having left their part to crude, untrained minds. A glance at the patents list, with its hundreds of ridiculous, foolish specifications for insulating wire, presents evidence of uneducated dabbling.

In conjunction with this neglect of the engineer there has been an almost entire absence of co-operation of the business man with the technical.

As a consequence of the striving of the one for cheap wire and the other for high quality, and no attention to the intervening details, the market has run into two channels : On the one hand a cheap and very poor insulation, and on the other, a high quality and very high price. There are a few grades between, of insignificant amounts, which do not affect the argument.

A friendly association on the part of these men, with a little more regard to the importance of the details involved, would in all probability have made a market for a fair-priced medium wire. For an instance, a wire is needed for interior and hidden work, to go on a 52-volt circuit alternating current from a 1,000-volt main line, or on a 125-volt circuit direct, constant current. For this a wire is demanded having an insulation resistance of from 800 to 1200 megohms per mile, and unless the engineer and underwriters are hoodwinked, a high-priced rubber-covered wire is put in which is capable of withstanding without rupture the shock of from 5,000 to 10,-000 volts alternating. This appears an excessively large factor for safety, but with their present knowledge those on whom the responsibility lies cannot accept anything less costly, and take the risk of perhaps an early breakdown of their insulation. They have a general knowledge that an insulation compound made up with a large percentage of pure rubber will resist water and not be short-lived, and that a more attenuated compound, or another compound, may be good or rubbish. They cannot tell without the test of time, and not being familiar with its manufacture, they will not accept risks on another's ipse dixit. They therefore must stick to an extra superfine where an ordinary wire would do.

This ordinary wire, by which is meant a fair insulation resistance sufficiently long-lived, at a medium price, would undoubtedly be forthcoming if the market demanded it. In this regard, however, when the engineer does not demand anything more than an insulated wire, the purchasing agent has an opportunity to get something cheap, and the lowest tender gets the contract. This is the place where very poor stuff masquerades as electrically-insulated wire. One very prominent kind, which is literally a whited sepulchre, is a wire covered with a braid or wrap of cotton, or other fibrous material, very hygroscopic, which is saturated with pitch, or some much vaunted insulating paint, to render it non-hygroscopic, which it does not; the whole then receives a plaster of whiting and fish-glue, or similar compound, to render it fire-proof, which it does not.

Then again, where the high cost wire is put in, the ends on the cut-outs, rosettes, etc., are often left bare or worse by being insulated with sticking tape, made of cotton (hygroscopic) and poor rubber compound which quickly oxidises, and against such weak spots a wire having an insulating resistance of 200 megohms per mile should be more than ample. In addition to this the flexible drop cords to the lamps have simply been called for in specifications as rubber-covered lamp cord, and the purchasing agent buys the cheapest article which can legally be labelled "rubber-covered." If a drop of salt water be dropped on this cord when the circuit is closed its quality will probably show up in a very bright way.

The insulation called "weatherproof," which is not weatherproof, however, serves the purpose for which it is generally used very well. It is generally used as a line wire on currents at a low voltage, and providing that the pole insulators are good, it simply serves as a separator, preventing a dead short circuit if stray wires of low voltage touch it. In choosing this wire, if the choice were made more with regard to its usefulness and life, and not so much to the highly polished surface, a saving in renewals might be effected without an extra first outlay, in that the money saved by foregoing the extra work in fancy finishing could be put into the material, by having a heavier, stronger covering.

It must be remembered that this covering only acts as a separator, that it soaks up water almost as a sponge, and that the end to be gained is that the covering required be strong enough to resist the rough usage it gets from the kerb-stones, posts, trees, road-gravel, etc., when the linemen are stringing it, and from the sun and rain afterwards. To this end it is necessary to put on two or more strong jute or cotton braids, and to saturate them with a compound which will stick the braids to the wire and preserve them from rotting. There can be very little of the polish left when the wire is stretched on the poles after this handling.

The above somewhat short and imperfect remarks, if they succeed in calling attention to a very backward branch of electrical work, may suggest many ideas for improvement.

(1). It may be suggested that some of our prominent consulting engineers (men above financial interest in any particular manufacture) make a special study of this subject, in order to have more than a mere general knowledge of it.

(2). That the leading fire insurance companies together engage a man thoroughly experienced in wire insulating as a permanent inspector of insulations. (N. B. They already do something abortive in this direction.) That they fit him up with a laboratory, and have a sample of all wires tested and put on record before they are allowed to be strung, or

(3). That the government appoint this official and

(4). That a law be passed making it a misdemeanor for anyone to string wires a sample of which has not been officially accepted by the inspector.

(5). That our colleges give some open lectures on the chemistry, etc., of caoutchouc, resins, cotton, silk, waxes, and so forth, which are or may be useful as dielectrics or protectors to dielectrics.

A course on such lines as this would spread the gospel of insulation, and would make for progress, in spite of the stumbling blocks our rule of thumb men prove themselves to be, with their shellac and resin or such like compounds, of which they make such a "dark and bloody mystery."

## THE UTILITY OF ELECTRIC CLUBS.

ELECTRICITY is now recognized as one of the greatest factors in the commercial world, and the number of new enterprises coming into existence, and for which the services of competent electricians, engineers, etc., are required, emphasizes the necessity of intending applicants such positions thoroughly fitting themselves for the same.

The formation of Electric Clubs in the different cities is probably one of the best means of education and improvement. The opportunity is thus afforded for the interchange of ideas and the presentation of papers on practical subjects of interest. The Montreal Electric Club enjoyed a period of usefulness during its existence, but owing to the removal from Montreal of some of its members, and the fact that the work was left largely to a few it has ceased to exist. We are pleased to learn, however, that an effort will be made this fall to revive the Club. It is a significant fact that many of the prominent members of this Club, which was composed largely of the younger electricians, have secured responsible and lucrative positions.

In the city of Toronto there is also a good field for the organization of a similar club. The number of electricians, engineers, students, etc., in Toronto, should be sufficient to ensure a fair membership. During the winter months meetings could be held, say twice a month, at which papers should be presented and discussions held upon subjects relating to the various departments of electrical work. It is hoped that steps in this direction may be taken before the season is too far advanced.

## MODERN PRACTICE IN INTERIOR WIRING.

IN the course of his paper on the "Evolution of Interior Conduits from the Electrical Standpoint," before the National Electric Light Association at New York, Luther Stieringer made the following statement :

The best experience in the past fifteen years in interior wiring has demonstrated the following facts :

First—Indiscriminate wiring with staples is universally condemned.

Second—Cleat wiring is admissible in exposed work where the circumstances admit, but not in concealed work.

Third—Wires imbedded in plaster, depending on the insulation only for protection, are condemned.

Fourth—Lead-covered wires are also condemned, except where protected in a conduit.

Fifth—Wires in mouldings do not afford mechanical or chemical protection, and are only admissible in surface work.

Sixth—Wires carried in plaster, and covered with split or zinc tubes to prevent injury by trowels, are condemned.

Seventh—Glass or porcelain insulators can only be utilized in special cases of exposed work.

Eighth—Paper tubes do not afford absolute mechanical and chemical protection.

Ninth—Insulated tubes covered with a thin coating of brass or other metals do not afford absolute mechanical and chemical protection, but in exposed work they are to a certain extent admissible.

Tenth—Woven fabric conduit does not afford absolute chemical protection.

Eleventh.—Heavy insulating covering, integral with the insulation offers no absolute protection against mechanical and chemical injury, and is analagous to rubber tubing for gas distribution installed throughout a building.

Twelfth—Concentric wiring is practiced in England with satisfactory results, but it is not in use in the United States. It offers many possibilities in the direction of a solid and fixed system.

Thirteenth—Paper-lined iron or steel pipes, known as "iron-armored conduit," "builders' tube," "armorite," "Clifton," and plain iron or steel pipe, are the only conduits that can afford absolute security against mechanical and chemical injury and assure permanence.

## CANADIAN ASSOCIATION OF STATIONARY ENGINEERS.

NOTE—Secretaries of Associations are requested to forward matter for publication in this Department not later than the 25th of each month.

### TORONTO, NO. I.

THE regular meeting of Toronto No. 1 was held on the 15th of July. A pleasing feature of the meeting was the presentation of a family rocking chair each to Bro. T. Eversfield, engineer at Toronto University, and Bro. Wm. Butler, engineer at Nordheimer's piano factory. The presentation was made by the President, Bro. J. Fox, on behalf of the Association.

### HAMILTON NO. 2.

AT the regular meeting on the 3rd of July, the newly-elected officers were installed by Bro. A. E. Edkins, after which he gave a brief address in connection with the approaching annual convention. Bro. Pettigrew also spoke along the same line.

### BRANTFORD NO. 4.

The following is a list of officers of the above association for the term ending June 30th, 1897 ; President, J. B. Forsyth ; Vice-President, Jos. Ogle ; Secretary, Thos. Pilgrim, Continental Cordage Co.; Treasurer, L. Fordham ; Conductor, F. Temperance ; Door-Keeper, A. McKinnon.

### PETERBORO NO. 14.

At the regular meeting of Peterboro Branch No. 14, held in Engineer's Hall, the following officers were elected : President, W. L. Outhwaite ; Vice-President, W. Forster ; Secretary, A. E. McCallum ; Treasurer, W. Taylor ; Conductor, G. Pogue ; Door-Keeper, P. Milloy. Mr. Outhwaite was appointed as the representative to the annual convention in Kingston.

### BROCKVILLE NO. 15.

James Aikens, Recording Secretary of the above branch, writes : On the sixth of July we met in our rooms for the purpose of electing officers for the ensuing year. The following was the result : President, Archibald Franklin ; Vice-President, John Grundy ; Recording Secretary, James Aikins ; Treasurer, John McCaw ; Financial Secretary, Wm. Robinson ; Conductor, Fred.

Andrews ; Door-Keeper, John Boyd ; Trustees, Ernest Carr, Edward Devine, James McRitchie.

Immediately after the election, the Past President, Bro. W. F. Chapman, proceeded to instruct the new officers in the discharge of their duties, his remarks being well received. The next part of the programme was a speech by Bro. Albert E. Henry, on the benefits which he received in the way of technical knowledge by joining the C. A. S. E. We all hope to see No. 15 prosper in the future as it has done in the past two years, when Bro. Chapman was leader, and no doubt the President's chair will be ably filled by Bro. Franklin, the veteran, and chief engineer of the water works in this town.

THE ANNUAL CONVENTION.

The local association at Kingston have made every arrangement for the entertainment of the delegates to the annual convention to be held in that city on the 18th and 19th inst. It is expected that about one hundred delegates will be present, and as Kingston is favorably situated for a summer meeting, a pleasant as well as a profitable time is assured.

Mayor Elliott has consented to deliver an address of welcome in the Council Chamber, after which a business session will be held. On the second day the delegates and their friends will sail down the river by special steamer, visiting some of the most picturesque islands of the St. Lawrence. Among other social features will be a drive to Fort Henry and around the Kingston Mills, a visit to the penitentiary and other places of interest, a lawn party at Ontario Park, and a banquet at the Hotel Frontenac on the evening of the last day, by the courtesy of the local association.

The business programme was not finally arranged at time of going to press, but it is expected that some interesting papers on engineering subjects will be presented, and other questions of interest to the association brought up for discussion.

The members of the Kingston association are working faithfully to ensure the success of the convention, and it is hoped their efforts will be rewarded by a large attendance of members.

## A NOVEL INSTALLATION.

THE Royal Electric Company recently installed at Peterboro one of their synchronous motors to operate a stone crusher, used by Messrs. Corry & Laverdure, contractors for the construction of the Trent Valley Canal, to crush all stone required for that section of the canal.

The Peterboro Light and Power Company furnish the current operating this motor from their 180 kilowatt " S.K.C." generator recently obtained from the Royal Electric Company.

This plant is interesting because it is, we believe, the first of this kind and the only one in commercial operation in Canada, and because it indicates a useful, profitable and practical direction in which central stations can employ their plants during the period of the day when lighting is not required.

Mr. J. F. H. Wyse, representative of the Royal Electric Company, who directed the installation, spoke of it as follows : "The current to operate the stone crusher is transmitted from the station of the Peterboro Light and Power Company, a mile distant. The motor plant consists of a fifty kilowatt alternating current synchronous motor, with its exciter, and a five horse power starting motor. The alternating current is taken

by the motor directly from the transmission line at 1,000 volts. The stone crusher is belted to one end of the shaft of the synchronous motor, to the other end being belted in tandem the exciter and starting motor.

It was intended at first to use a shifting device or clutch arrangement to connect and put into operation the stone crusher after the synchronous motor had attained the required speed. This plan was changed, however, in order to simplify the arrangement, and the five horse power starting motor relied upon to bring up to speed the synchronous motor with stone crusher attached, as well as to drive the exciter.

Although this demanded more power than the rated capacity of the motor, it did the work with ease, and readily brought the synchronous motor to above the required speed.

To indicate to the attendant on the stone-crushing plant the proper time to connect the synchronous motor with the alternating current transmission line from the station of the Peterboro Electric Light and Power Company, a regular " S.K.C." synchronizer, as made by the Royal Electric Company, is used, which consists simply of two of their " 2 C " Stanley transformers, so connected that when the synchronous motor is at the required speed and in step with the generator, a mile away, a lamp connected with these transformers goes out, giving positive indication to the attendant when to connect the motor with the transmission line.

The plant was started June 17th and the stone crusher has been successfully doing its work every day. It has been put to the utmost test ; the greatest possible loads have been put on ; the crusher has been jammed full of the hardest stones obtainable ; the greatest variations possible in load, from nothing to the extreme capacity of the crusher, have taken place rapidly, but no variation in speed occurred, the synchronous motor meeting every demand upon it without change.

Messrs. Corry & Laverdure express themselves as more than pleased at the operation of the plant. They have also bought another motor from the Royal Electric Company to operate a pile driver.

## THE VALUE OF ADVERTISING.

ONE of the largest advertisers in London says : "We once hit upon a novel expedient for ascertaining over what area our advertisements were read. We published a couple of half-column ads. in which we purposely misstated half a dozen historical facts. In less than a week we received between 300 and 400 letters from all parts of the country, from people wishing to know why on earth we kept such a consummate idiot, who knew so little about English history. The letters kept pouring in for three or four weeks. It was one of the best paying ads. we ever printed, but we did not repeat our experiment, because the one I refer to served its purpose. Our letters came from school-boys, girls, professors, clergymen, school-teachers and, in two instances, from eminent men who have a world-wide reputation. I was more impressed with the value of advertising from those two advertisements than I should have been by volumes of theories."—Exchange.

A second edition of the Inventor's Guide has been issued by Messrs. Ridout & Maybee, patent solicitors, Toronto. It has been considerably enlarged, and contains, in addition to other interesting features, a table containing economic statistics of the population, area, industries, etc., of the different countries of the world.

In re appeal of the New Westminster and Burrard Inlet Telephone Company, the Supreme Court of British Columbia held that telephone wires, whether carried above or underneath the soil of the highway, are liable to be taxed by the city of Vancouver. A switchboard is not a fixture and therefore not liable to be taxed.

## SPARKS.

The Brantford Electric Light Company propose putting in a new plant.

Only one tender was received by the Toronto City Council for the telephone franchise for the city.

The employees of the Bell Telephone Company, Montreal, held their annual pic-nic on the 25th of July.

The town of Goderich has passed a by-law to introduce the incandescent system of electric lighting.

The Citizens' Light & Power Co., Cote St. Paul, Montreal, will build an addition to their engine house.

Mackay & Guest, of Renfrew, Ont., intend erecting an isolated water power plant and putting in another machine.

The City Council of St. Thomas, Ont., will submit a by-law to the ratepayers for the establishment of an electric light plant.

The Lachine Rapids Hydraulic and Land Company have been granted permission to increase their capital stock to $2,000,000.

Judge Thos. Deacon has decided that the Bell Telephone Company at Arnprior must pay taxes on $1,000 worth of real property.

A Chicago lawyer has defined a promoter as follows : " One who sells nothing for something to a man who thinks he is getting something for nothing."

Mr. Nicola Tesla is announced to have discovered a method of successfully transmitting electricity, upon a commercial basis, over a distance of at least 500 miles.

The Board of Governors of the Hamilton general hospital are considering the question of installing an electric light plant. An estimate for a plant of 250 lights places the cost at $2,500.

It is said that the Lake Superior Power Company, of Sault Ste. Marie, Ont., will go extensively into the production of calcium carbide, the substance from which the new acetylene gas is manufactured.

Arthur Gagnon, a Bell telephone lineman, while working on one of the company's poles on McGill street, Montreal, came in contact with a live wire and fell forty feet to the ground, being instantly killed.

The city of Vancouver, B. C., recently made a contract for lighting the city at a cost of 27½ cents per night per lamp of 2,000 candle power. This is stated to be the lowest rate yet obtained by any city of less than 30,000 population.

Said the maiden, archly smiling :
" Why all this cathodic fuss ?
Men should know we've long seen through them,
But they'll never see through us."
—San Francisco Examiner.

The town of Trenton, Ont., has moved for an interim injunction to restrain the Trenton Electric Light Company from supplying electricity to persons outside the town and from using poles planted in the streets of the town for that purpose. The case will be heard at Cobourg in September.

Incorporation is announced of the Little Salmon River Telephone Company, for the purpose of constructing a telephone line between Sussex, Clover Hill, Waterford and Havelock, N. B. The promoters are Messrs. S. H. White, W. J. Mills, A. L. Price, C. J. Armstrong and H. B. Price, of Sussex.

The Toronto Street Railway Company have under consideration the construction of an electric road from Hamilton to Toronto. The road will, in all probability, be an extension of the present line to Long Branch. Mr. McCulloch, electrical engineer for the company, is making a survey of the route.

The Auburn Light and Power Company has been organized at Peterborough, Ont., and is applying for incorporation. The personnel of the company is Messrs. James McKendry, M.P., John Carnegie and W. H. Meldrum, manager. The object is stated to be to supply electricity for power and lighting purposes.

The town of Peterboro, Ont., recently asked tenders for electric street lighting. The Peterboro' Light and Power Company tendered at the following figures for 300 nights in the year : Two year's term, $65.00 per light ; three years, $62.50 ; five years, $57.50. The Auburn Light and Power Company tendered at $75.00, $73.50 and $72.00 respectively. No action has as yet been taken by the council.

The Lachine Rapids Hydraulic & Land Co., of Montreal, having acquired the rights from the Standard Light & Power Company to place wires under ground in the city, have applied to the city council for permission to proceed with the work. The company propose to use cement lined iron tubes, similar to those used in many cities in the United States, and vitrified tile after the style of the Niagara Cataract Construction Company's plan, and for sub-mains Edison's tubing, filled in with asphalt by hydraulic pressure.

A despatch from Chicago, dated July 27th, says : A combine has been formed for the purpose of maintaining prices, by the leading manufacturers of incandescent lamps in the United States. This agreement will practically put an end to the war in prices, which has virtually done away with all the profits in this line of business. The factories and corporations under the new combine are : The General Electric, the Bryan-Marsh Company, Columbia, Packard, Westinghouse, Buckeye, Sunbeam, Adams-Bagnall, Perkins, Bernstein, Beacon and Warren.

## PERSONAL.

Mr. Wm. MacKenzie, President of the Toronto Railway Co., has just returned from England.

Mr. William Ahearn, jr., has been appointed manager of the Ottawa Porcelain and Carbon Co., vice Mr. J. W. Taylor, resigned.

M1. L. B. McFarlane, of Montreal, has been appointed General Superintendent of the Bell Telephone Company of Canada.

Mr. H. B. Spencer, managing director of the Hull and Aylmer electric railway, has taken an office in the Central Chambers, Ottawa.

Mr. C. R. Hosmer, president and general manager of the C. P. R. Telegraph Company, has recently returned from a trip to England.

Mr. Harry Nuttall, financial secretary of Montreal No. 1 C.A.S.E., is at present in England paying a visit to his friends in his native land.

Mr. Geo. W. Sadler, of the well-known firm of Robin, Sadler & Haworth, has been elected as alderman to represent St. Antoine ward, Montreal.

Mr. De Hart, superintendent of the London, Ont., street railway, is said to have received an offer to manage a street railway in a large city in New Mexico.

Mr. C. J. Mullen, formerly electrician of the Ottawa Electric Railway Company, and who has recently been in South America, has arrived at Burban, South Africa.

Mr. Chas. Aird, inspector, and Mr. Geo. M. Seguin, cashier, of the Ottawa Electric Railway Co., have been appointed train master and accountant respectively for the Hull Electric Railway.

The death is announced in London on July 7th of Sir John Pender, one of the original promoters of the Atlantic cable, and prominently connected for 40 years past with sub-marine cable companies and enterprises. Sir John Pender had reached the advanced age of 80 years.

Mr. H. W. Kent, manager of the New Westminster and Burrard Inlet Telephone Company, was married on the 8th of July to Miss Florence Emily Findley, of Charlottetown, P. E. I. Mr. Kent is well known throughout the western province, having been connected with the establishment of a number of telephone systems.

Mr. Gordon J. Henderson, of Montreal, has been appointed manager of the Hamilton Electric Light and Power Co. Mr. Henderson is a brother of Mr. C. W. Henderson, electrical contractor, of Montreal. Mr. J. J. Wright, of Toronto, under whose management the company has been for some time past, has been appointed on the Board of Directors.

## TRADE NOTES.

The W. A. Johnson Electric Co. have been awarded the contract for a 40 arc light dynamo for the electric plant at Toronto Junction.

The St. Johns Electric Co., St. Johns, Newfoundland, are making extensive additions to their plant. E. Leonard & Son, of London, are supplying the steam plant.

The Le Roi Mining Co., of Rossland, B. C., have placed an order with the Ingersoll Rock Drill Co., of Montreal, for a large direct-acting winding roll, 24 x 40, and for three 125 h.p. boilers.

The following is a partial list of motors installed by the Kay Electrical Mfg. Co. during the last month : Messrs. Buntin & Reid, Toronto, 4 h.p. motor ; Central Press Agency Co., Toronto, one electrotyping dynamo ; Linden Creamery Co., Toronto, one 5 h.p. motor ; Wherle Brush Co., Toronto, one 5 h.p. motor ; Mr. Hutchison, wood yard, Toronto, 10 h.p. motor ; Steel-Clad Bath & Metal Co., Toronto, one electroplating dynamo and one 1 h.p. 4-pole motor ; Mr. A. Moore, Toronto, one 2 h.p. motor ; Mr. Woods, printer, Toronto, one 2 h.p. motor ; Kemp Mfg. Co., Toronto, two 6 h.p. motors ; McLean Publishing Co., Toronto, one 6 h.p. motor ; Mr. B. Lindman, Toronto, one 2 h.p. motor ; Mr. H. R. Cuddon, St. Catharines, one 3 h.p. motor ; Mr. G. C. Hinton, Victoria, B. C., one 3 h.p. motor and one 6 h.p. motor.

# ELECTRIC RAILWAY DEPARTMENT.

## NEW RAILROAD MOTOR.

A TEST of a new electric motor, the invention of Mr. Nicola Tesla, will shortly be made at the works of the Westinghouse Company in Pittsburgh. ·The motor is destined for use on the elevated railways in Boston, and is a polyphase or induction motor, applying an alternating current, which is said to be preferable for long-distance transmission. Its distinctive characteristic is the utilization of the rotating magnetic field. It does away with the commutator and the brush, necessary to the use of the direct currents in action. Mr. Tesla states that the discarding of these makes his motor less costly —an important consideration—more reliable, easier to handle, and less perilous to those who handle it.

## THE HURON AND ONTARIO ELECTRIC RAILWAY.

THE Huron and Ontario Electric Railway Company are slowly but steadily completing arrangements for the construction of the road. According to the act of incorporation, the capital stock of the company is to be two million dollars. Mr. N. McNamara, of Walkerton, is president, Dr. Rollston, of Shelburne, vice-president, and Mr. A. McK. Cameron, of Meaford, secretary. The road will extend from Port Perry to Kincardine, with two branches, one running north from Priceville, through Meaford, Owen Sound, Tiverton, etc., around to Kincardine, and the other extending from Walkerton, through Mildmay, Teeswater, and Lucknow to Goderich, with a connection between Lucknow and Kincardine through Ripley. The entire length of the road will be something over 300 miles, and motive power for its operation will be supplied from stations at Eugenia, Glen Roden, Southampton and Thompsonville.

The company is authorized to issue bonds to the extent of $10,000 per mile for construction purposes, and $6,000 additional for each mile double-tracked. At a meeting of the shareholders held in Toronto recently an offer for construction was received from a New York firm. It was stated that most of the municipalities interested had passed resolutions adopting the by-laws and agreements with the company. A survey of the route is now being made by engineers. This will occupy about two months, after which track-laying will be proceeded with.

In Chemnitz, Saxony, no poles are used for operating the electric street railway. The method of stringing wires is by means of ornamental rosettes fastened into the woodwork or walls of houses, having projecting hooks to which the wires are attached. Those hooks are firmly fastened and are tested with seven times the weight they are called upon to bear. The railway tracks are level with the pavements, and accidents are rare. The cars run at a rate of 220 yards a minute in the centre of the city. No conductors are employed, the motorman being the only person on board who represents the company. By doing away with conductors the company saves 44,000 marks annually. The fare is only 10 pfennigs, or a trifle less than 2½ cents, on all routes, including transfers. Should 150,000 persons evade payment in 12 months, the loss would be only 15,000 marks. It would take 450,000 evasions in fare to offset the company's savings by dispensing with conductor's salaries. Fare boxes are attached to both ends of the car.

## SPARKS.

George Beattie was killed by a trolley car on the Hull & Aylmer electric railway.

A. W. Prestine, a carpenter of Hespeler, Ont., was killed on the Galt, Preston & Hespeler street railway by falling between the motor car and trailer.

An exhibition of Reynold's self-loading electric car was given in Montreal recently under the supervision of Mr. St. George, City Surveyor. The work done was satisfactory.

In lieu of privileges granted by the city of Hull, Que., the Hull and Aylmer Electric Railway Company have agreed to light the city for five years with thirty-two candle-power lamps.

Arrangements are being made in St. Johns, Nfld., for the construction of an electric railway, to operate between the city and suburban villages within a distance of twenty miles. The plant will be driven by water power.

The Canadian Electric Railway and Power Co. is seeking power from the Dominion Government to build an electric railway from Cobourg via Port Hope, Bowmanville, Oshawa, Whitby, Toronto, Oakville and Hamilton to Suspension Bridge and Niagara Falls.

The Vancouver, Victoria and Eastern Railway and Navigation Company is applying for incorporation to construct telegraph and telephone systems along the line of a proposed railway from Vancouver, B. C., through Manitoba, Ontario and Quebec to the Atlantic seaboard. The solicitor for the company is Donald G. Macdonell, of Vancouver.

Experiments have recently been conducted in New York by the New York Central Railroad Company, with a new hot water motor. The hot water, under enormous pressure, is stored in supply boilers and then charged into the battery cylinders of the motor. The great merit of the motor is said to be its cheapness. The cars can be operated upon any track, all that is required being a number of boiler houses along the road.

An effort is being made by the citizens of Cote des Neiges to compel the Montreal Street Railway Company to extend their line along Grey street and up Cote des Neiges Hill. The company object to building the line up the hill on the grounds that there is little traffic and the danger to life would be very great. The matter has been referred to the city attorney, with the object of learning who is the competent authority to determine where lines should be built in accordance with the terms of the franchise.

The belt line railway around Toronto which was constructed some years ago by the Grand Trunk Railway Company did not prove a paying investment, and has not been operated for some time. A company is now being formed, to be known as the Toronto Radial Railway Company, to acquire the property and franchise of the said railway, with the object of electrifying the road, and with the privilege of making extensions within a radius of fifty miles. Messrs. Dewart & Raney, Toronto, are solicitors for the company.

The directors of the proposed Carp, Almonte and Lanark Railway held a meeting on Wednesday last, when it was decided to begin preliminary surveys at once. It is proposed to have the line run from Carp to Bridgewater, a distance of some 68 miles, passing through Almonte and Lanark. From Bridgewater the line will connect with the Central Ontario R. R. and the Grand Trunk. Among the promoters are Mr. T. W. Rains, president, and Messrs. W. H. Stafford, D. M. Fraser, D. Shaw, Dr. Groves and J. W. McElroy.

The Ottawa Electric Railway Company provide amusements for their patrons at the parks owned by the company adjacent to the city. On the 22nd ultimo an exhibition of Edison's latest invention, the Vitascope, was given at " West End " park. The Vitascope is an improvement on the Kinetoscope, and instead of objects being reproduced in miniature in a cabinet, they are thrown in life size on a large screen, just the same as lime-light views. A view of Prospect Park, Brooklyn, showing foot passengers, bicyclists and horses passing, was an interesting feature, as was also the breakwater at Coney Island.

### SPARKS.

The Toronto and Suburban street railway has been extended to Lambton Mills.

The Galt, Preston & Hespeler electric railway carried 23,000 passengers in June.

The Hamilton Radial Railway Company are actively engaged in the construction of their line.

A Cleveland syndicate is said to be desirous of purchasing from Col. Stacey the street railway franchise for St. Thomas.

The shareholders of the Hamilton and Dundas Railway have approved of the conversion of the road into an electric line.

Contracts will be awarded this week for materials required for the extension of the Hamilton, Grimsby and Beamsville Railway to Beamsville.

Miss Lizzie Cole was killed on Queen street, Toronto, by being struck by a trolley car. The jury brought in a verdict exonerating the employees on the car from blame, but recommended that the speed of cars be regulated by city authorities and that more efficient fenders be used.

The projectors of the Hamilton, Ancaster and Alberton Radial Railway have requested that a right of way be granted for their proposed road from Hamilton to Ancaster. Among the promoters of the scheme are Messrs. W. H. Wardrope, W. F. Walker, Q.C., F. G. Beckett and Major Snider.

The city council of Hamilton have released the $20,000 of bonds of the Hamilton, Grimsby and Beamsville Railway, held by the city as security for the continued operation of the road. This will enable the company to extend the road to Beamsville at once, arrangements for which are now being made.

The Montreal Street Railway Company will hereafter issue certificates of merit to their employees. Any man who has been in the company's service for five years will be entitled to wear one gold band; ten years' service will give two gold bands, and an additional gold band for every extra five years' service.

Mr. E. A. C. Pew, of the Hamilton and Lake Erie Power Co., has written to the Mayor of Hamilton offering to furnish power to run the pumping machinery for the waterworks for $10,000 a year. He states that arrangements have been made with capitalists to have the canal and plant constructed by November next.

Suit is said to have been entered on behalf of Dr. Rolston, of Shelburne, one of the promoters of the Huron & Ontario Electric Railway, to recover $3,000 for services, and an injunction to prevent the payment out of the moneys which the plaintiff claims have, through his efforts, found their way into the company's treasury. The company and E. A. C. Pew are defendants in the action.

The Hull Electric Railway desire to run a short spur line in front of their office in Hull, and this is opposed by the City Council on the ground that the company have no right to any part of the street other than that occupied by the main track. The question will probably be settled by law, and as the act under which the company operates gives them the authority to operate an electric road on the streets of Hull, the decision will be interesting.

The power house of the Montreal Park and Island Railway Company, at Mile End, Montreal, was completely destroyed by fire on the 30th of July. Besides the electric power plant the company lost twelve open cars, four trailers, two electric sweepers, two tar waggons and a small steam locomotive. The loss is in the neighborhood of $40,000, and is covered by insurance. The service on the road was continued as usual the following day, the Montreal Street Railway Company providing the power.

The Elektrotechnische Zeitschrift gives the following description of the arrangement recently introduced in Stockholm for the automatic control of the operation of the exchange: The exchanges are fitted for night service, and for that purpose the annunciators are provided with contact devices, which in falling close a circuit and cause a bell to ring. This arrangement is now employed during the daytime, but a call-indicator is put in place of the bell. This call-indicator consists of an electromagnet, the armature of which moves a signal behind a little window. The ring-off indicators are provided with the same device, the only difference being a different color of the signal. Besides facilitating the control of the operators the arrangement assists the operator to a large extent, as it saves the continuous and tiring observation of the annunciator board. The operator has only to watch the call-indicator and, when it signals, to look at the annunciator board. The new exchange at Christiania, just opened, has a similar device.

# CANADIAN GENERAL ELECTRIC CO.
## (LIMITED)

# Railway Apparatus

We invite attention to the commanding position the Company has attained in the Electric Railway field.

Our Apparatus is now in use on almost, without exception, every independent road in the Dominion, and the fact that we have been awarded practically all the contracts placed during the past year for equipment either for new or existing roads, is the strongest possible testimonial both as to the superiority of the Apparatus itself, and the fair and liberal basis on which the business of the Company is conducted. Amongst the roads from which orders for apparatus have recently been received may be mentioned the following :

| | |
|---|---|
| Hull & Aylmer Electric Ry. Co. | Vancouver & Westminster Tramway Co. |
| Moncton Electric Street Railway, Heat & Power Co. | City and Suburban Street Ry. |
| | Guelph Electric Street Ry. |
| Hamilton adial RElectric Ry. Co. | Berlin & Waterloo Street Ry. |
| Cornwall Electric Street Ry. | Port Dalhousie, St. Catharines & Thorold Street |
| Halifax Electric Tramway Co. | Railway Co. |
| St. John Street Railway Co. | Brantford Street Railway Co. |
| Montreal Street Railway. | London Street Railway Co. |
| Toronto Street Railway. | Kingston, Portsmouth & Cataraqui Railway. |
| Chateaugay & Northern Ry. | |

# Lighting and Power
# Transmission Apparatus

In considering the development of our systems of apparatus for lighting and power transmission, we have kept in view the important fact, that varying conditions of service require varying methods to meet them.

We have in our **Edison Direct Current Three-Wire System,** our **500 Volt Direct Current System,** our **Single-Phase Alternating Current System,** our **Monocyclic System** and our **Three-Phase System,** a series of methods, each superior to all others, for the service to which it is adapted. We are not confined to one system only, but cover the whole range of Direct Current, Single-Phase and Multi-Phase Alternating Apparatus. Our interest in each case, therefore, lies in using the most suitable system, since we manufacture all ; not in twisting the conditions to suit one particular system, however ill-adapted to the particular case.

**Our recent sales of Lighting and Power Apparatus have exceeded all previous records and include the sale to the Lachine Rapids Hydraulic and Land Co'y, of twelve three-phase generators, each of 1000 h.p. capacity.**

CANADIAN

# ELECTRICAL NEWS

AND

## STEAM ENGINEERING JOURNAL.

| VOL. VI. | SEPTEMBER, 1896 | No. 9. |

### TANDEM COMPOUND STEAM ENGINE.

THOSE who have observed the trend of steam engine designing during the past few years will have noticed that there is a tendency towards a short, compact, heavy built frame, with strong simple parts, suited to the severe and incessant work imposed upon power plants by street railway and other heavy work. Corliss and other types of long stroke engines have been shortened and strengthened in order to meet these conditions and to occupy less room, and there is also a tendency to increase the speed to suit direct driven dynamos and give better regulation. In fact, there seems a tendency for the advocates of high and low speed to meet half

is of the "Sweet" or "Straight Line" pattern, used in all engines made by the Robb Co., is of the simplest and most sensitive form and directly connected to the valves. The high pressure cylinder is placed next to the frame, low pressure in rear, and so arranged that the cylinder head and pistons may be removed without disturbing the cylinders, valves or other parts. The valves are of the "Porter" type, consisting of a flat plate balanced by a pressure plate, which have proved so successful in the "Porter-Allen," "Straight Line" and other engines, their greatest merit being simplicity and freedom from wear. Both high and low pressure valves are attached to the governor in such a way as to

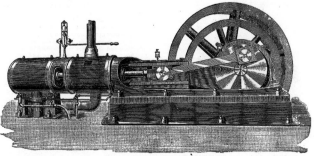

TANDEM COMPOUND STEAM ENGINE.

way in a type of engine which will embody the best points of each.

As an example of what is being done in this way, we give an illustration of a tandem compound engine, built by the Robb Engineering Co., of Amherst, Nova Scotia. The cut is from one of four engines of 300 h. p. each recently installed for the Halifax Electric Tramway Co., for railway and lighting purposes, and it represents a type of engine designed with a view to combine the best points of long and short stroke engines.

The design of frame and general proportion of parts is similar to recent types of long and medium stroke engines designed for railway work. The shaft bearings, crank and crosshead pins are much larger than usual, to insure cool running under stress of overloading or irregular work. The guides are cylindrical, allowing the crosshead free alignment. The disc crank contains sufficient metal to permit the crank pin and shaft to be forced in under heavy hydraulic pressure, and is balanced. The main journal has quarter boxes with adjustment at top and sides. The governor, which

divide the load exactly between the high and low pressure cylinders. This system is new and peculiar to the Robb engines and is found to give better economy with variable loads, such as are found in railway work.

The manufacturers are now building a full line of these engines, in simple, tandem and cross-compounds, up to 700 h. p., having a medium length of stroke, speed from 150 to 200 revolutions per minute ; and as the parts are massive, and bearings unusually large, parts simple and strong, they are splendidly adapted for direct connection to electric generators or other variable work.

The extension of the Hamilton, Grimsby and Beamsville railway to Beamsville will shortly be completed.

The number of miles possible to be ridden in the United States on a street car for five cents is said to range from 8¼ miles in Jersey City up to 18 miles in Brooklyn, the average of ten cities being 13 miles. At Chicago a ride of 21 miles can be had for this small sum on an ordinary railroad.

## THE HAMILTON ELECTRIC LIGHT AND POWER COMPANY.

ONE of the best equipped electric light and power plants of Ontario is that of the above company. The plant began operations in April, 1892. It occupies two buildings. The dynamo room was designed and constructed under the direction of Mr. D. Thomson, the late manager. The placing of the machinery was directed by Mr. Dickinson, chief engineer. The dynamo building, on Main St., is 135 x 70 ft. with hip roof, supported by wooden beams covered with corrugated iron ; a cupalo surmounts the top. The stone foundation is laid in Portland cement and sharp sand.

Two pair of "Brown" (Polson Works) engines of 700 h. p. each, 78 revolutions per minute, 22 inch cylinder, 50 inch stroke, with fly wheel 16 x 4 ft., and one high speed compound engine of 150 h. p., 250 revolutions per minute, all non-condensing, drive the 120 feet

MR. GORDON J. HENDERSON,
Manager Hamilton Electric Light and Power Company.

of line shafting, 5½ inches in diameter, and 7 inches at driven pulleys, which runs down the centre of the building between the engines and the machines. A Leonard-Ball engine is also in place, but is not used. These engines are belted to the shafting by 4 foot 3 ply Robin, Sadler & Haworth belting. A Bain & Coville clutch coupling is in the centre of the shafting. The fly wheels are boxed in casings next the passage between them and the wall. At the rear wall are two Northey pumps, 8 inch cylinder, 12 inch stroke, and 5 inch rams, pumping water at 208° into the boilers from a heater built by Bain & Coville from a design by the chief engineer. The different compartments of the heater are full of hay. The hot water passes through this hay, which extracts the lime and magnesia from the water, which forms in a hard lime-stoney substance on top of the hay.

Entering the rear wall on a level with the cross beams are two 10 inch wrought iron steam mains conveying steam to the several engines. The branch pipes are 6 inches. Where the mains enter the building are three 10-inch valves and a pipe connection between the mains. The valves work so that both mains can supply the engines, or all the engines run from one main.

The floor is concrete except one spot, which is of wood and underneath which is a large tank containing 7,000 gallons of water, to be used in case the water mains give out.

The machines consist of arc and incandescent lighters and electric power generators. There are two "Royal" alternators with exciters, (one of 2,000 lights and the other of 1,250 lights), one "C. G. E." 2,000 lighter and exciter, and one Westinghouse 1,800 lighter and exciter. Fifteen machines of varying power supply arc light, fourteen of which are from the Royal Company ; the other, of 35 lights, being a Toronto Electric Light machine. The Royal arcs are as follows : four 35 lighters, eight 50 lighters, and two 12 lighters. An "Edison" 75 K. W. generator and a "Royal" 100 K. W. multipolar generator generate power for motors of from ½ to 15 h. p.

The machines are set on stone foundations, 5 feet

deep, and run very smoothly. Mr. Martin, the chief electrician, has his office to the right ; behind this is the store room, where are all kinds of supplies and repairs for arc or incandescent lighting and installing of same. They use Packard lamps and Royal arcs, and Ottawa and C. G. E. ⅝ carbons. The repair shop and lamp testing room is 20 x 25 feet. A full equipment for all repairs is in this shop, except for heavy machines, which are wound by the Toronto Electric Light Co. Along the wall of the store room is the large slate switch board with full complement of instruments for the alternators and generators. On the arc switch board are 14 circuits and on each circuit is an automatic pilot light. In case of the opening of a circuit or trouble at the machine, the pilot light supplied from an incandescent circuit lights up and the operator can see at a glance what circuit or machine is in trouble. This is the invention of Mr. Martin, the chief electrician.

On the arc poles the lamps are hung on hinged arms, no drop ropes being used. Each trimmer is given 90 lamps a day to trim and he covers his circuit with a horse and two wheeled "chariot," which was designed by the chief electrician.

On King street is a three-storey structure in front, combining the offices of the company, and several other offices. The rear is taken up by the boiler room. There are three batteries, the first comprising five 60-inch Osborne-Killey tubulars, 75 h. p. each, and the second two, 66 inch Goldie & McCulloch tubulars, 90 h. p. each ; and two Polson water tube boilers, 200 h. p. each. All these boilers are connected to a square brick smoke stack 125 feet high. The tubular boilers are connected to one main and the water tube boilers to the other.

The fuel used under these boilers is hard and soft coal screenings, three-fifths being soft coal. Between the two buildings is a space of about ten feet, and the coal as it passes through is weighed on scales operated from the chief engineer's office, which is in the rear corner of the dynamo room. The engines in the dynamo room exhaust under the floor through a 12-inch pipe, which passes into the rear building 3 feet underground

MR. T. W. MARTIN,
Chief Electrician Hamilton Electric Light and Power Company.

in a box of sawdust. The steam mains carried between the buildings are covered with 12 inches of mineral wool with box castings.

The city use 375 arc lights, and 75 arcs are in commercial service. The capacity is 500 lights. Over 9000 incandescent lamps are installed and on an average 300 new lamps are installed each month.

The directorate of the above company and of the Toronto Electric Light Company being largely the same, Mr. J. J. Wright, manager of the latter company, assumed the management upon the resignation of Mr. Thomson about two years ago. Since the amalgamation of the Toronto Electric Light Company and the Incandescent Light Company in Toronto, however, the duties of Mr. Wright have been so onerous that it has been found necessary to appoint a separate manager for the Hamilton company. The appointment has been given to Mr. Gordon J. Henderson, of Montreal.

Mr. Gordon J. Henderson, who has recently been appointed manager of the company, and whose portrait is herewith presented, was born in the city of Montreal in the year 1872. He is a son of Mr. David H. Henderson, a prominent lumber merchant. For some years he has been connected with his brother, Mr. C. W. Henderson, the well-known electrical contractor of that city. He is quite prominent in Montreal's society, and holds a commission as Captain in the 6th Battalion Fusiliers, having the honor of turning out the best drilled company in his battalion. Mr. Henderson is a business man of considerable ability, and under his supervision the company will no doubt enjoy a marked degree of prosperity.

Mr. T. W. Martin, chief electrician, was born in London, England, 26 years ago. His family came to Canada, and at the age of fourteen years he entered the employ of the Toronto Electric Light Company, under Mr. Wright, in the old Sherbourne street plant. He was removed to Hamilton two years ago. He is a

MR. R. DICKENSON,
Chief Engineer Hamilton Electric Light and Power Company.

clever electrician and fills his position in a creditable manner.

Mr. R. Dickenson, chief engineer, was born in Dover, Kent, in 1841. He entered the Royal Navy in 1858, in which he served for some years, leaving it for merchant vessels. In 1874 he came to Canada, filling different positions, and eleven years ago he entered the employ of the Hamilton Electric Light Co.

## THE PARAGON OF EXHIBITIONS.

THE major part of the entries having now been made for Toronto's big exhibition, which is to be held from August 31st to September 12th, it is possible to state definitely that the scale of the exhibition will really be greater than ever. Never before did the exhibits cover such a wide range as they will this year. It almost looks as if every province had striven to do its best to make the exhibition worthy of the country. At the forthcoming exhibition in Toronto there will be seen food products of Prince Edward Island ; food products, manufactures, fruit and live stock, of Nova Scotia and New Brunswick ; an extensive display of horses and cattle, manufactures and minerals, from Quebec ; the products of forests, waters, mines, gardens, farms, studs, workshops and art studios of Ontario ; the grain, minerals and horses of Manitoba ; the grain and minerals of the North-West ; and cereals, fish and minerals of British Columbia. The governments of Ontario, the Dominion and British Columbia will make special exhibits of the wealth of the earth, while the Canadian Pacific Railway will supplement these displays by showing cereals, vegetables and minerals from many points on their lines, to the extent of double what the company has shown in other years. In art especially will the exhibition be strong, with the three pictures painted by F. M. Bell-Smith, illustrating incidents connected with the death of Sir John Thompson, at Windsor Castle, for one of which pictures Her Majesty the Queen, Princess Beatrice and members of the Royal household gave special sittings. There will be Edison's wonderful Eidoloscope, an electric theatre ; Ontario Trotting Horse Breeders' stake races ; Lockhart's performing elephants ; the magnificent historical spectacle, entitled the "Feast of Nations" and commemorating the "Taking of the Bastile," and a thousand and one other things ; while in consideration of the cattle being on show the first week the railways have agreed to grant one fare for the round trip for the entire exhibition from all points in Canada, and to run a special cheap excursion the first week, on Sept. 3rd, and two the second week.

## TRANSFORMERS.
### BY G. W. F.

THE really distinctive feature of an alternating current system is the transformer. Without it the alternating current would possess no advantages over the direct, and the transmission of power for lighting or motor purposes would be impracticable except at the cost of very large conductors. The use of the alternating current in connection with station transformers arose out of the practical limitations imposed on direct current apparatus, and in so far was an improvement in the art. In direct current working the pressure generated by the dynamo is maintained throughout the entire system ; a 220 volt machine will cause a pressure of 220 volts (less the "drop" of course) between the positive and negative wires at all points ; a 500 volt machine gives a pressure of 500 volts everywhere, and so on. In order to distribute current over an extensive area, it is evidently necessary to use either heavy, and therefore expensive, feeders with a 220 volt pressure, or to use a higher pressure and so allow of smaller feeders. But as it is not at all desirable to introduce a high pressure into lamps placed in private buildings, where they have to be handled constantly, and where the wires are frequently exposed to risk of grounding, it is evident that a limit of pressure is soon reached, and that any extension of business must be met by an additional expenditure for feeder copper. In a district where there is a large amount of lighting this may be commercially possible, but it is quite easy to imagine conditions where the additional amount of lighting would not actually justify the necessary feeder expense. It is easily seen that any method which permits of the use of a high pressure for transmission, and at the same time of a low pressure for utilization, meets the conditions of economical supply and safe use. The static transformer renders possible an advantage beyond the power of the direct current.

It would be strange if a piece of apparatus possessing such great importance were not worth capable study, and in fact the electrical principles governing its action, and the electrical, magnetic, and mechanical features entering into and influencing its design and construction are not merely of great interest, but a thorough comprehension of them is necessary before the constructing or operating electrician can be considered conversant with alternate current working. To the casual observer a transformer is merely a quantity of insulated copper wire wound in two separate coils round an iron core ; the whole placed inside a box and what goes on inside that box when the current is turned on is of no more interest to them than the mechanism of a musical box—you turn the handle and grind out music ; you turn on the current and you get light—somehow. It is thought by those whose interest in electrical matters leads them no furthur than the study of how to pay the least money for plant—that once a transformer is hung up on a pole and connected into circuit there is the end of it ; that the worst thing that can happen to it is to have one of its fuses blow, or lightning get into it and burn it up. As to its being a source of expense all the time, as to its capacity for wasting current, the matter not only does not occur to them, but they actually smile when it is suggested to them, How can a transformer be a source of expense ? How can it waste current ? It isn't doing anything ; it isn't moving or revolving ; there's no friction about it—it doesn't need oiling—might as well suggest that a glass insulator is a source of expense. A little investigation, however, will show that the transformer is not the simple thing it is popularly supposed to be, and that careful study and educated thought were just as necessary in its evolution as introducing the high class modern dynamo. The basis of transformer action is the same as that of dynamo action —induction. If a closed conductor be placed in a magnetic field, the intensity of which is rapidly varying, an E. M. F. is set up in that conductor, the direction of the E. M. F. will depend on whether the intensity is increasing or decreasing its strength on the rate of variation.

N S are two poles, the space between them being a

magnetic field as indicated by arrows. C is a closed conducting ring capable of being revolved on A, as axis. It is understood that A is really at right angles to the direction N S, and that the plane of the ring C is perpendicular to the direction of the lines of force from N

Diagram I.

to S.   Now if the strength of the magnetic field N S is always the same (as it generally is in a dynamo) and if the ring C be held stationary in any position, it will be evident that nothing is varying, and consequently there will be no current set up in C.   But if we now revolve the axis A (in either direction) and with it the ring C, it will be seen that, although the same amount of lines of force will always flow from N to S, the ring will in some positions hold less of them than it will in others. In the diagram No. 1 the ring is perpendicular to the field and will contain say X lines of force.   In diagram No. 2, having now revolved it through a quarter of a

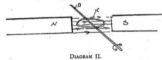

Diagram II.

circle, the plane of the ring is parallel to the lines of force, and contains none of them at all.
So that in the course of a quarter revolution we have varied the lines of force contained by the ring from a certain maximum down to nothing, and this is the condition necessary for the setting up in the ring of an electromotive force.   Turning C through another quarter circle would again vary the field with respect to the ring from nothing up to the same maximum as before ; the third quarter turn would bring it back to nothing ; the fourth raise it again to the first position. Thus, revolving the ring in a constant magnetic field, causes a variation with respect to the ring which sets up an E. M. F. in it.   The same result would be obtained by holding the ring stationary and causing the field to revolve, as indicated in diagram No. 3.

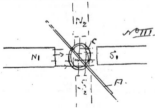

Diagram III.

The ring C is perpendicular to field N₁ S₁, and if these poles be shifted to positions N₂ S₂, the ring being not moved ; the ring will be parallel to the new field N₂ S₂, hence during the shifting of the positions of the poles an E. M. F. will have been set up in C.   It is therefore evident that so long as there is relative motion, it does not matter whether the ring be moved or the field. The necessary condition being a varying of the lines of force passing through the ring, a third method is possible which will attain that object without revolving either the ring or the poles.   If in the above diagrams the poles N and S are supposed to be electro magnets, (that is iron bars which are made magnets by the passage around them of a current), and we have some means of varying the current passing through the wire,

either by means of a rheostat or other equivalent means, then, remembering that the strength of an electro-magnet (the strength of its field) varies within certain limits, in the same proportion as the current producing it, it will be evident that the field can be varied up or down by simply turning the rheostat, leaving both poles and ring stationary.   And from this last method it is but a step to the energizing of the poles by an alternating current which will cause an even greater variation of the field than the rheostat can accomplish with a direct current.   This will be plain when it is considered that the current in an alternating circuit begins at nothing, grows rapidly to a certain maximum, diminishes again down to nothing, then actually changes its direction and grows to a negative maximum, and then decreases again to nothing.   Turning to diagram No. 1, we will suppose ring C to be one of the coils of an alternating dynamo.   In the position where C is perpendicular to the lines of force N S, any very slight revolving of A will not vary the amount of them contained by C much ; in fact in this position a slight revolving will really generate no E. M. F. at all, but as A is revolved (counter clockwise) ring C will hold less and less lines of force until it reaches the position in diagram 2, when the rate at which C is decreasing is greatest, and as the E. M. F. generated depends on the rate of variation, the highest E. M. F. generated in C will be when it is passing through the position in diagram 2, and a proportionate E. M. F. will be generated in C at any intermediate position.   So that the E. M. F. will grow during a quarter revolution from o to a maximum.   When C has been revolved through a half turn, conditions will be as they were in diagram No. 1, and at this point no

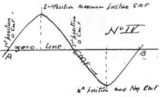

No IV.

E. M. F. will be generated in C, it having decreased from the position of maximum E. M. F. in diagram 2. If C be revolved through and then quarter turn, then everything will be the same as in diagram No. 2, and the E. M. F. will again be at a maximum, except that the direction of the E. M. F. has reversed, and instead of being from right to left is now from left to right ; or, if we call the first direction positive we can call the new one negative.   From the third quarter revolution to the fourth brings C back to the first position.   All these changes can be put into a diagram form as in No. 4, where the curved line shows how the E. M. F. in the ring varies between a positive and a negative maximum. It will be understood that whereas the direction of the E. M. F. is from right to left in the upper part of the diagram, it becomes from left to right in the lower ; the strength of the E. M. F. at any point in the revolution being indicated by the height of the curve above the zero line and the distance A B representing one revolution of the axis.
Now suppose we have two magnets M, P, energized by an alternating current as above, so that they shall be of opposite polarities and a ring C.   From position 1 to position 2 (diagram 4) M will have a north pole, and P a south pole, constantly increasing in strength, and from position 2 to position 3 the polarities will be the same, but the strength of the field will diminish constantly back to nothing.   At position 3, however, the direction of the E. M. F. changes so that, from 3 to 4, M will become a south pole (instead of a north), and P will now be a north pole (instead of a south), and the strength of this reversed field will constantly increase to a maximum at 4, and thence down to nothing at position 1, when the E. M. F. again changes its direction,

and the poles consequently again change sign. During these variations and reversals, the ring C will (although stationary) have been placed in a varying field produced by the alternating electromagnets, and consequently an alternating E. M. F. will have been induced in C. This is the simple theory on which is based the action of the transformer, but its application gives rise to phenomena which introduce new and less simple considerations. A transformer could be constructed on the plan shown in the diagrams, but a more convenient, and in every way

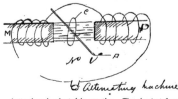

better form is adopted in practice. The simplest form may be shown in diagram where A is a bar of iron, P is an insulated wire wound round and carrying an alternating current from the generator G, S being another insulated wire also wound round A and P, (but insulated from both) and leading to say the lamp L. On passing the alternating current from G round A, the bar is at once made an electromagnet, the poles of which reverse their sign as the direction of the alternating current reverses. The whole space surrounding A becomes a magnetic field, the lines of force radiating in the manner

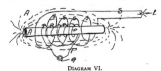

DIAGRAM VI.

indicated from one pole to the other. It is evident that under these conditions the coil S is just as much placed in a varying magnetic field as it was in diagram No. 5 ; hence a current will be induced in it. All transformers are built in this way : two coils wound together round a magnetic circuit, and insulated both from it and from each other, the one carrying the energizing current called the primary, the other in which current is induced called the secondary. The feature of special importance in this induction is that, no matter what may be the voltage in the primary coil, we can get what voltage we desire in the secondary, so that we can run our alternator at 1,000 or 2,000 or5,000 volts, and still have only 52, or 104, or any other desired voltage in our secondary coils. This, of course, permits of the use of high voltage for distribution and low voltage at lamps, obtaining both economy and safety. The difference in voltage between primary and secondary wires depends directly on the proportion between the number of primary and secondary coils. If there be ten turns of the primary to each secondary turn, then the secondary voltage will be only one-tenth of the primary, and so on.

The action of the transformer is, as described above, that when the primary circuit is closed round the bar the alternating current transforms it into an electromagnet, with rapidly reversing polarity, and the varying and reversing field induces an alternating current of the same periodity in the secondary wire. This appears to be so simple a process that the person who does not examine it more closely will not easily believe when told that a transformer wastes coal. As a fact, however, every transformer built—even the very best—necessarily wastes energy ; good transformers waste less than second-rate ones. These wastes have been located and can be calculated. They result as follows :

On closing the circuit in the primary the core becomes

an electromagnet, whose polarity reverses at the same times as the current reverses. It has been observed that subjecting iron to an alternating magnetomotive force raises the temperature of the iron—and this phenomenon has been accounted for by the following theory : Consider a bar of iron M. Pass a current round it from the source K in the direction of the arrow. Instantly the one end becomes a north pole N, and the other a south pole S. Now we may consider the bar to be made up of an infinite number of small atoms of iron pivoted at their centre points, each of which becomes a little atomic magnet, their N poles all pointing to the N end of bar, and their S poles all pointing toward the S end. Now reverse the direction of the current from K. Instantly everything is changed. The old north pole now becomes a south pole ; the old south pole is now a north pole. Every little atomic magnet has swung round, and is now pointing in the opposite direction to what it did before. Plainly they cannot have done all

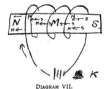

DIAGRAM VII.

this swinging without rubbing against each other and getting warm, and this friction requires a little expenditure of energy to overcome it. We can easily see, therefore, that a certain amount of energy is expended in the iron core of the transformer itself in producing the necessary alternating magnetism ; and this energy will be greater in proportion as the number of reversals of the current becomes greater. It is also evident that as the number of atomic magnets increases—that is, as the total mass of the bar increases—a greater power must be expended in overcoming their friction ; it will take twice as much power to swing 2,000 atomic magnets as to swing 1,000. Once more, it is evident that as the strength with which each atomic magnet points in one direction increases, i.e., as the strength of the magnet increases so will it take more and more power to force it to point in the opposite direction. Consequently, it is plain that a certain amount of power must necessarily be expended in the transformer itself in producing the alternating magnetism, and the actual amount of power so expended depends first on the number of times the current alternates ; next on the number of atomic magnets to be reversed (that is the size of the whole iron bar), and third on the strength of the magnet (the amount of magnetic flux). It depends on the transformer itself whether the total amount of energy expended in this way is greater or smaller, but in any case it has to be supplied by the primary current and hence by the coal pile. It therefore follows that any means of reducing it is an advantage. The amount so expended depends, we have seen, on the number of reversals, the total mass, and the magnetic induction. It would therefore be an advantage to use a lower alternation, but this is limited by considerations outside the purpose of this article. It certainly will be of advantage to use a smaller bar of iron, but as in order to construct a transformer capable of giving a certain secondary voltage, we have to use a proportionate magnetic induction, the only way he can reduce the size of the iron is by using a better class metal whose permeability is higher. It is necessary to remember that bars of different kinds and qualities of iron will not give the same magnetic strength for the same current, but that the poorer the iron the less need be the magnetising current to give a desired strength. Consequently we can only reduce the mass by using a better quality of iron. Better quality means higher price—it is no economy to select the cheapest transformer.

*To be Continued.*)

PUBLISHED ON THE FIFTH OF EVERY MONTH BY

## CHAS. H. MORTIMER.

OFFICE : CONFEDERATION LIFE BUILDING,
Corner Yonge and Richmond Streets,

TORONTO, - - - CANADA.
Telephone 2362.

NEW YORK LIFE INSURANCE BUILDING, MONTREAL.
Bell Telephone 2299.

*ADVERTISEMENTS.*

Advertising rates sent promptly on application. Orders for advertising should reach the office of publication not later than the 24th day of the month immediately preceding date of issue. Changes in advertisements will be made whenever desired, without cost to the advertiser, but to insure proper compliance with the instructions of the advertiser, requests for change should reach the office as early as the 2nd day of the month.

*SUBSCRIPTIONS.*

The ELECTRICAL NEWS will be mailed to subscribers in the Dominion, or the United States, post free, for $1.00 per annum, 50 cents for six months. The price of subscription should be remitted by currency, registered letter, or postal order payable to C. H. Mortimer. Please do not send cheques on local banks unless 5 cents is added for cost of discount. Money sent in unregistered letters will be at senders' risk. Subscriptions from foreign countries embraced in the General Postal Union $1.50 per annum. Subscriptions are payable in advance. The paper will be discontinued at expiration of term paid for if so stipulated by the subscriber, but where no such understanding exists, will be continued until instructions to discontinue are received and all arrearages paid.
Subscribers may, have the mailing address changed as often as desired. When ordering change, always give the old as well as the new address.
The Publisher should be notified of the failure of subscribers to receive their paper promptly and regularly.

*EDITOR'S ANNOUNCEMENTS.*

Correspondence is invited upon all topics legitimately coming within the scope of this journal.

The "Canadian Electrical News" has been appointed the official paper of the Canadian Electrical Association.

## CANADIAN ELECTRICAL ASSOCIATION.

### OFFICERS :
PRESIDENT :
JOHN YULE, Manager Guelph Light & Power Company, Guelph, Ont.

1ST VICE-PRESIDENT :
L. B. McFARLANE, Manager Eastern Department, Bell Telephone Company, Montreal.

2ND VICE-PRESIDENT :
E. C. BREITHAUPT, Electrical Engineer, Berlin, Ont.

SECRETARY-TREASURER :
C. H. MORTIMER, Publisher ELECTRICAL NEWS, Toronto.

### EXECUTIVE COMMITTEE :
GEO. BLACK, G. N. W. Telegraph Co., Hamilton.
J. A. KAMMERER, General Agent, Royal Electric Co., Toronto.
K. J. DUNSTAN, Local Manager Bell Telephone Company, Toronto.
A. M. WICKENS, Electrician Parliament Buildings, Toronto.
I. J. WRIGHT, Manager Toronto Electric Light Company, Toronto.
ROSS MACKENZIE, Manager Niagara Falls Park and River Railway, Niagara Falls, Ont.
A. B. SMITH, Superintendent G. N. W. Telegraph Co., Toronto.
JOHN CARROLL, Sec.-Treas. Eugene F. Phillips Electrical Works, Montreal.
C. B. HUNT, London Electric Company, London.
F. C. ARMSTRONG, Canadian General Electric Co., Toronto.

## CANADIAN ASSOCIATION OF STATIONARY ENGINEERS.

President, JAMES DEVLIN, · · Kingston, Ont.
Vice-President, E. J. PHILIP, · · 11 Cumberland St., Toronto.
Secretary, W. F. CHAPMAN, · · Brockville, Ont.
Treasurer, R. C. PETTIGREW, · · Hamilton, Ont.
Conductor, J. MURPHY, · · Montreal, Que.
Door Keeper, F. J. MERRILL, · · Ottawa, Ont.

TORONTO BRANCH NO. 1.—Meets 1st and 3rd Wednesday each month in Engineers' Hall, 61 Victoria street. John Fox, President ; Chas. Moseley, Vice-President ; T. Eversfield, Recording Secretary, University Crescent.
MONTREAL BRANCH NO. 1.—Meets 1st and 3rd Thursday each month, in Engineers' Hall, Craig street. President, John Murphy ; 1st Vice-President, J. E. Huntington ; and Vice-President, Wm. Smyth ; Secretary, B. Archibald York ; Treasurer, Peter McNaughton.
ST. LAURENT BRANCH NO. 2.—Meets every Monday evening at 43 Bonsecours street, Montreal R. Drouin, President ; Alfred L. tour, Secretary, 306 Delisle street, St. Cunegonde.
BRANDON, MAN. BRANCH NO. 1.—Meets 1st and 3rd Friday each month, in City Hall A. R. Crawford, President ; Arthur Fleming, Secretary.
HAMILTON BRANCH NO. 2.—Meets 1st and 3rd Friday each month in Maccabee's Hall. Wm. Norris, President ; E. Teeter, Vice-President ; Jas. Ironsides, Corresponding Secretary.
STRATFORD BRANCH NO. 3.—John Hoy, President ; Samuel H. Weir, Secretary.
BRANTFORD BRANCH NO. 4.—Meets 2nd and 4th Friday each month. J. B. Forsyth, President ; Jos. Ogle, Vice-President ; T. Pilgrim, Continental Cordage Co., Secretary.

---

LONDON BRANCH NO. 5.—Meets once a month in the Huron and Erie Loan Savings Co.'s block. Robert Simmie, President ; E. Kidner, Vice-President ; Wm. Meaden, Secretary Treasurer, 533 Richmond street.
GUELPH BRANCH NO. 6.—Meets 1st and 3rd Wednesday each month at 7.30 p. m. H. Geary, President ; Thos. Anderson, Vice-President ; H. Flewelling, Rec.-Secretary ; P. Ryan, Fin.-Secretary ; Treasurer, C. F. Jordan.
OTTAWA BRANCH NO. 7.—Meet every second and fourth Saturday in each month, in Borbridge's hall, Rideau street ; Frank Robert, President ; F. Merrill, Secretary, 352 Wellington street.
DRESDEN BRANCH NO. 8.—Meets 1st and Thursday in each month. Thos Steeper, Secretary.
BERLIN BRANCH NO. 9.—Meets 2nd and 4th Saturday each month at 8 p.m. J. R. Utley, President ; G. Steinmetz, Vice-President ; Secretary and Treasurer, W. J. Rhodes, Berlin, Ont.
KINGSTON BRANCH NO. 10.—Meets 1st and 3rd Tuesday in each month in Fraser Hall, King street, at 8 p. m. President, S. Donnelly ; Vice-President, Henry Hopkins ; Secretary, J. W. Tandvin.
WINNIPEG BRANCH NO. 11.—President, G. M. Hazlett ; Rec.-Secretary, J. Sutherland ; Financial Secretary, A. B. Jones.
KINCARDINE BRANCH NO. 12.—Meets every Tuesday at 8 o'clock, in McKibbon's block. President, Daniel Bennett ; Vice President, Joseph Lighthall ; Secretary, Percy C. Walker, Waterworks.
WIARTON BRANCH NO. 13.—President, Wm. Craddock ; Rec.-Secretary, Ed. Dunham.

**The C.A.S.E.** IN the present number will be found a report of the proceedings of the annual convention of the Canadian Association of Stationary Engineers held at Kingston. A perusal of the report leads to the conviction that in point of attendance and also as regards the importance of the discussions and business transacted, this convention suffers by comparison with those of previous years. Our truest friends are those who sometimes call us to account for our shortcomings, as well as commend us for what is meritorious in our conduct. We trust, therefore, that the Association will not take it amiss if we give expression to a few opinions with regard to its policy and work. The avowed object of the Association, viz., to educate its members up to a higher standard of efficiency, thereby fitting them to improve their social and financial standing, is one which must commend itself to everyone. It appears to us, however, that this object is in some degree being lost sight of, as witness the fact that at the recent convention only one paper, and that upon a subject not intimately connected with engineering practice, was presented. There was practically no discussion whatever upon engineering practice—the subject above all others in which members of the Association are interested, and on which they need enlightenment. A large proportion of the time of the delegates was taken up with sight-seeing, and most of the remainder in considering ways and means of raising the revenue, which appears to be on the decline. Might not the energy which is being dissipated on publishing schemes, which are entirely without the province of the Association, and not calculated to enhance the respect in which it should be held by manufacturers and the public generally, be more profitably employed in the collection and dissemination of engineering knowledge of a character which should result in permanently increasing the membership and the finances ? However this may be, the ultimate success of the Association will depend on the extent to which the desire for a pleasant outing is subordinated to a determination to promote the education and welfare of every member.

**Operating Lamps at High Pressure.** THERE is a growing feeling in favor of operating lamps at a pressure of 220 volts instead of, as heretofore, at 110. Reports from Europe show that this method of distribution is rapidly gaining favor, and in the States there are quite a number of plants adopting the improvement. It will be obvious that the advantages of this are that a very considerably greater area can be served from the same station with the same loss, and that the percentage of variation of voltage will be very much less than it was with the 110 volts. This again reacts on the lamps, so that their average life is greatly increased. The only matter that seems to retard the full development of the 220 volt distribution system seems to be the difficulty of producing good commercial lamps to suit the high pressure, and this seems to be in a fair way to being overcome. We recommend central stations to keep their eyes on this development, with the view of adopting it ultimately.

**Designing an Electric Plant.** No part of the designing engineer's duty, when laying out a power house for electric lighting or railway purposes, is more important, or requires greater care and experience, than the general proportioning of the various pieces of apparatus and machinery, so that they may work together with the highest ultimate efficiency. Nothing is more apparent in the large proportion of lighting stations in the Dominion, than the complete absence of any continuous, coherent scheme, binding together and running through the entire plant, and we are sorry to say that nothing could be more unanimous than the complaint from those owning such plants, that electric lighting is not a very lucrative business. And yet very little consideration will show that these unfortunate results are but a necessary consequence of the policy—or rather the lack of policy—adopted by owners. It is too usual to consider a power house as consisting of two separate portions—steam plant and electric plant—and to consider them without much reference to each other. The purchaser is told that, generally speaking, it takes 10 lamps to a horse power, and on this very approximate and unsatisfactory basis he proceeds to make his own arrangements for steam engine, without having any idea as to the efficiency of the dynamo he proposes to try, or as to the most economical voltage drop, whether 5% or 10%, or as to the many other data which would all greatly influence the proper power of the engine. As a matter of fact he places himself entirely in the hands of the engine builders, who certainly cannot be expected to be extremely well posted on electrical matters, and as both he and they are very insufficiently informed as to what power would be actually required, the chances are that between them they decide—"in order to be certain"—on an engine 25% larger than there was any necessity for. Now it is well to be on the safe side, undoubtedly ; but then what is enough is enough—any more is a superfluity ; and there is great likelihood that had advice been obtained from some independent engineer of competence the saving in the size of the engine would more than pay his fees. If it were only the question of saving a few dollars on the price of the engine, this would be not of sufficient importance to warrant any great extra expense, but such unnecessary extra horse power means a continual yearly extra and unnecessary expense for fuel, over and above what would be necessary with an engine of proper size. This will be perfectly evident when one considers that it takes an appreciable percentage of the power of engine to merely turn itself over without any load. This percentage is frequently placed at 10%; so that a 100 h. p. engine would take 10 h. p. to turn it over. Now on the supposition that an engine has been purchased that is larger than necessary, it is very easily demonstrated that each horse power of such unnecessary extra use will cost the central station, on the average, one-half ton of coal per year more than necessary, assuming such a low coal consumption as

3 lbs. per h. p. h. This may seem a small amount, but then it is unnecessary, and capitalized at 5% per annum, it represents a sum of $40 per h. p. Then there is the further consideration of the less average efficiency of the larger engine working on a load only sufficient for a smaller one. What would be a full load for a 100 h. p. engine is only 80% of a full load for one of 125 h. p., and as the percentage efficiency of steam engines falls off rapidly as the proportion of load decreases, it is plainly seen that a too large engine is by no means a prudent precaution.

Another matter on which a word of caution is in season is the proper size of generators to use. The size selected is too often a matter of purely arbitrary choice on the part of the purchaser, who does not take sufficiently into account such very important factors as population of town, class of inhabitants, number of churches, halls, etc. Here again, it is no economy to base one's ideas on one's own inexperience, instead of calling in professional independent advice, and so profiting by the accumulated experience of the electrical profession. The problems presented are—to get enough ; to not get too much, and to arrange the generators in units of such size as that such, and as many as may be operating at any moment, may be operating at their maximum efficiency. This efficiency question is one which plays a very much more important part in the operation of electric plants than most of their owners are aware of, and if more attention were paid to it there would be less complaint about the unprofitableness of electric lighting business. The main trouble seems to arise out of the injudicious selection of sizes of generators ; they being, as a rule, so selected that for the very large proportion of the time, they are operating at much less than half load. For instance, a machine of 1000 capacity may be installed in a town of 2000 inhabitants ; now only for about one hour per night during the depth of winter, will that machine be called on for 1000 lights ; at all other times it will be supplying less than that number, and for the greater portion of the time (from about 10 p. m. to 5 a. m.) it will be running on loads of from 900 down to 200, with the smaller number predominating. Now it is quite plain that the machine, under these conditions, will be running at full efficiency for only about 10 per cent. of time, or probably less, and that its average load will be considerably less than half. Consequently it will be running principally on half load efficiency, or even less. To the thinking mind this consideration will at once lead to the inevitable conclusion that the 1000 light generator is too large as a unit, and that one of more like 500 lights would be better to use under the circumstances. Two units of this size operating together, will supply the 1000 lights ; their full load efficiency will be nearly equal to that of the 1000 light machine. When the load comes down to 500 lights then one machine will take care of it, at the same efficiency and at the lowest point of the load, viz., 200 lights. The proportion that this load bears to the 500 light full load is just double what it bears to the 1000 light full load machine. Consequently the average efficiency throughout is higher in the latter case than it was in the former, and the economy greater. It cannot be too strongly impressed on central station owners that the division of their generating plant—both steam and electric—into economical units is a most important matter.

## BOILER FEED PUMPS.

By WM. THOMPSON, Montreal West.

A RECENT visit to a small plant, operated by a somewhat young engineer who was constantly having trouble with his boiler feed pump, owing to want of knowledge as to the principles of construction and operation, must be my excuse for again troubling my fellow engineers. At the outset allow me to say that it is not my intention to discuss the use of feed pumps from an economical standpoint, but as far as my ability will allow me to endeavor to explain the principles of operation of the ordinary feed pump to be found in use in many of our engine rooms at the present day.

In a few instances I have found pumps connected direct to the boiler and pumping direct thereto, delivering the feed water to the boiler at the same temperature as it left the water cylinder of the pump. In some cases no attempt had been made to heat the water, and in others a tank filled with water was heated by means of a jet of steam, taken usually from the exhaust main, and the temperature of the water raised as high as they could get it, or as high as they could get the pump to handle it. Most modern plants are now, however, fitted with some kind of apparatus for heating the water after it leaves the pump and before it enters the boiler. In non-condensing plants this is usually done by means of a tubular heater constructed on the principle of distributing a large amount of heating surface to the cold water, or inversely to the principles of the surface condenser. This heater is commonly situated at some point in the exhaust main, and makes at once an admirable means of heating the boiler feed water at the smallest possible cost. In condensing plants a great many means are adopted, but apparently the engineers of to-day prefer the use of "economizers," through which the water is forced, and the waste gases from the furnace are utilized for the purposes of heating, thus enabling the engineer to utilize full value of his fuel to the last possible moment.

There are so many forms of boiler feed pumps working under so many different conditions of service, that I shall not attempt to describe any of them, except to take for an example the simplest form of a pump operated from some part of the engine or its reciprocating parts, and commonly known as a single acting plunger pump.

The principles of action of this pump may be explained from the accompanying diagram, representing a single acting plunger pump shown in section, and with the suction embedded in water, the

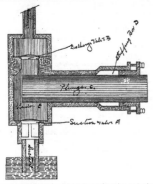

pump being empty, valve A being the suction valve, and valve B being the discharge valve, the plunger C being operated from some part of the machinery giving the necessary motion.

The water has the pressure of the atmosphere resting upon its surface, and the pump being also filled with air at atmospherical pressure, the inner face of the water within the suction pipe is also under atmospherical pressure, and consequently in a state of equilibrium.

Now, suppose that the stuffing box D has been securely packed to prevent the admission of air, and that plunger C has been moved to the right, as no more air can get into the pump, that already within it will expand and as a consequence will become lighter, therefore the pressure on the inner face of the suction

valve A will have been reduced, and as a result the water will rise up in the pipe, raising suction valve A in its passage and into pump chamber E. Let me here say that this act is very frequently misunderstood by young engineers and is the cause of a great deal of his troubles ; he imagining that the moving of the plunger to the right drew or sucked the water into the pump chamber E, while as a matter of fact the water rises in the suction pipe owing to the pressure on the inner face of valve A having been reduced, as a result of the expansion of the air previously mentioned ; therefore, the pressure of the atmosphere exerted on the surface of the water forces the water up into the pump chamber. To obtain this result the engineer will notice that it is compulsory that the admission of air to pump chamber must be prevented, or expansion cannot be effected, or in other words a vacuum will not be created, and the respective weights between the two points will not have been in any way changed, and water will as a consequence remain stationery.

The water inside the pipe will rise above that outside in proportion to the amount to which it is relieved of the pressure of the air, and that if the first stroke of the plunger to the right reduces the pressure from 15 pounds per square inch (atmospheric pressure) to 14 lbs., the water will be forced up the suction pipe a distance of about 2¼ feet, because a column of water one square inch in section and 2¼ feet high is equal to one pound in weight.

When the plunger has completed its travel to the right, the suction valve will fall to its seat and enclose the water in the pump chamber ; but as soon as the plunger moves back to the left and enters the pump chamber it will compress the water and force it to raise the discharge valve (B), and expell from the pump a volume of water or air equal in volume to the cubical contents of that part of the plunger that enters the pump chamber and displaces water. To prevent the plunger from forcing the water in the pump chamber back to the suction pipe the suction valve must first close and remain closed until the plunger has completed its stroke to the left. And if when the plunger was at the end of its stroke to the right the pump was partly filled with air, this air will be expelled from the pump before any water is ; but if the pump was filled with water, then water only will be delivered.

Now let us suppose that the plunger during its first stroke reduced the pressure within the pump chamber from 15 to 14 lbs. per square inch, and that the second and each subsequent stroke of the plunger reduced the pressure in the suction pipe one pound each stroke, the water in the suction pipe will rise 2¼ feet for each stroke of the plunger, until the weight of the column of water within the suction pipe is equal in weight to the pressure of the atmosphere bearing on the surface of the water ; and thus to ascertain how far a pump of this kind will cause the water to rise, will be found by calculation to be equal to a column of water nearly 34 feet high. Consequently it must always be borne in mind that no matter how high the pump may be set above the level of the water, it is impossible for the water to rise more than 34 feet up the suction pipe, no matter how perfect a vacuum can be got, because the force that propels the water is a fixed quantity of about 15 lbs. to the square inch, and it cannot raise a column of water greater in weight than itself. It is considered excellent practice when a pump will create a vacuum sufficiently good to raise water 30 feet.

This principle of operation is applied to all feed pumps with, however, many different mechanical appliances, to suit different purposes and conditions of service.

When this pump, or rather this style of pump, is applied to the purpose indicated, it will be observed that the pressure within the pump chamber when the plunger is discharging is at all times equal to the pressure contained in the boiler, and that to secure the proper performance of the pump for feed purposes, the following methods of construction and operation must be observed, viz :

1st : That the vertical distance between the top of the pump chamber E and the surface of the water must not be more than 30 feet, and that all pipes and connections must be perfectly air tight to prevent the admission of air between the valve and the water.

2nd : That the suction valve must weigh less than 15 lbs. per square inch of cross section. It will be borne in mind that the weight of the valve acts directly on, and against the pressure of the atmosphere on the surface of the water, and reduces the height to which the water will rise directly as the pressure required to be exerted per square inch on the valve to raise it off its seat. An instance of this occurs to my mind, where an engineer of my acquaintance purchased at second-hand a duplex steam pump, which my friend set up to pump water from a tank to his boiler. Much to his surprise he found it would not work, although he knew it had been doing excellent service where last in use. On examination he found that the springs on the suction valves had been adjusted at about 30 lbs. per square inch to suit former service, where water was pumped direct from town mains. As soon as proper adjustment had been made the pump performed quite satisfactorily.

3rd : That all air must be excluded from pump cylinder or chamber, and that all flanges and stuffing boxes must be kept tight, not only to prevent the admission of air, but to prevent leakage of water while pump is in operation.

4th : The discharge valve and pipe must also be clear, and all check valves, stop valves, etc., in proper working order, so that the plunger, or piston of the pump, will not be subject to any greater pressure than that within the boiler.

It will be unnecessary for me to add that water sufficiently hot to form steam at atmospherical pressure cannot be pumped owing to the destruction of the vacuum by the vapor. Nor will it be necessary to enumerate the various disorders to which pumps are subjected, as all minor troubles can invariably be traced to some of the causes already discussed.

## THE MEASUREMENT OF RESISTANCE.

SINCE the resistance of no two metals is the same, it was necessary to select the resistance of some accurately defined substance as a standard of measurement. The unit adopted by the international electrical congress in 1893 and called the ohm, after the discoverer of what is called Ohm's law, is " the resistance offered to any unvarying electric current by a column of mercury at the temperature of melting ice, 14,4521 grammes in mass of a constant cross sectional area and of a length of 106.3 centimeters." From this is obtained the standard unit of resistance, but for practical purposes wires of known resistance or resistance coils are used.

The resistance coils require great accuracy in their measurement, in the insulation of the wire and in the mounting of the coils. The wires must be carefully selected and tested. The insulation must be such as will withstand the highest temperature to which it is subjected without change. Silk thread is extensively used for the insulation. The wire is usually wound on spools or in coils so as to occupy as little room as possible, and are mounted in a box, which protects them from injury and places them in a convenient form to be carried. The ends of the coils are connected to plates or binding posts in the cover. This, also, must be carefully constructed so that the resistance at the point of contact will be as low as possible. A single coil is sometimes placed in an ebony case, or any number, according as the work for which it is to be used seems to require. When a large number is placed in one box the ends of the wires are usually connected to metal blocks, placed at such a distance apart that a metal plug will make a good connection between any two.

The resistance coils being uniform in size, the entire resistance or any part may be used. This is one of several styles of resistance boxes which are manufactured by instrument makers, and is the one commonly used. In measuring the resistance of an electric circuit, we cannot take our standard of measurement as we would take a foot measure to obtain the length of a piece of timber, but we can use it in another way, which will be explained with the Wheatstone bridge. If that of which we wish to measure the resistance is carrying a current and we have a voltmeter and ammeter so we may obtain the difference of potential and amount of current, the resistance is easily obtained by means of Ohm's law, the resistance equaling the electromotive force divided by the current.

## ELECTRICAL ITEMS WORTH REMEMBERING.

DROPPING a steel magnet, or vibrating it in other ways, diminishes its magnetism.

It is said that steel containing 12 per cent. of manganese cannot be magnetised.

Flames and currents of very hot air are good conductors of electricity. An electrified body placed near a flame soon loses its charge.

In changing a secondary battery, the charging electro-motive force should not exceed the electro-motive force of the battery more than 5 per cent.

The resistance of copper rises about 0.21 per cent. for each degree Cent.

A lightning rod is the seat of a continuous current, so long as the earth at its base and the air at its apex are of different potentials.

The rate of transmission on the Atlantic cables is eighteen words of five letters each per minute. With the "duplex" this rate of transmission is nearly doubled.

The effect of age and of strong currents on German silver is to render it brittle. A similar change takes place in an alloy of gold and silver.

To obtain the number of turns of wire in an electro-magnet, multiply the thickness of the coils by the length, and divide by the diameter of the wire squared.

A test for the porosity of porous cells consists in filling the cell with clean water and taking the per cent. of leakage. The correct amount of leakage is 15 per cent. in 24 hours.

If the air had been as good a conductor of electricity as copper, says Prof. Alfred Daniell, we would probably never have known anything about electricity, for our attention would never have been directed to any electrical phenomena.

For resistance coils, for moderately heavy currents, hoop iron, bent into zigzag shape, answers very well. One yard of hoop iron $\frac{1}{2}$ inch wide and 1-32 inch thick measures about 1-100 of an ohm.

The voltage of a secondary battery must always be equal to or slightly in excess of the voltage of the lamp to be burned. For example, a 20 volt lamp will require 10 secondary cells, but ten cells will supply more than 20 lamps.

Compression of air increases its dielectric strength. Cailletet found dry air compressed to a pressure of 40 or 50 atmospheres resisted the passage through it of a spark from a powerful induction coil, while the discharge points were only 0.05 centimeter apart.

An accumulator with 17 plates, 10 by 12 inches, is reckoned, in horse-power hours, equal to about one horse-power hour. Taking this as a basis, it will require 6 cells for one horse-power for 6 hours, or 30 cells for 5 horse-power for the same length of time.

To obtain the length of wire on an electro-magnet, add the thickness of the coils to the diameter of the core outside of the insulation, multipy by 3.14, again by the length, and again by the thickness of the coils, and divide by the diameter of the wire squared.

Blotting paper, saturated with a solution of iodide of potassium to which a little starch paste has been added, forms a chemical test paper for testing weak currents. When the paper (slightly damp) is placed between the terminals of a battery, a blue stain appears at the anode, or wire connected with the carbon or positive pole of the battery.—Scientific American.

# CANADIAN ASSOCIATION OF STATIONARY ENGINEERS

### SEVENTH ANNUAL CONVENTION.

THE limestone city of Kingston was honored this year as the seat of the seventh annual convention of the Canadian Association of Stationary Engineers, which was called for the 17th, 18th and 19th of August. The duties of entertainment therefore devolved upon Kingston Branch No. 10, and the manner in which the delegates were received proved conclusively their qualifications in this respect. While the number in attendance was not as large as desired, the convention throughout was extremely interesting and enjoyable.

By the kindness of the city authorities the Council chamber was placed at the disposal of the association, and at 10 a. m. the delegates convened in session. The Executive Committee was fully represented as follows: W. G. Blackgrove, Toronto, president; J. Devlin, Kingston, vice-president; E. J. Philip, Toronto, secretary; R. C. Pettigrew, Hamilton, treasurer; W. F. Chapman, Brockville, conductor; F. G. Johnston, Ottawa, doorkeeper.

Mr. Jas. Devlin, Kingston, President.

The Executive president, Mr. Blackgrove, occupied the chair, and on his left sat vice-president Devlin.

The delegates and visitors from the various places were as below :

Toronto—A. E. Edkins, John Fox, W. Selby, J. Huggett, R. Pink, J. G. Bain, C. Moseley, A. M. Wickens, W. G. Blackgrove, Geo. Grant, Wm. McKay.

Montreal—B. A. York, John Murphy, Wilbur Ware, O. E. Grandberg, J. J. York, Frank J. Greene.

Hamilton—R. C. Pettigrew, W. Norris.

Ottawa—F. G. Johnston, F. Robert, F. J. Merrill.

Guelph—C. J. Jorden.

Wiarton—F. J. Cody.

Brockville—W. F. Chapman, J. McCaw.

Carleton Place—J. McKay.

Nearly all the members of the local association were present at some time during the session, prominent among whom were Sandford Donnelly, president; John Tandvin, secretary; Charles Selby, treasurer; Daniel Reeves, John McDonald, Charles Asselstine, Thomas Burns, Fred Simmonds.

The president presented the delegates to the Mayor, who gave a brief address of welcome. He expressed himself as being assured that the subjects coming before the convention would receive that consideration which their importance demanded. Being informed that the constitution of the association very properly provides that the organization shall not be used as a means to encourage strikes or interference in any way between its members and their employers, he thought this fact a

matter for congratulation. "In coming from all parts of the Dominion to assist in educating the minds of others in your calling," he said, "Your mission is as noble as your calling is responsible. We all must recognize the importance of having responsible and reliable men placed in positions where human lives are placed at their mercy. It is, therefore, a personal pleasure for me to welcome a representative body of brother mechanics to our city; men who thoroughly understand the subjects they discuss and who can practice what they preach; men possessing a thorough and practical knowledge of their calling. I trust your deliberations while in our city will be beneficial to your order and the public in general." During their leisure hours the Mayor requested the delegates to visit the various public buildings and places of interest throughout the city.

The president replied that it was equally pleasant for him to thank His Worship for the kind and hearty welcome. The reception was thoroughly appreciated by the delegates, who had come to the beautiful city for both business and pleasure—business to discuss matters whereby both manufacturer and steam user may profit by their experience. He extended to the Mayor and Council a cordial invitation to visit the convention at any time. The delegates were confident they would be right royally entertained. The association's objects were purely educational, believing it is never too late to learn.

### PRESIDENT'S ADDRESS.

In addressing the convention the president stated his gratification at seeing so many familiar faces present, and he felt in good company. He made a touching reference to the late Bro. Duncan Robertson, of Hamilton, whose death occurred shortly after the last convention in Ottawa. A faithful officer, a true friend and kind husband, his death was deeply regretted. He asked the new members to join heartily in the work to be done. The most important question to come up would be the changing of the name of the association, and he hoped the brothers would weigh their thoughts before expressing their views on the matter. Another important question would be the holding of the convention every two years, instead of annually as in the past.

Mr. E. J. Philip, Toronto, Vice-President.

The compulsory issuing of certificates of membership and other topics of interest would also be considered. He stated that Stratford No. 3 had been reorganized, and an application had been received from Waterloo, where it was desired to inaugurate a branch association. He called the attention of the members to the programme that was laid before them, and thought it was the best the association had ever had. From the correspondence received from Brothers Tandvin and Devlin he felt perfectly satisfied that the members of Kingston No. 10

were a whole-souled and hard-working lot of men. Nothing had been left undone in the way of making this meeting both successful and entertaining to the delegates.

The secretary then read the minutes of the last convention, which were adopted.

Standing committees were appointed as follows :—

Auditing Committee—Bros. F. G. Johnston, W. Selby, J. G. Bain.

Constitution and By-Laws—Bros. J. J. York, chairman ; A. M. Wickens, W. Norris, C. J. Jorden, S. Donnelly.

Educational and Good of Order—Bros. A. M. Wickens, chairman ; J. J. York, J. Devlin.

Mileage — R. C. Pettigrew, chairman; C. Moseley, J. Murphy, J. F. Cody, F. J. Merrill.

Credentials—Bros. O. E. Granberg, chairman ; J. Huggett, W. F. Chapman.

MR. W. F. CHAPMAN, Brockville, Secretary.

Bro. Devlin asked what had been done in connection with securing reduced rates from the Correspondence School of Scranton, Pa., to which the Secretary replied that he did not understand their terms, as a scholarship would cost a member of the association the same as an outsider. Mr. H. S. Robertson, who represented the school, gave a statement of the facts, which showed that the Canadian association could obtain the same privileges as the International association. By these privileges members were not compelled to take their educational course in full, but could take up any branch desired.

The secretary was asked what had been done towards securing a reduced insurance rate for members of the association. He stated that several insurance companies had promised to give the association reduced rates by deducting the agents' commission. In his opinion the death rate had been increased by the encroachment of scientific inventions.

The convention then adjourned, to meet again in business session in the evening.

At 2 o'clock in the afternoon about four hundred delegates and their friends boarded the steamer Hero for a sail on the St. Lawrence among the Thousand Islands. Although attended with occasional showers of rain, the trip was thoroughly enjoyed, and to many was quite a revelation. On the return trip the city was reached about nine o'clock.

EVENING SESSION.

Reassembling in convention, Bro. John Fox, chief engineer at O'Keefe's brewery, Toronto, read the following interesting paper :—

ICE MAKING MACHINERY.

For some time I have looked forward to the preparation of a short paper on cold storage and refrigeration, and in presenting it I will endeavor to be as practical in my few remarks as the subject will admit. I will therefore dwell principally on that system of refrigeration which is now under my charge at the O'Keefe Brewing Co.'s works, namely, the " Delevergne," or direct expansion system. The substance used in this system is anhydrous ammonia. We are told that ammonia is a combination of nitrogen and hydrogen, expressed by the formula $NH^3$, which means that an atom of nitrogen (representing 14 parts by weight) is combined with three atoms of hydrogen (representing 3 parts by weight), at ordinary temperatures. The ammonia, or anhydrous ammonia, as it is called in its natural condition, is a gas or vapor, at the temperature of 30° F. It becomes a liquid at the ordinary pressure of the atmosphere, and at higher temperatures also, if higher pressures are employed. The anhydrous ammonia dissolves in water in different proportions, forming what is known as ammonia water, liquid ammonia, aqua ammonia, etc. At a temperature of 900° F. ammonia dissociates, that is, it is decomposed into its constituents, nitrogen and hydrogen. The latter being a combustible gas, it appears that partial decomposition takes place at lower temperatures, but probably not to the extent frequently supposed.

Ammonia is not combustible at the ordinary temperatures, and a flame is extinguished if plunged into the gas, but if ammonia be mixed with oxygen, the mixed gases may be ignited and will burn with a pale yellow flame. Such mixtures may be termed explosive in a sense. If a flame sufficiently hot is applied to a jet of ammonia, it (or rather the hydrogen of the same) burns as long as the flame is applied, furnishing the heat for the decomposition of the ammonia.

Ammonia is not explosive, but when stored in drums with insufficient space left for it to expand, with a high temperature, the drums will burst, as has happened in hot seasons.

Ammonia vapor is highly suffocating and for that reason persons employed in rooms charged with ammonia gas must protect their respiration properly.

With the direct expansion system, the liquid ammonia is directly conducted to the place where heat shall be absorbed, or, we might say, into the rooms which are to be cooled. The gas is then drawn back to the machines or compressors, where it is again compressed and discharged into a pressure tank, and from there to the condenser, where it is again liquefied. In liquefying the gas, cold water is allowed to trickle over the condenser, or we might call it the condensing coils, thereby cooling the ammonia. The liquid then passes on to the separating tank and if any oil should get into the liquid it is caught there. The ammonia then goes on through expansion valves into cold storage rooms where the heat of room is absorbed, thereby cooling or lowering the temperature of same, completing its work thus to repeat its circulation over and over again.

Now let us see what we have to consider in the shape of mechanical work performed. As you may know the equivalent of a ton of ice is 284,000 heat units, or the amount of heat that is required to convert a ton of ice at 32° F. into a ton of water at 32° F.; or conversely, it is the amount of heat that must be extracted from a ton of water at 32° F. in order to convert it into a ton of ice at 32° F.

Let us take, for instance, a 50 ton plant. The latent heat of one pound of ice is 142 heat units; multiplying this by 2,000 gives us the number of heat units in one ton. Now, as we are considering a 50 ton plant, this will be 14,200,000 heat units in 24 hours of time, or in other words, a 50 ton plant in 24 hours will absorb this amount of heat units. I might say here that in speaking of a plant of so many tons capacity, it is always understood to mean for 24 hours of time.

MR. R. C. PETTIGREW, Hamilton, Treas.

The temperature of expanding ammonia would have to be about 10° F. lower than the temperature of a cold storage room, which we will take as 35° F., consequently by using latent heat of vaporization at that temperature, which is 35° F. – 10° F. = 25° F, we find it to be 540.03, which is refrigerating effect of 1 lb. of ammonia when the temperature of refrigeration is 25 deg. F., and that of condenser 70 deg., specific heat of the ammonia being 1 deg. F. The amount of ammonia to be evaporated, therefore, per minute of our 50 ton plant is (540.03 - (70 - 25) = 495.03 latent heat of ammonia at 25 deg. F. Omitting the decimals and taking this in round numbers, 495 × 60 = 29,700. This divided into 591658.33 = 19.92, which is the number of pounds of ammonia we require per minute for our 50 ton plant. We require about 20 lbs., and the volume of 1 lb. of ammonia vapor at 25 deg. F. is equal to 5.26

cubic feet, consequently compressor capacity per minute will have to be 105.20 cubic feet. If we add to this 20%, which is a fair allowance for losses by radiation, etc., we require an actual compressor capacity of 126.24 cubic feet per minute.

Let us see how the plant I operate compares with this theoretical calculation just made. The compressor cylinders are $11'' \times 22''$, which is equal to about 1 1/5 cubic feet capacity of each cylinder. Our engine makes 40 revolutions per minute and each is double acting. Diameter $11'' \times 11'' = 121 \times .7854 = 95.0334 \times 22 \div 1728 = 1.2099$. Consequently at each revolution of crank shaft each compressor discharges its contents twice, which gives us a total discharge of about 192 cubic feet per minute. If we deduct 20% from this for clearance, losses, etc., we get 154 cubic feet, or about 27 feet more than required by our theoretical calculation, which would be the amount allowed to come and go on, which I think close enough for all practical purposes.

Now comes the question of piping required for cold storage rooms. In piping cold storage rooms, from what information I can gather on this subject, it is usual to allow about one square foot of pipe surface for every 1,000 heat units to be absorbed. This is equal to about 1.6 running feet of 2 inch pipe. For a 50 ton plant, according to this rule, we will require a sufficient amount to absorb 14,200,000 heat units in 24 hours, which in round numbers will be $14,200,000 \times 1.6 \div 3,000 = 7,573$ running feet of 2 in. pipe. Of course you understand this estimate is approximate. If we were using 1 in. pipe instead of 2 in. pipe, and the same factor, namely, 3,000 heat units, to be absorbed in 24 hours per each square foot of pipe surface exposed, it would require about 2,833 running feet of pipe. The condensers are a system of pipes or coils into which the ammonia, after being compressed in compressors, is forced, where it is cooled by cold water trickling over the pipes. These are called atmospheric or surface condensers. The ammonia in passing through the condensers yields to the cooling water the heat which it has acquired in doing refrigerating duty by its evaporation and the heat it has acquired during compression, superheating being prevented by a liberal supply of oil in our case.

The mechanical work done during compression is converted into its equivalent of heat. This amount of heat is also equal to the latent heat of volatilization of the ammonia at the temperature of the condenser. The efficiency of the condenser determines in a great measure the economical working of the machine, and for this reason it is good policy to have as much condenser surface as practical consideration may permit. It is said for average conditions (incoming water 65 deg. F., outgoing 85 deg. F,) it will require 20 square feet of surface per ton, or for a 50 ton machine it will take 1,600 linear feet of 2 inch pipe. The main difference of outgoing and incoming water is 20 deg., $485.42 \times 20 \times 60 \div 20 \div 8.33 = 3496$, which is amount of water in gallons per hour.

GRIDIRWORKS FOR ECONOMIZING COOLING WATER.—Where cooling water is very scarce, and especially where atmospheric conditions—dryness of air, etc.—are favorable, the cooling water may be re-used by subjecting the spent water to an artificial cooling process by running the same over large surfaces exposed to the air in a fine spray. A device of this kind is described as being a chimney-like structure, built of boards. Its height is 25 feet, the other dimensions being $8' \times 8'$. Inside this structure are placed a number of partitions of thin boards, spaced 4 inches apart, extending to within 1 foot of the bottom of the structure; but the lower halves of these partitions are placed at right angles to the upper halves. This arrangement gives better results than unbroken partitions. The water to be cooled enters the structure at the top, where, by the use of a galvanized iron overflow gutter, it is spread evenly over the partitions and walls and flows downward in thin sheets. At the base of the structure air is introduced in such quantity that the upward current has a velocity of about 20 feet per second. The air meeting the downward flow of water absorbs the heat by contact and also by vaporizing during the passage, 20 deg. F.

The oil used for lubricating the compressors differs from ordinary lubricating oil in that it must not congeal at low temperatures, and must be free from vegetable or animal oil. For this reason only mineral oils can be used, and of these only such oils as will stand a low temperature without freezing, such as the best paraffin oil will do.

Bro. Edkins wished to know what a 50 ton ice plant was.

Bro. Fox replied that it absorbed the heat in a cold storage room with the same power as would 50 tons of ice kept at 32° F.

Bro. J. J. York said that only the previous week the Board of Trade of Montreal had met to consider the introduction of ice-making machinery on the steamships. It was a subject which the intelligent engineer would have to grapple with sooner or later.

A hearty vote of thanks was tendered to Bro. Fox for his paper.

Mr. J. M. Campbell, of Kingston, promised a paper on "Electrical Appliances," but was unavoidably absent from the city.

A visit was then paid to the electric light and gas works, under the direction of Mr. Simmons, the superintendent, who showed the delegates some experiments with acetylene gas.

## SECOND DAY.

The convention resumed at 10 a.m., the president in the chair.

Bro. E. J. Philip, Executive Secretary, presented his report, which showed that the total receipts for the year were $607.22 and the expenditure $505.65, leaving a balance of $101.57. The strength of the association had not been up to that of former years, neither numerically nor financially, but a large amount of good had been done by the association taking up the matters of education, insurance and certificates, and while none of these had been as successful as was anticipated, the probability was that during the following year they would be got into better shape. He suggested that the cost of certificates be lessened to the members and that they be made compulsory. He again reported at length on insurance and the Correspondence School, as also on a scheme that would make the Executive more of an educator, by establishing a Bureau of Information.

The Committee on Constitution and By-laws presented a report, which recommended that it be made compulsory to secure membership certificates, and that a reduction be made in the number of officers. This was necessary in order to meet expenses. The movement met with much opposition, some proposing raising the per capita tax, while others suggested meeting in convention every two years. The committee was requested to report again the following day.

The report of the Treasurer was then presented, in which it was stated that the Association had felt the effects of the commercial depression. There was a balance on hand of $101.57.

The report, which was certified correct by the auditors, was adopted, and the meeting adjourned for lunch.

At 2 o'clock, by the courtesy of Mr. B. W. Folger, manager of the street railway, a special car took the delegates and friends to the penitentiary. Mr. Devlin, chief engineer, escorted the visitors through the institution. They were shown a small hand engine built by the famous Percy, of Montreal, forty years ago, for the penitentiary, which cost $1,100. Over the door of the engine room was the lettering "Welcome C.A.S.E." surrounding the British coat of arms. On the wall was a crown of colored incandescent lights. Rockwood asylum was also visited, and in the evening the members had an outing at Lake Ontario Park.

## THIRD DAY.

At 9 a.m. on Thursday the business of the convention was again taken up.

The report of the Committee on Constitution and By-laws was adopted. It recommended the granting of a free certificate to every member of the association in good standing and the raising of the per capita tax 25c.

The report of the Committee on Education and Good of the Order recommended that members should take a course in the International Correspondence School. This report was also adopted.

Bro. Edkins suggested the advisability of taking steps to secure Dominion legislation compelling all engineers to hold certificates, instead of as at present, through the provincial government.

Bro. Granberg said Quebec would give every assistance. If there was a compulsory law passed by the Dominion government, the certificate holders of the Ontario association would readily join the Canadian association, and allow the former to lapse. This would bring in a membership of from 700 to 800.

Bro. Wickens said he had been through five legislative fights, and advised them to ask for enough, so that they would be able to get something.

It was moved by Bro. Edkins, and seconded by Bro. Philip, that a committee be appointed to co-operate with the executive board of the Ontario association with a view to securing Dominion legislation for the compulsory examination of stationary engineers. Carried.

It was resolved to publish a hand-book, giving a list of all the engineers in Canada, which number about 12,000.

The mileage report was then adopted.

Bro. Norris moved that a quarterly report from the Executive Secretary be sent to each branch giving its standing. Carried.

### ELECTION OF OFFICERS.

The next order of business was the election of officers. The president appointed Bro. Edkins returning officer and Bros. Robert and Tandvin scrutineers.

For president Bro. James Devlin, of Kingston, was elected by acclamation.

For vice-president four nominations were made, Bros. Philip, Pettigrew, Granberg and Chapman. Bro. E. J. Philip, of Toronto, was elected.

The contest for secretary was between Bros. W. F. Chapman, Brockville, and R. C. Pettigrew, Hamilton, the former being successful.

Bros. Pettigrew, B. A. York and Granberg were in the field for treasurer. Bro. Pettigrew was elected.

For conductor there were eight nominees, Bros. Huggett, Murphy, Bain, Wickens, B. A. York, Moseley, Johnson and Jorden. Bro. J. Murphy, of Montreal, was successful.

Bro. F. J. Merrill, of Ottawa, was elected doorkeeper. The other candidates were Bros. Huggett, Fox, Jorden, McKay, Johnston and Norris.

Past-president York installed the newly-elected officers, and the retiring president, Bro. Blackgrove, was presented with the customary jewel.

The officers elected thanked the convention for the honor conferred upon them and promised to endeavor to further the best interests of the C. A. S. E.

A vote of thanks was tendered the returning officers and scrutineers.

Votes of thanks were also tendered the Kingston Street Railway Company, the International Correspondence School, the Mayor and city council, the local association and the press.

Brockville and Hamilton both tendered for next year's convention, with the result that Brockville was chosen.

The president-elect appointed Bro. Granberg district-deputy for Quebec and Bro. Cody district-deputy for Ontario.

Bros. Wickens, Edkins and Norris were appointed a Committee on Legislation.

After conclusive remarks the convention closed by singing "God Save the Queen."

At 3 p. m. the members took carriages for a drive to Kingston Mills and Fort Henry, lunching at the former place.

### THE BANQUET.

A banquet at the British American Hotel on Thursday evening fittingly closed the convention. The chair was occupied by President S. Donnelly, of Kingston No. 10. On his right sat the Mayor, Aldermen Skinner, Ryan and Tait. On his left was Executive President Devlin, Past President Blackgrove and Past President Wickens.

The first toast on the list was "The Queen," which was honored by singing the National Anthem.

The Mayor and Ald. Skinner, Ryan and Tait responded to the toast of the "City of Kingston."

"The C.A.S.E., its Aims and Objects," brought replies from Bros. Wickens, J. J. York, Granberg and Norris. Bro. Wickens said that the C.A.S.E. was bound to succeed, for it was founded on the rock of knowledge. Intelligent engineers did not believe in accidents; explosions were due to carelessness or ignorance.

Bro. York said that if all the engineers were as proud as he was to be a member of the C. A. S. E. the membership would increase ten fold. Steam users contemplating remodelling their plants or installing new machines should consult the C.A.S.E.

MR. S. DONNELLY, President, Kingston No. 10.

Bros. Granberg and Norris spoke briefly of the advantages afforded by the association.

"Our Manufacturers" was acknowledged by Mr. Anderson, of the Imperial Oil Company. He spoke of the practical emigration policy required in Canada, and advocated the further extension of foreign trade.

Mr. Robertson, of the Correspondence School at Scranton, responded to the toast "Our Technical Educators."

"Our Visitors and Kindred Societies" brought replies from Bros. Blackgrove and Edkins. Bro. Edkins said that many thought that the O.A.S.E. was an antagonist to the C.A.S.E., which was not the case. As soon as Dominion legislation compelled qualified licensed engineers the O.A.S.E., with its 700 members, would amalgamate with the C.A.S.E.. He showed how dangerous it was to place boilers in the care of incompetent men, by giving the relative explosive forces of steam and dynamite.

"Kingston No. 10" was acknowledged by Bros. S. Donnelly and J. Devlin.

"The Executive Council" was replied to by Bros. Devlin, Philip, Pettigrew and Chapman; "The Ladies" by Bro. Granberg and Ald. Tait, and "The Press" by representatives of the Kingston Whig and News, and the Canadian Engineer and ELECTRICAL NEWS, of Toronto.

The entertainment was interersed with songs by Messrs. Grant, Blackgrove, Cochrane, Skinner, Murphy. and Rubert.

### MR. JAMES DEVLIN.

Mr. Devlin, president-elect of the C. A. S. E., was born in Kingston, and is the eldest son of the late P. Devlin, a veteran fireman. After serving his apprenticeship with the Canadian Locomotive and Engine Company, he worked for a time with D. Ewen & Sons, and in 1873 was appointed engineer of the Government waterworks. Two years after he was transferred to the penitentiary at St. Vincent de Paul, near Montreal, as chief engineer. In 1885 he received the appointment of chief engineer of Kingston penitentiary, which position he still occupies. His connection with the C.A.S.E. dates from the year 1892, and since that time he has been an active worker. He was largely instrumental in securing the formation of the Kingston branch, and at the annual convention at Ottawa last year he was unanimously elected to the Executive Committee as vice-president, and filled the position with such credit that his qualifications for the duties of president are unquestioned. He is also a member of the Board of Examiners of the Ontario Association of Stationary Engineers, and a strong advocate of compulsory examination of engineers.

### SPARKS.

An electric light plant will probably be installed in the asylum at Brockville, Ont.

The Sherbrooke Telephone Co., of Sherbrooke, Que., is building 100 miles of new lines.

A by-law has been carried by the ratepayers of Listowel, Ont., in favor of electric lighting.

The Prescott Electric Light Co., of Prescott, Ont., contemplate making additions to their plant.

Local parties will probably install an electric light plant for street lighting at Winchester, Ont.

The Ottawa Electric Light Co. intend putting in a 7,000 candle power machine in No. 2 power house.

It is said to be the intention of Mr. Comstock, of Brockville, to place an electric light plant in his yacht.

On the 17th of August the ratepayers of Huntsville, Ont., sanctioned a by-law providing for an electric light plant.

It is announced that a gentleman from Halifax proposes establishing an electric light plant at Shubenacadie, N. S.

The plant of the Owen Sound Electric Light Co. is being enlarged to supply incandescent as well as arc lights.

The Galt, Preston & Hespeler electric railway carried 35,000 passengers and 930 tons of freight during the month of July.

The Auer Light Co., with a capital of $30,000 has been organized in Ottawa. Mr. C. S. Taggart has been appointed manager.

The Listowel Gas Co., of Listowel, Ont., purpose installing an electric plant when the present contract for lighting the town expires.

The Trojan Car Coupler Co.'s branch at Smith's Falls, Ont., are supplying the Quebec and St. John electric road with car couplers.

Incorporation is being sought by the Amherstburg Electric Light, Heat & Power Co., of Amherstburg, Ont., with a capital stock of $20,000.

It is reported that Michael & Becker, owners of a patent telephone system, are prepared to make an offer for the telephone franchise of Toronto.

The city council of Windsor, Ont., have purchased the machinery in the electric lighting station from E. Leonard & Sons, of London, at the price of $2,685.

Mrs. R. McLaughlin, of Summerville, has entered a suit against the Toronto & Mimico Electric Railway Co., to recover $20,000 damages for the death of her husband, who was killed by one of the company's cars.

Owing to the destruction of the power house of the Montreal Park & Island Railway Co., two new power houses are being erected, one at Lachine and the other at St. Laurent.

The Electric Railway Co. at Sherbrooke, Que., is said to be negotiating for water power at Brompton Falls, and if successful will commence the construction of the road at an early date.

Dr. Harrison Chamberlain has presented to the city council of Buffalo a petition asking for permission to construct and maintain a telephone system in that city. It is proposed to organize a new company.

The Hamilton, Chedoke & Ancaster Electric Railway Co. have secured almost all the right of way necessary for their proposed line, and steps are now being taken to organize the company. It is the intention to have the road in operation early next summer.

In the decision of the Privy-Council in the case of the Toronto Street Railway Co. re the duty on steel rails, several other railway companies are interested, as follows : London Street Railway Co., $12,000 to $15,000 ; Hamilton Street Railway Co., $18,000 ; Windsor Street Railway Co., $6,000. The Winnipeg Street Railway Co. is also said to be interested.

Incorporation has been granted to the Callendar Telephone Exchange Co., with head office in Toronto. The object of the company is to deal in telephone patents, and to build and operate telephone lines throughout Canada. The promoters are : Romaine Callander and Edward H. Hart, of Brantford, and J. Enoch Thompson, of Toronto. The capital stock is $100,000.

The corporation of the town of Markham, Ont., have leased their lighting plant to the Markham Electric Light Company, who have extended and improved the system. They have installed a 20 K.W. alternating plant furnished them by the Royal Electric Company, and about 350 lamps. The street lighting has also been changed from arc to incandescent lights, which are giving every satisfaction.

At the annual meeting of the Ottawa Car Co., held recently, it was stated that a dividend of 8 per cent. had been paid to the shareholders and a sum equal to 4 per cent. placed to the reserve account of the company. Directors were elected as follows : Thomas Ahearn, president and managing director ; James D. Fraser, secretary-treasurer ; W. W. Wiley, J. W. McRae, W. Y. Soper and Wm. Scott.

The addition to the power house of the Montreal Street Railway Co. is nearing completion. A new boiler house with chimney 250 feet in height is being constructed, in which will be placed three Babcock & Wilcox water tube steam boilers and a direct-connected engine of nearly three thousand horse power. This addition has been rendered necessary in order to cope with the increasing passenger traffic.

R. Anderson, Ottawa, is applying for a patent on an electric switch to turn on one, two or three lights, or any number, in rotation. It is useful in high chandeliers on occasions where only one or two lights are required. He is also applying for a patent for an invalid's push hanging switch. One push lights the lamp and the next puts it out. This is a very simple and useful contrivance. It will be made of porcelain.

The Richmond County Electric Co. have decided to entirely remodel their plant outside the station. All their old type transformers will be discarded and be replaced by Stanley type in large units, and all their primary circuits will be transformed to 2,000 volts. By these changes they will secure twenty per cent. increase in capacity and be able to take care of their increased business without changing their present dynamos. The order for transformers and supplies has been placed with the Royal Electric Company.

At the recent carnival at Halifax, N. S., an attraction of much interest was the electrical illuminations of the war-ships and the illuminated procession of steamers, yachts and boats. The flagship " Crescent " was decorated with incandescent lamps, while the English and French war-ships and the cable ships " Mackay Bennett " were beautifully displayed with search lights and blue fire. On the steamer " Annie " was placed a special engine and dynamo, with strings of incandescent lamps covered with fancy shades and Wheeler reflectors. This installation was the work of John Starr, Son & Co., of Halifax, and reflected credit upon the firm.

## TO AVOID FIRES FROM ELECTRICAL APPLIANCES.

The National Board of Fire Underwriters of the United States has promulgated a series of rules referring to electrical appliances for light and power. It publishes the following cautions for the information of the public.

1. Have your wiring done by responsible parties, and make contract subject to the underwriter's rules. Cheap work and dangerous work usually go hand in hand.

2. Switch bases and cut-off blocks should be non-combustible (porcelain or glass).

3. Incandescent lamps get hot ; therefore all inflammable material should be kept away from them. Many fires have been caused by inflammable goods being placed in contact with incandescent lamp globes and sockets.

4. The use of flexible cord should be restricted to straight pendant drops, and should not be used in show windows.

5. Wires should be supported on glass or porcelain, and never on wooden cleats ; or else they should run in approved conduits.

6. Wires should not approach each other nearer than eight inches in arc, and two and one-half inches in incandescent lighting.

7. Wires should not come into contact with metal pipes.

8. Metal staples to fasten wires should not be used.

9. Wires should not come into contact with other substances than their designed insulating supports.

10. All joints and splices should be thoroughly soldered and carefully wrapped with tape.

11. Wires should be always protected with tubes of glass or porcelain where passing through wall, partitions, timbers, etc. Soft rubber tubes are especially dangerous.

12. All combination fixtures, such as gas fixtures with electric lamps and wires attached, should have approved insulating joints. The use of soft rubber or any material in such joints that will shrink or crack by variation of temperature is dangerous.

13. Electric gas lighting and electric lights on the same fixture always increase the hazard of fire, and should be avoided.

14. An electric arc light gives off sparks and embers. All arc lamps in vicinity of inflammable material should have wire nets surrounding the globe, and such spark-arresters reaching from globe to body of lamp as will prevent the escape of sparks, melted copper, and particals of carbon.

15. Arc light wires should never be concealed.

16. Current from street railway wires should never be used for lighting or power in any building, as it is extremely dangerous.

17. When possible, the current should be shut off by a switch where the wires enter the building, when the light and power are not in use.

18. Remember that "resistance boxes," "regulators," "rheostats," "reducers," and all such things, are sources of heat and should be treated like stoves. Any resistance introduced in an electric circuit, transforms electric energy into heat. Electric heaters are constructed on this principle. Do not use wooden cases made for these stoves nor mount them on wood work.

## AUTOMATIC COAL WEIGHING MACHINES FOR POWER STATIONS.

THE late Hon. Eckley B. Coxe, who strongly advocated the substitution of a continuous record of actual boiler performance for the prevailing system of occasional tests, once stated the matter very tersely, saying :

"I am not so much interested in knowing what some expert may be able to do with my boilers as to know what work my firemen are actually getting from them every day."

To know this, however, means the measuring of the intake and the output—means accounting for the entire supply and production, so that the necessary comparisons may be made for formulating the result. The automatic weighing machine supplies this requirement, automatically handling coal and water, much after the manner of an ordinary water meter, say, interposed in a water pipe, or a gas meter for that matter, giving a continuous and reliable record of what has passed through it.

In another way, too, may the automatic weighing machine serve a good purpose. Daniel Webster has been quoted for the way in which, in one of his speeches, he emphasized "the tremendous power of six per cent." Certainly the investor of to-day looks sharply enough. to the difference of one per cent. in the rate of interest chargeable against him. But does he look as closely to the other components of the "cost?" For instance, recent experiments indicate that the anthracite coal generally used for steam making will hold about four per cent. of water without much dripping ; and much of that coal is "watered" to this extent before delivery. If, now, the coal pile be replenished twice a year with wet coal, it is evident that the buyer pays the interest rate plus eight per cent. of the purchase price as the cost of the capital employed in "carrying" the fuel account.

Although the coal cannot, for obvious reasons, always be obtained dry, the drying may be readily effected in nearly all power stations before the coal reaches the bins by using heated air drawn from the upper part of the boiler rooms. Then, by weighing in through an automatic weigher and reweighing in the same way, first to the bins, next directly to the furnaces, all of the required facts are obtained. The first weighings, by showing the amounts taken and delivered to the bins indicate the evaporation, and a comparison of the records of the second and third readings will show, at any time, the amount held in storage in each bin, besides giving the amount chargeable to each set of boilers.—Francis H. Richards in Cassier's Magazine for August.

## ERRATA.

Mr. John Patterson, of the Power Company, Hamilton, writes ; "Will you kindly correct a couple of errors in your article on the Cataract Power Co., of Hamilton, in the August issue of your paper. The water flowing over the fall is more than 6 inches by 48 feet, instead of 5 inches by 18 feet, and the head is about 210 feet instead of 270. The flow of water is over eight thousand cubic feet per minute, which at this head gives something over the 2500 horse power."

Mr. P. G. Gossler, engineer for the Royal Electric Co., Montreal, is at present engaged on the reconstruction of the Montreal station. They are replacing 17 alternators by five 300 kilowatt generators, directly connected to three engines. They are also building an entirely new switch board for light and power from the two phase circuits. Their present system of 1000 volts is being changed to a 2000 volt system and their transformers are being replaced by more modern ones. Power will be obtained from Chambly Falls, fifteen miles distant.

# CORRESPONDENCE

To the Editor of the CANADIAN ELECTRICAL NEWS.

SIR :—I clipped the following letter from the August issue of Power, and with your permission will reproduce it here, as the writer evidently has discovered some new element of danger in our already dangerous but interesting occupation :

#### WHERE QUICK ACTION WAS NECESSARY.

Several years ago a gentleman who is now chief in a large water works in New York had charge of an engine in a saw mill in the then wild woods of Michigan. One morning he came to the mill, and found no steam. The watchman told him that he had been firing very heavy for several hours with dry fuel. An examination showed that the boiler, dome, and steam pipes were full of water, also that the boiler and arch were excessively hot. Here was a case of danger that required a superior knowledge of engineering, philosophy, good judgment, and prompt action. What would you do?

J. W. POWER.

Will my fellow Canadian professional brethren enlighten me as to what this writer intends to explain, whether the boiler in question had become so overheated as to have become dangerous from this cause, or simply whether the engineer was fearful of detaining the sawmill hand until such time as he could reduce his water level and make the steam, the watchman had been so long trying to get?

After reading the letter I am of the opinion that the writer intends to state that the boiler was overheated and in danger of collapse from this cause. He tells us that the boiler, dome and steam pipes were full of water, but does not say how they happened to be full ; quite evidently the water had fed into the boiler from some source after the engineer had shut down the previous night—a not altogether unknown occurrence, as most engineers can verify, especially when there happens to be a cold water pressure on the feed main. No matter how the thing happened, however, we are told that the apparatus in question was full, that the watchman had been firing for several hours, (apparently on this particular morning the watchman got at the fires earlier than usual) ; that he had no steam, that the boiler as a consequence of the untiring energy of the watchman was excessively hot, that immediate danger was so great that this model up-to-date engineer had to know so much and act so quickly under those trying circumstances that Mr. Power thinks it worthy of record as a brilliant engineering feat and wants to know what other engineers would have done under the circumstances.

Well, Mr. Power, for my part under such circumstances as these I should have kept exceedingly quiet and taken off my coat and made steam. In the first place, if, as Mr. Power tells us, the boiler, etc., was full of water, how was it that the almost incessant firing of the watchman did not impart some heat to this same water, particularly when the boiler shell was so overheated? Mr. Power no doubt knows that the circulation within the boiler was just the same under circumstances stated as it would be with dome full of steam, under equal pressure, and also as heat was imparted to the water it would in course of events follow the natural law and expand. But we are told everything was full and therefore water could not expand or increase in bulk, and increase of pressure on the boiler would follow as a natural consequence and would be recorded on the steam gauge just the same as though the boiler was making steam in the usual course of events. We are told, however, that when the engineer arrived there was no steam, or in other words no pressure was indicated on the gauge. If so, then it is quite evident that the heat from the fires had never reached the water within the boiler, and that overheating of boiler plates was caused rather from want of water than there being too much water in the boiler.

After reading Mr. Power's letter I am very much inclined to believe that a very large amount of danger was imaginary and that a great many of the circumstances are the result of an imaginative brain rather than an actual occurrence.

"ENGINEER."

---

The Lozier Manufacturing Co., of Toronto, have applied for incorporation to build motor vehicles, etc. The capital stock is $500,000.

Mr. Higman, head of the Government Electric Light Inspection Department, recently made a test of the electric light service supplied to the city of Toronto by the Toronto Electric Light Co. In his report to the City Board of Control he states that "voltage readings were made at 112 different lamps, which show an average pressure of 51 volts. If an average of the whole ampere readings is made we have 10.6 amperes as the current strength ; this multiplied by the mean voltage gives an average of 541 watts, or 25 per cent. in excess of what the contract called for." The report concludes with the statement that "not only is the company fulfilling its obligations, but is doing so generously, and the operation of the whole plant is decidedly creditable."

## BEARDSHAW'S "PROFILE" TOOL STEEL.

ON another page of this journal will be found an advertisement by Messrs. Winn & Holland, of Montreal, who have secured the sole agency for Canada for the sale of Beardshaw's Profile Tool Steel, which is now being largely introduced in Montreal. In introducing to the engineers and allied trades tool steel in a new form, little need be said, other than to point out its various applications, (which is done in the catalogues supplied by the agents) for to say, that the steel comes to the user in such form or profiles as to enable tools to be made by grinding only, is to indicate that a long-felt want has been met.

This steel is now rolled in six different sections or profiles, most required by engineers and machinists, and other profiles will be made as required. Tools for the lathe, planer, shaping, slotting and drilling machines, also chisels, rimers, taps, bits, broaches, gravers, etc., are made from this steel without forging, and there is, therefore, no wasting in the fire. The steel as it comes from the rolls is ready to cut to lengths and grind into tools, which is a great saving in time. There is also a saving over 50% in the weight of steel used.

The manufacturers claim that the quality of this steel is much finer than any ordinary tool steel put upon the market. Owing to it not being necessary to forge it, it is possible to supply a very much more durable steel, which if forged, would be liable to be spoiled by the blacksmith.

The adoption of this Profile Steel amongst the most up-to-date engineers is now an assured fact. An undoubted success of this steel has been made in Europe, and it is now being introduced in the colonies. The Profile Steel is also being produced in self-hard quality, so that no heating is necessary.

## PERSONAL.

Mr. A. A. Dion, superintendent of the Ottawa Electric Light & Power Co., returned a fortnight ago from a trip to New York.

Mr. George C. Peters, manager of the New Brunswick Telegraph Co., at Moncton, N. B., recently had the misfortune to break one of his arms.

Mr. J. C. Mullen, formerly with the Ottawa Electric Railway Co., is at present engaged on the construction of a government electric line in Durban, South Africa.

It is stated that Mr. James Devlin, chief engineer of the penitentiary at Kingston, is an applicant for the position of Superintendent of Public Works for the Dominion.

Mr. A. Porter, superintendent of the Cornwall Electric Street Railway, has resigned, to accept a position in Montreal. The employees of the company presented him with a gold watch on the eve of his departure.

Mr. Angus Grant, for many years superintendent of the Great Northwestern Telegraph Co., at Montreal, died at his home in Prescott on the 15th ult. Some time ago, owing to ill health, he was obliged to resign his position.

The death occurred in Toronto early in August of Mr. Thomas Northey, father of Mr. J. P. Northey, of the Northey Mfg. Co. Deceased was well known throughout Canada, and was in his 80th year. Fifty years ago he established a foundry in Hamilton, removing to Toronto in 1880.

Mr. W. E. Davis, formerly of the Toronto Street Railway Co., and latterly electrical engineer and purchasing agent of the Detroit Railway, has resigned his position and removed to Saginaw, where he becomes manager of the Bearinger Electric Railway between that city and Bay City.

Mr. J. J. Ashworth, for several years attached to the engineering department and latterly to the agency staff of the Canadian General Electric Co., has severed his connection with the company and will engage in the future in independent engineering and construction work. The termination of Mr. Ashworth's connection with the company was made the occasion of a pleasant expression of good will on part of the Toronto staff it the form of an address accompanying the presentation of a locket with monogram suitably engraved.

It is with deepest regret that we announce the death of Mr. A. W. Congdon, of the engineering staff of the Canadian General Electric Co., to whose illness reference was made in these columns recently. The hope then expressed that he would probably recover was unfortunately not well founded, the disease having taken deeper hold upon his system than had been supposed. It is believed that in the trip which Mr. Congdon made several years ago to Japan, his constitution was undermined.

## TRADE NOTES.

The Canadian General Electric Co. are installing a 500-light incandescent plant in the Winnipeg general hospital.

The Dodge Wood Split Pulley Company have removed their Toronto office from 86 King street to 74 York street.

W. A. Johnson, of Dresden, has ordered a 500 light single phase alternating plant from the Canadian General Electric Co.

The Vancouver Consolidated Ry. Co. have ordered a 150 k. w. "Monocyclic" generator from the Canadian General Electric Co.

The Niagara Falls Light & Power Co. recently started up the second of the 2,000-light single phase generators purchased from the Canadian General Electric Co.

The Hull Electric Co. have ordered a parlor car from the Canadian General Electric Co. to meet the demand for such a service on part of excursion parties from the capital city.

The Robb Engineering Co., of Amherst, N. S., are placing a 100 horse power engine and a 125 horse power Monarch boiler in the power house of the Moncton Street Railway Co.

Messrs. Hooper & Starr, who are constructing the Cornwall St. Ry., owing to the increasing traffic, have ordered additional C. G. E. 800 motors from the Canadian General Electric Co.

The Dominion Oil & Supply Co., of Montreal, is applying for incorporation, to manufacture engine and boiler supplies, etc. Among the applicants are Tancrede Hout and Paul Gailbart.

The Goldie & McCulloch Co., of Galt, have been given an order by the Galt Waterworks Department for a compound steam pump capable of pumping 1,000,000 imperial gallons per 24 hours.

The Paxton, Tait Co., of Port Perry, Ont., are seeking incorporation, with a capital stock of $99,000. The first directors are Hon. John Dryden, George William Dryden and William McGill.

Messrs. B. Bell & Sons, agricultural implement manufacturers, of St. George, Ont., are having their large works lighted throughout by electricity. The Royal Electric Company are installing the plant.

The Montreal Park and Island Railway Company have recently placed an order with the Royal Electric Company for ten (two 30 h. p. motor) car equipments, and one 250 k. w. railway generator. These are to replace their plant recently destroyed by fire.

The Berlin & Waterloo Electric St. Ry. Co. have placed an order for two closed motor car bodies with the Canadian General Electric Co. These cars will be of the company's standard vestibule type, somewhat modified to meet the views of the president of the road, Mr. E. Carl Breithaupt, and are intended to be models both in design and construction.

La Compagine Electrique has recently purchased a 30 h. p. S. K. C. two-phase motor from the Royal Electric Co. This motor will be used to operate a woollen mill. This installation is of special importance as showing the development of multiphase motor work in small plants, and how small plants can increase their earning capacity by operating in the day time as well as the short time of load at night.

The Corporation of the town of Newmarket have closed a contract with the Royal Electric Co. for one of their "S. K. C." alternating current dynamos, having a capacity of 1,000 16 candle power lamps; and also have ordered 800 light capacity in "S. K. C." transformers. They are re-building the old plant and are also installing a new arc machine and lamps, purchased from the Canadian General Electric Co.

The Royal Electric Company are installing for McMaster Bros., of Ridgetown, Ont., one of their 75 k.w. "S. K. C." transformers of 500 light capacity. The generator is direct belted to a 100 h.p. high speed Leonard Ball engine, making a very compact and modern plant. Their new brick power house is a model of neatness, and when their plant is completed they will have a fine modern and complete electric lighting plant.

Mr. Geo. E. Matthews, manager of the Electric Repair and Contracting Co., of Montreal, states that the results of the company's first year's business are very satisfactory, and they are entering upon the second year with fair prospects. The company are prepared to execute any class of new or repair work. The fact that Mr. Matthews was for ten years foreman of the repair department of the Royal Electric Co. should be a sufficient guarantee of competency.

The Montreal Cotton Co. are installing electrical power apparatus in their mill at Valleyfield. Mr. Louis Simpson, the general manager of the company, after examining the operation of similar plants in the cotton mills in the United States, has placed an order with the Canadian General Electric Co. for a 600 h. p. three phase generator, and for 350 h. p. capacity in induction motors. Some of the latter will be of the inverted type attached to the ceiling and direct coupled to the line shafts which they are to run.

The Chateauguay & Northern Ry. Co. are proceeding rapidly with the equipment of their road, which is really a branch of the Montreal Island Belt Line system. The contract for the entire electrical equipment and for the car bodies has been awarded to the Canadian General Electric Co. and consists initially of one 200 k. w. multipolar railway generator, and two open and four closed cars equipped with C. G. E. 800 motors, and type "K" controllers. The road is expected to be ready for operation not later than September 15th.

## UNSOLVED PROBLEMS IN THE MANUFAC-TURE OF LIGHT.

PROF. John Cox, in a recent lecture on the above-named theme, before the Royal Society of Canada, presented, in a very striking way the enormous percentage loss of energy in all attempts heretofore made to manufacture light through the agency of the steam engine.

To begin with he points out that in practice not more than from 7 to 16 per cent. of the energy of the fuel used can be realized through the engine, and theoretical considerations establish a limit at about 30 per cent., beyond which it would seem to be hopeless to expect to pass in any form of heat engine. This he terms one of the unsolved problems.

It is, however, not unsolvable if we can devise some means of extracting the energy of coal otherwise than by heat—say in some such manner as that in burning zinc in a voltaic battery. That this is not beyond the scope of our present scientific knowledge the recent experiments of Borchers and others bear strong evidence.

In the second stage of the operation of producing the electric light, the dynamo is already so nearly perfect that hardly any heat is lost in its conversion into current.

The third stage brings us to the lamp, with some 7 per cent. of the original energy still available. The only means thus far available for producing luminous energy is to heat the molecules of some substance, and in this operation we are compelled to waste the greater portion of our available energy in producing heat before we obtain the light rays.

"Here, then, is the second unsolved problem, since even in the incandescent lamp and the arc lamp not more than from 3 to 5 per cent. of the energy supplied is converted into light. Thus, of the original store in the coal less than three parts in a thousand ultimately become useful. In the last six years, however, some hint of means to overcome the difficulty has been obtained from the proof by Maxwell and Hertz that light is only, an electric radiation. Could we produce electric oscillations of a sufficient rapidity, we might discard the molecules of matter, and directly manufacture light without their intervention. To do this we must be able to produce oscillations at the rate of 400,000,000,000 per second. Tesla has produced them in thousands and millions per second, and Crookes has shown how, by means of high vacua, to raise many bodies to brilliant fluorescence at a small expense of energy. . . These are hints toward a solution of the problem, but give no solution as yet. Prof. Langley states that the Cuban firefly spends the whole of its energy upon the visual rays without wasting any upon heat, and is some four hundred times more efficient as a light producer than the electric arc, and even ten times more efficient than the sun in this respect. Thus, while at present we have no solution of these important problems, we have reason to hope that in the not distant future one may be obtained, and the human inventor may not be put to shame by his humble insect rival."

# ELECTRIC RAILWAY DEPARTMENT.

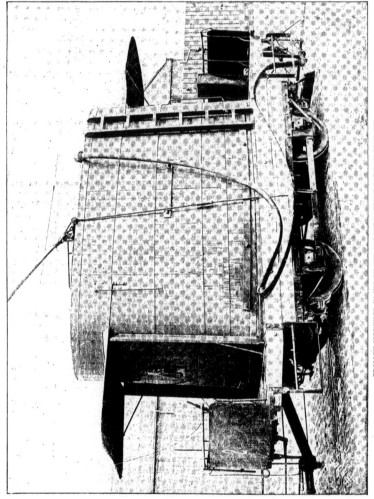

TROLLEY SPRINKLER IN USE BY THE TORONTO STREET RAILWAY COMPANY.

## A MONTREAL ELECTRIC LINE.

AN electric railway is nearing completion which will connect the city of Montreal with Bout de l'Isle, the extreme east end of the island of Montreal, about twelve miles from the city. The promoters of the road are the Chateauguay and Northern Electric Railway Company, which commenced construction in June last. It will extend from the Montreal Street Railway Company's line on Ontario street through Point Aux Trembles and Longue Point to the east end of the island opposite Charlemagne, connecting with that place by a steam ferry. The road is of the ordinary standard gauge, the right of way secured measuring eighty feet. When the line leaves Maissoneuve the route is perfectly straight for a distance of six miles, the sharpest curve on the road being one of two degrees, while it is practically level the entire distance. The trolley poles are of cedar and of the ordinary size.

The rolling stock is being furnished by the Canadian General Electric Company. Eight cars are now nearing completion, which are finished inside and out in mahogany. They will have a seating capacity of from 60 to 100 persons and will be lighted by means of electric globes.

The power house is being constructed at Point Aux Trembles, is 85x46 ft., with 18 ft. walls, in which will be placed a full complement of machinery. The engines, four in number, are being furnished by the Goldie & McCulloch Company, of Galt. The company have acquired a fine hardwood grove of twenty acres, at the end of the island, and where Riviere des Prairies empties into the St. Lawrence, and this in due time will be converted into a park. The return trip from the city to Bout de l'Isle will be made inside of an hour.

## HULL & AYLMER ROAD.

THE starting up of the Hull Electric Co.'s new system between Hull and Aylmer on July 1, was a marked success, both in point of the patronage which it seems likely to secure, and in the operation of the electric plant. We hope to be in a position in our next issue to place before our readers the detailed description of the various interesting features presented by this important addition to the electric railways of the Dominion.

The twenty-seventh annual general meeting of the shareholders of the Dominion Telegraph Co. was held in Toronto on the 14th of August. The report of the directors showed that the liabilities of the company were $1,015,972.70, and the assets $1,313,905.24. The report of the directors was unanimously adopted, and the following were elected directors for the ensuing year : Thos. Swinyard, president ; Sir Frank Smith, vice-president ; General— Thos. Eckert, Charles A. Tinker, A. G. Ramsay, Henry Pellatt, Hector MacKenzie, Thomas S. Clark and Thos. R. Wood. In recognition of his services, a presentation of a handsome service of silver plate was made to Mr. Swinyard, president.

## SPARKS.

Arrangements have not yet been completed for the conversion of the Hamilton and Dundas Railway into an electric road.

The time for beginning the construction of the Perth and Lanark electric railway has been extended for one year by the Perth town council, provided satisfactory guarantees of bona fides be furnished.

The city of Victoria, B. C., and the Consolidated Electric Railway Company have been made joint defendants in an action to recover $50,000 damages for the death of Mrs. Prevost, who was one of the victims of the Point Ellice bridge accident.

Frank Stevens, of Hamilton, who was injured by a fall from a pole while repairing the Hamilton, Grimsby and Beamsville railway has entered an action against the company for damages, alleging that the fall was due to an electric shock received as the result of a defective bell.

The Gravenhurst Electric Light and Power Co., of which Mr. E. F. G. Fletcher is manager, are re-modelling and re-building their plant. They are erecting a brick power house 30 x 40 on the shore of Gull lake where they have an abundance of water for condensing purposes and are easy of access to the Grand Trunk Ry. for their fuel. The order for an 80 h. p. Wheelock engine with condenser, boiler and shafting, was secured by the Goldie & McCulloch Company, Ltd., of Galt, Ont. This will give them ample power. They have also purchased from the Royal Electric Company a 50 k.w. "S. K. C." alternating generator and switch-board, and an 18-light 6½ ampere "T. H." Royal arc dynamo with lamps, which are now being installed. When completed, this will be one of the finest electric lighting plants in Canada, and speaks well for the enterprise and push of the new company.

THIS is to certify that MR. ROBERT ANDERSON, of the City of Ottawa, installed an Electric Light Plant in the village of Eganville for me. I am pleased to say that the work was done in a most satisfactory manner. Any person requiring his services will find him reliable and trustworthy.

JOHN D. McRAE.

EGANVILLE, May 4, 1896.

## SPARKS.

A syndicate of capitalists is reported to be in course of formation in Quebec to purchase the Lower Town street railway and Mr. Beecher's electric railway franchise.

The electric street railway at Moncton, N. B., was put in operation on the 12th of August. The road is over two miles in length, and construction occupied less than fifty days.

A description of a new electric telegraph has recently been received at the Department of State, Washington. The apparatus, it is said, makes it possible to communicate with a ship at a certain anchor ground without any direct line from land. An electric battery is placed on the shore with one pole in contact with water or moist earth, while the current from the other pole, through a telegraph key and a revolution interrupter, is conducted to a cable which is laid out to the anchor ground and placed around the latter in a coil having a diameter of 1,000 to 1,200 feet. On board the ship is a small solenoid with which a telephone is connected. When a message is sent from the land a bell sounds on the ship and the communication is sent by the telegraph key through the telephone instrument.

# ROBIN, SADLER & HAWORTH

Manufacturers of

## OAK-TANNED LEATHER BELTING

### MONTREAL AND TORONTO

Orders addressed either to our Toronto or Montreal Factory will have prompt care.
Goods will be forwarded same day as order is received.

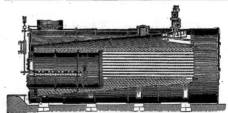

OLD SERIES, VOL. XV.—No. 6.
NEW SERIES, VOL. VI.—No. 10.

·OCTOBER, 1896

PRICE 10 CENTS
$1.00 PER YEAR.

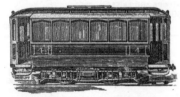

CANADIAN

# ELECTRICAL NEWS
### AND
## STEAM ENGINEERING JOURNAL.

| VOL. VI. | OCTOBER, 1896 | No. 10. |

## THE HORNSBY-AKROYD OIL ENGINE.

THE Northey Manufacturing Company, Ltd., well-known throughout the Dominion as builders of high class pumping machinery, have lately entered upon the manufacture of a most decided novelty in Canada, in the shape of the Hornsby-Akroyd oil engine. This engine, as will be seen from the accompanying cut, is a very compact and simple machine, and one which will prove most useful in a great many situations in which

4th. The expulsion of the spent gases by the piston.

In starting the oil engine, the small lamp, fed by the same oil as is used in the engine, is lighted and placed under the vaporiser, which is the part immediately behind the cylinder proper. In about ten minutes the vaporiser is hot and the engine ready to start. The fly-wheel is turned by hand a couple of revolutions, to draw air into the cylinder, and the engine then works automatically, giving out power in exact proportion to

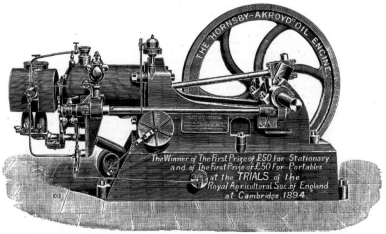

The Winner of The First Prize of £50 for Stationary and of The First Prize of £50 For Portables at the TRIALS of the Royal Agricultural Soc. of England at Cambridge, 1894.

the steam engine is neither so convenient or so economical.

In the oil engine the power is produced direct from a low grade of petroleum, by internal combustion, without the intervention of a boiler or steam in any form.

The oil engine which the Northey Mfg. Co. is about to place on the Canadian market, works on what is known as the "Otto cycle," which may be briefly explained as follows :—

1st. The admission of atmospheric air into the cylinder during the forward movement of the piston.

2nd. The compression of this air during the backward movement of the piston and its intimate intermixture with the oil vapor, previously introduced into the vaporiser.

3rd. The expansion by combustion of the mixture of gas and air in the cylinder.

the work to be done, and running evenly and quietly without further attention, so long as the supply of oil is maintained. The consumption of oil is less than one pint per horse power per hour, and a cheap gas oil, first distillation, is used, costing 7½ cents per gallon.

It will be noticed that the power in the oil engine is obtained from the expansion by combustion of a mixture of gas and air, and special attention is directed to the special safety from explosion or fire which the oil engine affords. The only fire while the engine is running is inside the cylinder, and the supply of oil is contained in a cast-iron receptable in the bed, secure from all danger. There are no sparks, no smoke and no ashes.

Attention is also called to the ease and quickness of starting of this engine, and its great economy and safety from fire hazard. It may be used advantageously wherever a steam engine can be used, and in many situations where a steam engine could not be used. For threshing it is specially useful, as no large supply of water is required, and the portable type is light and compact.

In combination with a pump it affords cheap and economical waterworks for towns and villages ; and the engine may be used with excellent results for driving dynamos for lighting and other purposes. In fact, the special applications of the Northey Mfg. Co's. oil engine cannot be enumerated, but will readily suggest themselves to parties requiring power.

Prices and catalogues may be obtained from the makers, the Northey Manufacturing Company, Ltd., King street subway, Toronto, who will be glad to furnish estimates and information to all interested.

## ROCKING GRATES.

IN view of the rather adverse conclusion arrived at by the American Boiler Manufacturers' Association with regard to the advantage to be secured by the use of shaking furnace grates, the ELECTRICAL NEWS solicited the opinion of several well-known engineers on the subject. The opinions received are printed below. This subject, like any other which affects the fuel account, should have a particular interest for owners and operators of steam plant. We would therefore be pleased to see subjects of this character discussed from time to time in our columns.

Mr. A. E. Edkins, Toronto, on the eve of his departure for England, writes briefly as follows : " Re decision of American Boiler Makers' Association, and rocking grates, I have never yet seen a fireman use them to cause waste of fuel, but on the contrary, as a general rule, I find they do not operate them often enough. If the fuel is suitable, I believe the rocking grate to be a good thing and conducive to economy."

Mr. E. J. Philip, 11 Cumberland St., Toronto, writes :

" In reference to the advantages and disadvantages of shaking grates, many things may be said, from the fact that different people look at the same thing from different standpoints. A shaking grate may save money if properly designed and managed, or it may waste coal and be a bill of expense in repairs.

" A good shaking grate will save coal, increase the capacity of the boiler, and will reduce the work of the fireman ; but to do this it must be properly designed for the work it has to do, and put in a properly proportioned furnace, and it must be carefully managed. If any of these requirements are neglected it will very likely fail to meet the expectations of the purchaser. Coal may not be saved, and yet the grates may be satisfactory if they accomplish what they were put in to do. I know of an instance in which a firm put in a shaking grate and their fuel consumption was increased, yet they were satisfied, because they wanted to be able to burn more coal and increase their capacity. In this case the grate was badly proportioned—the air space was very wide. In other places, under other conditions, this grate would have been condemned.

" In another place a new make of shaking grate was put in and an old stationary grate taken out. The new grate was to save 15%. When a test was made a loss of from 1 to 3% was shown. The grates are still being used and are satisfactory, in that they increase the capacity of the boiler and the work of firing is less.

" Against these cases may be cited two sets of shaking grates put in and the furnaces rebuilt and entirely altered, by which a saving of 11% was effected with the same coal (large egg) and a change to pea coal showed a saving in cost of 24%. A large percentage of this should be credited to the new furnaces. The old furnace and ash-pit was very low, and had far too large a grate area. The height of ash-pit and furnace was doubled and the area reduced nearly one-half. This is an exceptional case ; it was not a boiler furnace.

" There are a number of shaking grates in the city that are giving good satisfaction and have shown a good saving. Where there is only one boiler and it has ample capacity, a good stationary grate is the best under most conditions, but if a fireman has much steam to make and a number of boilers to fire, a shaking grate will be found of advantage.

" It must not be forgotten that there are shaking grates, and shaking and dumping grates, with all sorts of combinations on both. The dumping feature is as a rule dangerous to the coal pile, and is often expensive in repairs in the hands of many firemen. This type of grate, however, may be used with advantage and economy in large plants, or when in charge of an unusually careful fireman.

" Under all conditions a grate of any description, to be satisfactory, should have the air space properly proportioned for the particular fuel to be burned. If a shaking grate, it should be designed so that it will not be likely to get out of order, and that, if a part is broken, it can be easily replaced without disturbing the entire grate. It should be put in a properly proportioned furnace. Lastly, it must be carefully managed. The furnace should be proportioned to the grade of fuel used. If all these conditions are complied with, coal will be saved and the capacity of the boiler increased, and the work of the fireman made easier.

" There is not enough thought and care used in building in a boiler, designing the furnace and selecting the grate. Nearly every furnace has some arrangement to save coal, increase the capacity or reduce the work, and yet where is there to be found a furnace that is entirely satisfactory in every respect ? Discussions on labor and coal saving devices would bring out much information and would be of benefit to us all."

Mr. G. C. Mooring, Toronto, writes : " To my mind the main and best feature of the shaking grate bar was not touched upon by the American Boiler Makers' convention, (or at least was not reported), and that is the possibility of being able to clean the fire without having to open the furnace doors. If the fireman has too much steam he sometimes opens the furnace doors and the steam drops very suddenly. It takes from two to five minutes to clean a fire. Any thoughtful engineer or boiler maker knows what a great loss in fuel this causes, not to speak of the injury to the boiler. Throughout the whole discussion the main point against shaker bars is that the fireman does too much shaking. Is that the fault of the bars? I have seen great waste of coal from the same cause and from the coal being too fine for the mesh of the bars as well as from trying to burn hard coal dust without mixing some soft screenings with it, which latter method cokes and prevents much loss. I do not agree with Mr. Leonard, who says that shaker bars work best with poor fuel. If Mr. Leonard would try firing with the coal we get sometimes that melts and runs over the bars like iron—runs partly through the bars until the cold air trying to get through chills it and it sticks there. Let him try to shake under these conditions, and he will wish those shaker bars in a still hotter place than they are. Shaker bars work best with good coal, either hard or soft. Ask locomotive engineers how they would get along these days without shaker bars. I think that the shaker bar has an advantage over the straight bar; still I would not recommend any firm to change unless the straight bars were burnt out. I do not recommend any particular shaker bar, but whatever the make, it should have as much air space as possible."

Mr. Geo. C. Robb, Chief Engineer of the Boiler Insurance and Inspection Co., Toronto, writes :—

"The best method of burning coal in a steam boiler is still an unsettled question, and likely to remain so. One reason for this is, that it seems to be impossible to get the best results out of a given quantity of coal, and at the same time, get the greatest amount of work out of the boiler in which the coal is burned. To get the greatest amount of steam out of a boiler is often a far more important matter for the owner than to get steam with the least possible amount of coal. Another reason why the question is so difficult of settlement is that there are so many varieties of coal, each requiring to be used in some particular way in order to get best results. The amount of air which should pass into the furnace, how much of it should go up through the coal, and how much should enter above the fire, form points of detail upon which great differences of opinion are found to exist. It will repay any one interested in the subject to make a study of the theory of combustion and then try to carry the theory into practice, and carefully note the results.

"The argument used by the American Boiler Manufacturers' Association seems to be rather a poor one. Supposing it were true that sometimes some unburnt coal did fall through because the fireman shook the bars too vigorously; that is an evil which can be easily remedied, and it is an evil of much less magnitude than having the furnace doors kept wide open while the vigorous fireman is stirring up the fuel and the cold air is rushing in, cooling off the boiler, and developing rivet cracks at the seam over the bridge wall.

"Shaking or rocking grates enable a fireman to keep the whole surface in better condition for the proper passage of air than can be done by stirring with slice bar in the hand. The fire can be shaken up without the doors being opened, except for the actual admission of the fuel. Fuel is wasted and boilers are injured by sudden changes of temperature in the furnace, and as shaking bars diminish the time when the furnace doors must be kept open, it follows that they must if properly used, both save fuel and prevent injury to the boiler. It is quite possible by sufficient shaking to dump the whole fire into the ash pit, but that would not be a fair way to use them, and if a fault, it should be laid on the fireman rather than on the grates.

"It may be taken as proved, that economy in fuel in a steam boiler is promoted by burning the fuel at as high a temperature as possible, by keeping that high temperature as uniform as possible, and by having the rate of combustion as regular as possible. Mechanical stokers, rocking grate bars and other appliances help a fireman to keep a furnace in these conditions and hence, unless there be other objections to their use, it would seem that they should be more used than they are."

## LONG BURNING ARC LAMPS.

A RECENT innovation in arc lighting practice which has already attracted considerable attention from central station managers as well as the manufacing companies is the "long burning arc lamp." A recognized objection to the use of the arc lamp for general illumination has been the cost of the carbons and the daily expense involved in their renewal. An additional drawback has been the inadaptability of the existing arc lamp for candle powers lower than those which obtain for ordinary street lighting service.

A very simple, and it is claimed satisfactory arc lamp with enclosed arc for "long burning" service has been placed upon the market recently by the Canadian General Electric Company. Among the principal features of value claimed for a lamp of this type may be noted briefly the following : It requires very little attention, and therefore the expense of trimming is greatly reduced ; it is independent of other lamps on the same circuit, and may be cut in or out without affecting them ; it does not cast deep shadows ; it is artistic in appearance and compact in design, having a self contained resistance.

Two classes are being made at present, burning with one trimming 100 hours and 150 hours respectively. The former is about 37" and the latter about 46". Both are made with three different styles of finish as follows: Plain back japan finish ; ornamental dull black ebony finish, and ornamental polished brass finish.

The standard lamps are made for 5 amperes and can be adjusted to take from $4\frac{1}{4}$ to $4\frac{1}{2}$ amperes if desired. Lamps of smaller amperage can be furnished if so ordered. Lamps for 3 to $3\frac{1}{4}$ amperes are not considered impracticable, but small carbons should be used. All lamps are carefully adjusted and tested at 110 volts before shipment.

The mechanism is extremely simple, consisting of a pair of magnet coils, the armature of which carries the clutch and controls the feeding device, the clutch being perfectly positive, and at the same time feeding with the utmost delicacy. In order to meet the varying conditions of line voltage, an adjustment for the voltage at the arc is provided in the resistance at the top of the lamp. Changing this resistance varies only the length and potential of the arc and not the current strength. The method of securing the inner globe and lower carbon is very simple and effective, rendering it convenient for trimming and cleaning. The inner globe completely encloses the arc. This is designed to increase the life of the carbons, by excluding the air and thereby preventing combustion. The outer globe holder is a new, patented, self-locking device which is very convenient and perfectly secure. The globe is supported at all times from below and when lowered for trimming the top of the globe is level with the bottom of the frame, rendering the lower carbon holder accessible.

The use of high grade, solid carbons is necessary to prevent undue coating of inner globe to give satisfactory service. As there is more or less variation in the size of $\frac{1}{2}$" carbons, the opening of the cap of the inner globe is .525" diameter, and the carbons used should come within the following limits : .520" max. diam. ; .505 min. diam. The opening or space between the cap and carbon should be only sufficient to allow the free passage of the carbon as it feeds downward. If air is allowed to enter the inner globe the life of the carbons is greatly shortened. Attention should be given to polishing and cleaning the carbon rod at every trimming, to prevent its becoming sticky from atmospheric conditions. With solid "electra" carbons, which have been found to give the best results, a potential of 75 to 80 volts at the arc is required.

When these lamps are properly trimmed with correct lengths of carbons, more than the rated time of burning can be expected. They will not need further attention through an entire run, and will cut out properly when the carbons are consumed. In most cases the piece left in the upper holder is of correct length for the lower holder for the next full run.

A further desideratum is a lamp equally simple and effective adopted for use on alternating circuits. Such a lamp is promised and indeed assured by experimental work as a development of the immediate future.

## CANADIAN ASSOCIATION OF STATIONARY ENGINEERS.

NOTE.—Secretaries of Associations are requested to forward matter for publication in this Department not later than the 25th of each month.

### KINGSTON ASSOCIATION NO. 10.

At the last regular meeting of the above association the officers for the ensuing year were installed as follows: Past President, S. Donnelly; President, F. Simmons; Vice-President, J. Tandvin; Treasurer, C. Selby; Secretary, A. Macdonald; Doorkeeper, R. McDonald; Conductor, R. Bajus; Trustees, John L. Orr and S. Donnelly. Letters were read from delegates to the recent annual convention of the C.A.S.E., expressing their appreciation of the hospitality extended to them.

### BROCKVILLE ASSOCIATION NO. 15.

Mr. J. Aikens, Recording Secretary of this association, reports that since the Kingston convention some very interesting instruction meetings have been held, at which there has been a good attendance of earnest workers. At these meetings the blackboard has been in constant use for purposes of illustration. In the unavoidable absence on some occasions of the President, Bro. Franklin, the Past President, Bro. Chapman, gave the members the benefit of his assistance in solving the problems under discussion.

## ONTARIO ASSOCIATION STATIONARY ENGINEERS.

TORONTO, Sept. 11, 1896.

To the Editor of the ELECTRICAL NEWS.

SIR,—The following engineers have recently passed their examinations: Third class—Geo. H. Bull, Rosemeath; Chas. Kemp, Petrolea; A. Ritchie, Orillia; C. Labarge, Hull, P. Q.; J. Radmore, Buckingham, P.Q.; Fred. Nagle, Paris; J. Carol, Hamilton; Albert Martin, Toronto; F. C. Corrie, Stratford; D. Anderson, Mt. Forest; J. Wilson, Hamilton; Geo. E. Bower, Lucknow; Geo. H. Cooper, Oakville. Second class—D. H. Vincent, Belleville; B. Deo, St. Thomas; Thos. R. Seaton, Toronto. In all twenty engineers wrote for examination, four of whom failed, either through not having had the required experience or other cause.

Enquiries are coming in daily from all parts of the province concerning examinations, which goes to prove that the feeling is growing, viz., that all engineers should hold certificates.

The city council of Hamilton have decided that no engineer shall be employed by that corporation unless he hold an Ontario certificate. This is a move in the right direction and might well be followed by our city council in Toronto.

There are about forty certificate holders who have not paid their renewal fees so far this year, and I shall be glad if they will return old certificates, either with or without renewal fees, as the certificates are the Board's property and must be returned to this office when expired.

I shall be glad to send information regarding examinations to any engineer desiring same on receipt of post card giving name and address.

Yours truly,
A. E. EDKINS, Registrar.

Office, 88 Caroline St., Toronto.

## BURSTING STEAM PIPES.

The explosion of steam pipes has been occurring lately with such frequency as leads one to ask, Why? As a general thing a steam pipe is stronger for the pressure it has to carry than is a steam boiler, and yet they explode, showing that some force is at work which produces a weakening effect on the pipe. A long line of steam pipe is difficult to keep tight unless some special arrangement is employed that will allow, not only for expansion and contractions but other strains to which the pipe is subjected.

There are few engines that run so steadily but what they cause vibration of the steam pipe and in some cases the vibration becomes so great that it is necessary to use extra braces or stays to prevent its going beyond limits. Constant vibration of metal under strain is known to have a tendency towards producing crystalization, and this is probably what results in some steam pipes.

### KEEP AT IT.

If you expect to conquer
　In the battle of to-day,
You will have to blow your trumpet
　In a firm and steady way.
If you toot your little whistle
　And then lay aside the horn,
There's not a soul will ever know
　That such a man was born.

The man that owns his acres
　Is the man that plows all day;
And the man that keeps a humping
　Is the man that's here to stray.
But the man that advertises
　With a sort of sudden jerk,
Is the man that blames the printer
　Because it didn't work.

But the man that gets the business
　Uses brainy printers' ink,
Not a clatter and a sputter,
　But an ad. that makes you think;
And he plans his advertisements
　As he plans his well-bought stock,
And the future of his business
　Is as solid as a rock.

### ERRATUM.

NEW YORK, Sept. 22, 1896.

To the Editor of the CANADIAN ELECTRICAL NEWS.

DEAR SIR,—I beg to call attention to an error (probably a misprint) which appears in your paper for September. In the column headed "Electrical Items worth Remembering," there appears: "The resistance of copper rises about 0.21 per cent. for each degree centigrade," which should read "0.21° F."

Respectfully yours,
V. M. BENEDIKT, E. E.,
27 Thames st., New York.

## DEFINITIONS OF ELECTRICAL TERMS.

ACCUMULATOR.—Storage or secondary battery, in which electricity has been carried and has been converted into chemical energy, being retransformed into electricity when the battery is put to use for the purpose of furnishing energy or light.

AMPERE.—The unit of strength of the current per second. It represents, perhaps, the volume of electricity, and its value is the quantity of the fluid which flows per second through one ohm of resistance when impelled by one volt of electro-motive force.

ANODE.—The positive pole of a battery.

ARC.—The space between the points of the carbons in an electric light or lamp which is bridged by the current represented by the flame.

ARMATURE.—The revolving arm of an electric generator.

BATTERY.—A primary battery is one in which electricity is obtained through the decomposition of metals in chemical solutions. Zinc and copper may be the metals and sulphuric acid the chemical· Gold, silver, platinum, iron or tin may also be used as the metals and sal-ammoniac, bi-chromate of potash, nitric acid and sulphate of copper may also be used as the chemicals. The storage battery is a cell of acidulated water, containing, for example, plates of lead. This arrangement has an electric current directed into it, which it will give back in almost an equal quantity when the energy is wanted. There are various methods and ways of making both primary and secondary or storage batteries, but the above are the general principles governing their construction.

BRUSH.—The copper string which connects with the commutator of a dynamo and gathers the electricity for the conductors.

CANDLE.—Our unit of illuminating power.

CARBONS.—Rods of carbon are used in arc lights for first establishing the current, and then, when withdrawn, form the arc over which the electric flame leaps. They are made of powdered coke by a secret process.

CELL.—The vessel in which chemical action produces electricity.

CIRCUIT.—The path along which an electric current travels.

COMMUTATOR.—The collector of the electricity generated, and from which the fluid is taken by the brushes.

CONDENSER.—An arrangement for collecting a large quantity of electricity on a small surface.

CONDUCTIVITY.—The comparative ability of a substance to convey a current of electricity.

CONDUCTOR.—Conveyors of the electric current, silver being the best, and copper next, in conductivity.

CORE.—The iron that becomes magnetized in an electro-magnet. In helix, this iron is of the softest kind.

COULOMB.—The unit of dynamic quantity represented by one ampere of current.

CURRENT.—The flow of electricity along a conductor. Its strength in amperes is found by dividing the electromotive force in volts by the resistance in ohms.

## A WORD OF PRAISE.

MR. B. A. YORK, Secretary of Montreal Association of Stationary Engineers, writes the publisher of the ELECTRICAL NEWS as follows :—"At our regular meeting your paper received much praise for the way you had so ably and fully reported all that took place at our last convention, and I will take opportunity to thank you and wish your paper every success."

## THE ADVANTAGES OF VERTICAL ENGINES.

The great increase in the use of power for the generation of electricity in large quantities has served to develop large stationary engines, and as such plants are usually in thickly populated districts, where land or floor space is expensive, the vertical engine has received the preference to a great extent ; for a given power it occupies less floor space than any other type. For the same rotative speed and power the cost of building such engines is about equal, whether the vertical or horizontal type is used, but, as builders become used to designing the vertical engine, I think the first cost will be in favor of this type.

As to accessibility for repairs and care in running, there is little to choose between them, but with a properly rigged overhead travelling crane I think the matter of overhauling the vertical is the easier, whereas in running it is doubtless more convenient to have everything on one level.

In the matter of friction the vertical engine, too, has a great advantage, as the packing, besides its appropriate office of preventing steam leakage past the piston, has only to guide it also, whereas in the horizontal engine it must not only support the entire weight of the piston, but also the pressure of steam, as the "bull ring" generally fits the bottom half of the cylinder steam tight, but allows the steam to enter on top as far as the packing ring.

In the vertical design the weight of the cross-head does not increase the slide friction, which is not the case with the horizontal engine when running, with the crank passing the upper arc as the piston goes toward the shaft ; and when the reverse direction is used, although the slide is relieved of the weight of the cross-head, a worse trouble is introduced, namely, the slapping up and down of the cross-head at each end of the stroke.—Charles H. Manning in Cassier's for October.

## SPECIALIZATION IN ENGINEERING.

THE civil engineer of past generations, who was supposed to command a comprehensive knowledge of every branch of engineering then practiced, from the design of a steam engine or machine tool to that of a bridge or city drain system or complete waterworks plant, has virtually ceased to exist, and in his stead, says a writer in Cassier's Magazine, we find the steam engineer, the sanitary engineer, the bridge engineer and the engineer of various other subdivisions of the great field of engineering, each an expert in his particular line. It has been found impossible for one man to combine within himself the detail knowledge necessary to practice all these branches with entire success. One branch alone is almost sufficient to make a life study, and the engineering specialist of to-day finds himself busily enough occupied in keeping abreast of the times.

Messrs. John Starr, Son & Co., Halifax, have just installed a 50 light plant for the St. Croix Paper Mills Co., of Hartsville, N. S.

F. Stancliffe, of Flat Lands, N. B., has had a 50 light plant installed in his shingle mill. This plant was supplied and installed by John Starr, Son & Co., of Halifax, N. S.

Messrs. John Starr, Son & Co., Halifax, have recently installed a 200 light plant for Kilgour Shives, of Campbellton, N. B. This is used for lighting Mr. Shives' extensive lumber mills and yards

The "Unique" telephones as manufactured by John Starr, Son & Co., Halifax, are having a large sale. This firm have recently supplied a number of telephones and switchboards to Campbellton and Quebec, both of which orders were "repeats" which speaks well for the "Unique" telephones which have now been on the market for several years.

PUBLISHED ON THE FIFTH OF EVERY MONTH BY

## CHAS. H. MORTIMER,

OFFICE : CONFEDERATION LIFE BUILDING,

Corner Yonge and Richmond Streets,

### TORONTO, - - - CANADA.

Telephone 2362.

NEW YORK LIFE INSURANCE BUILDING, MONTREAL.

Bell Telephone 2299.

*ADVERTISEMENTS.*

Advertising rates sent promptly on application. Orders for advertising should reach the office of publication not later than the 26th day of the month immediately preceding date of issue. Changes in advertisements will be made whenever desired, without cost to the advertiser, but to insure proper compliance with the instructions of the advertiser, requests for change should reach the office as early as the 2 nd day of the month.

*SUBSCRIPTIONS.*

The ELECTRICAL NEWS will be mailed to subscribers in the Dominion, or the United States, post free, for $1.00 per annum. 50 cents for six months. The price of subscription should be remitted by currency, registered letter, or postal order payable to C. H. Mortimer. Please do not send cheques on local banks unless 5 cents is added for cost of discount. Money sent in unregistered letters will be at senders' risk. Subscriptions from foreign countries embraced in the General Postal Union $1.50 per annum. Subscriptions are payable in advance. The paper will be discontinued at expiration of term paid for if so stipulated by the subscriber, but where no such understanding exists, will be continued until instructions to discontinue are received and all arrearages paid.

Subscribers may have the mailing address changed as often as desired. When ordering change, always give the old as well as the new address.

The Publisher should be notified of the failure of subscribers to receive their paper promptly and regularly.

*EDITOR'S ANNOUNCEMENTS.*

Correspondence is invited upon all topics legitimately coming within the scope of this journal.

The "Canadian Electrical News" has been appointed the official paper of the Canadian Electrical Association.

## CANADIAN ELECTRICAL ASSOCIATION.

### OFFICERS:

PRESIDENT :

JOHN YULE, Manager Guelph Light & Power Company, Guelph, Ont.

1ST VICE-PRESIDENT :

L. B. McFARLANE, Manager Eastern Department, Bell Telephone Company, Montreal.

2ND VICE-PRESIDENT :

E. C. BREITHAUPT, Electrical Engineer, Berlin, Ont.

SECRETARY-TREASURER :

C. H. MORTIMER, Publisher ELECTRICAL NEWS, Toronto.

EXECUTIVE COMMITTEE :

GEO. BLACK, G. N. W. Telegraph Co., Hamilton.

J. A. KAMMERER, General Agent, Royal Electric Co., Toronto.

K. J. DUNSTAN, Local Manager Bell Telephone Company, Toronto.

A. M. WICKENS, Electrician Parliament Buildings, Toronto.

J. J. WRIGHT, Manager Toronto Electric Light Company, Toronto.

ROSS MACKENZIE, Manager Niagara Falls Park and River Railway, Niagara Falls, Ont.

A. B. SMITH, Superintendent G. N. W. Telegraph Co., Toronto.

JOHN CARROLL, Sec.-Treas. Eugene F. Phillips Electrical Works, Montreal.

C. B. HUNT, London Electric Company, London.

F. C. ARMSTRONG, Canadian General Electric Co., Toronto.

## CANADIAN ASSOCIATION OF STATIONARY ENGINEERS.

President, JAMES DEVLIN,    Kingston, Ont.
Vice-President, E. J. PHILIP,    11 Cumberland St., Toronto.
Secretary, W. F. CHAPMAN,    Brockville, Ont.
Treasurer, R. C. PETTIGREW,    Hamilton, Ont.
Conductor, J. MURPHY,    Montreal, Que.
Door Keeper, F. J. MERRILL,    Ottawa, Ont.

TORONTO BRANCH NO. 1.—Meets 1st and 3rd Wednesday each month in Engineers' Hall, 61 Victoria street. John Fox, President ; Chas. Moseley, Vice-President ; T. Eversfield, Recording Secretary, University Crescent.

MONTREAL BRANCH NO. 1.—Meets 1st and 3rd Thursday each month, in Engineers' Hall, Craig street. President, John Murphy ; 1st Vice-President, J. E. Huntington ; and Vice-President, Wm. Smyth ; Secretary, B. Archibald York ; Treasurer, Peter McNaughton.

ST. LAURENT BRANCH NO. 2.—Meets every Monday evening at 43 Bonsecours street, Montreal. R. Drouin, President ; Alfred Latour, Secretary, 306 Delisle street, St. Cunegonde.

BRANDON, MAN., BRANCH NO. 1.—Meets 1st and 3rd Friday each month, in City Hall A. R. Crawford, President ; Arthur Fleming, Secretary.

HAMILTON BRANCH NO. 2.—Meets 1st and 3rd Friday each month in Maccabee's Hall. Wm. Norris, President ; E. Teeter, Vice-President ; Jas. Ironsides, Corresponding Secretary.

STRATFORD BRANCH NO. 3.—John Hoy, President ; Samuel H. Weir, Secretary.

BRANTFORD BRANCH NO. 4.—Meets 2nd and 4th Friday each month. J. B. Forsyth, President ; Jos. Ogle, Vice-President ; T. Pilgrim, Continental Cordage Co., Secretary

LONDON BRANCH NO. 5.—Meets once a month in the Huron and Erie Loan Savings Co.'s block. Robert Simmie, President ; E. Kidner, Vice-President ; Wm. Meaden, Secretary Treasurer, 533 Richmond street.

GUELPH BRANCH NO. 6.—Meets 1st and 3rd Wednesday each month at 7.30 p. m. H. Geary, President ; Thos. Anderson, Vice-President ; H. Flewelling, Rec.-Secretary ; F. Ryan, Fin.-Secretary ; Treasurer, C. F. Jordan.

OTTAWA BRANCH NO. 7.—Meets every second and fourth Saturday in each month, in Borbridge's hall, Rideau street ; Frank Robert, President ; F. Merrill, Secretary, 352 Wellington street.

DRESDEN BRANCH NO. 8.—Meets 1st and Thursday in each month. Thos Steeper, Secretary.

BERLIN BRANCH NO. 9.—Meets 2nd and 4th Saturday each month at 8 p.m. J. R. Utley, President ; G. Steinmetz, Vice-President ; Secretary and Treasurer, W. J. Rhodes, Berlin, Ont.

KINGSTON BRANCH NO. 10.—Meets 1st and 3rd Tuesday in each month in Fraser Hall, King street, at 8 p. m. President, F. Simmons ; Vice-President, J. W. Tandvin ; Secretary, A. Macdonald.

WINNIPEG BRANCH NO. 11.—President, G. M. Haslett ; Rec.-Secretary, J. Sutherland ; Financial Secretary, A. B. Jones.

KINCARDINE BRANCH NO. 12.—Meets every Tuesday at 8 o'clock, in McKibbon's block. President, Daniel Bennett ; Vice-President, Joseph Lighthall; Secretary, Percy C. Walker, Waterworks.

WIARTON BRANCH NO. 13.—President, Wm. Craddock ; Rec.-Secretary, Ed. Dunham

PETERBOROUGH BRANCH NO. 14.—Meets 2nd and 4th Wednesday in each month W. L. Outhwaite, President ; W Forster, Vice-President ; A. E. McCallum, Secretary.

BROCKVILLE BRANCH NO. 15.—Meets every Monday and Friday evening. President, Archibald Franklin ; Vice-President, John Grundy ; Recordin, Secretary, James Aikins.

CARLETON PLACE BRANCH NO. 16.—Meets every Saturday evening. President, Jos. McKay ; Secretary, J. D. Armstrong.

## ONTARIO ASSOCIATION OF STATIONARY ENGINEERS.

### BOARD OF EXAMINERS.

President, A. AMES,    Brantford, Ont.
Vice-President, F. G. MITCHELL    London, Ont.
Registrar, A. E. EDKINS    88 Caroline St , Toronto.
Treasurer, R. MACKIE,    28 Napier st., Hamilton.
Solicitor, J. A. McANDREWS,    Toronto.

TORONTO—A. E. Edkins, A. M. Wickens, E. J. Phillips, F. Donaldson. J. Bain.
HAMILTON—R. Mackie T. Elliott.
BRANTFORD—A. Ames, care Patterson & Sons.
OTTAWA—Thomas Wesley.
KINGSTON—J. Devlin, (Chief Engineer Pen tentiary), J. Campbell.
LONDON—F. Mitchell
NIAGARA FALLS—W. Phillips.

Information regarding examinations will be furnished on application to any member of the Board.

THE electrical congress held at Geneva *The Geneva Congress.* in August, was poorly attended. Notwithstanding that representatives were not present from many of the leading scientific societies of the world, including the British and American Institutes of Electrical Engineers, the Congress felt no hesitation in rejecting the magnetic units sanctioned by the American Institute, and adopting a system of photometric units.

AT a convention of street lighting *Some Results of Municipal Lighting.* officials held recently at New Haven, a bad showing was made on behalf of municipal control of electric lighting plants. The statement was made that Wabash, Ind., purchased a plant for $18,000 and sold it for $30 ; Xenia, O., paid $35,000 for a plant and ten years later sold it for $10,000 ; Moline, Ill., bought a plant at $15,000 and four years after sold it for $8,000 ; Michigan City bought a $10,000 plant and sold it for $2,500.

THE agreement entered into five years *The Telephone Situation.* ago between the Bell Telephone Company and the City Council of Toronto, is about to expire. Under this agreement five per cent. of the receipts of the Toronto exchange were to be paid into the city exchequer and the yearly rental was decreased from $50 to $45 per instrument for commercial use. The Telephone Company have notified the Council that they will decline to renew the agreement, and it is said to be their intention to increase the rental of their instruments when the period of the present arrangement shall terminate. The Council have invited tenders for the franchise, but are understood to have had no offers. Representatives of the Strowger automatic telephone have, however, set up several of their instruments in the business part of the city with the purpose of demonstrating their utility. Appearances would seem to indicate that the Bell Telephone Co. are likely to remain in control of the situation in Toronto unless dislodged by the less expensive method which Prof. Bell is reported to have discovered of transmitting messages by means of a ray of light.

**Those Alleged Portraits.** MESSRS. A. M. Wickens, A. E. Edkins, and John Fox, have for many years been among the most intelligent and hard-working promoters of the prosperity of the Canadian Association of Stationary Engineers. Presumably in recognition of their self-sacrificing efforts, they find themselves depicted in the columns of the Canadian Engineer as "the villains in the play," or as a couple of cracksmen who have just finished a term in the strong institution at Kingston and are on the look-out for another job.

**The Feed Water of Water Tube Boilers.** IN the case of the recent explosion of a water tube boiler in England, the Board of Trade stated the cause to have been the closing of the down-take tubes by calcareous deposit. The stoppage of the circulation due to this deposit caused undue expansion of the horizontal tubes and placed such a severe strain on front and rear cast iron headers, as caused their fracture. Forged steel is now being substituted for cast iron in headers in some boilers of this description. Notwithstanding, there would still appear to be an element of danger where pure feed water cannot be obtained.

**Three-Cent Fares.** THE universal demand for cheapness has led to an agitation for a reduction of the street car fare to three cents. Mr. H. A. Everett, formerly of Toronto, now the principal owner of the new electric street railway at Detroit, Mich., was one of the few men in the business who believed that it would be to the advantage of the companies to offer a three-cent fare. He reduced the fare accordingly, but the results have not justified the wisdom of the step, and a return has been made to the former price. Especially in view of the serious inroads which the increasing use of the bicycle is making in the business of city roads, any reduction in the present fares is out of the question.

**Insurance Against Accident to Electrical Machinery.** THIS is the day of electrical enterprise. Every day new concerns are started, new machinery introduced, new methods invented; machinery and apparatus are being continually improved and cheapened, and the man who neglects to read neglects his own interests. Enterprises are being organized in every direction, which have for their object the lowering of prices of supplies, machines, etc., and the latest that we have heard of is one for the insuring of electrical machinery against loss by accident, and against repairs. This seems to be a most valuable business, for electrical men cannot tell when their engines, or armatures, transformers, or motors may break down and require expert attention. The fact is that there is a most unsatisfactory amount of old and out of date apparatus being operated in central stations, the repairs on which must amount annually to a considerable sum, and any arrangement which will permit of the owners being guaranteed against ruinous accidents ought to pay both the owners and the guarantors. Besides which, a company of experts who make it their business to keep plants in efficient working order will be a great boon to those whose acquaintance with electrical matters is limited.

**Electricity in Photography.** WE are constantly being astonished by the multitude and variety of the purposes to which electricity is being adapted. One of the latest and most wonderful is to be seen in connection with the cinematograph now on exhibition in Toronto. By means of this instrument, which is the production of a French inventor named Lemaire, pictures in which the activities of living creatures and of nature are reproduced with the utmost fidelity, are thrown upon the canvass. Electricity has made it possible to take a series of photographs of objects in motion with such lightning-like rapidity, that when the photographs are placed side by side together and passed through the cinematograph, there is presented to view a reproduction of the whole scene as it appeared to the eye of the original beholder.

**Shaking Furnace Grates.** THE American Boiler Manufacturers' Association, at its recent convention, discussed at some length the relative advantages and disadvantages of shaking grates. The conclusion arrived at was, that owing to the disposition of firemen to do too much shaking, an unduly large percentage of coal is dropped through the grate into the ash-pit. This would appear to be the fault of the fireman rather than of the grate. There are a great many costly fuel-saving devices on the market at the present day, for most of which large claims are made. It is highly desirable that those who have had practical experience in the use of any of these devices, should make known for the general welfare of owners and operators of steam plants, how far these claims are capable of being realized. We would take it as a favor if any of our readers who have had experience with shaking grates and suchlike modern devices, would write us their opinion of them.

**Steam Turbines.** A CLASS of steam plant is now forcing itself on the notice of the electrical operating interest that presents many most interesting points and is well worthy of careful investigation. We allude to the machines known as "steam turbines." There are two of these that are well, and we may say favorably known to those who keep a place on the advance guard of electrical progress. The first is the Parsons, and the second the DeLaval steam turbine. In both the principle is to take advantage of the tremendous impact force of steam escaping (under pressure) from the boiler, to turn what may be termed a wheel with little discs or fans placed on its periphery. In this aspect, steam turbines are analogous to the Pelton and other impact turbines that rely for their turning moment more on impact than on static pressure. The tests on these turbines shew a very high degree of efficiency, the comparison being somewhat as follows : A single cylinder non-condensing high pressure engine will require about 30 lbs. of steam to maintain a horse power. A compound (two expansion) engine of superior make will require about 21 lbs. condensing ; but some most extensive and apparently competent tests on a DeLaval steam turbine give a consumption of a little over 18 lbs. per horse power, which is 'an uncommonly good showing, and one worthy of attracting the attention of the' electrical profession. These machines revolve at a very high rate, and of course require most careful construction. Hitherto they have been connected to special dynamos through gearing, but in the near future, no doubt, they will be equally well adapted to belt connections. We strongly advise all electrical men to watch their development.

Montreal vs. Toronto. THE city of Montreal has announced its purpose to hold an. International Exhibition in 1897 or the year following. Toronto also gives notice of its intention to hold a Dominion Exhibition next year. Toronto claims it was first in the field, and says it doesn't want to undertake an International Exhibition, and asks Montreal to defer the larger enterprise for a year or two. Montreal replies that the holding of a Dominion Exhibition in Toronto next year would seriously impair the chances of an International Exhibition a year or two later. Both cities have applied to the Dominion and Provincial governments for aid. Both have admitted that without such aid they cannot hope to make their scheme a success. Therefore, the decision as to which of the enterprises shall go forward at the present time would appear to rest with the government, unless, as we trust will be the case, a satisfactory arrangement can be reached between the representatives of the two cities.

Efficiency of Transformers. IN his valuable paper on "Some Central Station Economies" presented to the Canadian Electrical Association, Mr. P. G. Gossler makes a very conclusive showing as to the amount of saving which it is possible to effect by substituting for old-style transformers modern high efficiency apparatus. He instances a case in which the saving thus effected was sufficient to pay the cost of the new transformers within a period of less than three years. Mr. Gossler is authority for the statement that the efficiency of transformers varies from 50 to 100 per cent. If this statement be correct, and we judge it to be so, then there is need of the exercise of greater knowledge and skill on the part of some of the manufacturers of transformers in order that their production may be brought nearer up to the standard of machines of the highest efficiency. In other classes of electrical apparatus such a wide variation in efficiency does not exist, nor should it be allowed to continue in an instrument with functions so important as those of the transformer.

Specifications. WE have had the advantage of seeing the specifications on which a number of electric lighting plants have recently been purchased, and have been struck by their laxity, and generally vague nature. In many cases—in most of them, in fact—it seemed as though special stress was laid on comparatively unimportant matters, whereas those points on which should really depend the selection of machinery were either not alluded to at all, or received only the most cursory notice. We have particularly in mind an arc plant specification which called for "a plant of 50 light nominal 2000 c. p. capacity with lamps, etc.," and then it went on to state that the candle power would be tested, and must be as specified.

Now, in the first place, what is the actual candle power of a nominal 2000 c. p. lamp? and is that actual candle power to be tested in the horizontal plane, or in any other plane making an angle with the horizontal? There was no efficiency requirements—no maximum temperature limit and the only really onerous condition was that the. plant would have to operate to the satisfaction of the engineer, who, by the way, knew just enough about electrical apparatus to carbon the lamps. In another case a 60 k. w. alternator was called for. There was no specification as to voltage, maximum line loss, temperature limits, efficiencies, or any important feature of a machine, but it was clearly stated that the machine would be required to carry its full rated load for 24 consecutive hours, without undue heating (sic) in any part. What is "undue heating"? May the limit be placed at 200° F or 50° F, or where—and who is to fix it? Furthermore, what man in his senses is going to run a machine, in a small town, for 24 hours, in ordinary practice? If the machine is ever required to carry its full load for more than four hours at a time, that is all that it would ever be called on to do. And yet this same specification that left the alternator and transformer to the mercy of the contractors in every important point, imposed the most rigorous and minute conditions as to how the poles were to be placed—their size, and how many times they were to be painted, and the exact color, finishing up consistently by neglecting to say how many were required. Is this the way to buy machinery? Persons who throw themselves on the mercy of contractors by making specifications of the above description are laying themselves open to all kinds of deception and trickery, and only deserve to be taken in. The electrical market is full of machinery, good and bad and medium, and of course a purchaser is entitled to choose which he prefers, but carelessly or ignorantly prepared specifications impose rigorous conditions only on those manufacturing companies that will not condescend to sell poor apparatus, and leave every loop-hole of escape to those second-rate concerns that trade upon the inexperience of a credulous and penurious public.

## EXPANSION OF BOILERS.

In a communication upon the above subject presented before the American Boiler Manufacturers' Association, by Mr. Fred Leonard, of London, Ont., the author said that during the last year an opportunity was offered to measure carefully the expansion of a stationary boiler bricked in and a small locomotive boiler mounted on skids, and it would appear that the expansion amounts to very little. The stationery boiler was 60 inches in diameter, 12 feet long, and stood three days, being cleaned and washed out. On the fourth day it was steamed up with a working pressure of 90 pounds, and a difference only of ¼ of an inch could be seen in length when standing cooled off and steam on. The locomotive boiler was 34 inches in diameter, 12 feet 9 inches long and carried 95 pounds steam, measured only ¼ inch less after having the water drawn off and standing 24 hours. From this it would appear that the plates and rollers under the brackets are unnecessary, as ¼ inch expansion in 12 feet amounts to practically nothing.

By request Mr. Leonard explained that the measurements were made on the return tubular boiler by means of a rod with a hook on the end which should be shoved through the tubes to the back connection.

## PEMBROKE ELECTRIC LIGHT COMPANY.

THE composition of the Pembroke Electric Light Company is as follows : President, Hon. P. White ; Vice-President, A. Foster ; Directors, Geo. Smith and Alex. Miller.

The building is an L shaped structure, 52 x 40 feet, in which the plant is situated, the boiler room being situated in the smaller part. The building is of brick, with a steep roof, and is situated on the banks of the Madawaska river, convenient for condensing purposes.

MR. J. A. THIBODEAU.
Manager Pembroke Electric Light Company.

The C. P. R. track passes within a few feet of it, and coal is easily handled.

In the dynamo room two Wheelock cut-off condensing engines operate the machinery ; one of them is a tandem compound of 110 h. p., the other of 128 h. p. These engines are belted to 40 feet of 4 in. shafting, from which is run a Royal alternator of 1000 lights, two Edison three-wire system generators and two 25 light Western arcs, all of modern design. The switch board is 11 x 18 feet, and is fully equipped with all necessary instruments. On the shafting are two Goldie & McCulloch clutch couplings, which permit the engines to run separately or together.

The boiler room is 22 x 30 feet, and contains two Goldie & McCulloch boilers of 100 and 70 h. p. respectively, fired by wood. Two Northey condensers in the dynamo room supply them through two Austin heaters.

The company was organized in 1889, and three years ago erected the building, an illustration of which appears on this page.

Mr. Thibodeau, the manager, whose portrait appears herewith, is a shrewd business man and is connected with many other enterprises in the town.

The plant is in charge of Mr. A. Cone, electrician, and Mr. Thos. Mackie, engineer.

---

The Hull and Aylmer Electric Railway Co. have purchased a park a mile and a half further up Deschene Lake than the present park at Aylmer. The park has a frontage of nearly half a mile on the lake. It is said to be the intention of the company to double track the road from Hull, and to purchase another locomotive and ten 40-foot trailers for handling excursion parties. A Ruggle's rotary snow plough has also been ordered.

## BY THE WAY.

A NOVEL cause of dispute has arisen between the City Council of Toronto and the Toronto Railway Company Under its agreement the Company pay mileage fees to the city on the pavement between their tracks. The point in dispute is whether curves and intersections should be included in the mileage pavements. The city argues yes, and the Company, no. The latter quote the opinion of Mr. W. T. Jennings, late City Engineer, who drafted the engineering clauses of the agreement, in support of their contention. If the curves and intersections are to be counted in, the Company will be required to pay $4,000 per year additional mileage.

× × × ×

THE city of Detroit rejected the offer of the Detroit Electric Light & Power Co., to furnish light at $102.20 per lamp per year, and went into the business as a municipal enterprise. The sum of $630,141.92 was invested in the plant. A report of the first nine months' operations has just been published, by which it is shown that it has cost the city $68.52 per lamp, exclusive of any allowance for depreciation, interest on investment, water, rent and insurance. If these items are taken into account, as they certainly ought to be, and counting in also the amount which would have been received in taxes from a private lighting company, the actual cost per lamp is shown to be upwards of $130 per year, or more than $25 per lamp per year in excess of what a private concern offered to supply the light for.

× × × ×

THE State of Ohio, following in the wake of New York state, has recently placed upon its statute books a law which makes electricity the instrument by which in future the death penalty is to be inflicted. The prison

PEMBROKE ELECTRIC LIGHT COMPANY—DYNAMO AND ENGINE ROOM.

official whose duty it was to purchase the required electrical apparatus for this purpose is said to have made the round of the electrical supply companies in Chicago and found that not one of them was willing to sell a dynamo to generate current to stop the current of human life. He had previously visited New York with the same result. The New York State authorities are said to have met with the same difficulty, and were finally obliged to buy their apparatus through a second party. The above circumstance would appear to indicate that there is not yet a complete divorce between business and sentiment. The electrical fraternity have, no doubt, also felt it to be their duty not to assist to accentuate the idea which the daily press had succeeded in instilling into the minds of the people that the use of electricity was attended with the greatest possible danger to property and life.

## THE "NIAGARA" INJECTOR.

BELOW is a sectional cut of the "Niagara Injector" an injector which is rapidly becoming popular among steam users. This boiler feeder is manufactured in St. John, N. B., by W. H. Stirling. The machine has only been on the market one year and is now in actual use in most of the cities and towns throughout Canada.

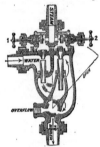

THE NIAGARA INJECTOR.

The machine is complete in itself requiring no valves as will be seen by the cut.

It can be throttled by means of valve No. 1 on suction side, so as to supply from full capacity down to required quantity, thus reducing the quantity of steam used, and delivering the water 90° hotter. The manufacturer states that this feature will save the price of the injector many times over in fuel alone, and that this fact has been demonstrated beyond doubt by the "Niagara" Injector being connected where other machines have been taken off.

Mr. Stirling has shipped these injectors to nearly every western city in Canada as far west as British Columbia.

The "Niagara" Injector is sold in Montreal by Samuel Fisher, 57 Sulpice street, and other dealers.

## SPEED OF PULLEYS.

The diameter of the driven being given, to find its number of revolutions : Rule—Multiply the diameter of the driver by its number of revolutions, and divide the product by the diameter of the driven ; the quotient will be the number of revolutions of the driven.

Ex.—24in. diameter of driver x 150, number of revolutions = 3,600 ÷ 12in. diameter of driven = 300.

The diameter and revolutions of the driver being given, to find the diameter of the driven, that shall make any given number of revolutions in the same time : Rule—Multiply the diameter of the driver by its number of revolutions, and divide the product by the number of required revolutions of the driven ; the quotient will be its diameter.

Ex.—Diameter of driver (as before) 24in. x revolutions 150 = 3,600. Number of revolutions of driven required = 300. Then 3,600 ÷ 300 = 12in.

The rules following are but changes of the same, and will be readily understood from the foregoing examples.

To ascertain the size of the driver : Rule—Multiply the diameter of the driver by the number of revolutions you wish to make, and divide the product by the required revolutions of the driver ; the quotient will be the size of the driver.

To ascertain the size of pulleys for given speed ; Rule—Multiply all the diameters of the drivers together and all the diameters of the driven together ; divide the drivers by the driven ; the answer multiply by the known revolutions of main shaft.

## THE NEW KAY MOTOR.

THIS motor was designed to meet the increasing demand for small power, and they are made in sizes from ½ h.p. up to 10 h.p., the object being to produce an efficient, durable and cheap machine. There is only one joint in the magnetic field, therefore the loss in the magnetic circuit is scarcely perceptible. The bearings are self aligning and self oiling, having a metal ring at each end to carry the oil on to the shaft from the collar below. These bearings are made of the best phosphor bronze that can be had, in fact all the material used in the construction of the machines is of the best and the the workmanship unsurpassed.

These machines have been tested in different places by expert electricians and they claim that they are as high in efficiency as any others that have tested and higher than a good many. The simplicity of their construction enables the firm to put them on the market at a very reasonable price. Every machine that is turned out is tested up to its full capacity and guaranteed against all electrical and mechanical defects for two years from the time they are started. The demand for these machines is so great that the company's factory is taxed to its utmost.

In the last few years electricity as a motive power has

THE NEW KAY MOTOR.

come so rapidly to the front that there is scarcely a village or factory where it is not employed extensively for lighting and power purposes. The Kay Electrical Mfg. Co. being among the pioneers in this line, have endeavoured to keep pace with the most advanced improvements and there is hardly a village, city or town from Quebec to Vancouver where there is not more or less of their machinery in operation. In the city of Toronto there are more than three hundred of their machines in use ; in Hamilton nine-tenths of the electric power is used through Kay motors. Guelph, Brantford, St. Catharines and Montreal are all extensive users of these machines.

## TRANSFORMERS.

### By G. W. F.

#### (Concluded from September Number.)

A COMPARISON of tests made on a number of transformers shows that between the best and the worst there was the difference of about 100 watts per hour in the amount of energy consumed in the above friction—which is called hysteresis. Reducing this to a question of coal consumed, the better transformer consumed on no load more than one ton less than the other. This meant a saving of about $3.50 per year per transformer, by using the better and more expensive one. This saving capitalized shows that the better transformer was worth at least $100 more than the other in point of hysteresis saving alone, not considering losses to be investigated later.

A second source of wasted energy is the generation, within the core itself, of Eddy currents which heat up the iron, and so consume power. It will be evident from an inspection of the diagram 6 that the passage of a current through the primary wire P will set up currents, not only in the secondary wire S, but also in the bar A, which is actually a conductor placed in a varying field. These currents will circulate through the bar in directions at right angles to its length, and any means of checking them or reducing them will be an advantage as tending to reduce the losses. A current is stopped by breaking a circuit, and this method is employed in the construction of transformer cores, which are built up of thin sheets of iron placed side by side with some form of insulation between them. Thus, instead of the bar being solid, and so constituting a metallic circuit for Eddy currents, it may be represented by the accompanying diagram No. 8, which shows it made

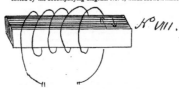

of sheets separated by other sheets of insulation. The direction of the Eddy currents would be across the length, as indicated by the arrows ; but they are evidently checked by the insulation, and so cannot flow in such great strength. The insulation does not interfere with the flow of the line of magnetic force, whose direction is along the bar. ' It is impossible to quite check or do away with Eddy currents altogether, because just as long as iron is subjected to a varying field, it must necessarily have currents set up in it. We can only minimize the evil by efficient design and construction. Lest it might be thought that these losses—from hysteresis and Eddy currents—are too insignificant to really take any account of, it may be here stated that results of a series of most carefully conducted tests, by persons whose competence was quite beyond question, showed that with transformers of superior make, the losses in very small sizes were sufficient to form about 10 per cent. of the capacity of the transformer, and in the larger sizes between 7 and 8 per cent. What the percentage would be in transformers of inferior make is impossible to estimate for all cases, but a test made on several different transformers by Prof. Jackson showed that, taking two for comparison, a central station using 100 transformers of the size considered would find a difference of $1,200 in operating expenses between the two makes— that is, the better type would cost less to operate than the other by $1,200 per year. Transformers are just like everything else— there are good ones and poor ones. A good transformer is the only one that a central station can afford to buy, and a good transformer costs money to build and is therefore expensive.

A third source of waste is the " magnetizing current," and this again can be minimized by careful design and good construction, but not entirely stopped.

This magnetizing current can be understood thus : P is a primary wire from the generator G. S is the secondary, and both are wound round a bar A. It is generally thought that if the secondary circuit is open—that is, when there are no lamps being lighted —that no current will be flowing in the primary circuit P. This is an error. There is a current flowing in P just as long as the generator is operating, whether the secondary is open or closed. The current, it is true, will be but small in the former case, and will increase as the load on the secondary becomes greater ; but

it is evident that no matter whether that secondary is open or not, the primary circuit is always connected right across the 1,000-volt mains, and must therefore carry some current. The reason it is so small at open secondary is because a counter electromotive force is set up in the primary by the alternating magnetism which it is itself the cause of in the core. This counter electromotive force is almost equal to the impressed E. M. F., and only the difference

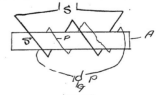

DIAGRAM IX.

between is available for setting up the flow of the small magnetizing current. Whenever the circuit through the secondary is closed, through a lamp or lamps, an E.M.F. will be set up in the secondary, which will indirectly assist the E.M.F. impressed on the primary. In this case the difference between the impressed and .the counter electromotive forces acting to force a current through the primary will be greater than at open secondary, and will set up a primary current which will increase as the secondary resistance becomes less by throwing in more lamps. But it must be clearly borne in mind that a current is flowing in the primary whether the secondary is open or not ; and further that the amount' of this magnetizing current depends on the construction of the transformer, being capable of reduction to a very small amount, or of being made to assume very uneconomical proportions. Two high-class transformers on the market to-day are guaranteed to have the following magnetizing currents : A has ·125 of an ampere ; B has ·0656 of an ampere in transformers of 6,000 watt capacities. Just as long as the generator is operating, A will consume ·125 of an ampere, and B ·0656 of an ampere, whether there are lamps burning or not. This is sometimes incorrectly called the leakage current. Take an installation of 1,000 lights, using ten of this size transformers. The magnetizing current of the lot will be, with A type 1¼ amperes, and with B type ·656 of an ampere. Reducing this to a matter of watts or horse power : A type will cause a necessary waste of one and seven-tenths of a horse power every hour the generator runs, while B will cause a necessary waste of only nine-tenths (9/10) of a horse power. Assuming that the plant operates for an average of eight hours for 365 nights during the year, and taking coal at $3.00 per short ton, and allowing 4 lbs. per h.p.h., the calculation is easy that the waste in magnetizing current only using A type transformers amounts to $30 worth of coal per year, and with B type to $15.75. This cost, being a constant yearly expenditure, should be capitalized, and at say 5% interest it shews that B type transformers of the above total capacity are worth $300 more than A type, or $30 each ; and consequently that to use A type instead of B type is a very marked extravagance, unless they can be bought for $30 less. It cannot be too strongly emphasized, that it is by careful attention to, and consideration of such details that central stations must look to their profits. There have been numerous instances where central stations have turned a yearly deficit into a satisfactory profit by scrapping all their old transformers—with their wasteful magnetizing currents, and heavy hysteresis and Eddy losses—and using instead transformers of the most modern type. In the former case the generator current was lost in the transformer primaries ; in the latter it was saved and available for sale. In the calculation made above, two really high class transformers were compared. What will be the results if transformers using half an ampere are taken as the basis ? And yet it may easily be verified that plenty of those now hanging on poles in provincial towns take all of that. From the above considerations it will be plain that the efficiency of a transformer is a matter to be seriously taken into account, and that to buy such apparatus on the basis of lowest cost is a most imprudent policy. The facts that such apparatus cannot be watched during operation should make purchasers all the more careful in selecting it.

A point of considerable importance, although not entailing any loss on the central station, is that of the pressure regulation. A transformer, having no means of automatically raising its voltage as the load becomes greater, necessarily allows a "drop" between no load and full load. A generator has some provision made, either by compounding its field or by hand regulation of the exciting current, for increasing its initial pressure as the current gets larger so that at all amperages the final pressure at the lamps will be constant. A transformer, however, cannot be compound wound ; nor is it convenient to have a man up the pole to work a rheostat periodically, so that the pressure at its primary terminals is all that is available for causing the flow of current from no load to full load. It is plain, therefore, that at full load the pressure on the lamps will be somewhat lower than at light load, and the difference between these two pressures depends— as indeed does all the data—on construction and design. In the above transformers A has a regulation of 2½ per cent., B, of 1¾. This means that at no load, on one lamp, the pressure will be 2½ volts higher than it will at full load or 100 lamps, with A, and 1¾ volts higher with one lamp than with 100 lamps with B. Thus B will subject its lamps to ¾ of a volt less variation than A, and as the life of lamps decreases about 15 per cent. with every 1% of excessive voltage, it requires no great ability to see that for the consumers' interests B is the better transformer. All the foregoing considerations shew conclusively that transformers are in their way just as important as generators ; that they are just as susceptible of careful and educated design and construction as any other apparatus ; that to build a thoroughly good transformer requires very superior material and equally superior workmanship; that consequently a good transformer necessarily costs money, and a cheap one bears prima facie evidence of inferiority ; that a cheap transformer is the most expensive piece of apparatus that can be bought ; and a high priced one the truest economy and the best investment. It is to be hoped also that, in the near future, central station men will come to understand that the more sharply and intelligently that they study their plants and business, and the more they try to keep abreast of the times the better for themselves.

## IMPROVED THERMO-ELECTRIC BATTERY.

### By James Asher

THE problem of how to transform heat economically into electricity is one of the most important that can be laid before the inventor. The electrical efficiency of the best thermo-electric battery is probably about one-twentieth only of that of a dynamo driven by a steam engine whose boiler uses an equal quantity of fuel.

We shall now consider where the great waste of heat occurs in the thermo-electric battery. In the first place there is an enormous waste of heat from the chimney. The quantity of heat which escapes from the chimney of a thermo-electric battery is perhaps about equal to that which escapes from the chimney of a stove consuming an equal quantity of fuel in an equal time. Nearly one-fourth of the total heat is radiated from a stove, and three-fourths passes up the chimney and does no useful work except the production of a draught of air for supplying the furnace.

The writer has invented several methods of securing better economy in those thermo-electric batteries which have chimneys and which use no water to cool the ends of the elements.

First Method.—All the air which feeds the furnace is caused to pass within a casing, along and in contact with the ends of the elements which need to be cooled, and then it enters the furnace at an elevated temperature. Thus the heat that would have otherwise been wasted is returned to the furnace, and part of the waste is thereby avoided. In fact, in this method I apply the regenerative principle to the thermo-electric battery. This method will not enable us to save all the heat which otherwise would have been wasted by radiation and convection from the exposed ends of the elements of the battery, because the products of combustion will leave the chimney at a higher temperature than they would otherwise.

Second Method.—An artificial draft is employed in the furnace. It is well known that an artificial draught can be maintained much more economically than can a natural draught in the furnaces of steam boilers. When we use a natural draught a high temperature in the escaping gases is necessary, otherwise the draught would be very feeble. But when we use a blower to force air into the furnace the products of combustion can be made to leave the chimney if required at a temperature but little higher than that of the atmosphere. Hence, the heat from the products of combustion may be used to elevate the temperature of the inner junctions of a thermo-electric battery, row after row, each row

receiving heat from the gases at a lower temperature than the preceding row. The gases part with nearly all their heat before entering the chimney. The first row in this method would naturally receive heat at the inner junctions, but this temperature would be too high for the metals to endure without either fusion or rapid oxidation. In order to overcome this difficulty the writer proposes to force air into the thermo-electric battery beyond the furnace, and so as to mingle with the products of combustion of very high temperature which proceed from the furnace. By this plan we shall have a large volume of mixed gases, at a temperature which will not be too high for the inner junctions of the battery to endure. A great many elements will be needed to absorb the heat from the large volume of mixed gases. The temperature of the last set of elements at their outer ends should preferably not greatly exceed the temperature of the atmosphere.

Third Method.—It is said that the range of temperature in thermo batteries is only about ninety degrees. There is no advantage in maintaining the heated ends at a temperature of more than ninety degrees higher than their cooler ends. This being the case, I propose to utilize the outer ends of the first set of elements as a source of heat for the inner or hotter ends of the second set of elements, the ends of which are nearly in contact therewith, then the outer ends of the second set as a source of heat for the inner or hotter ends of another or third set, and so on until the inner ends of the last set have a temperature of about ninety degrees above that of the atmosphere. Thus I might have about ten sets of elements, the outer end of each set serving as a source of heat for the next set. A portion of the heat which enters each set of elements is transformed into electricity, and therefore, as heat, it disappears.

It is probable that in a thermo-electric battery, constructed according to my plans, the second law of thermodynamics would approximately hold. Supposing that we maintain the inner junctions of the inner set of elements at a temperature of 960° Fahrenheit, and the ends of the last set at a temperature of 150°, then if the second law of thermodynamics holds good here we should have a theoretical efficiency of

$$\frac{(960° + 460°) - (150° + 460°)}{960° + 460°}$$

which is equal to fifty-seven per cent. This is a much greater efficiency than any which has ever been obtained from any steam or gas engine. It should be stated, however, that a deduction should be made for the power required to operate the blower when we use one.

Here, then, are several methods proposed for economizing heat in the thermo-electric battery. All these methods may be combined, then we shall obtain the highest efficiency.

## SPARKS.

The failure is announced of the Holmes Electric Co., of Montreal. The assets of the company were recently sold by auction.

It is reported that the first building in Canada to be lighted with acetylene gas will be the new Presbyterian church, Palmerston, Ont.

An employee of the Royal Electric Co., named Sabaouth was killed in the Company's factory recently by coming in contact with a large belt.

Following the example of the ladies of London, the members of the King's Daughters Society, Cornwall, acted as conductors on the electric cars on September 16th, with the object of raising funds for the general hospital in that town.

Messrs. R. J. McGowan, Secretary Fire Dept., Toronto, and Z. Benoit, Chief of the Montreal Fire Department, were among the promoters of the National Association of Police and Fire Telegraph Superintendents organized in Brooklyn, N. Y., on September 15th.

Messrs. Fregeau & Lecroix have recently purchased the electric light plant at Three Rivers formerly owned by the corporation. The purchasers propose to obtain power from the falls at Price's Mills on the Batiscan river, 14 miles distant from the city. It is said to be their intention to also supply light to the neighboring villages.

When 350 watts make one horse-power, when copper wire sells for five cents a ton, when six inches make one foot, when two feet make one yard, when one watt equals a kilowatt—then 53 cents will make one dollar, and the people of the United States will stand as the largest aggregation of dishonest repudiators in the history of the world.—New York Electrical Review.

## SPARKS.

Since the first of August the Montreal Street Railway Company have refused to accept payment of fares in American silver.

There was a collision on the Hamilton Radial Railway Co.'s line on Sept. 10th. Only two persons were injured, and these but slightly.

The Hamilton Street Railway Co. have notified their employees that a reduction of ten per cent. in salaries will be made commencing October 1st.

The Port Arthur Pulp Timber Co. is being incorporated to manufacture timber and to construct electric light and power works. The capital stock is $200,000.

Mr. John Patterson states that construction work will be commenced immediately on the plant of the Cataract Power Co., who propose to convey power from DeCew Falls to Hamilton.

The Armington & Sims' Engine Co., of Providence, Rhode Island, which suspended on the first of the month, had a contract to make a 600 horse-power engine for the London Street Railway Company.

The Telephone Company of St. Francois, Riviere du Sud, will seek for an amendment to its charter at the next session of the Provincial Legislature to enable it to prolong its line as far as Montmagny.

Negotiations are in progress between Toronto capitalists and the company who operate the horse car system at Niagara Falls, looking to the transfer of the line and its transformation into an electric road.

The management of the Toronto Technical School have added $100 to the salary of Mr. James Milne, Lecturer in Electricity, and have given him the supervision of the drafting room, mechanics, electricity, steam and the steam engine.

Application will shortly be made to Parliament for the incorporation of the Moto-Cycle Co., of Canada, Ltd., to manufacture and sell horseless vehicles. The headquarters of the company are to be at Montreal. The proposed capital is $150,000 in shares of $10.00 each.

At the annual meeting of the standard Light & Power Co., held in Montreal recently, the former Board of Directors was re-elected and the following officers appointed for the ensuing year : R. Wilson Smith, president ; W. McLea Walbank, vice-president and managing director ; E. Craig, secretary-treasurer.

The building and plant of the Palmerston Electric Light Co. was totally destroyed by fire last month, together with 100 cords of wood belonging to the company. The loss is a heavy one, the insurance being only $1,600. It is said to be the intention of the company to rebuild and instal a new plant immediately.

In order to get better service to points between Arnprior and Pembroke, the Bell Telephone Co. has constructed a new copper wire line from Ottawa to Arnprior. The Company have also a direct line from Ottawa to Brockville via Almonte and Carleton Place, and a new line is under construction from Ottawa to Morrisburg, via Metcalf and Winchester.

The courts of Montreal will decide in a day or two whether or not the Standard Light & Power Co. and the Bell Telephone Co. have power to open up the streets of the city and lay underground mains without the consent of the City Council, by virtue of the authority conferred upon them by the Quebec Legislature. Each company has commenced the construction of underground mains, but work has been stopped by the city authorities, pending a legal decision on the above point.

A bill is now under consideration in the Dominion Parliament for the incorporation of the Mather Bridge and Power Company. The object of the company is to bridge the Niagara river at Port Erie and to generate electrical power by means of an immense paddle wheel attached to the centre of the bridge. The bill is being opposed on the grounds that the wheel would be an obstruction to navigation, and that the country should receive a revenue from the utilization of the water power.

The streets of the town of Newmarket have been in darkness since last April, at which time the electric light service was discontinued, on the ground of unprofitableness. Ten thousand dollars have recently been invested by the municipality in a new plant, from which arc lighting for the streets and incandescent lighting for commercial and private use will be furnished. The new system is being rapidly put in working order, and is expected to go into operation within a week from date.

The International Trading Co. will submit to the City Council of Kaslo, B. C., a proposition to install an efficient lighting system, arc and incandescent, on condition that the city will contract at the price of $100 per month for twelve arc lights, for ten years, and exempt the property of the company from taxation. The company guarantee also to furnish light to private consumers at a reasonable figure, and to sell their plant to the city at a figure to be agreed upon at the expiration of the term of their contract.

The town of Peterboro' recently invited tenders for public lighting. No tenders, were however, received. A letter was read from the Peterboro' Light & Power Co., stating that they refrained from submitting a tender on the ground that the contract was too stringent. They objected specially to the clauses stipulating that the poles be painted and that a penalty of 75 cts. per lamp be imposed should the candle power be found to be at any time less than 2,000. The Council are now considering the question of purchasing a plant and operating it as a municipal enterprise.

Strained relations have existed for some time past between the City Council of Winnipeg, and the Electric Street Railway Company of that city. The Council contend that the company have forfeited their franchise by ignoring the terms of their charter as to character of service and condition of maintenance of the road. Mr. Wm. McKenzie, the president of the road, has just visited Winnipeg with the purpose of arriving at a settlement of the difficulty. He has offered to sell the road to the city, as he claims that it has been a source of trouble to the present owners ever since it was put in operation.

The formal opening of the Lachine Rapids Hydraulic and Power Company's works for the utilization and transmission to Montreal of the power of the Lachine Rapids, took place last month, and was attended by a number of prominent citizens. Mr. Burland, President of the company, reviewed the history of the enterprise, and stated that up to the present about $800,000 had been paid into the company, proof sufficient of confidence on the part of the directors and shareholders in the success of the scheme. Much credit was deservedly bestowed upon Messrs. W. McLea Walbank and E. T. Pringle, who were the original promoters and subsequently the engineers of the work.

The Chambly Water Power Co. have let the contract to Mr. Peter Lyall, of Montreal, for the construction of a dam across the Richelieu river at Chambly for generating electric power. A contract has also been given to the Stillwell-Bierce Co., of Dayton, Ohio, for the required machinery, while tenders have also been invited for sub-contracts amounting to upwards of half a million dollars. It is proposed to transmit electric power from Chambly to Montreal, a distance of 15 miles. It is expected that about 40,000 horse power can be generated. It is understood that the Royal Electric Co., of Montreal, who are shareholders in the Chambly Power Co., have contracted for a considerable portion of the available power.

In reply to Mayor Fleming, who called in question the impartiality of the tests of the electric light supplied in Toronto, Mr. Higman, Chief of the Inspection Department at Ottawa, declares his report to be an exact record of the conditions as he found them. He states further that no arrangement was made with either party as to the time when the test should be made, nor had either party any opportunity of "padding" the report or exerting any influence that would tend to bias the report for or against either party. Mr. Higman also points to the fact that arc lighting dynamos being constant current machines, it would be exceedingly difficult and inconvenient for the company to vary the output to any appreciable extent.

Some experiments on the effect of heat on insulating materials made by Mr. C. E. Skinner, are summarized in the following conclusions : 1. The insulation resistance of all ordinary fibrous insulating materials, such as paper, cloth, etc., decreases upon being heated up, and then increases again when the moisture is expelled. 2. Continued heating of 31 hours at 120 degrees centigrade does not lower the insulation resistance of paper. 3. The insulation resistance of completed apparatus shows the same characteristics as the insulation resistance of materials taken separately. 4. A low insulation resistance is not necessarily an indication of poor insulation, but probably an indication of the conditions of the apparatus in regard to moisture. 5. A high electromotive force should not be applied to apparatus when the insulation resistance is low. 6. Material which is badly deteriorated mechanically by heat may still have a high insulation resistance but very poor insulating qualities.

## SPARKS.

The City Council of St. Thomas have decided to grant a three years lighting contract to the gas company.

An effort is being made to induce the Council of North Bay to have the streets of that town lighted by electricity.

A bill has been introduced in the Dominion Parliament to in- corporate the Columbia Telephone & Telegraph Co.

The Gravenhurst Electric Light & Power Co. have succeeded to the business of the Gravenhurst Electric Light Co.

The St. Johns, Que., Electric Light Co., are negotiating for the necessary supply of power to operate their system successfully.

Wm. Simpson, an electrician with the Cortland Automatic Fire Alarm Co., won the first prize in the recent bicycle road race at Toronto.

The Consolidated Railway & Light Co., Victoria, B. C., will install an additional dynamo, weighing 8,000 lbs., purchased in England.

The Peoples' Heat & Light Co., composed of Boston capitalists, is reported to have purchased the franchise and works of the Halifax Gas Light Co.

The Nova Scotia Telephone Co. have just completed their new line between Glasgow, Pictou & Truro, in connection with their long distance line to Halifax.

Mr. J. W. Taylor, late manager of the Ottawa Porcelain & Carbon Co., is said to have purchased a valuable Feldspar mine suitable for the manufacture of porcelain ware and insulating ma- terial.

It is reported that an electric railway is to be immediately con- structed from Liverpool, N. S., to the pulp mill at Millton, N. S., to carry the product of the mill to the seaport, and also to carry passengers.

The Sussex, N. B., Water & Electric Light Co. has recently been organized, and is about to erect a station 28 x 50 ft. in size. The company expect to have their plant in operation before the close of the year.

The Ottawa Electric Railway Co. are equipping their cars with fenders. It is proposed to place a fender at one end of the car only, and to construct loops so that the car may be turned around at the end of the trip.

As a result of a recent visit of the president and directors of the St. John Railway Co., it was decided to remove the electric plant to the Company's new building on Smythe street, and to install additional machinery.

The Montreal Street Railway Co. have just completed the con- struction of an immense new chimney, the diameter of which is 54 feet at the base, and the height 225 feet. Two million bricks were used in its construction.

A bill is before Parliament to authorize the formation of the Canadian Electric Light & Power Co., with authority to build an electric railway from Cobourg, via Port Hope, Toronto and Hamilton, to the Suspension Bridge.

Li Hung Chang, the distinguished Chinaman who recently visited Canada, took his first ride in an electric car on the Niagara Falls Park & River Electric Railway, having previously refused to embark on the American Gorge road.

The Quebec Legislature will be asked at its next session to in- corporate the St. Hyacinthe City & Granby Railway Co. to con- struct a railway to be operated by steam, electricity or other motive power, from Brigham, Brome county, to St. Hyacinthe. Capital $100,000.

On the route of the extension of the Montreal Park and Island Railway to Lachine, there has been discovered a piece of swamp which will entail a great deal of expense in the way of filling up. About 2000 loads of slabs have already been used as filling, and the end is not yet.

The Richelieu Telephone Co.'s property has recently been pur- chased by the Pare & Pare Telephone Co. The company's lines run from St. Ramie to St. Guilleaume, Que., connecting with other lines, forming a system of 263 miles in length, connecting 47 towns and villages.

The corporation of St. Johns, Que., have recently entered into a new contract with the electric light company. In future only a few of the principal places in the town will be lighted with arc lights, other parts being lighted by 25 c. p. incandescent lights. The total cost to the corporation will be $1,200 per year.

Mr. W. H. Meldrum is at the head of a new company which has lately been formed in Peterboro' for the purpose of supplying electric power. An electric plant, including multiphase generat- ors and motors, costing in the neighborhood of $25,000, is being installed for the company, under the direction of Mr. J. M. Camp- bell, electrical engineer of Gananoque.

The following are the officers elect of the Halifax Tramway Co.: President, Henry M. Whitney; Vice-Presidents, John Y. Payzant and Hon. D. McKeen; Secretary, B. F. Pearson; Directors, John Y. Payzant, Adam Burns and Thomas Fysche. The company have taken over the road from Mr. Brown, the con- tractor, and it is in successful operation.

The Hull and Aylmer Electric Railway Co. are endeavoring to obtain from the Dominion Parliament the necessary legislation to allow them to change the name of the company to the " Hull and Aylmer Railway Co.," and to cross the Suspension bridge and land passengers in Ottawa. Their application is being strongly opposed by the Ottawa Electric Railway Co.

Notice is given that application will be made at the next session of the Dominion Parliament for the incorporation of a company to construct and operate a railway easterly from Vancouver through the North-West Territory and Manitoba and the province of Ontario to the Great Lakes, and to construct and operate tele- graph and telephone lines along the said railway.

A charter will be applied for on behalf of the St. Jerome Power & Electric Light Co., Ltd. The object of the company is to acquire the electric plant now in operation at St. Jerome, Que., and the water power and mill privileges by which the same is operated. The proposed capital stock is $50,000; the chief place of business to be at St. Jerome, and the head office at Montreal.

A company has been formed at Quebec to take over the fran- chise given by the City Council to Mr. Beemer for the construc- tion of an electric street railway. About a quarter of a million dollars has already been subscribed, and the construction of the road is to be commenced immediately. The power for the opera- tion of the road will be supplied by the Montmorency Electric Light & Power Co.

It is reported that owing to the unwillingness of the Dundas town Council to grant the assistance asked for, the Hamilton & Dundas Railway Co. have abandoned, for the present at least, their intention of reconstructing the road and adopting electricity as the motive power. Instead of so doing it is said to be the intention to purchase a new dummy engine and new cars, and to lay new track between Dundas and the limits of the city of Hamilton.

A report on the route of the proposed Huron & Ontario electric railway has been presented to the president and directors of the company by the engineer, Mr. A. Brunel. It is proposed to utilize a number of water powers in the county of Grey for power purposes. The entire length of the road will be 285 miles, and it is stated that the road will open up a new section of country with a population of 140,000, and secure trade which is now carried on by means of horses.

It is reported that the business of the Niagara Falls Park & River Railway Co., for the past season has been disappointing, and that steps are to be taken to reduce the cost of operation. In this connection it is also reported that Mr. Ross Mackenzie, the manager of the road, has tendered his resignation and has been succeeded by Mr. Phillips, late chief engineer, and that nego- tiations are in progress between the directors of this road and the Niagara Gorge road, looking to the amalgamation of the two systems.

The Lachine Rapids Hydraulic & Power Co. have commenced the erection of a receiving station at the corner of Seminary and McCord streets, from which wires will be strung on steel poles to the generating works at Lachine. On some streets, however, underground conduits will be laid, provided the dispute between the company and the city of Montreal regarding the laying of these conduits shall result favorably to the company. A contract for 507,000 lineal feet of concrete-lined iron conduit has been given to the National Conduit Co., of New York, at the price of $150,000.

The City Council of Victoria, B. C., recently passed a by-law imposing a number of restrictions upon the operation of the street car system of the Consolidated Railway Co., of that city. The railway company have obtained from the courts an order that the validity of this by-law shall be argued and legally decided. The railway company claim that the by-law is illegal, inasmuch that it seeks to impose conditions and restrictions which are at variance with the agreement of 1894 between the city and the Victoria Electric Railway & Lighting Company, to whose charter the Consolidated Railway Co. have succeeded.

## PRESERVATION OF WOODEN POSTS.

The conduction of the electric current for various purposes necessitates the use of an immense number of wooden posts as supports for the conducting wires and cables, and the preservation of these posts, which are set in the ground, is a question which has caused electric and other engineers a large amount of thought.

According to "La Nature" great interests are involved, for it is estimated that in Europe, alone, there are about 20,000,000 posts in use for carrying electric wires. The wood, where it is set in the ground, "betwixt wind and water," is very soon destroyed, and a number of posts have to be replaced each year. It is estimated that in Europe, alone, the maintenance of the posts costs nearly $4,000,000 per year.

Attempts have been made to prolong the life of the post by the injection of metallic salts, as sulphate of copper, or iron, or creosoting, and a certain measure of success has been obtained ; but in time rains dissolve and carry away these substances. It is now proposed to protect the weak point of a post by a stoneware cover. This has been tried and has given good results.

Numerous observations have shown that a post is attacked for a length of from ten to twelve inches from the surface of the ground, which is the depth to which rains usually penetrate. This distance is covered with two half cylinders of salt-glazed stoneware joined together, and the space between the stoneware and the post is filled with a damp-resisting cement, such as Portland cement with sand or gravel. A ring is fixed onto the post just above the level of the stoneware coat, and the top is made up of cement laid at an angle so that the rain will run off.

Very careful experiments go to show that this method of preserving a post will increase its life by more than five times, and the cost would be very slight in comparison to the benefits obtained.

## LEGAL.

Re Brantford Electric and Power Co. and Draper.—Judgment on appeal by the company from order of Falconbridge, J., referring an award back to the arbitrator. The company were the assignees of the lessor, and Draper was assignee of the lessees, mentioned in the lease of water power. The lessor had an option under the terms of the lease to refuse to renew the term, in which case he was to pay for the "building and erections" on the land, at a price to be ascertained by the arbitrator. The award in question made upon the lessor's election not to renew did not award to the lessee the value of certain fixed and moveable machinery. The order of Falconbridge, J., directed that the value of the machinery should be included. The court dismissed the appeal with costs.

The suit of the Royal Electric Co., of Montreal, against the town of Maissoneuve, and the Edison General Electric Co., intervening, was argued recently before Mr. Justice Charland in the Superior Court at Montreal. The Montreal Gazette prints the following particulars and decision : "The plaintiff alleged that on the 16th October, 1891, by deed before notary, it was agreed between the parties that the plaintiff should furnish the town of Maissoneuve with a complete system for the lighting of the town by electricity, the sum stipulated being $9,300 ; that the plaintiff immediately manufactured the necessary apparatus and prosecuted the work with diligence until stopped by an injunction and other legal proceedings. The plaintiff claimed the sum of 4,375.50, with interest, for work done and material furnished. The defendant called in the Edison General Electric Company in warranty, and the latter company took up the suit in behalf of defendant. The Court held that the plaintiff had proved its allegations, and that the intervening party had not proved its pleas, and judgment was given in favor of the plaintiff for the amount claimed, $4,861.17.'

## PERSONAL.

Mr. A. E. Edkins, of the Boiler Insurance and Inspection Co.'s staff, sailed a few days ago per steamer Lucania for England.

Mr. H. M. Whitney, formerly president of the West End Street Railway company, of Boston, has been elected president of the Halifax, N. S. Electric Tramway company.

The many friends of Mr. Ross McKenzie, manager of the Niagara Falls Park & River Railway, will be pleased to learn that he is recovering from the severe attack of typhoid fever, which at one time threatened to prove fatal.

Mr. W. E. Davis, formerly electrician of the Toronto Railway company, and more recently electrical engineer and purchasing agent of the Detroit Railway, has been appointed manager of the Bearinger electric road, operating between Saginaw and Bay City, Mich.

It is announced on the authority of Professor McCallum, who is at present in Europe in connection with arrangements for the meeting of the British Association next year, that Lord Kelvin, the celebrated electrician, will be among the scientists who will attend this meeting.

Mr. C. F. Medbury has resigned his position with Messrs. Ahearn & Soper, of Ottawa, and has accepted a position with the Western Electric Company, of Chicago, with headquarters in New York city. The removal of Mr. Medbury will be deeply regretted in electrical circles in Canada. He was acknowledged to be one of the brightest, most energetic and gentlemanly of the representatives of the manufacturing companies in the Dominion, and may be expected to give a good account of himself in whatever capacity he may be placed.

## TRADE NOTES.

The Beaverton Electric Light Co. are adding a 250 light Edison dynamo to their present plant.

The Consolidated Railway Co., of Vancouver, have installed a 150 kilowatt monocyclic generator.

P. McIntosh & Sons, Toronto, have purchased a 300 light plant from the Canadian General Electric Co.

The Canadian General Electric Co. have sold a 150 kilowatt monocyclic generator to the Hull Electric Co.

The O'Keefe Brewing Co., of Toronto, are installing a 500 light direct-connected unit. The Canadian General Electric Co. have the contract.

The New Glasgow Electric Light & Power Co. are installing a 75 kilowatt alternator of the Canadian General Electric Company's monocyclic type.

The firm of Ness, McLaren & Bate, electrical supplies, Montreal, has been dissolved. Mr. Norman W. McLaren will continue the business under the former name.

The Royal Electric Co. are installing in the asylum for insane at Mimico two direct current generators with a capacity of 500 lamps, to supplement the plant which was put in there some years ago.

Messrs. John Starr, Son & Co., of Halifax, have recently issued a very complete catalogue of 70 pages, and of convenient size, containing illustrations and prices of the various kinds of electrical apparatus which they handle.

The Weekes-Eldred Co., sole manufacturers for Canada of the Improved Jones Under-Feed Mechanical Stoker, have opened an office at No. 512 Board of Trade Building, Toronto, preparatory to introducing the invention throughout the Dominion.

The Colliery Engineer Co., proprietors of the International Correspondence Schools, at Scranton, Pa., were partially burned out on the morning of the 30th of August. They advise us that fortunately their printing plant was in another building, and they had reserves of all instruction and question papers, drawing plates and other supplies and stationery used in the schools in still another building, so that their business will not be seriously interfered with. They have secured new and more commodious offices and are prepared to enroll and instruct students as usual.

The Corporation of the town of Sudbury have closed a contract with the Royal Electric Co., for the installation of one of their 75 kilowatt "S. K. C." two phase alternating current generators ; from which they will operate 15 alternating current arc lamps for street lighting, about 1,000 incandescent lamps, and a number of motors. We are advised that this is the first alternating current plant in Ontario, other than experimental, and demonstrates the flexibility of the alternating two phase system. They can serve a night load as well as a day load, from the same dynamo, and only use one circuit, making it possible to run a lighting and power system with only one circuit and one machine.

# ELECTRIC RAILWAY DEPARTMENT.

## ELECTRICITY ON A STEAM ROAD.

THE latest development in Canadian electric railway work is the equipping of the Aylmer branch of the Canadian Pacific Railway with electric service. This line extends from Hull, a suburb of Ottawa, to Aylmer, where it connects with the Pontiac Pacific Junction Railway extending 60 or 70 miles up the north side of the Ottawa river. The section from Hull to Aylmer has been leased by the Hull Electric Co. for a term of 35 years, the understanding being that besides passenger and mail traffic they are to handle all through and local freight delivered to them by either the Canadian Pacific Railway or the Pontiac Pacific Junction Railway. As they are the only connecting link with the Pontiac

Ont., and operate under a head of 9 feet. Four 60 inch wheels are now installed and space is provided for two more.

The electrical equipment of the power house consists of two M. P. 4-200-425 generators built by the Canadian General Electric Company. For controlling the output of these machines there is a white marble switchboard consisting of two generator panels, two feeder panels and a total output panel, all of the General Electric standard type and supplied by the Canadian Co. Besides these there are three panels containing the "Barbour" water wheel regulator by which the current output of the generators is automatically kept constant by cutting in or out dead resistance

ELECTRIC LOCOMOTIVE—AYLMER BRANCH CANADIAN PACIFIC RAILWAY.

Pacific Junction Road it can readily be understood that the quantity of freight is considerable, amounting usually to 50 or 75 cars per day. This freight is mostly handled at night, leaving the road free during the day for passenger traffic.

At the Aylmer end of the line the company owns 60 acres situated on Deschesne Lake, a sheet of water three miles wide by 27 miles long; an ideal spot for sailing and boating, thus forming a strong attraction for the Ottawa citizens. Indeed the traffic has been far beyond expectations and the train service had to be increased until they are now running 36 regular trains each way per day besides special excursion trains.

The power is obtained from Deschesne Rapids, where the lake of the same name empties itself into the Ottawa River at a point midway between the termini of the road.

The wheels are of the "New American" type manufactured by Wm. Kennedy & Sons of Owen Sound,

as the load varies on the line. By this means the speed of the machines is kept constant and the variation in voltage is held within a very close limit.

The car sheds and repair shops are also at Deschesne and are fully equipped with all modern appliances for handling and inspecting the rolling stock which at present consists of five closed cars and five open cars, besides a mail, baggage and express car and a locomotive. All the cars are mounted on double trucks, and are each equipped with two G. E. 1200 motors with K.21 controllers. The closed cars are 42 feet long over all and finished in mahogany throughout, the outside sheeting being also solid mahogany finish, in the natural wood. These cars have extra large vestibules at each end provided with seats for the accommodation of smokers, and divided from the main part of the car by double sliding doors. The open cars have 13 benches with reversible backs and their finish and solidity are excellent. All these cars were built and equipped at the Canadian General Electric Co.'s

Peterboro factories from where they were shipped complete ready for delivery on the track.

The locomotive is of particular interest, being the first of the kind operated in Canada. It weighs something over 20 tons and is provided with double trucks, each axle being equipped with a motor. As all the wheels are driven full traction advantage is obtained from the total weight and a draw bar pull of 10,000 lbs. can therefore be exerted, equivalent to the power of the average 35 or 40 ton steam locomotive. This was also designed and built by the Canadian General Electric Co.

In equipping this road the Hull Electric Co. have evidently constantly kept before them the maxim that the best is the cheapest in the end, and will no doubt reap the advantage by long life in their apparatus and small repair bills.

The president is Mr. Alexander Fraser; vice-president, Mr. W. J. Conroy; secretary-treasurer, Mr. Jas. Gibson; and managing director, Mr. H. B. Spencer.

Besides operating the railway, the company have also exclusive privileges for both private and public lighting in the city of Hull and the town of Aylmer, and for the purpose there is installed at the power house a 150 K. W. monocyclic generator with a standard switchboard panel and equipment.

## A NEW DEPARTURE IN STREET RAILWAY PRACTICE.

THE proposition was seriously discussed, in connection with the opening of a new electric railway project in Eastern Ontario, recently, to employ good looking young women as conductors, as a means of popularizing and enhancing the receipts of the road. It remained, however, for Mr. C. E. A. Carr, manager of the London, Ont., Street Railway Co., to make a practical test of the idea.

With the object of raising funds to assist in furnishing the new Y. M. C. A. building, about eighty good looking and fashionable ladies of the city arranged with the street car company to act in the capacity of conductors on a certain day, trusting to their charms to swell the receipts and realize a surplus for the purpose mentioned. On the day preceding the one on which they were to enter upon their duties, the ladies took practice trips over the lines, and made careful mental notes of the manner in which the conductors performed their duties.

It was arranged that the ladies should divide themselves into detachments, each detachment remaining on duty for two hours at a time. Much to everybody's surprise, especially as the morning of the day fixed for the experiment proved to be a wet one, every lady conductor reported for duty at the early hour at which the cars begin running. More than half of the preceding night had been spent in decorating the cars with bunting, and when the rain came, it destroyed the results of all the labor bestowed in this direction. Instead of giving way to discouragement, the ladies soon had the interior of the cars charmingly decorated with cut flowers.

In order that the company might not violate the clause in their agreement with the city which provides that at least two men shall be in charge of each car, the manager writes us that the company's own conductors had charge as on other days, the ladies merely collecting fares with the fare box and issuing transfer tickets. One young woman, however, is credited with having

done all the work in conducting her car during several shifts. She collected fares, stopped and started the car to take on and let off passengers, registered the fares, made change and issued transfers, and also ran ahead of the car at the railway crossings.

The ladies are said to have refused to recognize passes, no matter by whom presented, and certain of the city officials who are accustomed to free transportation were told that they must either put up the amount of their fare in good coin of the realm or get off and walk. Having become unaccustomed to walking, they had recourse to the other alternative.

Among the many amusing incidents of the day, a local paper records the following : "One of the officials of the road saw a very funny thing on the Springbank line just before three o'clock in the afternoon. The conductor on the car in question (which was returning to the city) did not have a chaperon or any passengers on board, and the young woman was on the front platform taking instructions from the motorman. Seeing another car coming, and thinking that some of the officials might be aboard the motorman tried to get the young woman to leave the controller, and the switch. This she would not do, and the motorman, bound to be found on duty at all events, put his both arms about the girl and also held the mechanism governing the current. Passengers on the passing car caught a glimpse of the queer sight as the up car passed, the girl smiling saucily, and the motorman looking abashed at having to hold in his arms a bundle of charms in broad daylight."

Manager Carr informs us that notwithstanding the unfavorable weather the venture was on the whole a satisfactory one for the street railway.

## BERLIN AND WATERLOO ELECTRIC RAILWAY.

MR. E. Carl Briethaupt, President and Manager of the Berlin and Waterloo Electric Street Railway, is evidently determined to make this a thoroughly up-to-date road. When electricity was adopted as the motive power, light rails which were previously in use were retained. Last winter, however, proved them to be unsuitable, great difficulty being experienced from snow and ice. The old horse cars were also made to duty, after having been vestibuled. They too proved to be unsuited to the new order of things, and have been replaced by the most modern style of coaches. The necessary quantity of 60 lb. steel rails has now been purchased, to replace those at present in use ; new car barns are in process of construction, and the spring of 1897 will see the road in a position to offer its patrons first-class accommodation.

## CANADIAN VS. ENGLISH ENTERPRISE.

It has been recently stated on good authority, says London Lightning, that there are at work or under construction in Canada 36 electric street railroads, with a total length of nearly 600 miles. No less than 750 motor cars are in use or building, and the sum invested in the various undertakings is a little over four millions sterling. England in the meantime is just beginning to wake up, rub her eyes, and wonder whether the horse tram could really be improved upon, and whether ½d. a mile running costs, with speed, cleanliness and comfort, are, on the whole, preferable to 7d. a mile running costs, polluted streets, frowsy 'buses, the travelling powers of a gouty tortoise and the horrors of cruelty to animals. "But," says the dear old grandmother of nations, "we might hurt somebody, or frighten a dog ; and then, after all, we have got on without these things so far," and she dozes again. The Canadian company which recently purchased control of the Manchester tramway, will shortly show our British friends how to construct and operate an up-to-date road.

The town of Orillia is receiving tenders for the supply of a fire alarm system.

The village of Huntsville is receiving tenders for the installation of an electric lighting system.

A recently quite popular method of measuring the commercial efficiency of the boiler and furnace combined, is to determine the cost of evaporating from and at 212° F. 1,000 lbs of water.

Where the feed pipe between heater and boiler is exposed there is considerable loss of heat, amounting to sufficient to lower the temperature of the water 5 to 10 degrees and even more, in some cases.

Superheated steam cannot exist in contact with the water from which it was generated. Its temperature is higher, its specific volume greater, and its density less than saturated steam of the same pressure.

# The Royal Electric Co.

### MONTREAL, QUE.      Western Office— TORONTO, ONT.

MANUFACTURERS OF

# ELECTRICAL MACHINERY AND APPARATUS

WEST END MAIN MACHINE SHOP OF THE ROYAL ELECTRIC CO'Y.

The machine shop illustrated above is acknowledged by those experienced in factory design, arrangement and equipment, as equal in every respect to any on this continent and the best in the Dominion of Canada.

It is equipped completely and contains many machine tools larger than any of similar kind in Canada, and of the most modern type and highest class, which, with best arrangement and thorough shop organization, secures highest standard of accuracy and finish in product.

It is devoted exclusively to the manufacture of Electrical Machinery and has facilities for all kinds and all sizes up to the largest ever made anywhere.

In it are being made all the appliances included in the **largest single order ever given for Electrical Apparatus,** viz.: **EIGHT "S. K. C." TWO-PHASE GENERATORS,** each of **2650 Horse Power** capacity, or a total of **21,200 Horse Power,** for the **Chambly Mfg. Co'y,** together with Switchboards, Transformers, etc., etc.

With such a shop it is not necessary to have made outside of Canada any Electrical Machinery contracted for with . . . . .

## THE ROYAL ELECTRIC COMPANY

## SOME AIR RULES.

The volume of air under constant pressure increases directly as the absolute temperature increases.

The absolute temperature is the temperature shown by the thermometer, plus 461.

To find the volume of air at any higher temperature, multiply the volume at lower temperature by the higher absolute temperature and divide the product by the lower absolute temperature.

Example : If 50 cubic feet of air enters a heating coil at 20 degrees, what is its volume after it leaves coil at 100 degrees temperature?

(Volume) $50 \times 561$ (higher absolute temperature) $= 28,050 \div 481$ (lower absolute temperature) $= 58.3$ cubic feet.

A pound of air at atmospheric pressure will occupy 12.38 cubic feet at 32 degrees temperature.

By substituting, a constant for above value is obtained, 39.8, and with this constant the cubic feet of a pound of air can be found for any temperature by this rule :

Divide the absolute temperature of the air by 39.08. By doing this for two temperatures the difference will be the change in volume due to the difference in heat for each pound of air.

The length of an indicator diagram should be twice its height, in order to be proportionate in appearance.

You should keep before the people,
For they are very apt, you know,
To forget you are in business,
If you cease to tell them so.

## ROBERT A. ROSS, E.E.

(M.E. Grad. S P.S.; E.E. Degree Toronto University ;
Member A.I.E.E )
Late Chief Electrical Engineer Royal Electric Co ,
Montreal; Works Engineer Canadian General
Electric Co., and previously with
Engine Companies.

## ELECTRICAL AND ....
## MECHANICAL ENGINEER

Specifications, Plans, Supervision, Valuation and Advice on Standard Electric
Plants  Special Machinery Designed.

*Room 15, Hamilton Chambers*
*17 St. John St., MONTREAL*

## WANTED—BY ELECTRICIAN AND STEAM ENGINEER

Position on plant of 500 incandescent lights and upwards, with arc lighting, as general man in charge of plant (willing to learn). Nearly 10 years' experience on all kinds of construction; thoroughly posted in arc, motor, constant current s and 3 wire incandescent ; also transformer work, single and two phase, with 2 and 3 wires. Good practical knowledge of station working, steam or water. Thoroughly well up in incandescent lighting  A great hustler on inside wiring and pole line work.  First class references.  Salary low for steady employment  Address, C. O. F.,
ELECTRICAL NEWS Office.

## THE ENDURANCE OF ROTATING SHAFTS.

SOME of the results arrived at by recent tests made by the government authorities at the arsenal in Watertown, Mass., may be regarded as of special importance in relation to the endurance of rotating shafts. Thus, while it has been found that great improvements in tensile strength and elastic limit have been obtained, it has not been shown whether the limit of endurance under repeated strains has been increased. In the rotating tests of cylindrical shafts, alternate tensile and compressive strains are successfully applied, and under these conditions of loading, no steel has yet been experimented with which will endure a fibre stress of 40,000 per square inch without rupturing, and this result has been reached after a total number of repetitions of from four to seven millions for steels of high elastic limit and tensile strength.

Friction is very nearly proportional to pressure.

The velocity of a river, by reason of the friction of the banks, is greatest in mid-channels, a little below the surface, and least near the banks.

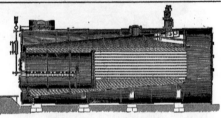

# CANADIAN ELECTRICAL NEWS

## STEAM ENGINEERING JOURNAL

OLD SERIES, VOL. XV.—No. 6.
NEW SERIES, VOL. VI.—No. 11.

NOVEMBER, 1896

PRICE 10 CENTS
$1.00 PER YEAR.

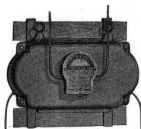

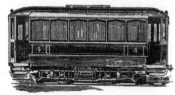

CANADIAN

# ELECTRICAL NEWS

AND

## STEAM ENGINEERING JOURNAL.

| VOL. VI. | NOVEMBER, 1896 | No. 11. |

### AN OLD FIRE ENGINE.

THROUGH the courtesy of Mr. James Devlin, Chief Engineer of the Kingston Penetentiary, we are enabled to present to our readers an illustration and some particulars of an old fire engine which has found a resting place in that institution, and which was an object of much interest to the delegates to the last annual convention of the C. A. S. E., on the occasion of their visit to the penetentiary. The engine was made by Mr. Perry of Montreal, brother of Mr. Alfred Perry of that city, and took first prize at an exhibition held in Montreal, the finest workmanship being displayed in its construction.

The engine was purchased by the government of Upper Canada upwards of forty years ago for the sum of $1,100, and is to-day in all its working parts quite as good as new.

The large number of torpedo boats now being built for the U. S. navy brings forth some features in machinery exceptionally interesting and novel. On some

AN OLD FIRE ENGINE—KINGSTON PENITENTIARY.

boats of this kind recently launched, the equipment of steam pumps, as well as the main engines, are run without the use of oil in the steam cylinders. While this is not a new idea so far as vertical steam engines are concerned, it has never been the practice to run steam pumps without oil. The pumps are arranged without any oil holes whatever, so that it is impossible to get oil into the steam cylinders. These pumps were given an exhaustive test for several days by the manufacturers and they operated with entire satisfaction, and without using a drop of oil. The doing away with the use of oil in the steam cylinder is a matter of considerable importance, as there is no necessity of carrying feed water filters and no anxiety about oil injuring the body.—Scientific Machinist.

The Canadian Customs authorities have lately given a decision to the effect that all electric bells imported into the country are subject to a duty of 25 per cent. ad valorem.

### IMPROVEMENTS BY THE TORONTO ELECTRIC LIGHT CO.

THE Toronto Electric Light Co. have erected a new incandescent and power station, adjoining their other stations on the Esplanade. The new station is constructed entirely of brick and iron, so that there is nothing for fire to feed upon. The work on this station has been considerably delayed owing to the failure of the contractors to furnish the galvanized iron roofing material as promptly as required. Inability to roof in the building has prevented the installation of the machinery.

An immense foundation of concrete has been put in to carry the required steam and electrical plant. A large vertical engine is now being put in position, and it is expected that the entire equipment of machinery for this new station will be in operation before the close of the year. In the large station adjoining there has just been put in a new condenser operating vertically and driven by a small direct connected vertical engine. This condenser was built under the direction of Mr. J. J. Wright, the manager. It is of much heavier construction than most of those we are accustomed to see, the object being to avoid the many break-downs which have been experienced from the constant operation of lighter machines. Mr. Wright is a believer in the wisdom of putting a sufficiency of cast iron into steam and electrical machinery for use in the electrical business, to give it the strength necessary to withstand the strain of constant operation.

The most economical point of cut off in an engine, says the Boston Journal of Commerce, is not that point that will expand the steam down to the atmospheric pressure, but is a point between three-tenths and one-third cut-off. A variation of a slight amount either side of this point makes but little difference, and we would not call an engine with eight pounds terminal under eighty pounds pressure as overloaded.

## THE LACHINE RAPIDS HYDRAULIC & LAND COMPANY'S WORKS.

It is barely a year since we first announced that a company was being formed, whose object it would be to harness the Lachine Rapids to utilize its dormant power for electric purposes.

Without a visit to the works in question it is impossible for anyone to conceive an idea of the magnitude of the undertaking and the amount of work already accomplished.

It was only at the end of August, 1895, that permission was asked to construct the dams, which permission was granted to Mr. McLea Walbank and Thomas Pringle, the former a civil and the latter a mechanical engineer, both of Montreal After this permission was obtained from the Dominion Government it was necessary to form a company for putting the same into execution, which took considerable time. Plans and specifications were prepared, estimates were made and tenders asked for carrying out the work.

The intention of the company was to develop water power and sell it to existing corporations and deliver it to them upon the jack shaft of the company's power house at the rapids. On this understanding a company, with a capital of one million dollars ($1,000,000) was formed. The contract for the excavation of head

and tail races and the construction of stone and crib dams was awarded to Messrs. Wm. Davis & Sons, of Ottawa, and the work actually got under way late last fall.

The directors immediately set about trying to dispose of their power, and having offered it at what they considered a reasonable figure to several large corporations, who declined to negotiate until the completion of the works in question, perhaps thinking that when completed so large a power would only be in the market for the very wealthy corporations, they made no

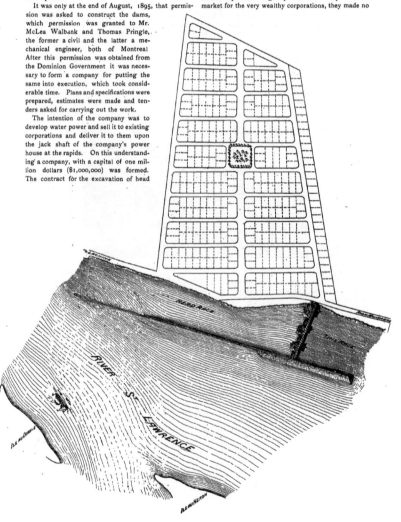

offer, doubtless figuring that "Possession is nine points of the law," and that a company without rights or franchises would have small chance of disposing of their power, and they waited for a bargain.

The directors of the Lachine Rapids Hydraulic & Land Company knew the value of their power, they had

MR. G. B. BURLAND, President.

faith in their undertaking, at least the promoters had, and at a meeting of the shareholders, held some time later, it was decided, that it the opinions of the company's engineers could be corroborated by outside experts, the necessary money would be forthcoming to convert the power into electricity, and sell it wholesale and retail directly to the customers themselves. Messrs. Walter Shanley and T. C. Keefer, two of our best known engineers, were consulted, and after hearing the explanations of the company's engineers on the question of frazil or anchor ice, back water, examining the plans and specifications of the works, their report more than

MR. T. PRINGLE, Vice-President and Engineer.

endorsed the statement already given, and it was therefore decided to increase the capital of the company to two million dollars, and to award contracts for electrical machinery.

The next question considered was the ingress into the city, and here the company met with considerable opposition, but having quietly secured the controlling interest in the stock of the Citizens' Light & Power Company, Ltd., whose works are situated at Cote St. Paul, a company having a notarial agreement with the city of Montreal for the erection of poles in all its streets, having contracts for street lighting for long terms of years with the towns of St. Henry, St. Cune-

gonde, Westmount and St. Louis de Mile End, and the lighting of the Montreal Harbor, also various contracts for private lighting, they seized the opportunity of obtaining an extraordinary entrance into the city of Montreal. Existing corporations did not like this step on the part of the Lachine Rapids Hydraulic & Land Company, and as soon as they commenced to avail themselves of their rights to erect poles, no less than two companies entered suit to prevent them doing so. This resulted in the company acquiring the charter of the Standard Light & Power Company, which was granted by the Legislature of the Province of Quebec, and which gives, without the city's consent, the

MR. W. McLEA WALBANK, Managing Director.

authority to lay underground conduits in the streets of the city of Montreal. It was however reserved in the charter for the city to prescribe the manner in which the streets might be opened. But the city treated the requests of the company with contempt, and on the company commencing operations, their men were forcibly restrained and arrested at the request of the Montreal City Council. The result was, that Mr. Walbank, the vice-president and managing director of the Standard Light & Power Company, applied to the superior court for an injunction to restrain the city police from interfering with the men in the carrying out

MR. A. PRINGLE, Secretary and Engineer.

of their work. The injunction was granted and the case argued and won by the company. The city went to appeal, and judgment in appeal was rendered quite recently, unanimously maintaining the company's charter, and declaring the injunction absolute and allowing the work to go on. Therefore the Lachine

Rapids Hydraulic & Land Company with its associated companies have now pole rights and underground rights throughout the whole city.

DESCRIPTION OF THE WORKS.

The plan of the works on a very small scale is shown, which gives the head race and wing dam, guard pier, tail race and main dam with power houses on it, and the land acquired for the future model high class suburb that the company propose to erect.

The proposed head race is one thousand feet wide at the main dam and is about four thousand feet long, and will have an average depth of water of thirteen feet. The bottom of the tail race will be some nine or ten feet lower than the bottom of the head race, and will be some twelve or fifteen hundred feet wide, allowing ample capacity for the carrying away of the water discharged from the wheels. The main dam is constructed of cut stone, about four feet wide and about fifty feet long, set to quarter inch joints and grouted in cement filled in with concrete between the stones. These stone piers rest upon concrete foundations which start at the wheel pit level and finish at the top of the flume bottom. There are two vertical sliding gates in each flume and system of stop logs at the lower end

MR. ALEX. FRASER, Director.

with a steel frame and iron rods secured to the bottom of the flumes, so that the weight of the water will have a tendency to assist in holding the stop logs in position. In front of these is an iron rack especially made to keep off all debris from the wheels. The wheels themselves are "Victor" make and have been tested at Holyoke and given over eighty per cent efficiency. The wheels are vertical and governed by an improved governor which will control the speed within 2% from no load to full load. The wheels will be coupled by bevel core gears to a horizontal shaft, six in series, and at the extremity of the shaft will be a 750 K.W. generator which is being made in Schenectady for the Canadian General Electric Company. The speed of these machines will be 175 revolutions, generating a current of 4400 volts. They will be the most modern three-phase machines and built to be easily moved for repairs.

The increase of temperature after a continuous run of twenty-four hours at rated speed must not exceed forty degrees centigrade above the temperature of the air at five feet distance from the shaft. The inherent regulation of the dynamo shaft shall be set with the drop of potential between 10% at rated load, and the non-inductive load shall not exceed 6%. The dynamos

are to operate regularly and parallel, having common bus bars. And in so operating, each dynamo must carry its portion of the total load. The insulation is to withstand successfully a potential ten times as great as that for which it is designed, and the armature to be tested for twenty thousand volts.

As already stated, the wheels shall be connected six to each machine, there being twelve machines, there

MR. PETER LYALL, Director.

will be in all seventy-two wheels. The power house will consist of a building about one thousand feet long with three centre portions divided off so as to form the dynamo house starting from the shore.

There will be on each line of shaft six wheels and a generator; then a generator and six wheels; then again six wheels and a generator and a generator and six wheels; then six wheels and a generator; then a generator and six wheels. This doubled will complete the electrical lay-out. The building is of steel construction, the dynamo portions being Laprairie pressed brick. The foundations and flooring of the dynamo houses are of concrete and steel beams, and the turbine sheds of steel work with three inch plank covered with corrugated iron.

A traveller, capable of carrying twenty-five tons with

MR. E. K. GREENE, Director.

thirty-nine feet span, will travel the full length of the dynamo house and turbine sheds moving any machinery that may be required and at the same time placed so as to lower or raise the head gates. The switch-boards will be of the latest type. The pole line starting from the power house is constructed of lattice steel poles and calculated to stand a wind pressure of six thousand pounds. It is fitted up with 6 x 6 cross-arms, made of

British Columbia fir with double petticoated porcelain insulators, and carrying No. o wires as far as the Lachine Canal, near the Curran bridge, when it will pass underground until it enters the sub-station located at the intersection of Seminary and McCord streets. The current will from this point be distributed throughout the city in underground conduits, which are being constructed by the Standard Light & Power Company.

Mr. W. P. Davis, Senior Member of the Firm of Wm. Davis & Sons, Contractors.

The names of the officers of the company and of the engineers and contractors of the works, are as follows :—

G. B. Burland, Montreal, President.
Thomas Pringle, Montreal, Vice-President.
W. McLea Walbank, B.A. Sc., Montreal, Managing Director.
Alexander Pringle, Secretary Pro-tem, Montreal,
Alexander Fraser, Ottawa, Director.
E. Kirk Greene, Montreal, Director.
Hugh Graham, Montreal, Director.
Peter Lyall, Montreal, Director.

ENGINEERS.

T. Pringle & Son and W. McLea Walbank, Montreal.

Mr. W. H. Davis, of Wm. Davis & Sons.

CONTRACTORS.

For dams and excavation, William Davis & Sons.
For iron work, Dominion Bridge Company.
For wood work, the James Shearer Company, Ltd.
For water wheels and hydraulic machinery, Stilwell-Bierce & Smith-Vaille Company.
For electrical machinery, the Canadian General Electric Co.

For wire work, the Dominion Wire Works.
Underground conduits, National Underground Conduit Co., New York.

We are pleased to be able to present to our readers, accompanying this article, portraits of some of the above named gentlemen.

## EXPANSION OF BOILERS BY HEAT.

A. C. KIRK, in a recent discussion on "Hard Firing in Boilers," gives it as his opinion that the expansion of the material of boilers, and the strains that frequently cause them to leak are largely due to irregular heating, and this irregular heating must be greatest in the largest boilers whose various surfaces are exposed to very different temperatures, says the Scientific Machinist. Thus, the temperature of the flues of a Lancashire boiler being much higher than that of the shell, its movement must necessarily be greater. This would be an argument in favor of boilers of compact form like those used in ships. It has often been observed that new boilers, that is to say, boilers having clean inner and outer surfaces give much better results in steaming than those that have been in service for months, and this is used as a reason for decrying the tests of new

Mr. Jas. D. Davis, Junior Member of Wm. Davis & Sons.

boilers as a standard of duty. Nevertheless the duty of a new boiler could be kept up till it was worn out if it were fed with pure water and fired with gas or other smokeless fuel. So the results of test should rather be taken as standards of efficiency to be worked up to. This is all the more feasible since various methods of forced combustion render it possible to burn slack, steam coal, gas coal, patent fuel and anthracite in the same furnace, and equally without smoke.

The problem of raising steam in a boiler begins with the burning of the fuel, which must be consumed at the required rate and burned completely so as to keep the heating surfaces as clean as possible. The heat has to be communicated to the water with the smallest loss on the way, and the steam has to be used in the least possible time after production, in order to work with economy.

Notice of application for the incorporation of the Willson Carbide Works of St. Catharines, Limited, has been given. The object is the manufacture of calcium carbide and any other metallurgical substances, the capital to be $200,000, The incorporators are : Thomas Leopold Willson, St. Catharines ; E. A. Neresheimer, New York ; A. M. Scott, Woodstock ; John Garry, St. Catharines, and R. G. Cox, St. Catharines. Messrs. Willson, Neresheimer, Scott and Garry will be the provisional directors.

PUBLISHED ON THE FIFTH OF EVERY MONTH BY

### CHAS. H. MORTIMER,

OFFICE : CONFEDERATION LIFE BUILDING,

Corner Yonge and Richmond Streets,

TORONTO, - - CANADA.

Telephone 9362.

NEW YORK LIFE INSURANCE BUILDING, MONTREAL.

Bell Telephone 2209.

---

The "Canadian Electrical News" has been appointed the official paper of the Canadian Electrical Association.

---

### CANADIAN ELECTRICAL ASSOCIATION.

---

### CANADIAN ASSOCIATION OF STATIONARY ENGINEERS.

---

### ONTARIO ASSOCIATION OF STATIONARY ENGINEERS.

---

BOILER explosions are principally due to two causes, viz., defective steam apparatus and ignorant or careless engineers and firemen. The extent to which these causes exist is shown by a tabulated statement of boiler explosions for the past sixteen years, which was submitted at the annual convention of the National Association of Stationary Engineers held recently in Buffalo. The explosions during this period aggregate 3,586, involving the loss of 4,508 lives, 6,348 cases of serious injury, and damage to property in 13½ per cent. of the explosions amounting to $8,288,370. These figures speak volumes in favor of the better education of the engineer and the periodical inspection of steam producing apparatus.

*Boiler Explosions.*

In connection with the revival of the Sundar car agitation, the City Council of Toronto have obtained from prominent legal gentlemen, including Sir Oliver Mowat, Minister of Justice, the opinion that as a condition of submitting the question of Sunday cars to a vote of the citizens, the Council have the power to impose such special conditions upon the company as they may deem advisable. One would suppose that the most important consideration, especially from the standpoint of those who hold that a Sunday service is objectionable, would be to secure a properly regulated and limited service. It is surprising to observe, however, that the special condition insisted upon by the anti-Sunday car element in the Council, is that the company shall be compelled to grant a three-cent fare on Sunday. It cannot be doubted that cheap fares are an inducement to the citizens to use the cars. That being the case, is it not highly inconsistent on the part of those who declare that Sunday street car traffic is demoralizing, to compel

*Sunday Cars.*

the street railway company to offer the citizens the greatest possible inducement to patronize the cars on Sunday.

**The Canadian Westinghouse Company.** A DISTINCT addition is to be made to the electrical manufacturing industries of Canada, by the establishment at Hamilton of a branch manufactory of the Westinghouse Electrical Manufacturing Company, of Pittsburgh, Pa. All the arrangements to this end, including the purchase of the McKechnie factory building, and the granting of special privileges by the city, have been concluded, and the works are expected to be in operation early in the new year. The Canadian company will be known as the Westinghouse Manufacturing Co., Limited, of Hamilton. The names of the incorporators are :— Messrs. George Westinghouse, Henry Herman Westinghouse and John Caldwell, all of Pittsburg, manufacturers, and the Hon. James M. Gibson and Mr. Archibald E. Malloch, of Hamilton. Messrs. H. H. Westinghouse, the Hon. James M. Gibson and Archibald E. Malloch are to be the provisional directors of the company.

**A Fair Method of Appointment.** IT has been usual in the past for a great deal of wire-pulling to be done to secure the position of engineer in large public buildings. In consequence it frequently happened that the man who was best equipped by knowledge and experience failed to get the position. In other words the man with the greatest amount of "pull" got the appointment over the heads of men who were better qualified, but were no good as wire-pullers. We are pleased to observe that these conditions are not to be allowed to prevail in connection with the new city buildings now approaching completion in Toronto. At a recent meeting of the City Council a resolution was adopted that the stationary engineer to be placed in charge of the boilers and heating of the buildings be appointed after a competitive examination. It begins to appear as if those engineers who aim to secure the good positions in the future must rely less upon their ability to pull wires and more upon their knowledge of engineering.

**The Power of Niagara.** THE company which is utilizing the power of Niagara Falls on the American side has paid the Ontario government the sum of $25,000 per year for several years past for an option to the exclusive privilege of utilizing the water power on the Canadian side also. The term of this option will expire in two years, and the government are being urged not to renew the arrangement, as Canadian capitalists are prepared to utilize the power of the Canadian Fall if the opportunity of doing so is given them. It is improbable that the American company will let so valuable a privilege slip through their fingers unless they should find ample scope for their capital and ambition in connection with their huge enterprise on the American side. But if they do forfeit it, the Ontario government should consider very carefully the terms on which such privileges should be granted in the future. The success which has attended the methods employed for the utilization of the power of the American cataract permits of a more accurate estimate of the possibilities in this direction and consequently of the value which should be placed upon exclusive rights to the use of the power. There is no

doubt whatever that before many years shall have passed the means will be found of transmitting power from Niagara to the towns and cities in and adjacent to the Niagara peninsula. If it is deemed wise to grant exclusive rights to one company, the period for which such rights are to be granted should be one of reasonable limit, and the revenue received by the province should be carefully proportioned to the prospective value of the privilege. As a method of disposing of the privilege to the best advantage the government might publicly invite bids for the franchise.

**A Long-Distance Power Transmission Scheme.** THE scheme for the transmission of electric power by the Keewatin Power Co., from Keewatin to Winnipeg, which was the subject of comment last year, is not dead as might be supposed, seeing that nothing has recently been heard about it. On the contrary we learn that the Power Company have employed a competent electrical engineer to go thoroughly into the possibilities of the enterprise and to formulate plans and prepare specifications upon which tenders for the required plant may be based. Not only has this been done, but we understand that tenders have actually been submitted by the Royal, Westinghouse and General Electric Companies, and their merits are now being considered. It is rumored that the line potential proposed is between 25,000 and 30,000 volts. The distance between the terminals of the line would be 120 miles. This is one of the most ambitious power transmission schemes which has yet assumed tangible form ; its further development is therefore a subject of wide interest.

**Qualification of Wiremen.** AT the present moment, the only qualification that a man must have, in order to apply for a wiring job, is the possession of a screw driver and hammer, and the cheapest man gets it. Whenever a town is being lighted, a host of perfectly irresponsible men settle there and go about canvassing for installations, which they put in without any reference to the general wiring plan of the town, and in many cases, in the most unworkmanlike manner. The consequence is, as has been already pointed out in these columns with reference to a particular case, that no coherent scheme is possible, and that in two adjacent houses lamps will be subjected to pressures differing at any moment, by so much as three volts. In one large hotel in a small provincial town, a test shewed that lamps in two adjacent rooms that formed the terminii of two branches, had six volts difference in their pressures. This is, it is to be hoped, an unique case, but it illustrates the evils of allowing anyone to do wiring. An examining board might be formed, including representatives from the Fire Underwriters, the Canadian Electrical Association, the larger manufacturing companies, and from some of the larger operating companies, who would examine candidates as to their knowledge of wiring rules and methods, their knowledge of practical electricity, and their acquaintance with central station economies. Certificates would follow of two or three classes, shewing that the examinee had been found competent by reading or experience to (a) entirely manage a central station plant, both alternating and direct current, steam or water power, both the plant and the business ; (b) merely look after the electrical machinery, as dynamo tender and electrician, to effect repairs and make extensions ; (c) do the wiring in houses or streets. B would include C. It is not to be expected that this proposal would immediately find acceptance. Old timers, who "know it all" already, would treat it with contempt, but there are many companies who would gladly avail themselves of the assurance afforded by such a certificate of the competence of their employees, and who would even hold out higher pay inducements to their staff, in order that they might qualify themselves for it. At first, of course, it would be no proof of incompetence if a man did not hold such a document, but it would always be strong argument in his favor if he did. A board of examiners could be selected, the personnel of which had such unquestioned standing in the electrical profession that their endorsation would carry weight and disarm criticism. We shall be glad to have expressions of opinion on this matter from central station men.

We are convinced that the general level of electrical knowledge would be raised by the inauguration of such an innovation. There would, in course of time be an emulation among the younger and more ambitious members of the profession, which would lead to more careful and thorough study, and a rapid dissemination of knowledge that could not be other than beneficial both to those who operate power houses and those who draw dividends from them. The opinion that now seems to obtain, that competent knowledge of electricity is to be acquired by "induction" so to speak—that a man who works around dynamos and lamps for a couple of years is a thoroughly well posted "electrician," would become less strong in proportion as study became more general ; for of no science is the saying more true that " the more a man studies the less he knows."

THE specifications for the plumbing

Wiring Specifications. work for the new municipal buildings in Toronto, have given rise to very unfavorable criticism at the hands of some prominent manufacturers of heating and sanitary apparatus ; and we should like to suggest that the electrical manufacturing industry has an equal cause for complaint, and on similar grounds. The plumbers' complaint is that the specifications, instead of calling for apparatus of "any" make, but imposing a certain particular make as a standard of excellence, seem to impose as a condition that none but one particular make shall be supplied, thus imposing limitations that practically withdraw the matter from open competition. In the electrical work also—which, by the way, in the interests of the public, we should like to know was being planned by an electrical specialist, certain particular conduits have been named in the specifications to the exclusion of others, equally excellent. Why should this be so? Specifications should, no doubt, be so drawn up as that none but suitable machinery and apparatus could be tendered for thereon, but it seems to be hardly in the public interests to expressly favor one class of manufacture at the expense of others, when the bona fides of all manufacturers can be insured by specifying appropriate tests. And one other matter we would draw attention to—why include electrical wiring with the plumbing and heating work? What obvious connection is there between a plumber and an incandescent wireman? Would it not have been as well to make a separate contract for this work ?

Status of the Electrical Engineer in Canada. GOOD engineering is the great want of the day in electrical matters, and if works that are distinctly electrical enterprises are carried on by members of other engineering specialties, it is not because there are no electrical engineers in Canada. McGill College turns out electrical specialists who are sought for to fill the most responsible positions in the United States ; the School of Practical Science in Toronto gives a most admirable technical training ; and there are electrical courses in Kingston R. M. C., the Toronto Technical School, and other institutions ; and yet it is doubtful whether the electrical profession has even a status in Canada. The engineering of electric lighting plants, power transmissions, and railways, is placed in the hands of architects, sanitary engineers, provincial land surveyors—anybody but electrical engineers, and the results are just what might be expected. Architects and civil engineers require all their time to keep themselves posted in their own business ; and it is not to be expected that they can keep themselves up to the times in electrical matters where improvement is so rapid along many differing lines, and where the text book knowledge of ten years ago is left behind in the dark ages. Special training is necessary in order to qualify anyone to follow along the march of electrical progress, without falling far to the rear ; and it is inevitable that specialists in other branches, even highly educated civil engineers, will merely voice the ideas of some particular electrical manufacturing company, to which they have referred for technical assistance.

## PERSONAL.

The employees of the Great North-Western Telegraph Co. presented Mr. W. D. Toye with a marble clock on the occasion of his recent marriage.

Mr. Wm. A. Sweet has resigned his position with the Hamilton, Grimsby & Beamsville Electric Railway to accept the position of chief engineer of the Hamilton Radial Electric Railway.

Mr. Robt. A. Ross, who was for some years chief electrical engineer with the Royal Electric Co., and previously occupied a similar position with the Canadian General Electric Co. has commenced business in Montreal as a consulting mechanical and electrical engineer.

Mr. G. F. Cummings, an eminent electrician, was recently in Toronto in the interests of the Conduit Electrical Company, of Detroit, who are considering the question of opening a branch in this city. Mr. Cummings is well known in Toronto, being at one time in charge of the engineering department of the Edison Electric Light Works.

Mr. D. H. Keeley, Superintendent of Government Telegraphs, has returned from an inspection trip in the east, after an absence of two months. Taking the government cable steamer Newfield, he repaired the cables between the Magdalen Islands and Cape North, and between St. Paul's Island and Cape North, also the Bay of Fundy cable to Whitehead Island.

We learn that the statement we published in our last issue that Mr. C. F. Medbury had accepted a position with the Western Electric Co., of Chicago, is incorrect. Mr. Medbury is at present acting as consulting engineer to Mr. Horace Beemer, promoter of the Quebec Electric Railway Co., and to the promoters of electrical enterprises in other parts of Canada. The wish of his numerous friends is that he may find sufficient employment for his talents to warrant him in again taking up his residence in Canada.

By a special car of the Winnipeg electric street railway, twenty-four Indian children from the Brandon Industrial Institute were recently conveyed to the power house, where Mr. Glenwright, superintendent of the railway, explained to them the operation of the machinery. The children are said to have evinced great interest in their visit.

## BY THE WAY.

As an illustration of the power-consuming qualities of a modern pulp mill, it may be mentioned that the mill at Grande Mere, Quebec, requires for its operation 6,000 horse power, or an amount equal to that at present consumed by all the industries of Hamilton. What a big coal pile would be required to keep the machinery of this mill in operation, if it were not located in a country of magnificent water powers.

x x x x

I NOTICED in passing the Toronto Industrial Exhibition Grounds the other day, that the rails used for the operation of the pioneer electric railway on the Van Depœle system, by means of which ten years ago visitors were carried from Strachan Avenue to the Fair grounds, are still in position. Are they being kept on exhibition as one of the curiosities of the Fair? If not, the wonder is that the management has not before this converted them into cash at the price of old iron.

x x x x

I AM told that the British Columbia mining boom has already proved to be a good thing for the Canadian telegraph companies, whose western business has expanded considerably since the gold fever set in. It is to be regretted that owing to the lack of railway communication, our eastern manufacturers and supply houses, do not secure for their goods the British Columbia market, which of right should be theirs, and which if open to them would greatly stimulate business in eastern Canada.

x x x x

PETROLIA, the headquarters of the Canadian petroleum industry, is now lighted by electricity. Naturally enough the shareholders in the oil companies opposed the introduction of the modern illuminant, no doubt on the ground that to do otherwise would be tacidly to admit that coal oil lighting is an out-of-date method. The battle between these interested champions of old-time methods and the progressive element of the community was waged for many months, before victory for progress was achieved. Following close on the adoption of modern lighting the town has set about the construction of an extensive and costly water works system, and seems to be determined to be in every respect a thoroughly up-to-date community.

x x x x

AN acquaintance of mine, a competent electrician, who has been endeavoring to secure a situation as superintendent of a lighting station, writes me in the following vigorous fashion : " I am entirely disgusted with the electrical business. Central station owners want a first-class man to work for nothing. They expect all the best requirements for about 35 to 40 dollars per month. In fact they want a "fireman and electrician." Anything that can use the coal shovel is good enough to run a central station, and as to wiring, so long as a man can get up a pole that is all that is required. The ability to shovel coal, climb a pole, put a pair of carbons in an arc lamp and sling oil around the power house are the essential requisites." I fear that there is more truth than poetry in the above complaint. Men buy expensive machinery and put it in charge of ignorant, careless attendants, because their services can be had for less money than would be required to pay competent men. The apparent saving in salary, is the point on which their attention is fixed ; with such men frequent break-downs, unnecessary wear and tear of machinery, inefficiency of production, waste of supplies, and the

hundred and one leaks, pass unnoticed. A little careful enquiry in these directions would show that the losses arising throughout the year from these causes would suffice to pay many times over the additional sum necessary to secure the services of a thoroughly competent superintendent. My friend's letter reveals I believe one of the principal reasons why many electric lighting plants do not pay, and the sooner the owners of such plants come to recognize that profits are contingent upon good management, the sooner will the electric lighting business take a step upward, and rank with other enterprises from which satisfactory profits are derived.

x x x x

I AM reminded by a conversation I had with a citizen of Dundas the other day, that it is an unwise policy on the part of holders of public franchises not to cultivate in every possible way the good will of the citizens of the municipalities upon which they must largely depend for support. This man, who claimed to voice the sentiments of the community, complained very strongly of the attitude of the management of the Hamilton and Dundas Railway Company towards the town of Dundas. He claimed that the Company had no intention of transforming the road into an electric road, but were trying to put the town in the position of being compelled to contribute by way of bonus to the funds of the Company a sum of about $2500 per year for all time. He complained of the action of the Company in making a charge of 10 cents for small parcels which passengers could carry on their knees, and for which formerly no charge was made, and said that the lack of consideration in such particulars had given rise to a feeling of antagonism on the part of the citizens towards the railway Company, as a result of which an effort would probably be made to induce the Hamilton Street Railway Company to extend their lines to Dundas by way of the main street of the town, over which and one or two other streets the Hamilton and Dundas Company does not control the right of way. There is no doubt that if a trolley line were in operation between Hamilton and Dundas, it would at once capture the traffic and render the present steam road a valueless piece of property. As to whether the complaints preferred against the present Company are well founded, I have no means of knowing, but it is at least unfortunate for the Company that citizens of the town should feel themselves called upon to give public expression to opinions which must tend to deprive it of the good wishes and cordial support of the community. In this connection I may mention that the Company are at present putting down new rails, and it is said to be their intention to put on some new cars

---

A new design in steam engines has been patented in Canada, United States and Europe, by Cleveland and Peterson, of Brandon, Man. One engine on this pattern has been built at Ritchie's machine works, Hamilton.

The Town of Chatham, Ont., is considering the question of installing an electric light plant, to be controlled by the town, and has secured the following statistics relative to the cost of operating similar plants in other towns. Chatham, 65 lamps, all night, 23½ cents each per lamp ; Brantford, 55 lights, all night, 23 cents ; Cobourg, 23 lights till midnight, 21 cents ; Port Hope, 33 lamps till midnight, 15 cents ; Peterboro, 25 cents per light all night ; Ingersoll, 36 lights till midnight, 20 cents ; Woodstock, 70 lights till midnight, 19 cents ; Belleville, 61 lights, all night, 24 cents ; Galt, 50 lamps, 22 cents till midnight ; Hamilton, 369 lights, all night, 25 cents ; Guelph, 90 lights, all night, 24½ cents ; Owen Sound, 30 lamps, all night, 30 cents ; London, 300 lamps, all night, 25 cents.

## COSTS OF MANUFACTURE.

### By H. W. Nelson.

Somebody has said that the mountain travailed and groaned and brought forth a mouse. This might be said of the superfluous energy expended in finding the costs of manufacture. Reams of paper and myriads of figures are collected, collated, and the result is one little most unreliable observation of the actual cost. The nearest actual knowledge of the amount spent in manufacture is gained from the balance sheet, but this cannot be called the absolute truth on account of the use of unreliable data in the valuation of the inventory of goods in stock, and in a state of partial completion. The uncertainty of the inventory valuation will nevertheless figure as a small percentage of error on the balance sheet, provided the finished stock from year to year is kept at an uniform amount and valued at an arbitrary figure, providing the same figure is used for each inventory, and that the quantity of the partially finished product is kept small. If such provisos are practicable, the result obtained in any case is a very unsatisfactory one, it simply shows the total profit or loss, i.e.:—the cost on the whole manufacture. Very few companies, however, limit themselves to one uniform article of manufacture—thus their balance sheet, although perhaps showing a profit, does not show whether the profit over the cost of one article is being absorbed by the loss on another. To overcome this, a Cost Department is instituted and a multiplicity of shop orders is issued. The shop orders are supposed to furnish the correct observations, and the Cost Department to collect and collate them. The department without doubt works hard and conscientiously and probably considers that it obtains the actual figures of cost. To see the fallacy of this one has only to follow the shop orders closely on their passage to see what a snow-ball like amount of error they are accumulating. To explain by an example :—Supposing that a 50 kilowatt dynamo, direct current, is to be built ; Shop Order No. 100 is issued on the factory for the machine and all material and labour expended on it is to be charged to order No. 100, and all returns to be credited of course.

The dynamo consists essentially of a bed-plate, two pillow-blocks, armature with shaft, commutator and brush-carriage, pole-pieces and field-coils. The bed-plate and pillow-blocks are received in the rough from the foundry, and are charged correctly against the shop order, from the foundry bill price ; labour handling them to the mechanics however goes against factory expense, etc; the labor planing, boring, drilling, is charged correctly ; then bolts being necessary, are ordered from the store-

ordered for the bearings ; only part is used, the rest remains in the melting-pot, and is used on another order.

Take the armature and its accessories :— The laborer's work for handling is charged to the general dumping order, viz ; Factory Expense, or worse still, the laborers make out time tickets on which they guess the apportionment of the time they have spent on different shop orders, to them meaning-

### PLATE I.

*50 Kilo-Watt 125 Volt Generator. Material*

| Description of parts | No. of pieces | Weight in lb. rough | Price per unit | Cartage handling etc | Cost lb machined | Weight finished | Credit scrap remelted |
|---|---|---|---|---|---|---|---|
| **Bed Plate** | 1 | 1210 | | | | | |
| Sub-plate | 2 | 300 | | | | | |
| Bolts | 4 | — | | | | | |
| **Pillow Blocks** | | | | | | | |
| Caps | 2 | 600 | | | | | |
| Cap bolts | 2 | 75 | | | | | |
| Oil rings | 8 | 5 | | | | | |
| Babbitt metal | 2 | 20 | | | | | |
| Bolts | 8 | | | | | | |
| **Armature** | | | | | | | |
| Front plate for body | 1 | 10 | | | | | |
| Back " | 1 | 10 | | | | | |
| Reten. bolts | 16 | — | | | | | |
| Discs for body | 360 | 800 | | | | | |
| Insulation | — | 10 | | | | | |
| Paper | — | 50 yards | | | | | |
| Cloth | — | 2 yards | | | | | |
| Varnish | — | 1 gal. | | | | | |
| Mica | — | — | | | | | |
| Wire | — | 800 | | | | | |
| Tape | | 5 | | | | | |
| Commutator | | | | | | | |
| Front ring | | 50 | | | | | |
| Back " | | 75 | | | | | |
| Cylinder | | 100 | | | | | |
| Bolts nuts | 32 | — | | | | | |
| Screws | 16 | — | | | | | |
| Bars | 50 | 65 | | | | | |
| Mica | | 15 | | | | | |
| **Shaft** | | | | | | | |
| Nuts & washer | | | | | | | |
| Washers | | | | | | | |
| Flanges | | | | | | | |
| **Pulley** | | | | | | | |
| Set screw | | | | | | | |
| Key | | | | | | | |

less numbers. The armature winder commences work with a full or possibly a superfluous equipment of material, which is charged up correctly ; his foreman is perhaps obliged to take him off this job temporarily (a frequent necessity in shops) to rush through another armature, the man is on piece-work and needs material in a hurry, uses some of that supplied on the first order and fails to report the matter. The first order thereby becomes charged at completion with a superfluity of material and the second one not charged sufficiently. To enumerate more errors of observation would make this article too long and tedious ; practical men will recall dozens and dozens of instances to illustrate the point. The argument does not hold good that these and similar faults show up bad foremen. To follow up all the details of all the shop orders in work, closely and accurately, would require a force of clerks in the factory, and it cannot be expected that foremen and workmen alone, who are not trained to clerical work, can follow or grasp the importance of the matter.

The suggestion occurs, why not take one individual piece of work of each kind and have a clerk familiar with shop work follow it through from start to finish ? Very good ; but this is only one observation and is liable to variation, and what is wanted is an average taken over a large number of pieces of one particular kind or size. This brings the matter back to the old shop orders which have been shown to be uncertain and unreliable, and which also demand a large amount of clerical labor continually.

A method of approximations is in force in one of the large American shops, which has all the appearance of being a good one both mathematically and practically, as will now be shown.

### PLATE II.

*Steel iron castings. Weight of rough pieces as shown on foundry bill.—*

| Name of casting | 1 Kw | 2 Kw | 5 Kw | 7½ Kw | 10 Kw | 11 Kw | 15 Kw | 20 Kw | 30 Kw | 40 Kw | 50 Kw | 80 Kw | 100 Kw | | | Date Noted |
|---|---|---|---|---|---|---|---|---|---|---|---|---|---|---|---|---|
| **Bed Plate** | | | | | | | | | | | 1205 | | | | | Noted |
| | | | | | | | | | | | 1190 | | | | | " 10 |
| | | | | | | | | | | | 1260 | | | | | " 11 |
| | | | | | | | | | | | 1199 | | | | | " 14 |
| | | | | | | | | | | | 1201 | | | | | " 16 |
| | | | | | | | | | | | 1209 | | | | | " 18 |
| *Average weight* | | | | | | | | | | | 7259 | | | | | |
| | | | | | | | | | | | 1209 | | | | | |
| **Pillow Blocks** | | | | | | | | | | | | | | | | |
| *Average weight* | | | | | | | | | | | | | | | | |

that different bolts are better, which are thereupon substituted, and the discarded ones are forgotten (a common error among the best of foremen) and remain with the oddments on the mechanics bench. Shop Order No. 100 is thereby overcharged. Later on possibly the discarded bolts apply on another job and shop order, and are not charged. Then again 25 lbs. of Babbitt's metal is room on the shop order ; the superintending engineer decides

It may be mentioned at once that the working out of such a scheme must be left entirely in the hands of the Works Manager, and must not be left in the hands of clerks, as they generally lack the judgment necessary in this regard, or even to accountants, however capable they may be as accountants. The latter from the nature of their training are wedded to exact balances, and would shudder at an approximation.

To start with, and to stick to the example shown above: the factory is engaged in building direct current dynamos and motors, and the cost of manufacturing is required. Lists of the constituent parts of the different sizes of machine, classified under

the headings of their essential parts must first be prepared (see Plate I and Plate II). The weights in the rough on Plate I must be taken from the best average that can be obtained from Plate II. It should be the duty of the receiving clerk to keep Plate II (which is a first-class birds-eye view for the manager, over the foundry weights) posted up to date.

The drawings and engineers' data will give the theoretical quantities of other material, and a few careful observations of the amounts actually used must be made under the personal supervision of the Works Manager. Space will not permit a more elaborate illustration of these details.

The labor is the next thing to consider : there is the labor which is strictly productive, and there is that which is so often dumped into a general expense account. In the first, if the whole shop is on piece work, finding the cost is more or less a simple matter ; if otherwise, reliable figures must be obtained from an average over many observations. Taking the latter case, it is advisable to commence with, to get one good observation on several different sizes of the same type, and plot them on cross-section paper as shown in Plate III. This gives a good basis on which to approximate the time taken on intermediate sizes, viz ; by simply following the horizontal lines to the curve A.B., and reading on the perpendicular lines the number of hours as shown. Experience has proven that an approximation such as this is not merely guess work, but differs very slightly (a fraction of a percentage) from the afterwards ascertained actual facts. Theoretically the curved line A.B. should be the dotted straight line A.B., but there are practical considerations which prevent it. This curve is only a preliminary one, and fresh observations must be made from time to time to check it, and to demonstrate the fact of its deviation from the dotted line A.B., being due to good practical reasons.

Next comes the labor which is so often confounded with expense. For instance, handling, painting, extra chippings on castings, helpers to operatives, etc. There seems to be no legitimate reason for not classing this in with the productive labor ; it is a labor cost on the machine, and can be accounted for directly. This labor can be approximated in the same manner as the purely productive labor as shown by Plate III. It is, however, a some-

what more variable quantity, and requires the average of several very close observations on the extreme, and two or three intermediate sizes before plotting the curve. Works Managers will at once see the strength of this caution, recognizing as they do, that this class of labor is one of the "little foxes" which play such sad havoc with the cost of production.

Finally the general expense of the factory is to be accounted for. In the Factory General Expense account care must be taken to EXCLUDE all expenses of the selling offices, advertising, and freight on finished product, in short all expenses outside the factory gates. Clearly understanding this, the method of apportioning to material and labor as above, the true proportion of factory general expense is very simple. It is understood, of course, that under the heading of material is included freight, cartage, duty and other dues ; under labor is included all operatives, and their helpers' pay, laborers handling the raw material or product in any form ; in brief, the pay of all persons handling the material inside the factory gates, otherwise than storekeepers. Factory General Expense includes managers, foremen, clerical help, storekeepers, expenses of light, heat and power, rent, insurance and other charges incidental with running the factory. It is readily seen by this that the factory general expense bill is practically a fixed quantity, whereas the Labor bill is a quantity varying with the busy and slack seasons. Expense bears to Labor a definite relation, and on the contrary has no relation to Material. To explain this more fully :—Factory General Expense is the salary of the personnel and the incidental running expenses as shown above, and can be lowered to a very slight extent, if to any, during slack seasons, while on the other hand, the number of productive employees has to be increased or decreased according to the work on hand. Thus in a busy season the factory general expense may bear the relation of five dollars expense to five dollars of operatives' wages, in a slack one the five dollars expense remains fixed, whereas the operatives' wages bill is only one dollar.

All that is necessary now is to get an exact statement from the Accounting Department of the charges to Factory General Expense

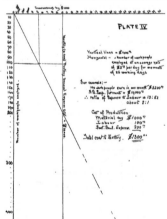

over a definite period, the longer the better, and make a diagram as in Plate IV.

Plate IV not only demonstrates the fact without figuring that the Factory General Expense piles up the actual cost of production considerably higher in slack than in busy seasons, but it gives the manager a simple and ready-reckoner in figuring against a close call from competing firms and saves the nerve-racking uncertainty of adding arbitrary percentages for the safety margins.

The above, although badly cramped from its necessary conciseness, should be a fairly lucid demonstration of a simple and cheap method of "cost making," which eliminates almost entirely the personal equation, viz : the arbitrary percentage margins, and unsystematic guess-work allowances so very much in vogue in many of the most up-to-date companies.

The figures herein shown are strictly systematic and scientific, and it cannot be contended that such approximations are in any way guesses.

## THE STEAM ENGINE.

### By JOHN C. GOUGH.

THOSE of us who have made for ourselves a toy engine (and I am inclined to believe that the number will be found not inconsiderable), and have watched it doing its whirling revolutions of some 2,000 or 3,000 per minute, might naturally be led to the conclusion that there was no limit to the velocity with which steam can travel, or to the speed with which it can enter and pass out of the cylinder, and its motion be changed or reversed. Such a notion would be, however, erroneous in the extreme, as the speed at which steam travels on its way to or from the cylinder of a steam engine is an important factor in determining the best size of the steam passages, ports, etc. It has been found by experiment that so long as the steam in a pipe is not required to travel at a higher rate of speed than 100 ft. per second, or 6,000 ft. per minute, the loss from friction, etc., in the pipes and passages is not a serious one, and the pressure at which it can be obtained on the piston is not much reduced from the boiler pressure. Beyond this speed, however, serious loss of pressure takes place, and hence the importance of having the steam ports so large that the velocity referred to may not be exceeded.

Those engineers who have had much experience in the over-hauling of steam engines, are aware that in a large number of old engines the ports and passages in the cylinders are very much too small. In a case which recently came under the notice of the writer, the area of the steam port was not more than one-thirtieth of the cylinder area, and the loss from wire-drawing, poor vacuum, etc., was in consequence very considerable, although the piston speed was not a high one.

When the steam ports and passages of an engine are too crippled, there is, in addition to the loss from wire-drawing, etc., a further loss of power from the increased resistance to the exit of the steam from the cylinder, causing what is known as "back-pressure."

In condensing engines, even though there may be a plentiful supply of cold water and good "vacuum" in the cylinder, when the exhaust passages are tortuous, and contracted, an exceedingly poor effect of the vacuum will be obtained in the cylinder.

Perhaps, as good an illustration of this as could be obtained is shown in the diagrams (Figs. 1 and 2) taken from a condensing

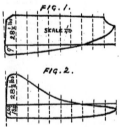

engine a few months ago. The engine was furnished with a cut-off valve, by means of which the steam could be cut off at any desired portion of the stroke. In Fig. 1, it will be observed, the steam is admitted during about nine-tenths of the piston's stroke, and when released, escapes with such difficulty through the contracted exhaust passages, that the back-pressure exceeds that of the atmosphere during three-tenths of the return stroke, and, at the termination of the stroke, when the full effect of the vacuum is usually obtained in the cylinder, there is an absolute back-pressure of something like 6 lbs.—that is to say, to use the common expression, the "vacuum" is only 9 lbs.

Fig. 2 was taken from the same engine and immediately after Fig. 1, when driving a much lighter load, and the steam cut off about ⅙th the stroke. The quantity of steam to be condensed is here so much less that a "vacuum" of 13½ lbs. is now shown in the cylinder, and although the effect of the restricted exhaust is

still seen at the commencement of the exhaust, a very much greater effect is obtained from the vacuum throughout the stroke than under the conditions shown in Fig. 1. Hence we see, that besides the loss of pressure in admission when the ports and passages are too small, an even greater loss is experienced from the back-pressure, resulting from the resistance to the egress of the steam.

Many rules have been given for determining the size of steam ports, etc., but none can be considered as other than empirical which does not take into account the fact above mentioned, viz., that the velocity of the steam in following a piston should not in the passages exceed 100 ft. per second, or 6,000 ft. per minute. Bearing this in mind, we shall see that the area of the steam ports should have the same ratio to the area of the cylinder that the piston speed in feet per minute has to reach 6,000, from which we easily deduce a rule, thus :

$$\text{Area of port} = \frac{\text{Area of Cylinder} \times \text{Piston speed in feet per minute}}{6,000}$$

A rule in very general use amongst engineers and to be found in most engineering pocket-books, is to make the area of steam port equal to 1/16th the area of the cylinders. Let us see how this corresponds with the rule we have just found. Take for example, say, an engine having 5 feet stroke running 50 revolutions per minute : that is to say, with a piston speed of 500 feet per minute. By our rule we have,

$$\text{Area of port} = \frac{\text{Area of Cylinder} \times 500}{6,000} = \frac{\text{Area of Cylinder}}{12}$$

so that we see for 500 ft. piston speed, the common rule for making the steam ports equal to 1/16th the cylinder area would give too small a passage, the port requiring to be not less than 1/12th the cylinder area, in order not to lose pressure and power from the entering steam having too high a velocity. If the piston speed were only 400 ft. the port might be made only 1/15th area of cylinder, because

$$\frac{\text{Area of Cylinder} \times 400 \text{ ft. per minute}}{6,000} = \frac{\text{Area of Cylinder}}{15}$$

and so on if the piston speed were still further reduced ; the common rule would give a larger steam port than is necessary. In order to avoid the great loss experienced by restriction of the exhaust, as shown in the above diagrams, it is customary to make the exhaust ports about half as large again as the steam ports, so that the egress may be as free as possible, and the back-pressure on the piston reduced to a minimum.

There being then so much advantage in having large passages to and from our cylinders, the consideration naturally arises as to what conditions determine the limits in the opposite direction, or can we really make the steam passages and steam ports too large? It is well known that too much "clearance" space at the ends of the cylinder means loss of steam at each stroke of the piston, and in the same way an excessive capacity of steam port or passage would mean waste, through more steam having to be thrown away in exhausting than is really necessary. A further consideration respecting the disadvantage of having too great width of area of ports is found in the cooling influence of the exhaust steam, the temperature of which, at or below the atmospheric pressure, is very much lower than that of the fresh steam at its entrance to the cylinder. The passage of the exhaust steam through the ports as the high-pressure steam has to pass through, abstracts a considerable amount of heat from every part of the metal with which it comes in contact, and this heat has to be supplied again by the entering steam at each stroke. The question, therefore, of having the steam ports of an engine a proper size, neither too small, so as to cause loss by wire-drawing, back-pressure, etc., nor too large, so as to give rise to loss from cooling, etc., it will be seen is one which has a material influence on economical working. The rule here given is a practical one of long personal experience and one which the writer can recommend.

The steam ports in Corliss engines are, in usual practice, 1/12th, and exhaust ports ⅙th of the piston area. The piston speeds, of course, govern the application of those rules in nearly every instance.

MONTREAL, QUE., 12th Oct., 1896.

The Spokane and British Columbia Telephone and Telegraph Co. propose constructing an international telephone line between Washington and British Columbia. Arrangements have been made for connection at Spokane with the Inland Telephone Company, and on the British Columbia side with the Vernon and Nelson Telephone Company's system.

## CANADIAN ASSOCIATION OF STATIONARY ENGINEERS.

NOTE.—Secretaries of Associations are requested to forward matter for publication in this Department not later than the 25th of each month

### TORONTO NO. 1.

TORONTO No. 1 held their regular fortnightly meeting in Engineer's Hall, 61 Victoria street, on Wednesday evening, 21st ultimo, at which one candidate was initiated. After the regular routine of business, Bro. Cross gave an interesting "chalk talk" on figures, and was presented with a hearty vote of thanks.

The Library Committee reported that the first of a series of book cases was on view in the library room, and that upwards of 100 volumes were already on the shelves. They wish to thank those who have been kind enough to donate to this work, and to state that they would be pleased to hear from the manufacturers or others who feel inclined to assist this branch of the Association's good work.

The Banquet Committee reported that they had made arrangements for the tenth annual banquet, to be held at the Palmer House on November 25th (Thanksgiving Eve). This hotel is centrally located, at the corner of King and York streets, and the dining hall has facilities for accommodating 300 guests. The secretary of the committee hereby extends a hearty invitation to all members of the C. A. S. E., and hopes that in case written invitations are not received all officers and members will consider the above invitation sufficient.

G. C. MOORING, Sec. of Com.
15 Charlotte street.

### QUESTIONS AND ANSWERS.

A. B. C., Ontario, writes : "Please give the horse power or friction load of a 100 horse power high speed engine belted direct to a slow speed alternator of 1000 lamps capacity ; also the horse power required to charge two miles of primary line (without any lamp on) and transformers of 1000 lamps capacity."

ANSWER.—It is impossible to give quantitative answers to your questions. The friction load of an engine belted to a generator, depends on many very variable quantities, such as condition of bearings, state of belt, make of engine and generator, etc. About ten horse power, however, should not be exceeded, on running everything empty in your size of plant. Your second question is even more impossible than the first. By the expression "charge" when you ask the horse power required to "charge" 2 miles of line and transformers, we think you mean, that if a primary line two miles long has transformer primaries connected up to a capacity of 1000 lights, and the secondaries are left open ; then how much current will the ammeter indicate? In other terms, you want to know the magnetizing current of transformers to a capacity of 1000 lamps. There is no rule for this ; the magnetizing current depends, in amount, on the design of the transformer, the materials of which it is constructed, its capacity ; a large transformer will generally require a smaller magnetizing current, in proportion, than will a smaller one of the same make ; and you will find that transformers of to-day, which embody the principles of design and construction which have been arrived at by careful scientific investigation, will prove considerably more economical on this point than are those which were made a few years ago, before transformers had received the study which has been devoted to them during the past two or three years. You will also find that the general idea that a transformer is the easiest thing in the world to make, is erroneous. We are sorry that our answers are necessarily so very indefinite, but the questions themselves do not admit of anything more so. The manufacturers of your transformers have probably made observations of the magnetizing current, and might give you information.

### ELECTRICAL DEVELOPMENT IN EUROPE.

L'INDUSTRIE Electrique publishes the following figures in regard to electrical development in Europe : There are 560 miles of electric roads in Europe, which is an increase of 125 miles in one year. The number of electric cars has increased from 1236 to 1747 in the same time. Germany has 250 miles of electric roads and 857 motor cars. France has 82 miles and 225 motor cars. Great Britain has 65 miles, with 168 cars, and Austria-Hungary has 45 miles, with 157 cars. Next comes Switzerland, Italy, Spain and Belgium, in the order given, while Russia has but one electric railroad, with 6 miles of track and 32 motor cars, and Portugal ends the list with 1⅞ miles. Of the 111 European lines 91 are overhead trolleys, of which there were 35 in Germany, 12 in Switzerland, 10 in France, and 7 each in England and Italy, and 6 in Austria-Hungary, etc. Of electric railroads with underground current there were but three at the beginning of this year, one each in England, Germany and Hungary. Nine lines are provided with an insulated central track, through which the current is conducted, eight of these railroads being in Great Britain and one in France. The remaining eight lines are provided with accumulators. Of these, four are in France and two in Austria, and one each in England and the Netherlands.

### SPARKS.

At the annual meeting of the Nanaimo, B. C., Telephone Company, held early in October, the following directors were elected ; Messrs. J. C. Armstrong, O. Plunkett, M. Bray, E. Pimbury and G. Norris. Subsequently G. Norris was chosen president ; J. C. Armstrong, vice-president ; W. K. Leighton, secretary and collector, and G. E. T. Pittendrigh, manager.

The Royal Electric Company are installing for the Sussex Water and Electric Company, Sussex, N. S., one of their 40 k. w. "S.K.C." two phase generators, with 360 16 c.p. lights capacity in transformers, and are wiring up the town. The Sussex Company are supplying both arc and incandescent lighting, as well as motors from the same dynamo and circuits. With the continued improvement in alternating arc lamps, it is now quite feasible to do this. A number of plants are now in operation furnishing arc and incandescent light, as well as motors from the same dynamo and circuits ; excellent results are reported. The use of motors makes it possible to run the plant the full 24 hours with increased capacity.

The shareholders of the Merchants' Telephone Company, Montreal, held their annual meeting on the 7th of October, Mr. F. X. Moisan, the president, in the chair. The annual report stated that two hundred and thirty-one new telephones had been placed during the past year, making a total of over 900 now in operation. The treasurer's report showed a surplus of $19,000. It was decided to negotiate a further loan of $45,000 to extend operations. The following gentlemen were elected directors : F. X. Moisan, L. E. Beauchamp, J. E. Beaudoin, A. S. Delisle, G. N. Ducharme, L. H. Henault, M. T. Lefebvre and F. Dagenais. At a subsequent meeting of the directors, Mr. F. X. Moisan was re-elected president, Mr. A. S. Hamelin, vice-president and L. E. Beauchamp, treasurer.

## GAS CYLINDER EXPLOSIONS.

A GOVERNMENT committee in England has made official inquiry, and found that, of nineteen cases of gas cylinder explosions in different parts of the world, four were due to carelessness, one from mixed gas or vapor due to improper compressing arrangements, four to bad cylinders, three either to bad cylinders or to an excessive pressure due to overcharging, one due to ignition from oil, and one for which no cause could be assigned. The committee recommends that in the case of cylinders of compressed gas—that is, oxygen, hydrogen or coal gas—and of lap-welded wrought iron, a greatest working pressure of 120 atmospheres, or 1,000 pounds to the square inch, and the stress due to working pressure not to exceed six and one-half tons to the square inch ; proof pressure in hydraulic test, after annealing, 224 atmospheres, or 3,360 pounds to the square inch ; permanent stretch in hydraulic test not to be more than 10 per cent. of the elastic stretch ; and one cylinder in fifty to be subjected to a statical bending test, and to stand crushing nearly to flatness between two rounded knife edges without cracking. In the case of lap-welded or seamless steam cylinders, the greatest working pressure is fixed by this committee at 120 atmospheres, or 1,800 pounds to the square inch, carbon in steel not to exceed 0.25 per cent., or iron to be less than 99 per cent.; tenacity of steel not to be less than 26, nor more than 33 tons to the square inch.

## BOILER SCALE AND STEAM EFFICIENCY.

DISCUSSING the subject of boilers and feed water recently, Professor F. B. Crocker made some terse remarks on the subject. The water used in steam boilers is obtained either from the regular city water supply or from some source such as a pond, river, or well. Which of these is best to employ depends upon the circumstances in each particular case, but in almost every instance the question of the purity of the water is an important matter. Almost any water available for use in boilers contains from 10 to 100 grains of solid material per gallon, and since a 100 horse-power boiler evaporates about 30,000 pounds of water per day of ten hours, or about 400 tons per month, the accumulation of this material becomes very considerable, assuming only half of it to be deposited. Impurities of water are of two distinct kinds : First, small particles of solid material mechanically held in suspension, the presence of which is perfectly evident to the eye, forming what is called, in plain language, muddy or dirty water. The other class of impurities are mineral substances dissolved in water, producing little or no change in its appearance or transparency. Impurities of the first kind can be removed by filtering, or by simply allowing the suspended particles to settle ; but impurities actually dissolved in the water can only be eliminated by some process of chemical or physical precipitation. The so-called " hard water " is simply water containing compounds of lime, magnesia, etc., in solution, which are particularly objectionable in water for boilers, since they are deposited as a scale or incrustation upon the interior, and seriously interfere with the transmission of heat through the metal, thereby reducing the efficiency of the boiler and also introducing a danger that it will become excessively heated and weakened. These deposits in boilers sometimes reach a thickness of half an inch or more, and are extremely troublesome and difficult to prevent or to remove after they have formed.

It is estimated that scale one-sixteenth of an inch thick necessitates the use of about 10 per cent. more fuel, one-fourth inch almost 40 per cent more, and half to three-quarter inch scale actually doubles the amount of fuel required to generate a given quantity of steam. These facts, and the greatly increased repairs and danger arising from scale in boilers, show the great importance of eliminating it.

## THE EFFECT UPON THE DIAGRAMS OF LONG PIPE-CONNECTIONS FOR STEAM ENGINE INDICATORS.

IF an indicator is to be relied upon to give a true record of the varying pressures and volumes within an engine cylinder, its connection therewith must be direct and very short.

Any pipe connection between an indicator and an engine cylinder is likely to effect the action of the indicator ; under ordinary conditions of speed and pressure, a very short length of pipe may produce a measurable effect in the diagram, and a length of three feet or more may be sufficient to render the cards valueless except for rough or approximate work.

In general, the effect of the pipe is to retard the pencil action of the indicator attached to it.

Other conditions being equal, the effects produced by a pipe between an indicator and an engine cylinder become more pronounced as the speed of the engine is increased.

Modifications in the form of the diagram resulting from the presence of a pipe are proportionately greater for short cut-off cards than for those of longer cut-off, other things being equal.

Events of the stroke (cut-off, release, beginning of compression) are recorded, by an indicator attached to a pipe, later than the actual occurrence of the events in the cylinder.

As recorded by an indicator attached to a pipe, pressures during the greater part of expansion are higher, and during compression are lower, than the actual pressures existing in the cylinder.

The area of diagrams made by an indicator attached to a pipe may be greater or less than the area of the true card, depending upon the length of the pipe ; for lengths such as are ordinarily used, the area of the pipe-cards will be greater than that of the true cards.

Within limits, the indicated power of the engine is increased by increasing the length of the indicator pipe.

Conclusions concerning the character of the expansion of compression curves, or concerning changes in the quality of the mixture in the cylinder during expansion or compression, are unreliable when based upon cards obtained from indicators attached to the cylinder through the medium of a pipe, even though the pipe is short.— W. F. M. Goss, in Scientific Machinist.

The attention of those of our readers who are desirous of becoming acquainted with the principles and applications of Roentgen rays and phenomena of the anode and cathode, is directed to a book on this subject by Edward P. Thompson, M. E., E. E., New York, and recently published by Messrs. D. VanNostrand & Co., of the same city. This book, consisting of 200 pages, reviews the history of investigations and experiments in connection with the electric discharge from the time of Faraday, Davy, Page and others, and treats of the variety of purposes to which our present knowledge of the subject may be applied. It is suggested that the study of the subject might well have a place in the curriculum of scientific schools. The book, which is illustrated with numerous engravings, sells at $1.50 per copy.

## SPARKS.

The Metropolitan street railway is being extended to Richmond Hill.

The London Street Railway Company are putting in a new 536 horse power engine in their power house on Bathurst street.

An electric light plant has been installed at Trail, B. C., consisting of an alternating dynamo of 1000 16 c.p. lights and an arc machine of 25 lights.

The dispute between the city of Winnipeg and the Street Railway Company has been finally settled, and arrangements are now being made to extend the line.

The Asbestos and Danville Railway Company, of Danville, Que., will apply to parliament for authority to build an electric railway from Danville to Asbestos.

George Hunt, proprietor of the St. Lawrence Machinery Supply Company of Montreal, is reported to be offering his creditors ten per cent. on claims amounting to $3,100.

The Montreal Park and Island Railway Company expect to have their line completed to Lachine this fall. The power houses at St. Laurent and Lachine are also nearing completion.

Judgment for $700 and costs was recovered by Mr. Nelles, ex-manager of the Hamilton, Grimsby and Beamsville Railway, in his suit against the company for alleged wrongful dismissal.

A company of Brantford capitalists are promoting a scheme for the construction of an electric railway from Brantford to Paris and Ayr, and probably to Galt. A charter for the road is held by the promoters.

The St. Hyacinthe and Granby Railway Co., of St. Hyacinthe Que., is seeking incorporation with a capital stock of $100,000, to build a steam or electric railway, between Bingham, Brome County and St. Hyacinthe.

There is now in course of construction for the Ottawa Street Railway Company an electric locomotive for hauling lumber from the yards of Messrs. W. C. Edwards & Co. to the Canada Atlantic Railway after the hours of the regular passenger service.

The Chateau & Northern Electric Railway Co. will shortly complete their electric road from the city of Montreal to Bout d'Isle, a distance of twelve miles. An initial trip over the section between Maisonneuve-and Point Aux Trembles, where the power house is located, was made a few days ago. Four cars have arrived from the Canadian General Electric Co., Peterboro', and two more are shortly expected.

The announcement has been made within the past few days that an international syndicate of capitalists has secured control of the largest tramway in London, England. Mr. Wm. McKenzie, president of the Toronto Street Railway Co., who some weeks ago was given a franchise for an electric railway in Birmingham, is at the head of the syndicate, and with him are associated Mr. James Ross, of Montreal, and several street railway capitalists of New York, St. Louis and Philadelphia.

The quarterly meeting of the shareholders of the Hamilton, Grimsby and Beamsville Electric Railway Company was held on the 2nd inst., when the statements of the secretary-treasurer, Mr. Adam Rutherford, for the six months ending September 30 were accepted. Mr. Rutherford was voted stock in the railway to the value of $2,000 for his services up to March, 1894. The Beamsville extension of the road has been completed, and was placed in operation a few days ago. The annual meeting of the company will be held on the fourth Monday in January.

Arrangements have been completed by the Westinghouse Air Brake Manufacturing Company, of Pittsburg, for the establishment of a branch factory in Hamilton, Ont., and a new Canadian company will be organized, to be known as the Westinghouse Manufacturing Co., Ltd., of Hamilton. The capital stock will be $500,000. The names of the applicants are: George Westinghouse, Henry Herman Westinghouse and John Caldwell of Pittsburg, and Hon. J. M. Gibson and A. E. Malloch, of Hamilton. It is proposed to manufacture electrical appliances and air brakes for railways, switches, etc.

Tenders for a telephone franchise for the city of Toronto, were opened a fortnight ago. Tender No. 1 was by George Mussol, who offered to establish the system described as automatic, giving absolute secret connection and a continuous service at nights. No. 2 tender proposed that the Citizens Telephone Company of Toronto (to be incorporated) would furnish The Wilhelm Telephone Company System of Buffalo. No. 3 was by Clark, Gowes & Co.,

on behalf of a client. Tender No. 4 was from Messrs. Beauchemin, Montreal, offering to furnish a service on the basis of the Merchants' Telephone Company of Montreal, at $25 each.

The paragraph which appeared in our October number relative to the proposed electric railway at Quebec, was slightly inaccurate. We are advised that Mr. Beemer, the original promotor and owner of the franchise of the railway, has concluded an arrangement with a company for the construction of the road, which when completed and put in operation, will be taken over by Mr. Beemer.

The following figures show the prices paid for electric lighting by a number of Ontario towns :

| | | | |
|---|---|---|---|
| Belleville | 2,000 candle power. | $127.75 per light. | |
| Brampton | 2,000 " | " . | 64.00 " |
| Chatham | 2,000 " | " . | 85.27 " |
| Clinton | 2,000 " | " . | 66.00 " |
| Cobourg | 2,000 " | " . | 62.50 " |
| Dunnville | 2,000 " | " . | 60.82 " |
| Galt | 2,000 " | " . | 80.30 " |
| Guelph | 2,000 " | " . | 89.42 " |
| Ingersoll | 2,000 " | " . | 73.00 " |
| London | 2,000 " | " . | 108.58 " |
| Meaford | 2,000 " | " . | 55.00 " |
| Niagara Falls | 2,000 " | " . | 85.00 " |
| Owen Sound | 2,000 " | " . | 90.00 " |
| Simcoe | 2,000 " | " . | 73.00 " |
| Toronto | 2,000 " | " . | 108.58 " |
| Wallaceburg | 2,000 " | " . | 75.00 " |
| Welland | 2,000 " | " . | 57.20 " |

A company has been organized in Peterboro to install and operate a plant for the supply of electric power. The members of the company, Messrs. W. H. Meldrum, John Carnegie and James Kendry are well known and enterprising citizens, and in their hands the undertaking should have results at once beneficial to the town in enabling it to offer to manufacturers the advantage of cheap electric power, and at the same time the returns in a financial way should be satisfactory. The power site is located at Auburn, giving a maximum transmission distance of four miles to the farthest point at which power is to be supplied. The initial installation consists of a 250 K.W. slow speed 3 phase generator of the Canadian General Electric Co's. latest type. This machine will run at the very slow speed of 200 revolutions, and will be direct coupled to the line shaft, thus saving the loss in a belt transmission. Current will be distributed at the transmission voltage, 2080 volts, directly to all the motors of more than 50 horse-power capacity. For smaller motors step down transformers will be used to reduce the pressure to 115 volts. Contracts have already been secured by the Company for some 250 or 300 horse power and it is considered likely that the installation of a second generator will be necessary in the near future. The plant is to be in operation on the 1st of January next.

The North Shore Power Company of Three Rivers, Quebec, have secured a franchise from the City of Three Rivers to supply incandescent and arc lamps as well as to pump the city water. The corporation of the City of Three Rivers installed a municipal lighting plant, but have turned it over to the North Shore Power Co., who are going to operate it with power generated on the Batiscan River at Batiscan Chute, and convey the same to Three Rivers, a distance of 16 miles. This Company have purchased from the Royal Electric Co., two of their S.K.C. 2 phase generators, with a capacity of 240 K.W. each. It is the intention to generate the current at their water power, using step-up transformers, bringing the pressure up to 11,000 volts, and at Three Rivers step-down transfromers will be used to reduce the pressure to a working pressure of about 1000 volts, where it will be connected to the present lighting circuits that were turned over to the North Shore Power Company by the corporation of Three Rivers. The flexibility of the system being put in by the North Shore Company is being well demonstrated by the fact that the step-down transformers are located in the old lighting station at Three Rivers, and that the present circuits for incandescent lighting will be directly connected to the step-down transformers, and that the expense in making the change in the Three Rivers station is practically nil. The transformers in use for about 3000 lights already installed are of the Royal type of 16000 alternations, and as this is also the periodicity of the 2 phase generators being installed, no change on their lines or transformers is necessary. The corporation of Three Rivers had in operation one arc dynamo of 50 lights, and one with 30 lights capacity. It is intended to drive these two arc machines with one of their single phase alternators which have been in use a number of years there, and which will be coupled in one side of the 2 phase circuit and driven as a synchronous motor.

## SPARKS.

Mr. John Davidson, of Smith's Falls, Ont., has invented an electric heater.

Improvements have recently been made to the electric light station at Sherbrooke, Que.

An electric plant for lighting the town of Hull is being installed. It will be located at Deschenes Mills.

The incorporation is announced of the Amherstburg Electric Light Company, with a capital of $20,000.

An eastern syndicate is endeavoring to secure control of the New Westminster, B. C., electric light plant.

It is rumored that a rival telephone company will shortly commence business at Halifax, but the report lacks confirmation.

Prospects are said to be favorable for the conversion at an early date of the Hamilton and Dundas Railway into an electric road.

Mrs. E. Bradley, of Lynchburg, has begun an action against the Hamilton Radial Electric Railway Co. for $5,000 damages for the death of her son.

The Northern Electric Railway Company is applying to the Quebec legislature for incorporation, to build an electric railway from Montreal to St. Jerome.

The Ottawa Car Co. is now constructing a combined passenger, baggage and express car for the Ottawa Electric Street Railway Co. It will be 40 feet in length 13 feet longer than the ordinary passenger cars, and will have accomodation for 36 passengers and run on eight wheels.

The new exchange of the Bell Telephone Company at Winnipeg was recently opened. It is said to be one of the most complete systems in the Dominion, and was designed by Mr. J. A. Baylis, the company's expert. Prior to leaving for Montreal, Mr. Baylis and his assistants were tendered a complimentary dinner.

The Peterboro' Town Council has accepted the offer of the Peterboro' Light and Power Co. to supply the town with eighty-five 2,000 candle power arc lamps for the sum of $65 per arc lamp per year, they to pay the sum of $400 per year as rental for the use of the streets. The new contract is to be for the term of seven years from January 1st, 1897. Additional lamps over eighty-five are to be charged at the rate of $60 per year.

Daniel McAuley, a young mechanical engineer of Port Morien, C. B., has patented an invention to prevent boiler explosions. It is a steam boiler pressure indicating alarm, which is set to go off when the pressure on the boiler has reached a point over which it ought not to go, much the same as the engineer sets his alarm clock for five in the morning. If the steam valve is out of order, as often happens, no explosion can occur, because this patent will give the alarm.                    →

Arrangements have been made for the development of the water power of the Pend d'Oreille river, in British Columbia. A power station will be located at the mouth of the river, about twelve miles from Rossland. The plan contemplates the construction of a dam, from which the water will be conducted in steel flumes and delivered to the water wheels. It is claimed that 10,000 h.p. can be developed, but it is proposed to install 2,000 h. p. to begin with. The total investment in connection with the enterprise will be in the vicinity of $250,000.

In pursuance of a certain indenture made between the Yorkshire Guarantee and Securities Company and the Consolidated Railway Company, of Vancouver, B. C., the assets of the latter company will be offered for sale on the 17th inst. The property consists of an electric street railway extending throughout the cities of Vancouver, Victoria and New Westminster, and also between Vancouver and New Westminster, and between Victoria and the town of Esquimalt and Oak Bay, including power plants, rolling stock, etc., also lighting plants, power houses, machinery, etc., in Vancouver and Victoria.

At the annual meeting of the Acetylene Light, Heat and Power Company, held in New York during the past month, President Adams, in his address, stated that the new illuminant had been favorably reported on by both the Philadelphia Fire Underwriters' Association and the New York Board. He also presented the names of thirty fire insurance companies which had approved the use of the automatic generators. This last statement was based principally on the fact that permission had been granted a certain large risk to use acetylene, but under a number of conditions, among which was one prohibiting the storing of the carbide on the premises. It was also required that the tank be placed outside the building.

"During his present visit to Peterboro," says the Review, "the electric light inspector of this division, Mr. Wm. Johnston, has found two electric light meters of the other kind which were being used in private houses where ten lights are generally burned but which had not registered any of the electricity passed through them for several months. The cause of this lamentable state of affairs—as viewed by the company—was apparent last night, when the inspector broke those very official looking seals which he places on every correct meter and disclosed "da niggah in da fence." Doubtless in that terrific thunderstorm in June last, when poles were struck on Macdonnel and other streets and the company's loss was counted by hundreds of dollars, some of the electricity which everybody gets gratis entered these meters and destroyed their usefulness to the company. Another case settled by Mr. Johnston was that of a local company whose manager complained that the meter they had was running too fast. The test of this meter was made in the presence of the manager and a representative of the Light and Power Co., and was found to be one per cent slow, or in favor of the consumer."

## TRADE NOTES.

The Royal Electric Company are installing a lighting and power plant for the Brookfield Mining Associates at North Brookfield, N. S.

The Corporation of Huntsville purchased a 1000 light alternator of their standard single phase type from the Canadian General Electric Co.

The Royal Electric Co. have just completed the installation of an incandescent lighting plant in the large woollen mills of A. W. Brodie, Hespeler, Ont.

E. H. Thomas & Co., Norwich, Ont., are lighting their factories by electric light. The Royal Electric Co. are furnishing and installing the apparatus.

Mr. G. A. Adams, Adamsville, P. Q., has recently installed lighting plant for illuminating his mill and residence. The apparatus was supplied by the Royal Electric Company.

Wenger Bros., of Ayton, Ont., are lighting up their mills and a portion of the town by electricity; they expect to install about 200 lamps. The dynamo, etc., is being furnished by the Royal Electric Company.

Mr. J. W. Easton has severed connection with the John Abell Co., of Toronto, and connected himself with the Stevens Manufacturing Co., of London, who will in future build his latest improved electrical apparatus.

Mr. C. W. Henderson, contractor and manufacturer, has recently installed in the Canada Life new building, Montreal, electric calls in the elevators, which system is something new and very novel, being designed and manufactured expressly for that company.

The T. H. Taylor Co., Ltd., of Chatham, Ont., are lighting their large mills by electricity, and have placed their order for a 200 light dynamo with the Royal Electric Co. They are also having installed by the same firm 150 lamps throughout their mills and store house.

The Welland Vale Manufacturing Company, St. Catharines, Ont., have completed a large addition to their factory; it is being lighted throughout with electricity. There will be about 500 lamps. The plant is being furnished and installed by the Royal Electric Company.

Mr. C. W. Henderson, contractor and manufacturer of electrical supplies, Montreal, has recently fitted up some of the largest buildings in that city, notably the Montreal Street Railway Co., Montreal Diocesan Theological College, Standard Shirt Co., Thompson Shoe Co., Montreal Steam Laundry.

Mr. J. W. Skinner, of Mitchell, Ont., Canadian representative of the National Electric Mfg. Co., of Eau Claire, Wis., reports having recently made the following sales: 1000 light dynamo to the town of Goderich; 350 light plant to J. L. Eidt, to light the village of Auburn; 1000 light plant to the Kensington Furniture Co., of Goderich.

Letters patent have been issued incorporating the Paxton-Tate Company, of Port Perry, Ont., to manufacture saw mill machinery, water-wheels, etc. The capital stock is $99,000, and the promoters are George W. Dryden, James Carnegie and William McGill, of Port Perry, Leonard Burnett, of Greenbank, Hon. John Dryden, of Toronto, and F. W. Hodson, of Guelph.

# ELECTRIC RAILWAY DEPARTMENT.

## AMERICAN STREET RAILWAY ASSOCIA-
## TION.

THE recent convention of the above association at St. Louis is described as having been one of the most successful in its history. The name of Mr. C. E. A. Carr, manager of the London Street Railway Co., appears in the register of attendants as the sole representative of Canada. The association declined to entertain a proposal for amalgamation with the National Electric Light Association. Captain McCulloch, vice-president and manager of the St. Louis street railway corporations, was elected president for the ensuing year. Niagara Falls was selected as the place of next meeting.

## MONTREAL STREET RAILWAY.

THE annual meeting of the Montreal Street Railway Company was held on the 4th inst. The chair was occupied by Mr. L. J. Forget, the president of the company, and there was a good attendance of both directors and stockholders. The annual statement showed that the net earnings for the year ending September 30 last were $1,253,183.14, as against $1,096,911.31 for the previous year. The gross earnings were $555,033.69. The net profits were $462,106.79, as against $351,349.13 in 1895. Of this amount two dividends of 4 per cent. each and a bonus of 1 per cent. were declared, amounting in all to $360,000, the balance of $102,106.79 being added to the surplus.

The cost of operating during the entire year was 56.48 per cent. of the entire receipts, as compared with 59.20 per cent. for the previous year.

The good results obtained from the conversion of the system to electricity are very apparent in the statistical statement. The net earnings for the year 1896 are nearly as large as the gross earnings for 1892, being $555,033.69, as compared with $564,406.57, and the operating expenses per cent. of earnings has fallen from 82.68 to 56.48.

The rapid growth of traffic during last winter necessitated additional power, rolling stock, etc., and an additional boiler house at the William street power station to supply steam to a new 2,500 horse power direct-connected engine and generator, was erected. Fifty open motor cars were constructed in the spring, and twenty-four closed cars are now nearing completion at the company's shops.

A resolution of condolence was passed at the death of Mr. Edward Lusher, for many years connected with the company.

The number of passengers carried in 1896 was 4,018,-713 in excess of 1895. The figures for the last five years were 29,896,471 in 1896, 25,877,758 in 1895, 20,569,013 in 1894, 17,177,952 in 1893, and 11,631,386 in 1892. The transfers given last year were 8,541,530, or 28½ per each hundred passengers.

The Board of Directors was re-elected as follows : Mr. L. J. Forget, Mr. James Ross, Mr. K. W. Blackwell, Mr. G. C. Cunningham, Col. F. C. Henshaw.

It is stated to be the intention of the company to issue an additional million dollars of stock, the funds being required for extensions, improvements, etc.

## ELECTRIC RAILWAY FOR QUEBEC.

AFTER negotiations extending over a long period, the construction of an electric railway for the city of Quebec seems now to be an assured fact. A meeting of the shareholders was held on the 10th ultimo for the purpose of organizing the company. A report was read by Andrew Thomson, president of the Union Bank of Canada, showing the steps which had been taken towards organizing the company, the nature of the proposed contract, and the agreement with the Montmorency Power Company for the furnishing of power. The subscribed capital was limited to $320,000, and this amount had all been taken up. The following directors were elected : Messrs. Wm. Shaw, Andrew Thomson, John Breakey, E. E. Webb, Judge Chauveux, H. Kennedy, E. W. Methol. On motion of the Hon. L. P. Pelletier, seconded by Mr. A. Thomson, a resolution was adopted empowering the Board of Directors to enter into a contract with the Montmorency & Charlevoix Railway Company and the Montmorency Electric Power Company for the construction of the road under the former's contract with the city, and for the supply of power by the latter. Mr. Beemer transferred his franchise to the company, but reserves the right to redeem the road up to the 1st of July, 1898, by paying interest on the capital at 6 per cent., and a commission of 10 per cent. on the amount expended by the company.

Since the meeting the four parties to the agreement for the construction of the road have signed the contract. These are, the city of Quebec, the Montmorency and Charlevoix Railway Company, the Quebec and Levis Electric Power Company, and the Quebec District Railway Company, the latter being the name under which the company will operate the road.

Large quantities of materials have already arrived for the work, and over two hundred men are employed in construction. Within a very short time the citizens of Quebec will enjoy all the advantages of a thoroughly-equipped electric railway.

## CORNWALL ELECTRIC STREET RAILWAY.

THE authorities of the town of Cornwall, fully alive to the importance of rapid transit, determined to have an electric street railway, and the enterprising firm of Hooper & Starr were given the franchise. These two gentlemen are well known throughout Canada and have had a wide experience in electrical and railway work.

Ground was broken on the 21st of April of this year, and by the 24th of May 3½ miles of track were laid. There are now 5 miles of single track in operation. The handling of freight was expected to be the main source of revenue, but the passenger traffic has been greater than was anticipated, and the park, which was opened for the benefit of the patrons of the road, has proven a great attraction for the summer months.

### THE ROAD BED.

The construction of the road bed was placed in charge of Mr. Bruce, C.E., who was for some time with the C. P. R. Where there was solid bottom, 9 inches of heavy boulders were laid, and on top of this were placed 4 inches of broken stone. The ties (standard) were then laid on with earth between. A coat of macadam was afterwards placed on top, covering the rails. A heavy steam roller was then run over this, giving it a smooth, level surface. The rails are 56 lbs., with strap fish plates bonded with oooo wire, with a malleable tapering thimble which is set in the rail. The wire is run through this and the thimble is hammered in tight. This is claimed to be a decided improvement over the soldering method.

On the portion of streets where there were sandy or boggy bottoms, cedars were laid to a width of eight or nine feet, on top of which four to five inches of macadam were laid, with ties, etc., on top. There is over a mile of this construction. Some cedars were 40 feet in length, the minimum being 16 feet. One place on their private property near the G. T. R. depot—a boggy place—was made solid by laying boards diagonally, with the boulders, crushed stone, etc., on top. Curves are laid very flat, and heavy freight cars are hauled easily round them.

Little can be said of the overhead construction, as there is no feed wire, the station being in the centre of the circuits. The trolley wire is oo hard drawn.

### ROLLING STOCK.

The cars comprise four motors, three trailers, and a locomotive, but the three trailers are being converted into motors. Two open car bodies and one closed car body, and the body of the locomotive were built by the Rathbun Company, and the balance by the Canadian General Electric Company. The locomotive is equipped with four C. G. E. 800 motors on double trucks, two motors on each truck, and weighs 15 tons. The motors are arranged on the double series system, which permits of regulating the speed according to the load. All the cars and the locomotive are mounted on steam car wheels, preventing that rocking motion incident to cars using light wheels. The trucks are made by the Canada Switch and Spring Co.

### THE POWER HOUSE.

The power house is a handsome brick structure, faced with pressed brick, and designed by Mr. H. Ross Hooper, who was the architect of the car barns. It is 125 x 35 ft., divided into a dynamo and a boiler room, the dimensions of which are 75 x 35 ft. and 48 x 35 ft. respectively. The roof is supported by iron trusses.

The dynamo room is well lighted and ventilated, and the ceiling is sheeted with corrugated iron. The floor is matched hardwood, and the foundations of the machines and engines are of stone capped with brick. A 250 h.p. Robb-Armstrong cross compound engine drives a 200 k. w. C. G. E. generator. The water of the St. Lawrence is used for condensing, and a Northey condenser is in operation, with a National (Robb-Armstrong) heater. A slate switchboard is mounted with full C. G. E. equipment for the generator. The chief engineer's office and a work bench and tools occupy part of one side, and there is sufficient room for a duplicate engine and generator.

The boiler room contains two "Monarch" boilers of 150 h.p. each. The furnaces are fed with fuel of hard pea coal, mixed with the soft run of the mine. Fire-proof doors separate the dynamo and boiler rooms.

### THE CAR BARN.

The car barn is a frame structure, sheeted on roof and sides with metallic shingles and siding. It is 95 x 60 ft., part of which—16 x 95 ft.—is used for a freight shed.

The roof is supported by three independent trusses from a 60 foot span. The capacity of the barn is nine cars. A 30 foot repair pit is used for all repairs on trucks, motors, etc., besides a repair shop, which is in one corner. On the freight shed side are double tracks and a platform 6 ft. wide running the full length of the building. The freight is unloaded from the cars to the platform, and then into the shed.

### THE ROUTES.

The system centres at the post-office on Main street, and cars meet all trains and boats. There are two lines and a spur line. One line is on the east side to the park, and the other extends from the station on west side and connects with the east side line. The spur line runs to the mills. The cars run from 5.30 a. m. to 12.30 p. m. No registers are used, the conductors being supplied with what are called shot boxes. In each box is a little shot receptacle, which will not upset as long as the conductor does not turn the fare box upside down to rifle its contents.

The park owned by the company comprises 15 acres, and is prettily situated and laid out for the enjoyment of the patrons of the road. A merry-go-round is operated by a C. G. E. 800 motor.

The directorate of the road is as follows :—President, H. Ross Hooper ; vice-president and managing director, D. A. Starr ; secretary-treasurer, F. M. Siddall ; A. J. Hooper and W. R. Hitchcock.

### LEGAL.

In the action brought by one Burns against the London Street Railway Company, to recover damages for the killing of plaintiff's dog, which ran across the track within ten or fifteen feet of an approaching car, the first Division Court of Middlesex held that the case came within Hay v. G. W. R. W. Co., 37 Q. B. 465, and the action of the dog was the cause of its death, and therefore the plaintiff could not recover.

The appeal of the Toronto Railway Company from the decision of the Court of Revision confirming an assessment of $537,137 upon their street equipment, was argued a week ago before the County Judges of York, Peel and Ontario. Messrs. B. B. Osler and Wm. Laidlaw argued the case for the Company, and the City Counsel for the city. Mr. Osler contended that the company's franchise being a limited one they stood in the relation of tenants of the city, and as such were exempt by law from taxation—the taxes being payable, except under special agreement, by the landlord. Mr. Osler advanced the further argument that the railway was a highway, the rails being part of the soil, and as such should be exempt. Counsel for the city interpreted the assessment act as placing the rails, poles and wires of the Company within the meaning of real estate, and as such liable to assessment. He quoted the words "purchaser" and "vendor" in the agreement between the Company and the city to show that the company does not stand in the relation of a tenant of the city. Judgment in the case has not yet been rendered.

---

The Winnipeg Street Railway Company employ for the conveyance of pic-nic and excursion parties, a motor car, attached to which are four trailers consisting of old horse cars fitted with new platforms and sills, a railing all round, and seats arranged across both sides and ends, with space for a passageway from the steps at either end. These trailers are lighted by lamps strung on wires supported on poles at either end of the car.

The Montreal Street Railway Company have recently had a system of interlocking safety devices placed at the Wellington Bridge crossing the Lachine canal. An electric motor is employed to turn the bridge. Before the bridge is opened, a derail, consisting of a tongue switch, is so set as to turn the car off the track at a distance of 80 feet from the bridge on either side, thus preventing the possibility of a plunge into the canal while the bridge is open.

## SPARKS.

An electrician named A. Lepitre was instantly killed by coming in contact with an electric current in the Montreal Light & Power Company's station.

It is rumored that as soon as the Chambly water power is developed, steps will be taken to construct an electric railway between Montreal and St. John.

It is announced that the Cataract Power Co., of Hamilton, which was recently organized, have closed contracts in Hamilton for power to the value of $50,000 a year.

The Sussex Water and Electric Light Co. have entered into an agreement with the town of Sussex, N. B., to install an electric light plant. Mr. F. E. Norton, of St. John, has charge of the construction work. A 500 light power plant will be installed.

At a convention of street lighting officials held recently at New Haven, a poor showing was made on behalf of municipal control of electric lighting plants. It was stated that Wabash, Ind., purchased a plant for $18,000 and sold it for $30; Xenia, O., paid $35,000 for a plant and ten years later sold it for $10,000; Moline, Ill., bought a plant at $15,000 and four years later sold it for $8,000; Michigan City bought a $10,000 plant and sold it for $2,500.

The two rival electric light companies at Renfrew, Ont., are about to amalgamate, under the name of the Electric Light and Power Company. A charter of incorporation is now being asked for, the applicants being W. A. Mackay, T. W. Guest, A. C. Mackay, A. A. Wright and Howard Wright. The erection of a new power house will be one of the first improvements, a site for which has been secured. It will be a large structure, and will contain two steam engines, one being 225 horse power.

The Owen Sound Electric Manufacturing and Illuminating Co. have decided to furnish incandescent light and power, as well as arc lighting to the Town of Owen Sound. Their water-power is situated three miles from the business centre of the city; this they have very lately improved, and now have a steady power of 200 horse or more the whole year round. The Royal Electric Company have secured the contract to furnish them with "S.K.C." two phase dynamos and apparatus, in order that they may be able to supply power as well as incandescent light, thus enabling them to use their water-power the entire 24 hours of the day— during the daytime furnishing power to the different manufacturers, and at night furnishing the arc and incandescent lights. This was one of the first companies in the electric lighting business in Canada. They have always been very conservative, and have been successful from the start.

# CANADIAN GENERAL ELECTRIC CO.
## (LIMITED)

Authorized Capital, $2,000,000.00
Paid up Capital,　$1,500,000.00

### HEAD OFFICE:
## 65 FRONT STREET WEST, - - TORONTO, ONT.

BRANCH OFFICES AND WARE-ROOMS:

| | | |
|---|---|---|
| 1802 Notre Dame St. - MONTREAL. | Main Street - - WINNIPEG. | |
| 138 Hollis Street - HALIFAX. | Granville Street - - VANCOUVER. | |

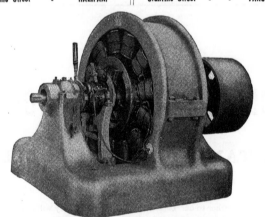

*150 K. W. MONOCYCLIC GENERATOR.*

# The Monocyclic System

has been established by the experience of the past year to be the only satisfactory system for the distribution of . . . . . .

# Light and Power

from the same generator and circuit. We invite attention to **its superior mechanical design and construction**; its absolute simplicity in distribution as compared with the complications of the polyphase systems; its perfect regulation secured by compounding to compensate for line loss; its freedom from unbalancing, the lighting circuit being single-phase; the perfect operation of our induction motors, which require no condensers.

## SPARKS.

Granby, Que., is agitating for the electric light.

The council of Dundas, Ont., have decided to have the town lighted by electricity.

The new building for the Brantford Electric Light Company will shortly be completed.

The Ontario Electric and Engineering Company, Toronto, is being incorporated with a capital stock of $10,000.

T. E. Bulwer has been registered proprietor of the firm of H. E. P. Bulwer, electrical supplies, Montreal.

Mr. James Noxon, Inspector of Ontario Asylums and Prisons, is experimenting with an appliance for burning tan bark for fuel for boilers, which, if successful, will be introduced in the prisons and asylums of Ontario.

By the bursting of a drive pulley at the electric light works, St. Thomas, Ont., the shafting and dynamo were badly damaged

The town of Kaslo, B. C., has rejected the proposal of Alexander & Retallack to put in an electric light plant, the by-law being defeated.

Some time ago the Western Electric Co., of Chicago, were negotiating to establish a plant in Vancouver, B. C., and deposited two bonds of $5,000 with the city.

The scheme has since been abandoned, and the city has returned the bonds.

It is said that much of the comparative comfort of the men accompanying Dr. Nansen on his North Pole venture was due to the electric current supplied by a dynamo driven by a windmill. Dr. Nansen's previous experiences in the Arctic regions led him to expect a continuous breeze in the level ice regions, and the compact windmill outfit which he took with him fulfilled all expectations. By means of the electric current thus obtained the ship was lighted by electricity, and it is also stated that the current was employed for purposes of heating.

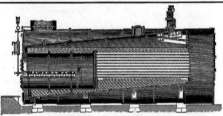

# Kay Electric Mfg. Co.

255 James St. N., HAMILTON, ONT.
58 Adelaide St. W., TORONTO, ONT.   Telephone 1214.

We are prepared to furnish——

*Dynamos of any capacity for any voltage, either compound or shunt wound.*
*Motors from 1-8 to 40 h. p., either series or compound wound.*
*Elevator Motors of all sizes.*
*Alternating Dynamos from 300 to 1000 lights.*
*Transformers of any capacity from 5 to 125 lights.*
*Electro-plating Dynamos, any capacity.*
*Electrical Experimenting in all its branches.*

WRITE FOR PARTICULARS AND ANY INFORMATION REQUIRED.

# GAS ENGINES

Of from 1 to 600 Brake Horse Power, for electrical Industrial and other purposes.

MANUFACTURED BY

FRIED. KRUPP GRUSONWERK, Magdeburg, Germany.

**JAS. W. PYKE & CO., Montreal, Que.** Representatives for the Dominion of Canada.

Particulars on Application.

# STEAM PUMPS

**DUPLEX**
**SINGLE**
**TRIPLEX**

## For All Duties

....

# NORTHEY MFG. CO., Ltd., TORONTO

The Laurie Engine Co., Monteal

SOLE AGENTS FOR PROVINCE OF QUEBEC

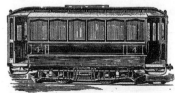

FINE - -

*ELECTRIC*

# Street Cars

.... OUR SPECIALTY ...

We also manufacture Horse and Trail Cars of every description.

**PATTERSON & CORBIN** ........ **ST. CATHARINES, ONT.**

# ~ TRANSFORMERS ~

WE have the largest and most thoroughly equipped factory in America devoted exclusively to the manufacture of Transformers. . .

THE "SIMPSON" TRANSFORMER has by long odds, by merit only, won the first place over all other transformers, and is in use by more companies than any other make in this country. . . . . . .

# G. T. SIMPSON · HAMILTON, ONT.

CANADIAN

# ELECTRICAL NEWS

AND

## STEAM ENGINEERING JOURNAL.

| ·VOL. VI. | DECEMBER, 1896 | No. 12. |

### THE ONTARIO ELECTRIC AND ENGINEER-ING COMPANY.

As a sign of the times and increasing prosperity in the industrial world, after a season of comparative inactivity, perhaps nothing is more encouraging than to note the appearance, from time to time, of new commercial enterprises springing up in spite of the blue ruinist's cry of hard times and keen competition.

It may be that, like ourselves, the promoters of these concerns hold their own opinions as to the time when best to launch out, and no doubt they have also the conviction that after a lengthened period of depression there must always come a revival in business.

Referring more particularly to the electrical industry, we note with pleasure the arrival into the Canadian field of the Ontario Electric and Engineering Co., Ltd., recently organized for the purpose of carrying on a general electric contracting, supply and repair business, with commodious headquarters at 77 to 81 Adelaide street west.

It is intended, we understand, to pay special attention to repair work, which feature will no doubt commend itself to central station men, who even with the best of good luck may sometimes require the quick co-operation of a well-equipped machine shop and competent engineers.

The secretary-treasurer, Mr. W. Heathcote, who for

60 K. W. SINGLE PHASE WARREN ALTERNATOR.

some years held a responsible position on the engineering staff of the Canadian General Electric Co., Ltd., is a gentleman of sound business experience and executive ability, who will doubtless perform his duties with credit to himself and profit to the company.

The position of chief engineer is held by Mr. Hazen Ritchie, A. I. E. E., a graduate of the Royal Military College, Kingston, who has had several years' experience with the larger companies both in England and on this continent.

The sales department will be in the hands of Mr. J. J. Ashworth, so well and favorably known to the electrical public as having been on the agency staff of the C. G. E. Co. since its inception, having only severed his connection to identify himself with the new enterprise.

We illustrate on this page a single phase alternating current dynamo which this company are now placing on the Canadian market. It is of the inductor type, with stationary armature, and, it is claimed, combines all the qualities of durability, slow speed (that of a 60 k.w. be-

ROTOR OF WARREN DYNAMO.

ing only 720 r. p. m.), good regulation, and high efficiency. The manufacturers are the Warren Electric Co., of Chicago, Ill., for whom the Ontario Electric and Engineering Co. are acting as sole agents for Canada. Sales are reported good, although the machine has been but a very few weeks before the public.

The company are also sole agents for the Eddy Electric Manufacturing Co., of Windsor, Conn., the well known makers of direct current machinery in all sizes.

The fact of being in a position to place apparatus of such high grade on the market, and having on its executive and engineering staff, men, each a specialist in his particular line, augurs well for the success of the company.

### QUESTIONS AND ANSWERS.

C. L. F., Parry Sound, Ont., writes : I enclose three pieces of wire, viz., No.'s 18 iron, 18 and 16 copper magnet, American wire gauge. Will you please tell me what size they are in B. & S. gauge ?

ANSWER.—American wire gauge is the same as B. & S. Guage. Of the three pieces of wire, the long brown one is .033 in. = B. & S. between No. 18 and 19 ; the very small white insulation is .005 in. = B. & S. No. 36 ; the short piece is .064 in. = B. & S. No. 14.

A correspondent in an Eastern Ontario city writes : " I believe that you can, better than anyone else, favor me with a definite opinion as to the outlook for college graduates in the field of electrical engineering, and from what I can judge it is the most promising of all professions at the present time, and I would like to know, with some degree of certainty, whether the field is already over supplied, as is claimed in some quarters, or whether the fault of non-success experienced by some technically qualified men is properly attributable to their own lack of energy, or, say, want of the exercise of common sense in casting about for employment?

ANSWER.—We scarcely know what opinion to express in reply to your enquiry. We have talked this subject over a number of times with men occupying leading positions in the electrical business, and the general opinion appears to be that the outlook for young men in this calling is not as promising as a great many people appear to imagine. If you look over the electrical field at the present time you will see that the number of really good positions, in this country at least, are very limited. We know of several qualified electricians formerly occupying good positions, who, having lost them, have found it impossible to secure others equally remunerative. We do not pretend to know all the circumstances in connection with these cases, and consequently are not able to say that these persons have not, in some degree, themselves to blame for the position in which they find themselves at present. The electrical business in this country appears to be at a point where it is very difficult indeed to estimate its future development, hence the difficulty of expressing an opinion on the subject of your enquiry. If the electric railroad continues to develop as it has done during the past five years, there should be a considerable number of openings for young men in that field. This applies also to the distribution of power by electricity over long distances. If the distribution of power in this manner is found to be commercially practicable and advantageous, it will probably lead to the establishing of quite a number of large power stations at certain points throughout the country, where water power is available, and in such power stations the services of one or two first class electricians will be indispensable. With regard to the electric lighting business, a great many of the men in charge of central stations at the present time have not had proper training for the position, and are consequently lacking in efficiency. Unfortunately, the owners of stations do not appear to realize, as they should, the necessity of employing properly qualified men and paying them satisfactory salaries. Until the owners of stations come to realize that a poorly qualified superintendent is dear at any price, there will be few openings in this direction for the services of properly qualified young men. We are not without hope, however, that the business will ultimately be placed on a proper footing and will be conducted more in accordance with the best known principles of business management. When that time arrives the number of openings for competent young men will be increased. This is the situation as it presents itself to us at the present time. What new developments in the use of electricity may be forthcoming in the near future it is impossible to know.

## POWER DEVELOPMENT AT NIAGARA FALLS.
### BY F. C. ARMSTRONG.

THE delivery in Buffalo on the 15th of November last, of the first thousand horse power out of eight thousand which the Cataract Construction Company are under contract to supply to the Buffalo Railway Co., marks the completion of an important stage in this notable enterprise. No undertaking in recent years has attracted the attention of the engineering and industrial world to so great a degree as the now accomplished "harnessing of Niagara"; and no undertaking of a certainty has had to win its way to a signal success in the face of greater difficulties and more discouraging and persistent prophesies of failure. Although so much has been written from time to time during the progress of the work that the

electrical public, at any rate, are pretty well conversant with its history, a brief recital of its main points may not be out of place at the present moment.

From the day when Father Hennepin in his Nouvelle Decouverte first published to the world a description and sketch of the mighty cataract, the Falls of Niagara have held their place as the great natural wonder of America, the main objective point on this continent of the globe trotter and the wedding tourist. It was not to be expected, of course, that the utilitarian spirit of recent years would be satisfied to find scenery alone in what was plainly meant for water power. Some early attempts at utilization were made, and the present Niagara Falls Hydraulic and Land Company is a development from the first hydraulic canal constructed between 1853 and 1861. In both Canada and the United States, however, a strong and wide-spread feeling existed against any further disfigurement of the naturally charming surroundings of the Falls which culminated in the nationalization for park purposes of the lands enclosing them on both sides.

In 1889 the Cataract Construction Company was organized to carry out the plans for power development worked out by Thomas Evershed. These embraced mainly the taking of the necessary water supply from the river by a short canal at a point one-and-one-half miles above the Falls, its delivery at this point, where the erection of the necessary buildings would not be objectionable from an æsthetic standpoint, into a wheel-pit 178 feet in depth, and its discharge through an underground tunnel into the river at a point directly below the upper Suspension Bridge, the capacity of the tunnel being fixed at 120,000 horse power. The personnel of the company, of which Mr. E. D. Adams was president, Mr. W. B. Rankine secretary, and Messrs. D. O. Mills, J. Pierrepont Morgan, W. K. Vanderbilt and J. J. Astor, members of the Board of Directors, was a sufficient guarantee that the capital necessary for an undertaking of such magnitude would be readily forthcoming.

As general consulting engineer the company retained Dr. Coleman Sellers, the hydraulic and electrical portions of the work being placed respectively in the hands of Mr. Clemens Herschel and Professor George Forbes, of London, England.

In 1893 the International Niagara Commission, composed of Sir William Thomson (Lord Kelvin), Dr. Sellers, Col. Theodore Turrettini, Professor Mascart and Professor William Unwin, were invited to examine existing methods and select plans for the detail apparatus required in the development and transmission of the power. For the turbines the design submitted by M. M. Faesch & Piccard, of Geneva, Switzerland, was selected. For the transmission, as might have been expected, electricity was finally adopted, though not without a careful examination into the merits of compressed air, hydraulic tubes and rope transmission.

Regarding the position taken by Lord Kelvin, Prof. Rowland and other authorities consulted, toward the particular electrical system and type of generator ultimately used, a somewhat acrimonious discussion has since been carried on. It seems fairly clear, however, that to Professor Forbes is due the credit of insisting on the employment of alternating instead of direct currents—a choice of which no one would to-day gainsay the wisdom in view of the different uses requiring widely varying voltages for which the current is now being required. A second point on which Professor Forbes was exposed to attack was his advocacy of a comparatively low frequency. Here again the advantage obtained of greatly lessened inductive loss on the long distance transmission lines, added to the much greater suitability of the low periodicity for rotary transformer work, has been amply sufficient to demonstrate the correctness of his judgment. The umbrella shaped type of generator adopted, with an external revolving field and stationary armature, which has proved itself admirably suited for the requirements of a large fly-wheel effect and light revolving weight, is substantially the design submitted by him as consulting engineer to the manufacturing companies. In this connection it may be added that whatever estimate is to be placed on Professor Forbes' work for the Cataract Company, he is certainly entitled to respect for the courage with which he has always been ready to defend his convictions. The Parthian dart which he discharged at his critics and detractors in his famous article in "Blackwoods," affords sufficient evidence on this point.

The first of the three five thousand horse-power generators forming the original order given to the Westinghouse Electric Manufacturing Company, was started up on the 5th of April, 1895, and shortly afterwards the regular supply of current to the amount of 2,000 h. p. to the Pittsburgh Reduction Company for the manufacture of aluminum was commenced. To the electro-chemical group of local users of the new power there has since been added the Carborundum Company, which produces in the electric furnace, from carbon, in one of its many metamorphic conditions, an abrasive claimed to be superior to emery.

A preliminary installation for the manufacture of carbide of calcium uses at present 1,000 horse-power. This amount will undoubtedly be greatly increased should acetylene gas in the future become something less of an ignis fatuus and more of a practical illuminant. Other local applications of the power are, with synchronous motors, to operate the generating plant of the Niagara Falls Electric Light Co., and with rotary transformers in supplying current at 500 volts for the Buffalo & Niagara Falls Railway.

It is as marking the satisfactory commencement of the second stage in the distribution of the power in which its successful transmission over considerable distances is the problem to be worked out, that the thousand horse-power already laid down in Buffalo becomes of the first importance. The difficulties to be overcome are not, of course, of an engineering nature, since several transmissions on a large scale over greater distances have been for some time in operation. The question has been whether electric power generated under the conditions which obtain in the Cataract Company's plant can be sold at a profit in Buffalo at prices as low as those at which steam power can be produced under the absolutely favorable conditions existing at that point. Comparative estimates made by the most capable engineers have differed regarding this all-important matter to a curious extent. The result, as indicated by the contract entered into with the Buffalo Railway Company for 8,000 horse-power delivered, at a price stated to be $36.00 per horse-power per annum, would seem to show that the Cataract Company's officials and one of the most important of their prospective customers, have been able to arrive at a mutually satisfactory basis of price for the transmitted electric power where the circumstances governing its previous production by steam were such as to render possible the very highest economy.

It seems reasonable to estimate the amount of power which will be disposed of in Buffalo within a year at not less than 15,000 horse-power, and in view of this and other increasing demands, an additional order has been placed with the manufacturers for five 5,000 horse-power generators, which will bring the total generating capacity of the plant up to 40,000 horse-power. For the transmission to Buffalo, which has been carried on under the plans of the General Electric Company, a line potential of 11,000 volts is now being used, but this will be doubled to 22,000 volts later on, in order to keep the copper cost and energy loss within reasonable commercial limits. The three-phase system is used for the transmission instead of the two-phase, on account of the very considerable saving effected in copper. At the sub-station in Buffalo the current stepped down to 2,000 volts is carried through underground cables to the Railway Company's power house, where, after further stepping-down, the General Electric Company's rotary converters change it from an alternating to a direct current at the standard railway voltage.

The commercial success, now practically assured, of the transmission to Buffalo, entails of course the extension of the company's field of operations in this direction over a wide area. Just where the commercial limitations which will govern in the matter will fix the point beyond which Niagara power cannot be profitably delivered, would be at this moment a very unsafe matter on which to hazard a definite opinion. It should be kept in mind, however, that the completed scheme of the Cataract Company involves the development of 200,000 horse-power on the American and 250,000 horse-power on the Canadian side of the Falls, to find a market for which will require a transmission radius considerably in excess of 100 miles.

The other important power development already referred to —the Niagara Falls Hydraulic and Land Company—will, along with various projects now under consideration on the Canadian side, be more fully considered in a subsequent paper.

The Minister of Education has promised to provide for the sustenance of the Toronto Technical School in case the city provides a permanent building therefor.

The Metropolitan Street Railway Company have extended their line to Richmond Hill, and are considering the further extension of the line to Lake Simcoe, in which case a new power house will probably be erected at Newmarket or Aurora. The present power house near Mount Pleasant will be improved.

## TRADE NOTES.

The Canadian General Electric Co. are supplying a 1,000-light standard single phase alternator to Victoriaville, P. Q.

A large engine for the St. Thomas Electric Light Works was recently supplied by Cowan & Company, of Galt, Ont.

The Ontario Electric and Engineering Co., Toronto, have sold a 500-light alternating plant for lighting the town of Newcastle, Ont.

The Canadian General Electric Co. have been awarded a contract for a 500-light incandescent plant for the town of Alvinston, Ont.

P. McIntosh & Sons, of Toronto, have installed in their factory a 300-light incandescent plant supplied by the Canadian General Electric Co.

The Almonte Electric Light Co. have added to their plant a 600-light incandescent generator manufactured by the Canadian General Electric Co.

The Canadian General Electric Co. are installing a 500-light single phase standard alternating plant for a local company recently organized in Embro, Ont.

The Toronto office of the J. C. McLaren Belting Company, of Montreal, has been removed to 69 Bay street. Craig, McArthur & Co. are the representatives in this city.

Messrs. Coristine & Co., of Montreal, have installed a 55 k. w. direct current incandescent generator of the Canadian General Electric Co.'s moderate speed multipolar type.

The Canadian General Electric Co. have closed a contract with the Canada Paper Co. for a 1,000-light incandescent generator of their latest multipolar steel type, with iron-clad armature.

The Electric Repair and Contracting Co., of Montreal, are at present busily engaged in rebuilding motors and generators damaged by fire which took place recently on the premises of the Montreal Park & Island Railway Co.

The St. Catharines Electric Light & Power Co. have placed an order for a 2,000-light standard single phase alternator with the Canadian General Electric Co. The 60 k.w. machine of the same type which they have been operating up to the present has proved insufficient in capacity to meet the growing demands of their business.

The Fraserville Co., Ltd., of which Mr. John MacFarlane, of the Canada Paper Co., Montreal, is president, are installing a complete 750-light alternating plant in the town of Fraserville, Que. The entire contract has been awarded to the Ontario Electric & Engineering Co., Ltd., who will install for the generating plant one of their 45 k.w. single phase "Warren" alternators.

The Berlin & Waterloo Railway Co. have just placed in service two new vestibuled cars, having a length over all of 27 feet 6 inches. These cars are exceedingly handsome in design and finish, solid mahogany being used throughout for the interior fittings, and embody important improvements in various details. They were constructed at the Peterboro shops of the Canadian General Electric Co.

Owing to the rapidly increasing demand for their goods, the Kay Electrical Mfg. Co., of Hamilton and Toronto, will shortly commence the building of an addition to their factory. The following is a partial list of their more recent sales :—Kemp Mfg. Co., Toronto, 2 motors ; H. R. Cuddon, St. Catharines, 1 motor ; M. Hutchinson, wood yard, Toronto, 1 motor ; A. Moore, Toronto, 1 motor ; Aylmer Electro Plating Co., 1 dynamo ; Steel Clad Bath & Metal Co., Toronto, 1 4-pole motor ; Wherle Brush Co., Toronto, 1 motor ; Leitch & Turnbull, Hamilton, 3 motors, for elevator purposes : A. R. Williams, Toronto, 3 motors ; Davis & Henderson, Toronto, 2 motors ; Mr. Garner, Toronto, 1 motor ; Mr. Enright, Toronto, 1 motor ; Mr. Bomberg, Toronto, 1 motor dynamo ; H. C. Hunter, Dundas, 1 4-pole 400-amp. dynamo ; Haskins Wine Co., Hamilton, 1 motor ; McPherson & Glassco, Hamilton, 1 motor ; Munderloh & Co., Montreal, 1 dynamo ; J. Turner & Son, Toronto, 1 motor ; Wm. Beers, Toronto, 1 motor ; T. Bell & Co., wood yard, Toronto, 1 motor ; Barber Bros., Georgetown, 1 30-h. p. 4-pole motor ; H. & F. Hoerr, Toronto, 1 motor, 15 h. p. ; Ontario Agricultural College, Guelph, plant for light and power ; Small & Fisher, Woodstock, N. B., 1 dynamo ; A. Laidlaw, Toronto, 1 motor ; Mr. L. Williams, Toronto, 1 motor ; John Forman, Montreal, 3 motors ; Wilson Pub. Co., Toronto, lighting plant ; T. E. Brandon, Toronto, 1 motor ; Davison & Holmes, Toronto, 1 motor ; Bennett & Wright, Toronto, 2 4-pole motors ; Diamond Machine & Tool Co., Toronto, 1 electro plating dynamo. This firm have also sent 10 electric machines to the North-west and British Columbia.

## THE YOUNG MAN'S CHANCES IN THE ELECTRICAL FIELD.

IN view of the opinion which seems largely to prevail that electricity is the thing to which young men should now turn their attention with the best hope of reaping satisfactory results from their labors, the editor of the ELECTRICAL NEWS deemed it advisable to solicit opinions on the subject. For this purpose the following letter was recently addressed to a few persons prominently identified with the electrical interests :

DEAR SIR,—To assist me in answering frequent enquiries as to the possibilities for qualified young men in the various departments of electrical work, I have thought it advisable to endea or to obtain an expression of opinion from a number of persons qualified to advise on the subject.

The enquiry may be briefly put thus :—"What are the chances of the young man who graduates as an Electrical Engineer in comparison with the young man who enters any of the other professions or commercial life?" I would esteem it a favor if you would kindly give me an expression of your views on this matter in time for publication in the ELECTRICAL NEWS for December

C. H. MORTIMER.

We trust the appended replies will be of assistance to parents and young men who find themselves face to face with the problem of choosing in what direction life's efforts should be expended :

Mr. Granville C. Cunningham, manager and chief engineer of the Montreal Street Railway Company, writes :—"At present there seems to be more opening in electrical engineering than in any other professions in this country. Of course the success of a man largely depends upon himself. There is little doubt, I think, but that electricity, during the coming years, will have large developments in this country.".

The manager of another important electrical company, who requests that his name be omitted, writes :— "Replying to enquiry contained in yours of 28th inst., it is common knowledge that every profession, trade, and calling is overcrowded, but that there is room at the top for persons of exceptional ability, is well known, and any person of even more than average ability will succeed fairly well whether he be on a farm, in commerce, or in professional life. What then are the chances of a young man of more than average ability who graduates as an electrical engineer, in comparison with those of a young man of equal ability who enters one of the other professions, say law or medicine? Let us see how the matter stands in Toronto. There are in round figures 500 lawyers. We will not be far out in saying that the number who possess more than average ability and who have established a practice is about 150, and these have incomes of $1,000 a year and upwards. Are there ten electrical engineers in Toronto earning this amount?

"There are lawyers in Canada making eight and ten thousand dollars per year and some as high as fifteen and twenty thousand. How many electrical engineers in the country are making half of the lowest figure?

"What is true in law holds equally so in medicine. There are about 400 doctors in Toronto, and judging by the houses they inhabit and the style of their living, the average income of an established doctor of more than average ability must at least be as great as that of his legal brother.

"The man of less than average ability has neither room nor place in any profession. He may graduate as an electrical engineer, but will end up in attending a dynamo or stringing wire at forty or fifty dollars a month. The time spent at college would have been better employed in getting a practical mechanical education or a sound business training.

"I have no desire to discourage persons from going into a business employing electricity. The prospects of a bright intelligent young man would be at least as good as they would be in any non-electric business, but I feel that our schools and colleges are turning out a hundred electrical engineers for every vacant position in the country. What is to become of them? Electricity does not spell any royal road to fortune."

Mr. Wm. H. Browne, general manager of the Royal Electric Co., Montreal, writes : "In reply to your enquiry as to what are the chances as an electrical engineer, compared with other professions or commercial life, I presume the answer would be that on the average the electrical engineer would be likely to do as well as the average man in other professions or commercial life.

"In electrical work, as in all other work, the most room is at the top, but it is quite likely that for some time to come the electrical engineer who can be at the top may not be as financially successful as his corresponding member of the legal or medical profession or the commercial man.

"The field for opportunity for clients is necessarily, at present, much more restricted in the electrical line than in the other professions or commercial life, because the industry is new, but there is no doubt that the growth of the electrical industry, by reason of the increase of the application of electric power, will very largely increase, and within a few years will require the talents of the best members of the profession, and those who may be capable of meeting these requirements will, no doubt, do as well as the best members of other professions.

"In my judgment, one of the greatest needs of the electrical business of this country to-day is the employment in all operating electrical plants, of thoroughly well qualified young men, graduated as electrical engineers.

"I have frequent applications in our business here, from parents of young boys, sixteen to eighteen years of age, to take them into our shops and teach them the electrical business.

"The impression appears to prevail, that this is all that is necessary to make competent electrical engineers.

"I am obliged to refuse all such applications and advise such parents that if their sons have special aptitude and inclination for mechanics, that they be sent to some good college to receive a thorough complete course in electrical and mechanical engineering, for the two are almost necessarily bound together, and after graduation, to seek occupation practically, either in the operation of an electrical plant or in a manufacturing establishment.

"The electrical engineer requires special qualifications to fit him for his profession and there have been many who have graduated as such who have probably made a grave mistake, by reason of not possessing the special aptitude and talents."

Prof. Galbraith, Principal of the School of Practical Science, Toronto, writes : "Your question is not an easy one to answer. It seems to me that it is well to assume that all money-making occupations, businesses and professions are full. This being the case, success will depend largely on the special fitness of the candidate for his chosen vocation. Natural capacity for one's work, supplemented by education and training ought, other things being equal, to ensure a reasonable amount of success. There is always room at the top ; to get there, however, requires special qualifications as well as opportunities. The man who takes an interest in his work for its own sake and not simply for the money which he may make from it, will not be discouraged by hard times, and will in all probability work his way through life more cheerfully than the man who values his occupation simply by the dollars and cents he may make out of it. A young man ought not to select his profession simply because at present business in it is good, nor ought he to reject a profession for the opposite reason. He ought to remember that his choice is not only for the immediate future, but for life, and that during his life ups and downs may be many and not far between."

Mr. G. J. H., manager of an Electric Light and Power Co., writes : " Complying with your request of the 28th November, will say that your enquiry covers quite a lengthy opinion.

" The comparison between a young man graduating as an electrical engineer with a man entering a commercial life, can be made as follows : The man entering a college course to qualify for an electrical engineer has before him, I think, a four years course. He enters at the age of 18, and say he gets plucked two years out of his course, which would bring him to the age of 24 when he qualified, he then really has to make a start in life, or in other words hang out his shingle that he is ready for business, unless he happens to be fortunate enough to secure a position with some reliable firm. If not, he may plod along for a couple of years, very often receiving smaller wages than the ordinary mechanic who has served his time at the bench. In this connection there comes to mind the cases of two personal friends of mine ; the first graduated as an electrical engineer from McGill about a year and a half ago ; he went to one of the largest cities in the States, and at the present time is drawing the heavy salary of $1.50 a day. The other, now out of college some time, secured the appointment of Construction Superintendent on an electric road, and after giving the company the benefit of his college education as an electrical engineer in overcoming technical difficulties and systematizing the whole road, was politely dismissed, to be replaced by a man that could never know as much as this engineer had forgotten, but it was a question of a few dollars a year in salary. As a rule you will find that college graduates expect to start their professional career at very large salaries. This is one of the greatest mistakes these graduates could make. When it comes to closing an engagement they prefer to hold off for several months, than close at a fair salary. As a consequence you will find college graduates filling commercial positions, for which purpose their college education is of very little use, to say nothing of the four to six years of their life that has been to all commercial purposes lost. I do not refer particularly to electrical graduates, as I could record several similar instances as applied to civil engineers. As you are well aware college education can never do a young man any harm, provided he can afford to take a course and spend the required time.

" As a rule a young man starting a commercial life would be about 15 years old, and would have from 15 to 24 to make a mark for himself, the ability to do which must naturally depend largely on himself. Provided he starts with a reliable firm, displays any ability,

or is at all industrious, he is almost certain to secure advancement, and in time, no doubt, will be given a position of trust, and by the time his friend had graduated at 24, the commercial man would have better prospects than the graduate.

" This is the age of development in electricity, and I think if I had a boy of 15 or 17 I would prepare him to take a course to qualify as an electrical engineer, but as we all know there are so many different opinions on the bringing up of boys, that it is a matter that would take hours of discussion."

Mr. E. Carl Breithaupt, Consulting Electrical Engineer, Berlin, Ont., writes : " Replying to your enquiry of the 28th ult., as to the relative chances of a young man who graduates as an electrical engineer as compared with one who enters any other profession or commercial life, it seems to me that such a comparison is not altogether a proper one to make ; a man must have a very particular fitness to make a success in any profession, and especially do I think this is the case in the three Engineering professions, the Civil, Mechanical and Electrical. If a boy shows aptitude and fondness for engineering work, is willing to work very diligently, and willing to don a suit of overalls and perform heavy manual labor at any time he may be called upon, either day or night, I think his chances as an electrical engineer are as good as those of any other calling in life. There is one thing, however, that must be remembered, viz., that very few engineers in any one of the three branches named have become very wealthy in the practice of their profession. Engineering work must be considered more as a labor of love than one for financial gain."

Messrs. Ahearn & Soper, Ottawa, write : "Replying to your favor of the 28th ultimo, asking what are the chances of the young man who graduates as an electrical engineer in comparison with the young man who enters any of the other professions or commercial life, we think his chances are now about equal. A few years ago his opportunity for obtaining employment might have been better, but electrical engineering to-day, like other professions, seems to have been overdone."

. Mr. George White Fraser, Consulting Electrical Engineer, writes : " In answer to your enquiry of date 30th, now as to the prospects of young men entering the electrical engineering profession in Canada, I would say : At the present moment there is practically no electrical engineering in Canada. When persons are contemplating an enterprise involving the use of electricity for lighting, power or railway purposes, the last man they think of consulting is an electrical specialist. This is due apparently to the fact that, first, the general public seem to think that they know enough about it to do without advice ; second, the manufacturing companies naturally do all they can to discourage the idea of consulting competent engineers in independent practice, and offer to do all engineering themselves free of charge. The general public accept this seemingly generous offer, shutting their eyes to the rather obvious consideration that this engineering has got to be paid for somehow, whether done by an independent person, or one employed by a manufacturing company, and that the engineering of the latter is necessarily biased in favor of the " system " exploited by his company ; third, there have been no competent electrical engineers doing business until quite recently—the only persons in that line having been more entitled to the name of ' electrical mechanics ' than of ' electrical engineers '— being able merely to make repairs and do small wiring jobs. Of course we have many instances of civil, hydraulic and mechanical engineers, and architects, and even land surveyors, who, without the slightest right to do so, have called themselves electrical engineers, and freely advertise their specialties as being electric railways, electric lighting, etc., and actually get work in those lines which they simply hand over bodily to the manufacturing company of their choice ; and fourth, that a great deal of such electrical work as there has been, has been more or less of a pettifogging character—municipal deals, small lighting plants, and so on. I think, however, that a different notion is taking hold of the public, that is rather encouraging to the independent engineer. In the first place, happily, these small plants are about all sold now, and people are getting a little less confident as to their electrical attainments. The evolution of machinery from the old D. C. or single phase alternating type to the latest polyphase development, with all the latest storage battery, inductor type, direct connected side issues, has rather brought electrical engineering, as such, to the front ; and as the public begin to read a little more, and hear a little more, and find out that electricity is not "in its infancy, " nor yet a matter of unspeakable mystery, but a science to be studied and understood ; a profession clearly distinct from civil or mechanical, or hydraulic engineering, and vastly different to architectural or land surveying, so do they think more of obtaining advice from electrical men—more especially as the number and variety of different types of machinery offered to them increase to their great perplexity.

"Briefly—I think that most of the small work is done. During the next several years large works will be promoted—large railways, power schemes, electrolytic plants, etc ; the men interested in them are business men who will not submit to the dictation of any manufacturing company, but retain outside independent engineers, knowing very well that electrical specialists can attain better results than the most experienced general practitioner. Therefore I think there is plenty work to do for electrical engineers who will vigorously insist on recognition, who will keep themselves absolutely free of the influence of any manufacturing company and who will keep themselves abreast of the times. It will be a hard fight, for we have many antagonists—we have the inertia of an ignorant public, the animosity of powerful manufacturing companies, who, in my own experience, will go to any length to persuade customers against calling in independent advice ; and we have the jealousy of the other branches of the great engineering profession, who do not care to see electricity defined as a specialty for which they are not professionally qualified. I, personally, shall be glad to welcome any accession to the ranks of the independent electrical engineering profession in Canada, and think that success is a matter of determined effort and co-operation."

Mr. C. E. A. Carr, manager of the London Street Railway Co., writes : "In reply to your inquiry of the 28th November, I should think the chances of success in the electrical field were much better than in any other, for the reason that the uses of electricity are daily becoming broader, which is not the case, in so marked a degree, in any other profession."

Mr. R. A. Ross, mechanical and electrical consulting engineer, Montreal, writes : "Replying to your enquiry as to what are the chances of young men who graduate as electrical engineers in comparison with those who enter other professions or commercial life, I should say, that without doubt at the present time electrical engineering is overcrowded, and will probably always remain so for the following reasons :

"To a new profession there is always a rush, and in this case the influx has been particularly large, because of the rapid expansion of electrical enterprise, necessitating a large amount of engineering supervision, which has become unnecessary as the enterprise settles down to a rigidly economical basis.

"Again, civil engineering has long been recognized as an overcrowded profession, and the tendency of those contemplating entering the engineering field has been to avoid the civil and enter the new and rapidly expanding electrical field. This result has obtained in spite of the fact that although there is room for a civil engineer or two in every county, there is not room for an electrical engineer in a dozen counties. Further, electrical engineering will always attract to itself more than its legitimate share of students because of its novel attractiveness, and will tend to remain crowded. A glance at the list of students now entered in electrical engineering at our colleges will give eloquent testimony to above opinions."

Mr. James Milne, Lecturer in Electricity, Toronto Technical School, writes : "The great trouble in these days, I think, is in giving the young man the impression that if he receives a university training and graduates as an electrical engineer, that his services will be in demand, and that he will be looked up to by every one in the business, while the man who has been less fortunate as regards his education, but serves an apprenticeship to some trade, will be inferior in every respect.

"I believe in giving a fair education to all, but after that education has been attained the best thing that can be done is to learn a trade, and in learning that trade care should be exercised in the selection of the proper place.

"A young man who serves his apprenticeship in a small place, that is in a place where there is a scarcity of tools, etc., will in most cases turn out a better workman than the one who serves his time in a very large concern. In the smaller place ingenuity has to be exercised to get the various job done with the tools that are at hand, while in the large place special tools are ready made for almost everything. Therefore, in this respect the proper place to serve an apprenticeship is where a turn at everything may be got, such as patternmaking, fitting, turning, armature winding, etc., etc., and finishing up with the drawing office. This is what a complete apprenticeship should comprise. In these large manufacturing concerns where premiums are paid for instruction, the chances of knowing something at the end of the time are very slim indeed. There is one good thing about the arrangement, however—the money is generally thrown away by those who can afford it, and benefits the electrical concern, but whether or not it benefits the other party is a secondary consideration.

"In the smaller place the young man gets a fair insight into everything, and gets accustomed to the use of tools, and by and by is sent out to do various jobs and gain valuable knowledge and experience, and in a comparatively short time becomes a first-class practical man.

"Our learned brother, the electrical engineer, who has just graduated, finds that before he can be of much use he must gain practical knowledge, and to do this he has to get into some shop. Now here is where the sticker is ; he has been led to believe that he will not have to soil his hands, and that his brains will do it all. He never made a greater mistake in all his life.

"For some unknown reasons, parties in charge of shops or branches of any manufacturing concern will almost invariably refuse to employ these graduates, even although their services are offered gratis, and it is right here where our premium system comes in. They pay the money for instruction, and simply put in the necessary time, and that is about the end of it.

"Our man who has served his apprenticeship in the small concern and spent his spare time in reading up, sees an advertisement which reads something after the following : ' Wanted—a good man to take charge of an electric light plant—apply at so and so.' He, of course, applies, and in his application he states his experience, etc., together with all the rest of his redeeming qualities. For the situation we have 100 applications, 99 of which are from electrical engineers, graduates of some university. The parties to whom the applications have been sent read all the applications and comparing all their good points decide to give the situation to our practical man. This, I think, is pretty nearly the universal experience.

"When we bear in mind that what might be termed ' good jobs ' are very scarce, and in Canada there probably are about a dozen of them, which at present are all filled, the chances of an opening are very slim indeed, unless by some unforeseen calamity, such as a death, and the chances of our electrical engineer dying are probably about the same.

"There can be no doubt, however, that exceptionally smart men, no matter what profession—be it electrical, mechanical or otherwise—will make their mark, but this does not mean that our fortune favored electrical engineer will in very many cases knock out our man with the more practical ideas.

"It appears to me that if this education business is forced much higher that premiums will soon have to be offered for ordinary working men. If some manufacturing concern were to start up, say in the electrical business, and several foremen were required, I know I am correct when I state that every foreman would be selected from the ranks of those who have graduated (?) in the shop, and should it happen that say a superintendent or engineer was required to supervise the whole engineering part, you will find that the man appointed will be also a thoroughly practical man, with, however, a good technical education.

"To finish up with, I would like to state that if any young man is an aspirant for some fine job—nothing to do and big money for it—let him keep out of the electrical business, and more especially the lighting and power part of it. Without exception it is the most aggravating, most tantalizing and thankless of all, and an eternal source of worry from early morn to late at night, which accounts for those in this part of the business being old men long before their time."

Mr. J. J. Wright, manager Toronto Electric Light Co., writes : "At the risk of being considered somewhat of a pessimist, I am compelled to take a view of the question that I am afraid will not suit the sanguine enthusiast who considers that "electricity is only in its infancy" and that there are unbounded possibilities in the business for those who study electricity. The difficulty is that the field—in Canada, at least—is extremely limited. There are perhaps a dozen positions in the entire Dominion that it would pay the enterprising young electrical engineer to aspire to, and, unlike a mercantile business or the profession of physician or lawyer, they are not likely to materially increase. For instance, in each large city we have one street railway company and one, or at most, two, electric light companies. As these cities increase in size and population they will require, or rather, there will be room for; more lawyers and more doctors to maintain the present ratio, but when they are twice the size there will be but the one street railway and the one electric light company—if, indeed, these have not amalgamated and still further reduced the meagre opportunities. In the mercantile business or the legal or medical profession, or in the many branches of trade or commerce, an enterprising young man who has gained the requisite knowledge and saved a few dollars may enter and may hold his own with the best, and there is something lacking if he does not make a mark. At any rate, he has an equal chance with the rest to reach the top of the tree. But, no matter how smart he is as an electrician, if he feels the irksomeness and limitation of a subordinate position, he cannot start an electric light company or a street railway company of his own—at least, not often, and if he waits for the manager, or the superintendent, or the electrician of his local company to die or to hang himself,· he has at least the satisfaction of knowing that·the chance of survival is about equal, if indeed he does not starve to death in the meanwhile. There are but two manufacturing concerns in the country of any size who make a specialty of electrical apparatus. Let us say that one or two more come into existence. They have each their staff of "electricians," who supervise the construction and installation of apparatus, and these concerns are already capable of considerable increase in their output without any further skilled help as 'electrical engineers.' As electrical installations increase, as no doubt they will, especially in railway work, it follows that the increase in men employed will consist of the rank and file of intelligent mechanics and laborers, whose functions will consist simply in handling the apparatus put in

their hands—mechanics who will be perfectly competent to repair and operate the machinery, and laborers, firemen and oilers, whose wages will run from a dollar to two and a half dollars a day. The 'electrical engineer' does not appear to come in it at all. When I mention lawyers and doctors I do so, not that I advise a young man to take up these professions—because they are admittedly overcrowded—but as an illustration ; simply to show that in similar professions there are opportunities for the clever man to rise to the top if he has the qualifications ; whereas it matters not how great the qualifications if there are no opportunities or openings for their exercise. As a matter of fact, most of the professions appear to be overcrowded. The mistake of parents generally arises from their desire to see their sons do better than they themselves have—at least in appearance, and to have them rise in the social scale. The money they are expending in over-educating their boys for a struggle in an unremunerative profession has been made by every-day plain and prosaic hard work, but work that has borne excellent fruit. There is many a farm acquired in this way that to-day is mortgaged to the hilt to pay for the university education of a boy who was considered too good for his surroundings and who was not satisfied with plain honest work like his father did, though that work would have brought him independence and comfort instead of the worry and strife that is necessary even to wrest a moderate living in the midst of the fierce competition of professional life. I believe that if a tithe of the same training and ability, method and scientific knowledge required for this were applied to the operations of the farm, that the results in wealth, comfort and happiness would far transcend the best that can be gleaned in the care-strewn paths of professional life.

"I am afraid I am making this letter somewhat long, and getting a little off the track of the 'Electrical Engineer,' but holding the position I do, and being brought into contact as I am every day with many who have the idea that electricity is the coming thing and who want to learn electricity with a view of being as it were 'in the swim,' I cannot refrain from giving expression to my views. In one of your articles in the November issue of your paper you voice the complaint of an electrical man who kicks about this very thing. Because the country plant wants only a skilled laborer to look after its meagre apparatus, and cannot afford a school-taught electrician with a stand-up collar, he considers it a grievance. The fact is that the lot of the electrical operator is not a happy one, and not nearly as desirable as the ordinary observer who sees no further than the outside ·glamour is apt to think it is. The farm-laborer has a picnic in comparison. If he has to rise at day dawn, at any rate he, gets his sleep at night—night with the electrical man is his time of greatest tension. There is no let up Sunday or week day, holidays as well ; life, as Mr. Mantalini would say,. is 'one deuced horrid grind.' The running of a high tension station is a hot, dirty, and to some, extent, dangerous job, and the most of the work is done while the rest of the world are enjoying their relaxation from toil. Competition with other methods of lighting compel the station manager to. exercise the most rigid economy. Therefore his work is subdivided and specialized. It is a 'one man one job' business. He wants a man to do one thing AND TO STAY AT IT. Therefore the young man who wants to learn the whole of it stands little chance. A modern electrical installation cannot be a training school or there would be an end of efficiency. It is often a matter of wonder to me what is to become of the comparatively large number of graduates in electricity and electrical engineering that are yearly turned out from our educational institutions. There will undoubtedly be a few who will drop into positions that will from time to time be available. More who will be compelled to work with the rank and file whose numbers can be recruited equally well from our Schools of Science, or from the grease pot and wiping rag, but the majority will be compelled to remain in a state of 'innocuous desuetude' and vegetate upon the wealth of their parents till some other career opens out to them."

PUBLISHED ON THE FIFTH OF EVERY MONTH BY

## CHAS. H. MORTIMER,

OFFICE : CONFEDERATION LIFE BUILDING,

Corner Yonge and Richmond Streets,

TORONTO,    -    -    CANADA.

Telephone 2362.

NEW YORK LIFE INSURANCE BUILDING, MONTREAL.

Bell Telephone 2209.

***ADVERTISEMENTS.***

Advertising rates sent promptly on application. Orders for advertising should reach the office of publication not later than the 24th day of the month immediately preceding date of issue. Changes in advertisements will be made whenever desired, without cost to the advertiser, but to insure proper compliance with the instructions of the advertiser, requests for change should reach the office as early as the 2nd day of the month.

***SUBSCRIPTIONS.***

The ELECTRICAL NEWS will be mailed to subscribers in the Dominion, or the United States, post free, for $1.00 per annum, 50 cents for six months. The price of subscription should be remitted by currency, registered letter, or postal order payable to C. H. Mortimer. Please do not send cheques on local banks unless 5 cents is added for cost of discount. Money sent in unregistered letters will be at senders' risk. Subscriptions from foreign countries embraced in the General Postal Union $1.50 per annum. Subscriptions are payable in advance. The paper will be discontinued at expiration of term paid for if so stipulated by the subscriber, but where no such under-standing exists, will be continued until instructions to discontinue are received and all arrearages paid.

Subscribers may have the mailing address changed as often as desired. When ordering change, always give the old as well as the new address.

The Publisher should be notified of the failure of subscribers to receive their paper promptly and regularly.

***EDITOR'S ANNOUNCEMENTS.***

Correspondence is invited upon all topics legitimately coming within the scope of this journal.

The "Canadian Electrical News" has been appointed the official paper of the Canadian Electrical Association.

## CANADIAN ELECTRICAL ASSOCIATION.

### OFFICERS:

PRESIDENT :

JOHN YULE, Manager Guelph Light & Power Company, Guelph, Ont.

1ST VICE-PRESIDENT :

L. B. McFARLANE, Manager Eastern Department, Bell Telephone Company, Montreal.

2ND VICE-PRESIDENT :

E. C. BREITHAUPT, Electrical Engineer, Berlin, Ont.

SECRETARY-TREASURER :

C. H. MORTIMER, Publisher ELECTRICAL NEWS, Toronto.

EXECUTIVE COMMITTEE :

GEO. BLACK, G. N. W. Telegraph Co., Hamilton.

J. A. KAMMERER, General Agent, Royal Electric Co., Toronto.

K. J. DUNSTAN, Local Manager Bell Telephone Company, Toronto.

A. M. WICKENS, Electrician Parliament Buildings, Toronto.

J. J. WRIGHT, Manager Toronto Electric Light Company, Toronto.

ROSS MACKENZIE, Manager Niagara Falls Park and River Railway, Niagara Falls, Ont.

A. B. SMITH, Superintendent G. N. W. Telegraph Co., Toronto.

JOHN CARROLL, Sec.-Treas. Eugene F. Phillips Electrical Works, Montreal.

C. B. HUNT, London Electric Company, London.

F. C. ARMSTRONG, Canadian General Electric Co., Toronto.

## CANADIAN ASSOCIATION OF STATIONARY ENGINEERS.

President, JAMES DEVLIN,          Kingston, Ont.
Vice-President, E. J. PHILIP,     11 Cumberland St., Toronto.
Secretary, W. F. CHAPMAN,         Brockville, Ont.
Treasurer, R. C. PETTIGREW,       Hamilton, Ont.
Conductor, J. MURPHY,             Montreal, Que
Door Keeper, F. J. MERRILL,       Ottawa, Ont.

TORONTO BRANCH NO. 1.—Meets 1st and 3rd Wednesday each month in Engineers' Hall, 61 Victoria street. John Fox, President ; Chas. Moseley Vice-President ; T. Eversfield, Recording Secretary, University Crescent.

MONTREAL BRANCH NO. 1.—Meets 1st and 3rd Thursday each month, in Engineers' Hall, Craig street. President, John Murphy ; 1st Vice-President, J. E. Huntington ; 2nd Vice-President, Wm. Smyth ; Secretary, B. Archibald York ; Treasurer, Peter McNaugton.

ST. LAURENT BRANCH NO. 2.—Meets every Monday evening at 43 Bonsecours street, Mont eal. R. Drouin, President ; Alfred Latour, Secretary, 306 Delisle eet, St. Cunegonde.

BRANDON, MAN., BRANCH NO. 1.—Meets 1st and 3rd Friday each month in City Hall. A. R. Crawford, President ; Arthur Fleming, Secretary.

HAMILTON BRANCH NO. 2.—Meets 1st and 3rd Friday each month in Maccabee's Hall. Wm. Norris, President ; E. Teeter, Vice-President ; Jos. Ironside, Corresponding Secretary, Markland St.

STRATFORD BRANCH NO. 3.—John Hoy, President ; Samuel H. Weir, Secretary.

BRANTFORD BRANCH NO. 4.—Meets 2nd and 4th Friday each month. J. B. Forsyth, President ; Jos. Ogle, Vice-President ; T. Pilgrim, Continental Cordage Co., Secretary.

LONDON BRANCH NO. 5.—Meets on the first and third Thursday in each month in Sherwood Hall. G. B. Risler, President ; D. Campbell, Vice-President ; Wm. Meaden, Secretary-Treasurer, 533 Richmond street.

GUELPH BRANCH NO. 6.—Meets 1st and 3rd Wednesday each month at 7.30 p. m.   H. Geary, President ; Thos. Anderson, Vice-President ; H. Flewelling, Rec.-Secretary ; P. Ryan, Fin.-Secretary ; Treasurer, C. F. Jordan.

OTTAWA BRANCH NO. 7.—Meets every second and fourth Saturday in each month, in Borbridge's hall, Rideau streett ; Frank Robert, President ; F. Merrill, Secretary, 352 Wellington street.

DRESDEN BRANCH NO. 8.—Meets 1st and Thursday in each month.  Thos Steeper, Secretary.

BERLIN BRANCH NO. 9.—Meets 2nd and 4th Saturday e ch month at 8 p.m. J. R. Utley, President ; G. Steinmetz, Vice-President ; Secretary and Treasurer, W. J. Rhodes, Berlin, Ont.

KINGSTON BRANCH NO. 10.—Meets 1st and 3rd Thursday in each month in Fraser Hall, King street, at 8 p. m.   President, F. Simmons ; Vice-President, J. W. Tandvin ; Secretary, A. Macdonald.

WINNIPEG BRANCH NO. 11.—President, G. M. Hazlett ; Rec.-Secretary, J. Sutherland ; Financial Secretary, A. B. Jones.

KINCARDINE BRANCH NO. 12.—Meets every Tuesday at 8 o'clock, in McKibbon's block.   President, Daniel Bennett ; Vice-President, Joseph Lighthall ; Secretary, Percy C. Walker, Waterworks.

WIARTON BRANCH NO. 13.—President, Wm. Craddock ; Rec.-Secretary, Ed. Dunham

PETERBOROUGH BRANCH NO. 14.—Meets 2nd and 4th Wednesday in each month   W. L. Outhwaite, President ; W Forster, Vice-President ; A. E. McCallum, Secretary.

BROCKVILLE BRANCH NO. 15.—Meets every Monday and Friday evening. President, Archibald Franklin ; Vice-President, John Grundy ; Recording Secretary, James Aikins.

CARLETON PLACE BRANCH NO. 16.—Meets every Saturday evening. President, Jos. McKay ; Secretary, J. D. Armstrong.

## ONTARIO ASSOCIATION OF STATIONARY ENGINEERS.

President, A. AMES,         Brantford, Ont.
Vice-President. F. G. MITCHELL     London, Ont.
Registrar, A. E. EDKINS    88 Caroline St , Toronto.
Treasurer, R. MACKIE,      28 Napier st., Hamilton.
Solicitor, J. A. McANDREWS,     Toronto.

### BOARD OF EXAMINERS.

TORONTO—A. E. Edkins, A. M. Wickens, E. J. Phillips, F. Donaldson. J. Bain.

HAMILTON—R. Mackie  T. Elliott.

BRANTFORD—A. Ames, care Patterson & Sons.

OTTAWA—Thomas Wesley.

KINGSTON—J. Devlin, (Chief Engineer Penitentiary), J. Campbell.

LONDON—F. Mitchell.

NIAGARA FALLS—W. Phillips.

Information regarding examinations will be furnished on application to any member of the Board.

**Central Station Insurance.**   CENTRAL stations are regarded as very poor risks, generally, by the fire insurance companies ; and no wonder. In very many cases an electric lighting business is undertaken as collateral to a planing or saw mill, and the building containing the electrical machinery is added in the form of a small shed, to the mill. The leads from the machine to the outside lines are looped about anywhere, and very inefficient precautions are taken against bad grounds on the machine or switchboard. Now conditions that increase the fire risk, are very favorable for expensive leaks and wastes of current, so that any precautions taken to lessen the risk are of double benefit as tending to lower the operating costs as well. Insurance people are apt to be very cautious and conservative when dealing with electricity, and often impose conditions that seem to be unnecessarily strict ; but then they are in a position to more or less dictate terms ; and as a matter of fact, the method of making installations in smaller towns is apt to be very lax.

**Polyphase Electric Railways.**   THE application of alternating currents to street railway purposes has long been a problem that now seems to be in a fair way of being solved. The advantages of high voltage transmission are too obvious to require any mention, and would be of special importance in the development of many country railway systems, the length of whose routes, however, would necessitate a very large expenditure for feeder copper on the usual 500 volt system, apart from the complication introduced by the double trolleys and overhead work required by any polyphase motor when used for traction purposes. The fact that induction motors, as now constructed, are built for only one efficient speed has been regarded as an insuperable obstacle to their use on cars. This, however, is not so very serious a difficulty after all, for on country routes where stops are infrequent, and at certain definite places only, it is not required to vary the speed, except, of course, on starting, and perhaps on very sharp curves and crossings of roads. This condition has actually been met in practice, by the expedient of so constructing the motor as that certain poles can be cut in or out, thus causing the rotor to travel at a less or a greater angular speed. Rumors are current of satisfactory results having been reached by experimenters in this field; and we may reasonably expect within the next very few years to have polyphase railways running with commercial success.

**The Steam Plant in Central Stations.** WE have lately presented in our columns, a paper on the subject of steam engine indicators and the advantages to be derived from their frequent use. The steam part of an electric power house is one that receives but little attention very often from the owners of such plants, and their ill success is due, in great part, to this laxity. We have lately had occasion to critically examine indicator cards taken from a number of engines of different makes that have been continuously in use for a number of years, and if all our readers had the same opportunity of seeing the results as we have, we believe the use of the indicator would be much more frequent than it is. There seems to be an opinion among the smaller steam users, that once an engine has been put in—there is an end of it. They conclude that iron and steel last for ever; that there is no wear and tear; and that if the engine is properly oiled up by a $1.00 a day mechanic, it requires no further attention. We desire to most emphatically protest against this idea, and urge our readers to keep a constant watch over the performance of their engines. Valves will inevitably wear, pistons have an unfortunate habit of abrading, and the result is some inefficiency of steam use that consumes more fuel than would otherwise be necessary. There is no more useful—we might say essential—instrument in a power house than an indicator, nor is there one that is more seldom found. An intelligent man with an indicator and voltmeter and ammeter can, in the course of a month, learn more about the economical generation of electricity, and the manufacture of dividends, than a whole year of study in mathematical works for which his technical education is not sufficient.

**Improved Methods.** IT is exceedingly interesting to observe the growth of electrical enterprises in Canada during the last few months. Not only is this evident in the number of new enterprises inaugurated, but in their varied character; and more especially in the fact that considerably more attention is being paid to the preliminary engineering than used to be the case. It seems to be more generally recognized that electric lighting is a business by itself, and deserves careful attention as such. The "survival of the fittest" principle has also been exemplified in the fact that there really seems to be less poor machinery on the market, and more satisfactory apparatus. The purchasing public are becoming more awake to the fact that machinery should not be selected on the basis of its price only, but also in consideration of its inherent electrical and mechanical excellence, and that there is sufficient difference between good and poor machinery, to more than

counterbalance a considerable difference in first cost. We have also observed that there is a greater variety in the types of machines—both steam and electrical—purchased than used to be; and the reason for this seems to be that purchasers are beginning to rely less on their own ideas as to what is best, and to advise with independent authorities. The first result of this course has been that a better all-round class of construction has been undertaken; and the second that purchasers have not been limited to one or two makes of machine, but have felt less hesitation in purchasing in the open market.

**Electrical Progress.** CONCENTRATION and improvement are the most marked evidences of electrical progress. Within the last few months many electrical companies have been making great steps in those directions, and, we are glad to learn, always with satisfactory results. It is very interesting and instructive to note the particular lines along which such new dispositions seem to be principally made, and their perfect similarity should give good forethought to electrical men who desire to keep their places in the front rank of intelligent operators. Investigation into the operating economies seems to have had a large influence in suggesting changes; first, in an entire reconstruction of distributing systems, lines and transformers; next, in the substitution of a few large generating units of modern make, for many small units of a type which, although representing the best that could be made some years ago, has been proved wanting. Next we have observed an encouraging tendency to build new power houses and to pay considerable attention to their designing for convenience and efficiency; and lastly, and, we think, best of all, there is an increasing demand for a better educated and more capable class of operator. We take some credit for having in a measure influenced this improvement. During the past eighteen months we have repeatedly called attention, both editorially and through papers by competent authorities, to the many important subjects for investigation by enterprising electrical men. It would be greatly to the interest of the whole electrical generating industry, if they could arrange to tackle central station problems as an association instead of as individuals. The suggestion has been made that there might with advantage be established under competent direction, a central station laboratory, in which accurate tests could be made of the quality of lamps, transformers, carbons, wire and electrical supplies of all kinds. We would be glad to have opinions on this subject.

### PERSONAL.

Mr. A. S. Colpitt has been appointed city electrician for Halifax, N. S.

Mr. A. E. Edkins, Registrar of the O. A. S. E., has recently returned from a trip to England.

Hon. Louis Tourville, who was largely interested in the Tourville Electric Light Co., of Louisville, Que., died in Montreal early in November.

On the 25th of November, Mr. Adam Rutherford, secretary-treasurer of the Hamilton, Grimsby and Beamsville Railway, was married at Grimsby to Miss Marie Nelles, of that town. Rev. C. R. Lee officiated.

It is stated that Mr. Romaine Callender has gone to England to commence a telegraphic business between that country, France, Germany, etc., and the United States. Mr. Callender is said to have invented a new system of telegraphy, making it possible to turn out over half a million words in less than 28 seconds.

## SOME ELECTRICAL INDUSTRIES OF ST. CATHARINES.

SITUATED on the old Welland canal, the city of St. Catharines is provided with a water power such as few cities in Ontario can boast of. Each lock is harnessed to furnish power to some industry, and flumes run in all directions to convey the water to turn the wheels of commerce. In electrical enterprises the city occupies a prominent position, having a first-class electric street railway, lighting station, and a number of other electrical industries. The street railway is known as the

### PORT DALHOUSIE, ST. CATHARINES AND THOROLD ELECTRIC STREET RAILWAY.

Although not yet reaching Port Dalhousie, its extension to that point is now under way. It performs the functions of a freight and passenger service, the freight being carried by flat cars. The line extends from St. Catharines to Thorold, passing through Merritton. A branch also runs to the cemetery outside of St. Catharines.

Fifteen years ago Dr. T. S. Oille (now president of the Niagara Central Railway) organized a company to operate a horse car railway between St. Catharines and Thorold. This was operated successfully until 1887 when Mr. E. E. Smith gained control of the road, and converted it into the first electrical road in Canada. It was of the Van Depoele system, the trolley travelling on top of double trolley wires, and being connected to car by a flexible rubber covered cable.

During the past spring the road was converted into an up-to-date electrical railway, under the supervision of the present owners, Messrs. Dawson & Symmes, who are practical electricians. The conversion of the road from the old Van Depoele system to the modern equipment has placed St. Catharines alongside of her sister cities.

The machines in the power house were overhauled and a new C. G. E. generator added. A new electric water governor of Mr. Symmes' design does admirable service, as the route is so hilly and tortuous that a governor is indispensable.

The track, eight miles in length, is single, and is laid with 56 and 66 pound rails. The overhead construction is No. o trolley wire and ooo feed wire. There are no rigid brackets.

The rolling stock consists of eight Patterson & Corbin cars. Three are closed and three open motor cars, the other two being trailers. The motor cars are equipped by the Canadian General Electric Co.

The offices and barns are on the main street of the city. The barns are well equipped with tools, by means of which the company are enabled to make nearly all their repairs.

Like many other street railway companies, a park is controlled by the company. It consists of six acres, well situated and nicely wooded.

The road has improved wonderfully under the management of Messrs. Dawson & Symmes, and is one of the best equipped in the province.

### ST. CATHARINES ELECTRIC COMPANY.

The city is supplied with electric light by the St. Catharines Electric Company, the superintendent of which is Mr. McNaugh, who has a wide experience in electrical matters. Over a thousand incandescent lights are in use, also 75 arcs, 62 on the streets and 10 in business houses, three being used for lighting the plant itself.

The plant is situated on the old canal, a considerable distance from any other buildings, thus lessening the fire risk. It is in charge of Mr. Chas. Steel, who has been with the company for many years.

A heavy head of water operates two water wheels of 135 h.p. and 100 h.p. respectively. These wheels each turn a shaft 30 feet in length and five inches in diameter. One shaft drives the C. G. E. 1000 light alternator and exciter and a Minneapolis water wheel governor, while the other drives the three 35 light Royal arc machines. The current of the alternator is controlled from a skeleton C. G. E. switch-board, with its complement of instruments, etc. On the wall near the arc machines are the arc instruments, complete with lightning arrestors, etc.

Over the wheel room, which is in the annex, is the repair shop and lamp testing room, well supplied with requisites.

One interesting feature noticed was an alarm bell, which is rung by the person approaching the front door. As he approaches within a few feet of the door he necessarily steps on the platform, underneath which are metallic plates which come in contact and form a circuit on which is a bell in the dynamo room.

### COOK & SONS' ELECTRIC POWER PLANT.

Within the last few years, the use of electric motors for operating small plants has become so important that the above named firm, who owned a valuable water site and building on the old canal, decided to put in a generator. The head of water of 12 feet turns two Little Giant turbines of 96 h.p. each, which together operate a shaft 30 feet long by 4½ inches in diameter, on which is a fly-wheel 7 feet in diameter. A belt from this fly-wheel drives a 100 K. W. C. G. E. generator. The current is directed by a frame switch-board, on which are C. G. E. instruments, except a direct reading Weston volt meter. The instruments consist of an ammeter, a circuit breaker, a volt meter, a rheostat, and two lightning arresters. The wheels are controlled by a hand regulator, which is a greater economizer than an electric governor, as the latter uses up too much current for a small machine. Over twenty-five patrons are supplied with power from motors of from 1½ K.W. to 24 K. W. The noted street car builders, Messrs. Patterson & Corbin, use a 12 K. W. motor, and Cook & Son use a large motor in their planing mill.

Messrs. Cook & Son intend catering for lighting patronage, and a 1000 light alternator will be in operation at an early date.

### THOROLD ELECTRIC LIGHT COMPANY.

The suburban town of Thorold, connected by the street railway, is lighted by the Thorold Electric Light Co., whose plant is also located on the old canal. From an 11 feet head two wheels are turned, one a Little Giant and one a Leffel of 60 h. p. each, which in turn operate a 25 foot shaft. From this shaft are driven a 600 light C. G. E. alternator and exciter and two 25 Ball arc machines. A C. G. E. skeleton switch-board with its full equipment of C. G. E. instruments is connected to the alternator. To the Royal arc machines are connected Ball arc instruments, which are fastened on the wall.

Below the dynamo room in the basement are the wheel pits, and down there Mr. Jas. McGill, jr., the electrician, has a dark room where he finishes his photographs, for Mr. McGill is an enthusiastic amateur photographer. In the rear of the dynamo room is a well equipped repair shop. In this shop are piled, ready for shipment, boxes of Mr. McGill's patent wire connector, which was described in the columns of this journal some months ago.

The proprietors of the plant are Messrs. John McGill, sr., and John Battle, the latter being also the proprietor of the Thorold Cement Works. Their business has been quite successful, and larger machines may shortly be put in.

Prominent among the other electrical concerns of St. Catharines may be mentioned the Packard Electric Co., Patterson & Corbin, and T. L. Wilson's acetylene gas works, descriptions of which will probably be given in a future issue.

## THE ALLISTON MILLING COMPANY.

The above company some time ago secured control of the Alliston flour mills, in which was located the electric lighting plant. Finding that the dust arising from the manufacture of the flour had a bad effect on the successful working of the electrical machines, they decided to build a separate building for the electric plant. This building is a red brick cottage structure, 32 x 40 feet. It is divided into three parts, one for the boiler, one for the engine, and one for the dynamos. Through the wall next the mill runs a shaft, which, when water is high, can run both the mill and the dynamos, or when the water

THE ALLISTON MILLING COMPANY'S BUILDING.

is low the engine can run both the mill and the dynamo.

In the mill are two Little Giant water wheels, one of 25 h.p. and the other of 40 h.p. These wheels can be connected to the 12 foot shaft entering into the dynamo room when required.

In the boiler room is an 80 h.p Osborne-Killey boiler, supplying steam to a 75 h.p. Osborne-Killey engine. The engine and a hot water heater occupy one room. The water goes to the boiler at 205° Fahrenheit. A plunger pump is connected to the fly wheel.

The machines in the dynamo room are, first, a 500 light C. G. E. alternator with exciter, 700 lamps being installed throughout the town ; second, a Reliance arc machine of thirty-five lights capacity, twenty-two of which are in use—eleven commercial and eleven on the streets. On the C. G. E. skeleton switch-board are the instruments of both machines.

The dynamo room is lighted by an arc light, and the other two rooms and the mill by incandescent lights.

It is probable that waterworks machinery will be placed under the building, which, if installed, will necessitate an addition to the building.

## RECENT CANADIAN PATENTS.

Mr. Ferdinand de Camp, of Berlin, Germany, has been granted a patent, No. 53,449, for a furnace and apparatus for burning coal dust, which consists of the combination with a coal dust feeding device of a fan so arranged as to propel the coal dust together with air into the furnace, and a rotary shifting cylinder to uniformly distribute the same, also of an arrangement which closes the issue of the coal dust hopper in such a manner that the coal dust taken up by the conical portion of the worm from the hopper is conveyed to the cylindrical enlarged portion in a loose condition for further conveyance.

A patent has been granted to Joseph Hardill, Benson French, and R. T. Harding, all of Stratford, Ont., for a steam engine, with cylinder provided with two pistons and rods, and a suitably operated valve whereby steam is directed against the outer ends of faces of the pistons to force them inwardly, and at the end of the inward stroke is admitted into the cylinder between the pistons to force them outwardly. It is claimed to consist of a combination of the cylinder, the piston and piston heads having movement therein, the steam chest, elongated parts connecting the central portion of the chest to the cylinder, the valve provided with double ports designed to co-act with a central port and elongated ports, the exhaust port in the valve designed to connect with the central port, etc.

The Bell Telephone Company, of Montreal, have been granted a patent, No. 53,605, for a telephone key-board, also for a keyboard apparatus for telephone switchboards.

A patent for converting simple into polyphase alternating circuits has been granted to Charles S. Bradley, of Avon, N. Y. It consists in the combination with a simple alternating current circuit, of a plurality of transformers supplied thereby, means for creating a difference of phase in the several transformers, and interconnections for combining the displaced phases to produce a resultant phase, and a plurality of coils in inductive relation to the several phases, said coils being connected in series relation.

The longest commercial distance at which the long-distance telephone is now operated is from Boston to St. Louis, a distance of 1,400 miles. The line is almost twice as long as any European telephone line.

It is reported that the construction of the Huron and Ontario Electric Railway is to be commenced at once. The road will extend from Kincardine and Goderich via Walkerton to Eugenia, the junction town, thence north to Meaford and south to Port Perry. A meeting of the provisional directors will be held in a few days to ratify the agreement with the contractor.

## CANADIAN ASSOCIATION OF STATIONARY ENGINEERS.

NOTE.—Secretaries of Associations are requested to forward matter for publication in this Department not later than the 24th of each month.

### ANNUAL DINNER OF TORONTO NO. I.

THE annual dinners of Toronto Association No. 1 have become to be looked forward to by the members and their friends with much anticipation and interest, and each year their growing popularity is shown by an increase in attendance. This year it was the tenth annual banquet, and was held on Thanksgiving eve. The scene of the festive gathering was the Palmer House, and the spacious dining-room afforded ample accommodation for the guests, who numbered about 200. The tables permitted of all being comfortably seated at one time, which was a marked improvement over former years. The ten large electric light chandeliers, each decorated with four union jacks, presented an attractive appearance. The arrangements were complete and well carried out.

The duties of Chairman devolved upon Mr. John Fox, President of the Association. To his right were Mr. John Yule, President of the Canadian Electrical Association, Mr. E. H. Keating, City Engineer, Ald. McMurrich, and Prof. Galbraith, Principal of the School of Practical Science. On his left were Dr. Orr and Past-Presidents Wickens and Lewis.

The visitors from outside places were Messrs. R. C. Pettigrew, Treasurer of the Executive Committee, and R. Mackie, Treasurer of the Ontario Association, both of Hamilton; G. M. Hazlett, President of Winnipeg No. 11, and W. L. Oathwaite, President of Peterboro' No. 14.

The Chairman read letters of regret from the following persons: Messrs. R. J. Fleming, Mayor of Toronto; O. P. St. John, President Marine Engineers' Association; A. Ames, President Ontario Association; James Devlin, President Executive Council; John Galt, C. E., Mechanical Engineer; and J. C. Robb, of the Boiler Inspection and Insurance Company.

The supper provided was of excellent quality, and such as would satisfy the appetites of the most ravenous. The menu was as follows:

### MENU

OYSTERS.
New York Counts (Raw au Lemon).

SOUP.
Cream of Oyster.

FISH.
Boiled Sea Salmon, Hollandeuse Sauce.     Hors d'Oeuvres.     Pomme de Terre.
Parisienne.     Celery.     Pickled Beets.     Red Cabbage.

BOILED.
Sugar Cured Ham, with Spinach.     Leg Southdown Mutton, Caper Sauce.

ROASTS.
Sirloin Beef, Yorkshire Pudding.     Young Turkey, Stuffed, Cranberry Sauce.
Spring Duck, Apple Sauce     Haunch Venison, Red Currant Jelley.

COLD MEATS.
Tongue.     Pigs' Feet.     Lambs' Tongues.     Head Cheese.

VEGETABLES.
Tomatoes.     Green Peas.     Boiled and Mashed Potatoes.

SALADS.
Chicken.     Celery.     Bohemian.

PUDDINGS.
English Plum, Brandy Sauce.     Baked Coccanut, a la Creme.

PASTRY.
Apple.     Mince.     Lemon.

DESSERT.
Charlotte Russe.     Lemon Sponge.

JELLIES.
Champagne.     Strawberry.     Sherry Wine.

FRUIT.
Florida Oranges.     Bananas.     Snow Apples.
Green Tea.     Black Tea,     French Coffee.
Crackers and Cheese.

Shortly after ten o'clock, after an hour or so had been spent in disposing of the viands, the Chairman addressed a few words of welcome to the guests, and proposed the toast of "The Queen," which was responded to by the singing of the National Anthem.

After a song by Mr. Grant, Mr. A. M. Wickens was called upon to respond to the toast, "Canada, Our Home." He referred to the large number of men of stability to be found in Canada, who, he said, were proud to call themselves Canadians wherever they went. He hoped that, instead of having five million people in Canada, we would shortly have twenty millions.

Ald. McMurrich, in the absence of the Mayor, acknowledged the toast of the city of Toronto. He came to the city over 52 years ago and had enjoyed every day since. He was interested in the success of the Stationary Engineers, and fully recognized the benefits to be derived from such an association. Their positions were among the most responsible which any person could occupy.

A song by Mr. Blackgrove was followed by the toast, "The Manufacturers," coupled with which were the names of Mr. John Main, of the Polson Company, and Mr. Weeks, of the Weeks-Eldred Company. Each spoke of the kindly feeling which existed between the manufacturers and the engineers. Mr. Main thought the prospects for the coming season were promising, and hoped soon to observe the return of good times.

A song by Ald. McMurrich was well received, after which the toast of the "Educational Interests" was proposed by the Chairman, to which Prof. Galbraith, of the School of Practical Science, and Dr. Orr, of the Technical School, responded.

Prof. Galbraith was pleased with the success of the Engineers' Association. It had begun in a small way, and for a time was not particularly prosperous, but was now assured of success. He stated that of late much attention had been given to the relative merits of low speed and high speed engines, but the problem had as yet never been satisfactorily settled. Electricians were now making engines half way between high speed and low speed. He referred to the recent experiments of Dr. Jakes, of Boston, who had endeavored to produce electricity by means of ordinary combustion, doing away with both the steam engine and the boiler. He stated that the experiments were not considered successful. In his opinion the only hopeful way by which the steam engine could be done away with in operating electrical machinery was by the use of the steam turbine, which was as yet only in its infancy. Late developments along this line strengthened this conviction, and so far had met with a moderate degree of success. By this method the steam from the boilers was blown into a turbine, thereby causing the turbine to revolve. He could not understand how it would be possible to do away with the boilers also. In his closing remarks he stated that we required a little more imagination among our technical teachers. We had always looked up too much to other countries.

Mr. Charles Palmer then voluntarily favored the company with a song of his own composition, entitled, "My Own Irish Love," which received a very hearty encore, as did also his response.

Dr. Orr was called upon, and spoke especially in regard to the Toronto Technical School, which, he said, had been an unprecedented success. In five years the number of pupils in attendance had grown from 246 to

1310. This in itself proved conclusively that the educational system of Ontario had not met the requirements of the country. Mechanics must be educated in this country as well as elsewhere. Germany had superseded England in manufacturing industries as the result of her thorough technical schools, and he believed we should have primary technical schools in every manufacturing centre in Ontario. He was pleased to learn that it was proposed to introduce manual training in our public schools. At present our children were educated only for one course—a non-productive course, and no one could make a good mechanic without a thorough knowledge of mathematics. The Toronto Technical School had of late been the subject of much adverse criticism, of which he thought it was entirely undeserving. Taking the statistics in connection with education in Toronto, it was shown that 34,000 pupils were attending the public schools, 1,300 the High schools, and 1,310 the Technical School. The cost per pupil in the public schools was $19, in the High schools $31, and in the Technical School $6. He thought this clearly showed that the money given to the Technical School was well and profitably spent. This school, he said, had been established largely as the result of efforts on the part of members of the Canadian Association of Stationary Engineers, in conjunction with Mr. John Galt, C. E. He wished the Association continued success.

A song by Mr. Ferrier followed.

The names of Mr. E. J. Philip, Vice-President, and Mr. R. C. Pettigrew, Treasurer, were coupled with the toast of "The Executive." Mr. Philip said that the Executive had brought up many new schemes, and were extending the field of the association. Two new branches had been started during the year. He hoped that they would shortly succeed in obtaining a license law, which he considered an absolute necessity. A thorough engineer required as much knowledge as any other profession, yet they had no law. There were plants in Toronto where hundreds of people were working above the boilers, which were in charge of incompetent and unreliable men. Their efforts to secure legislation had been defeated in the Ontario House, and now they proposed to endeavor to get an act through the Dominion Parliament.

Mr. Pettigrew referred briefly to the advantages to be gained by employing a licensed engineer.

A song by Mr. Blackgrove, entitled "Remember You Have Children of Your Own," was much appreciated.

The next toast was "Sister Societies."

The first speaker was Mr. John Yule, President of the Canadian Electrical Association. He congratulated the engineers upon the apparent success of their association, and thought their object was a worthy one and deserving of support.

A recitation was given by Mr. Post, and Mr. George Mills, President of the Brotherhood of Locomotive Engineers, spoke on behalf of that organization, which, he said, had been in existence for 33 years. It was first started in Rochester with twelve men, but now had a membership of 35,000. They had a Legislative Board both for Ontario and the Dominion, of which he was chairman.

Mr. A. E. Edkins represented the Ontario Association of Stationary Engineers, which, he said, was the outcome of the labors of the Canadian Association. Four

years ago a Board of Examiners was appointed to issue certificates, and he had just had the pleasure the previous evening of issuing the 650th certificate. Much credit was due the association for raising the standard of steam engineering over what it was fifteen years ago. Many engineers had been better fitted to fill their positions by the efforts put forth to qualify themselves to pass the examinations. During his visit to England he had been struck with the technical schools there, and had the pleasure of visiting the Birmingham Technical and Art School. In their efforts to secure legislation, he said the engineers should receive the support of steam users, and pointed to the advantages which would accrue. The English law did not require an engineer to hold a certificate, but steam users were compelled to employ a competent person and to have their plants inspected once a year. The owner is held responsible for any accident, and is therefore interested in preventing the same.

At this stage Dr. Orr proposed the toast of Toronto No. 1, to which Mr. James Huggett and Mr. Wickens responded. Mr. Huggett referred to educational matters and to the new library which had been commenced, while Mr. Wickens, in showing the advantages to be derived from the association, pointed out that many engineers had been thus enabled to double their salaries in a few years.

After a response to the toast, "Visiting Brethren," by Mr. Robert Mackie, of Hamilton, another song was rendered by Mr. Grant.

Mr. Hazlett, of Winnipeg, was also called upon. He stated that in Winnipeg they had a law relating to the inspection of boilers, and had tried to get a bill passed licensing engineers.

The hearty reception given to the toast of Mr. Fox, President of Toronto No. 1, was acknowledged in a few well-chosen remarks, and after "The Press" had been duly honored, and responded to on behalf of the Canadian Engineer and the ELECTRICAL NEWS AND STEAM ENGINEERING JOURNAL, the singing of "God Save the Queen" closed the evening's programme.

To the committee in charge of the arrangements is largely due the success of the banquet. Messrs. Thos. Eversfield, G. C. Mooring, A. M. Wickens, John Fox, J. Marr and J. Bain were the members thereof. Mr. R. G. Stapells presided at the piano.

---

### TORONTO NO. 1.

At the regular meeting of the above association, held on Wednesday, the 2nd inst., one candidate was initiated. After the transaction of routine business, Bro. E. J. Philip gave the first of a number of short talks on "Natural Philosophy," which was greatly appreciated, and for which he received a hearty vote of thanks from the members. Another talk will be given at the next regular meeting, at which it is hoped a large attendance will be present.

---

### LONDON NO. 5.

Our association met on the 19th ultimo and re-organized, with the following members as officers: G. B. Risler, Advertiser office, president; D. Campbell, Pottersburg, vice-president; W. Meaden, secretary-treasurer, re-elected; Duncan McKinley, recording secretary; Wm. McLean, guard. It was decided at the meeting to permit engineers to join before the first of January next at a lower rate. Some consideration was given to the Scranton Correspondence School. Our meetings are held on the first and third Thursdays in each month in Sherwood Hall.

D. McKINLEY, Rec.-Secretary,
292 Ridout street.

### HAMILTON NO. 2.

At the last regular meeting a paper on "Heat," which will be found printed in this number of the NEWS, was read by Mr, James Gill, B. A., of the Collegiate Institute. As the result of an interview by the officers of the association with the School Board, that body has agreed to stipulate that in future engineers in steam heated schools must hold at least a 3rd class certificate. Quite a number of new members have been received into the association recently.

JOSEPH IRONSIDE, Secretary.

### KINGSTON NO. 10.

At the last meeting of Kingston Branch No. 10, it was decided (by a standing vote of the members present) to change the meeting nights from the 1st and 3rd Tuesdays to the 1st and 3rd Thursdays of each month, the next meeting night occurring on Thursday evening, December 3rd, 1896.

Very truly yours,
JOHN McDONALD, Secretary.
98 Clergy Street.

### BROCKVILLE NO. 15.

SIR,—Since our last report we have removed to new rooms more suitable for our association. When any of our brother engineers come to Brockville they will find our rooms on the second storey of Richard's Block, on King St. The attendance at the meetings is good and one new member has been initiated.

JAS. AIKINS, Recording Secretary.

## PRESENT STATUS OF THE DISTRIBUTION AND TRANSMISSION OF ELEC-TRICAL ENERGY.

MR. Louis Duncan, President of the American Institute of Electrical Engineers, concludes a paper on the above subject, with the following summary of conclusions :—

My conclusions, subject always to the influence of local conditions, are as follows :

1. In both direct-current lighting and traction systems, where the power is generated in or near the area of distribution, it is best to use one station situated at the most economical point for producing power.

2. In the case of the traction systems, when the economical area of direct distribution is passed, boosters should be employed directly or in connection with batteries, to a distance of 10 or 12 miles from a station, and beyond this rotary transformers, whether with or without batteries, should be used.

3. In the case of direct-current lighting systems, the energy should be transmitted to storage batteries situated at centres of consumption either directly or by means of a rotary transformer and distributed from them.

4. Where batteries are used, it is best to place them at the end of feeder wires to obtain the advantage of a constant load on the wire.

5. The best system for the long-distance transmission of energy, for general purposes, is the three-phase alternating system.

6. Commercial transmissions are in successful operation for distances of 35 miles and for voltages as high as 15,000 volts.

Experience with these plants shows that the transmission to 50 miles, with a pressure of 20,000 volts, is practicable ; beyond these limits the transmission would be more or less experimental.

## MR. O. E. GRANBERG,

DISTRICT DEPUTY FOR QUEBEC, C. A. S. E.

In the portrait appearing below is presented the countenance of Mr. O. E. Granberg, District Deputy for Quebec for the Canadian Association of Stationary Engineers. Mr. Granberg was born in Norway, Europe, in 1852, and came to Canada in 1860. At 12 years of age he entered a blacksmith shop, where he remained for three years, and later served three years in a machine shop. After working some time in the foundry and pattern shops, he went to New York and worked for some years in different engine and boiler works, returning to Canada in 1875. He was then employed in erecting engines, boilers and machinery in

MR. O. E. GRANBERG,
District Deputy for Quebec C.A.S.E.

mines, and shafting, gearing and machinery in manufacturing establishments. After having fitted up machinery in a cotton mill, he was employed as chief engineer and master mechanic for some years, and was subsequently made manager of the mill. He gave up that position for the one he now occupies, that of Inspector of the Boiler Inspection and Insurance Company of Canada, which he has occupied for over six years.

Mr. Granberg holds first-class engineers' papers, and is a qualified and authorized boiler inspector and examiner of engineers and firemen, as per Industrial Establishment Act of the Province of Quebec. He received the appointment of Examiner of Boiler Inspectors from the Lieutenant-Governor of the Province of Quebec in 1894, and has been a member of the Canadian Association of Stationary Engineers for five years, in which organization he is very popular.

## WORDS OF APPRECIATION FROM THE FAR NORTH.

MR. E. B. Congdon, Fort Macleod, N. W. T., in renewing his subscription to the ELECTRICAL NEWS, writes : "I enjoy the paper so much that I would not like to have it discontinued."

## HEAT. *

### By James Gill, B A.

It was with some diffidence that we agreed to read this paper before you, knowing, as we do, that you are all practical men and that our knowledge for the most part is but theoretical. However, we will go on the assumption that most teachers take, that you know nothing about the subject.

Our first question with regard to heat is, what is it? In past time it was considered a material substance that entered into a body, and by its presence there rendered the body warmer; its absence left the body cold. There was this difficulty, however, in supposing heat to be a material substance, in that the body when warm weighed no more than when cold. Sir Humphrey Davy melted two blocks of ice by rubbing them together, and concluded that heat was not a material substance, but a form of motion. Heat is generally understood at the present time to be due to the motion of the molecules of a body. These molecules are in constant motion, and when their motion is quickened the body becomes warmer; when their motion is retarded the body becomes colder.

In the next place let us inquire into the ways of producing heat. We will place down six ways of obtaining heat:

1st. From MECHANICAL ACTION as shown in friction. You are all acquainted with the result of rubbing a button of brass on your coat-sleeve. It used to be a common trick with school boys to rub the button for some time and then place it on the back of a playmate's hand. It had about the same effect as the sun's rays through a lens. Also the Savage of the Isles of the Sea was accustomed to produce fire by rubbing two dry sticks together.

2nd. PERCUSSION—As shown in placing a piece of lead on an anvil and hammering it. It soon becomes quite hot. The lead bullet after striking the metal target is too hot to pick up.

3rd. COMPRESSION—As shown in placing a piece of tinder in a tube in which a tube moves up and down. The mere shoving of the piston downwards is enough to ignite the tinder.

4th. CHEMICAL ACTION—Wherever chemical action goes on heat results. Pour some sulphuric acid into a vessel of water and then place your hand against the outside, you will find that the vessel is warm. Again the heat in the human body is maintained by chemical action.

5th. HEAT FROM THE ELECTRIC CURRENT—If you take several cells and connect for battery purposes, and then hold in your hand the two terminals from the positive and negative poles, you will soon find them too hot to hold. You have no doubt heard of a whole meal being cooked in Ottawa by means of heat obtained from the current.

6th. RADIANT HEAT—As obtained from the sun. The sun radiates heat on all sides, and this is borne to us through the ether which is supposed to fill all space.

The first three of these classes may be placed under the one head of "mechanical action."

Then let us notice the effects of heat applied:

1st. EXPANSION—As shown in a bar of metal placed rigidly between two fixed supports and heated. The bar bends and twists out of the straight.

2nd. CHANGE OF STATE—As shown in a block of ice to which heat is applied. It is first converted into water, and then if sufficient heat be applied, into steam.

3rd. CHANGE OF TEMPERATURE—Which we measure by means of the common thermometer.

We would like you to notice here the difference between temperature and quantity of heat. A cup of water and a painful of water may be at the same temperature, but the painful has the greater quantity of heat because it has the greater amount of mass. Again, we would notice that there is always present a tendency to equalization of temperatures. This takes place in three ways:

1st. RADIATION.—If I light a fire in the stove here it soon makes itself felt throughout the room, by radiating heat in all directions.

2nd. CONDUCTION.—Place in the fire one end of an iron bar and it will not be long before you are unwilling to keep hold of the other end. This is due to the molecules of the bar conducting the heat from the end in the fire to the end held in the hand.

3rd. CONNECTION.—This is the warming of a room or house by the bodily movement of a heated substance, such as is shown in the warming of buildings by hot air. The air is heated at the furnace and moves bodily from there to the rooms of the building.

Physicists are in the habit of using certain units in which to express amount of heat. One of these units is the amount of heat needed to raise one pound of water through one degree Fahrenheit. By means of these units a relation between heat and work can be expressed. First a definition of work: If one pound of matter be raised vertically against gravity through one foot, one foot-pound of work is said to be done, or if a body be drawn through one foot against a resistance from friction of one pound, one foot pound of work is said to be done. It is found from careful experiments that one of the above heat units is equivalent to 772 foot pounds of work. You are also acquainted with the unit used in expressing rate of doing work, viz., the horse power. One horse power is equivalent to 33,000 foot pounds of work per minute.

Just here we might give the method of finding the horse power of an engine: Find the area of the piston head in square inches and multiply by the length of stroke doubled and by the number of revolutions per minute, and also by the pressure in lbs., which product divide by 33,000, and the answer is in horse power. Thus, if effective pressure of steam be 60 lbs., diameter of piston 14 inches, length of stroke 2½ feet, and revolutions 70 per minute, then the horse power of engine will equal

$$\frac{(14 \times 14 \times .7854) \times (2\frac{1}{2} \times 2) \times 70 \times 60}{33,000}$$

But the all important point with the engineer is the conversion of heat into work. Where heat is applied to water it confers upon the steam which is produced the power of doing work, such as driving the piston from one end of the cylinder to the other against resistance. For example, the heat energy of the boiler in the engine is transferred into mechanical motion. The steam is admitted to the cylinder, and by means of its expansive force drives the piston to the other end, then by a special movement of slide valves caused by the eccentrics, the steam is allowed in at the other end of cylinder and the piston moves in the other direction, and so the motion is maintained. Work is done by the steam during its admission into the cylinder, and also by expansion after its admission.

Steam in its expansion obeys the well known law of Boyle, viz., that if the temperature be kept constant the volume of a given body of gas varies inversely as pressure, density and elastic force. If the steam be allowed to enter at full pressure of 80 lbs. for say one fourth the stroke, and is then cut off, the piston will have to be forced to the other end by the steam working expansively.

What is known as back pressure must be taken into consideration in finding the work done. The back pressure is usually fifteen pounds to the square inch in a non-condensing engine, so that the steam in cylinder must not be allowed to expand so far as to bring its pressure down to that amount. The relation between pressure and volume in a given body of gas may be very easily shown to the eye by a graphic representation by taking horizontal lines to represent volumes and vertical lengths to represent pressure, but it seems to us that you are better acquainted with what is called technically the "indicator diagram" than we are.

Up to this point we have been reasonably sure of our ground; it appears to us that so far as the practical working of a steam engine is concerned, we have more reason to learn from you than you to learn from us.

## GOOD NEWS FOR MACHINE WORKMEN.

The Royal Electric Company, of Montreal, has recently closed several extensive contracts for large electrical machinery, which will keep their factory occupied night and day for more than a year, and necessitate a large increase in the number of their employees. They are advertising for a number of mechanics, to whom good wages will be paid.

They have added to their machinery equipment recently some of the largest tools of their kind in Canada, such as planers, boring mills and drills, and need men experienced on such tools. This is a hopeful evidence of improvement in Canadian manufacture and industry, the greatest part of their work being required for water power developments and railroad purposes.—Evening News, Nov. 25th, 1896.

The Kay Electric Co., of Hamilton, have requested that the 25 per cent. duty on soft copper wire be removed.

Messrs. James Whitcomb, of Toledo, and James A. Bailey, of Detroit, are looking around for a suitable site in Canada for the establishment of carbon works.

* Paper read before the Hamilton Association C. A. S. E.

## ELECTRIC LIGHT INSPECTION STATISTICS.

The annual report of the Commissioner of Inland Revenue for the Dominion, for the fiscal year ending June 30th, 1896, contains the first statistics compiled relating to the inspection of electric light meters under the provisions of the Electric Light Inspection Act, which went into operation in June of last year. The report states that offices for testing purposes have been fitted with the necessary apparatus at Windsor, London, Toronto, Hamilton, Belleville and Ottawa, in Ontario, at Montreal, Quebec and Sherbrooke, in Quebec, and at St. John, N. B., and Halifax, N. S. A set of Lord Kelvin's absolute standard apparatus, both for the measurement of current and potential, is being placed in position in the standards branch at Ottawa by Mr. O. Higman, electrical engineer of the department, which, it is expected, will prove of great value to electric lighting companies as a convenient means of standardizing their measuring instruments.

The total revenue during the year from registration and inspection of meters was $8,681.25, and the expenses of the department $6,693.23, a large portion of the latter being for permanent equipment.

It will be observed by the accompanying table that 3,705 meters were inspected, of which number only 110 were rejected, while four were verified after the first rejection. In the city of Ottawa there are the largest number of meters, 938, while Montreal has 626, Hamilton 537, and Toronto 345.

STATEMENT SHOWING THE NUMBER OF ELECTRIC LIGHT METERS VERIFIED, REJECTED AND VERIFIED AFTER REJECTION.

| Districts. | Number. | Verified as coming within the error tolerated by law. | | | Rejected | | | Verified after first rejection as coming within the error tolerated by law. | | |
|---|---|---|---|---|---|---|---|---|---|---|
| | | Correct. | Fast. | Slow. | Unsound. | Fast. | Slow. | Correct. | Fast. | Slow. |
| Belleville | 233 | 56 | 125 | 52 | ..... | ..... | .... | ..... | ..... | ..... |
| Hamilton | 537 | 149 | 98 | 285 | ..... | ..... | 5 | ..... | ..... | ..... |
| London | 361 | 118 | 158 | 85 | ..... | ..... | .... | ..... | ..... | ..... |
| Ottawa | 938 | 166 | 3:2 | 349 | 19 | 33 | 19 | ..... | ..... | ..... |
| Toronto | 345 | 4 | 319 | 22 | ..... | ..... | ..... | ..... | ..... | ..... |
| Montreal | 626 | 83 | 232 | 301 | 1 | 1 | 8 | ..... | ..... | ..... |
| Quebec | 216 | 20 | 99 | 87 | 4 | 1 | 2 | 1 | 2 | ..... |
| Sherbrooke | 6 | 1 | 1 | 4 | ..... | ..... | ..... | ..... | ..... | ..... |
| St. John | 166 | 35 | 71 | 78 | ..... | 1 | 1 | ..... | ..... | ..... |
| Halifax | 257 | 140 | 58 | 47 | 8 | 3 | ..... | 1 | ..... | ..... |
| Totals | 3,705 | 772 | 1,513 | 1,320 | 32 | 39 | 35 | 2 | 2 | ..... |

The following statement shows the electric light companies registered under the Act during the year, together with the number of lamps operated. Each arc lamp is reckoned as equal to ten incandescents:

STATEMENT SHOWING THE ELECTRIC LIGHT COMPANIES REGISTERED UNDER THE ELECTRIC LIGHT INSPECTION ACT DURING THE YEAR ENDED 30TH JUNE, 1896.

| FROM WHOM COLLECTED. | NUMBER OF LAMPS. | | |
|---|---|---|---|
| | Arc. | Incandescent. | Totals. |
| Trenton Electric Company | 53 | 103 | 633 |
| R. R. Casement & Co., Madoc | .... | .... | ........ |
| W. H. Pearson & Co., Belleville | 40 | 400 | 800 |
| Corporation, Town of Picton | 21 | 300 | 510 |
| Stormont Electric Light and Power Company | .... | 1,300 | 1,300 |
| Vankleek Hill Electric Works | .... | 500 | 500 |
| Village of Alexandria | .... | 400 | 400 |
| Kingston Light, Heat and Power Company | 105 | 2,000 | 3,050 |
| Napanee Water and Electric Light Company | 33 | 31 | 361 |
| Light, Heat and Power Company, Lindsay | 100 | 4,000 | 5,000 |
| Port Hope Electric Light and Power Company | 37 | 450 | 820 |
| Bowmanville Electric Light Company | 46 | ..... | 460 |
| Peterborough Light and Power Company | 118 | 3,000 | 4,180 |
| Corporation of Campbellford | 22 | 409 | 629 |
| Lakefield Electric Light Company | 20 | ..... | 200 |
| Fenelon Falls Electric Light Company | 5 | 270 | 320 |
| Village of Colborne Electric Light | 27 | 17 | 287 |
| Cobourg Electric Light and Power Company | 39 | 600 | 990 |
| W. J. & H. W. Foulds, Electric Light, Hastings | 8 | 132 | 212 |
| W. C. Harrison, Electric Light, Norwood | 25 | 119 | 369 |
| Millbrook Electric Light Company | 7 | 400 | 470 |
| Brockville Electric Light and Power Company | 34 | 1,600 | 1,940 |
| Morrisburg Electric Light (A. H. Merkly) | ...... | 600 | 600 |
| Gananoque Electric Light and Water Supply Company | 16 | 1,600 | 1,760 |
| Kemptville Electric Light Company | ..... | 380 | 380 |
| Merrickville Electric Company | ...... | 300 | 300 |
| Prescott Electric Light Company | 16 | 750 | 9 0 |
| Ingersoll Electric Power and Light Company | 60 | 550 | 1,150 |
| Brantford Electric Street Railway Company | .. | 2,531 | 2,531 |
| Woodstock Electric Light, Power and Street Railway Co | 68 | 550 | 1,230 |
| Brantford Electric and Power Company | 56 | 2,200 | 2,760 |
| Gas and Water Company, Simcoe | 36 | 225 | 585 |
| Paris Electric Light Company | 45 | 400 | 850 |
| Norwich Electric Light | 15 | 32 | 182 |
| Port Rowan Electric Light Supply | 20 | 24 | 424 |
| Tilsonburg Electric Light (F. J. Barkey) | 33 | 142 | 472 |
| Port Dover Electric Light Syndicate | 8 | 80 | 160 |
| Hamilton Electric Light Company | 450 | 8,000 | 12,500 |
| Dunnville Electric Light Company | 35 | 38 | 388 |
| Niagara Falls Electric Light, Heat and Power Company | 50 | 1,440 | 1,940 |
| Hagersville Electric Light Company | 20 | 5 | 205 |
| Welland Electric Light Company | 30 | 500 | 800 |
| St. Catharines Electric Light and Power Company | 68 | 900 | 1,580 |
| Cayuga Electric Light and Power Company | 9 | 359 | 449 |
| Thorold Electric Light Plant (Jas. McGill) | 36 | 375 | 735 |
| J. W. VanDyke Electric Light Plant, Grimsby | 8 | 250 | 330 |
| Corporation of the Town of Niagara | .... | 800 | 800 |
| London Electric Light Company | 350 | 5,155 | 8,655 |
| Sarnia Gas and Electric Light Company | 53 | 240 | 770 |
| St. Thomas Gas Co., supplying Electric Power and Light. | 55 | 137 | 687 |
| Fitzgerald & Sauermann Electric Light Company, Watford | 14 | 50 | 190 |
| Freeman N. Saylor, Strathroy | 29 | 51 | 341 |
| Petrolea Electric Light, Heat and Power Company | 30 | 770 | 1,070 |
| Hamilton & Prout, Forest | 17 | 50 | 220 |
| W. W. Gordon, Glencoe | 25 | 50 | 300 |
| H. C. Baird & Son, Parkhill | 26 | 75 | 335 |
| Aylmer Electric Light Company | 38 | 52 | 432 |
| Stratford Gas Company | 140 | 300 | 1,700 |
| Clinton Electric Light Company | 31 | 500 | 810 |
| Cook Brothers Electric Light Company, Hensall | .. | 240 | 240 |
| Corporation of the Town of Mitchell | 40 | 430 | 830 |
| Seaforth Electric Light, Heat and Power Company | 60 | 500 | 1,100 |
| Palmerston Electric Light Company | 19 | 340 | 530 |
| Exeter Electric Light Company | 10 | 200 | 300 |
| Town of Goderich | 84 | ..... | 840 |
| St. Mary's Electric Light Company | 33 | 650 | 980 |
| Wingham Electric Light Company | 50 | 600 | 1 00 |
| Blyth Electric Light Plant (John B. Kelly) | 17 | 140 | 310 |
| Brussels Electric Light Company | 21 | 10 | 220 |
| People's Electric Company, Windsor | 4 | 2,580 | 2,6 0 |
| Hiram Walker & Sons, Walkerville | ...... | 1,200 | 1,200 |
| W. A. Johnson & Co., Dresden | 26 | 27 | 287 |
| Smith & Anderson, Blenheim | 19 | 79 | 269 |
| J. & W. McMaster, Ridgetown | 44 | 10 | 450 |
| Wm. Laing, Essex | 35 | 12 | 362 |
| Tilbury Electric Light | 30 | 15 | 345 |
| Electric Company, Amherstburg | .. | 500 | 500 |
| Chatham Gas Company | 63 | 775 | 1,405 |
| Wallaceburg Electric Light Company | 40 | 12 | 452 |
| Leamington Electric Light Company | 21 | 206 | 416 |
| Geo. Munro, Thamesville | 7 | 175 | 245 |
| Ottawa Electric Company | 500 | 50,000 | 55,000 |
| Albert MacLaren, Buckingham | .. | 550 | 550 |
| A. W. Wright & Co., Renfrew | 20 | 800 | 1,000 |
| Mackay & Guest, Renfrew | .. | 1,140 | 1,140 |
| Citizens' Electric Light Company, Smith's Falls | ..... | 1,000 | 1,000 |
| Smith's Falls Electric Light Company | 20 | 2,000 | 2 200 |
| Carleton Place Electric Company | 40 | 700 | 1,100 |
| Almonte Electric Light Company | 23 | 800 | 1,030 |
| Pembroke Electric Light Company | 28 | 2,200 | 2,480 |
| Arnprior Electric Light and Power Company | 30 | 2,500 | 2,800 |
| Perth Electric Light Company | 47 | ........ | 470 |
| Tay Electric Light Company, of Perth | .. | 1,200 | 1,200 |
| Electric Light Company, of Eganville | ...... | 300 | 300 |
| Star Electric Light Company, of Eganville | .. | 300 | 300 |
| Guelph Light and Power Company | 100 | 1,400 | 2,400 |
| Galt Gas Light Company | 70 | 25 | 725 |
| Berlin Gas Light Company | 36 | 156 | 516 |
| Waterloo Electric Company | 24 | 81 | 321 |
| Howes & Leighton, Harriston | 30 | 20 | 320 |
| Jacob Morley, New Hamburg | 22 | 50 | 270 |
| John Shearer, Villages of Blair and Preston | ..... | 300 | 300 |
| A. Groves, Fergus | 35 | 700 | 1,050 |
| Corley & Collins, Mount Forest | 13 | 400 | 530 |
| James Fenwick, Preston | 30 | 20 | 320 |
| James Fenwick, Hespeler | 24 | 35 | 275 |
| Incandescent Light Company, of Toronto | .. | 30,000 | 30,000 |
| Toronto Electric Light Company | 1,500 | 50 | 15,050 |
| Milton Electric Light and Power Company | 19 | 236 | 426 |
| Barrie Electric Light Company | 47 | 2,500 | 2,970 |
| Town of Orillia | 43 | 2,000 | 2,430 |
| Penetanguishene & Midland St. Railr'd, Light & Power Co. | 14 | 1,200 | 1,340 |
| Midland Electric Company | 35 | 500 | 850 |
| Creemore Electric Light Company | ...... | 135 | 135 |
| Joseph Knox, Stayner | 4 | 400 | 440 |
| Glenwilliams Electric Light Company | 45 | ........ | 450 |
| W. J. Fletcher, Electric Light and Power Station, Alliston | 20 | 350 | 550 |
| Light, Heat and Power Company, of Newmarket | 15 | 400 | 550 |
| Hutton Electric Light Company, Brampton | 26 | 500 | 760 |
| Huntsville Electric Light Company | 1 | 550 | 560 |
| Nicholas Egan, Tottenham | ...... | 150 | 150 |
| Mattawa Electric Light and Power Company | ...... | 900 | 900 |
| Town of Bracebridge | ...... | 1,700 | 1,700 |
| John Bourke, Mattawa | ...... | 703 | 703 |
| Robert McGowan, Oakville | 35 | 800 | 1,150 |
| Lakefield and Whitby Electric Light Company | 40 | ...... | 400 |
| H. A. Train, Burk's Falls | ...... | 200 | 200 |
| Port Perry Illuminating Company | 20 | 213 | 413 |
| Tagoma Water and Light Company | 50 | 440 | 940 |
| Gravenhurst Electric and Trading Company | 11 | 500 | 610 |
| Appleyard Electric Light Company, Grand Valley | 7 | 75 | 145 |
| Orangeville Electric Light and Power Company | 26 | 850 | 1,110 |
| Corporation of the Village of Markham | 18 | 50 | 230 |
| Robertson, Rowland & Company, Walkerton | 30 | 800 | 1,100 |
| Owen Sound Electric Illuminating and Man'f'g Company | 55 | ....... | 550 |
| Corporation of the Town of Collingwood | 35 | 1,050 | 1,400 |
| Town of Kincardine | 19 | 525 | 715 |
| Wiarton Electric Light Company | 26 | 40 | 300 |
| Ramage Bros. Electric Light Company, Chesley | .. ... | 400 | 400 |
| Donald McIntyre Electric Light Company, Paisley | 7 | 231 | 301 |
| Wm. Moore & Sons, Meaford | 250 | 340 | 2,840 |
| T. & J. R. Andrews, Thornbury | 19 | 20 | 210 |
| Southampton Electric Light Company | 20 | 18 | 218 |
| John Bearman, Chesley | 22 | 210 | 430 |
| Daniel Knechtel, Hanover | 12 | 298 | 410 |
| Durham Electric Light Company | 22 | 250 | 470 |
| La Corporation de la Ville de Joliette | 33 | 1,262 | 1,592 |
| Royal Electric Company | 1,450 | 54,000 | 68,500 |

| FROM WHOM COLLECTED. | NUMBER OF LAMPS. | | |
|---|---|---|---|
| | Arc. | Incandescent. | Totals. |
| Citizens' Light and Power Company | 800 | 800 | 8,800 |
| La Compagnie Electrique St. Jean Baptiste | | 3,447 | 3,447 |
| La Ville de Maisonneuve | 26 | 319 | 579 |
| Corporation of the Town of Lachine | 42 | 1,315 | 1,435 |
| Temple Electric Company | 20 | 1,750 | 1,950 |
| I. B. Robert, Beauharnois | | 86 | 86 |
| Valleyfield Electric Company | | 775 | 775 |
| Electric Light Company of Terrebonne | | 337 | 337 |
| Magloire Ouimet, St. Jerome | | 497 | 497 |
| Montmorency Electric Power Company | 406 | 10,500 | 14,500 |
| Sherbrooke Gas and Water Company | 70 | 2,300 | 3,000 |
| Richmond County Electric Company | | 877 | 877 |
| Stanstead Electric Light Company | 25 | 900 | 1,150 |
| Coaticook Electric Light Company | 28 | 650 | 930 |
| Parker & Howe, Dixville | | 125 | 125 |
| French Bros., Sawyerville | | 80 | 80 |
| La Compagnie des Pouvoirs Hydrauliques de St. Hyacinthe | 2 | 3,000 | 3,020 |
| La Compagnie pour l'Eclairage au Gaz de St. Hyacinthe | 30 | 150 | 450 |
| Granby Electric Light Company | 35 | | 350 |
| Carleton Electric Light Company | 3 | 375 | 675 |
| St. John Railway Company | 401 | 6,110 | 10,120 |
| Fredericton Gas Light Company | 80 | 900 | 1,000 |
| Woodstock Electric Light Company | | 1,800 | 1,800 |
| Sackville Electric Light and Telephone Company | | 405 | 405 |
| Small & Fisher Company, Woodstock | | 500 | 500 |
| St. Stephen Electric Light Company | 40 | 183 | 583 |
| City of Moncton Light and Water Department | 98 | 975 | 1,895 |
| Prince Edward Island Electric Company | 80 | 2,000 | 2,800 |
| Full Electric Light Company, Charlottetown | | 490 | 490 |
| Halifax Gas Light Company | 75 | 2,000 | 2,950 |
| Halifax Illumination and Power Company | 220 | 5,000 | 7,200 |
| Dartmouth Gas, Electric Light, Heating and Power Co | | 850 | 850 |
| Windsor Electric Light and Power Company | 12 | 1,350 | 1,470 |
| "Chambers" Electric Light Company | 65 | 2,900 | 3,550 |
| Kentville Electric Light and Power Company | | 520 | 520 |
| Acadia Edison Electric Company, Wolfville | | 500 | 500 |
| Edison Electric Light and Power Company, Springhill | | 575 | 575 |
| Lunenburg Gas Company | | 600 | 600 |
| Bridgewater Electric Light and Water Power Company | | 475 | 475 |
| Canada Electric Light Company, Amherst | 26 | 1,500 | 1,760 |
| North Sydney Electric Company | | 700 | 700 |
| Sydney Gas and Electric Light Company | | 1,400 | 1,400 |
| New Glasgow Electric Company | 16 | 2,000 | 2,160 |
| Digby Electric Light Plant | | 300 | 300 |
| Bridgetown Electric Light Plant | | 200 | 200 |
| Annapolis Electric Light Company | | 450 | 450 |
| Citizens' Telephone and Electric Company, Rat Portage | | 3,500 | 3,500 |

## SPARKS.

The town of Paris, Ont., is said to be considering the purchase of an electric plant.

An effort is being made to secure the installation of an electric lighting plant at Embro, Ont.

It is announced that another electric lighting company is likely to be started at St. John, N. B.

The electric light plant of the Hull Electric Company at Hull, Que., has been put in operation.

John Norwood, of Alvinston, Ont., has secured the contract for lighting that town by electricity.

It is rumored that an Ottawa firm propose putting in an opposition electric plant at Arnprior, Ont.

Mr. Tache is making arrangements to install an electric light plant for street lighting at Huntingdon, Que.

A dividend of 12 per cent. per annum has been declared by the Nelson Electric Light Company, of Nelson, B. C.

The conduits laid in the streets of Montreal by the Lachine Rapids Hydraulic & Land Company cover a distance of 80 miles.

A small addition is now being erected to the power house of the Hamilton Electric Light & Power Company at Hamilton, Ont.

The new electric light plant at St. Marys, Ont., will shortly be put in operation, the erection of the power house being nearly completed.

An electric light plant will be established at Shawville, Que., a company having been given twenty years' exemption from taxation.

In the city of Halifax, N. S., an agitation has been commenced towards having all electric light and telephone wires placed underground.

M. F. Beach & Company have been granted permission by the village council of Winchester, Ont., to erect poles for electric street lighting.

The Cocoa Matting Co., of Cobourg, Ont., have installed an isolated plant in their factory. The Canadian General Electric Co. secured the contract.

The new plant of the Sussex Electric Light Company was recently put in successful operation at Sussex, N. B. Over 400 lights have already been taken.

A despatch from New York states that the Commercial Cable Company have secured control of the Postal Telegraph Company,

and propose amalgamating it with their own system, issuing $20,000,000 of debenture stocks.

Mr. Graham has made application to the village council of Norwich, Ont., for permission to erect poles and install a plant for incandescent and arc lighting.

The Toronto Electric Motor Company have requested that the present duty of 30 per cent. on magnet wire entering Canada from the United States be removed.

At the municipal elections in January a by-law will be voted on by the ratepayers of Bridgeburg, Ont., granting an electric light franchise for fifteen years to D. A. Coste.

The town council of Magog, Que., have in view the purchase of a dam on the property of the B. A. Land Company in order to secure power for electric lighting purposes.

The municipality of St. Louis du Mile End, Que., have awarded the contract to the Citizens' Light and Power Co., for lighting the streets for the next thirteen years, also for residential lighting.

Horseless carriages are to be manufactured in Montreal, a company with a capital of $140,000 having been organized for that purpose. It will be known as the Moto-Cycle Company of Canada, Limited.

The Minden & Northwestern Railway Company, which proposes constructing a railway from Irondale Junction to a point on the Georgian Bay district, ask for authority to use either steam or electricity as the motive power.

The British Columbia Electric Company, with head office at Tacoma, Washington, and a capital stock of $10,000, has been registered to do business in British Columbia. Its purpose is to equip fire and burglar alarm systems.

With a view to increasing the steam capacity of the locomotives, the Grand Trunk Railway are enlarging the boilers of some of their passenger locomotives by the addition of a square top over the boiler next to the cab engine.

A test of horseless vehicles was recently made at London, England, in which fifty-four vehicles were entered, including several German inventions, the two Duryea carriages from the United States, and a large number of English manufacture.

Bothwell & Irving, solicitors, of Victoria, B. C., are asking for an extension of time for the commencement of the works contemplated by the Kootenay Power Company's Construction Act. One of the objects of this company is the production of electricity.

At a meeting of the Standard Light & Power Company, Limited, held in Montreal on the 27th ultimo, Mr. W. McLea Walbank was elected president and Mr. Jeffrey H. Burland vice-president. A retiring director was replaced by the election of Mr. W. S. Evans.

The city council of Winnipeg have now under consideration the question of purchasing an electric light plant to be controlled by the city. The scheme has been sanctioned by the council, and the ratepayers will be requested to authorize the expenditure of $75,000 for the purpose.

At a recent meeting of the shareholders of the Parry Sound Electric Light Company, of Parry Sound, Ont., the chief business was the adoption of the auditor's report and the disposition of the surplus earning over cost of operation. The statement presented was considered satisfactory.

A despatch from St. Paul, Minn., states that a small rotary engine of novel design has been invented by Grant Brambel, of Sleepy Eye, Minn., for the patent of which Mr. H. F. Allen, of London, is said to have offered $1,600,000. The engine does away entirely with the crank motion of the steam engine, and uses its own plunger for a cut-off. It is steam tight and requires no ring packing.

Letters patent have been issued incorporating the St. Jerome Power and Electric Light Company, of St. Jerome, Que. The capital stock is $30,000, and the objects are to purchase the electric light plant now in operation at that town and to develop electric and water power for commercial purposes. Among the incorporators are C. L. Higgins and J. J. Westgate, of Montreal, and James Pearson, barrister, of Toronto.

As the result of a disagreement between the town authorities and the electric light company of Welland, Ont., the lights were recently shut off. Some time ago the council passed a resolution compelling the electric light company to raise the street lights at their own expense, which the manager of the company refused to do, and it is probable that tenders for a franchise for electric treet lighting will be asked for in consequence.

# ELECTRIC RAILWAY DEPARTMENT.

### GUELPH STREET RAILWAY.

ON the 17th of May, 1895, the first spike was driven on the above railway, by the president's amiable wife, in front of their residence, and four months later five miles of the road were in operation. Early this year another mile was completed, forming a belt line on which is located the baseball grounds.

It is claimed—and justly too—that it is the smoothest road in Canada. The rails on the new part are 65 lbs., while on the other part they are 56 lbs. It is a single track, with suspended trolley wires in the crowded parts of the city, while the rigid bracket is used in the less crowded streets.

A twenty minute service is provided on all the lines, and connections are made to and from all trains. There are three lines and a belt line, each starting from

of the power house. This saves the cost and operation of pumps. A large C. G. E. generator generates the electric current for the operation of the cars. The power house is sufficiently large to operate a twelve mile road.

The car barn is the same size as the power house and is similarly built; it has accommodation for about twelve cars.

The company is composed of the Sleeman family. Mr. Geo. Sleeman, sr., is the president, and Mr. E. Sleeman secretary and superintendent. The other offices are held by other members of the family.

The company own a pretty park near Waterloo ave., which they are laying out for the benefit of their patrons. It is probable that in the near future the road may be extended south west to Hespeler, in which event

GUELPH STREET RAILWAY.

the C. P. R. depot. One runs to the dairy building beyond the Ontario Agricultural College, climbing a steep hill on the way, and another runs down Waterloo ave., to the city limits in the south west, terminating at the Silver Creek Brewery. Both these lines run through the market place. The other line goes up Wyndham street—the main street—through St. George's Square, up Woolwich street to Elora Rd., and thence to the north west city limits. The belt line connects with the Waterloo ave. line at Edinburgh Rd., and with the Woolwick line at Suffolk street. On this line is the baseball grounds and the Collegiate Institute.

The rolling stock consists of five motor cars and a trailer, built by the Canadian General Electric Co. They are neatly furnished and upholstered.

The power house on Waterloo ave. is 100 x 40 feet in dimensions, with a ceiling 20 feet high, which is sheeted with Georgia pine. The two 100 h.p. Goldie & McCulloch boilers generate steam for a 150 h.p. cross compound Wheelock engine. There is space for another boiler and engine of the same power. In the basement is a large Northey condenser.

Water is supplied to the boiler by gravitation, there being a creek near by which is dammed 100 yards back

connection will be made with the large manufacturing town of Galt.

Application will be made to Parliament for the incorporation of a company to build an electric railway from Wabigoon, Ont., on the C. P. R., through the Manitou country to Rainy River, and touching nearly all the chief mining camps. Mr. E. A. C. Pew, of Toronto, is said to be one of the promoters.

Mr. Black, manager of the street railway at Niagara Falls, Ont., returned from New York a fortnight ago, where he interviewed the shareholders of the company with the object of converting the road to an electrical system. Nothing, however, can be done until next spring, and in the meantime present circumstances will have to alter considerably to secure the carrying out of the scheme, as the attitude of the council is said to be unfavorable thereto.

The Judicial Committee of the Privy Council has allowed an appeal, with costs, to the Edison Electric Company against the Westminster and Vancouver Tramway Companies, the Bank of British Columbia, and others. The Edison Company appealed against the decision of the Supreme Court of British Columbia granting a judgment in favor of the Bank of British Columbia against the tramway companies, to the prejudice of the Edison Company, who are the creditors of the tramways. The decision of the Judicial Committee of the Privy Council declares the judgment of the Supreme Court of British Columbia against the tramways to be null.

## SPARKS.

The Canadian General Electric Co. are installing an isolated plant at Thetford Mines, P. Q.

The Canadian General Electric Co. have installed an isolated incandescent plant at Kearney, Ont.

There is talk of building an electric railway from Hurdman's Bridge to Morrisburg, Ont., via Metcalfe.

The Nova Scotia Steel Co., New Glasgow, have placed an order for an isolated incandescent plant with the Canadian General Electric Co.

The Metropolitan Street Railway Company, of Toronto, will increase their capital stock to $500,000 by the issue of 5,000 new shares of $100 each.

It is learned that the recently organized Peterborough Electric Power Co. have entered into a contract to supply 200 h. p. to the Canadian General Electric Company's works.

During the six months ending October 31 the Hamilton, Grimsby and Beamsville Railway carried 147,000 passengers, 545,000 pounds of fruit, and considerable other freight.

According to the report of the Canadian consul, concessionary powers have been granted to build 100 miles of electric tramway at Johannesburg, South Africa, with a capital of £2,000,000.

The Toronto Street Railway Company are building a passenger car for the Hamilton and Dundas Railway. It will be a handsome car of cherry wood, 60 feet in length, and superior in design and finish to anything the company have yet turned out.

D. Knechtel, of Hanover, has extended his monocyclic circuit to Neustadt, a distance of six miles, and to Caresruhe, a distance of 8½ miles from his generating station, and is now supplying a regular lighting service to both of these towns, as well as to Hanover.

It is probable that the construction of an electric railway from Brantford to Galt by way of Paris will be carried out at an early date. A charter for this road was granted about two years ago, some of the promoters being Mayor Elliott, H. McKay Wilson and Dr. Secord, of Brantford, and Henry Stroud, of Paris.

Wm. Shearer, of Blair, Ont., recently had a narrow escape from death. He was engaged oiling the machinery in the power house when the spout of the oil can came in contact with one of the bearings, directing the current to his left hand. Fortunately, his other hand being free, the circuit was broken, but his hand was badly burned.

An order-in-council has been issued incorporating the Hamilton, Chedoke and Ancaster Electric Railway Company, with a capital stock of $100,000, divided into 10,000 shares of $10 each. The names of incorporators are Edward Henderson, Ancaster; Harry Maxey, F. G. Beckett and Frederick Snider, of Hamilton, and Henry Beckett, of the township of Barton. The promoters expect to commence operations in the early spring.

The Fraserville Company, Limited, of Fraserville, Que., which is seeking incorporation, will engage in the manufacture of pulp and the manufacture, sale and distribution of electrical machinery, electric power, etc. The capital stock is $50,000, and among the promoters are Geo. White-Fraser, of Toronto, and David Cooke, of Fraserville.

The Montreal Street Railway Company have decided to grant bonuses to their employees who have been in the service of the company for five years and upwards, as follows : Men in the service five years, $10 a year ; ten years, $20 ; fifteen years, $30 ; twenty years and upwards, $40. A fortnight ago the bonuses for this year were paid to the men.

The Hamilton Radial Electric Railway Co. have added to the winter equipment of their road a Ruggles rotary snow plow, which will be operated with four G. E. 1,200 motors supplied by the Canadian General Electric Co. This will be the first instance in Canada of the use of the electric rotary, and the powerful equipment provided should be an absolute guarantee against any interruption of the company's service even during the severest snow fall.

C. C. Udell, of Los Angeles, Ont., met death from an electric shock which he received while using a telephone. Udell was a conductor on the street railway, and was in the act of telephoning to the car house when he was thrown to the ground in an unconscious condition, 1,500 volts of electricity having passed through his body from an electric light wire which had become crossed with the private telephone line of the railroad company.

The O'Keefe Brewing Co. are installing a 500 light isolated generating plant, and the contract has been awarded to the Goldie & McCulloch Co. for an "Ideal" engine, and to the Canadian General Electric Co. for a 40 k.w. direct connected generator of their slow-speed multipolar type. The surplus capacity of the generator will be utilized to furnish current, from which will be operated motors installed at various points throughout the buildings, in place of the small steam engines at present in use.

### FINDS IT INTERESTING.

Mr. A. W. McMaugh, St. Catharines, in remitting his subscription for the ELECTRICAL NEWS, writes : "It is a very interesting monthly, and am pleased with some of the able articles appearing from time to time in its columns."

### CAN'T DO WITHOUT IT.

Mr. N. Smith, engineer, in charge of the Alexandria, Ont., water works, in remitting his subscription, writes : "I find your NEWS so interesting that I can't do without it."

In the United States Circuit Court Judge Showalter issued an order recently of importance to the telephone industry, which practically puts an end to the alleged right of exclusive manufacture held by the Western Electric Company. The order made in the case is to the effect that the Western telephone switch patent expired July 30, 1895, by virtue of its having been taken out in Canada by the inventor June 30, 1890. The discovery that a Canadian patent had been issued on the invention was made only recently by the Western Telephone Construction Company, which concern was the principal defendant in the suit brought by the Bell and Western Electric combination. The court has had the case in which the Western Electric Company charged infringement under advisement since last summer, and, in the meantime, the Western Telephone Construction Company discovered this new evidence. Aside from ordering the admission of this testimony Judge Showalter could do nothing further until he renders his decision. Under the decision in the Bates refrigerator case, the discovery of the Canadian patent shortens the time of the Watson concession five years, it having been ruled that a patent taken out in a foreign country acts against a later patent issued in other countries.

The Montmorency Electric Power Co., of Montmorency Falls, have placed an order with the Royal Electric Co., of Montreal, for two 600 k.w. "S. K. C." two phase alternating current generators and switchboards complete. One of these generators is to be placed at the Montmorency Falls, where it will be driven by a water power, and the current carried 9 miles into the city of Quebec to the sub-station of the Montmorency Electric and Power Co., where the second 600 k.w. alternating current generator will be located. The generator at the Falls will deliver 5,000 volts direct to the line, and the generator in the sub-station will be used as a synchronous motor and receive the current at 5,000 volts direct into the machine, so that no step-up or step-down transformers or any intermediate apparatus is required. The synchronous motor is to be direct-connected to a direct current railway generator, which is to furnish power to the new Quebec Street Railway, which will be built early the coming year.

# ROBIN, SADLER & HAWORTH

Manufacturers of

## OAK-TANNED LEATHER BELTING

### MONTREAL AND TORONTO

Orders addressed either to our Toronto or Montreal Factory will have prompt care.
Goods will be forwarded same day as order is received.

Please mention the CANADIAN ELECTRICAL NEWS when corresponding with Advertisers.

## SPARKS.

The Montreal Cotton Co. have placed an order for additional induction motors for their Valleyfield plant with the Canadian General Electric Co.

The Toronto Electric Light Co. will shortly be ready to start up the new vertical engine driving the 450 k. w. Canadian General generator in their new power house.

The City Council of Chatham, Ont., will shortly consider the question of submitting a by-law to the ratepayers providing for the purchase of an electric lighting plant.

The Amherstburg Electric Light and Power Co. have increased their plant by adding a 1,000 light single phase alternator of the Canadian General Electric Co.'s standard type.

The corporation of the city of Guelph has given notice of its intention to submit a by-law to the ratepayers to authorize the purchase of an electric lighting plant to be operated by the city.

Canadian patents Nos. 53,456, 53,457, 53,458, 53,550, 53,551 and 53,552, covering improvements on telephone switchboards, signalling apparatus, etc., have been granted to the Bell Telephone Co. of Canada.

The Keewatin Power Company, of which Mr. John Mathers, of Ottawa, is manager, propose to utilize the water power obtained from its dam at Keewatin by transmitting electric energy of 5,000 horse power to Winnipeg. The distance between the two points is 120 miles.

A Russian electrician is said to have perfected a telephone which practically disregards distance. At a recent test between Moscow and Rostoff, a distance of 890 miles, talking, singing and instrumental music at one end of the line were distinctly heard by listeners at the other. An experiment is to be made by land wires and Atlantic cable in talking between London and New York.

A plant for the manufacture of acetylene gas was recently put in operation at Niagara Falls, N. Y., by the Acetylene Heat, Light and Power Company. In the process of manufacture 75 lbs. of pure lime and 100 lbs. of pure coke are mixed together and put through a crusher, then through the rolling mill, the coke passing through what is termed a 50 mesh sieve. After being thoroughly crushed the mixture is dumped into a porcelain iron mixer with round pebbles in it, where the compound is shaken and mixed. It is then carried to the furnaces, where it is dumped into the crucibles on the furnace, of which there are four, each holding 800 pounds of the mixture. The crucibles are clamped by jaws, which are worked by wedges drawn in and out by endless screws. In each crucible is placed a few inches of ground coke, and then the carbon is let down on top of the mass, thus completing the connection; in short, circuiting the transformer. The carbon is afterwards gradually drawn up, all the while the mixture is being dumped in on each side. The chemical combination immediately begins to take place, and changes the coke and lime into calcium carbide. From two to two and one quarter tons of the mixture makes one ton of carbide.

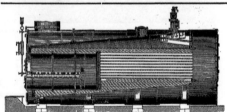

Lightning Source UK Ltd.
Milton Keynes UK
UKHW022135011218
333024UK00011BA/1833/P